This online teaching and l... **integrates the** entire digita... **most effective instructor** a... **to fit every learning style.**

With **WileyPLUS**:

- Students achieve concept mastery in a rich, structured environment that's available 24/7

- Instructors personalize and manage their course more effectively with assessment, assignments, grade tracking, and more

- manage time better
- study smarter
- save money

From multiple study paths, to self-assessment, to a wealth of interactive visual and audio resources, *WileyPLUS* gives you everything you need to personalize the teaching and learning experience.

»Find out how to MAKE IT YOURS»

www.wiley**plus**.com

10TH EDITION

Elementary Linear Algebra

with Supplemental Applications

International Student Version

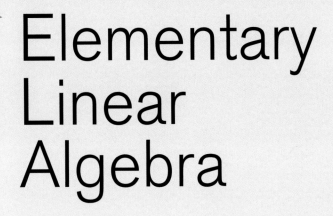

HOWARD ANTON

Professor Emeritus, Drexel University

CHRIS RORRES

University of Pennsylvania

WILEY

Copyright © 2011 John Wiley & Sons (Asia) Pte Ltd

Cover image from © iStockphoto

Contributing Subject Matter Expert: Dr. Heather Hulett, University of Wisconsin-La Crosse

All rights reserved. **This book is authorized for sale in Europe, Asia, Africa and the Middle East only and may not be exported outside of these territories.** Exportation from or importation of this book to another region without the Publisher's authorization is illegal and is a violation of the Publisher's rights. The Publisher may take legal action to enforce its rights. The Publisher may recover damages and costs, including but not limited to lost profits and attorney's fees, in the event legal action is required.

No part of this publication may be reproduced, stored in a retrieval system, or transmitted in any form or by any means, electronic, mechanical, photocopying, recording, scanning, or otherwise, except as permitted under Section 107 or 108 of the 1976 United States Copyright Act, without either the prior written permission of the Publisher or authorization through payment of the appropriate per-copy fee to the Copyright Clearance Center, Inc., 222 Rosewood Drive, Danvers, MA 01923, website www.copyright.com. Requests to the Publisher for permission should be addressed to the Permissions Department, John Wiley & Sons, Inc., 111 River Street, Hoboken, NJ 07030, (201) 748-6011, fax (201) 748-6008, website http://www.wiley.com/go/permissions.

ISBN: 978-0-470-56157-7

Printed in Asia

10 9 8 7 6 5 4 3 2

To
My wife, Pat
My children, Brian, David, and Lauren
My parents, Shirley and Benjamin
My benefactor, Stephen Girard (1750–1831),
   whose philanthropy changed my life

Howard Anton

To
Billie

Chris Rorres

ABOUT THE AUTHORS

Howard Anton obtained his B.A. from Lehigh University, his M.A. from the University of Illinois, and his Ph.D. from the Polytechnic University of Brooklyn, all in mathematics. In the early 1960s he worked for Burroughs Corporation and Avco Corporation at Cape Canaveral, Florida, where he was involved with the manned space program. In 1968 he joined the Mathematics Department at Drexel University, where he taught full time until 1983. Since then he has devoted the majority of his time to textbook writing and activities for mathematical associations. Dr. Anton was president of the EPADEL Section of the Mathematical Association of America (MAA), served on the Board of Governors of that organization, and guided the creation of the Student Chapters of the MAA. In addition to various pedagogical articles, he has published numerous research papers in functional analysis, approximation theory, and topology. He is best known for his textbooks in mathematics, which are among the most widely used in the world. There are currently more than 150 versions of his books, including translations into Spanish, Arabic, Portuguese, Italian, Indonesian, French, Japanese, Chinese, Hebrew, and German. For relaxation, Dr. Anton enjoys travel and photography.

Chris Rorres earned his B.S. degree from Drexel University and his Ph.D. from the Courant Institute of New York University. He was a faculty member of the Department of Mathematics at Drexel University for more than 30 years where, in addition to teaching, he did applied research in solar engineering, acoustic scattering, population dynamics, computer system reliability, geometry of archaeological sites, optimal animal harvesting policies, and decision theory. He retired from Drexel in 2001 as a Professor Emeritus of Mathematics and is now a mathematical consultant. He also has a research position at the School of Veterinary Medicine at the University of Pennsylvania where he does mathematical modeling of animal epidemics. Dr. Rorres is a recognized expert on the life and work of Archimedes and has appeared in various television documentaries on that subject. His highly acclaimed website on Archimedes (http://www.math.nyu.edu/~crorres/Archimedes/contents.html) is a virtual book that has become an important teaching tool in mathematical history for students around the world.

PREFACE

This edition of *Elementary Linear Algebra with Supplemental Applications*, 10th edition, gives an introductory treatment of linear algebra that is suitable for a first undergraduate course. Its aim is to present the fundamentals of linear algebra in the clearest possible way—sound pedagogy is the main consideration. Although calculus is not a prerequisite, there is some optional material that is clearly marked for students with a calculus background. If desired, that material can be omitted without loss of continuity.

Technology is not required to use this text, but for instructors who would like to use MATLAB, *Mathematica*, Maple, or calculators with linear algebra capabilities, we have posted some supporting material that can be accessed at the following Web site:

www.wiley.com/go/global/anton

Summary of Changes in this Edition

This edition is a major revision of its predecessor. In addition to including some new material, some of the old material has been streamlined to ensure that the major topics can all be covered in a standard course. These are the most significant changes:

- **Vectors in 2-space, 3-space, and n-space** Chapters 3 and 4 of the previous edition have been combined into a single chapter. This has enabled us to eliminate some duplicate exposition and to juxtapose concepts in n-space with those in 2-space and 3-space, thereby conveying more clearly how n-space ideas generalize those already familiar to the student.

- **New Pedagogical Elements** Each section now ends with a *Concept Review* and a *Skills* mastery that provide the student a convenient reference to the main ideas in that section.

- **New Exercises** Many new exercises have been added, including a set of True/False exercises at the end of most sections.

- **Earlier Coverage of Eigenvalues and Eigenvectors** The chapter on eigenvalues and eigenvectors, which was Chapter 7 in the previous edition, is Chapter 5 in this edition.

- **Complex Vector Spaces** The chapter entitled *Complex Vector Spaces* in the previous edition has been completely revised. The most important ideas are now covered in Sections 5.3 and 7.5 in the context of matrix diagonalization. A brief review of complex numbers is included in the Appendix.

- **Quadratic Forms** This material has been extensively rewritten and streamlined to focus more precisely on the most important ideas.

- **New Chapter on Numerical Methods** In the previous edition an assortment of topics appeared in the last chapter. That chapter has been replaced by a new chapter that focuses *exclusively* on numerical methods of linear algebra. We achieved this by moving those topics not concerned with numerical methods elsewhere in the text.

- **Singular-Value Decomposition** In recognition of its growing importance, a new section on *Singular-Value Decomposition* has been added to the chapter on numerical methods.

- **Internet Search and the Power Method** A new section on the *Power Method* and its application to Internet search engines has been added to the chapter on numerical methods.

Hallmark Features

- **Relationships Among Concepts** One of our main pedagogical goals is to convey to the student that linear algebra is a cohesive subject and not simply a collection of isolated definitions and techniques. One way in which we do this is by using a crescendo of *Equivalent Statements* theorems that continually revisit relationships among systems of equations, matrices, determinants, vectors, linear transformations, and eigenvalues. To get a general sense of how we use this technique see Theorems 1.5.3, 1.6.4, 2.2.8, 4.8.10, 4.10.4 and then Theorem 5.1.6, for example.

- **Smooth Transition to Abstraction** Because the transition from R^n to general vector spaces is difficult for many students, considerable effort is devoted to explaining the purpose of abstraction and helping the student to "visualize" abstract ideas by drawing analogies to familiar geometric ideas.

- **Mathematical Precision** When reasonable, we try to be mathematically precise. In keeping with the level of student audience, proofs are presented in a patient style that is tailored for beginners. There is a brief section in the Appendix on how to read proof statements, and there are various exercises in which students are guided through the steps of a proof and asked for justification.

- **Suitability for a Diverse Audience** This text is designed to serve the needs of students in engineering, computer science, biology, physics, business, and economics as well as those majoring in mathematics.

- **Historical Notes** To give the students a sense of mathematical history and to convey that real people created the mathematical theorems and equations they are studying, we have included numerous *Historical Notes* that put the topic being studied in historical perspective.

About the Exercises

- **Graded Exercise Sets** Each exercise set begins with routine drill problems and progresses to problems with more substance.

- **True/False Exercises** Most exercise sets end with a set of True/False exercises that are designed to check conceptual understanding and logical reasoning. To avoid pure guessing, the students are required to justify their responses in some way.

- **Supplementary Exercise Sets** Most chapters end with a set of supplementary exercises that tend to be more challenging and force the student to draw on ideas from the *entire* chapter rather than a specific section.

Supplementary Materials for Students

- **Technology Exercises and Data Files** The technology exercises that appeared in the previous edition have been moved to the Web site that accompanies this text. Those exercises are designed to be solved using MATLAB, *Mathematica*, or Maple and are accompanied by data files in all three formats. The exercises and data can be downloaded from the following Web site:

  www.wiley.com/go/global/anton

Supplementary Materials for Instructors

- **Instructor's Solutions Manual** This online supplement provides worked-out solutions to most exercises in the text.

- **WileyPLUS™** This is Wiley's proprietary online teaching and learning environment that integrates a digital version of this textbook with instructor and student resources to fit a variety of teaching and learning styles. WileyPLUS will help your students master concepts in a rich and structured environment that is available to them 24/7. It will also help you to personalize and manage your course more effectively with student assessments, assignments, grade tracking, and other useful tools.

  - Your students will receive timely access to resources that address their individual needs and will receive immediate feedback and remediation resources when needed.

- There are also self-assessment tools that are linked to the relevant portions of the text that will enable your students to take control of their own learning and practice.
- WileyPLUS will help you to identify those students who are falling behind and to intervene in a timely manner without waiting for scheduled office hours.

More information about WileyPLUS can be obtained from your Wiley representative.

A Guide for the Instructor

Although linear algebra courses vary widely in content and philosophy, most courses fall into two categories—those with about 35–40 lectures and those with about 25–30 lectures. Accordingly, we have created long and short templates as possible starting points for constructing a course outline. Of course, these are just guides, and you will certainly want to customize them to fit your local interests and requirements. Neither of these sample templates includes applications. Those can be added, if desired, as time permits.

	Long Template	Short Template
Chapter 1: Systems of Linear Equations and Matrices	7 lectures	6 lectures
Chapter 2: Determinants	3 lectures	2 lectures
Chapter 3: Euclidean Vector Spaces	4 lectures	3 lectures
Chapter 4: General Vector Spaces	10 lectures	10 lectures
Chapter 5: Eigenvalues and Eigenvectors	3 lectures	3 lectures
Chapter 6: Inner Product Spaces	3 lectures	1 lecture
Chapter 7: Diagonalization and Quadratic Forms	4 lectures	3 lectures
Chapter 8: Linear Transformations	3 lectures	2 lectures
Total:	**37 lectures**	**30 lectures**

An Applications-Oriented Course

Once the necessary core material is covered, the instructor can choose applications from the first nine chapters or from Chapter 10. The following table classifies each of the 20 sections in Chapter 10 according to difficulty:

Easy: The average student who has met the stated prerequisites should be able to read the material with no help from the instructor.

Moderate: The average student wo has met the stated prerequisites may require a little help from the instructor.

More Difficult: The average student who has met the stated prerequisites will probably need help from the instructor.

	1	2	3	4	5	6	7	8	9	10	11	12	13	14	15	16	17	18	19	20
EASY	•		•																	
MODERATE		•			•	•	•	•	•	•					•				•	•
MORE DIFFICULT				•									•	•		•	•	•		

Preface

Acknowledgements

We would like to express our appreciation to the following people whose helpful guidance has greatly improved the text:

Reviewers and Contributors

Don Allen, *Texas A&M University*
John Alongi, *Northwestern University*
John Beachy, *Northern Illinois University*
Przemslaw Bogacki, *Old Dominion University*
Robert Buchanan, *Millersville University of Pennsylvania*
Ralph Byers, *University of Kansas*
Evangelos A. Coutsias, *University of New Mexico*
Joshua Du, *Kennesaw State University*
Fatemeh Emdad, *Michigan Technological University*
Vincent Ervin, *Clemson University*
Anda Gadidov, *Kennesaw State University*
Guillermo Goldsztein, *Georgia Institute of Technology*
Tracy Hamilton, *California State University, Sacramento*
Amanda Hattway, *Wentworth Institute of Technology*
Heather Hulett, *University of Wisconsin–La Crosse*
David Hyeon, *Northern Illinois University*
Matt Insall, *Missouri University of Science and Technology*
Mic Jackson, *Earlham College*
Anton Kaul, *California Polytechnic Institute, San Luis Obispo*
Harihar Khanal, *Embry-Riddle University*
Hendrik Kuiper, *Arizona State University*
Kouok Law, *Georgia Perimeter College*
James McKinney, *California State University, Pomona*
Eric Schmutz, *Drexel University*
Qin Sheng, *Baylor University*
Adam Sikora, *State University of New York at Buffalo*
Allan Silberger, *Cleveland State University*
Dana Williams, *Dartmouth College*

Mathematical Advisors

Special thanks are due to a number of talented teachers and mathematicians who provided pedagogical guidance, provided help with answers and exercises, or provided detailed checking or proofreading:

John Alongi, *Northwestern University*
Scott Annin, *California State University, Fullerton*
Anton Kaul, *California Polytechnic State University*
Sarah Streett
Cindy Trimble, *C Trimble and Associates*
Brad Davis, *C Trimble and Associates*

The Wiley Support Team

David Dietz, Senior Acquisitions Editor
Jeff Benson, Assistant Editor
Pamela Lashbrook, Senior Editorial Assistant
Janet Foxman, Production Editor
Maddy Lesure, Senior Designer
Laurie Rosatone, Vice President and Publisher
Sarah Davis, Senior Marketing Manager
Diana Smith, Marketing Assistant
Melissa Edwards, Media Editor

Lisa Sabatini, Media Project Manager
Sheena Goldstein, Photo Editor
Carol Sawyer, Production Manager
Lilian Brady, Copyeditor

Special Contributions The talents and dedication of many individuals are required to produce a book such as this, and I am fortunate to have benefited from the expertise of the following people:

David Dietz – our editor, for his attention to detail, his sound judgment, his faith in us.

Jeff Benson – our assistant editor, who did an unbelievable job in organizing and coordinating the many threads required to make this edition a reality.

Carol Sawyer – of *The Perfect Proof*, who coordinated the myriad of details in the production process.

Dan Kirschenbaum – of *The Art of Arlene and Dan Kirschenbaum*, whose artistic and technical expertise resolved some difficult and critical illustration issues.

Bill Tuohy – who read parts of the manuscript and whose critical eye for detail had an important influence on the evolution of the text.

Pat Anton – who proofread manuscript, when needed.

Maddy Lesure – our text and cover designer, whose unerring sense of elegant design is apparent in the pages of this book.

Rena Lam – of *Techsetters, Inc.*, who did an absolutely amazing job of wading through a nightmare of author edits, scribbles, and last-minute changes to produce a beautiful book.

John Rogosich – of *Techsetters, Inc.*, who skillfully programmed the design elements of the book and resolved numerous thorny typesetting issues.

Lilian Brady – our copyeditor of many years, whose eye for typography and whose knowledge of language is amazing.

The Wiley Team – There are many other people at Wiley who worked behind the scenes and to whom I owe a debt of gratitude: Laurie Rosatone, Ann Berlin, Dorothy Sinclair, Janet Foxman, Sarah Davis, Harry Nolan, Sheena Goldstein, Melissa Edwards, and Norm Christiansen. Thanks to you all.

Wiley Publishers would also like to thank Dr. Heather Hulett, University of Wisconsin-La Crosse, for her contributions to the International Student Version.

CONTENTS

CHAPTER 1 Systems of Linear Equations and Matrices 1

1.1 Introduction to Systems of Linear Equations 2
1.2 Gaussian Elimination 11
1.3 Matrices and Matrix Operations 25
1.4 Inverses; Algebraic Properties of Matrices 38
1.5 Elementary Matrices and a Method for Finding A^{-1} 51
1.6 More on Linear Systems and Invertible Matrices 60
1.7 Diagonal, Triangular, and Symmetric Matrices 66
1.8 **Application:** Applications of Linear Systems 73
1.9 **Application:** Leontief Input-Output Models 85

CHAPTER 2 Determinants 93

2.1 Determinants by Cofactor Expansion 93
2.2 Evaluating Determinants by Row Reduction 100
2.3 Properties of Determinants; Cramer's Rule 106

CHAPTER 3 Euclidean Vector Spaces 119

3.1 Vectors in 2-Space, 3-Space, and n-Space 119
3.2 Norm, Dot Product, and Distance in R^n 130
3.3 Orthogonality 143
3.4 The Geometry of Linear Systems 152
3.5 Cross Product 161

CHAPTER 4 General Vector Spaces 171

4.1 Real Vector Spaces 171
4.2 Subspaces 179
4.3 Linear Independence 190
4.4 Coordinates and Basis 200
4.5 Dimension 209
4.6 Change of Basis 217
4.7 Row Space, Column Space, and Null Space 225
4.8 Rank, Nullity, and the Fundamental Matrix Spaces 237
4.9 Matrix Transformations from R^n to R^m 247
4.10 Properties of Matrix Transformations 263
4.11 **Application:** Geometry of Matrix Operators on R^2 273
4.12 **Application:** Dynamical Systems and Markov Chains 282

CHAPTER 5 Eigenvalues and Eigenvectors 295

5.1 Eigenvalues and Eigenvectors 295
5.2 Diagonalization 305

5.3 Complex Vector Spaces 315
5.4 **Application:** Differential Equations 327

CHAPTER 6 Inner Product Spaces 335

6.1 Inner Products 335
6.2 Angle and Orthogonality in Inner Product Spaces 345
6.3 Gram–Schmidt Process; QR-Decomposition 352
6.4 Best Approximation; Least Squares 366
6.5 **Application:** Least Squares Fitting to Data 376
6.6 **Application:** Function Approximation; Fourier Series 382

CHAPTER 7 Diagonalization and Quadratic Forms 389

7.1 Orthogonal Matrices 389
7.2 Orthogonal Diagonalization 397
7.3 Quadratic Forms 405
7.4 Optimization Using Quadratic Forms 417
7.5 Hermitian, Unitary, and Normal Matrices 424

CHAPTER 8 Linear Transformations 433

8.1 General Linear Transformations 433
8.2 Isomorphism 445
8.3 Compositions and Inverse Transformations 452
8.4 Matrices for General Linear Transformations 458
8.5 Similarity 468

CHAPTER 9 Numerical Methods 477

9.1 LU-Decompositions 477
9.2 The Power Method 487
9.3 **Application:** Internet Search Engines 496
9.4 Comparison of Procedures for Solving Linear Systems 501
9.5 Singular Value Decomposition 506
9.6 **Application:** Data Compression Using Singular Value Decomposition 514

CHAPTER 10 Applications of Linear Algebra 519

10.1 Constructing Curves and Surfaces Through Specified Points 520
10.2 Geometric Linear Programming 525
10.3 The Earliest Applications of Linear Algebra 536
10.4 Cubic Spline Interpolation 543
10.5 Markov Chains 553
10.6 Graph Theory 563
10.7 Games of Strategy 572
10.8 Leontief Economic Models 581
10.9 Forest Management 590

10.10 Computer Graphics 597
10.11 Equilibrium Temperature Distributions 605
10.12 Computed Tomography 615
10.13 Fractals 626
10.14 Chaos 641
10.15 Cryptography 654
10.16 Genetics 665
10.17 Age-Specific Population Growth 676
10.18 Harvesting of Animal Populations 686
10.19 A Least Squares Model for Human Hearing 693
10.20 Warps and Morphs 700

APPENDIX A How to Read Theorems 711

APPENDIX B Complex Numbers 713

Answers to Exercises 721

Index 765

CHAPTER 1

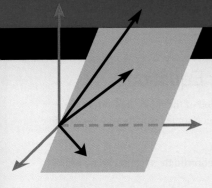

Systems of Linear Equations and Matrices

CHAPTER CONTENTS

1.1 Introduction to Systems of Linear Equations 2
1.2 Gaussian Elimination 11
1.3 Matrices and Matrix Operations 25
1.4 Inverses; Algebraic Properties of Matrices 38
1.5 Elementary Matrices and a Method for Finding A^{-1} 51
1.6 More on Linear Systems and Invertible Matrices 60
1.7 Diagonal, Triangular, and Symmetric Matrices 66
1.8 Applications of Linear Systems 73
  - Network Analysis (Traffic Flow) 73
  - Electrical Circuits 76
  - Balancing Chemical Equations 78
  - Polynomial Interpolation 80
1.9 Leontief Input-Output Models 85

INTRODUCTION Information in science, business, and mathematics is often organized into rows and columns to form rectangular arrays called "matrices" (plural of "matrix"). Matrices often appear as tables of numerical data that arise from physical observations, but they occur in various mathematical contexts as well. For example, we will see in this chapter that all of the information required to solve a system of equations such as

$$5x + y = 3$$
$$2x - y = 4$$

is embodied in the matrix

$$\begin{bmatrix} 5 & 1 & 3 \\ 2 & -1 & 4 \end{bmatrix}$$

and that the solution of the system can be obtained by performing appropriate operations on this matrix. This is particularly important in developing computer programs for solving systems of equations because computers are well suited for manipulating arrays of numerical information. However, matrices are not simply a notational tool for solving systems of equations; they can be viewed as mathematical objects in their own right, and there is a rich and important theory associated with them that has a multitude of practical applications. It is the study of matrices and related topics that forms the mathematical field that we call "linear algebra." In this chapter we will begin our study of matrices.

1.1 Introduction to Systems of Linear Equations

Systems of linear equations and their solutions constitute one of the major topics that we will study in this course. In this first section we will introduce some basic terminology and discuss a method for solving such systems.

Linear Equations Recall that in two dimensions a line in a rectangular xy-coordinate system can be represented by an equation of the form

$$ax + by = c \quad (a, b \text{ not both } 0)$$

and in three dimensions a plane in a rectangular xyz-coordinate system can be represented by an equation of the form

$$ax + by + cz = d \quad (a, b, c \text{ not all } 0)$$

These are examples of "linear equations," the first being a linear equation in the variables x and y and the second a linear equation in the variables x, y, and z. More generally, we define a **linear equation** in the n variables $x_1, x_2, \ldots, x_n$ to be one that can be expressed in the form

$$a_1 x_1 + a_2 x_2 + \cdots + a_n x_n = b \tag{1}$$

where $a_1, a_2, \ldots, a_n$ and b are constants, and the a's are not all zero. In the special cases where $n = 2$ or $n = 3$, we will often use variables without subscripts and write linear equations as

$$a_1 x + a_2 y = b \quad (a_1, a_2 \text{ not both } 0) \tag{2}$$

$$a_1 x + a_2 y + a_3 z = b \quad (a_1, a_2, a_3 \text{ not all } 0) \tag{3}$$

In the special case where $b = 0$, Equation (1) has the form

$$a_1 x_1 + a_2 x_2 + \cdots + a_n x_n = 0 \tag{4}$$

which is called a **homogeneous linear equation** in the variables $x_1, x_2, \ldots, x_n$.

▶ **EXAMPLE 1 Linear Equations**

Observe that a linear equation does not involve any products or roots of variables. All variables occur only to the first power and do not appear, for example, as arguments of trigonometric, logarithmic, or exponential functions. The following are linear equations:

$$x + 3y = 7 \qquad\qquad x_1 - 2x_2 - 3x_3 + x_4 = 0$$
$$\tfrac{1}{2}x - y + 3z = -1 \qquad\qquad x_1 + x_2 + \cdots + x_n = 1$$

The following are not linear equations:

$$x + 3y^2 = 4 \qquad\qquad 3x + 2y - xy = 5$$
$$\sin x + y = 0 \qquad\qquad \sqrt{x_1} + 2x_2 + x_3 = 1 \blacktriangleleft$$

A finite set of linear equations is called a **system of linear equations** or, more briefly, a **linear system**. The variables are called **unknowns**. For example, system (5) that follows has unknowns x and y, and system (6) has unknowns x_1, x_2, and x_3.

$$\begin{array}{ll} 5x + y = 3 & \quad 4x_1 - x_2 + 3x_3 = -1 \\ 2x - y = 4 & \quad 3x_1 + x_2 + 9x_3 = -4 \end{array} \tag{5–6}$$

A general linear system of m equations in the n unknowns $x_1, x_2, \ldots, x_n$ can be written as

$$\begin{aligned} a_{11}x_1 + a_{12}x_2 + \cdots + a_{1n}x_n &= b_1 \\ a_{21}x_1 + a_{22}x_2 + \cdots + a_{2n}x_n &= b_2 \\ \vdots \qquad \vdots \qquad\qquad \vdots \qquad &\;\; \vdots \\ a_{m1}x_1 + a_{m2}x_2 + \cdots + a_{mn}x_n &= b_m \end{aligned} \qquad (7)$$

> The double subscripting on the coefficients a_{ij} of the unknowns gives their location in the system—the first subscript indicates the equation in which the coefficient occurs, and the second indicates which unknown it multiplies. Thus, a_{12} is in the first equation and multiplies x_2.

A **solution** of a linear system in n unknowns $x_1, x_2, \ldots, x_n$ is a sequence of n numbers $s_1, s_2, \ldots, s_n$ for which the substitution

$$x_1 = s_1, \quad x_2 = s_2, \ldots, \quad x_n = s_n$$

makes each equation a true statement. For example, the system in (5) has the solution

$$x = 1, \quad y = -2$$

and the system in (6) has the solution

$$x_1 = 1, \quad x_2 = 2, \quad x_3 = -1$$

These solutions can be written more succinctly as

$$(1, -2) \quad \text{and} \quad (1, 2, -1)$$

in which the names of the variables are omitted. This notation allows us to interpret these solutions geometrically as points in two-dimensional and three-dimensional space. More generally, a solution

$$x_1 = s_1, \quad x_2 = s_2, \ldots, \quad x_n = s_n$$

of a linear system in n unknowns can be written as

$$(s_1, s_2, \ldots, s_n)$$

which is called an **ordered n-tuple**. With this notation it is understood that all variables appear in the same order in each equation. If $n = 2$, then the n-tuple is called an **ordered pair**, and if $n = 3$, then it is called an **ordered triple**.

Linear Systems with Two and Three Unknowns

Linear systems in two unknowns arise in connection with intersections of lines. For example, consider the linear system

$$\begin{aligned} a_1 x + b_1 y &= c_1 \\ a_2 x + b_2 y &= c_2 \end{aligned}$$

in which the graphs of the equations are lines in the xy-plane. Each solution (x, y) of this system corresponds to a point of intersection of the lines, so there are three possibilities (Figure 1.1.1):

1. The lines may be parallel and distinct, in which case there is no intersection and consequently no solution.
2. The lines may intersect at only one point, in which case the system has exactly one solution.
3. The lines may coincide, in which case there are infinitely many points of intersection (the points on the common line) and consequently infinitely many solutions.

In general, we say that a linear system is **consistent** if it has at least one solution and **inconsistent** if it has no solutions. Thus, a *consistent* linear system of two equations in

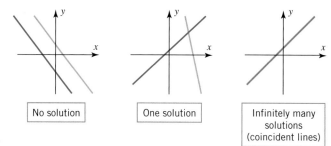

Figure 1.1.1

two unknowns has either one solution or infinitely many solutions—there are no other possibilities. The same is true for a linear system of three equations in three unknowns

$$a_1 x + b_1 y + c_1 z = d_1$$
$$a_2 x + b_2 y + c_2 z = d_2$$
$$a_3 x + b_3 y + c_3 z = d_3$$

in which the graphs of the equations are planes. The solutions of the system, if any, correspond to points where all three planes intersect, so again we see that there are only three possibilities—no solutions, one solution, or infinitely many solutions (Figure 1.1.2).

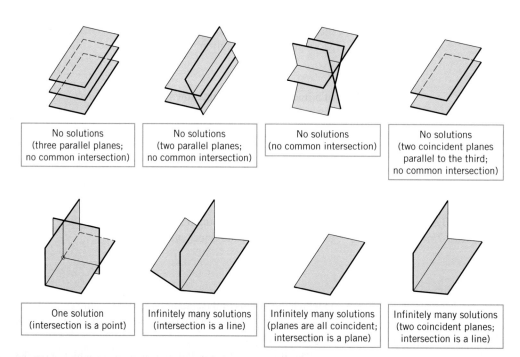

Figure 1.1.2

We will prove later that our observations about the number of solutions of linear systems of two equations in two unknowns and linear systems of three equations in three unknowns actually hold for *all* linear systems. That is:

Every system of linear equations has zero, one, or infinitely many solutions. There are no other possibilities.

1.1 Introduction to Systems of Linear Equations

▶ **EXAMPLE 2** **A Linear System with One Solution**

Solve the linear system
$$x - y = 1$$
$$2x + y = 6$$

Solution We can eliminate x from the second equation by adding -2 times the first equation to the second. This yields the simplified system
$$x - y = 1$$
$$3y = 4$$

From the second equation we obtain $y = \frac{4}{3}$, and on substituting this value in the first equation we obtain $x = 1 + y = \frac{7}{3}$. Thus, the system has the unique solution
$$x = \tfrac{7}{3}, \quad y = \tfrac{4}{3}$$

Geometrically, this means that the lines represented by the equations in the system intersect at the single point $\left(\tfrac{7}{3}, \tfrac{4}{3}\right)$. We leave it for you to check this by graphing the lines.

▶ **EXAMPLE 3** **A Linear System with No Solutions**

Solve the linear system
$$x + y = 4$$
$$3x + 3y = 6$$

Solution We can eliminate x from the second equation by adding -3 times the first equation to the second equation. This yields the simplified system
$$x + y = 4$$
$$0 = -6$$

The second equation is contradictory, so the given system has no solution. Geometrically, this means that the lines corresponding to the equations in the original system are parallel and distinct. We leave it for you to check this by graphing the lines or by showing that they have the same slope but different y-intercepts.

▶ **EXAMPLE 4** **A Linear System with Infinitely Many Solutions**

Solve the linear system
$$4x - 2y = 1$$
$$16x - 8y = 4$$

Solution We can eliminate x from the second equation by adding -4 times the first equation to the second. This yields the simplified system
$$4x - 2y = 1$$
$$0 = 0$$

The second equation does not impose any restrictions on x and y and hence can be omitted. Thus, the solutions of the system are those values of x and y that satisfy the single equation
$$4x - 2y = 1 \tag{8}$$

Geometrically, this means the lines corresponding to the two equations in the original system coincide. One way to describe the solution set is to solve this equation for x in terms of y to obtain $x = \tfrac{1}{4} + \tfrac{1}{2}y$ and then assign an arbitrary value t (called a ***parameter***)

In Example 4 we could have also obtained parametric equations for the solutions by solving (8) for y in terms of x, and letting $x = t$ be the parameter. The resulting parametric equations would look different but would define the same solution set.

to y. This allows us to express the solution by the pair of equations (called *parametric equations*)

$$x = \tfrac{1}{4} + \tfrac{1}{2}t, \quad y = t$$

We can obtain specific numerical solutions from these equations by substituting numerical values for the parameter. For example, $t = 0$ yields the solution $\left(\tfrac{1}{4}, 0\right)$, $t = 1$ yields the solution $\left(\tfrac{3}{4}, 1\right)$, and $t = -1$ yields the solution $\left(-\tfrac{1}{4}, -1\right)$. You can confirm that these are solutions by substituting the coordinates into the given equations.

▶ **EXAMPLE 5 A Linear System with Infinitely Many Solutions**

Solve the linear system

$$\begin{aligned} x - y + 2z &= 5 \\ 2x - 2y + 4z &= 10 \\ 3x - 3y + 6z &= 15 \end{aligned}$$

Solution This system can be solved by inspection, since the second and third equations are multiples of the first. Geometrically, this means that the three planes coincide and that those values of x, y, and z that satisfy the equation

$$x - y + 2z = 5 \qquad (9)$$

automatically satisfy all three equations. Thus, it suffices to find the solutions of (9). We can do this by first solving (9) for x in terms of y and z, then assigning arbitrary values r and s (parameters) to these two variables, and then expressing the solution by the three parametric equations

$$x = 5 + r - 2s, \quad y = r, \quad z = s$$

Specific solutions can be obtained by choosing numerical values for the parameters r and s. For example, taking $r = 1$ and $s = 0$ yields the solution $(6, 1, 0)$. ◀

Augmented Matrices and Elementary Row Operations

As the number of equations and unknowns in a linear system increases, so does the complexity of the algebra involved in finding solutions. The required computations can be made more manageable by simplifying notation and standardizing procedures. For example, by mentally keeping track of the location of the $+$'s, the x's, and the $=$'s in the linear system

$$\begin{aligned} a_{11}x_1 + a_{12}x_2 + \cdots + a_{1n}x_n &= b_1 \\ a_{21}x_1 + a_{22}x_2 + \cdots + a_{2n}x_n &= b_2 \\ \vdots \qquad \vdots \qquad \qquad \vdots \qquad &\vdots \\ a_{m1}x_1 + a_{m2}x_2 + \cdots + a_{mn}x_n &= b_m \end{aligned}$$

we can abbreviate the system by writing only the rectangular array of numbers

$$\begin{bmatrix} a_{11} & a_{12} & \cdots & a_{1n} & b_1 \\ a_{21} & a_{22} & \cdots & a_{2n} & b_2 \\ \vdots & \vdots & & \vdots & \vdots \\ a_{m1} & a_{m2} & \cdots & a_{mn} & b_m \end{bmatrix}$$

As noted in the introduction to this chapter, the term "matrix" is used in mathematics to denote a rectangular array of numbers. In a later section we will study matrices in detail, but for now we will only be concerned with augmented matrices for linear systems.

This is called the *augmented matrix* for the system. For example, the augmented matrix for the system of equations

$$\begin{aligned} x_1 + x_2 + 2x_3 &= 9 \\ 2x_1 + 4x_2 - 3x_3 &= 1 \\ 3x_1 + 6x_2 - 5x_3 &= 0 \end{aligned} \quad \text{is} \quad \begin{bmatrix} 1 & 1 & 2 & 9 \\ 2 & 4 & -3 & 1 \\ 3 & 6 & -5 & 0 \end{bmatrix}$$

The basic method for solving a linear system is to perform algebraic operations on the system that do not alter the solution set and that produce a succession of increasingly

1.1 Introduction to Systems of Linear Equations

simpler systems, until a point is reached where it can be ascertained whether the system is consistent, and if so, what its solutions are. Typically, the algebraic operations are as follows:

1. Multiply an equation through by a nonzero constant.
2. Interchange two equations.
3. Add a constant times one equation to another.

Since the rows (horizontal lines) of an augmented matrix correspond to the equations in the associated system, these three operations correspond to the following operations on the rows of the augmented matrix:

1. Multiply a row through by a nonzero constant.
2. Interchange two rows.
3. Add a constant times one row to another.

These are called ***elementary row operations*** on a matrix.

In the following example we will illustrate how to use elementary row operations and an augmented matrix to solve a linear system in three unknowns. Since a systematic procedure for solving linear systems will be developed in the next section, do not worry about how the steps in the example were chosen. Your objective here should be simply to understand the computations.

▶ **EXAMPLE 6** Using Elementary Row Operations

In the left column we solve a system of linear equations by operating on the equations in the system, and in the right column we solve the same system by operating on the rows of the augmented matrix.

$$\begin{aligned} x + y + 2z &= 9 \\ 2x + 4y - 3z &= 1 \\ 3x + 6y - 5z &= 0 \end{aligned} \qquad \begin{bmatrix} 1 & 1 & 2 & 9 \\ 2 & 4 & -3 & 1 \\ 3 & 6 & -5 & 0 \end{bmatrix}$$

Add -2 times the first equation to the second to obtain

Add -2 times the first row to the second to obtain

$$\begin{aligned} x + y + 2z &= 9 \\ 2y - 7z &= -17 \\ 3x + 6y - 5z &= 0 \end{aligned} \qquad \begin{bmatrix} 1 & 1 & 2 & 9 \\ 0 & 2 & -7 & -17 \\ 3 & 6 & -5 & 0 \end{bmatrix}$$

Maxime Bôcher
(1867–1918)

Historical Note The first known use of augmented matrices appeared between 200 B.C. and 100 B.C. in a Chinese manuscript entitled *Nine Chapters of Mathematical Art*. The coefficients were arranged in columns rather than in rows, as today, but remarkably the system was solved by performing a succession of operations on the columns. The actual use of the term *augmented matrix* appears to have been introduced by the American mathematician Maxime Bôcher in his book *Introduction to Higher Algebra*, published in 1907. In addition to being an outstanding research mathematician and an expert in Latin, chemistry, philosophy, zoology, geography, meteorology, art, and music, Bôcher was an outstanding expositor of mathematics whose elementary textbooks were greatly appreciated by students and are still in demand today.

[*Image: Courtesy of the American Mathematical Society*]

Chapter 1 Systems of Linear Equations and Matrices

Add -3 times the first equation to the third to obtain

$$x + y + 2z = 9$$
$$2y - 7z = -17$$
$$3y - 11z = -27$$

Add -3 times the first row to the third to obtain

$$\begin{bmatrix} 1 & 1 & 2 & 9 \\ 0 & 2 & -7 & -17 \\ 0 & 3 & -11 & -27 \end{bmatrix}$$

Multiply the second equation by $\frac{1}{2}$ to obtain

$$x + y + 2z = 9$$
$$y - \tfrac{7}{2}z = -\tfrac{17}{2}$$
$$3y - 11z = -27$$

Multiply the second row by $\frac{1}{2}$ to obtain

$$\begin{bmatrix} 1 & 1 & 2 & 9 \\ 0 & 1 & -\tfrac{7}{2} & -\tfrac{17}{2} \\ 0 & 3 & -11 & -27 \end{bmatrix}$$

Add -3 times the second equation to the third to obtain

$$x + y + 2z = 9$$
$$y - \tfrac{7}{2}z = -\tfrac{17}{2}$$
$$-\tfrac{1}{2}z = -\tfrac{3}{2}$$

Add -3 times the second row to the third to obtain

$$\begin{bmatrix} 1 & 1 & 2 & 9 \\ 0 & 1 & -\tfrac{7}{2} & -\tfrac{17}{2} \\ 0 & 0 & -\tfrac{1}{2} & -\tfrac{3}{2} \end{bmatrix}$$

Multiply the third equation by -2 to obtain

$$x + y + 2z = 9$$
$$y - \tfrac{7}{2}z = -\tfrac{17}{2}$$
$$z = 3$$

Multiply the third row by -2 to obtain

$$\begin{bmatrix} 1 & 1 & 2 & 9 \\ 0 & 1 & -\tfrac{7}{2} & -\tfrac{17}{2} \\ 0 & 0 & 1 & 3 \end{bmatrix}$$

Add -1 times the second equation to the first to obtain

$$x \quad + \tfrac{11}{2}z = \tfrac{35}{2}$$
$$y - \tfrac{7}{2}z = -\tfrac{17}{2}$$
$$z = 3$$

Add -1 times the second row to the first to obtain

$$\begin{bmatrix} 1 & 0 & \tfrac{11}{2} & \tfrac{35}{2} \\ 0 & 1 & -\tfrac{7}{2} & -\tfrac{17}{2} \\ 0 & 0 & 1 & 3 \end{bmatrix}$$

Add $-\tfrac{11}{2}$ times the third equation to the first and $\tfrac{7}{2}$ times the third equation to the second to obtain

$$x = 1$$
$$y = 2$$
$$z = 3$$

Add $-\tfrac{11}{2}$ times the third row to the first and $\tfrac{7}{2}$ times the third row to the second to obtain

$$\begin{bmatrix} 1 & 0 & 0 & 1 \\ 0 & 1 & 0 & 2 \\ 0 & 0 & 1 & 3 \end{bmatrix}$$

The solution $x = 1$, $y = 2$, $z = 3$ is now evident. ◀

Concept Review

- Linear equation
- Homogeneous linear equation
- System of linear equations
- Solution of a linear system
- Ordered n-tuple
- Consistent linear system
- Inconsistent linear system
- Parameter
- Parametric equations
- Augmented matrix
- Elementary row operations

Skills

- Determine whether a given equation is linear.
- Determine whether a given n-tuple is a solution of a linear system.
- Find the augmented matrix of a linear system.
- Find the linear system corresponding to a given augmented matrix.
- Perform elementary row operations on a linear system and on its corresponding augmented matrix.
- Determine whether a linear system is consistent or inconsistent.
- Find the set of solutions to a consistent linear system.

Exercise Set 1.1

1. In each part, determine whether the equation is linear in x_1, x_2, and x_3.

   (a) $x_1 + 5x_2 - \sqrt{2}x_3 = 1$ (b) $x_1 + 3x_2 + x_1x_3 = 2$

   (c) $x_1 = -7x_2 + 3x_3$ (d) $x_1^{-2} + x_2 + 8x_3 = 5$

   (e) $x_1^{3/5} - 2x_2 + x_3 = 4$

   (f) $\pi x_1 - \sqrt{2}x_2 + \frac{1}{3}x_3 = 7^{1/3}$

2. In each part, determine whether the equations form a linear system.

   (a) $-2x + 4y + z = 2$
   $3x - \dfrac{2}{y} = 0$

   (b) $x = 4$
   $2x = 8$

   (c) $4x - y + 2z = -1$
   $-x + (\ln 2)y - 3z = 0$

   (d) $3z + x = -4$
   $y + 5z = 1$
   $6x + 2z = 3$
   $-x - y - z = 4$

3. In each part, determine whether the equations form a linear system.

   (a) $x_1 - x_2 + x_3 = \cos(\pi)$
   $3x_1 - x_2 \quad\; x_3 = 2$

   (b) $\quad\quad 5y + w = 1$
   $2x + 5y - 4z + w = 1$

   (c) $7x_1 - \;\; x_2 + 2x_3 = \quad 0$
   $2x_1 + \;\; x_2 - x_3x_4 = \quad 3$
   $-x_1 + 5x_2 - \quad\; x_4 = -1$

   (d) $x_1 + x_2 = x_3 + x_4$

4. For each system in Exercise 2 that is linear, determine whether it is consistent.

5. For each system in Exercise 3 that is linear, determine whether it is consistent.

6. Write a system of linear equations consisting of three equations in three unknowns with

   (a) no solutions.

   (b) exactly one solution.

   (c) infinitely many solutions.

7. In each part, determine whether the given vector is a solution of the linear system

$$3x_1 + 2x_2 - 2x_3 = \;\;\, 1$$
$$2x_1 - \;\; x_2 + \;\; x_3 = \;\;\, 2$$
$$x_1 + 3x_2 - 3x_3 = -1$$

   (a) $(5, -4, 0)$ (b) $\left(\frac{5}{7}, \frac{-4}{7}, 0\right)$ (c) $(3, -2, 2)$

   (d) $\left(\frac{5}{7}, \frac{3}{7}, 1\right)$ (e) $(-3, 0, -5)$

8. In each part, determine whether the given vector is a solution of the linear system

$$x_1 + 2x_2 - 2x_3 = 3$$
$$3x_1 - \;\; x_2 + \;\; x_3 = 1$$
$$-x_1 + 5x_2 - 5x_3 = 5$$

   (a) $\left(\frac{5}{7}, \frac{8}{7}, 1\right)$ (b) $\left(\frac{5}{7}, \frac{8}{7}, 0\right)$ (c) $(5, 8, 1)$

   (d) $\left(\frac{5}{7}, \frac{10}{7}, \frac{2}{7}\right)$ (e) $\left(\frac{5}{7}, \frac{22}{7}, 2\right)$

9. In each part, find the solution set of the linear equation by using parameters as necessary.

   (a) $2x + 4y = 3$

   (b) $3x_1 - 5x_2 + x_3 + 4x_4 = 9$

10. In each part, find the solution set of the linear equation by using parameters as necessary.

    (a) $3x_1 - 5x_2 + 4x_3 = 7$

    (b) $3v - 8w + 2x - y + 4z = 0$

11. In each part, find a system of linear equations corresponding to the given augmented matrix.

(a) $\begin{bmatrix} 2 & 5 & 6 \\ 0 & 1 & 2 \\ -1 & 0 & 0 \end{bmatrix}$ (b) $\begin{bmatrix} 3 & 0 & -2 & 5 \\ 7 & 1 & 4 & -3 \\ 0 & -2 & 1 & 7 \end{bmatrix}$

(c) $\begin{bmatrix} 1 & 5 & 7 & -1 & 3 \\ 2 & 2 & 1 & 1 & 0 \end{bmatrix}$

(d) $\begin{bmatrix} 1 & 0 & 0 & 0 & 7 \\ 0 & 1 & 0 & 0 & -2 \\ 0 & 0 & 1 & 0 & 3 \\ 0 & 0 & 0 & 1 & 4 \end{bmatrix}$

12. In each part, find a system of linear equations corresponding to the given augmented matrix.

(a) $\begin{bmatrix} 2 & -1 \\ -4 & -6 \\ 1 & -1 \\ 3 & 0 \end{bmatrix}$ (b) $\begin{bmatrix} 0 & 3 & -1 & -1 & -1 \\ 5 & 2 & 0 & -3 & -6 \end{bmatrix}$

(c) $\begin{bmatrix} 1 & 2 & 3 & 4 \\ -4 & -3 & -2 & -1 \\ 5 & -6 & 1 & 1 \\ -8 & 0 & 0 & 3 \end{bmatrix}$

(d) $\begin{bmatrix} 3 & 0 & 1 & -4 & 3 \\ -4 & 0 & 4 & 1 & -3 \\ -1 & 3 & 0 & -2 & -9 \\ 0 & 0 & 0 & -1 & -2 \end{bmatrix}$

13. In each part, find the augmented matrix for the given system of linear equations.

(a) $-2x_1 = 6$
$3x_1 = 8$
$9x_1 = -3$

(b) $3x_1 \quad - x_3 + 6x_4 = 0$
$2x_2 - x_3 - 5x_4 = -2$

(c) $2x_2 \quad - 3x_4 + x_5 = 0$
$-3x_1 - x_2 + x_3 \quad = -1$
$6x_1 + 2x_2 - x_3 + 2x_4 - 3x_5 = 6$

(d) $x_1 - x_3 = 4$
$x_2 + x_4 = 9$

14. In each part, find the augmented matrix for the given system of linear equations.

(a) $3x_1 - 2x_2 = -1$
$4x_1 + 5x_2 = 3$
$7x_1 + 3x_2 = 2$

(b) $2x_1 \quad + 2x_3 = 1$
$3x_1 - x_2 + 4x_3 = 7$
$6x_1 + x_2 - x_3 = 0$

(c) $x_1 + 2x_2 \quad - x_4 + x_5 = 1$
$3x_2 + x_3 \quad - x_5 = 2$
$x_3 + 7x_4 \quad = 1$

(d) $x_1 \quad = 1$
$x_2 \quad = 2$
$x_3 = 3$

15. The curve $y = ax^2 + bx + c$ shown in the accompanying figure passes through the points (x_1, y_1), (x_2, y_2), and (x_3, y_3).

Show that the coefficients a, b, and c are a solution of the system of linear equations whose augmented matrix is

$$\begin{bmatrix} x_1^2 & x_1 & 1 & y_1 \\ x_2^2 & x_2 & 1 & y_2 \\ x_3^2 & x_3 & 1 & y_3 \end{bmatrix}$$

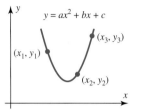

◀ Figure Ex-15

16. Explain why each of the three elementary row operations does not affect the solution set of a linear system.

17. Show that if the linear equations

$$x_1 + kx_2 = c \quad \text{and} \quad x_1 + lx_2 = d$$

have the same solution set, then the two equations are identical (i.e., $k = l$ and $c = d$).

True-False Exercises

In parts (a)–(h) determine whether the statement is true or false, and justify your answer.

(a) A linear system whose equations are all homogeneous must have a unique solution.

(b) Multiplying a linear equation through by zero is an acceptable elementary row operation.

(c) The linear system

$$x - y = 3$$
$$2x - 2y = k$$

cannot have a unique solution, regardless of the value of k.

(d) A single linear equation with two or more unknowns must always have infinitely many solutions.

(e) If the number of unknowns in a linear system exceeds the number of equations, then the system must be consistent.

(f) If each equation in a consistent linear system is multiplied through by a constant c, then all solutions to the new system can be obtained by multiplying solutions from the original system by c.

(g) Elementary row operations permit one equation in a linear system to be subtracted from another.

(h) The linear system with corresponding augmented matrix

$$\begin{bmatrix} 2 & -1 & 4 \\ 0 & 0 & -1 \end{bmatrix}$$

is consistent.

1.2 Gaussian Elimination

In this section we will develop a systematic procedure for solving systems of linear equations. The procedure is based on the idea of performing certain operations on the rows of the augmented matrix that simplifies it to a form from which the solution of the system can be ascertained by inspection.

Considerations in Solving Linear Systems

When considering methods for solving systems of linear equations, it is important to distinguish between large systems that must be solved by computer and small systems that can be solved by hand. For example, there are many applications that lead to linear systems in thousands or even millions of unknowns. Large systems require special techniques to deal with issues of memory size, roundoff errors, solution time, and so forth. Such techniques are studied in the field of **numerical analysis** and will only be touched on in this text. However, almost all of the methods that are used for large systems are based on the ideas that we will develop in this section.

Echelon Forms

In Example 6 of the last section, we solved a linear system in the unknowns x, y, and z by reducing the augmented matrix to the form

$$\begin{bmatrix} 1 & 0 & 0 & 1 \\ 0 & 1 & 0 & 2 \\ 0 & 0 & 1 & 3 \end{bmatrix}$$

from which the solution $x = 1$, $y = 2$, $z = 3$ became evident. This is an example of a matrix that is in **reduced row echelon form**. To be of this form, a matrix must have the following properties:

1. If a row does not consist entirely of zeros, then the first nonzero number in the row is a 1. We call this a **leading 1**.
2. If there are any rows that consist entirely of zeros, then they are grouped together at the bottom of the matrix.
3. In any two successive rows that do not consist entirely of zeros, the leading 1 in the lower row occurs farther to the right than the leading 1 in the higher row.
4. Each column that contains a leading 1 has zeros everywhere else in that column.

A matrix that has the first three properties is said to be in **row echelon form**. (Thus, a matrix in reduced row echelon form is of necessity in row echelon form, but not conversely.)

▶ **EXAMPLE 1 Row Echelon and Reduced Row Echelon Form**

The following matrices are in reduced row echelon form.

$$\begin{bmatrix} 1 & 0 & 0 & 4 \\ 0 & 1 & 0 & 7 \\ 0 & 0 & 1 & -1 \end{bmatrix}, \begin{bmatrix} 1 & 0 & 0 \\ 0 & 1 & 0 \\ 0 & 0 & 1 \end{bmatrix}, \begin{bmatrix} 0 & 1 & -2 & 0 & 1 \\ 0 & 0 & 0 & 1 & 3 \\ 0 & 0 & 0 & 0 & 0 \\ 0 & 0 & 0 & 0 & 0 \end{bmatrix}, \begin{bmatrix} 0 & 0 \\ 0 & 0 \end{bmatrix}$$

The following matrices are in row echelon form but not reduced row echelon form.

$$\begin{bmatrix} 1 & 4 & -3 & 7 \\ 0 & 1 & 6 & 2 \\ 0 & 0 & 1 & 5 \end{bmatrix}, \begin{bmatrix} 1 & 1 & 0 \\ 0 & 1 & 0 \\ 0 & 0 & 0 \end{bmatrix}, \begin{bmatrix} 0 & 1 & 2 & 6 & 0 \\ 0 & 0 & 1 & -1 & 0 \\ 0 & 0 & 0 & 0 & 1 \end{bmatrix}$$

EXAMPLE 2 More on Row Echelon and Reduced Row Echelon Form

As Example 1 illustrates, a matrix in row echelon form has zeros below each leading 1, whereas a matrix in reduced row echelon form has zeros below *and above* each leading 1. Thus, with any real numbers substituted for the *'s, all matrices of the following types are in row echelon form:

$$\begin{bmatrix} 1 & * & * & * \\ 0 & 1 & * & * \\ 0 & 0 & 1 & * \\ 0 & 0 & 0 & 1 \end{bmatrix}, \begin{bmatrix} 1 & * & * & * \\ 0 & 1 & * & * \\ 0 & 0 & 1 & * \\ 0 & 0 & 0 & 0 \end{bmatrix}, \begin{bmatrix} 1 & * & * & * \\ 0 & 1 & * & * \\ 0 & 0 & 0 & 0 \\ 0 & 0 & 0 & 0 \end{bmatrix}, \begin{bmatrix} 0 & 1 & * & * & * & * & * & * & * \\ 0 & 0 & 0 & 1 & * & * & * & * & * \\ 0 & 0 & 0 & 0 & 1 & * & * & * & * \\ 0 & 0 & 0 & 0 & 0 & 1 & * & * & * \\ 0 & 0 & 0 & 0 & 0 & 0 & 0 & 1 & * \end{bmatrix}$$

All matrices of the following types are in reduced row echelon form:

$$\begin{bmatrix} 1 & 0 & 0 & 0 \\ 0 & 1 & 0 & 0 \\ 0 & 0 & 1 & 0 \\ 0 & 0 & 0 & 1 \end{bmatrix}, \begin{bmatrix} 1 & 0 & 0 & * \\ 0 & 1 & 0 & * \\ 0 & 0 & 1 & * \\ 0 & 0 & 0 & 0 \end{bmatrix}, \begin{bmatrix} 1 & 0 & * & * \\ 0 & 1 & * & * \\ 0 & 0 & 0 & 0 \\ 0 & 0 & 0 & 0 \end{bmatrix}, \begin{bmatrix} 0 & 1 & * & 0 & 0 & 0 & * & * & 0 & * \\ 0 & 0 & 0 & 1 & 0 & 0 & * & * & 0 & * \\ 0 & 0 & 0 & 0 & 1 & 0 & * & * & 0 & * \\ 0 & 0 & 0 & 0 & 0 & 1 & * & * & 0 & * \\ 0 & 0 & 0 & 0 & 0 & 0 & 0 & 0 & 1 & * \end{bmatrix} \blacktriangleleft$$

If, by a sequence of elementary row operations, the augmented matrix for a system of linear equations is put in *reduced* row echelon form, then the solution set can be obtained either by inspection or by converting certain linear equations to parametric form. Here are some examples.

► EXAMPLE 3 Unique Solution

Suppose that the augmented matrix for a linear system in the unknowns x_1, x_2, x_3, and x_4 has been reduced by elementary row operations to

$$\begin{bmatrix} 1 & 0 & 0 & 0 & 3 \\ 0 & 1 & 0 & 0 & -1 \\ 0 & 0 & 1 & 0 & 0 \\ 0 & 0 & 0 & 1 & 5 \end{bmatrix}$$

This matrix is in reduced row echelon form and corresponds to the equations

$$\begin{aligned} x_1 &= 3 \\ x_2 &= -1 \\ x_3 &= 0 \\ x_4 &= 5 \end{aligned}$$

Thus, the system has a unique solution, namely, $x_1 = 3$, $x_2 = -1$, $x_3 = 0$, $x_4 = 5$.

> In Example 3 we could, if desired, express the solution more succinctly as the 4-tuple $(3, -1, 0, 5)$.

► EXAMPLE 4 Linear Systems in Three Unknowns

In each part, suppose that the augmented matrix for a linear system in the unknowns x, y, and z has been reduced by elementary row operations to the given reduced row echelon form. Solve the system.

(a) $\begin{bmatrix} 1 & 0 & 0 & 0 \\ 0 & 1 & 2 & 0 \\ 0 & 0 & 0 & 1 \end{bmatrix}$ (b) $\begin{bmatrix} 1 & 0 & 3 & -1 \\ 0 & 1 & -4 & 2 \\ 0 & 0 & 0 & 0 \end{bmatrix}$ (c) $\begin{bmatrix} 1 & -5 & 1 & 4 \\ 0 & 0 & 0 & 0 \\ 0 & 0 & 0 & 0 \end{bmatrix}$

Solution (a) The equation that corresponds to the last row of the augmented matrix is

$$0x + 0y + 0z = 1$$

Since this equation is not satisfied by any values of x, y, and z, the system is inconsistent.

Solution (b) The equation that corresponds to the last row of the augmented matrix is

$$0x + 0y + 0z = 0$$

This equation can be omitted since it imposes no restrictions on x, y, and z; hence, the linear system corresponding to the augmented matrix is

$$\begin{aligned} x \phantom{{}+y} + 3z &= -1 \\ y - 4z &= 2 \end{aligned}$$

Since x and y correspond to the leading 1's in the augmented matrix, we call these the ***leading variables***. The remaining variables (in this case z) are called ***free variables***. Solving for the leading variables in terms of the free variables gives

$$\begin{aligned} x &= -1 - 3z \\ y &= 2 + 4z \end{aligned}$$

From these equations we see that the free variable z can be treated as a parameter and assigned an arbitrary value t, which then determines values for x and y. Thus, the solution set can be represented by the parametric equations

$$x = -1 - 3t, \quad y = 2 + 4t, \quad z = t$$

By substituting various values for t in these equations we can obtain various solutions of the system. For example, setting $t = 0$ yields the solution

$$x = -1, \quad y = 2, \quad z = 0$$

and setting $t = 1$ yields the solution

$$x = -4, \quad y = 6, \quad z = 1$$

Solution (c) As explained in part (b), we can omit the equations corresponding to the zero rows, in which case the linear system associated with the augmented matrix consists of the single equation

$$x - 5y + z = 4 \tag{1}$$

from which we see that the solution set is a plane in three-dimensional space. Although (1) is a valid form of the solution set, there are many applications in which it is preferable to express the solution set in parametric form. We can convert (1) to parametric form by solving for the leading variable x in terms of the free variables y and z to obtain

$$x = 4 + 5y - z$$

From this equation we see that the free variables can be assigned arbitrary values, say $y = s$ and $z = t$, which then determine the value of x. Thus, the solution set can be expressed parametrically as

$$x = 4 + 5s - t, \quad y = s, \quad z = t \blacktriangleleft \tag{2}$$

Formulas, such as (2), that express the solution set of a linear system parametrically have some associated terminology.

We will usually denote parameters in a general solution by the letters $r, s, t, \ldots$, but any letters that do not conflict with the names of the unknowns can be used. For systems with more than three unknowns, subscripted letters such as $t_1, t_2, t_3, \ldots$ are convenient.

DEFINITION 1 If a linear system has infinitely many solutions, then a set of parametric equations from which all solutions can be obtained by assigning numerical values to the parameters is called a ***general solution*** of the system.

Chapter 1 Systems of Linear Equations and Matrices

Elimination Methods

We have just seen how easy it is to solve a system of linear equations once its augmented matrix is in reduced row echelon form. Now we will give a step-by-step *elimination procedure* that can be used to reduce any matrix to reduced row echelon form. As we state each step in the procedure, we illustrate the idea by reducing the following matrix to reduced row echelon form.

$$\begin{bmatrix} 0 & 0 & -2 & 0 & 7 & 12 \\ 2 & 4 & -10 & 6 & 12 & 28 \\ 2 & 4 & -5 & 6 & -5 & -1 \end{bmatrix}$$

Step 1. Locate the leftmost column that does not consist entirely of zeros.

$$\begin{bmatrix} 0 & 0 & -2 & 0 & 7 & 12 \\ 2 & 4 & -10 & 6 & 12 & 28 \\ 2 & 4 & -5 & 6 & -5 & -1 \end{bmatrix}$$
↑
└── Leftmost nonzero column

Step 2. Interchange the top row with another row, if necessary, to bring a nonzero entry to the top of the column found in Step 1.

$$\begin{bmatrix} 2 & 4 & -10 & 6 & 12 & 28 \\ 0 & 0 & -2 & 0 & 7 & 12 \\ 2 & 4 & -5 & 6 & -5 & -1 \end{bmatrix}$$ ⟵ The first and second rows in the preceding matrix were interchanged.

Step 3. If the entry that is now at the top of the column found in Step 1 is a, multiply the first row by $1/a$ in order to introduce a leading 1.

$$\begin{bmatrix} 1 & 2 & -5 & 3 & 6 & 14 \\ 0 & 0 & -2 & 0 & 7 & 12 \\ 2 & 4 & -5 & 6 & -5 & -1 \end{bmatrix}$$ ⟵ The first row of the preceding matrix was multiplied by $\frac{1}{2}$.

Step 4. Add suitable multiples of the top row to the rows below so that all entries below the leading 1 become zeros.

$$\begin{bmatrix} 1 & 2 & -5 & 3 & 6 & 14 \\ 0 & 0 & -2 & 0 & 7 & 12 \\ 0 & 0 & 5 & 0 & -17 & -29 \end{bmatrix}$$ ⟵ -2 times the first row of the preceding matrix was added to the third row.

Step 5. Now cover the top row in the matrix and begin again with Step 1 applied to the submatrix that remains. Continue in this way until the *entire* matrix is in row echelon form.

$$\begin{bmatrix} 1 & 2 & -5 & 3 & 6 & 14 \\ 0 & 0 & -2 & 0 & 7 & 12 \\ 0 & 0 & 5 & 0 & -17 & -29 \end{bmatrix}$$
↑
└── Leftmost nonzero column
in the submatrix

$$\begin{bmatrix} 1 & 2 & -5 & 3 & 6 & 14 \\ 0 & 0 & 1 & 0 & -\frac{7}{2} & -6 \\ 0 & 0 & 5 & 0 & -17 & -29 \end{bmatrix}$$ ⟵ The first row in the submatrix was multiplied by $-\frac{1}{2}$ to introduce a leading 1.

1.2 Gaussian Elimination

$$\begin{bmatrix} 1 & 2 & -5 & 3 & 6 & 14 \\ 0 & 0 & 1 & 0 & -\frac{7}{2} & -6 \\ 0 & 0 & 0 & 0 & \frac{1}{2} & 1 \end{bmatrix}$$ ← −5 times the first row of the submatrix was added to the second row of the submatrix to introduce a zero below the leading 1.

$$\begin{bmatrix} 1 & 2 & -5 & 3 & 6 & 14 \\ 0 & 0 & 1 & 0 & -\frac{7}{2} & -6 \\ 0 & 0 & 0 & 0 & \frac{1}{2} & 1 \end{bmatrix}$$ ← The top row in the submatrix was covered, and we returned again to Step 1.

↑
Leftmost nonzero column in the new submatrix

$$\begin{bmatrix} 1 & 2 & -5 & 3 & 6 & 14 \\ 0 & 0 & 1 & 0 & -\frac{7}{2} & -6 \\ 0 & 0 & 0 & 0 & 1 & 2 \end{bmatrix}$$ ← The first (and only) row in the new submatrix was multiplied by 2 to introduce a leading 1.

The *entire* matrix is now in row echelon form. To find the reduced row echelon form we need the following additional step.

Step 6. Beginning with the last nonzero row and working upward, add suitable multiples of each row to the rows above to introduce zeros above the leading 1's.

$$\begin{bmatrix} 1 & 2 & -5 & 3 & 6 & 14 \\ 0 & 0 & 1 & 0 & 0 & 1 \\ 0 & 0 & 0 & 0 & 1 & 2 \end{bmatrix}$$ ← $\frac{7}{2}$ times the third row of the preceding matrix was added to the second row.

$$\begin{bmatrix} 1 & 2 & -5 & 3 & 0 & 2 \\ 0 & 0 & 1 & 0 & 0 & 1 \\ 0 & 0 & 0 & 0 & 1 & 2 \end{bmatrix}$$ ← −6 times the third row was added to the first row.

$$\begin{bmatrix} 1 & 2 & 0 & 3 & 0 & 7 \\ 0 & 0 & 1 & 0 & 0 & 1 \\ 0 & 0 & 0 & 0 & 1 & 2 \end{bmatrix}$$ ← 5 times the second row was added to the first row.

The last matrix is in reduced row echelon form.

The procedure (or algorithm) we have just described for reducing a matrix to reduced row echelon form is called ***Gauss–Jordan elimination***. This algorithm consists of two parts, a ***forward phase*** in which zeros are introduced below the leading 1's and a ***backward phase*** in which zeros are introduced above the leading 1's. If only the forward phase

Carl Friedrich Gauss (1777–1855)

Wilhelm Jordan (1842–1899)

Historical Note Although versions of Gaussian elimination were known much earlier, the power of the method was not recognized until the great German mathematician Carl Friedrich Gauss used it to compute the orbit of the asteroid Ceres from limited data. What happened was this: On January 1, 1801 the Sicilian astronomer Giuseppe Piazzi (1746–1826) noticed a dim celestial object that he believed might be a "missing planet." He named the object Ceres and made a limited number of positional observations but then lost the object as it neared the Sun. Gauss undertook the problem of computing the orbit from the limited data and the procedure that we now call Gaussian elimination. The work of Gauss caused a sensation when Ceres reappeared a year later in the constellation Virgo at almost the precise position that Gauss predicted! The method was further popularized by the German engineer Wilhelm Jordan in his handbook on geodesy (the science of measuring Earth shapes) entitled *Handbuch der Vermessungskunde* and published in 1888.
[*Images: Granger Collection (Gauss); wikipedia (Jordan)*]

is used, then the procedure produces a row echelon form only and is called *Gaussian elimination*. For example, in the preceding computations a row echelon form was obtained at the end of Step 5.

▶ **EXAMPLE 5** Gauss–Jordan Elimination

Solve by Gauss–Jordan elimination.

$$\begin{aligned} x_1 + 3x_2 - 2x_3 + 2x_5 &= 0 \\ 2x_1 + 6x_2 - 5x_3 - 2x_4 + 4x_5 - 3x_6 &= -1 \\ 5x_3 + 10x_4 + 15x_6 &= 5 \\ 2x_1 + 6x_2 + 8x_4 + 4x_5 + 18x_6 &= 6 \end{aligned}$$

Solution The augmented matrix for the system is

$$\begin{bmatrix} 1 & 3 & -2 & 0 & 2 & 0 & 0 \\ 2 & 6 & -5 & -2 & 4 & -3 & -1 \\ 0 & 0 & 5 & 10 & 0 & 15 & 5 \\ 2 & 6 & 0 & 8 & 4 & 18 & 6 \end{bmatrix}$$

Adding -2 times the first row to the second and fourth rows gives

$$\begin{bmatrix} 1 & 3 & -2 & 0 & 2 & 0 & 0 \\ 0 & 0 & -1 & -2 & 0 & -3 & -1 \\ 0 & 0 & 5 & 10 & 0 & 15 & 5 \\ 0 & 0 & 4 & 8 & 0 & 18 & 6 \end{bmatrix}$$

Multiplying the second row by -1 and then adding -5 times the new second row to the third row and -4 times the new second row to the fourth row gives

$$\begin{bmatrix} 1 & 3 & -2 & 0 & 2 & 0 & 0 \\ 0 & 0 & 1 & 2 & 0 & 3 & 1 \\ 0 & 0 & 0 & 0 & 0 & 0 & 0 \\ 0 & 0 & 0 & 0 & 0 & 6 & 2 \end{bmatrix}$$

Interchanging the third and fourth rows and then multiplying the third row of the resulting matrix by $\frac{1}{6}$ gives the row echelon form

$$\begin{bmatrix} 1 & 3 & -2 & 0 & 2 & 0 & 0 \\ 0 & 0 & 1 & 2 & 0 & 3 & 1 \\ 0 & 0 & 0 & 0 & 0 & 1 & \frac{1}{3} \\ 0 & 0 & 0 & 0 & 0 & 0 & 0 \end{bmatrix}$$ This completes the forward phase since there are zeros below the leading 1's.

Adding -3 times the third row to the second row and then adding 2 times the second row of the resulting matrix to the first row yields the reduced row echelon form

$$\begin{bmatrix} 1 & 3 & 0 & 4 & 2 & 0 & 0 \\ 0 & 0 & 1 & 2 & 0 & 0 & 0 \\ 0 & 0 & 0 & 0 & 0 & 1 & \frac{1}{3} \\ 0 & 0 & 0 & 0 & 0 & 0 & 0 \end{bmatrix}$$ This completes the backward phase since there are zeros above the leading 1's.

> Note that in constructing the linear system in (3) we ignored the row of zeros in the corresponding augmented matrix. Why is this justified?

The corresponding system of equations is

$$\begin{aligned} x_1 + 3x_2 + 4x_4 + 2x_5 &= 0 \\ x_3 + 2x_4 &= 0 \\ x_6 &= \tfrac{1}{3} \end{aligned} \qquad (3)$$

Solving for the leading variables we obtain

$$\begin{aligned} x_1 &= -3x_2 - 4x_4 - 2x_5 \\ x_3 &= -2x_4 \\ x_6 &= \tfrac{1}{3} \end{aligned}$$

Finally, we express the general solution of the system parametrically by assigning the free variables x_2, x_4, and x_5 arbitrary values r, s, and t, respectively. This yields

$$x_1 = -3r - 4s - 2t, \quad x_2 = r, \quad x_3 = -2s, \quad x_4 = s, \quad x_5 = t, \quad x_6 = \tfrac{1}{3} \blacktriangleleft$$

Homogeneous Linear Systems

A system of linear equations is said to be **homogeneous** if the constant terms are all zero; that is, the system has the form

$$\begin{aligned} a_{11}x_1 + a_{12}x_2 + \cdots + a_{1n}x_n &= 0 \\ a_{21}x_1 + a_{22}x_2 + \cdots + a_{2n}x_n &= 0 \\ &\vdots \\ a_{m1}x_1 + a_{m2}x_2 + \cdots + a_{mn}x_n &= 0 \end{aligned}$$

Every homogeneous system of linear equations is consistent because all such systems have $x_1 = 0$, $x_2 = 0$, ..., $x_n = 0$ as a solution. This solution is called the ***trivial solution***; if there are other solutions, they are called ***nontrivial solutions***.

Because a homogeneous linear system always has the trivial solution, there are only two possibilities for its solutions:

- The system has only the trivial solution.
- The system has infinitely many solutions in addition to the trivial solution.

In the special case of a homogeneous linear system of two equations in two unknowns, say

$$\begin{aligned} a_1 x + b_1 y &= 0 \quad (a_1, b_1 \text{ not both zero}) \\ a_2 x + b_2 y &= 0 \quad (a_2, b_2 \text{ not both zero}) \end{aligned}$$

the graphs of the equations are lines through the origin, and the trivial solution corresponds to the point of intersection at the origin (Figure 1.2.1).

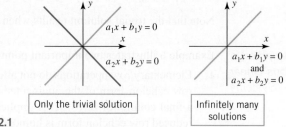

▶ Figure 1.2.1

There is one case in which a homogeneous system is assured of having nontrivial solutions—namely, whenever the system involves more unknowns than equations. To see why, consider the following example of four equations in six unknowns.

▶ EXAMPLE 6 A Homogeneous System

Use Gauss–Jordan elimination to solve the homogeneous linear system

$$\begin{aligned} x_1 + 3x_2 - 2x_3 \quad\quad\quad + 2x_5 \quad\quad\quad &= 0 \\ 2x_1 + 6x_2 - 5x_3 - 2x_4 + 4x_5 - 3x_6 &= 0 \\ 5x_3 + 10x_4 \quad\quad\quad + 15x_6 &= 0 \\ 2x_1 + 6x_2 \quad\quad\quad + 8x_4 + 4x_5 + 18x_6 &= 0 \end{aligned} \quad (4)$$

Solution Observe first that the coefficients of the unknowns in this system are the same as those in Example 5; that is, the two systems differ only in the constants on the right side. The augmented matrix for the given homogeneous system is

$$\begin{bmatrix} 1 & 3 & -2 & 0 & 2 & 0 & 0 \\ 2 & 6 & -5 & -2 & 4 & -3 & 0 \\ 0 & 0 & 5 & 10 & 0 & 15 & 0 \\ 2 & 6 & 0 & 8 & 4 & 18 & 0 \end{bmatrix} \quad (5)$$

which is the same as the augmented matrix for the system in Example 5, except for zeros in the last column. Thus, the reduced row echelon form of this matrix will be the same as that of the augmented matrix in Example 5, except for the last column. However, a moment's reflection will make it evident that a column of zeros is not changed by an elementary row operation, so the reduced row echelon form of (5) is

$$\begin{bmatrix} 1 & 3 & 0 & 4 & 2 & 0 & 0 \\ 0 & 0 & 1 & 2 & 0 & 0 & 0 \\ 0 & 0 & 0 & 0 & 0 & 1 & 0 \\ 0 & 0 & 0 & 0 & 0 & 0 & 0 \end{bmatrix} \quad (6)$$

The corresponding system of equations is

$$\begin{aligned} x_1 + 3x_2 \quad\quad + 4x_4 + 2x_5 \quad\quad &= 0 \\ x_3 + 2x_4 \quad\quad\quad &= 0 \\ x_6 &= 0 \end{aligned}$$

Solving for the leading variables we obtain

$$\begin{aligned} x_1 &= -3x_2 - 4x_4 - 2x_5 \\ x_3 &= -2x_4 \\ x_6 &= 0 \end{aligned} \quad (7)$$

If we now assign the free variables x_2, x_4, and x_5 arbitrary values r, s, and t, respectively, then we can express the solution set parametrically as

$$x_1 = -3r - 4s - 2t, \quad x_2 = r, \quad x_3 = -2s, \quad x_4 = s, \quad x_5 = t, \quad x_6 = 0$$

Note that the trivial solution results when $r = s = t = 0$. ◀

Free Variables in Homogeneous Linear Systems

Example 6 illustrates two important points about solving homogeneous linear systems:

1. Elementary row operations do not alter columns of zeros in a matrix, so the reduced row echelon form of the augmented matrix for a homogeneous linear system has a final column of zeros. This implies that the linear system corresponding to the reduced row echelon form is homogeneous, just like the original system.

2. When we constructed the homogeneous linear system corresponding to augmented matrix (6), we ignored the row of zeros because the corresponding equation

$$0x_1 + 0x_2 + 0x_3 + 0x_4 + 0x_5 + 0x_6 = 0$$

does not impose any conditions on the unknowns. Thus, depending on whether or not the reduced row echelon form of the augmented matrix for a homogeneous linear system has any rows of zero, the linear system corresponding to that reduced row echelon form will either have the same number of equations as the original system or it will have fewer.

Now consider a general homogeneous linear system with n unknowns, and suppose that the reduced row echelon form of the augmented matrix has r nonzero rows. Since each nonzero row has a leading 1, and since each leading 1 corresponds to a leading variable, the homogeneous system corresponding to the reduced row echelon form of the augmented matrix must have r leading variables and $n - r$ free variables. Thus, this system is of the form

$$
\begin{aligned}
x_{k_1} + \sum(\;) = 0 \\
x_{k_2} + \sum(\;) = 0 \\
\ddots \vdots \\
x_{k_r} + \sum(\;) = 0
\end{aligned}
\tag{8}
$$

where in each equation the expression $\sum(\;)$ denotes a sum that involves the free variables, if any [see (7), for example]. In summary, we have the following result.

> **THEOREM 1.2.1** *Free Variable Theorem for Homogeneous Systems*
>
> *If a homogeneous linear system has n unknowns, and if the reduced row echelon form of its augmented matrix has r nonzero rows, then the system has $n - r$ free variables.*

Note that Theorem 1.2.2 applies only to homogeneous systems—a *nonhomogeneous* system with more unknowns than equations need not be consistent. However, we will prove later that if a nonhomogeneous system with more unknowns then equations is consistent, then it has infinitely many solutions.

Theorem 1.2.1 has an important implication for homogeneous linear systems with more unknowns than equations. Specifically, if a homogeneous linear system has m equations in n unknowns, and if $m < n$, then it must also be true that $r < n$ (why?). This being the case, the theorem implies that there is at least one free variable, and this implies that the system has infinitely many solutions. Thus, we have the following result.

> **THEOREM 1.2.2** *A homogeneous linear system with more unknowns than equations has infinitely many solutions.*

In retrospect, we could have anticipated that the homogeneous system in Example 6 would have infinitely many solutions since it has four equations in six unknowns.

Gaussian Elimination and Back-Substitution

For small linear systems that are solved by hand (such as most of those in this text), Gauss–Jordan elimination (reduction to reduced row echelon form) is a good procedure to use. However, for large linear systems that require a computer solution, it is generally more efficient to use Gaussian elimination (reduction to row echelon form) followed by a technique known as **back-substitution** to complete the process of solving the system. The next example illustrates this technique.

▶ **EXAMPLE 7** *Example 5 Solved by Back-Substitution*

From the computations in Example 5, a row echelon form of the augmented matrix is

$$
\begin{bmatrix}
1 & 3 & -2 & 0 & 2 & 0 & 0 \\
0 & 0 & 1 & 2 & 0 & 3 & 1 \\
0 & 0 & 0 & 0 & 0 & 1 & \frac{1}{3} \\
0 & 0 & 0 & 0 & 0 & 0 & 0
\end{bmatrix}
$$

To solve the corresponding system of equations

$$x_1 + 3x_2 - 2x_3 \quad\quad + 2x_5 \quad\quad = 0$$
$$x_3 + 2x_4 \quad\quad + 3x_6 = 1$$
$$x_6 = \tfrac{1}{3}$$

we proceed as follows:

Step 1. Solve the equations for the leading variables.

$$x_1 = -3x_2 + 2x_3 - 2x_5$$
$$x_3 = 1 - 2x_4 - 3x_6$$
$$x_6 = \tfrac{1}{3}$$

Step 2. Beginning with the bottom equation and working upward, successively substitute each equation into all the equations above it.

Substituting $x_6 = \tfrac{1}{3}$ into the second equation yield

$$x_1 = -3x_2 + 2x_3 - 2x_5$$
$$x_3 = -2x_4$$
$$x_6 = \tfrac{1}{3}$$

Substituting $x_3 = -2x_4$ into the first equation yields

$$x_1 = -3x_2 - 4x_4 - 2x_5$$
$$x_3 = -2x_4$$
$$x_6 = \tfrac{1}{3}$$

Step 3. Assign arbitrary values to the free variables, if any.

If we now assign x_2, x_4, and x_5 the arbitrary values r, s, and t, respectively, the general solution is given by the formulas

$$x_1 = -3r - 4s - 2t, \quad x_2 = r, \quad x_3 = -2s, \quad x_4 = s, \quad x_5 = t, \quad x_6 = \tfrac{1}{3}$$

This agrees with the solution obtained in Example 5.

▶ **EXAMPLE 8**

Suppose that the matrices below are augmented matrices for linear systems in the unknowns x_1, x_2, x_3, and x_4. These matrices are all in row echelon form but not reduced row echelon form. Discuss the existence and uniqueness of solutions to the corresponding linear systems

(a) $\begin{bmatrix} 1 & -3 & 7 & 2 & 5 \\ 0 & 1 & 2 & -4 & 1 \\ 0 & 0 & 1 & 6 & 9 \\ 0 & 0 & 0 & 0 & 1 \end{bmatrix}$
(b) $\begin{bmatrix} 1 & -3 & 7 & 2 & 5 \\ 0 & 1 & 2 & -4 & 1 \\ 0 & 0 & 1 & 6 & 9 \\ 0 & 0 & 0 & 0 & 0 \end{bmatrix}$
(c) $\begin{bmatrix} 1 & -3 & 7 & 2 & 5 \\ 0 & 1 & 2 & -4 & 1 \\ 0 & 0 & 1 & 6 & 9 \\ 0 & 0 & 0 & 1 & 0 \end{bmatrix}$

Solution (a) The last row corresponds to the equation

$$0x_1 + 0x_2 + 0x_3 + 0x_4 = 1$$

from which it is evident that the system is inconsistent.

Solution (b) The last row corresponds to the equation

$$0x_1 + 0x_2 + 0x_3 + 0x_4 = 0$$

which has no effect on the solution set. In the remaining three equations the variables x_1, x_2, and x_3 correspond to leading 1's and hence are leading variables. The variable x_4

is a free variable. With a little algebra, the leading variables can be expressed in terms of the free variable, and the free variable can be assigned an arbitrary value. Thus, the system must have infinitely many solutions.

Solution (c) The last row corresponds to the equation

$$x_4 = 0$$

which gives us a numerical value for x_4. If we substitute this value into the third equation, namely,

$$x_3 + 6x_4 = 9$$

we obtain $x_3 = 9$. You should now be able to see that if we continue this process and substitute the known values of x_3 and x_4 into the equation corresponding to the second row, we will obtain a unique numerical value for x_2; and if, finally, we substitute the known values of x_4, x_3, and x_2 into the equation corresponding to the first row, we will produce a unique numerical value for x_1. Thus, the system has a unique solution. ◀

Some Facts About Echelon Forms

There are three facts about row echelon forms and reduced row echelon forms that are important to know but we will not prove:

1. Every matrix has a unique reduced row echelon form; that is, regardless of whether you use Gauss–Jordan elimination or some other sequence of elementary row operations, the same reduced row echelon form will result in the end.[*]

2. Row echelon forms are not unique; that is, different sequences of elementary row operations can result in different row echelon forms.

3. Although row echelon forms are not unique, all row echelon forms of a matrix A have the same number of zero rows, and the leading 1's always occur in the same positions in the row echelon forms of A. Those are called the **pivot positions** of A. A column that contains a pivot position is called a **pivot column** of A.

▶ **EXAMPLE 9** **Pivot Positions and Columns**

Earlier in this section (immediately after Definition 1) we found a row echelon form of

$$A = \begin{bmatrix} 0 & 0 & -2 & 0 & 7 & 12 \\ 2 & 4 & -10 & 6 & 12 & 28 \\ 2 & 4 & -5 & 6 & -5 & -1 \end{bmatrix}$$

to be

$$\begin{bmatrix} 1 & 2 & -5 & 3 & 6 & 14 \\ 0 & 0 & 1 & 0 & -\frac{7}{2} & -6 \\ 0 & 0 & 0 & 0 & 1 & 2 \end{bmatrix}$$

The leading 1's occur in positions (row 1, column 1), (row 2, column 3), and (row 3, column 5). These are the pivot positions. The pivot columns are columns 1, 3, and 5. ◀

Roundoff Error and Instability

There is often a gap between mathematical theory and its practical implementation—Gauss–Jordan elimination and Gaussian elimination being good examples. The problem is that computers generally approximate numbers, thereby introducing **roundoff** errors,

---

[*]A proof of this result can be found in the article "The Reduced Row Echelon Form of a Matrix Is Unique: A Simple Proof," by Thomas Yuster, *Mathematics Magazine*, Vol. 57, No. 2, 1984, pp. 93–94.

so unless precautions are taken, successive calculations may degrade an answer to a degree that makes it useless. Algorithms (procedures) in which this happens are called ***unstable***. There are various techniques for minimizing roundoff error and instability. For example, it can be shown that for large linear systems Gauss–Jordan elimination involves roughly 50% more operations than Gaussian elimination, so most computer algorithms are based on the latter method. Some of these matters will be considered in Chapter 9.

Concept Review

- Reduced row echelon form
- Row echelon form
- Leading 1
- Leading variables
- Free variables
- General solution to a linear system
- Gaussian elimination
- Gauss–Jordan elimination
- Forward phase
- Backward phase
- Homogeneous linear system
- Trivial solution
- Nontrivial solution
- Dimension Theorem for Homogeneous Systems
- Back-substitution

Skills

- Recognize whether a given matrix is in row echelon form, reduced row echelon form, or neither.
- Construct solutions to linear systems whose corresponding augmented matrices that are in row echelon form or reduced row echelon form.
- Use Gaussian elimination to find the general solution of a linear system.
- Use Gauss–Jordan elimination to find the general solution of a linear system.
- Analyze homogeneous linear systems using the Free Variable Theorem for Homogeneous Systems.

Exercise Set 1.2

1. In each part, determine whether the matrix is in row echelon form, reduced row echelon form, both, or neither.

(a) $\begin{bmatrix} 1 & 0 & 0 \\ 0 & 1 & 0 \\ 0 & 0 & 1 \end{bmatrix}$
(b) $\begin{bmatrix} 1 & 0 & 0 \\ 0 & 1 & 0 \\ 0 & 0 & 0 \end{bmatrix}$
(c) $\begin{bmatrix} 0 & 1 & 0 \\ 0 & 0 & 1 \\ 0 & 0 & 0 \end{bmatrix}$
(d) $\begin{bmatrix} 1 & 2 & 3 & 1 \\ 0 & 0 & 0 & 0 \\ 0 & 0 & 0 & 1 \end{bmatrix}$
(e) $\begin{bmatrix} 1 & -2 & 2 & 0 \\ 0 & 0 & 0 & 1 \end{bmatrix}$

(d) $\begin{bmatrix} 1 & 0 & 3 & 1 \\ 0 & 1 & 2 & 4 \end{bmatrix}$
(e) $\begin{bmatrix} 1 & 2 & 0 & 3 & 0 \\ 0 & 0 & 1 & 1 & 0 \\ 0 & 0 & 0 & 0 & 1 \\ 0 & 0 & 0 & 0 & 0 \end{bmatrix}$
(f) $\begin{bmatrix} 2 & 0 \\ 0 & 1 \end{bmatrix}$
(g) $\begin{bmatrix} 1 & 2 & 4 & 0 & 1 \\ 0 & 0 & 0 & 1 & 2 \end{bmatrix}$

(f) $\begin{bmatrix} 0 & 0 \\ 0 & 0 \\ 0 & 0 \end{bmatrix}$
(g) $\begin{bmatrix} 1 & -7 & 5 & 5 \\ 0 & 1 & 3 & 2 \end{bmatrix}$

3. In each part, suppose that the augmented matrix for a system of linear equations has been reduced by row operations to the given row echelon form. Solve the system.

(a) $\begin{bmatrix} 1 & -3 & 4 & 7 \\ 0 & 1 & 2 & 2 \\ 0 & 0 & 1 & 5 \end{bmatrix}$

(b) $\begin{bmatrix} 1 & 0 & 8 & -5 & 6 \\ 0 & 1 & 4 & -9 & 3 \\ 0 & 0 & 1 & 1 & 2 \end{bmatrix}$

2. In each part, determine whether the matrix is in row echelon form, reduced row echelon form, both, or neither.

(a) $\begin{bmatrix} 1 & 1 & 2 \\ 0 & 1 & 1 \\ 0 & 0 & 1 \end{bmatrix}$
(b) $\begin{bmatrix} 0 & 0 & 1 \\ 0 & 1 & 1 \\ 1 & 0 & 1 \end{bmatrix}$
(c) $\begin{bmatrix} 1 & 0 & 0 \\ 0 & 0 & 1 \\ 0 & 0 & 0 \end{bmatrix}$

(c) $\begin{bmatrix} 1 & 7 & -2 & 0 & -8 & -3 \\ 0 & 0 & 1 & 1 & 6 & 5 \\ 0 & 0 & 0 & 1 & 3 & 9 \\ 0 & 0 & 0 & 0 & 0 & 0 \end{bmatrix}$

(d) $\begin{bmatrix} 1 & -3 & 7 & 1 \\ 0 & 1 & 4 & 0 \\ 0 & 0 & 0 & 1 \end{bmatrix}$

4. In each part, suppose that the augmented matrix for a system of linear equations has been reduced by row operations to the given row echelon form. Solve the system.

(a) $\begin{bmatrix} 1 & 0 & 0 & 0 \\ 0 & 1 & 0 & -2 \\ 0 & 0 & 1 & 4 \end{bmatrix}$

(b) $\begin{bmatrix} 1 & 0 & 0 & 3 & 2 \\ 0 & 1 & 0 & 1 & 0 \\ 0 & 0 & 1 & -4 & 1 \end{bmatrix}$

(c) $\begin{bmatrix} 1 & 0 & 0 & 0 & 2 & -2 \\ 0 & 1 & -2 & 0 & 0 & 1 \\ 0 & 0 & 0 & 1 & 7 & 0 \\ 0 & 0 & 0 & 0 & 0 & 0 \end{bmatrix}$

(d) $\begin{bmatrix} 1 & 2 & 0 & 0 \\ 0 & 0 & 1 & 0 \\ 0 & 0 & 0 & 1 \end{bmatrix}$

▶ In Exercises 5–8, solve the linear system by Gauss–Jordan elimination. ◀

5. $x_1 + 2x_2 - 3x_3 = 6$
$2x_1 - x_2 + 4x_3 = 1$
$x_1 - x_2 + x_3 = 3$

6. $2x_1 + 2x_2 + 2x_3 = 0$
$-2x_1 + 5x_2 + 2x_3 = 1$
$8x_1 + x_2 + 4x_3 = -1$

7. $3x - y + z + 7w = 13$
$-2x + y - z - 3w = -9$
$-2x + y \quad\;\; - 7w = -8$

8. $\quad\;\; - 2b + 3c = 1$
$3a + 6b - 3c = -2$
$6a + 6b + 3c = 5$

▶ In Exercises 9–12, solve the linear system by Gaussian elimination. ◀

9. Exercise 5
10. Exercise 6
11. Exercise 7
12. Exercise 8

▶ In Exercises 13–16, determine whether the homogeneous system has nontrivial solutions by inspection (without pencil and paper). ◀

13. $2x_1 - 3x_2 + 4x_3 - x_4 = 0$
$7x_1 + x_2 - 8x_3 + 9x_4 = 0$
$2x_1 + 8x_2 + x_3 - x_4 = 0$

14. $x_1 + 3x_2 - x_3 = 0$
$x_2 - 8x_3 = 0$
$4x_3 = 0$

15. $a_{11}x_1 + a_{12}x_2 + a_{13}x_3 = 0$
$a_{21}x_1 + a_{22}x_2 + a_{23}x_3 = 0$

16. $3x_1 - 2x_2 = 0$
$6x_1 - 4x_2 = 0$

▶ In Exercises 17–24, solve the given linear system by any method. ◀

17. $2x + y + 4z = 0$
$3x + y + 6z = 0$
$4x + y + 9z = 0$

18. $2x - y - 3z = 0$
$-x + 2y - 3z = 0$
$x + y + 4z = 0$

19. $x_1 - x_2 + 7x_3 + x_4 = 0$
$x_1 + 2x_2 - 6x_3 - x_4 = 0$

20. $\quad\;\; v + 3w - 2x = 0$
$2u + v - 4w + 3x = 0$
$2u + 3v + 2w - x = 0$
$-4u - 3v + 5w - 4x = 0$

21. $\quad\;\; 2x + 2y + 4z = 0$
$w \quad\;\; - y - 3z = 0$
$2w + 3x + y + z = 0$
$-2w + x + 3y - 2z = 0$

22. $x_1 + 3x_2 \quad\;\; + x_4 = 0$
$x_1 + 4x_2 + 2x_3 \quad\;\; = 0$
$\quad\;\; - 2x_2 - 2x_3 - x_4 = 0$
$2x_1 - 4x_2 + x_3 + x_4 = 0$
$x_1 - 2x_2 - x_3 + x_4 = 0$

23. $2I_1 - I_2 + 3I_3 + 4I_4 = 9$
$I_1 \quad\;\; - 2I_3 + 7I_4 = 11$
$3I_1 - 3I_2 + I_3 + 5I_4 = 8$
$2I_1 + I_2 + 4I_3 + 4I_4 = 10$

24. $\quad\;\; Z_3 + Z_4 + Z_5 = 0$
$-Z_1 - Z_2 + 2Z_3 - 3Z_4 + Z_5 = 0$
$Z_1 + Z_2 - 2Z_3 \quad\;\; - Z_5 = 0$
$2Z_1 + 2Z_2 - Z_3 \quad\;\; + Z_5 = 0$

▶ In Exercises 25–28, determine the values of a for which the system has no solutions, exactly one solution, or infinitely many solutions. ◀

25. $x + 2y + z = 2$
$2x - 2y + 3z = 1$
$x + 2y - az = a$

26. $x + 2y + z = 2$
$2x - 2y + 3z = 1$
$x + 2y - (a^2 - 3)z = a$

27. $x + 2y - 3z = 4$
$3x - y + 5z = 2$
$4x + y + (a^2 - 2)z = a + 4$

28. $x + y + 7z = -7$
$2x + 3y + 17z = -16$
$x + 2y + (a^2 + 1)z = 3a$

In Exercises 29–30, solve the following systems, where a, b, and c are constants.

29. $2x + y = a$
$3x + 6y = b$

30. $x_1 + x_2 + x_3 = a$
$2x_1 + 2x_3 = b$
$ 3x_2 + 3x_3 = c$

31. Find two different row echelon forms of

$$\begin{bmatrix} 1 & 3 \\ 2 & 7 \end{bmatrix}$$

This exercise shows that a matrix can have multiple row echelon forms.

32. Reduce

$$\begin{bmatrix} 2 & 1 & 3 \\ 0 & -2 & -29 \\ 3 & 4 & 5 \end{bmatrix}$$

to reduced row echelon form without introducing fractions at any intermediate stage.

33. Show that the following nonlinear system has 18 solutions if $0 \le \alpha \le 2\pi$, $0 \le \beta \le 2\pi$, and $0 \le \gamma \le 2\pi$.

$$\sin\alpha + 2\cos\beta + 3\tan\gamma = 0$$
$$2\sin\alpha + 5\cos\beta + 3\tan\gamma = 0$$
$$-\sin\alpha - 5\cos\beta + 5\tan\gamma = 0$$

[*Hint:* Begin by making the substitutions $x = \sin\alpha$, $y = \cos\beta$, and $z = \tan\gamma$.]

34. Solve the following system of nonlinear equations for the unknown angles α, β, and γ, where $0 \le \alpha \le 2\pi$, $0 \le \beta \le 2\pi$, and $0 \le \gamma < \pi$.

$$2\sin\alpha - \cos\beta + 3\tan\gamma = 3$$
$$4\sin\alpha + 2\cos\beta - 2\tan\gamma = 2$$
$$6\sin\alpha - 3\cos\beta + \tan\gamma = 9$$

35. Solve the following system of nonlinear equations for x, y, and z.

$$x^2 + y^2 + z^2 = 6$$
$$x^2 - y^2 + 2z^2 = 2$$
$$2x^2 + y^2 - z^2 = 3$$

[*Hint:* Begin by making the substitutions $X = x^2$, $Y = y^2$, $Z = z^2$.]

36. Solve the following system for x, y, and z.

$$\frac{1}{x} + \frac{2}{y} - \frac{4}{z} = 1$$
$$\frac{2}{x} + \frac{3}{y} + \frac{8}{z} = 0$$
$$-\frac{1}{x} + \frac{9}{y} + \frac{10}{z} = 5$$

37. Find the coefficients a, b, c, and d so that the curve shown in the accompanying figure is the graph of the equation $y = ax^3 + bx^2 + cx + d$.

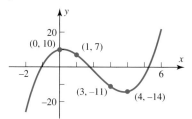

◀ Figure Ex-37

38. Find the coefficients a, b, c, and d so that the curve shown in the accompanying figure is given by the equation $ax^2 + ay^2 + bx + cy + d = 0$.

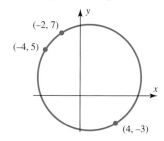

◀ Figure Ex-38

39. If the linear system

$$a_1x + b_1y + c_1z = 0$$
$$a_2x - b_2y + c_2z = 0$$
$$a_3x + b_3y - c_3z = 0$$

has only the trivial solution, what can be said about the solutions of the following system?

$$a_1x + b_1y + c_1z = 3$$
$$a_2x - b_2y + c_2z = 7$$
$$a_3x + b_3y - c_3z = 11$$

40. (a) If A is a 3×5 matrix, then what is the maximum possible number of leading 1's in its reduced row echelon form?

(b) If B is a 3×6 matrix whose last column has all zeros, then what is the maximum possible number of parameters in the general solution of the linear system with augmented matrix B?

(c) If C is a 5×3 matrix, then what is the minimum possible number of rows of zeros in any row echelon form of C?

41. (a) Prove that if $ad - bc \ne 0$, then the reduced row echelon form of

$$\begin{bmatrix} a & b \\ c & d \end{bmatrix} \text{ is } \begin{bmatrix} 1 & 0 \\ 0 & 1 \end{bmatrix}$$

(b) Use the result in part (a) to prove that if $ad - bc \ne 0$, then the linear system

$$ax + by = k$$
$$cx + dy = l$$

has exactly one solution.

42. Consider the system of equations
$$ax + by = 0$$
$$cx + dy = 0$$
$$ex + fy = 0$$
Discuss the relative positions of the lines $ax + by = 0$, $cx + dy = 0$, and $ex + fy = 0$ when (a) the system has only the trivial solution, and (b) the system has nontrivial solutions.

43. Describe all possible reduced row echelon forms of

(a) $\begin{bmatrix} a & b & c \\ d & e & f \\ g & h & i \end{bmatrix}$ (b) $\begin{bmatrix} a & b & c & d \\ e & f & g & h \\ i & j & k & l \\ m & n & p & q \end{bmatrix}$

True-False Exercises

In parts (a)–(i) determine whether the statement is true or false, and justify your answer.

(a) If a matrix is in reduced row echelon form, then it is also in row echelon form.

(b) If an elementary row operation is applied to a matrix that is in row echelon form, the resulting matrix will still be in row echelon form.

(c) Every matrix has a unique row echelon form.

(d) A homogeneous linear system in n unknowns whose corresponding augmented matrix has a reduced row echelon form with r leading 1's has $n - r$ free variables.

(e) All leading 1's in a matrix in row echelon form must occur in different columns.

(f) If every column of a matrix in row echelon form has a leading 1 then all entries that are not leading 1's are zero.

(g) If a homogeneous linear system of n equations in n unknowns has a corresponding augmented matrix with a reduced row echelon form containing n leading 1's, then the linear system has only the trivial solution.

(h) If the reduced row echelon form of the augmented matrix for a linear system has a row of zeros, then the system must have infinitely many solutions.

(i) If a linear system has more unknowns than equations, then it must have infinitely many solutions.

1.3 Matrices and Matrix Operations

Rectangular arrays of real numbers arise in contexts other than as augmented matrices for linear systems. In this section we will begin to study matrices as objects in their own right by defining operations of addition, subtraction, and multiplication on them.

Matrix Notation and Terminology

In Section 1.2 we used rectangular arrays of numbers, called *augmented matrices*, to abbreviate systems of linear equations. However, rectangular arrays of numbers occur in other contexts as well. For example, the following rectangular array with three rows and seven columns might describe the number of hours that a student spent studying three subjects during a certain week:

	Mon.	Tues.	Wed.	Thurs.	Fri.	Sat.	Sun.
Math	2	3	2	4	1	4	2
History	0	3	1	4	3	2	2
Language	4	1	3	1	0	0	2

If we suppress the headings, then we are left with the following rectangular array of numbers with three rows and seven columns, called a "matrix":

$$\begin{bmatrix} 2 & 3 & 2 & 4 & 1 & 4 & 2 \\ 0 & 3 & 1 & 4 & 3 & 2 & 2 \\ 4 & 1 & 3 & 1 & 0 & 0 & 2 \end{bmatrix}$$

More generally, we make the following definition.

26 Chapter 1 Systems of Linear Equations and Matrices

> **DEFINITION 1** A *matrix* is a rectangular array of numbers. The numbers in the array are called the *entries* in the matrix.

▶ **EXAMPLE 1 Examples of Matrices**

Some examples of matrices are

$$\begin{bmatrix} 1 & 2 \\ 3 & 0 \\ -1 & 4 \end{bmatrix}, \quad [2 \ 1 \ 0 \ -3], \quad \begin{bmatrix} e & \pi & -\sqrt{2} \\ 0 & \frac{1}{2} & 1 \\ 0 & 0 & 0 \end{bmatrix}, \quad \begin{bmatrix} 1 \\ 3 \end{bmatrix}, \quad [4] \quad ◀$$

A matrix with only one column is called a *column vector* or a *column matrix*, and a matrix with only one row is called a *row vector* or a *row matrix*. In Example 1, the 2×1 matrix is a column vector, the 1×4 matrix is a row vector, and the 1×1 matrix is both a row vector and a column vector.

The *size* of a matrix is described in terms of the number of rows (horizontal lines) and columns (vertical lines) it contains. For example, the first matrix in Example 1 has three rows and two columns, so its size is 3 by 2 (written 3×2). In a size description, the first number always denotes the number of rows, and the second denotes the number of columns. The remaining matrices in Example 1 have sizes 1×4, 3×3, 2×1, and 1×1, respectively.

We will use capital letters to denote matrices and lowercase letters to denote numerical quantities; thus we might write

$$A = \begin{bmatrix} 2 & 1 & 7 \\ 3 & 4 & 2 \end{bmatrix} \quad \text{or} \quad C = \begin{bmatrix} a & b & c \\ d & e & f \end{bmatrix}$$

When discussing matrices, it is common to refer to numerical quantities as *scalars*. Unless stated otherwise, *scalars will be real numbers*; complex scalars will be considered later in the text.

The entry that occurs in row i and column j of a matrix A will be denoted by a_{ij}. Thus a general 3×4 matrix might be written as

$$A = \begin{bmatrix} a_{11} & a_{12} & a_{13} & a_{14} \\ a_{21} & a_{22} & a_{23} & a_{24} \\ a_{31} & a_{32} & a_{33} & a_{34} \end{bmatrix}$$

Matrix brackets are often omitted from 1×1 matrices, making it impossible to tell, for example, whether the symbol 4 denotes the number "four" or the matrix [4]. This rarely causes problems because it is usually possible to tell which is meant from the context.

and a general $m \times n$ matrix as

$$A = \begin{bmatrix} a_{11} & a_{12} & \cdots & a_{1n} \\ a_{21} & a_{22} & \cdots & a_{2n} \\ \vdots & \vdots & & \vdots \\ a_{m1} & a_{m2} & \cdots & a_{mn} \end{bmatrix} \quad (1)$$

When a compact notation is desired, the preceding matrix can be written as

$$[a_{ij}]_{m \times n} \quad \text{or} \quad [a_{ij}]$$

the first notation being used when it is important in the discussion to know the size, and the second when the size need not be emphasized. Usually, we will match the letter denoting a matrix with the letter denoting its entries; thus, for a matrix B we would generally use b_{ij} for the entry in row i and column j, and for a matrix C we would use the notation c_{ij}.

The entry in row i and column j of a matrix A is also commonly denoted by the symbol $(A)_{ij}$. Thus, for matrix (1) above, we have

$$(A)_{ij} = a_{ij}$$

and for the matrix
$$A = \begin{bmatrix} 2 & -3 \\ 7 & 0 \end{bmatrix}$$
we have $(A)_{11} = 2$, $(A)_{12} = -3$, $(A)_{21} = 7$, and $(A)_{22} = 0$.

Row and column vectors are of special importance, and it is common practice to denote them by boldface lowercase letters rather than capital letters. For such matrices, double subscripting of the entries is unnecessary. Thus a general $1 \times n$ row vector **a** and a general $m \times 1$ column vector **b** would be written as

$$\mathbf{a} = \begin{bmatrix} a_1 & a_2 & \cdots & a_n \end{bmatrix} \quad \text{and} \quad \mathbf{b} = \begin{bmatrix} b_1 \\ b_2 \\ \vdots \\ b_m \end{bmatrix}$$

A matrix A with n rows and n columns is called a **square matrix of order n**, and the shaded entries $a_{11}, a_{22}, \ldots, a_{nn}$ in (2) are said to be on the **main diagonal** of A.

$$\begin{bmatrix} a_{11} & a_{12} & \cdots & a_{1n} \\ a_{21} & a_{22} & \cdots & a_{2n} \\ \vdots & \vdots & & \vdots \\ a_{n1} & a_{n2} & \cdots & a_{nn} \end{bmatrix} \qquad (2)$$

Operations on Matrices

So far, we have used matrices to abbreviate the work in solving systems of linear equations. For other applications, however, it is desirable to develop an "arithmetic of matrices" in which matrices can be added, subtracted, and multiplied in a useful way. The remainder of this section will be devoted to developing this arithmetic.

> **DEFINITION 2** Two matrices are defined to be *equal* if they have the same size and their corresponding entries are equal.

The equality of two matrices

$$A = [a_{ij}] \quad \text{and} \quad B = [b_{ij}]$$

of the same size can be expressed either by writing

$$(A)_{ij} = (B)_{ij}$$

or by writing

$$a_{ij} = b_{ij}$$

where it is understood that the equalities hold for all values of i and j.

▶ **EXAMPLE 2 Equality of Matrices**

Consider the matrices

$$A = \begin{bmatrix} 2 & 1 \\ 3 & x \end{bmatrix}, \quad B = \begin{bmatrix} 2 & 1 \\ 3 & 5 \end{bmatrix}, \quad C = \begin{bmatrix} 2 & 1 & 0 \\ 3 & 4 & 0 \end{bmatrix}$$

If $x = 5$, then $A = B$, but for all other values of x the matrices A and B are not equal, since not all of their corresponding entries are equal. There is no value of x for which $A = C$ since A and C have different sizes. ◀

> **DEFINITION 3** If A and B are matrices of the same size, then the *sum* $A + B$ is the matrix obtained by adding the entries of B to the corresponding entries of A, and the *difference* $A - B$ is the matrix obtained by subtracting the entries of B from the corresponding entries of A. Matrices of different sizes cannot be added or subtracted.

In matrix notation, if $A = [a_{ij}]$ and $B = [b_{ij}]$ have the same size, then

$$(A + B)_{ij} = (A)_{ij} + (B)_{ij} = a_{ij} + b_{ij} \quad \text{and} \quad (A - B)_{ij} = (A)_{ij} - (B)_{ij} = a_{ij} - b_{ij}$$

EXAMPLE 3 Addition and Subtraction

Consider the matrices

$$A = \begin{bmatrix} 2 & 1 & 0 & 3 \\ -1 & 0 & 2 & 4 \\ 4 & -2 & 7 & 0 \end{bmatrix}, \quad B = \begin{bmatrix} -4 & 3 & 5 & 1 \\ 2 & 2 & 0 & -1 \\ 3 & 2 & -4 & 5 \end{bmatrix}, \quad C = \begin{bmatrix} 1 & 1 \\ 2 & 2 \end{bmatrix}$$

Then

$$A + B = \begin{bmatrix} -2 & 4 & 5 & 4 \\ 1 & 2 & 2 & 3 \\ 7 & 0 & 3 & 5 \end{bmatrix} \text{ and } A - B = \begin{bmatrix} 6 & -2 & -5 & 2 \\ -3 & -2 & 2 & 5 \\ 1 & -4 & 11 & -5 \end{bmatrix}$$

The expressions $A + C$, $B + C$, $A - C$, and $B - C$ are undefined. ◄

> **DEFINITION 4** If A is any matrix and c is any scalar, then the **product** cA is the matrix obtained by multiplying each entry of the matrix A by c. The matrix cA is said to be a **scalar multiple** of A.

In matrix notation, if $A = [a_{ij}]$, then

$$(cA)_{ij} = c(A)_{ij} = ca_{ij}$$

EXAMPLE 4 Scalar Multiples

For the matrices

$$A = \begin{bmatrix} 2 & 3 & 4 \\ 1 & 3 & 1 \end{bmatrix}, \quad B = \begin{bmatrix} 0 & 2 & 7 \\ -1 & 3 & -5 \end{bmatrix}, \quad C = \begin{bmatrix} 9 & -6 & 3 \\ 3 & 0 & 12 \end{bmatrix}$$

we have

$$2A = \begin{bmatrix} 4 & 6 & 8 \\ 2 & 6 & 2 \end{bmatrix}, \quad (-1)B = \begin{bmatrix} 0 & -2 & -7 \\ 1 & -3 & 5 \end{bmatrix}, \quad \tfrac{1}{3}C = \begin{bmatrix} 3 & -2 & 1 \\ 1 & 0 & 4 \end{bmatrix}$$

It is common practice to denote $(-1)B$ by $-B$. ◄

Thus far we have defined multiplication of a matrix by a scalar but not the multiplication of two matrices. Since matrices are added by adding corresponding entries and subtracted by subtracting corresponding entries, it would seem natural to define multiplication of matrices by multiplying corresponding entries. However, it turns out that such a definition would not be very useful for most problems. Experience has led mathematicians to the following more useful definition of matrix multiplication.

> **DEFINITION 5** If A is an $m \times r$ matrix and B is an $r \times n$ matrix, then the **product** AB is the $m \times n$ matrix whose entries are determined as follows: To find the entry in row i and column j of AB, single out row i from the matrix A and column j from the matrix B. Multiply the corresponding entries from the row and column together, and then add up the resulting products.

EXAMPLE 5 Multiplying Matrices

Consider the matrices

$$A = \begin{bmatrix} 1 & 2 & 4 \\ 2 & 6 & 0 \end{bmatrix}, \quad B = \begin{bmatrix} 4 & 1 & 4 & 3 \\ 0 & -1 & 3 & 1 \\ 2 & 7 & 5 & 2 \end{bmatrix}$$

1.3 Matrices and Matrix Operations

Since A is a 2×3 matrix and B is a 3×4 matrix, the product AB is a 2×4 matrix. To determine, for example, the entry in row 2 and column 3 of AB, we single out row 2 from A and column 3 from B. Then, as illustrated below, we multiply corresponding entries together and add up these products.

$$\begin{bmatrix} 1 & 2 & 4 \\ 2 & 6 & 0 \end{bmatrix} \begin{bmatrix} 4 & 1 & 4 & 3 \\ 0 & -1 & 3 & 1 \\ 2 & 7 & 5 & 2 \end{bmatrix} = \begin{bmatrix} \square & \square & \square & \square \\ \square & \square & 26 & \square \end{bmatrix}$$

$$(2 \cdot 4) + (6 \cdot 3) + (0 \cdot 5) = 26$$

The entry in row 1 and column 4 of AB is computed as follows:

$$\begin{bmatrix} 1 & 2 & 4 \\ 2 & 6 & 0 \end{bmatrix} \begin{bmatrix} 4 & 1 & 4 & 3 \\ 0 & -1 & 3 & 1 \\ 2 & 7 & 5 & 2 \end{bmatrix} = \begin{bmatrix} \square & \square & \square & 13 \\ \square & \square & \square & \square \end{bmatrix}$$

$$(1 \cdot 3) + (2 \cdot 1) + (4 \cdot 2) = 13$$

The computations for the remaining entries are

$$(1 \cdot 4) + (2 \cdot 0) + (4 \cdot 2) = 12$$
$$(1 \cdot 1) - (2 \cdot 1) + (4 \cdot 7) = 27$$
$$(1 \cdot 4) + (2 \cdot 3) + (4 \cdot 5) = 30$$
$$(2 \cdot 4) + (6 \cdot 0) + (0 \cdot 2) = 8$$
$$(2 \cdot 1) - (6 \cdot 1) + (0 \cdot 7) = -4$$
$$(2 \cdot 3) + (6 \cdot 1) + (0 \cdot 2) = 12$$

$$AB = \begin{bmatrix} 12 & 27 & 30 & 13 \\ 8 & -4 & 26 & 12 \end{bmatrix} \blacktriangleleft$$

The definition of matrix multiplication requires that the number of columns of the first factor A be the same as the number of rows of the second factor B in order to form the product AB. If this condition is not satisfied, the product is undefined. A convenient way to determine whether a product of two matrices is defined is to write down the size of the first factor and, to the right of it, write down the size of the second factor. If, as in (3), the inside numbers are the same, then the product is defined. The outside numbers then give the size of the product.

$$\underset{m \times r}{A} \quad \underset{r \times n}{B} = \underset{m \times n}{AB} \tag{3}$$

Inside / Outside

Gotthold Eisenstein (1823–1852)

Historical Note The concept of matrix multiplication is due to the German mathematician Gotthold Eisenstein, who introduced the idea around 1844 to simplify the process of making substitutions in linear systems. The idea was then expanded on and formalized by Cayley in his *Memoir on the Theory of Matrices* that was published in 1858. Eisenstein was a pupil of Gauss, who ranked him as the equal of Isaac Newton and Archimedes. However, Eisenstein, suffering from bad health his entire life, died at age 30, so his potential was never realized.

[Image: wikipedia]

▶ EXAMPLE 6 Determining Whether a Product Is Defined

Suppose that A, B, and C are matrices with the following sizes:

$$\begin{array}{ccc} A & B & C \\ 3 \times 4 & 4 \times 7 & 7 \times 3 \end{array}$$

Then by (3), AB is defined and is a 3×7 matrix; BC is defined and is a 4×3 matrix; and CA is defined and is a 7×4 matrix. The products AC, CB, and BA are all undefined. ◀

In general, if $A = [a_{ij}]$ is an $m \times r$ matrix and $B = [b_{ij}]$ is an $r \times n$ matrix, then, as illustrated by the shading in (4),

$$AB = \begin{bmatrix} a_{11} & a_{12} & \cdots & a_{1r} \\ a_{21} & a_{22} & \cdots & a_{2r} \\ \vdots & \vdots & & \vdots \\ a_{i1} & a_{i2} & \cdots & a_{ir} \\ \vdots & \vdots & & \vdots \\ a_{m1} & a_{m2} & \cdots & a_{mr} \end{bmatrix} \begin{bmatrix} b_{11} & b_{12} & \cdots & b_{1j} & \cdots & b_{1n} \\ b_{21} & b_{22} & \cdots & b_{2j} & \cdots & b_{2n} \\ \vdots & \vdots & & \vdots & & \vdots \\ b_{r1} & b_{r2} & \cdots & b_{rj} & \cdots & b_{rn} \end{bmatrix} \qquad (4)$$

the entry $(AB)_{ij}$ in row i and column j of AB is given by

$$(AB)_{ij} = a_{i1}b_{1j} + a_{i2}b_{2j} + a_{i3}b_{3j} + \cdots + a_{ir}b_{rj} \qquad (5)$$

Partitioned Matrices A matrix can be subdivided or **partitioned** into smaller matrices by inserting horizontal and vertical rules between selected rows and columns. For example, the following are three possible partitions of a general 3×4 matrix A—the first is a partition of A into four **submatrices** A_{11}, A_{12}, A_{21}, and A_{22}; the second is a partition of A into its row vectors $\mathbf{r}_1$, $\mathbf{r}_2$, and $\mathbf{r}_3$; and the third is a partition of A into its column vectors $\mathbf{c}_1$, $\mathbf{c}_2$, $\mathbf{c}_3$, and $\mathbf{c}_4$:

$$A = \left[\begin{array}{ccc|c} a_{11} & a_{12} & a_{13} & a_{14} \\ a_{21} & a_{22} & a_{23} & a_{24} \\ \hline a_{31} & a_{32} & a_{33} & a_{34} \end{array}\right] = \begin{bmatrix} A_{11} & A_{12} \\ A_{21} & A_{22} \end{bmatrix}$$

$$A = \left[\begin{array}{cccc} a_{11} & a_{12} & a_{13} & a_{14} \\ \hline a_{21} & a_{22} & a_{23} & a_{24} \\ \hline a_{31} & a_{32} & a_{33} & a_{34} \end{array}\right] = \begin{bmatrix} \mathbf{r}_1 \\ \mathbf{r}_2 \\ \mathbf{r}_3 \end{bmatrix}$$

$$A = \left[\begin{array}{c|c|c|c} a_{11} & a_{12} & a_{13} & a_{14} \\ a_{21} & a_{22} & a_{23} & a_{24} \\ a_{31} & a_{32} & a_{33} & a_{34} \end{array}\right] = [\mathbf{c}_1 \quad \mathbf{c}_2 \quad \mathbf{c}_3 \quad \mathbf{c}_4]$$

Matrix Multiplication by Columns and by Rows Partitioning has many uses, one of which is for finding particular rows or columns of a matrix product AB without computing the entire product. Specifically, the following formulas, whose proofs are left as exercises, show how individual column vectors of AB can be obtained by partitioning B into column vectors and how individual row vectors of AB can be obtained by partitioning A into row vectors.

$$AB = A[\mathbf{b}_1 \quad \mathbf{b}_2 \quad \cdots \quad \mathbf{b}_n] = [A\mathbf{b}_1 \quad A\mathbf{b}_2 \quad \cdots \quad A\mathbf{b}_n] \quad (6)$$

(AB computed column by column)

$$AB = \begin{bmatrix} \mathbf{a}_1 \\ \mathbf{a}_2 \\ \vdots \\ \mathbf{a}_m \end{bmatrix} B = \begin{bmatrix} \mathbf{a}_1 B \\ \mathbf{a}_2 B \\ \vdots \\ \mathbf{a}_m B \end{bmatrix} \quad (7)$$

(AB computed row by row)

In words, these formulas state that

$$j\text{th column vector of } AB = A[\,j\text{th column vector of } B] \quad (8)$$

$$i\text{th row vector of } AB = [i\text{th row vector of } A]B \quad (9)$$

▶ **EXAMPLE 7** **Example 5 Revisited**

If A and B are the matrices in Example 5, then from (8) the second column vector of AB can be obtained by the computation

$$\begin{bmatrix} 1 & 2 & 4 \\ 2 & 6 & 0 \end{bmatrix} \begin{bmatrix} 1 \\ -1 \\ 7 \end{bmatrix} = \begin{bmatrix} 27 \\ -4 \end{bmatrix}$$

Second column of B Second column of AB

and from (9) the first row vector of AB can be obtained by the computation

$$[1 \quad 2 \quad 4] \begin{bmatrix} 4 & 1 & 4 & 3 \\ 0 & -1 & 3 & 1 \\ 2 & 7 & 5 & 2 \end{bmatrix} = [12 \quad 27 \quad 30 \quad 13]$$

First row of A First row of AB ◀

Matrix Products as Linear Combinations

We have discussed three methods for computing a matrix product AB—entry by entry, column by column, and row by row. The following definition provides yet another way of thinking about matrix multiplication.

DEFINITION 6 If $A_1, A_2, \ldots, A_r$ are matrices of the same size, and if $c_1, c_2, \ldots, c_r$ are scalars, then an expression of the form

$$c_1 A_1 + c_2 A_2 + \cdots + c_r A_r$$

is called a **linear combination** of $A_1, A_2, \ldots, A_r$ with **coefficients** $c_1, c_2, \ldots, c_r$.

To see how matrix products can be viewed as linear combinations, let A be an $m \times n$ matrix and $\mathbf{x}$ an $n \times 1$ column vector, say

$$A = \begin{bmatrix} a_{11} & a_{12} & \cdots & a_{1n} \\ a_{21} & a_{22} & \cdots & a_{2n} \\ \vdots & \vdots & & \vdots \\ a_{m1} & a_{m2} & \cdots & a_{mn} \end{bmatrix} \quad \text{and} \quad \mathbf{x} = \begin{bmatrix} x_1 \\ x_2 \\ \vdots \\ x_n \end{bmatrix}$$

Then

$$A\mathbf{x} = \begin{bmatrix} a_{11}x_1 + a_{12}x_2 + \cdots + a_{1n}x_n \\ a_{21}x_1 + a_{22}x_2 + \cdots + a_{2n}x_n \\ \vdots & \vdots & & \vdots \\ a_{m1}x_1 + a_{m2}x_2 + \cdots + a_{mn}x_n \end{bmatrix} = x_1 \begin{bmatrix} a_{11} \\ a_{21} \\ \vdots \\ a_{m1} \end{bmatrix} + x_2 \begin{bmatrix} a_{12} \\ a_{22} \\ \vdots \\ a_{m2} \end{bmatrix} + \cdots + x_n \begin{bmatrix} a_{1n} \\ a_{2n} \\ \vdots \\ a_{mn} \end{bmatrix}$$

(10)

This proves the following theorem.

THEOREM 1.3.1 *If A is an $m \times n$ matrix, and if $\mathbf{x}$ is an $n \times 1$ column vector, then the product $A\mathbf{x}$ can be expressed as a linear combination of the column vectors of A in which the coefficients are the entries of $\mathbf{x}$.*

▶ **EXAMPLE 8** Matrix Products as Linear Combinations

The matrix product

$$\begin{bmatrix} -1 & 3 & 2 \\ 1 & 2 & -3 \\ 2 & 1 & -2 \end{bmatrix} \begin{bmatrix} 2 \\ -1 \\ 3 \end{bmatrix} = \begin{bmatrix} 1 \\ -9 \\ -3 \end{bmatrix}$$

can be written as the following linear combination of column vectors

$$2\begin{bmatrix} -1 \\ 1 \\ 2 \end{bmatrix} - 1\begin{bmatrix} 3 \\ 2 \\ 1 \end{bmatrix} + 3\begin{bmatrix} 2 \\ -3 \\ -2 \end{bmatrix} = \begin{bmatrix} 1 \\ -9 \\ -3 \end{bmatrix}$$

▶ **EXAMPLE 9** Columns of a Product AB as Linear Combinations

We showed in Example 5 that

$$AB = \begin{bmatrix} 1 & 2 & 4 \\ 2 & 6 & 0 \end{bmatrix} \begin{bmatrix} 4 & 1 & 4 & 3 \\ 0 & -1 & 3 & 1 \\ 2 & 7 & 5 & 2 \end{bmatrix} = \begin{bmatrix} 12 & 27 & 30 & 13 \\ 8 & -4 & 26 & 12 \end{bmatrix}$$

It follows from Formula (6) and Theorem 1.3.1 that the jth column vector of AB can be expressed as a linear combination of the column vectors of A in which the coefficients in the linear combination are the entries from the jth column of B. The computations are as follows:

$$\begin{bmatrix} 12 \\ 8 \end{bmatrix} = 4\begin{bmatrix} 1 \\ 2 \end{bmatrix} + 0\begin{bmatrix} 2 \\ 6 \end{bmatrix} + 2\begin{bmatrix} 4 \\ 0 \end{bmatrix}$$

$$\begin{bmatrix} 27 \\ -4 \end{bmatrix} = \begin{bmatrix} 1 \\ 2 \end{bmatrix} - \begin{bmatrix} 2 \\ 6 \end{bmatrix} + 7\begin{bmatrix} 4 \\ 0 \end{bmatrix}$$

$$\begin{bmatrix} 30 \\ 26 \end{bmatrix} = 4\begin{bmatrix} 1 \\ 2 \end{bmatrix} + 3\begin{bmatrix} 2 \\ 6 \end{bmatrix} + 5\begin{bmatrix} 4 \\ 0 \end{bmatrix}$$

$$\begin{bmatrix} 13 \\ 12 \end{bmatrix} = 3\begin{bmatrix} 1 \\ 2 \end{bmatrix} + \begin{bmatrix} 2 \\ 6 \end{bmatrix} + 2\begin{bmatrix} 4 \\ 0 \end{bmatrix} \blacktriangleleft$$

Matrix Form of a Linear System

Matrix multiplication has an important application to systems of linear equations. Consider a system of m linear equations in n unknowns:

$$\begin{aligned} a_{11}x_1 + a_{12}x_2 + \cdots + a_{1n}x_n &= b_1 \\ a_{21}x_1 + a_{22}x_2 + \cdots + a_{2n}x_n &= b_2 \\ &\vdots \\ a_{m1}x_1 + a_{m2}x_2 + \cdots + a_{mn}x_n &= b_m \end{aligned}$$

Since two matrices are equal if and only if their corresponding entries are equal, we can replace the m equations in this system by the single matrix equation

$$\begin{bmatrix} a_{11}x_1 + a_{12}x_2 + \cdots + a_{1n}x_n \\ a_{21}x_1 + a_{22}x_2 + \cdots + a_{2n}x_n \\ \vdots \\ a_{m1}x_1 + a_{m2}x_2 + \cdots + a_{mn}x_n \end{bmatrix} = \begin{bmatrix} b_1 \\ b_2 \\ \vdots \\ b_m \end{bmatrix}$$

The $m \times 1$ matrix on the left side of this equation can be written as a product to give

$$\begin{bmatrix} a_{11} & a_{12} & \cdots & a_{1n} \\ a_{21} & a_{22} & \cdots & a_{2n} \\ \vdots & \vdots & & \vdots \\ a_{m1} & a_{m2} & \cdots & a_{mn} \end{bmatrix} \begin{bmatrix} x_1 \\ x_2 \\ \vdots \\ x_n \end{bmatrix} = \begin{bmatrix} b_1 \\ b_2 \\ \vdots \\ b_m \end{bmatrix}$$

If we designate these matrices by A, $\mathbf{x}$, and $\mathbf{b}$, respectively, then we can replace the original system of m equations in n unknowns by the single matrix equation

$$A\mathbf{x} = \mathbf{b}$$

The matrix A in this equation is called the ***coefficient matrix*** of the system. The ***augmented matrix*** for the system is obtained by adjoining $\mathbf{b}$ to A as the last column; thus the augmented matrix is

> The vertical bar in $[A \mid \mathbf{b}]$ is a convenient way to separate A from $\mathbf{b}$ visually; it has no mathematical significance.

$$[A \mid \mathbf{b}] = \begin{bmatrix} a_{11} & a_{12} & \cdots & a_{1n} & \bigm| & b_1 \\ a_{21} & a_{22} & \cdots & a_{2n} & \bigm| & b_2 \\ \vdots & \vdots & & \vdots & & \vdots \\ a_{m1} & a_{m2} & \cdots & a_{mn} & \bigm| & b_m \end{bmatrix}$$

Transpose of a Matrix

We conclude this section by defining two matrix operations that have no analogs in the arithmetic of real numbers.

DEFINITION 7 If A is any $m \times n$ matrix, then the ***transpose of* A**, denoted by A^T, is defined to be the $n \times m$ matrix that results by interchanging the rows and columns of A; that is, the first column of A^T is the first row of A, the second column of A^T is the second row of A, and so forth.

▶ **EXAMPLE 10 Some Transposes**

The following are some examples of matrices and their transposes.

$$A = \begin{bmatrix} a_{11} & a_{12} & a_{13} & a_{14} \\ a_{21} & a_{22} & a_{23} & a_{24} \\ a_{31} & a_{32} & a_{33} & a_{34} \end{bmatrix}, \quad B = \begin{bmatrix} 2 & 3 \\ 1 & 4 \\ 5 & 6 \end{bmatrix}, \quad C = [1 \ 3 \ 5], \quad D = [4]$$

$$A^T = \begin{bmatrix} a_{11} & a_{21} & a_{31} \\ a_{12} & a_{22} & a_{32} \\ a_{13} & a_{23} & a_{33} \\ a_{14} & a_{24} & a_{34} \end{bmatrix}, \quad B^T = \begin{bmatrix} 2 & 1 & 5 \\ 3 & 4 & 6 \end{bmatrix}, \quad C^T = \begin{bmatrix} 1 \\ 3 \\ 5 \end{bmatrix}, \quad D^T = [4] \blacktriangleleft$$

Observe that not only are the columns of A^T the rows of A, but the rows of A^T are the columns of A. Thus the entry in row i and column j of A^T is the entry in row j and column i of A; that is,

$$(A^T)_{ij} = (A)_{ji} \tag{11}$$

Note the reversal of the subscripts.

In the special case where A is a square matrix, the transpose of A can be obtained by interchanging entries that are symmetrically positioned about the main diagonal. In (12) we see that A^T can also be obtained by "reflecting" A about its main diagonal.

$$A = \begin{bmatrix} 1 & -2 & 4 \\ 3 & 7 & 0 \\ -5 & 8 & 6 \end{bmatrix} \rightarrow \begin{bmatrix} 1 & -2 & 4 \\ 3 & 7 & 0 \\ -5 & 8 & 6 \end{bmatrix} \rightarrow A^T = \begin{bmatrix} 1 & 3 & -5 \\ -2 & 7 & 8 \\ 4 & 0 & 6 \end{bmatrix} \tag{12}$$

Interchange entries that are symmetrically positioned about the main diagonal.

James Sylvester
(1814–1897)

Arthur Cayley
(1821–1895)

Historical Note The term *matrix* was first used by the English mathematician James Sylvester, who defined the term in 1850 to be an "oblong arrangement of terms." Sylvester communicated his work on matrices to a fellow English mathematician and lawyer named Arthur Cayley, who then introduced some of the basic operations on matrices in a book entitled *Memoir on the Theory of Matrices* that was published in 1858. As a matter of interest, Sylvester, who was Jewish, did not get his college degree because he refused to sign a required oath to the Church of England. He was appointed to a chair at the University of Virginia in the United States but resigned after swatting a student with a stick because he was reading a newspaper in class. Sylvester, thinking he had killed the student, fled back to England on the first available ship. Fortunately, the student was not dead, just in shock!

[*Images: The Granger Collection, New York*]

DEFINITION 8 If A is a square matrix, then the *trace of* A, denoted by $\text{tr}(A)$, is defined to be the sum of the entries on the main diagonal of A. The trace of A is undefined if A is not a square matrix.

▶ **EXAMPLE 11** Trace of a Matrix

The following are examples of matrices and their traces.

$$A = \begin{bmatrix} a_{11} & a_{12} & a_{13} \\ a_{21} & a_{22} & a_{23} \\ a_{31} & a_{32} & a_{33} \end{bmatrix}, \quad B = \begin{bmatrix} -1 & 2 & 7 & 0 \\ 3 & 5 & -8 & 4 \\ 1 & 2 & 7 & -3 \\ 4 & -2 & 1 & 0 \end{bmatrix}$$

$$\text{tr}(A) = a_{11} + a_{22} + a_{33} \qquad \text{tr}(B) = -1 + 5 + 7 + 0 = 11 \quad ◀$$

In the exercises you will have some practice working with the transpose and trace operations.

Concept Review

- Matrix
- Entries
- Column vector (or column matrix)
- Row vector (or row matrix)
- Square matrix
- Main diagonal
- Equal matrices
- Matrix operations: sum, difference, scalar multiplication
- Linear combination of matrices
- Product of matrices (matrix multiplication)
- Partitioned matrices
- Submatrices
- Row-column method
- Column method
- Row method
- Coefficient matrix of a linear system
- Transpose
- Trace

Skills

- Determine the size of a given matrix.
- Identify the row vectors and column vectors of a given matrix.
- Perform the arithmetic operations of matrix addition, subtraction, scalar multiplication, and multiplication.
- Determine whether the product of two given matrices is defined.
- Compute matrix products using the row-column method, the column method, and the row method.
- Express the product of a matrix and a column vector as a linear combination of the columns of the matrix.
- Express a linear system as a matrix equation, and identify the coefficient matrix.
- Compute the transpose of a matrix.
- Compute the trace of a square matrix.

Exercise Set 1.3

1. Suppose that A, B, C, D, and E are matrices with the following sizes:

A	B	C	D	E
(4×5)	(4×5)	(5×2)	(4×2)	(5×4)

   In each part, determine whether the given matrix expression is defined. For those that are defined, give the size of the resulting matrix.

   (a) BA (b) $AC + D$ (c) $AE + B$
   (d) $AB + B$ (e) $E(A + B)$ (f) $E(AC)$
   (g) $E^T A$ (h) $(A^T + E)D$

2. Suppose that A, B, C, D, and E are matrices with the following sizes:

A	B	C	D	E
(3×1)	(3×6)	(6×2)	(2×6)	(1×3)

In each part, determine whether the given matrix expression is defined. For those that are defined, give the size of the resulting matrix.

(a) EA (b) AB^T (c) $B^T(A + E^T)$

(d) $2A + C$ (e) $(C^T + D)B^T$ (f) $CD + B^T E^T$

(g) $(BD^T)C^T$ (h) $DC + EA$

3. Consider the matrices

$$A = \begin{bmatrix} 2 & 0 \\ -4 & 6 \end{bmatrix}, \quad B = \begin{bmatrix} 1 & -7 & 2 \\ 5 & 3 & 0 \end{bmatrix}, \quad C = \begin{bmatrix} 4 & 9 \\ -3 & 0 \\ 2 & 1 \end{bmatrix},$$

$$D = \begin{bmatrix} -2 & 1 & 8 \\ 3 & 0 & 2 \\ 4 & -6 & 3 \end{bmatrix}, \quad E = \begin{bmatrix} 0 & 3 & 0 \\ -5 & 1 & 1 \\ 7 & 6 & 2 \end{bmatrix}$$

In each part, compute the given expression (where possible).

(a) $D + E$ (b) $D - E$ (c) $5A$

(d) $-9D$ (e) $2B - C$ (f) $7E - 3D$

(g) $2(D + 5E)$ (h) $B - B$ (i) $\text{tr}(D)$

(j) $\text{tr}(D - E)$ (k) $2\,\text{tr}(4B)$ (l) $\text{tr}(A)$

4. Using the matrices in Exercise 3, in each part compute the given expression (where possible).

(a) $2A^T + C$ (b) $D^T - E^T$ (c) $(D - E)^T$

(d) $B^T + 5C^T$ (e) $\tfrac{1}{2}C^T - \tfrac{1}{4}A$ (f) $B - B^T$

(g) $2E^T - 3D^T$ (h) $(2E^T - 3D^T)^T$ (i) $(CD)E$

(j) $C(BA)$ (k) $\text{tr}(DE^T)$ (l) $\text{tr}(BC)$

5. Using the matrices in Exercise 3, in each part compute the given expression (where possible).

(a) AB (b) BA (c) $(3E)D$

(d) $(AB)C$ (e) $A(BC)$ (f) CC^T

(g) $(DC)^T$ (h) $(C^T B)A^T$ (i) $\text{tr}(DD^T)$

(j) $\text{tr}(4E^T - D)$ (k) $\text{tr}(A^T C^T + 2E^T)$ (l) $\text{tr}((E^T C)B)$

6. Using the matrices in Exercise 3, in each part compute the given expression (where possible).

(a) $(2D^T - E)A$ (b) $(4B)C + 2B$

(c) $(-AC)^T + 5D^T$ (d) $(BA^T - 2C)^T$

(e) $B^T(CC^T - A^T A)$ (f) $D^T E^T - (ED)^T$

7. Let

$$A = \begin{bmatrix} 3 & -2 & 7 \\ 6 & 5 & 4 \\ 0 & 4 & 9 \end{bmatrix} \quad \text{and} \quad B = \begin{bmatrix} 6 & -2 & 4 \\ 0 & 1 & 3 \\ 7 & 7 & 5 \end{bmatrix}$$

Use the row method or column method (as appropriate) to find

(a) the first row of AB. (b) the third row of AB.

(c) the second column of AB. (d) the first column of BA.

(e) the third row of AA. (f) the third column of AA.

8. Referring to the matrices in Exercise 7, use the row method or column method (as appropriate) to find

(a) the first column of AB. (b) the third column of BB.

(c) the second row of BB. (d) the first column of AA.

(e) the third column of AB. (f) the first row of BA.

9. Referring to the matrices in Exercise 7 and Example 9,

(a) express each column vector of AA as a linear combination of the column vectors of A.

(b) express each column vector of BB as a linear combination of the column vectors of B.

10. Referring to the matrices in Exercise 7 and Example 9,

(a) express each column vector of AB as a linear combination of the column vectors of A.

(b) express each column vector of BA as a linear combination of the column vectors of B.

11. In each part, find matrices A, $\mathbf{x}$, and $\mathbf{b}$ that express the given system of linear equations as a single matrix equation $A\mathbf{x} = \mathbf{b}$, and write out this matrix equation.

(a) $\begin{aligned} 5x + y + z &= 2 \\ 2x \quad\quad + 3z &= 1 \\ x + 2y \quad\quad &= 0 \end{aligned}$

(b) $\begin{aligned} x_1 + x_2 - x_3 - 7x_4 &= 6 \\ -x_2 + 4x_3 + x_4 &= 1 \\ 4x_1 + 2x_2 + x_3 + 8x_4 &= 0 \end{aligned}$

12. In each part, find matrices A, $\mathbf{x}$, and $\mathbf{b}$ that express the given system of linear equations as a single matrix equation $A\mathbf{x} = \mathbf{b}$, and write out this matrix equation.

(a) $\begin{aligned} x_1 - 2x_2 + 3x_3 &= -3 \\ 2x_1 + x_2 \quad\quad &= 0 \\ -3x_2 + 4x_3 &= 1 \\ x_1 \quad\quad + x_3 &= 5 \end{aligned}$

(b) $\begin{aligned} 3x_1 + 3x_2 + 3x_3 &= -3 \\ -x_1 - 5x_2 - 2x_3 &= 3 \\ -4x_2 + x_3 &= 0 \end{aligned}$

13. In each part, express the matrix equation as a system of linear equations.

(a) $\begin{bmatrix} 5 & 6 & -7 \\ -1 & -2 & 3 \\ 0 & 4 & -1 \end{bmatrix} \begin{bmatrix} x_1 \\ x_2 \\ x_3 \end{bmatrix} = \begin{bmatrix} 2 \\ 0 \\ 3 \end{bmatrix}$

(b) $\begin{bmatrix} 1 & 1 & 1 \\ 2 & 3 & 0 \\ 5 & -3 & -6 \end{bmatrix} \begin{bmatrix} x_1 \\ x_2 \\ x_3 \end{bmatrix} = \begin{bmatrix} 2 \\ 2 \\ -9 \end{bmatrix}$

14. In each part, express the matrix equation as a system of linear equations.

(a) $\begin{bmatrix} 3 & -1 & 2 \\ 4 & 3 & 7 \\ -2 & 1 & 5 \end{bmatrix} \begin{bmatrix} x_1 \\ x_2 \\ x_3 \end{bmatrix} = \begin{bmatrix} 2 \\ -1 \\ 4 \end{bmatrix}$

(b) $\begin{bmatrix} 3 & -2 & 0 & 1 \\ 5 & 0 & 2 & -2 \\ 3 & 1 & 4 & 7 \\ -2 & 5 & 1 & 6 \end{bmatrix} \begin{bmatrix} w \\ x \\ y \\ z \end{bmatrix} = \begin{bmatrix} 0 \\ 0 \\ 0 \\ 0 \end{bmatrix}$

▶ In Exercises 15–16, find all values of k, if any, that satisfy the equation. ◀

15. $\begin{bmatrix} k & 1 & 1 \end{bmatrix} \begin{bmatrix} 1 & 1 & 0 \\ 1 & 0 & 2 \\ 0 & 2 & -3 \end{bmatrix} \begin{bmatrix} k \\ 1 \\ 1 \end{bmatrix} = 0$

16. $\begin{bmatrix} 2 & 2 & k \end{bmatrix} \begin{bmatrix} 1 & 2 & 0 \\ 2 & 0 & 3 \\ 0 & 3 & 1 \end{bmatrix} \begin{bmatrix} 2 \\ 2 \\ k \end{bmatrix} = 0$

▶ In Exercises 17–18, solve the matrix equation for $a, b, c,$ and d.

17. $\begin{bmatrix} 3 & a \\ 1 & a+b \end{bmatrix} = \begin{bmatrix} b & c-2d \\ c+2d & 0 \end{bmatrix}$

18. $\begin{bmatrix} a-b & b+a \\ 3d+c & 2d-c \end{bmatrix} = \begin{bmatrix} 8 & 1 \\ 7 & 6 \end{bmatrix}$

19. Let A be any $m \times n$ matrix and let 0 be the $m \times n$ matrix each of whose entries is zero. Show that if $kA = 0$, either $k = 0$ or $A = 0$.

20. (a) Show that if AB and BA are both defined, then AB and BA are square matrices.

(b) Show that if A is an $m \times n$ matrix and $A(BA)$ is defined, then B is an $n \times m$ matrix.

21. Prove: If A and B are $n \times n$ matrices, then

$$\text{tr}(A+B) = \text{tr}(A) + \text{tr}(B)$$

22. (a) Show that if B is any matrix with a column of zeros and A is any matrix for which AB is defined, then AB also has a column of zeros.

(b) Find a similar result involving a row of zeros.

23. In each part, find a 6×6 matrix $[a_{ij}]$ that satisfies the stated condition. Make your answers as general as possible by using letters rather than specific numbers for the nonzero entries.

(a) $a_{ij} = 0$ if $i \neq j$ (b) $a_{ij} = 0$ if $i > j$

(c) $a_{ij} = 0$ if $i < j$

(d) $a_{ij} = 0$ if $|i - j| > 1$

24. Find the 4×4 matrix $A = [a_{ij}]$ whose entries satisfy the stated condition.

(a) $a_{ij} = i - j$ (b) $a_{ij} = (-1)^1 ij$

(c) $a_{ij} = \begin{cases} 0 & |i-j| \geq 1 \\ -1 & |i-j| < 1 \end{cases}$

25. Consider the function $y = f(x)$ defined for 2×1 matrices x by $y = Ax$, where

$$A = \begin{bmatrix} 1 & 1 \\ 0 & 1 \end{bmatrix}$$

Plot $f(x)$ together with x in each case below. How would you describe the action of f?

(a) $x = \begin{pmatrix} 1 \\ 1 \end{pmatrix}$ (b) $x = \begin{pmatrix} 2 \\ 0 \end{pmatrix}$

(c) $x = \begin{pmatrix} 4 \\ 3 \end{pmatrix}$ (d) $x = \begin{pmatrix} 2 \\ -2 \end{pmatrix}$

26. Let I be the $n \times n$ matrix whose entry in row i and column j is

$$\begin{cases} 1 & \text{if } i = j \\ 0 & \text{if } i \neq j \end{cases}$$

Show that $AI = IA = A$ for every $n \times n$ matrix A.

27. How many 3×3 matrices A can you find such that

$$A \begin{bmatrix} x \\ y \\ z \end{bmatrix} = \begin{bmatrix} x+y \\ x-y \\ 0 \end{bmatrix}$$

for all choices of $x, y,$ and z?

28. How many 3×3 matrices A can you find such that

$$A \begin{bmatrix} x \\ y \\ z \end{bmatrix} = \begin{bmatrix} xy \\ 0 \\ 0 \end{bmatrix}$$

for all choices of $x, y,$ and z?

29. A matrix B is said to be a *square root* of a matrix A if $BB = A$.

(a) Find two square roots of $A = \begin{bmatrix} 2 & 2 \\ 2 & 2 \end{bmatrix}$.

(b) How many different square roots can you find of $A = \begin{bmatrix} 5 & 0 \\ 0 & 9 \end{bmatrix}$?

(c) Do you think that every 2×2 matrix has at least one square root? Explain your reasoning.

30. Let 0 denote a 2×2 matrix, each of whose entries is zero.

(a) Is there a 2×2 matrix A such that $A \neq 0$ and $AA = 0$? Justify your answer.

(b) Is there a 2×2 matrix A such that $A \neq 0$ and $AA = A$? Justify your answer.

True-False Exercises

In parts (a)–(o) determine whether the statement is true or false, and justify your answer.

(a) The matrix $\begin{bmatrix} 1 & 2 & 3 \\ 4 & 5 & 6 \end{bmatrix}$ has no main diagonal.

(b) An $m \times n$ matrix has m column vectors and n row vectors.

(c) If $AB = BA$, then A must equal B.

(d) If $A\mathbf{x} = \mathbf{b}$, then $\mathbf{b}$ must be a linear combination of the columns of A.

(e) For every matrix A, it is true that $(A^T)^T = A$.

(f) If A and B are square matrices of the same order, then $\text{tr}(AB) = \text{tr}(A)\text{tr}(B)$.

(g) If A and B are square matrices of the same order, then $(AB)^T = A^T B^T$.

(h) For every square matrix A, it is true that $\text{tr}(A^T) = \text{tr}(A)$.

(i) If A is a 6×4 matrix and B is an $m \times n$ matrix such that $B^T A^T$ is a 2×6 matrix, then $m = 4$ and $n = 2$.

(j) If A is an $n \times n$ matrix and c is a scalar, then $\text{tr}(cA) = c\,\text{tr}(A)$.

(k) If A, B, and C are matrices of the same size such that $A - C = B - C$, then $A = B$.

(l) If A, B, and C are square matrices of the same order such that $AC = BC$, then $A = B$.

(m) If $AB + BA$ is defined, then A and B are square matrices of the same size.

(n) If B has a column of zeros, then so does AB if this product is defined.

(o) If B has a column of zeros, then so does BA if this product is defined.

1.4 Inverses; Algebraic Properties of Matrices

In this section we will discuss some of the algebraic properties of matrix operations. We will see that many of the basic rules of arithmetic for real numbers hold for matrices, but we will also see that some do not.

Properties of Matrix Addition and Scalar Multiplication

The following theorem lists the basic algebraic properties of the matrix operations.

THEOREM 1.4.1 *Properties of Matrix Arithmetic*

Assuming that the sizes of the matrices are such that the indicated operations can be performed, the following rules of matrix arithmetic are valid.

(a) $A + B = B + A$ (Commutative law for addition)
(b) $A + (B + C) = (A + B) + C$ (Associative law for addition)
(c) $A(BC) = (AB)C$ (Associative law for multiplication)
(d) $A(B + C) = AB + AC$ (Left distributive law)
(e) $(B + C)A = BA + CA$ (Right distributive law)
(f) $A(B - C) = AB - AC$
(g) $(B - C)A = BA - CA$
(h) $a(B + C) = aB + aC$
(i) $a(B - C) = aB - aC$
(j) $(a + b)C = aC + bC$
(k) $(a - b)C = aC - bC$
(l) $a(bC) = (ab)C$
(m) $a(BC) = (aB)C = B(aC)$

To prove any of the equalities in this theorem we must show that the matrix on the left side has the same size as that on the right and that the corresponding entries on the two sides are the same. Most of the proofs follow the same pattern, so we will prove part (*d*) as a sample. The proof of the associative law for multiplication is more complicated than the rest and is outlined in the exercises.

Proof (d) We must show that $A(B + C)$ and $AB + AC$ have the same size and that corresponding entries are equal. To form $A(B + C)$, the matrices B and C must have the same size, say $m \times n$, and the matrix A must then have m columns, so its size must be of the form $r \times m$. This makes $A(B + C)$ an $r \times n$ matrix. It follows that $AB + AC$ is also an $r \times n$ matrix and, consequently, $A(B + C)$ and $AB + AC$ have the same size.

Suppose that $A = [a_{ij}]$, $B = [b_{ij}]$, and $C = [c_{ij}]$. We want to show that corresponding entries of $A(B + C)$ and $AB + AC$ are equal; that is,

$$[A(B + C)]_{ij} = [AB + AC]_{ij}$$

for all values of i and j. But from the definitions of matrix addition and matrix multiplication, we have

$$[A(B+C)]_{ij} = a_{i1}(b_{1j} + c_{1j}) + a_{i2}(b_{2j} + c_{2j}) + \cdots + a_{im}(b_{mj} + c_{mj})$$
$$= (a_{i1}b_{1j} + a_{i2}b_{2j} + \cdots + a_{im}b_{mj}) + (a_{i1}c_{1j} + a_{i2}c_{2j} + \cdots + a_{im}c_{mj})$$
$$= [AB]_{ij} + [AC]_{ij} = [AB + AC]_{ij} \blacktriangleleft$$

> There are three basic ways to prove that two matrices of the same size are equal—prove that corresponding entries are the same, prove that corresponding row vectors are the same, or prove that corresponding column vectors are the same.

Remark Although the operations of matrix addition and matrix multiplication were defined for pairs of matrices, associative laws (*b*) and (*c*) enable us to denote sums and products of three matrices as $A + B + C$ and ABC without inserting any parentheses. This is justified by the fact that no matter how parentheses are inserted, the associative laws guarantee that the same end result will be obtained. In general, *given any sum or any product of matrices, pairs of parentheses can be inserted or deleted anywhere within the expression without affecting the end result.*

▶ **EXAMPLE 1 Associativity of Matrix Multiplication**

As an illustration of the associative law for matrix multiplication, consider

$$A = \begin{bmatrix} 1 & 2 \\ 3 & 4 \\ 0 & 1 \end{bmatrix}, \quad B = \begin{bmatrix} 4 & 3 \\ 2 & 1 \end{bmatrix}, \quad C = \begin{bmatrix} 1 & 0 \\ 2 & 3 \end{bmatrix}$$

Then

$$AB = \begin{bmatrix} 1 & 2 \\ 3 & 4 \\ 0 & 1 \end{bmatrix} \begin{bmatrix} 4 & 3 \\ 2 & 1 \end{bmatrix} = \begin{bmatrix} 8 & 5 \\ 20 & 13 \\ 2 & 1 \end{bmatrix} \quad \text{and} \quad BC = \begin{bmatrix} 4 & 3 \\ 2 & 1 \end{bmatrix} \begin{bmatrix} 1 & 0 \\ 2 & 3 \end{bmatrix} = \begin{bmatrix} 10 & 9 \\ 4 & 3 \end{bmatrix}$$

Thus

$$(AB)C = \begin{bmatrix} 8 & 5 \\ 20 & 13 \\ 2 & 1 \end{bmatrix} \begin{bmatrix} 1 & 0 \\ 2 & 3 \end{bmatrix} = \begin{bmatrix} 18 & 15 \\ 46 & 39 \\ 4 & 3 \end{bmatrix}$$

and

$$A(BC) = \begin{bmatrix} 1 & 2 \\ 3 & 4 \\ 0 & 1 \end{bmatrix} \begin{bmatrix} 10 & 9 \\ 4 & 3 \end{bmatrix} = \begin{bmatrix} 18 & 15 \\ 46 & 39 \\ 4 & 3 \end{bmatrix}$$

so $(AB)C = A(BC)$, as guaranteed by Theorem 1.4.1(*c*). ◀

Properties of Matrix Multiplication

Do not let Theorem 1.4.1 lull you into believing that *all* laws of real arithmetic carry over to matrix arithmetic. For example, you know that in real arithmetic it is always true that $ab = ba$, which is called the *commutative law for multiplication*. In matrix arithmetic, however, the equality of AB and BA can fail for three possible reasons:

1. AB may be defined and BA may not (for example, if A is 2×3 and B is 3×4).
2. AB and BA may both be defined, but they may have different sizes (for example, if A is 2×3 and B is 3×2).
3. AB and BA may both be defined and have the same size, but the two matrices may be different (as illustrated in the next example).

▶ **EXAMPLE 2 Order Matters in Matrix Multiplication**

Consider the matrices

$$A = \begin{bmatrix} -1 & 0 \\ 2 & 3 \end{bmatrix} \quad \text{and} \quad B = \begin{bmatrix} 1 & 2 \\ 3 & 0 \end{bmatrix}$$

Multiplying gives

$$AB = \begin{bmatrix} -1 & -2 \\ 11 & 4 \end{bmatrix} \quad \text{and} \quad BA = \begin{bmatrix} 3 & 6 \\ -3 & 0 \end{bmatrix}$$

Thus, $AB \neq BA$. ◀

> Do not read too much into Example 2—it does not rule out the possibility that AB and BA may be equal in *certain* cases, just that they are not equal in *all* cases. If it so happens that $AB = BA$, then we say that AB and BA **commute**.

Zero Matrices

A matrix whose entries are all zero is called a ***zero matrix***. Some examples are

$$\begin{bmatrix} 0 & 0 \\ 0 & 0 \end{bmatrix}, \quad \begin{bmatrix} 0 & 0 & 0 \\ 0 & 0 & 0 \\ 0 & 0 & 0 \end{bmatrix}, \quad \begin{bmatrix} 0 & 0 & 0 & 0 \\ 0 & 0 & 0 & 0 \end{bmatrix}, \quad \begin{bmatrix} 0 \\ 0 \\ 0 \\ 0 \end{bmatrix}, \quad [0]$$

We will denote a zero matrix by 0 unless it is important to specify its size, in which case we will denote the $m \times n$ zero matrix by $0_{m \times n}$.

It should be evident that if A and 0 are matrices with the same size, then

$$A + 0 = 0 + A = A$$

Thus, 0 plays the same role in this matrix equation that the number 0 plays in the numerical equation $a + 0 = 0 + a = a$.

The following theorem lists the basic properties of zero matrices. Since the results should be self-evident, we will omit the formal proofs.

THEOREM 1.4.2 *Properties of Zero Matrices*

If c is a scalar, and if the sizes of the matrices are such that the operations can be perfomed, then:

(a) $A + 0 = 0 + A = A$
(b) $A - 0 = A$
(c) $A - A = A + (-A) = 0$
(d) $0A = 0$
(e) *If $cA = 0$, then $c = 0$ or $A = 0$.*

Since we know that the commutative law of real arithmetic is not valid in matrix arithmetic, it should not be surprising that there are other rules that fail as well. For example, consider the following two laws of real arithmetic:

- If $ab = bc$ and $a \neq 0$, then $b = c$. **[The cancellation law]**
- If $ab = 0$, then at least one of the factors on the left is 0.

The next two examples show that these laws are not universally true in matrix arithmetic.

▶ **EXAMPLE 3** **Failure of the Cancellation Law**

Consider the matrices

$$A = \begin{bmatrix} 0 & 1 \\ 0 & 2 \end{bmatrix}, \quad B = \begin{bmatrix} 1 & 1 \\ 3 & 4 \end{bmatrix}, \quad C = \begin{bmatrix} 2 & 5 \\ 3 & 4 \end{bmatrix}$$

We leave it for you to confirm that

$$AB = AC = \begin{bmatrix} 3 & 4 \\ 6 & 8 \end{bmatrix}$$

Although $A \neq 0$, canceling A from both sides of the equation $AB = AC$ would lead to the incorrect conclusion that $B = C$. Thus, the cancellation law does not hold, in general, for matrix multiplication.

▶ **EXAMPLE 4** **A Zero Product with Nonzero Factors**

Here are two matrices for which $AB = 0$, but $A \neq 0$ and $B \neq 0$:

$$A = \begin{bmatrix} 0 & 1 \\ 0 & 2 \end{bmatrix}, \quad B = \begin{bmatrix} 3 & 7 \\ 0 & 0 \end{bmatrix} \blacktriangleleft$$

Identity Matrices A square matrix with 1's on the main diagonal and zeros elsewhere is called an **identity matrix**. Some examples are

$$[1], \quad \begin{bmatrix} 1 & 0 \\ 0 & 1 \end{bmatrix}, \quad \begin{bmatrix} 1 & 0 & 0 \\ 0 & 1 & 0 \\ 0 & 0 & 1 \end{bmatrix}, \quad \begin{bmatrix} 1 & 0 & 0 & 0 \\ 0 & 1 & 0 & 0 \\ 0 & 0 & 1 & 0 \\ 0 & 0 & 0 & 1 \end{bmatrix}$$

An identity matrix is denoted by the letter I. If it is important to emphasize the size, we will write I_n for the $n \times n$ identity matrix.

To explain the role of identity matrices in matrix arithmetic, let us consider the effect of multiplying a general 2×3 matrix A on each side by an identity matrix. Multiplying on the right by the 3×3 identity matrix yields

$$AI_3 = \begin{bmatrix} a_{11} & a_{12} & a_{13} \\ a_{21} & a_{22} & a_{23} \end{bmatrix} \begin{bmatrix} 1 & 0 & 0 \\ 0 & 1 & 0 \\ 0 & 0 & 1 \end{bmatrix} = \begin{bmatrix} a_{11} & a_{12} & a_{13} \\ a_{21} & a_{22} & a_{23} \end{bmatrix} = A$$

and multiplying on the left by the 2×2 identity matrix yields

$$I_2 A = \begin{bmatrix} 1 & 0 \\ 0 & 1 \end{bmatrix} \begin{bmatrix} a_{11} & a_{12} & a_{13} \\ a_{21} & a_{22} & a_{23} \end{bmatrix} = \begin{bmatrix} a_{11} & a_{12} & a_{13} \\ a_{21} & a_{22} & a_{23} \end{bmatrix} = A$$

The same result holds in general; that is, if A is any $m \times n$ matrix, then

$$AI_n = A \quad \text{and} \quad I_m A = A$$

42 Chapter 1 Systems of Linear Equations and Matrices

Thus, the identity matrices play the same role in these matrix equations that the number 1 plays in the numerical equation $a \cdot 1 = 1 \cdot a = a$.

As the next theorem shows, identity matrices arise naturally in studying reduced row echelon forms of *square* matrices.

> **THEOREM 1.4.3** *If R is the reduced row echelon form of an $n \times n$ matrix A, then either R has a row of zeros or R is the identity matrix I_n.*

Proof Suppose that the reduced row echelon form of A is

$$R = \begin{bmatrix} r_{11} & r_{12} & \cdots & r_{1n} \\ r_{21} & r_{22} & \cdots & r_{2n} \\ \vdots & \vdots & & \vdots \\ r_{n1} & r_{n2} & \cdots & r_{nn} \end{bmatrix}$$

Either the last row in this matrix consists entirely of zeros or it does not. If not, the matrix contains no zero rows, and consequently each of the n rows has a leading entry of 1. Since these leading 1's occur progressively farther to the right as we move down the matrix, each of these 1's must occur on the main diagonal. Since the other entries in the same column as one of these 1's are zero, R must be I_n. Thus, either R has a row of zeros or $R = I_n$. ◂

Inverse of a Matrix

In real arithmetic every nonzero number a has a reciprocal $a^{-1}(=1/a)$ with the property

$$a \cdot a^{-1} = a^{-1} \cdot a = 1$$

The number a^{-1} is sometimes called the *multiplicative inverse* of a. Our next objective is to develop an analog of this result for matrix arithmetic. For this purpose we make the following definition.

> **DEFINITION 1** If A is a square matrix, and if a matrix B of the same size can be found such that $AB = BA = I$, then A is said to be ***invertible*** (or ***nonsingular***) and B is called an ***inverse*** of A. If no such matrix B can be found, then A is said to be ***singular***.

Remark The relationship $AB = BA = I$ is not changed by interchanging A and B, so if A is invertible and B is an inverse of A, then it is also true that B is invertible, and A is an inverse of B. Thus, when

$$AB = BA = I$$

we say that A and B are *inverses of one another*.

▶ **EXAMPLE 5 An Invertible Matrix**

Let

$$A = \begin{bmatrix} 2 & -5 \\ -1 & 3 \end{bmatrix} \quad \text{and} \quad B = \begin{bmatrix} 3 & 5 \\ 1 & 2 \end{bmatrix}$$

Then

$$AB = \begin{bmatrix} 2 & -5 \\ -1 & 3 \end{bmatrix} \begin{bmatrix} 3 & 5 \\ 1 & 2 \end{bmatrix} = \begin{bmatrix} 1 & 0 \\ 0 & 1 \end{bmatrix} = I$$

$$BA = \begin{bmatrix} 3 & 5 \\ 1 & 2 \end{bmatrix} \begin{bmatrix} 2 & -5 \\ -1 & 3 \end{bmatrix} = \begin{bmatrix} 1 & 0 \\ 0 & 1 \end{bmatrix} = I$$

Thus, A and B are invertible and each is an inverse of the other. ◂

▶ **EXAMPLE 6 A Class of Singular Matrices**

In general, a square matrix with a row or column of zeros is singular. To help understand why this is so, consider the matrix

$$A = \begin{bmatrix} 1 & 4 & 0 \\ 2 & 5 & 0 \\ 3 & 6 & 0 \end{bmatrix}$$

To prove that A is singular we must show that there is no 3×3 matrix B such that $AB = BA = I$. For this purpose let $\mathbf{c}_1, \mathbf{c}_2, \mathbf{0}$ be the column vectors of A. Thus, for any 3×3 matrix B we can express the product BA as

$$BA = B[\mathbf{c}_1 \quad \mathbf{c}_2 \quad \mathbf{0}] = [B\mathbf{c}_1 \quad B\mathbf{c}_2 \quad \mathbf{0}] \quad \text{[Formula (6) of Section 1.3]}$$

The column of zeros shows that $BA \neq I$ and hence that A is singular. ◀

Properties of Inverses

It is reasonable to ask whether an invertible matrix can have more than one inverse. The next theorem shows that the answer is no—*an invertible matrix has exactly one inverse.*

THEOREM 1.4.4 *If B and C are both inverses of the matrix A, then $B = C$.*

Proof Since B is an inverse of A, we have $BA = I$. Multiplying both sides on the right by C gives $(BA)C = IC = C$. But it is also true that $(BA)C = B(AC) = BI = B$, so $C = B$. ◀

As a consequence of this important result, we can now speak of "the" inverse of an invertible matrix. If A is invertible, then its inverse will be denoted by the symbol A^{-1}. Thus,

$$AA^{-1} = I \quad \text{and} \quad A^{-1}A = I \tag{1}$$

The inverse of A plays much the same role in matrix arithmetic that the reciprocal a^{-1} plays in the numerical relationships $aa^{-1} = 1$ and $a^{-1}a = 1$.

In the next section we will develop a method for computing the inverse of an invertible matrix of any size. For now we give the following theorem that specifies conditions under which a 2×2 matrix is invertible and provides a simple formula for its inverse.

THEOREM 1.4.5 *The matrix*

$$A = \begin{bmatrix} a & b \\ c & d \end{bmatrix}$$

is invertible if and only if $ad - bc \neq 0$, in which case the inverse is given by the formula

$$A^{-1} = \frac{1}{ad - bc} \begin{bmatrix} d & -b \\ -c & a \end{bmatrix} \tag{2}$$

The quantity $ad - bc$ in Theorem 1.4.5 is called the **determinant** of the 2×2 matrix A and is denoted by

$$\det(A) = ad - bc$$

or alternatively by

$$\begin{vmatrix} a & b \\ c & d \end{vmatrix} = ad - bc$$

We will omit the proof, because we will study a more general version of this theorem later. For now, you should at least confirm the validity of Formula (2) by showing that $AA^{-1} = A^{-1}A = I$.

Historical Note The formula for A^{-1} given in Theorem 1.4.5 first appeared (in a more general form) in Arthur Cayley's 1858 *Memoir on the Theory of Matrices*. The more general result that Cayley discovered will be studied later.

$$\det(A) = \begin{vmatrix} a & b \\ c & d \end{vmatrix} = ad - bc$$

▲ Figure 1.4.1

Remark Figure 1.4.1 illustrates that the determinant of a 2×2 matrix A is the product of the entries on its main diagonal minus the product of the entries *off* its main diagonal. In words, Theorem 1.4.5 states that a 2×2 matrix A is invertible if and only if its determinant is nonzero, and if invertible, then its inverse can be obtained by interchanging its diagonal entries, reversing the signs of its off-diagonal entries, and multiplying the entries by the reciprocal of the determinant of A.

▶ **EXAMPLE 7** **Calculating the Inverse of a 2 × 2 Matrix**

In each part, determine whether the matrix is invertible. If so, find its inverse.

$$\text{(a) } A = \begin{bmatrix} 6 & 1 \\ 5 & 2 \end{bmatrix} \qquad \text{(b) } A = \begin{bmatrix} -1 & 2 \\ 3 & -6 \end{bmatrix}$$

Solution (a) The determinant of A is $\det(A) = (6)(2) - (1)(5) = 7$, which is nonzero. Thus, A is invertible, and its inverse is

$$A^{-1} = \frac{1}{7}\begin{bmatrix} 2 & -1 \\ -5 & 6 \end{bmatrix} = \begin{bmatrix} \frac{2}{7} & -\frac{1}{7} \\ -\frac{5}{7} & \frac{6}{7} \end{bmatrix}$$

We leave it for you to confirm that $AA^{-1} = A^{-1}A = I$.

Solution (b) The matrix is not invertible since $\det(A) = (-1)(-6) - (2)(3) = 0$.

▶ **EXAMPLE 8** **Solution of a Linear System by Matrix Inversion**

A problem that arises in many applications is to solve a pair of equations of the form

$$u = ax + by$$
$$v = cx + dy$$

for x and y in terms of u and v. One approach is to treat this as a linear system of two equations in the unknowns x and y and use Gauss–Jordan elimination to solve for x and y. However, because the coefficients of the unknowns are *literal* rather than *numerical*, this procedure is a little clumsy. As an alternative approach, let us replace the two equations by the single matrix equation

$$\begin{bmatrix} u \\ v \end{bmatrix} = \begin{bmatrix} ax + by \\ cx + dy \end{bmatrix}$$

which we can rewrite as

$$\begin{bmatrix} u \\ v \end{bmatrix} = \begin{bmatrix} a & b \\ c & d \end{bmatrix}\begin{bmatrix} x \\ y \end{bmatrix}$$

If we assume that the 2×2 matrix is invertible (i.e., $ad - bc \neq 0$), then we can multiply through on the left by the inverse and rewrite the equation as

$$\begin{bmatrix} a & b \\ c & d \end{bmatrix}^{-1}\begin{bmatrix} u \\ v \end{bmatrix} = \begin{bmatrix} a & b \\ c & d \end{bmatrix}^{-1}\begin{bmatrix} a & b \\ c & d \end{bmatrix}\begin{bmatrix} x \\ y \end{bmatrix}$$

which simplifies to

$$\begin{bmatrix} a & b \\ c & d \end{bmatrix}^{-1}\begin{bmatrix} u \\ v \end{bmatrix} = \begin{bmatrix} x \\ y \end{bmatrix}$$

Using Theorem 1.4.5, we can rewrite this equation as

$$\frac{1}{ad - bc}\begin{bmatrix} d & -b \\ -c & a \end{bmatrix}\begin{bmatrix} u \\ v \end{bmatrix} = \begin{bmatrix} x \\ y \end{bmatrix}$$

from which we obtain
$$x = \frac{du - bv}{ad - bc}, \quad y = \frac{av - cu}{ad - bc}$$

The next theorem is concerned with inverses of matrix products.

THEOREM 1.4.6 *If A and B are invertible matrices with the same size, then AB is invertible and*
$$(AB)^{-1} = B^{-1}A^{-1}$$

Proof We can establish the invertibility and obtain the stated formula at the same time by showing that
$$(AB)(B^{-1}A^{-1}) = (B^{-1}A^{-1})(AB) = I$$
But
$$(AB)(B^{-1}A^{-1}) = A(BB^{-1})A^{-1} = AIA^{-1} = AA^{-1} = I$$
and similarly, $(B^{-1}A^{-1})(AB) = I$. ◄

Although we will not prove it, this result can be extended to three or more factors:

A product of any number of invertible matrices is invertible, and the inverse of the product is the product of the inverses in the reverse order.

► **EXAMPLE 9** **The Inverse of a Product**

Consider the matrices
$$A = \begin{bmatrix} 1 & 2 \\ 1 & 3 \end{bmatrix}, \quad B = \begin{bmatrix} 3 & 2 \\ 2 & 2 \end{bmatrix}$$

We leave it for you to show that

If a product of matrices is singular, then at least one of the factors must be singular. Why?

$$AB = \begin{bmatrix} 7 & 6 \\ 9 & 8 \end{bmatrix}, \quad (AB)^{-1} = \begin{bmatrix} 4 & -3 \\ -\frac{9}{2} & \frac{7}{2} \end{bmatrix}$$

and also that
$$A^{-1} = \begin{bmatrix} 3 & -2 \\ -1 & 1 \end{bmatrix}, \quad B^{-1} = \begin{bmatrix} 1 & -1 \\ -1 & \frac{3}{2} \end{bmatrix}, \quad B^{-1}A^{-1} = \begin{bmatrix} 1 & -1 \\ -1 & \frac{3}{2} \end{bmatrix}\begin{bmatrix} 3 & -2 \\ -1 & 1 \end{bmatrix} = \begin{bmatrix} 4 & -3 \\ -\frac{9}{2} & \frac{7}{2} \end{bmatrix}$$

Thus, $(AB)^{-1} = B^{-1}A^{-1}$ as guaranteed by Theorem 1.4.6. ◄

Powers of a Matrix   If A is a *square* matrix, then we define the nonnegative integer powers of A to be
$$A^0 = I \quad \text{and} \quad A^n = AA \cdots A \quad \text{[n factors]}$$
and if A is invertible, then we define the negative integer powers of A to be
$$A^{-n} = (A^{-1})^n = A^{-1}A^{-1} \cdots A^{-1} \quad \text{[n factors]}$$

Because these definitions parallel those for real numbers, the usual laws of nonnegative exponents hold; for example,
$$A^r A^s = A^{r+s} \quad \text{and} \quad (A^r)^s = A^{rs}$$

In addition, we have the following properties of negative exponents.

> **THEOREM 1.4.7** *If A is invertible and n is a nonnegative integer, then:*
> (a) A^{-1} *is invertible and* $(A^{-1})^{-1} = A$.
> (b) A^n *is invertible and* $(A^n)^{-1} = A^{-n} = (A^{-1})^n$.
> (c) kA *is invertible for any nonzero scalar k, and* $(kA)^{-1} = k^{-1}A^{-1}$.

We will prove part (c) and leave the proofs of parts (a) and (b) as exercises.

Proof (c) Property (c) in Theorem 1.4.1 and property (f) in Theorem 1.4.2 imply that
$$(kA)(k^{-1}A^{-1}) = k^{-1}(kA)A^{-1} = (k^{-1}k)AA^{-1} = (1)I = I$$
and similarly, $(k^{-1}A^{-1})(kA) = I$. Thus, kA is invertible and $(kA)^{-1} = k^{-1}A^{-1}$. ◄

▶ **EXAMPLE 10** **Properties of Exponents**

Let A and A^{-1} be the matrices in Example 9; that is,
$$A = \begin{bmatrix} 1 & 2 \\ 1 & 3 \end{bmatrix} \quad \text{and} \quad A^{-1} = \begin{bmatrix} 3 & -2 \\ -1 & 1 \end{bmatrix}$$

Then
$$A^{-3} = (A^{-1})^3 = \begin{bmatrix} 3 & -2 \\ -1 & 1 \end{bmatrix}\begin{bmatrix} 3 & -2 \\ -1 & 1 \end{bmatrix}\begin{bmatrix} 3 & -2 \\ -1 & 1 \end{bmatrix} = \begin{bmatrix} 41 & -30 \\ -15 & 11 \end{bmatrix}$$

Also,
$$A^3 = \begin{bmatrix} 1 & 2 \\ 1 & 3 \end{bmatrix}\begin{bmatrix} 1 & 2 \\ 1 & 3 \end{bmatrix}\begin{bmatrix} 1 & 2 \\ 1 & 3 \end{bmatrix} = \begin{bmatrix} 11 & 30 \\ 15 & 41 \end{bmatrix}$$

so, as expected from Theorem 1.4.7(b),
$$(A^3)^{-1} = \frac{1}{(11)(41) - (30)(15)}\begin{bmatrix} 41 & -30 \\ -15 & 11 \end{bmatrix} = \begin{bmatrix} 41 & -30 \\ -15 & 11 \end{bmatrix} = (A^{-1})^3$$

▶ **EXAMPLE 11** **The Square of a Matrix Sum**

In real arithmetic, where we have a commutative law for multiplication, we can write
$$(a + b)^2 = a^2 + ab + ba + b^2 = a^2 + ab + ab + b^2 = a^2 + 2ab + b^2$$
However, in matrix arithmetic, where we have no commutative law for multiplication, the best we can do is to write
$$(A + B)^2 = A^2 + AB + BA + B^2$$
It is only in the special case where A and B *commute* (i.e., $AB = BA$) that we can go a step further and write
$$(A + B)^2 = A^2 + 2AB + B^2 \quad ◄$$

Matrix Polynomials

If A is a square matrix, say $n \times n$, and if
$$p(x) = a_0 + a_1 x + a_2 x^2 + \cdots + a_m x^m$$
is any polynomial, then we define the $n \times n$ matrix $p(A)$ to be
$$p(A) = a_0 I + a_1 A + a_2 A^2 + \cdots + a_m A^m \tag{3}$$

where I is the $n \times n$ identity matrix; that is, $p(A)$ is obtained by substituting A for x and replacing the constant term a_0 by the matrix $a_0 I$. An expression of form (3) is called a **matrix polynomial in A**.

▶ **EXAMPLE 12** A Matrix Polynomial

Find $p(A)$ for
$$p(x) = x^2 - 2x - 3 \quad \text{and} \quad A = \begin{bmatrix} -1 & 2 \\ 0 & 3 \end{bmatrix}$$

Solution
$$p(A) = A^2 - 2A - 3I$$
$$= \begin{bmatrix} -1 & 2 \\ 0 & 3 \end{bmatrix}^2 - 2\begin{bmatrix} -1 & 2 \\ 0 & 3 \end{bmatrix} - 3\begin{bmatrix} 1 & 0 \\ 0 & 1 \end{bmatrix}$$
$$= \begin{bmatrix} 1 & 4 \\ 0 & 9 \end{bmatrix} - \begin{bmatrix} -2 & 4 \\ 0 & 6 \end{bmatrix} - \begin{bmatrix} 3 & 0 \\ 0 & 3 \end{bmatrix} = \begin{bmatrix} 0 & 0 \\ 0 & 0 \end{bmatrix}$$

or more briefly, $p(A) = 0$. ◀

Remark It follows from the fact that $A^r A^s = A^{r+s} = A^{s+r} = A^s A^r$ that powers of a square matrix commute, and since a matrix polynomial in A is built up from powers of A, any two matrix polynomials in A also commute; that is, for any polynomials p_1 and p_2 we have
$$p_1(A) p_2(A) = p_2(A) p_1(A) \tag{4}$$

Properties of the Transpose

The following theorem lists the main properties of the transpose.

THEOREM 1.4.8 *If the sizes of the matrices are such that the stated operations can be performed, then*:

(a) $(A^T)^T = A$
(b) $(A + B)^T = A^T + B^T$
(c) $(A - B)^T = A^T - B^T$
(d) $(kA)^T = kA^T$
(e) $(AB)^T = B^T A^T$

If you keep in mind that transposing a matrix interchanges its rows and columns, then you should have little trouble visualizing the results in parts (a)–(d). For example, part (a) states the obvious fact that interchanging rows and columns twice leaves a matrix unchanged; and part (b) states that adding two matrices and then interchanging the rows and columns produces the same result as interchanging the rows and columns before adding. We will omit the formal proofs. Part (e) is a less obvious, but for brevity we will omit its proof as well. The result in that part can be extended to three or more factors and restated as:

The transpose of a product of any number of matrices is the product of the transposes in the reverse order.

The following theorem establishes a relationship between the inverse of a matrix and the inverse of its transpose.

THEOREM 1.4.9 *If A is an invertible matrix, then A^T is also invertible and*
$$(A^T)^{-1} = (A^{-1})^T$$

Proof We can establish the invertibility and obtain the formula at the same time by showing that
$$A^T(A^{-1})^T = (A^{-1})^T A^T = I$$
But from part (*e*) of Theorem 1.4.8 and the fact that $I^T = I$, we have
$$A^T(A^{-1})^T = (A^{-1}A)^T = I^T = I$$
$$(A^{-1})^T A^T = (AA^{-1})^T = I^T = I$$
which completes the proof. ◂

▶ **EXAMPLE 13** **Inverse of a Transpose**

Consider a general 2×2 invertible matrix and its transpose:
$$A = \begin{bmatrix} a & b \\ c & d \end{bmatrix} \quad \text{and} \quad A^T = \begin{bmatrix} a & c \\ b & d \end{bmatrix}$$

Since A is invertible, its determinant $ad - bc$ is nonzero. But the determinant of A^T is also $ad - bc$ (verify), so A^T is also invertible. It follows from Theorem 1.4.5 that
$$(A^T)^{-1} = \begin{bmatrix} \dfrac{d}{ad-bc} & -\dfrac{c}{ad-bc} \\ -\dfrac{b}{ad-bc} & \dfrac{a}{ad-bc} \end{bmatrix}$$
which is the same matrix that results if A^{-1} is transposed (verify). Thus,
$$(A^T)^{-1} = (A^{-1})^T$$
as guaranteed by Theorem 1.4.9. ◂

Concept Review

- Commutative law for matrix addition
- Associative law for matrix addition
- Associative law for matrix multiplication
- Left and right distributive laws
- Zero matrix
- Identity matrix
- Inverse of a matrix
- Invertible matrix
- Nonsingular matrix
- Singular matrix
- Determinant
- Power of a matrix
- Matrix polynomial

Skills

- Know the arithmetic properties of matrix operations.
- Be able to prove arithmetic properties of matrices.
- Know the properties of zero matrices.
- Know the properties of identity matrices.
- Be able to recognize when two square matrices are inverses of each other.
- Be able to determine whether a 2×2 matrix is invertible.
- Be able to solve a linear system of two equations in two unknowns whose coefficient matrix is invertible.
- Be able to prove basic properties involving invertible matrices.
- Know the properties of the matrix transpose and its relationship with invertible matrices.

Exercise Set 1.4

1. Let
$$A = \begin{bmatrix} 2 & -1 & 3 \\ 0 & 4 & 5 \\ -2 & 1 & 4 \end{bmatrix}, \quad B = \begin{bmatrix} 8 & -3 & -5 \\ 0 & 1 & 2 \\ 4 & -7 & 6 \end{bmatrix},$$
$$C = \begin{bmatrix} 0 & -2 & 3 \\ 1 & 7 & 4 \\ 3 & 5 & 9 \end{bmatrix}, \quad a = 4, \quad b = -7$$

Show that

(a) $A + (B + C) = (A + B) + C$

(b) $(AB)C = A(BC)$ (c) $(a + b)C = aC + bC$

(d) $a(B - C) = aB - aC$

2. Using the matrices and scalars in Exercise 1, verify that

(a) $a(BC) = (aB)C = B(aC)$

(b) $A(B - C) = AB - AC$ (c) $(B + C)A = BA + CA$

(d) $a(bC) = (ab)C$

3. Using the matrices and scalars in Exercise 1, verify that

(a) $(B^T)^T = B$ (b) $(A + C)^T = A^T + C^T$

(c) $(bA)^T = bA^T$ (d) $(CA)^T = A^T C^T$

▶ In Exercises 4–7, use Theorem 1.4.5 to compute the inverses of the following matrices. ◀

4. $A = \begin{bmatrix} 3 & 1 \\ 5 & 2 \end{bmatrix}$ **5.** $B = \begin{bmatrix} 6 & 3 \\ -5 & -2 \end{bmatrix}$

6. $C = \begin{bmatrix} 6 & 4 \\ -2 & -1 \end{bmatrix}$ **7.** $D = \begin{bmatrix} 0 & -3 \\ 7 & 2 \end{bmatrix}$

8. Find the inverse of
$$\begin{bmatrix} \cos\theta & \sin\theta \\ -\sin\theta & \cos\theta \end{bmatrix}$$

9. Find the inverse of
$$\begin{bmatrix} \frac{1}{2}(e^x + e^{-x}) & \frac{1}{2}(e^x - e^{-x}) \\ \frac{1}{2}(e^x - e^{-x}) & \frac{1}{2}(e^x + e^{-x}) \end{bmatrix}$$

10. Use the matrix A in Exercise 4 to verify that $(A^T)^{-1} = (A^{-1})^T$.

11. Use the matrix B in Exercise 5 to verify that $(B^T)^{-1} = (B^{-1})^T$.

12. Use the matrices A and B in Exercises 4 and 5 to verify that $(AB)^{-1} = B^{-1}A^{-1}$.

13. Use the matrices A, B, and C in Exercises 4–6 to verify that $(ABC)^{-1} = C^{-1}B^{-1}A^{-1}$.

▶ In Exercises 14–17, use the given information to find A. ◀

14. $A^{-1} = \begin{bmatrix} 2 & -1 \\ 3 & 5 \end{bmatrix}$ **15.** $(5A)^{-1} = \begin{bmatrix} 4 & 2 \\ 1 & 3 \end{bmatrix}$

16. $(5A^T)^{-1} = \begin{bmatrix} -3 & -1 \\ 5 & 2 \end{bmatrix}$ **17.** $(I + 2A)^{-1} = \begin{bmatrix} -1 & 2 \\ 4 & 5 \end{bmatrix}$

18. Let A be the matrix
$$\begin{bmatrix} 2 & 0 \\ 4 & 1 \end{bmatrix}$$
In each part, compute the given quantity.

(a) A^3 (b) A^{-3} (c) $A^2 - 2A + I$

(d) $p(A)$, where $p(x) = x - 2$

(e) $p(A)$, where $p(x) = 2x^2 - x + 1$

(f) $p(A)$, where $p(x) = x^3 - 2x + 4$

19. Repeat Exercise 18 for the matrix
$$A = \begin{bmatrix} 1 & -1 \\ -2 & 3 \end{bmatrix}$$

20. Repeat Exercise 18 for the matrix
$$A = \begin{bmatrix} 3 & 0 & -1 \\ 0 & -2 & 0 \\ 5 & 0 & 2 \end{bmatrix}$$

21. Repeat Exercise 18 for the matrix
$$A = \begin{bmatrix} 3 & 0 & 0 \\ 0 & -1 & 3 \\ 0 & -3 & -1 \end{bmatrix}$$

▶ In Exercises 22–24, let $p_1(x) = x^2 - 9$, $p_2(x) = x + 3$, and $p_3(x) = x - 3$. Show that $p_1(A) = p_2(A)p_3(A)$ for the given matrix. ◀

22. The matrix A in Exercise 18.

23. The matrix A in Exercise 21.

24. An arbitrary square matrix A.

25. Show that if $p(x) = x^2 - (a + d)x + (ad - bc)$ and
$$A = \begin{bmatrix} a & b \\ c & d \end{bmatrix}$$
then $p(A) = 0$.

26. Show that if $p(x) = x^3 - (a + b + c)x^2 + (ab + ae + be - cd)x - a(be - cd)$ and
$$A = \begin{bmatrix} a & 0 & 0 \\ 0 & b & c \\ 0 & d & e \end{bmatrix}$$
then $p(A) = 0$.

27. Consider the matrix

$$A = \begin{bmatrix} a_{11} & 0 & \cdots & 0 \\ 0 & a_{22} & \cdots & 0 \\ \vdots & \vdots & & \vdots \\ 0 & 0 & \cdots & a_{nn} \end{bmatrix}$$

where $a_{11} a_{22} \cdots a_{nn} \neq 0$. Show that A is invertible and find its inverse.

28. Show that if a square matrix A satisfies the equation $A^2 + 5A - 2I = 0$, then $A^{-1} = \frac{1}{2}(A + 5I)$.

29. (a) Show that a matrix with a row of zeros cannot have an inverse.

   (b) Show that a matrix with a column of zeros cannot have an inverse.

30. Assuming that all matrices are $n \times n$ and invertible, solve for D.
$$ABC^T DBA^T C = AB^T$$

31. Assuming that all matrices are $n \times n$ and invertible, solve for D.
$$C^T B^{-1} A^2 BAC^{-1} DA^{-2} B^T C^{-2} = C^T$$

32. If A is a square matrix and n is a positive integer, is it true that $(A^n)^T = (A^T)^n$? Justify your answer.

33. Simplify:
$$D^{-1}CBA(BA)^{-1}C^{-1}(C^{-1}D)^{-1}$$

34. Simplify:
$$(AC^{-1})^{-1}(AC^{-1})(AC^{-1})^{-1}AD^{-1}$$

▶ In Exercises 35–37, determine whether A is invertible, and if so, find the inverse. [Hint: Solve $AX = I$ for X by equating corresponding entries on the two sides.] ◀

35. $A = \begin{bmatrix} 1 & 0 & 1 \\ 1 & 1 & 0 \\ 0 & 1 & 1 \end{bmatrix}$

36. $A = \begin{bmatrix} 1 & 1 & 1 \\ 1 & 0 & 0 \\ 0 & 1 & 1 \end{bmatrix}$

37. $A = \begin{bmatrix} 1 & 0 & 1 \\ 0 & 1 & 0 \\ 1 & 0 & -1 \end{bmatrix}$

38. Prove Theorem 1.4.2.

▶ In Exercises 39–42, use the method of Example 8 to find the unique solution of the given linear system. ◀

39. $3x_1 - 2x_2 = -1$
    $4x_1 + 5x_2 = 3$

40. $-x_1 + 5x_2 = 4$
    $-x_1 - 3x_2 = 1$

41. $7x_1 + 2x_2 = 3$
    $3x_1 + x_2 = 0$

42. $2x_1 - 2x_2 = 4$
    $x_1 + 4x_2 = 4$

43. Prove part (a) of Theorem 1.4.1.

44. Prove part (c) of Theorem 1.4.1.

45. Prove part (f) of Theorem 1.4.1.

46. Prove part (b) of Theorem 1.4.2.

47. Prove part (c) of Theorem 1.4.2.

48. Verify Formula (4) in the text by a direct calculation.

49. Prove part (d) of Theorem 1.4.8.

50. Prove part (e) of Theorem 1.4.8.

51. (a) Show that if A is invertible and $AB = AC$, then $B = C$.

   (b) Explain why part (a) and Example 3 do not contradict one another.

52. Show that if A is invertible and k is any nonzero scalar, then $(kA)^n = k^n A^n$ for all integer values of n.

53. (a) Show that if A, B, and $A + B$ are invertible matrices with the same size, then
$$A(A^{-1} + B^{-1})B(A + B)^{-1} = I$$

   (b) What does the result in part (a) tell you about the matrix $A^{-1} + B^{-1}$?

54. A square matrix A is said to be *idempotent* if $A^2 = A$.

   (a) Show that if A is idempotent, then so is $I - A$.

   (b) Show that if A is idempotent, then $2A - I$ is invertible and is its own inverse.

55. Show that if A is a square matrix such that $A^k = 0$ for some positive integer k, then the matrix A is invertible and
$$(I - A)^{-1} = I + A + A^2 + \cdots + A^{k-1}$$

True-False Exercises

In parts (a)–(k) determine whether the statement is true or false, and justify your answer.

(a) Two $n \times n$ matrices, A and B, are inverses of one another if and only if $AB = BA = 0$.

(b) For all square matrices A and B of the same size, it is true that $(A + B)^2 = A^2 + 2AB + B^2$.

(c) Give an example of a 2×2 matrix A that is idempotent but is not the zero matrix or the identity matrix.

(d) If A and B are invertible matrices of the same size, then AB is invertible and $(AB)^{-1} = A^{-1}B^{-1}$.

(e) If A and B are matrices such that AB is defined, then it is true that $(AB)^T = A^T B^T$.

(f) The matrix
$$A = \begin{bmatrix} a & b \\ c & d \end{bmatrix}$$
is invertible if and only if $ad - bc \neq 0$.

(g) If A and B are matrices of the same size and k is a constant, then $(kA + B)^T = kA^T + B^T$.

(h) If A is an invertible matrix, then so is A^T.

(i) If $p(x) = a_0 + a_1 x + a_2 x^2 + \cdots + a_m x^m$ and I is an identity matrix, then $p(I) = a_0 + a_1 + a_2 + \cdots + a_m$.

(j) A square matrix containing a row or column of zeros cannot be invertible.

(k) The sum of two invertible matrices of the same size must be invertible.

1.5 Elementary Matrices and a Method for Finding A^{-1}

In this section we will develop an algorithm for finding the inverse of a matrix, and we will discuss some of the basic properties of invertible matrices.

In Section 1.1 we defined three elementary row operations on a matrix A:

1. Multiply a row by a nonzero constant c.
2. Interchange two rows.
3. Add a constant c times one row to another.

It should be evident that if we let B be the matrix that results from A by performing one of the operations in this list, then the matrix A can be recovered from B by performing the corresponding operation in the following list:

1. Multiply the same row by $1/c$.
2. Interchange the same two rows.
3. If B resulted by adding c times row r_1 of A to row r_2, then add $-c$ times r_1 to row r_2.

It follows that if B is obtained from A by performing a sequence of elementary row operations, then there is a second sequence of elementary row operations, which when applied to B recovers A (Exercise 43). Accordingly, we make the following definition.

> **DEFINITION 1** Matrices A and B are said to be *row equivalent* if either (hence each) can be obtained from the other by a sequence of elementary row operations.

Our next goal is to show how matrix multiplication can be used to carry out an elementary row operation.

> **DEFINITION 2** An $n \times n$ matrix is called an *elementary matrix* if it can be obtained from the $n \times n$ identity matrix I_n by performing a *single* elementary row operation.

▶ **EXAMPLE 1** Elementary Matrices and Row Operations

Listed below are four elementary matrices and the operations that produce them.

$$\begin{bmatrix} 1 & 0 \\ 0 & -3 \end{bmatrix} \quad \begin{bmatrix} 1 & 0 & 0 & 0 \\ 0 & 0 & 0 & 1 \\ 0 & 0 & 1 & 0 \\ 0 & 1 & 0 & 0 \end{bmatrix} \quad \begin{bmatrix} 1 & 0 & 3 \\ 0 & 1 & 0 \\ 0 & 0 & 1 \end{bmatrix} \quad \begin{bmatrix} 1 & 0 & 0 \\ 0 & 1 & 0 \\ 0 & 0 & 1 \end{bmatrix}$$

Multiply the second row of I_2 by -3.

Interchange the second and fourth rows of I_4.

Add 3 times the third row of I_3 to the first row.

Multiply the first row of I_3 by 1. ◀

The following theorem, whose proof is left as an exercise, shows that when a matrix A is multiplied on the *left* by an elementary matrix E, the effect is to perform an elementary row operation on A.

THEOREM 1.5.1 *Row Operations by Matrix Multiplication*

If the elementary matrix E results from performing a certain row operation on I_m and if A is an $m \times n$ matrix, then the product EA is the matrix that results when this same row operation is performed on A.

▶ **EXAMPLE 2** **Using Elementary Matrices**

Consider the matrix
$$A = \begin{bmatrix} 1 & 0 & 2 & 3 \\ 2 & -1 & 3 & 6 \\ 1 & 4 & 4 & 0 \end{bmatrix}$$
and consider the elementary matrix
$$E = \begin{bmatrix} 1 & 0 & 0 \\ 0 & 1 & 0 \\ 3 & 0 & 1 \end{bmatrix}$$
which results from adding 3 times the first row of I_3 to the third row. The product EA is
$$EA = \begin{bmatrix} 1 & 0 & 2 & 3 \\ 2 & -1 & 3 & 6 \\ 4 & 4 & 10 & 9 \end{bmatrix}$$
which is precisely the matrix that results when we add 3 times the first row of A to the third row. ◀

Theorem 1.5.1 will be a useful tool for developing new results about matrices, but as a practical matter it is usually preferable to perform row operations directly.

We know from the discussion at the beginning of this section that if E is an elementary matrix that results from performing an elementary row operation on an identity matrix I, then there is a second elementary row operation, which when applied to E, produces I back again. Table 1 lists these operations. The operations on the right side of the table are called the ***inverse operations*** of the corresponding operations on the left.

Table 1

Row Operation on I That Produces E	Row Operation on E That Reproduces I
Multiply row i by $c \neq 0$	Multiply row i by $1/c$
Interchange rows i and j	Interchange rows i and j
Add c times row i to row j	Add $-c$ times row i to row j

▶ **EXAMPLE 3** **Row Operations and Inverse Row Operations**

In each of the following, an elementary row operation is applied to the 2×2 identity matrix to obtain an elementary matrix E, then E is restored to the identity matrix by applying the inverse row operation.

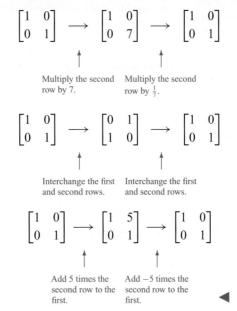

The next theorem is a key result about invertibility of elementary matrices. It will be a building block for many results that follow.

THEOREM 1.5.2 *Every elementary matrix is invertible, and the inverse is also an elementary matrix.*

Proof If E is an elementary matrix, then E results by performing some row operation on I. Let E_0 be the matrix that results when the inverse of this operation is performed on I. Applying Theorem 1.5.1 and using the fact that inverse row operations cancel the effect of each other, it follows that
$$E_0 E = I \quad \text{and} \quad E E_0 = I$$
Thus, the elementary matrix E_0 is the inverse of E. ◄

Equivalence Theorem

One of our objectives as we progress through this text is to show how seemingly diverse ideas in linear algebra are related. The following theorem, which relates results we have obtained about invertibility of matrices, homogeneous linear systems, reduced row echelon forms, and elementary matrices, is our first step in that direction. As we study new topics, more statements will be added to this theorem.

THEOREM 1.5.3 **Equivalent Statements**

If A is an $n \times n$ matrix, then the following statements are equivalent, that is, all true or all false.

(a) *A is invertible.*
(b) *$A\mathbf{x} = \mathbf{0}$ has only the trivial solution.*
(c) *The reduced row echelon form of A is I_n.*
(d) *A is expressible as a product of elementary matrices.*

It may make the logic of our proof of Theorem 1.5.3 more apparent by writing the implications

$(a) \Rightarrow (b) \Rightarrow (c) \Rightarrow (d) \Rightarrow (a)$

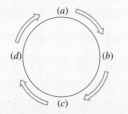

This makes it evident visually that the validity of any one statement implies the validity of all the others, and hence that the falsity of any one implies the falsity of the others.

Proof We will prove the equivalence by establishing the chain of implications: $(a) \Rightarrow (b) \Rightarrow (c) \Rightarrow (d) \Rightarrow (a)$.

(a) $\Rightarrow$ (b) Assume A is invertible and let $\mathbf{x}_0$ be any solution of $A\mathbf{x} = \mathbf{0}$. Multiplying both sides of this equation by the matrix A^{-1} gives $A^{-1}(A\mathbf{x}_0) = A^{-1}\mathbf{0}$, or $(A^{-1}A)\mathbf{x}_0 = \mathbf{0}$, or $I\mathbf{x}_0 = \mathbf{0}$, or $\mathbf{x}_0 = \mathbf{0}$. Thus, $A\mathbf{x} = \mathbf{0}$ has only the trivial solution.

(b) $\Rightarrow$ (c) Let $A\mathbf{x} = \mathbf{0}$ be the matrix form of the system

$$\begin{aligned} a_{11}x_1 + a_{12}x_2 + \cdots + a_{1n}x_n &= 0 \\ a_{21}x_1 + a_{22}x_2 + \cdots + a_{2n}x_n &= 0 \\ \vdots \quad \vdots \quad \vdots \quad \vdots & \\ a_{n1}x_1 + a_{n2}x_2 + \cdots + a_{nn}x_n &= 0 \end{aligned} \quad (1)$$

and assume that the system has only the trivial solution. If we solve by Gauss–Jordan elimination, then the system of equations corresponding to the reduced row echelon form of the augmented matrix will be

$$\begin{aligned} x_1 \qquad\qquad\qquad &= 0 \\ x_2 \qquad\qquad &= 0 \\ \ddots \qquad & \\ x_n &= 0 \end{aligned} \quad (2)$$

Thus the augmented matrix

$$\begin{bmatrix} a_{11} & a_{12} & \cdots & a_{1n} & 0 \\ a_{21} & a_{22} & \cdots & a_{2n} & 0 \\ \vdots & \vdots & & \vdots & \vdots \\ a_{n1} & a_{n2} & \cdots & a_{nn} & 0 \end{bmatrix}$$

for (1) can be reduced to the augmented matrix

$$\begin{bmatrix} 1 & 0 & 0 & \cdots & 0 & 0 \\ 0 & 1 & 0 & \cdots & 0 & 0 \\ 0 & 0 & 1 & \cdots & 0 & 0 \\ \vdots & \vdots & \vdots & & \vdots & \vdots \\ 0 & 0 & 0 & \cdots & 1 & 0 \end{bmatrix}$$

for (2) by a sequence of elementary row operations. If we disregard the last column (all zeros) in each of these matrices, we can conclude that the reduced row echelon form of A is I_n.

(c) $\Rightarrow$ (d) Assume that the reduced row echelon form of A is I_n, so that A can be reduced to I_n by a finite sequence of elementary row operations. By Theorem 1.5.1, each of these operations can be accomplished by multiplying on the left by an appropriate elementary matrix. Thus we can find elementary matrices $E_1, E_2, \ldots, E_k$ such that

$$E_k \cdots E_2 E_1 A = I_n \quad (3)$$

By Theorem 1.5.2, $E_1, E_2, \ldots, E_k$ are invertible. Multiplying both sides of Equation (3) on the left successively by $E_k^{-1}, \ldots, E_2^{-1}, E_1^{-1}$ we obtain

$$A = E_1^{-1} E_2^{-1} \cdots E_k^{-1} I_n = E_1^{-1} E_2^{-1} \cdots E_k^{-1} \qquad (4)$$

By Theorem 1.5.2, this equation expresses A as a product of elementary matrices.

(d) $\Rightarrow$ (a) If A is a product of elementary matrices, then from Theorems 1.4.7 and 1.5.2, the matrix A is a product of invertible matrices and hence is invertible. ◄

A Method for Inverting Matrices

As a first application of Theorem 1.5.3, we will develop a procedure (or algorithm) that can be used to tell whether a given matrix is invertible, and if so, produce its inverse. To derive this algorithm, assume for the moment, that A is an invertible $n \times n$ matrix. In Equation (3), the elementary matrices execute a sequence of row operations that reduce A to I_n. If we multiply both sides of this equation on the right by A^{-1} and simplify, we obtain

$$A^{-1} = E_k \cdots E_2 E_1 I_n$$

But this equation tells us that *the same sequence of row operations that reduces A to I_n will transform I_n to A^{-1}*. Thus, we have established the following result.

> **Inversion Algorithm** To find the inverse of an invertible matrix A, find a sequence of elementary row operations that reduces A to the identity and then perform that same sequence of operations on I_n to obtain A^{-1}.

A simple method for carrying out this procedure is given in the following example.

▶ **EXAMPLE 4** Using Row Operations to Find A^{-1}

Find the inverse of

$$A = \begin{bmatrix} 1 & 2 & 3 \\ 2 & 5 & 3 \\ 1 & 0 & 8 \end{bmatrix}$$

Solution We want to reduce A to the identity matrix by row operations and simultaneously apply these operations to I to produce A^{-1}. To accomplish this we will adjoin the identity matrix to the right side of A, thereby producing a partitioned matrix of the form

$$[A \mid I]$$

Then we will apply row operations to this matrix until the left side is reduced to I; these operations will convert the right side to A^{-1}, so the final matrix will have the form

$$[I \mid A^{-1}]$$

The computations are as follows:

$$\begin{bmatrix} 1 & 2 & 3 & | & 1 & 0 & 0 \\ 2 & 5 & 3 & | & 0 & 1 & 0 \\ 1 & 0 & 8 & | & 0 & 0 & 1 \end{bmatrix}$$

$$\begin{bmatrix} 1 & 2 & 3 & | & 1 & 0 & 0 \\ 0 & 1 & -3 & | & -2 & 1 & 0 \\ 0 & -2 & 5 & | & -1 & 0 & 1 \end{bmatrix}$$ ← We added -2 times the first row to the second and -1 times the first row to the third.

$$\begin{bmatrix} 1 & 2 & 3 & | & 1 & 0 & 0 \\ 0 & 1 & -3 & | & -2 & 1 & 0 \\ 0 & 0 & -1 & | & -5 & 2 & 1 \end{bmatrix}$$ ← We added 2 times the second row to the third.

$$\begin{bmatrix} 1 & 2 & 3 & | & 1 & 0 & 0 \\ 0 & 1 & -3 & | & -2 & 1 & 0 \\ 0 & 0 & 1 & | & 5 & -2 & -1 \end{bmatrix}$$ ← We multiplied the third row by -1.

$$\begin{bmatrix} 1 & 2 & 0 & | & -14 & 6 & 3 \\ 0 & 1 & 0 & | & 13 & -5 & -3 \\ 0 & 0 & 1 & | & 5 & -2 & -1 \end{bmatrix}$$ ← We added 3 times the third row to the second and -3 times the third row to the first.

$$\begin{bmatrix} 1 & 0 & 0 & | & -40 & 16 & 9 \\ 0 & 1 & 0 & | & 13 & -5 & -3 \\ 0 & 0 & 1 & | & 5 & -2 & -1 \end{bmatrix}$$ ← We added -2 times the second row to the first.

Thus,

$$A^{-1} = \begin{bmatrix} -40 & 16 & 9 \\ 13 & -5 & -3 \\ 5 & -2 & -1 \end{bmatrix}$$ ◀

Often it will not be known in advance if a given $n \times n$ matrix A is invertible. However, if it is not, then by parts (*a*) and (*c*) of Theorem 1.5.3 it will be impossible to reduce A to I_n by elementary row operations. This will be signaled by a row of zeros appearing on the *left side* of the partition at some stage of the inversion algorithm. If this occurs, then you can stop the computations and conclude that A is not invertible.

▶ **EXAMPLE 5 Showing That a Matrix Is Not Invertible**

Consider the matrix

$$A = \begin{bmatrix} 1 & 6 & 4 \\ 2 & 4 & -1 \\ -1 & 2 & 5 \end{bmatrix}$$

Applying the procedure of Example 4 yields

$$\begin{bmatrix} 1 & 6 & 4 & | & 1 & 0 & 0 \\ 2 & 4 & -1 & | & 0 & 1 & 0 \\ -1 & 2 & 5 & | & 0 & 0 & 1 \end{bmatrix}$$

$$\begin{bmatrix} 1 & 6 & 4 & | & 1 & 0 & 0 \\ 0 & -8 & -9 & | & -2 & 1 & 0 \\ 0 & 8 & 9 & | & 1 & 0 & 1 \end{bmatrix} \longleftarrow \text{We added } -2 \text{ times the first row to the second and added the first row to the third.}$$

$$\begin{bmatrix} 1 & 6 & 4 & | & 1 & 0 & 0 \\ 0 & -8 & -9 & | & -2 & 1 & 0 \\ 0 & 0 & 0 & | & -1 & 1 & 1 \end{bmatrix} \longleftarrow \text{We added the second row to the third.}$$

Since we have obtained a row of zeros on the left side, A is not invertible.

▶ **EXAMPLE 6** **Analyzing Homogeneous Systems**

Use Theorem 1.5.3 to determine whether the given homogeneous system has nontrivial solutions.

(a) $\begin{aligned} x_1 + 2x_2 + 3x_3 &= 0 \\ 2x_1 + 5x_2 + 3x_3 &= 0 \\ x_1 \phantom{{}+2x_2} + 8x_3 &= 0 \end{aligned}$ (b) $\begin{aligned} x_1 + 6x_2 + 4x_3 &= 0 \\ 2x_1 + 4x_2 - x_3 &= 0 \\ -x_1 + 2x_2 + 5x_3 &= 0 \end{aligned}$

Solution From parts (*a*) and (*b*) of Theorem 1.5.3 a homogeneous linear system has only the trivial solution if and only if its coefficient matrix is invertible. From Examples 4 and 5 the coefficient matrix of system (a) is invertible and that of system (b) is not. Thus, system (a) has only the trivial solution whereas system (b) has nontrivial solutions.

◀

Concept Review

- Row equivalent matrices
- Elementary matrix
- Inverse operations
- Inversion algorithm

Skills

- Determine whether a given square matrix is an elementary.
- Determine whether two square matrices are row equivalent.
- Apply the inverse of a given elementary row operation to a matrix.
- Apply elementary row operations to reduce a given square matrix to the identity matrix.
- Understand the relationships between statements that are equivalent to the invertibility of a square matrix (Theorem 1.5.3).
- Use the inversion algorithm to find the inverse of an invertible matrix.
- Express an invertible matrix as a product of elementary matrices.

Exercise Set 1.5

1. Decide whether each matrix below is an elementary matrix.

(a) $\begin{bmatrix} 1 & 0 \\ -5 & 1 \end{bmatrix}$

(b) $\begin{bmatrix} -5 & 1 \\ 1 & 0 \end{bmatrix}$

(c) $\begin{bmatrix} 1 & 1 & 0 \\ 0 & 0 & 1 \\ 0 & 0 & 0 \end{bmatrix}$

(d) $\begin{bmatrix} 2 & 0 & 0 & 2 \\ 0 & 1 & 0 & 0 \\ 0 & 0 & 1 & 0 \\ 0 & 0 & 0 & 1 \end{bmatrix}$

2. Decide whether each matrix below is an elementary matrix.

(a) $\begin{bmatrix} 1 & 0 \\ 0 & \sqrt{3} \end{bmatrix}$

(b) $\begin{bmatrix} 0 & 0 & 1 \\ 0 & 1 & 0 \\ 1 & 0 & 0 \end{bmatrix}$

(c) $\begin{bmatrix} 1 & 0 & 0 \\ 0 & 1 & 9 \\ 0 & 0 & 1 \end{bmatrix}$

(d) $\begin{bmatrix} -1 & 0 & 0 \\ 0 & 0 & 1 \\ 0 & 1 & 0 \end{bmatrix}$

3. Find a row operation and the corresponding elementary matrix that will restore the given elementary matrix to the identity matrix.

(a) $\begin{bmatrix} 1 & 0 \\ 0 & -4 \end{bmatrix}$

(b) $\begin{bmatrix} 1 & 9 & 0 \\ 0 & 1 & 0 \\ 0 & 0 & 1 \end{bmatrix}$

(c) $\begin{bmatrix} 1 & 0 & 0 \\ 0 & 0 & 1 \\ 0 & 1 & 0 \end{bmatrix}$

(d) $\begin{bmatrix} 1 & 0 & 0 & 0 \\ 0 & 1 & 0 & -1 \\ 0 & 0 & 1 & 0 \\ 0 & 0 & 0 & 1 \end{bmatrix}$

4. Find a row operation and the corresponding elementary matrix that will restore the given elementary matrix to the identity matrix.

(a) $\begin{bmatrix} 1 & 0 \\ -3 & 1 \end{bmatrix}$

(b) $\begin{bmatrix} 1 & 0 & 0 \\ 0 & 1 & 0 \\ 0 & 0 & 3 \end{bmatrix}$

(c) $\begin{bmatrix} 0 & 0 & 0 & 1 \\ 0 & 1 & 0 & 0 \\ 0 & 0 & 1 & 0 \\ 1 & 0 & 0 & 0 \end{bmatrix}$

(d) $\begin{bmatrix} 1 & 0 & -\frac{1}{7} & 0 \\ 0 & 1 & 0 & 0 \\ 0 & 0 & 1 & 0 \\ 0 & 0 & 0 & 1 \end{bmatrix}$

5. In each part, an elementary matrix E and a matrix A are given. Write down the row operation corresponding to E and show that the product EA results from applying the row operation to A.

(a) $E = \begin{bmatrix} 0 & 1 \\ 1 & 0 \end{bmatrix}$, $A = \begin{bmatrix} -1 & -2 & 5 & -1 \\ 3 & -6 & -6 & -6 \end{bmatrix}$

(b) $E = \begin{bmatrix} 1 & 0 & 0 \\ 0 & 1 & 0 \\ 0 & -3 & 1 \end{bmatrix}$, $A = \begin{bmatrix} 2 & -1 & 0 & -4 & -4 \\ 1 & -3 & -1 & 5 & 3 \\ 2 & 0 & 1 & 3 & -1 \end{bmatrix}$

(c) $E = \begin{bmatrix} 1 & 0 & 4 \\ 0 & 1 & 0 \\ 0 & 0 & 1 \end{bmatrix}$, $A = \begin{bmatrix} 1 & 4 \\ 2 & 5 \\ 3 & 6 \end{bmatrix}$

6. In each part, an elementary matrix E and a matrix A are given. Write down the row operation corresponding to E and show that the product EA results from applying the row operation to A.

(a) $E = \begin{bmatrix} -6 & 0 \\ 0 & 1 \end{bmatrix}$, $A = \begin{bmatrix} -1 & -2 & 5 & -1 \\ 3 & -6 & -6 & -6 \end{bmatrix}$

(b) $E = \begin{bmatrix} 1 & 0 & 0 \\ -4 & 1 & 0 \\ 0 & 0 & 1 \end{bmatrix}$, $A = \begin{bmatrix} 2 & -1 & 0 & -4 & -4 \\ 1 & -3 & -1 & 5 & 3 \\ 2 & 0 & 1 & 3 & -1 \end{bmatrix}$

(c) $E = \begin{bmatrix} 1 & 0 & 0 \\ 0 & 5 & 0 \\ 0 & 0 & 1 \end{bmatrix}$, $A = \begin{bmatrix} 1 & 4 \\ 2 & 5 \\ 3 & 6 \end{bmatrix}$

▶ In Exercises 7–8, use the following matrices.

$A = \begin{bmatrix} 2 & 6 & -8 \\ 0 & 5 & 3 \\ 4 & 7 & 9 \end{bmatrix}$, $B = \begin{bmatrix} 2 & 6 & -8 \\ 0 & 5 & 3 \\ 0 & -5 & 25 \end{bmatrix}$

$C = \begin{bmatrix} 1 & 3 & -4 \\ 0 & 5 & 3 \\ 4 & 7 & 9 \end{bmatrix}$, $D = \begin{bmatrix} 2 & 6 & -8 \\ 0 & 0 & 28 \\ 0 & -5 & 25 \end{bmatrix}$

$F = \begin{bmatrix} 2 & 6 & -8 \\ 6 & 23 & -21 \\ 0 & -5 & 25 \end{bmatrix}$ ◀

7. Find an elementary matrix E that satisfies the equation.

(a) $EA = B$ (b) $EB = A$

(c) $EA = C$ (d) $EC = A$

8. Find an elementary matrix E that satisfies the equation.

(a) $EB = D$ (b) $ED = B$

(c) $EB = F$ (d) $EF = B$

▶ In Exercises 9–24, use the inversion algorithm to find the inverse of the given matrix, if the inverse exists. ◀

9. $\begin{bmatrix} 1 & 4 \\ 2 & 7 \end{bmatrix}$

10. $\begin{bmatrix} -3 & 6 \\ 4 & 5 \end{bmatrix}$

11. $\begin{bmatrix} 4 & 6 \\ 2 & 3 \end{bmatrix}$

12. $\begin{bmatrix} 6 & -4 \\ -3 & 2 \end{bmatrix}$

13. $\begin{bmatrix} 2 & 1 & -1 \\ 0 & 6 & 4 \\ 0 & -2 & 2 \end{bmatrix}$

14. $\begin{bmatrix} 1 & 2 & 0 \\ 2 & 1 & 2 \\ 0 & 2 & 1 \end{bmatrix}$

15. $\begin{bmatrix} -1 & 3 & -4 \\ 2 & 4 & 1 \\ -4 & 2 & -9 \end{bmatrix}$

16. $\begin{bmatrix} \frac{1}{5} & \frac{1}{5} & -\frac{2}{5} \\ \frac{1}{5} & \frac{1}{5} & \frac{1}{10} \\ \frac{1}{5} & -\frac{4}{5} & \frac{1}{10} \end{bmatrix}$

17. $\begin{bmatrix} 1 & 0 & 1 \\ 0 & 1 & 1 \\ 1 & 1 & 0 \end{bmatrix}$

18. $\begin{bmatrix} \sqrt{2} & 3\sqrt{2} & 0 \\ -4\sqrt{2} & \sqrt{2} & 0 \\ 0 & 0 & 1 \end{bmatrix}$

19. $\begin{bmatrix} 1 & 4 & 4 \\ 1 & 2 & 4 \\ 1 & 3 & 2 \end{bmatrix}$

20. $\begin{bmatrix} 1 & 0 & 0 & 0 \\ 1 & 3 & 0 & 0 \\ 1 & 3 & 5 & 0 \\ 1 & 3 & 5 & 7 \end{bmatrix}$

21. $\begin{bmatrix} 2 & -4 & 0 & 0 \\ 1 & 2 & 12 & 0 \\ 0 & 0 & 2 & 0 \\ 0 & -1 & -4 & -5 \end{bmatrix}$

22. $\begin{bmatrix} -8 & 17 & 2 & \frac{1}{3} \\ 4 & 0 & \frac{2}{5} & -9 \\ 0 & 0 & 0 & 0 \\ -1 & 13 & 4 & 2 \end{bmatrix}$

23. $\begin{bmatrix} -1 & 0 & 1 & 0 \\ 2 & 3 & -2 & 6 \\ 0 & -1 & 2 & 0 \\ 0 & 0 & 1 & 5 \end{bmatrix}$

24. $\begin{bmatrix} 0 & 0 & 2 & 0 \\ 1 & 0 & 0 & 1 \\ 0 & -1 & 3 & 0 \\ 2 & 1 & 5 & -3 \end{bmatrix}$

▶ In Exercises 25–26, find the inverse of each of the following 4×4 matrices, where k_1, k_2, k_3, k_4, and k are all nonzero. ◀

25. (a) $\begin{bmatrix} k_1 & 0 & 0 & 0 \\ 0 & k_2 & 0 & 0 \\ 0 & 0 & k_3 & 0 \\ 0 & 0 & 0 & k_4 \end{bmatrix}$ (b) $\begin{bmatrix} k & 1 & 0 & 0 \\ 0 & 1 & 0 & 0 \\ 0 & 0 & k & 1 \\ 0 & 0 & 0 & 1 \end{bmatrix}$

26. (a) $\begin{bmatrix} 0 & 0 & 0 & k_1 \\ 0 & 0 & k_2 & 0 \\ 0 & k_3 & 0 & 0 \\ k_4 & 0 & 0 & 0 \end{bmatrix}$ (b) $\begin{bmatrix} k & 0 & 0 & 0 \\ 1 & k & 0 & 0 \\ 0 & 1 & k & 0 \\ 0 & 0 & 1 & k \end{bmatrix}$

▶ In Exercises 27–28, find all values of c, if any, for which the given matrix is invertible. ◀

27. $\begin{bmatrix} c & -c & c \\ 1 & c & 1 \\ 0 & 0 & c \end{bmatrix}$

28. $\begin{bmatrix} c & 1 & 0 \\ 1 & c & 1 \\ 0 & 1 & c \end{bmatrix}$

▶ In Exercises 29–32, write the given matrix as a product of elementary matrices. ◀

29. $\begin{bmatrix} -2 & 3 \\ 1 & 0 \end{bmatrix}$

30. $\begin{bmatrix} 1 & 0 \\ -5 & 2 \end{bmatrix}$

31. $\begin{bmatrix} 1 & 0 & -2 \\ 0 & 4 & 3 \\ 0 & 0 & 1 \end{bmatrix}$

32. $\begin{bmatrix} 1 & 1 & 0 \\ 1 & 1 & 1 \\ 0 & 1 & 1 \end{bmatrix}$

▶ In Exercises 33–36, write the *inverse* of the given matrix as a product of elementary matrices. ◀

33. The matrix in Exercise 29.

34. The matrix in Exercise 30.

35. The matrix in Exercise 31.

36. The matrix in Exercise 32.

▶ In Exercises 37–38, show that the given matrices A and B are row equivalent, and find a sequence of elementary row operations that produces B from A. ◀

37. $A = \begin{bmatrix} 7 & 1 & -2 \\ -1 & 3 & 4 \\ 5 & 6 & 8 \end{bmatrix}$, $B = \begin{bmatrix} -3 & -11 & -18 \\ 5 & 6 & 8 \\ -1 & 3 & 4 \end{bmatrix}$

38. $A = \begin{bmatrix} 2 & 1 & 0 \\ -1 & 1 & 0 \\ 3 & 0 & -1 \end{bmatrix}$, $B = \begin{bmatrix} 6 & 9 & 4 \\ -5 & -1 & 0 \\ -1 & -2 & -1 \end{bmatrix}$

39. Show that if
$$A = \begin{bmatrix} 1 & 0 & 0 \\ 0 & 1 & 0 \\ a & b & c \end{bmatrix}$$
is an elementary matrix, then at least one entry in the third row must be zero.

40. Show that
$$A = \begin{bmatrix} 0 & a & 0 & 0 & 0 \\ b & 0 & c & 0 & 0 \\ 0 & d & 0 & e & 0 \\ 0 & 0 & f & 0 & g \\ 0 & 0 & 0 & h & 0 \end{bmatrix}$$
is not invertible for any values of the entries.

41. Prove that if A and B are $m \times n$ matrices, then A and B are row equivalent if and only if A and B have the same reduced row echelon form.

42. Prove that if A is an invertible matrix and B is row equivalent to A, then B is also invertible.

43. Show that if B is obtained from A by performing a sequence of elementary row operations, then there is a second sequence of elementary row operations, which when applied to B recovers A.

True-False Exercises

In parts (a)–(g) determine whether the statement is true or false, and justify your answer.

(a) The product of two elementary matrices of the same size must be an elementary matrix.

(b) Every elementary matrix is invertible.

(c) If A and B are row equivalent, and if B and C are row equivalent, then A and C are row equivalent.

(d) If A is an $n \times n$ matrix that is not invertible, then the linear system $A\mathbf{x} = \mathbf{0}$ has infinitely many solutions.

(e) If A is an $n \times n$ matrix that is not invertible, then the matrix obtained by interchanging two rows of A cannot be invertible.

(f) If A is invertible and a multiple of the first row of A is added to the second row, then the resulting matrix is invertible.

(g) An expression of the invertible matrix A as a product of elementary matrices is unique.

1.6 More on Linear Systems and Invertible Matrices

In this section we will show how the inverse of a matrix can be used to solve a linear system and we will develop some more results about invertible matrices.

Number of Solutions of a Linear System

In Section 1.1 we made the statement (based on Figures 1.1.1 and 1.1.2) that every linear system has either no solutions, has exactly one solution, or has infinitely many solutions. We are now in a position to prove this fundamental result.

> **THEOREM 1.6.1** *A system of linear equations has zero, one, or infinitely many solutions. There are no other possibilities.*

Proof If $A\mathbf{x} = \mathbf{b}$ is a system of linear equations, exactly one of the following is true: (a) the system has no solutions, (b) the system has exactly one solution, or (c) the system has more than one solution. The proof will be complete if we can show that the system has infinitely many solutions in case (c).

Assume that $A\mathbf{x} = \mathbf{b}$ has more than one solution, and let $\mathbf{x}_0 = \mathbf{x}_1 - \mathbf{x}_2$, where $\mathbf{x}_1$ and $\mathbf{x}_2$ are any two distinct solutions. Because $\mathbf{x}_1$ and $\mathbf{x}_2$ are distinct, the matrix $\mathbf{x}_0$ is nonzero; moreover,

$$A\mathbf{x}_0 = A(\mathbf{x}_1 - \mathbf{x}_2) = A\mathbf{x}_1 - A\mathbf{x}_2 = \mathbf{b} - \mathbf{b} = \mathbf{0}$$

If we now let k be any scalar, then

$$A(\mathbf{x}_1 + k\mathbf{x}_0) = A\mathbf{x}_1 + A(k\mathbf{x}_0) = A\mathbf{x}_1 + k(A\mathbf{x}_0)$$
$$= \mathbf{b} + k\mathbf{0} = \mathbf{b} + \mathbf{0} = \mathbf{b}$$

But this says that $\mathbf{x}_1 + k\mathbf{x}_0$ is a solution of $A\mathbf{x} = \mathbf{b}$. Since $\mathbf{x}_0$ is nonzero and there are infinitely many choices for k, the system $A\mathbf{x} = \mathbf{b}$ has infinitely many solutions. ◀

Solving Linear Systems by Matrix Inversion

Thus far we have studied two *procedures* for solving linear systems—Gauss–Jordan elimination and Gaussian elimination. The following theorem provides an actual *formula* for the solution of a linear system of n equations in n unknowns in the case where the coefficient matrix is invertible.

> **THEOREM 1.6.2** *If A is an invertible $n \times n$ matrix, then for each $n \times 1$ matrix $\mathbf{b}$, the system of equations $A\mathbf{x} = \mathbf{b}$ has exactly one solution, namely, $\mathbf{x} = A^{-1}\mathbf{b}$.*

Proof Since $A(A^{-1}\mathbf{b}) = \mathbf{b}$, it follows that $\mathbf{x} = A^{-1}\mathbf{b}$ is a solution of $A\mathbf{x} = \mathbf{b}$. To show that this is the only solution, we will assume that $\mathbf{x}_0$ is an arbitrary solution and then show that $\mathbf{x}_0$ must be the solution $A^{-1}\mathbf{b}$.

If $\mathbf{x}_0$ is any solution of $A\mathbf{x} = \mathbf{b}$, then $A\mathbf{x}_0 = \mathbf{b}$. Multiplying both sides of this equation by A^{-1}, we obtain $\mathbf{x}_0 = A^{-1}\mathbf{b}$. ◂

▶ **EXAMPLE 1** **Solution of a Linear System Using A^{-1}**

Consider the system of linear equations

$$x_1 + 2x_2 + 3x_3 = 5$$
$$2x_1 + 5x_2 + 3x_3 = 3$$
$$x_1 \qquad\quad + 8x_3 = 17$$

In matrix form this system can be written as $A\mathbf{x} = \mathbf{b}$, where

$$A = \begin{bmatrix} 1 & 2 & 3 \\ 2 & 5 & 3 \\ 1 & 0 & 8 \end{bmatrix}, \quad \mathbf{x} = \begin{bmatrix} x_1 \\ x_2 \\ x_3 \end{bmatrix}, \quad \mathbf{b} = \begin{bmatrix} 5 \\ 3 \\ 17 \end{bmatrix}$$

In Example 4 of the preceding section, we showed that A is invertible and

$$A^{-1} = \begin{bmatrix} -40 & 16 & 9 \\ 13 & -5 & -3 \\ 5 & -2 & -1 \end{bmatrix}$$

By Theorem 1.6.2, the solution of the system is

$$\mathbf{x} = A^{-1}\mathbf{b} = \begin{bmatrix} -40 & 16 & 9 \\ 13 & -5 & -3 \\ 5 & -2 & -1 \end{bmatrix} \begin{bmatrix} 5 \\ 3 \\ 17 \end{bmatrix} = \begin{bmatrix} 1 \\ -1 \\ 2 \end{bmatrix}$$

or $x_1 = 1, x_2 = -1, x_3 = 2$. ◂

> Keep in mind that the method of Example 1 only applies when the system has as many equations as unknowns and the coefficient matrix is invertible.

Linear Systems with a Common Coefficient Matrix

Frequently, one is concerned with solving a sequence of systems

$$A\mathbf{x} = \mathbf{b}_1, \quad A\mathbf{x} = \mathbf{b}_2, \quad A\mathbf{x} = \mathbf{b}_3, \ldots, \quad A\mathbf{x} = \mathbf{b}_k$$

each of which has the same square coefficient matrix A. If A is invertible, then the solutions

$$\mathbf{x}_1 = A^{-1}\mathbf{b}_1, \quad \mathbf{x}_2 = A^{-1}\mathbf{b}_2, \quad \mathbf{x}_3 = A^{-1}\mathbf{b}_3, \ldots, \quad \mathbf{x}_k = A^{-1}\mathbf{b}_k$$

can be obtained with one matrix inversion and k matrix multiplications. An efficient way to do this is to form the partitioned matrix

$$[A \mid \mathbf{b}_1 \mid \mathbf{b}_2 \mid \cdots \mid \mathbf{b}_k] \qquad (1)$$

in which the coefficient matrix A is "augmented" by all k of the matrices $\mathbf{b}_1, \mathbf{b}_2, \ldots, \mathbf{b}_k$, and then reduce (1) to reduced row echelon form by Gauss–Jordan elimination. In this way we can solve all k systems at once. This method has the added advantage that it applies even when A is not invertible.

EXAMPLE 2 Solving Two Linear Systems at Once

Solve the systems

(a) $x_1 + 2x_2 + 3x_3 = 4$
$\quad 2x_1 + 5x_2 + 3x_3 = 5$
$\quad x_1 + 8x_3 = 9$

(b) $x_1 + 2x_2 + 3x_3 = 1$
$\quad 2x_1 + 5x_2 + 3x_3 = 6$
$\quad x_1 + 8x_3 = -6$

Solution The two systems have the same coefficient matrix. If we augment this coefficient matrix with the columns of constants on the right sides of these systems, we obtain

$$\begin{bmatrix} 1 & 2 & 3 & 4 & 1 \\ 2 & 5 & 3 & 5 & 6 \\ 1 & 0 & 8 & 9 & -6 \end{bmatrix}$$

Reducing this matrix to reduced row echelon form yields (verify)

$$\begin{bmatrix} 1 & 0 & 0 & 1 & 2 \\ 0 & 1 & 0 & 0 & 1 \\ 0 & 0 & 1 & 1 & -1 \end{bmatrix}$$

It follows from the last two columns that the solution of system (a) is $x_1 = 1$, $x_2 = 0$, $x_3 = 1$ and the solution of system (b) is $x_1 = 2$, $x_2 = 1$, $x_3 = -1$. ◀

Properties of Invertible Matrices

Up to now, to show that an $n \times n$ matrix A is invertible, it has been necessary to find an $n \times n$ matrix B such that

$$AB = I \quad \text{and} \quad BA = I$$

The next theorem shows that if we produce an $n \times n$ matrix B satisfying *either* condition, then the other condition holds automatically.

THEOREM 1.6.3 *Let A be a square matrix.*

(a) *If B is a square matrix satisfying $BA = I$, then $B = A^{-1}$.*

(b) *If B is a square matrix satisfying $AB = I$, then $B = A^{-1}$.*

We will prove part (*a*) and leave part (*b*) as an exercise.

Proof (a) Assume that $BA = I$. If we can show that A is invertible, the proof can be completed by multiplying $BA = I$ on both sides by A^{-1} to obtain

$$BAA^{-1} = IA^{-1} \quad \text{or} \quad BI = IA^{-1} \quad \text{or} \quad B = A^{-1}$$

To show that A is invertible, it suffices to show that the system $A\mathbf{x} = \mathbf{0}$ has only the trivial solution (see Theorem 1.5.3). Let $\mathbf{x}_0$ be any solution of this system. If we multiply both sides of $A\mathbf{x}_0 = \mathbf{0}$ on the left by B, we obtain $BA\mathbf{x}_0 = B\mathbf{0}$ or $I\mathbf{x}_0 = \mathbf{0}$ or $\mathbf{x}_0 = \mathbf{0}$. Thus, the system of equations $A\mathbf{x} = \mathbf{0}$ has only the trivial solution. ◀

Equivalence Theorem

We are now in a position to add two more statements to the four given in Theorem 1.5.3.

THEOREM 1.6.4 Equivalent Statements

If A is an $n \times n$ matrix, then the following are equivalent.

(a) A is invertible.
(b) $A\mathbf{x} = \mathbf{0}$ has only the trivial solution.
(c) The reduced row echelon form of A is I_n.
(d) A is expressible as a product of elementary matrices.
(e) $A\mathbf{x} = \mathbf{b}$ is consistent for every $n \times 1$ matrix $\mathbf{b}$.
(f) $A\mathbf{x} = \mathbf{b}$ has exactly one solution for every $n \times 1$ matrix $\mathbf{b}$.

Proof Since we proved in Theorem 1.5.3 that (a), (b), (c), and (d) are equivalent, it will be sufficient to prove that $(a) \Rightarrow (f) \Rightarrow (e) \Rightarrow (a)$.

(a) $\Rightarrow$ (f) This was already proved in Theorem 1.6.2.

(f) $\Rightarrow$ (e) This is almost self-evident, for if $A\mathbf{x} = \mathbf{b}$ has exactly one solution for every $n \times 1$ matrix $\mathbf{b}$, then $A\mathbf{x} = \mathbf{b}$ is consistent for every $n \times 1$ matrix $\mathbf{b}$.

(e) $\Rightarrow$ (a) If the system $A\mathbf{x} = \mathbf{b}$ is consistent for every $n \times 1$ matrix $\mathbf{b}$, then, in particular, this is so for the systems

$$A\mathbf{x} = \begin{bmatrix} 1 \\ 0 \\ 0 \\ \vdots \\ 0 \end{bmatrix}, \quad A\mathbf{x} = \begin{bmatrix} 0 \\ 1 \\ 0 \\ \vdots \\ 0 \end{bmatrix}, \ldots, \quad A\mathbf{x} = \begin{bmatrix} 0 \\ 0 \\ 0 \\ \vdots \\ 1 \end{bmatrix}$$

Let $\mathbf{x}_1, \mathbf{x}_2, \ldots, \mathbf{x}_n$ be solutions of the respective systems, and let us form an $n \times n$ matrix C having these solutions as columns. Thus C has the form

$$C = [\mathbf{x}_1 \mid \mathbf{x}_2 \mid \cdots \mid \mathbf{x}_n]$$

As discussed in Section 1.3, the successive columns of the product AC will be

$$A\mathbf{x}_1, A\mathbf{x}_2, \ldots, A\mathbf{x}_n$$

[see Formula (8) of Section 1.3]. Thus,

$$AC = [A\mathbf{x}_1 \mid A\mathbf{x}_2 \mid \cdots \mid A\mathbf{x}_n] = \begin{bmatrix} 1 & 0 & \cdots & 0 \\ 0 & 1 & \cdots & 0 \\ 0 & 0 & \cdots & 0 \\ \vdots & \vdots & & \vdots \\ 0 & 0 & \cdots & 1 \end{bmatrix} = I$$

> It follows from the equivalency of parts (e) and (f) that if you can show that $A\mathbf{x} = \mathbf{b}$ has at *least one* solution for every $n \times 1$ matrix $\mathbf{b}$, then you can conclude that it has *exactly one* solution for every $n \times 1$ matrix $\mathbf{b}$.

By part (b) of Theorem 1.6.3, it follows that $C = A^{-1}$. Thus, A is invertible. ◀

We know from earlier work that invertible matrix factors produce an invertible product. Conversely, the following theorem shows that if the product of square matrices is invertible, then the factors themselves must be invertible.

THEOREM 1.6.5 Let A and B be square matrices of the same size. If AB is invertible, then A and B must also be invertible.

In our later work the following fundamental problem will occur frequently in various contexts.

> **A Fundamental Problem** Let A be a fixed $m \times n$ matrix. Find all $m \times 1$ matrices $\mathbf{b}$ such that the system of equations $A\mathbf{x} = \mathbf{b}$ is consistent.

If A is an invertible matrix, Theorem 1.6.2 completely solves this problem by asserting that for *every* $m \times 1$ matrix $\mathbf{b}$, the linear system $A\mathbf{x} = \mathbf{b}$ has the unique solution $\mathbf{x} = A^{-1}\mathbf{b}$. If A is not square, or if A is square but not invertible, then Theorem 1.6.2 does not apply. In these cases the matrix $\mathbf{b}$ must usually satisfy certain conditions in order for $A\mathbf{x} = \mathbf{b}$ to be consistent. The following example illustrates how the methods of Section 1.2 can be used to determine such conditions.

▶ **EXAMPLE 3 Determining Consistency by Elimination**

What conditions must b_1, b_2, and b_3 satisfy in order for the system of equations

$$\begin{aligned} x_1 + x_2 + 2x_3 &= b_1 \\ x_1 + x_3 &= b_2 \\ 2x_1 + x_2 + 3x_3 &= b_3 \end{aligned}$$

to be consistent?

Solution The augmented matrix is

$$\begin{bmatrix} 1 & 1 & 2 & b_1 \\ 1 & 0 & 1 & b_2 \\ 2 & 1 & 3 & b_3 \end{bmatrix}$$

which can be reduced to row echelon form as follows:

$$\begin{bmatrix} 1 & 1 & 2 & b_1 \\ 0 & -1 & -1 & b_2 - b_1 \\ 0 & -1 & -1 & b_3 - 2b_1 \end{bmatrix} \quad \longleftarrow \text{-1 times the first row was added to the second and -2 times the first row was added to the third.}$$

$$\begin{bmatrix} 1 & 1 & 2 & b_1 \\ 0 & 1 & 1 & b_1 - b_2 \\ 0 & -1 & -1 & b_3 - 2b_1 \end{bmatrix} \quad \longleftarrow \text{The second row was multiplied by -1.}$$

$$\begin{bmatrix} 1 & 1 & 2 & b_1 \\ 0 & 1 & 1 & b_1 - b_2 \\ 0 & 0 & 0 & b_3 - b_2 - b_1 \end{bmatrix} \quad \longleftarrow \text{The second row was added to the third.}$$

It is now evident from the third row in the matrix that the system has a solution if and only if b_1, b_2, and b_3 satisfy the condition

$$b_3 - b_2 - b_1 = 0 \quad \text{or} \quad b_3 = b_1 + b_2$$

To express this condition another way, $A\mathbf{x} = \mathbf{b}$ is consistent if and only if $\mathbf{b}$ is a matrix of the form

$$\mathbf{b} = \begin{bmatrix} b_1 \\ b_2 \\ b_1 + b_2 \end{bmatrix}$$

where b_1 and b_2 are arbitrary.

► EXAMPLE 4 Determining Consistency by Elimination

What conditions must b_1, b_2, and b_3 satisfy in order for the system of equations

$$x_1 + 2x_2 + 3x_3 = b_1$$
$$2x_1 + 5x_2 + 3x_3 = b_2$$
$$x_1 \quad\quad + 8x_3 = b_3$$

to be consistent?

Solution The augmented matrix is

$$\begin{bmatrix} 1 & 2 & 3 & b_1 \\ 2 & 5 & 3 & b_2 \\ 1 & 0 & 8 & b_3 \end{bmatrix}$$

Reducing this to reduced row echelon form yields (verify)

$$\begin{bmatrix} 1 & 0 & 0 & -40b_1 + 16b_2 + 9b_3 \\ 0 & 1 & 0 & 13b_1 - 5b_2 - 3b_3 \\ 0 & 0 & 1 & 5b_1 - 2b_2 - b_3 \end{bmatrix} \quad (2)$$

In this case there are no restrictions on b_1, b_2, and b_3, so the system has the unique solution

$$x_1 = -40b_1 + 16b_2 + 9b_3, \quad x_2 = 13b_1 - 5b_2 - 3b_3, \quad x_3 = 5b_1 - 2b_2 - b_3 \quad (3)$$

for all values of b_1, b_2, and b_3. ◄

What does the result in Example 4 tell you about the coefficient matrix of the system?

Skills

- Determine whether a linear system of equations has no solutions, exactly one solution, or infinitely many solutions.
- Solve linear systems by inverting its coefficient matrix.
- Solve multiple linear systems with the same coefficient matrix simultaneously.
- Be familiar with the additional conditions of invertibility stated in the Equivalence Theorem.

Exercise Set 1.6

► In Exercises 1–8, solve the system by inverting the coefficient matrix and using Theorem 1.6.2. ◄

1. $3x_1 + 5x_2 = -2$
$\quad x_1 + 2x_2 = 3$

2. $4x_1 - 3x_2 = -3$
$\quad 2x_1 - 5x_2 = 9$

3. $x_1 + 3x_2 + x_3 = 4$
$\quad 2x_1 + 2x_2 + x_3 = -1$
$\quad 2x_1 + 3x_2 + x_3 = 3$

4. $5x_1 + 3x_2 + 2x_3 = 4$
$\quad 3x_1 + 3x_2 + 2x_3 = 2$
$\quad\quad\quad x_2 + x_3 = 5$

5. $x_1 \quad\quad - x_3 = 6$
$\quad x_1 + x_2 + x_3 = -3$
$\quad -x_1 + x_2 \quad = 12$

6. $\quad\quad -x - 2y - 3z = 0$
$\quad w + x + 4y + 4z = 7$
$\quad w + 3x + 7y + 9z = 4$
$\quad -w - 2x - 4y - 6z = 6$

7. $x_1 + x_2 = b_1$
$\quad 5x_1 + 6x_2 = b_2$

8. $x_1 + 2x_2 + 3x_3 = b_1$
$\quad 2x_1 + 5x_2 + 5x_3 = b_2$
$\quad 3x_1 + 5x_2 + 8x_3 = b_3$

► In Exercises 9–12, solve the linear systems together by reducing the appropriate augmented matrix. ◄

9. $x_1 - 5x_2 = b_1$
$\quad 3x_1 + 2x_2 = b_2$
(i) $b_1 = 1$, $b_2 = 4$ (ii) $b_1 = -2$, $b_2 = 5$

10. $-x_1 + 4x_2 + x_3 = b_1$
$\quad x_1 + 9x_2 - 2x_3 = b_2$
$\quad 6x_1 + 4x_2 - 8x_3 = b_3$
(i) $b_1 = 0$, $b_2 = 1$, $b_3 = 0$
(ii) $b_1 = -3$, $b_2 = 4$, $b_3 = -5$

11. $6x_1 + 5x_2 = b_1$
$\quad 5x_1 + 4x_2 = b_2$
(i) $b_1 = 0$, $b_2 = 1$ (ii) $b_1 = -4$, $b_2 = 6$
(iii) $b_1 = -1$, $b_2 = 3$ (iv) $b_1 = -5$, $b_2 = 1$

12.
$$x_1 + 3x_2 + 5x_3 = b_1$$
$$-x_1 - 2x_2 = b_2$$
$$2x_1 + 5x_2 + 4x_3 = b_3$$

(i) $b_1 = 1, \ b_2 = 0, \ b_3 = -1$
(ii) $b_1 = 0, \ b_2 = 1, \ b_3 = 1$
(iii) $b_1 = -1, \ b_2 = -1, \ b_3 = 0$

▶ In Exercises 13–17, determine conditions on the b_i's, if any, in order to guarantee that the linear system is consistent. ◀

13.
$$x_1 - 3x_2 = b_1$$
$$4x_1 - 12x_2 = b_2$$

14.
$$2x_1 - 5x_2 = b_1$$
$$3x_1 + 6x_2 = b_2$$

15.
$$x_1 - 2x_2 + 5x_3 = b_1$$
$$4x_1 - 5x_2 + 8x_3 = b_2$$
$$-3x_1 + 3x_2 - 3x_3 = b_3$$

16.
$$x_1 - 2x_2 - x_3 = b_1$$
$$-4x_1 + 5x_2 + 2x_3 = b_2$$
$$-4x_1 + 7x_2 + 4x_3 = b_3$$

17.
$$x_1 + 3x_2 - x_3 + 2x_4 = b_1$$
$$-2x_1 + x_2 + 5x_3 + x_4 = b_2$$
$$3x_1 - 2x_2 - 2x_3 + x_4 = b_3$$
$$5x_1 - 7x_2 - 3x_3 = b_4$$

18. Consider the matrices

$$A = \begin{bmatrix} 2 & 1 & 2 \\ 2 & 2 & -2 \\ 3 & 1 & 1 \end{bmatrix} \quad \text{and} \quad \mathbf{x} = \begin{bmatrix} x_1 \\ x_2 \\ x_3 \end{bmatrix}$$

(a) Show that the equation $A\mathbf{x} = \mathbf{x}$ can be rewritten as $(A - I)\mathbf{x} = \mathbf{0}$ and use this result to solve $A\mathbf{x} = \mathbf{x}$ for $\mathbf{x}$.

(b) Solve $A\mathbf{x} = 4\mathbf{x}$.

▶ In Exercises 19–20, solve the given matrix equation for X. ◀

19. $\begin{bmatrix} 1 & 2 & 3 \\ 3 & 7 & 6 \\ 1 & 0 & 8 \end{bmatrix} X = \begin{bmatrix} 1 & 4 & -2 & 0 & 3 \\ 0 & -1 & 5 & 2 & 7 \\ -3 & 6 & 8 & 9 & 0 \end{bmatrix}$

20. $\begin{bmatrix} -2 & 0 & 1 \\ 0 & -1 & -1 \\ 1 & 1 & -4 \end{bmatrix} X = \begin{bmatrix} 4 & 3 & 2 & 1 \\ 6 & 7 & 8 & 9 \\ 1 & 3 & 7 & 9 \end{bmatrix}$

21. Let $A\mathbf{x} = \mathbf{0}$ be a homogeneous system of n linear equations in n unknowns that has only the trivial solution. Show that if k is any positive integer, then the system $A^k\mathbf{x} = \mathbf{0}$ also has only the trivial solution.

22. Let $A\mathbf{x} = \mathbf{0}$ be a homogeneous system of n linear equations in n unknowns, and let Q be an invertible $n \times n$ matrix. Show that $A\mathbf{x} = \mathbf{0}$ has just the trivial solution if and only if $(QA)\mathbf{x} = \mathbf{0}$ has just the trivial solution.

23. Let $A\mathbf{x} = \mathbf{b}$ be any consistent system of linear equations, and let $\mathbf{x}_1$ be a fixed solution. Show that every solution to the system can be written in the form $\mathbf{x} = \mathbf{x}_1 + \mathbf{x}_0$, where $\mathbf{x}_0$ is a solution to $A\mathbf{x} = \mathbf{0}$. Show also that every matrix of this form is a solution.

24. Use part (a) of Theorem 1.6.3 to prove part (b).

True-False Exercises

In parts (a)–(g) determine whether the statement is true or false, and justify your answer.

(a) It is impossible for a system of linear equations to have exactly two solutions.

(b) If the linear system $A\mathbf{x} = \mathbf{b}$ has a unique solution, then the linear system $A\mathbf{x} = \mathbf{c}$ also must have a unique solution.

(c) If A and B are $n \times n$ matrices such that $AB = I_n$, then $BA = I_n$.

(d) If A and B are row equivalent matrices, then the linear systems $A\mathbf{x} = \mathbf{0}$ and $B\mathbf{x} = \mathbf{0}$ have the same solution set.

(e) If A is an $n \times n$ matrix and S is an $n \times n$ invertible matrix, then if $\mathbf{x}$ is a solution to the linear system $(S^{-1}AS)\mathbf{x} = \mathbf{b}$, then $S\mathbf{x}$ is a solution to the linear system $A\mathbf{y} = S\mathbf{b}$.

(f) Let A be an $n \times n$ matrix. The linear system $A\mathbf{x} = 4\mathbf{x}$ has a unique solution if and only if $A - 4I$ is an invertible matrix.

(g) Let A and B be $n \times n$ matrices. If AB is invertible, then both A and B must be invertible.

1.7 Diagonal, Triangular, and Symmetric Matrices

In this section we will discuss matrices that have various special forms. These matrices arise in a wide variety of applications and will play an important role in our subsequent work.

Diagonal Matrices A square matrix in which all the entries off the main diagonal are zero is called a ***diagonal matrix***. Here are some examples:

$$\begin{bmatrix} 0 & 0 \\ 0 & 0 \end{bmatrix}, \quad \begin{bmatrix} 2 & 0 \\ 0 & -5 \end{bmatrix}, \quad \begin{bmatrix} 1 & 0 & 0 \\ 0 & 1 & 0 \\ 0 & 0 & 1 \end{bmatrix}, \quad \begin{bmatrix} 6 & 0 & 0 & 0 \\ 0 & -4 & 0 & 0 \\ 0 & 0 & 0 & 0 \\ 0 & 0 & 0 & 8 \end{bmatrix}$$

1.7 Diagonal, Triangular, and Symmetric Matrices

A general $n \times n$ diagonal matrix D can be written as

$$D = \begin{bmatrix} d_1 & 0 & \cdots & 0 \\ 0 & d_2 & \cdots & 0 \\ \vdots & \vdots & & \vdots \\ 0 & 0 & \cdots & d_n \end{bmatrix} \quad (1)$$

> Confirm Formula (2) by showing that
> $$DD^{-1} = D^{-1}D = I$$

A diagonal matrix is invertible if and only if all of its diagonal entries are nonzero; in this case the inverse of (1) is

$$D^{-1} = \begin{bmatrix} 1/d_1 & 0 & \cdots & 0 \\ 0 & 1/d_2 & \cdots & 0 \\ \vdots & \vdots & & \vdots \\ 0 & 0 & \cdots & 1/d_n \end{bmatrix} \quad (2)$$

Powers of diagonal matrices are easy to compute; we leave it for you to verify that if D is the diagonal matrix (1) and k is a positive integer, then

$$D^k = \begin{bmatrix} d_1^k & 0 & \cdots & 0 \\ 0 & d_2^k & \cdots & 0 \\ \vdots & \vdots & & \vdots \\ 0 & 0 & \cdots & d_n^k \end{bmatrix} \quad (3)$$

▶ **EXAMPLE 1** Inverses and Powers of Diagonal Matrices

If

$$A = \begin{bmatrix} 1 & 0 & 0 \\ 0 & -3 & 0 \\ 0 & 0 & 2 \end{bmatrix}$$

then

$$A^{-1} = \begin{bmatrix} 1 & 0 & 0 \\ 0 & -\frac{1}{3} & 0 \\ 0 & 0 & \frac{1}{2} \end{bmatrix}, \quad A^5 = \begin{bmatrix} 1 & 0 & 0 \\ 0 & -243 & 0 \\ 0 & 0 & 32 \end{bmatrix}, \quad A^{-5} = \begin{bmatrix} 1 & 0 & 0 \\ 0 & -\frac{1}{243} & 0 \\ 0 & 0 & \frac{1}{32} \end{bmatrix}$$

◀

Matrix products that involve diagonal factors are especially easy to compute. For example,

$$\begin{bmatrix} d_1 & 0 & 0 \\ 0 & d_2 & 0 \\ 0 & 0 & d_3 \end{bmatrix} \begin{bmatrix} a_{11} & a_{12} & a_{13} & a_{14} \\ a_{21} & a_{22} & a_{23} & a_{24} \\ a_{31} & a_{32} & a_{33} & a_{34} \end{bmatrix} = \begin{bmatrix} d_1 a_{11} & d_1 a_{12} & d_1 a_{13} & d_1 a_{14} \\ d_2 a_{21} & d_2 a_{22} & d_2 a_{23} & d_2 a_{24} \\ d_3 a_{31} & d_3 a_{32} & d_3 a_{33} & d_3 a_{34} \end{bmatrix}$$

$$\begin{bmatrix} a_{11} & a_{12} & a_{13} \\ a_{21} & a_{22} & a_{23} \\ a_{31} & a_{32} & a_{33} \\ a_{41} & a_{42} & a_{43} \end{bmatrix} \begin{bmatrix} d_1 & 0 & 0 \\ 0 & d_2 & 0 \\ 0 & 0 & d_3 \end{bmatrix} = \begin{bmatrix} d_1 a_{11} & d_2 a_{12} & d_3 a_{13} \\ d_1 a_{21} & d_2 a_{22} & d_3 a_{23} \\ d_1 a_{31} & d_2 a_{32} & d_3 a_{33} \\ d_1 a_{41} & d_2 a_{42} & d_3 a_{43} \end{bmatrix}$$

> In words, *to multiply a matrix A on the left by a diagonal matrix D, one can multiply successive rows of A by the successive diagonal entries of D, and to multiply A on the right by D, one can multiply successive columns of A by the successive diagonal entries of D.*

Triangular Matrices

A square matrix in which all the entries above the main diagonal are zero is called **lower triangular**, and a square matrix in which all the entries below the main diagonal are zero is called **upper triangular**. A matrix that is either upper triangular or lower triangular is called **triangular**.

▶ EXAMPLE 2 Upper and Lower Triangular Matrices

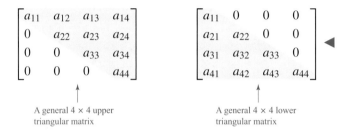

Remark Observe that diagonal matrices are both upper triangular and lower triangular since they have zeros below and above the main diagonal. Observe also that a *square* matrix in row echelon form is upper triangular since it has zeros below the main diagonal.

Properties of Triangular Matrices

Example 2 illustrates the following four facts about triangular matrices that we will state without formal proof:

- A square matrix $A = [a_{ij}]$ is upper triangular if and only if all entries to the left of the main diagonal are zero; that is, $a_{ij} = 0$ if $i > j$ (Figure 1.7.1).
- A square matrix $A = [a_{ij}]$ is lower triangular if and only if all entries to the right of the main diagonal are zero; that is, $a_{ij} = 0$ if $i < j$ (Figure 1.7.1).
- A square matrix $A = [a_{ij}]$ is upper triangular if and only if the ith row starts with at least $i - 1$ zeros for every i.
- A square matrix $A = [a_{ij}]$ is lower triangular if and only if the jth column starts with at least $j - 1$ zeros for every j.

▲ Figure 1.7.1

The following theorem lists some of the basic properties of triangular matrices.

THEOREM 1.7.1

(a) *The transpose of a lower triangular matrix is upper triangular, and the transpose of an upper triangular matrix is lower triangular.*

(b) *The product of lower triangular matrices is lower triangular, and the product of upper triangular matrices is upper triangular.*

(c) *A triangular matrix is invertible if and only if its diagonal entries are all nonzero.*

(d) *The inverse of an invertible lower triangular matrix is lower triangular, and the inverse of an invertible upper triangular matrix is upper triangular.*

Part (*a*) is evident from the fact that transposing a square matrix can be accomplished by reflecting the entries about the main diagonal; we omit the formal proof. We will prove (*b*), but we will defer the proofs of (*c*) and (*d*) to the next chapter, where we will have the tools to prove those results more efficiently.

1.7 Diagonal, Triangular, and Symmetric Matrices

Proof (b) We will prove the result for lower triangular matrices; the proof for upper triangular matrices is similar. Let $A = [a_{ij}]$ and $B = [b_{ij}]$ be lower triangular $n \times n$ matrices, and let $C = [c_{ij}]$ be the product $C = AB$. We can prove that C is lower triangular by showing that $c_{ij} = 0$ for $i < j$. But from the definition of matrix multiplication,

$$c_{ij} = a_{i1}b_{1j} + a_{i2}b_{2j} + \cdots + a_{in}b_{nj}$$

If we assume that $i < j$, then the terms in this expression can be grouped as follows:

$$c_{ij} = \underbrace{a_{i1}b_{1j} + a_{i2}b_{2j} + \cdots + a_{i(j-1)}b_{(j-1)j}}_{\substack{\text{Terms in which the row} \\ \text{number of } b \text{ is less than} \\ \text{the column number of } b}} + \underbrace{a_{ij}b_j + \cdots + a_{in}b_{nj}}_{\substack{\text{Terms in which the row} \\ \text{number of } a \text{ is less than} \\ \text{the column number of } a}}$$

In the first grouping all of the b factors are zero since B is lower triangular, and in the second grouping all of the a factors are zero since A is lower triangular. Thus, $c_{ij} = 0$, which is what we wanted to prove. ◄

▶ **EXAMPLE 3** **Computations with Triangular Matrices**

Consider the upper triangular matrices

$$A = \begin{bmatrix} 1 & 3 & -1 \\ 0 & 2 & 4 \\ 0 & 0 & 5 \end{bmatrix}, \quad B = \begin{bmatrix} 3 & -2 & 2 \\ 0 & 0 & -1 \\ 0 & 0 & 1 \end{bmatrix}$$

It follows from part (c) of Theorem 1.7.1 that the matrix A is invertible but the matrix B is not. Moreover, the theorem also tells us that A^{-1}, AB, and BA must be upper triangular. We leave it for you to confirm these three statements by showing that

$$A^{-1} = \begin{bmatrix} 1 & -\frac{3}{2} & \frac{7}{5} \\ 0 & \frac{1}{2} & -\frac{2}{5} \\ 0 & 0 & \frac{1}{5} \end{bmatrix}, \quad AB = \begin{bmatrix} 3 & -2 & -2 \\ 0 & 0 & 2 \\ 0 & 0 & 5 \end{bmatrix}, \quad BA = \begin{bmatrix} 3 & 5 & -1 \\ 0 & 0 & -5 \\ 0 & 0 & 5 \end{bmatrix} \blacktriangleleft$$

Symmetric Matrices

DEFINITION 1 A square matrix A is said to be ***symmetric*** if $A = A^T$.

It is easy to recognize a symmetric matrix by inspection: The entries on the main diagonal have no restrictions, but mirror images of entries *across* the main diagonal must be equal. Here is a picture using the second matrix in Example 4:

$$\begin{bmatrix} 1 & 4 & 5 \\ 4 & -3 & 0 \\ 5 & 0 & 7 \end{bmatrix}$$

All diagonal matrices, such as the third matrix in Example 4, have this property.

▶ **EXAMPLE 4** **Symmetric Matrices**

The following matrices are symmetric, since each is equal to its own transpose (verify).

$$\begin{bmatrix} 7 & -3 \\ -3 & 5 \end{bmatrix}, \quad \begin{bmatrix} 1 & 4 & 5 \\ 4 & -3 & 0 \\ 5 & 0 & 7 \end{bmatrix}, \quad \begin{bmatrix} d_1 & 0 & 0 & 0 \\ 0 & d_2 & 0 & 0 \\ 0 & 0 & d_3 & 0 \\ 0 & 0 & 0 & d_4 \end{bmatrix} \blacktriangleleft$$

Remark It follows from Formula (11) of Section 1.3 that a square matrix $A = [a_{ij}]$ is symmetric if and only if

$$(A)_{ij} = (A)_{ji} \tag{4}$$

for all values of i and j.

The following theorem lists the main algebraic properties of symmetric matrices. The proofs are direct consequences of Theorem 1.4.8 and are omitted.

THEOREM 1.7.2 *If A and B are symmetric matrices with the same size, and if k is any scalar, then:*

(a) A^T *is symmetric.*

(b) $A + B$ *and* $A - B$ *are symmetric.*

(c) kA *is symmetric.*

It is not true, in general, that the product of symmetric matrices is symmetric. To see why this is so, let A and B be symmetric matrices with the same size. Then it follows from part (e) of Theorem 1.4.8 and the symmetry of A and B that

$$(AB)^T = B^T A^T = BA$$

Thus, $(AB)^T = AB$ if and only if $AB = BA$, that is, if and only if A and B commute. In summary, we have the following result.

> Use the result in Theorem 1.4.5 to confirm Theorem 1.7.3 in the case where the symmetric matrix
> $$\begin{bmatrix} a & b \\ b & d \end{bmatrix}$$
> is invertible.

THEOREM 1.7.3 *The product of two symmetric matrices is symmetric if and only if the matrices commute.*

▶ **EXAMPLE 5 Products of Symmetric Matrices**

The first of the following equations shows a product of symmetric matrices that *is not* symmetric, and the second shows a product of symmetric matrices that *is* symmetric. We conclude that the factors in the first equation do not commute, but those in the second equation do. We leave it for you to verify that this is so.

$$\begin{bmatrix} 1 & 2 \\ 2 & 3 \end{bmatrix} \begin{bmatrix} -4 & 1 \\ 1 & 0 \end{bmatrix} = \begin{bmatrix} -2 & 1 \\ -5 & 2 \end{bmatrix}$$

$$\begin{bmatrix} 1 & 2 \\ 2 & 3 \end{bmatrix} \begin{bmatrix} -4 & 3 \\ 3 & -1 \end{bmatrix} = \begin{bmatrix} 2 & 1 \\ 1 & 3 \end{bmatrix} \blacktriangleleft$$

Invertibility of Symmetric Matrices

In general, a symmetric matrix need not be invertible. For example, a diagonal matrix with a zero on the main diagonal is symmetric but not invertible. However, the following theorem shows that if a symmetric matrix happens to be invertible, then its inverse must also be symmetric.

THEOREM 1.7.4 *If A is an invertible symmetric matrix, then* A^{-1} *is symmetric.*

Proof Assume that A is symmetric and invertible. From Theorem 1.4.9 and the fact that $A = A^T$, we have

$$(A^{-1})^T = (A^T)^{-1} = A^{-1}$$

which proves that A^{-1} is symmetric. ◀

Products AA^T and A^TA

Matrix products of the form AA^T and A^TA arise in a variety of applications. If A is an $m \times n$ matrix, then A^T is an $n \times m$ matrix, so the products AA^T and A^TA are both square matrices—the matrix AA^T has size $m \times m$, and the matrix A^TA has size $n \times n$. Such products are always symmetric since

$$(AA^T)^T = (A^T)^T A^T = AA^T \quad \text{and} \quad (A^TA)^T = A^T(A^T)^T = A^TA$$

▶ **EXAMPLE 6** **The Product of a Matrix and Its Transpose Is Symmetric**

Let A be the 2×3 matrix

$$A = \begin{bmatrix} 1 & -2 & 4 \\ 3 & 0 & -5 \end{bmatrix}$$

Then

$$A^T A = \begin{bmatrix} 1 & 3 \\ -2 & 0 \\ 4 & -5 \end{bmatrix} \begin{bmatrix} 1 & -2 & 4 \\ 3 & 0 & -5 \end{bmatrix} = \begin{bmatrix} 10 & -2 & -11 \\ -2 & 4 & -8 \\ -11 & -8 & 41 \end{bmatrix}$$

$$A A^T = \begin{bmatrix} 1 & -2 & 4 \\ 3 & 0 & -5 \end{bmatrix} \begin{bmatrix} 1 & 3 \\ -2 & 0 \\ 4 & -5 \end{bmatrix} = \begin{bmatrix} 21 & -17 \\ -17 & 34 \end{bmatrix}$$

Observe that $A^T A$ and $A A^T$ are symmetric as expected. ◀

Later in this text, we will obtain general conditions on A under which AA^T and $A^T A$ are invertible. However, in the special case where A is *square*, we have the following result.

THEOREM 1.7.5 *If A is an invertible matrix, then AA^T and $A^T A$ are also invertible.*

Proof Since A is invertible, so is A^T by Theorem 1.4.9. Thus AA^T and $A^T A$ are invertible, since they are the products of invertible matrices. ◀

Concept Review

- Diagonal matrix
- Lower triangular matrix
- Upper triangular matrix
- Triangular matrix
- Symmetric matrix

Skills

- Determine whether a diagonal matrix is invertible with no computations.
- Compute matrix products involving diagonal matrices by inspection.
- Determine whether a matrix is triangular.
- Understand how the transpose operation affects diagonal and triangular matrices.
- Understand how inversion affects diagonal and triangular matrices.
- Determine whether a matrix is a symmetric matrix.

Exercise Set 1.7

▶ In Exercises **1–4**, determine whether the given matrix is invertible. ◀

▶ In Exercises **5–8**, determine the product by inspection. ◀

1. $\begin{bmatrix} 4 & 0 \\ 0 & -2 \end{bmatrix}$

2. $\begin{bmatrix} 5 & 0 & 0 & 0 \\ 0 & 6 & 0 & 0 \\ 0 & 0 & 0 & 0 \\ 0 & 0 & 0 & 1 \end{bmatrix}$

3. $\begin{bmatrix} -1 & 0 & 0 \\ 0 & 2 & 0 \\ 0 & 0 & \frac{1}{3} \end{bmatrix}$

4. $\begin{bmatrix} -1 & 0 & 0 & 0 \\ 0 & 3 & 0 & 0 \\ 0 & 0 & -3 & 0 \\ 0 & 0 & 0 & -2 \end{bmatrix}$

5. $\begin{bmatrix} 3 & 0 & 0 \\ 0 & -1 & 0 \\ 0 & 0 & 2 \end{bmatrix} \begin{bmatrix} 2 & 1 \\ -4 & 1 \\ 2 & 5 \end{bmatrix}$

6. $\begin{bmatrix} -3 & 2 & 8 \\ 4 & 1 & 6 \end{bmatrix} \begin{bmatrix} 7 & 0 & 0 \\ 0 & -1 & 0 \\ 0 & 0 & \frac{1}{2} \end{bmatrix}$

7. $\begin{bmatrix} 5 & 0 & 0 \\ 0 & 2 & 0 \\ 0 & 0 & -3 \end{bmatrix} \begin{bmatrix} -3 & 2 & 0 & 4 & -4 \\ 1 & -5 & 3 & 0 & 3 \\ -6 & 2 & 2 & 2 & 2 \end{bmatrix}$

8. $\begin{bmatrix} 2 & 0 & 0 \\ 0 & -1 & 0 \\ 0 & 0 & 4 \end{bmatrix} \begin{bmatrix} 4 & -1 & 3 \\ 1 & 2 & 0 \\ -5 & 1 & -2 \end{bmatrix} \begin{bmatrix} -3 & 0 & 0 \\ 0 & 5 & 0 \\ 0 & 0 & 2 \end{bmatrix}$

▶ In Exercises 9–12, find A^2, A^{-2}, and A^{-k} (where k is any integer) by inspection. ◀

9. $A = \begin{bmatrix} 2 & 0 \\ 0 & -1 \end{bmatrix}$

10. $A = \begin{bmatrix} -6 & 0 & 0 \\ 0 & 3 & 0 \\ 0 & 0 & 5 \end{bmatrix}$

11. $A = \begin{bmatrix} 3 & 0 & 0 \\ 0 & \frac{1}{2} & 0 \\ 0 & 0 & \frac{1}{5} \end{bmatrix}$

12. $A = \begin{bmatrix} -2 & 0 & 0 & 0 \\ 0 & -4 & 0 & 0 \\ 0 & 0 & -3 & 0 \\ 0 & 0 & 0 & 2 \end{bmatrix}$

▶ In Exercises 13–19, decide whether the given matrix is symmetric. ◀

13. $\begin{bmatrix} 1 & 2 \\ 2 & -1 \end{bmatrix}$
14. $\begin{bmatrix} 2 & 0 \\ 1 & 2 \end{bmatrix}$
15. $\begin{bmatrix} 0 & -7 \\ -7 & 7 \end{bmatrix}$

16. $\begin{bmatrix} 3 & 4 \\ 4 & 0 \end{bmatrix}$
17. $\begin{bmatrix} 2 & 3a & 0 \\ 3a & 1 & b \\ 0 & b & 0 \end{bmatrix}$

18. $\begin{bmatrix} 2 & -1 & 3 \\ -1 & 5 & 1 \\ 3 & 1 & 7 \end{bmatrix}$
19. $\begin{bmatrix} 0 & 0 & 1 \\ 0 & 2 & 0 \\ 3 & 0 & 0 \end{bmatrix}$

▶ In Exercises 20–22, decide by inspection whether the given matrix is invertible. ◀

20. $\begin{bmatrix} 2 & 6 & 0 \\ 0 & 2 & 1 \\ 0 & 0 & 0 \end{bmatrix}$
21. $\begin{bmatrix} 9 & 0 & 0 & 1 \\ 0 & 2 & 0 & 3 \\ 0 & 0 & 4 & -5 \\ 0 & 0 & 0 & 6 \end{bmatrix}$

22. $\begin{bmatrix} 2 & 0 & 0 & 0 \\ -3 & -1 & 0 & 0 \\ -4 & -6 & 0 & 0 \\ 0 & 3 & 8 & -5 \end{bmatrix}$

▶ In Exercises 23–24, find all values of the unknown constant(s) in order for A to be symmetric. ◀

23. $A = \begin{bmatrix} -3 & a^2 \\ 4 & 0 \end{bmatrix}$

24. $A = \begin{bmatrix} 2 & a-2b+2c & 2a+b+c \\ 3 & 5 & a+c \\ 0 & -2 & 7 \end{bmatrix}$

▶ In Exercises 25–26, find all values of x in order for A to be invertible. ◀

25. $A = \begin{bmatrix} 2-x & 5 & x^2 \\ 0 & x+3 & x-1 \\ 0 & 0 & x \end{bmatrix}$

26. $A = \begin{bmatrix} x-\frac{1}{2} & 0 & 0 \\ x & x-\frac{1}{3} & 0 \\ x^2 & x^3 & x+\frac{1}{4} \end{bmatrix}$

▶ In Exercises 27–28, find a diagonal matrix A that satisfies the given condition. ◀

27. $A^5 = \begin{bmatrix} 1 & 0 & 0 \\ 0 & -1 & 0 \\ 0 & 0 & -1 \end{bmatrix}$
28. $A^{-2} = \begin{bmatrix} 9 & 0 & 0 \\ 0 & 4 & 0 \\ 0 & 0 & 1 \end{bmatrix}$

29. Verify Theorem 1.7.1(b) for the product AB, where
$$A = \begin{bmatrix} 2 & 0 & 0 \\ 1 & 1 & 0 \\ -3 & 4 & 5 \end{bmatrix}, \quad B = \begin{bmatrix} -1 & 0 & 0 \\ 6 & 2 & 0 \\ 1 & 1 & 5 \end{bmatrix}$$

30. Verify Theorem 1.7.1(d) for the matrices A and B in Exercise 29.

31. Verify Theorem 1.7.4 for the given matrix A.

(a) $A = \begin{bmatrix} 2 & -1 \\ -1 & 3 \end{bmatrix}$
(b) $A = \begin{bmatrix} 1 & -2 & 3 \\ -2 & 1 & -7 \\ 3 & -7 & 4 \end{bmatrix}$

32. Let A be an $n \times n$ symmetric matrix.
   (a) Show that A^2 is symmetric.
   (b) Show that $2A^2 - 3A + I$ is symmetric.

33. Prove: If $A^T A = A$, then A is symmetric and $A = A^2$.

34. Find all 3×3 diagonal matrices A that satisfy $A^2 - 3A - 4I = 0$.

35. Let $A = [a_{ij}]$ be an $n \times n$ matrix. Determine whether A is symmetric.
   (a) $a_{ij} = i^2 + j^2$
   (b) $a_{ij} = i^2 - j^2$
   (c) $a_{ij} = 2i + 2j$
   (d) $a_{ij} = 2i^2 + 2j^3$

36. On the basis of your experience with Exercise 35, devise a general test that can be applied to a formula for a_{ij} to determine whether $A = [a_{ij}]$ is symmetric.

37. A square matrix A is called **skew-symmetric** if $A^T = -A$. Prove:
   (a) If A is an invertible skew-symmetric matrix, then A^{-1} is skew-symmetric.
   (b) If A and B are skew-symmetric matrices, then so are A^T, $A + B$, $A - B$, and kA for any scalar k.
   (c) Every square matrix A can be expressed as the sum of a symmetric matrix and a skew-symmetric matrix. [*Hint:* Note the identity $A = \frac{1}{2}(A + A^T) + \frac{1}{2}(A - A^T)$.]

▶ In Exercises 38–39, fill in the missing entries (marked with ×) to produce a skew-symmetric matrix. ◀

38. $A = \begin{bmatrix} \times & \times & 4 \\ 0 & \times & \times \\ \times & -1 & \times \end{bmatrix}$

39. $A = \begin{bmatrix} \times & 0 & \times \\ \times & \times & -4 \\ 8 & \times & \times \end{bmatrix}$

40. Find all values of $a, b, c,$ and d for which A is skew-symmetric.

$$A = \begin{bmatrix} 0 & 2a - 3b + c & 3a - 5b + 5c \\ -2 & 0 & 5a - 8b + 6c \\ -3 & -5 & d \end{bmatrix}$$

41. We showed in the text that the product of symmetric matrices is symmetric if and only if the matrices commute. Is the product of commuting skew-symmetric matrices skew-symmetric? Explain. [*Note:* See Exercise 37 for the definition of *skew-symmetric*.]

42. If the $n \times n$ matrix A can be expressed as $A = LU$, where L is a lower triangular matrix and U is an upper triangular matrix, then the linear system $A\mathbf{x} = \mathbf{b}$ can be expressed as $LU\mathbf{x} = \mathbf{b}$ and can be solved in two steps:

Step 1. Let $U\mathbf{x} = \mathbf{y}$, so that $LU\mathbf{x} = \mathbf{b}$ can be expressed as $L\mathbf{y} = \mathbf{b}$. Solve this system.

Step 2. Solve the system $U\mathbf{x} = \mathbf{y}$ for $\mathbf{x}$.

In each part, use this two-step method to solve the given system.

(a) $\begin{bmatrix} 1 & 0 & 0 \\ -2 & 3 & 0 \\ 2 & 4 & 1 \end{bmatrix} \begin{bmatrix} 2 & -1 & 3 \\ 0 & 1 & 2 \\ 0 & 0 & 4 \end{bmatrix} \begin{bmatrix} x_1 \\ x_2 \\ x_3 \end{bmatrix} = \begin{bmatrix} 1 \\ -2 \\ 0 \end{bmatrix}$

(b) $\begin{bmatrix} 2 & 0 & 0 \\ 4 & 1 & 0 \\ -3 & -2 & 3 \end{bmatrix} \begin{bmatrix} 3 & -5 & 2 \\ 0 & 4 & 1 \\ 0 & 0 & 2 \end{bmatrix} \begin{bmatrix} x_1 \\ x_2 \\ x_3 \end{bmatrix} = \begin{bmatrix} 4 \\ -5 \\ 2 \end{bmatrix}$

43. Find a lower triangular matrix that satisfies

$$A^3 = \begin{bmatrix} 8 & 0 \\ 9 & -1 \end{bmatrix}$$

True-False Exercises

In parts (a)–(m) determine whether the statement is true or false, and justify your answer.

(a) The transpose of a diagonal matrix is a diagonal matrix.

(b) The transpose of an upper triangular matrix is an upper triangular matrix.

(c) The sum of an upper triangular matrix and a lower triangular matrix is a diagonal matrix.

(d) All entries of a symmetric matrix are determined by the entries occurring on and above the main diagonal.

(e) All entries of an upper triangular matrix are determined by the entries occurring on and above the main diagonal.

(f) The inverse of an invertible lower triangular matrix is an upper triangular matrix.

(g) A diagonal matrix is invertible if and only if all of its diagonal entries are positive.

(h) The sum of a diagonal matrix and a lower triangular matrix is a lower triangular matrix.

(i) A matrix that is both symmetric and upper triangular must be a diagonal matrix.

(j) If A and B are $n \times n$ matrices such that $A + B$ is symmetric, then A and B are symmetric.

(k) If A and B are $n \times n$ matrices such that $A + B$ is upper triangular, then A and B are upper triangular.

(l) If A^2 is a symmetric matrix, then A is a symmetric matrix.

(m) If kA is a symmetric matrix for some $k \neq 0$, then A is a symmetric matrix.

1.8 Applications of Linear Systems

In this section we will discuss some brief applications of linear systems. These are but a small sample of the wide variety of real-world problems to which our study of linear systems is applicable.

Network Analysis The concept of a **network** appears in a variety of applications. Loosely stated, a **network** is a set of **branches** through which something "flows." For example, the branches might be electrical wires through which electricity flows, pipes through which water or oil flows, traffic lanes through which vehicular traffic flows, or economic linkages through which money flows, to name a few possibilities.

In most networks, the branches meet at points, called **nodes** or **junctions**, where the flow divides. For example, in an electrical network, nodes occur where three or more wires join, in a traffic network they occur at street intersections, and in a financial network they occur at banking centers where incoming money is distributed to individuals or other institutions.

In the study of networks, there is generally some numerical measure of the rate at which the medium flows through a branch. For example, the flow rate of electricity is often measured in amperes, the flow rate of water or oil in gallons per minute, the flow rate of traffic in vehicles per hour, and the flow rate of European currency in millions of Euros per day. We will restrict our attention to networks in which there is **flow conservation** at each node, by which we mean that *the rate of flow into any node is equal to the rate of flow out of that node.* This ensures that the flow medium does not build up at the nodes and block the free movement of the medium through the network.

A common problem in network analysis is to use known flow rates in certain branches to find the flow rates in all of the branches. Here is an example.

▶ **EXAMPLE 1** Network Analysis Using Linear Systems

Figure 1.8.1 shows a network with four nodes in which the flow rate and direction of flow in certain branches are known. Find the flow rates and directions of flow in the remaining branches.

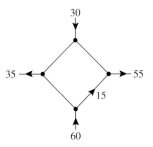

▲ Figure 1.8.1

Solution As illustrated in Figure 1.8.2, we have assigned arbitrary directions to the unknown flow rates x_1, x_2, and x_3. We need not be concerned if some of the directions are incorrect, since an incorrect direction will be signaled by a negative value for the flow rate when we solve for the unknowns.

It follows from the conservation of flow at node A that

$$x_1 + x_2 = 30$$

Similarly, at the other nodes we have

$$x_2 + x_3 = 35 \quad (\text{node } B)$$
$$x_3 + 15 = 60 \quad (\text{node } C)$$
$$x_1 + 15 = 55 \quad (\text{node } D)$$

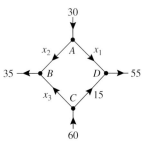

▲ Figure 1.8.2

These four conditions produce the linear system

$$\begin{aligned} x_1 + x_2 \phantom{{}+x_3} &= 30 \\ x_2 + x_3 &= 35 \\ x_3 &= 45 \\ x_1 \phantom{{}+x_2+x_3} &= 40 \end{aligned}$$

which we can now try to solve for the unknown flow rates. In this particular case the system is sufficiently simple that it can be solved by inspection (work from the bottom up). We leave it for you to confirm that the solution is

$$x_1 = 40, \quad x_2 = -10, \quad x_3 = 45$$

The fact that x_2 is negative tells us that the direction assigned to that flow in Figure 1.8.2 is incorrect; that is, the flow in that branch is *into* node A.

▶ **EXAMPLE 2** Design of Traffic Patterns

The network in Figure 1.8.3 shows a proposed plan for the traffic flow around a new park that will house the Liberty Bell in Philadelphia, Pennsylvania. The plan calls for a computerized traffic light at the north exit on Fifth Street, and the diagram indicates the average number of vehicles per hour that are expected to flow in and out of the streets that border the complex. All streets are one-way.

(a) How many vehicles per hour should the traffic light let through to ensure that the average number of vehicles per hour flowing into the complex is the same as the average number of vehicles flowing out?

(b) Assuming that the traffic light has been set to balance the total flow in and out of the complex, what can you say about the average number of vehicles per hour that will flow along the streets that border the complex?

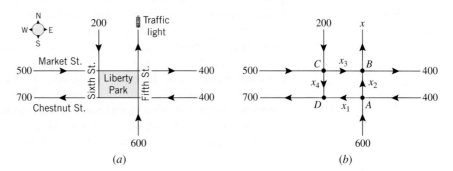

▶ Figure 1.8.3

Solution (a) If, as indicated in Figure 1.8.3b, we let x denote the number of vehicles per hour that the traffic light must let through, then the total number of vehicles per hour that flow in and out of the complex will be

Flowing in: $500 + 400 + 600 + 200 = 1700$
Flowing out: $x + 700 + 400$

Equating the flows in and out shows that the traffic light should let $x = 600$ vehicles per hour pass through.

Solution (b) To avoid traffic congestion, the flow in must equal the flow out at each intersection. For this to happen, the following conditions must be satisfied:

Intersection	Flow In		Flow Out
A	$400 + 600$	=	$x_1 + x_2$
B	$x_2 + x_3$	=	$400 + x$
C	$500 + 200$	=	$x_3 + x_4$
D	$x_1 + x_4$	=	700

Thus, with $x = 600$, as computed in part (a), we obtain the following linear system:

$$\begin{aligned} x_1 + x_2 &= 1000 \\ x_2 + x_3 &= 1000 \\ x_3 + x_4 &= 700 \\ x_1 \phantom{{}+x_2} + x_4 &= 700 \end{aligned}$$

We leave it for you to show that the system has infinitely many solutions and that these are given by the parametric equations

$$x_1 = 700 - t, \quad x_2 = 300 + t, \quad x_3 = 700 - t, \quad x_4 = t \qquad (1)$$

However, the parameter t is not completely arbitrary here, since there are physical constraints to be considered. For example, the average flow rates must be nonnegative since we have assumed the streets to be one-way, and a negative flow rate would indicate a flow in the wrong direction. This being the case, we see from (1) that t can be any real number that satisfies $0 \leq t \leq 700$, which implies that the average flow rates along the streets will fall in the ranges

$$0 \leq x_1 \leq 700, \quad 300 \leq x_2 \leq 1000, \quad 0 \leq x_3 \leq 700, \quad 0 \leq x_4 \leq 700 \blacktriangleleft$$

76 Chapter 1 Systems of Linear Equations and Matrices

Electrical Circuits

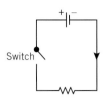

▲ Figure 1.8.4

Next, we will show how network analysis can be used to analyze electrical circuits consisting of batteries and resistors. A **battery** is a source of electric energy, and a **resistor**, such as a lightbulb, is an element that dissipates electric energy. Figure 1.8.4 shows a schematic diagram of a circuit with one battery (represented by the symbol ⊣⊢), one resistor (represented by the symbol ⌐⌐⌐), and a switch. The battery has a **positive pole** (+) and a **negative pole** (−). When the switch is closed, electrical current is considered to flow from the positive pole of the battery, through the resistor, and back to the negative pole (indicated by the arrowhead in the figure).

Electrical current, which is a flow of electrons through wires, behaves much like the flow of water through pipes. A battery acts like a pump that creates "electrical pressure" to increase the flow rate of electrons, and a resistor acts like a restriction in a pipe that reduces the flow rate of electrons. The technical term for electrical pressure is **electrical potential**; it is commonly measured in **volts** (V). The degree to which a resistor reduces the electrical potential is called its **resistance** and is commonly measured in **ohms** (Ω). The rate of flow of electrons in a wire is called **current** and is commonly measured in **amperes** (also called **amps**) (A). The precise effect of a resistor is given by the following law:

> **Ohm's Law** If a current of I amperes passes through a resistor with a resistance of R ohms, then there is a resulting drop of E volts in electrical potential that is the product of the current and resistance; that is,
>
> $$E = IR$$

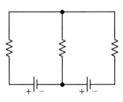

▲ Figure 1.8.5

A typical electrical network will have multiple batteries and resistors joined by some configuration of wires. A point at which three or more wires in a network are joined is called a **node** (or **junction point**). A **branch** is a wire connecting two nodes, and a **closed loop** is a succession of connected branches that begin and end at the same node. For example, the electrical network in Figure 1.8.5 has two nodes and three closed loops—two inner loops and one outer loop. As current flows through an electrical network, it undergoes increases and decreases in electrical potential, called **voltage rises** and **voltage drops**, respectively. The behavior of the current at the nodes and around closed loops is governed by two fundamental laws:

> **Kirchhoff's Current Law** The sum of the currents flowing into any node is equal to the sum of the currents flowing out.

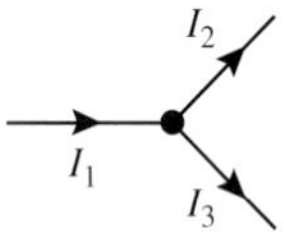

▲ Figure 1.8.6

> **Kirchhoff's Voltage Law** In one traversal of any closed loop, the sum of the voltage rises equals the sum of the voltage drops.

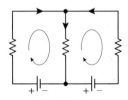

Clockwise closed-loop convention with arbitrary direction assignments to currents in the branches

▲ Figure 1.8.7

Kirchhoff's current law is a restatement of the principle of flow conservation at a node that was stated for general networks. Thus, for example, the currents at the top node in Figure 1.8.6 satisfy the equation $I_1 = I_2 + I_3$.

In circuits with multiple loops and batteries there is usually no way to tell in advance which way the currents are flowing, so the usual procedure in circuit analysis is to assign *arbitrary* directions to the current flows in the branches and let the mathematical computations determine whether the assignments are correct. In addition to assigning directions to the current flows, Kirchhoff's voltage law requires a direction of travel for each closed loop. The choice is arbitrary, but for consistency we will always take this direction to be *clockwise* (Figure 1.8.7). We also make the following conventions:

- A voltage drop occurs at a resistor if the direction assigned to the current through the resistor is the same as the direction assigned to the loop, and a voltage rise occurs at a resistor if the direction assigned to the current through the resistor is the opposite to that assigned to the loop.
- A voltage rise occurs at a battery if the direction assigned to the loop is from $-$ to $+$ through the battery, and a voltage drop occurs at a battery if the direction assigned to the loop is from $+$ to $-$ through the battery.

If you follow these conventions when calculating currents, then those currents whose directions were assigned correctly will have positive values and those whose directions were assigned incorrectly will have negative values.

▶ **EXAMPLE 3** **A Circuit with One Closed Loop**

Determine the current I in the circuit shown in Figure 1.8.8.

Solution Since the direction assigned to the current through the resistor is the same as the direction of the loop, there is a voltage drop at the resistor. By Ohm's law this voltage drop is $E = IR = 3I$. Also, since the direction assigned to the loop is from $-$ to $+$ through the battery, there is a voltage rise of 6 volts at the battery. Thus, it follows from Kirchhoff's voltage law that

$$3I = 6$$

from which we conclude that the current is $I = 2$ A. Since I is positive, the direction assigned to the current flow is correct.

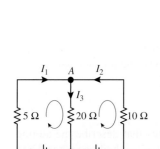

▲ Figure 1.8.8

▶ **EXAMPLE 4** **A Circuit with Three Closed Loops**

Determine the currents I_1, I_2, and I_3 in the circuit shown in Figure 1.8.9.

Solution Using the assigned directions for the currents, Kirchhoff's current law provides one equation for each node:

Node	Current In		Current Out
A	$I_1 + I_2$	=	I_3
B	I_3	=	$I_1 + I_2$

However, these equations are really the same, since both can be expressed as

$$I_1 + I_2 - I_3 = 0 \tag{2}$$

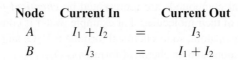

▲ Figure 1.8.9

Gustav Kirchhoff (1824–1887)

Historical Note The German physicist Gustav Kirchhoff was a student of Gauss. His work on Kirchhoff's laws, announced in 1854, was a major advance in the calculation of currents, voltages, and resistances of electrical circuits. Kirchhoff was severely disabled and spent most of his life on crutches or in a wheelchair.
[Image: ©SSPL/The Image Works]

To find unique values for the currents we will need two more equations, which we will obtain from Kirchhoff's voltage law. We can see from the network diagram that there are three closed loops, a left inner loop containing the 50 V battery, a right inner loop containing the 30 V battery, and an outer loop that contains both batteries. Thus, Kirchhoff's voltage law will actually produce three equations. With a clockwise traversal of the loops, the voltage rises and drops in these loops are as follows:

	Voltage Rises	Voltage Drops
Left Inside Loop	50	$5I_1 + 20I_3$
Right Inside Loop	$30 + 10I_2 + 20I_3$	0
Outside Loop	$30 + 50 + 10I_2$	$5I_1$

These conditions can be rewritten as

$$\begin{aligned} 5I_1 + 20I_3 &= 50 \\ 10I_2 + 20I_3 &= -30 \\ 5I_1 - 10I_2 &= 80 \end{aligned} \quad (3)$$

However, the last equation is superfluous, since it is the difference of the first two. Thus, if we combine (2) and the first two equations in (3), we obtain the following linear system of three equations in the three unknown currents:

$$\begin{aligned} I_1 + I_2 - I_3 &= 0 \\ 5I_1 + 20I_3 &= 50 \\ 10I_2 + 20I_3 &= -30 \end{aligned}$$

We leave it for you to solve this system and show that $I_1 = 6$ A, $I_2 = -5$ A, and $I_3 = 1$ A. The fact that I_2 is negative tells us that the direction of this current is opposite to that indicated in Figure 1.8.9. ◀

Balancing Chemical Equations

Chemical compounds are represented by **chemical formulas** that describe the atomic makeup of their molecules. For example, water is composed of two hydrogen atoms and one oxygen atom, so its chemical formula is H_2O; and stable oxygen is composed of two oxygen atoms, so its chemical formula is O_2.

When chemical compounds are combined under the right conditions, the atoms in their molecules rearrange to form new compounds. For example, when methane burns, the methane (CH_4) and stable oxygen (O_2) react to form carbon dioxide (CO_2) and water (H_2O). This is indicated by the **chemical equation**

$$CH_4 + O_2 \longrightarrow CO_2 + H_2O \quad (4)$$

The molecules to the left of the arrow are called the **reactants** and those to the right the **products**. In this equation the plus signs serve to separate the molecules and are not intended as algebraic operations. However, this equation does not tell the whole story, since it fails to account for the proportions of molecules required for a **complete reaction** (no reactants left over). For example, we can see from the right side of (4) that to produce one molecule of carbon dioxide and one molecule of water, one needs *three* oxygen atoms for each carbon atom. However, from the left side of (4) we see that one molecule of methane and one molecule of stable oxygen have only *two* oxygen atoms for each carbon atom. Thus, on the reactant side the ratio of methane to stable oxygen cannot be one-to-one in a complete reaction.

A chemical equation is said to be **balanced** if for each type of atom in the reaction, the same number of atoms appears on each side of the arrow. For example, the balanced version of Equation (4) is

$$CH_4 + 2O_2 \longrightarrow CO_2 + 2H_2O \tag{5}$$

by which we mean that one methane molecule combines with two stable oxygen molecules to produce one carbon dioxide molecule and two water molecules. In theory, one could multiply this equation through by any positive integer. For example, multiplying through by 2 yields the balanced chemical equation

$$2CH_4 + 4O_2 \longrightarrow 2CO_2 + 4H_2O$$

However, the standard convention is to use the smallest positive integers that will balance the equation.

Equation (4) is sufficiently simple that it could have been balanced by trial and error, but for more complicated chemical equations we will need a systematic method. There are various methods that can be used, but we will give one that uses systems of linear equations. To illustrate the method let us reexamine Equation (4). To balance this equation we must find positive integers, x_1, x_2, x_3, and x_4 such that

$$x_1 \,(CH_4) + x_2 \,(O_2) \longrightarrow x_3 \,(CO_2) + x_4 \,(H_2O) \tag{6}$$

For each of the atoms in the equation, the number of atoms on the left must be equal to the number of atoms on the right. Expressing this in tabular form we have

	Left Side		Right Side
Carbon	x_1	=	x_3
Hydrogen	$4x_1$	=	$2x_4$
Oxygen	$2x_2$	=	$2x_3 + x_4$

from which we obtain the homogeneous linear system

$$\begin{aligned} x_1 - x_3 &= 0 \\ 4x_1 - 2x_4 &= 0 \\ 2x_2 - 2x_3 - x_4 &= 0 \end{aligned}$$

The augmented matrix for this system is

$$\begin{bmatrix} 1 & 0 & -1 & 0 & 0 \\ 4 & 0 & 0 & -2 & 0 \\ 0 & 2 & -2 & -1 & 0 \end{bmatrix}$$

We leave it for you to show that the reduced row echelon form of this matrix is

$$\begin{bmatrix} 1 & 0 & 0 & -\frac{1}{2} & 0 \\ 0 & 1 & 0 & -1 & 0 \\ 0 & 0 & 1 & -\frac{1}{2} & 0 \end{bmatrix}$$

from which we conclude that the general solution of the system is

$$x_1 = t/2, \quad x_2 = t, \quad x_3 = t/2, \quad x_4 = t$$

where t is arbitrary. The smallest positive integer values for the unknowns occur when we let $t = 2$, so the equation can be balanced by letting $x_1 = 1$, $x_2 = 2$, $x_3 = 1$, $x_4 = 2$. This agrees with our earlier conclusions, since substituting these values into Equation (6) yields Equation (5).

► EXAMPLE 5 Balancing Chemical Equations Using Linear Systems

Balance the chemical equation

$$HCl + Na_3PO_4 \longrightarrow H_3PO_4 + NaCl$$

[hydrochloric acid] + [sodium phosphate] $\longrightarrow$ [phosphoric acid] + [sodium chloride]

Solution Let x_1, x_2, x_3, and x_4 be positive integers that balance the equation

$$x_1\,(HCl) + x_2\,(Na_3PO_4) \longrightarrow x_3\,(H_3PO_4) + x_4\,(NaCl) \tag{7}$$

Equating the number of atoms of each type on the two sides yields

$$\begin{aligned} 1x_1 &= 3x_3 & \text{Hydrogen (H)} \\ 1x_1 &= 1x_4 & \text{Chlorine (Cl)} \\ 3x_2 &= 1x_4 & \text{Sodium (Na)} \\ 1x_2 &= 1x_3 & \text{Phosphorous (P)} \\ 4x_2 &= 4x_3 & \text{Oxygen (O)} \end{aligned}$$

from which we obtain the homogeneous linear system

$$\begin{aligned} x_1 \phantom{{}+3x_2} - 3x_3 \phantom{{}-x_4} &= 0 \\ x_1 \phantom{{}+3x_2-3x_3} - x_4 &= 0 \\ 3x_2 \phantom{{}-3x_3} - x_4 &= 0 \\ x_2 - x_3 \phantom{{}-x_4} &= 0 \\ 4x_2 - 4x_3 \phantom{{}-x_4} &= 0 \end{aligned}$$

We leave it for you to show that the reduced row echelon form of the augmented matrix for this system is

$$\begin{bmatrix} 1 & 0 & 0 & -1 & 0 \\ 0 & 1 & 0 & -\tfrac{1}{3} & 0 \\ 0 & 0 & 1 & -\tfrac{1}{3} & 0 \\ 0 & 0 & 0 & 0 & 0 \\ 0 & 0 & 0 & 0 & 0 \end{bmatrix}$$

from which we conclude that the general solution of the system is

$$x_1 = t, \quad x_2 = t/3, \quad x_3 = t/3, \quad x_4 = t$$

where t is arbitrary. To obtain the smallest positive integers that balance the equation, we let $t = 3$, in which case we obtain $x_1 = 3$, $x_2 = 1$, $x_3 = 1$, and $x_4 = 3$. Substituting these values in (7) produces the balanced equation

$$3HCl + Na_3PO_4 \longrightarrow H_3PO_4 + 3NaCl \quad \blacktriangleleft$$

Polynomial Interpolation   An important problem in various applications is to find a polynomial whose graph passes through a specified set of points in the plane; this is called an ***interpolating polynomial*** for the points. The simplest example of such a problem is to find a linear polynomial

$$p(x) = ax + b \tag{8}$$

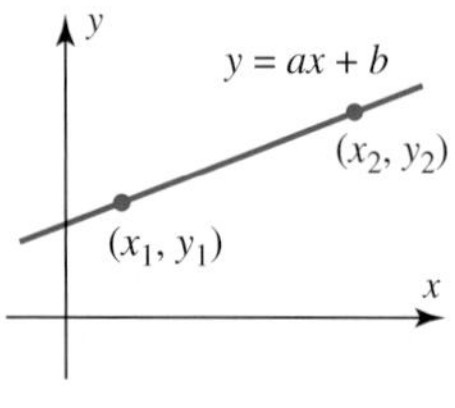

whose graph passes through two known distinct points, (x_1, y_1) and (x_2, y_2), in the xy-plane (Figure 1.8.10). You have probably encountered various methods in analytic geometry for finding the equation of a line through two points, but here we will give a method based on linear systems that can be adapted to general polynomial interpolation.

▲ Figure 1.8.10

The graph of (8) is the line $y = ax + b$, and for this line to pass through the points (x_1, y_1) and (x_2, y_2), we must have

$$y_1 = ax_1 + b \quad \text{and} \quad y_2 = ax_2 + b$$

Therefore, the unknown coefficients a and b can be obtained by solving the linear system

$$ax_1 + b = y_1$$
$$ax_2 + b = y_2$$

We don't need any fancy methods to solve this system—the value of a can be obtained by subtracting the equations to eliminate b, and then the value of a can be substituted into either equation to find b. We leave it as an exercise for you to find a and b and then show that they can be expressed in the form

$$a = \frac{y_2 - y_1}{x_2 - x_1} \quad \text{and} \quad b = \frac{y_1 x_2 - y_2 x_1}{x_2 - x_1} \tag{9}$$

provided $x_1 \neq x_2$. Thus, for example, the line $y = ax + b$ that passes through the points

$$(2, 1) \quad \text{and} \quad (5, 4)$$

can be obtained by taking $(x_1, y_1) = (2, 1)$ and $(x_2, y_2) = (5, 4)$, in which case (9) yields

$$a = \frac{4 - 1}{5 - 2} = 1 \quad \text{and} \quad b = \frac{(1)(5) - (4)(2)}{5 - 2} = -1$$

Therefore, the equation of the line is

$$y = x - 1$$

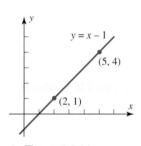

▲ Figure 1.8.11

(Figure 1.8.11).

Now let us consider the more general problem of finding a polynomial whose graph passes through n points with distinct x-coordinates

$$(x_1, y_1), \quad (x_2, y_2), \quad (x_3, y_3), \ldots, \quad (x_n, y_n) \tag{10}$$

Since there are n conditions to be satisfied, intuition suggests that we should begin by looking for a polynomial of the form

$$p(x) = a_0 + a_1 x + a_2 x^2 + \cdots + a_{n-1} x^{n-1} \tag{11}$$

since a polynomial of this form has n coefficients that are at our disposal to satisfy the n conditions. However, we want to allow for cases where the points may lie on a line or have some other configuration that would make it possible to use a polynomial whose degree is less than $n - 1$; thus, we allow for the possibility that a_{n-1} and other coefficients in (11) may be zero.

The following theorem, which we will prove later in the text, is the basic result on polynomial interpolation.

> **THEOREM 1.8.1 Polynomial Interpolation**
>
> *Given any n points in the xy-plane that have distinct x-coordinates, there is a unique polynomial of degree $n - 1$ or less whose graph passes through those points.*

Let us now consider how we might go about finding the interpolating polynomial (11) whose graph passes through the points in (10). Since the graph of this polynomial is the graph of the equation

$$y = a_0 + a_1 x + a_2 x^2 + \cdots + a_{n-1} x^{n-1} \tag{12}$$

it follows that the coordinates of the points must satisfy

$$a_0 + a_1 x_1 + a_2 x_1^2 + \cdots + a_{n-1} x_1^{n-1} = y_1$$
$$a_0 + a_1 x_2 + a_2 x_2^2 + \cdots + a_{n-1} x_2^{n-1} = y_2$$
$$\vdots \qquad \vdots \qquad \vdots \qquad \qquad \vdots \qquad \vdots \qquad (13)$$
$$a_0 + a_1 x_n + a_2 x_n^2 + \cdots + a_{n-1} x_n^{n-1} = y_n$$

In these equations the values of x's and y's are assumed to be known, so we can view this as a linear system in the unknowns $a_0, a_1, \ldots, a_{n-1}$. From this point of view the augmented matrix for the system is

$$\begin{bmatrix} 1 & x_1 & x_1^2 & \cdots & x_1^{n-1} & y_1 \\ 1 & x_2 & x_2^2 & \cdots & x_2^{n-1} & y_2 \\ \vdots & \vdots & \vdots & & \vdots & \vdots \\ 1 & x_n & x_n^2 & \cdots & x_n^{n-1} & y_n \end{bmatrix} \qquad (14)$$

and hence the interpolating polynomial can be found by reducing this matrix to reduced row echelon form (Gauss–Jordan elimination).

▶ **EXAMPLE 6** **Polynomial Interpolation by Gauss–Jordan Elimination**

Find a cubic polynomial whose graph passes through the points

$$(1, 3), \quad (2, -2), \quad (3, -5), \quad (4, 0)$$

Solution Since there are four points, we will use an interpolating polynomial of degree $n = 3$. Denote this polynomial by

$$p(x) = a_0 + a_1 x + a_2 x^2 + a_3 x^3$$

and denote the x- and y-coordinates of the given points by

$$x_1 = 1, \quad x_2 = 2, \quad x_3 = 3, \quad x_4 = 4 \quad \text{and} \quad y_1 = 3, \quad y_2 = -2, \quad y_3 = -5, \quad y_4 = 0$$

Thus, it follows from (14) that the augmented matrix for the linear system in the unknowns $a_0, a_1, a_2,$ and a_3 is

$$\begin{bmatrix} 1 & x_1 & x_1^2 & x_1^3 & y_1 \\ 1 & x_2 & x_2^2 & x_2^3 & y_2 \\ 1 & x_3 & x_3^2 & x_3^3 & y_3 \\ 1 & x_4 & x_4^2 & x_4^3 & y_4 \end{bmatrix} = \begin{bmatrix} 1 & 1 & 1 & 1 & 3 \\ 1 & 2 & 4 & 8 & -2 \\ 1 & 3 & 9 & 27 & -5 \\ 1 & 4 & 16 & 64 & 0 \end{bmatrix}$$

We leave it for you to confirm that the reduced row echelon form of this matrix is

$$\begin{bmatrix} 1 & 0 & 0 & 0 & 4 \\ 0 & 1 & 0 & 0 & 3 \\ 0 & 0 & 1 & 0 & -5 \\ 0 & 0 & 0 & 1 & 1 \end{bmatrix}$$

from which it follows that $a_0 = 4, a_1 = 3, a_2 = -5, a_3 = 1$. Thus, the interpolating polynomial is

$$p(x) = 4 + 3x - 5x^2 + x^3$$

The graph of this polynomial and the given points are shown in Figure 1.8.12. ◀

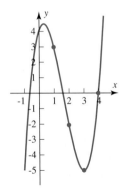

▲ Figure 1.8.12

Remark Later we will give a more efficient method for finding interpolating polynomials that is better suited for problems in which the number of data points is large.

CALCULUS AND CALCULATING UTILITY REQUIRED

▶ **EXAMPLE 7 Approximate Integration**

There is no way to evaluate the integral

$$\int_0^1 \sin\left(\frac{\pi x^2}{2}\right) dx$$

directly since there is no way to express an antiderivative of the integrand in terms of elementary functions. This integral could be approximated by Simpson's rule or some comparable method, but an alternative approach is to approximate the integrand by an interpolating polynomial and integrate the approximating polynomial. For example, let us consider the five points

$$x_0 = 0, \quad x_1 = 0.25, \quad x_2 = 0.5, \quad x_3 = 0.75, \quad x_4 = 1$$

that divide the interval $[0, 1]$ into four equally spaced subintervals. The values of

$$f(x) = \sin\left(\frac{\pi x^2}{2}\right)$$

at these points are approximately

$$f(0) = 0, \quad f(0.25) = 0.098017, \quad f(0.5) = 0.382683,$$
$$f(0.75) = 0.77301, \quad f(1) = 1$$

The interpolating polynomial is (verify)

$$p(x) = 0.098796x + 0.762356x^2 + 2.14429x^3 - 2.00544x^4 \tag{15}$$

and

$$\int_0^1 p(x)\,dx \approx 0.438501 \tag{16}$$

As shown in Figure 1.8.13, the graphs of f and p match very closely over the interval $[0, 1]$, so the approximation is quite good. ◀

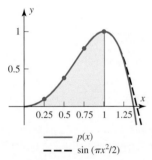

▲ Figure 1.8.13

Concept Review

- Network
- Branches
- Nodes
- Flow conservation
- Electrical circuits: battery, resistor, poles (positive and negative), electrical potential, Ohm's law, Kirchhoff's current law, Kirchhoff's voltage law
- Chemical equations: reactants, products, balanced equation
- Interpolating polynomial

Skills

- Find the flow rates and directions of flow in branches of a network.
- Find the amount of current flowing through parts of an electrical circuit.
- Write a balanced chemical equation for a given chemical reaction.
- Find an interpolating polynomial for a graph passing through a given collection of points.

Exercise Set 1.8

1. The accompanying figure shows a network in which the flow rate and direction of flow in certain branches are known. Find the flow rates and directions of flow in the remaining branches.

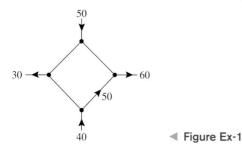

◀ Figure Ex-1

2. The accompanying figure shows known flow rates of hydrocarbons into and out of a network of pipes at an oil refinery.
   (a) Set up a linear system whose solution provides the unknown flow rates.
   (b) Solve the system for the unknown flow rates.
   (c) Find the flow rates and directions of flow if $x_4 = 50$ and $x_6 = 0$.

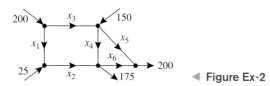

◀ Figure Ex-2

3. The accompanying figure shows a network of one-way streets with traffic flowing in the directions indicated. The flow rates along the streets are measured as the average number of vehicles per hour.
   (a) Set up a linear system whose solution provides the unknown flow rates.
   (b) Solve the system for the unknown flow rates.
   (c) If the flow along the road from A to B must be reduced for construction, what is the minimum flow that is required to keep traffic flowing on all roads?

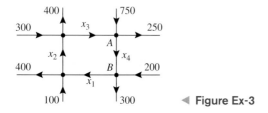

◀ Figure Ex-3

4. The accompanying figure shows a network of one-way streets with traffic flowing in the directions indicated. The flow rates along the streets are measured as the average number of vehicles per hour.
   (a) Set up a linear system whose solution provides the unknown flow rates.
   (b) Solve the system for the unknown flow rates.
   (c) Is it possible to close the road from A to B for construction and keep traffic flowing on the other streets? Explain.

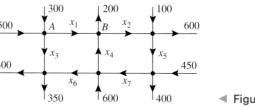

◀ Figure Ex-4

▶ In Exercises 5–8, analyze the given electrical circuits by finding the unknown currents. ◀

5.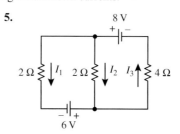

6.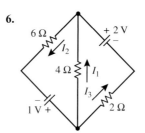

7.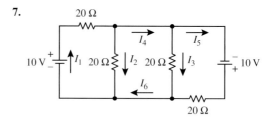

8.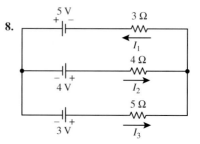

▶ In Exercises 9–12, write a balanced equation for the given chemical reaction. ◀

9. $C_3H_8 + O_2 \rightarrow CO_2 + H_2O$ (propane combustion)

10. $C_6H_{12}O_6 \rightarrow CO_2 + C_2H_5OH$ (fermentation of sugar)

11. $CH_3COF + H_2O \rightarrow CH_3COOH + HF$

12. $CO_2 + H_2O \rightarrow C_6H_{12}O_6 + O_2$ (photosynthesis)

13. Find the quadratic polynomial whose graph passes through the points $(0, -1)$, $(1, 2)$, and $(-1, 0)$.

14. Find the quadratic polynomial whose graph passes through the points $(0, 0)$, $(-1, 1)$, and $(1, 1)$.

15. Find the cubic polynomial whose graph passes through the points $(-1, -1)$, $(0, 1)$, $(1, 3)$, $(4, -1)$.

16. The accompanying figure shows the graph of a cubic polynomial. Find the polynomial.

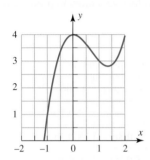

◀ Figure Ex-16

17. (a) Find an equation that represents the family of all second-degree polynomials that pass through the points $(0, 1)$ and $(1, 2)$. [*Hint:* The equation will involve one arbitrary parameter that produces the members of the family when varied.]

    (b) By hand, or with the help of a graphing utility, sketch four curves in the family.

18. In this section we have selected only a few applications of linear systems. Using the Internet as a search tool, try to find some more real-world applications of such systems. Select one that is of interest to you, and write a paragraph about it.

True-False Exercises

In parts (a)–(e) determine whether the statement is true or false, and justify your answer.

(a) In any network, the sum of the flows out of a node must equal the sum of the flows into a node.

(b) When a current passes through a resistor, there is an increase in the electrical potential in a circuit.

(c) Kirchhoff's current law states that the sum of the currents flowing into a node equals the sum of the currents flowing out of the node.

(d) A chemical equation is called balanced if the total number of atoms on each side of the equation is the same.

(e) Given any n points in the xy-plane, there is a unique polynomial of degree $n - 1$ or less whose graph passes through those points.

1.9 Leontief Input-Output Models

In 1973 the economist Wassily Leontief was awarded the Nobel prize for his work on economic modeling in which he used matrix methods to study the relationships between different sectors in an economy. In this section we will discuss some of the ideas developed by Leontief.

Inputs and Outputs in an Economy

One way to analyze an economy is to divide it into **sectors** and study how the sectors interact with one another. For example, a simple economy might be divided into three sectors—manufacturing, agriculture, and utilities. Typically, a sector will produce certain **outputs** but will require **inputs** from the other sectors and itself. For example, the agricultural sector may produce wheat as an output but will require inputs of farm machinery from the manufacturing sector, electrical power from the utilities sector, and food from its own sector to feed its workers. Thus, we can imagine an economy to be a network in which inputs and outputs flow in and out of the sectors; the study of such flows is called **input-output analysis**. Inputs and outputs are commonly measured in monetary units (dollars or millions of dollars, for example) but other units of measurement are also possible.

The flows between sectors of a real economy are not always obvious. For example, in World War II the United States had a demand for 50,000 new airplanes that required the construction of many new aluminum manufacturing plants. This produced an unexpectedly large demand for certain copper electrical components, which in turn produced a copper shortage. The problem was eventually resolved by using silver borrowed from Fort Knox as a copper substitute. In all likelihood modern input-output analysis would have anticipated the copper shortage.

Most sectors of an economy will produce outputs, but there may exist sectors that consume outputs without producing anything themselves (the consumer market, for example). Those sectors that do not produce outputs are called *open sectors*. Economies with no open sectors are called *closed economies*, and economies with one or more open sectors are called *open economies* (Figure 1.9.1). In this section we will be concerned with economies with one open sector, and our primary goal will be to determine the output levels that are required for the productive sectors to sustain themselves and satisfy the demand of the open sector.

Leontief Model of an Open Economy

Let us consider a simple open economy with one open sector and three product-producing sectors: manufacturing, agriculture, and utilities. Assume that inputs and outputs are measured in dollars and that the inputs required by the productive sectors to produce one dollar's worth of output are in accordance with Table 1.

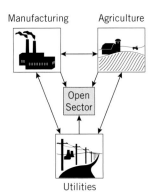

▲ Figure 1.9.1

Table 1

		Input Required per Dollar Output		
		Manufacturing	Agriculture	Utilities
Provider	Manufacturing	$ 0.50	$ 0.10	$ 0.10
	Agriculture	$ 0.20	$ 0.50	$ 0.30
	Utilities	$ 0.10	$ 0.30	$ 0.40

Usually, one would suppress the labeling and express this matrix as

$$C = \begin{bmatrix} 0.5 & 0.1 & 0.1 \\ 0.2 & 0.5 & 0.3 \\ 0.1 & 0.3 & 0.4 \end{bmatrix} \tag{1}$$

This is called the *consumption matrix* (or sometimes the *technology matrix*) for the economy. The column vectors

$$\mathbf{c}_1 = \begin{bmatrix} 0.5 \\ 0.2 \\ 0.1 \end{bmatrix}, \quad \mathbf{c}_2 = \begin{bmatrix} 0.1 \\ 0.5 \\ 0.3 \end{bmatrix}, \quad \mathbf{c}_3 = \begin{bmatrix} 0.1 \\ 0.3 \\ 0.4 \end{bmatrix}$$

**Wassily Leontief
(1906–1999)**

Historical Note It is somewhat ironic that it was the Russian-born Wassily Leontief who won the Nobel prize in 1973 for pioneering the modern methods for analyzing free-market economies. Leontief was a precocious student who entered the University of Leningrad at age 15. Bothered by the intellectual restrictions of the Soviet system, he was put in jail for anti-Communist activities, after which he headed for the University of Berlin, receiving his Ph.D. there in 1928. He came to the United States in 1931, where he held professorships at Harvard and then New York University.

[*Image:* ©Bettmann/©Corbis]

1.9 Leontief Input-Output Models

> What is the economic significance of the row sums of the consumption matrix?

in C list the inputs required by the manufacturing, agricultural, and utilities sectors, respectively, to produce $1.00 worth of output. These are called the ***consumption vectors*** of the sectors. For example, $\mathbf{c}_1$ tells us that to produce $1.00 worth of output the manufacturing sector needs $0.50 worth of manufacturing output, $0.20 worth of agricultural output, and $0.10 worth of utilities output.

Continuing with the above example, suppose that the open sector wants the economy to supply it manufactured goods, agricultural products, and utilities with dollar values:

d_1 dollars of manufactured goods
d_2 dollars of agricultural products
d_3 dollars of utilities

The column vector $\mathbf{d}$ that has these numbers as successive components is called the ***outside demand vector***. Since the product-producing sectors consume some of their own output, the dollar value of their output must cover their own needs plus the outside demand. Suppose that the dollar values required to do this are

x_1 dollars of manufactured goods
x_2 dollars of agricultural products
x_3 dollars of utilities

The column vector $\mathbf{x}$ that has these numbers as successive components is called the ***production vector*** for the economy. For the economy with consumption matrix (1), that portion of the production vector $\mathbf{x}$ that will be consumed by the three productive sectors is

$$x_1 \begin{bmatrix} 0.5 \\ 0.2 \\ 0.1 \end{bmatrix} + x_2 \begin{bmatrix} 0.1 \\ 0.5 \\ 0.3 \end{bmatrix} + x_3 \begin{bmatrix} 0.1 \\ 0.3 \\ 0.4 \end{bmatrix} = \begin{bmatrix} 0.5 & 0.1 & 0.1 \\ 0.2 & 0.5 & 0.3 \\ 0.1 & 0.3 & 0.4 \end{bmatrix} \begin{bmatrix} x_1 \\ x_2 \\ x_3 \end{bmatrix} = C\mathbf{x}$$

(Fractions consumed by manufacturing) (Fractions consumed by agriculture) (Fractions consumed by utilities)

The vector $C\mathbf{x}$ is called the ***intermediate demand vector*** for the economy. Once the intermediate demand is met, the portion of the production that is left to satisfy the outside demand is $\mathbf{x} - C\mathbf{x}$. Thus, if the outside demand vector is $\mathbf{d}$, then $\mathbf{x}$ must satisfy the equation

$$\underbrace{\mathbf{x}}_{\text{Amount produced}} - \underbrace{C\mathbf{x}}_{\text{Intermediate demand}} = \underbrace{\mathbf{d}}_{\text{Outside demand}}$$

which we will find convenient to rewrite as

$$(I - C)\mathbf{x} = \mathbf{d} \qquad (2)$$

The matrix $I - C$ is called the ***Leontief matrix*** and (2) is called the ***Leontief equation***.

▶ **EXAMPLE 1** Satisfying Outside Demand

Consider the economy described in Table 1. Suppose that the open sector has a demand for $7900 worth of manufacturing products, $3950 worth of agricultural products, and $1975 worth of utilities.

(a) Can the economy meet this demand?
(b) If so, find a production vector $\mathbf{x}$ that will meet it exactly.

Solution The consumption matrix, production vector, and outside demand vector are

$$C = \begin{bmatrix} 0.5 & 0.1 & 0.1 \\ 0.2 & 0.5 & 0.3 \\ 0.1 & 0.3 & 0.4 \end{bmatrix}, \quad \mathbf{x} = \begin{bmatrix} x_1 \\ x_2 \\ x_3 \end{bmatrix}, \quad \mathbf{d} = \begin{bmatrix} 7900 \\ 3950 \\ 1975 \end{bmatrix} \quad (3)$$

To meet the outside demand, the vector $\mathbf{x}$ must satisfy the Leontief equation (2), so the problem reduces to solving the linear system

$$\underbrace{\begin{bmatrix} 0.5 & -0.1 & -0.1 \\ -0.2 & 0.5 & -0.3 \\ -0.1 & -0.3 & 0.6 \end{bmatrix}}_{I-C} \underbrace{\begin{bmatrix} x_1 \\ x_2 \\ x_3 \end{bmatrix}}_{\mathbf{x}} = \underbrace{\begin{bmatrix} 7900 \\ 3950 \\ 1975 \end{bmatrix}}_{\mathbf{d}} \quad (4)$$

(if consistent). We leave it for you to show that the reduced row echelon form of the augmented matrix for this system is

$$\begin{bmatrix} 1 & 0 & 0 & | & 27{,}500 \\ 0 & 1 & 0 & | & 33{,}750 \\ 0 & 0 & 1 & | & 24{,}750 \end{bmatrix}$$

This tells us that (4) is consistent, and the economy can satisfy the demand of the open sector exactly by producing $27,500 worth of manufacturing output, $33,750 worth of agricultural output, and $24,750 worth of utilities output. ◀

Productive Open Economies

In the preceding discussion we considered an open economy with three product-producing sectors; the same ideas apply to an open economy with n product-producing sectors. In this case, the consumption matrix, production vector, and outside demand vector have the form

$$C = \begin{bmatrix} c_{11} & c_{12} & \cdots & c_{1n} \\ c_{21} & c_{22} & \cdots & c_{2n} \\ \vdots & \vdots & & \vdots \\ c_{n1} & c_{n2} & \cdots & c_{nn} \end{bmatrix}, \quad \mathbf{x} = \begin{bmatrix} x_1 \\ x_2 \\ \vdots \\ x_n \end{bmatrix}, \quad \mathbf{d} = \begin{bmatrix} d_1 \\ d_2 \\ \vdots \\ d_n \end{bmatrix}$$

where all entries are nonnegative and

$c_{ij} = $ the monetary value of the output of the ith sector that is needed by the jth sector to produce one unit of output

$x_i = $ the monetary value of the output of the ith sector

$d_i = $ the monetary value of the output of the ith sector that is required to meet the demand of the open sector

Remark Note that the jth column vector of C contains the monetary values that the jth sector requires of the other sectors to produce one monetary unit of output, and the ith row vector of C contains the monetary values required of the ith sector by the other sectors for each of them to produce one monetary unit of output.

As discussed in our example above, a production vector $\mathbf{x}$ that meets the demand $\mathbf{d}$ of the outside sector must satisfy the Leontief equation

$$(I - C)\mathbf{x} = \mathbf{d}$$

If the matrix $I - C$ is invertible, then this equation has the unique solution

$$\mathbf{x} = (I - C)^{-1}\mathbf{d} \quad (5)$$

for every demand vector **d**. However, for **x** to be a valid production vector it must have nonnegative entries, so the problem of importance in economics is to determine conditions under which the Leontief equation has a solution with nonnegative entries.

It is evident from the form of (5) that if $I - C$ is invertible, and if $(I - C)^{-1}$ has nonnegative entries, then for every demand vector **d** the corresponding **x** will also have nonnegative entries, and hence will be a valid production vector for the economy. Economies for which $(I - C)^{-1}$ has nonnegative entries are said to be ***productive***. Such economies are desirable because demand can always be met by some level of production. The following theorem, whose proof can be found in many books on economics, gives conditions under which open economies are productive.

THEOREM 1.9.1 *If C is the consumption matrix for an open economy, and if all of the column sums are less than 1, then the matrix $I - C$ is invertible, the entries of $(I - C)^{-1}$ are nonnegative, and the economy is productive.*

Remark The jth column sum of C represents the total dollar value of input that the jth sector requires to produce $1 of output, so if the jth column sum is less than 1, then the jth sector requires less than $1 of input to produce $1 of output; in this case we say that the jth sector is ***profitable***. Thus, Theorem 1.9.1 states that if all product-producing sectors of an open economy are profitable, then the economy is productive. In the exercises we will ask you to show that an open economy is productive if all of the row sums of C are less than 1 (Exercise 11). Thus, an open economy is productive if *either* all of the column sums or all of the row sums of C are less than 1.

▶ **EXAMPLE 2** **An Open Economy Whose Sectors Are All Profitable**

The column sums of the consumption matrix C in (1) are less than 1, so $(I - C)^{-1}$ exists and has nonnegative entries. Use a calculating utility to confirm this, and use this inverse to solve Equation (4) in Example 1.

Solution We leave it for you to show that

$$(I - C)^{-1} \approx \begin{bmatrix} 2.65823 & 1.13924 & 1.01266 \\ 1.89873 & 3.67089 & 2.15190 \\ 1.39241 & 2.02532 & 2.91139 \end{bmatrix}$$

This matrix has nonnegative entries, and

$$\mathbf{x} = (I - C)^{-1}\mathbf{d} \approx \begin{bmatrix} 2.65823 & 1.13924 & 1.01266 \\ 1.89873 & 3.67089 & 2.15190 \\ 1.39241 & 2.02532 & 2.91139 \end{bmatrix} \begin{bmatrix} 7900 \\ 3950 \\ 1975 \end{bmatrix} \approx \begin{bmatrix} 27,500 \\ 33,750 \\ 24,750 \end{bmatrix}$$

which is consistent with the solution in Example 1. ◀

Concept Review

- Sectors
- Inputs
- Outputs
- Input-output analysis
- Open sector
- Economies: open, closed
- Consumption (technology) matrix
- Consumption vector
- Outside demand vector
- Production vector
- Intermediate demand vector
- Leontief matrix
- Leontief equation

Skills

- Construct a consumption matrix for an economy.
- Understand the relationships among the vectors of a sector of an economy: consumption, outside demand, production, and intermediate demand.

Exercise Set 1.9

1. A copy store (C) and a paper company (P) use each other's services. For each $1.00 of business that C does, it uses $0.10 of its own services and $0.30 of P's services, and for each $1.00 of business that P does, it uses $0.20 of its own services and $0.10 of C's services.
   (a) Construct a consumption matrix for this economy.
   (b) How much must C and P each produce to provide customers with $9000 worth of copy services and $6000 of paper supplies?

2. A simple economy produces food (F) and housing (H). The production of $1.00 worth of food requires $0.30 worth of food and $0.10 worth of housing, and the production of $1.00 worth of housing requires $0.20 worth of food and $0.60 worth of housing.
   (a) Construct a consumption matrix for this economy.
   (b) What dollar value of food and housing must be produced for the economy to provide consumers $130,000 worth of food and $130,000 worth of housing?

3. Consider the open economy described by the accompanying table, where the input is in dollars needed for $1.00 of output.
   (a) Find the consumption matrix for the economy.
   (b) Suppose that the open sector has a demand for $6000 worth of housing, $3000 worth of food, and $2000 worth of utilities. Use row reduction to find a production vector that will meet this demand exactly.

Table Ex-3

Input Required per Dollar Output

		Housing	Food	Utilities
Provider	Housing	$ 0.20	$ 0.60	$ 0.50
	Food	$ 0.40	$ 0.20	$ 0.20
	Utilities	$ 0.20	$ 0.10	$ 0.20

4. A company produces Web design, software, and networking services. View the company as an open economy described by the accompanying table, where input is in dollars needed for $1.00 of output.
   (a) Find the consumption matrix for the company.
   (b) Suppose that the customers (the open sector) have a demand for $5400 worth of Web design, $2700 worth of software, and $900 worth of networking. Use row reduction to find a production vector that will meet this demand exactly.

Table Ex-4

Input Required per Dollar Output

		Web Design	Software	Networking
Provider	Web Design	$ 0.40	$ 0.20	$ 0.45
	Software	$ 0.30	$ 0.35	$ 0.30
	Networking	$ 0.15	$ 0.10	$ 0.20

▶ In Exercises 5–6, use matrix inversion to find the production vector **x** that meets the demand **d** for the consumption matrix C.

5. $C = \begin{bmatrix} 0.1 & 0.3 \\ 0.5 & 0.4 \end{bmatrix}$; $\mathbf{d} = \begin{bmatrix} 50 \\ 60 \end{bmatrix}$

6. $C = \begin{bmatrix} 0.3 & 0.1 \\ 0.3 & 0.7 \end{bmatrix}$; $\mathbf{d} = \begin{bmatrix} 22 \\ 14 \end{bmatrix}$

7. In each part, show that the open economy with consumption matrix C is productive:

(a) $C = \begin{bmatrix} 0.4 & 0.3 & 0.5 \\ 0.2 & 0.5 & 0.2 \\ 0.2 & 0.1 & 0.1 \end{bmatrix}$

(b) $C = \begin{bmatrix} 0.4 & 0.1 & 0.2 \\ 0.3 & 0.1 & 0.5 \\ 0.4 & 0.3 & 0.2 \end{bmatrix}$

8. Consider an open economy with consumption matrix

$$C = \begin{bmatrix} \frac{1}{2} & \frac{1}{4} & \frac{1}{4} \\ \frac{1}{2} & \frac{1}{8} & \frac{1}{4} \\ \frac{1}{2} & \frac{1}{4} & \frac{1}{8} \end{bmatrix}$$

If the open sector demands the same dollar value from each product-producing sector, which such sector must produce the greatest dollar value to meet the demand?

9. Consider an open economy with consumption matrix

$$C = \begin{bmatrix} c_{11} & c_{12} \\ c_{21} & 0 \end{bmatrix}$$

Show that the Leontief equation $\mathbf{x} - C\mathbf{x} = \mathbf{d}$ has a unique solution for every demand vector $\mathbf{d}$ if $c_{21}c_{12} < 1 - c_{11}$.

10. (a) Consider an open economy with a consumption matrix C whose column sums are less than 1, and let $\mathbf{x}$ be the production vector that satisfies an outside demand $\mathbf{d}$; that is, $(I - C)^{-1}\mathbf{d} = \mathbf{x}$. Let $\mathbf{d}_j$ be the demand vector that is obtained by increasing the jth entry of $\mathbf{d}$ by 1 and leaving the other entries fixed. Prove that the production vector $\mathbf{x}_j$ that meets this demand is

$$\mathbf{x}_j = \mathbf{x} + j\text{th column vector of } (I - C)^{-1}$$

(b) In words, what is the economic significance of the jth column vector of $(I - C)^{-1}$? [Hint: Look at $\mathbf{x}_j - \mathbf{x}$.]

11. Prove: If C is an $n \times n$ matrix whose entries are nonnegative and whose row sums are less than 1, then $I - C$ is invertible and has nonnegative entries. [Hint: $(A^T)^{-1} = (A^{-1})^T$ for any invertible matrix A.]

True-False Exercises

In parts (a)–(e) determine whether the statement is true or false, and justify your answer.

(a) Sectors of an economy that produce outputs are called open sectors.

(b) A closed economy is an economy that has no open sectors.

(c) The rows of a consumption matrix represent the outputs in a sector of an economy.

(d) If the column sums of the consumption matrix are all less than 1, then the Leontif matrix is invertible.

(e) The Leontif equation relates the production vector for an economy to the outside demand vector.

Chapter 1 Supplementary Exercises

▶ In Exercises 1–4 the given matrix represents an augmented matrix for a linear system. Write the corresponding set of linear equations for the system, and use Gaussian elimination to solve the linear system. Introduce free parameters as necessary. ◀

1. $\begin{bmatrix} 3 & -1 & 0 & 4 & 1 \\ 2 & 0 & 3 & 3 & -1 \end{bmatrix}$

2. $\begin{bmatrix} 1 & 4 & -1 \\ -2 & -8 & 2 \\ 3 & 12 & -3 \\ 0 & 0 & 0 \end{bmatrix}$

3. $\begin{bmatrix} 2 & -4 & 1 & 6 \\ -4 & 0 & 3 & -1 \\ 0 & 1 & -1 & 3 \end{bmatrix}$

4. $\begin{bmatrix} 3 & 1 & -2 \\ -9 & -3 & 6 \\ 6 & 2 & 1 \end{bmatrix}$

5. Use Gauss–Jordan elimination to solve for x' and y' in terms of x and y.

$$x = \tfrac{3}{5}x' - \tfrac{4}{5}y'$$
$$y = \tfrac{4}{5}x' + \tfrac{3}{5}y'$$

6. Use Gauss–Jordan elimination to solve for x' and y' in terms of x and y.

$$x = x' \cos\theta - y' \sin\theta$$
$$y = x' \sin\theta + y' \cos\theta$$

7. Find positive integers that satisfy

$$x + y + z = 9$$
$$x + 5y + 10z = 44$$

8. A box containing pennies, nickels, and dimes has 13 coins with a total value of 83 cents. How many coins of each type are in the box?

9. Let

$$\begin{bmatrix} a & 0 & b & 2 \\ a & a & 4 & 4 \\ 0 & a & 2 & b \end{bmatrix}$$

be the augmented matrix for a linear system. Find for what values of a and b the system has

(a) a unique solution.

(b) a one-parameter solution.

(c) a two-parameter solution.   (d) no solution.

10. For which value(s) of a does the following system have zero solutions? One solution? Infinitely many solutions?

$$x_1 + x_2 + x_3 = 4$$
$$x_3 = 2$$
$$(a^2 - 4)x_3 = a - 2$$

11. Find a matrix K such that $AKB = C$ given that
$$A = \begin{bmatrix} 1 & 4 \\ -2 & 3 \\ 1 & -2 \end{bmatrix}, \quad B = \begin{bmatrix} 2 & 0 & 0 \\ 0 & 1 & -1 \end{bmatrix},$$
$$C = \begin{bmatrix} 8 & 6 & -6 \\ 6 & -1 & 1 \\ -4 & 0 & 0 \end{bmatrix}.$$

12. How should the coefficients a, b, and c be chosen so that the system
$$ax + by - 3z = -3$$
$$-2x - by + cz = -1$$
$$ax + 3y - cz = -3$$
has the solution $x = 1$, $y = -1$, and $z = 2$?

13. In each part, solve the matrix equation for X.

(a) $X \begin{bmatrix} -1 & 0 & 1 \\ 1 & 1 & 0 \\ 3 & 1 & -1 \end{bmatrix} = \begin{bmatrix} 1 & 2 & 0 \\ -3 & 1 & 5 \end{bmatrix}$

(b) $X \begin{bmatrix} 1 & -1 & 2 \\ 3 & 0 & 1 \end{bmatrix} = \begin{bmatrix} -5 & -1 & 0 \\ 6 & -3 & 7 \end{bmatrix}$

(c) $\begin{bmatrix} 3 & 1 \\ -1 & 2 \end{bmatrix} X - X \begin{bmatrix} 1 & 4 \\ 2 & 0 \end{bmatrix} = \begin{bmatrix} 2 & -2 \\ 5 & 4 \end{bmatrix}$

14. Let A be a square matrix.

(a) Show that $(I - A)^{-1} = I + A + A^2 + A^3$ if $A^4 = 0$.

(b) Show that
$$(I - A)^{-1} = I + A + A^2 + \cdots + A^n$$
if $A^{n+1} = 0$.

15. Find values of a, b, and c such that the graph of the polynomial $p(x) = ax^2 + bx + c$ passes through the points $(1, 2)$, $(-1, 6)$, and $(2, 3)$.

16. (*Calculus required*) Find values of a, b, and c such that the graph of $p(x) = ax^2 + bx + c$ passes through the point $(-1, 0)$ and has a horizontal tangent at $(2, -9)$.

17. Let J_n be the $n \times n$ matrix each of whose entries is 1. Show that if $n > 1$, then
$$(I - J_n)^{-1} = I - \frac{1}{n-1} J_n$$

18. Show that if a square matrix A satisfies
$$A^3 + 4A^2 - 2A + 7I = 0$$
then so does A^T.

19. Prove: If B is invertible, then $AB^{-1} = B^{-1}A$ if and only if $AB = BA$.

20. Prove: If A is invertible, then $A + B$ and $I + BA^{-1}$ are both invertible or both not invertible.

21. Prove: If A is an $m \times n$ matrix and B is the $n \times 1$ matrix each of whose entries is $1/n$, then
$$AB = \begin{bmatrix} \bar{r}_1 \\ \bar{r}_2 \\ \vdots \\ \bar{r}_m \end{bmatrix}$$
where $\bar{r}_i$ is the average of the entries in the ith row of A.

22. (*Calculus required*) If the entries of the matrix
$$C = \begin{bmatrix} c_{11}(x) & c_{12}(x) & \cdots & c_{1n}(x) \\ c_{21}(x) & c_{22}(x) & \cdots & c_{2n}(x) \\ \vdots & \vdots & & \vdots \\ c_{m1}(x) & c_{m2}(x) & \cdots & c_{mn}(x) \end{bmatrix}$$
are differentiable functions of x, then we define
$$\frac{dC}{dx} = \begin{bmatrix} c'_{11}(x) & c'_{12}(x) & \cdots & c'_{1n}(x) \\ c'_{21}(x) & c'_{22}(x) & \cdots & c'_{2n}(x) \\ \vdots & \vdots & & \vdots \\ c'_{m1}(x) & c'_{m2}(x) & \cdots & c'_{mn}(x) \end{bmatrix}$$

Show that if the entries in A and B are differentiable functions of x and the sizes of the matrices are such that the stated operations can be performed, then

(a) $\dfrac{d}{dx}(kA) = k\dfrac{dA}{dx}$

(b) $\dfrac{d}{dx}(A + B) = \dfrac{dA}{dx} + \dfrac{dB}{dx}$

(c) $\dfrac{d}{dx}(AB) = \dfrac{dA}{dx}B + A\dfrac{dB}{dx}$

23. (*Calculus required*) Use part (c) of Exercise 22 to show that
$$\frac{dA^{-1}}{dx} = -A^{-1}\frac{dA}{dx}A^{-1}$$
State all the assumptions you make in obtaining this formula.

24. Assuming that the stated inverses exist, prove the following equalities.

(a) $(C^{-1} + D^{-1})^{-1} = C(C + D)^{-1}D$

(b) $(I + CD)^{-1}C = C(I + DC)^{-1}$

(c) $(C + DD^T)^{-1}D = C^{-1}D(I + D^T C^{-1}D)^{-1}$

CHAPTER 2

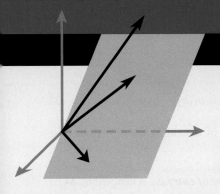

Determinants

CHAPTER CONTENTS
2.1 Determinants by Cofactor Expansion 93
2.2 Evaluating Determinants by Row Reduction 100
2.3 Properties of Determinants; Cramer's Rule 106

INTRODUCTION In this chapter we will study "determinants" or, more precisely, "determinant functions." Unlike real-valued functions, such as $f(x) = x^2$, that assign a real number to a real variable x, determinant functions assign a real number $f(A)$ to a matrix variable A. Although determinants first arose in the context of solving systems of linear equations, they are no longer used for that purpose in real-world applications. Although they can be useful for solving very small linear systems (say two or three unknowns), our main interest in them stems from the fact that they link together various concepts in linear algebra and provide a useful formula for the inverse of a matrix.

2.1 Determinants by Cofactor Expansion

In this section we will define the notion of a "determinant." This will enable us to give a specific formula for the inverse of an invertible matrix, whereas up to now we have had only a computational procedure for finding it. This, in turn, will eventually provide us with a formula for solutions of certain kinds of linear systems.

Recall from Theorem 1.4.5 that the 2×2 matrix

$$A = \begin{bmatrix} a & b \\ c & d \end{bmatrix}$$

WARNING It is important to keep in mind that $\det(A)$ is a *number*, whereas A is a *matrix*.

is invertible if and only if $ad - bc \neq 0$ and that the expression $ad - bc$ is called the ***determinant*** of the matrix A. Recall also that this determinant is denoted by writing

$$\det(A) = ad - bc \quad \text{or} \quad \begin{vmatrix} a & b \\ c & d \end{vmatrix} = ad - bc \tag{1}$$

and that the inverse of A can be expressed in terms of the determinant as

$$A^{-1} = \frac{1}{\det(A)} \begin{bmatrix} d & -b \\ -c & a \end{bmatrix} \tag{2}$$

Minors and Cofactors   One of our main goals in this chapter is to obtain an analog of Formula (2) that is applicable to square matrices of *all orders*. For this purpose we will find it convenient to use subscripted entries when writing matrices or determinants. Thus, if we denote a 2×2 matrix as

$$A = \begin{bmatrix} a_{11} & a_{12} \\ a_{21} & a_{22} \end{bmatrix}$$

Chapter 2 Determinants

We define the determinant of a 1×1 matrix $A = [a_{11}]$ as
$$\det[A] = \det[a_{11}] = a_{11}$$

then the two equations in (1) take the form

$$\det(A) = \begin{vmatrix} a_{11} & a_{12} \\ a_{21} & a_{22} \end{vmatrix} = a_{11}a_{22} - a_{12}a_{21} \quad (3)$$

The following definition will be key to our goal of extending the definition of a determinant to higher order matrices.

> **DEFINITION 1** If A is a square matrix, then the ***minor of entry*** a_{ij} is denoted by M_{ij} and is defined to be the determinant of the submatrix that remains after the ith row and jth column are deleted from A. The number $(-1)^{i+j}M_{ij}$ is denoted by C_{ij} and is called the ***cofactor of entry*** a_{ij}.

▶ **EXAMPLE 1** Finding Minors and Cofactors

Let
$$A = \begin{bmatrix} 3 & 1 & -4 \\ 2 & 5 & 6 \\ 1 & 4 & 8 \end{bmatrix}$$

WARNING We have followed the standard convention of using capital letters to denote minors and cofactors even though they are numbers, not matrices.

The minor of entry a_{11} is

$$M_{11} = \begin{vmatrix} 3 & 1 & -4 \\ 2 & 5 & 6 \\ 1 & 4 & 8 \end{vmatrix} = \begin{vmatrix} 5 & 6 \\ 4 & 8 \end{vmatrix} = 16$$

The cofactor of a_{11} is
$$C_{11} = (-1)^{1+1}M_{11} = M_{11} = 16$$

Similarly, the minor of entry a_{32} is

$$M_{32} = \begin{vmatrix} 3 & 1 & -4 \\ 2 & 5 & 6 \\ 1 & 4 & 8 \end{vmatrix} = \begin{vmatrix} 3 & -4 \\ 2 & 6 \end{vmatrix} = 26$$

The cofactor of a_{32} is
$$C_{32} = (-1)^{3+2}M_{32} = -M_{32} = -26 \quad ◀$$

Historical Note The term *determinant* was first introduced by the German mathematician Carl Friedrich Gauss in 1801 (see p. 15), who used them to "determine" properties of certain kinds of functions. Interestingly, the term *matrix* is derived from a Latin word for "womb" because it was viewed as a container of determinants.

Historical Note The term *minor* is apparently due to the English mathematician James Sylvester (see p. 34), who wrote the following in a paper published in 1850: "Now conceive any one line and any one column be struck out, we get...a square, one term less in breadth and depth than the original square; and by varying in every possible selection of the line and column excluded, we obtain, supposing the original square to consist of n lines and n columns, n^2 such minor squares, each of which will represent what I term a 'First Minor Determinant' relative to the principal or complete determinant."

2.1 Determinants by Cofactor Expansion

Remark Note that a minor M_{ij} and its corresponding cofactor C_{ij} are either the same or negatives of each other and that the relating sign $(-1)^{i+j}$ is either $+1$ or -1 in accordance with the pattern in the "checkerboard" array

$$\begin{bmatrix} + & - & + & - & + & \cdots \\ - & + & - & + & - & \cdots \\ + & - & + & - & + & \cdots \\ - & + & - & + & - & \cdots \\ \vdots & \vdots & \vdots & \vdots & \vdots & \end{bmatrix}$$

For example,
$$C_{11} = M_{11}, \quad C_{21} = -M_{21}, \quad C_{22} = M_{22}$$

and so forth. Thus, it is never really necessary to calculate $(-1)^{i+j}$ to calculate C_{ij}—you can simply compute the minor M_{ij} and then adjust the sign in accordance with the checkerboard pattern. Try this in Example 1.

▶ **EXAMPLE 2** Cofactor Expansions of a 2 × 2 Matrix

The checkerboard pattern for a 2×2 matrix $A = [a_{ij}]$ is

$$\begin{bmatrix} + & - \\ - & + \end{bmatrix}$$

so that
$$C_{11} = M_{11} = a_{22} \qquad C_{12} = -M_{12} = -a_{21}$$
$$C_{21} = -M_{21} = -a_{12} \qquad C_{22} = M_{22} = a_{11}$$

We leave it for you to use Formula (3) to verify that $\det(A)$ can be expressed in terms of cofactors in the following four ways:

$$\det(A) = \begin{vmatrix} a_{11} & a_{12} \\ a_{21} & a_{22} \end{vmatrix}$$
$$= a_{11}C_{11} + a_{12}C_{12}$$
$$= a_{21}C_{21} + a_{22}C_{22} \qquad (4)$$
$$= a_{11}C_{11} + a_{21}C_{21}$$
$$= a_{12}C_{12} + a_{22}C_{22}$$

Each of last four equations is called a *cofactor expansion* of $\det[A]$. In each cofactor expansion the entries and cofactors all come from the same row or same column of A. For example, in the first equation the entries and cofactors all come from the first row of A, in the second they all come from the second row of A, in the third they all come from the first column of A, and in the fourth they all come from the second column of A. ◀

Definition of a General Determinant

Formula (4) is a special case of the following general result, which we will state without proof.

> **THEOREM 2.1.1** *If A is an $n \times n$ matrix, then regardless of which row or column of A is chosen, the number obtained by multiplying the entries in that row or column by the corresponding cofactors and adding the resulting products is always the same.*

This result allows us to make the following definition.

Chapter 2 Determinants

DEFINITION 2 If A is an $n \times n$ matrix, then the number obtained by multiplying the entries in any row or column of A by the corresponding cofactors and adding the resulting products is called the ***determinant of A***, and the sums themselves are called ***cofactor expansions of A***. That is,

$$\det(A) = a_{1j}C_{1j} + a_{2j}C_{2j} + \cdots + a_{nj}C_{nj} \quad (5)$$
[cofactor expansion along the jth column]

and

$$\det(A) = a_{i1}C_{i1} + a_{i2}C_{i2} + \cdots + a_{in}C_{in} \quad (6)$$
[cofactor expansion along the ith row]

▶ EXAMPLE 3 Cofactor Expansion Along the First Row

Find the determinant of the matrix

$$A = \begin{bmatrix} 3 & 1 & 0 \\ -2 & -4 & 3 \\ 5 & 4 & -2 \end{bmatrix}$$

by cofactor expansion along the first row.

Solution

$$\det(A) = \begin{vmatrix} 3 & 1 & 0 \\ -2 & -4 & 3 \\ 5 & 4 & -2 \end{vmatrix} = 3\begin{vmatrix} -4 & 3 \\ 4 & -2 \end{vmatrix} - 1\begin{vmatrix} -2 & 3 \\ 5 & -2 \end{vmatrix} + 0\begin{vmatrix} -2 & -4 \\ 5 & 4 \end{vmatrix}$$

$$= 3(-4) - (1)(-11) + 0 = -1$$

▶ EXAMPLE 4 Cofactor Expansion Along the First Column

Let A be the matrix in Example 3, and evaluate $\det(A)$ by cofactor expansion along the first column of A.

Solution

$$\det(A) = \begin{vmatrix} 3 & 1 & 0 \\ -2 & -4 & 3 \\ 5 & 4 & -2 \end{vmatrix} = 3\begin{vmatrix} -4 & 3 \\ 4 & -2 \end{vmatrix} - (-2)\begin{vmatrix} 1 & 0 \\ 4 & -2 \end{vmatrix} + 5\begin{vmatrix} 1 & 0 \\ -4 & 3 \end{vmatrix}$$

$$= 3(-4) - (-2)(-2) + 5(3) = -1$$

This agrees with the result obtained in Example 3.

Note that in Example 4 we had to compute three cofactors, whereas in Example 3 only two were needed because the third was multiplied by zero. As a rule, the best strategy for cofactor expansion is to expand along a row or column with the most zeros.

**Charles Lutwidge Dodgson
(Lewis Carroll)
(1832–1898)**

Historical Note Cofactor expansion is not the only method for expressing the determinant of a matrix in terms of determinants of lower order. For example, although it is not well known, the English mathematician Charles Dodgson, who was the author of *Alice's Adventures in Wonderland* and *Through the Looking Glass* under the pen name of Lewis Carroll, invented such a method, called "*condensation*." That method has recently been resurrected from obscurity because of its suitability for parallel processing on computers.

[*Image: Time & Life Pictures/Getty Images, Inc.*]

2.1 Determinants by Cofactor Expansion

▶ **EXAMPLE 5** Smart Choice of Row or Column

If A is the 4×4 matrix

$$A = \begin{bmatrix} 1 & 0 & 0 & -1 \\ 3 & 1 & 2 & 2 \\ 1 & 0 & -2 & 1 \\ 2 & 0 & 0 & 1 \end{bmatrix}$$

then to find $\det(A)$ it will be easiest to use cofactor expansion along the second column, since it has the most zeros:

$$\det(A) = 1 \cdot \begin{vmatrix} 1 & 0 & -1 \\ 1 & -2 & 1 \\ 2 & 0 & 1 \end{vmatrix}$$

For the 3×3 determinant, it will be easiest to use cofactor expansion along its second column, since it has the most zeros:

$$\det(A) = 1 \cdot -2 \cdot \begin{vmatrix} 1 & -1 \\ 2 & 1 \end{vmatrix}$$
$$= -2(1+2)$$
$$= -6$$

▶ **EXAMPLE 6** Determinant of an Upper Triangular Matrix

The following computation shows that the determinant of a 4×4 upper triangular matrix is the product of its diagonal entries. Each part of the computation uses a cofactor expansion along the first row.

$$\begin{vmatrix} a_{11} & 0 & 0 & 0 \\ a_{21} & a_{22} & 0 & 0 \\ a_{31} & a_{32} & a_{33} & 0 \\ a_{41} & a_{42} & a_{43} & a_{44} \end{vmatrix} = a_{11} \begin{vmatrix} a_{22} & 0 & 0 \\ a_{32} & a_{33} & 0 \\ a_{42} & a_{43} & a_{44} \end{vmatrix}$$

$$= a_{11}a_{22} \begin{vmatrix} a_{33} & 0 \\ a_{43} & a_{44} \end{vmatrix}$$

$$= a_{11}a_{22}a_{33}|a_{44}| = a_{11}a_{22}a_{33}a_{44} \blacktriangleleft$$

The method illustrated in Example 6 can be easily adapted to prove the following general result.

THEOREM 2.1.2 *If A is an $n \times n$ triangular matrix (upper triangular, lower triangular, or diagonal), then $\det(A)$ is the product of the entries on the main diagonal of the matrix; that is, $\det(A) = a_{11}a_{22} \cdots a_{nn}$.*

A Useful Technique for Evaluating 2×2 and 3×3 Determinants

Determinants of 2×2 and 3×3 matrices can be evaluated very efficiently using the pattern suggested in Figure 2.1.1.

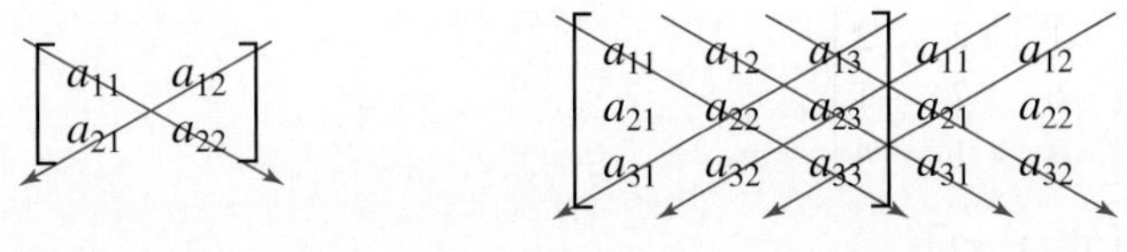

▶ **Figure 2.1.1**

In the 2×2 case, the determinant can be computed by forming the product of the entries on the rightward arrow and subtracting the product of the entries on the leftward arrow. In the 3×3 case we first recopy the first and second columns as shown in the figure, after

WARNING The arrow technique only works for determinants of 2×2 and 3×3 matrices.

which we can compute the determinant by summing the products of the entries on the rightward arrows and subtracting the products on the leftward arrows. These procedures execute the computations

$$\begin{vmatrix} a_{11} & a_{12} \\ a_{21} & a_{22} \end{vmatrix} = a_{11}a_{22} - a_{12}a_{21}$$

$$\begin{vmatrix} a_{11} & a_{12} & a_{13} \\ a_{21} & a_{22} & a_{23} \\ a_{31} & a_{32} & a_{33} \end{vmatrix} = a_{11}\begin{vmatrix} a_{22} & a_{23} \\ a_{32} & a_{33} \end{vmatrix} - a_{12}\begin{vmatrix} a_{21} & a_{23} \\ a_{31} & a_{33} \end{vmatrix} + a_{13}\begin{vmatrix} a_{21} & a_{22} \\ a_{31} & a_{32} \end{vmatrix}$$

$$= a_{11}(a_{22}a_{33} - a_{23}a_{32}) - a_{12}(a_{21}a_{33} - a_{23}a_{31}) + a_{13}(a_{21}a_{32} - a_{22}a_{31})$$

$$= a_{11}a_{22}a_{33} + a_{12}a_{23}a_{31} + a_{13}a_{21}a_{32} - a_{13}a_{22}a_{31} - a_{12}a_{21}a_{33} - a_{11}a_{23}a_{32}$$

which agrees with the cofactor expansions along the first row.

▶ **EXAMPLE 7** A Technique for Evaluating 2 × 2 and 3 × 3 Determinants

$$\begin{vmatrix} 3 & 1 \\ 4 & -2 \end{vmatrix} = (3)(-2) - (1)(4) = -10$$

$$\begin{vmatrix} 1 & 2 & 3 \\ -4 & 5 & 6 \\ 7 & -8 & 9 \end{vmatrix} = [45 + 84 + 96] - [105 - 48 - 72] = 240 \quad \blacktriangleleft$$

Concept Review

- Determinant
- Minor
- Cofactor
- Cofactor expansion

Skills

- Find the minors and cofactors of a square matrix.
- Use cofactor expansion to evaluate the determinant of a square matrix.
- Use the arrow technique to evaluate the determinant of a 2×2 or 3×3 matrix.
- Use the determinant of a 2×2 invertible matrix to find the inverse of that matrix.
- Find the determinant of an upper triangular, lower triangular, or diagonal matrix by inspection.

Exercise Set 2.1

▶ In Exercises 1–2, find all the minors and cofactors of the matrix A. ◀

1. $A = \begin{bmatrix} 1 & -2 & 3 \\ 6 & 7 & -1 \\ -3 & 1 & 4 \end{bmatrix}$

2. $A = \begin{bmatrix} 1 & 1 & 2 \\ 3 & 3 & 6 \\ 0 & 1 & 4 \end{bmatrix}$

3. Let

$$A = \begin{bmatrix} 4 & -1 & 1 & 6 \\ 0 & 0 & -3 & 3 \\ 4 & 1 & 0 & 14 \\ 4 & 1 & 3 & 2 \end{bmatrix}$$

Find

(a) M_{11} and C_{11}.

(b) M_{32} and C_{32}.

(c) M_{12} and C_{12}.

(d) M_{43} and C_{43}.

4. Let
$$A = \begin{bmatrix} 2 & 3 & -1 & 1 \\ -3 & 2 & 0 & 3 \\ 3 & -2 & 1 & 0 \\ 3 & -2 & 1 & 4 \end{bmatrix}$$

Find

(a) M_{32} and C_{32}. (b) M_{44} and C_{44}.

(c) M_{41} and C_{41}. (d) M_{24} and C_{24}.

▶ In Exercises 5–8, evaluate the determinant of the given matrix. If the matrix is invertible, use Equation (2) to find its inverse. ◀

5. $\begin{bmatrix} 4 & 4 \\ 2 & 3 \end{bmatrix}$ **6.** $\begin{bmatrix} 4 & 1 \\ 8 & 2 \end{bmatrix}$

7. $\begin{bmatrix} -5 & 7 \\ -7 & -2 \end{bmatrix}$ **8.** $\begin{bmatrix} \sqrt{2} & \sqrt{6} \\ 4 & \sqrt{3} \end{bmatrix}$

▶ In Exercises 9–14, use the arrow technique to evaluate the determinant of the given matrix. ◀

9. $\begin{bmatrix} a-3 & 5 \\ -3 & a-2 \end{bmatrix}$ **10.** $\begin{bmatrix} -2 & 7 & 6 \\ 5 & 1 & -2 \\ 3 & 8 & 4 \end{bmatrix}$

11. $\begin{bmatrix} 1 & 2 & 4 \\ -3 & 3 & 5 \\ 7 & 0 & 6 \end{bmatrix}$ **12.** $\begin{bmatrix} -1 & 1 & 2 \\ 3 & 0 & -5 \\ 1 & 7 & 2 \end{bmatrix}$

13. $\begin{bmatrix} 3 & 0 & 0 \\ 2 & -1 & 5 \\ 1 & 9 & -4 \end{bmatrix}$ **14.** $\begin{bmatrix} c & -4 & 3 \\ 2 & 1 & c^2 \\ 4 & c-1 & 2 \end{bmatrix}$

▶ In Exercises 15–18, find all values of λ for which $\det(A) = 0$.

15. $A = \begin{bmatrix} \lambda+4 & 0 \\ 4 & \lambda+2 \end{bmatrix}$ **16.** $A = \begin{bmatrix} \lambda-4 & 0 & 0 \\ 0 & \lambda & 2 \\ 0 & 3 & \lambda-1 \end{bmatrix}$

17. $A = \begin{bmatrix} \lambda-3 & 1 \\ -1 & \lambda+3 \end{bmatrix}$ **18.** $A = \begin{bmatrix} \lambda-4 & 4 & 0 \\ -1 & \lambda & 0 \\ 0 & 0 & \lambda-5 \end{bmatrix}$

19. Evaluate the determinant of the matrix in Exercise 13 by a cofactor expansion along

(a) the first row. (b) the first column.

(c) the second row. (d) the second column.

(e) the third row. (f) the third column.

20. Evaluate the determinant of the matrix in Exercise 12 by a cofactor expansion along

(a) the first row. (b) the first column.

(c) the second row. (d) the second column.

(e) the third row. (f) the third column.

▶ In Exercises 21–26, evaluate $\det(A)$ by a cofactor expansion along a row or column of your choice. ◀

21. $A = \begin{bmatrix} -3 & 0 & 7 \\ 2 & 5 & 1 \\ -1 & 0 & 5 \end{bmatrix}$ **22.** $A = \begin{bmatrix} 3 & 3 & 1 \\ 1 & 0 & -4 \\ 1 & -3 & 5 \end{bmatrix}$

23. $A = \begin{bmatrix} 1 & k & k^2 \\ 1 & k & k^2 \\ 1 & k & k^2 \end{bmatrix}$ **24.** $A = \begin{bmatrix} k+1 & k-1 & 7 \\ 2 & k-3 & 4 \\ 5 & k+1 & k \end{bmatrix}$

25. $A = \begin{bmatrix} 2 & 2 & 1 & 0 \\ -1 & 0 & 3 & 0 \\ 4 & 9 & 3 & 1 \\ 0 & -1 & 5 & 7 \end{bmatrix}$

26. $A = \begin{bmatrix} 4 & 0 & 0 & 1 & 0 \\ 3 & 3 & 3 & -1 & 0 \\ 1 & 2 & 4 & 2 & 3 \\ 9 & 4 & 6 & 2 & 3 \\ 2 & 2 & 4 & 2 & 3 \end{bmatrix}$

▶ In Exercises 27–32, evaluate the determinant of the given matrix by inspection. ◀

27. $\begin{bmatrix} 1 & 0 & 0 \\ 0 & -1 & 0 \\ 0 & 0 & 1 \end{bmatrix}$ **28.** $\begin{bmatrix} 2 & 0 & 0 \\ 0 & 2 & 0 \\ 0 & 0 & 2 \end{bmatrix}$

29. $\begin{bmatrix} 0 & 0 & 0 & 0 \\ 1 & 2 & 0 & 0 \\ 0 & 4 & 3 & 0 \\ 1 & 2 & 3 & 8 \end{bmatrix}$ **30.** $\begin{bmatrix} 1 & 1 & 1 & 1 \\ 0 & 2 & 2 & 2 \\ 0 & 0 & 3 & 3 \\ 0 & 0 & 0 & 4 \end{bmatrix}$

31. $\begin{bmatrix} 1 & 2 & 7 & -3 \\ 0 & 1 & -4 & 1 \\ 0 & 0 & 2 & 7 \\ 0 & 0 & 0 & 3 \end{bmatrix}$ **32.** $\begin{bmatrix} -3 & 0 & 0 & 0 \\ 1 & 2 & 0 & 0 \\ 40 & 10 & -1 & 0 \\ 100 & 200 & -23 & 3 \end{bmatrix}$

33. Show that the value of the following determinant is independent of θ.

$$\begin{vmatrix} \sin(\theta) & \cos(\theta) & 0 \\ -\cos(\theta) & \sin(\theta) & 0 \\ \sin(\theta) - \cos(\theta) & \sin(\theta) + \cos(\theta) & 1 \end{vmatrix}$$

34. Show that the matrices

$$A = \begin{bmatrix} a & b \\ 0 & c \end{bmatrix} \quad \text{and} \quad B = \begin{bmatrix} d & e \\ 0 & f \end{bmatrix}$$

commute if and only if

$$\begin{vmatrix} b & a-c \\ e & d-f \end{vmatrix} = 0$$

35. By inspection, what is the relationship between the following determinants?

$$d_1 = \begin{vmatrix} a & b & c \\ d & 1 & f \\ g & 0 & 1 \end{vmatrix} \quad \text{and} \quad d_2 = \begin{vmatrix} a+\lambda & b & c \\ d & 1 & f \\ g & 0 & 1 \end{vmatrix}$$

36. Show that

$$\det(A) = \frac{1}{2} \begin{vmatrix} \operatorname{tr}(A) & 1 \\ \operatorname{tr}(A^2) & \operatorname{tr}(A) \end{vmatrix}$$

for every 2×2 matrix A.

37. What can you say about an nth-order determinant all of whose entries are 1? Explain your reasoning.

38. What is the maximum number of zeros that a 3×3 matrix can have without having a zero determinant? Explain your reasoning.

39. What is the maximum number of zeros that a 4×4 matrix can have without having a zero determinant? Explain your reasoning.

40. Prove that (x_1, y_1), (x_2, y_2), and (x_3, y_3) are collinear points if and only if

$$\begin{vmatrix} x_1 & y_1 & 1 \\ x_2 & y_2 & 1 \\ x_3 & y_3 & 1 \end{vmatrix} = 0$$

41. Prove that the equation of the line through the distinct points (a_1, b_1) and (a_2, b_2) can be written as

$$\begin{vmatrix} x & y & 1 \\ a_1 & b_1 & 1 \\ a_2 & b_2 & 1 \end{vmatrix} = 0$$

42. Prove that if A is upper triangular and B_{ij} is the matrix that results when the ith row and jth column of A are deleted, then B_{ij} is upper triangular if $i < j$.

True-False Exercises

In parts (a)–(j) determine whether the statement is true or false, and justify your answer.

(a) The determinant of the 2×2 matrix $\begin{bmatrix} a & b \\ c & d \end{bmatrix}$ is $ad + bc$.

(b) Two square matrices A and B can have the same determinant only if they are the same size.

(c) The minor M_{ij} is the same as the cofactor C_{ij} if and only if $i + j$ is even.

(d) If A is a 3×3 symmetric matrix, then $C_{ij} = C_{ji}$ for all i and j.

(e) The value of a cofactor expansion of a matrix A is independent of the row or column chosen for the expansion.

(f) The determinant of a lower triangular matrix is the sum of the entries along its main diagonal.

(g) For every square matrix A and every scalar c, we have $\det(cA) = c \det(A)$.

(h) For all square matrices A and B, we have $\det(A + B) = \det(A) + \det(B)$.

(i) For every 2×2 matrix A, we have $\det(A^2) = (\det(A))^2$.

2.2 Evaluating Determinants by Row Reduction

In this section we will show how to evaluate a determinant by reducing the associated matrix to row echelon form. In general, this method requires less computation than cofactor expansion and hence is the method of choice for large matrices.

A Basic Theorem We begin with a fundamental theorem that will lead us to an efficient procedure for evaluating the determinant of a square matrix of any size.

> **THEOREM 2.2.1** *Let A be a square matrix. If A has a row of zeros or a column of zeros, then $\det(A) = 0$.*

Proof Since the determinant of A can be found by a cofactor expansion along any row or column, we can use the row or column of zeros. Thus, if we let $C_1, C_2, \ldots, C_n$ denote the cofactors of A along that row or column, then it follows from Formula (5) or (6) in Section 2.1 that

$$\det(A) = 0 \cdot C_1 + 0 \cdot C_2 + \cdots + 0 \cdot C_n = 0 \quad \blacktriangleleft$$

2.2 Evaluating Determinants by Row Reduction

The following useful theorem relates the determinant of a matrix and the determinant of its transpose.

> Because transposing a matrix changes its columns to rows and its rows to columns, almost every theorem about the rows of a determinant has a companion version about columns, and vice versa.

THEOREM 2.2.2 *Let A be a square matrix. Then* $\det(A) = \det(A^T)$.

Proof Since transposing a matrix changes its columns to rows and its rows to columns, the cofactor expansion of A along any row is the same as the cofactor expansion of A^T along the corresponding column. Thus, both have the same determinant. ◂

Elementary Row Operations

The next theorem shows how an elementary row operation on a square matrix affects the value of its determinant. In place of a formal proof we have provided a table to illustrate the ideas in the 3×3 case (see Table 1).

THEOREM 2.2.3 *Let A be an $n \times n$ matrix.*

(a) *If B is the matrix that results when a single row or single column of A is multiplied by a scalar k, then* $\det(B) = k \det(A)$.

(b) *If B is the matrix that results when two rows or two columns of A are interchanged, then* $\det(B) = -\det(A)$.

(c) *If B is the matrix that results when a multiple of one row of A is added to another row or when a multiple of one column is added to another column, then* $\det(B) = \det(A)$.

> The first panel of Table 1 shows that you can bring a common factor from any row (column) of a determinant through the determinant sign. This is a slightly different way of thinking about part (a) of Theorem 2.2.3.

Table 1

Relationship	Operation
$\begin{vmatrix} ka_{11} & ka_{12} & ka_{13} \\ a_{21} & a_{22} & a_{23} \\ a_{31} & a_{32} & a_{33} \end{vmatrix} = k \begin{vmatrix} a_{11} & a_{12} & a_{13} \\ a_{21} & a_{22} & a_{23} \\ a_{31} & a_{32} & a_{33} \end{vmatrix}$ $\det(B) = k\det(A)$	The first row of A is multiplied by k.
$\begin{vmatrix} a_{21} & a_{22} & a_{23} \\ a_{11} & a_{12} & a_{13} \\ a_{31} & a_{32} & a_{33} \end{vmatrix} = - \begin{vmatrix} a_{11} & a_{12} & a_{13} \\ a_{21} & a_{22} & a_{23} \\ a_{31} & a_{32} & a_{33} \end{vmatrix}$ $\det(B) = -\det(A)$	The first and second rows of A are interchanged.
$\begin{vmatrix} a_{11}+ka_{21} & a_{12}+ka_{22} & a_{13}+ka_{23} \\ a_{21} & a_{22} & a_{23} \\ a_{31} & a_{32} & a_{33} \end{vmatrix} = \begin{vmatrix} a_{11} & a_{12} & a_{13} \\ a_{21} & a_{22} & a_{23} \\ a_{31} & a_{32} & a_{33} \end{vmatrix}$ $\det(B) = \det(A)$	A multiple of the second row of A is added to the first row.

We will verify the first equation in Table 1 and leave the other two for you. To start, note that the determinants on the two sides of the equation differ only in the first row, so these determinants have the same cofactors, C_{11}, C_{12}, C_{13}, along that row (since those cofactors depend only on the entries in the *second* two rows). Thus, expanding the left side by cofactors along the first row yields

$$\begin{vmatrix} ka_{11} & ka_{12} & ka_{13} \\ a_{21} & a_{22} & a_{23} \\ a_{31} & a_{32} & a_{33} \end{vmatrix} = ka_{11}C_{11} + ka_{12}C_{12} + ka_{33}C_{13}$$

$$= k(a_{11}C_{11} + a_{12}C_{12} + a_{33}C_{13})$$

$$= k \begin{vmatrix} a_{11} & a_{12} & a_{13} \\ a_{21} & a_{22} & a_{23} \\ a_{31} & a_{32} & a_{33} \end{vmatrix}$$

Elementary Matrices   It will be useful to consider the special case of Theorem 2.2.3 in which $A = I_n$ is the $n \times n$ identity matrix and E (rather than B) denotes the elementary matrix that results when the row operation is performed on I_n. In this special case Theorem 2.2.3 implies the following result.

THEOREM 2.2.4 *Let E be an $n \times n$ elementary matrix.*
(a) *If E results from multiplying a row of I_n by a nonzero number k, then $\det(E) = k$.*
(b) *If E results from interchanging two rows of I_n, then $\det(E) = -1$.*
(c) *If E results from adding a multiple of one row of I_n to another, then $\det(E) = 1$.*

▶ **EXAMPLE 1** **Determinants of Elementary Matrices**

The following determinants of elementary matrices, which are evaluated by inspection, illustrate Theorem 2.2.4.

Observe that the determinant of an elementary matrix cannot be zero.

$$\begin{vmatrix} 1 & 0 & 0 & 0 \\ 0 & 3 & 0 & 0 \\ 0 & 0 & 1 & 0 \\ 0 & 0 & 0 & 1 \end{vmatrix} = 3, \quad \begin{vmatrix} 0 & 0 & 0 & 1 \\ 0 & 1 & 0 & 0 \\ 0 & 0 & 1 & 0 \\ 1 & 0 & 0 & 0 \end{vmatrix} = -1, \quad \begin{vmatrix} 1 & 0 & 0 & 7 \\ 0 & 1 & 0 & 0 \\ 0 & 0 & 1 & 0 \\ 0 & 0 & 0 & 1 \end{vmatrix} = 1 \blacktriangleleft$$

The second row of I_4 was multiplied by 3. | The first and last rows of I_4 were interchanged. | 7 times the last row of I_4 was added to the first row.

Matrices with Proportional Rows or Columns   If a square matrix A has two proportional rows, then a row of zeros can be introduced by adding a suitable multiple of one of the rows to the other. Similarly for columns. But adding a multiple of one row or column to another does not change the determinant, so from Theorem 2.2.1, we must have $\det(A) = 0$. This proves the following theorem.

THEOREM 2.2.5 *If A is a square matrix with two proportional rows or two proportional columns, then $\det(A) = 0$.*

▶ **EXAMPLE 2** **Introducing Zero Rows**

The following computation shows how to introduce a row of zeros when there are two proportional rows.

$$\begin{vmatrix} 1 & 3 & -2 & 4 \\ 2 & 6 & -4 & 8 \\ 3 & 9 & 1 & 5 \\ 1 & 1 & 4 & 8 \end{vmatrix} = \begin{vmatrix} 1 & 3 & -2 & 4 \\ 0 & 0 & 0 & 0 \\ 3 & 9 & 1 & 5 \\ 1 & 1 & 4 & 8 \end{vmatrix} = 0 \quad \longleftarrow \text{The second row is 2 times the first, so we added } -2 \text{ times the first row to the second to introduce a row of zeros.}$$

2.2 Evaluating Determinants by Row Reduction

Each of the following matrices has two proportional rows or columns; thus, each has a determinant of zero.

$$\begin{bmatrix} -1 & 4 \\ -2 & 8 \end{bmatrix}, \quad \begin{bmatrix} 1 & -2 & 7 \\ -4 & 8 & 5 \\ 2 & -4 & 3 \end{bmatrix}, \quad \begin{bmatrix} 3 & -1 & 4 & -5 \\ 6 & -2 & 5 & 2 \\ 5 & 8 & 1 & 4 \\ -9 & 3 & -12 & 15 \end{bmatrix} \blacktriangleleft$$

Evaluating Determinants by Row Reduction

We will now give a method for evaluating determinants that involves substantially less computation than cofactor expansion. The idea of the method is to reduce the given matrix to upper triangular form by elementary row operations, then compute the determinant of the upper triangular matrix (an easy computation), and then relate that determinant to that of the original matrix. Here is an example.

▶ **EXAMPLE 3** Using Row Reduction to Evaluate a Determinant

Evaluate det(A) where

$$A = \begin{bmatrix} 0 & 1 & 5 \\ 3 & -6 & 9 \\ 2 & 6 & 1 \end{bmatrix}$$

Solution We will reduce A to row echelon form (which is upper triangular) and then apply Theorem 2.1.2.

$$\det(A) = \begin{vmatrix} 0 & 1 & 5 \\ 3 & -6 & 9 \\ 2 & 6 & 1 \end{vmatrix} = - \begin{vmatrix} 3 & -6 & 9 \\ 0 & 1 & 5 \\ 2 & 6 & 1 \end{vmatrix} \quad \longleftarrow \text{The first and second rows of } A \text{ were interchanged.}$$

$$= -3 \begin{vmatrix} 1 & -2 & 3 \\ 0 & 1 & 5 \\ 2 & 6 & 1 \end{vmatrix} \quad \longleftarrow \text{A common factor of 3 from the first row was taken through the determinant sign.}$$

$$= -3 \begin{vmatrix} 1 & -2 & 3 \\ 0 & 1 & 5 \\ 0 & 10 & -5 \end{vmatrix} \quad \longleftarrow -2 \text{ times the first row was added to the third row.}$$

$$= -3 \begin{vmatrix} 1 & -2 & 3 \\ 0 & 1 & 5 \\ 0 & 0 & -55 \end{vmatrix} \quad \longleftarrow -10 \text{ times the second row was added to the third row.}$$

$$= (-3)(-55) \begin{vmatrix} 1 & -2 & 3 \\ 0 & 1 & 5 \\ 0 & 0 & 1 \end{vmatrix} \quad \longleftarrow \text{A common factor of } -55 \text{ from the last row was taken through the determinant sign.}$$

$$= (-3)(-55)(1) = 165$$

> Even with today's fastest computers it would take millions of years to calculate a 25 × 25 determinant by cofactor expansion, so methods based on row reduction are often used for large determinants. For determinants of small size (such as those in this text), cofactor expansion is often a reasonable choice.

▶ **EXAMPLE 4** Using Column Operations to Evaluate a Determinant

Compute the determinant of

$$A = \begin{bmatrix} 1 & 0 & 0 & 3 \\ 2 & 7 & 0 & 6 \\ 0 & 6 & 3 & 0 \\ 7 & 3 & 1 & -5 \end{bmatrix}$$

Example 4 points out that it is always wise to keep an eye open for column operations that can shorten computations.

Solution This determinant could be computed as above by using elementary row operations to reduce A to row echelon form, but we can put A in lower triangular form in one step by adding -3 times the first column to the fourth to obtain

$$\det(A) = \det \begin{bmatrix} 1 & 0 & 0 & 0 \\ 2 & 7 & 0 & 0 \\ 0 & 6 & 3 & 0 \\ 7 & 3 & 1 & -26 \end{bmatrix} = (1)(7)(3)(-26) = -546 \blacktriangleleft$$

Cofactor expansion and row or column operations can sometimes be used in combination to provide an effective method for evaluating determinants. The following example illustrates this idea.

▶ **EXAMPLE 5** **Row Operations and Cofactor Expansion**

Evaluate $\det(A)$ where

$$A = \begin{bmatrix} 3 & 5 & -2 & 6 \\ 1 & 2 & -1 & 1 \\ 2 & 4 & 1 & 5 \\ 3 & 7 & 5 & 3 \end{bmatrix}$$

Solution By adding suitable multiples of the second row to the remaining rows, we obtain

$$\det(A) = \begin{vmatrix} 0 & -1 & 1 & 3 \\ 1 & 2 & -1 & 1 \\ 0 & 0 & 3 & 3 \\ 0 & 1 & 8 & 0 \end{vmatrix}$$

$$= - \begin{vmatrix} -1 & 1 & 3 \\ 0 & 3 & 3 \\ 1 & 8 & 0 \end{vmatrix} \quad \longleftarrow \text{Cofactor expansion along the first column}$$

$$= - \begin{vmatrix} -1 & 1 & 3 \\ 0 & 3 & 3 \\ 0 & 9 & 3 \end{vmatrix} \quad \longleftarrow \text{We added the first row to the third row.}$$

$$= -(-1) \begin{vmatrix} 3 & 3 \\ 9 & 3 \end{vmatrix} \quad \longleftarrow \text{Cofactor expansion along the first column}$$

$$= -18 \blacktriangleleft$$

Skills

- Know the effect of elementary row operations on the value of a determinant.
- Know the determinants of the three types of elementary matrices.
- Know how to introduce zeros into the rows or columns of a matrix to facilitate the evaluation of its determinant.
- Use row reduction to evaluate the determinant of a matrix.
- Use column operations to evaluate the determinant of a matrix.
- Combine the use of row reduction and cofactor expansion to evaluate the determinant of a matrix.

Exercise Set 2.2

▶ In Exercises 1–4, verify that $\det(A) = \det(A^T)$. ◀

1. $A = \begin{bmatrix} -2 & 3 \\ 1 & 4 \end{bmatrix}$

2. $A = \begin{bmatrix} -6 & 1 \\ 2 & -2 \end{bmatrix}$

3. $A = \begin{bmatrix} 3 & 1 & -2 \\ 1 & 0 & 4 \\ 5 & -3 & 6 \end{bmatrix}$

4. $A = \begin{bmatrix} 4 & 2 & -1 \\ 0 & 2 & -3 \\ -1 & 1 & 5 \end{bmatrix}$

▶ In Exercises 5–9, find the determinant of the given elementary matrix by inspection. ◀

5. $\begin{bmatrix} 1 & 0 & 0 & 0 \\ 0 & 1 & 0 & -5 \\ 0 & 0 & 1 & 0 \\ 0 & 0 & 0 & 1 \end{bmatrix}$

6. $\begin{bmatrix} 1 & 0 & 0 \\ 0 & 1 & 0 \\ -5 & 0 & 1 \end{bmatrix}$

7. $\begin{bmatrix} 1 & 0 & 0 \\ 0 & -2 & 0 \\ 0 & 0 & 1 \end{bmatrix}$

8. $\begin{bmatrix} 1 & 0 & 0 & 0 \\ 0 & -\frac{1}{3} & 0 & 0 \\ 0 & 0 & 1 & 0 \\ 0 & 0 & 0 & 1 \end{bmatrix}$

9. $\begin{bmatrix} 1 & 0 & 0 & 0 \\ 0 & 1 & 0 & -9 \\ 0 & 0 & 1 & 0 \\ 0 & 0 & 0 & 1 \end{bmatrix}$

▶ In Exercises 10–17, evaluate the determinant of the given matrix by reducing the matrix to row echelon form. ◀

10. $\begin{bmatrix} 3 & 6 & -9 \\ 0 & 0 & -2 \\ -2 & 1 & 5 \end{bmatrix}$

11. $\begin{bmatrix} 0 & 3 & 1 \\ 1 & 1 & 2 \\ 3 & 2 & 4 \end{bmatrix}$

12. $\begin{bmatrix} 1 & -3 & 0 \\ -2 & 4 & 1 \\ 5 & -2 & 2 \end{bmatrix}$

13. $\begin{bmatrix} 3 & -6 & 9 \\ -2 & 7 & -2 \\ 0 & 1 & 5 \end{bmatrix}$

14. $\begin{bmatrix} 1 & -2 & 3 & 1 \\ 5 & -9 & 6 & 3 \\ -1 & 2 & -6 & -2 \\ 2 & 8 & 6 & 1 \end{bmatrix}$

15. $\begin{bmatrix} 2 & 1 & 3 & 1 \\ 1 & 0 & 1 & 1 \\ 0 & 2 & 1 & 0 \\ 0 & 1 & 2 & 3 \end{bmatrix}$

16. $\begin{bmatrix} 0 & 1 & 1 & 1 \\ \frac{1}{2} & \frac{1}{2} & 1 & \frac{1}{2} \\ \frac{2}{3} & \frac{1}{3} & \frac{1}{3} & 0 \\ -\frac{1}{3} & \frac{2}{3} & 0 & 0 \end{bmatrix}$

17. $\begin{bmatrix} 1 & 3 & 1 & 5 & 3 \\ -2 & -7 & 0 & -4 & 2 \\ 0 & 0 & 1 & 0 & 1 \\ 0 & 0 & 2 & 1 & 1 \\ 0 & 0 & 0 & 1 & 1 \end{bmatrix}$

18. Repeat Exercises 11–13 by using a combination of row operations and cofactor expansion.

19. Repeat Exercises 14–17 by using a combination of row operations and cofactor expansion.

▶ In Exercises 20–27, evaluate the determinant, given that

$$\begin{vmatrix} a & b & c \\ d & e & f \\ g & h & i \end{vmatrix} = -6 \quad ◀$$

20. $\begin{vmatrix} g & h & i \\ d & e & f \\ a & b & c \end{vmatrix}$

21. $\begin{vmatrix} d & e & f \\ g & h & i \\ a & b & c \end{vmatrix}$

22. $\begin{vmatrix} a & b & c \\ d & e & f \\ 2a & 2b & 2c \end{vmatrix}$

23. $\begin{vmatrix} -a & -b & -c \\ 2d & 2e & 2f \\ 5g & 5h & 5i \end{vmatrix}$

24. $\begin{vmatrix} a+d & b+e & c+f \\ -d & -e & -f \\ g & h & i \end{vmatrix}$

25. $\begin{vmatrix} a+g & b+h & c+i \\ d & e & f \\ g & h & i \end{vmatrix}$

26. $\begin{vmatrix} a & b & c \\ 2d & 2e & 2f \\ g+3a & h+3b & i+3c \end{vmatrix}$

27. $\begin{vmatrix} 3g & 3h & 3i \\ 2a+d & 2b+e & 2c+f \\ d & e & f \end{vmatrix}$

28. Show that

(a) $\det \begin{bmatrix} 0 & 0 & a_{13} \\ 0 & a_{22} & a_{23} \\ a_{31} & a_{32} & a_{33} \end{bmatrix} = -a_{13}a_{22}a_{31}$

(b) $\det \begin{bmatrix} 0 & 0 & 0 & a_{14} \\ 0 & 0 & a_{23} & a_{24} \\ 0 & a_{32} & a_{33} & a_{34} \\ a_{41} & a_{42} & a_{43} & a_{44} \end{bmatrix} = a_{14}a_{23}a_{32}a_{41}$

29. Use row reduction to show that

$$\begin{vmatrix} 1 & 1 & 1 \\ a & b & c \\ a^2 & b^2 & c^2 \end{vmatrix} = (b-a)(c-a)(c-b)$$

▶ In Exercises 30–33, confirm the identities without evaluating the determinants directly. ◀

30. $\begin{vmatrix} a_1 + b_1 t & a_2 + b_2 t & a_3 + b_3 t \\ a_1 t + b_1 & a_2 t + b_2 & a_3 t + b_3 \\ c_1 & c_2 & c_3 \end{vmatrix} = (1 - t^2) \begin{vmatrix} a_1 & a_2 & a_3 \\ b_1 & b_2 & b_3 \\ c_1 & c_2 & c_3 \end{vmatrix}$

31. $\begin{vmatrix} a_1 & b_1 & a_1 + b_1 + c_1 \\ a_2 & b_2 & a_2 + b_2 + c_2 \\ a_3 & b_3 & a_3 + b_3 + c_3 \end{vmatrix} = \begin{vmatrix} a_1 & b_1 & c_1 \\ a_2 & b_2 & c_2 \\ a_3 & b_3 & c_3 \end{vmatrix}$

32. $\begin{vmatrix} a_1 & b_1+ta_1 & c_1+rb_1+sa_1 \\ a_2 & b_2+ta_2 & c_2+rb_2+sa_2 \\ a_3 & b_3+ta_3 & c_3+rb_3+sa_3 \end{vmatrix} = \begin{vmatrix} a_1 & a_2 & a_3 \\ b_1 & b_2 & b_3 \\ c_1 & c_2 & c_3 \end{vmatrix}$

33. $\begin{vmatrix} a_1+b_1 & a_1-b_1 & c_1 \\ a_2+b_2 & a_2-b_2 & c_2 \\ a_3+b_3 & a_3-b_3 & c_3 \end{vmatrix} = -2\begin{vmatrix} a_1 & b_1 & c_1 \\ a_2 & b_2 & c_2 \\ a_3 & b_3 & c_3 \end{vmatrix}$

34. Find the determinant of the following matrix.

$$\begin{bmatrix} a & b & b & b \\ b & a & b & b \\ b & b & a & b \\ b & b & b & a \end{bmatrix}$$

▶ In Exercises 35–36, show that $\det(A) = 0$ without directly evaluating the determinant. ◀

35. $A = \begin{bmatrix} 2 & 0 & -1 & 3 \\ 1 & 3 & 5 & 7 \\ -3 & -3 & -4 & -10 \\ 5 & 1 & 0 & 6 \end{bmatrix}$

36. $A = \begin{bmatrix} -4 & 1 & 1 & 1 & 1 \\ 1 & -4 & 1 & 1 & 1 \\ 1 & 1 & -4 & 1 & 1 \\ 1 & 1 & 1 & -4 & 1 \\ 1 & 1 & 1 & 1 & -4 \end{bmatrix}$

True-False Exercises

In parts (a)–(f) determine whether the statement is true or false, and justify your answer.

(a) If A is a 4×4 matrix and B is obtained from A by interchanging the first two rows and then interchanging the last two rows, then $\det(B) = \det(A)$.

(b) If A is a 3×3 matrix and B is obtained from A by multiplying the first column by 4 and multiplying the third column by $\frac{3}{4}$, then $\det(B) = 3\det(A)$.

(c) If A is a 3×3 matrix and B is obtained from A by adding 5 times the first row to each of the second and third rows, then $\det(B) = 25\det(A)$.

(d) If A is an $n \times n$ matrix and B is obtained from A by multiplying each row of A by its row number, then
$$\det(B) = \frac{n(n+1)}{2}\det(A)$$

(e) If A is a square matrix with two identical columns, then $\det(A) = 0$.

(f) If the sum of the second and fourth row vectors of a 6×6 matrix A is equal to the last row vector, then $\det(A) = 0$.

2.3 Properties of Determinants; Cramer's Rule

In this section we will develop some fundamental properties of matrices, and we will use these results to derive a formula for the inverse of an invertible matrix and formulas for the solutions of certain kinds of linear systems.

Basic Properties of Determinants

Suppose that A and B are $n \times n$ matrices and k is any scalar. We begin by considering possible relationships between $\det(A)$, $\det(B)$, and

$$\det(kA), \quad \det(A+B), \quad \text{and} \quad \det(AB)$$

Since a common factor of any row of a matrix can be moved through the determinant sign, and since each of the n rows in kA has a common factor of k, it follows that

$$\det(kA) = k^n \det(A) \tag{1}$$

For example,

$$\begin{vmatrix} ka_{11} & ka_{12} & ka_{13} \\ ka_{21} & ka_{22} & ka_{23} \\ ka_{31} & ka_{32} & ka_{33} \end{vmatrix} = k^3 \begin{vmatrix} a_{11} & a_{12} & a_{13} \\ a_{21} & a_{22} & a_{23} \\ a_{31} & a_{32} & a_{33} \end{vmatrix}$$

Unfortunately, no simple relationship exists among $\det(A)$, $\det(B)$, and $\det(A+B)$. In particular, we emphasize that $\det(A+B)$ will usually *not* be equal to $\det(A) + \det(B)$. The following example illustrates this fact.

2.3 Properties of Determinants; Cramer's Rule

▶ **EXAMPLE 1** $\det(A + B) \neq \det(A) + \det(B)$

Consider
$$A = \begin{bmatrix} 1 & 2 \\ 2 & 5 \end{bmatrix}, \quad B = \begin{bmatrix} 3 & 1 \\ 1 & 3 \end{bmatrix}, \quad A + B = \begin{bmatrix} 4 & 3 \\ 3 & 8 \end{bmatrix}$$

We have $\det(A) = 1$, $\det(B) = 8$, and $\det(A + B) = 23$; thus
$$\det(A + B) \neq \det(A) + \det(B) \quad \blacktriangleleft$$

In spite of the previous example, there is a useful relationship concerning sums of determinants that is applicable when the matrices involved are the same except for *one* row (column). For example, consider the following two matrices that differ only in the second row:
$$A = \begin{bmatrix} a_{11} & a_{12} \\ a_{21} & a_{22} \end{bmatrix} \quad \text{and} \quad B = \begin{bmatrix} a_{11} & a_{12} \\ b_{21} & b_{22} \end{bmatrix}$$

Calculating the determinants of A and B we obtain
$$\det(A) + \det(B) = (a_{11}a_{22} - a_{12}a_{21}) + (a_{11}b_{22} - a_{12}b_{21})$$
$$= a_{11}(a_{22} + b_{22}) - a_{12}(a_{21} + b_{21})$$
$$= \det \begin{bmatrix} a_{11} & a_{12} \\ a_{21} + b_{21} & a_{22} + b_{22} \end{bmatrix}$$

Thus
$$\det \begin{bmatrix} a_{11} & a_{12} \\ a_{21} & a_{22} \end{bmatrix} + \det \begin{bmatrix} a_{11} & a_{12} \\ b_{21} & b_{22} \end{bmatrix} = \det \begin{bmatrix} a_{11} & a_{12} \\ a_{21} + b_{21} & a_{22} + b_{22} \end{bmatrix}$$

This is a special case of the following general result.

THEOREM 2.3.1 *Let A, B, and C be $n \times n$ matrices that differ only in a single row, say the rth, and assume that the rth row of C can be obtained by adding corresponding entries in the rth rows of A and B. Then*
$$\det(C) = \det(A) + \det(B)$$
The same result holds for columns.

▶ **EXAMPLE 2 Sums of Determinants**

We leave it to you to confirm the following equality by evaluating the determinants.
$$\det \begin{bmatrix} 1 & 7 & 5 \\ 2 & 0 & 3 \\ 1+0 & 4+1 & 7+(-1) \end{bmatrix} = \det \begin{bmatrix} 1 & 7 & 5 \\ 2 & 0 & 3 \\ 1 & 4 & 7 \end{bmatrix} + \det \begin{bmatrix} 1 & 7 & 5 \\ 2 & 0 & 3 \\ 0 & 1 & -1 \end{bmatrix} \quad \blacktriangleleft$$

Determinant of a Matrix Product

Considering the complexity of the formulas for determinants and matrix multiplication, it would seem unlikely that a simple relationship should exist between them. This is what makes the simplicity of our next result so surprising. We will show that if A and B are square matrices of the same size, then

$$\det(AB) = \det(A)\det(B) \qquad (2)$$

The proof of this theorem is fairly intricate, so we will have to develop some preliminary results first. We begin with the special case of (2) in which A is an elementary matrix. Because this special case is only a prelude to (2), we call it a lemma.

LEMMA 2.3.2 *If B is an $n \times n$ matrix and E is an $n \times n$ elementary matrix, then*
$$\det(EB) = \det(E)\det(B)$$

Proof We will consider three cases, each in accordance with the row operation that produces the matrix E.

Case 1 If E results from multiplying a row of I_n by k, then by Theorem 1.5.1, EB results from B by multiplying the corresponding row by k; so from Theorem 2.2.3(a) we have
$$\det(EB) = k \det(B)$$
But from Theorem 2.2.4(a) we have $\det(E) = k$, so
$$\det(EB) = \det(E)\det(B)$$

Cases 2 and 3 The proofs of the cases where E results from interchanging two rows of I_n or from adding a multiple of one row to another follow the same pattern as Case 1 and are left as exercises. ◄

Remark It follows by repeated applications of Lemma 2.3.2 that if B is an $n \times n$ matrix and $E_1, E_2, \ldots, E_r$ are $n \times n$ elementary matrices, then
$$\det(E_1 E_2 \cdots E_r B) = \det(E_1)\det(E_2) \cdots \det(E_r)\det(B) \qquad (3)$$

Determinant Test for Invertibility

Our next theorem provides an important criterion for determining whether a matrix is invertible. It also takes us a step closer to establishing Formula (2).

THEOREM 2.3.3 *A square matrix A is invertible if and only if $\det(A) \neq 0$.*

Proof Let R be the reduced row echelon form of A. As a preliminary step, we will show that $\det(A)$ and $\det(R)$ are both zero or both nonzero: Let $E_1, E_2, \ldots, E_r$ be the elementary matrices that correspond to the elementary row operations that produce R from A. Thus
$$R = E_r \cdots E_2 E_1 A$$
and from (3),
$$\det(R) = \det(E_r) \cdots \det(E_2)\det(E_1)\det(A) \qquad (4)$$

We pointed out in the margin note that accompanies Theorem 2.2.4 that the determinant of an elementary matrix is nonzero. Thus, it follows from Formula (4) that $\det(A)$ and $\det(R)$ are either both zero or both nonzero, which sets the stage for the main part of the proof. If we assume first that A is invertible, then it follows from Theorem 1.6.4 that $R = I$ and hence that $\det(R) = 1 \, (\neq 0)$. This, in turn, implies that $\det(A) \neq 0$, which is what we wanted to show.

Conversely, assume that $\det(A) \neq 0$. It follows from this that $\det(R) \neq 0$, which tells us that R cannot have a row of zeros. Thus, it follows from Theorem 1.4.3 that $R = I$ and hence that A is invertible by Theorem 1.6.4. ◄

It follows from Theorems 2.3.3 and 2.2.5 that a square matrix with two proportional rows or two proportional columns is not invertible.

▶ **EXAMPLE 3** **Determinant Test for Invertibility**

Since the first and third rows of

$$A = \begin{bmatrix} 1 & 2 & 3 \\ 1 & 0 & 1 \\ 2 & 4 & 6 \end{bmatrix}$$

are proportional, $\det(A) = 0$. Thus A is not invertible. ◀

We are now ready for the main result concerning products of matrices.

THEOREM 2.3.4 *If A and B are square matrices of the same size, then*

$$\det(AB) = \det(A)\det(B)$$

Proof We divide the proof into two cases that depend on whether or not A is invertible. If the matrix A is not invertible, then by Theorem 1.6.5 neither is the product AB. Thus, from Theorem 2.3.3, we have $\det(AB) = 0$ and $\det(A) = 0$, so it follows that $\det(AB) = \det(A)\det(B)$.

Now assume that A is invertible. By Theorem 1.6.4, the matrix A is expressible as a product of elementary matrices, say

$$A = E_1 E_2 \cdots E_r \qquad (5)$$

so

$$AB = E_1 E_2 \cdots E_r B$$

Applying (3) to this equation yields

$$\det(AB) = \det(E_1)\det(E_2) \cdots \det(E_r)\det(B)$$

and applying (3) again yields

$$\det(AB) = \det(E_1 E_2 \cdots E_r)\det(B)$$

which, from (5), can be written as $\det(AB) = \det(A)\det(B)$. ◀

▶ **EXAMPLE 4** **Verifying That det(AB) = det(A) det(B)**

Consider the matrices

$$A = \begin{bmatrix} 3 & 1 \\ 2 & 1 \end{bmatrix}, \quad B = \begin{bmatrix} -1 & 3 \\ 5 & 8 \end{bmatrix}, \quad AB = \begin{bmatrix} 2 & 17 \\ 3 & 14 \end{bmatrix}$$

We leave it for you to verify that

$$\det(A) = 1, \quad \det(B) = -23, \quad \text{and} \quad \det(AB) = -23$$

Thus $\det(AB) = \det(A)\det(B)$, as guaranteed by Theorem 2.3.4. ◀

Augustin Louis Cauchy
(1789–1857)

Historical Note In 1815 the great French mathematician Augustin Cauchy published a landmark paper in which he gave the first systematic and modern treatment of determinants. It was in that paper that Theorem 2.3.4 was stated and proved in full generality for the first time. Special cases of the theorem had been stated and proved earlier, but it was Cauchy who made the final jump.
[*Image: The Granger Collection, New York*]

The following theorem gives a useful relationship between the determinant of an invertible matrix and the determinant of its inverse.

THEOREM 2.3.5 *If A is invertible, then*
$$\det(A^{-1}) = \frac{1}{\det(A)}$$

Proof Since $A^{-1}A = I$, it follows that $\det(A^{-1}A) = \det(I)$. Therefore, we must have $\det(A^{-1})\det(A) = 1$. Since $\det(A) \neq 0$, the proof can be completed by dividing through by $\det(A)$. ◄

Adjoint of a Matrix

In a cofactor expansion we compute $\det(A)$ by multiplying the entries in a row or column by their cofactors and adding the resulting products. It turns out that if one multiplies the entries in any row by the corresponding cofactors from a *different* row, the sum of these products is always zero. (This result also holds for columns.) Although we omit the general proof, the next example illustrates the idea of the proof in a special case.

It follows from Theorems 2.3.5 and 2.1.2 that
$$\det(A^{-1}) = \frac{1}{a_{11}}\frac{1}{a_{22}}\cdots\frac{1}{a_{nn}}$$
Moreover, by using the adjoint formula it is possible to show that
$$\frac{1}{a_{11}}, \frac{1}{a_{22}}, \ldots, \frac{1}{a_{nn}}$$
are actually the successive diagonal entries of A^{-1} (compare A and A^{-1} in Example 3 of Section 1.7).

► **EXAMPLE 5** **Entries and Cofactors from Different Rows**

Let
$$A = \begin{bmatrix} a_{11} & a_{12} & a_{13} \\ a_{21} & a_{22} & a_{23} \\ a_{31} & a_{32} & a_{33} \end{bmatrix}$$

Consider the quantity
$$a_{11}C_{31} + a_{12}C_{32} + a_{13}C_{33}$$
that is formed by multiplying the entries in the first row by the cofactors of the corresponding entries in the third row and adding the resulting products. We can show that this quantity is equal to zero by the following trick: Construct a new matrix A' by replacing the third row of A with another copy of the first row. That is,
$$A' = \begin{bmatrix} a_{11} & a_{12} & a_{13} \\ a_{21} & a_{22} & a_{23} \\ a_{11} & a_{12} & a_{13} \end{bmatrix}$$

Let $C'_{31}, C'_{32}, C'_{33}$ be the cofactors of the entries in the third row of A'. Since the first two rows of A and A' are the same, and since the computations of $C_{31}, C_{32}, C_{33}, C'_{31}, C'_{32}$, and C'_{33} involve only entries from the first two rows of A and A', it follows that
$$C_{31} = C'_{31}, \quad C_{32} = C'_{32}, \quad C_{33} = C'_{33}$$

Since A' has two identical rows, it follows from (3) that
$$\det(A') = 0 \tag{6}$$

On the other hand, evaluating $\det(A')$ by cofactor expansion along the third row gives
$$\det(A') = a_{11}C'_{31} + a_{12}C'_{32} + a_{13}C'_{33} = a_{11}C_{31} + a_{12}C_{32} + a_{13}C_{33} \tag{7}$$

From (6) and (7) we obtain
$$a_{11}C_{31} + a_{12}C_{32} + a_{13}C_{33} = 0 \quad ◄$$

2.3 Properties of Determinants; Cramer's Rule

DEFINITION 1 If A is any $n \times n$ matrix and C_{ij} is the cofactor of a_{ij}, then the matrix

$$\begin{bmatrix} C_{11} & C_{12} & \cdots & C_{1n} \\ C_{21} & C_{22} & \cdots & C_{2n} \\ \vdots & \vdots & & \vdots \\ C_{n1} & C_{n2} & \cdots & C_{nn} \end{bmatrix}$$

is called the **matrix of cofactors from** A. The transpose of this matrix is called the **adjoint of** A and is denoted by $\text{adj}(A)$.

Leonard Eugene Dickson (1874–1954)

Historical Note The use of the term *adjoint* for the transpose of the matrix of cofactors appears to have been introduced by the American mathematician L. E. Dickson in a research paper that he published in 1902.
[*Image: Courtesy of the American Mathematical Society*]

▶ **EXAMPLE 6** Adjoint of a 3 × 3 Matrix

Let
$$A = \begin{bmatrix} 3 & 2 & -1 \\ 1 & 6 & 3 \\ 2 & -4 & 0 \end{bmatrix}$$

The cofactors of A are

$$C_{11} = 12 \quad C_{12} = 6 \quad C_{13} = -16$$
$$C_{21} = 4 \quad C_{22} = 2 \quad C_{23} = 16$$
$$C_{31} = 12 \quad C_{32} = -10 \quad C_{33} = 16$$

so the matrix of cofactors is

$$\begin{bmatrix} 12 & 6 & -16 \\ 4 & 2 & 16 \\ 12 & -10 & 16 \end{bmatrix}$$

and the adjoint of A is

$$\text{adj}(A) = \begin{bmatrix} 12 & 4 & 12 \\ 6 & 2 & -10 \\ -16 & 16 & 16 \end{bmatrix} \blacktriangleleft$$

In Theorem 1.4.5 we gave a formula for the inverse of a 2 × 2 invertible matrix. Our next theorem extends that result to $n \times n$ invertible matrices.

THEOREM 2.3.6 *Inverse of a Matrix Using Its Adjoint*

If A is an invertible matrix, then

$$A^{-1} = \frac{1}{\det(A)} \text{adj}(A) \tag{8}$$

Proof We show first that
$$A \, \text{adj}(A) = \det(A) I$$

Consider the product

$$A \, \text{adj}(A) = \begin{bmatrix} a_{11} & a_{12} & \cdots & a_{1n} \\ a_{21} & a_{22} & \cdots & a_{2n} \\ \vdots & \vdots & & \vdots \\ a_{i1} & a_{i2} & \cdots & a_{in} \\ \vdots & \vdots & & \vdots \\ a_{n1} & a_{n2} & \cdots & a_{nn} \end{bmatrix} \begin{bmatrix} C_{11} & C_{21} & \cdots & C_{j1} & \cdots & C_{n1} \\ C_{12} & C_{22} & \cdots & C_{j2} & \cdots & C_{n2} \\ \vdots & \vdots & & \vdots & & \vdots \\ C_{1n} & C_{2n} & \cdots & C_{jn} & \cdots & C_{nn} \end{bmatrix}$$

The entry in the ith row and jth column of the product $A \operatorname{adj}(A)$ is

$$a_{i1}C_{j1} + a_{i2}C_{j2} + \cdots + a_{in}C_{jn} \tag{9}$$

(see the shaded lines above).

If $i = j$, then (9) is the cofactor expansion of $\det(A)$ along the ith row of A (Theorem 2.1.1), and if $i \neq j$, then the a's and the cofactors come from different rows of A, so the value of (9) is zero. Therefore,

$$A \operatorname{adj}(A) = \begin{bmatrix} \det(A) & 0 & \cdots & 0 \\ 0 & \det(A) & \cdots & 0 \\ \vdots & \vdots & & \vdots \\ 0 & 0 & \cdots & \det(A) \end{bmatrix} = \det(A) I \tag{10}$$

Since A is invertible, $\det(A) \neq 0$. Therefore, Equation (10) can be rewritten as

$$\frac{1}{\det(A)}[A \operatorname{adj}(A)] = I \quad \text{or} \quad A\left[\frac{1}{\det(A)} \operatorname{adj}(A)\right] = I$$

Multiplying both sides on the left by A^{-1} yields

$$A^{-1} = \frac{1}{\det(A)} \operatorname{adj}(A) \blacktriangleleft$$

▶ **EXAMPLE 7 Using the Adjoint to Find an Inverse Matrix**

Use (8) to find the inverse of the matrix A in Example 6.

Solution We leave it for you to check that $\det(A) = 64$. Thus

$$A^{-1} = \frac{1}{\det(A)} \operatorname{adj}(A) = \frac{1}{64} \begin{bmatrix} 12 & 4 & 12 \\ 6 & 2 & -10 \\ -16 & 16 & 16 \end{bmatrix} = \begin{bmatrix} \frac{12}{64} & \frac{4}{64} & \frac{12}{64} \\ \frac{6}{64} & \frac{2}{64} & -\frac{10}{64} \\ -\frac{16}{64} & \frac{16}{64} & \frac{16}{64} \end{bmatrix} \blacktriangleleft$$

Cramer's Rule

Our next theorem uses the formula for the inverse of an invertible matrix to produce a formula, called **Cramer's rule**, for the solution of a linear system $A\mathbf{x} = \mathbf{b}$ of n equations in n unknowns in the case where the coefficient matrix A is invertible (or, equivalently, when $\det(A) \neq 0$).

THEOREM 2.3.7 Cramer's Rule

If $A\mathbf{x} = \mathbf{b}$ is a system of n linear equations in n unknowns such that $\det(A) \neq 0$, then the system has a unique solution. This solution is

$$x_1 = \frac{\det(A_1)}{\det(A)}, \quad x_2 = \frac{\det(A_2)}{\det(A)}, \ldots, \quad x_n = \frac{\det(A_n)}{\det(A)}$$

where A_j is the matrix obtained by replacing the entries in the jth column of A by the entries in the matrix

$$\mathbf{b} = \begin{bmatrix} b_1 \\ b_2 \\ \vdots \\ b_n \end{bmatrix}$$

2.3 Properties of Determinants; Cramer's Rule

Proof If $\det(A) \neq 0$, then A is invertible, and by Theorem 1.6.2, $\mathbf{x} = A^{-1}\mathbf{b}$ is the unique solution of $A\mathbf{x} = \mathbf{b}$. Therefore, by Theorem 2.3.6 we have

$$\mathbf{x} = A^{-1}\mathbf{b} = \frac{1}{\det(A)}\text{adj}(A)\mathbf{b} = \frac{1}{\det(A)}\begin{bmatrix} C_{11} & C_{21} & \cdots & C_{n1} \\ C_{12} & C_{22} & \cdots & C_{n2} \\ \vdots & \vdots & & \vdots \\ C_{1n} & C_{2n} & \cdots & C_{nn} \end{bmatrix} \begin{bmatrix} b_1 \\ b_2 \\ \vdots \\ b_n \end{bmatrix}$$

Multiplying the matrices out gives

$$\mathbf{x} = \frac{1}{\det(A)}\begin{bmatrix} b_1 C_{11} + b_2 C_{21} + \cdots + b_n C_{n1} \\ b_1 C_{12} + b_2 C_{22} + \cdots + b_n C_{n2} \\ \vdots \\ b_1 C_{1n} + b_2 C_{2n} + \cdots + b_n C_{nn} \end{bmatrix}$$

The entry in the jth row of $\mathbf{x}$ is therefore

$$x_j = \frac{b_1 C_{1j} + b_2 C_{2j} + \cdots + b_n C_{nj}}{\det(A)} \tag{11}$$

Now let

$$A_j = \begin{bmatrix} a_{11} & a_{12} & \cdots & a_{1j-1} & b_1 & a_{1j+1} & \cdots & a_{1n} \\ a_{21} & a_{22} & \cdots & a_{2j-1} & b_2 & a_{2j+1} & \cdots & a_{2n} \\ \vdots & \vdots & & \vdots & \vdots & \vdots & & \vdots \\ a_{n1} & a_{n2} & \cdots & a_{nj-1} & b_n & a_{nj+1} & \cdots & a_{nn} \end{bmatrix}$$

Since A_j differs from A only in the jth column, it follows that the cofactors of entries $b_1, b_2, \ldots, b_n$ in A_j are the same as the cofactors of the corresponding entries in the jth column of A. The cofactor expansion of $\det(A_j)$ along the jth column is therefore

$$\det(A_j) = b_1 C_{1j} + b_2 C_{2j} + \cdots + b_n C_{nj}$$

Substituting this result in (11) gives

$$x_j = \frac{\det(A_j)}{\det(A)} \blacktriangleleft$$

Gabriel Cramer
(1704–1752)

Historical Note Variations of Cramer's rule were fairly well known before the Swiss mathematician discussed it in work he published in 1750. It was Cramer's superior notation that popularized the method and led mathematicians to attach his name to it.
[Image: Granger Collection]

▶ **EXAMPLE 8** Using Cramer's Rule to Solve a Linear System

Use Cramer's rule to solve

$$\begin{aligned} x_1 \phantom{{}+4x_2} + 2x_3 &= 6 \\ -3x_1 + 4x_2 + 6x_3 &= 30 \\ -x_1 - 2x_2 + 3x_3 &= 8 \end{aligned}$$

Solution

$$A = \begin{bmatrix} 1 & 0 & 2 \\ -3 & 4 & 6 \\ -1 & -2 & 3 \end{bmatrix}, \quad A_1 = \begin{bmatrix} 6 & 0 & 2 \\ 30 & 4 & 6 \\ 8 & -2 & 3 \end{bmatrix},$$

$$A_2 = \begin{bmatrix} 1 & 6 & 2 \\ -3 & 30 & 6 \\ -1 & 8 & 3 \end{bmatrix}, \quad A_3 = \begin{bmatrix} 1 & 0 & 6 \\ -3 & 4 & 30 \\ -1 & -2 & 8 \end{bmatrix}$$

For $n > 3$, it is usually more efficient to solve a linear system with n equations in n unknowns by Gauss–Jordan elimination than by Cramer's rule. Its main use is for obtaining properties of solutions of a linear system without actually solving the system.

Therefore,

$$x_1 = \frac{\det(A_1)}{\det(A)} = \frac{-40}{44} = \frac{-10}{11}, \quad x_2 = \frac{\det(A_2)}{\det(A)} = \frac{72}{44} = \frac{18}{11},$$

$$x_3 = \frac{\det(A_3)}{\det(A)} = \frac{152}{44} = \frac{38}{11} \blacktriangleleft$$

Equivalence Theorem In Theorem 1.6.4 we listed five results that are equivalent to the invertibility of a matrix A. We conclude this section by merging Theorem 2.3.3 with that list to produce the following theorem that relates all of the major topics we have studied thus far.

> **THEOREM 2.3.8 Equivalent Statements**
>
> *If A is an $n \times n$ matrix, then the following statements are equivalent.*
>
> (a) A is invertible.
>
> (b) $A\mathbf{x} = \mathbf{0}$ has only the trivial solution.
>
> (c) The reduced row echelon form of A is I_n.
>
> (d) A can be expressed as a product of elementary matrices.
>
> (e) $A\mathbf{x} = \mathbf{b}$ is consistent for every $n \times 1$ matrix $\mathbf{b}$.
>
> (f) $A\mathbf{x} = \mathbf{b}$ has exactly one solution for every $n \times 1$ matrix $\mathbf{b}$.
>
> (g) $\det(A) \neq 0$.

OPTIONAL

We now have all of the machinery necessary to prove the following two results, which we stated without proof in Theorem 1.7.1:

- **Theorem 1.7.1(c)** A triangular matrix is invertible if and only if its diagonal entries are all nonzero.

- **Theorem 1.7.1(d)** The inverse of an invertible lower triangular matrix is lower triangular, and the inverse of an invertible upper triangular matrix is upper triangular.

Proof of Theorem 1.7.1(c) Let $A = [a_{ij}]$ be a triangular matrix, so that its diagonal entries are

$$a_{11}, a_{22}, \ldots, a_{nn}$$

From Theorem 2.1.2, the matrix A is invertible if and only if

$$\det(A) = a_{11}a_{22} \cdots a_{nn}$$

is nonzero, which is true if and only if the diagonal entries are all nonzero.

Proof of Theorem 1.7.1(d) We will prove the result for upper triangular matrices and leave the lower triangular case for you. Assume that A is upper triangular and invertible. Since

$$A^{-1} = \frac{1}{\det(A)} \text{adj}(A)$$

we can prove that A^{-1} is upper triangular by showing that $\text{adj}(A)$ is upper triangular or, equivalently, that the matrix of cofactors is lower triangular. We can do this by showing that every cofactor C_{ij} with $i < j$ (i.e., above the main diagonal) is zero. Since

$$C_{ij} = (-1)^{i+j} M_{ij}$$

it suffices to show that each minor M_{ij} with $i < j$ is zero. For this purpose, let B_{ij} be the matrix that results when the ith row and jth column of A are deleted, so

$$M_{ij} = \det(B_{ij}) \tag{12}$$

From the assumption that $i < j$, it follows that B_{ij} is upper triangular (see Figure 1.7.1). Since A is upper triangular, its $(i+1)$-st row begins with at least i zeros. But the ith row of B_{ij} is the $(i+1)$-st row of A with the entry in the jth column removed. Since $i < j$, none of the first i zeros is removed by deleting the jth column; thus the ith row of B_{ij} starts with at least i zeros, which implies that this row has a zero on the main diagonal. It now follows from Theorem 2.1.2 that $\det(B_{ij}) = 0$ and from (12) that $M_{ij} = 0$. ◄

2.3 Properties of Determinants; Cramer's Rule

Concept Review
- Determinant test for invertibility
- Matrix of cofactors
- Adjoint of a matrix
- Cramer's rule
- Equivalent statements about an invertible matrix

Skills
- Know how determinants behave with respect to basic arithmetic operations, as given in Equation (1), Theorem 2.3.1, Lemma 2.3.2, and Theorem 2.3.4.
- Use the determinant to test a matrix for invertibility.
- Know how $\det(A)$ and $\det(A^{-1})$ are related.
- Compute the matrix of cofactors for a square matrix A.
- Compute $\text{adj}(A)$ for a square matrix A.
- Use the adjoint of an invertible matrix to find its inverse.
- Use Cramer's rule to solve linear systems of equations.
- Know the equivalent characterizations of an invertible matrix given in Theorem 2.3.8.

Exercise Set 2.3

▶ In Exercises 1–4, verify that $\det(kA) = k^n \det(A)$. ◀

1. $A = \begin{bmatrix} 3 & 5 \\ -2 & -4 \end{bmatrix}$; $k = 3$

2. $A = \begin{bmatrix} 2 & 2 \\ 5 & -2 \end{bmatrix}$; $k = -4$

3. $A = \begin{bmatrix} 2 & -1 & 3 \\ 3 & 2 & 1 \\ 1 & 4 & 5 \end{bmatrix}$; $k = -2$

4. $A = \begin{bmatrix} 1 & 1 & 1 \\ 0 & 2 & 3 \\ 0 & 1 & -2 \end{bmatrix}$; $k = 3$

▶ In Exercises 5–6, verify that $\det(AB) = \det(BA)$ and determine whether the equality $\det(A + B) = \det(A) + \det(B)$ holds. ◀

5. $A = \begin{bmatrix} 2 & 0 & -1 \\ 3 & 0 & 5 \\ 0 & 4 & 0 \end{bmatrix}$ and $B = \begin{bmatrix} 4 & 0 & 1 \\ 6 & -2 & 1 \\ -3 & 5 & 2 \end{bmatrix}$

6. $A = \begin{bmatrix} -1 & 8 & 2 \\ 1 & 0 & -1 \\ -2 & 2 & 2 \end{bmatrix}$ and $B = \begin{bmatrix} 2 & -1 & -4 \\ 1 & 1 & 3 \\ 0 & 3 & -1 \end{bmatrix}$

▶ In Exercises 7–14, use determinants to decide whether the given matrix is invertible. ◀

7. $A = \begin{bmatrix} 3 & 6 & 1 \\ 0 & 2 & -4 \\ 0 & 0 & 1 \end{bmatrix}$

8. $A = \begin{bmatrix} 2 & 0 & 3 \\ 0 & 3 & 2 \\ -2 & 0 & -4 \end{bmatrix}$

9. $A = \begin{bmatrix} 1 & 4 & 5 \\ 1 & 3 & 3 \\ 2 & 4 & 3 \end{bmatrix}$

10. $A = \begin{bmatrix} -3 & 0 & 1 \\ 5 & 0 & 6 \\ 8 & 0 & 3 \end{bmatrix}$

11. $A = \begin{bmatrix} 4 & 2 & 8 \\ -2 & 1 & -4 \\ 3 & 1 & 6 \end{bmatrix}$

12. $A = \begin{bmatrix} 1 & 0 & -1 \\ 9 & -1 & 4 \\ 8 & 9 & -1 \end{bmatrix}$

13. $A = \begin{bmatrix} 2 & 0 & 0 \\ 8 & 1 & 0 \\ -5 & 3 & 6 \end{bmatrix}$

14. $A = \begin{bmatrix} \sqrt{2} & -\sqrt{7} & 0 \\ 3\sqrt{2} & -3\sqrt{7} & 0 \\ 5 & -9 & 0 \end{bmatrix}$

▶ In Exercises 15–18, find the values of k for which A is invertible. ◀

15. $A = \begin{bmatrix} k-3 & -2 \\ -2 & k-2 \end{bmatrix}$

16. $A = \begin{bmatrix} k & 2 \\ 2 & k \end{bmatrix}$

17. $A = \begin{bmatrix} 1 & 3 & k \\ 2 & 1 & 3 \\ 4 & 6 & 2 \end{bmatrix}$

18. $A = \begin{bmatrix} 2 & 1 & 0 \\ k & 2 & k \\ 2 & 4 & 2 \end{bmatrix}$

▶ In Exercises 19–23, decide whether the given matrix is invertible, and if so, use the adjoint method to find its inverse. ◀

19. $A = \begin{bmatrix} 2 & 5 & 5 \\ -1 & -1 & 0 \\ 2 & 4 & 3 \end{bmatrix}$

20. $A = \begin{bmatrix} 2 & 0 & 3 \\ 0 & 3 & 2 \\ -2 & 0 & -4 \end{bmatrix}$

21. $A = \begin{bmatrix} 2 & -3 & 5 \\ 0 & 1 & -3 \\ 0 & 0 & 2 \end{bmatrix}$

22. $A = \begin{bmatrix} 2 & 0 & 0 \\ 8 & 1 & 0 \\ -5 & 3 & 6 \end{bmatrix}$

23. $A = \begin{bmatrix} 1 & 3 & 1 & 1 \\ 2 & 5 & 2 & 2 \\ 1 & 3 & 8 & 9 \\ 1 & 3 & 2 & 2 \end{bmatrix}$

▶ In Exercises 24–29, solve by Cramer's rule, where it applies. ◀

24. $7x_1 - 2x_2 = 3$
$3x_1 + x_2 = 5$

25. $3x_1 + 5x_2 = 7$
$6x_1 + 2x_2 + 4x_3 = 10$
$-x_1 + 4x_2 - 3x_3 = 0$

26. $x - 4y + z = 6$
$4x - y + 2z = -1$
$2x + 2y - 3z = -20$

27. $x_1 - 3x_2 + x_3 = 4$
$2x_1 - x_2 = -2$
$4x_1 - 3x_3 = 0$

28.
$$-x_1 - 4x_2 + 2x_3 + x_4 = -32$$
$$2x_1 - x_2 + 7x_3 + 9x_4 = 14$$
$$-x_1 + x_2 + 3x_3 + x_4 = 11$$
$$x_1 - 2x_2 + x_3 - 4x_4 = -4$$

29.
$$3x_1 - x_2 + x_3 = 4$$
$$-x_1 + 7x_2 - 2x_3 = 1$$
$$2x_1 + 6x_2 - x_3 = 5$$

30. Show that the matrix

$$A = \begin{bmatrix} \cos\theta & \sin\theta & 0 \\ -\sin\theta & \cos\theta & 0 \\ 0 & 0 & 1 \end{bmatrix}$$

is invertible for all values of θ; then find A^{-1} using Theorem 2.3.6.

31. Use Cramer's rule to solve for y without solving for the unknowns x, z, and w.

$$4x + y + z + w = 6$$
$$3x + 7y - z + w = 1$$
$$7x + 3y - 5z + 8w = -3$$
$$x + y + z + 2w = 3$$

32. Let $A\mathbf{x} = \mathbf{b}$ be the system in Exercise 31.

(a) Solve by Cramer's rule.

(b) Solve by Gauss–Jordan elimination.

(c) Which method involves fewer computations?

33. Prove that if $\det(A) = 1$ and all the entries in A are integers, then all the entries in A^{-1} are integers.

34. Let $A\mathbf{x} = \mathbf{b}$ be a system of n linear equations in n unknowns with integer coefficients and integer constants. Prove that if $\det(A) = 1$, the solution $\mathbf{x}$ has integer entries.

35. Let

$$A = \begin{bmatrix} a & b & c \\ d & e & f \\ g & h & i \end{bmatrix}$$

Assuming that $\det(A) = -5$, find

(a) $\det(-4A)$ (b) $\det(A^{-1})$ (c) $\det(3A^{-1})$

(d) $\det((3A)^{-1})$ (e) $\det \begin{bmatrix} a & g & d \\ b & h & e \\ c & i & f \end{bmatrix}$

36. In each part, find the determinant given that A is a 4×4 matrix for which $\det(A) = -2$.

(a) $\det(-A)$ (b) $\det(A^{-1})$ (c) $\det(2A^T)$ (d) $\det(A^3)$

37. In each part, find the determinant given that A is a 3×3 matrix for which $\det(A) = 7$.

(a) $\det(3A)$ (b) $\det(A^{-1})$

(c) $\det(2A^{-1})$ (d) $\det((2A)^{-1})$

38. Prove that a square matrix A is invertible if and only if A^TA is invertible.

39. Show that if A is a square matrix, then $\det(A^TA) = \det(AA^T)$.

True-False Exercises

In parts (a)–(l) determine whether the statement is true or false, and justify your answer.

(a) If A is a 3×3 matrix, then $\det(2A) = 2\det(A)$.

(b) If A and B are square matrices of the same size such that $\det(A) = \det(B)$, then $\det(A + B) = 2\det(A)$.

(c) If A and B are square matrices of the same size and A is invertible, then

$$\det(A^{-1}BA) = \det(B)$$

(d) A square matrix A is invertible if and only if $\det(A) = 0$.

(e) The matrix of cofactors of A is precisely $[\mathrm{adj}(A)]^T$.

(f) For every $n \times n$ matrix A, we have

$$A \cdot \mathrm{adj}(A) = (\det(A))I_n$$

(g) If A is a square matrix and the linear system $A\mathbf{x} = \mathbf{0}$ has multiple solutions for $\mathbf{x}$, then $\det(A) = 0$.

(h) If A is an $n \times n$ matrix and there exists an $n \times 1$ matrix $\mathbf{b}$ such that the linear system $A\mathbf{x} = \mathbf{b}$ has no solutions, then the reduced row echelon form of A cannot be I_n.

(i) If E is an elementary matrix, then $E\mathbf{x} = \mathbf{0}$ has only the trivial solution.

(j) If A is an invertible matrix, then the linear system $A\mathbf{x} = \mathbf{0}$ has only the trivial solution if and only if the linear system $A^{-1}\mathbf{x} = \mathbf{0}$ has only the trivial solution.

(k) If A is invertible, then $\mathrm{adj}(A)$ must also be invertible.

(l) If A has a row of zeros, then so does $\mathrm{adj}(A)$.

Chapter 2 Supplementary Exercises

In Exercises **1–8**, evaluate the determinant of the given matrix by (a) cofactor expansion and (b) using elementary row operations to introduce zeros into the matrix.

1. $\begin{bmatrix} -4 & 2 \\ 3 & 3 \end{bmatrix}$

2. $\begin{bmatrix} 7 & -1 \\ -2 & -6 \end{bmatrix}$

3. $\begin{bmatrix} -1 & 5 & 2 \\ 0 & 2 & -1 \\ -3 & 1 & 1 \end{bmatrix}$

4. $\begin{bmatrix} -1 & -2 & -3 \\ -4 & -5 & -6 \\ -7 & -8 & -9 \end{bmatrix}$

5. $\begin{bmatrix} 3 & 0 & -1 \\ 1 & 1 & 1 \\ 0 & 4 & 2 \end{bmatrix}$

6. $\begin{bmatrix} -5 & 1 & 4 \\ 3 & 0 & 2 \\ 1 & -2 & 2 \end{bmatrix}$

7. $\begin{bmatrix} 3 & 6 & 0 & 1 \\ -2 & 3 & 1 & 4 \\ 1 & 0 & -1 & 1 \\ -9 & 2 & -2 & 2 \end{bmatrix}$

8. $\begin{bmatrix} -1 & -2 & -3 & -4 \\ 4 & 3 & 2 & 1 \\ 1 & 2 & 3 & 4 \\ -4 & -3 & -2 & -1 \end{bmatrix}$

9. Evaluate the determinants in Exercises 3–6 by using the arrow technique (see Example 7 in Section 2.1).

10. (a) Construct a 4×4 matrix whose determinant is easy to compute using cofactor expansion but hard to evaluate using elementary row operations.
    (b) Construct a 4×4 matrix whose determinant is easy to compute using elementary row operations but hard to evaluate using cofactor expansion.

11. Use the determinant to decide whether the matrices in Exercises 1–4 are invertible.

12. Use the determinant to decide whether the matrices in Exercises 5–8 are invertible.

In Exercises **13–15**, find the determinant of the given matrix by any method.

13. $\begin{vmatrix} 5 & b-3 \\ b-2 & -3 \end{vmatrix}$

14. $\begin{vmatrix} 3 & -4 & a \\ a^2 & 1 & 2 \\ 2 & a-1 & 4 \end{vmatrix}$

15. $\begin{vmatrix} 0 & 0 & 0 & 0 & -3 \\ 0 & 0 & 0 & -4 & 0 \\ 0 & 0 & -1 & 0 & 0 \\ 0 & 2 & 0 & 0 & 0 \\ 5 & 0 & 0 & 0 & 0 \end{vmatrix}$

16. Solve for x.
$$\begin{vmatrix} x & -1 \\ 3 & 1-x \end{vmatrix} = \begin{vmatrix} 1 & 0 & -3 \\ 2 & x & -6 \\ 1 & 3 & x-5 \end{vmatrix}$$

In Exercises **17–24**, use the adjoint method (Theorem 2.3.6) to find the inverse of the given matrix, if it exists.

17. The matrix in Exercise 1.   18. The matrix in Exercise 2.
19. The matrix in Exercise 3.   20. The matrix in Exercise 4.
21. The matrix in Exercise 5.   22. The matrix in Exercise 6.
23. The matrix in Exercise 7.   24. The matrix in Exercise 8.

25. Use Cramer's rule to solve for x' and y' in terms of x and y.
$$x = \tfrac{3}{5}x' - \tfrac{4}{5}y'$$
$$y = \tfrac{4}{5}x' + \tfrac{3}{5}y'$$

26. Use Cramer's rule to solve for x' and y' in terms of x and y.
$$x = x'\cos\theta - y'\sin\theta$$
$$y = x'\sin\theta + y'\cos\theta$$

27. By examining the determinant of the coefficient matrix, show that the following system has a nontrivial solution if and only if $\alpha = \beta$.
$$\begin{aligned} x + y + \alpha z &= 0 \\ x + y + \beta z &= 0 \\ \alpha x + \beta y + z &= 0 \end{aligned}$$

28. Let A be a 3×3 matrix, each of whose entries is 1 or 0. What is the largest possible value for $\det(A)$?

29. (a) For the triangle in the accompanying figure, use trigonometry to show that
$$b\cos\gamma + c\cos\beta = a$$
$$c\cos\alpha + a\cos\gamma = b$$
$$a\cos\beta + b\cos\alpha = c$$
and then apply Cramer's rule to show that
$$\cos\alpha = \frac{b^2 + c^2 - a^2}{2bc}$$
(b) Use Cramer's rule to obtain similar formulas for $\cos\beta$ and $\cos\gamma$.

◀ Figure Ex-29

30. Use determinants to show that for all real values of λ, the only solution of
$$x - 2y = \lambda x$$
$$x - y = \lambda y$$
is $x = 0, y = 0$.

31. Prove: If A is invertible, then $\mathrm{adj}(A)$ is invertible and
$$[\mathrm{adj}(A)]^{-1} = \frac{1}{\det(A)}A = \mathrm{adj}(A^{-1})$$

32. Prove: If A is an $n \times n$ matrix, then

$$\det[\operatorname{adj}(A)] = [\det(A)]^{n-1}$$

33. Prove: If the entries in each row of an $n \times n$ matrix A add up to zero, then the determinant of A is zero. [*Hint:* Consider the product $A\mathbf{x}$, where $\mathbf{x}$ is the $n \times 1$ matrix, each of whose entries is one.]

34. (a) In the accompanying figure, the area of the triangle ABC can be expressed as

area ABC = area $ADEC$ + area $CEFB$ − area $ADFB$

Use this and the fact that the area of a trapezoid equals $\frac{1}{2}$ the altitude times the sum of the parallel sides to show that

$$\text{area } ABC = \frac{1}{2} \begin{vmatrix} x_1 & y_1 & 1 \\ x_2 & y_2 & 1 \\ x_3 & y_3 & 1 \end{vmatrix}$$

[*Note:* In the derivation of this formula, the vertices are labeled such that the triangle is traced counterclockwise proceeding from (x_1, y_1) to (x_2, y_2) to (x_3, y_3). For a clockwise orientation, the determinant above yields the *negative* of the area.]

(b) Use the result in (a) to find the area of the triangle with vertices $(3, 3)$, $(4, 0)$, $(-2, -1)$.

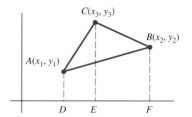

◀ **Figure Ex-34**

35. Use the fact that 21,375, 38,798, 34,162, 40,223, and 79,154 are all divisible by 19 to show that

$$\begin{vmatrix} 2 & 1 & 3 & 7 & 5 \\ 3 & 8 & 7 & 9 & 8 \\ 3 & 4 & 1 & 6 & 2 \\ 4 & 0 & 2 & 2 & 3 \\ 7 & 9 & 1 & 5 & 4 \end{vmatrix}$$

is divisible by 19 without directly evaluating the determinant.

36. Without directly evaluating the determinant, show that

$$\begin{vmatrix} \sin \alpha & \cos \alpha & \sin(\alpha + \delta) \\ \sin \beta & \cos \beta & \sin(\beta + \delta) \\ \sin \gamma & \cos \gamma & \sin(\gamma + \delta) \end{vmatrix} = 0$$

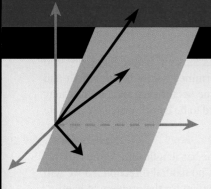

CHAPTER 3

Euclidean Vector Spaces

CHAPTER CONTENTS
3.1 Vectors in 2-Space, 3-Space, and n-Space 119
3.2 Norm, Dot Product, and Distance in R^n 130
3.3 Orthogonality 143
3.4 The Geometry of Linear Systems 152
3.5 Cross Product 161

INTRODUCTION Engineers and physicists distinguish between two types of physical quantities—*scalars*, which are quantities that can be described by a numerical value alone, and *vectors*, which are quantities that require both a number and a direction for their complete physical description. For example, temperature, length, and speed are scalars because they can be fully described by a number that tells "how much"—a temperature of 20°C, a length of 5 cm, or a speed of 75 km/h. In contrast, velocity and force are vectors because they require a number that tells "how much" and a direction that tells "which way"—say, a boat moving at 10 knots in a direction 45° northeast, or a force of 100 lb acting vertically. Although the notions of vectors and scalars that we will study in this text have their origins in physics and engineering, we will be more concerned with using them to build mathematical structures and then applying those structures to such diverse fields as genetics, computer science, economics, telecommunications, and environmental science.

3.1 Vectors in 2-Space, 3-Space, and n-Space

Linear algebra is concerned with two kinds of mathematical objects, "matrices" and "vectors." We are already familiar with the basic ideas about matrices, so in this section we will introduce some of the basic ideas about vectors. As we progress through this text we will see that vectors and matrices are closely related and that much of linear algebra is concerned with that relationship.

Geometric Vectors

Engineers and physicists represent vectors in two dimensions (also called **2-space**) or in three dimensions (also called **3-space**) by arrows. The direction of the arrowhead specifies the *direction* of the vector and the *length* of the arrow specifies the magnitude. Mathematicians call these **geometric vectors**. The tail of the arrow is called the *initial point* of the vector and the tip the *terminal point* (Figure 3.1.1).

In this text we will denote vectors in boldface type such as **a**, **b**, **v**, **w**, and **x**, and we will denote scalars in lowercase italic type such as a, k, v, w, and x. When we want to indicate that a vector **v** has initial point A and terminal point B, then, as shown in Figure 3.1.2, we will write

$$\mathbf{v} = \overrightarrow{AB}$$

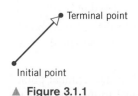
▲ Figure 3.1.1

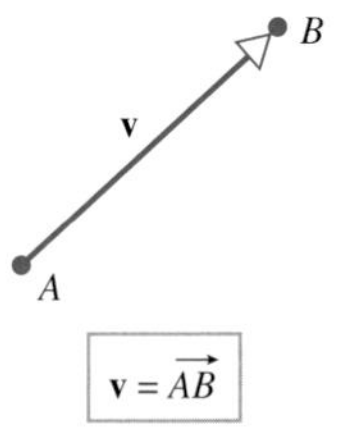

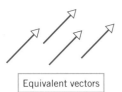

Figure 3.1.2

Vectors with the same length and direction, such as those in Figure 3.1.3, are said to be **equivalent**. Since we want a vector to be determined solely by its length and direction, equivalent vectors are regarded to be the same vector even though they may be in different positions. Equivalent vectors are also said to be **equal**, which we indicate by writing

$$\mathbf{v} = \mathbf{w}$$

The vector whose initial and terminal points coincide has length zero, so we call this the **zero vector** and denote it by **0**. The zero vector has no natural direction, so we will agree that it can be assigned any direction that is convenient for the problem at hand.

Vector Addition

There are a number of important algebraic operations on vectors, all of which have their origin in laws of physics.

Figure 3.1.3

Parallelogram Rule for Vector Addition If **v** and **w** are vectors in 2-space or 3-space that are positioned so their initial points coincide, then the two vectors form adjacent sides of a parallelogram, and the **sum v + w** is the vector represented by the arrow from the common initial point of **v** and **w** to the opposite vertex of the parallelogram (Figure 3.1.4a).

Here is another way to form the sum of two vectors.

Triangle Rule for Vector Addition If **v** and **w** are vectors in 2-space or 3-space that are positioned so the initial point of **w** is at the terminal point of **v**, then the **sum v + w** is represented by the arrow from the initial point of **v** to the terminal point of **w** (Figure 3.1.4b).

In Figure 3.1.4c we have constructed the sums **v + w** and **w + v** by the triangle rule. This construction makes it evident that

$$\mathbf{v} + \mathbf{w} = \mathbf{w} + \mathbf{v} \tag{1}$$

and that the sum obtained by the triangle rule is the same as the sum obtained by the parallelogram rule.

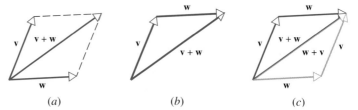

▶ Figure 3.1.4

Vector addition can also be viewed as a process of translating points.

Vector Addition Viewed as Translation If **v**, **w**, and **v + w** are positioned so their initial points coincide, then the terminal point of **v + w** can be viewed in two ways:

1. The terminal point of **v + w** is the point that results when the terminal point of **v** is translated in the direction of **w** by a distance equal to the length of **w** (Figure 3.1.5a).

2. The terminal point of **v + w** is the point that results when the terminal point of **w** is translated in the direction of **v** by a distance equal to the length of **v** (Figure 3.1.5b).

Accordingly, we say that **v + w** is the **translation of v by w** or, alternatively, the **translation of w by v**.

3.1 Vectors in 2-Space, 3-Space, and n-Space

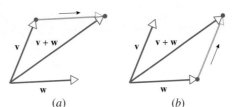

Figure 3.1.5 (a) (b)

Vector Subtraction In ordinary arithmetic we can write $a - b = a + (-b)$, which expresses subtraction in terms of addition. There is an analogous idea in vector arithmetic.

> **Vector Subtraction** The *negative* of a vector **v**, denoted by $-\mathbf{v}$, is the vector that has the same length as **v** but is oppositely directed (Figure 3.1.6a), and the *difference* of **v** from **w**, denoted by $\mathbf{w} - \mathbf{v}$, is taken to be the sum
> $$\mathbf{w} - \mathbf{v} = \mathbf{w} + (-\mathbf{v}) \tag{2}$$

The difference of **v** from **w** can be obtained geometrically by the parallelogram method shown in Figure 3.1.6b, or more directly by positioning **w** and **v** so their initial points coincide and drawing the vector from the terminal point of **v** to the terminal point of **w** (Figure 3.1.6c).

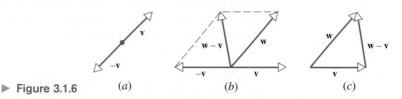

Figure 3.1.6 (a) (b) (c)

Scalar Multiplication Sometimes there is a need to change the length of a vector or change its length and reverse its direction. This is accomplished by a type of multiplication in which vectors are multiplied by scalars. As an example, the product $2\mathbf{v}$ denotes the vector that has the same direction as **v** but twice the length, and the product $-2\mathbf{v}$ denotes the vector that is oppositely directed to **v** and has twice the length. Here is the general result.

> **Scalar Multiplication** If **v** is a nonzero vector in 2-space or 3-space, and if k is a nonzero scalar, then we define the *scalar product of* **v** *by* k to be the vector whose length is $|k|$ times the length of **v** and whose direction is the same as that of **v** if k is positive and opposite to that of **v** if k is negative. If $k = 0$ or $\mathbf{v} = \mathbf{0}$, then we define $k\mathbf{v}$ to be $\mathbf{0}$.

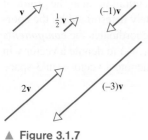

Figure 3.1.7 shows the geometric relationship between a vector **v** and some of its scalar multiples. In particular, observe that $(-1)\mathbf{v}$ has the same length as **v** but is oppositely directed; therefore,
$$(-1)\mathbf{v} = -\mathbf{v} \tag{3}$$

Figure 3.1.7

Parallel and Collinear Vectors Suppose that **v** and **w** are vectors in 2-space or 3-space with a common initial point. If one of the vectors is a scalar multiple of the other, then the vectors lie on a common line, so it is reasonable to say that they are *collinear* (Figure 3.1.8a). However, if we translate one of the vectors, as indicated in Figure 3.1.8b, then the vectors are *parallel* but no longer collinear. This creates a linguistic problem because translating a vector does not change it. The only way to resolve this problem is to agree that the terms *parallel* and

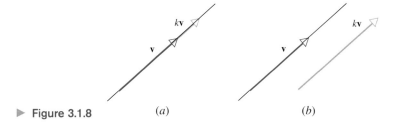

Figure 3.1.8 (a) (b)

collinear mean the same thing when applied to vectors. Although the vector **0** has no clearly defined direction, we will regard it to be parallel to all vectors when convenient.

Sums of Three or More Vectors

Vector addition satisfies the **associative law for addition**, meaning that when we add three vectors, say **u**, **v**, and **w**, it does not matter which two we add first; that is,

$$\mathbf{u} + (\mathbf{v} + \mathbf{w}) = (\mathbf{u} + \mathbf{v}) + \mathbf{w}$$

It follows from this that there is no ambiguity in the expression $\mathbf{u} + \mathbf{v} + \mathbf{w}$ because the same result is obtained no matter how the vectors are grouped.

A simple way to construct $\mathbf{u} + \mathbf{v} + \mathbf{w}$ is to place the vectors "tip to tail" in succession and then draw the vector from the initial point of **u** to the terminal point of **w** (Figure 3.1.9a). The tip-to-tail method also works for four or more vectors (Figure 3.1.9b). The tip-to-tail method also makes it evident that if **u**, **v**, and **w** are vectors in 3-space with a *common initial point*, then $\mathbf{u} + \mathbf{v} + \mathbf{w}$ is the diagonal of the parallelepiped that has the three vectors as adjacent sides (Figure 3.1.9c).

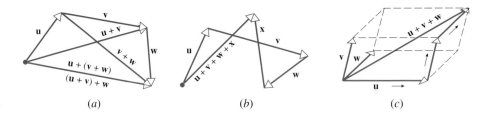

Figure 3.1.9 (a) (b) (c)

Vectors in Coordinate Systems

Up until now we have discussed vectors without reference to a coordinate system. However, as we will soon see, computations with vectors are much simpler to perform if a coordinate system is present to work with.

If a vector **v** in 2-space or 3-space is positioned with its initial point at the origin of a rectangular coordinate system, then the vector is completely determined by the coordinates of its terminal point (Figure 3.1.10). We call these coordinates the **components** of **v** relative to the coordinate system. We will write $\mathbf{v} = (v_1, v_2)$ to denote a vector **v** in 2-space with components (v_1, v_2), and $\mathbf{v} = (v_1, v_2, v_3)$ to denote a vector **v** in 3-space with components (v_1, v_2, v_3).

The component forms of the zero vector are $\mathbf{0} = (0, 0)$ in 2-space and $\mathbf{0} = (0, 0, 0)$ in 3-space.

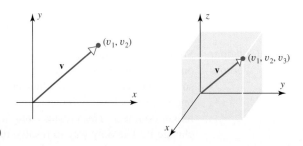

Figure 3.1.10

It should be evident geometrically that two vectors in 2-space or 3-space are equivalent if and only if they have the same terminal point when their initial points are at the origin. Algebraically, this means that two vectors are equivalent if and only if their corresponding components are equal. Thus, for example, the vectors

$$\mathbf{v} = (v_1, v_2, v_3) \quad \text{and} \quad \mathbf{w} = (w_1, w_2, w_3)$$

in 3-space are equivalent if and only if

$$v_1 = w_1, \quad v_2 = w_2, \quad v_3 = w_3$$

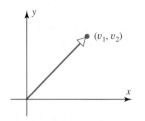

▲ Figure 3.1.11 The ordered pair (v_1, v_2) can represent a point or a vector.

Remark It may have occurred to you that an ordered pair (v_1, v_2) can represent either a vector with *components* v_1 and v_2 or a point with *components* v_1 and v_2 (and similarly for ordered triples). Both are valid geometric interpretations, so the appropriate choice will depend on the geometric viewpoint that we want to emphasize (Figure 3.1.11).

Vectors Whose Initial Point Is Not at the Origin

It is sometimes necessary to consider vectors whose initial points are not at the origin. If $\overrightarrow{P_1 P_2}$ denotes the vector with initial point $P_1(x_1, y_1)$ and terminal point $P_2(x_2, y_2)$, then the components of this vector are given by the formula

$$\overrightarrow{P_1 P_2} = (x_2 - x_1, y_2 - y_1) \tag{4}$$

That is, the components of $\overrightarrow{P_1 P_2}$ are obtained by subtracting the coordinates of the initial point from the coordinates of the terminal point. For example, in Figure 3.1.12 the vector $\overrightarrow{P_1 P_2}$ is the difference of vectors $\overrightarrow{OP_2}$ and $\overrightarrow{OP_1}$, so

$$\overrightarrow{P_1 P_2} = \overrightarrow{OP_2} - \overrightarrow{OP_1} = (x_2, y_2) - (x_1, y_1) = (x_2 - x_1, y_2 - y_1)$$

As you might expect, the components of a vector in 3-space that has initial point $P_1(x_1, y_1, z_1)$ and terminal point $P_2(x_2, y_2, z_2)$ are given by

$$\overrightarrow{P_1 P_2} = (x_2 - x_1, y_2 - y_1, z_2 - z_1) \tag{5}$$

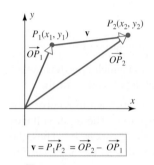

▲ Figure 3.1.12

▶ **EXAMPLE 1** Finding the Components of a Vector

The components of the vector $\mathbf{v} = \overrightarrow{P_1 P_2}$ with initial point $P_1(2, -1, 4)$ and terminal point $P_2(7, 5, -8)$ are

$$\mathbf{v} = (7 - 2, 5 - (-1), (-8) - 4) = (5, 6, -12) \quad \blacktriangleleft$$

n-Space

The idea of using ordered pairs and triples of real numbers to represent points in two-dimensional space and three-dimensional space was well known in the eighteenth and nineteenth centuries. By the dawn of the twentieth century, mathematicians and physicists were exploring the use of "higher-dimensional" spaces in mathematics and physics. Today, even the layman is familiar with the notion of time as a fourth dimension, an idea used by Albert Einstein in developing the general theory of relativity. Today, physicists working in the field of "string theory" commonly use 11-dimensional space in their quest for a unified theory that will explain how the fundamental forces of nature work. Much of the remaining work in this section is concerned with extending the notion of space to n-dimensions.

To explore these ideas further, we start with some terminology and notation. The set of all real numbers can be viewed geometrically as a line. It is called the **real line** and is denoted by R or R^1. The superscript reinforces the intuitive idea that a line is one-dimensional. The set of all ordered pairs of real numbers (called **2-tuples**) and the set of all ordered triples of real numbers (called **3-tuples**) are denoted by R^2 and

R^3, respectively. The superscript reinforces the idea that the ordered pairs correspond to points in the plane (two-dimensional) and ordered triples to points in space (three-dimensional). The following definition extends this idea.

> **DEFINITION 1** If n is a positive integer, then an ***ordered n-tuple*** is a sequence of n real numbers $(v_1, v_2, \ldots, v_n)$. The set of all ordered n-tuples is called ***n-space*** and is denoted by R^n.

Remark You can think of the numbers in an n-tuple $(v_1, v_2, \ldots, v_n)$ as either the coordinates of a *generalized point* or the components of a *generalized vector*, depending on the geometric image you want to bring to mind—the choice makes no difference mathematically, since it is the algebraic properties of n-tuples that are of concern.

Here are some typical applications that lead to n-tuples.

- **Experimental Data**—A scientist performs an experiment and makes n numerical measurements each time the experiment is performed. The result of each experiment can be regarded as a vector $\mathbf{y} = (y_1, y_2, \ldots, y_n)$ in R^n in which $y_1, y_2, \ldots, y_n$ are the measured values.

- **Storage and Warehousing**—A national trucking company has 15 depots for storing and servicing its trucks. At each point in time the distribution of trucks in the service depots can be described by a 15-tuple $\mathbf{x} = (x_1, x_2, \ldots, x_{15})$ in which x_1 is the number of trucks in the first depot, x_2 is the number in the second depot, and so forth.

- **Electrical Circuits**—A certain kind of processing chip is designed to receive four input voltages and produces three output voltages in response. The input voltages can be regarded as vectors in R^4 and the output voltages as vectors in R^3. Thus, the chip can be viewed as a device that transforms an input vector $\mathbf{v} = (v_1, v_2, v_3, v_4)$ in R^4 into an output vector $\mathbf{w} = (w_1, w_2, w_3)$ in R^3.

- **Graphical Images**—One way in which color images are created on computer screens is by assigning each pixel (an addressable point on the screen) three numbers that describe the **hue**, **saturation**, and **brightness** of the pixel. Thus, a complete color image can be viewed as a set of 5-tuples of the form $\mathbf{v} = (x, y, h, s, b)$ in which x and y are the screen coordinates of a pixel and h, s, and b are its hue, saturation, and brightness.

- **Economics**—One approach to economic analysis is to divide an economy into sectors (manufacturing, services, utilities, and so forth) and measure the output of each sector by a dollar value. Thus, in an economy with 10 sectors the economic output of the entire economy can be represented by a 10-tuple $\mathbf{s} = (s_1, s_2, \ldots, s_{10})$ in which the numbers $s_1, s_2, \ldots, s_{10}$ are the outputs of the individual sectors.

Albert Einstein (1879–1955)

Historical Note The German-born physicist Albert Einstein immigrated to the United States in 1935, where he settled at Princeton University. Einstein spent the last three decades of his life working unsuccessfully at producing a *unified field theory* that would establish an underlying link between the forces of gravity and electromagnetism. Recently, physicists have made progress on the problem using a framework known as *string theory*. In this theory the smallest, indivisible components of the Universe are not particles but loops that behave like vibrating strings. Whereas Einstein's space-time universe was four-dimensional, strings reside in an 11-dimensional world that is the focus of current research.

[*Image:* ©Bettmann/©Corbis]

3.1 Vectors in 2-Space, 3-Space, and n-Space

- **Mechanical Systems**—Suppose that six particles move along the same coordinate line so that at time t their coordinates are $x_1, x_2, \ldots, x_6$ and their velocities are $v_1, v_2, \ldots, v_6$, respectively. This information can be represented by the vector

$$\mathbf{v} = (x_1, x_2, x_3, x_4, x_5, x_6, v_1, v_2, v_3, v_4, v_5, v_6, t)$$

in R^{13}. This vector is called the ***state*** of the particle system at time t.

Operations on Vectors in R^n

Our next goal is to define useful operations on vectors in R^n. These operations will all be natural extensions of the familiar operations on vectors in R^2 and R^3. We will denote a vector $\mathbf{v}$ in R^n using the notation

$$\mathbf{v} = (v_1, v_2, \ldots, v_n)$$

and we will call $\mathbf{0} = (0, 0, \ldots, 0)$ the ***zero vector***.

We noted earlier that in R^2 and R^3 two vectors are equivalent (equal) if and only if their corresponding components are the same. Thus, we make the following definition.

> **DEFINITION 2** Vectors $\mathbf{v} = (v_1, v_2, \ldots, v_n)$ and $\mathbf{w} = (w_1, w_2, \ldots, w_n)$ in R^n are said to be ***equivalent*** (also called ***equal***) if
>
> $$v_1 = w_1, \quad v_2 = w_2, \ldots, \quad v_n = w_n$$
>
> We indicate this by writing $\mathbf{v} = \mathbf{w}$.

▶ **EXAMPLE 2** Equality of Vectors

$$(a, b, c, d) = (1, -4, 2, 7)$$

if and only if $a = 1, b = -4, c = 2$, and $d = 7$. ◀

Our next objective is to define the operations of addition, subtraction, and scalar multiplication for vectors in R^n. To motivate these ideas, we will consider how these operations can be performed on vectors in R^2 using components. By studying Figure 3.1.13 you should be able to deduce that if $\mathbf{v} = (v_1, v_2)$ and $\mathbf{w} = (w_1, w_2)$, then

$$\mathbf{v} + \mathbf{w} = (v_1 + w_1, v_2 + w_2) \tag{6}$$

$$k\mathbf{v} = (kv_1, kv_2) \tag{7}$$

In particular, it follows from (7) that

$$-\mathbf{v} = (-1)\mathbf{v} = (-v_1, -v_2) \tag{8}$$

and hence that

$$\mathbf{w} - \mathbf{v} = \mathbf{w} + (-\mathbf{v}) = (w_1 - v_1, w_2 - v_2) \tag{9}$$

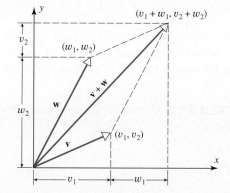
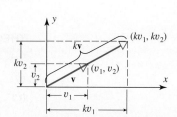

▶ Figure 3.1.13

Motivated by Formulas (6)–(9), we make the following definition.

> **DEFINITION 3** If $\mathbf{v} = (v_1, v_2, \ldots, v_n)$ and $\mathbf{w} = (w_1, w_2, \ldots, w_n)$ are vectors in R^n, and if k is any scalar, then we define
> $$\mathbf{v} + \mathbf{w} = (v_1 + w_1, v_2 + w_2, \ldots, v_n + w_n) \quad (10)$$
> $$k\mathbf{v} = (kv_1, kv_2, \ldots, kv_n) \quad (11)$$
> $$-\mathbf{v} = (-v_1, -v_2, \ldots, -v_n) \quad (12)$$
> $$\mathbf{w} - \mathbf{v} = \mathbf{w} + (-\mathbf{v}) = (w_1 - v_1, w_2 - v_2, \ldots, w_n - v_n) \quad (13)$$

In words, vectors are added (or subtracted) by adding (or subtracting) their corresponding components, and a vector is multiplied by a scalar by multiplying each component by that scalar.

▶ **EXAMPLE 3** Algebraic Operations Using Components

If $\mathbf{v} = (1, -3, 2)$ and $\mathbf{w} = (4, 2, 1)$, then
$$\mathbf{v} + \mathbf{w} = (5, -1, 3), \quad 2\mathbf{v} = (2, -6, 4)$$
$$-\mathbf{w} = (-4, -2, -1), \quad \mathbf{v} - \mathbf{w} = \mathbf{v} + (-\mathbf{w}) = (-3, -5, 1) \blacktriangleleft$$

The following theorem summarizes the most important properties of vector operations.

> **THEOREM 3.1.1** If $\mathbf{u}, \mathbf{v}$, and $\mathbf{w}$ are vectors in R^n, and if k and m are scalars, then:
> (a) $\mathbf{u} + \mathbf{v} = \mathbf{v} + \mathbf{u}$
> (b) $(\mathbf{u} + \mathbf{v}) + \mathbf{w} = \mathbf{u} + (\mathbf{v} + \mathbf{w})$
> (c) $\mathbf{u} + \mathbf{0} = \mathbf{0} + \mathbf{u} = \mathbf{u}$
> (d) $\mathbf{u} + (-\mathbf{u}) = \mathbf{0}$
> (e) $k(\mathbf{u} + \mathbf{v}) = k\mathbf{u} + k\mathbf{v}$
> (f) $(k + m)\mathbf{u} = k\mathbf{u} + m\mathbf{u}$
> (g) $k(m\mathbf{u}) = (km)\mathbf{u}$
> (h) $1\mathbf{u} = \mathbf{u}$

We will prove part (b) and leave some of the other proofs as exercises.

Proof (b) Let $\mathbf{u} = (u_1, u_2, \ldots, u_n)$, $\mathbf{v} = (v_1, v_2, \ldots, v_n)$, and $\mathbf{w} = (w_1, w_2, \ldots, w_n)$. Then
$$(\mathbf{u} + \mathbf{v}) + \mathbf{w} = ((u_1, u_2, \ldots, u_n) + (v_1, v_2, \ldots, v_n)) + (w_1, w_2, \ldots, w_n)$$
$$= (u_1 + v_1, u_2 + v_2, \ldots, u_n + v_n) + (w_1, w_2, \ldots, w_n) \quad \text{[Vector addition]}$$
$$= ((u_1 + v_1) + w_1, (u_2 + v_2) + w_2, \ldots, (u_n + v_n) + w_n) \quad \text{[Vector addition]}$$
$$= (u_1 + (v_1 + w_1), u_2 + (v_2 + w_2), \ldots, u_n + (v_n + w_n)) \quad \text{[Regroup]}$$
$$= (u_1, u_2, \ldots, u_n) + (v_1 + w_1, v_2 + w_2, \ldots, v_n + w_n) \quad \text{[Vector addition]}$$
$$= \mathbf{u} + (\mathbf{v} + \mathbf{w}) \blacktriangleleft$$

The following additional properties of vectors in R^n can be deduced easily by expressing the vectors in terms of components (verify).

> **THEOREM 3.1.2** If $\mathbf{v}$ is a vector in R^n and k is a scalar, then:
> (a) $0\mathbf{v} = \mathbf{0}$
> (b) $k\mathbf{0} = \mathbf{0}$
> (c) $(-1)\mathbf{v} = -\mathbf{v}$

Calculating Without Components

One of the powerful consequences of Theorems 3.1.1 and 3.1.2 is that they allow calculations to be performed without expressing the vectors in terms of components. For example, suppose that $\mathbf{x}$, $\mathbf{a}$, and $\mathbf{b}$ are vectors in R^n, and we want to solve the vector equation $\mathbf{x} + \mathbf{a} = \mathbf{b}$ for the vector $\mathbf{x}$ without using components. We could proceed as follows:

$$\mathbf{x} + \mathbf{a} = \mathbf{b} \quad \text{[Given]}$$
$$(\mathbf{x} + \mathbf{a}) + (-\mathbf{a}) = \mathbf{b} + (-\mathbf{a}) \quad \text{[Add the negative of a to both sides]}$$
$$\mathbf{x} + (\mathbf{a} + (-\mathbf{a})) = \mathbf{b} - \mathbf{a} \quad \text{[Part (b) of Theorem 3.1.1]}$$
$$\mathbf{x} + \mathbf{0} = \mathbf{b} - \mathbf{a} \quad \text{[Part (d) of Theorem 3.1.1]}$$
$$\mathbf{x} = \mathbf{b} - \mathbf{a} \quad \text{[Part (c) of Theorem 3.1.1]}$$

While this method is obviously more cumbersome than computing with components in R^n, it will become important later in the text where we will encounter more general kinds of vectors.

Linear Combinations

Addition, subtraction, and scalar multiplication are frequently used in combination to form new vectors. For example, if $\mathbf{v}_1$, $\mathbf{v}_2$, and $\mathbf{v}_3$ are vectors in R^n, then the vectors

$$\mathbf{u} = 2\mathbf{v}_1 + 3\mathbf{v}_2 + \mathbf{v}_3 \quad \text{and} \quad \mathbf{w} = 7\mathbf{v}_1 - 6\mathbf{v}_2 + 8\mathbf{v}_3$$

are formed in this way. In general, we make the following definition.

> **DEFINITION 4** If $\mathbf{w}$ is a vector in R^n, then $\mathbf{w}$ is said to be a ***linear combination*** of the vectors $\mathbf{v}_1, \mathbf{v}_2, \ldots, \mathbf{v}_r$ in R^n if it can be expressed in the form
> $$\mathbf{w} = k_1\mathbf{v}_1 + k_2\mathbf{v}_2 + \cdots + k_r\mathbf{v}_r \tag{14}$$
> where $k_1, k_2, \ldots, k_r$ are scalars. These scalars are called the ***coefficients*** of the linear combination. In the case where $r = 1$, Formula (14) becomes $\mathbf{w} = k_1\mathbf{v}_1$, so that a linear combination of a single vector is just a scalar muliple of that vector.

Note that this definition of a linear combination is consistent with that given in the context of matrices (see Definition 6 in Section 1.3).

Application of Linear Combinations to Color Models

Colors on computer monitors are commonly based on what is called the **RGB** *color model*. Colors in this system are created by adding together percentages of the primary colors red (R), green (G), and blue (B). One way to do this is to identify the primary colors with the vectors

$$\mathbf{r} = (1, 0, 0) \quad \text{(pure red)},$$
$$\mathbf{g} = (0, 1, 0) \quad \text{(pure green)},$$
$$\mathbf{b} = (0, 0, 1) \quad \text{(pure blue)}$$

in R^3 and to create all other colors by forming linear combinations of $\mathbf{r}$, $\mathbf{g}$, and $\mathbf{b}$ using coefficients between 0 and 1, inclusive; these coefficients represent the percentage of each pure color in the mix.

The set of all such color vectors is called **RGB** *space* or the **RGB** *color cube* (Figure 3.1.14). Thus, each color vector $\mathbf{c}$ in this cube is expressible as a linear combination of the form

$$\mathbf{c} = k_1\mathbf{r} + k_2\mathbf{g} + k_3\mathbf{b}$$
$$= k_1(1, 0, 0) + k_2(0, 1, 0) + k_3(0, 0, 1)$$
$$= (k_1, k_2, k_3)$$

where $0 \leq k_i \leq 1$. As indicated in the figure, the corners of the cube represent the pure primary colors together with the colors black, white, magenta, cyan, and yellow. The vectors along the diagonal running from black to white correspond to shades of gray.

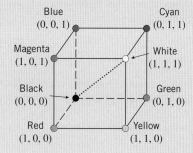

▶ Figure 3.1.14

Alternative Notations for Vectors

Up to now we have been writing vectors in R^n using the notation

$$\mathbf{v} = (v_1, v_2, \ldots, v_n) \tag{15}$$

We call this the **comma-delimited** form. However, since a vector in R^n is just a list of its n components in a specific order, any notation that displays those components in the correct order is a valid way of representing the vector. For example, the vector in (15) can be written as

$$\mathbf{v} = [v_1 \quad v_2 \quad \cdots \quad v_n] \tag{16}$$

which is called *row-matrix* form, or as

$$\mathbf{v} = \begin{bmatrix} v_1 \\ v_2 \\ \vdots \\ v_n \end{bmatrix} \tag{17}$$

which is called **column-matrix** form. The choice of notation is often a matter of taste or convenience, but sometimes the nature of a problem will suggest a preferred notation. Notations (15), (16), and (17) will all be used at various places in this text.

Concept Review

- Geometric vector
- Direction
- Length
- Initial point
- Terminal point
- Equivalent vectors
- Zero vector
- Vector addition: parallelogram rule and triangle rule
- Vector subtraction
- Negative of a vector
- Scalar multiplication
- Collinear (i.e., parallel) vectors
- Components of a vector
- Coordinates of a point
- n-tuple
- n-space
- Vector operations in n-space: addition, subtraction, scalar multiplication
- Linear combination of vectors

Skills

- Perform geometric operations on vectors: addition, subtraction, and scalar multiplication.
- Perform algebraic operations on vectors: addition, subtraction, and scalar multiplication.
- Determine whether two vectors are equivalent.
- Determine whether two vectors are collinear.
- Sketch vectors whose initial and terminal points are given.
- Find components of a vector whose initial and terminal points are given.
- Prove basic algebraic properties of vectors (Theorems 3.1.1 and 3.1.2).

Exercise Set 3.1

▶ In Exercises 1–2, draw a coordinate system (as in Figure 3.1.10) and locate the points whose coordinates are given. ◀

1. (a) $(3, 4, 5)$ (b) $(-3, 4, 5)$ (c) $(3, -4, 5)$
   (d) $(3, 4, -5)$ (e) $(-3, -4, 5)$ (f) $(-3, 4, -5)$

2. (a) $(0, 3, -3)$ (b) $(3, -3, 0)$ (c) $(-3, 0, 0)$
   (d) $(3, 0, 3)$ (e) $(0, 0, -3)$ (f) $(0, 3, 0)$

▶ In Exercises 3–4, sketch the following vectors with the initial points located at the origin. ◀

3. (a) $\mathbf{v}_1 = (2, 5)$ (b) $\mathbf{v}_2 = (-3, 2)$
   (c) $\mathbf{v}_3 = (-4, -3)$ (d) $\mathbf{v}_4 = (3, 4, 5)$
   (e) $\mathbf{v}_5 = (3, 3, 0)$ (f) $\mathbf{v}_6 = (-1, 0, 2)$

4. (a) $\mathbf{v}_1 = (5, -4)$ (b) $\mathbf{v}_2 = (3, 0)$
   (c) $\mathbf{v}_3 = (0, -7)$ (d) $\mathbf{v}_4 = (0, 0, -3)$
   (e) $\mathbf{v}_5 = (0, 4, -1)$ (f) $\mathbf{v}_6 = (2, 2, 2)$

▶ In Exercises 5–6, sketch the following vectors with the initial points located at the origin. ◀

5. (a) $P_1(-3, 5)$, $P_2(2, 3)$
   (b) $P_1(4, 1)$, $P_2(6, 4)$
   (c) $P_1(3, -7, 2)$, $P_2(-2, 5, -4)$

6. (a) $P_1(-5, 0)$, $P_2(-3, 1)$
   (b) $P_1(0, 0)$, $P_2(3, 4)$
   (c) $P_1(-1, 0, 2)$, $P_2(0, -1, 0)$
   (d) $P_1(2, 2, 2)$, $P_2(0, 0, 0)$

▶ In Exercises 7–8, find the components of the vector $\overrightarrow{P_1 P_2}$. ◀

7. (a) $P_1(3, 5)$, $P_2(2, 8)$
   (b) $P_1(5, -2, 1)$, $P_2(2, 4, 2)$

8. (a) $P_1(-6, 2)$, $P_2(-4, -1)$
   (b) $P_1(0, 0, 0)$, $P_2(-1, 6, 1)$

9. (a) Find the terminal point of the vector that is equivalent to $\mathbf{u} = (5, 2)$ and whose initial point is $A(3, 2)$.
   (b) Find the terminal point of the vector that is equivalent to $\mathbf{u} = (1, 2, 2)$ and whose initial point is $B(3, -1, 0)$.

10. (a) Find the initial point of the vector that is equivalent to $\mathbf{u} = (1, 2)$ and whose terminal point is $B(2, 0)$.
    (b) Find the terminal point of the vector that is equivalent to $\mathbf{u} = (1, 1, 3)$ and whose initial point is $A(0, 2, 0)$.

11. Find a nonzero vector $\mathbf{u}$ with terminal point $Q(3, 0, -5)$ such that
    (a) $\mathbf{u}$ has the same direction as $\mathbf{v} = (4, -2, -1)$.
    (b) $\mathbf{u}$ is oppositely directed to $\mathbf{v} = (4, -2, -1)$.

12. Find a nonzero vector $\mathbf{u}$ with initial point $P(-1, 3, -5)$ such that
    (a) $\mathbf{u}$ has the same direction as $\mathbf{v} = (6, 7, -3)$.
    (b) $\mathbf{u}$ is oppositely directed to $\mathbf{v} = (6, 7, -3)$.

13. Let $\mathbf{u} = (3, -2)$, $\mathbf{v} = (1, 0)$, and $\mathbf{w} = (-2, 4)$. Find the components of
    (a) $\mathbf{u} + \mathbf{w}$
    (b) $\mathbf{v} - 3\mathbf{u}$
    (c) $2(\mathbf{u} - 5\mathbf{w})$
    (d) $3\mathbf{v} - 2(\mathbf{u} + 2\mathbf{w})$
    (e) $-3(\mathbf{w} - 2\mathbf{u} + \mathbf{v})$
    (f) $(-2\mathbf{u} - \mathbf{v}) - 5(\mathbf{v} + 3\mathbf{w})$

14. Let $\mathbf{u} = (-3, 1, 2)$, $\mathbf{v} = (4, 0, -8)$, and $\mathbf{w} = (6, -1, -4)$. Find the components of
    (a) $\mathbf{v} - \mathbf{w}$
    (b) $6\mathbf{u} + 2\mathbf{v}$
    (c) $-\mathbf{v} + \mathbf{u}$
    (d) $5(\mathbf{v} - 4\mathbf{u})$
    (e) $-3(\mathbf{v} - 8\mathbf{w})$
    (f) $(2\mathbf{u} - 7\mathbf{w}) - (8\mathbf{v} + \mathbf{u})$

15. Let $\mathbf{u} = (-3, 2, 1, 0)$, $\mathbf{v} = (4, 7, -3, 2)$, and $\mathbf{w} = (5, -2, 8, 1)$. Find the components of
    (a) $\mathbf{v} - \mathbf{w}$
    (b) $2\mathbf{u} + 7\mathbf{v}$
    (c) $-\mathbf{u} + (\mathbf{v} - 4\mathbf{w})$
    (d) $6(\mathbf{u} - 3\mathbf{v})$
    (e) $-\mathbf{v} - \mathbf{w}$
    (f) $(6\mathbf{v} - \mathbf{w}) - (4\mathbf{u} + \mathbf{v})$

16. Let $\mathbf{u}$, $\mathbf{v}$, and $\mathbf{w}$ be the vectors in Exercise 15. Find the vector $\mathbf{x}$ that satisfies $5\mathbf{x} - 2\mathbf{v} = 2(\mathbf{w} - 5\mathbf{x})$.

17. Let $\mathbf{u} = (5, -1, 0, 3, -3)$, $\mathbf{v} = (-1, -1, 7, 2, 0)$, and $\mathbf{w} = (-4, 2, -3, -5, 2)$. Find the components of
    (a) $\mathbf{u} - \mathbf{v}$
    (b) $3\mathbf{v} - 2\mathbf{w}$
    (c) $5(\mathbf{u} + 2\mathbf{w}) - 3\mathbf{v}$
    (d) $5(-\mathbf{v} + 4\mathbf{u} - \mathbf{w})$
    (e) $-2(3\mathbf{w} + \mathbf{v}) + (2\mathbf{u} + \mathbf{w})$
    (f) $\frac{1}{2}(\mathbf{w} - 5\mathbf{v} + 2\mathbf{u}) + \mathbf{v}$

18. Let $\mathbf{u} = (1, 2, -3, 5, 0)$, $\mathbf{v} = (0, 4, -1, 1, 2)$, and $\mathbf{w} = (7, 1, -4, -2, 3)$. Find the components of
    (a) $\mathbf{v} + \mathbf{w}$
    (b) $3(2\mathbf{u} - \mathbf{v})$
    (c) $(3\mathbf{u} - \mathbf{v}) - (2\mathbf{u} + 4\mathbf{w})$

19. Let $\mathbf{u} = (-3, 1, 2, 4, 4)$, $\mathbf{v} = (4, 0, -8, 1, 2)$, and $\mathbf{w} = (6, -1, -4, 3, -5)$. Find the components of
    (a) $\mathbf{u} - \mathbf{v}$
    (b) $2\mathbf{v} + 3\mathbf{w}$
    (c) $(3\mathbf{u} + 4\mathbf{v}) - (7\mathbf{w} + 3\mathbf{u})$

20. Let $\mathbf{u}$, $\mathbf{v}$, and $\mathbf{w}$ be the vectors in Exercise 18. Find the components of the vector $\mathbf{x}$ that satisfies the equation $3\mathbf{u} + \mathbf{v} - 2\mathbf{w} = 3\mathbf{x} + 2\mathbf{w}$.

21. Let $\mathbf{u}$, $\mathbf{v}$, and $\mathbf{w}$ be the vectors in Exercise 19. Find the components of the vector $\mathbf{x}$ that satisfies the equation $2\mathbf{u} + \mathbf{v} + \mathbf{x} = 6\mathbf{x} + \mathbf{w}$.

22. For what value(s) of t, if any, is the given vector parallel to $\mathbf{u} = (4, -1)$?
    (a) $(8t, -2)$
    (b) $(8t, 2t)$
    (c) $(1, t^2)$

23. Which of the following vectors in R^6 are parallel to $\mathbf{u} = (-2, 1, 0, 3, 5, 1)$?
    (a) $(4, 2, 0, 6, 10, 2)$
    (b) $(4, -2, 0, -6, -10, -2)$
    (c) $(0, 0, 0, 0, 0, 0)$

24. Let $\mathbf{u} = (2, 1, 0, 1, -1)$ and $\mathbf{v} = (-2, 3, 1, 0, 2)$. Find scalars a and b so that $a\mathbf{u} + b\mathbf{v} = (-8, 8, 3, -1, 7)$.

25. Let $\mathbf{u} = (3, 1, -1, 5)$ and $\mathbf{v} = (0, 2, 1, -3)$. Find scalars a and b so that $a\mathbf{u} + b\mathbf{v} = (3, -3, -3, 11)$.

26. Find all scalars c_1, c_2, and c_3 such that
$$c_1(1, 2, 0) + c_2(2, 1, 1) + c_3(0, 3, 1) = (0, 0, 0)$$

27. Find all scalars c_1, c_2, and c_3 such that
$$c_1(1, -1, 0) + c_2(3, 2, 1) + c_3(0, 1, 4) = (-1, 1, 19)$$

28. Find all scalars c_1, c_2, and c_3 such that
$$c_1(-1, 0, 2) + c_2(2, 2, -2) + c_3(1, -2, 1) = (-6, 12, 4)$$

130 Chapter 3 Euclidean Vector Spaces

29. Let $\mathbf{u}_1 = (4, 7, -3, 1)$, $\mathbf{u}_2 = (0, -1, 1, 0)$, $\mathbf{u}_3 = (-1, 2, 1, 0)$, and $\mathbf{u}_4 = (0, 1, 0, 1)$. Find scalars c_1, c_2, c_3, and c_4 such that $c_1\mathbf{u}_1 + c_2\mathbf{u}_2 + c_3\mathbf{u}_3 + c_4\mathbf{u}_4 = (6, 14, -5, 5)$.

30. Show that there do not exist scalars c_1, c_2, and c_3 such that
$$c_1(1, 0, 1, 0) + c_2(1, 0, -2, 1) + c_3(2, 0, 1, 2) = (1, -2, 2, 3)$$

31. Show that there do not exist scalars c_1, c_2, and c_3 such that
$$c_1(-2, 9, 6) + c_2(-3, 2, 1) + c_3(1, 7, 5) = (0, 5, 4)$$

32. Consider Figure 3.1.12. Discuss a geometric interpretation of the vector
$$\mathbf{u} = \overrightarrow{OP_1} + \tfrac{1}{2}(\overrightarrow{OP_2} - \overrightarrow{OP_1})$$

33. Let P be the point $(2, 3, -2)$ and Q the point $(7, -4, 1)$.
    (a) Find the midpoint of the line segment connecting P and Q.
    (b) Find the point on the line segment connecting P and Q that is $\tfrac{3}{4}$ of the way from P to Q.

34. Let P be the point $(1, 3, 7)$. If the point $(4, 0, -6)$ is the midpoint of the line segment connecting P and Q, what is Q?

35. Prove parts (a), (c), and (d) of Theorem 3.1.1.

36. Prove parts (e)–(h) of Theorem 3.1.1.

37. Prove parts (a)–(c) of Theorem 3.1.2.

True-False Exercises

In parts (a)–(k) determine whether the statement is true or false, and justify your answer.

(a) Two equivalent vectors must have the same initial point.

(b) The vectors (a, b) and $(a, b, 0)$ are equivalent.

(c) If k is a scalar and $\mathbf{v}$ is a vector, then $\mathbf{v}$ and $k\mathbf{v}$ are parallel if and only if $k \geq 0$.

(d) The vectors $\mathbf{v} + (\mathbf{u} + \mathbf{w})$ and $(\mathbf{w} + \mathbf{v}) + \mathbf{u}$ are the same.

(e) If $\mathbf{u} + \mathbf{v} = \mathbf{u} + \mathbf{w}$, then $\mathbf{v} = \mathbf{w}$.

(f) If a and b are scalars such that $a\mathbf{u} + b\mathbf{v} = \mathbf{0}$, then $\mathbf{u}$ and $\mathbf{v}$ are parallel vectors.

(g) Collinear vectors with the same length are equal.

(h) If $(a, b, c) + (x, y, z) = (x, y, z)$, then (a, b, c) must be the zero vector.

(i) If k and m are scalars and $\mathbf{u}$ and $\mathbf{v}$ are vectors, then
$$(k + m)(\mathbf{u} + \mathbf{v}) = k\mathbf{u} + m\mathbf{v}$$

(j) If the vectors $\mathbf{v}$ and $\mathbf{w}$ are given, then the vector equation
$$3(2\mathbf{v} - \mathbf{x}) = 5\mathbf{x} - 4\mathbf{w} + \mathbf{v}$$
can be solved for $\mathbf{x}$.

(k) The linear combinations $a_1\mathbf{v}_1 + a_2\mathbf{v}_2$ and $b_1\mathbf{v}_1 + b_2\mathbf{v}_2$ can only be equal if $a_1 = b_1$ and $a_2 = b_2$.

3.2 Norm, Dot Product, and Distance in R^n

In this section we will be concerned with the notions of length and distance as they relate to vectors. We will first discuss these ideas in R^2 and R^3 and then extend them algebraically to R^n.

Norm of a Vector   In this text we will denote the length of a vector $\mathbf{v}$ by the symbol $\|\mathbf{v}\|$, which is read as the *norm* of $\mathbf{v}$, the *length* of $\mathbf{v}$, or the *magnitude* of $\mathbf{v}$ (the term "norm" being a common mathematical synonym for length). As suggested in Figure 3.2.1a, it follows from the Theorem of Pythagoras that the norm of a vector (v_1, v_2) in R^2 is

$$\|\mathbf{v}\| = \sqrt{v_1^2 + v_2^2} \tag{1}$$

Similarly, for a vector (v_1, v_2, v_3) in R^3, it follows from Figure 3.2.1b and two applications of the Theorem of Pythagoras that

$$\|\mathbf{v}\|^2 = (OR)^2 + (RP)^2 = (OQ)^2 + (QR)^2 + (RP)^2 = v_1^2 + v_2^2 + v_3^2$$

and hence that

$$\|\mathbf{v}\| = \sqrt{v_1^2 + v_2^2 + v_3^2} \tag{2}$$

Motivated by the pattern of Formulas (1) and (2) we make the following definition.

3.2 Norm, Dot Product, and Distance in R^n

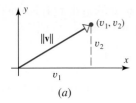

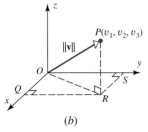

▲ Figure 3.2.1

DEFINITION 1 If $\mathbf{v} = (v_1, v_2, \ldots, v_n)$ is a vector in R^n, then the **norm** of $\mathbf{v}$ (also called the **length** of $\mathbf{v}$ or the **magnitude** of $\mathbf{v}$) is denoted by $\|\mathbf{v}\|$, and is defined by the formula

$$\|\mathbf{v}\| = \sqrt{v_1^2 + v_2^2 + v_3^2 + \cdots + v_n^2} \tag{3}$$

▶ **EXAMPLE 1 Calculating Norms**

It follows from Formula (2) that the norm of the vector $\mathbf{v} = (-3, 2, 1)$ in R^3 is

$$\|\mathbf{v}\| = \sqrt{(-3)^2 + 2^2 + 1^2} = \sqrt{14}$$

and it follows from Formula (3) that the norm of the vector $\mathbf{v} = (2, -1, 3, -5)$ in R^4 is

$$\|\mathbf{v}\| = \sqrt{2^2 + (-1)^2 + 3^2 + (-5)^2} = \sqrt{39} \; ◀$$

Our first theorem in this section will generalize to R^n the following three familiar facts about vectors in R^2 and R^3:

- Distances are nonnegative.
- The zero vector is the only vector of length zero.
- Multiplying a vector by a scalar multiplies its length by the absolute value of that scalar.

It is important to recognize that just because these results hold in R^2 and R^3 does not guarantee that they hold in R^n—their validity in R^n must be *proved* using algebraic properties of n-tuples.

THEOREM 3.2.1 *If $\mathbf{v}$ is a vector in R^n, and if k is any scalar, then:*
(a) $\|\mathbf{v}\| \geq 0$
(b) $\|\mathbf{v}\| = 0$ *if and only if* $\mathbf{v} = \mathbf{0}$
(c) $\|k\mathbf{v}\| = |k|\|\mathbf{v}\|$

We will prove part (c) and leave (a) and (b) as exercises.

Proof (c) If $\mathbf{v} = (v_1, v_2, \ldots, v_n)$, then $k\mathbf{v} = (kv_1, kv_2, \ldots, kv_n)$, so

$$\|k\mathbf{v}\| = \sqrt{(kv_1)^2 + (kv_2)^2 + \cdots + (kv_n)^2}$$
$$= \sqrt{(k^2)(v_1^2 + v_2^2 + \cdots + v_n^2)}$$
$$= |k|\sqrt{v_1^2 + v_2^2 + \cdots + v_n^2}$$
$$= |k|\|\mathbf{v}\| \; ◀$$

Unit Vectors A vector of norm 1 is called a **unit vector**. Such vectors are useful for specifying a direction when length is not relevant to the problem at hand. You can obtain a unit vector in a desired direction by choosing any *nonzero* vector $\mathbf{v}$ in that direction and multiplying $\mathbf{v}$ by the reciprocal of its length. For example, if $\mathbf{v}$ is a vector of length 2 in R^2 or R^3, then $\frac{1}{2}\mathbf{v}$ is a unit vector in the same direction as $\mathbf{v}$. More generally, if $\mathbf{v}$ is any nonzero vector in R^n, then

$$\mathbf{u} = \frac{1}{\|\mathbf{v}\|}\mathbf{v} \tag{4}$$

WARNING Sometimes you will see Formula (4) expressed as
$$\mathbf{u} = \frac{\mathbf{v}}{\|\mathbf{v}\|}$$
This is just a more compact way of writing that formula and is *not* intended to convey that **v** is being divided by $\|\mathbf{v}\|$.

defines a unit vector that is in the same direction as **v**. We can confirm that (4) is a unit vector by applying part (c) of Theorem 3.2.1 with $k = 1/\|\mathbf{v}\|$ to obtain

$$\|\mathbf{u}\| = \|k\mathbf{v}\| = |k|\|\mathbf{v}\| = k\|\mathbf{v}\| = \frac{1}{\|\mathbf{v}\|}\|\mathbf{v}\| = 1$$

The process of multiplying a nonzero vector by the reciprocal of its length to obtain a unit vector is called ***normalizing* v**.

▶ **EXAMPLE 2** **Normalizing a Vector**

Find the unit vector **u** that has the same direction as $\mathbf{v} = (2, 2, -1)$.

Solution The vector **v** has length

$$\|\mathbf{v}\| = \sqrt{2^2 + 2^2 + (-1)^2} = 3$$

Thus, from (4)

$$\mathbf{u} = \tfrac{1}{3}(2, 2, -1) = \left(\tfrac{2}{3}, \tfrac{2}{3}, -\tfrac{1}{3}\right)$$

As a check, you may want to confirm that $\|\mathbf{u}\| = 1$. ◀

The Standard Unit Vectors

When a rectangular coordinate system is introduced in R^2 or R^3, the unit vectors in the positive directions of the coordinate axes are called the ***standard unit vectors***. In R^2 these vectors are denoted by

$$\mathbf{i} = (1, 0) \quad \text{and} \quad \mathbf{j} = (0, 1)$$

and in R^3 by

$$\mathbf{i} = (1, 0, 0), \quad \mathbf{j} = (0, 1, 0), \quad \text{and} \quad \mathbf{k} = (0, 0, 1)$$

(Figure 3.2.2). Every vector $\mathbf{v} = (v_1, v_2)$ in R^2 and every vector $\mathbf{v} = (v_1, v_2, v_3)$ in R^3 can be expressed as a linear combination of standard unit vectors by writing

$$\mathbf{v} = (v_1, v_2) = v_1(1, 0) + v_2(0, 1) = v_1\mathbf{i} + v_2\mathbf{j} \tag{5}$$

$$\mathbf{v} = (v_1, v_2, v_3) = v_1(1, 0, 0) + v_2(0, 1, 0) + v_3(0, 0, 1) = v_1\mathbf{i} + v_2\mathbf{j} + v_3\mathbf{k} \tag{6}$$

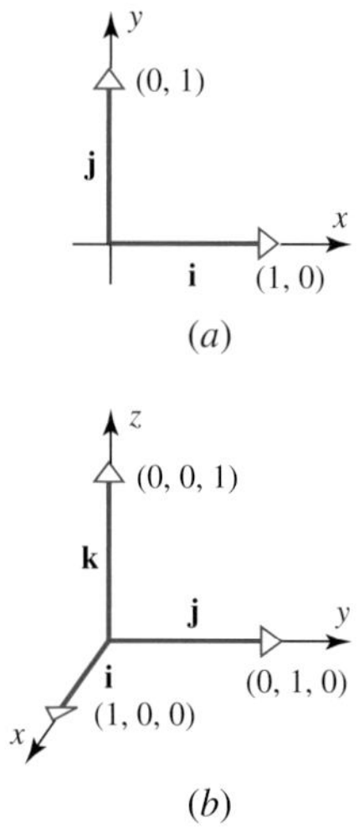

Moreover, we can generalize these formulas to R^n by defining the ***standard unit vectors in R^n*** to be

$$\mathbf{e}_1 = (1, 0, 0, \ldots, 0), \quad \mathbf{e}_2 = (0, 1, 0, \ldots, 0), \ldots, \quad \mathbf{e}_n = (0, 0, 0, \ldots, 1) \tag{7}$$

in which case every vector $\mathbf{v} = (v_1, v_2, \ldots, v_n)$ in R^n can be expressed as

$$\mathbf{v} = (v_1, v_2, \ldots, v_n) = v_1\mathbf{e}_1 + v_2\mathbf{e}_2 + \cdots + v_n\mathbf{e}_n \tag{8}$$

▲ Figure 3.2.2

▶ **EXAMPLE 3** **Linear Combinations of Standard Unit Vectors**

$$(2, -3, 4) = 2\mathbf{i} - 3\mathbf{j} + 4\mathbf{k}$$
$$(7, 3, -4, 5) = 7\mathbf{e}_1 + 3\mathbf{e}_2 - 4\mathbf{e}_3 + 5\mathbf{e}_4 \quad ◀$$

Distance in R^n

If P_1 and P_2 are points in R^2 or R^3, then the length of the vector $\overrightarrow{P_1P_2}$ is equal to the distance d between the two points (Figure 3.2.3). Specifically, if $P_1(x_1, y_1)$ and $P_2(x_2, y_2)$ are points in R^2, then Formula (4) of Section 3.1 implies that

$$d = \|\overrightarrow{P_1P_2}\| = \sqrt{(x_2 - x_1)^2 + (y_2 - y_1)^2} \tag{9}$$

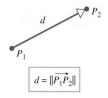

▲ Figure 3.2.3

We noted in the previous section that n-tuples can be viewed either as vectors or points in R^n. In Definition 2 we chose to describe them as points, as that seemed the more natural interpretation.

This is the familiar distance formula from analytic geometry. Similarly, the distance between the points $P_1(x_1, y_1, z_1)$ and $P_2(x_2, y_2, z_2)$ in 3-space is

$$d(\mathbf{u}, \mathbf{v}) = \|\overrightarrow{P_1 P_2}\| = \sqrt{(x_2 - x_1)^2 + (y_2 - y_1)^2 + (z_2 - z_1)^2} \quad (10)$$

Motivated by Formulas (9) and (10), we make the following definition.

DEFINITION 2 If $\mathbf{u} = (u_1, u_2, \ldots, u_n)$ and $\mathbf{v} = (v_1, v_2, \ldots, v_n)$ are points in R^n, then we denote the ***distance*** between $\mathbf{u}$ and $\mathbf{v}$ by $d(\mathbf{u}, \mathbf{v})$ and define it to be

$$d(\mathbf{u}, \mathbf{v}) = \|\mathbf{u} - \mathbf{v}\| = \sqrt{(u_1 - v_1)^2 + (u_2 - v_2)^2 + \cdots + (u_n - v_n)^2} \quad (11)$$

▶ **EXAMPLE 4** Calculating Distance in R^n

If
$$\mathbf{u} = (1, 3, -2, 7) \quad \text{and} \quad \mathbf{v} = (0, 7, 2, 2)$$

then the distance between $\mathbf{u}$ and $\mathbf{v}$ is

$$d(\mathbf{u}, \mathbf{v}) = \sqrt{(1-0)^2 + (3-7)^2 + (-2-2)^2 + (7-2)^2} = \sqrt{58} \quad \blacktriangleleft$$

Dot Product

Our next objective is to define a useful multiplication operation on vectors in R^2 and R^3 and then extend that operation to R^n. To do this we will first need to define exactly what we mean by the "angle" between two vectors in R^2 or R^3. For this purpose, let $\mathbf{u}$ and $\mathbf{v}$ be nonzero vectors in R^2 or R^3 that have been positioned so that their initial points coincide. We define the ***angle between*** $\mathbf{u}$ ***and*** $\mathbf{v}$ to be the angle θ determined by $\mathbf{u}$ and $\mathbf{v}$ that satisfies the inequalities $0 \leq \theta \leq \pi$ (Figure 3.2.4).

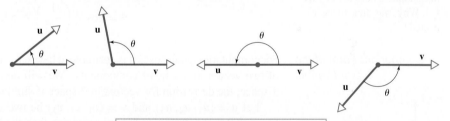

▶ Figure 3.2.4

The angle θ between $\mathbf{u}$ and $\mathbf{v}$ satisfies $0 \leq \theta \leq \pi$.

DEFINITION 3 If $\mathbf{u}$ and $\mathbf{v}$ are nonzero vectors in R^2 or R^3, and if θ is the angle between $\mathbf{u}$ and $\mathbf{v}$, then the ***dot product*** (also called the ***Euclidean inner product***) of $\mathbf{u}$ and $\mathbf{v}$ is denoted by $\mathbf{u} \cdot \mathbf{v}$ and is defined as

$$\mathbf{u} \cdot \mathbf{v} = \|\mathbf{u}\| \|\mathbf{v}\| \cos \theta \quad (12)$$

If $\mathbf{u} = \mathbf{0}$ or $\mathbf{v} = \mathbf{0}$, then we define $\mathbf{u} \cdot \mathbf{v}$ to be 0.

The sign of the dot product reveals information about the angle θ that we can obtain by rewriting Formula (12) as

$$\cos \theta = \frac{\mathbf{u} \cdot \mathbf{v}}{\|\mathbf{u}\| \|\mathbf{v}\|} \quad (13)$$

Since $0 \leq \theta \leq \pi$, it follows from Formula (13) and properties of the cosine function studied in trigonometry that

- θ is acute if $\mathbf{u} \cdot \mathbf{v} > 0$.
- θ is obtuse if $\mathbf{u} \cdot \mathbf{v} < 0$.
- $\theta = \pi/2$ if $\mathbf{u} \cdot \mathbf{v} = 0$.

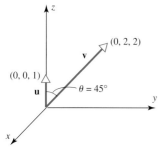

▲ Figure 3.2.5

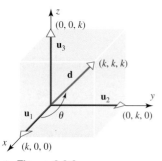

▲ Figure 3.2.6

Note that the angle θ obtained in Example 6 does not involve k. Why was this to be expected?

Component Form of the Dot Product

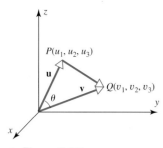

▲ Figure 3.2.7

▶ **EXAMPLE 5** **Dot Product**

Find the dot product of the vectors shown in Figure 3.2.5.

Solution The lengths of the vectors are

$$\|\mathbf{u}\| = 1 \quad \text{and} \quad \|\mathbf{v}\| = \sqrt{8} = 2\sqrt{2}$$

and the cosine of the angle θ between them is

$$\cos(45°) = 1/\sqrt{2}$$

Thus, it follows from Formula (12) that

$$\mathbf{u} \cdot \mathbf{v} = \|\mathbf{u}\|\|\mathbf{v}\| \cos \theta = (1)(2\sqrt{2})(1/\sqrt{2}) = 2$$

▶ **EXAMPLE 6** **A Geometry Problem Solved Using Dot Product**

Find the angle between a diagonal of a cube and one of its edges.

Solution Let k be the length of an edge and introduce a coordinate system as shown in Figure 3.2.6. If we let $\mathbf{u}_1 = (k, 0, 0)$, $\mathbf{u}_2 = (0, k, 0)$, and $\mathbf{u}_3 = (0, 0, k)$, then the vector

$$\mathbf{d} = (k, k, k) = \mathbf{u}_1 + \mathbf{u}_2 + \mathbf{u}_3$$

is a diagonal of the cube. It follows from Formula (13) that the angle θ between $\mathbf{d}$ and the edge $\mathbf{u}_1$ satisfies

$$\cos \theta = \frac{\mathbf{u}_1 \cdot \mathbf{d}}{\|\mathbf{u}_1\|\|\mathbf{d}\|} = \frac{k^2}{(k)(\sqrt{3k^2})} = \frac{1}{\sqrt{3}}$$

With the help of a calculator we obtain

$$\theta = \cos^{-1}\left(\frac{1}{\sqrt{3}}\right) \approx 54.74° \quad ◀$$

For computational purposes it is desirable to have a formula that expresses the dot product of two vectors in terms of components. We will derive such a formula for vectors in 3-space; the derivation for vectors in 2-space is similar.

Let $\mathbf{u} = (u_1, u_2, u_3)$ and $\mathbf{v} = (v_1, v_2, v_3)$ be two nonzero vectors. If, as shown in Figure 3.2.7, θ is the angle between $\mathbf{u}$ and $\mathbf{v}$, then the law of cosines yields

$$\|\overrightarrow{PQ}\|^2 = \|\mathbf{u}\|^2 + \|\mathbf{v}\|^2 - 2\|\mathbf{u}\|\|\mathbf{v}\| \cos \theta \tag{14}$$

Josiah Willard Gibbs
(1839–1903)

Historical Note The dot product notation was first introduced by the American physicist and mathematician J. Willard Gibbs in a pamphlet distributed to his students at Yale University in the 1880s. The product was originally written on the baseline, rather than centered as today, and was referred to as the *direct product*. Gibbs's pamphlet was eventually incorporated into a book entitled *Vector Analysis* that was published in 1901 and coauthored with one of his students. Gibbs made major contributions to the fields of thermodynamics and electromagnetic theory and is generally regarded as the greatest American physicist of the nineteenth century.

[Image: The Granger Collection, New York]

Since $\overrightarrow{PQ} = \mathbf{v} - \mathbf{u}$, we can rewrite (14) as

$$\|\mathbf{u}\|\|\mathbf{v}\|\cos\theta = \tfrac{1}{2}(\|\mathbf{u}\|^2 + \|\mathbf{v}\|^2 - \|\mathbf{v} - \mathbf{u}\|^2)$$

or

$$\mathbf{u} \cdot \mathbf{v} = \tfrac{1}{2}(\|\mathbf{u}\|^2 + \|\mathbf{v}\|^2 - \|\mathbf{v} - \mathbf{u}\|^2)$$

Substituting

$$\|\mathbf{u}\|^2 = u_1^2 + u_2^2 + u_3^2, \qquad \|\mathbf{v}\|^2 = v_1^2 + v_2^2 + v_3^2$$

and

$$\|\mathbf{v} - \mathbf{u}\|^2 = (v_1 - u_1)^2 + (v_2 - u_2)^2 + (v_3 - u_3)^2$$

we obtain, after simplifying,

$$\mathbf{u} \cdot \mathbf{v} = u_1 v_1 + u_2 v_2 + u_3 v_3 \tag{15}$$

The companion formula for vectors in 2-space is

$$\mathbf{u} \cdot \mathbf{v} = u_1 v_1 + u_2 v_2 \tag{16}$$

Motivated by the pattern in Formulas (15) and (16), we make the following definition.

> Although we derived Formula (15) and its 2-space companion under the assumption that **u** and **v** are nonzero, it turned out that these formulas are also applicable if $\mathbf{u} = \mathbf{0}$ or $\mathbf{v} = \mathbf{0}$ (verify).

> In words, *to calculate the dot product (Euclidean inner product) multiply corresponding components and add the resulting products.*

DEFINITION 4 If $\mathbf{u} = (u_1, u_2, \ldots, u_n)$ and $\mathbf{v} = (v_1, v_2, \ldots, v_n)$ are vectors in R^n, then the ***dot product*** (also called the ***Euclidean inner product***) of **u** and **v** is denoted by $\mathbf{u} \cdot \mathbf{v}$ and is defined by

$$\mathbf{u} \cdot \mathbf{v} = u_1 v_1 + u_2 v_2 + \cdots + u_n v_n \tag{17}$$

▶ **EXAMPLE 7** **Calculating Dot Products Using Components**

(a) Use Formula (15) to compute the dot product of the vectors **u** and **v** in Example 5.

(b) Calculate $\mathbf{u} \cdot \mathbf{v}$ for the following vectors in R^4:

$$\mathbf{u} = (-1, 3, 5, 7), \quad \mathbf{v} = (-3, -4, 1, 0)$$

Solution (a) The component forms of the vectors are $\mathbf{u} = (0, 0, 1)$ and $\mathbf{v} = (0, 2, 2)$. Thus,

$$\mathbf{u} \cdot \mathbf{v} = (0)(0) + (0)(2) + (1)(2) = 2$$

which agrees with the result obtained geometrically in Example 5.

Solution (b)

$$\mathbf{u} \cdot \mathbf{v} = (-1)(-3) + (3)(-4) + (5)(1) + (7)(0) = -4 \quad \blacktriangleleft$$

Algebraic Properties of the Dot Product

In the special case where $\mathbf{u} = \mathbf{v}$ in Definition 4, we obtain the relationship

$$\mathbf{v} \cdot \mathbf{v} = v_1^2 + v_2^2 + \cdots + v_n^2 = \|\mathbf{v}\|^2 \tag{18}$$

This yields the following formula for expressing the length of a vector in terms of a dot product:

$$\|\mathbf{v}\| = \sqrt{\mathbf{v} \cdot \mathbf{v}} \tag{19}$$

Dot products have many of the same algebraic properties as products of real numbers.

THEOREM 3.2.2 *If* **u**, **v**, *and* **w** *are vectors in* R^n, *and if* k *is a scalar, then*:
(a) $\mathbf{u} \cdot \mathbf{v} = \mathbf{v} \cdot \mathbf{u}$ [Symmetry property]
(b) $\mathbf{u} \cdot (\mathbf{v} + \mathbf{w}) = \mathbf{u} \cdot \mathbf{v} + \mathbf{u} \cdot \mathbf{w}$ [Distributive property]
(c) $k(\mathbf{u} \cdot \mathbf{v}) = (k\mathbf{u}) \cdot \mathbf{v}$ [Homogeneity property]
(d) $\mathbf{v} \cdot \mathbf{v} \geq 0$ *and* $\mathbf{v} \cdot \mathbf{v} = 0$ *if and only if* $\mathbf{v} = \mathbf{0}$ [Positivity property]

We will prove parts (c) and (d) and leave the other proofs as exercises.

Proof (c) Let $\mathbf{u} = (u_1, u_2, \ldots, u_n)$ and $\mathbf{v} = (v_1, v_2, \ldots, v_n)$. Then
$$k(\mathbf{u} \cdot \mathbf{v}) = k(u_1 v_1 + u_2 v_2 + \cdots + u_n v_n)$$
$$= (ku_1)v_1 + (ku_2)v_2 + \cdots + (ku_n)v_n = (k\mathbf{u}) \cdot \mathbf{v}$$

Proof (d) The result follows from parts (a) and (b) of Theorem 3.2.1 and the fact that
$$\mathbf{v} \cdot \mathbf{v} = v_1 v_1 + v_2 v_2 + \cdots + v_n v_n = v_1^2 + v_2^2 + \cdots + v_n^2 = \|\mathbf{v}\|^2 \blacktriangleleft$$

The next theorem gives additional properties of dot products. The proofs can be obtained either by expressing the vectors in terms of components or by using the algebraic properties established in Theorem 3.2.2.

THEOREM 3.2.3 *If* **u**, **v**, *and* **w** *are vectors in* R^n, *and if* k *is a scalar, then*:
(a) $\mathbf{0} \cdot \mathbf{v} = \mathbf{v} \cdot \mathbf{0} = 0$
(b) $(\mathbf{u} + \mathbf{v}) \cdot \mathbf{w} = \mathbf{u} \cdot \mathbf{w} + \mathbf{v} \cdot \mathbf{w}$
(c) $\mathbf{u} \cdot (\mathbf{v} - \mathbf{w}) = \mathbf{u} \cdot \mathbf{v} - \mathbf{u} \cdot \mathbf{w}$
(d) $(\mathbf{u} - \mathbf{v}) \cdot \mathbf{w} = \mathbf{u} \cdot \mathbf{w} - \mathbf{v} \cdot \mathbf{w}$
(e) $k(\mathbf{u} \cdot \mathbf{v}) = \mathbf{u} \cdot (k\mathbf{v})$

We will show how Theorem 3.2.2 can be used to prove part (b) without breaking the vectors into components. The other proofs are left as exercises.

Proof (b)
$$(\mathbf{u} + \mathbf{v}) \cdot \mathbf{w} = \mathbf{w} \cdot (\mathbf{u} + \mathbf{v}) \quad \text{[By symmetry]}$$
$$= \mathbf{w} \cdot \mathbf{u} + \mathbf{w} \cdot \mathbf{v} \quad \text{[By distributivity]}$$
$$= \mathbf{u} \cdot \mathbf{w} + \mathbf{v} \cdot \mathbf{w} \quad \text{[By symmetry]} \blacktriangleleft$$

Formulas (18) and (19) together with Theorems 3.2.2 and 3.2.3 make it possible to manipulate expressions involving dot products using familiar algebraic techniques.

▶ **EXAMPLE 8 Calculating with Dot Products**

$$(\mathbf{u} - 2\mathbf{v}) \cdot (3\mathbf{u} + 4\mathbf{v}) = \mathbf{u} \cdot (3\mathbf{u} + 4\mathbf{v}) - 2\mathbf{v} \cdot (3\mathbf{u} + 4\mathbf{v})$$
$$= 3(\mathbf{u} \cdot \mathbf{u}) + 4(\mathbf{u} \cdot \mathbf{v}) - 6(\mathbf{v} \cdot \mathbf{u}) - 8(\mathbf{v} \cdot \mathbf{v})$$
$$= 3\|\mathbf{u}\|^2 - 2(\mathbf{u} \cdot \mathbf{v}) - 8\|\mathbf{v}\|^2 \blacktriangleleft$$

Cauchy–Schwarz Inequality and Angles in R^n

Our next objective is to extend to R^n the notion of "angle" between nonzero vectors **u** and **v**. We will do this by starting with the formula

$$\theta = \cos^{-1}\left(\frac{\mathbf{u} \cdot \mathbf{v}}{\|\mathbf{u}\|\|\mathbf{v}\|}\right) \quad (20)$$

which we previously derived for nonzero vectors in R^2 and R^3. Since dot products and norms have been defined for vectors in R^n, it would seem that this formula has all the ingredients to serve as a *definition* of the angle θ between two vectors, **u** and **v**, in R^n. However, there is a fly in the ointment, the problem being that the inverse cosine in Formula (20) is not defined unless its argument satisfies the inequalities

$$-1 \leq \frac{\mathbf{u} \cdot \mathbf{v}}{\|\mathbf{u}\|\|\mathbf{v}\|} \leq 1 \quad (21)$$

Fortunately, these inequalities *do* hold for all nonzero vectors in R^n as a result of the following fundamental result known as the ***Cauchy–Schwarz inequality***.

THEOREM 3.2.4 *Cauchy–Schwarz Inequality*

If $\mathbf{u} = (u_1, u_2, \ldots, u_n)$ *and* $\mathbf{v} = (v_1, v_2, \ldots, v_n)$ *are vectors in* R^n, *then*

$$|\mathbf{u} \cdot \mathbf{v}| \leq \|\mathbf{u}\|\|\mathbf{v}\| \quad (22)$$

or in terms of components

$$|u_1 v_1 + u_2 v_2 + \cdots + u_n v_n| \leq (u_1^2 + u_2^2 + \cdots + u_n^2)^{1/2} (v_1^2 + v_2^2 + \cdots + v_n^2)^{1/2} \quad (23)$$

We will omit the proof of this theorem because later in the text we will prove a more general version of which this will be a special case. Our goal for now will be to use this theorem to prove that the inequalities in (21) hold for all nonzero vectors in R^n. Once that is done we will have established all the results required to use Formula (20) as our *definition* of the angle between nonzero vectors **u** and **v** in R^n.

To prove that the inequalities in (21) hold for all nonzero vectors in R^n, divide both sides of Formula (22) by the product $\|\mathbf{u}\|\|\mathbf{v}\|$ to obtain

$$\frac{|\mathbf{u} \cdot \mathbf{v}|}{\|\mathbf{u}\|\|\mathbf{v}\|} \leq 1 \quad \text{or equivalently} \quad \left|\frac{\mathbf{u} \cdot \mathbf{v}}{\|\mathbf{u}\|\|\mathbf{v}\|}\right| \leq 1$$

from which (21) follows.

Hermann Amandus
Schwarz
(1843–1921)

Viktor Yakovlevich
Bunyakovsky
(1804–1889)

Historical Note The Cauchy–Schwarz inequality is named in honor of the French mathematician Augustin Cauchy (see p. 109) and the German mathematician Hermann Schwarz. Variations of this inequality occur in many different settings and under various names. Depending on the context in which the inequality occurs, you may find it called Cauchy's inequality, the Schwarz inequality, or sometimes even the Bunyakovsky inequality, in recognition of the Russian mathematician who published his version of the inequality in 1859, about 25 years before Schwarz.
[*Images: wikipedia (Schwarz); wikipedia (Bunyakovsky)*]

138 Chapter 3 Euclidean Vector Spaces

Geometry in R^n

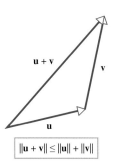

Figure 3.2.8

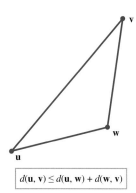

Figure 3.2.9

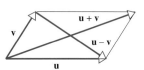

Figure 3.2.10

Earlier in this section we extended various concepts to R^n with the idea that familiar results that we can visualize in R^2 and R^3 might be valid in R^n as well. Here are two fundamental theorems from plane geometry whose validity extends to R^n:

- The sum of the lengths of two side of a triangle is at least as large as the third (Figure 3.2.8).
- The shortest distance between two points is a straight line (Figure 3.2.9).

The following theorem generalizes these theorems to R^n.

THEOREM 3.2.5 *If $\mathbf{u}$, $\mathbf{v}$, and $\mathbf{w}$ are vectors in R^n, and if k is any scalar, then:*

(a) $\|\mathbf{u} + \mathbf{v}\| \leq \|\mathbf{u}\| + \|\mathbf{v}\|$ [Triangle inequality for vectors]

(b) $d(\mathbf{u}, \mathbf{v}) \leq d(\mathbf{u}, \mathbf{w}) + d(\mathbf{w}, \mathbf{v})$ [Triangle inequality for distances]

Proof (a)

$$\begin{aligned}\|\mathbf{u} + \mathbf{v}\|^2 &= (\mathbf{u} + \mathbf{v}) \cdot (\mathbf{u} + \mathbf{v}) = (\mathbf{u} \cdot \mathbf{u}) + 2(\mathbf{u} \cdot \mathbf{v}) + (\mathbf{v} \cdot \mathbf{v}) \\ &= \|\mathbf{u}\|^2 + 2(\mathbf{u} \cdot \mathbf{v}) + \|\mathbf{v}\|^2 \\ &\leq \|\mathbf{u}\|^2 + 2|\mathbf{u} \cdot \mathbf{v}| + \|\mathbf{v}\|^2 \quad \longleftarrow \text{Property of absolute value} \\ &\leq \|\mathbf{u}\|^2 + 2\|\mathbf{u}\|\|\mathbf{v}\| + \|\mathbf{v}\|^2 \quad \longleftarrow \text{Cauchy–Schwarz inequality} \\ &= (\|\mathbf{u}\| + \|\mathbf{v}\|)^2\end{aligned}$$

Proof (b) It follows from part (a) and Formula (11) that

$$\begin{aligned}d(\mathbf{u}, \mathbf{v}) &= \|\mathbf{u} - \mathbf{v}\| = \|(\mathbf{u} - \mathbf{w}) + (\mathbf{w} - \mathbf{v})\| \\ &\leq \|\mathbf{u} - \mathbf{w}\| + \|\mathbf{w} - \mathbf{v}\| = d(\mathbf{u}, \mathbf{w}) + d(\mathbf{w}, \mathbf{v}) \quad \blacktriangleleft\end{aligned}$$

It is proved in plane geometry that for any parallelogram the sum of the squares of the diagonals is equal to the sum of the squares of the four sides (Figure 3.2.10). The following theorem generalizes that result to R^n.

THEOREM 3.2.6 *Parallelogram Equation for Vectors*

If $\mathbf{u}$ and $\mathbf{v}$ are vectors in R^n, then

$$\|\mathbf{u} + \mathbf{v}\|^2 + \|\mathbf{u} - \mathbf{v}\|^2 = 2\left(\|\mathbf{u}\|^2 + \|\mathbf{v}\|^2\right) \tag{24}$$

Proof

$$\begin{aligned}\|\mathbf{u} + \mathbf{v}\|^2 + \|\mathbf{u} - \mathbf{v}\|^2 &= (\mathbf{u} + \mathbf{v}) \cdot (\mathbf{u} + \mathbf{v}) + (\mathbf{u} - \mathbf{v}) \cdot (\mathbf{u} - \mathbf{v}) \\ &= 2(\mathbf{u} \cdot \mathbf{u}) + 2(\mathbf{v} \cdot \mathbf{v}) \\ &= 2\left(\|\mathbf{u}\|^2 + \|\mathbf{v}\|^2\right) \quad \blacktriangleleft\end{aligned}$$

We could state and prove many more theorems from plane geometry that generalize to R^n, but the ones already given should suffice to convince you that R^n is not so different from R^2 and R^3 even though we cannot visualize it directly. The next theorem establishes a fundamental relationship between the dot product and norm in R^n.

3.2 Norm, Dot Product, and Distance in R^n

THEOREM 3.2.7 *If $\mathbf{u}$ and $\mathbf{v}$ are vectors in R^n with the Euclidean inner product, then*

$$\mathbf{u} \cdot \mathbf{v} = \tfrac{1}{4}\|\mathbf{u} + \mathbf{v}\|^2 - \tfrac{1}{4}\|\mathbf{u} - \mathbf{v}\|^2 \qquad (25)$$

Note that Formula (25) expresses the dot product in terms of norms.

Proof
$$\|\mathbf{u} + \mathbf{v}\|^2 = (\mathbf{u} + \mathbf{v}) \cdot (\mathbf{u} + \mathbf{v}) = \|\mathbf{u}\|^2 + 2(\mathbf{u} \cdot \mathbf{v}) + \|\mathbf{v}\|^2$$
$$\|\mathbf{u} - \mathbf{v}\|^2 = (\mathbf{u} - \mathbf{v}) \cdot (\mathbf{u} - \mathbf{v}) = \|\mathbf{u}\|^2 - 2(\mathbf{u} \cdot \mathbf{v}) + \|\mathbf{v}\|^2$$

from which (25) follows by simple algebra. ◄

Dot Products as Matrix Multiplication

There are various ways to express the dot product of vectors using matrix notation. The formulas depend on whether the vectors are expressed as row matrices or column matrices. Here are the possibilities.

Table 1

Form	Dot Product	Example
$\mathbf{u}$ a column matrix and $\mathbf{v}$ a column matrix	$\mathbf{u} \cdot \mathbf{v} = \mathbf{u}^T\mathbf{v} = \mathbf{v}^T\mathbf{u}$	$\mathbf{u} = \begin{bmatrix} 1 \\ -3 \\ 5 \end{bmatrix}$ $\mathbf{v} = \begin{bmatrix} 5 \\ 4 \\ 0 \end{bmatrix}$ $\mathbf{u}^T\mathbf{v} = \begin{bmatrix} 1 & -3 & 5 \end{bmatrix}\begin{bmatrix} 5 \\ 4 \\ 0 \end{bmatrix} = -7$ $\mathbf{v}^T\mathbf{u} = \begin{bmatrix} 5 & 4 & 0 \end{bmatrix}\begin{bmatrix} 1 \\ -3 \\ 5 \end{bmatrix} = -7$
$\mathbf{u}$ a row matrix and $\mathbf{v}$ a column matrix	$\mathbf{u} \cdot \mathbf{v} = \mathbf{u}\mathbf{v} = \mathbf{v}^T\mathbf{u}^T$	$\mathbf{u} = \begin{bmatrix} 1 & -3 & 5 \end{bmatrix}$ $\mathbf{v} = \begin{bmatrix} 5 \\ 4 \\ 0 \end{bmatrix}$ $\mathbf{u}\mathbf{v} = \begin{bmatrix} 1 & -3 & 5 \end{bmatrix}\begin{bmatrix} 5 \\ 4 \\ 0 \end{bmatrix} = -7$ $\mathbf{v}^T\mathbf{u}^T = \begin{bmatrix} 5 & 4 & 0 \end{bmatrix}\begin{bmatrix} 1 \\ -3 \\ 5 \end{bmatrix} = -7$
$\mathbf{u}$ a column matrix and $\mathbf{v}$ a row matrix	$\mathbf{u} \cdot \mathbf{v} = \mathbf{v}\mathbf{u} = \mathbf{u}^T\mathbf{v}^T$	$\mathbf{u} = \begin{bmatrix} 1 \\ -3 \\ 5 \end{bmatrix}$ $\mathbf{v} = \begin{bmatrix} 5 & 4 & 0 \end{bmatrix}$ $\mathbf{v}\mathbf{u} = \begin{bmatrix} 5 & 4 & 0 \end{bmatrix}\begin{bmatrix} 1 \\ -3 \\ 5 \end{bmatrix} = -7$ $\mathbf{u}^T\mathbf{v}^T = \begin{bmatrix} 1 & -3 & 5 \end{bmatrix}\begin{bmatrix} 5 \\ 4 \\ 0 \end{bmatrix} = -7$
$\mathbf{u}$ a row matrix and $\mathbf{v}$ a row matrix	$\mathbf{u} \cdot \mathbf{v} = \mathbf{u}\mathbf{v}^T = \mathbf{v}\mathbf{u}^T$	$\mathbf{u} = \begin{bmatrix} 1 & -3 & 5 \end{bmatrix}$ $\mathbf{v} = \begin{bmatrix} 5 & 4 & 0 \end{bmatrix}$ $\mathbf{u}\mathbf{v}^T = \begin{bmatrix} 1 & -3 & 5 \end{bmatrix}\begin{bmatrix} 5 \\ 4 \\ 0 \end{bmatrix} = -7$ $\mathbf{v}\mathbf{u}^T = \begin{bmatrix} 5 & 4 & 0 \end{bmatrix}\begin{bmatrix} 1 \\ -3 \\ 5 \end{bmatrix} = -7$

If A is an $n \times n$ matrix and $\mathbf{u}$ and $\mathbf{v}$ are $n \times 1$ matrices, then it follows from the first row in Table 1 and properties of the transpose that

$$A\mathbf{u} \cdot \mathbf{v} = \mathbf{v}^T(A\mathbf{u}) = (\mathbf{v}^T A)\mathbf{u} = (A^T\mathbf{v})^T\mathbf{u} = \mathbf{u} \cdot A^T\mathbf{v}$$
$$\mathbf{u} \cdot A\mathbf{v} = (A\mathbf{v})^T\mathbf{u} = (\mathbf{v}^T A^T)\mathbf{u} = \mathbf{v}^T(A^T\mathbf{u}) = A^T\mathbf{u} \cdot \mathbf{v}$$

The resulting formulas

$$A\mathbf{u} \cdot \mathbf{v} = \mathbf{u} \cdot A^T\mathbf{v} \qquad (26)$$

$$\mathbf{u} \cdot A\mathbf{v} = A^T\mathbf{u} \cdot \mathbf{v} \qquad (27)$$

provide an important link between multiplication by an $n \times n$ matrix A and multiplication by A^T.

▶ **EXAMPLE 9** Verifying That $A\mathbf{u} \cdot \mathbf{v} = \mathbf{u} \cdot A^T\mathbf{v}$

Suppose that

$$A = \begin{bmatrix} 1 & -2 & 3 \\ 2 & 4 & 1 \\ -1 & 0 & 1 \end{bmatrix}, \quad \mathbf{u} = \begin{bmatrix} -1 \\ 2 \\ 4 \end{bmatrix}, \quad \mathbf{v} = \begin{bmatrix} -2 \\ 0 \\ 5 \end{bmatrix}$$

Then

$$A\mathbf{u} = \begin{bmatrix} 1 & -2 & 3 \\ 2 & 4 & 1 \\ -1 & 0 & 1 \end{bmatrix} \begin{bmatrix} -1 \\ 2 \\ 4 \end{bmatrix} = \begin{bmatrix} 7 \\ 10 \\ 5 \end{bmatrix}$$

$$A^T\mathbf{v} = \begin{bmatrix} 1 & 2 & -1 \\ -2 & 4 & 0 \\ 3 & 1 & 1 \end{bmatrix} \begin{bmatrix} -2 \\ 0 \\ 5 \end{bmatrix} = \begin{bmatrix} -7 \\ 4 \\ -1 \end{bmatrix}$$

from which we obtain

$$A\mathbf{u} \cdot \mathbf{v} = 7(-2) + 10(0) + 5(5) = 11$$
$$\mathbf{u} \cdot A^T\mathbf{v} = (-1)(-7) + 2(4) + 4(-1) = 11$$

Thus, $A\mathbf{u} \cdot \mathbf{v} = \mathbf{u} \cdot A^T\mathbf{v}$ as guaranteed by Formula (8). We leave it for you to verify that Formula (27) also holds. ◀

A Dot Product View of Matrix Multiplication

Dot products provide another way of thinking about matrix multiplication. Recall that if $A = [a_{ij}]$ is an $m \times r$ matrix and $B = [b_{ij}]$ is an $r \times n$ matrix, then the ijth entry of AB is

$$a_{i1}b_{1j} + a_{i2}b_{2j} + \cdots + a_{ir}b_{rj}$$

which is the dot product of the ith row vector of A

$$[a_{i1} \quad a_{i2} \quad \cdots \quad a_{ir}]$$

and the jth column vector of B

$$\begin{bmatrix} b_{1j} \\ b_{2j} \\ \vdots \\ b_{rj} \end{bmatrix}$$

Thus, if the row vectors of A are $\mathbf{r}_1, \mathbf{r}_2, \ldots, \mathbf{r}_m$ and the column vectors of B are $\mathbf{c}_1, \mathbf{c}_2, \ldots, \mathbf{c}_n$, then the matrix product AB can be expressed as

$$AB = \begin{bmatrix} \mathbf{r}_1 \cdot \mathbf{c}_1 & \mathbf{r}_1 \cdot \mathbf{c}_2 & \cdots & \mathbf{r}_1 \cdot \mathbf{c}_n \\ \mathbf{r}_2 \cdot \mathbf{c}_1 & \mathbf{r}_2 \cdot \mathbf{c}_2 & \cdots & \mathbf{r}_2 \cdot \mathbf{c}_n \\ \vdots & \vdots & & \vdots \\ \mathbf{r}_m \cdot \mathbf{c}_1 & \mathbf{r}_m \cdot \mathbf{c}_2 & \cdots & \mathbf{r}_m \cdot \mathbf{c}_n \end{bmatrix} \qquad (28)$$

Application of Dot Products to ISBN Numbers

Although the system has recently changed, most books published in the last 25 years have been assigned a unique 10-digit number called an *International Standard Book Number* or ISBN. The first nine digits of this number are split into three groups—the first group representing the country or group of countries in which the book originates, the second identifying the publisher, and the third assigned to the book title itself. The tenth and final digit, called a *check digit*, is computed from the first nine digits and is used to ensure that an electronic transmission of the ISBN, say over the Internet, occurs without error.

To explain how this is done, regard the first nine digits of the ISBN as a vector $\mathbf{b}$ in R^9, and let $\mathbf{a}$ be the vector

$$\mathbf{a} = (1, 2, 3, 4, 5, 6, 7, 8, 9)$$

Then the check digit c is computed using the following procedure:

1. Form the dot product $\mathbf{a} \cdot \mathbf{b}$.
2. Divide $\mathbf{a} \cdot \mathbf{b}$ by 11, thereby producing a remainder c that is an integer between 0 and 10, inclusive. The check digit is taken to be c, with the proviso that $c = 10$ is written as X to avoid double digits.

For example, the ISBN of the brief edition of *Calculus*, sixth edition, by Howard Anton is

$$0\text{-}471\text{-}15307\text{-}9$$

which has a check digit of 9. This is consistent with the first nine digits of the ISBN, since

$$\mathbf{a} \cdot \mathbf{b} = (1, 2, 3, 4, 5, 6, 7, 8, 9) \cdot (0, 4, 7, 1, 1, 5, 3, 0, 7) = 152$$

Dividing 152 by 11 produces a quotient of 13 and a remainder of 9, so the check digit is $c = 9$. If an electronic order is placed for a book with a certain ISBN, then the warehouse can use the above procedure to verify that the check digit is consistent with the first nine digits, thereby reducing the possibility of a costly shipping error.

Concept Review

- Norm (or length or magnitude) of a vector
- Unit vector
- Normalized vector
- Standard unit vectors
- Distance between points in R^n
- Angle between two vectors in R^n
- Dot product (or Euclidean inner product) of two vectors in R^n
- Cauchy–Schwarz inequality
- Triangle inequality
- Parallelogram equation for vectors

Skills

- Compute the norm of a vector in R^n.
- Determine whether a given vector in R^n is a unit vector.
- Normalize a nonzero vector in R^n.
- Determine the distance between two vectors in R^n.
- Compute the dot product of two vectors in R^n.
- Compute the angle between two nonzero vectors in R^n.
- Prove basic properties pertaining to norms and dot products (Theorems 3.2.1–3.2.3 and 3.2.5–3.2.7).

Exercise Set 3.2

▶ In Exercises 1–2, find the norm of $\mathbf{v}$, a unit vector that has the same direction as $\mathbf{v}$, and a unit vector that is oppositely directed to $\mathbf{v}$. ◀

1. (a) $\mathbf{v} = (3, -5)$ (b) $\mathbf{v} = (3, 3, 1)$
   (c) $\mathbf{v} = (0, 1, -1, 2, 6)$

2. (a) $\mathbf{v} = (-5, 12)$ (b) $\mathbf{v} = (1, -1, 2)$
   (c) $\mathbf{v} = (-2, 3, 3, -1)$

▶ In Exercises 3–4, evaluate the given expression with $\mathbf{u} = (2, -2, 3)$, $\mathbf{v} = (1, -3, 4)$, and $\mathbf{w} = (3, 6, -4)$. ◀

3. (a) $\|\mathbf{u} - \mathbf{v}\|$ (b) $\|\mathbf{u}\| - \|\mathbf{v}\|$
   (c) $\|3\mathbf{u} + 3\mathbf{w}\|$ (d) $\|2\mathbf{u} - 4\mathbf{v} + \mathbf{w}\|$

4. (a) $\|\mathbf{u} + \mathbf{v} + \mathbf{w}\|$ (b) $\|\mathbf{u} - \mathbf{v}\|$
   (c) $\|3\mathbf{v}\| - 3\|\mathbf{v}\|$ (d) $\|\mathbf{u}\| - \|\mathbf{v}\|$

▶ In Exercises 5–6, evaluate the given expression with $\mathbf{u} = (-2, -1, 4, 5)$, $\mathbf{v} = (3, 1, -5, 7)$, and $\mathbf{w} = (-6, 2, 1, 1)$. ◀

5. (a) $\|3\mathbf{u} - 5\mathbf{v} + \mathbf{w}\|$ (b) $\|3\mathbf{u}\| - 5\|\mathbf{v}\| + \|\mathbf{w}\|$
   (c) $\|-\|\mathbf{u}\|\mathbf{v}\|$

6. (a) $\|\mathbf{u}\| - 2\|\mathbf{v}\| - 3\|\mathbf{w}\|$ (b) $\|\mathbf{u}\| + \|-2\mathbf{v}\| + \|-3\mathbf{w}\|$
   (c) $\|\|\mathbf{u} - \mathbf{v}\|\mathbf{w}\|$

7. Let $\mathbf{v} = (0, 2, -6, 3)$. Find all scalars k such that $\|k\mathbf{v}\| = 14$.

8. Let $\mathbf{v} = (1, 1, 2, -3, 1)$. Find all scalars k such that $\|k\mathbf{v}\| = 4$.

142 Chapter 3 Euclidean Vector Spaces

▶ In Exercises 9–10, find $\mathbf{u} \cdot \mathbf{v}$, $\mathbf{u} \cdot \mathbf{u}$, and $\mathbf{v} \cdot \mathbf{v}$. ◀

9. (a) $\mathbf{u} = (1, 2, -3)$, $\mathbf{v} = (3, -3, 5)$
   (b) $\mathbf{u} = (2, 1, -2, 4)$, $\mathbf{v} = (0, -1, -3, 1)$

10. (a) $\mathbf{u} = (1, 1, -2, 3)$, $\mathbf{v} = (-1, 0, 5, 1)$
    (b) $\mathbf{u} = (2, -1, 1, 0, -2)$, $\mathbf{v} = (1, 2, 2, 2, 1)$

▶ In Exercises 11–12, find the Euclidean distance between $\mathbf{u}$ and $\mathbf{v}$. ◀

11. (a) $\mathbf{u} = (3, 3, 3)$, $\mathbf{v} = (1, 0, 4)$
    (b) $\mathbf{u} = (0, -2, -1, 1)$, $\mathbf{v} = (-3, 2, 4, 4)$
    (c) $\mathbf{u} = (3, -3, -2, 0, -3, 13, 5)$,
        $\mathbf{v} = (-4, 1, -1, 5, 0, -11, 4)$

12. (a) $\mathbf{u} = (1, 2, -3, 0)$, $\mathbf{v} = (5, 1, 2, -2)$
    (b) $\mathbf{u} = (2, -1, -4, 1, 0, 6, -3, 1)$,
        $\mathbf{v} = (-2, -1, 0, 3, 7, 2, -5, 1)$
    (c) $\mathbf{u} = (0, 1, 1, 1, 2)$, $\mathbf{v} = (2, 1, 0, -1, 3)$

13. Find the cosine of the angle between the vectors in each part of Exercise 11, and then state whether the angle is acute, obtuse, or 90°.

14. Find the cosine of the angle between the vectors in each part of Exercise 12, and then state whether the angle is acute, obtuse, or 90°.

15. A vector $\mathbf{a}$ in the xy-plane has a length of 4 units and points in the positive x-direction, and a vector $\mathbf{b}$ in that plane has a length of 3 units and points in a direction that is 60° counterclockwise from the the positive x-direction. Find $\mathbf{a} \cdot \mathbf{b}$.

16. Suppose that a vector $\mathbf{a}$ in the xy-plane points in a direction that is 47° counterclockwise from the positive x-axis, and a vector $\mathbf{b}$ in that plane points in a direction that is 43° clockwise from the positive x-axis. What can you say about the value of $\mathbf{a} \cdot \mathbf{b}$?

▶ In Exercises 17–18, determine whether the expression makes sense mathematically. If not, explain why. ◀

17. (a) $\mathbf{u} \cdot (\mathbf{v} \cdot \mathbf{w})$ (b) $\mathbf{u} \cdot (\mathbf{v} + \mathbf{w})$
    (c) $\|\mathbf{u} \cdot \mathbf{v}\|$ (d) $(\mathbf{u} \cdot \mathbf{v}) - \|\mathbf{u}\|$

18. (a) $\|\mathbf{u}\| \cdot \|\mathbf{v}\|$ (b) $(\mathbf{u} \cdot \mathbf{v}) - \mathbf{w}$
    (c) $(\mathbf{u} \cdot \mathbf{v}) - k$ (d) $k \cdot \mathbf{u}$

19. Find a unit vector that has the same direction as the given vector.
    (a) $(3, -4)$ (b) $(3, 3)$
    (c) $(3, 6, -2)$ (d) $(1, 3, 5, -2)$

20. Find a unit vector that is oppositely directed to the given vector.
    (a) $(-12, -5)$ (b) $(3, -3, -3)$
    (c) $(-6, 8)$ (d) $(-3, 1, \sqrt{6}, 3)$

21. State a procedure for finding a vector of a specified length m that points in the same direction as a given vector $\mathbf{v}$.

22. If $\|\mathbf{v}\| = 2$ and $\|\mathbf{w}\| = 3$, what are the largest and smallest values possible for $\|\mathbf{v} - \mathbf{w}\|$? Give a geometric explanation of your results.

23. Find the cosine of the angle θ between $\mathbf{u}$ and $\mathbf{v}$.
    (a) $\mathbf{u} = (1, 1)$, $\mathbf{v} = (3, 0)$
    (b) $\mathbf{u} = (-3, 5)$, $\mathbf{v} = (2, 7)$
    (c) $\mathbf{u} = (2, 1, 3)$, $\mathbf{v} = (1, 2, -4)$
    (d) $\mathbf{u} = (2, 0, 1, -2)$, $\mathbf{v} = (1, 5, -3, 2)$

24. Find the radian measure of the angle θ (with $0 \leq \theta \leq \pi$) between $\mathbf{u}$ and $\mathbf{v}$.
    (a) $(1, -7)$ and $(21, 3)$ (b) $(0, 2)$ and $(3, -3)$
    (c) $(-1, 1, 0)$ and $(0, -1, 1)$ (d) $(1, -1, 0)$ and $(1, 0, 0)$

▶ In Exercises 25–26, verify that the Cauchy–Schwarz inequality holds. ◀

25. (a) $\mathbf{u} = (2, 3)$, $\mathbf{v} = (5, -7)$
    (b) $\mathbf{u} = (1, -5, 4)$, $\mathbf{v} = (3, 3, 3)$
    (c) $\mathbf{u} = (0, 2, 2, 1)$, $\mathbf{v} = (1, 1, 1, 1)$

26. (a) $\mathbf{u} = (4, 1, 1)$, $\mathbf{v} = (1, 2, 3)$
    (b) $\mathbf{u} = (1, 2, 1, 2, 3)$, $\mathbf{v} = (0, 1, 1, 5, -2)$
    (c) $\mathbf{u} = (1, 3, 5, 2, 0, 1)$, $\mathbf{v} = (0, 2, 4, 1, 3, 5)$

27. Let $\mathbf{p}_0 = (x_0, y_0, z_0)$ and $\mathbf{p} = (x, y, z)$. Describe the set of all points (x, y, z) for which $\|\mathbf{p} - \mathbf{p}_0\| = 1$.

28. (a) Show that the components of the vector $\mathbf{v} = (v_1, v_2)$ in Figure Ex-28a are $v_1 = \|\mathbf{v}\| \cos \theta$ and $v_2 = \|\mathbf{v}\| \sin \theta$.
    (b) Let $\mathbf{u}$ and $\mathbf{v}$ be the vectors in Figure Ex-28b. Use the result in part (a) to find the components of $4\mathbf{u} - 5\mathbf{v}$.

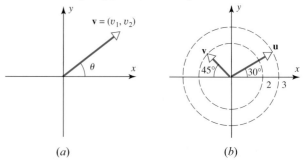

▲ Figure Ex-28

29. Prove parts (*a*) and (*b*) of Theorem 3.2.1.

30. Prove parts (*a*) and (*c*) of Theorem 3.2.3.

31. Prove parts (*d*) and (*e*) of Theorem 3.2.3.

32. Under what conditions will the triangle inequality (Theorem 3.2.5a) be an equality? Explain your answer geometrically.

33. What can you say about two nonzero vectors, **u** and **v**, that satisfy the equation $\|\mathbf{u}+\mathbf{v}\| = \|\mathbf{u}\| + \|\mathbf{v}\|$?

34. (a) What relationship must hold for the point $\mathbf{p} = (a,b,c)$ to be equidistant from the origin and the xz-plane? Make sure that the relationship you state is valid for positive and negative values of a, b, and c.

    (b) What relationship must hold for the point $\mathbf{p} = (a,b,c)$ to be farther from the origin than from the xz-plane? Make sure that the relationship you state is valid for positive and negative values of a, b, and c

True-False Exercises

In parts (a)–(j) determine whether the statement is true or false, and justify your answer.

(a) If each component of a vector in R^3 is doubled, the norm of that vector is doubled.

(b) In R^2, the vectors of norm 5 whose initial points are at the origin have terminal points lying on a circle of radius 5 centered at the origin.

(c) Every vector in R^n has a positive norm.

(d) If **v** is a nonzero vector in R^n, there are exactly two unit vectors that are parallel to **v**.

(e) If $\|\mathbf{u}\| = 2$, $\|\mathbf{v}\| = 1$, and $\mathbf{u} \cdot \mathbf{v} = 1$, then the angle between **u** and **v** is $\pi/3$ radians.

(f) The expressions $(\mathbf{u} \cdot \mathbf{v}) + \mathbf{w}$ and $\mathbf{u} \cdot (\mathbf{v} + \mathbf{w})$ are both meaningful and equal to each other.

(g) If $\mathbf{u} \cdot \mathbf{v} = \mathbf{u} \cdot \mathbf{w}$, then $\mathbf{v} = \mathbf{w}$.

(h) If $\mathbf{u} \cdot \mathbf{v} = 0$, then either $\mathbf{u} = \mathbf{0}$ or $\mathbf{v} = \mathbf{0}$.

(i) In R^2, if **u** lies in the first quadrant and **v** lies in the third quadrant, then $\mathbf{u} \cdot \mathbf{v}$ cannot be positive.

(j) For all vectors **u**, **v**, and **w** in R^n, we have
$$\|\mathbf{u}+\mathbf{v}+\mathbf{w}\| \le \|\mathbf{u}\| + \|\mathbf{v}\| + \|\mathbf{w}\|$$

3.3 Orthogonality

In the last section we defined the notion of "angle" between vectors in R^n. In this section we will focus on the notion of "perpendicularity." Perpendicular vectors in R^n play an important role in a wide variety of applications.

Orthogonal Vectors Recall from Formula (20) in the previous section that the angle θ between two *nonzero* vectors **u** and **v** in R^n is defned by the formula

$$\theta = \cos^{-1}\left(\frac{\mathbf{u} \cdot \mathbf{v}}{\|\mathbf{u}\|\|\mathbf{v}\|}\right)$$

It follows from this that $\theta = \pi/2$ if and only if $\mathbf{u} \cdot \mathbf{v} = 0$. Thus, we make the following definition.

> **DEFINITION 1** Two nonzero vectors **u** and **v** in R^n are said to be **orthogonal** (or **perpendicular**) if $\mathbf{u} \cdot \mathbf{v} = 0$. We will also agree that the zero vector in R^n is orthogonal to *every* vector in R^n. A nonempty set of vectors in R^n is called an **orthogonal set** if all pairs of distinct vectors in the set are orthogonal. An orthogonal set of unit vectors is called an **orthonormal set**.

▶ **EXAMPLE 1 Orthogonal Vectors**

(a) Show that $\mathbf{u} = (-2, 3, 1, 4)$ and $\mathbf{v} = (1, 2, 0, -1)$ are orthogonal vectors in R^4.

(b) Show that the set $S = \{\mathbf{i}, \mathbf{j}, \mathbf{k}\}$ of standard unit vectors is an orthogonal set in R^3.

Solution (a) The vectors are orthogonal since

$$\mathbf{u} \cdot \mathbf{v} = (-2)(1) + (3)(2) + (1)(0) + (4)(-1) = 0$$

144 Chapter 3 Euclidean Vector Spaces

In Example 1 there is no need to check that

$$j \cdot i = k \cdot i = k \cdot j = 0$$

since this follows from computations in the example and the symmetry property of the dot product.

Solution (b) We must show that all pairs of distinct vectors are orthogonal, that is,

$$i \cdot j = i \cdot k = j \cdot k = 0$$

This is evident geometrically (Figure 3.2.2), but it can be seen as well from the computations

$$i \cdot j = (1, 0, 0) \cdot (0, 1, 0) = 0$$
$$i \cdot k = (1, 0, 0) \cdot (0, 0, 1) = 0$$
$$j \cdot k = (0, 1, 0) \cdot (0, 0, 1) = 0 \blacktriangleleft$$

Lines and Planes Determined by Points and Normals

One learns in analytic geometry that a line in R^2 is determined uniquely by its slope and one of its points, and that a plane in R^3 is determined uniquely by its "inclination" and one of its points. One way of specifying slope and inclination is to use a *nonzero* vector **n**, called a ***normal***, that is orthogonal to the line or plane in question. For example, Figure 3.3.1 shows the line through the point $P_0(x_0, y_0)$ that has normal $\mathbf{n} = (a, b)$ and the plane through the point $P_0(x_0, y_0, z_0)$ that has normal $\mathbf{n} = (a, b, c)$. Both the line and the plane are represented by the vector equation

$$\mathbf{n} \cdot \overrightarrow{P_0P} = 0 \tag{1}$$

where P is either an arbitrary point (x, y) on the line or an arbitrary point (x, y, z) in the plane. The vector $\overrightarrow{P_0P}$ can be expressed in terms of components as

$$\overrightarrow{P_0P} = (x - x_0, y - y_0) \quad \text{[line]}$$
$$\overrightarrow{P_0P} = (x - x_0, y - y_0, z - z_0) \quad \text{[plane]}$$

Thus, Equation (1) can be written as

$$a(x - x_0) + b(y - y_0) = 0 \quad \text{[line]} \tag{2}$$

$$a(x - x_0) + b(y - y_0) + c(z - z_0) = 0 \quad \text{[plane]} \tag{3}$$

These are called the ***point-normal*** equations of the line and plane.

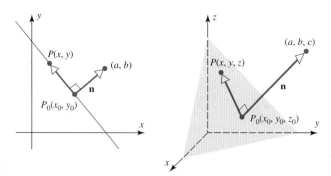

▶ Figure 3.3.1

▶ **EXAMPLE 2 Point-Normal Equations**

It follows from (2) that in R^2 the equation

$$6(x - 3) + (y + 7) = 0$$

represents the line through the point $(3, -7)$ with normal $\mathbf{n} = (6, 1)$; and it follows from (3) that in R^3 the equation

$$4(x - 3) + 2y - 5(z - 7) = 0$$

represents the plane through the point $(3, 0, 7)$ with normal $\mathbf{n} = (4, 2, -5)$. ◀

When convenient, the terms in Equations (2) and (3) can be multiplied out and the constants combined. This leads to the following theorem.

THEOREM 3.3.1

(a) *If a and b are constants that are not both zero, then an equation of the form*

$$ax + by + c = 0 \qquad (4)$$

represents a line in R^2 with normal $\mathbf{n} = (a, b)$.

(b) *If a, b, and c are constants that are not all zero, then an equation of the form*

$$ax + by + cz + d = 0 \qquad (5)$$

represents a plane in R^3 with normal $\mathbf{n} = (a, b, c)$.

▶ **EXAMPLE 3** **Vectors Orthogonal to Lines and Planes Through the Origin**

(a) The equation $ax + by = 0$ represents a line through the origin in R^2. Show that the vector $\mathbf{n}_1 = (a, b)$ formed from the coefficients of the equation is orthogonal to the line, that is, orthogonal to every vector along the line.

(b) The equation $ax + by + cz = 0$ represents a plane through the origin in R^3. Show that the vector $\mathbf{n}_2 = (a, b, c)$ formed from the coefficients of the equation is orthogonal to the plane, that is, orthogonal to every vector that lies in the plane.

Solution We will solve both problems together. The two equations can be written as

$$(a, b) \cdot (x, y) = 0 \quad \text{and} \quad (a, b, c) \cdot (x, y, z) = 0$$

or, alternatively, as

$$\mathbf{n}_1 \cdot (x, y) = 0 \quad \text{and} \quad \mathbf{n}_2 \cdot (x, y, z) = 0$$

These equations show that $\mathbf{n}_1$ is orthogonal to every vector (x, y) on the line and that $\mathbf{n}_2$ is orthogonal to every vector (x, y, z) in the plane (Figure 3.3.1). ◀

Recall that

$$ax + by = 0 \quad \text{and} \quad ax + by + cz = 0$$

are called *homogeneous equations*. Example 3 illustrates that homogeneous equations in two or three unknowns can be written in the vector form

$$\mathbf{n} \cdot \mathbf{x} = 0 \qquad (6)$$

where $\mathbf{n}$ is the vector of coefficients and $\mathbf{x}$ is the vector of unknowns. In R^2 this is called the ***vector form of a line*** through the origin, and in R^3 it is called the ***vector form of a plane*** through the origin.

Referring to Table 1 of Section 3.2, in what other ways can you write (6) if $\mathbf{n}$ and $\mathbf{x}$ are expressed in matrix form?

Orthogonal Projections

In many applications it is necessary to "decompose" a vector **u** into a sum of two terms, one term being a scalar multiple of a specified nonzero vector **a** and the other term being orthogonal to **a**. For example, if **u** and **a** are vectors in R^2 that are positioned so their initial points coincide at a point Q, then we can create such a decomposition as follows (Figure 3.3.2):

- Drop a perpendicular from the tip of **u** to the line through **a**.
- Construct the vector $\mathbf{w}_1$ from Q to the foot of the perpendicular.
- Construct the vector $\mathbf{w}_2 = \mathbf{u} - \mathbf{w}_1$.

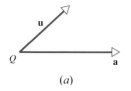(a) 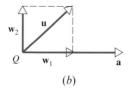(b) 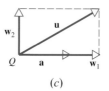(c) 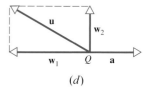(d)

▲ Figure 3.3.2 In parts (b) through (d), $\mathbf{u} = \mathbf{w}_1 + \mathbf{w}_2$, where $\mathbf{w}_1$ is parallel to **a** and $\mathbf{w}_2$ is orthogonal to **a**.

Since

$$\mathbf{w}_1 + \mathbf{w}_2 = \mathbf{w}_1 + (\mathbf{u} - \mathbf{w}_1) = \mathbf{u}$$

we have decomposed **u** into a sum of two orthogonal vectors, the first term being a scalar multiple of **a** and the second being orthogonal to **a**.

The following theorem shows that the foregoing results, which we illustrated using vectors in R^2, apply as well in R^n.

THEOREM 3.3.2 **Projection Theorem**

*If **u** and **a** are vectors in R^n, and if $\mathbf{a} \neq 0$, then **u** can be expressed in exactly one way in the form $\mathbf{u} = \mathbf{w}_1 + \mathbf{w}_2$, where $\mathbf{w}_1$ is a scalar multiple of **a** and $\mathbf{w}_2$ is orthogonal to **a**.*

Proof Since the vector $\mathbf{w}_1$ is to be a scalar multiple of **a**, it must have the form

$$\mathbf{w}_1 = k\mathbf{a} \qquad (7)$$

Our goal is to find a value of the scalar k and a vector $\mathbf{w}_2$ that is orthogonal to **a** such that

$$\mathbf{u} = \mathbf{w}_1 + \mathbf{w}_2 \qquad (8)$$

We can determine k by using (7) to rewrite (8) as

$$\mathbf{u} = \mathbf{w}_1 + \mathbf{w}_2 = k\mathbf{a} + \mathbf{w}_2$$

and then applying Theorems 3.2.2 and 3.2.3 to obtain

$$\mathbf{u} \cdot \mathbf{a} = (k\mathbf{a} + \mathbf{w}_2) \cdot \mathbf{a} = k\|\mathbf{a}\|^2 + (\mathbf{w}_2 \cdot \mathbf{a}) \qquad (9)$$

Since $\mathbf{w}_2$ is to be orthogonal to **a**, the last term in (9) must be 0, and hence k must satisfy the equation

$$\mathbf{u} \cdot \mathbf{a} = k\|\mathbf{a}\|^2$$

from which we obtain

$$k = \frac{\mathbf{u} \cdot \mathbf{a}}{\|\mathbf{a}\|^2}$$

as the only possible value for k. The proof can be completed by rewriting (8) as

$$\mathbf{w}_2 = \mathbf{u} - \mathbf{w}_1 = \mathbf{u} - k\mathbf{a} = \mathbf{u} - \frac{\mathbf{u} \cdot \mathbf{a}}{\|\mathbf{a}\|^2}\mathbf{a}$$

and then confirming that $\mathbf{w}_2$ is orthogonal to $\mathbf{a}$ by showing that $\mathbf{w}_2 \cdot \mathbf{a} = 0$ (we leave the details for you). ◀

The vectors $\mathbf{w}_1$ and $\mathbf{w}_2$ in the Projection Theorem have associated names—the vector $\mathbf{w}_1$ is called the *orthogonal projection of* $\mathbf{u}$ *on* $\mathbf{a}$ or sometimes *the vector component of* $\mathbf{u}$ *along* $\mathbf{a}$, and the vector $\mathbf{w}_2$ is called the vector *component of* $\mathbf{u}$ *orthogonal to* $\mathbf{a}$. The vector $\mathbf{w}_1$ is commonly denoted by the symbol $\text{proj}_{\mathbf{a}}\mathbf{u}$, in which case it follows from (8) that $\mathbf{w}_2 = \mathbf{u} - \text{proj}_{\mathbf{a}}\mathbf{u}$. In summary,

$$\text{proj}_{\mathbf{a}}\mathbf{u} = \frac{\mathbf{u} \cdot \mathbf{a}}{\|\mathbf{a}\|^2}\mathbf{a} \quad (\textit{vector component of } \mathbf{u} \textit{ along } \mathbf{a}) \tag{10}$$

$$\mathbf{u} - \text{proj}_{\mathbf{a}}\mathbf{u} = \mathbf{u} - \frac{\mathbf{u} \cdot \mathbf{a}}{\|\mathbf{a}\|^2}\mathbf{a} \quad (\textit{vector component of } \mathbf{u} \textit{ orthogonal to } \mathbf{a}) \tag{11}$$

▶ **EXAMPLE 4** **Orthogonal Projection on a Line**

Find the orthogonal projections of the vectors $\mathbf{e}_1 = (1, 0)$ and $\mathbf{e}_2 = (0, 1)$ on the line L that makes an angle θ with the positive x-axis in R^2.

Solution As illustrated in Figure 3.3.3, $\mathbf{a} = (\cos\theta, \sin\theta)$ is a unit vector along the line L, so our first problem is to find the orthogonal projection of $\mathbf{e}_1$ along $\mathbf{a}$. Since

$$\|\mathbf{a}\| = \sqrt{\sin^2\theta + \cos^2\theta} = 1 \quad \text{and} \quad \mathbf{e}_1 \cdot \mathbf{a} = (1, 0) \cdot (\cos\theta, \sin\theta) = \cos\theta$$

it follows from Formula (10) that this projection is

$$\text{proj}_{\mathbf{a}}\mathbf{e}_1 = \frac{\mathbf{e}_1 \cdot \mathbf{a}}{\|\mathbf{a}\|^2}\mathbf{a} = (\cos\theta)(\cos\theta, \sin\theta) = (\cos^2\theta, \sin\theta\cos\theta)$$

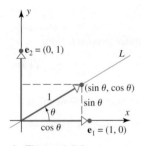

▲ Figure 3.3.3

Similarly, since $\mathbf{e}_2 \cdot \mathbf{a} = (0, 1) \cdot (\cos\theta, \sin\theta) = \sin\theta$, it follows from Formula (10) that

$$\text{proj}_{\mathbf{a}}\mathbf{e}_2 = \frac{\mathbf{e}_2 \cdot \mathbf{a}}{\|\mathbf{a}\|^2}\mathbf{a} = (\sin\theta)(\cos\theta, \sin\theta) = (\sin\theta\cos\theta, \sin^2\theta)$$

▶ **EXAMPLE 5** **Vector Component of u Along a**

Let $\mathbf{u} = (2, -1, 3)$ and $\mathbf{a} = (4, -1, 2)$. Find the vector component of $\mathbf{u}$ along $\mathbf{a}$ and the vector component of $\mathbf{u}$ orthogonal to $\mathbf{a}$.

Solution

$$\mathbf{u} \cdot \mathbf{a} = (2)(4) + (-1)(-1) + (3)(2) = 15$$
$$\|\mathbf{a}\|^2 = 4^2 + (-1)^2 + 2^2 = 21$$

Thus the vector component of $\mathbf{u}$ along $\mathbf{a}$ is

$$\text{proj}_{\mathbf{a}}\mathbf{u} = \frac{\mathbf{u} \cdot \mathbf{a}}{\|\mathbf{a}\|^2}\mathbf{a} = \tfrac{15}{21}(4, -1, 2) = \left(\tfrac{20}{7}, -\tfrac{5}{7}, \tfrac{10}{7}\right)$$

and the vector component of $\mathbf{u}$ orthogonal to $\mathbf{a}$ is

$$\mathbf{u} - \text{proj}_{\mathbf{a}}\mathbf{u} = (2, -1, 3) - \left(\tfrac{20}{7}, -\tfrac{5}{7}, \tfrac{10}{7}\right) = \left(-\tfrac{6}{7}, -\tfrac{2}{7}, \tfrac{11}{7}\right)$$

As a check, you may wish to verify that the vectors $\mathbf{u} - \text{proj}_{\mathbf{a}}\mathbf{u}$ and $\mathbf{a}$ are perpendicular by showing that their dot product is zero. ◀

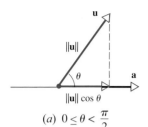

(a) $0 \leq \theta < \frac{\pi}{2}$

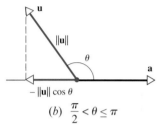

(b) $\frac{\pi}{2} < \theta \leq \pi$

▲ Figure 3.3.4

Sometimes we will be more interested in the *norm* of the vector component of **u** along **a** than in the vector component itself. A formula for this norm can be derived as follows:

$$\|\text{proj}_\mathbf{a}\mathbf{u}\| = \left\| \frac{\mathbf{u} \cdot \mathbf{a}}{\|\mathbf{a}\|^2}\mathbf{a} \right\| = \left| \frac{\mathbf{u} \cdot \mathbf{a}}{\|\mathbf{a}\|^2} \right| \|\mathbf{a}\| = \frac{|\mathbf{u} \cdot \mathbf{a}|}{\|\mathbf{a}\|^2} \|\mathbf{a}\|$$

where the second equality follows from part (*c*) of Theorem 3.2.1 and the third from the fact that $\|\mathbf{a}\|^2 > 0$. Thus,

$$\|\text{proj}_\mathbf{a}\mathbf{u}\| = \frac{|\mathbf{u} \cdot \mathbf{a}|}{\|\mathbf{a}\|} \tag{12}$$

If θ denotes the angle between **u** and **a**, then $\mathbf{u} \cdot \mathbf{a} = \|\mathbf{u}\|\|\mathbf{a}\|\cos\theta$, so (12) can also be written as

$$\|\text{proj}_\mathbf{a}\mathbf{u}\| = \|\mathbf{u}\||\cos\theta| \tag{13}$$

(Verify.) A geometric interpretation of this result is given in Figure 3.3.4.

The Theorem of Pythagoras

In Section 3.2 we found that many theorems about vectors in R^2 and R^3 also hold in R^n. Another example of this is the following generalization of the Theorem of Pythagoras (Figure 3.3.5).

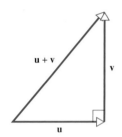

▲ Figure 3.3.5

THEOREM 3.3.3 *Theorem of Pythagoras in R^n*

If **u** *and* **v** *are orthogonal vectors in R^n with the Euclidean inner product, then*

$$\|\mathbf{u} + \mathbf{v}\|^2 = \|\mathbf{u}\|^2 + \|\mathbf{v}\|^2 \tag{14}$$

Proof Since **u** and **v** are orthogonal, we have $\mathbf{u} \cdot \mathbf{v} = 0$, from which it follows that

$$\|\mathbf{u} + \mathbf{v}\|^2 = (\mathbf{u} + \mathbf{v}) \cdot (\mathbf{u} + \mathbf{v}) = \|\mathbf{u}\|^2 + 2(\mathbf{u} \cdot \mathbf{v}) + \|\mathbf{v}\|^2 = \|\mathbf{u}\|^2 + \|\mathbf{v}\|^2 \blacktriangleleft$$

▶ **EXAMPLE 6** Theorem of Pythagoras in R^4

We showed in Example 1 that the vectors

$$\mathbf{u} = (-2, 3, 1, 4) \quad \text{and} \quad \mathbf{v} = (1, 2, 0, -1)$$

are orthogonal. Verify the Theorem of Pythagoras for these vectors.

Solution We leave it for you to confirm that

$$\mathbf{u} + \mathbf{v} = (-1, 5, 1, 3)$$
$$\|\mathbf{u} + \mathbf{v}\|^2 = 36$$
$$\|\mathbf{u}\|^2 + \|\mathbf{v}\|^2 = 30 + 6$$

Thus, $\|\mathbf{u} + \mathbf{v}\|^2 = \|\mathbf{u}\|^2 + \|\mathbf{v}\|^2$ ◀

OPTIONAL
Distance Problems

We will now show how orthogonal projections can be used to solve the following three distance problems:

Problem 1. Find the distance between a point and a line in R^2.

Problem 2. Find the distance between a point and a plane in R^3.

Problem 3. Find the distance between two parallel planes in R^3.

3.3 Orthogonality

A method for solving the first two problems is provided by the next theorem. Since the proofs of the two parts are similar, we will prove part (*b*) and leave part (*a*) as an exercise.

THEOREM 3.3.4

(a) *In R^2 the distance D between the point $P_0(x_0, y_0)$ and the line $ax + by + c = 0$ is*

$$D = \frac{|ax_0 + by_0 + c|}{\sqrt{a^2 + b^2}} \quad (15)$$

(b) *In R^3 the distance D between the point $P_0(x_0, y_0, z_0)$ and the plane $ax + by + cz + d = 0$ is*

$$D = \frac{|ax_0 + by_0 + cz_0 + d|}{\sqrt{a^2 + b^2 + c^2}} \quad (16)$$

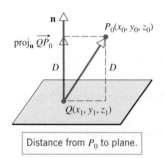

Distance from P_0 to plane.

▲ Figure 3.3.6

Proof (b) Let $Q(x_1, y_1, z_1)$ be any point in the plane. Position the normal $\mathbf{n} = (a, b, c)$ so that its initial point is at Q. As illustrated in Figure 3.3.6, the distance D is equal to the length of the orthogonal projection of $\overrightarrow{QP_0}$ on $\mathbf{n}$. Thus, it follows from Formula (16) that

$$D = \|\text{proj}_\mathbf{n} \overrightarrow{QP_0}\| = \frac{|\overrightarrow{QP_0} \cdot \mathbf{n}|}{\|\mathbf{n}\|}$$

But

$$\overrightarrow{QP_0} = (x_0 - x_1, y_0 - y_1, z_0 - z_1)$$
$$\overrightarrow{QP_0} \cdot \mathbf{n} = a(x_0 - x_1) + b(y_0 - y_1) + c(z_0 - z_1)$$
$$\|\mathbf{n}\| = \sqrt{a^2 + b^2 + c^2}$$

Thus

$$D = \frac{|a(x_0 - x_1) + b(y_0 - y_1) + c(z_0 - z_1)|}{\sqrt{a^2 + b^2 + c^2}} \quad (17)$$

Since the point $Q(x_1, y_1, z_1)$ lies in the given plane, its coordinates satisfy the equation of that plane; thus

$$ax_1 + by_1 + cz_1 + d = 0$$

or

$$d = -ax_1 - by_1 - cz_1$$

Substituting this expression in (17) yields (16). ◄

▶ **EXAMPLE 7** **Distance Between a Point and a Plane**

Find the distance D between the point $(1, -4, -3)$ and the plane $2x - 3y + 6z = -1$.

Solution Since the distance formulas in Theorem 3.3.4 require that the equations of the line and plane be written with zero on the right side, we first need to rewrite the equation of the plane as

$$2x - 3y + 6z + 1 = 0$$

from which we obtain

$$D = \frac{|2(1) + (-3)(-4) + 6(-3) + 1|}{\sqrt{2^2 + (-3)^2 + 6^2}} = \frac{|-3|}{7} = \frac{3}{7} \quad ◄$$

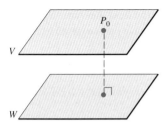

▲ Figure 3.3.7 The distance between the parallel planes V and W is equal to the distance between P_0 and W.

The third distance problem posed above is to find the distance between two parallel planes in R^3. As suggested in Figure 3.3.7, the distance between a plane V and a plane W can be obtained by finding any point P_0 in one of the planes, and computing the distance between that point and the other plane. Here is an example.

▶ **EXAMPLE 8** Distance Between Parallel Planes

The planes
$$x + 2y - 2z = 3 \quad \text{and} \quad 2x + 4y - 4z = 7$$
are parallel since their normals, $(1, 2, -2)$ and $(2, 4, -4)$, are parallel vectors. Find the distance between these planes.

Solution To find the distance D between the planes, we can select an arbitrary point in one of the planes and compute its distance to the other plane. By setting $y = z = 0$ in the equation $x + 2y - 2z = 3$, we obtain the point $P_0(3, 0, 0)$ in this plane. From (16), the distance between P_0 and the plane $2x + 4y - 4z = 7$ is
$$D = \frac{|2(3) + 4(0) + (-4)(0) - 7|}{\sqrt{2^2 + 4^2 + (-4)^2}} = \frac{1}{6} \quad \blacktriangleleft$$

Concept Review

- Orthogonal (perpendicular) vectors
- Orthogonal set of vectors
- Normal to a line
- Normal to a plane
- Point-normal equations
- Vector form of a line
- Vector form of a plane
- Orthogonal projection of **u** on **a**
- Vector component of **u** along **a**
- Vector component of **u** orthogonal to **a**
- Theorem of Pythagoras

Skills

- Determine whether two vectors are orthogonal.
- Determine whether a given set of vectors forms an orthogonal set.
- Find equations for lines (or planes) by using a normal vector and a point on the line (or plane).
- Find the vector form of a line or plane through the origin.
- Compute the vector component of **u** along **a** and orthogonal to **a**.
- Find the distance between a point and a line in R^2 or R^3.
- Find the distance between two parallel planes in R^3.
- Find the distance between a point and a plane.

Exercise Set 3.3

▶ In Exercises 1–2, determine whether **u** and **v** are orthogonal vectors. ◀

1. (a) $\mathbf{u} = (1, 3, -2)$, $\mathbf{v} = (-5, 3, 2)$
   (b) $\mathbf{u} = (0, 1, 0)$, $\mathbf{v} = (-1, 1, 0)$
   (c) $\mathbf{u} = (6, 2, -2)$, $\mathbf{v} = (-2, 3, -3)$
   (d) $\mathbf{u} = (2, 4, 5)$, $\mathbf{v} = (-5, 4, -1)$

2. (a) $\mathbf{u} = (2, 3)$, $\mathbf{v} = (5, -7)$
   (b) $\mathbf{u} = (-6, -2)$, $\mathbf{v} = (4, 0)$
   (c) $\mathbf{u} = (1, -5, 4)$, $\mathbf{v} = (3, 3, 3)$
   (d) $\mathbf{u} = (-2, 2, 3)$, $\mathbf{v} = (1, 7, -4)$

▶ In Exercises 3–4, determine whether the vectors form an orthogonal set. ◀

3. (a) $\mathbf{v}_1 = (1, 2)$, $\mathbf{v}_2 = (-2, 1)$
   (b) $\mathbf{v}_1 = (2, 4)$, $\mathbf{v}_2 = (-1, 2)$
   (c) $\mathbf{v}_1 = (1, 3, -1)$, $\mathbf{v}_2 = (-2, 2, 4)$, $\mathbf{v}_3 = (14, -2, 8)$
   (d) $\mathbf{v}_1 = (5, -1, 2)$, $\mathbf{v}_2 = (-1, 2, 3)$, $\mathbf{v}_3 = (4, -1, 2)$

4. (a) $\mathbf{v}_1 = (2, 3)$, $\mathbf{v}_2 = (-3, 2)$
   (b) $\mathbf{v}_1 = (1, -2)$, $\mathbf{v}_2 = (-2, 1)$
   (c) $\mathbf{v}_1 = (1, 0, 1)$, $\mathbf{v}_2 = (1, 1, 1)$, $\mathbf{v}_3 = (-1, 0, 1)$
   (d) $\mathbf{v}_1 = (2, -2, 1)$, $\mathbf{v}_2 = (2, 1, -2)$, $\mathbf{v}_3 = (1, 2, 2)$

5. Find a unit vector that is orthogonal to both $\mathbf{u} = (1, 1, 0)$ and $\mathbf{v} = (-1, 0, 1)$.

6. (a) Show that $\mathbf{v} = (a, b)$ and $\mathbf{w} = (-b, a)$ are orthogonal vectors.

(b) Use the result in part (a) to find two vectors that are orthogonal to $\mathbf{v} = (2, -3)$.

(c) Find two unit vectors that are orthogonal to $(-3, 4)$.

7. Do the points $A(1, 1, 1), B(-2, 0, 3)$, and $C(-3, -1, 1)$ form the vertices of a right triangle? Explain your answer.

8. Repeat Exercise 7 for the points $A(3, 0, 2), B(4, 3, 0)$, and $C(8, 1, -1)$.

▷ In Exercises 9–12, find a point-normal form of the equation of the plane passing through P and having $\mathbf{n}$ as a normal. ◁

9. $P(2, 3, -4); \; \mathbf{n} = (1, -1, 2)$

10. $P(1, 1, 4); \; \mathbf{n} = (1, 9, 8)$ **11.** $P(1, 1, 1); \; \mathbf{n} = (2, 0, 0)$

12. $P(0, 0, 0); \; \mathbf{n} = (1, 2, 3)$

▷ In Exercises 13–16, determine whether the given planes are parallel. ◁

13. $3x - 2y + z = 6$ and $2x - y + 4z = 0$

14. $x - 4y - 3z - 2 = 0$ and $3x - 12y - 9z - 7 = 0$

15. $2y = 8x - 4z + 5$ and $x = \frac{1}{2}z + \frac{1}{4}y$

16. $(-4, 1, 2) \cdot (x, y, z) = 0$ and $(8, -2, -4) \cdot (x, y, z) = 0$

▷ In Exercises 17–18, determine whether the given planes are perpendicular. ◁

17. $3x + y - 2z - 6 = 0$ and $2x - 4y + z - 5 = 0$

18. $x - 2y + 3z = 4, \; -2x + 5y + 4z = -1$

▷ In Exercises 19–20, find $\|\text{proj}_\mathbf{a} \mathbf{u}\|$. ◁

19. (a) $\mathbf{u} = (2, 4), \; \mathbf{a} = (1, 1)$
(b) $\mathbf{u} = (1, -1, 0), \; \mathbf{a} = (2, 0, 1)$

20. (a) $\mathbf{u} = (5, 6), \; \mathbf{a} = (2, -1)$
(b) $\mathbf{u} = (3, -2, 6), \; \mathbf{a} = (1, 2, -7)$

▷ In Exercises 21–28, find the vector component of $\mathbf{u}$ along $\mathbf{a}$ and the vector component of $\mathbf{u}$ orthogonal to $\mathbf{a}$. ◁

21. $\mathbf{u} = (6, 2), \; \mathbf{a} = (3, -9)$ **22.** $\mathbf{u} = (-1, -2), \; \mathbf{a} = (-2, 3)$

23. $\mathbf{u} = (3, -1, 2), \; \mathbf{a} = (1, 3, 0)$

24. $\mathbf{u} = (1, 0, 0), \; \mathbf{a} = (4, 3, 8)$

25. $\mathbf{u} = (1, 1, 1), \; \mathbf{a} = (0, 2, -1)$

26. $\mathbf{u} = (2, 0, 1), \; \mathbf{a} = (1, 2, 3)$

27. $\mathbf{u} = (2, 1, 1, 2), \; \mathbf{a} = (4, -4, 2, -2)$

28. $\mathbf{u} = (5, 0, -3, 7), \; \mathbf{a} = (2, 1, -1, -1)$

▷ In Exercises 29–32, find the distance between the point and the line. ◁

29. $(-3, 1); \; 4x + 3y + 4 = 0$

30. $(-1, 4); \; x - 3y + 2 = 0$

31. $(2, -5); \; y = -4x + 2$

32. $(1, 8); \; 3x + y = 5$

▷ In Exercises 33–36, find the distance between the point and the plane. ◁

33. $(2, 1, -3); \; 2x - y - 2z = 6$

34. $(-1, -1, 2); \; 2x + 5y - 6z = 4$

35. $(-1, 2, 1); \; 2x + 3y - 4z = 1$

36. $(0, 3, -2); \; x - y - z = 3$

▷ In Exercises 37–40, find the distance between the given parallel planes. ◁

37. $2x - y - z = 5$ and $-4x + 2y + 2z = 12$

38. $3x - 4y + z = 1$ and $6x - 8y + 2z = 3$

39. $-4x + y - 3z = 0$ and $8x - 2y + 6z = 0$

40. $2x - y + z = 1$ and $2x - y + z = -1$

41. Let $\mathbf{i}, \mathbf{j}$, and $\mathbf{k}$ be unit vectors along the positive x, y, and z axes of a rectangular coordinate system in 3-space. If $\mathbf{v} = (a, b, c)$ is a nonzero vector, then the angles α, β, and γ between $\mathbf{v}$ and the vectors $\mathbf{i}, \mathbf{j}$, and $\mathbf{k}$, respectively, are called the *direction angles* of $\mathbf{v}$ (Figure Ex-41), and the numbers $\cos \alpha, \cos \beta$, and $\cos \gamma$ are called the *direction cosines* of $\mathbf{v}$.

(a) Show that $\cos \alpha = a/\|\mathbf{v}\|$.

(b) Find $\cos \beta$ and $\cos \gamma$.

(c) Show that $\mathbf{v}/\|\mathbf{v}\| = (\cos \alpha, \cos \beta, \cos \gamma)$.

(d) Show that $\cos^2 \alpha + \cos^2 \beta + \cos^2 \gamma = 1$.

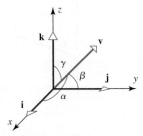

◁ Figure Ex-41

42. Use the result in Exercise 41 to estimate, to the nearest degree, the angles that a diagonal of a box with dimensions $10 \text{ cm} \times 15 \text{ cm} \times 25 \text{ cm}$ makes with the edges of the box.

43. Show that if $\mathbf{v}$ is orthogonal to both $\mathbf{w}_1$ and $\mathbf{w}_2$, then $\mathbf{v}$ is orthogonal to $k_1 \mathbf{w}_1 + k_2 \mathbf{w}_2$ for all scalars k_1 and k_2.

44. Let $\mathbf{u}$ and $\mathbf{v}$ be nonzero vectors in 2- or 3-space, and let $k = \|\mathbf{u}\|$ and $l = \|\mathbf{v}\|$. Show that the vector $\mathbf{w} = l\mathbf{u} + k\mathbf{v}$ bisects the angle between $\mathbf{u}$ and $\mathbf{v}$.

45. Prove part (*a*) of Theorem 3.3.4.

46. Is it possible to have

$$\text{proj}_a \mathbf{u} = \text{proj}_u \mathbf{a}?$$

Explain your reasoning.

True-False Exercises

In parts (a)–(g) determine whether the statement is true or false, and justify your answer.

(a) The vectors $(3, -1, 2)$ and $(0, 0, 0)$ are orthogonal.

(b) If $\mathbf{u}$ and $\mathbf{v}$ are orthogonal vectors, then for all nonzero scalars k and m, $k\mathbf{u}$ and $m\mathbf{v}$ are orthogonal vectors.

(c) The orthogonal projection of $\mathbf{u}$ on $\mathbf{a}$ is perpendicular to the vector component of $\mathbf{u}$ orthogonal to $\mathbf{a}$.

(d) If $\mathbf{a}$ and $\mathbf{b}$ are orthogonal vectors, then for every nonzero vector $\mathbf{u}$, we have

$$\text{proj}_a(\text{proj}_b(\mathbf{u})) = \mathbf{0}$$

(e) If $\mathbf{a}$ and $\mathbf{u}$ are nonzero vectors, then

$$\text{proj}_a(\text{proj}_a(\mathbf{u})) = \text{proj}_a(\mathbf{u})$$

(f) If the relationship

$$\text{proj}_a \mathbf{u} = \text{proj}_a \mathbf{v}$$

holds for some nonzero vector $\mathbf{a}$, then $\mathbf{u} = \mathbf{v}$.

(g) For all vectors $\mathbf{u}$ and $\mathbf{v}$, it is true that

$$\|\mathbf{u} + \mathbf{v}\| = \|\mathbf{u}\| + \|\mathbf{v}\|$$

3.4 The Geometry of Linear Systems

In this section we will use parametric and vector methods to study general systems of linear equations. This work will enable us to interpret solution sets of linear systems with n unknowns as geometric objects in R^n just as we interpreted solution sets of linear systems with two and three unknowns as points, lines, and planes in R^2 and R^3.

Vector and Parametric Equations of Lines in R^2 and R^3

In the last section we derived equations of lines and planes that are determined by a point and a normal vector. However, there are other useful ways of specifying lines and planes. For example, a unique line in R^2 or R^3 is determined by a point $\mathbf{x}_0$ on the line and a nonzero vector $\mathbf{v}$ parallel to the line, and a unique plane in R^3 is determined by a point $\mathbf{x}_0$ in the plane and two noncollinear vectors $\mathbf{v}_1$ and $\mathbf{v}_2$ parallel to the plane. The best way to visualize this is to translate the vectors so their initial points are at $\mathbf{x}_0$ (Figure 3.4.1).

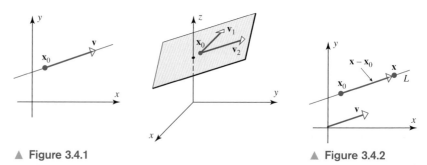

▲ Figure 3.4.1   ▲ Figure 3.4.2

Let us begin by deriving an equation for the line L that contains the point $\mathbf{x}_0$ and is parallel to $\mathbf{v}$. If $\mathbf{x}$ is a general point on such a line, then, as illustrated in Figure 3.4.2, the vector $\mathbf{x} - \mathbf{x}_0$ will be some scalar multiple of $\mathbf{v}$, say

$$\mathbf{x} - \mathbf{x}_0 = t\mathbf{v} \quad \text{or equivalently} \quad \mathbf{x} = \mathbf{x}_0 + t\mathbf{v}$$

As the variable t (called a ***parameter***) varies from $-\infty$ to ∞, the point $\mathbf{x}$ traces out the line L. Accordingly, we have the following result.

3.4 The Geometry of Linear Systems

Although it is not stated explicitly, it is understood in Formulas (1) and (2) that the parameter t varies from $-\infty$ to ∞. This applies to all vector and parametric equations in this text except where stated otherwise.

THEOREM 3.4.1 *Let L be the line in R^2 or R^3 that contains the point $\mathbf{x}_0$ and is parallel to the nonzero vector $\mathbf{v}$. Then the equation of the line through $\mathbf{x}_0$ that is parallel to $\mathbf{v}$ is*

$$\mathbf{x} = \mathbf{x}_0 + t\mathbf{v} \quad (1)$$

If $\mathbf{x}_0 = \mathbf{0}$, then the line passes through the origin and the equation has the form

$$\mathbf{x} = t\mathbf{v} \quad (2)$$

Vector and Parametric Equations of Planes in R^3

Next we will derive an equation for the plane W that contains the point $\mathbf{x}_0$ and is parallel to the noncollinear vectors $\mathbf{v}_1$ and $\mathbf{v}_2$. As shown in Figure 3.4.3, if $\mathbf{x}$ is any point in the plane, then by forming suitable scalar multiples of $\mathbf{v}_1$ and $\mathbf{v}_2$, say $t_1\mathbf{v}_1$ and $t_2\mathbf{v}_2$, we can create a parallelogram with diagonal $\mathbf{x} - \mathbf{x}_0$ and adjacent sides $t_1\mathbf{v}_1$ and $t_2\mathbf{v}_2$. Thus, we have

$$\mathbf{x} - \mathbf{x}_0 = t_1\mathbf{v}_1 + t_2\mathbf{v}_2 \quad \text{or equivalently} \quad \mathbf{x} = \mathbf{x}_0 + t_1\mathbf{v}_1 + t_2\mathbf{v}_2$$

As the variables t_1 and t_2 (called *parameters*) vary independently from $-\infty$ to ∞, the point $\mathbf{x}$ varies over the entire plane W. Accordingly, we make the following definition.

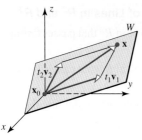

▲ Figure 3.4.3

THEOREM 3.4.2 *Let W be the plane in R^3 that contains the point $\mathbf{x}_0$ and is parallel to the noncollinear vectors $\mathbf{v}_1$ and $\mathbf{v}_2$. Then an equation of the plane through $\mathbf{x}_0$ that is parallel to $\mathbf{v}_1$ and $\mathbf{v}_2$ is given by*

$$\mathbf{x} = \mathbf{x}_0 + t_1\mathbf{v}_1 + t_2\mathbf{v}_2 \quad (3)$$

If $\mathbf{x}_0 = \mathbf{0}$, then the plane passes through the origin and the equation has the form

$$\mathbf{x} = t_1\mathbf{v}_1 + t_2\mathbf{v}_2 \quad (4)$$

Remark Observe that the line through $\mathbf{x}_0$ represented by Equation (1) is the translation by $\mathbf{x}_0$ of the line through the origin represented by Equation (2) and that the plane through $\mathbf{x}_0$ represented by Equation (3) is the translation by $\mathbf{x}_0$ of the plane through the origin represented by Equation (4) (Figure 3.4.4).

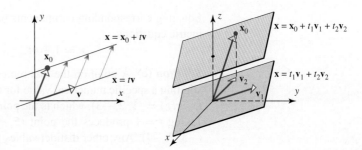

▶ Figure 3.4.4

Motivated by the forms of Formulas (1) to (4), we can extend the notions of line and plane to R^n by making the following definitions.

DEFINITION 1 If $\mathbf{x}_0$ and $\mathbf{v}$ are vectors in R^n, and if $\mathbf{v}$ is nonzero, then the equation

$$\mathbf{x} = \mathbf{x}_0 + t\mathbf{v} \quad (5)$$

defines the *line through $\mathbf{x}_0$ that is parallel to $\mathbf{v}$*. In the special case where $\mathbf{x}_0 = \mathbf{0}$, the line is said to *pass through the origin*.

DEFINITION 2 If $\mathbf{x}_0$, $\mathbf{v}_1$, and $\mathbf{v}_2$ are vectors in R^n, and if $\mathbf{v}_1$ and $\mathbf{v}_2$ are not collinear, then the equation

$$\mathbf{x} = \mathbf{x}_0 + t_1\mathbf{v}_1 + t_2\mathbf{v}_2 \tag{6}$$

defines the *plane through $\mathbf{x}_0$ that is parallel to $\mathbf{v}_1$ and $\mathbf{v}_2$*. In the special case where $\mathbf{x}_0 = \mathbf{0}$, the plane is said to *pass through the origin*.

Equations (5) and (6) are called *vector forms* of a line and plane in R^n. If the vectors in these equations are expressed in terms of their components and the corresponding components on each side are equated, then the resulting equations are called *parametric equations* of the line and plane. Here are some examples.

▶ **EXAMPLE 1** Vector and Parametric Equations of Lines in R^2 and R^3

(a) Find a vector equation and parametric equations of the line in R^2 that passes through the origin and is parallel to the vector $\mathbf{v} = (-2, 3)$.

(b) Find a vector equation and parametric equations of the line in R^3 that passes through the point $P_0(1, 2, -3)$ and is parallel to the vector $\mathbf{v} = (4, -5, 1)$.

(c) Use the vector equation obtained in part (b) to find two points on the line that are different from P_0.

Solution (a) It follows from (5) with $\mathbf{x}_0 = \mathbf{0}$ that a vector equation of the line is $\mathbf{x} = t\mathbf{v}$. If we let $\mathbf{x} = (x, y)$, then this equation can be expressed in vector form as

$$(x, y) = t(-2, 3)$$

Equating corresponding components on the two sides of this equation yields the parametric equations

$$x = -2t, \quad y = 3t$$

Solution (b) It follows from (5) that a vector equation of the line is $\mathbf{x} = \mathbf{x}_0 + t\mathbf{v}$. If we let $\mathbf{x} = (x, y, z)$, and if we take $\mathbf{x}_0 = (1, 2, -3)$, then this equation can be expressed in vector form as

$$(x, y, z) = (1, 2, -3) + t(4, -5, 1) \tag{7}$$

Equating corresponding components on the two sides of this equation yields the parametric equations

$$x = 1 + 4t, \quad y = 2 - 5t, \quad z = -3 + t$$

Solution (c) A point on the line represented by Equation (7) can be obtained by substituting a specific numerical value for the parameter t. However, since $t = 0$ produces $(x, y, z) = (1, 2, -3)$, which is the point P_0, this value of t does not serve our purpose. Taking $t = 1$ produces the point $(5, -3, -2)$ and taking $t = -1$ produces the point $(-3, 7, -4)$. Any other distinct values for t (except $t = 0$) would work just as well. ◀

▶ **EXAMPLE 2** Vector and Parametric Equations of a Plane in R^3

Find vector and parametric equations of the plane $x - y + 2z = 5$.

Solution We will find the parametric equations first. We can do this by solving the equation for any one of the variables in terms of the other two and then using those two variables as parameters. For example, solving for x in terms of y and z yields

$$x = 5 + y - 2z \tag{8}$$

We would have obtained different parametric and vector equations in Example 2 had we solved (8) for y or z rather than x. However, one can show the same plane results in all three cases as the parameters vary from $-\infty$ to ∞.

and then using y and z as parameters t_1 and t_2, respectively, yields the parametric equations

$$x = 5 + t_1 - 2t_2, \quad y = t_1, \quad z = t_2$$

To obtain a vector equation of the plane we rewrite these parametric equations as

$$(x, y, z) = (5 + t_1 - 2t_2, t_1, t_2)$$

or, equivalently, as

$$(x, y, z) = (5, 0, 0) + t_1(1, 1, 0) + t_2(-2, 0, 1)$$

▶ **EXAMPLE 3** Vector and Parametric Equations of Lines and Planes in R^4

(a) Find vector and parametric equations of the line through the origin of R^4 that is parallel to the vector $\mathbf{v} = (5, -3, 6, 1)$.

(b) Find vector and parametric equations of the plane in R^4 that passes through the point $\mathbf{x}_0 = (2, -1, 0, 3)$ and is parallel to both $\mathbf{v}_1 = (1, 5, 2, -4)$ and $\mathbf{v}_2 = (0, 7, -8, 6)$.

Solution (a) If we let $\mathbf{x} = (x_1, x_2, x_3, x_4)$, then the vector equation $\mathbf{x} = t\mathbf{v}$ can be expressed as

$$(x_1, x_2, x_3, x_4) = t(5, -3, 6, 1)$$

Equating corresponding components yields the parametric equations

$$x_1 = 5t, \quad x_2 = -3t, \quad x_3 = 6t, \quad x_4 = t$$

Solution (b) The vector equation $\mathbf{x} = \mathbf{x}_0 + t_1\mathbf{v}_1 + t_2\mathbf{v}_2$ can be expressed as

$$(x_1, x_2, x_3, x_4) = (2, -1, 0, 3) + t_1(1, 5, 2, -4) + t_2(0, 7, -8, 6)$$

which yields the parametric equations

$$x_1 = 2 + t_1$$
$$x_2 = -1 + 5t_1 + 7t_2$$
$$x_3 = 2t_1 - 8t_2$$
$$x_4 = 3 - 4t_1 + 6t_2 \quad \blacktriangleleft$$

Lines Through Two Points in R^n

If $\mathbf{x}_0$ and $\mathbf{x}_1$ are distinct points in R^n, then the line determined by these points is parallel to the vector $\mathbf{v} = \mathbf{x}_1 - \mathbf{x}_0$ (Figure 3.4.5), so it follows from (5) that the line can be expressed in vector form as

$$\mathbf{x} = \mathbf{x}_0 + t(\mathbf{x}_1 - \mathbf{x}_0) \tag{9}$$

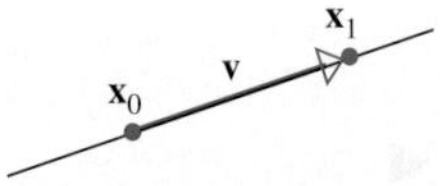

or, equivalently, as

$$\mathbf{x} = (1-t)\mathbf{x}_0 + t\mathbf{x}_1 \tag{10}$$

▲ Figure 3.4.5

These are called the ***two-point vector equations*** of a line in R^n.

▶ **EXAMPLE 4** A Line Through Two Points in R^2

Find vector and parametric equations for the line in R^2 that passes through the points $P(0, 7)$ and $Q(5, 0)$.

Solution We will see below that it does not matter which point we take to be $\mathbf{x}_0$ and which we take to be $\mathbf{x}_1$, so let us choose $\mathbf{x}_0 = (0, 7)$ and $\mathbf{x}_1 = (5, 0)$. It follows that $\mathbf{x}_1 - \mathbf{x}_0 = (5, -7)$ and hence that

$$(x, y) = (0, 7) + t(5, -7) \tag{11}$$

which we can rewrite in parametric form as

$$x = 5t, \quad y = 7 - 7t$$

Had we reversed our choices and taken $\mathbf{x}_0 = (5, 0)$ and $\mathbf{x}_1 = (0, 7)$, then the resulting vector equation would have been

$$(x, y) = (5, 0) + t(-5, 7) \tag{12}$$

and the parametric equations would have been

$$x = 5 - 5t, \quad y = 7t$$

(verify). Although (11) and (12) look different, they both represent the line whose equation in rectangular coordinates is

$$7x + 5y = 35$$

(Figure 3.4.6). This can be seen by eliminating the parameter t from the parametric equations (verify). ◀

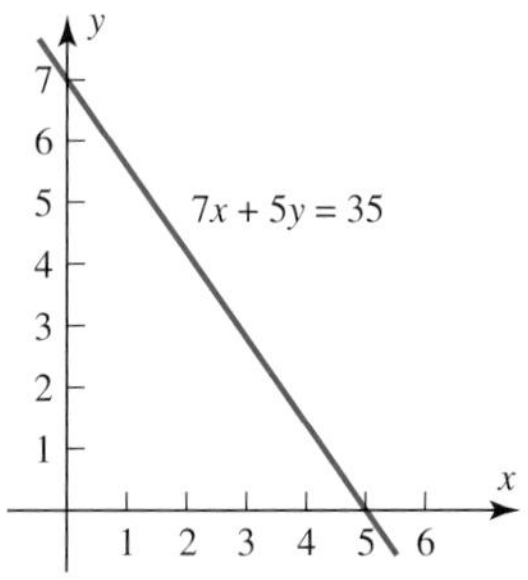

▲ Figure 3.4.6

The point $\mathbf{x} = (x, y)$ in Equations (10) and (11) traces an entire line in R^2 as the parameter t varies over the interval $(-\infty, \infty)$. If, however, we restrict the parameter to vary from $t = 0$ to $t = 1$, then $\mathbf{x}$ will not trace the entire line but rather just the *line segment* joining the points $\mathbf{x}_0$ and $\mathbf{x}_1$. The point $\mathbf{x}$ will start at $\mathbf{x}_0$ when $t = 0$ and end at $\mathbf{x}_1$ when $t = 1$. Accordingly, we make the following definition.

DEFINITION 3 If $\mathbf{x}_0$ and $\mathbf{x}_1$ are vectors in R^n, then the equation

$$\mathbf{x} = \mathbf{x}_0 + t(\mathbf{x}_1 - \mathbf{x}_0) \quad (0 \leq t \leq 1) \tag{13}$$

defines the **line segment from $\mathbf{x}_0$ to $\mathbf{x}_1$**. When convenient, Equation (13) can be written as

$$\mathbf{x} = (1 - t)\mathbf{x}_0 + t\mathbf{x}_1 \quad (0 \leq t \leq 1) \tag{14}$$

▶ **EXAMPLE 5** A Line Segment from One Point to Another in R^2

It follows from (13) and (14) that the line segment in R^2 from $\mathbf{x}_0 = (1, -3)$ to $\mathbf{x}_1 = (5, 6)$ can be represented either by the equation

$$\mathbf{x} = (1, -3) + t(4, 9) \quad (0 \leq t \leq 1)$$

or by

$$\mathbf{x} = (1 - t)(1, -3) + t(5, 6) \quad (0 \leq t \leq 1) \blacktriangleleft$$

Dot Product Form of a Linear System

Our next objective is to show how to express linear equations and linear systems in dot product notation. This will lead us to some important results about orthogonality and linear systems.

Recall that a *linear equation* in the variables $x_1, x_2, \ldots, x_n$ has the form

$$a_1 x_1 + a_2 x_2 + \cdots + a_n x_n = b \quad (a_1, a_2, \ldots, a_n \text{ not all zero}) \tag{15}$$

and that the corresponding *homogeneous* equation is

$$a_1 x_1 + a_2 x_2 + \cdots + a_n x_n = 0 \quad (a_1, a_2, \ldots, a_n \text{ not all zero}) \tag{16}$$

These equations can be rewritten in vector form by letting

$$\mathbf{a} = (a_1, a_2, \ldots, a_n) \quad \text{and} \quad \mathbf{x} = (x_1, x_2, \ldots, x_n)$$

in which case Formula (15) can be written as

$$\mathbf{a} \cdot \mathbf{x} = b \tag{17}$$

and Formula (16) as

$$\mathbf{a} \cdot \mathbf{x} = 0 \tag{18}$$

Except for a notational change from **n** to **a**, Formula (18) is the extension to R^n of Formula (6) in Section 3.3. This equation reveals that *each solution vector* **x** *of a homogeneous equation is orthogonal to the coefficient vector* **a**. To take this geometric observation a step further, consider the homogeneous system

$$\begin{aligned} a_{11}x_1 + a_{12}x_2 + \cdots + a_{1n}x_n &= 0 \\ a_{21}x_1 + a_{22}x_2 + \cdots + a_{2n}x_n &= 0 \\ \vdots \qquad \vdots \qquad \vdots \qquad \vdots \\ a_{m1}x_1 + a_{m2}x_2 + \cdots + a_{mn}x_n &= 0 \end{aligned}$$

If we denote the successive row vectors of the coefficient matrix by $\mathbf{r}_1, \mathbf{r}_2, \ldots, \mathbf{r}_m$, then we can rewrite this system in dot product form as

$$\begin{aligned} \mathbf{r}_1 \cdot \mathbf{x} &= \mathbf{0} \\ \mathbf{r}_2 \cdot \mathbf{x} &= \mathbf{0} \\ \vdots \qquad \vdots \\ \mathbf{r}_m \cdot \mathbf{x} &= \mathbf{0} \end{aligned} \tag{19}$$

from which we see that every solution vector **x** is orthogonal to every row vector of the coefficient matrix. In summary, we have the following result.

THEOREM 3.4.3 *If A is an m × n matrix, then the solution set of the homogeneous linear system* $A\mathbf{x} = \mathbf{0}$ *consists of all vectors in R^n that are orthogonal to every row vector of A.*

▶ **EXAMPLE 6** Orthogonality of Row Vectors and Solution Vectors

We showed in Example 6 of Section 1.2 that the general solution of the homogeneous linear system

$$\begin{bmatrix} 1 & 3 & -2 & 0 & 2 & 0 \\ 2 & 6 & -5 & -2 & 4 & -3 \\ 0 & 0 & 5 & 10 & 0 & 15 \\ 2 & 6 & 0 & 8 & 4 & 18 \end{bmatrix} \begin{bmatrix} x_1 \\ x_2 \\ x_3 \\ x_4 \\ x_5 \\ x_6 \end{bmatrix} = \begin{bmatrix} 0 \\ 0 \\ 0 \\ 0 \end{bmatrix}$$

is

$$x_1 = -3r - 4s - 2t, \quad x_2 = r, \quad x_3 = -2s, \quad x_4 = s, \quad x_5 = t, \quad x_6 = 0$$

which we can rewrite in vector form as

$$\mathbf{x} = (-3r - 4s - 2t, r, -2s, s, t, 0)$$

According to Theorem 3.4.3, the vector **x** must be orthogonal to each of the row vectors

$$\begin{aligned} \mathbf{r}_1 &= (1, 3, -2, 0, 2, 0) \\ \mathbf{r}_2 &= (2, 6, -5, -2, 4, -3) \\ \mathbf{r}_3 &= (0, 0, 5, 10, 0, 15) \\ \mathbf{r}_4 &= (2, 6, 0, 8, 4, 18) \end{aligned}$$

Chapter 3 Euclidean Vector Spaces

We will confirm that $\mathbf{x}$ is orthogonal to $\mathbf{r}_1$, and leave it for you to verify that $\mathbf{x}$ is orthogonal to the other three row vectors as well. The dot product of $\mathbf{r}_1$ and $\mathbf{x}$ is

$$\mathbf{r}_1 \cdot \mathbf{x} = 1(-3r - 4s - 2t) + 3(r) + (-2)(-2s) + 0(s) + 2(t) + 0(0) = 0$$

which establishes the orthogonality. ◀

The Relationship Between $A\mathbf{x} = \mathbf{0}$ and $A\mathbf{x} = \mathbf{b}$

We will conclude this section by exploring the relationship between the solutions of a homogeneous linear system $A\mathbf{x} = \mathbf{0}$ and the solutions (if any) of a nonhomogeneous linear system $A\mathbf{x} = \mathbf{b}$ that has the same coefficient matrix. These are called *corresponding linear systems*.

To motivate the result we are seeking, let us compare the solutions of the corresponding linear systems

$$\begin{bmatrix} 1 & 3 & -2 & 0 & 2 & 0 \\ 2 & 6 & -5 & -2 & 4 & -3 \\ 0 & 0 & 5 & 10 & 0 & 15 \\ 2 & 6 & 0 & 8 & 4 & 18 \end{bmatrix} \begin{bmatrix} x_1 \\ x_2 \\ x_3 \\ x_4 \\ x_5 \\ x_6 \end{bmatrix} = \begin{bmatrix} 0 \\ 0 \\ 0 \\ 0 \end{bmatrix} \quad \text{and} \quad \begin{bmatrix} 1 & 3 & -2 & 0 & 2 & 0 \\ 2 & 6 & -5 & -2 & 4 & -3 \\ 0 & 0 & 5 & 10 & 0 & 15 \\ 2 & 6 & 0 & 8 & 4 & 18 \end{bmatrix} \begin{bmatrix} x_1 \\ x_2 \\ x_3 \\ x_4 \\ x_5 \\ x_6 \end{bmatrix} = \begin{bmatrix} 0 \\ -1 \\ 5 \\ 6 \end{bmatrix}$$

We showed in Examples 5 and 6 of Section 1.2 that the general solutions of these linear systems can be written in parametric form as

homogeneous ⟶ $x_1 = -3r - 4s - 2t, \quad x_2 = r, \quad x_3 = -2s, \quad x_4 = s, \quad x_5 = t, \quad x_6 = 0$

nonhomogeneous ⟶ $x_1 = -3r - 4s - 2t, \quad x_2 = r, \quad x_3 = -2s, \quad x_4 = s, \quad x_5 = t, \quad x_6 = \frac{1}{3}$

which we can then rewrite in vector form as

homogeneous ⟶ $(x_1, x_2, x_3, x_4, x_5) = (-3r - 4s - 2t, r, -2s, s, t, 0)$

nonhomogeneous ⟶ $(x_1, x_2, x_3, x_4, x_5) = \left(-3r - 4s - 2t, r, -2s, s, t, \frac{1}{3}\right)$

By splitting the vectors on the right apart and collecting terms with like parameters, we can rewrite these equations as

homogeneous ⟶ $(x_1, x_2, x_3, x_4, x_5) = r(-3, 1, 0, 0, 0) + s(-4, 0, -2, 1, 0, 0) + t(-2, 0, 0, 0, 1, 0)$ (20)

nonhomogeneous ⟶ $(x_1, x_2, x_3, x_4, x_5) = r(-3, 1, 0, 0, 0) + s(-4, 0, -2, 1, 0, 0)$
$\qquad\qquad\qquad\qquad\qquad\qquad + t(-2, 0, 0, 0, 1, 0) + \left(0, 0, 0, 0, 0, \frac{1}{3}\right)$ (21)

Formulas (20) and (21) reveal that each solution of the nonhomogeneous system can be obtained by adding the fixed vector $\left(0, 0, 0, 0, 0, \frac{1}{3}\right)$ to the corresponding solution of the homogeneous system. This is a special case of the following general result.

THEOREM 3.4.4 *The general solution of a consistent linear system $A\mathbf{x} = \mathbf{b}$ can be obtained by adding any specific solution of $A\mathbf{x} = \mathbf{b}$ to the general solution of $A\mathbf{x} = \mathbf{0}$.*

Proof Let $\mathbf{x}_0$ be any specific solution of $A\mathbf{x} = \mathbf{b}$, let W denote the solution set of $A\mathbf{x} = \mathbf{0}$, and let $\mathbf{x}_0 + W$ denote the set of all vectors that result by adding $\mathbf{x}_0$ to each vector in W. We must show that if $\mathbf{x}$ is a vector in $\mathbf{x}_0 + W$, then $\mathbf{x}$ is a solution of $A\mathbf{x} = \mathbf{b}$, and conversely, that every solution of $A\mathbf{x} = \mathbf{b}$ is in the set $\mathbf{x}_0 + W$.

Assume first that $\mathbf{x}$ is a vector in $\mathbf{x}_0 + W$. This implies that $\mathbf{x}$ is expressible in the form $\mathbf{x} = \mathbf{x}_0 + \mathbf{w}$, where $A\mathbf{x}_0 = \mathbf{b}$ and $A\mathbf{w} = \mathbf{0}$. Thus,

$$A\mathbf{x} = A(\mathbf{x}_0 + \mathbf{w}) = A\mathbf{x}_0 + A\mathbf{w} = \mathbf{b} + \mathbf{0} = \mathbf{b}$$

which shows that $\mathbf{x}$ is a solution of $A\mathbf{x} = \mathbf{b}$.

Conversely, let $\mathbf{x}$ be any solution of $A\mathbf{x} = \mathbf{b}$. To show that $\mathbf{x}$ is in the set $\mathbf{x}_0 + W$ we must show that $\mathbf{x}$ is expressible in the form

$$\mathbf{x} = \mathbf{x}_0 + \mathbf{w} \tag{22}$$

where $\mathbf{w}$ is in W (i.e., $A\mathbf{w} = \mathbf{0}$). We can do this by taking $\mathbf{w} = \mathbf{x} - \mathbf{x}_0$. This vector obviously satisfies (22), and it is in W since

$$A\mathbf{w} = A(\mathbf{x} - \mathbf{x}_0) = A\mathbf{x} - A\mathbf{x}_0 = \mathbf{b} - \mathbf{b} = \mathbf{0} \blacktriangleleft$$

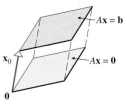

▲ **Figure 3.4.7** The solution set of $A\mathbf{x} = \mathbf{b}$ is a translation of the solution space of $A\mathbf{x} = \mathbf{0}$.

Remark Theorem 3.4.4 has a useful geometric interpretation that is illustrated in Figure 3.4.7. If, as discussed in Section 3.1, we interpret vector addition as translation, then the theorem states that if $\mathbf{x}_0$ is *any* specific solution of $A\mathbf{x} = \mathbf{b}$, then the *entire* solution set of $A\mathbf{x} = \mathbf{b}$ can be obtained by translating the solution set of $A\mathbf{x} = \mathbf{0}$ by the vector $\mathbf{x}_0$.

Concept Review

- Parameters
- Parametric equations of lines
- Parametric equations of planes
- Two-point vector equations of a line
- Vector equation of a line
- Vector equation of a plane

Skills

- Express the equations of lines in R^2 and R^3 using either vector or parametric equations.
- Express the equations of planes in R^n using either vector or parametric equations.
- Express the equation of a line containing two given points in R^2 or R^3 using either vector or parametric equations.
- Find equations of a line and a line segment.
- Verify the orthogonality of the row vectors of a linear system of equations and a solution vector.
- Use a specific solution to the nonhomogeneous linear system $A\mathbf{x} = \mathbf{b}$ and the general solution of the corresponding linear system $A\mathbf{x} = \mathbf{0}$ to obtain the general solution to $A\mathbf{x} = \mathbf{b}$.

Exercise Set 3.4

▶ In Exercises **1–4**, find vector and parametric equations of the line containing the point and parallel to the vector. ◀

1. Point: $(3, 2)$; parallel vector: $(-1, 0)$

2. Point: $(2, -1)$; vector: $\mathbf{v} = (-4, -2)$

3. Point: $(0, 0, 0)$; vector: $\mathbf{v} = (-3, 0, 1)$

4. Point: $(-6, 2, 5)$; parallel vector: $(2, 1, 4)$

▶ In Exercises **5–8**, use the given equation of a line to find a point on the line and a vector parallel to the line. ◀

5. $\mathbf{x} = (-2 + 4t, 3 - t)$ **6.** $(x, y, z) = (4t, 7, 4 + 3t)$

7. $\mathbf{x} = t(1, 4) + (1 - t)(2, -2)$

8. $\mathbf{x} = (1 - t)(0, -5, 1)$

▶ In Exercises **9–12**, find vector and parametric equations of the plane containing the given point and parallel vectors. ◀

9. Point: $(1, -2, 0)$; parallel vectors: $(0, 2, 4)$ and $(-1, 3, 2)$

10. Point: $(0, 6, -2)$; vectors: $\mathbf{v}_1 = (0, 9, -1)$ and $\mathbf{v}_2 = (0, -3, 0)$

11. Point: $(-1, 1, 4)$; vectors: $\mathbf{v}_1 = (6, -1, 0)$ and $\mathbf{v}_2 = (-1, 3, 1)$

12. Point: $(0, 5, -4)$; vectors: $\mathbf{v}_1 = (0, 0, -5)$ and $\mathbf{v}_2 = (1, -3, -2)$

▶ In Exercises **13–14**, find vector and parametric equations of the line in R^2 that passes through the origin and is orthogonal to $\mathbf{v}$.

13. $\mathbf{v} = (3, -1)$ **14.** $\mathbf{v} = (1, -4)$

▶ In Exercises **15–16**, find vector and parametric equations of the plane in R^3 that passes through the origin and is orthogonal to $\mathbf{v}$. ◀

15. $\mathbf{v} = (4, 0, -5)$ [*Hint:* Construct two nonparallel vectors orthogonal to $\mathbf{v}$ in R^3].

16. $\mathbf{v} = (3, 1, -6)$

▶ In Exercises **17–20**, find the general solution to the linear system and confirm that the row vectors of the coefficient matrix are orthogonal to the solution vectors. ◀

160 Chapter 3 Euclidean Vector Spaces

17. $2x_1 + x_2 - x_3 = 0$
    $4x_1 + 2x_2 - 2x_3 = 0$
    $x_1 + 3x_2 - 3x_3 = 0$

18. $x_1 + 3x_2 - 4x_3 = 0$
    $2x_1 + 6x_2 - 8x_3 = 0$

19. $x_1 + 5x_2 + x_3 + 2x_4 - x_5 = 0$
    $x_1 - 2x_2 - x_3 + 3x_4 + 2x_5 = 0$

20. $x_1 + 3x_2 - 4x_3 = 0$
    $x_1 + 2x_2 + 3x_3 = 0$

21. (a) The equation $x + y + z = 1$ can be viewed as a linear system of one equation in three unknowns. Express a general solution of this equation as a particular solution plus a general solution of the associated homogeneous system.

    (b) Give a geometric interpretation of the result in part (a).

22. (a) The equation $x + y = 1$ can be viewed as a linear system of one equation in two unknowns. Express a general solution of this equation as a particular solution plus a general solution of the associated homogeneous system.

    (b) Give a geometric interpretation of the result in part (a).

23. (a) Find a homogeneous linear system of two equations in three unknowns whose solution space consists of those vectors in R^3 that are orthogonal to $\mathbf{a} = (1, 1, 1)$ and $\mathbf{b} = (-2, 3, 0)$.

    (b) What kind of geometric object is the solution space?

    (c) Find a general solution of the system obtained in part (a), and confirm that Theorem 3.4.3 holds.

24. (a) Find a homogeneous linear system of two equations in three unknowns whose solution space consists of those vectors in R^3 that are orthogonal to $\mathbf{a} = (-3, 2, -1)$ and $\mathbf{b} = (0, -2, -2)$.

    (b) What kind of geometric object is the solution space?

    (c) Find a general solution of the system obtained in part (a), and confirm that Theorem 3.4.3 holds.

25. Consider the linear systems

$$\begin{bmatrix} 2 & 1 & -3 \\ 6 & 3 & -9 \\ -2 & -1 & 3 \end{bmatrix} \begin{bmatrix} x_1 \\ x_2 \\ x_3 \end{bmatrix} = \begin{bmatrix} 0 \\ 0 \\ 0 \end{bmatrix}$$

and

$$\begin{bmatrix} 2 & 1 & -3 \\ 6 & 3 & -9 \\ -2 & -1 & 3 \end{bmatrix} \begin{bmatrix} x_1 \\ x_2 \\ x_3 \end{bmatrix} = \begin{bmatrix} -3 \\ -9 \\ 3 \end{bmatrix}$$

(a) Find a general solution of the homogeneous system.

(b) Confirm that $x_1 = 1$, $x_2 = -2$, $x_3 = 1$ is a solution of the nonhomogeneous system.

(c) Use the results in parts (a) and (b) to find a general solution of the nonhomogeneous system.

(d) Check your result in part (c) by solving the nonhomogeneous system directly.

26. Consider the linear systems

$$\begin{bmatrix} 1 & -2 & 3 \\ 2 & 1 & 4 \\ 1 & -7 & 5 \end{bmatrix} \begin{bmatrix} x_1 \\ x_2 \\ x_3 \end{bmatrix} = \begin{bmatrix} 0 \\ 0 \\ 0 \end{bmatrix}$$

and

$$\begin{bmatrix} 1 & -2 & 3 \\ 2 & 1 & 4 \\ 1 & -7 & 5 \end{bmatrix} \begin{bmatrix} x_1 \\ x_2 \\ x_3 \end{bmatrix} = \begin{bmatrix} 2 \\ 7 \\ -1 \end{bmatrix}$$

(a) Find a general solution of the homogeneous system.

(b) Confirm that $x_1 = 1$, $x_2 = 1$, $x_3 = 1$ is a solution of the nonhomogeneous system.

(c) Use the results in parts (a) and (b) to find a general solution of the nonhomogeneous system.

(d) Check your result in part (c) by solving the nonhomogeneous system directly.

▶ In Exercises 27–28, find a general solution of the system, and use that solution to find a general solution of the associated homogeneous system and a particular solution of the given system.

27. $\begin{bmatrix} 4 & 3 & 2 & 1 \\ 12 & 9 & 3 & 4 \\ -4 & -3 & -2 & 4 \end{bmatrix} \begin{bmatrix} x_1 \\ x_2 \\ x_3 \\ x_4 \end{bmatrix} = \begin{bmatrix} 1 \\ 10 \\ 4 \end{bmatrix}$

28. $\begin{bmatrix} 9 & -3 & 5 & 6 \\ 6 & -2 & 3 & 1 \\ 3 & -1 & 3 & 14 \end{bmatrix} \begin{bmatrix} x_1 \\ x_2 \\ x_3 \\ x_4 \end{bmatrix} = \begin{bmatrix} 4 \\ 5 \\ -8 \end{bmatrix}$

True-False Exercises

In parts (a)–(f) determine whether the statement is true or false, and justify your answer.

(a) The vector equation of a line can be determined from any point lying on the line and a nonzero vector parallel to the line.

(b) The vector equation of a plane can be determined from any point lying in the plane and a nonzero vector parallel to the plane.

(c) The points lying on a line through the origin in R^2 or R^3 are all scalar multiples of any nonzero vector on the line.

(d) All solution vectors of the linear system $A\mathbf{x} = \mathbf{b}$ are orthogonal to the row vectors of the matrix A if and only if $\mathbf{b} = \mathbf{0}$.

(e) The general solution of the nonhomogeneous linear system $A\mathbf{x} = \mathbf{b}$ can be obtained by adding $\mathbf{b}$ to the general solution of the homogeneous linear system $A\mathbf{x} = \mathbf{0}$.

(f) If $\mathbf{x}_1$ and $\mathbf{x}_2$ are two solutions of the nonhomogeneous linear system $A\mathbf{x} = \mathbf{b}$, then $\mathbf{x}_1 - \mathbf{x}_2$ is a solution of the corresponding homogeneous linear system.

3.5 Cross Product

This optional section is concerned with properties of vectors in 3-space that are important to physicists and engineers. It can be omitted, if desired, since subsequent sections do not depend on its content. Among other things, we define an operation that provides a way of constructing a vector in 3-space that is perpendicular to two given vectors, and we give a geometric interpretation of 3×3 determinants.

Cross Product of Vectors In Section 3.2 we defined the dot product of two vectors **u** and **v** in n-space. That operation produced a *scalar* as its result. We will now define a type of vector multiplication that produces a *vector* as the result but which is applicable only to vectors in 3-space.

> **DEFINITION 1** If $\mathbf{u} = (u_1, u_2, u_3)$ and $\mathbf{v} = (v_1, v_2, v_3)$ are vectors in 3-space, then the *cross product* $\mathbf{u} \times \mathbf{v}$ is the vector defined by
>
> $$\mathbf{u} \times \mathbf{v} = (u_2 v_3 - u_3 v_2, \, u_3 v_1 - u_1 v_3, \, u_1 v_2 - u_2 v_1)$$
>
> or, in determinant notation,
>
> $$\mathbf{u} \times \mathbf{v} = \left(\begin{vmatrix} u_2 & u_3 \\ v_2 & v_3 \end{vmatrix}, \, -\begin{vmatrix} u_1 & u_3 \\ v_1 & v_3 \end{vmatrix}, \, \begin{vmatrix} u_1 & u_2 \\ v_1 & v_2 \end{vmatrix} \right) \tag{1}$$

Remark Instead of memorizing (1), you can obtain the components of $\mathbf{u} \times \mathbf{v}$ as follows:

- Form the 2×3 matrix $\begin{bmatrix} u_1 & u_2 & u_3 \\ v_1 & v_2 & v_3 \end{bmatrix}$ whose first row contains the components of **u** and whose second row contains the components of **v**.
- To find the first component of $\mathbf{u} \times \mathbf{v}$, delete the first column and take the determinant; to find the second component, delete the second column and take the negative of the determinant; and to find the third component, delete the third column and take the determinant.

▶ **EXAMPLE 1 Calculating a Cross Product**

Find $\mathbf{u} \times \mathbf{v}$, where $\mathbf{u} = (1, 2, -2)$ and $\mathbf{v} = (3, 0, 1)$.

Solution From either (1) or the mnemonic in the preceding remark, we have

$$\mathbf{u} \times \mathbf{v} = \left(\begin{vmatrix} 2 & -2 \\ 0 & 1 \end{vmatrix}, \, -\begin{vmatrix} 1 & -2 \\ 3 & 1 \end{vmatrix}, \, \begin{vmatrix} 1 & 2 \\ 3 & 0 \end{vmatrix} \right)$$

$$= (2, -7, -6) \blacktriangleleft$$

The following theorem gives some important relationships between the dot product and cross product and also shows that $\mathbf{u} \times \mathbf{v}$ is orthogonal to both **u** and **v**.

Historical Note The cross product notation $A \times B$ was introduced by the American physicist and mathematician J. Willard Gibbs, (see p. 134) in a series of unpublished lecture notes for his students at Yale University. It appeared in a published work for the first time in the second edition of the book *Vector Analysis*, by Edwin Wilson (1879–1964), a student of Gibbs. Gibbs originally referred to $A \times B$ as the "skew product."

162 Chapter 3 Euclidean Vector Spaces

> **THEOREM 3.5.1** **Relationships Involving Cross Product and Dot Product**
>
> *If* **u**, **v**, *and* **w** *are vectors in 3-space, then*
>
> (a) $\mathbf{u} \cdot (\mathbf{u} \times \mathbf{v}) = 0$ ($\mathbf{u} \times \mathbf{v}$ *is orthogonal to* $\mathbf{u}$)
>
> (b) $\mathbf{v} \cdot (\mathbf{u} \times \mathbf{v}) = 0$ ($\mathbf{u} \times \mathbf{v}$ *is orthogonal to* $\mathbf{v}$)
>
> (c) $\|\mathbf{u} \times \mathbf{v}\|^2 = \|\mathbf{u}\|^2 \|\mathbf{v}\|^2 - (\mathbf{u} \cdot \mathbf{v})^2$ (*Lagrange's identity*)
>
> (d) $\mathbf{u} \times (\mathbf{v} \times \mathbf{w}) = (\mathbf{u} \cdot \mathbf{w})\mathbf{v} - (\mathbf{u} \cdot \mathbf{v})\mathbf{w}$ (*relationship between cross and dot products*)
>
> (e) $(\mathbf{u} \times \mathbf{v}) \times \mathbf{w} = (\mathbf{u} \cdot \mathbf{w})\mathbf{v} - (\mathbf{v} \cdot \mathbf{w})\mathbf{u}$ (*relationship between cross and dot products*)

Proof (a)   Let $\mathbf{u} = (u_1, u_2, u_3)$ and $\mathbf{v} = (v_1, v_2, v_3)$. Then

$$\mathbf{u} \cdot (\mathbf{u} \times \mathbf{v}) = (u_1, u_2, u_3) \cdot (u_2 v_3 - u_3 v_2, u_3 v_1 - u_1 v_3, u_1 v_2 - u_2 v_1)$$
$$= u_1(u_2 v_3 - u_3 v_2) + u_2(u_3 v_1 - u_1 v_3) + u_3(u_1 v_2 - u_2 v_1) = 0$$

Proof (b)   Similar to (*a*).

Proof (c)   Since

$$\|\mathbf{u} \times \mathbf{v}\|^2 = (u_2 v_3 - u_3 v_2)^2 + (u_3 v_1 - u_1 v_3)^2 + (u_1 v_2 - u_2 v_1)^2 \qquad (2)$$

and

$$\|\mathbf{u}\|^2 \|\mathbf{v}\|^2 - (\mathbf{u} \cdot \mathbf{v})^2 = (u_1^2 + u_2^2 + u_3^2)(v_1^2 + v_2^2 + v_3^2) - (u_1 v_1 + u_2 v_2 + u_3 v_3)^2 \qquad (3)$$

the proof can be completed by "multiplying out" the right sides of (2) and (3) and verifying their equality.

Proof (d) and (e)   See Exercises 38 and 39. ◀

▶ **EXAMPLE 2** **u × v Is Perpendicular to u and to v**

Consider the vectors

$$\mathbf{u} = (1, 2, -2) \quad \text{and} \quad \mathbf{v} = (3, 0, 1)$$

In Example 1 we showed that

$$\mathbf{u} \times \mathbf{v} = (2, -7, -6)$$

Since

$$\mathbf{u} \cdot (\mathbf{u} \times \mathbf{v}) = (1)(2) + (2)(-7) + (-2)(-6) = 0$$

and

$$\mathbf{v} \cdot (\mathbf{u} \times \mathbf{v}) = (3)(2) + (0)(-7) + (1)(-6) = 0$$

$\mathbf{u} \times \mathbf{v}$ is orthogonal to both **u** and **v**, as guaranteed by Theorem 3.5.1. ◀

Joseph Louis Lagrange
(1736–1813)

Historical Note   Joseph Louis Lagrange was a French-Italian mathematician and astronomer. Although his father wanted him to become a lawyer, Lagrange was attracted to mathematics and astronomy after reading a memoir by the astronomer Halley. At age 16 he began to study mathematics on his own and by age 19 was appointed to a professorship at the Royal Artillery School in Turin. The following year he solved some famous problems using new methods that eventually blossomed into a branch of mathematics called the *calculus of variations*. These methods and Lagrange's applications of them to problems in celestial mechanics were so monumental that by age 25 he was regarded by many of his contemporaries as the greatest living mathematician. One of Lagrange's most famous works is a memoir, *Mécanique Analytique*, in which he reduced the theory of mechanics to a few general formulas from which all other necessary equations could be derived. Napoleon was a great admirer of Lagrange and showered him with many honors. In spite of his fame, Lagrange was a shy and modest man. On his death, he was buried with honor in the Pantheon.

[*Image:* ©SSPL/The Image Works]

The main arithmetic properties of the cross product are listed in the next theorem.

THEOREM 3.5.2 **Properties of Cross Product**

*If **u**, **v**, and **w** are any vectors in 3-space and k is any scalar, then*:

(a) $\mathbf{u} \times \mathbf{v} = -(\mathbf{v} \times \mathbf{u})$
(b) $\mathbf{u} \times (\mathbf{v} + \mathbf{w}) = (\mathbf{u} \times \mathbf{v}) + (\mathbf{u} \times \mathbf{w})$
(c) $(\mathbf{u} + \mathbf{v}) \times \mathbf{w} = (\mathbf{u} \times \mathbf{w}) + (\mathbf{v} \times \mathbf{w})$
(d) $k(\mathbf{u} \times \mathbf{v}) = (k\mathbf{u}) \times \mathbf{v} = \mathbf{u} \times (k\mathbf{v})$
(e) $\mathbf{u} \times \mathbf{0} = \mathbf{0} \times \mathbf{u} = \mathbf{0}$
(f) $\mathbf{u} \times \mathbf{u} = \mathbf{0}$

The proofs follow immediately from Formula (1) and properties of determinants; for example, part (*a*) can be proved as follows.

Proof (a)   Interchanging **u** and **v** in (1) interchanges the rows of the three determinants on the right side of (1) and hence changes the sign of each component in the cross product. Thus $\mathbf{u} \times \mathbf{v} = -(\mathbf{v} \times \mathbf{u})$. ◂

The proofs of the remaining parts are left as exercises.

▶ **EXAMPLE 3** **Standard Unit Vectors**

Consider the vectors

$$\mathbf{i} = (1, 0, 0), \quad \mathbf{j} = (0, 1, 0), \quad \mathbf{k} = (0, 0, 1)$$

These vectors each have length 1 and lie along the coordinate axes (Figure 3.5.1). They are called the ***standard unit vectors*** in 3-space. Every vector $\mathbf{v} = (v_1, v_2, v_3)$ in 3-space is expressible in terms of **i**, **j**, and **k** since we can write

$$\mathbf{v} = (v_1, v_2, v_3) = v_1(1, 0, 0) + v_2(0, 1, 0) + v_3(0, 0, 1) = v_1\mathbf{i} + v_2\mathbf{j} + v_3\mathbf{k}$$

For example,

$$(2, -3, 4) = 2\mathbf{i} - 3\mathbf{j} + 4\mathbf{k}$$

From (1) we obtain

$$\mathbf{i} \times \mathbf{j} = \left(\begin{vmatrix} 0 & 0 \\ 1 & 0 \end{vmatrix}, -\begin{vmatrix} 1 & 0 \\ 0 & 0 \end{vmatrix}, \begin{vmatrix} 1 & 0 \\ 0 & 1 \end{vmatrix} \right) = (0, 0, 1) = \mathbf{k} \quad ◂$$

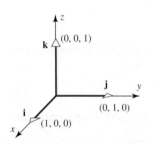

▲ **Figure 3.5.1** The standard unit vectors.

You should have no trouble obtaining the following results:

$$\mathbf{i} \times \mathbf{i} = \mathbf{0} \qquad \mathbf{j} \times \mathbf{j} = \mathbf{0} \qquad \mathbf{k} \times \mathbf{k} = \mathbf{0}$$
$$\mathbf{i} \times \mathbf{j} = \mathbf{k} \qquad \mathbf{j} \times \mathbf{k} = \mathbf{i} \qquad \mathbf{k} \times \mathbf{i} = \mathbf{j}$$
$$\mathbf{j} \times \mathbf{i} = -\mathbf{k} \qquad \mathbf{k} \times \mathbf{j} = -\mathbf{i} \qquad \mathbf{i} \times \mathbf{k} = -\mathbf{j}$$

▲ **Figure 3.5.2**

Figure 3.5.2 is helpful for remembering these results. Referring to this diagram, the cross product of two consecutive vectors going clockwise is the next vector around, and the cross product of two consecutive vectors going counterclockwise is the negative of the next vector around.

Determinant Form of Cross Product

It is also worth noting that a cross product can be represented symbolically in the form

$$\mathbf{u} \times \mathbf{v} = \begin{vmatrix} \mathbf{i} & \mathbf{j} & \mathbf{k} \\ u_1 & u_2 & u_3 \\ v_1 & v_2 & v_3 \end{vmatrix} = \begin{vmatrix} u_2 & u_3 \\ v_2 & v_3 \end{vmatrix} \mathbf{i} - \begin{vmatrix} u_1 & u_3 \\ v_1 & v_3 \end{vmatrix} \mathbf{j} + \begin{vmatrix} u_1 & u_2 \\ v_1 & v_2 \end{vmatrix} \mathbf{k} \quad (4)$$

For example, if $\mathbf{u} = (1, 2, -2)$ and $\mathbf{v} = (3, 0, 1)$, then

$$\mathbf{u} \times \mathbf{v} = \begin{vmatrix} \mathbf{i} & \mathbf{j} & \mathbf{k} \\ 1 & 2 & -2 \\ 3 & 0 & 1 \end{vmatrix} = 2\mathbf{i} - 7\mathbf{j} - 6\mathbf{k}$$

which agrees with the result obtained in Example 1.

WARNING It is not true in general that $\mathbf{u} \times (\mathbf{v} \times \mathbf{w}) = (\mathbf{u} \times \mathbf{v}) \times \mathbf{w}$. For example,

$$\mathbf{i} \times (\mathbf{j} \times \mathbf{j}) = \mathbf{i} \times \mathbf{0} = \mathbf{0}$$

and

$$(\mathbf{i} \times \mathbf{j}) \times \mathbf{j} = \mathbf{k} \times \mathbf{j} = -\mathbf{i}$$

so

$$\mathbf{i} \times (\mathbf{j} \times \mathbf{j}) \neq (\mathbf{i} \times \mathbf{j}) \times \mathbf{j}$$

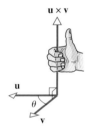

▲ Figure 3.5.3

We know from Theorem 3.5.1 that $\mathbf{u} \times \mathbf{v}$ is orthogonal to both $\mathbf{u}$ and $\mathbf{v}$. If $\mathbf{u}$ and $\mathbf{v}$ are nonzero vectors, it can be shown that the direction of $\mathbf{u} \times \mathbf{v}$ can be determined using the following "right-hand rule" (Figure 3.5.3): Let θ be the angle between $\mathbf{u}$ and $\mathbf{v}$, and suppose $\mathbf{u}$ is rotated through the angle θ until it coincides with $\mathbf{v}$. If the fingers of the right hand are cupped so that they point in the direction of rotation, then the thumb indicates (roughly) the direction of $\mathbf{u} \times \mathbf{v}$.

You may find it instructive to practice this rule with the products

$$\mathbf{i} \times \mathbf{j} = \mathbf{k}, \quad \mathbf{j} \times \mathbf{k} = \mathbf{i}, \quad \mathbf{k} \times \mathbf{i} = \mathbf{j}$$

Geometric Interpretation of Cross Product

If $\mathbf{u}$ and $\mathbf{v}$ are vectors in 3-space, then the norm of $\mathbf{u} \times \mathbf{v}$ has a useful geometric interpretation. Lagrange's identity, given in Theorem 3.5.1, states that

$$\|\mathbf{u} \times \mathbf{v}\|^2 = \|\mathbf{u}\|^2 \|\mathbf{v}\|^2 - (\mathbf{u} \cdot \mathbf{v})^2 \quad (5)$$

If θ denotes the angle between $\mathbf{u}$ and $\mathbf{v}$, then $\mathbf{u} \cdot \mathbf{v} = \|\mathbf{u}\| \|\mathbf{v}\| \cos \theta$, so (5) can be rewritten as

$$\|\mathbf{u} \times \mathbf{v}\|^2 = \|\mathbf{u}\|^2 \|\mathbf{v}\|^2 - \|\mathbf{u}\|^2 \|\mathbf{v}\|^2 \cos^2 \theta$$
$$= \|\mathbf{u}\|^2 \|\mathbf{v}\|^2 (1 - \cos^2 \theta)$$
$$= \|\mathbf{u}\|^2 \|\mathbf{v}\|^2 \sin^2 \theta$$

Since $0 \leq \theta \leq \pi$, it follows that $\sin \theta \geq 0$, so this can be rewritten as

$$\|\mathbf{u} \times \mathbf{v}\| = \|\mathbf{u}\| \|\mathbf{v}\| \sin \theta \quad (6)$$

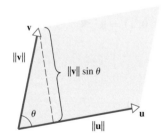

▲ Figure 3.5.4

But $\|\mathbf{v}\| \sin \theta$ is the altitude of the parallelogram determined by $\mathbf{u}$ and $\mathbf{v}$ (Figure 3.5.4). Thus, from (6), the area A of this parallelogram is given by

$$A = \text{(base)(altitude)} = \|\mathbf{u}\| \|\mathbf{v}\| \sin \theta = \|\mathbf{u} \times \mathbf{v}\|$$

This result is even correct if **u** and **v** are collinear, since the parallelogram determined by **u** and **v** has zero area and from (6) we have $\mathbf{u} \times \mathbf{v} = \mathbf{0}$ because $\theta = 0$ in this case. Thus we have the following theorem.

> **THEOREM 3.5.3** **Area of a Parallelogram**
>
> *If **u** and **v** are vectors in 3-space, then* $\|\mathbf{u} \times \mathbf{v}\|$ *is equal to the area of the parallelogram determined by **u** and **v**.*

▶ **EXAMPLE 4** **Area of a Triangle**

Find the area of the triangle determined by the points $P_1(2, 2, 0)$, $P_2(-1, 0, 2)$, and $P_3(0, 4, 3)$.

Solution The area A of the triangle is $\frac{1}{2}$ the area of the parallelogram determined by the vectors $\overrightarrow{P_1 P_2}$ and $\overrightarrow{P_1 P_3}$ (Figure 3.5.5). Using the method discussed in Example 1 of Section 3.1, $\overrightarrow{P_1 P_2} = (-3, -2, 2)$ and $\overrightarrow{P_1 P_3} = (-2, 2, 3)$. It follows that

$$\overrightarrow{P_1 P_2} \times \overrightarrow{P_1 P_3} = (-10, 5, -10)$$

(verify) and consequently that

$$A = \tfrac{1}{2}\|\overrightarrow{P_1 P_2} \times \overrightarrow{P_1 P_3}\| = \tfrac{1}{2}(15) = \tfrac{15}{2} \blacktriangleleft$$

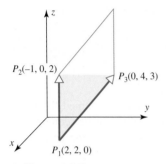

▲ Figure 3.5.5

> **DEFINITION 2** If **u**, **v**, and **w** are vectors in 3-space, then
>
> $$\mathbf{u} \cdot (\mathbf{v} \times \mathbf{w})$$
>
> is called the ***scalar triple product*** of **u**, **v**, and **w**.

The scalar triple product of $\mathbf{u} = (u_1, u_2, u_3)$, $\mathbf{v} = (v_1, v_2, v_3)$, and $\mathbf{w} = (w_1, w_2, w_3)$ can be calculated from the formula

$$\mathbf{u} \cdot (\mathbf{v} \times \mathbf{w}) = \begin{vmatrix} u_1 & u_2 & u_3 \\ v_1 & v_2 & v_3 \\ w_1 & w_2 & w_3 \end{vmatrix} \quad (7)$$

This follows from Formula (4) since

$$\mathbf{u} \cdot (\mathbf{v} \times \mathbf{w}) = \mathbf{u} \cdot \left(\begin{vmatrix} v_2 & v_3 \\ w_2 & w_3 \end{vmatrix} \mathbf{i} - \begin{vmatrix} v_1 & v_3 \\ w_1 & w_3 \end{vmatrix} \mathbf{j} + \begin{vmatrix} v_1 & v_2 \\ w_1 & w_2 \end{vmatrix} \mathbf{k} \right)$$

$$= \begin{vmatrix} v_2 & v_3 \\ w_2 & w_3 \end{vmatrix} u_1 - \begin{vmatrix} v_1 & v_3 \\ w_1 & w_3 \end{vmatrix} u_2 + \begin{vmatrix} v_1 & v_2 \\ w_1 & w_2 \end{vmatrix} u_3$$

$$= \begin{vmatrix} u_1 & u_2 & u_3 \\ v_1 & v_2 & v_3 \\ w_1 & w_2 & w_3 \end{vmatrix}$$

▶ **EXAMPLE 5** **Calculating a Scalar Triple Product**

Calculate the scalar triple product $\mathbf{u} \cdot (\mathbf{v} \times \mathbf{w})$ of the vectors

$$\mathbf{u} = 3\mathbf{i} - 2\mathbf{j} - 5\mathbf{k}, \quad \mathbf{v} = \mathbf{i} + 4\mathbf{j} - 4\mathbf{k}, \quad \mathbf{w} = 3\mathbf{j} + 2\mathbf{k}$$

Solution From (7),

$$\mathbf{u} \cdot (\mathbf{v} \times \mathbf{w}) = \begin{vmatrix} 3 & -2 & -5 \\ 1 & 4 & -4 \\ 0 & 3 & 2 \end{vmatrix}$$

$$= 3\begin{vmatrix} 4 & -4 \\ 3 & 2 \end{vmatrix} - (-2)\begin{vmatrix} 1 & -4 \\ 0 & 2 \end{vmatrix} + (-5)\begin{vmatrix} 1 & 4 \\ 0 & 3 \end{vmatrix}$$

$$= 60 + 4 - 15 = 49 \quad \blacktriangleleft$$

Remark The symbol $(\mathbf{u} \cdot \mathbf{v}) \times \mathbf{w}$ makes no sense because we cannot form the cross product of a scalar and a vector. Thus, no ambiguity arises if we write $\mathbf{u} \cdot \mathbf{v} \times \mathbf{w}$ rather than $\mathbf{u} \cdot (\mathbf{v} \times \mathbf{w})$. However, for clarity we will usually keep the parentheses.

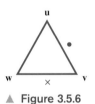

▲ Figure 3.5.6

It follows from (7) that

$$\mathbf{u} \cdot (\mathbf{v} \times \mathbf{w}) = \mathbf{w} \cdot (\mathbf{u} \times \mathbf{v}) = \mathbf{v} \cdot (\mathbf{w} \times \mathbf{u})$$

since the 3×3 determinants that represent these products can be obtained from one another by *two* row interchanges. (Verify.) These relationships can be remembered by moving the vectors $\mathbf{u}$, $\mathbf{v}$, and $\mathbf{w}$ clockwise around the vertices of the triangle in Figure 3.5.6.

Geometric Interpretation of Determinants

The next theorem provides a useful geometric interpretation of 2×2 and 3×3 determinants.

THEOREM 3.5.4

(a) The absolute value of the determinant

$$\det\begin{bmatrix} u_1 & u_2 \\ v_1 & v_2 \end{bmatrix}$$

is equal to the area of the parallelogram in 2-space determined by the vectors $\mathbf{u} = (u_1, u_2)$ and $\mathbf{v} = (v_1, v_2)$. (See Figure 3.5.7a.)

(b) The absolute value of the determinant

$$\det\begin{bmatrix} u_1 & u_2 & u_3 \\ v_1 & v_2 & v_3 \\ w_1 & w_2 & w_3 \end{bmatrix}$$

is equal to the volume of the parallelepiped in 3-space determined by the vectors $\mathbf{u} = (u_1, u_2, u_3)$, $\mathbf{v} = (v_1, v_2, v_3)$, and $\mathbf{w} = (w_1, w_2, w_3)$. (See Figure 3.5.7b.)

Proof (a) The key to the proof is to use Theorem 3.5.3. However, that theorem applies to vectors in 3-space, whereas $\mathbf{u} = (u_1, u_2)$ and $\mathbf{v} = (v_1, v_2)$ are vectors in 2-space. To circumvent this "dimension problem," we will view $\mathbf{u}$ and $\mathbf{v}$ as vectors in the xy-plane of an xyz-coordinate system (Figure 3.5.7c), in which case these vectors are expressed as $\mathbf{u} = (u_1, u_2, 0)$ and $\mathbf{v} = (v_1, v_2, 0)$. Thus

$$\mathbf{u} \times \mathbf{v} = \begin{vmatrix} \mathbf{i} & \mathbf{j} & \mathbf{k} \\ u_1 & u_2 & 0 \\ v_1 & v_2 & 0 \end{vmatrix} = \begin{vmatrix} u_1 & u_2 \\ v_1 & v_2 \end{vmatrix} \mathbf{k} = \det\begin{bmatrix} u_1 & u_2 \\ v_1 & v_2 \end{bmatrix} \mathbf{k}$$

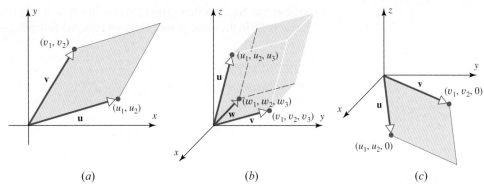

▲ Figure 3.5.7

It now follows from Theorem 3.5.3 and the fact that $\|\mathbf{k}\| = 1$ that the area A of the parallelogram determined by $\mathbf{u}$ and $\mathbf{v}$ is

$$A = \|\mathbf{u} \times \mathbf{v}\| = \left\| \det \begin{bmatrix} u_1 & u_2 \\ v_1 & v_2 \end{bmatrix} \mathbf{k} \right\| = \left| \det \begin{bmatrix} u_1 & u_2 \\ v_1 & v_2 \end{bmatrix} \right| \|\mathbf{k}\| = \left| \det \begin{bmatrix} u_1 & u_2 \\ v_1 & v_2 \end{bmatrix} \right|$$

which completes the proof.

Proof (b) As shown in Figure 3.5.8, take the base of the parallelepiped determined by $\mathbf{u}$, $\mathbf{v}$, and $\mathbf{w}$ to be the parallelogram determined by $\mathbf{v}$ and $\mathbf{w}$. It follows from Theorem 3.5.3 that the area of the base is $\|\mathbf{v} \times \mathbf{w}\|$ and, as illustrated in Figure 3.5.8, the height h of the parallelepiped is the length of the orthogonal projection of $\mathbf{u}$ on $\mathbf{v} \times \mathbf{w}$. Therefore, by Formula (12) of Section 3.3,

$$h = \|\text{proj}_{\mathbf{v} \times \mathbf{w}} \mathbf{u}\| = \frac{|\mathbf{u} \cdot (\mathbf{v} \times \mathbf{w})|}{\|\mathbf{v} \times \mathbf{w}\|}$$

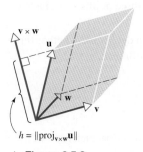

▲ Figure 3.5.8

It follows that the volume V of the parallelepiped is

$$V = (\text{area of base}) \cdot \text{height} = \|\mathbf{v} \times \mathbf{w}\| \frac{|\mathbf{u} \cdot (\mathbf{v} \times \mathbf{w})|}{\|\mathbf{v} \times \mathbf{w}\|} = |\mathbf{u} \cdot (\mathbf{v} \times \mathbf{w})|$$

so from (7),

$$V = \left| \det \begin{bmatrix} u_1 & u_2 & u_3 \\ v_1 & v_2 & v_3 \\ w_1 & w_2 & w_3 \end{bmatrix} \right| \qquad (8)$$

which completes the proof. ◀

Remark If V denotes the volume of the parallelepiped determined by vectors $\mathbf{u}$, $\mathbf{v}$, and $\mathbf{w}$, then it follows from Formulas (7) and (8) that

$$V = \begin{bmatrix} \text{volume of parallelepiped} \\ \text{determined by } \mathbf{u}, \mathbf{v}, \text{ and } \mathbf{w} \end{bmatrix} = |\mathbf{u} \cdot (\mathbf{v} \times \mathbf{w})| \qquad (9)$$

From this result and the discussion immediately following Definition 3 of Section 3.2, we can conclude that

$$\mathbf{u} \cdot (\mathbf{v} \times \mathbf{w}) = \pm V$$

where the $+$ or $-$ results depending on whether $\mathbf{u}$ makes an acute or an obtuse angle with $\mathbf{v} \times \mathbf{w}$.

Formula (9) leads to a useful test for ascertaining whether three given vectors lie in the same plane. Since three vectors not in the same plane determine a parallelepiped of positive volume, it follows from (9) that $|\mathbf{u} \cdot (\mathbf{v} \times \mathbf{w})| = 0$ if and only if the vectors $\mathbf{u}$, $\mathbf{v}$, and $\mathbf{w}$ lie in the same plane. Thus we have the following result.

> **THEOREM 3.5.5** *If the vectors* $\mathbf{u} = (u_1, u_2, u_3)$, $\mathbf{v} = (v_1, v_2, v_3)$, *and* $\mathbf{w} = (w_1, w_2, w_3)$ *have the same initial point, then they lie in the same plane if and only if*
> $$\mathbf{u} \cdot (\mathbf{v} \times \mathbf{w}) = \begin{vmatrix} u_1 & u_2 & u_3 \\ v_1 & v_2 & v_3 \\ w_1 & w_2 & w_3 \end{vmatrix} = 0$$

Concept Review

- Cross product of two vectors
- Determinant form of cross product
- Scalar triple product

Skills

- Compute the cross product of two vectors $\mathbf{u}$ and $\mathbf{v}$ in R^3.
- Know the geometric relationship between $\mathbf{u} \times \mathbf{v}$ to $\mathbf{u}$ and $\mathbf{v}$.
- Know the properties of the cross product (listed in Theorem 3.5.2).
- Compute the scalar triple product of three vectors in 3-space.
- Know the geometric interpretation of the scalar triple product.
- Compute the areas of triangles and parallelograms determined by two vectors or three points in 2-space or 3-space.
- Use the scalar triple product to determine whether three given vectors in 3-space are collinear.

Exercise Set 3.5

In Exercises 1–2, let $\mathbf{u} = (3, 2, -1)$, $\mathbf{v} = (0, 2, -3)$, and $\mathbf{w} = (2, 6, 7)$. Compute the indicated vectors.

1. (a) $\mathbf{u} \times \mathbf{v}$ (b) $(\mathbf{v} \times \mathbf{u}) \times \mathbf{w}$ (c) $\mathbf{v} \times (\mathbf{u} \times \mathbf{w})$

2. (a) $(\mathbf{u} \times \mathbf{v}) \times (\mathbf{v} \times \mathbf{w})$ (b) $\mathbf{u} \times (\mathbf{v} - 2\mathbf{w})$
   (c) $(\mathbf{u} \times \mathbf{v}) - 2\mathbf{w}$

In Exercises 3–6, use the cross product to find a vector that is orthogonal to both $\mathbf{u}$ and $\mathbf{v}$.

3. $\mathbf{u} = (2, 3, -1)$, $\mathbf{v} = (4, 1, 3)$
4. $\mathbf{u} = (1, 1, -2)$, $\mathbf{v} = (2, -1, 2)$
5. $\mathbf{u} = (0, 2, -2)$, $\mathbf{v} = (1, 3, 0)$
6. $\mathbf{u} = (3, 3, 1)$, $\mathbf{v} = (0, 4, 2)$

In Exercises 7–10, find the area of the parallelogram determined by the given vectors $\mathbf{u}$ and $\mathbf{v}$.

7. $\mathbf{u} = (1, 3, 4)$, $\mathbf{v} = (5, 1, 2)$
8. $\mathbf{u} = (3, -1, 4)$, $\mathbf{v} = (6, -2, 8)$
9. $\mathbf{u} = (2, 3, 0)$, $\mathbf{v} = (-1, 2, -2)$
10. $\mathbf{u} = (1, 1, 1)$, $\mathbf{v} = (3, 2, -5)$

In Exercises 11–12, find the area of the parallelogram with the given vertices.

11. $P_1(2, 3)$, $P_2(1, 4)$, $P_3(5, 2)$, $P_4(4, 3)$
12. $P_1(3, 2)$, $P_2(5, 4)$, $P_3(9, 4)$, $P_4(7, 2)$

In Exercises 13–14, find the area of the triangle with the given vertices.

13. $A(0, 3)$, $B(1, -2)$, $C(2, 2)$
14. $A(1, 1)$, $B(2, 2)$, $C(3, -3)$

In Exercises 15–16, find the area of the triangle in 3-space that has the given vertices.

15. $P_1(2, 6, -1)$, $P_2(1, 1, 1)$, $P_3(4, 6, 2)$
16. $P(1, -1, 2)$, $Q(0, 3, 4)$, $R(6, 1, 8)$

In Exercises 17–18, find the volume of the parallelepiped with sides $\mathbf{u}$, $\mathbf{v}$, and $\mathbf{w}$.

17. $\mathbf{u} = (0, 2, -2)$, $\mathbf{v} = (1, 2, 0)$, $\mathbf{w} = (-2, 3, 1)$
18. $\mathbf{u} = (3, 1, 2)$, $\mathbf{v} = (4, 5, 1)$, $\mathbf{w} = (1, 2, 4)$

In Exercises 19–20, determine whether $\mathbf{u}$, $\mathbf{v}$, and $\mathbf{w}$ lie in the same plane when positioned so that their initial points coincide.

19. $\mathbf{u} = (0, 1, -1)$, $\mathbf{v} = (2, 2, 0)$, $\mathbf{w} = (4, 1, 2)$

20. $\mathbf{u} = (5, -2, 1)$, $\mathbf{v} = (4, -1, 1)$, $\mathbf{w} = (1, -1, 0)$

▶ In Exercises 21–24, compute the scalar triple product $\mathbf{u} \cdot (\mathbf{v} \times \mathbf{w})$. ◀

21. $\mathbf{u} = (5, 1, 0)$, $\mathbf{v} = (6, 2, 0)$, $\mathbf{w} = (4, 2, 2)$

22. $\mathbf{u} = (-1, 2, 4)$, $\mathbf{v} = (3, 4, -2)$, $\mathbf{w} = (-1, 2, 5)$

23. $\mathbf{u} = (a, 0, 0)$, $\mathbf{v} = (0, b, 0)$, $\mathbf{w} = (0, 0, c)$

24. $\mathbf{u} = (3, -1, 6)$, $\mathbf{v} = (2, 4, 3)$, $\mathbf{w} = (5, -1, 2)$

▶ In Exercises 25–26, suppose that $\mathbf{u} \cdot (\mathbf{v} \times \mathbf{w}) = -4$. Find

25. (a) $\mathbf{u} \cdot (\mathbf{w} \times \mathbf{v})$ (b) $(\mathbf{v} \times \mathbf{w}) \cdot \mathbf{u}$ (c) $\mathbf{w} \cdot (\mathbf{u} \times \mathbf{v})$

26. (a) $\mathbf{v} \cdot (\mathbf{u} \times \mathbf{w})$ (b) $(\mathbf{u} \times \mathbf{w}) \cdot \mathbf{v}$ (c) $\mathbf{v} \cdot (\mathbf{w} \times \mathbf{u})$

27. (a) Find the area of the triangle having vertices $A(1, 0, 1)$, $B(0, 2, 3)$, and $C(2, 1, 0)$.

    (b) Use the result of part (a) to find the length of the altitude from vertex C to side AB.

28. Use the cross product to find the sine of the angle between the vectors $\mathbf{u} = (2, 3, -6)$ and $\mathbf{v} = (2, 3, 6)$.

29. Simplify $(\mathbf{u} + \mathbf{v}) \times (\mathbf{u} - \mathbf{v})$.

30. Let $\mathbf{a} = (a_1, a_2, a_3)$, $\mathbf{b} = (b_1, b_2, b_3)$, $\mathbf{c} = (c_1, c_2, c_3)$, and $\mathbf{d} = (d_1, d_2, d_3)$. Show that

$$(\mathbf{a} + \mathbf{d}) \cdot (\mathbf{b} \times \mathbf{c}) = \mathbf{a} \cdot (\mathbf{b} \times \mathbf{c}) + \mathbf{d} \cdot (\mathbf{b} \times \mathbf{c})$$

31. Let $\mathbf{u}$, $\mathbf{v}$, and $\mathbf{w}$ be nonzero vectors in 3-space with the same initial point, but such that no two of them are collinear. Show that

    (a) $\mathbf{u} \times (\mathbf{v} \times \mathbf{w})$ lies in the plane determined by $\mathbf{v}$ and $\mathbf{w}$.

    (b) $(\mathbf{u} \times \mathbf{v}) \times \mathbf{w}$ lies in the plane determined by $\mathbf{u}$ and $\mathbf{v}$.

32. Prove the following identities.

    (a) $(\mathbf{u} + k\mathbf{v}) \times \mathbf{v} = \mathbf{u} \times \mathbf{v}$

    (b) $\mathbf{u} \cdot (\mathbf{v} \times \mathbf{z}) = -(\mathbf{u} \times \mathbf{z}) \cdot \mathbf{v}$

33. Prove: If $\mathbf{a}$, $\mathbf{b}$, $\mathbf{c}$, and $\mathbf{d}$ lie in the same plane, then $(\mathbf{a} \times \mathbf{b}) \times (\mathbf{c} \times \mathbf{d}) = \mathbf{0}$.

34. Prove: If θ is the angle between $\mathbf{u}$ and $\mathbf{v}$ and $\mathbf{u} \cdot \mathbf{v} \neq 0$, then $\tan \theta = \|\mathbf{u} \times \mathbf{v}\|/(\mathbf{u} \cdot \mathbf{v})$.

35. Show that if $\mathbf{u}$, $\mathbf{v}$, and $\mathbf{w}$ are vectors in R^3, no two of which are collinear, then $\mathbf{u} \times (\mathbf{v} \times \mathbf{w})$ lies in the plane determined by $\mathbf{v}$ and $\mathbf{w}$.

36. It is a theorem of solid geometry that the volume of a tetrahedron is $\frac{1}{3}$(area of base) · (height). Use this result to prove that the volume of a tetrahedron whose sides are the vectors $\mathbf{a}$, $\mathbf{b}$, and $\mathbf{c}$ is $\frac{1}{6}|\mathbf{a} \cdot (\mathbf{b} \times \mathbf{c})|$ (see the accompanying figure).

◀ Figure Ex-36

37. Use the result of Exercise 36 to find the volume of the tetrahedron with vertices P, Q, R, S.

    (a) $P(-1, 2, 0)$, $Q(2, 1, -3)$, $R(1, 1, 1)$, $S(3, -2, 3)$

    (b) $P(0, 0, 0)$, $Q(1, 2, -1)$, $R(3, 4, 0)$, $S(-1, -3, 4)$

38. Prove part (d) of Theorem 3.5.1. [Hint: First prove the result in the case where $\mathbf{w} = \mathbf{i} = (1, 0, 0)$, then when $\mathbf{w} = \mathbf{j} = (0, 1, 0)$, and then when $\mathbf{w} = \mathbf{k} = (0, 0, 1)$. Finally, prove it for an arbitrary vector $\mathbf{w} = (w_1, w_2, w_3)$ by writing $\mathbf{w} = w_1\mathbf{i} + w_2\mathbf{j} + w_3\mathbf{k}$.]

39. Prove part (e) of Theorem 3.5.1. [Hint: Apply part (a) of Theorem 3.5.2 to the result in part (d) of Theorem 3.5.1.]

40. Prove:

    (a) Prove (b) of Theorem 3.5.2.

    (b) Prove (c) of Theorem 3.5.2.

    (c) Prove (d) of Theorem 3.5.2.

    (d) Prove (e) of Theorem 3.5.2.

    (e) Prove (f) of Theorem 3.5.2.

True-False Exercises

In parts (a)–(f) determine whether the statement is true or false, and justify your answer.

(a) The cross product of two nonzero vectors $\mathbf{u}$ and $\mathbf{v}$ is a nonzero vector if and only if $\mathbf{u}$ and $\mathbf{v}$ are not parallel.

(b) A normal vector to a plane can be obtained by taking the cross product of two nonzero and noncollinear vectors lying in the plane.

(c) The scalar triple product of $\mathbf{u}$, $\mathbf{v}$, and $\mathbf{w}$ determines a vector whose length is equal to the volume of the parallelepiped determined by $\mathbf{u}$, $\mathbf{v}$, and $\mathbf{w}$.

(d) If $\mathbf{u}$ and $\mathbf{v}$ are vectors in 3-space, then $\|\mathbf{v} \times \mathbf{u}\|$ is equal to the area of the parallelogram determined by $\mathbf{u}$ and $\mathbf{v}$.

(e) For all vectors $\mathbf{u}$, $\mathbf{v}$, and $\mathbf{w}$ in 3-space, the vectors $(\mathbf{u} \times \mathbf{v}) \times \mathbf{w}$ and $\mathbf{u} \times (\mathbf{v} \times \mathbf{w})$ are the same.

(f) If $\mathbf{u}$, $\mathbf{v}$, and $\mathbf{w}$ are vectors in R^3, where $\mathbf{u}$ is nonzero and $\mathbf{u} \times \mathbf{v} = \mathbf{u} \times \mathbf{w}$, then $\mathbf{v} = \mathbf{w}$.

Chapter 3   Supplementary Exercises

1. Let $\mathbf{u} = (-2, 0, 4)$, $\mathbf{v} = (3, -1, 6)$, and $\mathbf{w} = (2, -5, -5)$. Compute

   (a) $3\mathbf{v} - 2\mathbf{u}$
   (b) $\|\mathbf{u} + \mathbf{v} + \mathbf{w}\|$
   (c) the distance between $-3\mathbf{u}$ and $\mathbf{v} + 5\mathbf{w}$
   (d) $\text{proj}_{\mathbf{w}} \mathbf{u}$
   (e) $\mathbf{u} \cdot (\mathbf{v} \times \mathbf{w})$
   (f) $(-5\mathbf{v} + \mathbf{w}) \times ((\mathbf{u} \cdot \mathbf{v})\mathbf{w})$

2. Repeat Exercise 1 for the vectors $\mathbf{u} = 3\mathbf{i} - 5\mathbf{j} + \mathbf{k}$, $\mathbf{v} = -2\mathbf{i} + 2\mathbf{k}$, and $\mathbf{w} = -\mathbf{j} + 4\mathbf{k}$.

3. Repeat parts (a)–(d) of Exercise 1 for the vectors $\mathbf{u} = (-2, 6, 2, 1)$, $\mathbf{v} = (-3, 0, 8, 0)$, and $\mathbf{w} = (9, 1, -6, -6)$.

4. Repeat parts (a)–(d) of Exercise 1 for the vectors $\mathbf{u} = (0, 5, 0, -1, -2)$, $\mathbf{v} = (1, -1, 6, -2, 0)$, and $\mathbf{w} = (-4, -1, 4, 0, 2)$.

▶ In Exercises 5–6, determine whether the given set of vectors forms an orthogonal set. If so, normalize each vector to form an orthonormal set. ◀

5. $(-32, -1, 19)$, $(3, -1, 5)$, $(1, 6, 2)$

6. $(-2, 0, 1)$, $(1, 1, 2)$, $(1, -5, 2)$

7. (a) The set of all vectors in R^2 that are orthogonal to a nonzero vector is what kind of geometric object?
   (b) The set of all vectors in R^3 that are orthogonal to a nonzero vector is what kind of geometric object?
   (c) The set of all vectors in R^2 that are orthogonal to two noncollinear vectors is what kind of geometric object?
   (d) The set of all vectors in R^3 that are orthogonal to two noncollinear vectors is what kind of geometric object?

8. Show that $\mathbf{v}_1 = \left(\frac{2}{3}, \frac{1}{3}, \frac{2}{3}\right)$ and $\mathbf{v}_2 = \left(\frac{1}{3}, \frac{2}{3}, -\frac{2}{3}\right)$ are orthonormal vectors, and find a third vector $\mathbf{v}_3$ for which $\{\mathbf{v}_1, \mathbf{v}_2, \mathbf{v}_3\}$ is an orthonormal set.

9. *True or False:* If $\mathbf{u}$ and $\mathbf{v}$ are nonzero vectors such that $\|\mathbf{u} + \mathbf{v}\|^2 = \|\mathbf{u}\|^2 + \|\mathbf{v}\|^2$, then $\mathbf{u}$ and $\mathbf{v}$ are orthogonal.

10. *True or False:* If $\mathbf{u}$ is orthogonal to $\mathbf{v} + \mathbf{w}$, then $\mathbf{u}$ is orthogonal to $\mathbf{v}$ and $\mathbf{w}$.

11. Consider the points $P(3, -1, 4)$, $Q(6, 0, 2)$, and $R(5, 1, 1)$. Find the point S in R^3 whose first component is -1 and such that $\overrightarrow{PQ}$ is parallel to $\overrightarrow{RS}$.

12. Consider the points $P(-3, 1, 0, 6)$, $Q(0, 5, 1, -2)$, and $R(-4, 1, 4, 0)$. Find the point S in R^4 whose third component is 6 and such that $\overrightarrow{PQ}$ is parallel to $\overrightarrow{RS}$.

13. Using the points in Exercise 11, find the cosine of the angle between the vectors $\overrightarrow{PQ}$ and $\overrightarrow{PR}$.

14. Using the points in Exercise 12, find the cosine of the angle between the vectors $\overrightarrow{PQ}$ and $\overrightarrow{PR}$.

15. Find the distance between the point $P(-3, 1, 3)$ and the plane $5x + z = 3y - 4$.

16. Show that the planes $3x - y + 6z = 7$ and $-6x + 2y - 12z = 1$ are parallel, and find the distance between the planes.

▶ In Exercises 17–22, find vector and parametric equations for the line or plane in question. ◀

17. The plane in R^3 that contains the points $P(-2, 1, 3)$, $Q(-1, -1, 1)$, and $R(3, 0, -2)$.

18. The line in R^3 that contains the point $P(-1, 6, 0)$ and is orthogonal to the plane $4x - z = 5$.

19. The line in R^2 that is parallel to the vector $\mathbf{v} = (8, -1)$ and contains the point $P(0, -3)$.

20. The plane in R^3 that contains the point $P(-2, 1, 0)$ and is parallel to the plane $-8x + 6y - z = 4$.

21. The line in R^2 with equation $y = 3x - 5$.

22. The plane in R^3 with equation $2x - 6y + 3z = 5$.

▶ In Exercises 23–25, find a point-normal equation for the given plane. ◀

23. The plane that is represented by the vector equation $(x, y, z) = (-1, 5, 6) + t_1(0, -1, 3) + t_2(2, -1, 0)$.

24. The plane that contains the point $P(-5, 1, 0)$ and is orthogonal to the line with parametric equations $x = 3 - 5t$, $y = 2t$, and $z = 7$.

25. The plane that passes through the points $P(9, 0, 4)$, $Q(-1, 4, 3)$, and $R(0, 6, -2)$.

26. Suppose that $\{\mathbf{v}_1, \mathbf{v}_2, \mathbf{v}_3\}$ and $\{\mathbf{w}_1, \mathbf{w}_2\}$ are two sets of vectors such that $\mathbf{v}_i$ and $\mathbf{w}_j$ are orthogonal for all i and j. Prove that if a_1, a_2, a_3, b_1, b_2 are any scalars, then the vectors $\mathbf{v} = a_1\mathbf{v}_1 + a_2\mathbf{v}_2 + a_3\mathbf{v}_3$ and $\mathbf{w} = b_1\mathbf{w}_1 + b_2\mathbf{w}_2$ are orthogonal.

27. Prove that if two vectors $\mathbf{u}$ and $\mathbf{v}$ in R^2 are orthogonal to a nonzero vector $\mathbf{w}$ in R^2, then $\mathbf{u}$ and $\mathbf{v}$ are scalar multiples of each other.

28. Prove that $\|\mathbf{u} + \mathbf{v}\| = \|\mathbf{u}\| + \|\mathbf{v}\|$ if and only if $\mathbf{u}$ and $\mathbf{v}$ are parallel vectors.

29. The equation $Ax + By = 0$ represents a line through the origin in R^2 if A and B are not both zero. What does this equation represent in R^3 if you think of it as $Ax + By + 0z = 0$? Explain.

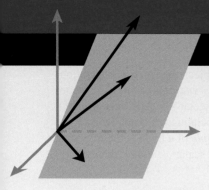

CHAPTER 4

General Vector Spaces

CHAPTER CONTENTS
4.1 Real Vector Spaces 171
4.2 Subspaces 179
4.3 Linear Independence 190
4.4 Coordinates and Basis 200
4.5 Dimension 209
4.6 Change of Basis 217
4.7 Row Space, Column Space, and Null Space 225
4.8 Rank, Nullity, and the Fundamental Matrix Spaces 237
4.9 Matrix Transformations from R^n to R^m 247
4.10 Properties of Matrix Transformations 263
4.11 Geometry of Matrix Operators on R^2 273
4.12 Dynamical Systems and Markov Chains 282

INTRODUCTION Recall that we began our study of vectors by viewing them as directed line segments (arrows). We then extended this idea by introducing rectangular coordinate systems, which enabled us to view vectors as ordered pairs and ordered triples of real numbers. As we developed properties of these vectors we noticed patterns in various formulas that enabled us to extend the notion of a vector to an n-tuple of real numbers. Although n-tuples took us outside the realm of our "visual experience," it gave us a valuable tool for understanding and studying systems of linear equations. In this chapter we will extend the concept of a vector yet again by using the most important algebraic properties of vectors in R^n as axioms. These axioms, if satisfied by a set of objects, will enable us to think of those objects as vectors.

4.1 Real Vector Spaces

In this section we will extend the concept of a vector by using the basic properties of vectors in R^n as axioms, which if satisfied by a set of objects, guarantee that those objects behave like familiar vectors.

Vector Space Axioms The following definition consists of ten axioms, eight of which are properties of vectors in R^n that were stated in Theorem 3.1.1. It is important to keep in mind that one does not *prove* axioms; rather, they are assumptions that serve as the starting point for proving theorems.

172 Chapter 4 General Vector Spaces

Vector space scalars can be real numbers or complex numbers. Vector spaces with real scalars are called *real vector spaces* and those with complex scalars are called *complex vector spaces*. For now we will be concerned exclusively with real vector spaces. We will consider complex vector spaces later.

DEFINITION 1 Let V be an arbitrary nonempty set of objects on which two operations are defined: addition, and multiplication by scalars. By *addition* we mean a rule for associating with each pair of objects $\mathbf{u}$ and $\mathbf{v}$ in V an object $\mathbf{u} + \mathbf{v}$, called the *sum* of $\mathbf{u}$ and $\mathbf{v}$; by *scalar multiplication* we mean a rule for associating with each scalar k and each object $\mathbf{u}$ in V an object $k\mathbf{u}$, called the *scalar multiple* of $\mathbf{u}$ by k. If the following axioms are satisfied by all objects $\mathbf{u}, \mathbf{v}, \mathbf{w}$ in V and all scalars k and m, then we call V a *vector space* and we call the objects in V *vectors*.

1. If $\mathbf{u}$ and $\mathbf{v}$ are objects in V, then $\mathbf{u} + \mathbf{v}$ is in V.
2. $\mathbf{u} + \mathbf{v} = \mathbf{v} + \mathbf{u}$
3. $\mathbf{u} + (\mathbf{v} + \mathbf{w}) = (\mathbf{u} + \mathbf{v}) + \mathbf{w}$
4. There is an object $\mathbf{0}$ in V, called a *zero vector* for V, such that $\mathbf{0} + \mathbf{u} = \mathbf{u} + \mathbf{0} = \mathbf{u}$ for all $\mathbf{u}$ in V.
5. For each $\mathbf{u}$ in V, there is an object $-\mathbf{u}$ in V, called a *negative* of $\mathbf{u}$, such that $\mathbf{u} + (-\mathbf{u}) = (-\mathbf{u}) + \mathbf{u} = \mathbf{0}$.
6. If k is any scalar and $\mathbf{u}$ is any object in V, then $k\mathbf{u}$ is in V.
7. $k(\mathbf{u} + \mathbf{v}) = k\mathbf{u} + k\mathbf{v}$
8. $(k + m)\mathbf{u} = k\mathbf{u} + m\mathbf{u}$
9. $k(m\mathbf{u}) = (km)(\mathbf{u})$
10. $1\mathbf{u} = \mathbf{u}$

Observe that the definition of a vector space does not specify the nature of the vectors or the operations. Any kind of object can be a vector, and the operations of addition and scalar multiplication need not have any relationship to those on R^n. The only requirement is that the ten vector space axioms be satisfied. In the examples that follow we will use four basic steps to show that a set with two operations is a vector space.

To Show that a Set with Two Operations is a Vector Space

Step 1. Identify the set V of objects that will become vectors.

Step 2. Identify the addition and scalar multiplication operations on V.

Step 3. Verify Axioms 1 and 6; that is, adding two vectors in V produces a vector in V, and multiplying a vector in V by a scalar also produces a vector in V. Axiom 1 is called *closure under addition*, and Axiom 6 is called *closure under scalar multiplication*.

Step 4. Confirm that Axioms 2, 3, 4, 5, 7, 8, 9, and 10 hold.

Hermann Günther Grassmann (1809–1877)

Historical Note The notion of an "abstract vector space" evolved over many years and had many contributors. The idea crystallized with the work of the German mathematician H. G. Grassmann, who published a paper in 1862 in which he considered abstract systems of unspecified elements on which he defined formal operations of addition and scalar multiplication. Grassmann's work was controversial, and others, including Augustin Cauchy (p. 137), laid reasonable claim to the idea.

[*Image:* ©Sueddeutsche Zeitung Photo/The Image Works]

Our first example is the simplest of all vector spaces in that it contains only one object. Since Axiom 4 requires that every vector space contain a zero vector, the object will have to be that vector.

▶ **EXAMPLE 1** **The Zero Vector Space**

Let V consist of a single object, which we denote by $\mathbf{0}$, and define
$$\mathbf{0} + \mathbf{0} = \mathbf{0} \quad \text{and} \quad k\mathbf{0} = \mathbf{0}$$
for all scalars k. It is easy to check that all the vector space axioms are satisfied. We call this the *zero vector space*. ◀

Our second example is one of the most important of all vector spaces—the familiar space R^n. It should not be surprising that the operations on R^n satisfy the vector space axioms because those axioms were based on known properties of operations on R^n.

▶ **EXAMPLE 2** *R^n Is a Vector Space*

Let $V = R^n$, and define the vector space operations on V to be the usual operations of addition and scalar multiplication of n-tuples; that is,
$$\mathbf{u} + \mathbf{v} = (u_1, u_2, \ldots, u_n) + (v_1, v_2, \ldots, v_n) = (u_1 + v_1, u_2 + v_2, \ldots, u_n + v_n)$$
$$k\mathbf{u} = (ku_1, ku_2, \ldots, ku_n)$$

The set $V = R^n$ is closed under addition and scalar multiplication because the foregoing operations produce n-tuples as their end result, and these operations satisfy Axioms 2, 3, 4, 5, 7, 8, 9, and 10 by virtue of Theorem 3.1.1. ◀

Our next example is a generalization of R^n in which we allow vectors to have infinitely many components.

▶ **EXAMPLE 3** **The Vector Space of Infinite Sequences of Real Numbers**

Let V consist of objects of the form
$$\mathbf{u} = (u_1, u_2, \ldots, u_n, \ldots)$$
in which $u_1, u_2, \ldots, u_n, \ldots$ is an infinite sequence of real numbers. We define two infinite sequences to be *equal* if their corresponding components are equal, and we define addition and scalar multiplication componentwise by
$$\mathbf{u} + \mathbf{v} = (u_1, u_2, \ldots, u_n, \ldots) + (v_1, v_2, \ldots, v_n, \ldots)$$
$$= (u_1 + v_1, u_2 + v_2, \ldots, u_n + v_n, \ldots)$$
$$k\mathbf{u} = (ku_1, ku_2, \ldots, ku_n, \ldots)$$

We leave it as an exercise to confirm that V with these operations is a vector space. We will denote this vector space by the symbol R^∞. ◀

In the next example our vectors will be matrices. This may be a little confusing at first because matrices are composed of rows and columns, which are themselves vectors (row vectors and column vectors). However, here we will not be concerned with the individual rows and columns but rather with the properties of the matrix operations as they relate to the matrix as a whole.

> **EXAMPLE 4** **A Vector Space of 2 × 2 Matrices**
>
> Let V be the set of 2×2 matrices with real entries, and take the vector space operations on V to be the usual operations of matrix addition and scalar multiplication; that is,
>
> $$\mathbf{u} + \mathbf{v} = \begin{bmatrix} u_{11} & u_{12} \\ u_{21} & u_{22} \end{bmatrix} + \begin{bmatrix} v_{11} & v_{12} \\ v_{21} & v_{22} \end{bmatrix} = \begin{bmatrix} u_{11} + v_{11} & u_{12} + v_{12} \\ u_{21} + v_{21} & u_{22} + v_{22} \end{bmatrix} \quad (1)$$
>
> $$k\mathbf{u} = k \begin{bmatrix} u_{11} & u_{12} \\ u_{21} & u_{22} \end{bmatrix} = \begin{bmatrix} ku_{11} & ku_{12} \\ ku_{21} & ku_{22} \end{bmatrix}$$

Note that Equation (1) involves three *different addition operations: the addition operation on vectors, the addition operation on matrices, and the addition operation on real numbers.*

> The set V is closed under addition and scalar multiplication because the foregoing operations produce 2×2 matrices as the end result. Thus, it remains to confirm that Axioms 2, 3, 4, 5, 7, 8, 9, and 10 hold. Some of these are standard properties of matrix operations. For example, Axiom 2 follows from Theorem 1.4.1a since
>
> $$\mathbf{u} + \mathbf{v} = \begin{bmatrix} u_{11} & u_{12} \\ u_{21} & u_{22} \end{bmatrix} + \begin{bmatrix} v_{11} & v_{12} \\ v_{21} & v_{22} \end{bmatrix} = \begin{bmatrix} v_{11} & v_{12} \\ v_{21} & v_{22} \end{bmatrix} + \begin{bmatrix} u_{11} & u_{12} \\ u_{21} & u_{22} \end{bmatrix} = \mathbf{v} + \mathbf{u}$$
>
> Similarly, Axioms 3, 7, 8, and 9 follow from parts (b), (h), (j), and (e), respectively, of that theorem (verify). This leaves Axioms 4, 5, and 10 that remain to be verified.
>
> To confirm that Axiom 4 is satisfied, we must find a 2×2 matrix $\mathbf{0}$ in V for which $\mathbf{u} + \mathbf{0} = \mathbf{0} + \mathbf{u}$ for all 2×2 matrices in V. We can do this by taking
>
> $$\mathbf{0} = \begin{bmatrix} 0 & 0 \\ 0 & 0 \end{bmatrix}$$
>
> With this definition,
>
> $$\mathbf{0} + \mathbf{u} = \begin{bmatrix} 0 & 0 \\ 0 & 0 \end{bmatrix} + \begin{bmatrix} u_{11} & u_{12} \\ u_{21} & u_{22} \end{bmatrix} = \begin{bmatrix} u_{11} & u_{12} \\ u_{21} & u_{22} \end{bmatrix} = \mathbf{u}$$
>
> and similarly $\mathbf{u} + \mathbf{0} = \mathbf{u}$. To verify that Axiom 5 holds we must show that each object $\mathbf{u}$ in V has a negative $-\mathbf{u}$ in V such that $\mathbf{u} + (-\mathbf{u}) = \mathbf{0}$ and $(-\mathbf{u}) + \mathbf{u} = \mathbf{0}$. This can be done by defining the negative of $\mathbf{u}$ to be
>
> $$-\mathbf{u} = \begin{bmatrix} -u_{11} & -u_{12} \\ -u_{21} & -u_{22} \end{bmatrix}$$
>
> With this definition,
>
> $$\mathbf{u} + (-\mathbf{u}) = \begin{bmatrix} u_{11} & u_{12} \\ u_{21} & u_{22} \end{bmatrix} + \begin{bmatrix} -u_{11} & -u_{12} \\ -u_{21} & -u_{22} \end{bmatrix} = \begin{bmatrix} 0 & 0 \\ 0 & 0 \end{bmatrix} = \mathbf{0}$$
>
> and similarly $(-\mathbf{u}) + \mathbf{u} = \mathbf{0}$. Finally, Axiom 10 holds because
>
> $$1\mathbf{u} = 1 \begin{bmatrix} u_{11} & u_{12} \\ u_{21} & u_{22} \end{bmatrix} = \begin{bmatrix} u_{11} & u_{12} \\ u_{21} & u_{22} \end{bmatrix} = \mathbf{u}$$

> **EXAMPLE 5** **The Vector Space of $m \times n$ Matrices**
>
> Example 4 is a special case of a more general class of vector spaces. You should have no trouble adapting the argument used in that example to show that the set V of all $m \times n$ matrices with the usual matrix operations of addition and scalar multiplication is a vector space. We will denote this vector space by the symbol M_{mn}. Thus, for example, the vector space in Example 4 is denoted as M_{22}.

▶ **EXAMPLE 6** **The Vector Space of Real-Valued Functions**

Let V be the set of real-valued functions that are defined at each x in the interval $(-\infty, \infty)$. If $\mathbf{f} = f(x)$ and $\mathbf{g} = g(x)$ are two functions in V and if k is any scalar, then define the operations of addition and scalar multiplication by

$$(\mathbf{f} + \mathbf{g})(x) = f(x) + g(x) \tag{2}$$

$$(k\mathbf{f})(x) = kf(x) \tag{3}$$

One way to think about these operations is to view the numbers $f(x)$ and $g(x)$ as "components" of $\mathbf{f}$ and $\mathbf{g}$ at the point x, in which case Equations (2) and (3) state that two functions are added by adding corresponding components, and a function is multiplied by a scalar by multiplying each component by that scalar—exactly as in R^n and R^∞. This idea is illustrated in parts (a) and (b) of Figure 4.1.1. The set V with these operations is denoted by the symbol $F(-\infty, \infty)$. We can prove that this is a vector space as follows:

Axioms 1 and 6: These closure axioms require that if we add two functions that are defined at each x in the interval $(-\infty, \infty)$, then sums and scalar multiples of those functions are also defined at each x in the interval $(-\infty, \infty)$. This follows from Formulas (2) and (3).

Axiom 4: This axiom requires that there exists a function $\mathbf{0}$ in $F(-\infty, \infty)$, which when added to any other function $\mathbf{f}$ in $F(-\infty, \infty)$ produces $\mathbf{f}$ back again as the result. The function, whose value at every point x in the interval $(-\infty, \infty)$ is zero, has this property. Geometrically, the graph of the function $\mathbf{0}$ is the line that coincides with the x-axis.

Axiom 5: This axiom requires that for each function $\mathbf{f}$ in $F(-\infty, \infty)$ there exists a function $-\mathbf{f}$ in $F(-\infty, \infty)$, which when added to $\mathbf{f}$ produces the function $\mathbf{0}$. The function defined by $-\mathbf{f}(x) = -f(x)$ has this property. The graph of $-\mathbf{f}$ can be obtained by reflecting the graph of $\mathbf{f}$ about the x-axis (Figure 4.1.1c).

Axioms 2, 3, 7, 8, 9, 10: The validity of each of these axioms follows from properties of real numbers. For example, if $\mathbf{f}$ and $\mathbf{g}$ are functions in $F(-\infty, \infty)$, then Axiom 2 requires that $\mathbf{f} + \mathbf{g} = \mathbf{g} + \mathbf{f}$. This follows from the computation

$$(\mathbf{f} + \mathbf{g})(x) = \mathbf{f}(x) + \mathbf{g}(x) = \mathbf{g}(x) + \mathbf{f}(x) = (\mathbf{g} + \mathbf{f})(x)$$

in which the first and last equalities follow from (2), and the middle equality is a property of real numbers. We will leave the proofs of the remaining parts as exercises. ◀

> In Example 6 the functions were defined on the entire interval $(-\infty, \infty)$. However, the arguments used in that example apply as well on all subintervals of $(-\infty, \infty)$, such as a closed interval $[a, b]$ or an open interval (a, b). We will denote the vector spaces of functions on these intervals by $F[a, b]$ and $F(a, b)$, respectively.

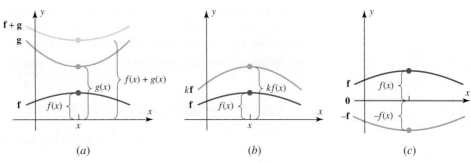

▲ Figure 4.1.1

It is important to recognize that you cannot impose any two operations on any set V and expect the vector space axioms to hold. For example, if V is the set of n-tuples with *positive* components, and if the standard operations from R^n are used, then V is not closed under scalar multiplication, because if $\mathbf{u}$ is a nonzero n-tuple in V, then $(-1)\mathbf{u}$ has at least one negative component and hence is not in V. The following is a less obvious example in which only one of the ten vector space axioms fails to hold.

▶ EXAMPLE 7 A Set That Is Not a Vector Space

Let $V = R^2$ and define addition and scalar multiplication operations as follows: If $\mathbf{u} = (u_1, u_2)$ and $\mathbf{v} = (v_1, v_2)$, then define

$$\mathbf{u} + \mathbf{v} = (u_1 + v_1, u_2 + v_2)$$

and if k is any real number, then define

$$k\mathbf{u} = (ku_1, 0)$$

For example, if $\mathbf{u} = (2, 4)$, $\mathbf{v} = (-3, 5)$, and $k = 7$, then

$$\mathbf{u} + \mathbf{v} = (2 + (-3), 4 + 5) = (-1, 9)$$
$$k\mathbf{u} = 7\mathbf{u} = (7 \cdot 2, 0) = (14, 0)$$

The addition operation is the standard one from R^2, but the scalar multiplication is not. In the exercises we will ask you to show that the first nine vector space axioms are satisfied. However, Axiom 10 fails to hold for certain vectors. For example, if $\mathbf{u} = (u_1, u_2)$ is such that $u_2 \neq 0$, then

$$1\mathbf{u} = 1(u_1, u_2) = (1 \cdot u_1, 0) = (u_1, 0) \neq \mathbf{u}$$

Thus, V is not a vector space with the stated operations. ◀

Our final example will be an unusual vector space that we have included to illustrate how varied vector spaces can be. Since the objects in this space will be real numbers, it will be important for you to keep track of which operations are intended as vector operations and which ones as ordinary operations on real numbers.

▶ EXAMPLE 8 An Unusual Vector Space

Let V be the set of positive real numbers, and define the operations on V to be

$$u + v = uv \quad \text{[Vector addition is numerical multiplication.]}$$
$$ku = u^k \quad \text{[Scalar multiplication is numerical exponentiation.]}$$

Thus, for example, $1 + 1 = 1$ and $(2)(1) = 1^2 = 1$—strange indeed, but nevertheless the set V with these operations satisfies the 10 vector space axioms and hence is a vector space. We will confirm Axioms 4, 5, and 7, and leave the others as exercises.

- Axiom 4—The zero vector in this space is the number 1 (i.e., $\mathbf{0} = 1$) since

$$u + 1 = u \cdot 1 = u$$

- Axiom 5—The negative of a vector u is its reciprocal (i.e., $-u = 1/u$) since

$$u + \frac{1}{u} = u\left(\frac{1}{u}\right) = 1 \,(= \mathbf{0})$$

- Axiom 7—$k(u + v) = (uv)^k = u^k v^k = (ku) + (kv)$. ◀

Some Properties of Vectors

The following is our first theorem about general vector spaces. As you will see, its proof is very formal with each step being justified by a vector space axiom or a known property of real numbers. There will not be many rigidly formal proofs of this type in the text, but we have included these to reinforce the idea that the familiar properties of vectors can all be derived from the vector space axioms.

THEOREM 4.1.1 *Let V be a vector space, **u** a vector in V, and k a scalar; then:*
(a) $0\mathbf{u} = \mathbf{0}$
(b) $k\mathbf{0} = \mathbf{0}$
(c) $(-1)\mathbf{u} = -\mathbf{u}$
(d) *If* $k\mathbf{u} = \mathbf{0}$, *then* $k = 0$ *or* $\mathbf{u} = \mathbf{0}$.

We will prove parts (*a*) and (*c*) and leave proofs of the remaining parts as exercises.

Proof (a) We can write
$$0\mathbf{u} + 0\mathbf{u} = (0 + 0)\mathbf{u} \quad \text{[Axiom 8]}$$
$$= 0\mathbf{u} \quad \text{[Property of the number 0]}$$

By Axiom 5 the vector $0\mathbf{u}$ has a negative, $-0\mathbf{u}$. Adding this negative to both sides above yields
$$[0\mathbf{u} + 0\mathbf{u}] + (-0\mathbf{u}) = 0\mathbf{u} + (-0\mathbf{u})$$
or
$$0\mathbf{u} + [0\mathbf{u} + (-0\mathbf{u})] = 0\mathbf{u} + (-0\mathbf{u}) \quad \text{[Axiom 3]}$$
$$0\mathbf{u} + \mathbf{0} = \mathbf{0} \quad \text{[Axiom 5]}$$
$$0\mathbf{u} = \mathbf{0} \quad \text{[Axiom 4]}$$

Proof (c) To prove that $(-1)\mathbf{u} = -\mathbf{u}$, we must show that $\mathbf{u} + (-1)\mathbf{u} = \mathbf{0}$. The proof is as follows:
$$\mathbf{u} + (-1)\mathbf{u} = 1\mathbf{u} + (-1)\mathbf{u} \quad \text{[Axiom 10]}$$
$$= (1 + (-1))\mathbf{u} \quad \text{[Axiom 8]}$$
$$= 0\mathbf{u} \quad \text{[Property of numbers]}$$
$$= \mathbf{0} \quad \text{[Part (}a\text{) of this theorem]} \blacktriangleleft$$

A Closing Observation This section of the text is very important to the overall plan of linear algebra in that it establishes a common thread between such diverse mathematical objects as geometric vectors, vectors in R^n, infinite sequences, matrices, and real-valued functions, to name a few. As a result, whenever we discover a new theorem about general vector spaces, we will at the same time be discovering a theorem about geometric vectors, vectors in R^n, sequences, matrices, real-valued functions, and about any new kinds of vectors that we might discover.

To illustrate this idea, consider what the rather innocent-looking result in part (*a*) of Theorem 4.1.1 says about the vector space in Example 8. Keeping in mind that the vectors in that space are positive real numbers, that scalar multiplication means numerical exponentiation, and that the zero vector is the number 1, the equation
$$0\mathbf{u} = \mathbf{0}$$
is a statement of the fact that if u is a positive real number, then
$$u^0 = 1$$

Concept Review

- Vector space
- Closure under addition
- Closure under scalar multiplication
- Examples of vector spaces

Skills

- Determine whether a given set with two operations is a vector space.
- Show that a set with two operations is not a vector space by demonstrating that at least one of the vector space axioms fails.

Exercise Set 4.1

1. Let V be the set of all ordered pairs of real numbers, and consider the following addition and scalar multiplication operations on $\mathbf{u} = (u_1, u_2)$ and $\mathbf{v} = (v_1, v_2)$:

   $$\mathbf{u} + \mathbf{v} = (u_1 + v_1, u_2 + v_2), \quad k\mathbf{u} = (0, ku_2)$$

   (a) Compute $\mathbf{u} + \mathbf{v}$ and $k\mathbf{u}$ for $\mathbf{u} = (2, 4)$, $\mathbf{v} = (1, -3)$, and $k = 5$.

   (b) In words, explain why V is closed under addition and scalar multiplication.

   (c) Since addition on V is the standard addition operation on R^2, certain vector space axioms hold for V because they are known to hold for R^2. Which axioms are they?

   (d) Show that Axioms 7, 8, and 9 hold.

   (e) Show that Axiom 10 fails and hence that V is not a vector space under the given operations.

2. Let V be the set of all ordered pairs of real numbers, and consider the following addition and scalar multiplication operations on $\mathbf{u} = (u_1, u_2)$ and $\mathbf{v} = (v_1, v_2)$:

   $$\mathbf{u} + \mathbf{v} = (u_1 + v_1 - 1, u_2 + v_2 - 1), \quad k\mathbf{u} = (ku_1, ku_2)$$

   (a) Compute $\mathbf{u} + \mathbf{v}$ and $k\mathbf{u}$ for $\mathbf{u} = (1, -2)$, $\mathbf{v} = (2, 0)$, and $k = 3$.

   (b) Show that $(0, 0) \neq \mathbf{0}$.

   (c) Show that $(1, 1) = \mathbf{0}$.

   (d) Show that Axiom 5 holds by producing an ordered pair $-\mathbf{u}$ such that $\mathbf{u} + (-\mathbf{u}) = \mathbf{0}$ for $\mathbf{u} = (u_1, u_2)$.

   (e) Find two vector space axioms that fail to hold.

▶ In Exercises 3–12, determine whether each set equipped with the given operations is a vector space. For those that are not vector spaces identify the vector space axioms that fail. ◀

3. The set of all real numbers with the standard operations of addition and multiplication.

4. The set of all pairs of real numbers of the form $(0, y)$, with the standard operations on R^2.

5. The set of all pairs of real numbers of the form (x, y), where $x \geq 0$, with the standard operations on R^2.

6. The set of all n-tuples of real numbers that have the form $(x, x, \ldots, x)$ with the standard operations on R^n.

7. The set of all triples of real numbers with the standard vector addition but with scalar multiplication defined by

   $$k(x, y, z) = (k^2 x, k^2 y, k^2 z)$$

8. The set of all 2×2 invertible matrices with the standard matrix addition and scalar multiplication.

9. The set of all 2×2 matrices of the form

   $$\begin{bmatrix} a & 0 \\ 0 & b \end{bmatrix}$$

   with the standard matrix addition and scalar multiplication.

10. The set of all real-valued functions f defined everywhere on the real line and such that $f(1) = 0$ with the operations used in Example 6.

11. The set of all pairs of real numbers of the form $(x, 1)$ with the operations

    $$(x, 1) + (x', 1) = (x + x', 1) \quad \text{and} \quad k(x, 1) = (k^2 x, 1)$$

12. The set of polynomials of the form $a_0 + a_1 x$ with the operations

    $$(a_0 + a_1 x) + (b_0 + b_1 x) = (a_0 + b_0) + (a_1 + b_1)x$$

    and

    $$k(a_0 + a_1 x) = (ka_0) + (ka_1)x$$

13. Verify Axioms 3, 7, 8, and 9 for the vector space given in Example 4.

14. Verify Axioms 1, 2, 3, 7, 8, 9, and 10 for the vector space given in Example 6.

15. With the addition and scalar multiplication operations defined in Example 7, show that $V = R^2$ satisfies Axioms 1–9.

16. Verify Axioms 1, 2, 3, 6, 8, 9, and 10 for the vector space given in Example 8.

17. Show that the set of all points in R^2 lying on a line is a vector space with respect to the standard operations of vector addition and scalar multiplication if and only if the line passes through the origin.

18. Show that the set of all points in R^3 lying in a plane is a vector space with respect to the standard operations of vector addition and scalar multiplication if and only if the plane passes through the origin.

▶ In Exercises 19–21, prove that the given set with the stated operations is a vector space. ◀

19. The set $V = \{\mathbf{0}\}$ with the operations of addition and scalar multiplication given in Example 1.

20. The set R^∞ of all infinite sequences of real numbers with the operations of addition and scalar multiplication given in Example 3.

21. The set M_{mn} of all $m \times n$ matrices with the usual operations of addition and scalar multiplication.

22. Prove part (d) of Theorem 4.1.1.

23. The argument that follows proves that if $\mathbf{u}$, $\mathbf{v}$, and $\mathbf{w}$ are vectors in a vector space V such that $\mathbf{u} + \mathbf{w} = \mathbf{v} + \mathbf{w}$, then $\mathbf{u} = \mathbf{v}$ (the *cancellation law* for vector addition). As illustrated, justify the steps by filling in the blanks.

$\mathbf{u} + \mathbf{w} = \mathbf{v} + \mathbf{w}$	Hypothesis
$(\mathbf{u} + \mathbf{w}) + (-\mathbf{w}) = (\mathbf{v} + \mathbf{w}) + (-\mathbf{w})$	Add $-\mathbf{w}$ to both sides.
$\mathbf{u} + [\mathbf{w} + (-\mathbf{w})] = \mathbf{v} + [\mathbf{w} + (-\mathbf{w})]$	_____
$\mathbf{u} + \mathbf{0} = \mathbf{v} + \mathbf{0}$	_____
$\mathbf{u} = \mathbf{v}$	_____

24. Let $\mathbf{v}$ be any vector in a vector space V. Prove that $0\mathbf{v} = \mathbf{0}$.

25. Below is a seven-step proof of part (b) of Theorem 4.1.1. Justify each step either by stating that it is true by *hypothesis* or by specifying which of the ten vector space axioms applies.

Hypothesis: Let $\mathbf{u}$ be any vector in a vector space V, let $\mathbf{0}$ be the zero vector in V, and let k be a scalar.

Conclusion: Then $k\mathbf{0} = \mathbf{0}$.

Proof:
(1) $k\mathbf{0} + k\mathbf{u} = k(\mathbf{0} + \mathbf{u})$
(2) $\phantom{k\mathbf{0} + k\mathbf{u}} = k\mathbf{u}$
(3) Since $k\mathbf{u}$ is in V, $-k\mathbf{u}$ is in V.
(4) Therefore, $(k\mathbf{0} + k\mathbf{u}) + (-k\mathbf{u}) = k\mathbf{u} + (-k\mathbf{u})$.
(5) $ k\mathbf{0} + (k\mathbf{u} + (-k\mathbf{u})) = k\mathbf{u} + (-k\mathbf{u})$
(6) $ k\mathbf{0} + \mathbf{0} = \mathbf{0}$
(7) $ k\mathbf{0} = \mathbf{0}$

26. Let $\mathbf{v}$ be any vector in a vector space V. Prove that $-\mathbf{v} = (-1)\mathbf{v}$.

27. Prove: If $\mathbf{u}$ is a vector in a vector space V and k a scalar such that $k\mathbf{u} = \mathbf{0}$, then either $k = 0$ or $\mathbf{u} = \mathbf{0}$. [*Suggestion:* Show that if $k\mathbf{u} = \mathbf{0}$ and $k \neq 0$, then $\mathbf{u} = \mathbf{0}$. The result then follows as a logical consequence of this.]

True-False Exercises

In parts (a)–(e) determine whether the statement is true or false, and justify your answer.

(a) A vector is a directed line segment (an arrow).

(b) A vector is an n-tuple of real numbers.

(c) A vector is any element of a vector space.

(d) There is a vector space consisting of exactly two distinct vectors.

(e) The set of polynomials with degree exactly 1 is a vector space under the operations defined in Exercise 12.

4.2 Subspaces

It is possible for one vector space to be contained within another. We will explore this idea in this section, we will discuss how to recognize such vector spaces, and we will give a variety of examples that will be used in our later work.

We will begin with some terminology.

DEFINITION 1 A subset W of a vector space V is called a ***subspace*** of V if W is itself a vector space under the addition and scalar multiplication defined on V.

In general, to show that a nonempty set W with two operations is a vector space one must verify the ten vector space axioms. However, if W is a subspace of a known vector space V, then certain axioms need not be verified because they are "inherited" from V. For example, it is *not* necessary to verify that $\mathbf{u} + \mathbf{v} = \mathbf{v} + \mathbf{u}$ holds in W because it holds for all vectors in V including those in W. On the other hand, it *is* necessary to verify that W is closed under addition and scalar multiplication since it is possible that adding two vectors in W or multiplying a vector in W by a scalar produces a vector in V that is outside of W (Figure 4.2.1).

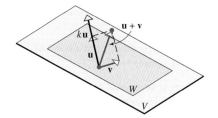

▶ **Figure 4.2.1** The vectors **u** and **v** are in W, but the vectors $\mathbf{u} + \mathbf{v}$ and $k\mathbf{u}$ are not.

Those axioms that are *not* inherited by W are

Axiom 1—Closure of W under addition

Axiom 4—Existence of a zero vector in W

Axiom 5—Existence of a negative in W for every vector in W

Axiom 6—Closure of W under scalar multiplication

so these must be verified to prove that it is a subspace of V. However, the following theorem shows that if Axiom 1 and Axiom 6 hold in W, then Axioms 4 and 5 hold in W as a consequence and hence need not be verified.

> **THEOREM 4.2.1** *If W is a set of one or more vectors in a vector space V, then W is a subspace of V if and only if the following conditions hold.*
>
> (a) *If $\mathbf{u}$ and $\mathbf{v}$ are vectors in W, then $\mathbf{u} + \mathbf{v}$ is in W.*
>
> (b) *If k is any scalar and $\mathbf{u}$ is any vector in W, then $k\mathbf{u}$ is in W.*

Proof If W is a subspace of V, then all the vector space axioms hold in W, including Axioms 1 and 6, which are precisely conditions (*a*) and (*b*).

Conversely, assume that conditions (*a*) and (*b*) hold. Since these are Axioms 1 and 6, and since Axioms 2, 3, 7, 8, 9, and 10 are inherited from V, we only need to show that Axioms 4 and 5 hold in W. For this purpose, let $\mathbf{u}$ be any vector in W. It follows from condition (*b*) that $k\mathbf{u}$ is a vector in W for every scalar k. In particular, $0\mathbf{u} = \mathbf{0}$ and $(-1)\mathbf{u} = -\mathbf{u}$ are in W, which shows that Axioms 4 and 5 hold in W. ◀

In words, Theorem 4.2.1 states that W is a subspace of V if and only if it is closed under addition and scalar multiplication.

▶ **EXAMPLE 1** **The Zero Subspace**

If V is any vector space, and if $W = \{\mathbf{0}\}$ is the subset of V that consists of the zero vector only, then W is closed under addition and scalar multiplication since

$$\mathbf{0} + \mathbf{0} = \mathbf{0} \quad \text{and} \quad k\mathbf{0} = \mathbf{0}$$

for any scalar k. We call W the ***zero subspace*** of V.

Note that every vector space has at least two subspaces, itself and its zero subspace.

▶ **EXAMPLE 2** **Lines Through the Origin Are Subspaces of R^2 and of R^3**

If W is a line through the origin of either R^2 or R^3, then adding two vectors on the line W or multiplying a vector on the line W by a scalar produces another vector on the line W, so W is closed under addition and scalar multiplication (see Figure 4.2.2 for an illustration in R^3).

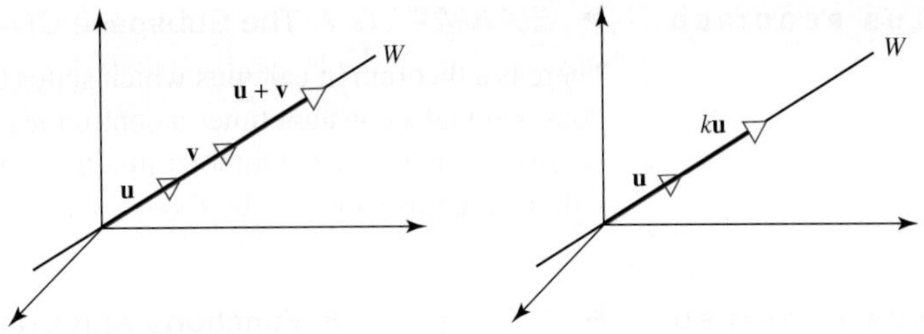

4.2 Subspaces

(a) W is closed under addition.

(b) W is closed under scalar multiplication.

► Figure 4.2.2

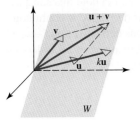

▲ Figure 4.2.3 The vectors $\mathbf{u} + \mathbf{v}$ and $k\mathbf{u}$ both lie in the same plane as $\mathbf{u}$ and $\mathbf{v}$.

► **EXAMPLE 3** Planes Through the Origin Are Subspaces of R^3

If $\mathbf{u}$ and $\mathbf{v}$ are vectors in a plane W through the origin of R^3, then it is evident geometrically that $\mathbf{u} + \mathbf{v}$ and $k\mathbf{u}$ lie in the same plane W for any scalar k (Figure 4.2.3). Thus W is closed under addition and scalar multiplication. ◄

Table 1 that follows gives a list of subspaces of R^2 and of R^3 that we have encountered thus far. We will see later that these are the only subspaces of R^2 and of R^3.

Table 1

Subspaces of R^2	Subspaces of R^3
• $\{\mathbf{0}\}$	• $\{\mathbf{0}\}$
• Lines through the origin	• Lines through the origin
• R^2	• Planes through the origin
	• R^3

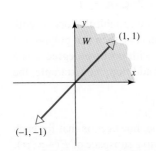

▲ Figure 4.2.4 W is not closed under scalar multiplication.

► **EXAMPLE 4** A Subset of R^2 That Is Not a Subspace

Let W be the set of all points (x, y) in R^2 for which $x \geq 0$ and $y \geq 0$ (the shaded region in Figure 4.2.4). This set is not a subspace of R^2 because it is not closed under scalar multiplication. For example, $\mathbf{v} = (1, 1)$ is a vector in W, but $(-1)\mathbf{v} = (-1, -1)$ is not.

► **EXAMPLE 5** Subspaces of M_{nn}

We know from Theorem 1.7.2 that the sum of two symmetric $n \times n$ matrices is symmetric and that a scalar multiple of a symmetric $n \times n$ matrix is symmetric. Thus, the set of symmetric $n \times n$ matrices is closed under addition and scalar multiplication and hence is a subspace of M_{nn}. Similarly, the sets of upper triangular matrices, lower triangular matrices, and diagonal matrices are subspaces of M_{nn}.

► **EXAMPLE 6** A Subset of M_{nn} That Is Not a Subspace

The set W of invertible $n \times n$ matrices is not a subspace of M_{nn}, failing on two counts—it is not closed under addition and not closed under scalar multiplication. We will illustrate this with an example in M_{22} that you can readily adapt to M_{nn}. Consider the matrices

$$U = \begin{bmatrix} 1 & 2 \\ 2 & 5 \end{bmatrix} \quad \text{and} \quad V = \begin{bmatrix} -1 & 2 \\ -2 & 5 \end{bmatrix}$$

The matrix $0U$ is the 2×2 zero matrix and hence is not invertible, and the matrix $U + V$ has a column of zeros, so it also is not invertible.

182 Chapter 4 General Vector Spaces

CALCULUS REQUIRED

▶ **EXAMPLE 7** **The Subspace** $C(-\infty, \infty)$

There is a theorem in calculus which states that a sum of continuous functions is continuous and that a constant times a continuous function is continuous. Rephrased in vector language, the set of continuous functions on $(-\infty, \infty)$ is a subspace of $F(-\infty, \infty)$. We will denote this subspace by $C(-\infty, \infty)$.

CALCULUS REQUIRED

▶ **EXAMPLE 8** **Functions with Continuous Derivatives**

A function with a continuous derivative is said to be *continuously differentiable*. There is a theorem in calculus which states that the sum of two continuously differentiable functions is continuously differentiable and that a constant times a continuously differentiable function is continuously differentiable. Thus, the functions that are continuously differentiable on $(-\infty, \infty)$ form a subspace of $F(-\infty, \infty)$. We will denote this subspace by $C^1(-\infty, \infty)$, where the superscript emphasizes that the *first* derivative is continuous. To take this a step further, the set of functions with m continuous derivatives on $(-\infty, \infty)$ is a subspace of $F(-\infty, \infty)$ as is the set of functions with derivatives of all orders on $(-\infty, \infty)$. We will denote these subspaces by $C^m(-\infty, \infty)$ and $C^\infty(-\infty, \infty)$, respectively.

▶ **EXAMPLE 9** **The Subspace of All Polynomials**

Recall that a ***polynomial*** is a function that can be expressed in the form

$$p(x) = a_0 + a_1 x + \cdots + a_n x^n \tag{1}$$

where $a_0, a_1, \ldots, a_n$ are constants. It is evident that the sum of two polynomials is a polynomial and that a constant times a polynomial is a polynomial. Thus, the set W of all polynomials is closed under addition and scalar multiplication and hence is a subspace of $F(-\infty, \infty)$. We will denote this space by P_∞.

In this text we regard all constants to be polynomials of degree zero. Be aware, however, that some authors do not assign a degree to the constant 0.

▶ **EXAMPLE 10** **The Subspace of Polynomials of Degree** $\leq n$

Recall that the ***degree*** of a polynomial is the highest power of the variable that occurs with a nonzero coefficient. Thus, for example, if $a_n \neq 0$ in Formula (1), then that polynomial has degree n. It is *not* true that the set W of polynomials with positive degree n is a subspace of $F(-\infty, \infty)$ because that set is not closed under addition. For example, the polynomials

$$1 + 2x + 3x^2 \quad \text{and} \quad 5 + 7x - 3x^2$$

both have degree 2, but their sum has degree 1. What *is* true, however, is that for each nonnegative integer n the polynomials of degree n *or less* form a subspace of $F(-\infty, \infty)$. We will denote this space by P_n. ◀

The Hierarchy of Function Spaces

It is proved in calculus that polynomials are continuous functions and have continuous derivatives of all orders on $(-\infty, \infty)$. Thus, it follows that P_∞ is not only a subspace of $F(-\infty, \infty)$, as previously observed, but is also a subspace of $C^\infty(-\infty, \infty)$. We leave it for you to convince yourself that the vector spaces discussed in Examples 7 to 10 are "nested" one inside the other as illustrated in Figure 4.2.5.

Remark In our previous examples, and as illustrated in Figure 4.2.5, we have only considered functions that are defined at all points of the interval $(-\infty, \infty)$. Sometimes we will want to consider functions that are only defined on some subinterval of $(-\infty, \infty)$, say the closed interval $[a, b]$ or the open interval (a, b). In such cases we will make an appropriate notation change. For example, $C[a, b]$ is the space of continuous functions on $[a, b]$ and $C(a, b)$ is the space of continuous functions on (a, b).

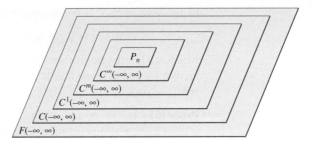
▶ Figure 4.2.5

Building Subspaces

The following theorem provides a useful way of creating a new subspace from known subspaces.

THEOREM 4.2.2 *If $W_1, W_2, \ldots, W_r$ are subspaces of a vector space V, then the intersection of these subspaces is also a subspace of V.*

Proof Let W be the intersection of the subspaces $W_1, W_2, \ldots, W_r$. This set is not empty because each of these subspaces contains the zero vector of V, and hence so does their intersection. Thus, it remains to show that W is closed under addition and scalar multiplication.

To prove closure under addition, let $\mathbf{u}$ and $\mathbf{v}$ be vectors in W. Since W is the intersection of $W_1, W_2, \ldots, W_r$, it follows that $\mathbf{u}$ and $\mathbf{v}$ also lie in each of these subspaces. Since these subspaces are all closed under addition, they all contain the vector $\mathbf{u} + \mathbf{v}$ and hence so does their intersection W. This proves that W is closed under addition. We leave the proof that W is closed under scalar multiplication to you. ◀

> Note that the first step in proving Theorem 4.2.2 was to establish that W contained at least one vector. This is important, for otherwise the subsequent argument might be logically correct but meaningless.

Sometimes we will want to find the "smallest" subspace of a vector space V that contains all of the vectors in some set of interest. The following definition, which generalizes Definition 4 of Section 3.1, will help us to do that.

DEFINITION 2 If $\mathbf{w}$ is a vector in a vector space V, then $\mathbf{w}$ is said to be a *linear combination* of the vectors $\mathbf{v}_1, \mathbf{v}_2, \ldots, \mathbf{v}_r$ in V if $\mathbf{w}$ can be expressed in the form

$$\mathbf{w} = k_1\mathbf{v}_1 + k_2\mathbf{v}_2 + \cdots + k_r\mathbf{v}_r \qquad (2)$$

where $k_1, k_2, \ldots, k_r$ are scalars. These scalars are called the *coefficients* of the linear combination.

> If $k = 1$, then Equation (2) has the form $\mathbf{w} = k_1\mathbf{v}_1$, in which case the linear combination is just a scalar multiple of $\mathbf{v}_1$.

THEOREM 4.2.3 *If $S = \{\mathbf{w}_1, \mathbf{w}_2, \ldots, \mathbf{w}_r\}$ is a nonempty set of vectors in a vector space V, then*:

(a) *The set W of all possible linear combinations of the vectors in S is a subspace of V.*

(b) *The set W in part (a) is the "smallest" subspace of V that contains all of the vectors in S in the sense that any other subspace that contains those vectors contains W.*

Proof (a) Let W be the set of all possible linear combinations of the vectors in S. We must show that S is closed under addition and scalar multiplication. To prove closure under addition, let

$$\mathbf{u} = c_1\mathbf{w}_1 + c_2\mathbf{w}_2 + \cdots + c_r\mathbf{w}_r \quad \text{and} \quad \mathbf{v} = k_1\mathbf{w}_1 + k_2\mathbf{w}_2 + \cdots + k_r\mathbf{w}_r$$

be two vectors in S. It follows that their sum can be written as

$$\mathbf{u} + \mathbf{v} = (c_1 + k_1)\mathbf{w}_1 + (c_2 + k_2)\mathbf{w}_2 + \cdots + (c_r + k_r)\mathbf{w}_r$$

which is a linear combination of the vectors in S. Thus, W is closed under addition. We leave it for you to prove that W is also closed under scalar multiplication and hence is a subspace of V.

Proof (b) Let W' be any subspace of V that contains all of the vectors in S. Since W' is closed under addition and scalar multiplication, it contains all linear combinations of the vectors in S and hence contains W. ◄

The following definition gives some important notation and terminology related to Theorem 4.2.3.

DEFINITION 3 The subspace of a vector space V that is formed from all possible linear combinations of the vectors in a nonempty set S is called the ***span of S***, and we say that the vectors in S ***span*** that subspace. If $S = \{\mathbf{w}_1, \mathbf{w}_2, \ldots, \mathbf{w}_r\}$, then we denote the span of S by

$$\mathrm{span}\{\mathbf{w}_1, \mathbf{w}_2, \ldots, \mathbf{w}_r\} \quad \text{or} \quad \mathrm{span}(S)$$

▶ **EXAMPLE 11** **The Standard Unit Vectors Span R^n**

Recall that the standard unit vectors in R^n are

$$\mathbf{e}_1 = (1, 0, 0, \ldots, 0), \quad \mathbf{e}_2 = (0, 1, 0, \ldots, 0), \ldots, \quad \mathbf{e}_n = (0, 0, 0, \ldots, 1)$$

These vectors span R^n since every vector $\mathbf{v} = (v_1, v_2, \ldots, v_n)$ in R^n can be expressed as

$$\mathbf{v} = v_1\mathbf{e}_1 + v_2\mathbf{e}_2 + \cdots + v_n\mathbf{e}_n$$

which is a linear combination of $\mathbf{e}_1, \mathbf{e}_2, \ldots, \mathbf{e}_n$. Thus, for example, the vectors

$$\mathbf{i} = (1, 0, 0), \quad \mathbf{j} = (0, 1, 0), \quad \mathbf{k} = (0, 0, 1)$$

span R^3 since every vector $\mathbf{v} = (a, b, c)$ in this space can be expressed as

$$\mathbf{v} = (a, b, c) = a(1, 0, 0) + b(0, 1, 0) + c(0, 0, 1) = a\mathbf{i} + b\mathbf{j} + c\mathbf{k}.$$

▶ **EXAMPLE 12** **A Geometric View of Spanning in R^2 and R^3**

(a) If $\mathbf{v}$ is a nonzero vector in R^2 or R^3 that has its initial point at the origin, then span$\{\mathbf{v}\}$, which is the set of all scalar multiples of $\mathbf{v}$, is the line through the origin determined by $\mathbf{v}$. You should be able to visualize this from Figure 4.2.6a by observing that the

George William Hill (1838–1914)

Historical Note The terms *linearly independent* and *linearly dependent* were introduced by Maxime Bôcher (see p. 7) in his book *Introduction to Higher Algebra*, published in 1907. The term *linear combination* is due to the American mathematician G. W. Hill, who introduced it in a research paper on planetary motion published in 1900. Hill was a "loner" who preferred to work out of his home in West Nyack, New York, rather than in academia, though he did try lecturing at Columbia University for a few years. Interestingly, he apparently returned the teaching salary, indicating that he did not need the money and did not want to be bothered looking after it. Although technically a mathematician, Hill had little interest in modern developments of mathematics and worked almost entirely on the theory of planetary orbits.

[*Image: Courtesy of the American Mathematical Society*]

tip of the vector $k\mathbf{v}$ can be made to fall at any point on the line by choosing the value of k appropriately.

(b) If $\mathbf{v}_1$ and $\mathbf{v}_2$ are nonzero vectors in R^3 that have their initial points at the origin, then span$\{\mathbf{v}_1, \mathbf{v}_2\}$, which consists of all linear combinations of $\mathbf{v}_1$ and $\mathbf{v}_2$, is the plane through the origin determined by these two vectors. You should be able to visualize this from Figure 4.2.6b by observing that the tip of the vector $k_1\mathbf{v}_1 + k_2\mathbf{v}_2$ can be made to fall at any point in the plane by adjusting the scalars k_1 and k_2 to lengthen, shorten, or reverse the directions of the vectors $k_1\mathbf{v}_1$ and $k_2\mathbf{v}_2$ appropriately.

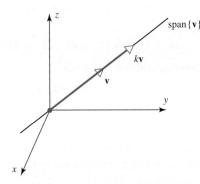
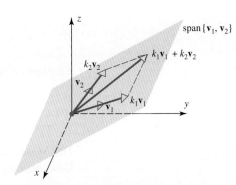

▶ Figure 4.2.6

(a) Span$\{\mathbf{v}\}$ is the line through the origin determined by $\mathbf{v}$.

(b) Span$\{\mathbf{v}_1, \mathbf{v}_2\}$ is the plane through the origin determined by $\mathbf{v}_1$ and $\mathbf{v}_2$.

▶ **EXAMPLE 13 A Spanning Set for P_n**

The polynomials $1, x, x^2, \ldots, x^n$ span the vector space P_n defined in Example 10 since each polynomial $\mathbf{p}$ in P_n can be written as

$$\mathbf{p} = a_0 + a_1 x + \cdots + a_n x^n$$

which is a linear combination of $1, x, x^2, \ldots, x^n$. We can denote this by writing

$$P_n = \text{span}\{1, x, x^2, \ldots, x^n\} \quad \blacktriangleleft$$

The next two examples are concerned with two important types of problems:

- Given a set S of vectors in R^n and a vector $\mathbf{v}$ in R^n, determine whether $\mathbf{v}$ is a linear combination of the vectors in S.
- Given a set S of vectors in R^n, determine whether the vectors span R^n.

▶ **EXAMPLE 14 Linear Combinations**

Consider the vectors $\mathbf{u} = (1, 2, -1)$ and $\mathbf{v} = (6, 4, 2)$ in R^3. Show that $\mathbf{w} = (9, 2, 7)$ is a linear combination of $\mathbf{u}$ and $\mathbf{v}$ and that $\mathbf{w}' = (4, -1, 8)$ is *not* a linear combination of $\mathbf{u}$ and $\mathbf{v}$.

Solution In order for $\mathbf{w}$ to be a linear combination of $\mathbf{u}$ and $\mathbf{v}$, there must be scalars k_1 and k_2 such that $\mathbf{w} = k_1\mathbf{u} + k_2\mathbf{v}$; that is,

$$(9, 2, 7) = k_1(1, 2, -1) + k_2(6, 4, 2)$$

or

$$(9, 2, 7) = (k_1 + 6k_2, 2k_1 + 4k_2, -k_1 + 2k_2)$$

Equating corresponding components gives

$$k_1 + 6k_2 = 9$$
$$2k_1 + 4k_2 = 2$$
$$-k_1 + 2k_2 = 7$$

Solving this system using Gaussian elimination yields $k_1 = -3$, $k_2 = 2$, so

$$\mathbf{w} = -3\mathbf{u} + 2\mathbf{v}$$

Similarly, for $\mathbf{w}'$ to be a linear combination of $\mathbf{u}$ and $\mathbf{v}$, there must be scalars k_1 and k_2 such that $\mathbf{w}' = k_1\mathbf{u} + k_2\mathbf{v}$; that is,

$$(4, -1, 8) = k_1(1, 2, -1) + k_2(6, 4, 2)$$

or

$$(4, -1, 8) = (k_1 + 6k_2, 2k_1 + 4k_2, -k_1 + 2k_2)$$

Equating corresponding components gives

$$k_1 + 6k_2 = 4$$
$$2k_1 + 4k_2 = -1$$
$$-k_1 + 2k_2 = 8$$

This system of equations is inconsistent (verify), so no such scalars k_1 and k_2 exist. Consequently, $\mathbf{w}'$ is not a linear combination of $\mathbf{u}$ and $\mathbf{v}$.

▶ **EXAMPLE 15 Testing for Spanning**

Determine whether $\mathbf{v}_1 = (1, 1, 2)$, $\mathbf{v}_2 = (1, 0, 1)$, and $\mathbf{v}_3 = (2, 1, 3)$ span the vector space R^3.

Solution We must determine whether an arbitrary vector $\mathbf{b} = (b_1, b_2, b_3)$ in R^3 can be expressed as a linear combination

$$\mathbf{b} = k_1\mathbf{v}_1 + k_2\mathbf{v}_2 + k_3\mathbf{v}_3$$

of the vectors $\mathbf{v}_1$, $\mathbf{v}_2$, and $\mathbf{v}_3$. Expressing this equation in terms of components gives

$$(b_1, b_2, b_3) = k_1(1, 1, 2) + k_2(1, 0, 1) + k_3(2, 1, 3)$$

or

$$(b_1, b_2, b_3) = (k_1 + k_2 + 2k_3, k_1 + k_3, 2k_1 + k_2 + 3k_3)$$

or

$$k_1 + k_2 + 2k_3 = b_1$$
$$k_1 \quad\quad\; + k_3 = b_2$$
$$2k_1 + k_2 + 3k_3 = b_3$$

Thus, our problem reduces to ascertaining whether this system is consistent for all values of b_1, b_2, and b_3. One way of doing this is to use parts (e) and (g) of Theorem 2.3.8, which state that the system is consistent if and only if its coefficient matrix

$$A = \begin{bmatrix} 1 & 1 & 2 \\ 1 & 0 & 1 \\ 2 & 1 & 3 \end{bmatrix}$$

has a nonzero determinant. But this is *not* the case here; we leave it for you to confirm that $\det(A) = 0$, so $\mathbf{v}_1$, $\mathbf{v}_2$, and $\mathbf{v}_3$ do not span R^3. ◀

Solution Spaces of Homogeneous Systems The solutions of a homogeneous linear system $A\mathbf{x} = \mathbf{0}$ of m equations in n unknowns can be viewed as vectors in R^n. The following theorem provides a useful insight into the geometric structure of the solution set.

4.2 Subspaces

THEOREM 4.2.4 *The solution set of a homogeneous linear system* $A\mathbf{x} = \mathbf{0}$ *in n unknowns is a subspace of* R^n.

Proof Let W be the solution set for the system. The set W is not empty because it contains at least the trivial solution $\mathbf{x} = \mathbf{0}$.

To show that W is a subspace of R^n, we must show that it is closed under addition and scalar multiplication. To do this, let $\mathbf{x}_1$ and $\mathbf{x}_2$ be vectors in W. Since these vectors are solutions of $A\mathbf{x} = \mathbf{0}$, we have

$$A\mathbf{x}_1 = \mathbf{0} \quad \text{and} \quad A\mathbf{x}_2 = \mathbf{0}$$

It follows from these equations and the distributive property of matrix multiplication that

$$A(\mathbf{x}_1 + \mathbf{x}_2) = A\mathbf{x}_1 + A\mathbf{x}_2 = \mathbf{0} + \mathbf{0} = \mathbf{0}$$

so W is closed under addition. Similarly, if k is any scalar then

$$A(k\mathbf{x}_1) = kA\mathbf{x}_1 = k\mathbf{0} = \mathbf{0}$$

so W is also closed under scalar multiplication. ◄

*Because the solution set of a homogeneous system in n unknowns is actually a subspace of R^n, we will generally refer to it as the **solution space** of the system.*

► **EXAMPLE 16 Solution Spaces of Homogeneous Systems**

Consider the linear systems

(a) $\begin{bmatrix} 1 & -2 & 3 \\ 2 & -4 & 6 \\ 3 & -6 & 9 \end{bmatrix} \begin{bmatrix} x \\ y \\ z \end{bmatrix} = \begin{bmatrix} 0 \\ 0 \\ 0 \end{bmatrix}$ (b) $\begin{bmatrix} 1 & -2 & 3 \\ -3 & 7 & -8 \\ -2 & 4 & -6 \end{bmatrix} \begin{bmatrix} x \\ y \\ z \end{bmatrix} = \begin{bmatrix} 0 \\ 0 \\ 0 \end{bmatrix}$

(c) $\begin{bmatrix} 1 & -2 & 3 \\ -3 & 7 & -8 \\ 4 & 1 & 2 \end{bmatrix} \begin{bmatrix} x \\ y \\ z \end{bmatrix} = \begin{bmatrix} 0 \\ 0 \\ 0 \end{bmatrix}$ (d) $\begin{bmatrix} 0 & 0 & 0 \\ 0 & 0 & 0 \\ 0 & 0 & 0 \end{bmatrix} \begin{bmatrix} x \\ y \\ z \end{bmatrix} = \begin{bmatrix} 0 \\ 0 \\ 0 \end{bmatrix}$

Solution

(a) We leave it for you to verify that the solutions are

$$x = 2s - 3t, \quad y = s, \quad z = t$$

from which it follows that

$$x = 2y - 3z \quad \text{or} \quad x - 2y + 3z = 0$$

This is the equation of a plane through the origin that has $\mathbf{n} = (1, -2, 3)$ as a normal.

(b) We leave it for you to verify that the solutions are

$$x = -5t, \quad y = -t, \quad z = t$$

which are parametric equations for the line through the origin that is parallel to the vector $\mathbf{v} = (-5, -1, 1)$.

(c) We leave it for you to verify that the only solution is $x = 0$, $y = 0$, $z = 0$, so the solution space is $\{\mathbf{0}\}$.

(d) This linear system is satisfied by all real values of x, y, and z, so the solution space is all of R^3. ◄

Remark Whereas the solution set of every *homogeneous* system of m equations in n unknowns is a subspace of R^n, it is *never* true that the solution set of a *nonhomogeneous* system of m equations in n unknowns is a subspace of R^n. There are two possible scenarios: first, the system may not have any solutions at all, and second, if there are solutions, then the solution set will not be closed under either addition or under scalar multiplication (Exercise 18).

A Concluding Observation

It is important to recognize that spanning sets are not unique. For example, any nonzero vector on the line in Figure 4.2.6a will span that line, and any two noncollinear vectors in the plane in Figure 4.2.6b will span that plane. The following theorem, whose proof we leave as an exercise, states conditions under which two sets of vectors will span the same space.

THEOREM 4.2.5 *If $S = \{\mathbf{v}_1, \mathbf{v}_2, \ldots, \mathbf{v}_r\}$ and $S' = \{\mathbf{w}_1, \mathbf{w}_2, \ldots, \mathbf{w}_k\}$ are nonempty sets of vectors in a vector space V, then*

$$\text{span}\{\mathbf{v}_1, \mathbf{v}_2, \ldots, \mathbf{v}_r\} = \text{span}\{\mathbf{w}_1, \mathbf{w}_2, \ldots, \mathbf{w}_k\}$$

if and only if each vector in S is a linear combination of those in S', and each vector in S' is a linear combination of those in S.

Concept Review

- Subspace
- Zero subspace
- Examples of subspaces
- Linear combination
- Span
- Solution space

Skills

- Determine whether a subset of a vector space is a subspace.
- Show that a subset of a vector space is a subspace.
- Show that a nonempty subset of a vector space is not a subspace by demonstrating that the set is either not closed under addition or not closed under scalar multiplication.
- Given a set S of vectors in R^n and a vector $\mathbf{v}$ in R^n, determine whether $\mathbf{v}$ is a linear combination of the vectors in S.
- Given a set S of vectors in R^n, determine whether the vectors in S span R^n.
- Determine whether two nonempty sets of vectors in a vector space V span the same subspace of V.

Exercise Set 4.2

1. Use Theorem 4.2.1 to determine which of the following are subspaces of R^3.

   (a) All vectors of the form $(a, 0, 0)$.

   (b) All vectors of the form $(a, 1, 0)$.

   (c) All vectors of the form (a, b, c), where $b = a + c$.

   (d) All vectors of the form (a, b, c), where $c = a - b$.

   (e) All vectors of the form $(a, -a, 0)$.

2. Use Theorem 4.2.1 to determine which of the following are subspaces of M_{nn}.

   (a) The set of all diagonal $n \times n$ matrices.

   (b) The set of all $n \times n$ matrices A such that $\det(A) = 0$.

   (c) The set of all $n \times n$ matrices A such that $\text{tr}(A) = 0$.

   (d) The set of all symmetric $n \times n$ matrices.

   (e) The set of all $n \times n$ matrices A such that $A^T = -A$.

   (f) The set of all $n \times n$ matrices A for which $A\mathbf{x} = \mathbf{0}$ has only the trivial solution.

   (g) The set of all $n \times n$ matrices A such that $AB = BA$ for some fixed $n \times n$ matrix B.

3. Use Theorem 4.2.1 to determine which of the following are subspaces of P_3.

   (a) All polynomials of the form $a_0 + a_1 x + a_2 x^2 + a_3 x^3$ where $a_1 = a_2$.

   (b) All polynomials of the form $a_0 + a_1 x + a_2 x^2 + a_3 x^3$ where $a_0 = 0$.

(c) All polynomials of the form $a_0 + a_1x + a_2x^2 + a_3x^3$ in which $a_0, a_1, a_2,$ and a_3 are integers.

(d) All polynomials of the form $a_0 + a_1x$, where a_0 and a_1 are real numbers.

4. Which of the following are subspaces of $F(-\infty, \infty)$?

(a) All functions f in $F(-\infty, \infty)$ for which $f(0) = 0$.

(b) All functions f in $F(-\infty, \infty)$ for which $f(0) = 1$.

(c) All functions f in $F(-\infty, \infty)$ for which $f(-x) = f(x)$.

(d) All polynomials of degree 2.

5. Which of the following are subspaces of R^∞?

(a) All sequences $\mathbf{v}$ in R^∞ of the form
$\mathbf{v} = (v, 0, v, 0, v, 0, \ldots)$.

(b) All sequences $\mathbf{v}$ in R^∞ of the form
$\mathbf{v} = (v, 1, v, 1, v, 1, \ldots)$.

(c) All sequences $\mathbf{v}$ in R^∞ of the form
$\mathbf{v} = (v, 2v, 4v, 8v, 16v, \ldots)$.

(d) All sequences in R^∞ whose components are 0 from some point on.

6. A line L through the origin in R^3 can be represented by parametric equations of the form $x = at$, $y = bt$, and $z = ct$. Use these equations to show that L is a subspace of R^3 by showing that if $\mathbf{v}_1 = (x_1, y_1, z_1)$ and $\mathbf{v}_2 = (x_2, y_2, z_2)$ are points on L and k is any real number, then $k\mathbf{v}_1$ and $\mathbf{v}_1 + \mathbf{v}_2$ are also points on L.

7. Which of the following are linear combinations of $\mathbf{u} = (1, -3, 2)$ and $\mathbf{v} = (1, 0, -4)$?

(a) $(0, -3, 6)$ (b) $(3, -9, -2)$

(c) $(0, 0, 0)$ (d) $(1, 6, -16)$

8. Express the following as linear combinations of $\mathbf{u} = (2, 1, 4)$, $\mathbf{v} = (1, -1, 3)$, and $\mathbf{w} = (3, 2, 5)$.

(a) $(-9, -7, -15)$ (b) $(6, 11, 6)$

(c) $(0, 0, 0)$ (d) $(7, 8, 9)$

9. Which of the following are linear combinations of
$$A = \begin{bmatrix} 3 & 2 \\ 0 & 1 \end{bmatrix}, \quad B = \begin{bmatrix} 0 & 2 \\ -2 & 4 \end{bmatrix}, \quad C = \begin{bmatrix} 1 & 1 \\ -2 & 5 \end{bmatrix}?$$

(a) $\begin{bmatrix} 2 & 5 \\ -2 & 4 \end{bmatrix}$ (b) $\begin{bmatrix} 4 & 5 \\ -2 & 10 \end{bmatrix}$

(c) $\begin{bmatrix} 1 & 3 \\ -4 & 1 \end{bmatrix}$ (d) $\begin{bmatrix} 9 & 9 \\ -8 & 21 \end{bmatrix}$

10. In each part express the vector as a linear combination of $\mathbf{p}_1 = 2 + x + 4x^2$, $\mathbf{p}_2 = 1 - x + 3x^2$, and $\mathbf{p}_3 = 3 + 2x + 5x^2$.

(a) $-9 - 7x - 15x^2$ (b) $6 + 11x + 6x^2$

(c) 0 (d) $7 + 8x + 9x^2$

11. In each part, determine whether the given vectors span R^3.

(a) $\mathbf{v}_1 = (1, 2, 3)$, $\mathbf{v}_2 = (2, 0, 0)$, $\mathbf{v}_3 = (-2, 1, 0)$

(b) $\mathbf{v}_1 = (2, -1, 2)$, $\mathbf{v}_2 = (4, 1, 3)$, $\mathbf{v}_3 = (2, 2, 1)$

(c) $\mathbf{v}_1 = (-1, 5, 2)$, $\mathbf{v}_2 = (3, 1, 1)$, $\mathbf{v}_3 = (2, 0, -2)$, $\mathbf{v}_4 = (4, 1, 0)$

(d) $\mathbf{v}_1 = (3, 2, 4)$, $\mathbf{v}_2 = (-3, -1, 0)$, $\mathbf{v}_3 = (0, 1, 4)$, $\mathbf{v}_4 = (0, 2, 8)$

12. Suppose that $\mathbf{v}_1 = (2, 1, 0, 3)$, $\mathbf{v}_2 = (3, -1, 5, 2)$, and $\mathbf{v}_3 = (-1, 0, 2, 1)$. Which of the following vectors are in span$\{\mathbf{v}_1, \mathbf{v}_2, \mathbf{v}_3\}$?

(a) $(2, 3, -7, 3)$ (b) $(0, 0, 0, 0)$

(c) $(1, 1, 1, 1)$ (d) $(-4, 6, -13, 4)$

13. Determine whether the following polynomials span P_2.
$$\mathbf{p}_1 = 1 - x + 2x^2, \quad \mathbf{p}_2 = 3 + x,$$
$$\mathbf{p}_3 = 5 - x + 4x^2, \quad \mathbf{p}_4 = -2 - 2x + 2x^2$$

14. Let $\mathbf{f} = \cos^2 x$ and $\mathbf{g} = \sin^2 x$. Which of the following lie in the space spanned by $\mathbf{f}$ and $\mathbf{g}$?

(a) $\cos 2x$ (b) $3 + x^2$ (c) 1 (d) $\sin x$ (e) 0

15. Determine whether the solution space of the system $A\mathbf{x} = \mathbf{0}$ is a line through the origin, a plane through the origin, or the origin only. If it is a plane, find an equation for it. If it is a line, find parametric equations for it.

(a) $A = \begin{bmatrix} 1 & -2 & 6 \\ 3 & -6 & 18 \\ -7 & 14 & -42 \end{bmatrix}$ (b) $A = \begin{bmatrix} 1 & -2 & 3 \\ -3 & 6 & 9 \\ -2 & 4 & -6 \end{bmatrix}$

(c) $A = \begin{bmatrix} 1 & 0 & 0 \\ 9 & -11 & 3 \\ 3 & -4 & 1 \end{bmatrix}$ (d) $A = \begin{bmatrix} 1 & 2 & -6 \\ 1 & 4 & 4 \\ 3 & 10 & 6 \end{bmatrix}$

(e) $A = \begin{bmatrix} 1 & -4 & 0 \\ -2 & 8 & 1 \\ 4 & -16 & 0 \end{bmatrix}$ (f) $A = \begin{bmatrix} 1 & -3 & 1 \\ 2 & -6 & 2 \\ 3 & -9 & 3 \end{bmatrix}$

16. (*Calculus required*) Show that the following sets of functions are subspaces of $F(-\infty, \infty)$.

(a) All continuous functions on $(-\infty, \infty)$.

(b) All differentiable functions on $(-\infty, \infty)$.

(c) All differentiable functions on $(-\infty, \infty)$ that satisfy $\mathbf{f}' + 2\mathbf{f} = \mathbf{0}$.

17. (*Calculus required*) Show that the set of continuous functions $\mathbf{f} = f(x)$ on $[a, b]$ such that
$$\int_a^b f(x)\,dx = 0$$
is a subspace of $C[a, b]$.

18. Show that the solution vectors of a consistent nonhomogeneous system of m linear equations in n unknowns do not form a subspace of R^n.

19. Prove Theorem 4.2.5.

20. Use Theorem 4.2.5 to show that the vectors $\mathbf{v}_1 = (1, 6, 4)$, $\mathbf{v}_2 = (2, 4, -1)$, $\mathbf{v}_3 = (-1, 2, 5)$, and the vectors $\mathbf{w}_1 = (1, -2, -5)$, $\mathbf{w}_2 = (0, 8, 9)$ span the same subspace of R^3.

True-False Exercises

In parts (a)–(k) determine whether the statement is true or false, and justify your answer.

(a) Every subspace of a vector space is itself a vector space.

(b) Every vector space is a subspace of itself.

(c) Every subset of a vector space V that contains the zero vector in V is a subspace of V.

(d) The set R^2 is a subspace of R^3.

(e) The solution set of a consistent linear system $A\mathbf{x} = \mathbf{b}$ of m equations in n unknowns is a subspace of R^n.

(f) The span of any finite set of vectors in a vector space is closed under addition and scalar multiplication.

(g) The intersection of any two subspaces of a vector space V is a subspace of V.

(h) The union of any two subspaces of a vector space V is a subspace of V.

(i) Two subsets of a vector space V that span the same subspace of V must be equal.

(j) The set of upper triangular $n \times n$ matrices is a subspace of the vector space of all $n \times n$ matrices.

(k) The polynomials $x - 1$, $(x - 1)^2$, and $(x - 1)^3$ span P_3.

4.3 Linear Independence

In this section we will consider the question of whether the vectors in a given set are interrelated in the sense that one or more of them can be expressed as a linear combination of the others. This is important to know in applications because the existence of such relationships often signals that some kind of complication is likely to occur.

Extraneous Vectors

In a rectangular xy-coordinate system every vector in the plane can be expressed in exactly one way as a linear combination of the standard unit vectors. For example, the only way to express the vector $(3, 2)$ as a linear combination of $\mathbf{i} = (1, 0)$ and $\mathbf{j} = (0, 1)$ is

$$(3, 2) = 3(1, 0) + 2(0, 1) = 3\mathbf{i} + 2\mathbf{j} \tag{1}$$

(Figure 4.3.1). Suppose, however, that we were to introduce a third coordinate axis that makes an angle of 45° with the x-axis. Call it the w-axis. As illustrated in Figure 4.3.2, the unit vector along the w-axis is

$$\mathbf{w} = \left(\frac{1}{\sqrt{2}}, \frac{1}{\sqrt{2}}\right)$$

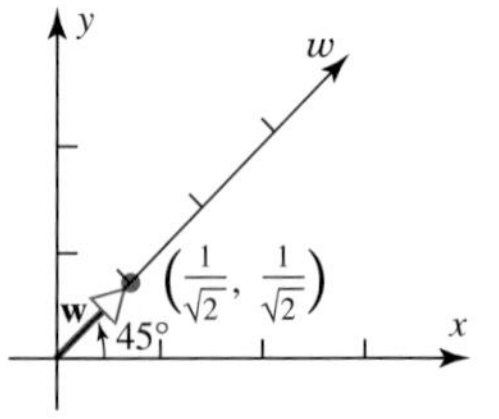

Whereas Formula (1) shows the only way to express the vector $(3, 2)$ as a linear combination of $\mathbf{i}$ and $\mathbf{j}$, there are infinitely many ways to express this vector as a linear combination of $\mathbf{i}$, $\mathbf{j}$, and $\mathbf{w}$. Three possibilities are

$$(3, 2) = 3(1, 0) + 2(0, 1) + 0\left(\frac{1}{\sqrt{2}}, \frac{1}{\sqrt{2}}\right) = 3\mathbf{i} + 2\mathbf{j} + 0\mathbf{w}$$

$$(3, 2) = 2(1, 0) + (0, 1) + \sqrt{2}\left(\frac{1}{\sqrt{2}}, \frac{1}{\sqrt{2}}\right) = 3\mathbf{i} + \mathbf{j} + \sqrt{2}\mathbf{w}$$

$$(3, 2) = 4(1, 0) + 3(0, 1) - \sqrt{2}\left(\frac{1}{\sqrt{2}}, \frac{1}{\sqrt{2}}\right) = 4\mathbf{i} + 3\mathbf{j} - \sqrt{2}\mathbf{w}$$

In short, by introducing a superfluous axis we created the complication of having multiple ways of assigning coordinates to points in the plane. What makes the vector $\mathbf{w}$ superfluous is the fact that it can be expressed as a linear combination of the vectors $\mathbf{i}$ and $\mathbf{j}$, namely,

$$\mathbf{w} = \left(\frac{1}{\sqrt{2}}, \frac{1}{\sqrt{2}}\right) = \frac{1}{\sqrt{2}}\mathbf{i} + \frac{1}{\sqrt{2}}\mathbf{j}$$

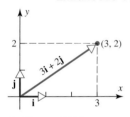

▲ Figure 4.3.1

▲ Figure 4.3.2

4.3 Linear Independence

Thus, one of our main tasks in this section will be to develop ways of ascertaining whether one vector in a set S is a linear combination of other vectors in S.

Linear Independence and Dependence

DEFINITION 1 If $S = \{\mathbf{v}_1, \mathbf{v}_2, \ldots, \mathbf{v}_r\}$ is a nonempty set of vectors in a vector space V, then the vector equation

$$k_1\mathbf{v}_1 + k_2\mathbf{v}_2 + \cdots + k_r\mathbf{v}_r = \mathbf{0}$$

has at least one solution, namely,

$$k_1 = 0, \quad k_2 = 0, \ldots, \quad k_r = 0$$

We call this the ***trivial solution***. If this is the only solution, then S is said to be a ***linearly independent set***. If there are solutions in addition to the trivial solution, then S is said to be a ***linearly dependent set***.

> We will often apply the terms *linearly independent* and *linearly dependent* to the vectors themselves rather than to the set.

▶ **EXAMPLE 1** Linear Independence of the Standard Unit Vectors in R^n

The most basic linearly independent set in R^n is the set of standard unit vectors

$$\mathbf{e}_1 = (1, 0, 0, \ldots, 0), \quad \mathbf{e}_2 = (0, 1, 0, \ldots, 0), \ldots, \quad \mathbf{e}_n = (0, 0, 0, \ldots, 1)$$

For notational simplicity, we will prove the linear independence in R^3 of

$$\mathbf{i} = (1, 0, 0), \quad \mathbf{j} = (0, 1, 0), \quad \mathbf{k} = (0, 0, 1)$$

The linear independence or linear dependence of these vectors is determined by whether there exist nontrivial solutions of the vector equation

$$k_1\mathbf{i} + k_2\mathbf{j} + k_3\mathbf{k} = \mathbf{0} \tag{2}$$

Since the component form of this equation is

$$(k_1, k_2, k_3) = (0, 0, 0)$$

it follows that $k_1 = k_2 = k_3 = 0$. This implies that (2) has only the trivial solution and hence that the vectors are linearly independent. ◀

▶ **EXAMPLE 2** Linear Independence in R^3

Determine whether the vectors

$$\mathbf{v}_1 = (1, -2, 3), \quad \mathbf{v}_2 = (5, 6, -1), \quad \mathbf{v}_3 = (3, 2, 1)$$

are linearly independent or linearly dependent in R^3.

Solution The linear independence or linear dependence of these vectors is determined by whether there exist nontrivial solutions of the vector equation

$$k_1\mathbf{v}_1 + k_2\mathbf{v}_2 + k_3\mathbf{v}_3 = \mathbf{0} \tag{3}$$

or, equivalently, of

$$k_1(1, -2, 3) + k_2(5, 6, -1) + k_3(3, 2, 1) = (0, 0, 0)$$

Equating corresponding components on the two sides yields the homogeneous linear system

$$\begin{aligned} k_1 + 5k_2 + 3k_3 &= 0 \\ -2k_1 + 6k_2 + 2k_3 &= 0 \\ 3k_1 - k_2 + k_3 &= 0 \end{aligned} \tag{4}$$

Thus, our problem reduces to determining whether this system has nontrivial solutions. There are various ways to do this; one possibility is to simply solve the system, which yields

$$k_1 = -\tfrac{1}{2}t, \quad k_2 = -\tfrac{1}{2}t, \quad k_3 = t$$

(we omit the details). This shows that the system has nontrivial solutions and hence that the vectors are linearly dependent. A second method for obtaining the same result is to compute the determinant of the coefficient matrix

$$A = \begin{bmatrix} 1 & 5 & 3 \\ -2 & 6 & 2 \\ 3 & -1 & 1 \end{bmatrix}$$

> In Example 2, what relationship do you see between the components of $\mathbf{v}_1$, $\mathbf{v}_2$, and $\mathbf{v}_3$ and the columns of the coefficient matrix A?

and use parts (b) and (g) of Theorem 2.3.8. We leave it for you to verify that $\det(A) = 0$, from which it follows (3) has nontrivial solutions and the vectors are linearly dependent.

▶ **EXAMPLE 3** **Linear Independence in R^4**

Determine whether the vectors

$$\mathbf{v}_1 = (1, 2, 2, -1), \quad \mathbf{v}_2 = (4, 9, 9, -4), \quad \mathbf{v}_3 = (5, 8, 9, -5)$$

in R^4 are linearly dependent or linearly independent.

Solution The linear independence or linear dependence of these vectors is determined by whether there exist nontrivial solutions of the vector equation

$$k_1 \mathbf{v}_1 + k_2 \mathbf{v}_2 + k_3 \mathbf{v}_3 = \mathbf{0}$$

or, equivalently, of

$$k_1(1, 2, 2, -1) + k_2(4, 9, 9, -4) + k_3(5, 8, 9, -5) = (0, 0, 0, 0)$$

Equating corresponding components on the two sides yields the homogeneous linear system

$$\begin{aligned} k_1 + 4k_2 + 5k_3 &= 0 \\ 2k_1 + 9k_2 + 8k_3 &= 0 \\ 2k_1 + 9k_2 + 9k_3 &= 0 \\ -k_1 - 4k_2 - 5k_3 &= 0 \end{aligned}$$

We leave it for you to show that this system has only the trivial solution

$$k_1 = 0, \quad k_2 = 0, \quad k_3 = 0$$

from which you can conclude that $\mathbf{v}_1$, $\mathbf{v}_2$, and $\mathbf{v}_3$ are linearly independent.

▶ **EXAMPLE 4** **An Important Linearly Independent Set in P_n**

Show that the polynomials

$$1, \quad x, \quad x^2, \ldots, \quad x^n$$

form a linearly independent set in P_n.

Solution For convenience, let us denote the polynomials as

$$\mathbf{p}_0 = 1, \quad \mathbf{p}_1 = x, \quad \mathbf{p}_2 = x^2, \ldots, \quad \mathbf{p}_n = x^n$$

We must show that the vector equation

$$a_0 \mathbf{p}_0 + a_1 \mathbf{p}_1 + a_2 \mathbf{p}_2 + \cdots + a_n \mathbf{p}_n = \mathbf{0} \qquad (5)$$

has only the trivial solution

$$a_0 = a_1 = a_2 = \cdots = a_n = 0$$

But (5) is equivalent to the statement that

$$a_0 + a_1 x + a_2 x^2 + \cdots + a_n x^n = 0 \tag{6}$$

for all x in $(-\infty, \infty)$, so we must show that this holds if and only if each coefficient in (6) is zero. To see that this is so, recall from algebra that a nonzero polynomial of degree n has at most n distinct roots. That being the case, each coefficient in (6) must be zero, for otherwise the left side of the equation would be a nonzero polynomial with infinitely many roots. Thus, (5) has only the trivial solution. ◀

The following example shows that the problem of determining whether a given set of vectors in P_n is linearly independent or linearly dependent can be reduced to determining whether a certain set of vectors in R^n is linearly dependent or independent.

▶ **EXAMPLE 5** Linear Independence of Polynomials

Determine whether the polynomials

$$\mathbf{p}_1 = 1 - x, \quad \mathbf{p}_2 = 5 + 3x - 2x^2, \quad \mathbf{p}_3 = 1 + 3x - x^2$$

are linearly dependent or linearly independent in P_2.

Solution The linear independence or linear dependence of these vectors is determined by whether there exist nontrivial solutions of the vector equation

$$k_1 \mathbf{p}_1 + k_2 \mathbf{p}_2 + k_3 \mathbf{p}_3 = \mathbf{0} \tag{7}$$

This equation can be written as

$$k_1(1 - x) + k_2(5 + 3x - 2x^2) + k_3(1 + 3x - x^2) = 0 \tag{8}$$

or, equivalently, as

$$(k_1 + 5k_2 + k_3) + (-k_1 + 3k_2 + 3k_3)x + (-2k_2 - k_3)x^2 = 0$$

Since this equation must be satisfied by all x in $(-\infty, \infty)$, each coefficient must be zero (as explained in the previous example). Thus, the linear dependence or independence of the given polynomials hinges on whether the following linear system has a nontrivial solution:

$$\begin{aligned} k_1 + 5k_2 + k_3 &= 0 \\ -k_1 + 3k_2 + 3k_3 &= 0 \\ -2k_2 - k_3 &= 0 \end{aligned} \tag{9}$$

In Example 5, what relationship do you see between the coefficients of the given polynomials and the column vectors of the coefficient matrix of system (9)?

We leave it for you to show that this linear system has a nontrivial solutions either by solving it directly or by showing that the coefficient matrix has determinant zero. Thus, the set $\{\mathbf{p}_1, \mathbf{p}_2, \mathbf{p}_3\}$ is linearly dependent. ◀

An Alternative Interpretation of Linear Independence

The terms *linearly dependent* and *linearly independent* are intended to indicate whether the vectors in a given set are interrelated in some way. The following theorem, whose proof is deferred to the end of this section, makes this idea more precise.

THEOREM 4.3.1 *A set S with two or more vectors is*

(a) *Linearly dependent if and only if at least one of the vectors in S is expressible as a linear combination of the other vectors in S.*

(b) *Linearly independent if and only if no vector in S is expressible as a linear combination of the other vectors in S.*

▶ **EXAMPLE 6** Example 1 Revisited

In Example 1 we showed that the standard unit vectors in R^n are linearly independent. Thus, it follows from Theorem 4.3.1 that none of these vectors is expressible as a linear combination of the other two. To illustrate this in R^3, suppose, for example, that

$$\mathbf{k} = k_1 \mathbf{i} + k_2 \mathbf{j}$$

or in terms of components that

$$(0, 0, 1) = (k_1, k_2, 0)$$

Since this equation cannot be satisfied by any values of k_1 and k_2, there is no way to express $\mathbf{k}$ as a linear combination of $\mathbf{i}$ and $\mathbf{j}$. Similarly, $\mathbf{i}$ is not expressible as a linear combination of $\mathbf{j}$ and $\mathbf{k}$, and $\mathbf{j}$ is not expressible as a linear combination of $\mathbf{i}$ and $\mathbf{k}$.

▶ **EXAMPLE 7** Example 2 Revisited

In Example 2 we saw that the vectors

$$\mathbf{v}_1 = (1, -2, 3), \quad \mathbf{v}_2 = (5, 6, -1), \quad \mathbf{v}_3 = (3, 2, 1)$$

are linearly dependent. Thus, it follows from Theorem 4.3.1 that at least one of these vectors is expressible as a linear combination of the other two. We leave it for you to confirm that these vectors satisfy the equation

$$\tfrac{1}{2}\mathbf{v}_1 + \tfrac{1}{2}\mathbf{v}_2 - \mathbf{v}_3 = \mathbf{0}$$

from which it follows, for example, that

$$\mathbf{v}_3 = \tfrac{1}{2}\mathbf{v}_1 + \tfrac{1}{2}\mathbf{v}_2 \quad ◀$$

Sets with One or Two Vectors

The following basic theorem is concerned with the linear independence and linear dependence of sets with one or two vectors and sets that contain the zero vector.

THEOREM 4.3.2

(a) *A finite set that contains* $\mathbf{0}$ *is linearly dependent.*

(b) *A set with exactly one vector is linearly independent if and only if that vector is not* $\mathbf{0}$.

(c) *A set with exactly two vectors is linearly independent if and only if neither vector is a scalar multiple of the other.*

Józef Hoëné de Wroński (1778–1853)

Historical Note The Polish-French mathematician Józef Hoëné de Wroński was born Józef Hoëné and adopted the name Wroński after he married. Wroński's life was fraught with controversy and conflict, which some say was due to his psychopathic tendencies and his exaggeration of the importance of his own work. Although Wroński's work was dismissed as rubbish for many years, and much of it was indeed erroneous, some of his ideas contained hidden brilliance and have survived. Among other things, Wroński designed a caterpillar vehicle to compete with trains (though it was never manufactured) and did research on the famous problem of determining the longitude of a ship at sea. His final years were spent in poverty.

[*Image*: wikipedia]

We will prove part (*a*) and leave the rest as exercises.

Proof (a) For any vectors $\mathbf{v}_1, \mathbf{v}_2, \ldots, \mathbf{v}_r$, the set $S = \{\mathbf{v}_1, \mathbf{v}_2, \ldots, \mathbf{v}_r, \mathbf{0}\}$ is linearly dependent since the equation

$$0\mathbf{v}_1 + 0\mathbf{v}_2 + \cdots + 0\mathbf{v}_r + 1(\mathbf{0}) = \mathbf{0}$$

expresses $\mathbf{0}$ as a linear combination of the vectors in S with coefficients that are not all zero. ◄

▶ **EXAMPLE 8** Linear Independence of Two Functions

The functions $\mathbf{f}_1 = x$ and $\mathbf{f}_2 = \sin x$ are linearly independent vectors in $F(-\infty, \infty)$ since neither function is a scalar multiple of the other. On the other hand, the two functions $\mathbf{g}_1 = \sin 2x$ and $\mathbf{g}_2 = \sin x \cos x$ are linearly dependent because the trigonometric identity $\sin 2x = 2 \sin x \cos x$ reveals that $\mathbf{g}_1$ and $\mathbf{g}_2$ are scalar multiples of each other. ◄

A Geometric Interpretation of Linear Independence

Linear independence has the following useful geometric interpretations in R^2 and R^3:

- Two vectors in R^2 or R^3 are linearly independent if and only if they do not lie on the same line when they have their initial points at the origin. Otherwise one would be a scalar multiple of the other (Figure 4.3.3).

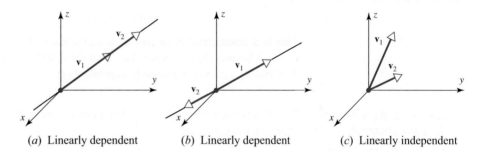

▶ **Figure 4.3.3**   (*a*) Linearly dependent   (*b*) Linearly dependent   (*c*) Linearly independent

- Three vectors in R^3 are linearly independent if and only if they do not lie in the same plane when they have their initial points at the origin. Otherwise at least one would be a linear combination of the other two (Figure 4.3.4).

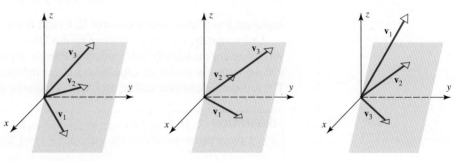

▶ **Figure 4.3.4**   (*a*) Linearly dependent   (*b*) Linearly dependent   (*c*) Linearly independent

At the beginning of this section we observed that a third coordinate axis in R^2 is superfluous by showing that a unit vector along such an axis would have to be expressible as a linear combination of unit vectors along the positive x- and y-axis. That result is a consequence of the next theorem, which shows that there can be at most n vectors in any linearly independent set R^n.

THEOREM 4.3.3 *Let $S = \{\mathbf{v}_1, \mathbf{v}_2, \ldots, \mathbf{v}_r\}$ be a set of vectors in R^n. If $r > n$, then S is linearly dependent.*

Proof Suppose that
$$\mathbf{v}_1 = (v_{11}, v_{12}, \ldots, v_{1n})$$
$$\mathbf{v}_2 = (v_{21}, v_{22}, \ldots, v_{2n})$$
$$\vdots \qquad \vdots$$
$$\mathbf{v}_r = (v_{r1}, v_{r2}, \ldots, v_{rn})$$

and consider the equation
$$k_1\mathbf{v}_1 + k_2\mathbf{v}_2 + \cdots + k_r\mathbf{v}_r = \mathbf{0}$$

If we express both sides of this equation in terms of components and then equate the corresponding components, we obtain the system

$$v_{11}k_1 + v_{21}k_2 + \cdots + v_{r1}k_r = 0$$
$$v_{12}k_1 + v_{22}k_2 + \cdots + v_{r2}k_r = 0$$
$$\vdots \qquad \vdots \qquad \qquad \vdots$$
$$v_{1n}k_1 + v_{2n}k_2 + \cdots + v_{rn}k_r = 0$$

It follows from Theorem 4.3.3, for example, that a set in R^2 with more than two vectors is linearly dependent and a set in R^3 with more than three vectors is linearly dependent.

This is a homogeneous system of n equations in the r unknowns $k_1, \ldots, k_r$. Since $r > n$, it follows from Theorem 1.2.2 that the system has nontrivial solutions. Therefore, $S = \{\mathbf{v}_1, \mathbf{v}_2, \ldots, \mathbf{v}_r\}$ is a linearly dependent set. ◄

CALCULUS REQUIRED
Linear Independence of Functions

Sometimes linear dependence of functions can be deduced from known identities. For example, the functions
$$\mathbf{f}_1 = \sin^2 x, \quad \mathbf{f}_2 = \cos^2 x, \quad \text{and} \quad \mathbf{f}_3 = 5$$
form a linearly dependent set in $F(-\infty, \infty)$, since the equation
$$5\mathbf{f}_1 + 5\mathbf{f}_2 - \mathbf{f}_3 = 5\sin^2 x + 5\cos^2 x - 5$$
$$= 5(\sin^2 x + \cos^2 x) - 5 = \mathbf{0}$$
expresses $\mathbf{0}$ as a linear combination of $\mathbf{f}_1$, $\mathbf{f}_2$, and $\mathbf{f}_3$ with coefficients that are not all zero.

Unfortunately, there is no *general* method that can be used to determine whether a set of functions is linearly independent or linearly dependent. However, there does exist a theorem that is useful for establishing linear independence in certain circumstances. The following definition will be useful for discussing that theorem.

DEFINITION 2 If $\mathbf{f}_1 = f_1(x), \mathbf{f}_2 = f_2(x), \ldots, \mathbf{f}_n = f_n(x)$ are functions that are $n - 1$ times differentiable on the interval $(-\infty, \infty)$, then the determinant

$$W(x) = \begin{vmatrix} f_1(x) & f_2(x) & \cdots & f_n(x) \\ f_1'(x) & f_2'(x) & \cdots & f_n'(x) \\ \vdots & \vdots & & \vdots \\ f_1^{(n-1)}(x) & f_2^{(n-1)}(x) & \cdots & f_n^{(n-1)}(x) \end{vmatrix}$$

is called the **Wronskian** of $f_1, f_2, \ldots, f_n$.

Suppose for the moment that $\mathbf{f}_1 = f_1(x), \mathbf{f}_2 = f_2(x), \ldots, \mathbf{f}_n = f_n(x)$ are *linearly dependent* vectors in $C^{(n-1)}(-\infty, \infty)$. This implies that for certain values of the coefficients the vector equation

$$k_1 \mathbf{f}_1 + k_2 \mathbf{f}_2 + \cdots + k_n \mathbf{f}_n = \mathbf{0}$$

has a nontrivial solution, or equivalently that the equation

$$k_1 f_1(x) + k_2 f_2(x) + \cdots + k_n f_n(x) = 0$$

is satisfied for all x in $(-\infty, \infty)$. Using this equation together with those that result by differentiating it $n-1$ times yields the linear system

$$\begin{aligned} k_1 f_1(x) & + k_2 f_2(x) & + \cdots + k_n f_n(x) & = 0 \\ k_1 f_1'(x) & + k_2 f_2'(x) & + \cdots + k_n f_n'(x) & = 0 \\ \vdots & \quad \vdots & \quad \vdots & \quad \vdots \\ k_1 f_1^{(n-1)}(x) & + k_2 f_2^{(n-1)}(x) & + \cdots + k_n f_n^{(n-1)}(x) & = 0 \end{aligned}$$

Thus, the linear dependence of $\mathbf{f}_1, \mathbf{f}_2, \ldots, \mathbf{f}_n$ implies that the linear system

$$\begin{bmatrix} f_1(x) & f_2(x) & \cdots & f_n(x) \\ f_1'(x) & f_2'(x) & \cdots & f_n'(x) \\ \vdots & \vdots & & \vdots \\ f_1^{(n-1)}(x) & f_2^{(n-1)}(x) & \cdots & f_n^{(n-1)}(x) \end{bmatrix} \begin{bmatrix} k_1 \\ k_2 \\ \vdots \\ k_n \end{bmatrix} = \begin{bmatrix} 0 \\ 0 \\ \vdots \\ 0 \end{bmatrix} \quad (10)$$

has a nontrivial solution. But this implies that the determinant of the coefficient matrix of (10) is zero for every such x. Since this determinant is the Wronskian of $f_1, f_2, \ldots, f_n$, we have established the following result.

THEOREM 4.3.4 *If the functions* $\mathbf{f}_1, \mathbf{f}_2, \ldots, \mathbf{f}_n$ *have* $n-1$ *continuous derivatives on the interval* $(-\infty, \infty)$, *and if the Wronskian of these functions is not identically zero on* $(-\infty, \infty)$, *then these functions form a linearly independent set of vectors in* $C^{(n-1)}(-\infty, \infty)$.

In Example 8 we showed that x and $\sin x$ are linearly independent functions by observing that neither is a scalar multiple of the other. The following example shows how to obtain the same result using the Wronskian (though it is a more complicated procedure in this particular case).

▶ **EXAMPLE 9** Linear Independence Using the Wronskian

Use the Wronskian to show that $\mathbf{f}_1 = x$ and $\mathbf{f}_2 = \sin x$ are linearly independent.

Solution The Wronskian is

$$W(x) = \begin{vmatrix} x & \sin x \\ 1 & \cos x \end{vmatrix} = x \cos x - \sin x$$

This function is not identically zero on the interval $(-\infty, \infty)$ since, for example,

$$W\left(\frac{\pi}{2}\right) = \frac{\pi}{2} \cos\left(\frac{\pi}{2}\right) - \sin\left(\frac{\pi}{2}\right) = \frac{\pi}{2}$$

Thus, the functions are linearly independent.

WARNING The converse of Theorem 4.3.4 is false. If the Wronskian of $\mathbf{f}_1, \mathbf{f}_2, \ldots, \mathbf{f}_n$ is identically zero on $(-\infty, \infty)$, then no conclusion can be reached about the linear independence of $\{\mathbf{f}_1, \mathbf{f}_2, \ldots, \mathbf{f}_n\}$—this set of vectors may be linearly independent or linearly dependent.

▶ **EXAMPLE 10** Linear Independence Using the Wronskian

Use the Wronskian to show that $\mathbf{f}_1 = 1$, $\mathbf{f}_2 = e^x$, and $\mathbf{f}_3 = e^{2x}$ are linearly independent.

Solution The Wronskian is

$$W(x) = \begin{vmatrix} 1 & e^x & e^{2x} \\ 0 & e^x & 2e^{2x} \\ 0 & e^x & 4e^{2x} \end{vmatrix} = 2e^{3x}$$

This function is obviously not identically zero on $(-\infty, \infty)$, so $\mathbf{f}_1$, $\mathbf{f}_2$, and $\mathbf{f}_3$ form a linearly independent set. ◀

OPTIONAL

We will close this section by proving part (*a*) of Theorem 4.3.1. We will leave the proof of part (*b*) as an exercise.

Proof of Theorem 4.3.1(a) Let $S = \{\mathbf{v}_1, \mathbf{v}_2, \ldots, \mathbf{v}_r\}$ be a set with two or more vectors. If we assume that S is linearly dependent, then there are scalars $k_1, k_2, \ldots, k_r$, not all zero, such that

$$k_1 \mathbf{v}_1 + k_2 \mathbf{v}_2 + \cdots + k_r \mathbf{v}_r = \mathbf{0} \qquad (11)$$

To be specific, suppose that $k_1 \neq 0$. Then (11) can be rewritten as

$$\mathbf{v}_1 = \left(-\frac{k_2}{k_1}\right) \mathbf{v}_2 + \cdots + \left(-\frac{k_r}{k_1}\right) \mathbf{v}_r$$

which expresses $\mathbf{v}_1$ as a linear combination of the other vectors in S. Similarly, if $k_j \neq 0$ in (11) for some $j = 2, 3, \ldots, r$, then $\mathbf{v}_j$ is expressible as a linear combination of the other vectors in S.

Conversely, let us assume that at least one of the vectors in S is expressible as a linear combination of the other vectors. To be specific, suppose that

$$\mathbf{v}_1 = c_2 \mathbf{v}_2 + c_3 \mathbf{v}_3 + \cdots + c_r \mathbf{v}_r$$

so

$$\mathbf{v}_1 - c_2 \mathbf{v}_2 - c_3 \mathbf{v}_3 - \cdots - c_r \mathbf{v}_r = \mathbf{0}$$

It follows that S is linearly dependent since the equation

$$k_1 \mathbf{v}_1 + k_2 \mathbf{v}_2 + \cdots + k_r \mathbf{v}_r = \mathbf{0}$$

is satisfied by

$$k_1 = 1, \quad k_2 = -c_2, \ldots, \quad k_r = -c_r$$

which are not all zero. The proof in the case where some vector other than $\mathbf{v}_1$ is expressible as a linear combination of the other vectors in S is similar. ◀

Concept Review

- Trivial solution
- Linearly independent set
- Linearly dependent set
- Wronskian

Skills

- Determine whether a set of vectors is linearly independent or linearly dependent.
- Express one vector in a linearly dependent set as a linear combination of the other vectors in the set.
- Use the Wronskian to show that a set of functions is linearly independent.

Exercise Set 4.3

1. Explain why the following are linearly dependent sets of vectors. (Solve this problem by inspection.)
   (a) $\mathbf{u}_1 = (3, -1)$ and $\mathbf{u}_2 = (6, -2)$ in R^2
   (b) $\mathbf{u}_1 = (-2, 0, 1)$, $\mathbf{u}_2 = (4, -2, 0)$, $\mathbf{u}_3 = (6, -6, 3)$ in R^3
   (c) $A = \begin{bmatrix} 0 & 1 \\ 2 & 3 \end{bmatrix}$ and $B = \begin{bmatrix} 0 & -1 \\ -2 & -3 \end{bmatrix}$ in M_{22}
   (d) $\mathbf{p}_1 = 2 + x - 3x^2$ and $\mathbf{p}_2 = -4 - 2x + 6x^2$ in P_2

2. Which of the following sets of vectors in R^3 are linearly dependent?
   (a) $(4, -1, 2)$, $(-4, 10, 2)$
   (b) $(-3, 0, 4)$, $(5, -1, 2)$, $(1, 1, 3)$
   (c) $(8, -1, 3)$, $(4, 0, 1)$
   (d) $(-2, 0, 1)$, $(3, 2, 5)$, $(6, -1, 1)$, $(7, 0, -2)$

3. Which of the following sets of vectors in R^4 are linearly dependent?
   (a) $(1, 2, -2, 1)$, $(3, 6, -6, 3)$, $(4, -2, 4, 1)$,
   (b) $(5, 2, 0, -1)$, $(0, -3, 0, 1)$, $(1, 0, -1, 2)$, $(3, 1, 0, 1)$
   (c) $(2, 1, 1, -4)$, $(2, -8, 9, -2)$, $(0, 3, -1, 5)$, $(0, -1, 2, 4)$
   (d) $(1, 0, -6, 3)$, $(0, 1, 3, 0)$, $(0, 2, 7, 0)$, $(0, 2, 0, 1)$

4. Which of the following sets of vectors in P_2 are linearly dependent?
   (a) $2 - x + 4x^2$, $3 + 6x + 2x^2$, $2 + 10x - 4x^2$
   (b) $3 + x + x^2$, $2 - x + 5x^2$, $4 - 3x^2$
   (c) $6 - x^2$, $1 + x + 4x^2$
   (d) $1 + 3x + 3x^2$, $x + 4x^2$, $5 + 6x + 3x^2$, $7 + 2x - x^2$

5. Assume that $\mathbf{v}_1$, $\mathbf{v}_2$, and $\mathbf{v}_3$ are vectors in R^3 that have their initial points at the origin. In each part, determine whether the three vectors lie in a plane.
   (a) $\mathbf{v}_1 = (3, 4, 5)$, $\mathbf{v}_2 = (1, -1, 0)$, $\mathbf{v}_3 = (2, 1, 0)$
   (b) $\mathbf{v}_1 = (2, 7, -6)$, $\mathbf{v}_2 = (1, 2, -4)$, $\mathbf{v}_3 = (-1, 1, 6)$

6. Assume that $\mathbf{v}_1$, $\mathbf{v}_2$, and $\mathbf{v}_3$ are vectors in R^3 that have their initial points at the origin. In each part, determine whether the three vectors lie on the same line.
   (a) $\mathbf{v}_1 = (-1, 2, 3)$, $\mathbf{v}_2 = (2, -4, -6)$, $\mathbf{v}_3 = (-3, 6, 0)$
   (b) $\mathbf{v}_1 = (2, -1, 4)$, $\mathbf{v}_2 = (4, 2, 3)$, $\mathbf{v}_3 = (2, 7, -6)$
   (c) $\mathbf{v}_1 = (4, 6, 8)$, $\mathbf{v}_2 = (2, 3, 4)$, $\mathbf{v}_3 = (-2, -3, -4)$

7. (a) Show that the vectors $\mathbf{v}_1 = (2, 0, -2, 1)$, $\mathbf{v}_2 = (3, 1, -5, 0)$, and $\mathbf{v}_3 = (2, 2, -6, -2)$ form a linearly dependent set in R^4.
   (b) Express each vector in part (a) as a linear combination of the other two.

8. (a) Show that the three vectors $\mathbf{v}_1 = (1, 2, 3, 4)$, $\mathbf{v}_2 = (0, 1, 0, -1)$, and $\mathbf{v}_3 = (1, 3, 3, 3)$ form a linearly dependent set in R^4.
   (b) Express each vector in part (a) as a linear combination of the other two.

9. For which real values of λ do the following vectors form a linearly dependent set in R^3?
   $$\mathbf{v}_1 = \left(\lambda, -\tfrac{1}{2}, -\tfrac{1}{2}\right), \quad \mathbf{v}_2 = \left(-\tfrac{1}{2}, \lambda, -\tfrac{1}{2}\right), \quad \mathbf{v}_3 = \left(-\tfrac{1}{2}, -\tfrac{1}{2}, \lambda\right)$$

10. Show that if $\{\mathbf{v}_1, \mathbf{v}_2, \mathbf{v}_3\}$ is a linearly independent set of vectors, then so are $\{\mathbf{v}_1, \mathbf{v}_2\}$, $\{\mathbf{v}_1, \mathbf{v}_3\}$, $\{\mathbf{v}_2, \mathbf{v}_3\}$, $\{\mathbf{v}_1\}$, $\{\mathbf{v}_2\}$, and $\{\mathbf{v}_3\}$.

11. Show that if $S = \{\mathbf{v}_1, \mathbf{v}_2, \ldots, \mathbf{v}_r\}$ is a linearly independent set of vectors, then so is every nonempty subset of S.

12. Show that if $S = \{\mathbf{v}_1, \mathbf{v}_2, \mathbf{v}_3\}$ is a linearly dependent set of vectors in a vector space V, and $\mathbf{v}_4$ is any vector in V that is not in S, then $\{\mathbf{v}_1, \mathbf{v}_2, \mathbf{v}_3, \mathbf{v}_4\}$ is also linearly dependent.

13. Show that if $S = \{\mathbf{v}_1, \mathbf{v}_2, \ldots, \mathbf{v}_r\}$ is a linearly dependent set of vectors in a vector space V, and if $\mathbf{v}_{r+1}, \ldots, \mathbf{v}_n$ are any vectors in V that are not in S, then $\{\mathbf{v}_1, \mathbf{v}_2, \ldots, \mathbf{v}_r, \mathbf{v}_{r+1}, \ldots, \mathbf{v}_n\}$ is also linearly dependent.

14. Show that in P_2 every set with more than three vectors is linearly dependent.

15. Show that if $\{\mathbf{v}_1, \mathbf{v}_2\}$ is linearly independent and $\mathbf{v}_3$ does not lie in span$\{\mathbf{v}_1, \mathbf{v}_2\}$, then $\{\mathbf{v}_1, \mathbf{v}_2, \mathbf{v}_3\}$ is linearly independent.

16. Prove: For any vectors $\mathbf{u}$, $\mathbf{v}$, and $\mathbf{w}$ in a vector space V, the vectors $\mathbf{u} - \mathbf{v}$, $\mathbf{v} - \mathbf{w}$, and $\mathbf{w} - \mathbf{u}$ form a linearly dependent set.

17. Prove: The space spanned by two vectors in R^3 is a line through the origin, a plane through the origin, or the origin itself.

18. Under what conditions is a set with one vector linearly independent?

19. Are the vectors $\mathbf{v}_1$, $\mathbf{v}_2$, and $\mathbf{v}_3$ in part (*a*) of the accompanying figure linearly independent? What about those in part (*b*)? Explain.

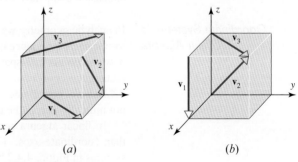

▲ Figure Ex-19

20. By using appropriate identities, where required, determine which of the following sets of vectors in $F(-\infty, \infty)$ are linearly dependent.

(a) $6,\ 3\sin^2 x,\ 2\cos^2 x$ (b) $x,\ \cos x$
(c) $1,\ \sin x,\ \sin 2x$ (d) $\cos 2x,\ \sin^2 x,\ \cos^2 x$
(e) $(3-x)^2,\ x^2 - 6x,\ 5$ (f) $0,\ \cos^3 \pi x,\ \sin^5 3\pi x$

21. The functions $f_1(x) = x$ and $f_2(x) = \cos x$ are linearly independent in $F(-\infty, \infty)$ because neither function is a scalar multiple of the other. Confirm the linear independence using Wroński's test.

22. The functions $f_1(x) = \sin x$ and $f_2(x) = \cos x$ are linearly independent in $F(-\infty, \infty)$ because neither function is a scalar multiple of the other. Confirm the linear independence using Wroński's test.

23. *(Calculus required)* Use the Wronskian to show that the following sets of vectors are linearly independent.

(a) $1,\ x,\ e^x$ (b) $1,\ x,\ x^2$

24. Use Wroński's test to show that the functions $f_1(x) = e^x$, $f_2(x) = xe^x$, and $f_3(x) = x^2 e^x$ are linearly independent vectors in $F(-\infty, \infty)$.

25. Use Wroński's test to show that the functions $f_1(x) = \sin x$, $f_2(x) = \cos x$, and $f_3(x) = x \cos x$ are linearly independent vectors in $F(-\infty, \infty)$.

26. Use part (a) of Theorem 4.3.1 to prove part (b).

27. Prove part (b) of Theorem 4.3.2.

28. (a) In Example 1 we showed that the mutually orthogonal vectors **i**, **j**, and **k** form a linearly independent set of vectors in R^3. Do you think that every set of three nonzero mutually orthogonal vectors in R^3 is linearly independent? Justify your conclusion with a geometric argument.

(b) Justify your conclusion with an algebraic argument. [*Hint:* Use dot products.]

True-False Exercises

In parts (a)–(h) determine whether the statement is true or false, and justify your answer.

(a) A set containing a single vector is linearly independent.

(b) The set of vectors $\{\mathbf{v}, k\mathbf{v}\}$ is linearly dependent for every scalar k.

(c) Every linearly dependent set contains the zero vector.

(d) If the set of vectors $\{\mathbf{v}_1, \mathbf{v}_2, \mathbf{v}_3\}$ is linearly independent, then $\{k\mathbf{v}_1, k\mathbf{v}_2, k\mathbf{v}_3\}$ is also linearly independent for every nonzero scalar k.

(e) If $\mathbf{v}_1, \ldots, \mathbf{v}_n$ are linearly dependent nonzero vectors, then at least one vector $\mathbf{v}_k$ is a unique linear combination of $\mathbf{v}_1, \ldots, \mathbf{v}_{k-1}$.

(f) The set of 2×2 matrices that contain exactly two 1's and two 0's is a linearly independent set in M_{22}.

(g) The three polynomials $(x-1)(x+2)$, $x(x+2)$, and $x(x-1)$ are linearly independent.

(h) The functions f_1 and f_2 are linearly dependent if there is a real number x so that $k_1 f_1(x) + k_2 f_2(x) = 0$ for some scalars k_1 and k_2.

4.4 Coordinates and Basis

We usually think of a line as being one-dimensional, a plane as two-dimensional, and the space around us as three-dimensional. It is the primary goal of this section and the next to make this intuitive notion of dimension precise. In this section we will discuss coordinate systems in general vector spaces and lay the groundwork for a precise definition of dimension in the next section.

Coordinate Systems in Linear Algebra

In analytic geometry we learned to use *rectangular* coordinate systems to create a one-to-one correspondence between points in 2-space and ordered pairs of real numbers and between points in 3-space and ordered triples of real numbers (Figure 4.4.1). Although rectangular coordinate systems are common, they are not essential. For example, Figure 4.4.2 shows coordinate systems in 2-space and 3-space in which the coordinate axes are not mutually perpendicular.

In linear algebra coordinate systems are commonly specified using vectors rather than coordinate axes. For example, in Figure 4.4.3 we have recreated the coordinate systems in Figure 4.4.2 by using unit vectors to identify the positive directions and then attaching coordinates to a point P using the scalar coefficients in the equations

$$\overrightarrow{OP} = a\mathbf{u}_1 + b\mathbf{u}_2 \quad \text{and} \quad \overrightarrow{OP} = a\mathbf{u}_1 + b\mathbf{u}_2 + c\mathbf{u}_3$$

4.4 Coordinates and Basis

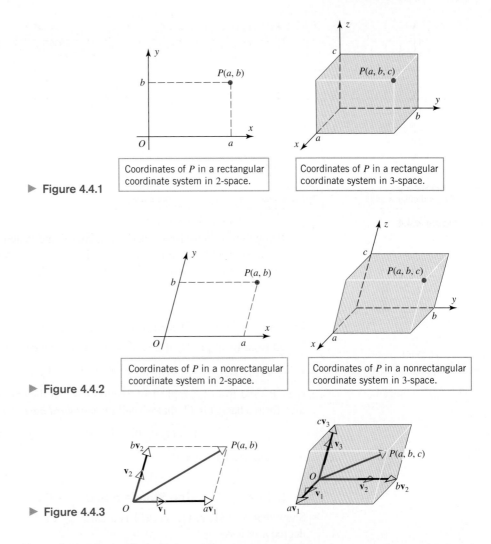

▶ Figure 4.4.1 Coordinates of P in a rectangular coordinate system in 2-space. / Coordinates of P in a rectangular coordinate system in 3-space.

▶ Figure 4.4.2 Coordinates of P in a nonrectangular coordinate system in 2-space. / Coordinates of P in a nonrectangular coordinate system in 3-space.

▶ Figure 4.4.3

Units of measurement are essential ingredients of any coordinate system. In geometry problems one tries to use the same unit of measurement on all axes to avoid distorting the shapes of figures. This is less important in applications where coordinates represent physical quantities with diverse units (for example, time in seconds on one axis and temperature in degrees Celsius on another axis). To allow for this level of generality, we will relax the requirement that *unit* vectors be used to identify the positive directions and require only that those vectors be linearly independent. We will refer to these as the "basis vectors" for the coordinate system. In summary, it is the directions of the basis vectors that establish the positive directions, and it is the lengths of the basis vectors that establish the spacing between the integer points on the axes (Figure 4.4.4).

Basis for a Vector Space

The following definition will make the preceding ideas more precise and will enable us to extend the concept of a coordinate system to general vector spaces.

Note that in Definition 1 we have required a basis to have finitely many vectors. Some authors call this a ***finite basis***, but we will not use this terminology.

DEFINITION 1 If V is any vector space and $S = \{\mathbf{v}_1, \mathbf{v}_2, \ldots, \mathbf{v}_n\}$ is a finite set of vectors in V, then S is called a ***basis*** for V if the following two conditions hold:

(a) S is linearly independent.
(b) S spans V.

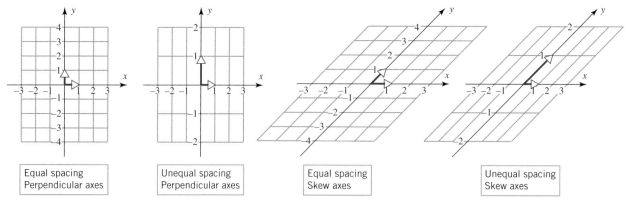

▲ Figure 4.4.4

If you think of a basis as describing a coordinate system for a vector space in V, then part (a) of this definition guarantees that there is no interrelationship between the basis vectors, and part (b) guarantees that there are enough basis vectors to provide coordinates for all vectors in V. Here are some examples.

▶ **EXAMPLE 1** The Standard Basis for R^n

Recall from Example 11 of Section 4.2 that the standard unit vectors

$$\mathbf{e}_1 = (1, 0, 0, \ldots, 0), \quad \mathbf{e}_2 = (0, 1, 0, \ldots, 0), \ldots, \quad \mathbf{e}_n = (0, 0, 0, \ldots, 1)$$

span R^n and from Example 1 of Section 4.3 that they are linearly independent. Thus, they form a basis for R^n that we call the ***standard basis for R^n***. In particular,

$$\mathbf{i} = (1, 0, 0), \quad \mathbf{j} = (0, 1, 0), \quad \mathbf{k} = (0, 0, 1)$$

is the standard basis for R^3.

▶ **EXAMPLE 2** The Standard Basis for P_n

Show that $S = \{1, x, x^2, \ldots, x^n\}$ is a basis for the vector space P_n of polynomials of degree n or less.

Solution We must show that the polynomials in S are linearly independent and span P_n. Let us denote these polynomials by

$$\mathbf{p}_0 = 1, \quad \mathbf{p}_1 = x, \quad \mathbf{p}_2 = x^2, \ldots, \quad \mathbf{p}_n = x^n$$

We showed in Example 13 of Section 4.2 that these vectors span P_n and in Example 4 of Section 4.3 that they are linearly independent. Thus, they form a basis for P_n that we call the ***standard basis for P_n***.

▶ **EXAMPLE 3** Another Basis for R^3

Show that the vectors $\mathbf{v}_1 = (1, 2, 1)$, $\mathbf{v}_2 = (2, 9, 0)$, and $\mathbf{v}_3 = (3, 3, 4)$ form a basis for R^3.

Solution We must show that these vectors are linearly independent and span R^3. To prove linear independence we must show that the vector equation

$$c_1 \mathbf{v}_1 + c_2 \mathbf{v}_2 + c_3 \mathbf{v}_3 = \mathbf{0} \tag{1}$$

has only the trivial solution; and to prove that the vectors span R^3 we must show that every vector $\mathbf{b} = (b_1, b_2, b_3)$ in R^3 can be expressed as

$$c_1 \mathbf{v}_1 + c_2 \mathbf{v}_2 + c_3 \mathbf{v}_3 = \mathbf{b} \tag{2}$$

By equating corresponding components on the two sides, these two equations can be expressed as the linear systems

$$\begin{array}{ll} c_1 + 2c_2 + 3c_3 = 0 & c_1 + 2c_2 + 3c_3 = b_1 \\ 2c_1 + 9c_2 + 3c_3 = 0 \quad \text{and} \quad & 2c_1 + 9c_2 + 3c_3 = b_2 \\ c_1 + 4c_3 = 0 & c_1 + 4c_3 = b_3 \end{array} \qquad (3)$$

(verify). Thus, we have reduced the problem to showing that in (3) the homogeneous system has only the trivial solution and that the nonhomogeneous system is consistent for all values of b_1, b_2, and b_3. But the two systems have the same coefficient matrix

$$A = \begin{bmatrix} 1 & 2 & 3 \\ 2 & 9 & 3 \\ 1 & 0 & 4 \end{bmatrix}$$

so it follows from parts (b), (e), and (g) of Theorem 2.3.8 that we can prove both results at the same time by showing that $\det(A) \neq 0$. We leave it for you to confirm that $\det(A) = -1$, which proves that the vectors $\mathbf{v}_1$, $\mathbf{v}_2$, and $\mathbf{v}_3$ form a basis for R^3.

▶ **EXAMPLE 4 The Standard Basis for M_{mn}**

Show that the matrices

$$M_1 = \begin{bmatrix} 1 & 0 \\ 0 & 0 \end{bmatrix}, \quad M_2 = \begin{bmatrix} 0 & 1 \\ 0 & 0 \end{bmatrix}, \quad M_3 = \begin{bmatrix} 0 & 0 \\ 1 & 0 \end{bmatrix}, \quad M_4 = \begin{bmatrix} 0 & 0 \\ 0 & 1 \end{bmatrix}$$

form a basis for the vector space M_{22} of 2×2 matrices.

Solution We must show that the matrices are linearly independent and span M_{22}. To prove linear independence we must show that the equation

$$c_1 M_1 + c_2 M_2 + c_3 M_3 + c_4 M_4 = \mathbf{0} \qquad (4)$$

has only the trivial solution, where $\mathbf{0}$ is the 2×2 zero matrix; and to prove that the matrices span M_{22} we must show that every 2×2 matrix

$$B = \begin{bmatrix} a & b \\ c & d \end{bmatrix}$$

can be expressed as

$$c_1 M_1 + c_2 M_2 + c_3 M_3 + c_4 M_4 = B \qquad (5)$$

The matrix forms of Equations (4) and (5) are

$$c_1 \begin{bmatrix} 1 & 0 \\ 0 & 0 \end{bmatrix} + c_2 \begin{bmatrix} 0 & 1 \\ 0 & 0 \end{bmatrix} + c_3 \begin{bmatrix} 0 & 0 \\ 1 & 0 \end{bmatrix} + c_4 \begin{bmatrix} 0 & 0 \\ 0 & 1 \end{bmatrix} = \begin{bmatrix} 0 & 0 \\ 0 & 0 \end{bmatrix}$$

and

$$c_1 \begin{bmatrix} 1 & 0 \\ 0 & 0 \end{bmatrix} + c_2 \begin{bmatrix} 0 & 1 \\ 0 & 0 \end{bmatrix} + c_3 \begin{bmatrix} 0 & 0 \\ 1 & 0 \end{bmatrix} + c_4 \begin{bmatrix} 0 & 0 \\ 0 & 1 \end{bmatrix} = \begin{bmatrix} a & b \\ c & d \end{bmatrix}$$

which can be rewritten as

$$\begin{bmatrix} c_1 & c_2 \\ c_3 & c_4 \end{bmatrix} = \begin{bmatrix} 0 & 0 \\ 0 & 0 \end{bmatrix} \quad \text{and} \quad \begin{bmatrix} c_1 & c_2 \\ c_3 & c_4 \end{bmatrix} = \begin{bmatrix} a & b \\ c & d \end{bmatrix}$$

Since the first equation has only the trivial solution

$$c_1 = c_2 = c_3 = c_4 = 0$$

the matrices are linearly independent, and since the second equation has the solution

$$c_1 = a, \quad c_2 = b, \quad c_3 = c, \quad c_4 = d$$

the matrices span M_{22}. This proves that the matrices M_1, M_2, M_3, M_4 form a basis for M_{22}. More generally, the mn different matrices whose entries are zero except for a single entry of 1 form a basis for M_{mn} called the **standard basis for M_{mn}**. ◀

> Some writers define the empty set to be a basis for the zero vector space, but we will not do so.

It is not true that every vector space has a basis in the sense of Definition 1. The simplest example is the zero vector space, which contains no linearly independent sets and hence no basis. The following is an example of a nonzero vector space that has no basis in the sense of Definition 1 because it cannot be spanned by finitely many vectors.

▶ **EXAMPLE 5 A Vector Space That Has No Finite Spanning Set**

Show that the vector space of P_∞ of all polynomials with real coefficients has no finite spanning set.

Solution If there were a finite spanning set, say $S = \{\mathbf{p}_1, \mathbf{p}_2, \ldots, \mathbf{p}_r\}$, then the degrees of the polynomials in S would have a maximum value, say n; and this in turn would imply that any linear combination of the polynomials in S would have degree at most n. Thus, there would be no way to express the polynomial x^{n+1} as a linear combination of the polynomials in S, contradicting the fact that the vectors in S span P_∞. ◀

For reasons that will become clear shortly, a vector space that cannot be spanned by finitely many vectors is said to be **infinite-dimensional**, whereas those that can are said to be **finite-dimensional**.

▶ **EXAMPLE 6 Some Finite- and Infinite-Dimensional Spaces**

In Examples 1, 2, and 4 we found bases for R^n, P_n, and M_{mn}, so these vector spaces are finite-dimensional. We showed in Example 5 that the vector space P_∞ is not spanned by finitely many vectors and hence is infinite-dimensional. In the exercises of this section and the next we will ask you to show that the vector spaces R^∞, $F(-\infty, \infty)$, $C(-\infty, \infty)$, $C^m(-\infty, \infty)$, and $C^\infty(-\infty, \infty)$ are infinite-dimensional. ◀

Coordinates Relative to a Basis

Earlier in this section we drew an informal analogy between basis vectors and coordinate systems. Our next goal is to make this informal idea precise by defining the notion of a coordinate system in a general vector space. The following theorem will be our first step in that direction.

THEOREM 4.4.1 **Uniqueness of Basis Representation**

If $S = \{\mathbf{v}_1, \mathbf{v}_2, \ldots, \mathbf{v}_n\}$ is a basis for a vector space V, then every vector $\mathbf{v}$ in V can be expressed in the form $\mathbf{v} = c_1\mathbf{v}_1 + c_2\mathbf{v}_2 + \cdots + c_n\mathbf{v}_n$ in exactly one way.

Proof Since S spans V, it follows from the definition of a spanning set that every vector in V is expressible as a linear combination of the vectors in S. To see that there is only *one* way to express a vector as a linear combination of the vectors in S, suppose that some vector $\mathbf{v}$ can be written as

$$\mathbf{v} = c_1\mathbf{v}_1 + c_2\mathbf{v}_2 + \cdots + c_n\mathbf{v}_n$$

and also as

$$\mathbf{v} = k_1\mathbf{v}_1 + k_2\mathbf{v}_2 + \cdots + k_n\mathbf{v}_n$$

Subtracting the second equation from the first gives

$$\mathbf{0} = (c_1 - k_1)\mathbf{v}_1 + (c_2 - k_2)\mathbf{v}_2 + \cdots + (c_n - k_n)\mathbf{v}_n$$

Since the right side of this equation is a linear combination of vectors in S, the linear independence of S implies that

$$c_1 - k_1 = 0, \quad c_2 - k_2 = 0, \ldots, \quad c_n - k_n = 0$$

that is,

$$c_1 = k_1, \quad c_2 = k_2, \ldots, \quad c_n = k_n$$

Thus, the two expressions for $\mathbf{v}$ are the same. ◀

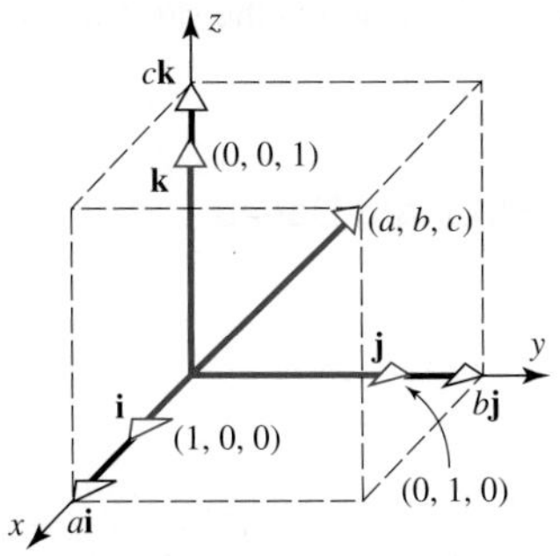

We now have all of the ingredients required to define the notion of "coordinates" in a general vector space V. For motivation, observe that in R^3, for example, the coordinates (a, b, c) of a vector $\mathbf{v}$ are precisely the coefficients in the formula

$$\mathbf{v} = a\mathbf{i} + b\mathbf{j} + c\mathbf{k}$$

that expresses $\mathbf{v}$ as a linear combination of the standard basis vectors for R^3 (see Figure 4.4.5). The following definition generalizes this idea.

▲ Figure 4.4.5

DEFINITION 2 If $S = \{\mathbf{v}_1, \mathbf{v}_2, \ldots, \mathbf{v}_n\}$ is a basis for a vector space V, and

$$\mathbf{v} = c_1\mathbf{v}_1 + c_2\mathbf{v}_2 + \cdots + c_n\mathbf{v}_n$$

is the expression for a vector $\mathbf{v}$ in terms of the basis S, then the scalars $c_1, c_2, \ldots, c_n$ are called the ***coordinates*** of $\mathbf{v}$ relative to the basis S. The vector $(c_1, c_2, \ldots, c_n)$ in R^n constructed from these coordinates is called the ***coordinate vector of $\mathbf{v}$ relative to S***; it is denoted by

$$(\mathbf{v})_S = (c_1, c_2, \ldots, c_n) \tag{6}$$

Sometimes it will be desirable to write a coordinate vector as a column matrix, in which case we will denote it using square brackets as

$$[\mathbf{v}]_S = \begin{bmatrix} c_1 \\ c_2 \\ \vdots \\ c_n \end{bmatrix}$$

We will refer to $[\mathbf{v}]_S$ as a ***coordinate matrix*** and reserve the terminology ***coordinate vector*** for the comma delimited form $(\mathbf{v})_S$.

Remark Recall that two *sets* are considered to be the same if they have the same members, even if those members are written in a different order. However, if $S = \{\mathbf{v}_1, \mathbf{v}_2, \ldots, \mathbf{v}_n\}$ is a set of *basis vectors*, then changing the order in which the vectors are written would change the order of the entries in $(\mathbf{v})_S$, possibly producing a different coordinate vector. To avoid this complication, we will make the convention that in any discussion involving a basis S the order of the vectors in S remains fixed. Some authors call a set of basis vectors with this restriction an ***ordered basis***. However, we will use this terminology only when emphasis on the order is required for clarity.

Observe that $(\mathbf{v})_S$ is a vector in R^n, so that once basis S is given for a vector space V, Theorem 4.4.1 establishes a one-to-one correspondence between vectors in V and vectors in R^n (Figure 4.4.6).

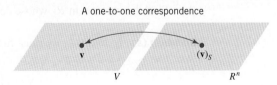

▶ Figure 4.4.6

▶ **EXAMPLE 7 Coordinates Relative to the Standard Basis for R^n**

In the special case where $V = R^n$ and S is the *standard basis*, the coordinate vector $(\mathbf{v})_S$ and the vector $\mathbf{v}$ are the same; that is,

$$\mathbf{v} = (\mathbf{v})_S$$

For example, in R^3 the representation of a vector $\mathbf{v} = (a, b, c)$ as a linear combination of the vectors in the standard basis $S = \{\mathbf{i}, \mathbf{j}, \mathbf{k}\}$ is

$$\mathbf{v} = a\mathbf{i} + b\mathbf{j} + c\mathbf{k}$$

so the coordinate vector relative to this basis is $(\mathbf{v})_S = (a, b, c)$, which is the same as the vector $\mathbf{v}$.

▶ **EXAMPLE 8** Coordinate Vectors Relative to Standard Bases

(a) Find the coordinate vector for the polynomial

$$\mathbf{p}(x) = c_0 + c_1 x + c_2 x^2 + \cdots + c_n x^n$$

relative to the standard basis for the vector space P_n.

(b) Find the coordinate vector of

$$B = \begin{bmatrix} a & b \\ c & d \end{bmatrix}$$

relative to the standard basis for M_{22}.

Solution (a) The given formula for $\mathbf{p}(x)$ expresses this polynomial as a linear combination of the standard basis vectors $S = \{1, x, x^2, \ldots, x^n\}$. Thus, the coordinate vector for $\mathbf{p}$ relative to S is

$$(\mathbf{p})_S = (c_0, c_1, c_2, \ldots, c_n)$$

Solution (b) We showed in Example 4 that the representation of a vector

$$B = \begin{bmatrix} a & b \\ c & d \end{bmatrix}$$

as a linear combination of the standard basis vectors is

$$B = \begin{bmatrix} a & b \\ c & d \end{bmatrix} = a\begin{bmatrix} 1 & 0 \\ 0 & 0 \end{bmatrix} + b\begin{bmatrix} 0 & 1 \\ 0 & 0 \end{bmatrix} + c\begin{bmatrix} 0 & 0 \\ 1 & 0 \end{bmatrix} + d\begin{bmatrix} 0 & 0 \\ 0 & 1 \end{bmatrix}$$

so the coordinate vector of B relative to S is

$$(B)_S = (a, b, c, d)$$

▶ **EXAMPLE 9** Coordinates in R^3

(a) We showed in Example 3 that the vectors

$$\mathbf{v}_1 = (1, 2, 1), \quad \mathbf{v}_2 = (2, 9, 0), \quad \mathbf{v}_3 = (3, 3, 4)$$

form a basis for R^3. Find the coordinate vector of $\mathbf{v} = (5, -1, 9)$ relative to the basis $S = \{\mathbf{v}_1, \mathbf{v}_2, \mathbf{v}_3\}$.

(b) Find the vector $\mathbf{v}$ in R^3 whose coordinate vector relative to S is $(\mathbf{v})_S = (-1, 3, 2)$.

Solution (a) To find $(\mathbf{v})_S$ we must first express $\mathbf{v}$ as a linear combination of the vectors in S; that is, we must find values of c_1, c_2, and c_3 such that

$$\mathbf{v} = c_1 \mathbf{v}_1 + c_2 \mathbf{v}_2 + c_3 \mathbf{v}_3$$

or, in terms of components,

$$(5, -1, 9) = c_1(1, 2, 1) + c_2(2, 9, 0) + c_3(3, 3, 4)$$

Equating corresponding components gives

$$\begin{aligned} c_1 + 2c_2 + 3c_3 &= 5 \\ 2c_1 + 9c_2 + 3c_3 &= -1 \\ c_1 + 4c_3 &= 9 \end{aligned}$$

Solving this system we obtain $c_1 = 1, c_2 = -1, c_3 = 2$ (verify). Therefore,
$$(\mathbf{v})_S = (1, -1, 2)$$

Solution (b) Using the definition of $(\mathbf{v})_S$, we obtain
$$\mathbf{v} = (-1)\mathbf{v}_1 + 3\mathbf{v}_2 + 2\mathbf{v}_3$$
$$= (-1)(1, 2, 1) + 3(2, 9, 0) + 2(3, 3, 4) = (11, 31, 7) \blacktriangleleft$$

Concept Review

- Basis
- Standard bases for R^n, P_n, M_{mn}
- Finite-dimensional
- Infinite-dimensional
- Coordinates
- Coordinate vector

Skills

- Show that a set of vectors is a basis for a vector space.
- Find the coordinates of a vector relative to a basis.
- Find the coordinate vector of a vector relative to a basis.

Exercise Set 4.4

1. In words, explain why the following sets of vectors are *not* bases for the indicated vector spaces.

(a) $\mathbf{u}_1 = (3, 2, 1)$, $\mathbf{u}_2 = (-2, 1, 0)$, $\mathbf{u}_3 = (5, 1, 1)$ for R^3

(b) $\mathbf{u}_1 = (1, 1)$, $\mathbf{u}_2 = (3, 5)$, $\mathbf{u}_3 = (4, 2)$ for R^2

(c) $\mathbf{p}_1 = 1 + x$, $\mathbf{p}_2 = 2x - x^2$ for P_2

(d) $A = \begin{bmatrix} 1 & 0 \\ 0 & 2 \end{bmatrix}$, $B = \begin{bmatrix} 0 & 3 \\ -5 & 1 \end{bmatrix}$, $C = \begin{bmatrix} 4 & -2 \\ 1 & 6 \end{bmatrix}$,

$D = \begin{bmatrix} 5 & 1 \\ 4 & 2 \end{bmatrix}$, $E = \begin{bmatrix} 7 & 1 \\ 2 & 9 \end{bmatrix}$ for M_{22}

2. Which of the following sets of vectors are bases for R^2?

(a) $\{(3, 1), (0, 0)\}$

(b) $\{(4, 1), (-7, -8)\}$

(c) $\{(5, 2), (-1, 3)\}$

(d) $\{(3, 9), (-4, -12)\}$

3. Which of the following sets of vectors are bases for R^3?

(a) $\{(1, 0, 0), (2, 2, 0), (3, 3, 3)\}$

(b) $\{(3, 1, -4), (2, 5, 6), (1, 4, 8)\}$

(c) $\{(2, -3, 1), (4, 1, 1), (0, -7, 1)\}$

(d) $\{(1, 6, 4), (2, 4, -1), (-1, 2, 5)\}$

4. Which of the following form bases for P_2?

(a) $2 - 4x + x^2$, $3 + 2x - x^2$, $1 + 6x - 2x^2$

(b) $3 + 2x - x^2$, $x + 5x^2$, $2 - 4x + x^2$

(c) $1 + x + x^2$, $x + x^2$, x^2

(d) $-4 + x + 3x^2$, $6 + 5x + 2x^2$, $8 + 4x + x^2$

5. Show that the following matrices form a basis for M_{22}.

$\begin{bmatrix} 3 & 6 \\ 3 & -6 \end{bmatrix}$, $\begin{bmatrix} 0 & -1 \\ -1 & 0 \end{bmatrix}$, $\begin{bmatrix} 0 & -8 \\ -12 & -4 \end{bmatrix}$, $\begin{bmatrix} 1 & 0 \\ -1 & 2 \end{bmatrix}$

6. Let V be the space spanned by $\mathbf{v}_1 = \cos^2 x$, $\mathbf{v}_2 = \sin^2 x$, $\mathbf{v}_3 = \cos 2x$.

(a) Show that $S = \{\mathbf{v}_1, \mathbf{v}_2, \mathbf{v}_3\}$ is not a basis for V.

(b) Find a basis for V.

7. Find the coordinate vector of $\mathbf{w}$ relative to the basis $S = \{\mathbf{u}_1, \mathbf{u}_2\}$ for R^2.

(a) $\mathbf{u}_1 = (0, 1)$, $\mathbf{u}_2 = (1, 0)$; $\mathbf{w} = (5, -3)$

(b) $\mathbf{u}_1 = (3, 8)$, $\mathbf{u}_2 = (1, 1)$; $\mathbf{w} = (1, 0)$

(c) $\mathbf{u}_1 = (1, 1)$, $\mathbf{u}_2 = (0, 2)$; $\mathbf{w} = (a, b)$

8. Find the coordinate vector of $\mathbf{w}$ relative to the basis $S = \{\mathbf{u}_1, \mathbf{u}_2\}$ of R^2.

(a) $\mathbf{u}_1 = (1, -1)$, $\mathbf{u}_2 = (1, 1)$; $\mathbf{w} = (1, 0)$

(b) $\mathbf{u}_1 = (1, -1)$, $\mathbf{u}_2 = (1, 1)$; $\mathbf{w} = (0, 1)$

(c) $\mathbf{u}_1 = (1, -1)$, $\mathbf{u}_2 = (1, 1)$; $\mathbf{w} = (1, 1)$

9. Find the coordinate vector of $\mathbf{v}$ relative to the basis $S = \{\mathbf{v}_1, \mathbf{v}_2, \mathbf{v}_3\}$.

(a) $\mathbf{v} = (3, 4, 3)$; $\mathbf{v}_1 = (3, 2, 1)$, $\mathbf{v}_2 = (-2, 1, 0)$, $\mathbf{v}_3 = (5, 0, 0)$

(b) $\mathbf{v} = (5, -12, 3)$; $\mathbf{v}_1 = (1, 2, 3)$, $\mathbf{v}_2 = (-4, 5, 6)$, $\mathbf{v}_3 = (7, -8, 9)$

10. Find the coordinate vector of $\mathbf{p}$ relative to the basis $S = \{\mathbf{p}_1, \mathbf{p}_2, \mathbf{p}_3\}$.

(a) $\mathbf{p} = 4 - 3x + x^2$; $\mathbf{p}_1 = 1$, $\mathbf{p}_2 = x$, $\mathbf{p}_3 = x^2$

(b) $\mathbf{p} = 2 - x + x^2$; $\mathbf{p}_1 = 1 + x$, $\mathbf{p}_2 = 1 + x^2$, $\mathbf{p}_3 = x + x^2$

208 Chapter 4 General Vector Spaces

11. Find the coordinate vector of A relative to the basis $S = \{A_1, A_2, A_3, A_4\}$.

$$A = \begin{bmatrix} 3 & -2 \\ 0 & 1 \end{bmatrix}; \quad A_1 = \begin{bmatrix} 1 & -1 \\ 0 & 0 \end{bmatrix}, \quad A_2 = \begin{bmatrix} 0 & 1 \\ 1 & 0 \end{bmatrix},$$

$$A_3 = \begin{bmatrix} 1 & 0 \\ 0 & 0 \end{bmatrix}, \quad A_4 = \begin{bmatrix} 0 & 1 \\ 0 & 1 \end{bmatrix}$$

▶ In Exercises 12–13, show that $\{A_1, A_2, A_3, A_4\}$ is a basis for M_{22}, and express A as a linear combination of the basis vectors.

12. $A_1 = \begin{bmatrix} 1 & 0 \\ 1 & 0 \end{bmatrix}, \quad A_2 = \begin{bmatrix} 1 & 1 \\ 0 & 0 \end{bmatrix}, \quad A_3 = \begin{bmatrix} 1 & 0 \\ 0 & 1 \end{bmatrix},$

$A_4 = \begin{bmatrix} 0 & 0 \\ 1 & 0 \end{bmatrix}; \quad A = \begin{bmatrix} 6 & 2 \\ 5 & 3 \end{bmatrix}$

13. $A_1 = \begin{bmatrix} 1 & 1 \\ 1 & 1 \end{bmatrix}, \quad A_2 = \begin{bmatrix} 0 & 1 \\ 1 & 1 \end{bmatrix}, \quad A_3 = \begin{bmatrix} 0 & 0 \\ 1 & 1 \end{bmatrix},$

$A_4 = \begin{bmatrix} 0 & 0 \\ 0 & 1 \end{bmatrix}; \quad A = \begin{bmatrix} 1 & 0 \\ 1 & 0 \end{bmatrix}$

▶ In Exercises 14–15, show that $\{\mathbf{p}_1, \mathbf{p}_2, \mathbf{p}_3\}$ is a basis for P_2, and express $\mathbf{p}$ as a linear combination of the basis vectors. ◀

14. $\mathbf{p}_1 = 1 + 2x + x^2, \quad \mathbf{p}_2 = 2 + 9x, \quad \mathbf{p}_3 = 3 + 3x + 4x^2;$
$\mathbf{p} = 2 + 17x - 3x^2$

15. $\mathbf{p}_1 = 1 + x + x^2, \quad \mathbf{p}_2 = x + x^2, \quad \mathbf{p}_3 = x^2;$
$\mathbf{p} = 7 - x + 2x^2$

16. The accompanying figure shows a rectangular xy-coordinate system and an $x'y'$-coordinate system with skewed axes. Assuming that 1-unit scales are used on all the axes, find the $x'y'$-coordinates of the points whose xy-coordinates are given.

(a) $(1, 1)$ (b) $(1, 0)$ (c) $(0, 1)$ (d) (a, b)

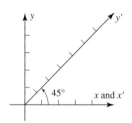

◀ Figure Ex-16

17. The accompanying figure shows a rectangular xy-coordinate system determined by the unit basis vectors $\mathbf{i}$ and $\mathbf{j}$ and an $x'y'$-coordinate system determined by unit basis vectors $\mathbf{u}_1$ and $\mathbf{u}_2$. Find the $x'y'$-coordinates of the points whose xy-coordinates are given.

(a) $(\sqrt{3}, 1)$ (b) $(1, 0)$ (c) $(0, 1)$ (d) (a, b)

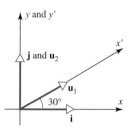

◀ Figure Ex-17

18. The basis that we gave for M_{22} in Example 4 consisted of noninvertible matrices. Do you think that there is a basis for M_{22} consisting of invertible matrices? Justify your answer.

19. Prove that R^∞ is infinite-dimensional.

True-False Exercises

In parts (a)–(e) determine whether the statement is true or false, and justify your answer.

(a) If $V = \text{span}\{\mathbf{v}_1, \ldots, \mathbf{v}_n\}$, then $\{\mathbf{v}_1, \ldots, \mathbf{v}_n\}$ is a basis for V.

(b) Every linearly independent subset of a vector space V is a basis for V.

(c) If $\{\mathbf{v}_1, \mathbf{v}_2, \ldots, \mathbf{v}_n\}$ is a basis for a vector space V, then every vector in V can be expressed as a linear combination of $\mathbf{v}_1, \mathbf{v}_2, \ldots, \mathbf{v}_n$.

(d) The coordinate vector of a vector $\mathbf{x}$ in R^n relative to the standard basis for R^n is $\mathbf{x}$.

(e) Every basis of P_4 contains at least one polynomial of degree 3 or less.

4.5 Dimension

We showed in the previous section that the standard basis R^n has n vectors and hence that the standard basis for R^3 has three vectors, the standard basis for R^2 has two vectors, and the standard basis for $R^1 (= R)$ has one vector. Since we think of space as three dimensional, a plane as two dimensional, and a line as one dimensional, there seems to be a link between the number of vectors in a basis and the dimension of a vector space. We will develop this idea in this section.

Number of Vectors in a Basis

Our first goal in this section is to establish the following fundamental theorem.

> **THEOREM 4.5.1** *All bases for a finite-dimensional vector space have the same number of vectors.*

To prove this theorem we will need the following preliminary result, whose proof is deferred to the end of the section.

> **THEOREM 4.5.2** *Let V be a finite-dimensional vector space, and let* $\{\mathbf{v}_1, \mathbf{v}_2, \ldots, \mathbf{v}_n\}$ *be any basis.*
> *(a) If a set has more than n vectors, then it is linearly dependent.*
> *(b) If a set has fewer than n vectors, then it does not span V.*

We can now see rather easily why Theorem 4.5.1 is true; for if

$$S = \{\mathbf{v}_1, \mathbf{v}_2, \ldots, \mathbf{v}_n\}$$

is an *arbitrary* basis for V, then the linear independence of S implies that any set in V with more than n vectors is linearly dependent and any set in V with fewer than n vectors does not span V. Thus, unless a set in V has exactly n vectors it cannot be a basis.

We noted in the introduction to this section that for certain familiar vector spaces the intuitive notion of dimension coincides with the number of vectors in a basis. The following definition makes this idea precise.

Some writers regard the empty set to be a basis for the zero vector space. This is consistent with our definition of dimension, since the empty set has no vectors and the zero vector space has dimension zero.

> **DEFINITION 1** The *dimension* of a finite-dimensional vector space V is denoted by $\dim(V)$ and is defined to be the number of vectors in a basis for V. In addition, the zero vector space is defined to have dimension zero.

*Engineers often use the term **degrees of freedom** as a synonym for dimension.*

▶ **EXAMPLE 1** Dimensions of Some Familiar Vector Spaces

$\dim(R^n) = n$ The standard basis has n vectors.

$\dim(P_n) = n + 1$ The standard basis has $n + 1$ vectors.

$\dim(M_{mn}) = mn$ The standard basis has mn vectors.

▶ **EXAMPLE 2** Dimension of Span(S)

If $S = \{\mathbf{v}_1, \mathbf{v}_2, \ldots, \mathbf{v}_r\}$ is a *linearly independent* set in a vector space V, then S is automatically a basis for span(S) (why?), and this implies that

$$\dim[\text{span}(S)] = r$$

In words, the dimension of the space spanned by a linearly independent set of vectors is equal to the number of vectors in that set.

▶ **EXAMPLE 3** **Dimension of a Solution Space**

Find a basis for and the dimension of the solution space of the homogeneous system

$$
\begin{aligned}
2x_1 + 2x_2 - x_3 + x_5 &= 0 \\
-x_1 - x_2 + 2x_3 - 3x_4 + x_5 &= 0 \\
x_1 + x_2 - 2x_3 - x_5 &= 0 \\
 x_3 + x_4 + x_5 &= 0
\end{aligned}
$$

Solution We leave it for you to solve this system by Gauss–Jordan elimination and show that its general solution is

$$x_1 = -s - t, \quad x_2 = s, \quad x_3 = -t, \quad x_4 = 0, \quad x_5 = t$$

which can be written in vector form as

$$(x_1, x_2, x_3, x_4, x_5) = (-s - t, s, -t, 0, t)$$

or, alternatively, as

$$(x_1, x_2, x_3, x_4, x_5) = s(-1, 1, 0, 0, 0) + t(-1, 0, -1, 0, 1)$$

This shows that the vectors $\mathbf{v}_1 = (-1, 1, 0, 0, 0)$ and $\mathbf{v}_2 = (-1, 0, -1, 0, 1)$ span the solution space. Since neither vector is a scalar multiple of the other, they are linearly independent and hence form a basis for the solution space. Thus, the solution space has dimension 2.

▶ **EXAMPLE 4** **Dimension of a Solution Space**

Find a basis for and the dimension of the solution space of the homogeneous system

$$
\begin{aligned}
x_1 + 3x_2 - 2x_3 + 2x_5 &= 0 \\
2x_1 + 6x_2 - 5x_3 - 2x_4 + 4x_5 - 3x_6 &= 0 \\
 5x_3 + 10x_4 + 15x_6 &= 0 \\
2x_1 + 6x_2 + 8x_4 + 4x_5 + 18x_6 &= 0
\end{aligned}
$$

Solution In Example 6 of Section 1.2 we found the solution of this system to be

$$x_1 = -3r - 4s - 2t, \quad x_2 = r, \quad x_3 = -2s, \quad x_4 = s, \quad x_5 = t, \quad x_6 = 0$$

which can be written in vector form as

$$(x_1, x_2, x_3, x_4, x_5, x_6) = (-3r - 4s - 2t, r, -2s, s, t, 0)$$

or, alternatively, as

$$(x_1, x_2, x_3, x_4, x_5) = r(-3, 1, 0, 0, 0, 0) + s(-4, 0, -2, 1, 0, 0) + t(-2, 0, 0, 0, 1, 0)$$

This shows that the vectors

$$\mathbf{v}_1 = (-3, 1, 0, 0, 0, 0), \quad \mathbf{v}_2 = (-4, 0, -2, 1, 0, 0), \quad \mathbf{v}_3 = (-2, 0, 0, 0, 1, 0)$$

span the solution space. We leave it for you to check that these vectors are linearly independent by showing that none of them is a linear combination of the other two (but see the remark that follows). Thus, the solution space has dimension 3. ◀

Remark It can be shown that for a homogeneous linear system, the method of the last example *always* produces a basis for the solution space of the system. We omit the formal proof.

Some Fundamental Theorems

We will devote the remainder of this section to a series of theorems that reveal the subtle interrelationships among the concepts of linear independence, basis, and dimension. These theorems are not simply exercises in mathematical theory—they are essential to the understanding of vector spaces and the applications that build on them.

We will start with a theorem (proved at the end of this section) that is concerned with the effect on linear independence and spanning if a vector is added to or removed from a given nonempty set of vectors. Informally stated, if you start with a linearly independent set S and adjoin to it a vector that is not a linear combination of those in S, then the enlarged set will still be linearly independent. Also, if you start with a set S of two or more vectors in which one of the vectors is a linear combination of the others, then that vector can be removed from S without affecting span(S) (Figure 4.5.1).

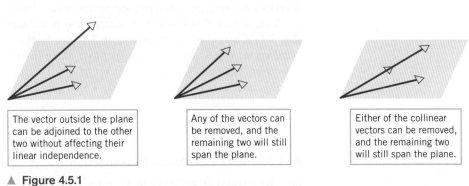

The vector outside the plane can be adjoined to the other two without affecting their linear independence.

Any of the vectors can be removed, and the remaining two will still span the plane.

Either of the collinear vectors can be removed, and the remaining two will still span the plane.

▲ Figure 4.5.1

THEOREM 4.5.3 **Plus/Minus Theorem**

Let S be a nonempty set of vectors in a vector space V.

(a) *If S is a linearly independent set, and if $\mathbf{v}$ is a vector in V that is outside of* span(S), *then the set $S \cup \{\mathbf{v}\}$ that results by inserting $\mathbf{v}$ into S is still linearly independent.*

(b) *If $\mathbf{v}$ is a vector in S that is expressible as a linear combination of other vectors in S, and if $S - \{\mathbf{v}\}$ denotes the set obtained by removing $\mathbf{v}$ from S, then S and $S - \{\mathbf{v}\}$ span the same space; that is,*

$$\text{span}(S) = \text{span}(S - \{\mathbf{v}\})$$

▶ **EXAMPLE 5** **Applying the Plus/Minus Theorem**

Show that $\mathbf{p}_1 = 1 - x^2$, $\mathbf{p}_2 = 2 - x^2$, and $\mathbf{p}_3 = x^3$ are linearly independent vectors.

Solution The set $S = \{\mathbf{p}_1, \mathbf{p}_2\}$ is linearly independent, since neither vector in S is a scalar multiple of the other. Since the vector $\mathbf{p}_3$ cannot be expressed as a linear combination of the vectors in S (why?), it can be adjoined to S to produce a linearly independent set $S' = \{\mathbf{p}_1, \mathbf{p}_2, \mathbf{p}_3\}$. ◀

In general, to show that a set of vectors $\{\mathbf{v}_1, \mathbf{v}_2, \ldots, \mathbf{v}_n\}$ is a basis for a vector space V, we must show that the vectors are linearly independent and span V. However, if we happen to know that V has dimension n (so that $\{\mathbf{v}_1, \mathbf{v}_2, \ldots, \mathbf{v}_n\}$ contains the right number of vectors for a basis), then it suffices to check *either* linear independence *or* spanning—the remaining condition will hold automatically. This is the content of the following theorem.

THEOREM 4.5.4 *Let V be an n-dimensional vector space, and let S be a set in V with exactly n vectors. Then S is a basis for V if and only if S spans V or S is linearly independent.*

Proof Assume that S has exactly n vectors and spans V. To prove that S is a basis, we must show that S is a linearly independent set. But if this is not so, then some vector $\mathbf{v}$ in S is a linear combination of the remaining vectors. If we remove this vector from S, then it follows from Theorem 4.5.3b that the remaining set of $n-1$ vectors still spans V. But this is impossible, since it follows from Theorem 4.5.2b that no set with fewer than n vectors can span an n-dimensional vector space. Thus S is linearly independent.

Assume that S has exactly n vectors and is a linearly independent set. To prove that S is a basis, we must show that S spans V. But if this is not so, then there is some vector $\mathbf{v}$ in V that is not in span(S). If we insert this vector into S, then it follows from Theorem 4.5.3a that this set of $n+1$ vectors is still linearly independent. But this is impossible, since Theorem 4.5.2a states that no set with more than n vectors in an n-dimensional vector space can be linearly independent. Thus S spans V. ◂

▶ **EXAMPLE 6 Bases by Inspection**

(a) By inspection, explain why $\mathbf{v}_1 = (-3, 7)$ and $\mathbf{v}_2 = (5, 5)$ form a basis for R^2.

(b) By inspection, explain why $\mathbf{v}_1 = (2, 0, -1)$, $\mathbf{v}_2 = (4, 0, 7)$, and $\mathbf{v}_3 = (-1, 1, 4)$ form a basis for R^3.

Solution (a) Since neither vector is a scalar multiple of the other, the two vectors form a linearly independent set in the two-dimensional space R^2, and hence they form a basis by Theorem 4.5.4.

Solution (b) The vectors $\mathbf{v}_1$ and $\mathbf{v}_2$ form a linearly independent set in the xz-plane (why?). The vector $\mathbf{v}_3$ is outside of the xz-plane, so the set $\{\mathbf{v}_1, \mathbf{v}_2, \mathbf{v}_3\}$ is also linearly independent. Since R^3 is three-dimensional, Theorem 4.5.4 implies that $\{\mathbf{v}_1, \mathbf{v}_2, \mathbf{v}_3\}$ is a basis for R^3. ◂

The next theorem (whose proof is deferred to the end of this section) reveals two important facts about the vectors in a finite-dimensional vector space V:

1. Every spanning set for a subspace is either a basis for that subspace or has a basis as a subset.
2. Every linearly independent set in a subspace is either a basis for that subspace or can be extended to a basis for it.

THEOREM 4.5.5 *Let S be a finite set of vectors in a finite-dimensional vector space V.*

(a) If S spans V but is not a basis for V, then S can be reduced to a basis for V by removing appropriate vectors from S.

(b) If S is a linearly independent set that is not already a basis for V, then S can be enlarged to a basis for V by inserting appropriate vectors into S.

We conclude this section with a theorem that relates the dimension of a vector space to the dimensions of its subspaces.

THEOREM 4.5.6 *If W is a subspace of a finite-dimensional vector space V, then*:

(a) *W is finite-dimensional.*

(b) $\dim(W) \leq \dim(V)$.

(c) $W = V$ *if and only if* $\dim(W) = \dim(V)$.

Proof (a) We will leave the proof of this part for the exercises.

Proof (b) Part (a) shows that W is finite-dimensional, so it has a basis
$$S = \{\mathbf{w}_1, \mathbf{w}_2, \ldots, \mathbf{w}_m\}$$
Either S is also a basis for V or it is not. If so, then $\dim(V) = m$, which means that $\dim(V) = \dim(W)$. If not, then because S is a linearly independent set it can be enlarged to a basis for V by part (b) of Theorem 4.5.5. But this implies that $\dim(W) < \dim(V)$, so we have shown that $\dim(W) \leq \dim(V)$ in all cases.

Proof (c) Assume that $\dim(W) = \dim(V)$ and that
$$S = \{\mathbf{w}_1, \mathbf{w}_2, \ldots, \mathbf{w}_m\}$$
is a basis for W. If S is not also a basis for V, then being linearly independent S can be extended to a basis for V by part (b) of Theorem 4.5.5. But this would mean that $\dim(V) > \dim(W)$, which contradicts our hypothesis. Thus S must also be a basis for V, which means that $\dim(W) = \dim(V)$. ◄

Figure 4.5.2 illustrates the geometric relationship between the subspaces of R^3 in order of increasing dimension.

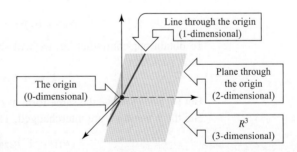

▶ Figure 4.5.2

OPTIONAL

We conclude this section with optional proofs of Theorems 4.5.2, 4.5.3, and 4.5.5.

Proof of Theorem 4.5.2(a) Let $S' = \{\mathbf{w}_1, \mathbf{w}_2, \ldots, \mathbf{w}_m\}$ be any set of m vectors in V, where $m > n$. We want to show that S' is linearly dependent. Since $S = \{\mathbf{v}_1, \mathbf{v}_2, \ldots, \mathbf{v}_n\}$ is a basis, each $\mathbf{w}_i$ can be expressed as a linear combination of the vectors in S, say

$$\begin{aligned}
\mathbf{w}_1 &= a_{11}\mathbf{v}_1 + a_{21}\mathbf{v}_2 + \cdots + a_{n1}\mathbf{v}_n \\
\mathbf{w}_2 &= a_{12}\mathbf{v}_1 + a_{22}\mathbf{v}_2 + \cdots + a_{n2}\mathbf{v}_n \\
&\vdots \qquad \vdots \qquad \vdots \qquad \qquad \vdots \\
\mathbf{w}_m &= a_{1m}\mathbf{v}_1 + a_{2m}\mathbf{v}_2 + \cdots + a_{nm}\mathbf{v}_n
\end{aligned} \qquad (1)$$

To show that S' is linearly dependent, we must find scalars $k_1, k_2, \ldots, k_m$, not all zero, such that

$$k_1\mathbf{w}_1 + k_2\mathbf{w}_2 + \cdots + k_m\mathbf{w}_m = \mathbf{0} \tag{2}$$

Using the equations in (1), we can rewrite (2) as

$$(k_1a_{11} + k_2a_{12} + \cdots + k_ma_{1m})\mathbf{v}_1$$
$$+ (k_1a_{21} + k_2a_{22} + \cdots + k_ma_{2m})\mathbf{v}_2$$
$$\ddots$$
$$+ (k_1a_{n1} + k_2a_{n2} + \cdots + k_ma_{nm})\mathbf{v}_n = \mathbf{0}$$

Thus, from the linear independence of S, the problem of proving that S' is a linearly dependent set reduces to showing there are scalars $k_1, k_2, \ldots, k_m$, not all zero, that satisfy

$$\begin{aligned} a_{11}k_1 + a_{12}k_2 + \cdots + a_{1m}k_m &= 0 \\ a_{21}k_1 + a_{22}k_2 + \cdots + a_{2m}k_m &= 0 \\ \vdots \qquad \vdots \qquad \quad \vdots \qquad \vdots & \\ a_{n1}k_1 + a_{n2}k_2 + \cdots + a_{nm}k_m &= 0 \end{aligned} \tag{3}$$

But (3) has more unknowns than equations, so the proof is complete since Theorem 1.2.2 guarantees the existence of nontrivial solutions.

Proof of Theorem 4.5.2(b) Let $S' = \{\mathbf{w}_1, \mathbf{w}_2, \ldots, \mathbf{w}_m\}$ be any set of m vectors in V, where $m < n$. We want to show that S' does not span V. We will do this by showing that the assumption that S' spans V leads to a contradiction of the linear independence of $\{\mathbf{v}_1, \mathbf{v}_2, \ldots, \mathbf{v}_n\}$.

If S' spans V, then every vector in V is a linear combination of the vectors in S'. In particular, each basis vector $\mathbf{v}_i$ is a linear combination of the vectors in S', say

$$\begin{aligned} \mathbf{v}_1 &= a_{11}\mathbf{w}_1 + a_{21}\mathbf{w}_2 + \cdots + a_{m1}\mathbf{w}_m \\ \mathbf{v}_2 &= a_{12}\mathbf{w}_1 + a_{22}\mathbf{w}_2 + \cdots + a_{m2}\mathbf{w}_m \\ \vdots \quad & \vdots \qquad \vdots \qquad \quad \vdots \\ \mathbf{v}_n &= a_{1n}\mathbf{w}_1 + a_{2n}\mathbf{w}_2 + \cdots + a_{mn}\mathbf{w}_m \end{aligned} \tag{4}$$

To obtain our contradiction, we will show that there are scalars $k_1, k_2, \ldots, k_n$, not all zero, such that

$$k_1\mathbf{v}_1 + k_2\mathbf{v}_2 + \cdots + k_n\mathbf{v}_n = \mathbf{0} \tag{5}$$

But (4) and (5) have the same form as (1) and (2) except that m and n are interchanged and the $\mathbf{w}$'s and $\mathbf{v}$'s are interchanged. Thus, the computations that led to (3) now yield

$$\begin{aligned} a_{11}k_1 + a_{12}k_2 + \cdots + a_{1n}k_n &= 0 \\ a_{21}k_1 + a_{22}k_2 + \cdots + a_{2n}k_n &= 0 \\ \vdots \qquad \vdots \qquad \quad \vdots \qquad \vdots & \\ a_{m1}k_1 + a_{m2}k_2 + \cdots + a_{mn}k_n &= 0 \end{aligned}$$

This linear system has more unknowns than equations and hence has nontrivial solutions by Theorem 1.2.2.

Proof of Theorem 4.5.3(a) Assume that $S = \{\mathbf{v}_1, \mathbf{v}_2, \ldots, \mathbf{v}_r\}$ is a linearly independent set of vectors in V, and $\mathbf{v}$ is a vector in V outside of span(S). To show that $S' = \{\mathbf{v}_1, \mathbf{v}_2, \ldots, \mathbf{v}_r, \mathbf{v}\}$ is a linearly independent set, we must show that the only scalars that satisfy

$$k_1\mathbf{v}_1 + k_2\mathbf{v}_2 + \cdots + k_r\mathbf{v}_r + k_{r+1}\mathbf{v} = \mathbf{0} \tag{6}$$

are $k_1 = k_2 = \cdots = k_r = k_{r+1} = 0$. But it must be true that $k_{r+1} = 0$ for otherwise we could solve (6) for $\mathbf{v}$ as a linear combination of $\mathbf{v}_1, \mathbf{v}_2, \ldots, \mathbf{v}_r$, contradicting the assumption that $\mathbf{v}$ is outside of span(S). Thus, (6) simplifies to

$$k_1\mathbf{v}_1 + k_2\mathbf{v}_2 + \cdots + k_r\mathbf{v}_r = \mathbf{0} \tag{7}$$

which, by the linear independence of $\{\mathbf{v}_1, \mathbf{v}_2, \ldots, \mathbf{v}_r\}$, implies that

$$k_1 = k_2 = \cdots = k_r = 0$$

Proof Theorem 4.5.3(b) Assume that $S = \{\mathbf{v}_1, \mathbf{v}_2, \ldots, \mathbf{v}_r\}$ is a set of vectors in V, and (to be specific) suppose that $\mathbf{v}_r$ is a linear combination of $\mathbf{v}_1, \mathbf{v}_2, \ldots, \mathbf{v}_{r-1}$, say

$$\mathbf{v}_r = c_1\mathbf{v}_1 + c_2\mathbf{v}_2 + \cdots + c_{r-1}\mathbf{v}_{r-1} \tag{8}$$

We want to show that if $\mathbf{v}_r$ is removed from S, then the remaining set of vectors $\{\mathbf{v}_1, \mathbf{v}_2, \ldots, \mathbf{v}_{r-1}\}$ still spans S; that is, we must show that every vector $\mathbf{w}$ in span(S) is expressible as a linear combination of $\{\mathbf{v}_1, \mathbf{v}_2, \ldots, \mathbf{v}_{r-1}\}$. But if $\mathbf{w}$ is in span(S), then $\mathbf{w}$ is expressible in the form

$$\mathbf{w} = k_1\mathbf{v}_1 + k_2\mathbf{v}_2 + \cdots + k_{r-1}\mathbf{v}_{r-1} + k_r\mathbf{v}_r$$

or, on substituting (8),

$$\mathbf{w} = k_1\mathbf{v}_1 + k_2\mathbf{v}_2 + \cdots + k_{r-1}\mathbf{v}_{r-1} + k_r(c_1\mathbf{v}_1 + c_2\mathbf{v}_2 + \cdots + c_{r-1}\mathbf{v}_{r-1})$$

which expresses $\mathbf{w}$ as a linear combination of $\mathbf{v}_1, \mathbf{v}_2, \ldots, \mathbf{v}_{r-1}$.

Proof of Theorem 4.5.5(a) If S is a set of vectors that spans V but is not a basis for V, then S is a linearly dependent set. Thus some vector $\mathbf{v}$ in S is expressible as a linear combination of the other vectors in S. By the Plus/Minus Theorem (4.5.3b), we can remove $\mathbf{v}$ from S, and the resulting set S' will still span V. If S' is linearly independent, then S' is a basis for V, and we are done. If S' is linearly dependent, then we can remove some appropriate vector from S' to produce a set S'' that still spans V. We can continue removing vectors in this way until we finally arrive at a set of vectors in S that is linearly independent and spans V. This subset of S is a basis for V.

Proof of Theorem 4.5.5(b) Suppose that $\dim(V) = n$. If S is a linearly independent set that is not already a basis for V, then S fails to span V, so there is some vector $\mathbf{v}$ in V that is not in span(S). By the Plus/Minus Theorem (4.5.3a), we can insert $\mathbf{v}$ into S, and the resulting set S' will still be linearly independent. If S' spans V, then S' is a basis for V, and we are finished. If S' does not span V, then we can insert an appropriate vector into S' to produce a set S'' that is still linearly independent. We can continue inserting vectors in this way until we reach a set with n linearly independent vectors in V. This set will be a basis for V by Theorem 4.5.4. ◂

Concept Review
- Dimension
- Relationships among the concepts of linear independence, basis, and dimension

Skills
- Find a basis for and the dimension of the solution space of a homogeneous linear system.
- Use dimension to determine whether a set of vectors is a basis for a finite-dimensional vector space.
- Extend a linearly independent set to a basis.

Exercise Set 4.5

▶ In Exercises 1–6, find a basis for the solution space of the homogeneous linear system, and find the dimension of that space.

1. $2x_1 - x_2 + x_3 = 0$
   $x_1 + x_2 = 0$
   $-2x_1 - x_2 + x_3 = 0$

2. $3x_1 + x_2 + x_3 + x_4 = 0$
   $5x_1 - x_2 + x_3 - x_4 = 0$

3. $3x_1 - x_2 + 2x_3 + x_4 = 0$
   $6x_1 - 2x_2 - 4x_3 = 0$

4. $x_1 - 3x_2 + x_3 = 0$
   $2x_1 - 6x_2 + 2x_3 = 0$
   $3x_1 - 9x_2 + 3x_3 = 0$

5. $2x_1 + x_2 + 3x_3 = 0$
   $x_1 + 5x_3 = 0$
   $ x_2 + x_3 = 0$

6. $x + y + z = 0$
   $3x + 2y - 2z = 0$
   $4x + 3y - z = 0$
   $6x + 5y + z = 0$

7. Find bases for the following subspaces of R^3.
   (a) The plane $2x + 4y - 3z = 0$.
   (b) The plane $y + z = 0$.
   (c) The line $x = 4t$, $y = 2t$, $z = -t$.
   (d) All vectors of the form (a, b, c), where $c = a - b$.

8. Find the dimensions of the following subspaces of R^4.
   (a) All vectors of the form $(a, b, c, 0)$.
   (b) All vectors of the form (a, b, c, d), where $d = a + b$ and $c = a - b$.
   (c) All vectors of the form (a, b, c, d), where $a = b = c = d$.

9. Find the dimension of each of the following vector spaces.
   (a) The vector space of all diagonal $n \times n$ matrices.
   (b) The vector space of all symmetric $n \times n$ matrices.
   (c) The vector space of all upper triangular $n \times n$ matrices.

10. Find the dimension of the subspace of P_3 consisting of all polynomials $a_0 + a_1x + a_2x^2 + a_3x^3$ for which $a_0 = 0$.

11. (a) Show that the set W of all polynomials in P_2 such that $p(1) = 0$ is a subspace of P_2.
    (b) Make a conjecture about the dimension of W.
    (c) Confirm your conjecture by finding a basis for W.

12. Find a standard basis vector for R^3 that can be added to the set $\{\mathbf{v}_1, \mathbf{v}_2\}$ to produce a basis for R^3.
    (a) $\mathbf{v}_1 = (1, 1, 1)$, $\mathbf{v}_2 = (2, -1, 3)$
    (b) $\mathbf{v}_1 = (5, 3, 0)$, $\mathbf{v}_2 = (1, -1, 2)$

13. Find standard basis vectors for R^4 that can be added to the set $\{\mathbf{v}_1, \mathbf{v}_2\}$ to produce a basis for R^4.
    $$\mathbf{v}_1 = (1, -4, 2, -3), \quad \mathbf{v}_2 = (-3, 8, -4, 6)$$

14. Let $\{\mathbf{v}_1, \mathbf{v}_2, \mathbf{v}_3\}$ be a basis for a vector space V. Show that $\{\mathbf{u}_1, \mathbf{u}_2, \mathbf{u}_3\}$ is also a basis, where $\mathbf{u}_1 = \mathbf{v}_1$, $\mathbf{u}_2 = \mathbf{v}_1 + \mathbf{v}_2$, and $\mathbf{u}_3 = \mathbf{v}_1 + \mathbf{v}_2 + \mathbf{v}_3$.

15. The vectors $\mathbf{v}_1 = (1, -2, 3)$ and $\mathbf{v}_2 = (0, 5, -3)$ are linearly independent. Enlarge $\{\mathbf{v}_1, \mathbf{v}_2\}$ to a basis for R^3.

16. The vectors $\mathbf{v}_1 = (1, -2, 3, -5)$ and $\mathbf{v}_2 = (0, -1, 2, -3)$ are linearly independent. Enlarge $\{\mathbf{v}_1, \mathbf{v}_2\}$ to a basis for R^4.

17. (a) Show that for every positive integer n, one can find $n + 1$ linearly independent vectors in $F(-\infty, \infty)$. [*Hint:* Look for polynomials.]
    (b) Use the result in part (a) to prove that $F(-\infty, \infty)$ is infinite-dimensional.
    (c) Prove that $C(-\infty, \infty)$, $C^m(-\infty, \infty)$, and $C^\infty(-\infty, \infty)$ are infinite-dimensional vector spaces.

18. Let S be a basis for an n-dimensional vector space V. Show that if $\mathbf{v}_1, \mathbf{v}_2, \ldots, \mathbf{v}_r$ form a linearly independent set of vectors in V, then the coordinate vectors $(\mathbf{v}_1)_S, (\mathbf{v}_2)_S, \ldots, (\mathbf{v}_r)_S$ form a linearly independent set in R^n, and conversely.

19. Using the notation from Exercise 18, show that if the vectors $\mathbf{v}_1, \mathbf{v}_2, \ldots, \mathbf{v}_r$ span V, then the coordinate vectors $(\mathbf{v}_1)_S, (\mathbf{v}_2)_S, \ldots, (\mathbf{v}_r)_S$ span R^n, and conversely.

20. Find a basis for the subspace of P_2 spanned by the given vectors.
    (a) $-1 + x - 2x^2$, $3 + 3x + 6x^2$, 9

(b) $1 + x$, x^2, $-2 + 2x^2$, $-3x$

(c) $1 + x - 3x^2$, $2 + 2x - 6x^2$, $3 + 3x - 9x^2$

[*Hint:* Let S be the standard basis for P_2, and work with the coordinate vectors relative to S as in Exercises 18 and 19.]

21. Prove: A subspace of a finite-dimensional vector space is finite-dimensional.

22. State the two parts of Theorem 4.5.2 in contrapositive form.

True-False Exercises

In parts (a)–(j) determine whether the statement is true or false, and justify your answer.

(a) The zero vector space has dimension zero.

(b) There is a set of 17 linearly independent vectors in R^{17}.

(c) There is a set of 11 vectors that span R^{17}.

(d) Every linearly independent set of five vectors in R^5 is a basis for R^5.

(e) Every set of five vectors that spans R^5 is a basis for R^5.

(f) Every set of vectors that spans R^n contains a basis for R^n.

(g) Every linearly independent set of vectors in R^n is contained in some basis for R^n.

(h) There is a basis for M_{22} consisting of invertible matrices.

(i) If A has size $n \times n$ and $I_n, A, A^2, \ldots, A^{n^2}$ are distinct matrices, then $\{I_n, A, A^2, \ldots, A^{n^2}\}$ is linearly dependent.

(j) There are at least two distinct three-dimensional subspaces of P_2.

4.6 Change of Basis

A basis that is suitable for one problem may not be suitable for another, so it is a common process in the study of vector spaces to change from one basis to another. Because a basis is the vector space generalization of a coordinate system, changing bases is akin to changing coordinate axes in R^2 and R^3. In this section we will study problems related to change of basis.

Coordinate Maps

If $S = \{\mathbf{v}_1, \mathbf{v}_2, \ldots, \mathbf{v}_n\}$ is a basis for a finite-dimensional vector space V, and if

$$(\mathbf{v})_S = (c_1, c_2, \ldots, c_n)$$

is the coordinate vector of $\mathbf{v}$ relative to S, then, as observed in Section 4.4, the mapping

$$\mathbf{v} \to (\mathbf{v})_S \quad (1)$$

creates a connection (a one-to-one correspondence) between vectors in the *general* vector space V and vectors in the *familiar* vector space R^n. We call (1) the **coordinate map** from V to R^n. In this section we will find it convenient to express coordinate vectors in the matrix form

$$[\mathbf{v}]_S = \begin{bmatrix} c_1 \\ c_2 \\ \vdots \\ c_n \end{bmatrix} \quad (2)$$

where the square brackets emphasize the matrix notation (Figure 4.6.1).

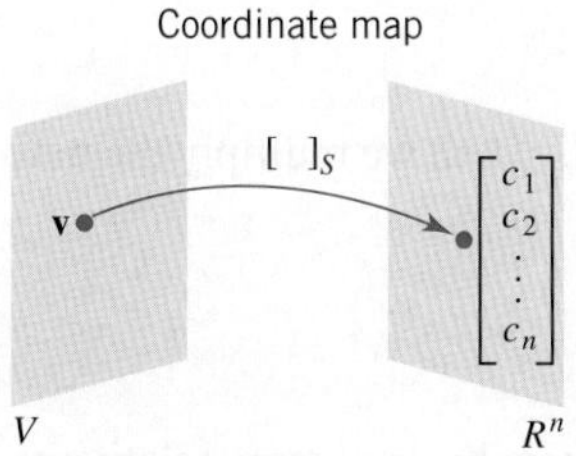

▲ Figure 4.6.1

Change of Basis

There are many applications in which it is necessary to work with more than one coordinate system. In such cases it becomes important to know how the coordinates of a fixed vector relative to each coordinate system are related. This leads to the following problem.

The Change-of-Basis Problem If $\mathbf{v}$ is a vector in a finite-dimensional vector space V, and if we change the basis for V from a basis B to a basis B', how are the coordinate vectors $[\mathbf{v}]_B$ and $[\mathbf{v}]_{B'}$ related?

Remark To solve this problem, it will be convenient to refer to B as the "old basis" and B' as the "new basis." Thus, our objective is to find a relationship between the old and new coordinates of a fixed vector $\mathbf{v}$ in V.

For simplicity, we will solve this problem for two-dimensional spaces. The solution for n-dimensional spaces is similar. Let

$$B = \{\mathbf{u}_1, \mathbf{u}_2\} \quad \text{and} \quad B' = \{\mathbf{u}'_1, \mathbf{u}'_2\}$$

be the old and new bases, respectively. We will need the coordinate vectors for the new basis vectors relative to the old basis. Suppose they are

$$[\mathbf{u}'_1]_B = \begin{bmatrix} a \\ b \end{bmatrix} \quad \text{and} \quad [\mathbf{u}'_2]_B = \begin{bmatrix} c \\ d \end{bmatrix} \tag{3}$$

That is,

$$\begin{aligned} \mathbf{u}'_1 &= a\mathbf{u}_1 + b\mathbf{u}_2 \\ \mathbf{u}'_2 &= c\mathbf{u}_1 + d\mathbf{u}_2 \end{aligned} \tag{4}$$

Now let $\mathbf{v}$ be any vector in V, and let

$$[\mathbf{v}]_{B'} = \begin{bmatrix} k_1 \\ k_2 \end{bmatrix} \tag{5}$$

be the new coordinate vector, so that

$$\mathbf{v} = k_1 \mathbf{u}'_1 + k_2 \mathbf{u}'_2 \tag{6}$$

In order to find the old coordinates of $\mathbf{v}$, we must express $\mathbf{v}$ in terms of the old basis B. To do this, we substitute (4) into (6). This yields

$$\mathbf{v} = k_1(a\mathbf{u}_1 + b\mathbf{u}_2) + k_2(c\mathbf{u}_1 + d\mathbf{u}_2)$$

or

$$\mathbf{v} = (k_1 a + k_2 c)\mathbf{u}_1 + (k_1 b + k_2 d)\mathbf{u}_2$$

Thus, the old coordinate vector for $\mathbf{v}$ is

$$[\mathbf{v}]_B = \begin{bmatrix} k_1 a + k_2 c \\ k_1 b + k_2 d \end{bmatrix}$$

which, by using (5), can be written as

$$[\mathbf{v}]_B = \begin{bmatrix} a & c \\ b & d \end{bmatrix} \begin{bmatrix} k_1 \\ k_2 \end{bmatrix} = \begin{bmatrix} a & c \\ b & d \end{bmatrix} [\mathbf{v}]_{B'}$$

This equation states that the old coordinate vector $[\mathbf{v}]_B$ results when we multiply the new coordinate vector $[\mathbf{v}]_{B'}$ on the left by the matrix

$$P = \begin{bmatrix} a & c \\ b & d \end{bmatrix}$$

Since the columns of this matrix are the coordinates of the new basis vectors relative to the old basis [see (3)] we have the following solution of the change-of-basis problem.

Solution of the Change-of-Basis Problem If we change the basis for a vector space V from an old basis $B = \{\mathbf{u}_1, \mathbf{u}_2, \ldots, \mathbf{u}_n\}$ to a new basis $B' = \{\mathbf{u}'_1, \mathbf{u}'_2, \ldots, \mathbf{u}'_n\}$, then for each vector $\mathbf{v}$ in V, the old coordinate vector $[\mathbf{v}]_B$ is related to the new coordinate vector $[\mathbf{v}]_{B'}$ by the equation

$$[\mathbf{v}]_B = P[\mathbf{v}]_{B'} \tag{7}$$

where the columns of P are the coordinate vectors of the new basis vectors relative to the old basis; that is, the column vectors of P are

$$[\mathbf{u}'_1]_B, \quad [\mathbf{u}'_2]_B, \ldots, \quad [\mathbf{u}'_n]_B \tag{8}$$

4.6 Change of Basis

Transition Matrices

The matrix P in Equation (7) is called the ***transition matrix*** from B' to B. For emphasis, we will often denote it by $P_{B' \to B}$. It follows from (8) that this matrix can be expressed in terms of its column vectors as

$$P_{B' \to B} = \left[[\mathbf{u}'_1]_B \mid [\mathbf{u}'_2]_B \mid \cdots \mid [\mathbf{u}'_n]_B \right] \qquad (9)$$

Similarly, the transition matrix from B to B' can be expressed in terms of its column vectors as

$$P_{B \to B'} = \left[[\mathbf{u}_1]_{B'} \mid [\mathbf{u}_2]_{B'} \mid \cdots \mid [\mathbf{u}_n]_{B'} \right] \qquad (10)$$

Remark There is a simple way to remember both of these formulas using the terms "old basis" and "new basis" defined earlier in this section: In Formula (9) the old basis is B' and the new basis is B, whereas in Formula (10) the old basis is B and the new basis is B'. Thus, both formulas can be restated as follows:

> *The columns of the transition matrix from an old basis to a new basis are the coordinate vectors of the old basis relative to the new basis.*

▶ **EXAMPLE 1 Finding Transition Matrices**

Consider the bases $B = \{\mathbf{u}_1, \mathbf{u}_2\}$ and $B' = \{\mathbf{u}'_1, \mathbf{u}'_2\}$ for R^2, where

$$\mathbf{u}_1 = (1, 0), \quad \mathbf{u}_2 = (0, 1), \quad \mathbf{u}'_1 = (1, 1), \quad \mathbf{u}'_2 = (2, 1)$$

(a) Find the transition matrix $P_{B' \to B}$ from B' to B.
(b) Find the transition matrix $P_{B \to B'}$ from B to B'.

Solution (a) Here the old basis vectors are $\mathbf{u}'_1$ and $\mathbf{u}'_2$ and the new basis vectors are $\mathbf{u}_1$ and $\mathbf{u}_2$. We want to find the coordinate matrices of the old basis vectors $\mathbf{u}'_1$ and $\mathbf{u}'_2$ relative to the new basis vectors $\mathbf{u}_1$ and $\mathbf{u}_2$. To do this, first we observe that

$$\mathbf{u}'_1 = \mathbf{u}_1 + \mathbf{u}_2$$
$$\mathbf{u}'_2 = 2\mathbf{u}_1 + \mathbf{u}_2$$

from which it follows that

$$[\mathbf{u}'_1]_B = \begin{bmatrix} 1 \\ 1 \end{bmatrix} \quad \text{and} \quad [\mathbf{u}'_2]_B = \begin{bmatrix} 2 \\ 1 \end{bmatrix}$$

and hence that

$$P_{B' \to B} = \begin{bmatrix} 1 & 2 \\ 1 & 1 \end{bmatrix}$$

Solution (b) Here the old basis vectors are $\mathbf{u}_1$ and $\mathbf{u}_2$ and the new basis vectors are $\mathbf{u}'_1$ and $\mathbf{u}'_2$. As in part (a), we want to find the coordinate matrices of the old basis vectors $\mathbf{u}'_1$ and $\mathbf{u}'_2$ relative to the new basis vectors $\mathbf{u}_1$ and $\mathbf{u}_2$. To do this, observe that

$$\mathbf{u}_1 = -\mathbf{u}'_1 + \mathbf{u}'_2$$
$$\mathbf{u}_2 = 2\mathbf{u}'_1 - \mathbf{u}'_2$$

from which it follows that

$$[\mathbf{u}_1]_{B'} = \begin{bmatrix} -1 \\ 1 \end{bmatrix} \quad \text{and} \quad [\mathbf{u}_2]_{B'} = \begin{bmatrix} 2 \\ -1 \end{bmatrix}$$

and hence that

$$P_{B \to B'} = \begin{bmatrix} -1 & 2 \\ 1 & -1 \end{bmatrix} \quad ◀$$

Suppose now that B and B' are bases for a finite-dimensional vector space V. Since multiplication by $P_{B' \to B}$ maps coordinate vectors relative to the basis B' into coordinate vectors relative to a basis B, and $P_{B \to B'}$ maps coordinate vectors relative to B into coordinate vectors relative to B', it follows that for every vector $\mathbf{v}$ in V we have

$$[\mathbf{v}]_B = P_{B' \to B} [\mathbf{v}]_{B'} \tag{11}$$

$$[\mathbf{v}]_{B'} = P_{B \to B'} [\mathbf{v}]_B \tag{12}$$

▶ **EXAMPLE 2 Computing Coordinate Vectors**

Let B and B' be the bases in Example 1. Use an appropriate formula to find $[\mathbf{v}]_B$ given that

$$[\mathbf{v}]_{B'} = \begin{bmatrix} -3 \\ 5 \end{bmatrix}$$

Solution To find $[\mathbf{v}]_B$ we need to make the transition from B' to B. It follows from Formula (11) and part (a) of Example 1 that

$$[\mathbf{v}]_B = P_{B' \to B} [\mathbf{v}]_{B'} = \begin{bmatrix} 1 & 2 \\ 1 & 1 \end{bmatrix} \begin{bmatrix} -3 \\ 5 \end{bmatrix} = \begin{bmatrix} 7 \\ 2 \end{bmatrix} \blacktriangleleft$$

Invertibility of Transition Matrices

If B and B' are bases for a finite-dimensional vector space V, then

$$(P_{B' \to B})(P_{B \to B'}) = P_{B \to B}$$

because multiplication by $(P_{B' \to B})(P_{B \to B'})$ first maps B-coordinates of a vector into B'-coordinates, and then maps those B'-coordinates back into the original B-coordinates. Since the net effect of the two operations is to leave each coordinate vector unchanged, we are led to conclude that $P_{B \to B}$ must be the identity matrix, that is,

$$(P_{B' \to B})(P_{B \to B'}) = I \tag{13}$$

(we omit the formal proof). For example, for the transition matrices obtained in Example 1 we have

$$(P_{B' \to B})(P_{B \to B'}) = \begin{bmatrix} 1 & 2 \\ 1 & 1 \end{bmatrix} \begin{bmatrix} -1 & 2 \\ 1 & -1 \end{bmatrix} = \begin{bmatrix} 1 & 0 \\ 0 & 1 \end{bmatrix} = I$$

It follows from (13) that $P_{B' \to B}$ is invertible and that its inverse is $P_{B \to B'}$. Thus, we have the following theorem.

THEOREM 4.6.1 *If P is the transition matrix from a basis B' to a basis B for a finite-dimensional vector space V, then P is invertible and P^{-1} is the transition matrix from B to B'.*

An Efficient Method for Computing Transition Matrices for R^n

Our next objective is to develop an efficient procedure for computing transition matrices *between bases for R^n*. As illustrated in Example 1, the first step in computing a transition matrix is to express each new basis vector as a linear combination of the old basis vectors. For R^n this involves solving n linear systems of n equations in n unknowns, each of which has the same coefficient matrix (why?). An efficient way to do this is by the method illustrated in Example 2 of Section 1.6, which is as follows:

A Procedure for Computing $P_{B \to B'}$

Step 1. Form the matrix $[B' \mid B]$.

Step 2. Use elementary row operations to reduce the matrix in Step 1 to reduced row echelon form.

Step 3. The resulting matrix will be $[I \mid P_{B \to B'}]$.

Step 4. Extract the matrix $P_{B \to B'}$ from the right side of the matrix in Step 3.

This procedure is captured in the following diagram.

$$[\text{new basis} \mid \text{old basis}] \xrightarrow{\text{row operations}} [I \mid \text{transition from old to new}] \quad (14)$$

▶ EXAMPLE 3 Example 1 Revisited

In Example 1 we considered the bases $B = \{\mathbf{u}_1, \mathbf{u}_2\}$ and $B' = \{\mathbf{u}'_1, \mathbf{u}'_2\}$ for R^2, where

$$\mathbf{u}_1 = (1, 0), \quad \mathbf{u}_2 = (0, 1), \quad \mathbf{u}'_1 = (1, 1), \quad \mathbf{u}'_2 = (2, 1)$$

(a) Use Formula (14) to find the transition matrix from B' to B.

(b) Use Formula (14) to find the transition matrix from B to B'.

Solution (a) Here B' is the old basis and B is the new basis, so

$$[\text{new basis} \mid \text{old basis}] = \begin{bmatrix} 1 & 0 & | & 1 & 2 \\ 0 & 1 & | & 1 & 1 \end{bmatrix}$$

Since the left side is already the identity matrix, no reduction is needed. We see by inspection that the transition matrix is

$$P_{B' \to B} = \begin{bmatrix} 1 & 2 \\ 1 & 1 \end{bmatrix}$$

which agrees with the result in Example 1.

Solution (b) Here B is the old basis and B' is the new basis, so

$$[\text{new basis} \mid \text{old basis}] = \begin{bmatrix} 1 & 2 & | & 1 & 0 \\ 1 & 1 & | & 0 & 1 \end{bmatrix}$$

By reducing this matrix, so the left side becomes the identity we obtain (verify)

$$[I \mid \text{transition from old to new}] = \begin{bmatrix} 1 & 0 & | & -1 & 2 \\ 0 & 1 & | & 1 & -1 \end{bmatrix}$$

so the transition matrix is

$$P_{B \to B'} = \begin{bmatrix} -1 & 2 \\ 1 & -1 \end{bmatrix}$$

which also agrees with the result in Example 1. ◀

Transition to the Standard Basis for R^n

Note that in part (a) of the last example the column vectors of the matrix that made the transition from the basis B' to the standard basis turned out to be the vectors in B' written in column form. This illustrates the following general result.

THEOREM 4.6.2 Let $B' = \{\mathbf{u}_1, \mathbf{u}_2, \ldots, \mathbf{u}_n\}$ be any basis for the vector space R^n and let $S = \{\mathbf{e}_1, \mathbf{e}_2, \ldots, \mathbf{e}_n\}$ be the standard basis for R^n. If the vectors in these bases are written in column form, then

$$P_{B' \to S} = [\mathbf{u}_1 \mid \mathbf{u}_2 \mid \cdots \mid \mathbf{u}_n] \tag{15}$$

It follows from this theorem that if

$$A = [\mathbf{u}_1 \mid \mathbf{u}_2 \mid \cdots \mid \mathbf{u}_n]$$

is *any* invertible $n \times n$ matrix, then A can be viewed as the transition matrix from the basis $\{\mathbf{u}_1, \mathbf{u}_2, \ldots, \mathbf{u}_n\}$ for R^n to the standard basis for R^n. Thus, for example, the matrix

$$A = \begin{bmatrix} 1 & 2 & 3 \\ 2 & 5 & 3 \\ 1 & 0 & 8 \end{bmatrix}$$

which was shown to be invertible in Example 4 of Section 1.5, is the transition matrix from the basis

$$\mathbf{u}_1 = (1, 2, 1), \quad \mathbf{u}_2 = (2, 5, 0), \quad \mathbf{u}_3 = (3, 3, 8)$$

to the basis

$$\mathbf{e}_1 = (1, 0, 0), \quad \mathbf{e}_2 = (0, 1, 0), \quad \mathbf{e}_3 = (0, 0, 1)$$

Concept Review
- Coordinate map
- Change-of-basis problem
- Transition matrix

Skills
- Find coordinate vectors relative to a given basis directly.
- Find the transition matrix from one basis to another.
- Use the transition matrix to compute coordinate vectors.

Exercise Set 4.6

1. Find the coordinate vector for $\mathbf{w}$ relative to the basis $S = \{\mathbf{u}_1, \mathbf{u}_2\}$ for R^2.
   (a) $\mathbf{u}_1 = (1, 0), \ \mathbf{u}_2 = (0, 1); \ \mathbf{w} = (-4, 3)$
   (b) $\mathbf{u}_1 = (1, -1), \ \mathbf{u}_2 = (2, 5); \ \mathbf{w} = (3, 7)$
   (c) $\mathbf{u}_1 = (1, 2), \ \mathbf{u}_2 = (-2, 1); \ \mathbf{w} = (a, b)$

2. Find the coordinate vector for $\mathbf{v}$ relative to the basis $S = \{\mathbf{v}_1, \mathbf{v}_2, \mathbf{v}_3\}$ for R^3.
   (a) $\mathbf{v} = (2, -1, 3); \ \mathbf{v}_1 = (1, 0, 0), \ \mathbf{v}_2 = (2, 2, 0),$
   $\mathbf{v}_3 = (3, 3, 3)$
   (b) $\mathbf{v} = (5, -12, 3); \ \mathbf{v}_1 = (1, 2, 3), \ \mathbf{v}_2 = (-4, 5, 6),$
   $\mathbf{v}_3 = (7, -8, 9)$

3. Find the coordinate vector for $\mathbf{p}$ relative to the basis $S = \{\mathbf{p}_1, \mathbf{p}_2, \mathbf{p}_3\}$ for P_2.
   (a) $\mathbf{p} = 5 - x + 3x^2; \ \mathbf{p}_1 = 5, \ \mathbf{p}_2 = x, \ \mathbf{p}_3 = x^2$
   (b) $\mathbf{p} = 2 + x - x^2; \ \mathbf{p}_1 = 1 - x, \ \mathbf{p}_2 = x + x^2,$
   $\mathbf{p}_3 = 1 - x^2$

4. Find the coordinate vector for A relative to the basis $S = \{A_1, A_2, A_3, A_4\}$ for M_{22}.

   $$A = \begin{bmatrix} 2 & 0 \\ -1 & 3 \end{bmatrix}, \quad A_1 = \begin{bmatrix} -1 & 1 \\ 0 & 0 \end{bmatrix}, \quad A_2 = \begin{bmatrix} 1 & 1 \\ 0 & 0 \end{bmatrix},$$
   $$A_3 = \begin{bmatrix} 0 & 0 \\ 1 & 0 \end{bmatrix}, \quad A_4 = \begin{bmatrix} 0 & 0 \\ 0 & 1 \end{bmatrix}$$

5. Consider the coordinate vectors

   $$[\mathbf{w}]_S = \begin{bmatrix} 6 \\ -1 \\ 4 \end{bmatrix}, \quad [\mathbf{q}]_S = \begin{bmatrix} 3 \\ 0 \\ 4 \end{bmatrix}, \quad [B]_S = \begin{bmatrix} -8 \\ 7 \\ 6 \\ 3 \end{bmatrix}$$

   (a) Find $\mathbf{w}$ if S is the basis in Exercise 2(a).
   (b) Find $\mathbf{q}$ if S is the basis in Exercise 3(a).
   (c) Find B if S is the basis in Exercise 4.

4.6 Change of Basis

6. Consider the bases $B = \{\mathbf{u}_1, \mathbf{u}_2\}$ and $B' = \{\mathbf{u}'_1, \mathbf{u}'_2\}$ for R^2, where

$$\mathbf{u}_1 = \begin{bmatrix} 1 \\ 0 \end{bmatrix}, \quad \mathbf{u}_2 = \begin{bmatrix} 0 \\ 1 \end{bmatrix}, \quad \mathbf{u}'_1 = \begin{bmatrix} 2 \\ 1 \end{bmatrix}, \quad \mathbf{u}'_2 = \begin{bmatrix} -3 \\ 4 \end{bmatrix}$$

(a) Find the transition matrix from B' to B.
(b) Find the transition matrix from B to B'.
(c) Compute the coordinate vector $[\mathbf{w}]_B$, where

$$\mathbf{w} = \begin{bmatrix} 3 \\ -5 \end{bmatrix}$$

and use (12) to compute $[\mathbf{w}]_{B'}$.
(d) Check your work by computing $[\mathbf{w}]_{B'}$ directly.

7. Repeat the directions of Exercise 6 with the same vector $\mathbf{w}$ but with

$$\mathbf{u}_1 = \begin{bmatrix} 4 \\ 1 \end{bmatrix}, \quad \mathbf{u}_2 = \begin{bmatrix} 3 \\ 1 \end{bmatrix}, \quad \mathbf{v}_1 = \begin{bmatrix} -1 \\ -2 \end{bmatrix}, \quad \mathbf{v}_2 = \begin{bmatrix} 2 \\ 3 \end{bmatrix}$$

8. Consider the bases $B = \{\mathbf{u}_1, \mathbf{u}_2, \mathbf{u}_3\}$ and $B' = \{\mathbf{u}'_1, \mathbf{u}'_2, \mathbf{u}'_3\}$ for R^3, where

$$\mathbf{u}_1 = \begin{bmatrix} -3 \\ 0 \\ -3 \end{bmatrix}, \quad \mathbf{u}_2 = \begin{bmatrix} -3 \\ 2 \\ -1 \end{bmatrix}, \quad \mathbf{u}_3 = \begin{bmatrix} 1 \\ 6 \\ -1 \end{bmatrix}$$

$$\mathbf{u}'_1 = \begin{bmatrix} -6 \\ -6 \\ 0 \end{bmatrix}, \quad \mathbf{u}'_2 = \begin{bmatrix} -2 \\ -6 \\ 4 \end{bmatrix}, \quad \mathbf{u}'_3 = \begin{bmatrix} -2 \\ -3 \\ 7 \end{bmatrix}$$

(a) Find the transition matrix from B to B'.
(b) Compute the coordinate vector $[\mathbf{w}]_B$, where

$$\mathbf{w} = \begin{bmatrix} -5 \\ 8 \\ -5 \end{bmatrix}$$

and use (12) to compute $[\mathbf{w}]_{B'}$.
(c) Check your work by computing $[\mathbf{w}]_{B'}$ directly.

9. Repeat the directions of Exercise 8 with the same vector $\mathbf{w}$, but with

$$\mathbf{u}_1 = \begin{bmatrix} 2 \\ 1 \\ 1 \end{bmatrix}, \quad \mathbf{u}_2 = \begin{bmatrix} 2 \\ -1 \\ 1 \end{bmatrix}, \quad \mathbf{u}_3 = \begin{bmatrix} 1 \\ 2 \\ 1 \end{bmatrix}$$

$$\mathbf{u}'_1 = \begin{bmatrix} 3 \\ 1 \\ -5 \end{bmatrix}, \quad \mathbf{u}'_2 = \begin{bmatrix} 1 \\ 1 \\ -3 \end{bmatrix}, \quad \mathbf{u}'_3 = \begin{bmatrix} -1 \\ 0 \\ 2 \end{bmatrix}$$

10. Consider the bases $B = \{\mathbf{p}_1, \mathbf{p}_2\}$ and $B' = \{\mathbf{q}_1, \mathbf{q}_2\}$ for P_1, where

$$\mathbf{p}_1 = 1 + 2x, \quad \mathbf{p}_2 = 3 - x, \quad \mathbf{q}_1 = 2 - 2x, \quad \mathbf{q}_2 = 4 + 3x$$

(a) Find the transition matrix from B' to B.
(b) Find the transition matrix from B to B'.

(c) Compute the coordinate vector $[\mathbf{p}]_B$, where $\mathbf{p} = 5 - x$, and use (12) to compute $[\mathbf{p}]_{B'}$.
(d) Check your work by computing $[\mathbf{p}]_{B'}$ directly.

11. Let V be the space spanned by $\mathbf{f}_1 = \sin x$ and $\mathbf{f}_2 = \cos x$.
(a) Show that $\mathbf{g}_1 = 2\sin x + \cos x$ and $\mathbf{g}_2 = 3\cos x$ form a basis for V.
(b) Find the transition matrix from $B' = \{\mathbf{g}_1, \mathbf{g}_2\}$ to $B = \{\mathbf{f}_1, \mathbf{f}_2\}$.
(c) Find the transition matrix from B to B'.
(d) Compute the coordinate vector $[\mathbf{h}]_B$, where $\mathbf{h} = 2\sin x - 5\cos x$, and use (12) to obtain $[\mathbf{h}]_{B'}$.
(e) Check your work by computing $[\mathbf{h}]_{B'}$ directly.

12. Let S be the standard basis for R^2, and let $B = \{\mathbf{v}_1, \mathbf{v}_2\}$ be the basis in which $\mathbf{v}_1 = (2, 1)$ and $\mathbf{v}_2 = (-3, 4)$.
(a) Find the transition matrix $P_{B \to S}$ by inspection.
(b) Use Formula (14) to find the transition matrix $P_{S \to B}$.
(c) Confirm that $P_{B \to S}$ and $P_{S \to B}$ are inverses of one another.
(d) Let $\mathbf{w} = (5, -3)$. Find $[\mathbf{w}]_B$ and then use Formula (11) to compute $[\mathbf{w}]_S$.
(e) Let $\mathbf{w} = (3, -5)$. Find $[\mathbf{w}]_S$ and then use Formula (12) to compute $[\mathbf{w}]_B$.

13. Let S be the standard basis for R^3, and let $B = \{\mathbf{v}_1, \mathbf{v}_2, \mathbf{v}_3\}$ be the basis in which $\mathbf{v}_1 = (1, 2, 1)$, $\mathbf{v}_2 = (2, 5, 0)$, and $\mathbf{v}_3 = (3, 3, 8)$.
(a) Find the transition matrix $P_{B \to S}$ by inspection.
(b) Use Formula (14) to find the transition matrix $P_{S \to B}$.
(c) Confirm that $P_{B \to S}$ and $P_{S \to B}$ are inverses of one another.
(d) Let $\mathbf{w} = (5, -3, 1)$. Find $[\mathbf{w}]_B$ and then use Formula (11) to compute $[\mathbf{w}]_S$.
(e) Let $\mathbf{w} = (3, -5, 0)$. Find $[\mathbf{w}]_S$ and then use Formula (12) to compute $[\mathbf{w}]_B$.

14. Let $B_1 = \{\mathbf{u}_1, \mathbf{u}_2\}$ and $B_2 = \{\mathbf{v}_1, \mathbf{v}_2\}$ be the bases for R^2 in which $\mathbf{u}_1 = (2, 2)$, $\mathbf{u}_2 = (4, -1)$, $\mathbf{v}_1 = (1, 3)$, and $\mathbf{v}_2 = (-1, -1)$.
(a) Use Formula (14) to find the transition matrix $P_{B_2 \to B_1}$.
(b) Use Formula (14) to find the transition matrix $P_{B_1 \to B_2}$.
(c) Confirm that $P_{B_2 \to B_1}$ and $P_{B_1 \to B_2}$ are inverses of one another.
(d) Let $\mathbf{w} = (5, -3)$. Find $[\mathbf{w}]_{B_1}$ and then use the matrix $P_{B_1 \to B_2}$ to compute $[\mathbf{w}]_{B_2}$ from $[\mathbf{w}]_{B_1}$.
(e) Let $\mathbf{w} = (3, -5)$. Find $[\mathbf{w}]_{B_2}$ and then use the matrix $P_{B_2 \to B_1}$ to compute $[\mathbf{w}]_{B_1}$ from $[\mathbf{w}]_{B_2}$.

15. Let $B_1 = \{\mathbf{u}_1, \mathbf{u}_2\}$ and $B_2 = \{\mathbf{v}_1, \mathbf{v}_2\}$ be the bases for R^2 in which $\mathbf{u}_1 = (1, 2)$, $\mathbf{u}_2 = (2, 3)$, $\mathbf{v}_1 = (1, 3)$, and $\mathbf{v}_2 = (1, 4)$.
(a) Use Formula (14) to find the transition matrix $P_{B_2 \to B_1}$.

(b) Use Formula (14) to find the transition matrix $P_{B_1 \to B_2}$.

(c) Confirm that $P_{B_2 \to B_1}$ and $P_{B_1 \to B_2}$ are inverses of one another.

(d) Let $\mathbf{w} = (1, 0)$. Find $[\mathbf{w}]_{B_1}$ and then use the matrix $P_{B_1 \to B_2}$ to compute $[\mathbf{w}]_{B_2}$ from $[\mathbf{w}]_{B_1}$.

(e) Let $\mathbf{w} = (3, -3)$. Find $[\mathbf{w}]_{B_2}$ and then use the matrix $P_{B_2 \to B_1}$ to compute $[\mathbf{w}]_{B_1}$ from $[\mathbf{w}]_{B_2}$.

16. Let $B_1 = \{\mathbf{u}_1, \mathbf{u}_2, \mathbf{u}_3\}$ and $B_2 = \{\mathbf{v}_1, \mathbf{v}_2, \mathbf{v}_3\}$ be the bases for R^3 in which $\mathbf{u}_1 = (-3, 0, -3)$, $\mathbf{u}_2 = (-3, 2, -1)$, $\mathbf{u}_3 = (1, 6, -1)$, $\mathbf{v}_1 = (-6, -6, 0)$, $\mathbf{v}_2 = (-2, -6, 4)$, and $\mathbf{v}_3 = (-2, -3, 7)$.

(a) Find the transition matrix $P_{B_1 \to B_2}$.

(b) Let $\mathbf{w} = (-5, 8, -5)$. Find $[\mathbf{w}]_{B_1}$ and then use the transition matrix obtained in part (a) to compute $[\mathbf{w}]_{B_2}$ by matrix multiplication.

(c) Check the result in part (b) by computing $[\mathbf{w}]_{B_2}$ directly.

17. Follow the directions of Exercise 16 with the same vector $\mathbf{w}$ but with $\mathbf{u}_1 = (2, 1, 1)$, $\mathbf{u}_2 = (2, -1, 1)$, $\mathbf{u}_3 = (1, 2, 1)$, $\mathbf{v}_1 = (3, 1, -5)$, $\mathbf{v}_2 = (1, 1, -3)$, and $\mathbf{v}_3 = (-1, 0, 2)$.

18. Let $S = \{\mathbf{e}_1, \mathbf{e}_2\}$ be the standard basis for R^2, and let $B = \{\mathbf{v}_1, \mathbf{v}_2\}$ be the basis that results when the vectors in S are reflected about the line $y = x$.

(a) Find the transition matrix $P_{B \to S}$.

(b) Let $P = P_{B \to S}$ and show that $P^T = P_{S \to B}$.

19. Let $S = \{\mathbf{e}_1, \mathbf{e}_2\}$ be the standard basis for R^2, and let $B = \{\mathbf{v}_1, \mathbf{v}_2\}$ be the basis that results when the vectors in S are reflected about the line that makes an angle θ with the positive x-axis.

(a) Find the transition matrix $P_{B \to S}$.

(b) Let $P = P_{B \to S}$ and show that $P^T = P_{S \to B}$.

20. If B_1, B_2, and B_3 are bases for R^2, and if

$$P_{B_1 \to B_2} = \begin{bmatrix} 3 & 1 \\ 5 & 2 \end{bmatrix} \text{ and } P_{B_2 \to B_3} = \begin{bmatrix} 7 & 2 \\ 4 & -1 \end{bmatrix}$$

then $P_{B_3 \to B_1} = $ _____.

21. If P is the transition matrix from a basis B' to a basis B, and Q is the transition matrix from B to a basis C, what is the transition matrix from B' to C? What is the transition matrix from C to B'?

22. To write the coordinate vector for a vector, it is necessary to specify an order for the vectors in the basis. If P is the transition matrix from a basis B' to a basis B, what is the effect on P if we reverse the order of vectors in B from $\mathbf{v}_1, \ldots, \mathbf{v}_n$ to $\mathbf{v}_n, \ldots, \mathbf{v}_1$? What is the effect on P if we reverse the order of vectors in both B' and B?

23. Consider the matrix

$$P = \begin{bmatrix} 1 & 1 & 0 \\ 1 & 0 & 2 \\ 0 & 2 & 1 \end{bmatrix}$$

(a) P is the transition matrix from what basis B to the standard basis $S = \{\mathbf{e}_1, \mathbf{e}_2, \mathbf{e}_3\}$ for R^3?

(b) P is the transition matrix from the standard basis $S = \{\mathbf{e}_1, \mathbf{e}_2, \mathbf{e}_3\}$ to what basis B for R^3?

24. The matrix

$$P = \begin{bmatrix} 1 & 0 & 0 \\ 0 & 3 & 2 \\ 0 & 1 & 1 \end{bmatrix}$$

is the transition matrix from what basis B to the basis $\{(1, 1, 1), (1, 1, 0), (1, 0, 0)\}$ for R^3?

25. Let B be a basis for R^n. Prove that the vectors $\mathbf{v}_1, \mathbf{v}_2, \ldots, \mathbf{v}_k$ form a linearly independent set in R^n if and only if the vectors $[\mathbf{v}_1]_B, [\mathbf{v}_2]_B, \ldots, [\mathbf{v}_k]_B$ form a linearly independent set in R^n.

26. Let B be a basis for R^n. Prove that the vectors $\mathbf{v}_1, \mathbf{v}_2, \ldots, \mathbf{v}_k$ span R^n if and only if the vectors $[\mathbf{v}_1]_B, [\mathbf{v}_2]_B, \ldots, [\mathbf{v}_k]_B$ span R^n.

27. If $[\mathbf{w}]_B = \mathbf{w}$ holds for all vectors $\mathbf{w}$ in R^n, what can you say about the basis B?

True-False Exercises

In parts (a)–(f) determine whether the statement is true or false, and justify your answer.

(a) If B_1 and B_2 are bases for a vector space V, then there exists a transition matrix from B_1 to B_2.

(b) Transition matrices are invertible.

(c) If B is a basis for a vector space R^n, then $P_{B \to B}$ is the identity matrix.

(d) If $P_{B_1 \to B_2}$ is a diagonal matrix, then each vector in B_2 is a scalar multiple of some vector in B_1.

(e) If each vector in B_2 is a scalar multiple of some vector in B_1, then $P_{B_1 \to B_2}$ is a diagonal matrix.

(f) If A is a square matrix, then $A = P_{B_1 \to B_2}$ for some bases B_1 and B_2 for R^n.

4.7 Row Space, Column Space, and Null Space

In this section we will study some important vector spaces that are associated with matrices. Our work here will provide us with a deeper understanding of the relationships between the solutions of a linear system and properties of its coefficient matrix.

Row Space, Column Space, and Null Space

Recall that vectors can be written in comma-delimited form or in matrix form as either row vectors or column vectors. In this section we will use the latter two.

DEFINITION 1 For an $m \times n$ matrix

$$A = \begin{bmatrix} a_{11} & a_{12} & \cdots & a_{1n} \\ a_{21} & a_{22} & \cdots & a_{2n} \\ \vdots & \vdots & & \vdots \\ a_{m1} & a_{m2} & \cdots & a_{mn} \end{bmatrix}$$

the vectors

$$\mathbf{r}_1 = [a_{11} \quad a_{12} \quad \cdots \quad a_{1n}]$$
$$\mathbf{r}_2 = [a_{21} \quad a_{22} \quad \cdots \quad a_{2n}]$$
$$\vdots$$
$$\mathbf{r}_m = [a_{m1} \quad a_{m2} \quad \cdots \quad a_{mn}]$$

in R^n that are formed from the rows of A are called the **row vectors** of A, and the vectors

$$\mathbf{c}_1 = \begin{bmatrix} a_{11} \\ a_{21} \\ \vdots \\ a_{m1} \end{bmatrix}, \quad \mathbf{c}_2 = \begin{bmatrix} a_{12} \\ a_{22} \\ \vdots \\ a_{m2} \end{bmatrix}, \ldots, \quad \mathbf{c}_n = \begin{bmatrix} a_{1n} \\ a_{2n} \\ \vdots \\ a_{mn} \end{bmatrix}$$

in R^m formed from the columns of A are called the **column vectors** of A.

▶ **EXAMPLE 1** Row and Column Vectors of a 2 × 3 Matrix

Let

$$A = \begin{bmatrix} 2 & 1 & 0 \\ 3 & -1 & 4 \end{bmatrix}$$

The row vectors of A are

$$\mathbf{r}_1 = [2 \quad 1 \quad 0] \quad \text{and} \quad \mathbf{r}_2 = [3 \quad -1 \quad 4]$$

and the column vectors of A are

$$\mathbf{c}_1 = \begin{bmatrix} 2 \\ 3 \end{bmatrix}, \quad \mathbf{c}_2 = \begin{bmatrix} 1 \\ -1 \end{bmatrix}, \quad \text{and} \quad \mathbf{c}_3 = \begin{bmatrix} 0 \\ 4 \end{bmatrix} \quad ◀$$

The following definition defines three important vector spaces associated with a matrix.

DEFINITION 2 If A is an $m \times n$ matrix, then the subspace of R^n spanned by the row vectors of A is called the **row space** of A, and the subspace of R^m spanned by the column vectors of A is called the **column space** of A. The solution space of the homogeneous system of equations $A\mathbf{x} = \mathbf{0}$, which is a subspace of R^n, is called the **null space** of A.

In this section and the next we will be concerned with two general questions:

Question 1. What relationships exist among the solutions of a linear system $A\mathbf{x} = \mathbf{b}$ and the row space, column space, and null space of the coefficient matrix A?

Question 2. What relationships exist among the row space, column space, and null space of a matrix?

Starting with the first question, suppose that

$$A = \begin{bmatrix} a_{11} & a_{12} & \cdots & a_{1n} \\ a_{21} & a_{22} & \cdots & a_{2n} \\ \vdots & \vdots & & \vdots \\ a_{m1} & a_{m2} & \cdots & a_{mn} \end{bmatrix} \quad \text{and} \quad \mathbf{x} = \begin{bmatrix} x_1 \\ x_2 \\ \vdots \\ x_n \end{bmatrix}$$

It follows from Formula (10) of Section 1.3 that if $\mathbf{c}_1, \mathbf{c}_2, \ldots, \mathbf{c}_n$ denote the column vectors of A, then the product $A\mathbf{x}$ can be expressed as a linear combination of these vectors with coefficients from $\mathbf{x}$; that is,

$$A\mathbf{x} = x_1\mathbf{c}_1 + x_2\mathbf{c}_2 + \cdots + x_n\mathbf{c}_n \tag{1}$$

Thus, a linear system, $A\mathbf{x} = \mathbf{b}$, of m equations in n unknowns can be written as

$$x_1\mathbf{c}_1 + x_2\mathbf{c}_2 + \cdots + x_n\mathbf{c}_n = \mathbf{b} \tag{2}$$

from which we conclude that $A\mathbf{x} = \mathbf{b}$ is consistent if and only if $\mathbf{b}$ is expressible as a linear combination of the column vectors of A. This yields the following theorem.

THEOREM 4.7.1 *A system of linear equations $A\mathbf{x} = \mathbf{b}$ is consistent if and only if $\mathbf{b}$ is in the column space of A.*

▶ **EXAMPLE 2** **A Vector b in the Column Space of A**

Let $A\mathbf{x} = \mathbf{b}$ be the linear system

$$\begin{bmatrix} -1 & 3 & 2 \\ 1 & 2 & -3 \\ 2 & 1 & -2 \end{bmatrix} \begin{bmatrix} x_1 \\ x_2 \\ x_3 \end{bmatrix} = \begin{bmatrix} 1 \\ -9 \\ -3 \end{bmatrix}$$

Show that $\mathbf{b}$ is in the column space of A by expressing it as a linear combination of the column vectors of A.

Solution Solving the system by Gaussian elimination yields (verify)

$$x_1 = 2, \quad x_2 = -1, \quad x_3 = 3$$

It follows from this and Formula (2) that

$$2\begin{bmatrix} -1 \\ 1 \\ 2 \end{bmatrix} - \begin{bmatrix} 3 \\ 2 \\ 1 \end{bmatrix} + 3\begin{bmatrix} 2 \\ -3 \\ -2 \end{bmatrix} = \begin{bmatrix} 1 \\ -9 \\ -3 \end{bmatrix} \quad \blacktriangleleft$$

Recall from Theorem 3.4.4 that the general solution of a consistent linear system $A\mathbf{x} = \mathbf{b}$ can be obtained by adding any specific solution of this system to the general solution of the corresponding homogeneous system $A\mathbf{x} = \mathbf{0}$. Keeping in mind that the null space of A is the same as the solution space of $A\mathbf{x} = \mathbf{0}$, we can rephrase that theorem in the following vector form.

THEOREM 4.7.2 *If $\mathbf{x}_0$ is any solution of a consistent linear system $A\mathbf{x} = \mathbf{b}$, and if $S = \{\mathbf{v}_1, \mathbf{v}_2, \ldots, \mathbf{v}_k\}$ is a basis for the null space of A, then every solution of $A\mathbf{x} = \mathbf{b}$ can be expressed in the form*

$$\mathbf{x} = \mathbf{x}_0 + c_1\mathbf{v}_1 + c_2\mathbf{v}_2 + \cdots + c_k\mathbf{v}_k \tag{3}$$

Conversely, for all choices of scalars $c_1, c_2, \ldots, c_k$, the vector $\mathbf{x}$ in this formula is a solution of $A\mathbf{x} = \mathbf{b}$.

Equation (3) gives a formula for the ***general solution of*** $A\mathbf{x} = \mathbf{b}$. The vector $\mathbf{x}_0$ in that formula is called a ***particular solution of*** $A\mathbf{x} = \mathbf{b}$, and the remaining part of the formula is called the ***general solution of*** $A\mathbf{x} = \mathbf{0}$. In words, this formula tells us that:

The general solution of a consistent linear system can be expressed as the sum of a particular solution of that system and the general solution of the corresponding homogeneous system.

Geometrically, the solution set of $A\mathbf{x} = \mathbf{b}$ can be viewed as the translation by $\mathbf{x}_0$ of the solution space of $A\mathbf{x} = \mathbf{0}$ (Figure 4.7.1).

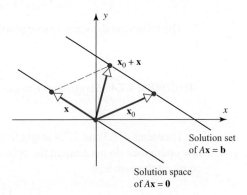

▶ **Figure 4.7.1**

▶ **EXAMPLE 3 General Solution of a Linear System** $A\mathbf{x} = \mathbf{b}$

In the concluding subsection of Section 3.4 we compared solutions of the linear systems

$$\begin{bmatrix} 1 & 3 & -2 & 0 & 2 & 0 \\ 2 & 6 & -5 & -2 & 4 & -3 \\ 0 & 0 & 5 & 10 & 0 & 15 \\ 2 & 6 & 0 & 8 & 4 & 18 \end{bmatrix} \begin{bmatrix} x_1 \\ x_2 \\ x_3 \\ x_4 \\ x_5 \\ x_6 \end{bmatrix} = \begin{bmatrix} 0 \\ 0 \\ 0 \\ 0 \end{bmatrix} \quad \text{and} \quad \begin{bmatrix} 1 & 3 & -2 & 0 & 2 & 0 \\ 2 & 6 & -5 & -2 & 4 & -3 \\ 0 & 0 & 5 & 10 & 0 & 15 \\ 2 & 6 & 0 & 8 & 4 & 18 \end{bmatrix} \begin{bmatrix} x_1 \\ x_2 \\ x_3 \\ x_4 \\ x_5 \\ x_6 \end{bmatrix} = \begin{bmatrix} 0 \\ -1 \\ 5 \\ 6 \end{bmatrix}$$

and deduced that the general solution $\mathbf{x}$ of the nonhomogeneous system and the general solution $\mathbf{x}_h$ of the corresponding homogeneous system (when written in column-vector form) are related by

$$\underbrace{\begin{bmatrix} x_1 \\ x_2 \\ x_3 \\ x_4 \\ x_5 \\ x_6 \end{bmatrix}}_{\mathbf{x}} = \begin{bmatrix} -3r-4s-2t \\ r \\ -2s \\ s \\ t \\ \frac{1}{3} \end{bmatrix} = \underbrace{\begin{bmatrix} 0 \\ 0 \\ 0 \\ 0 \\ 0 \\ \frac{1}{3} \end{bmatrix}}_{\mathbf{x}_0} + r\underbrace{\begin{bmatrix} -3 \\ 1 \\ 0 \\ 0 \\ 0 \\ 0 \end{bmatrix} + s\begin{bmatrix} -4 \\ 0 \\ -2 \\ 1 \\ 0 \\ 0 \end{bmatrix} + t\begin{bmatrix} -2 \\ 0 \\ 0 \\ 0 \\ 1 \\ 0 \end{bmatrix}}_{\mathbf{x}_h}$$

◀

Recall from the Remark following Example 4 of Section 4.5 that the vectors in $\mathbf{x}_h$ form a basis for the solution space of $A\mathbf{x} = \mathbf{0}$.

Bases for Row Spaces, Column Spaces, and Null Spaces

We first developed elementary row operations for the purpose of solving linear systems, and we know from that work that performing an elementary row operation on an augmented matrix does not change the solution set of the corresponding linear system. It follows that applying an elementary row operation to a matrix A does not change the solution set of the corresponding linear system $A\mathbf{x} = \mathbf{0}$, or, stated another way, it does not change the null space of A. Thus we have the following theorem.

THEOREM 4.7.3 *Elementary row operations do not change the null space of a matrix.*

The following theorem, whose proof is left as an exercise, is a companion to Theorem 4.7.3.

THEOREM 4.7.4 *Elementary row operations do not change the row space of a matrix.*

Theorems 4.7.3 and 4.7.4 might tempt you into *incorrectly* believing that elementary row operations do not change the column space of a matrix. To see why this is *not* true, compare the matrices

$$A = \begin{bmatrix} 1 & 3 \\ 2 & 6 \end{bmatrix} \quad \text{and} \quad B = \begin{bmatrix} 1 & 3 \\ 0 & 0 \end{bmatrix}$$

The matrix B can be obtained from A by adding -2 times the first row to the second. However, this operation has changed the column space of A, since that column space consists of all scalar multiples of

$$\begin{bmatrix} 1 \\ 2 \end{bmatrix}$$

whereas the column space of B consists of all scalar multiples of

$$\begin{bmatrix} 1 \\ 0 \end{bmatrix}$$

and the two are different spaces.

► EXAMPLE 4 Finding a Basis for the Null Space of a Matrix

Find a basis for the null space of the matrix

$$A = \begin{bmatrix} 1 & 3 & -2 & 0 & 2 & 0 \\ 2 & 6 & -5 & -2 & 4 & -3 \\ 0 & 0 & 5 & 10 & 0 & 15 \\ 2 & 6 & 0 & 8 & 4 & 18 \end{bmatrix}$$

Solution The null space of A is the solution space of the homogeneous linear system $A\mathbf{x} = \mathbf{0}$, which, as shown in Example 3, has the basis

$$\mathbf{v}_1 = \begin{bmatrix} -3 \\ 1 \\ 0 \\ 0 \\ 0 \\ 0 \end{bmatrix}, \quad \mathbf{v}_2 = \begin{bmatrix} -4 \\ 0 \\ -2 \\ 1 \\ 0 \\ 0 \end{bmatrix}, \quad \mathbf{v}_3 = \begin{bmatrix} -2 \\ 0 \\ 0 \\ 0 \\ 1 \\ 0 \end{bmatrix} \blacktriangleleft$$

Remark Observe that the basis vectors $\mathbf{v}_1$, $\mathbf{v}_2$, and $\mathbf{v}_3$ in the last example are the vectors that result by successively setting one of the parameters in the general solution equal to 1 and the others equal to 0.

The following theorem makes it possible to find bases for the row and column spaces of a matrix in row echelon form by inspection.

THEOREM 4.7.5 *If a matrix R is in row echelon form, then the row vectors with the leading 1's (the nonzero row vectors) form a basis for the row space of R, and the column vectors with the leading 1's of the row vectors form a basis for the column space of R.*

The proof involves little more than an analysis of the positions of the 0's and 1's of R. We omit the details.

► EXAMPLE 5 Bases for Row and Column Spaces

The matrix

$$R = \begin{bmatrix} 1 & -2 & 5 & 0 & 3 \\ 0 & 1 & 3 & 0 & 0 \\ 0 & 0 & 0 & 1 & 0 \\ 0 & 0 & 0 & 0 & 0 \end{bmatrix}$$

is in row echelon form. From Theorem 4.7.5, the vectors

$$\mathbf{r}_1 = [1 \quad -2 \quad 5 \quad 0 \quad 3]$$
$$\mathbf{r}_2 = [0 \quad 1 \quad 3 \quad 0 \quad 0]$$
$$\mathbf{r}_3 = [0 \quad 0 \quad 0 \quad 1 \quad 0]$$

form a basis for the row space of R, and the vectors

$$\mathbf{c}_1 = \begin{bmatrix} 1 \\ 0 \\ 0 \\ 0 \end{bmatrix}, \quad \mathbf{c}_2 = \begin{bmatrix} -2 \\ 1 \\ 0 \\ 0 \end{bmatrix}, \quad \mathbf{c}_4 = \begin{bmatrix} 0 \\ 0 \\ 1 \\ 0 \end{bmatrix}$$

form a basis for the column space of R.

▶ **EXAMPLE 6** **Basis for a Row Space by Row Reduction**

Find a basis for the row space of the matrix

$$A = \begin{bmatrix} 1 & -3 & 4 & -2 & 5 & 4 \\ 2 & -6 & 9 & -1 & 8 & 2 \\ 2 & -6 & 9 & -1 & 9 & 7 \\ -1 & 3 & -4 & 2 & -5 & -4 \end{bmatrix}$$

Solution Since elementary row operations do not change the row space of a matrix, we can find a basis for the row space of A by finding a basis for the row space of any row echelon form of A. Reducing A to row echelon form, we obtain (verify)

$$R = \begin{bmatrix} 1 & -3 & 4 & -2 & 5 & 4 \\ 0 & 0 & 1 & 3 & -2 & -6 \\ 0 & 0 & 0 & 0 & 1 & 5 \\ 0 & 0 & 0 & 0 & 0 & 0 \end{bmatrix}$$

By Theorem 4.7.5, the nonzero row vectors of R form a basis for the row space of R and hence form a basis for the row space of A. These basis vectors are

$$\mathbf{r}_1 = \begin{bmatrix} 1 & -3 & 4 & -2 & 5 & 4 \end{bmatrix}$$
$$\mathbf{r}_2 = \begin{bmatrix} 0 & 0 & 1 & 3 & -2 & -6 \end{bmatrix}$$
$$\mathbf{r}_3 = \begin{bmatrix} 0 & 0 & 0 & 0 & 1 & 5 \end{bmatrix} \blacktriangleleft$$

The problem of finding a basis for the column space of a matrix A in Example 6 is complicated by the fact that an elementary row operation can alter its column space. However, the good news is that *elementary row operations do not alter dependence relationships among the column vectors*. To make this more precise, suppose that $\mathbf{w}_1, \mathbf{w}_2, \ldots, \mathbf{w}_k$ are linearly dependent column vectors of A, so there are scalars $c_1, c_2, \ldots, c_k$ that are not all zero and such that

$$c_1 \mathbf{w}_1 + c_2 \mathbf{w}_2 + \cdots + c_k \mathbf{w}_k = \mathbf{0} \tag{4}$$

If we perform an elementary row operation on A, then these vectors will be changed into new column vectors $\mathbf{w}'_1, \mathbf{w}'_2, \ldots, \mathbf{w}'_k$. At first glance it would seem possible that the transformed vectors might be linearly independent. However, this is not so, since it can be proved that these new column vectors will be linear dependent and, in fact, related by an equation

$$c_1 \mathbf{w}'_1 + c_2 \mathbf{w}'_2 + \cdots + c_k \mathbf{w}'_k = \mathbf{0}$$

that has exactly the same coefficients as (4). It follows from the fact that elementary row operations are reversible that they also preserve linear independence among column vectors (why?). The following theorem summarizes all of these results.

4.7 Row Space, Column Space, and Null Space 231

THEOREM 4.7.6 *If A and B are row equivalent matrices, then*:

(a) *A given set of column vectors of A is linearly independent if and only if the corresponding column vectors of B are linearly independent.*

(b) *A given set of column vectors of A forms a basis for the column space of A if and only if the corresponding column vectors of B form a basis for the column space of B.*

▶ **EXAMPLE 7 Basis for a Column Space by Row Reduction**

Find a basis for the column space of the matrix

$$A = \begin{bmatrix} 1 & -3 & 4 & -2 & 5 & 4 \\ 2 & -6 & 9 & -1 & 8 & 2 \\ 2 & -6 & 9 & -1 & 9 & 7 \\ -1 & 3 & -4 & 2 & -5 & -4 \end{bmatrix}$$

Solution We observed in Example 6 that the matrix

$$R = \begin{bmatrix} 1 & -3 & 4 & -2 & 5 & 4 \\ 0 & 0 & 1 & 3 & -2 & -6 \\ 0 & 0 & 0 & 0 & 1 & 5 \\ 0 & 0 & 0 & 0 & 0 & 0 \end{bmatrix}$$

is a row echelon form of A. Keeping in mind that A and R can have different column spaces, we cannot find a basis for the column space of A directly from the column vectors of R. However, it follows from Theorem 4.7.6b that if we can find a set of column vectors of R that forms a basis for the column space of R, then the *corresponding* column vectors of A will form a basis for the column space of A.

Since the first, third, and fifth columns of R contain the leading 1's of the row vectors, the vectors

$$\mathbf{c}'_1 = \begin{bmatrix} 1 \\ 0 \\ 0 \\ 0 \end{bmatrix}, \quad \mathbf{c}'_3 = \begin{bmatrix} 4 \\ 1 \\ 0 \\ 0 \end{bmatrix}, \quad \mathbf{c}'_5 = \begin{bmatrix} 5 \\ -2 \\ 1 \\ 0 \end{bmatrix}$$

form a basis for the column space of R. Thus, the corresponding column vectors of A, which are

$$\mathbf{c}_1 = \begin{bmatrix} 1 \\ 2 \\ 2 \\ -1 \end{bmatrix}, \quad \mathbf{c}_3 = \begin{bmatrix} 4 \\ 9 \\ 9 \\ -4 \end{bmatrix}, \quad \mathbf{c}_5 = \begin{bmatrix} 5 \\ 8 \\ 9 \\ -5 \end{bmatrix}$$

form a basis for the column space of A. ◀

Up to now we have focused on methods for finding bases associated with matrices. Those methods can readily be adapted to the more general problem of finding a basis for the space spanned by a set of vectors in R^n.

▶ **EXAMPLE 8** **Basis for a Vector Space Using Row Operations**

Find a basis for the subspace of R^5 spanned by the vectors

$$\mathbf{v}_1 = (1, -2, 0, 0, 3), \quad \mathbf{v}_2 = (2, -5, -3, -2, 6),$$
$$\mathbf{v}_3 = (0, 5, 15, 10, 0), \quad \mathbf{v}_4 = (2, 6, 18, 8, 6)$$

Solution The space spanned by these vectors is the row space of the matrix

$$\begin{bmatrix} 1 & -2 & 0 & 0 & 3 \\ 2 & -5 & -3 & -2 & 6 \\ 0 & 5 & 15 & 10 & 0 \\ 2 & 6 & 18 & 8 & 6 \end{bmatrix}$$

Reducing this matrix to row echelon form, we obtain

$$\begin{bmatrix} 1 & -2 & 0 & 0 & 3 \\ 0 & 1 & 3 & 2 & 0 \\ 0 & 0 & 1 & 1 & 0 \\ 0 & 0 & 0 & 0 & 0 \end{bmatrix}$$

The nonzero row vectors in this matrix are

$$\mathbf{w}_1 = (1, -2, 0, 0, 3), \quad \mathbf{w}_2 = (0, 1, 3, 2, 0), \quad \mathbf{w}_3 = (0, 0, 1, 1, 0)$$

These vectors form a basis for the row space and consequently form a basis for the subspace of R^5 spanned by $\mathbf{v}_1, \mathbf{v}_2, \mathbf{v}_3$, and $\mathbf{v}_4$. ◀

Bases Formed from Row and Column Vectors of a Matrix

In all of the examples we have considered thus far we have looked for bases in which no restrictions were imposed on the individual vectors in the basis. We now want to focus on the problem of finding a basis for the row space of a matrix A consisting entirely of row vectors from A and a basis for the column space of A consisting entirely of column vectors of A.

Looking back on our earlier work, we see that the procedure followed in Example 7 did, in fact, produce a basis for the column space of A consisting of column vectors of A, whereas the procedure used in Example 6 produced a basis for the row space of A, but that basis did not consist of row vectors of A. The following example shows how to adapt the procedure from Example 7 to find a basis for the row space of a matrix that is formed from its row vectors.

▶ **EXAMPLE 9** **Basis for the Row Space of a Matrix**

Find a basis for the row space of

$$A = \begin{bmatrix} 1 & -2 & 0 & 0 & 3 \\ 2 & -5 & -3 & -2 & 6 \\ 0 & 5 & 15 & 10 & 0 \\ 2 & 6 & 18 & 8 & 6 \end{bmatrix}$$

consisting entirely of row vectors from A.

Solution We will transpose A, thereby converting the row space of A into the column space of A^T; then we will use the method of Example 7 to find a basis for the column space of A^T; and then we will transpose again to convert column vectors back to row vectors. Transposing A yields

$$A^T = \begin{bmatrix} 1 & 2 & 0 & 2 \\ -2 & -5 & 5 & 6 \\ 0 & -3 & 15 & 18 \\ 0 & -2 & 10 & 8 \\ 3 & 6 & 0 & 6 \end{bmatrix}$$

Reducing this matrix to row echelon form yields

$$\begin{bmatrix} 1 & 2 & 0 & 2 \\ 0 & 1 & -5 & -10 \\ 0 & 0 & 0 & 1 \\ 0 & 0 & 0 & 0 \\ 0 & 0 & 0 & 0 \end{bmatrix}$$

The first, second, and fourth columns contain the leading 1's, so the corresponding column vectors in A^T form a basis for the column space of A^T; these are

$$\mathbf{c}_1 = \begin{bmatrix} 1 \\ -2 \\ 0 \\ 0 \\ 3 \end{bmatrix}, \quad \mathbf{c}_2 = \begin{bmatrix} 2 \\ -5 \\ -3 \\ -2 \\ 6 \end{bmatrix}, \quad \text{and} \quad \mathbf{c}_4 = \begin{bmatrix} 2 \\ 6 \\ 18 \\ 8 \\ 6 \end{bmatrix}$$

Transposing again and adjusting the notation appropriately yields the basis vectors

$$\mathbf{r}_1 = \begin{bmatrix} 1 & -2 & 0 & 0 & 3 \end{bmatrix}, \quad \mathbf{r}_2 = \begin{bmatrix} 2 & -5 & -3 & -2 & 6 \end{bmatrix},$$

and

$$\mathbf{r}_4 = \begin{bmatrix} 2 & 6 & 18 & 8 & 6 \end{bmatrix}$$

for the row space of A. ◀

Next, we will give an example that adapts the methods we have developed above to solve the following general problem in R^n:

Problem Given a set of vectors $S = \{\mathbf{v}_1, \mathbf{v}_2, \ldots, \mathbf{v}_k\}$ in R^n, find a subset of these vectors that forms a basis for span(S), and express those vectors that are not in that basis as a linear combination of the basis vectors.

▶ **EXAMPLE 10 Basis and Linear Combinations**

(a) Find a subset of the vectors

$$\mathbf{v}_1 = (1, -2, 0, 3), \quad \mathbf{v}_2 = (2, -5, -3, 6),$$
$$\mathbf{v}_3 = (0, 1, 3, 0), \quad \mathbf{v}_4 = (2, -1, 4, -7), \quad \mathbf{v}_5 = (5, -8, 1, 2)$$

that forms a basis for the space spanned by these vectors.

(b) Express each vector not in the basis as a linear combination of the basis vectors.

Solution (a) We begin by constructing a matrix that has $v_1, v_2, \ldots, v_5$ as its column vectors:

$$\begin{bmatrix} 1 & 2 & 0 & 2 & 5 \\ -2 & -5 & 1 & -1 & -8 \\ 0 & -3 & 3 & 4 & 1 \\ 3 & 6 & 0 & -7 & 2 \end{bmatrix} \quad (5)$$
$$\uparrow \quad \uparrow \quad \uparrow \quad \uparrow \quad \uparrow$$
$$v_1 \quad v_2 \quad v_3 \quad v_4 \quad v_5$$

The first part of our problem can be solved by finding a basis for the column space of this matrix. Reducing the matrix to *reduced* row echelon form and denoting the column vectors of the resulting matrix by w_1, w_2, w_3, w_4, and w_5 yields

$$\begin{bmatrix} 1 & 0 & 2 & 0 & 1 \\ 0 & 1 & -1 & 0 & 1 \\ 0 & 0 & 0 & 1 & 1 \\ 0 & 0 & 0 & 0 & 0 \end{bmatrix} \quad (6)$$
$$\uparrow \quad \uparrow \quad \uparrow \quad \uparrow \quad \uparrow$$
$$w_1 \quad w_2 \quad w_3 \quad w_4 \quad w_5$$

The leading 1's occur in columns 1, 2, and 4, so by Theorem 4.7.5,

$$\{w_1, w_2, w_4\}$$

is a basis for the column space of (6), and consequently,

$$\{v_1, v_2, v_4\}$$

is a basis for the column space of (5).

Solution (b) We will start by expressing w_3 and w_5 as linear combinations of the basis vectors w_1, w_2, w_4. The simplest way of doing this is to express w_3 and w_5 in terms of basis vectors with smaller subscripts. Accordingly, we will express w_3 as a linear combination of w_1 and w_2, and we will express w_5 as a linear combination of w_1, w_2, and w_4. By inspection of (6), these linear combinations are

$$w_3 = 2w_1 - w_2$$
$$w_5 = w_1 + w_2 + w_4$$

We call these the ***dependency equations***. The corresponding relationships in (5) are

$$v_3 = 2v_1 - v_2$$
$$v_5 = v_1 + v_2 + v_4 \quad \blacktriangleleft$$

The following is a summary of the steps that we followed in our last example to solve the problem posed above.

Basis for Span(S)

Step 1. Form the matrix A having vectors in $S = \{v_1, v_2, \ldots, v_k\}$ as column vectors.

Step 2. Reduce the matrix A to reduced row echelon form R.

Step 3. Denote the column vectors of R by $w_1, w_2, \ldots, w_k$.

Step 4. Identify the columns of R that contain the leading 1's. The corresponding column vectors of A form a basis for span(S).

This completes the first part of the problem.

4.7 Row Space, Column Space, and Null Space

Step 5. Obtain a set of dependency equations by expressing each column vector of R that does not contain a leading 1 as a linear combination of preceding column vectors that do contain leading 1's.

Step 6. Replace the column vectors of R that appear in the dependency equations by the corresponding column vectors of A.

This completes the second part of the problem.

Concept Review

- Row vectors
- Column vectors
- Row space
- Column space
- Null space
- General solution
- Particular solution
- Relationships among linear systems and row spaces, column spaces, and null spaces
- Relationships among the row space, column space, and null space of a matrix
- Dependency equations

Skills

- Determine whether a given vector is in the column space of a matrix; if it is, express it as a linear combination of the column vectors of the matrix.
- Find a basis for the null space of a matrix.
- Find a basis for the row space of a matrix.
- Find a basis for the column space of a matrix.
- Find a basis for the span of a set of vectors in R^n.

Exercise Set 4.7

1. List the row vectors and column vectors of the matrix

$$\begin{bmatrix} 2 & -1 & 0 & 1 \\ 3 & 5 & 7 & -1 \\ 1 & 4 & 2 & 7 \end{bmatrix}$$

2. Express the product $A\mathbf{x}$ as a linear combination of the column vectors of A.

(a) $\begin{bmatrix} 3 & -1 \\ 1 & 4 \end{bmatrix} \begin{bmatrix} 5 \\ 2 \end{bmatrix}$

(b) $\begin{bmatrix} 4 & 0 & -1 \\ 3 & 6 & 2 \\ 0 & -1 & 4 \end{bmatrix} \begin{bmatrix} -2 \\ 3 \\ 5 \end{bmatrix}$

(c) $\begin{bmatrix} 5 & 2 & 6 \\ 1 & -1 & 3 \\ 0 & 1 & 7 \\ 2 & 1 & 3 \\ 4 & -2 & 1 \end{bmatrix} \begin{bmatrix} 4 \\ 6 \\ 9 \end{bmatrix}$

(d) $\begin{bmatrix} 2 & 1 & 5 \\ 6 & 3 & -8 \end{bmatrix} \begin{bmatrix} 3 \\ 0 \\ -5 \end{bmatrix}$

3. Determine whether $\mathbf{b}$ is in the column space of A, and if so, express $\mathbf{b}$ as a linear combination of the column vectors of A.

(a) $A = \begin{bmatrix} 5 & 1 \\ -1 & 5 \end{bmatrix}$; $\mathbf{b} = \begin{bmatrix} 1 \\ 0 \end{bmatrix}$

(b) $A = \begin{bmatrix} 0 & 1 & 4 \\ 2 & 1 & 1 \\ 2 & 2 & 5 \end{bmatrix}$; $\mathbf{b} = \begin{bmatrix} 1 \\ 0 \\ 2 \end{bmatrix}$

(c) $A = \begin{bmatrix} 1 & -1 & 1 \\ 9 & 3 & 1 \\ 1 & 1 & 1 \end{bmatrix}$; $\mathbf{b} = \begin{bmatrix} 5 \\ 1 \\ -1 \end{bmatrix}$

(d) $A = \begin{bmatrix} 1 & -1 & 1 \\ 1 & 1 & -1 \\ -1 & -1 & 1 \end{bmatrix}$; $\mathbf{b} = \begin{bmatrix} 2 \\ 0 \\ 0 \end{bmatrix}$

(e) $A = \begin{bmatrix} 1 & 2 & 0 & 1 \\ 0 & 1 & 2 & 1 \\ 1 & 2 & 1 & 3 \\ 0 & 1 & 2 & 2 \end{bmatrix}$; $\mathbf{b} = \begin{bmatrix} 4 \\ 3 \\ 5 \\ 7 \end{bmatrix}$

4. Suppose that $x_1 = -1$, $x_2 = 2$, $x_3 = 4$, $x_4 = -3$ is a solution of a nonhomogeneous linear system $A\mathbf{x} = \mathbf{b}$ and that the solution set of the homogeneous system $A\mathbf{x} = \mathbf{0}$ is given by the formulas

$$x_1 = -3r + 4s, \quad x_2 = r - s, \quad x_3 = r, \quad x_4 = s$$

(a) Find a vector form of the general solution of $A\mathbf{x} = \mathbf{0}$.

(b) Find a vector form of the general solution of $A\mathbf{x} = \mathbf{b}$.

5. In parts (a)–(d), find the vector form of the general solution of the given linear system $A\mathbf{x} = \mathbf{b}$; then use that result to find the vector form of the general solution of $A\mathbf{x} = \mathbf{0}$.

(a) $3x_1 + x_2 = 2$
$6x_1 + 2x_2 = 4$

(b) $x_1 + x_2 + 2x_3 = 5$
$x_1 + x_3 = -2$
$2x_1 + x_2 + 3x_3 = 3$

(c) $x_1 - 2x_2 + x_3 + x_4 = 1$
$-x_1 + x_2 - 2x_3 + x_4 = 2$
$ - 2x_2 - x_3 - x_4 = -2$
$x_1 - 3x_2 + 3x_4 = 4$

(d) $\begin{aligned} x_1 + 2x_2 - 3x_3 + x_4 &= 4 \\ -2x_1 + x_2 + 2x_3 + x_4 &= -1 \\ -x_1 + 3x_2 - x_3 + 2x_4 &= 3 \\ 4x_1 - 7x_2 - 5x_4 &= -5 \end{aligned}$

6. Find a basis for the null space of A.

(a) $A = \begin{bmatrix} 3 & -1 & 0 \\ 6 & -2 & 0 \\ 0 & 0 & 0 \end{bmatrix}$ (b) $A = \begin{bmatrix} 1 & -2 & 10 \\ 2 & -3 & 18 \\ 0 & -7 & 14 \end{bmatrix}$

(c) $A = \begin{bmatrix} 1 & 4 & 5 & 2 \\ 2 & 1 & 3 & 0 \\ -1 & 3 & 2 & 2 \end{bmatrix}$

(d) $A = \begin{bmatrix} 1 & 4 & 5 & 6 & 9 \\ 3 & -2 & 1 & 4 & -1 \\ -1 & 0 & -1 & -2 & -1 \\ 2 & 3 & 5 & 7 & 8 \end{bmatrix}$

(e) $A = \begin{bmatrix} 1 & -3 & 2 & 2 & 1 \\ 0 & 3 & 6 & 0 & -3 \\ 2 & -3 & -2 & 4 & 4 \\ 3 & -6 & 0 & 6 & 5 \\ -2 & 9 & 2 & -4 & -5 \end{bmatrix}$

7. In each part, a matrix in row echelon form is given. By inspection, find bases for the row and column spaces of A.

(a) $\begin{bmatrix} 1 & 0 & 2 \\ 0 & 0 & 1 \\ 0 & 0 & 0 \end{bmatrix}$ (b) $\begin{bmatrix} 1 & -3 & 0 & 0 \\ 0 & 1 & 0 & 0 \\ 0 & 0 & 0 & 0 \\ 0 & 0 & 0 & 0 \end{bmatrix}$

(c) $\begin{bmatrix} 1 & 2 & 4 & 5 \\ 0 & 1 & -3 & 0 \\ 0 & 0 & 1 & -3 \\ 0 & 0 & 0 & 1 \\ 0 & 0 & 0 & 0 \end{bmatrix}$ (d) $\begin{bmatrix} 1 & 2 & -1 & 5 \\ 0 & 1 & 4 & 3 \\ 0 & 0 & 1 & -7 \\ 0 & 0 & 0 & 1 \end{bmatrix}$

8. For the matrices in Exercise 6, find a basis for the row space of A by reducing the matrix to row echelon form.

9. By inspection, find a basis for the row space and a basis for the column space of each matrix.

(a) $\begin{bmatrix} 1 & 0 & 2 \\ 0 & 0 & 1 \\ 0 & 0 & 0 \end{bmatrix}$ (b) $\begin{bmatrix} 1 & -3 & 0 & 0 \\ 0 & 1 & 0 & 0 \\ 0 & 0 & 0 & 0 \\ 0 & 0 & 0 & 0 \end{bmatrix}$

(c) $\begin{bmatrix} 1 & 2 & 4 & 5 \\ 0 & 1 & -3 & 0 \\ 0 & 0 & 1 & -3 \\ 0 & 0 & 0 & 1 \\ 0 & 0 & 0 & 0 \end{bmatrix}$ (d) $\begin{bmatrix} 1 & 2 & -1 & 5 \\ 0 & 1 & 4 & 3 \\ 0 & 0 & 1 & -7 \\ 0 & 0 & 0 & 1 \end{bmatrix}$

10. For the matrices in Exercise 6, find a basis for the row space of A consisting entirely of row vectors of A.

11. Find a basis for the subspace of R^4 spanned by the given vectors.

(a) $(2, 4, -2, 3)$, $(-2, -2, 2, -4)$, $(1, 3, -1, 1)$

(b) $(-1, 1, -2, 0)$, $(3, 3, 6, 0)$, $(9, 0, 0, 3)$

(c) $(1, 1, 0, 0)$, $(0, 0, 1, 1)$, $(-2, 0, 2, 2)$, $(0, -3, 0, 3)$

12. Find a subset of the vectors that forms a basis for the space spanned by the vectors; then express each vector that is not in the basis as a linear combination of the basis vectors.

(a) $\mathbf{v}_1 = (1, -4, 1, 1)$, $\mathbf{v}_2 = (1, 0, -1, 1)$, $\mathbf{v}_3 = (1, 2, -2, 1)$, $\mathbf{v}_4 = (1, 2, 3, -2)$

(b) $\mathbf{v}_1 = (1, -2, 0, 3)$, $\mathbf{v}_2 = (2, -4, 0, 6)$, $\mathbf{v}_3 = (-1, 1, 2, 0)$, $\mathbf{v}_4 = (0, -1, 2, 3)$

(c) $\mathbf{v}_1 = (1, -1, 5, 2)$, $\mathbf{v}_2 = (-2, 3, 1, 0)$, $\mathbf{v}_3 = (4, -5, 9, 4)$, $\mathbf{v}_4 = (0, 4, 2, -3)$, $\mathbf{v}_5 = (-7, 18, 2, -8)$

13. Prove that the row vectors of an $n \times n$ invertible matrix A form a basis for R^n.

14. Construct a matrix whose null space consists of all linear combinations of the vectors

$$\mathbf{v}_1 = \begin{bmatrix} 1 \\ -1 \\ 3 \\ 2 \end{bmatrix} \quad \text{and} \quad \mathbf{v}_2 = \begin{bmatrix} 2 \\ 0 \\ -2 \\ 4 \end{bmatrix}$$

15. (a) Let

$$A = \begin{bmatrix} 0 & 1 & 0 \\ 1 & 0 & 0 \\ 0 & 0 & 0 \end{bmatrix}$$

Show that relative to an xyz-coordinate system in 3-space the null space of A consists of all points on the z-axis and that the column space consists of all points in the xy-plane (see the accompanying figure).

(b) Find a 3×3 matrix whose null space is the x-axis and whose column space is the yz-plane.

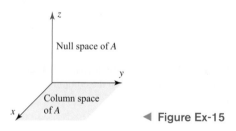

◀ Figure Ex-15

16. Find a 3×3 matrix whose null space is

(a) a point. (b) a line. (c) a plane.

17. (a) Find all 2×2 matrices whose null space is the line $x = 2t$, $y = -3t$.

(b) Sketch the null spaces of the following matrices:

$$A = \begin{bmatrix} 1 & 4 \\ 0 & 5 \end{bmatrix}, \quad B = \begin{bmatrix} 1 & 0 \\ 0 & 5 \end{bmatrix},$$

$$C = \begin{bmatrix} 6 & 2 \\ 3 & 1 \end{bmatrix}, \quad D = \begin{bmatrix} 0 & 0 \\ 0 & 0 \end{bmatrix}$$

18. The equation $x_1 + x_2 + x_3 = 1$ can be viewed as a linear system of one equation in three unknowns. Express its general solution as a particular solution plus the general solution of the corresponding homogeneous system. [*Suggestion:* Write the vectors in column form.]

19. Suppose that A and B are $n \times n$ matrices and A is invertible. Invent and prove a theorem that describes how the row spaces of AB and B are related.

True-False Exercises

In parts (a)–(j) determine whether the statement is true or false, and justify your answer.

(a) The span of $\mathbf{v}_1, \ldots, \mathbf{v}_n$ is the column space of the matrix whose column vectors are $\mathbf{v}_1, \ldots, \mathbf{v}_n$.

(b) The column space of a matrix A is the set of solutions of $A\mathbf{x} = \mathbf{b}$.

(c) If R is the reduced row echelon form of A, then those column vectors of R that contain the leading 1's form a basis for the column space of A.

(d) The set of nonzero row vectors of a matrix A is a basis for the row space of A.

(e) If A and B are $n \times n$ matrices that have the same row space, then A and B have the same column space.

(f) If E is an $m \times m$ elementary matrix and A is an $m \times n$ matrix, then the null space of EA is the same as the null space of A.

(g) If E is an $m \times m$ elementary matrix and A is an $m \times n$ matrix, then the row space of EA is the same as the row space of A.

(h) If E is an $m \times m$ elementary matrix and A is an $m \times n$ matrix, then the column space of EA is the same as the column space of A.

(i) The system $A\mathbf{x} = \mathbf{b}$ is inconsistent if and only if $\mathbf{b}$ is not in the column space of A.

(j) There is an invertible matrix A and a singular matrix B such that the row spaces of A and B are the same.

4.8 Rank, Nullity, and the Fundamental Matrix Spaces

In the last section we investigated relationships between a system of linear equations and the row space, column space, and null space of its coefficient matrix. In this section we will be concerned with the dimensions of those spaces. The results we obtain will provide a deeper insight into the relationship between a linear system and its coefficient matrix.

Row and Column Spaces Have Equal Dimensions

In Examples 6 and 7 of Section 4.7 we found that the row and column spaces of the matrix

$$A = \begin{bmatrix} 1 & -3 & 4 & -2 & 5 & 4 \\ 2 & -6 & 9 & -1 & 8 & 2 \\ 2 & -6 & 9 & -1 & 9 & 7 \\ -1 & 3 & -4 & 2 & -5 & -4 \end{bmatrix}$$

both have three basis vectors and hence are both three-dimensional. The fact that these spaces have the same dimension is not accidental, but rather a consequence of the following theorem.

THEOREM 4.8.1 *The row space and column space of a matrix A have the same dimension.*

Proof Let R be any row echelon form of A. It follows from Theorems 4.7.4 and 4.7.6b that

$$\dim(\text{row space of } A) = \dim(\text{row space of } R)$$
$$\dim(\text{column space of } A) = \dim(\text{column space of } R)$$

so it suffices to show that the row and column spaces of R have the same dimension. But the dimension of the row space of R is the number of nonzero rows, and by Theorem 4.7.5 the dimension of the column space of R is the number of leading 1's. Since these two numbers are the same, the row and column space have the same dimension. ◄

Rank and Nullity   The dimensions of the row space, column space, and null space of a matrix are such important numbers that there is some notation and terminology associated with them.

The proof of Theorem 4.8.1 shows that the rank of A can be interpreted as the number of leading 1's in any row echelon form of A.

DEFINITION 1 The common dimension of the row space and column space of a matrix A is called the **rank** of A and is denoted by rank(A); the dimension of the null space of A is called the **nullity** of A and is denoted by nullity(A).

▶ **EXAMPLE 1 Rank and Nullity of a 4 × 6 Matrix**

Find the rank and nullity of the matrix

$$A = \begin{bmatrix} -1 & 2 & 0 & 4 & 5 & -3 \\ 3 & -7 & 2 & 0 & 1 & 4 \\ 2 & -5 & 2 & 4 & 6 & 1 \\ 4 & -9 & 2 & -4 & -4 & 7 \end{bmatrix}$$

Solution The reduced row echelon form of A is

$$\begin{bmatrix} 1 & 0 & -4 & -28 & -37 & 13 \\ 0 & 1 & -2 & -12 & -16 & 5 \\ 0 & 0 & 0 & 0 & 0 & 0 \\ 0 & 0 & 0 & 0 & 0 & 0 \end{bmatrix} \quad (1)$$

(verify). Since this matrix has two leading 1's, its row and column spaces are two-dimensional and rank(A) = 2. To find the nullity of A, we must find the dimension of the solution space of the linear system $A\mathbf{x} = \mathbf{0}$. This system can be solved by reducing its augmented matrix to reduced row echelon form. The resulting matrix will be identical to (1), except that it will have an additional last column of zeros, and hence the corresponding system of equations will be

$$x_1 - 4x_3 - 28x_4 - 37x_5 + 13x_6 = 0$$
$$x_2 - 2x_3 - 12x_4 - 16x_5 + 5x_6 = 0$$

Solving these equations for the leading variables yields

$$x_1 = 4x_3 + 28x_4 + 37x_5 - 13x_6$$
$$x_2 = 2x_3 + 12x_4 + 16x_5 - 5x_6 \quad (2)$$

from which we obtain the general solution

$$x_1 = 4r + 28s + 37t - 13u$$
$$x_2 = 2r + 12s + 16t - 5u$$
$$x_3 = r$$
$$x_4 = s$$
$$x_5 = t$$
$$x_6 = u$$

or in column vector form

$$\begin{bmatrix} x_1 \\ x_2 \\ x_3 \\ x_4 \\ x_5 \\ x_6 \end{bmatrix} = r \begin{bmatrix} 4 \\ 2 \\ 1 \\ 0 \\ 0 \\ 0 \end{bmatrix} + s \begin{bmatrix} 28 \\ 12 \\ 0 \\ 1 \\ 0 \\ 0 \end{bmatrix} + t \begin{bmatrix} 37 \\ 16 \\ 0 \\ 0 \\ 1 \\ 0 \end{bmatrix} + u \begin{bmatrix} -13 \\ -5 \\ 0 \\ 0 \\ 0 \\ 1 \end{bmatrix} \qquad (3)$$

Because the four vectors on the right side of (3) form a basis for the solution space, nullity$(A) = 4$.

▶ **EXAMPLE 2** **Maximum Value for Rank**

What is the maximum possible rank of an $m \times n$ matrix A that is not square?

Solution Since the row vectors of A lie in R^n and the column vectors in R^m, the row space of A is at most n-dimensional and the column space is at most m-dimensional. Since the rank of A is the common dimension of its row and column space, it follows that the rank is at most the smaller of m and n. We denote this by writing

$$\text{rank}(A) \leq \min(m, n)$$

in which $\min(m, n)$ is the minimum of m and n. ◀

The following theorem establishes an important relationship between the rank and nullity of a matrix.

THEOREM 4.8.2 *Dimension Theorem for Matrices*

If A is a matrix with n columns, then

$$\text{rank}(A) + \text{nullity}(A) = n \qquad (4)$$

Proof Since A has n columns, the homogeneous linear system $A\mathbf{x} = \mathbf{0}$ has n unknowns (variables). These fall into two distinct categories: the leading variables and the free variables. Thus,

$$\begin{bmatrix} \text{number of leading} \\ \text{variables} \end{bmatrix} + \begin{bmatrix} \text{number of free} \\ \text{variables} \end{bmatrix} = n$$

But the number of leading variables is the same as the number of leading 1's in the reduced row echelon form of A, which is the rank of A; and the number of free variables is the same as the number of parameters in the general solution of $A\mathbf{x} = \mathbf{0}$, which is the nullity of A. This yields Formula (4). ◀

▶ **EXAMPLE 3** **The Sum of Rank and Nullity**

The matrix

$$A = \begin{bmatrix} -1 & 2 & 0 & 4 & 5 & -3 \\ 3 & -7 & 2 & 0 & 1 & 4 \\ 2 & -5 & 2 & 4 & 6 & 1 \\ 4 & -9 & 2 & -4 & -4 & 7 \end{bmatrix}$$

has 6 columns, so
$$\text{rank}(A) + \text{nullity}(A) = 6$$
This is consistent with Example 1, where we showed that
$$\text{rank}(A) = 2 \quad \text{and} \quad \text{nullity}(A) = 4 \blacktriangleleft$$

The following theorem, which summarizes results already obtained, interprets rank and nullity in the context of a homogeneous linear system.

THEOREM 4.8.3 *If A is an m × n matrix, then*
(a) rank(A) = the number of leading variables in the general solution of $A\mathbf{x} = \mathbf{0}$.
(b) nullity(A) = the number of parameters in the general solution of $A\mathbf{x} = \mathbf{0}$.

▶ **EXAMPLE 4 Number of Parameters in a General Solution**

Find the number of parameters in the general solution of $A\mathbf{x} = \mathbf{0}$ if A is a 5×7 matrix of rank 3.

Solution From (4),
$$\text{nullity}(A) = n - \text{rank}(A) = 7 - 3 = 4$$
Thus there are four parameters. ◀

Equivalence Theorem   In Theorem 2.3.8 we listed seven results that are equivalent to the invertibility of a square matrix A. We are now in a position to add eight more results to that list to produce a single theorem that summarizes most of the topics we have covered thus far.

THEOREM 4.8.4 Equivalent Statements
If A is an n × n matrix, then the following statements are equivalent.
(a)   A is invertible.
(b) $A\mathbf{x} = \mathbf{0}$ has only the trivial solution.
(c)   The reduced row echelon form of A is I_n.
(d)   A is expressible as a product of elementary matrices.
(e) $A\mathbf{x} = \mathbf{b}$ is consistent for every n × 1 matrix $\mathbf{b}$.
(f) $A\mathbf{x} = \mathbf{b}$ has exactly one solution for every n × 1 matrix $\mathbf{b}$.
(g) $\det(A) \neq 0$.
(h)   The column vectors of A are linearly independent.
(i)   The row vectors of A are linearly independent.
(j)   The column vectors of A span R^n.
(k)   The row vectors of A span R^n.
(l)   The column vectors of A form a basis for R^n.
(m) The row vectors of A form a basis for R^n.
(n)   A has rank n.
(o)   A has nullity 0.

Proof The equivalence of (*h*) through (*m*) follows from Theorem 4.5.4 (we omit the details). To complete the proof we will show that (*b*), (*n*), and (*o*) are equivalent by proving the chain of implications (*b*) $\Rightarrow$ (*o*) $\Rightarrow$ (*n*) $\Rightarrow$ (*b*).

(b) $\Rightarrow$ (o) If $A\mathbf{x} = \mathbf{0}$ has only the trivial solution, then there are no parameters in that solution, so nullity(A) = 0 by Theorem 4.8.3*b*.

(o) $\Rightarrow$ (n) Theorem 4.8.2.

(n) $\Rightarrow$ (b) If A has rank n, then Theorem 4.8.3*a* implies that there are n leading variables (hence no free variables) in the general solution of $A\mathbf{x} = \mathbf{0}$. This leaves the trivial solution as the only possibility. ◂

Overdetermined and Underdetermined Systems

In engineering and other applications, the occurrence of an overdetermined or underdetermined linear system often signals that one or more variables were omitted in formulating the problem or that extraneous variables were included. This often leads to some kind of undesirable physical result.

In many applications the equations in a linear system correspond to physical constraints or conditions that must be satisfied. In general, the most desirable systems are those that have the same number of constraints as unknowns, since such systems often have a unique solution. Unfortunately, it is not always possible to match the number of constraints and unknowns, so researchers are often faced with linear systems that have more constraints than unknowns, called ***overdetermined systems***, or with fewer constraints than unknowns, called ***underdetermined systems***. The following two theorems will help us to analyze both overdetermined and underdetermined systems.

THEOREM 4.8.5 *If $A\mathbf{x} = \mathbf{b}$ is a consistent linear system of m equations in n unknowns, and if A has rank r, then the general solution of the system contains $n - r$ parameters.*

Proof It follows from Theorem 4.7.2 that the number of parameters is equal to the nullity of A, which, by Theorem 4.8.2, is $n - r$. ◂

THEOREM 4.8.6 *Let A be an $m \times n$ matrix.*

(a) (***Overdetermined Case***). *If $m > n$, then the linear system $A\mathbf{x} = \mathbf{b}$ is inconsistent for at least one vector $\mathbf{b}$ in R^n.*

(b) (***Underdetermined Case***). *If $m < n$, then for each vector $\mathbf{b}$ in R^m the linear system $A\mathbf{x} = \mathbf{b}$ is either inconsistent or has infinitely many solutions.*

Proof (a) Assume that $m > n$, in which case the column vectors of A cannot span R^m (fewer vectors than the dimension of R^m). Thus, there is at least one vector $\mathbf{b}$ in R^m that is not in the column space of A, and for that $\mathbf{b}$ the system $A\mathbf{x} = \mathbf{b}$ is inconsistent by Theorem 4.7.1.

Proof (b) Assume that $m < n$. For each vector $\mathbf{b}$ in R^n there are two possibilities: either the system $A\mathbf{x} = \mathbf{b}$ is consistent or it is inconsistent. If it is inconsistent, then the proof is complete. If it is consistent, then Theorem 4.8.5 implies that the general solution has $n - r$ parameters, where $r = \text{rank}(A)$. But rank(A) is the smaller of m and n, so

$$n - r = n - m > 0$$

This means that the general solution has at least one parameter and hence there are infinitely many solutions. ◂

► EXAMPLE 5 Overdetermined and Underdetermined Systems

(a) What can you say about the solutions of an overdetermined system $A\mathbf{x} = \mathbf{b}$ of 7 equations in 5 unknowns in which A has rank $r = 4$?

(b) What can you say about the solutions of an underdetermined system $A\mathbf{x} = \mathbf{b}$ of 5 equations in 7 unknowns in which A has rank $r = 4$?

Solution (a) The system is consistent for some vector $\mathbf{b}$ in R^7, and for any such $\mathbf{b}$ the number of parameters in the general solution is $n - r = 5 - 4 = 1$.

Solution (b) The system may be consistent or inconsistent, but if it is consistent for the vector $\mathbf{b}$ in R^5, then the general solution has $n - r = 7 - 4 = 3$ parameters.

► EXAMPLE 6 An Overdetermined System

The linear system
$$\begin{aligned} x_1 - 2x_2 &= b_1 \\ x_1 - x_2 &= b_2 \\ x_1 + x_2 &= b_3 \\ x_1 + 2x_2 &= b_4 \\ x_1 + 3x_2 &= b_5 \end{aligned}$$

is overdetermined, so it cannot be consistent for all possible values of $b_1, b_2, b_3, b_4,$ and b_5. Exact conditions under which the system is consistent can be obtained by solving the linear system by Gauss–Jordan elimination. We leave it for you to show that the augmented matrix is row equivalent to

$$\begin{bmatrix} 1 & 0 & 2b_2 - b_1 \\ 0 & 1 & b_2 - b_1 \\ 0 & 0 & b_3 - 3b_2 + 2b_1 \\ 0 & 0 & b_4 - 4b_2 + 3b_1 \\ 0 & 0 & b_5 - 5b_2 + 4b_1 \end{bmatrix} \tag{5}$$

Thus, the system is consistent if and only if $b_1, b_2, b_3, b_4,$ and b_5 satisfy the conditions

$$\begin{aligned} 2b_1 - 3b_2 + b_3 &= 0 \\ 3b_1 - 4b_2 + b_4 &= 0 \\ 4b_1 - 5b_2 + b_5 &= 0 \end{aligned}$$

Solving this homogeneous linear system yields

$$b_1 = 5r - 4s, \quad b_2 = 4r - 3s, \quad b_3 = 2r - s, \quad b_4 = r, \quad b_5 = s$$

where r and s are arbitrary. ◄

Remark The coefficient matrix for the linear system in the last example has $n = 2$ columns, and it has rank $r = 2$ because there are two nonzero rows in its reduced row echelon form. This implies that when the system is consistent its general solution will contain $n - r = 0$ parameters; that is, the solution will be unique. With a moment's thought, you should be able to see that this is so from (5).

The Fundamental Spaces of a Matrix

There are six important vector spaces associated with a matrix A and its transpose A^T:

row space of A row space of A^T
column space of A column space of A^T
null space of A null space of A^T

4.8 Rank, Nullity, and the Fundamental Matrix Spaces

However, transposing a matrix converts row vectors into column vectors and conversely, so except for a difference in notation, the row space of A^T is the same as the column space of A, and the column space of A^T is the same as the row space of A. Thus, of the six spaces listed above, only the following four are distinct:

> If A is an $m \times n$ matrix, then the row space and null space of A are subspaces of R^n, and the column space of A and the null space of A^T are subspaces of R^m.

$$\text{row space of } A \qquad \text{column space of } A$$
$$\text{null space of } A \qquad \text{null space of } A^T$$

These are called the ***fundamental spaces*** of a matrix A. We will conclude this section by discussing how these four subspaces are related.

Let us focus for a moment on the matrix A^T. Since the row space and column space of a matrix have the same dimension, and since transposing a matrix converts its columns to rows and its rows to columns, the following result should not be surprising.

THEOREM 4.8.7 *If A is any matrix, then $\text{rank}(A) = \text{rank}(A^T)$.*

Proof

$$\text{rank}(A) = \dim(\text{row space of } A) = \dim(\text{column space of } A^T) = \text{rank}(A^T). \blacktriangleleft$$

This result has some important implications. For example, if A is an $m \times n$ matrix, then applying Formula (4) to the matrix A^T and using the fact that this matrix has m columns yields

$$\text{rank}(A^T) + \text{nullity}(A^T) = m$$

which, by virtue of Theorem 4.8.7, can be rewritten as

$$\text{rank}(A) + \text{nullity}(A^T) = m \qquad (6)$$

This alternative form of Formula (4) in Theorem 4.8.2 makes it possible to express the dimensions of all four fundamental spaces in terms of the size and rank of A. Specifically, if $\text{rank}(A) = r$, then

$$\dim[\text{row}(A)] = r \qquad \dim[\text{col}(A)] = r$$
$$\dim[\text{null}(A)] = n - r \qquad \dim[\text{null}(A^T)] = m - r \qquad (7)$$

The four formulas in (7) provide an *algebraic* relationship between the size of a matrix and the dimensions of its fundamental spaces. Our next objective is to find a *geometric* relationship between the fundamental spaces themselves. For this purpose recall from Theorem 3.4.3 that if A is an $m \times n$ matrix, then the null space of A consists of those vectors that are orthogonal to each of the row vectors of A. To develop that idea in more detail, we make the following definition.

DEFINITION 2 If W is a subspace of R^n, then the set of all vectors in R^n that are orthogonal to every vector in W is called the ***orthogonal complement*** of W and is denoted by the symbol $W^\perp$.

The following theorem lists three basic properties of orthogonal complements. We will omit the formal proof because a more general version of this theorem will be given later in the text.

THEOREM 4.8.8 *If W is a subspace of R^n, then*:
(a) $W^\perp$ *is a subspace of R^n.*
(b) *The only vector common to W and $W^\perp$ is* **0**.
(c) *The orthogonal complement of $W^\perp$ is W.*

▶ **EXAMPLE 7 Orthogonal Complements**

In R^2 the orthogonal complement of a line W through the origin is the line through the origin that is perpendicular to W (Figure 4.8.1a); and in R^3 the orthogonal complement of a plane W through the origin is the line through the origin that is perpendicular to that plane (Figure 4.8.1b). ◀

Explain why {**0**} and R^n are orthogonal complements.

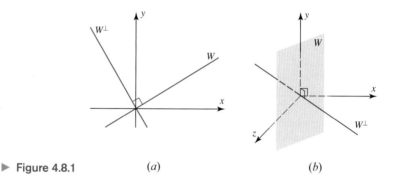

▶ Figure 4.8.1   (a)   (b)

A Geometric Link Between the Fundamental Spaces

The following theorem provides a geometric link between the fundamental spaces of a matrix. Part (a) is essentially a restatement of Theorem 3.4.3 in the language of orthogonal complements, and part (b), whose proof is left as an exercise, follows from part (a). The essential idea of the theorem is illustrated in Figure 4.8.2.

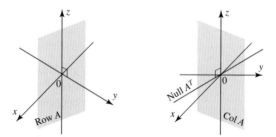

▶ Figure 4.8.2

THEOREM 4.8.9 *If A is an $m \times n$ matrix, then*:
(a) *The null space of A and the row space of A are orthogonal complements in R^n.*
(b) *The null space of A^T and the column space of A are orthogonal complements in R^m.*

More on the Equivalence Theorem

As our final result in this section, we will add two more statements to Theorem 4.8.4. We leave the proof that those statements are equivalent to the rest as an exercise.

> **THEOREM 4.8.10** Equivalent Statements
>
> *If A is an $n \times n$ matrix, then the following statements are equivalent.*
>
> (a) A is invertible.
> (b) $A\mathbf{x} = \mathbf{0}$ has only the trivial solution.
> (c) The reduced row echelon form of A is I_n.
> (d) A is expressible as a product of elementary matrices.
> (e) $A\mathbf{x} = \mathbf{b}$ is consistent for every $n \times 1$ matrix $\mathbf{b}$.
> (f) $A\mathbf{x} = \mathbf{b}$ has exactly one solution for every $n \times 1$ matrix $\mathbf{b}$.
> (g) $\det(A) \neq 0$.
> (h) The column vectors of A are linearly independent.
> (i) The row vectors of A are linearly independent.
> (j) The column vectors of A span R^n.
> (k) The row vectors of A span R^n.
> (l) The column vectors of A form a basis for R^n.
> (m) The row vectors of A form a basis for R^n.
> (n) A has rank n.
> (o) A has nullity 0.
> (p) The orthogonal complement of the null space of A is R^n.
> (q) The orthogonal complement of the row space of A is $\{\mathbf{0}\}$.

Applications of Rank

The advent of the Internet has stimulated research on finding efficient methods for transmitting large amounts of digital data over communications lines with limited bandwidths. Digital data are commonly stored in matrix form, and many techniques for improving transmission speed use the rank of a matrix in some way. Rank plays a role because it measures the "redundancy" in a matrix in the sense that if A is an $m \times n$ matrix of rank k, then $n - k$ of the column vectors and $m - k$ of the row vectors can be expressed in terms of k linearly independent column or row vectors. The essential idea in many data compression schemes is to approximate the original data set by a data set with smaller rank that conveys nearly the same information, then eliminate redundant vectors in the approximating set to speed up the transmission time.

Concept Review

- Rank
- Nullity
- Dimension Theorem
- Overdetermined system
- Underdetermined system
- Fundamental spaces of a matrix
- Relationships among the fundamental spaces
- Orthogonal complement
- Equivalent characterizations of invertible matrices

Skills

- Find the rank and nullity of a matrix.
- Find the dimension of the row space of a matrix.

Exercise Set 4.8

1. Verify that $\text{rank}(A) = \text{rank}(A^T)$.

$$A = \begin{bmatrix} 1 & 2 & 4 & 0 \\ -3 & 1 & 5 & 2 \\ -2 & 3 & 9 & 2 \end{bmatrix}$$

2. Find the rank and nullity of the matrix; then verify that the values obtained satisfy Formula (4) in the Dimension Theorem.

(a) $A = \begin{bmatrix} 2 & -1 & 3 \\ 4 & -2 & 1 \\ 2 & 1 & 0 \end{bmatrix}$ (b) $A = \begin{bmatrix} 2 & 0 & -1 \\ 4 & 0 & -2 \\ 0 & 0 & 0 \end{bmatrix}$

(c) $A = \begin{bmatrix} 1 & 3 & 1 & 4 \\ 2 & 4 & 2 & 0 \\ -1 & -3 & 0 & 5 \end{bmatrix}$

(d) $A = \begin{bmatrix} 1 & 4 & 5 & 6 & 9 \\ 3 & -2 & 1 & 4 & -1 \\ -1 & 0 & -1 & -2 & -1 \\ 2 & 3 & 5 & 7 & 8 \end{bmatrix}$

(e) $A = \begin{bmatrix} 1 & -3 & 2 & 2 & 1 \\ 0 & 3 & 6 & 0 & -3 \\ 2 & -3 & -2 & 4 & 4 \\ 3 & -6 & 0 & 6 & 5 \\ -2 & 9 & 2 & -4 & -5 \end{bmatrix}$

3. In each part of Exercise 2, use the results obtained to find the number of leading variables and the number of parameters in the solution of $A\mathbf{x} = \mathbf{0}$ without solving the system.

4. In each part, use the information in the table to find the dimension of the row space of A, column space of A, null space of A, and null space of A^T.

	(a)	(b)	(c)	(d)	(e)	(f)	(g)
Size of A	3×3	3×3	3×3	5×6	6×5	4×4	6×5
Rank(A)	3	2	1	2	2	0	5

5. In each part, find the largest possible value for the rank of A and the smallest possible value for the nullity of A.

(a) A is 4×6 (b) A is 5×5 (c) A is 6×4

6. If A is an $m \times n$ matrix, what is the largest possible value for its rank and the smallest possible value for its nullity?

7. In each part, use the information in the table to determine whether the linear system $A\mathbf{x} = \mathbf{b}$ is consistent. If so, state the number of parameters in its general solution.

	(a)	(b)	(c)	(d)	(e)	(f)	(g)
Size of A	3×3	3×3	3×3	5×9	5×9	4×4	6×2
Rank(A)	3	2	1	2	2	0	2
Rank$[A \mid \mathbf{b}]$	3	3	1	2	3	0	2

8. For each of the matrices in Exercise 7, find the nullity of A, and determine the number of parameters in the general solution of the homogeneous linear system $A\mathbf{x} = \mathbf{0}$.

9. What conditions must be satisfied by b_1, b_2, b_3, b_4, and b_5 for the overdetermined linear system

$$x_1 + x_2 = b_1$$
$$x_1 + 2x_2 = b_2$$
$$x_1 - x_2 = b_3$$
$$2x_1 + 4x_2 = b_4$$
$$x_1 + 3x_2 = b_5$$

to be consistent?

10. Let

$$A = \begin{bmatrix} a_{11} & a_{12} & a_{13} \\ a_{21} & a_{22} & a_{23} \end{bmatrix}$$

Show that A has rank 2 if and only if one or more of the determinants

$$\begin{vmatrix} a_{11} & a_{12} \\ a_{21} & a_{22} \end{vmatrix}, \quad \begin{vmatrix} a_{11} & a_{13} \\ a_{21} & a_{23} \end{vmatrix}, \quad \begin{vmatrix} a_{12} & a_{13} \\ a_{22} & a_{23} \end{vmatrix}$$

is nonzero.

11. Suppose that A is a 3×3 matrix whose null space is a line through the origin in 3-space. Can the row or column space of A also be a line through the origin? Explain.

12. Discuss how the rank of A varies with t.

(a) $A = \begin{bmatrix} 1 & -1 & t \\ 1 & t & -1 \\ t^2 & 1 & -1 \end{bmatrix}$ (b) $A = \begin{bmatrix} t & 3 & -1 \\ 3 & 6 & -2 \\ -1 & -3 & t \end{bmatrix}$

13. Are there values of r and s for which

$$\begin{bmatrix} 1 & 0 & 0 \\ 0 & r-2 & 2 \\ 0 & s-1 & r+2 \\ 0 & 0 & 3 \end{bmatrix}$$

has rank 1? Has rank 2? If so, find those values.

14. Use the result in Exercise 10 to show that the set of points (x, y, z) in R^3 for which the matrix

$$\begin{bmatrix} x & y & z \\ 1 & x & y \end{bmatrix}$$

has rank 1 is the curve with parametric equations $x = t$, $y = t^2$, $z = t^3$.

15. Prove: If $k \neq 0$, then A and kA have the same rank.

16. (a) Give an example of a 3×3 matrix whose column space is a plane through the origin in 3-space.

(b) What kind of geometric object is the null space of your matrix?

(c) What kind of geometric object is the row space of your matrix?

17. (a) If A is a 3×5 matrix, then the number of leading 1's in the reduced row echelon form of A is at most _____. Why?

(b) If A is a 3×5 matrix, then the number of parameters in the general solution of $A\mathbf{x} = \mathbf{0}$ is at most _____. Why?

(c) If A is a 5×3 matrix, then the number of leading 1's in the reduced row echelon form of A is at most _____. Why?

(d) If A is a 5×3 matrix, then the number of parameters in the general solution of $A\mathbf{x} = \mathbf{0}$ is at most _____. Why?

18. (a) If A is a 3×5 matrix, then the rank of A is at most _____. Why?

(b) If A is a 3×5 matrix, then the nullity of A is at most _____. Why?

(c) If A is a 3×5 matrix, then the rank of A^T is at most _____. Why?

(d) If A is a 3×5 matrix, then the nullity of A^T is at most _____. Why?

19. Find matrices A and B for which $\text{rank}(A) = \text{rank}(B)$, but $\text{rank}(A^2) \neq \text{rank}(B^2)$.

20. Prove: If a matrix A is not square, then either the row vectors or the column vectors of A are linearly dependent.

True-False Exercises

In parts (a)–(j) determine whether the statement is true or false, and justify your answer.

(a) Either the row vectors or the column vectors of a square matrix are linearly independent.

(b) A matrix with linearly independent row vectors and linearly independent column vectors is square.

(c) The nullity of a nonzero $m \times n$ matrix is at most m.

(d) Adding one additional column to a matrix increases its rank by one.

(e) The nullity of a square matrix with linearly dependent rows is at least one.

(f) If A is square and $A\mathbf{x} = \mathbf{b}$ is inconsistent for some vector $\mathbf{b}$, then the nullity of A is zero.

(g) If a matrix A has more rows than columns, then the dimension of the row space is greater than the dimension of the column space.

(h) If $\text{rank}(A^T) = \text{rank}(A)$, then A is square.

(i) There is no 3×3 matrix whose row space and null space are both lines in 3-space.

(j) If V is a subspace of R^n and W is a subspace of V, then $W^\perp$ is a subspace of $V^\perp$.

4.9 Matrix Transformations from R^n to R^m

In this section we will study functions of the form $\mathbf{w} = F(\mathbf{x})$, where the independent variable $\mathbf{x}$ is a vector in R^n and the dependent variable $\mathbf{w}$ is a vector in R^m. We will concentrate on a special class of such functions called "matrix transformations." Such transformations are fundamental in the study of linear algebra and have important applications in physics, engineering, social sciences, and various branches of mathematics.

Functions and Transformations

Recall that a ***function*** is a rule that associates with each element of a set A one and only one element in a set B. If f associates the element b with the element a, then we write

$$b = f(a)$$

and we say that b is the ***image*** of a under f or that $f(a)$ is the ***value*** of f at a. The set A is called the ***domain*** of f and the set B the ***codomain*** of f (Figure 4.9.1). The subset of the codomain that consists of all images of points in the domain is called the ***range*** of f.

For many common functions the domain and codomain are sets of real numbers, but in this text we will be concerned with functions for which the domain and codomain are vector spaces.

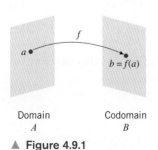

▲ Figure 4.9.1

DEFINITION 1 If V and W are vector spaces, and if f is a function with domain V and codomain W, then we say that f is a ***transformation*** from V to W or that f ***maps*** V to W, which we denote by writing

$$f : V \to W$$

In the special case where $V = W$, the transformation is also called an ***operator*** on V.

In this section we will be concerned exclusively with transformations from R^n to R^m; transformations of general vector spaces will be considered in a later section. To illustrate one way in which such transformations can arise, suppose that $f_1, f_2, \ldots, f_m$ are real-valued functions of n variables, say

$$\begin{aligned} w_1 &= f_1(x_1, x_2, \ldots, x_n) \\ w_2 &= f_2(x_1, x_2, \ldots, x_n) \\ &\vdots \\ w_m &= f_m(x_1, x_2, \ldots, x_n) \end{aligned} \qquad (1)$$

These m equations assign a unique point $(w_1, w_2, \ldots, w_m)$ in R^m to each point $(x_1, x_2, \ldots, x_n)$ in R^n and thus define a transformation from R^n to R^m. If we denote this transformation by T, then $T : R^n \to R^m$ and

$$T(x_1, x_2, \ldots, x_n) = (w_1, w_2, \ldots, w_m)$$

Matrix Transformations In the special case where the equations in (1) are linear, they can be expressed in the form

$$\begin{aligned} w_1 &= a_{11}x_1 + a_{12}x_2 + \cdots + a_{1n}x_n \\ w_2 &= a_{21}x_1 + a_{22}x_2 + \cdots + a_{2n}x_n \\ &\vdots \qquad \vdots \qquad \vdots \qquad \vdots \\ w_m &= a_{m1}x_1 + a_{m2}x_2 + \cdots + a_{mn}x_n \end{aligned} \qquad (2)$$

which we can write in matrix notation as

$$\begin{bmatrix} w_1 \\ w_2 \\ \vdots \\ w_m \end{bmatrix} = \begin{bmatrix} a_{11} & a_{12} & \cdots & a_{1n} \\ a_{21} & a_{22} & \cdots & a_{2n} \\ \vdots & \vdots & & \vdots \\ a_{m1} & a_{m2} & \cdots & a_{mn} \end{bmatrix} \begin{bmatrix} x_1 \\ x_2 \\ \vdots \\ x_n \end{bmatrix} \qquad (3)$$

or more briefly as

$$\mathbf{w} = A\mathbf{x} \qquad (4)$$

Although we could view this as a linear system, we will view it instead as a transformation that maps the column vector $\mathbf{x}$ in R^n into the column vector $\mathbf{w}$ in R^m by multiplying $\mathbf{x}$ on the left by A. We call this a ***matrix transformation*** (or ***matrix operator*** if $m = n$), and we denote it by $T_A : R^n \to R^m$. With this notation, Equation (4) can be expressed as

$$\mathbf{w} = T_A(\mathbf{x}) \qquad (5)$$

The matrix transformation T_A is called ***multiplication by*** A, and the matrix A is called the ***standard matrix*** for the transformation.

We will also find it convenient, on occasion, to express (5) in the schematic form

$$\mathbf{x} \xrightarrow{T_A} \mathbf{w} \qquad (6)$$

which is read "T_A maps $\mathbf{x}$ into $\mathbf{w}$."

▶ **EXAMPLE 1** **A Matrix Transformation from R^4 to R^3**

The matrix transformation $T: R^4 \to R^3$ defined by the equations

$$\begin{aligned} w_1 &= 2x_1 - 3x_2 + x_3 - 5x_4 \\ w_2 &= 4x_1 + x_2 - 2x_3 + x_4 \\ w_3 &= 5x_1 - x_2 + 4x_3 \end{aligned} \quad (7)$$

can be expressed in matrix form as

$$\begin{bmatrix} w_1 \\ w_2 \\ w_3 \end{bmatrix} = \begin{bmatrix} 2 & -3 & 1 & -5 \\ 4 & 1 & -2 & 1 \\ 5 & -1 & 4 & 0 \end{bmatrix} \begin{bmatrix} x_1 \\ x_2 \\ x_3 \\ x_4 \end{bmatrix} \quad (8)$$

so the standard matrix for T is

$$A = \begin{bmatrix} 2 & -3 & 1 & -5 \\ 4 & 1 & -2 & 1 \\ 5 & -1 & 4 & 0 \end{bmatrix}$$

The image of a point (x_1, x_2, x_3, x_4) can be computed directly from the defining equations (7) or from (8) by matrix multiplication. For example, if

$$(x_1, x_2, x_3, x_4) = (1, -3, 0, 2)$$

then substituting in (7) yields $w_1 = 1$, $w_2 = 3$, $w_3 = 8$ (verify), or alternatively from (8),

$$\begin{bmatrix} w_1 \\ w_2 \\ w_3 \end{bmatrix} = \begin{bmatrix} 2 & -3 & 1 & -5 \\ 4 & 1 & -2 & 1 \\ 5 & -1 & 4 & 0 \end{bmatrix} \begin{bmatrix} 1 \\ -3 \\ 0 \\ 2 \end{bmatrix} = \begin{bmatrix} 1 \\ 3 \\ 8 \end{bmatrix} \quad ◀$$

Some Notational Matters

Sometimes we will want to denote a matrix transformation without giving a name to the matrix itself. In such cases we will denote the standard matrix for $T: R^n \to R^m$ by the symbol $[T]$. Thus, the equation

$$T(\mathbf{x}) = [T]\mathbf{x} \quad (9)$$

is simply the statement that T is a matrix transformation with standard matrix $[T]$, and the image of $\mathbf{x}$ under this transformation is the product of the matrix $[T]$ and the column vector $\mathbf{x}$.

Properties of Matrix Transformations

The following theorem lists four basic properties of matrix transformations that follow from properties of matrix multiplication.

THEOREM 4.9.1 *For every matrix A the matrix transformation $T_A: R^n \to R^m$ has the following properties for all vectors $\mathbf{u}$ and $\mathbf{v}$ in R^n and for every scalar k:*

(a) $T_A(\mathbf{0}) = \mathbf{0}$
(b) $T_A(k\mathbf{u}) = kT_A(\mathbf{u})$ **[Homogeneity property]**
(c) $T_A(\mathbf{u} + \mathbf{v}) = T_A(\mathbf{u}) + T_A(\mathbf{v})$ **[Additivity property]**
(d) $T_A(\mathbf{u} - \mathbf{v}) = T_A(\mathbf{u}) - T_A(\mathbf{v})$

Proof All four parts are restatements of familiar properties of matrix multiplication:

$$A\mathbf{0} = \mathbf{0}, \quad A(k\mathbf{u}) = k(A\mathbf{u}), \quad A(\mathbf{u} + \mathbf{v}) = A\mathbf{u} + A\mathbf{v}, \quad A(\mathbf{u} - \mathbf{v}) = A\mathbf{u} - A\mathbf{v} \blacktriangleleft$$

It follows from Theorem 4.9.1 that a matrix transformation maps linear combinations of vectors in R^n into the corresponding linear combinations in R^m in the sense that

$$T_A(k_1\mathbf{u}_1 + k_2\mathbf{u}_2 + \cdots + k_r\mathbf{u}_r) = k_1 T_A(\mathbf{u}_1) + k_2 T_A(\mathbf{u}_2) + \cdots + k_r T_A(\mathbf{u}_r) \quad (10)$$

Depending on whether n-tuples and m-tuples are regarded as vectors or points, the geometric effect of a matrix transformation $T_A: R^n \to R^m$ is to map each vector (point) in R^n into a vector (point) in R^m (Figure 4.9.2).

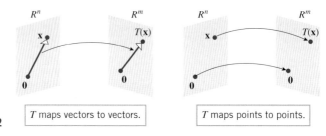

▶ **Figure 4.9.2**

T maps vectors to vectors. T maps points to points.

The following theorem states that if two matrix transformations from R^n to R^m have the same image at each point of R^n, then the matrices themselves must be the same.

THEOREM 4.9.2 *If $T_A: R^n \to R^m$ and $T_B: R^n \to R^m$ are matrix transformations, and if $T_A(\mathbf{x}) = T_B(\mathbf{x})$ for every vector $\mathbf{x}$ in R^n, then $A = B$.*

Proof To say that $T_A(\mathbf{x}) = T_B(\mathbf{x})$ for every vector in R^n is the same as saying that

$$A\mathbf{x} = B\mathbf{x}$$

for every vector $\mathbf{x}$ in R^n. This is true, in particular, if $\mathbf{x}$ is any of the standard basis vectors $\mathbf{e}_1, \mathbf{e}_2, \ldots, \mathbf{e}_n$ for R^n; that is,

$$A\mathbf{e}_j = B\mathbf{e}_j \quad (j = 1, 2, \ldots, n) \quad (11)$$

Since every entry of $\mathbf{e}_j$ is 0 except for the jth, which is 1, it follows from Theorem 1.3.1 that $A\mathbf{e}_j$ is the jth column of A and $B\mathbf{e}_j$ is the jth column of B. Thus, it follows from (11) that corresponding columns of A and B are the same, and hence that $A = B$. ◀

▶ **EXAMPLE 2 Zero Transformations**

If 0 is the $m \times n$ zero matrix, then

$$T_0(\mathbf{x}) = 0\mathbf{x} = \mathbf{0}$$

so multiplication by zero maps every vector in R^n into the zero vector in R^m. We call T_0 the ***zero transformation*** from R^n to R^m.

▶ **EXAMPLE 3** **Identity Operators**

If I is the $n \times n$ identity matrix, then

$$T_I(\mathbf{x}) = I\mathbf{x} = \mathbf{x}$$

so multiplication by I maps every vector in R^n into itself. We call T_I the *identity operator* on R^n. ◀

A Procedure for Finding Standard Matrices

There is a way of finding the standard matrix for a matrix transformation from R^n to R^m by considering the effect of that transformation on the standard basis vectors for R^n. To explain the idea, suppose that A is unknown and that

$$\mathbf{e}_1, \quad \mathbf{e}_2, \ldots, \quad \mathbf{e}_n$$

are the standard basis vectors for R^n. Suppose also that the images of these vectors under the transformation T_A are

$$T_A(\mathbf{e}_1) = A\mathbf{e}_1, \quad T_A(\mathbf{e}_2) = A\mathbf{e}_2, \ldots, \quad T_A(\mathbf{e}_n) = A\mathbf{e}_n$$

It follows from Theorem 1.3.1 that $A\mathbf{e}_j$ is a linear combination of the columns of A in which the successive coefficients are the entries of $\mathbf{e}_j$. But all entries of $\mathbf{e}_j$ are zero except the jth, so the product $A\mathbf{e}_j$ is just the jth column of the matrix A. Thus,

$$A = [T_A(\mathbf{e}_1) \mid T_A(\mathbf{e}_2) \mid \cdots \mid T_A(\mathbf{e}_n)] \tag{12}$$

In summary, we have the following procedure for finding the standard matrix for a matrix transformation:

Finding the Standard Matrix for a Matrix Transformation

Step 1. Find the images of the standard basis vectors $\mathbf{e}_1, \mathbf{e}_2, \ldots, \mathbf{e}_n$ for R^n in column form.

Step 2. Construct the matrix that has the images obtained in Step 1 as its successive columns. This matrix is the standard matrix for the transformation.

Reflection Operators

Some of the most basic matrix operators on R^2 and R^3 are those that map each point into its symmetric image about a fixed line or a fixed plane; these are called *reflection operators*. Table 1 shows the standard matrices for the reflections about the coordinate axes in R^2, and Table 2 shows the standard matrices for the reflections about the coordinate planes in R^3. In each case the standard matrix was obtained by finding the images of the standard basis vectors, converting those images to column vectors, and then using those column vectors as successive columns of the standard matrix.

Projection Operators

Matrix operators on R^2 and R^3 that map each point into its orthogonal projection on a fixed line or plane are called *projection operators* (or more precisely, *orthogonal projection* operators). Table 3 shows the standard matrices for the orthogonal projections on the coordinate axes in R^2, and Table 4 shows the standard matrices for the orthogonal projections on the coordinate planes in R^3.

Table 1

Operator	Illustration	Images of e_1 and e_2	Standard Matrix
Reflection about the y-axis $T(x, y) = (-x, y)$		$T(e_1) = T(1, 0) = (-1, 0)$ $T(e_2) = T(0, 1) = (0, 1)$	$\begin{bmatrix} -1 & 0 \\ 0 & 1 \end{bmatrix}$
Reflection about the x-axis $T(x, y) = (x, -y)$		$T(e_1) = T(1, 0) = (1, 0)$ $T(e_2) = T(0, 1) = (0, -1)$	$\begin{bmatrix} 1 & 0 \\ 0 & -1 \end{bmatrix}$
Reflection about the line $y = x$ $T(x, y) = (y, x)$		$T(e_1) = T(1, 0) = (0, 1)$ $T(e_2) = T(0, 1) = (1, 0)$	$\begin{bmatrix} 0 & 1 \\ 1 & 0 \end{bmatrix}$

Table 2

Operator	Illustration	e_1, e_2, e_3	Standard Matrix
Reflection about the xy-plane $T(x, y, z) = (x, y, -z)$		$T(e_1) = T(1, 0, 0) = (1, 0, 0)$ $T(e_2) = T(0, 1, 0) = (0, 1, 0)$ $T(e_3) = T(0, 0, 1) = (0, 0, -1)$	$\begin{bmatrix} 1 & 0 & 0 \\ 0 & 1 & 0 \\ 0 & 0 & -1 \end{bmatrix}$
Reflection about the xz-plane $T(x, y, z) = (x, -y, z)$		$T(e_1) = T(1, 0, 0) = (1, 0, 0)$ $T(e_2) = T(0, 1, 0) = (0, -1, 0)$ $T(e_3) = T(0, 0, 1) = (0, 0, 1)$	$\begin{bmatrix} 1 & 0 & 0 \\ 0 & -1 & 0 \\ 0 & 0 & 1 \end{bmatrix}$
Reflection about the yz-plane $T(x, y, z) = (-x, y, z)$		$T(e_1) = T(1, 0, 0) = (-1, 0, 0)$ $T(e_2) = T(0, 1, 0) = (0, 1, 0)$ $T(e_3) = T(0, 0, 1) = (0, 0, 1)$	$\begin{bmatrix} -1 & 0 & 0 \\ 0 & 1 & 0 \\ 0 & 0 & 1 \end{bmatrix}$

4.9 Matrix Transformations from R^n to R^m

Table 3

Operator	Illustration	Images of e_1 and e_2	Standard Matrix
Orthogonal projection on the x-axis $T(x, y) = (x, 0)$		$T(e_1) = T(1, 0) = (1, 0)$ $T(e_2) = T(0, 1) = (0, 0)$	$\begin{bmatrix} 1 & 0 \\ 0 & 0 \end{bmatrix}$
Orthogonal projection on the y-axis $T(x, y) = (0, y)$		$T(e_1) = T(1, 0) = (0, 0)$ $T(e_2) = T(0, 1) = (0, 1)$	$\begin{bmatrix} 0 & 0 \\ 0 & 1 \end{bmatrix}$

Table 4

Operator	Illustration	Images of e_1, e_2, e_3	Standard Matrix
Orthogonal projection on the xy-plane $T(x, y, z) = (x, y, 0)$		$T(e_1) = T(1, 0, 0) = (1, 0, 0)$ $T(e_2) = T(0, 1, 0) = (0, 1, 0)$ $T(e_3) = T(0, 0, 1) = (0, 0, 0)$	$\begin{bmatrix} 1 & 0 & 0 \\ 0 & 1 & 0 \\ 0 & 0 & 0 \end{bmatrix}$
Orthogonal projection on the xz-plane $T(x, y, z) = (x, 0, z)$		$T(e_1) = T(1, 0, 0) = (1, 0, 0)$ $T(e_2) = T(0, 1, 0) = (0, 0, 0)$ $T(e_3) = T(0, 0, 1) = (0, 0, 1)$	$\begin{bmatrix} 1 & 0 & 0 \\ 0 & 0 & 0 \\ 0 & 0 & 1 \end{bmatrix}$
Orthogonal projection on the yz-plane $T(x, y, z) = (0, y, z)$		$T(e_1) = T(1, 0, 0) = (0, 0, 0)$ $T(e_2) = T(0, 1, 0) = (0, 1, 0)$ $T(e_3) = T(0, 0, 1) = (0, 0, 1)$	$\begin{bmatrix} 0 & 0 & 0 \\ 0 & 1 & 0 \\ 0 & 0 & 1 \end{bmatrix}$

Rotation Operators   Matrix operators on R^2 and R^3 that move points along circular arcs are called **rotation operators**. Let us consider how to find the standard matrix for the rotation operator $T: R^2 \to R^2$ that moves points *counterclockwise* about the origin through an angle θ (Figure 4.9.3). As illustrated in Figure 4.9.3, the images of the standard basis vectors are

$$T(e_1) = T(1, 0) = (\cos\theta, \sin\theta) \quad \text{and} \quad T(e_2) = T(0, 1) = (-\sin\theta, \cos\theta)$$

so the standard matrix for T is

$$[T(e_1) \mid T(e_2)] = \begin{bmatrix} \cos\theta & -\sin\theta \\ \sin\theta & \cos\theta \end{bmatrix}$$

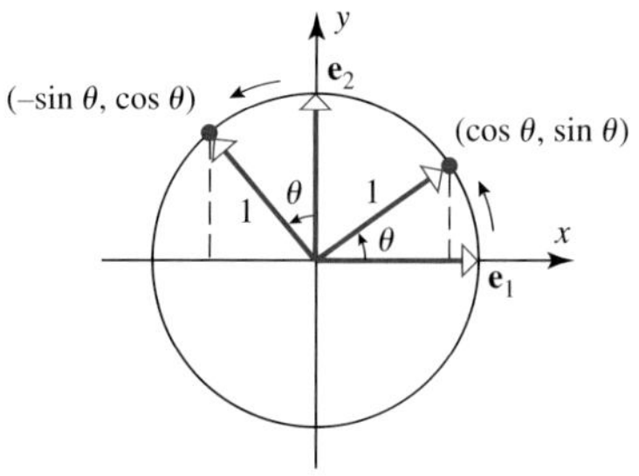

Figure 4.9.3

In keeping with common usage we will denote this operator by R_θ and call

$$R_\theta = \begin{bmatrix} \cos\theta & -\sin\theta \\ \sin\theta & \cos\theta \end{bmatrix} \quad (13)$$

the *rotation matrix* for R^2. If $\mathbf{x} = (x, y)$ is a vector in R^2, and if $\mathbf{w} = (w_1, w_2)$ is its image under the rotation, then the relationship $\mathbf{w} = R_\theta \mathbf{x}$ can be written in component form as

$$\begin{aligned} w_1 &= x\cos\theta - y\sin\theta \\ w_2 &= x\sin\theta + y\cos\theta \end{aligned} \quad (14)$$

These are called the *rotation equations* for R^2. These ideas are summarized in Table 5.

Table 5

Operator	Illustration	Rotation Equations	Standard Matrix
Rotation through an angle θ	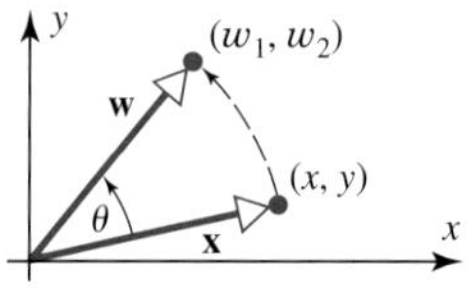	$\begin{aligned} w_1 &= x\cos\theta - y\sin\theta \\ w_2 &= x\sin\theta + y\cos\theta \end{aligned}$	$\begin{bmatrix} \cos\theta & -\sin\theta \\ \sin\theta & \cos\theta \end{bmatrix}$

In the plane, counterclockwise angles are positive and clockwise angles are negative. The rotation matrix for a *clockwise* rotation of $-\theta$ radians can be obtained by replacing θ by $-\theta$ in (12). After simplification this yields

$$R_{-\theta} = \begin{bmatrix} \cos\theta & \sin\theta \\ -\sin\theta & \cos\theta \end{bmatrix}$$

▶ **EXAMPLE 4 A Rotation Operator**

Find the image of $\mathbf{x} = (1, 1)$ under a rotation of $\pi/6$ radians ($= 30°$) about the origin.

Solution It follows from (13) with $\theta = \pi/6$ that

$$R_{\pi/6}\mathbf{x} = \begin{bmatrix} \frac{\sqrt{3}}{2} & -\frac{1}{2} \\ \frac{1}{2} & \frac{\sqrt{3}}{2} \end{bmatrix} \begin{bmatrix} 1 \\ 1 \end{bmatrix} = \begin{bmatrix} \frac{\sqrt{3}-1}{2} \\ \frac{1+\sqrt{3}}{2} \end{bmatrix} \approx \begin{bmatrix} 0.37 \\ 1.37 \end{bmatrix}$$

or in comma-delimited notation, $R_{\pi/6}(1, 1) = (0.37, 1.37)$. ◀

Rotations in R^3 A rotation of vectors in R^3 is usually described in relation to a ray emanating from the origin, called the *axis of rotation*. As a vector revolves around the axis of rotation, it sweeps out some portion of a cone (Figure 4.9.4a). The *angle of rotation*, which is measured in the base of the cone, is described as "clockwise" or "counterclockwise" in relation to a viewpoint that is along the axis of rotation *looking toward the origin*. For example, in Figure 4.9.4a the vector $\mathbf{w}$ results from rotating the vector $\mathbf{x}$ counterclockwise around the axis l through an angle θ. As in R^2, angles are *positive* if they are generated by counterclockwise rotations and *negative* if they are generated by clockwise rotations.

The most common way of describing a general axis of rotation is to specify a nonzero vector $\mathbf{u}$ that runs along the axis of rotation and has its initial point at the origin. The counterclockwise direction for a rotation about the axis can then be determined by a "right-hand rule" (Figure 4.9.4b): If the thumb of the right hand points in the direction of **u**, then the cupped fingers point in a counterclockwise direction.

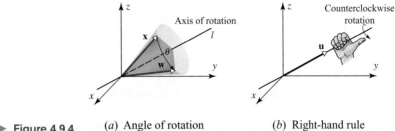

▶ Figure 4.9.4 (a) Angle of rotation (b) Right-hand rule

A *rotation operator* on R^3 is a matrix operator that rotates each vector in R^3 about some rotation axis through a fixed angle θ. In Table 6 we have described the rotation operators on R^3 whose axes of rotation are the positive coordinate axes. For each of these rotations one of the components is unchanged, and the relationships between the other components can be derived by the same procedure used to derive (14). For example, in the rotation about the z-axis, the z-components of $\mathbf{x}$ and $\mathbf{w} = T(\mathbf{x})$ are the same, and the x- and y-components are related as in (14). This yields the rotation equation shown in the last row of Table 6.

Table 6

Operator	Illustration	Rotation Equations	Standard Matrix
Counterclockwise rotation about the positive x-axis through an angle θ		$w_1 = x$ $w_2 = y \cos \theta - z \sin \theta$ $w_3 = y \sin \theta + z \cos \theta$	$\begin{bmatrix} 1 & 0 & 0 \\ 0 & \cos \theta & -\sin \theta \\ 0 & \sin \theta & \cos \theta \end{bmatrix}$
Counterclockwise rotation about the positive y-axis through an angle θ		$w_1 = x \cos \theta + z \sin \theta$ $w_2 = y$ $w_3 = -x \sin \theta + z \cos \theta$	$\begin{bmatrix} \cos \theta & 0 & \sin \theta \\ 0 & 1 & 0 \\ -\sin \theta & 0 & \cos \theta \end{bmatrix}$
Counterclockwise rotation about the positive z-axis through an angle θ		$w_1 = x \cos \theta - y \sin \theta$ $w_2 = x \sin \theta + y \cos \theta$ $w_3 = z$	$\begin{bmatrix} \cos \theta & -\sin \theta & 0 \\ \sin \theta & \cos \theta & 0 \\ 0 & 0 & 1 \end{bmatrix}$

For completeness, we note that the standard matrix for a counterclockwise rotation through an angle θ about an axis in R^3, which is determined by an arbitrary *unit vector* $\mathbf{u} = (a, b, c)$ that has its initial point at the origin, is

$$\begin{bmatrix} a^2(1-\cos\theta) + \cos\theta & ab(1-\cos\theta) - c\sin\theta & ac(1-\cos\theta) + b\sin\theta \\ ab(1-\cos\theta) + c\sin\theta & b^2(1-\cos\theta) + \cos\theta & bc(1-\cos\theta) - a\sin\theta \\ ac(1-\cos\theta) - b\sin\theta & bc(1-\cos\theta) + a\sin\theta & c^2(1-\cos\theta) + \cos\theta \end{bmatrix} \quad (15)$$

The derivation can be found in the book *Principles of Interactive Computer Graphics*, by W. M. Newman and R. F. Sproull (New York: McGraw-Hill, 1979). You may find it instructive to derive the results in Table 6 as special cases of this more general result.

Dilations and Contractions   If k is a nonnegative scalar, then the operator $T(\mathbf{x}) = k\mathbf{x}$ on R^2 or R^3 has the effect of increasing or decreasing the length of each vector by a factor of k. If $0 \leq k < 1$ the operator is called a **contraction** with factor k, and if $k > 1$ it is called a **dilation** with factor k (Figure 4.9.5). If $k = 1$, then T is the identity operator and can be regarded either as a contraction or a dilation. Tables 7 and 8 illustrate these operators.

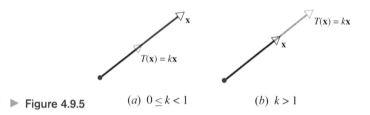

▶ **Figure 4.9.5**   (a) $0 \leq k < 1$ (b) $k > 1$

Table 7

Operator	Illustration $T(x, y) = (kx, ky)$	Effect on the Standard Basis		Standard Matrix
Contraction with factor k on R^2 ($0 \leq k < 1$)				$\begin{bmatrix} k & 0 \\ 0 & k \end{bmatrix}$
Dilation with factor k on R^2 ($k > 1$)				

Yaw, Pitch, and Roll

In aeronautics and astronautics, the orientation of an aircraft or space shuttle relative to an *xyz*-coordinate system is often described in terms of angles called *yaw*, *pitch*, and *roll*. If, for example, an aircraft is flying along the *y*-axis and the *xy*-plane defines the horizontal, then the aircraft's angle of rotation about the *z*-axis is called the *yaw*, its angle of rotation about the *x*-axis is called the *pitch*, and its angle of rotation about the *y*-axis is called the *roll*. A combination of yaw, pitch, and roll can be achieved by a single rotation about some axis through the origin. This is, in fact, how a space shuttle makes attitude adjustments—it doesn't perform each rotation separately; it calculates one axis, and rotates about that axis to get the correct orientation. Such rotation maneuvers are used to align an antenna, point the nose toward a celestial object, or position a payload bay for docking.

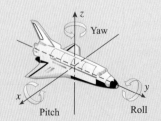

4.9 Matrix Transformations from R^n to R^m

Table 8

Operator	Illustration $T(x, y, z) = (kx, ky, kz)$	Standard Matrix
Contraction with factor k on R^3 ($0 \leq k \leq 1$)		$\begin{bmatrix} k & 0 & 0 \\ 0 & k & 0 \\ 0 & 0 & k \end{bmatrix}$
Dilation with factor k on R^3 ($k \geq 1$)		

Expansion and Compressions

In a dilation or contraction of R^2 or R^3, all coordinates are multiplied by a factor k. If only one of the coordinates is multiplied by k, then the resulting operator is called an ***expansion*** or ***compression*** with factor k. This is illustrated in Table 9 for R^2. You should have no trouble extending these results to R^3.

Table 9

Operator	Illustration $T(x, y) = (kx, y)$	Effect on the Standard Basis	Standard Matrix
Compression of R^2 in the x-direction with factor k ($0 \leq k < 1$)			$\begin{bmatrix} k & 0 \\ 0 & 1 \end{bmatrix}$
Expansion of R^2 in the x-direction with factor k ($k > 1$)			

Operator	Illustration $T(x, y) = (x, ky)$	Effect on the Standard Basis	Standard Matrix
Compression of R^2 in the y-direction with factor k ($0 \leq k < 1$)			$\begin{bmatrix} 1 & 0 \\ 0 & k \end{bmatrix}$
Expansion of R^2 in the y-direction with factor k ($k > 1$)			

Shears A matrix operator of the form $T(x, y) = (x + ky, y)$ translates a point (x, y) in the xy-plane parallel to the x-axis by an amount ky that is proportional to the y-coordinate of the point. This operator leaves the points on the x-axis fixed (since $y = 0$), but as we progress away from the x-axis, the translation distance increases. We call this operator the **shear in the x-direction with factor k**. Similarly, a matrix operator of the form $T(x, y) = (x, y + kx)$ is called the **shear in the y-direction with factor k**. Table 10 illustrates the basic information about shears in R^2.

Table 10

Operator	Effect on the Standard Basis	Standard Matrix
Shear of R^2 in the x-direction with factor k $T(x, y) = (x + ky, y)$	(diagrams showing unit square sheared for $k > 0$ and $k < 0$)	$\begin{bmatrix} 1 & k \\ 0 & 1 \end{bmatrix}$
Shear of R^2 in the y-direction with factor k $T(x, y) = (x, y + kx)$	(diagrams showing unit square sheared for $k > 0$ and $k < 0$)	$\begin{bmatrix} 1 & 0 \\ k & 1 \end{bmatrix}$

▶ **EXAMPLE 5** Some Basic Matrix Operators on R^2

In each part describe the matrix operator corresponding to A, and show its effect on the unit square.

(a) $A_1 = \begin{bmatrix} 1 & 2 \\ 0 & 1 \end{bmatrix}$ (b) $A_2 = \begin{bmatrix} 2 & 0 \\ 0 & 2 \end{bmatrix}$ (c) $A_3 = \begin{bmatrix} 2 & 0 \\ 0 & 1 \end{bmatrix}$

Solution By comparing the forms of these matrices to those in Tables 7, 9, and 10, we see that the matrix A_1 corresponds to a shear in the x-direction with factor 2, the matrix A_2 corresponds to a dilation with factor 2, and A_3 corresponds to an expansion in the x-direction with factor 2. The effects of these operators on the unit square are shown in Figure 4.9.6. ◀

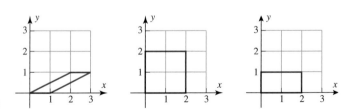

▶ Figure 4.9.6

OPTIONAL
Orthogonal Projections on Lines Through the Origin

In Table 3 we listed the standard matrices for the orthogonal projections on the coordinate axes in R^2. These are special cases of the more general operator $T: R^2 \to R^2$ that maps each point into its orthogonal projection on a line L through the origin that makes an

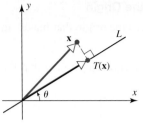

Figure 4.9.7

angle θ with the positive x-axis (Figure 4.9.7). In Example 4 of Section 3.3 we used Formula (10) of that section to find the orthogonal projections of the standard basis vectors for R^2 on that line. Expressed in matrix form, we found those projections to be

$$T(\mathbf{e}_1) = \begin{bmatrix} \cos^2\theta \\ \sin\theta\cos\theta \end{bmatrix} \quad \text{and} \quad T(\mathbf{e}_2) = \begin{bmatrix} \sin\theta\cos\theta \\ \sin^2\theta \end{bmatrix}$$

Thus, the standard matrix for T is

$$[T] = [T(\mathbf{e}_1) \mid T(\mathbf{e}_2)] = \begin{bmatrix} \cos^2\theta & \sin\theta\cos\theta \\ \sin\theta\cos\theta & \sin^2\theta \end{bmatrix} = \begin{bmatrix} \cos^2\theta & \tfrac{1}{2}\sin 2\theta \\ \tfrac{1}{2}\sin 2\theta & \sin^2\theta \end{bmatrix}$$

In keeping with common usage, we will denote this operator by

We have included two versions of Formula (16) because both are commonly used. Whereas the first version involves only the angle θ, the second involves both θ and 2θ.

$$P_\theta = \begin{bmatrix} \cos^2\theta & \sin\theta\cos\theta \\ \sin\theta\cos\theta & \sin^2\theta \end{bmatrix} = \begin{bmatrix} \cos^2\theta & \tfrac{1}{2}\sin 2\theta \\ \tfrac{1}{2}\sin 2\theta & \sin^2\theta \end{bmatrix} \quad (16)$$

▶ **EXAMPLE 6** Orthogonal Projection on a Line Through the Origin

Use Formula (16) to find the orthogonal projection of the vector $\mathbf{x} = (1, 5)$ on the line through the origin that makes an angle of $\pi/6 (= 30°)$ with the x-axis.

Solution Since $\sin(\pi/6) = 1/2$ and $\cos(\pi/6) = \sqrt{3}/2$, it follows from (16) that the standard matrix for this projection is

$$P_{\pi/6} = \begin{bmatrix} \cos^2(\pi/6) & \sin(\pi/6)\cos(\pi/6) \\ \sin(\pi/6)\cos(\pi/6) & \sin^2(\pi/6) \end{bmatrix} = \begin{bmatrix} \tfrac{3}{4} & \tfrac{\sqrt{3}}{4} \\ \tfrac{\sqrt{3}}{4} & \tfrac{1}{4} \end{bmatrix}$$

Thus,

$$P_{\pi/6}\mathbf{x} = \begin{bmatrix} \tfrac{3}{4} & \tfrac{\sqrt{3}}{4} \\ \tfrac{\sqrt{3}}{4} & \tfrac{1}{4} \end{bmatrix} \begin{bmatrix} 1 \\ 5 \end{bmatrix} = \begin{bmatrix} \tfrac{3+5\sqrt{3}}{4} \\ \tfrac{\sqrt{3}+5}{4} \end{bmatrix} \approx \begin{bmatrix} 2.91 \\ 1.68 \end{bmatrix}$$

or in comma-delimited notation, $P_{\pi/6}(1, 5) \approx (2.91, 1.68)$. ◀

Reflections About Lines Through the Origin

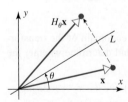

Figure 4.9.8

In Table 1 we listed the reflections about the coordinate axes in R^2. These are special cases of the more general operator $H_\theta: R^2 \to R^2$ that maps each point into its reflection about a line L through the origin that makes an angle θ with the positive x-axis (Figure 4.9.8). We could find the standard matrix for H_θ by finding the images of the standard basis vectors, but instead we will take advantage of our work on orthogonal projections by using the Formula (16) for P_θ to find a formula for H_θ.

You should be able to see from Figure 4.9.9 that for every vector $\mathbf{x}$ in R^n

$$P_\theta\mathbf{x} - \mathbf{x} = \tfrac{1}{2}(H_\theta\mathbf{x} - \mathbf{x}) \quad \text{or equivalently} \quad H_\theta\mathbf{x} = (2P_\theta - I)\mathbf{x}$$

Thus, it follows from Theorem 4.9.2 that

$$H_\theta = 2P_\theta - I \quad (17)$$

and hence from (16) that

$$H_\theta = \begin{bmatrix} \cos 2\theta & \sin 2\theta \\ \sin 2\theta & -\cos 2\theta \end{bmatrix} \quad (18)$$

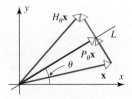

Figure 4.9.9

EXAMPLE 7 Reflection About a Line Through the Origin

Find the reflection of the vector $\mathbf{x} = (1, 5)$ on the line through the origin that makes an angle of $\pi/6 (= 30°)$ with the x-axis.

Solution Since $\sin(\pi/3) = \sqrt{3}/2$ and $\cos(\pi/3) = 1/2$, it follows from (18) that the standard matrix for this reflection is

$$H_{\pi/6} = \begin{bmatrix} \cos(\pi/3) & \sin(\pi/3) \\ \sin(\pi/3) & -\cos(\pi/3) \end{bmatrix} = \begin{bmatrix} \frac{1}{2} & \frac{\sqrt{3}}{2} \\ \frac{\sqrt{3}}{2} & -\frac{1}{2} \end{bmatrix}$$

Thus,

$$H_{\pi/6}\mathbf{x} = \begin{bmatrix} \frac{1}{2} & \frac{\sqrt{3}}{2} \\ \frac{\sqrt{3}}{2} & -\frac{1}{2} \end{bmatrix} \begin{bmatrix} 1 \\ 5 \end{bmatrix} = \begin{bmatrix} \frac{1+5\sqrt{3}}{2} \\ \frac{\sqrt{3}-5}{2} \end{bmatrix} \approx \begin{bmatrix} 4.83 \\ -1.63 \end{bmatrix}$$

or in comma-delimited notation, $H_{\pi/6}(1, 5) \approx (4.83, -1.63)$. ◀

> Show that the standard matrices in Tables 1 and 3 are special cases of (18) and (16).

Concept Review

- Function
- Image
- Value
- Domain
- Codomain
- Transformation
- Operator
- Matrix transformation
- Matrix operator
- Standard matrix
- Properties of matrix transformations
- Zero transformation
- Identity operator
- Reflection operator
- Projection operator
- Rotation operator
- Rotation matrix
- Rotation equations
- Axis of rotation in 3-space
- Angle of rotation in 3-space
- Expansion operator
- Compression operator
- Shear
- Dilation
- Contraction

Skills

- Find the domain and codomain of a transformation, and determine whether the transformation is linear.
- Find the standard matrix for a matrix transformation.
- Describe the effect of a matrix operator on the standard basis in R^n.

Exercise Set 4.9

▶ In Exercises 1–2, find the domain and codomain of the transformation $T_A(\mathbf{x}) = A\mathbf{x}$. ◀

1. (a) A has size 3×2.   (b) A has size 2×3.
    (c) A has size 3×3.   (d) A has size 1×6.

2. (a) A has size 4×5.   (b) A has size 5×4.
    (c) A has size 4×4.   (d) A has size 3×1.

3. If $T(x_1, x_2) = (x_1 - x_2, 2x_1, 3x_2 + x_1)$, then the domain of T is _____, the codomain of T is _____, and the image of $\mathbf{x} = (1, -2)$ under T is _____.

4. If $T(x_1, x_2, x_3) = (x_1 + 2x_2, x_1 - 2x_2)$, then the domain of T is _____, the codomain of T is _____, and the image of $\mathbf{x} = (0, -1, 4)$ under T is _____.

5. In each part, find the domain and codomain of the transformation defined by the equations, and determine whether the transformation is linear.

(a) $w_1 = 2x_1 - 3x_2 + 5x_3$
    $w_2 = 4x_1 - 6x_2 + 3x_3$

(b) $w_1 = x_1 - 3x_2x_3$
    $w_2 = 2x_1x_2 - 7x_3$

(c) $w_1 = 5x_1 - x_2 + x_3$
    $w_2 = -x_1 + x_2 + 7x_3$
    $w_3 = 2x_1 - 4x_2 - x_3$

(d) $w_1 = x_1^2 - 3x_2 + x_3 - 2x_4$
    $w_2 = 3x_1 - 4x_2 - x_3^2 + x_4$

6. In each part, determine whether T is a matrix transformation.

(a) $T(x, y) = (2x, y)$ (b) $T(x, y) = (-y, x)$
(c) $T(x, y) = (2x + y, x - y)$

(d) $T(x, y) = (x^2, y)$ (e) $T(x, y) = (x, y+1)$

7. In each part, determine whether T is a matrix transformation.
   (a) $T(x, y, z) = (0, 0)$ (b) $T(x, y, z, w) = (1, -1)$
   (c) $T(x, y, z) = (x - y + z, 0)$
   (d) $T(x, y, z) = (x, yz, x + y + z)$
   (e) $T(x, y, z) = (2y, x + z, -3y)$

8. Find the standard matrix for the transformation defined by the equations.
   (a) $w_1 = 2x_1 - 3x_2 + x_4$
   $w_2 = 3x_1 + 5x_2 - x_4$
   (b) $w_1 = 7x_1 + 2x_2 - 8x_3$
   $w_2 = - x_2 + 5x_3$
   $w_3 = 4x_1 + 7x_2 - x_3$
   (c) $w_1 = -x_1 + x_2$
   $w_2 = 3x_1 - 2x_2$
   $w_3 = 5x_1 - 7x_2$
   (d) $w_1 = x_1$
   $w_2 = x_1 + x_2$
   $w_3 = x_1 + x_2 + x_3$
   $w_4 = x_1 + x_2 + x_3 + x_4$

9. Find the standard matrix for the operator $T: R^3 \to R^3$ defined by
   $$w_1 = 4x_1 - 3x_2 + x_3$$
   $$w_2 = 2x_1 - x_2 + 5x_3$$
   $$w_3 = x_1 + 2x_2 - 2x_3$$
   and then calculate $T(-1, 2, 4)$ by directly substituting in the equations and also by matrix multiplication.

10. Find the standard matrix for the operator T defined by the formula.
    (a) $T(x_1, x_2) = (2x_1 - x_2, x_1 + x_2)$
    (b) $T(x_1, x_2) = (x_1, x_2)$
    (c) $T(x_1, x_2, x_3) = (x_1 + 2x_2 + x_3, x_1 + 5x_2, x_3)$
    (d) $T(x_1, x_2, x_3) = (4x_1, 7x_2, -8x_3)$

11. Find the standard matrix for the transformation T defined by the formula.
    (a) $T(x_1, x_2) = (x_2, -x_1, x_1 + 3x_2, x_1 - x_2)$
    (b) $T(x_1, x_2, x_3, x_4) = (7x_1 + 2x_2 - x_3 + x_4, x_2 + x_3, -x_1)$
    (c) $T(x_1, x_2, x_3) = (0, 0, 0, 0, 0)$
    (d) $T(x_1, x_2, x_3, x_4) = (x_4, x_1, x_3, x_2, x_1 - x_3)$

12. In each part, find $T(\mathbf{x})$, and express the answer in matrix form.
    (a) $[T] = \begin{bmatrix} 1 & 2 \\ 3 & 4 \end{bmatrix}$; $\mathbf{x} = \begin{bmatrix} 3 \\ -2 \end{bmatrix}$
    (b) $[T] = \begin{bmatrix} -1 & 2 & 0 \\ 3 & 1 & 5 \end{bmatrix}$; $\mathbf{x} = \begin{bmatrix} -1 \\ 1 \\ 3 \end{bmatrix}$
    (c) $[T] = \begin{bmatrix} -2 & 1 & 4 \\ 3 & 5 & 7 \\ 6 & 0 & -1 \end{bmatrix}$; $\mathbf{x} = \begin{bmatrix} x_1 \\ x_2 \\ x_3 \end{bmatrix}$
    (d) $[T] = \begin{bmatrix} -1 & 1 \\ 2 & 4 \\ 7 & 8 \end{bmatrix}$; $\mathbf{x} = \begin{bmatrix} x_1 \\ x_2 \end{bmatrix}$

13. In each part, use the standard matrix for T to find $T(\mathbf{x})$; then check the result by calculating $T(\mathbf{x})$ directly.
    (a) $T(x_1, x_2) = (-3x_1 + 4x_2, x_1)$; $\mathbf{x} = (-1, 4)$
    (b) $T(x_1, x_2, x_3) = (x_3, x_1 - 2x_2, 3x_1 + x_3)$; $\mathbf{x} = (2, 1, -3)$

14. Use matrix multiplication to find the reflection of $(-1, 2)$ about
    (a) the x-axis. (b) the y-axis. (c) the line $y = x$.

15. Use matrix multiplication to find the reflection of $(2, -5, 3)$ about
    (a) the xy-plane. (b) the xz-plane. (c) the yz-plane.

16. Use matrix multiplication to find the orthogonal projection of $(2, -5)$ on
    (a) the x-axis. (b) the y-axis.

17. Use matrix multiplication to find the orthogonal projection of $(-2, 1, 3)$ on
    (a) the xy-plane. (b) the xz-plane. (c) the yz-plane.

18. Use matrix multiplication to find the image of the vector $(3, -4)$ when it is rotated through an angle of
    (a) $\theta = 30°$. (b) $\theta = -60°$.
    (c) $\theta = 45°$. (d) $\theta = 90°$.

19. Use matrix multiplication to find the image of the vector $(-2, 1, 2)$ if it is rotated
    (a) $30°$ about the x-axis. (b) $45°$ about the y-axis.
    (c) $90°$ about the z-axis.

20. Find the standard matrix for the operator that rotates a vector in R^3 through an angle of $-60°$ about
    (a) the x-axis. (b) the y-axis. (c) the z-axis.

21. Use matrix multiplication to find the image of the vector $(-2, 1, 2)$ if it is rotated
    (a) $-30°$ about the x-axis. (b) $-45°$ about the y-axis.
    (c) $-90°$ about the z-axis.

22. In R^3 the **orthogonal projections** on the x-axis, y-axis, and z-axis are defined by
    $$T_1(x, y, z) = (x, 0, 0), \quad T_2(x, y, z) = (0, y, 0),$$
    $$T_3(x, y, z) = (0, 0, z)$$
    respectively.
    (a) Show that the orthogonal projections on the coordinate axes are matrix operators, and find their standard matrices.

(b) Show that if $T: R^3 \to R^3$ is an orthogonal projection on one of the coordinate axes, then for every vector $\mathbf{x}$ in R^3, the vectors $T(\mathbf{x})$ and $\mathbf{x} - T(\mathbf{x})$ are orthogonal.

(c) Make a sketch showing $\mathbf{x}$ and $\mathbf{x} - T(\mathbf{x})$ in the case where T is the orthogonal projection on the x-axis.

23. Use Formula (15) to derive the standard matrices for the rotations about the x-axis, y-axis, and z-axis in R^3.

24. Use Formula (15) to find the standard matrix for a rotation of $\pi/2$ radians about the axis determined by the vector $\mathbf{v} = (1, 1, 1)$. [*Note:* Formula (15) requires that the vector defining the axis of rotation have length 1.]

25. Use Formula (15) to find the standard matrix for a rotation of 180° about the axis determined by the vector $\mathbf{v} = (2, 2, 1)$. [*Note:* Formula (15) requires that the vector defining the axis of rotation have length 1.]

26. It can be proved that if A is a 2×2 matrix with orthonormal column vectors and for which $\det(A) = 1$, then multiplication by A is a rotation through some angle θ. Verify that

$$A = \begin{bmatrix} -\frac{1}{\sqrt{2}} & -\frac{1}{\sqrt{2}} \\ \frac{1}{\sqrt{2}} & -\frac{1}{\sqrt{2}} \end{bmatrix}$$

satisfies the stated conditions and find the angle of rotation.

27. The result stated in Exercise 26 can be extended to R^3; that is, it can be proved that if A is a 3×3 matrix with orthonormal column vectors and for which $\det(A) = 1$, then multiplication by A is a rotation about some axis through some angle θ. Use Formula (15) to show that the angle of rotation satisfies the equation

$$\cos \theta = \frac{\text{tr}(A) - 1}{2}$$

28. Let A be a 3×3 matrix (other than the identity matrix) satisfying the conditions stated in Exercise 27. It can be shown that if $\mathbf{x}$ is any nonzero vector in R^3, then the vector $\mathbf{u} = A\mathbf{x} + A^T\mathbf{x} + [1 - \text{tr}(A)]\mathbf{x}$ determines an axis of rotation when $\mathbf{u}$ is positioned with its initial point at the origin. [See "The Axis of Rotation: Analysis, Algebra, Geometry," by Dan Kalman, *Mathematics Magazine*, Vol. 62, No. 4, October 1989.]

(a) Show that multiplication by

$$A = \begin{bmatrix} \frac{1}{9} & -\frac{4}{9} & \frac{8}{9} \\ \frac{8}{9} & \frac{4}{9} & \frac{1}{9} \\ -\frac{4}{9} & \frac{7}{9} & \frac{4}{9} \end{bmatrix}$$

is a rotation.

(b) Find a vector of length 1 that defines an axis for the rotation.

(c) Use the result in Exercise 27 to find the angle of rotation about the axis obtained in part (b).

29. In words, describe the geometric effect of multiplying a vector $\mathbf{x}$ by the matrix A.

(a) $A = \begin{bmatrix} 0 & 0 \\ 0 & -3 \end{bmatrix}$ (b) $A = \begin{bmatrix} 1 & 0 \\ 0 & 4 \end{bmatrix}$

30. In words, describe the geometric effect of multiplying a vector $\mathbf{x}$ by the matrix A.

(a) $A = \begin{bmatrix} 2 & 0 \\ 0 & 3 \end{bmatrix}$ (b) $A = \begin{bmatrix} \frac{\sqrt{3}}{2} & -\frac{1}{2} \\ \frac{1}{2} & \frac{\sqrt{3}}{2} \end{bmatrix}$

31. In words, describe the geometric effect of multiplying a vector $\mathbf{x}$ by the matrix

$$A = \begin{bmatrix} \cos^2 \theta - \sin^2 \theta & -2 \sin \theta \cos \theta \\ 2 \sin \theta \cos \theta & \cos^2 \theta - \sin^2 \theta \end{bmatrix}$$

32. If multiplication by A rotates a vector $\mathbf{x}$ in the xy-plane through an angle θ, what is the effect of multiplying $\mathbf{x}$ by A^T? Explain your reasoning.

33. Let $\mathbf{x}_0$ be a nonzero column vector in R^2, and suppose that $T: R^2 \to R^2$ is the transformation defined by the formula $T(\mathbf{x}) = \mathbf{x}_0 + R_\theta \mathbf{x}$, where R_θ is the standard matrix of the rotation of R^2 about the origin through the angle θ. Give a geometric description of this transformation. Is it a matrix transformation? Explain.

34. A function of the form $f(x) = mx + b$ is commonly called a "linear function" because the graph of $y = mx + b$ is a line. Is f a matrix transformation on R?

35. Let $\mathbf{x} = \mathbf{x}_0 + t\mathbf{v}$ be a line in R^n, and let $T: R^n \to R^n$ be a matrix operator on R^n. What kind of geometric object is the image of this line under the operator T? Explain your reasoning.

True-False Exercises

In parts (a)–(i) determine whether the statement is true or false, and justify your answer.

(a) If A is a 2×3 matrix, then the domain of the transformation T_A is R^2.

(b) If A is an $m \times n$ matrix, then the codomain of the transformation T_A is R^n.

(c) If $T: R^n \to R^m$ and $T(\mathbf{0}) = \mathbf{0}$, then T is a matrix transformation.

(d) If $T: R^n \to R^m$ and $T(c_1 \mathbf{x} + c_2 \mathbf{y}) = c_1 T(\mathbf{x}) + c_2 T(\mathbf{y})$ for all scalars c_1 and c_2 and all vectors $\mathbf{x}$ and $\mathbf{y}$ in R^n, then T is a matrix transformation.

(e) There is only one matrix transformation $T: R^n \to R^m$ such that $T(-\mathbf{x}) = -T(\mathbf{x})$ for every vector $\mathbf{x}$ in R^n.

(f) There is only one matrix transformation $T: R^n \to R^m$ such that $T(\mathbf{x} + \mathbf{y}) = T(\mathbf{x} - \mathbf{y})$ for all vectors $\mathbf{x}$ and $\mathbf{y}$ in R^n.

(g) If **b** is a nonzero vector in R^n, then $T(\mathbf{x}) = \mathbf{x} + \mathbf{b}$ is a matrix operator on R^n.

(h) The matrix $\begin{bmatrix} \frac{1}{2} & -\frac{1}{2} \\ \frac{1}{2} & \frac{1}{2} \end{bmatrix}$ is the standard matrix for a rotation.

(i) The standard matrices of the reflections about the coordinate axes in 2-space have the form $\begin{bmatrix} a & 0 \\ 0 & -a \end{bmatrix}$, where $a = \pm 1$.

4.10 Properties of Matrix Transformations

In this section we will discuss properties of matrix transformations. We will show, for example, that if several matrix transformations are performed in succession, then the same result can be obtained by a single matrix transformation that is chosen appropriately. We will also explore the relationship between the invertibility of a matrix and properties of the corresponding transformation.

Compositions of Matrix Transformations

Suppose that T_A is a matrix transformation from R^n to R^k and T_B is a matrix transformation from R^k to R^m. If **x** is a vector in R^n, then T_A maps this vector into a vector $T_A(\mathbf{x})$ in R^k, and T_B, in turn, maps that vector into the vector $T_B(T_A(\mathbf{x}))$ in R^m. This process creates a transformation from R^n to R^m that we call the ***composition of T_B with T_A*** and denote by the symbol

$$T_B \circ T_A$$

which is read "T_B circle T_A". As illustrated in Figure 4.10.1, the transformation T_A in the formula is performed first; that is,

$$(T_B \circ T_A)(\mathbf{x}) = T_B(T_A(\mathbf{x})) \tag{1}$$

WARNING Just as it is *not* true, in general, that
$$AB = BA$$
so it is *not* true, in general, that
$$T_B \circ T_A = T_A \circ T_B$$
That is, *order matters when matrix transformations are composed.*

This composition is itself a matrix transformation since

$$(T_B \circ T_A)(\mathbf{x}) = T_B(T_A(\mathbf{x})) = B(T_A(\mathbf{x})) = B(A\mathbf{x}) = (BA)\mathbf{x}$$

which shows that it is multiplication by BA. This is expressed by the formula

$$T_B \circ T_A = T_{BA} \tag{2}$$

Compositions can be defined for any finite succession of matrix transformations whose domains and ranges have the appropriate dimensions. For example, to extend Formula (2) to three factors, consider the matrix transformations

$$T_A: R^n \to R^k, \quad T_B: R^k \to R^l, \quad T_C: R^l \to R^m$$

We define the composition $(T_C \circ T_B \circ T_A): R^n \to R^m$ by

$$(T_C \circ T_B \circ T_A)(\mathbf{x}) = T_C(T_B(T_A(\mathbf{x})))$$

As above, it can be shown that this is a matrix transformation whose standard matrix is CBA and that

$$T_C \circ T_B \circ T_A = T_{CBA} \tag{3}$$

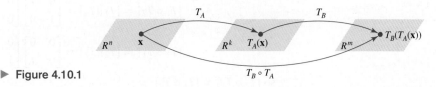

▶ Figure 4.10.1

As in Formula (9) of Section 4.9, we can use square brackets to denote a matrix transformation without referencing a specific matrix. Thus, for example, the formula

$$[T_2 \circ T_1] = [T_2][T_1] \tag{4}$$

is a restatement of Formula (2) which states that the standard matrix for a composition is the product of the standard matrices in the appropriate order. Similarly,

$$[T_3 \circ T_2 \circ T_1] = [T_3][T_2][T_1] \tag{5}$$

is a restatement of Formula (3).

▶ **EXAMPLE 1** **Composition of Two Rotations**

Let $T_1: R^2 \to R^2$ and $T_2: R^2 \to R^2$ be the matrix operators that rotate vectors through the angles θ_1 and θ_2, respectively. Thus the operation

$$(T_2 \circ T_1)(\mathbf{x}) = T_2(T_1(\mathbf{x}))$$

first rotates $\mathbf{x}$ through the angle θ_1, then rotates $T_1(\mathbf{x})$ through the angle θ_2. It follows that the net effect of $T_2 \circ T_1$ is to rotate each vector in R^2 through the angle $\theta_1 + \theta_2$ (Figure 4.10.2). Thus, the standard matrices for these matrix operators are

$$[T_1] = \begin{bmatrix} \cos\theta_1 & -\sin\theta_1 \\ \sin\theta_1 & \cos\theta_1 \end{bmatrix}, \quad [T_2] = \begin{bmatrix} \cos\theta_2 & -\sin\theta_2 \\ \sin\theta_2 & \cos\theta_2 \end{bmatrix},$$

$$[T_2 \circ T_1] = \begin{bmatrix} \cos(\theta_1 + \theta_2) & -\sin(\theta_1 + \theta_2) \\ \sin(\theta_1 + \theta_2) & \cos(\theta_1 + \theta_2) \end{bmatrix}$$

These matrices should satisfy (4). With the help of some basic trigonometric identities, we can confirm that this is so as follows:

$$[T_2][T_1] = \begin{bmatrix} \cos\theta_2 & -\sin\theta_2 \\ \sin\theta_2 & \cos\theta_2 \end{bmatrix} \begin{bmatrix} \cos\theta_1 & -\sin\theta_1 \\ \sin\theta_1 & \cos\theta_1 \end{bmatrix}$$

$$= \begin{bmatrix} \cos\theta_2 \cos\theta_1 - \sin\theta_2 \sin\theta_1 & -(\cos\theta_2 \sin\theta_1 + \sin\theta_2 \cos\theta_1) \\ \sin\theta_2 \cos\theta_1 + \cos\theta_2 \sin\theta_1 & -\sin\theta_2 \sin\theta_1 + \cos\theta_2 \cos\theta_1 \end{bmatrix}$$

$$= \begin{bmatrix} \cos(\theta_1 + \theta_2) & -\sin(\theta_1 + \theta_2) \\ \sin(\theta_1 + \theta_2) & \cos(\theta_1 + \theta_2) \end{bmatrix}$$

$$= [T_2 \circ T_1]$$

▶ **EXAMPLE 2** **Composition Is Not Commutative**

Let $T_1: R^2 \to R^2$ be the reflection about the line $y = x$, and let $T_2: R^2 \to R^2$ be the orthogonal projection on the y-axis. Figure 4.10.3 illustrates graphically that $T_1 \circ T_2$ and $T_2 \circ T_1$ have different effects on a vector $\mathbf{x}$. This same conclusion can be reached by showing that the standard matrices for T_1 and T_2 do not commute:

$$[T_1 \circ T_2] = [T_1][T_2] = \begin{bmatrix} 0 & 1 \\ 1 & 0 \end{bmatrix} \begin{bmatrix} 0 & 0 \\ 0 & 1 \end{bmatrix} = \begin{bmatrix} 0 & 1 \\ 0 & 0 \end{bmatrix}$$

$$[T_2 \circ T_1] = [T_2][T_1] = \begin{bmatrix} 0 & 0 \\ 0 & 1 \end{bmatrix} \begin{bmatrix} 0 & 1 \\ 1 & 0 \end{bmatrix} = \begin{bmatrix} 0 & 0 \\ 1 & 0 \end{bmatrix}$$

so $[T_2 \circ T_1] \neq [T_1 \circ T_2]$.

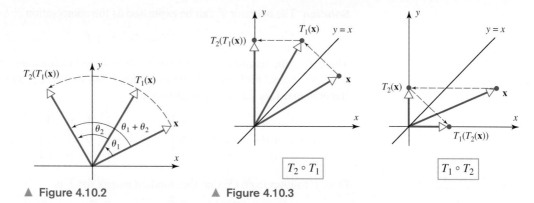

▲ Figure 4.10.2   ▲ Figure 4.10.3

▶ **EXAMPLE 3** **Composition of Two Reflections**

Let $T_1: R^2 \to R^2$ be the reflection about the y-axis, and let $T_2: R^2 \to R^2$ be the reflection about the x-axis. In this case $T_1 \circ T_2$ and $T_2 \circ T_1$ are the same; both map every vector $\mathbf{x} = (x, y)$ into its negative $-\mathbf{x} = (-x, -y)$ (Figure 4.10.4):

$$(T_1 \circ T_2)(x, y) = T_1(x, -y) = (-x, -y)$$
$$(T_2 \circ T_1)(x, y) = T_2(-x, y) = (-x, -y)$$

The equality of $T_1 \circ T_2$ and $T_2 \circ T_1$ can also be deduced by showing that the standard matrices for T_1 and T_2 commute:

$$[T_1 \circ T_2] = [T_1][T_2] = \begin{bmatrix} -1 & 0 \\ 0 & 1 \end{bmatrix} \begin{bmatrix} 1 & 0 \\ 0 & -1 \end{bmatrix} = \begin{bmatrix} -1 & 0 \\ 0 & -1 \end{bmatrix}$$

$$[T_2 \circ T_1] = [T_2][T_1] = \begin{bmatrix} 1 & 0 \\ 0 & -1 \end{bmatrix} \begin{bmatrix} -1 & 0 \\ 0 & 1 \end{bmatrix} = \begin{bmatrix} -1 & 0 \\ 0 & -1 \end{bmatrix}$$

The operator $T(\mathbf{x}) = -\mathbf{x}$ on R^2 or R^3 is called the ***reflection about the origin***. As the foregoing computations show, the standard matrix for this operator on R^2 is

$$[T] = \begin{bmatrix} -1 & 0 \\ 0 & -1 \end{bmatrix}$$

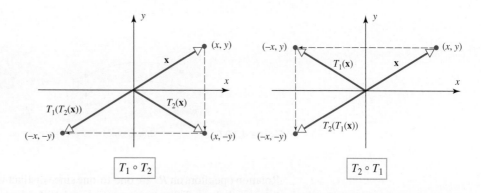

▶ Figure 4.10.4

▶ **EXAMPLE 4** **Composition of Three Transformations**

Find the standard matrix for the operator $T: R^3 \to R^3$ that first rotates a vector counterclockwise about the z-axis through an angle θ, then reflects the resulting vector about the yz-plane, and then projects that vector orthogonally onto the xy-plane.

Solution The operator T can be expressed as the composition

$$T = T_3 \circ T_2 \circ T_1$$

where T_1 is the rotation about the z-axis, T_2 is the reflection about the yz-plane, and T_3 is the orthogonal projection on the xy-plane. From Tables 6, 2, and 4 of Section 4.9, the standard matrices for these operators are

$$[T_1] = \begin{bmatrix} \cos\theta & -\sin\theta & 0 \\ \sin\theta & \cos\theta & 0 \\ 0 & 0 & 1 \end{bmatrix}, \quad [T_2] = \begin{bmatrix} -1 & 0 & 0 \\ 0 & 1 & 0 \\ 0 & 0 & 1 \end{bmatrix}, \quad [T_3] = \begin{bmatrix} 1 & 0 & 0 \\ 0 & 1 & 0 \\ 0 & 0 & 0 \end{bmatrix}$$

Thus, it follows from (5) that the standard matrix for T is

$$[T] = \begin{bmatrix} 1 & 0 & 0 \\ 0 & 1 & 0 \\ 0 & 0 & 0 \end{bmatrix} \begin{bmatrix} -1 & 0 & 0 \\ 0 & 1 & 0 \\ 0 & 0 & 1 \end{bmatrix} \begin{bmatrix} \cos\theta & -\sin\theta & 0 \\ \sin\theta & \cos\theta & 0 \\ 0 & 0 & 1 \end{bmatrix}$$

$$= \begin{bmatrix} -\cos\theta & \sin\theta & 0 \\ \sin\theta & \cos\theta & 0 \\ 0 & 0 & 0 \end{bmatrix} \blacktriangleleft$$

One-to-One Matrix Transformations

Our next objective is to establish a link between the invertibility of a matrix A and properties of the corresponding matrix transformation T_A.

> **DEFINITION 1** A matrix transformation $T_A: R^n \to R^m$ is said to be *one-to-one* if T_A maps distinct vectors (points) in R^n into distinct vectors (points) in R^m.

(See Figure 4.10.5). This idea can be expressed in various ways. For example, you should be able to see that the following are just restatements of Definition 1:

1. T_A is one-to-one if for each vector $\mathbf{b}$ in the range of A there is exactly one vector $\mathbf{x}$ in R^n such that $T_A \mathbf{x} = \mathbf{b}$.
2. T_A is one-to-one if the equality $T_A(\mathbf{u}) = T_A(\mathbf{v})$ implies that $\mathbf{u} = \mathbf{v}$.

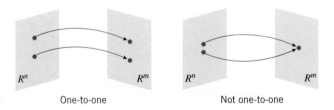

▶ Figure 4.10.5   One-to-one   Not one-to-one

Rotation operators on R^2 are one-to-one since distinct vectors that are rotated through the same angle have distinct images (Figure 4.10.6). In contrast, the orthogonal projection of R^3 on the xy-plane is not one-to-one because it maps distinct points on the same vertical line into the same point (Figure 4.10.7).

The following theorem establishes a fundamental relationship between the invertibility of a matrix and properties of the corresponding matrix transformation.

4.10 Properties of Matrix Transformations

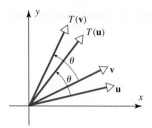

▲ **Figure 4.10.6** Distinct vectors **u** and **v** are rotated into distinct vectors $T(\mathbf{u})$ and $T(\mathbf{v})$.

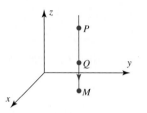

▲ **Figure 4.10.7** The distinct points P and Q are mapped into the same point M.

THEOREM 4.10.1 *If A is an $n \times n$ matrix and $T_A: R^n \to R^n$ is the corresponding matrix operator, then the following statements are equivalent.*

(a) *A is invertible.*

(b) *The range of T_A is R^n.*

(c) *T_A is one-to-one.*

Proof We will establish the chain of implications $(a) \Rightarrow (b) \Rightarrow (c) \Rightarrow (a)$.

$(a) \Rightarrow (b)$ Assume that A is invertible. By parts (a) and (e) of Theorem 4.8.10, the system $A\mathbf{x} = \mathbf{b}$ is consistent for every $n \times 1$ matrix $\mathbf{b}$ in R^n. This implies that T_A maps $\mathbf{x}$ into the arbitrary vector $\mathbf{b}$ in R^n, which in turn implies that the range of T_A is all of R^n.

$(b) \Rightarrow (c)$ Assume that the range of T_A is R^n. This implies that for every vector $\mathbf{b}$ in R^n there is some vector $\mathbf{x}$ in R^n for which $T_A(\mathbf{x}) = \mathbf{b}$ and hence that the linear system $A\mathbf{x} = \mathbf{b}$ is consistent for every vector $\mathbf{b}$ in R^n. But the equivalence of parts (e) and (f) of Theorem 4.8.10 implies that $A\mathbf{x} = \mathbf{b}$ has a unique solution for every vector $\mathbf{b}$ in R^n and hence that for every vector $\mathbf{b}$ in the range of T_A there is exactly one vector $\mathbf{x}$ in R^n such that $T_A\mathbf{x} = \mathbf{b}$.

$(c) \Rightarrow (a)$ Assume that T_A is one-to-one. Thus, if $\mathbf{b}$ is a vector in the range of T_A, there is a unique vector $\mathbf{x}$ in R^n for which $T_A(\mathbf{x}) = \mathbf{b}$. We leave it for you to complete the proof using Exercise 30. ◄

▶ **EXAMPLE 5 Properties of a Rotation Operator**

As indicated in Figure 4.10.6, the operator $T: R^n \to R^n$ that rotates vectors in R^2 through an angle θ is one-to-one. Confirm that $[T]$ is invertible in accordance with Theorem 4.10.1.

Solution From Table 5 of Section 4.9 the standard matrix for T is

$$[T] = \begin{bmatrix} \cos\theta & -\sin\theta \\ \sin\theta & \cos\theta \end{bmatrix}$$

This matrix is invertible because

$$\det[T] = \begin{vmatrix} \cos\theta & -\sin\theta \\ \sin\theta & \cos\theta \end{vmatrix} = \cos^2\theta + \sin^2\theta = 1 \neq 0$$

▶ **EXAMPLE 6 Properties of a Projection Operator**

As indicated in Figure 4.10.7, the operator $T: R^n \to R^n$ that projects each vector in R^3 orthogonally on the xy-plane is not one-to-one. Confirm that $[T]$ is not invertible in accordance with Theorem 4.10.1.

Solution From Table 4 of Section 4.9 the standard matrix for T is

$$[T] = \begin{bmatrix} 1 & 0 & 0 \\ 0 & 1 & 0 \\ 0 & 0 & 0 \end{bmatrix}$$

This matrix is not invertible since $\det[T] = 0$. ◄

Inverse of a One-to-One Matrix Operator

If $T_A: R^n \to R^n$ is a one-to-one matrix operator, then it follows from Theorem 4.10.1 that A is invertible. The matrix operator

$$T_{A^{-1}}: R^n \to R^n$$

that corresponds to A^{-1} is called the **inverse operator** or (more simply) the **inverse** of T_A. This terminology is appropriate because T_A and $T_{A^{-1}}$ cancel the effect of each other in the sense that if $\mathbf{x}$ is any vector in R^n, then

$$T_A(T_{A^{-1}}(\mathbf{x})) = AA^{-1}\mathbf{x} = I\mathbf{x} = \mathbf{x}$$
$$T_{A^{-1}}(T_A(\mathbf{x})) = A^{-1}A\mathbf{x} = I\mathbf{x} = \mathbf{x}$$

or, equivalently,

$$T_A \circ T_{A^{-1}} = T_{AA^{-1}} = T_I$$
$$T_{A^{-1}} \circ T_A = T_{A^{-1}A} = T_I$$

From a more geometric viewpoint, if $\mathbf{w}$ is the image of $\mathbf{x}$ under T_A, then $T_{A^{-1}}$ maps $\mathbf{w}$ back into $\mathbf{x}$, since

$$T_{A^{-1}}(\mathbf{w}) = T_{A^{-1}}(T_A(\mathbf{x})) = \mathbf{x}$$

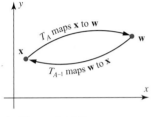

▲ Figure 4.10.8

(Figure 4.10.8).

Before considering examples, it will be helpful to touch on some notational matters. If $T_A: R^n \to R^n$ is a one-to-one matrix operator, and if $T_{A^{-1}}: R^n \to R^n$ is its inverse, then the standard matrices for these operators are related by the equation

$$T_{A^{-1}} = T_A^{-1} \tag{6}$$

In cases where it is preferable not to assign a name to the matrix, we will write this equation as

$$[T^{-1}] = [T]^{-1} \tag{7}$$

▶ **EXAMPLE 7 Standard Matrix for T^{-1}**

Let $T: R^2 \to R^2$ be the operator that rotates each vector in R^2 through the angle θ, so from Table 5 of Section 4.9,

$$[T] = \begin{bmatrix} \cos\theta & -\sin\theta \\ \sin\theta & \cos\theta \end{bmatrix} \tag{8}$$

It is evident geometrically that to undo the effect of T, one must rotate each vector in R^2 through the angle $-\theta$. But this is exactly what the operator T^{-1} does, since the standard matrix for T^{-1} is

$$[T^{-1}] = [T]^{-1} = \begin{bmatrix} \cos\theta & \sin\theta \\ -\sin\theta & \cos\theta \end{bmatrix} = \begin{bmatrix} \cos(-\theta) & -\sin(-\theta) \\ \sin(-\theta) & \cos(-\theta) \end{bmatrix}$$

(verify), which is the standard matrix for a rotation through the angle $-\theta$.

▶ **EXAMPLE 8 Finding T^{-1}**

Show that the operator $T: R^2 \to R^2$ defined by the equations

$$w_1 = 2x_1 + x_2$$
$$w_2 = 3x_1 + 4x_2$$

is one-to-one, and find $T^{-1}(w_1, w_2)$.

Solution The matrix form of these equations is

$$\begin{bmatrix} w_1 \\ w_2 \end{bmatrix} = \begin{bmatrix} 2 & 1 \\ 3 & 4 \end{bmatrix} \begin{bmatrix} x_1 \\ x_2 \end{bmatrix}$$

so the standard matrix for T is

$$[T] = \begin{bmatrix} 2 & 1 \\ 3 & 4 \end{bmatrix}$$

This matrix is invertible (so T is one-to-one) and the standard matrix for T^{-1} is

$$[T^{-1}] = [T]^{-1} = \begin{bmatrix} \frac{4}{5} & -\frac{1}{5} \\ -\frac{3}{5} & \frac{2}{5} \end{bmatrix}$$

Thus

$$[T^{-1}] \begin{bmatrix} w_1 \\ w_2 \end{bmatrix} = \begin{bmatrix} \frac{4}{5} & -\frac{1}{5} \\ -\frac{3}{5} & \frac{2}{5} \end{bmatrix} \begin{bmatrix} w_1 \\ w_2 \end{bmatrix} = \begin{bmatrix} \frac{4}{5}w_1 - \frac{1}{5}w_2 \\ -\frac{3}{5}w_1 + \frac{2}{5}w_2 \end{bmatrix}$$

from which we conclude that

$$T^{-1}(w_1, w_2) = \left(\tfrac{4}{5}w_1 - \tfrac{1}{5}w_2, -\tfrac{3}{5}w_1 + \tfrac{2}{5}w_2 \right) \blacktriangleleft$$

Linearity Properties

Up to now we have focused exclusively on matrix transformations from R^n to R^m. However, these are not the only kinds of transformations from R^n to R^m. For example, if $f_1, f_2, \ldots, f_m$ are any functions of the n variables $x_1, x_2, \ldots, x_n$, then the equations

$$w_1 = f_1(x_1, x_2, \ldots, x_n)$$
$$w_2 = f_2(x_1, x_2, \ldots, x_n)$$
$$\vdots$$
$$w_m = f_m(x_1, x_2, \ldots, x_n)$$

define a transformation $T: R^n \to R^m$ that maps the vector $\mathbf{x} = (x_1, x_2, \ldots, x_n)$ into the vector $(w_1, w_2, \ldots, w_m)$. But it is only in the case where these equations are *linear* that T is a matrix transformation. The question that we will now consider is this:

Question Are there algebraic properties of a transformation $T: R^n \to R^m$ that can be used to determine whether T is a matrix transformation?

The answer is provided by the following theorem.

THEOREM 4.10.2 *$T: R^n \to R^m$ is a matrix transformation if and only if the following relationships hold for all vectors $\mathbf{u}$ and $\mathbf{v}$ in R^n and for every scalar k:*

(i) $T(\mathbf{u} + \mathbf{v}) = T(\mathbf{u}) + T(\mathbf{v})$ [Additivity property]

(ii) $T(k\mathbf{u}) = kT(\mathbf{u})$ [Homogeneity property]

Proof If T is a matrix transformation, then properties (i) and (ii) follow respectively from parts (c) and (b) of Theorem 4.9.1.

Conversely, assume that properties (i) and (ii) hold. We must show that there exists an $m \times n$ matrix A such that

$$T(\mathbf{x}) = A\mathbf{x}$$

for every vector $\mathbf{x}$ in R^n. As a first step, recall from Formula (10) of Section 4.9 that the additivity and homogeneity properties imply that

$$T(k_1\mathbf{u}_1 + k_2\mathbf{u}_2 + \cdots + k_r\mathbf{u}_r) = k_1 T(\mathbf{u}_1) + k_2 T(\mathbf{u}_2) + \cdots + k_r T(\mathbf{u}_r) \quad (9)$$

for all scalars $k_1, k_2, \ldots, k_r$ and all vectors $\mathbf{u}_1, \mathbf{u}_2, \ldots, \mathbf{u}_r$ in R^n. Let A be the matrix

$$A = [T(\mathbf{e}_1) \mid T(\mathbf{e}_2) \mid \cdots \mid T(\mathbf{e}_n)]$$

in which $\mathbf{e}_1, \mathbf{e}_2, \ldots, \mathbf{e}_n$ are the standard basis vectors for R^n. It follows from Theorem 1.3.1 that $A\mathbf{x}$ is a linear combination of the columns of A in which the successive coefficients are the entries $x_1, x_2, \ldots, x_n$ of $\mathbf{x}$. That is,

$$A\mathbf{x} = x_1 T(\mathbf{e}_1) + x_2 T(\mathbf{e}_2) + \cdots + x_n T(\mathbf{e}_n)$$

Using (9) we can rewrite this as

$$A\mathbf{x} = T(x_1 \mathbf{e}_1 + x_2 \mathbf{e}_2 + \cdots + x_n \mathbf{e}_n) = T(\mathbf{x})$$

which completes the proof. ◄

The additivity and homogeneity properties in Theorem 4.10.2 are called *linearity conditions*, and a transformation that satisfies these conditions is called a *linear transformation*. Using this terminology Theorem 4.10.2 can be restated as follows.

THEOREM 4.10.3 *Every linear transformation from R^n to R^m is a matrix transformation, and conversely, every matrix transformation from R^n to R^m is a linear transformation.*

More on the Equivalence Theorem

As our final result in this section, we will add parts (b) and (c) of Theorem 4.10.1 to Theorem 4.8.10.

THEOREM 4.10.4 *Equivalent Statements*

If A is an $n \times n$ matrix, then the following statements are equivalent.

(a) A is invertible.
(b) $A\mathbf{x} = \mathbf{0}$ has only the trivial solution.
(c) The reduced row echelon form of A is I_n.
(d) A is expressible as a product of elementary matrices.
(e) $A\mathbf{x} = \mathbf{b}$ is consistent for every $n \times 1$ matrix $\mathbf{b}$.
(f) $A\mathbf{x} = \mathbf{b}$ has exactly one solution for every $n \times 1$ matrix $\mathbf{b}$.
(g) $\det(A) \neq 0$.
(h) The column vectors of A are linearly independent.
(i) The row vectors of A are linearly independent.
(j) The column vectors of A span R^n.
(k) The row vectors of A span R^n.
(l) The column vectors of A form a basis for R^n.
(m) The row vectors of A form a basis for R^n.
(n) A has rank n.
(o) A has nullity 0.
(p) The orthogonal complement of the null space of A is R^n.
(q) The orthogonal complement of the row space of A is $\{\mathbf{0}\}$.
(r) The range of T_A is R^n.
(s) T_A is one-to-one.

Concept Review

- Composition of matrix transformations
- Reflection about the origin
- One-to-one transformation
- Inverse of a matrix operator
- Linearity conditions
- Linear transformation
- Equivalent characterizations of invertible matrices

Skills

- Find the standard matrix for a composition of matrix transformations.
- Determine whether a matrix operator is one-to-one; if it is, then find the inverse operator.
- Determine whether a transformation is a linear transformation.

Exercise Set 4.10

▶ In Exercises 1–2, let T_A and T_B be the operators whose standard matrices are given. Find the standard matrices for $T_B \circ T_A$ and $T_A \circ T_B$. ◂

1. $A = \begin{bmatrix} 1 & -2 & 0 \\ 4 & 1 & -3 \\ 5 & 2 & 4 \end{bmatrix}$, $B = \begin{bmatrix} 2 & -3 & 3 \\ 5 & 0 & 1 \\ 6 & 1 & 7 \end{bmatrix}$

2. $A = \begin{bmatrix} 6 & 3 & -1 \\ 2 & 0 & 1 \\ 4 & -3 & 6 \end{bmatrix}$, $B = \begin{bmatrix} 4 & 0 & 4 \\ -1 & 5 & 2 \\ 2 & -3 & 8 \end{bmatrix}$

3. Let $T_1(x_1, x_2) = (x_1 + x_2, x_1 - x_2)$ and $T_2(x_1, x_2) = (3x_1, 2x_1 + 4x_2)$.
   (a) Find the standard matrices for T_1 and T_2.
   (b) Find the standard matrices for $T_2 \circ T_1$ and $T_1 \circ T_2$.
   (c) Use the matrices obtained in part (b) to find formulas for $T_1(T_2(x_1, x_2))$ and $T_2(T_1(x_1, x_2))$.

4. Let $T_1(x_1, x_2, x_3) = (4x_1, -2x_1 + x_2, -x_1 - 3x_2)$ and $T_2(x_1, x_2, x_3) = (x_1 + 2x_2, -x_3, 4x_1 - x_3)$.
   (a) Find the standard matrices for T_1 and T_2.
   (b) Find the standard matrices for $T_2 \circ T_1$ and $T_1 \circ T_2$.
   (c) Use the matrices obtained in part (b) to find formulas for $T_1(T_2(x_1, x_2, x_3))$ and $T_2(T_1(x_1, x_2, x_3))$.

5. Find the standard matrix for the stated composition in R^2.
   (a) A reflection about the line $y = x$, followed by a rotation of $60°$.
   (b) A contraction of a factor of $k = \frac{1}{3}$, followed by an orthogonal projection onto the y-axis.
   (c) A reflection about the y-axis, followed by a dilation with factor $k = 4$.

6. Find the standard matrix for the stated composition in R^2.
   (a) A rotation of $60°$, followed by an orthogonal projection on the x-axis, followed by a reflection about the line $y = x$.
   (b) A dilation with factor $k = 2$, followed by a rotation of $45°$, followed by a reflection about the y-axis.
   (c) A rotation of $15°$, followed by a rotation of $105°$, followed by a rotation of $60°$.

7. Find the standard matrix for the stated composition in R^3.
   (a) A reflection about the yz-plane, followed by an orthogonal projection on the xz-plane.
   (b) A rotation of $45°$ about the y-axis, followed by a dilation with factor $k = \sqrt{2}$.
   (c) An orthogonal projection on the xy-plane, followed by a reflection about the yz-plane.

8. Find the standard matrix for the stated composition in R^3.
   (a) A rotation of $30°$ about the x-axis, followed by a rotation of $30°$ about the z-axis, followed by a contraction with factor $k = \frac{1}{4}$.
   (b) A reflection about the xy-plane, followed by a reflection about the xz-plane, followed by an orthogonal projection on the yz-plane.
   (c) A rotation of $270°$ about the x-axis, followed by a rotation of $90°$ about the y-axis, followed by a rotation of $180°$ about the z-axis.

9. Determine whether $T_1 \circ T_2 = T_2 \circ T_1$.
   (a) $T_1: R^2 \to R^2$ is the orthogonal projection on the x-axis, and $T_2: R^2 \to R^2$ is the orthogonal projection on the y-axis.
   (b) $T_1: R^2 \to R^2$ is the rotation through an angle θ_1, and $T_2: R^2 \to R^2$ is the rotation through an angle θ_2.
   (c) $T_1: R^2 \to R^2$ is the orthogonal projection on the x-axis, and $T_2: R^2 \to R^2$ is the rotation through an angle θ.

10. Determine whether $T_1 \circ T_2 = T_2 \circ T_1$.
    (a) $T_1: R^3 \to R^3$ is a dilation by a factor k, and $T_2: R^3 \to R^3$ is the rotation about the z-axis through an angle θ.
    (b) $T_1: R^3 \to R^3$ is the rotation about the x-axis through an angle θ_1, and $T_2: R^3 \to R^3$ is the rotation about the z-axis through an angle θ_2.

11. By inspection, determine whether the matrix operator is one-to-one.
    (a) the orthogonal projection on the x-axis in R^2
    (b) the reflection about the y-axis in R^2
    (c) the reflection about the line $y = x$ in R^2
    (d) a contraction with factor $k > 0$ in R^2
    (e) a rotation about the z-axis in R^3
    (f) a reflection about the xy-plane in R^3
    (g) a dilation with factor $k > 0$ in R^3

12. Find the standard matrix for the matrix operator defined by the equations, and use Theorem 4.10.4 to determine whether the operator is one-to-one.
    (a) $w_1 = 8x_1 + 4x_2$
        $w_2 = 2x_1 + x_2$
    (b) $w_1 = 2x_1 - 3x_2$
        $w_2 = 5x_1 + x_2$
    (c) $w_1 = -x_1 + 3x_2 + 2x_3$
        $w_2 = 2x_1 \quad\quad + 4x_3$
        $w_3 = x_1 + 3x_2 + 6x_3$
    (d) $w_1 = x_1 + 2x_2 + 3x_3$
        $w_2 = 2x_1 + 5x_2 + 3x_3$
        $w_3 = x_1 \quad\quad + 8x_3$

13. Determine whether the matrix operator $T: R^2 \to R^2$ defined by the equations is one-to-one; if so, find the standard matrix for the inverse operator, and find $T^{-1}(w_1, w_2)$.
    (a) $w_1 = \quad\quad + 2x_2$
        $w_2 = -x_1$
    (b) $w_1 = 9x_1 + 5x_2$
        $w_2 = 2x_1 - 7x_2$
    (c) $w_1 = -x_2$
        $w_2 = -x_1$
    (d) $w_1 = 3x_1$
        $w_2 = -5x_1$

14. Determine whether the matrix operator $T: R^3 \to R^3$ defined by the equations is one-to-one; if so, find the standard matrix for the inverse operator, and find $T^{-1}(w_1, w_2, w_3)$.
    (a) $w_1 = x_1 - 2x_2 + 2x_3$
        $w_2 = 2x_1 + x_2 + x_3$
        $w_3 = x_1 + x_2$
    (b) $w_1 = x_1 - 3x_2 + 4x_3$
        $w_2 = -x_1 + x_2 + x_3$
        $w_3 = \quad\quad - 2x_2 + 5x_3$
    (c) $w_1 = x_1 + 4x_2 - x_3$
        $w_2 = 2x_1 + 7x_2 + x_3$
        $w_3 = x_1 + 3x_2$
    (d) $w_1 = x_1 + 2x_2 + x_3$
        $w_2 = -2x_1 + x_2 + 4x_3$
        $w_3 = 7x_1 + 4x_2 - 5x_3$

15. By inspection, find the inverse of the given one-to-one matrix operator.
    (a) The reflection about the x-axis in R^2.
    (b) The rotation through an angle of $\pi/4$ in R^2.
    (c) The dilation by a factor of 3 in R^2.
    (d) The reflection about the yz-plane in R^3.
    (e) The contraction by a factor of $\frac{1}{5}$ in R^3.

▶ In Exercises **16–17**, use Theorem 4.10.2 to determine whether $T: R^2 \to R^2$ is a matrix operator. ◀

16. (a) $T(x, y) = (2x, y)$
    (b) $T(x, y) = (x^2, y)$
    (c) $T(x, y) = (-y, x)$
    (d) $T(x, y) = (x, 0)$

17. (a) $T(x, y) = (\cos^2 x, \sin^2 y)$
    (b) $T(x, y) = (3x - y, 5y)$
    (c) $T(x, y) = (-x, -x)$
    (d) $T(x, y) = (x + y, x - y)$

▶ In Exercises **18–19**, use Theorem 4.10.2 to determine whether $T: R^3 \to R^2$ is a matrix transformation. ◀

18. (a) $T(x, y, z) = (x, x + y + z)$
    (b) $T(x, y, z) = (1, 1)$

19. (a) $T(x, y, z) = (0, 0)$
    (b) $T(x, y, z) = (6x + y, x - 6y)$

20. In each part, use Theorem 4.10.3 to find the standard matrix for the matrix operator from the images of the standard basis vectors.
    (a) The reflection operators on R^2 in Table 1 of Section 4.9.
    (b) The reflection operators on R^3 in Table 2 of Section 4.9.
    (c) The projection operators on R^2 in Table 3 of Section 4.9.
    (d) The projection operators on R^3 in Table 4 of Section 4.9.
    (e) The rotation operators on R^2 in Table 5 of Section 4.9.
    (f) The dilation and contraction operators on R^3 in Table 8 of Section 4.9.

21. Find the standard matrix for the given matrix operator.
    (a) $T: R^2 \to R^2$ projects a vector orthogonally onto the x-axis and then reflects that vector about the y-axis.
    (b) $T: R^2 \to R^2$ reflects a vector about the line $y = x$ and then reflects that vector about the x-axis.
    (c) $T: R^2 \to R^2$ dilates a vector by a factor of 3, then reflects that vector about the line $y = x$, and then projects that vector orthogonally onto the y-axis.

22. Find the standard matrix for the given matrix operator.
    (a) $T: R^3 \to R^3$ reflects a vector about the xz-plane and then contracts that vector by a factor of $\frac{1}{5}$.
    (b) $T: R^3 \to R^3$ projects a vector orthogonally onto the xz-plane and then projects that vector orthogonally onto the xy-plane.
    (c) $T: R^3 \to R^3$ reflects a vector about the xy-plane, then reflects that vector about the xz-plane, and then reflects that vector about the yz-plane.

23. Let $T_A: R^3 \to R^3$ be multiplication by
$$A = \begin{bmatrix} 4 & -1 & 2 \\ 5 & 1 & 2 \\ 3 & 6 & -4 \end{bmatrix}$$
and let $\mathbf{e}_1$, $\mathbf{e}_2$, and $\mathbf{e}_3$ be the standard basis vectors for R^3. Find the following vectors by inspection.
    (a) $T_A(\mathbf{e}_1)$, $T_A(\mathbf{e}_2)$, and $T_A(\mathbf{e}_3)$
    (b) $T_A(\mathbf{e}_1 + \mathbf{e}_2 + \mathbf{e}_3)$
    (c) $T_A(7\mathbf{e}_3)$

24. Determine whether multiplication by A is a one-to-one matrix transformation.

   (a) $A = \begin{bmatrix} 1 & -1 \\ 2 & 0 \\ 3 & -4 \end{bmatrix}$ (b) $A = \begin{bmatrix} 1 & 2 & 3 \\ -1 & 0 & -4 \end{bmatrix}$

   (c) $A = \begin{bmatrix} 1 & 2 & 1 \\ 0 & 1 & 1 \\ 1 & 1 & 0 \\ 1 & 0 & -1 \end{bmatrix}$

25. (a) Is a composition of one-to-one matrix transformations one-to-one? Justify your conclusion.

   (b) Can the composition of a one-to-one matrix transformation and a matrix transformation that is not one-to-one be one-to-one? Account for both possible orders of composition and justify your conclusion.

26. Show that $T(x, y) = (0, 0)$ defines a matrix operator on R^2 but $T(x, y) = (1, 1)$ does not.

27. (a) Prove: If $T: R^n \to R^m$ is a matrix transformation, then $T(\mathbf{0}) = \mathbf{0}$; that is, T maps the zero vector in R^n into the zero vector in R^m.

   (b) The converse of this is not true. Find an example of a function that satisfies $T(\mathbf{0}) = \mathbf{0}$ but is not a matrix transformation.

28. Prove: An $n \times n$ matrix A is invertible if and only if the linear system $A\mathbf{x} = \mathbf{w}$ has exactly one solution for every vector $\mathbf{w}$ in R^n for which the system is consistent.

29. Let A be an $n \times n$ matrix such that $\det(A) = 0$, and let $T: R^n \to R^n$ be multiplication by A.

   (a) What can you say about the range of the matrix operator T? Give an example that illustrates your conclusion.

   (b) What can you say about the number of vectors that T maps into $\mathbf{0}$?

30. Prove: If the matrix transformation $T_A: R^n \to R^n$ is one-to-one, then A is invertible.

True-False Exercises

In parts (a)–(f) determine whether the statement is true or false, and justify your answer.

(a) If $T: R^n \to R^m$ and $T(\mathbf{0}) = \mathbf{0}$, then T is a matrix transformation.

(b) If $T: R^n \to R^m$ and $T(c_1\mathbf{x} + c_2\mathbf{y}) = c_1 T(\mathbf{x}) + c_2 T(\mathbf{y})$ for all scalars c_1 and c_2 and all vectors $\mathbf{x}$ and $\mathbf{y}$ in R^n, then T is a matrix transformation.

(c) If $T: R^n \to R^m$ is a one-to-one matrix transformation, then there are no distinct vectors $\mathbf{x}$ and $\mathbf{y}$ for which $T(\mathbf{x} - \mathbf{y}) = \mathbf{0}$.

(d) If $T: R^n \to R^m$ is a matrix transformation and $m > n$, then T is one-to-one.

(e) If $T: R^n \to R^m$ is a matrix transformation and $m = n$, then T is one-to-one.

(f) If $T: R^n \to R^m$ is a matrix transformation and $m < n$, then T is one-to-one.

4.11 Geometry of Matrix Operators on R^2

In this optional section we will discuss matrix operators on R^2 in a little more depth. The ideas that we will develop here have important applications to computer graphics.

Transformations of Regions

In Section 4.9 we focused on the effect that a matrix operator has on individual vectors in R^2 and R^3. However, it is also important to understand how such operators affect the shapes of regions. For example, Figure 4.11.1 shows a famous picture of Albert Einstein and three computer-generated modifications of that image that result from matrix operators on R^2. The original picture was scanned and then digitized to decompose it into a rectangular array of pixels. The pixels were then transformed as follows:

- The program MATLAB was used to assign coordinates and a gray level to each pixel.
- The coordinates of the pixels were transformed by matrix multiplication.
- The pixels were then assigned their original gray levels to produce the transformed picture.

The overall effect of a matrix operator on R^2 can often be ascertained by graphing the images of the vertices $(0, 0)$, $(1, 0)$, $(0, 1)$, and $(1, 1)$ of the unit square (Figure 4.11.2).

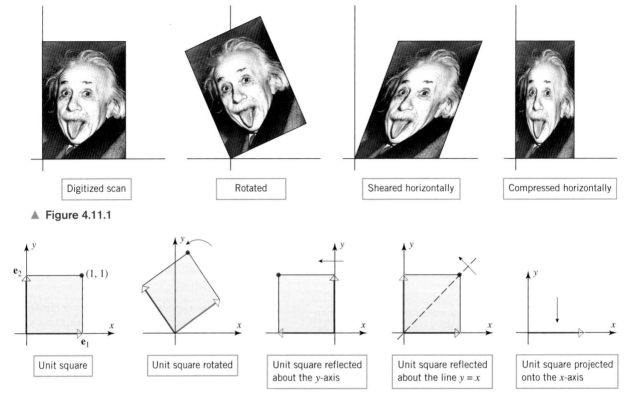

▲ Figure 4.11.1

▲ Figure 4.11.2

Table 1 shows the effect that some of the matrix operators studied in Section 4.9 have on the unit square. For clarity, we have shaded a portion of the original square and its corresponding image.

▶ **EXAMPLE 1 Transforming with Diagonal Matrices**

Suppose that the xy-plane first is compressed or expanded by a factor of k_1 in the x-direction and then is compressed or expanded by a factor of k_2 in the y-direction. Find a single matrix operator that performs both operations.

Solution The standard matrices for the two operations are

$$\begin{bmatrix} k_1 & 0 \\ 0 & 1 \end{bmatrix} \qquad \begin{bmatrix} 1 & 0 \\ 0 & k_2 \end{bmatrix}$$

x-compression (expansion) y-compression (expansion)

Thus, the standard matrix for the composition of the x-operation followed by the y-operation is

$$A = \begin{bmatrix} 1 & 0 \\ 0 & k_2 \end{bmatrix} \begin{bmatrix} k_1 & 0 \\ 0 & 1 \end{bmatrix} = \begin{bmatrix} k_1 & 0 \\ 0 & k_2 \end{bmatrix} \qquad (1)$$

This shows that multiplication by a diagonal 2×2 matrix compresses or expands the plane in the x-direction and also in the y-direction. In the special case where k_1 and k_2 are the same, say $k_1 = k_2 = k$, Formula (1) simplifies to

$$A = \begin{bmatrix} k & 0 \\ 0 & k \end{bmatrix}$$

which is a contraction or a dilation (Table 7 of Section 4.9). ◀

Table 1

Operator	Standard Matrix	Effect on the Unit Square
Reflection about the y-axis	$\begin{bmatrix} -1 & 0 \\ 0 & 1 \end{bmatrix}$	$(1,1) \rightarrow (-1,1)$
Reflection about the x-axis	$\begin{bmatrix} 1 & 0 \\ 0 & -1 \end{bmatrix}$	$(1,1) \rightarrow (1,-1)$
Reflection about the line $y = x$	$\begin{bmatrix} 0 & 1 \\ 1 & 0 \end{bmatrix}$	$(1,1) \rightarrow (1,1)$
Counterclockwise rotation through an angle θ	$\begin{bmatrix} \cos\theta & -\sin\theta \\ \sin\theta & \cos\theta \end{bmatrix}$	$(1,1) \rightarrow (\cos\theta - \sin\theta, \sin\theta + \cos\theta)$
Compression in the x-direction by a factor of k $(0 < k < 1)$	$\begin{bmatrix} k & 0 \\ 0 & 1 \end{bmatrix}$	$(1,1) \rightarrow (k,1)$
Expansion in the x-direction by a factor of k $(k > 1)$	$\begin{bmatrix} k & 0 \\ 0 & 1 \end{bmatrix}$	$(1,1) \rightarrow (k,1)$
Shear in the x-direction with factor $k > 0$	$\begin{bmatrix} 1 & k \\ 0 & 1 \end{bmatrix}$	$(1,1) \rightarrow (x+ky, y)$
Shear in the x-direction with factor $k < 0$	$\begin{bmatrix} 1 & k \\ 0 & 1 \end{bmatrix}$	$(1,1) \rightarrow (x+ky, y)$

EXAMPLE 2 Finding Matrix Operators

(a) Find the standard matrix for the operator on R^2 that first shears by a factor of 2 in the x-direction and then reflects the result about the line $y = x$. Sketch the image of the unit square under this operator.

(b) Find the standard matrix for the operator on R^2 that first reflects about $y = x$ and then shears by a factor of 2 in the x-direction. Sketch the image of the unit square under this operator.

(c) Confirm that the shear and the reflection in parts (a) and (b) do not commute.

Solution (a) The standard matrix for the shear is

$$A_1 = \begin{bmatrix} 1 & 2 \\ 0 & 1 \end{bmatrix}$$

and for the reflection is

$$A_2 = \begin{bmatrix} 0 & 1 \\ 1 & 0 \end{bmatrix}$$

Thus, the standard matrix for the shear followed by the reflection is

$$A_2 A_1 = \begin{bmatrix} 0 & 1 \\ 1 & 0 \end{bmatrix} \begin{bmatrix} 1 & 2 \\ 0 & 1 \end{bmatrix} = \begin{bmatrix} 0 & 1 \\ 1 & 2 \end{bmatrix}$$

Solution (b) The standard matrix for the reflection followed by the shear is

$$A_1 A_2 = \begin{bmatrix} 1 & 2 \\ 0 & 1 \end{bmatrix} \begin{bmatrix} 0 & 1 \\ 1 & 0 \end{bmatrix} = \begin{bmatrix} 2 & 1 \\ 1 & 0 \end{bmatrix}$$

Solution (c) The computations in Solutions (a) and (b) show that $A_1 A_2 \neq A_2 A_1$, so the standard matrices, and hence the operators, do not commute. The same conclusion follows from Figures 4.11.3 and 4.11.4, since the two operators produce different images of the unit square. ◀

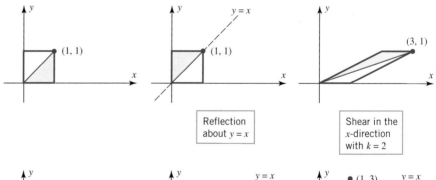

▶ Figure 4.11.3

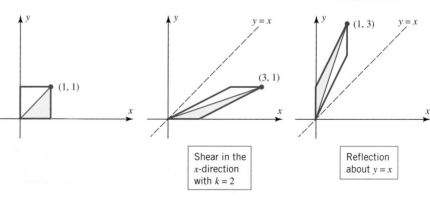

▶ Figure 4.11.4

4.11 Geometry of Matrix Operators on R^2

Geometry of One-to-One Matrix Operators

We will now turn our attention to one-to-one matrix operators on R^2, which are important because they map distinct points into distinct points. Recall from Theorem 4.10.4 (the Equivalence Theorem) that a matrix transformation T_A is one-to-one if and only if A can be expressed as a product of elementary matrices. Thus, we can analyze the effect of any one-to-one transformation T_A by first factoring the matrix A into a product of elementary matrices, say

$$A = E_1 E_2 \cdots E_r$$

and then expressing T_A as the composition

$$T_A = T_{E_1 E_2 \cdots E_r} = T_{E_1} \circ T_{E_2} \circ \cdots \circ T_{E_r} \qquad (2)$$

The following theorem explains the geometric effect of matrix operators corresponding to elementary matrices.

THEOREM 4.11.1 *If E is an elementary matrtix, then $T_E: R^2 \to R^2$ is one of the following*:
(a) *A shear along a coordinate axis.*
(b) *A reflection about $y = x$.*
(c) *A compression along a coordinate axis.*
(d) *An expansion along a coordinate axis.*
(e) *A reflection about a coordinate axis.*
(f) *A compression or expansion along a coordinate axis followed by a reflection about a coordinate axis.*

Proof Because a 2×2 elementary matrix results from performing a single elementary row operation on the 2×2 identity matrix, such a matrix must have one of the following forms (verify):

$$\begin{bmatrix} 1 & 0 \\ k & 1 \end{bmatrix}, \begin{bmatrix} 1 & k \\ 0 & 1 \end{bmatrix}, \begin{bmatrix} 0 & 1 \\ 1 & 0 \end{bmatrix}, \begin{bmatrix} k & 0 \\ 0 & 1 \end{bmatrix}, \begin{bmatrix} 1 & 0 \\ 0 & k \end{bmatrix}$$

The first two matrices represent shears along coordinate axes, and the third represents a reflection about $y = x$. If $k > 0$, the last two matrices represent compressions or expansions along coordinate axes, depending on whether $0 \le k < 1$ or $k > 1$. If $k < 0$, and if we express k in the form $k = -k_1$, where $k_1 > 0$, then the last two matrices can be written as

$$\begin{bmatrix} k & 0 \\ 0 & 1 \end{bmatrix} = \begin{bmatrix} -k_1 & 0 \\ 0 & 1 \end{bmatrix} = \begin{bmatrix} -1 & 0 \\ 0 & 1 \end{bmatrix} \begin{bmatrix} k_1 & 0 \\ 0 & 1 \end{bmatrix} \qquad (3)$$

$$\begin{bmatrix} 1 & 0 \\ 0 & k \end{bmatrix} = \begin{bmatrix} 1 & 0 \\ 0 & -k_1 \end{bmatrix} = \begin{bmatrix} 1 & 0 \\ 0 & -1 \end{bmatrix} \begin{bmatrix} 1 & 0 \\ 0 & k_1 \end{bmatrix} \qquad (4)$$

Since $k_1 > 0$, the product in (3) represents a compression or expansion along the x-axis followed by a reflection about the y-axis, and (4) represents a compression or expansion along the y-axis followed by a reflection about the x-axis. In the case where $k = -1$, transformations (3) and (4) are simply reflections about the y-axis and x-axis, respectively. ◀

Since every invertible matrix is a product of elementary matrices, the following result follows from Theorem 4.11.1 and Formula (2).

THEOREM 4.11.2 *If $T_A: R^2 \to R^2$ is multiplication by an invertible matrix A, then the geometric effect of T_A is the same as an appropriate succession of shears, compressions, expansions, and reflections.*

▶ **EXAMPLE 3** Analyzing the Geometric Effect of a Matrix Operator

Assuming that k_1 and k_2 are positive, express the diagonal matrix

$$A = \begin{bmatrix} k_1 & 0 \\ 0 & k_2 \end{bmatrix}$$

as a product of elementary matrices, and describe the geometric effect of multiplication by A in terms of compressions and expansions.

Solution From Example 1 we have

$$A = \begin{bmatrix} k_1 & 0 \\ 0 & k_2 \end{bmatrix} = \begin{bmatrix} 1 & 0 \\ 0 & k_2 \end{bmatrix} \begin{bmatrix} k_1 & 0 \\ 0 & 1 \end{bmatrix}$$

which shows that multiplication by A has the geometric effect of compressing or expanding by a factor of k_1 in the x-direction and then compressing or expanding by a factor of k_2 in the y-direction. ◀

▶ **EXAMPLE 4** Analyzing the Geometric Effect of a Matrix Operator

Express

$$A = \begin{bmatrix} 1 & 2 \\ 3 & 4 \end{bmatrix}$$

as a product of elementary matrices, and then describe the geometric effect of multiplication by A in terms of shears, compressions, expansions, and reflections.

Solution A can be reduced to I as follows:

$$\begin{bmatrix} 1 & 2 \\ 3 & 4 \end{bmatrix} \to \begin{bmatrix} 1 & 2 \\ 0 & -2 \end{bmatrix} \to \begin{bmatrix} 1 & 2 \\ 0 & 1 \end{bmatrix} \to \begin{bmatrix} 1 & 0 \\ 0 & 1 \end{bmatrix}$$

↑ Add -3 times the first row to the second. ↑ Multiply the second row by $-\frac{1}{2}$. ↑ Add -2 times the second row to the first.

The three successive row operations can be performed by multiplying A on the left successively by

$$E_1 = \begin{bmatrix} 1 & 0 \\ -3 & 1 \end{bmatrix}, \quad E_2 = \begin{bmatrix} 1 & 0 \\ 0 & -\frac{1}{2} \end{bmatrix}, \quad E_3 = \begin{bmatrix} 1 & -2 \\ 0 & 1 \end{bmatrix}$$

Inverting these matrices and using Formula (4) of Section 1.5 yields

$$A = E_1^{-1} E_2^{-1} E_3^{-1} = \begin{bmatrix} 1 & 0 \\ 3 & 1 \end{bmatrix} \begin{bmatrix} 1 & 0 \\ 0 & -2 \end{bmatrix} \begin{bmatrix} 1 & 2 \\ 0 & 1 \end{bmatrix}$$

Reading from right to left and noting that

$$\begin{bmatrix} 1 & 0 \\ 0 & -2 \end{bmatrix} = \begin{bmatrix} 1 & 0 \\ 0 & -1 \end{bmatrix} \begin{bmatrix} 1 & 0 \\ 0 & 2 \end{bmatrix}$$

it follows that the effect of multiplying by A is equivalent to

1. shearing by a factor of 2 in the x-direction,
2. then expanding by a factor of 2 in the y-direction,
3. then reflecting about the x-axis,
4. then shearing by a factor of 3 in the y-direction. ◀

Images of Lines Under Matrix Operators

Many images in computer graphics are constructed by connecting points with line segments. The following theorem, some of whose parts are proved in the exercises, is helpful for understanding how matrix operators transform such figures.

THEOREM 4.11.3 *If $T: R^2 \to R^2$ is multiplication by an invertible matrix, then:*

(a) *The image of a straight line is a straight line.*
(b) *The image of a straight line through the origin is a straight line through the origin.*
(c) *The images of parallel straight lines are parallel straight lines.*
(d) *The image of the line segment joining points P and Q is the line segment joining the images of P and Q.*
(e) *The images of three points lie on a line if and only if the points themselves lie on a line.*

Note that it follows from Theorem 4.11.3 that if A is an invertible 2×2 matrix, then multiplication by A maps triangles into triangles and parallelograms into parallelograms.

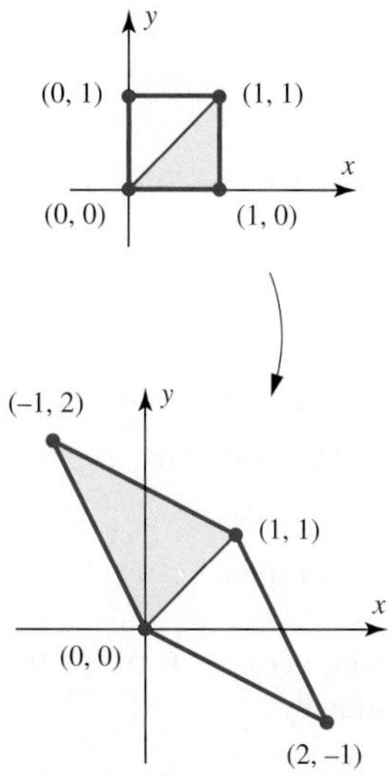

▲ **Figure 4.11.5**

▶ **EXAMPLE 5** **Image of a Square**

Sketch the image of the square with vertices $(0,0)$, $(1,0)$, $(1,1)$, and $(0,1)$ under multiplication by

$$A = \begin{bmatrix} -1 & 2 \\ 2 & -1 \end{bmatrix}$$

Solution Since

$$\begin{bmatrix} -1 & 2 \\ 2 & -1 \end{bmatrix} \begin{bmatrix} 0 \\ 0 \end{bmatrix} = \begin{bmatrix} 0 \\ 0 \end{bmatrix}, \quad \begin{bmatrix} -1 & 2 \\ 2 & -1 \end{bmatrix} \begin{bmatrix} 1 \\ 0 \end{bmatrix} = \begin{bmatrix} -1 \\ 2 \end{bmatrix}$$

$$\begin{bmatrix} -1 & 2 \\ 2 & -1 \end{bmatrix} \begin{bmatrix} 0 \\ 1 \end{bmatrix} = \begin{bmatrix} 2 \\ -1 \end{bmatrix}, \quad \begin{bmatrix} -1 & 2 \\ 2 & -1 \end{bmatrix} \begin{bmatrix} 1 \\ 1 \end{bmatrix} = \begin{bmatrix} 1 \\ 1 \end{bmatrix}$$

the image of the square is a parallelogram with vertices $(0,0)$, $(-1, 2)$, $(2, -1)$, and $(1, 1)$ (Figure 4.11.5).

▶ **EXAMPLE 6** **Image of a Line**

According to Theorem 4.11.3, the invertible matrix

$$A = \begin{bmatrix} 3 & 1 \\ 2 & 1 \end{bmatrix}$$

maps the line $y = 2x + 1$ into another line. Find its equation.

Solution Let (x, y) be a point on the line $y = 2x + 1$, and let (x', y') be its image under multiplication by A. Then

$$\begin{bmatrix} x' \\ y' \end{bmatrix} = \begin{bmatrix} 3 & 1 \\ 2 & 1 \end{bmatrix} \begin{bmatrix} x \\ y \end{bmatrix} \quad \text{and} \quad \begin{bmatrix} x \\ y \end{bmatrix} = \begin{bmatrix} 3 & 1 \\ 2 & 1 \end{bmatrix}^{-1} \begin{bmatrix} x' \\ y' \end{bmatrix} = \begin{bmatrix} 1 & -1 \\ -2 & 3 \end{bmatrix} \begin{bmatrix} x' \\ y' \end{bmatrix}$$

so
$$x = x' - y'$$
$$y = -2x' + 3y'$$

Substituting in $y = 2x + 1$ yields

$$-2x' + 3y' = 2(x' - y') + 1 \quad \text{or equivalently} \quad y' = \tfrac{4}{5}x' + \tfrac{1}{5}$$

Thus (x', y') satisfies

$$y = \tfrac{4}{5}x + \tfrac{1}{5}$$

which is the equation we want. ◀

Concept Review

- Effect of a matrix operator on the unit square
- Geometry of one-to-one matrix operators
- Images of lines under matrix operators

Skills

- Find standard matrices for geometric transformations of R^2.
- Describe the geometric effect of an invertible matrix operator.
- Find the image of the unit square under a matrix operator.
- Find the image of a line under a matrix operator.

Exercise Set 4.11

1. Find the standard matrix for the operator $T: R^2 \to R^2$ that maps a point (x, y) into
   (a) its reflection about the line $y = -x$.
   (b) its reflection through the origin.
   (c) its orthogonal projection on the x-axis.
   (d) its orthogonal projection on the y-axis.

2. For each part of Exercise 1, use the matrix you have obtained to compute $T(1, 3)$. Check your answers geometrically by plotting the points $(1, 3)$ and $T(1, 3)$.

3. Find the standard matrix for the operator $T: R^3 \to R^3$ that maps a point (x, y, z) into
   (a) its reflection through the xy-plane.
   (b) its reflection through the xz-plane.
   (c) its reflection through the yz-plane.

4. For each part of Exercise 3, use the matrix you have obtained to compute $T(1, 2, 1)$. Check your answers geometrically by plotting the points $(1, 2, 1)$ and $T(1, 2, 1)$.

5. Find the standard matrix for the operator $T: R^3 \to R^3$ that
   (a) rotates each vector $90°$ counterclockwise about the z-axis (looking along the positive z-axis toward the origin).
   (b) rotates each vector $90°$ counterclockwise about the x-axis (looking along the positive x-axis toward the origin).
   (c) rotates each vector $90°$ counterclockwise about the y-axis (looking along the positive y-axis toward the origin).

6. Sketch the image of the rectangle with vertices $(0, 0)$, $(1, 0)$, $(1, 2)$, and $(0, 2)$ under
   (a) a reflection about the x-axis.
   (b) a reflection about the y-axis.
   (c) a compression of factor $k = \tfrac{1}{4}$ in the y-direction.
   (d) an expansion of factor $k = 2$ in the x-direction.
   (e) a shear of factor $k = 3$ in the x-direction.
   (f) a shear of factor $k = 2$ in the y-direction.

7. Sketch the image of the square with vertices $(0, 0)$, $(1, 0)$, $(0, 1)$, and $(1, 1)$ under multiplication by

$$A = \begin{bmatrix} 0 & 2 \\ -1 & 0 \end{bmatrix}$$

8. Find the matrix that rotates a point (x, y) about the origin through
   (a) $45°$ (b) $90°$ (c) $180°$ (d) $270°$ (e) $-30°$

9. Find the matrix that shears by
   (a) a factor of $k = 4$ in the y-direction.
   (b) a factor of $k = -2$ in the x-direction.

10. Find the matrix that compresses or expands by
    (a) a factor of $\tfrac{1}{3}$ in the y-direction.
    (b) a factor of 6 in the x-direction.

11. In each part, describe the geometric effect on R^2 of multiplication by A.

   (a) $A = \begin{bmatrix} 1 & 0 \\ 0 & \frac{1}{3} \end{bmatrix}$ (b) $A = \begin{bmatrix} 6 & 0 \\ 0 & -1 \end{bmatrix}$ (c) $A = \begin{bmatrix} 2 & 1 \\ 4 & 0 \end{bmatrix}$

12. In each part, express the matrix as a product of elementary matrices, and then describe the effect on R^2 of multiplication by A in terms of compressions, expansions, reflections, and shears.

   (a) $A = \begin{bmatrix} 2 & 0 \\ 0 & 3 \end{bmatrix}$ (b) $A = \begin{bmatrix} 1 & 4 \\ 2 & 9 \end{bmatrix}$

   (c) $A = \begin{bmatrix} 0 & -2 \\ 4 & 0 \end{bmatrix}$ (d) $A = \begin{bmatrix} 1 & -3 \\ 4 & 6 \end{bmatrix}$

13. In each part, find a single matrix that performs the indicated succession of operations on R^2.

   (a) Compresses by a factor of $\frac{1}{2}$ in the x-direction, then expands by a factor of 5 in the y-direction.

   (b) Expands by a factor of 5 in the y-direction, then shears by a factor of 2 in the y-direction.

   (c) Reflects about $y = x$, then rotates through an angle of $180°$ about the origin.

14. In each part, find a single matrix that performs the indicated succession of operations on R^2.

   (a) Reflects about the y-axis, then expands by a factor of 5 in the x-direction, and then reflects about $y = x$.

   (b) Rotates through $30°$ about the origin, then shears by a factor of -2 in the y-direction, and then expands by a factor of 3 in the y-direction.

15. Use matrix inversion to show the following.

   (a) The inverse transformation for a reflection about $y = x$ is a reflection about $y = x$.

   (b) The inverse transformation for a compression along an axis is an expansion along that axis.

   (c) The inverse transformation for a reflection about a coordinate axis is a reflection about that axis.

   (d) The inverse transformation for a shear along a coordinate axis is a shear along that axis.

16. Find an equation of the image of the line $y = -4x + 3$ under multiplication by
$$A = \begin{bmatrix} 4 & -3 \\ 3 & -2 \end{bmatrix}$$

17. In parts (a) through (e), find an equation of the image of the line $y = 2x$ in R^2 under

   (a) a shear of factor 3 in the x-direction.

   (b) a compression of factor $\frac{1}{2}$ in the y-direction.

   (c) a reflection about $y = x$.

   (d) a reflection about the y-axis.

   (e) a rotation of $60°$ about the origin.

18. Find the matrix for a shear in the x-direction that transforms the triangle with vertices $(0, 0)$, $(2, 1)$, and $(3, 0)$ into a right triangle with the right angle at the origin.

19. (a) Show that multiplication by
$$A = \begin{bmatrix} 4 & 2 \\ 2 & 1 \end{bmatrix}$$
maps every point in the plane onto the line $y = \frac{1}{2}x$.

   (b) It follows from part (a) that the noncollinear points $(1, 0)$, $(0, 1)$, $(-1, 0)$ are mapped onto a line. Does this violate part (e) of Theorem 4.11.3?

20. Prove part (a) of Theorem 4.11.3. [Hint: A line in the plane has an equation of the form $Ax + By + C = 0$, where A and B are not both zero. Use the method of Example 6 to show that the image of this line under multiplication by the invertible matrix
$$\begin{bmatrix} a & b \\ c & d \end{bmatrix}$$
has the equation $A'x + B'y + C = 0$, where
$$A' = (dA - cB)/(ad - bc)$$
and
$$B' = (-bA + aB)/(ad - bc)$$
Then show that A' and B' are not both zero to conclude that the image is a line.]

21. Use the hint in Exercise 20 to prove parts (b) and (c) of Theorem 4.11.3.

22. In each part of the accompanying figure, find the standard matrix for the operator described.

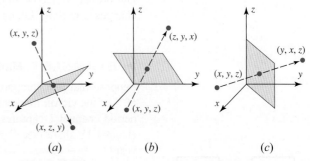

▲ Figure Ex-22

23. In R^3 the **shear in the xy-direction with factor k** is the matrix transformation that moves each point (x, y, z) parallel to the xy-plane to the new position $(x + kz, y + kz, z)$. (See the accompanying figure.)

   (a) Find the standard matrix for the shear in the xy-direction with factor k.

   (b) How would you define the shear in the xz-direction with factor k and the shear in the yz-direction with factor k?

Find the standard matrices for these matrix transformations.

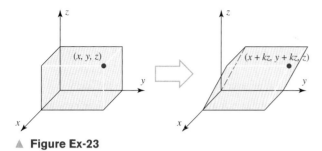
▲ Figure Ex-23

True-False Exercises

In parts (a)–(g) determine whether the statement is true or false, and justify your answer.

(a) The image of the unit square under a one-to-one matrix operator is a square.

(b) A 2×2 invertible matrix operator has the geometric effect of a succession of shears, compressions, expansions, and reflections.

(c) The image of a line under a one-to-one matrix operator is a line.

(d) Every reflection operator on R^2 is its own inverse.

(e) The matrix $\begin{bmatrix} 1 & 1 \\ 1 & -1 \end{bmatrix}$ represents reflection about a line.

(f) The matrix $\begin{bmatrix} 1 & -2 \\ 2 & 1 \end{bmatrix}$ represents a shear.

(g) The matrix $\begin{bmatrix} 1 & 0 \\ 0 & 3 \end{bmatrix}$ represents an expansion.

4.12 Dynamical Systems and Markov Chains

In this optional section we will show how matrix methods can be used to analyze the behavior of physical systems that evolve over time. The methods that we will study here have been applied to problems in business, ecology, demographics, sociology, and most of the physical sciences.

Dynamical Systems

A ***dynamical system*** is a finite set of variables whose values change with time. The value of a variable at a point in time is called the ***state of the variable*** at that time, and the vector formed from these states is called the ***state of the dynamical system*** at that time. Our primary objective in this section is to analyze how the state of a dynamical system changes with time. Let us begin with an example.

▶ **EXAMPLE 1** Market Share as a Dynamical System

Suppose that two competing television channels, channel 1 and channel 2, each have 50% of the viewer market at some initial point in time. Assume that over each one-year period channel 1 captures 10% of channel 2's share, and channel 2 captures 20% of channel 1's share (see Figure 4.12.1). What is each channel's market share after one year?

Solution Let us begin by introducing the time-dependent variables

$x_1(t)$ = fraction of the market held by channel 1 at time t
$x_2(t)$ = fraction of the market held by channel 2 at time t

and the column vector

$$\mathbf{x}(t) = \begin{bmatrix} x_1(t) \\ x_2(t) \end{bmatrix} \quad \begin{matrix} \leftarrow \text{Channel 1's fraction of the market at time } t \text{ in years} \\ \leftarrow \text{Channel 2's fraction of the market at time } t \text{ in years} \end{matrix}$$

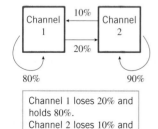
▲ Figure 4.12.1

The variables $x_1(t)$ and $x_2(t)$ form a dynamical system whose state at time t is the vector $\mathbf{x}(t)$. If we take $t = 0$ to be the starting point at which the two channels had 50% of the market, then the state of the system at that time is

$$\mathbf{x}(0) = \begin{bmatrix} x_1(0) \\ x_2(0) \end{bmatrix} = \begin{bmatrix} 0.5 \\ 0.5 \end{bmatrix} \quad \begin{matrix} \leftarrow \text{Channel 1's fraction of the market at time } t = 0 \\ \leftarrow \text{Channel 2's fraction of the market at time } t = 0 \end{matrix} \quad (1)$$

Now let us try to find the state of the system at time $t = 1$ (one year later). Over the one-year period, channel 1 retains 80% of its initial 50%, and it gains 10% of channel 2's initial 50%. Thus,

$$x_1(1) = 0.8(0.5) + 0.1(0.5) = 0.45 \quad (2)$$

Similarly, channel 2 gains 20% of channel 1's initial 50%, and retains 90% of its initial 50%. Thus,

$$x_2(1) = 0.2(0.5) + 0.9(0.5) = 0.55 \quad (3)$$

Therefore, the state of the system at time $t = 1$ is

$$\mathbf{x}(1) = \begin{bmatrix} x_1(1) \\ x_2(1) \end{bmatrix} = \begin{bmatrix} 0.45 \\ 0.55 \end{bmatrix} \quad \begin{matrix} \leftarrow \text{Channel 1's fraction of the market at time } t = 1 \\ \leftarrow \text{Channel 2's fraction of the market at time } t = 1 \end{matrix} \quad (4)$$

▶ **EXAMPLE 2** **Evolution of Market Share over Five Years**

Track the market shares of channels 1 and 2 in Example 1 over a five-year period.

Solution To solve this problem suppose that we have already computed the market share of each channel at time $t = k$ and we are interested in using the known values of $x_1(k)$ and $x_2(k)$ to compute the market shares $x_1(k + 1)$ and $x_2(k + 1)$ one year later. The analysis is exactly the same as that used to obtain Equations (2) and (3). Over the one-year period, channel 1 retains 80% of its starting fraction $x_1(k)$ and gains 10% of channel 2's starting fraction $x_2(k)$. Thus,

$$x_1(k + 1) = (0.8)x_1(k) + (0.1)x_2(k) \quad (5)$$

Similarly, channel 2 gains 20% of channel 1's starting fraction $x_1(k)$ and retains 90% of its own starting fraction $x_2(k)$. Thus,

$$x_2(k + 1) = (0.2)x_1(k) + (0.9)x_2(k) \quad (6)$$

Equations (5) and (6) can be expressed in matrix form as

$$\begin{bmatrix} x_1(k + 1) \\ x_2(k + 1) \end{bmatrix} = \begin{bmatrix} 0.8 & 0.1 \\ 0.2 & 0.9 \end{bmatrix} \begin{bmatrix} x_1(k) \\ x_2(k) \end{bmatrix} \quad (7)$$

which provides a way of using matrix multiplication to compute the state of the system at time $t = k + 1$ from the state at time $t = k$. For example, using (1) and (7) we obtain

$$\mathbf{x}(1) = \begin{bmatrix} 0.8 & 0.1 \\ 0.2 & 0.9 \end{bmatrix} \mathbf{x}(0) = \begin{bmatrix} 0.8 & 0.1 \\ 0.2 & 0.9 \end{bmatrix} \begin{bmatrix} 0.5 \\ 0.5 \end{bmatrix} = \begin{bmatrix} 0.45 \\ 0.55 \end{bmatrix}$$

which agrees with (4). Similarly,

$$\mathbf{x}(2) = \begin{bmatrix} 0.8 & 0.1 \\ 0.2 & 0.9 \end{bmatrix} \mathbf{x}(1) = \begin{bmatrix} 0.8 & 0.1 \\ 0.2 & 0.9 \end{bmatrix} \begin{bmatrix} 0.45 \\ 0.55 \end{bmatrix} = \begin{bmatrix} 0.415 \\ 0.585 \end{bmatrix}$$

We can now continue this process, using Formula (7) to compute $\mathbf{x}(3)$ from $\mathbf{x}(2)$, then $\mathbf{x}(4)$ from $\mathbf{x}(3)$, and so on. This yields (verify)

$$\mathbf{x}(3) = \begin{bmatrix} 0.3905 \\ 0.6095 \end{bmatrix}, \quad \mathbf{x}(4) = \begin{bmatrix} 0.37335 \\ 0.62665 \end{bmatrix}, \quad \mathbf{x}(5) = \begin{bmatrix} 0.361345 \\ 0.638655 \end{bmatrix} \quad (8)$$

Thus, after five years, channel 1 will hold about 36% of the market and channel 2 will hold about 64% of the market. ◀

If desired, we can continue the market analysis in the last example beyond the five-year period and explore what happens to the market share over the long term. We did so, using a computer, and obtained the following state vectors (rounded to six decimal places):

$$\mathbf{x}(10) \approx \begin{bmatrix} 0.338041 \\ 0.661959 \end{bmatrix}, \quad \mathbf{x}(20) \approx \begin{bmatrix} 0.333466 \\ 0.666534 \end{bmatrix}, \quad \mathbf{x}(40) \approx \begin{bmatrix} 0.333333 \\ 0.666667 \end{bmatrix} \quad (9)$$

All subsequent state vectors, when rounded to six decimal places, are the same as $\mathbf{x}(40)$, so we see that the market shares eventually stabilize with channel 1 holding about one-third of the market and channel 2 holding about two-thirds. Later in this section, we will explain why this stabilization occurs.

Markov Chains In many dynamical systems the states of the variables are not known with certainty but can be expressed as probabilities; such dynamical systems are called **stochastic processes** (from the Greek word *stokastikos*, meaning "proceeding by guesswork"). A detailed study of stochastic processes requires a precise definition of the term *probability*, which is outside the scope of this course. However, the following interpretation will suffice for our present purposes:

> *Stated informally, the **probability** that an experiment or observation will have a certain outcome is approximately the fraction of the time that the outcome would occur if the experiment were to be repeated many times under constant conditions—the greater the number of repetitions, the more accurately the probability describes the fraction of occurrences.*

For example, when we say that the probability of tossing heads with a fair coin is $\frac{1}{2}$, we mean that if the coin were tossed many times under constant conditions, then we would expect about half of the outcomes to be heads. Probabilities are often expressed as decimals or percentages. Thus, the probability of tossing heads with a fair coin can also be expressed as 0.5 or 50%.

If an experiment or observation has n possible outcomes, then the probabilities of those outcomes must be nonnegative fractions whose sum is 1. The probabilities are nonnegative because each describes the fraction of occurrences of an outcome over the long term, and the sum is 1 because they account for all possible outcomes. For example, if a box containing 10 balls has one red ball, three green balls, and six yellow balls, and if a ball is drawn at random from the box, then the probabilities of the various outcomes are

$$p_1 = \text{prob(red)} = 1/10 = 0.1$$
$$p_2 = \text{prob(green)} = 3/10 = 0.3$$
$$p_3 = \text{prob(yellow)} = 6/10 = 0.6$$

Each probability is a nonnegative fraction and

$$p_1 + p_2 + p_3 = 0.1 + 0.3 + 0.6 = 1$$

In a stochastic process with n possible states, the state vector at each time t has the form

$$\mathbf{x}(t) = \begin{bmatrix} x_1(t) \\ x_2(t) \\ \vdots \\ x_n(t) \end{bmatrix} \quad \begin{matrix} \text{Probability that the system is in state 1} \\ \text{Probability that the system is in state 2} \\ \vdots \\ \text{Probability that the system is in state } n \end{matrix}$$

The entries in this vector must add up to 1 since they account for all n possibilities. In general, a vector with nonnegative entries that add up to 1 is called a **probability vector**.

▶ **EXAMPLE 3** **Example 1 Revisited from the Probability Viewpoint**

Observe that the state vectors in Examples 1 and 2 are all probability vectors. This is to be expected since the entries in each state vector are the fractional market shares of the channels, and together they account for the entire market. In practice, it is preferable to interpret the entries in the state vectors as probabilities rather than exact market fractions, since market information is usually obtained by statistical sampling procedures with intrinsic uncertainties. Thus, for example, the state vector

$$\mathbf{x}(1) = \begin{bmatrix} x_1(1) \\ x_2(1) \end{bmatrix} = \begin{bmatrix} 0.45 \\ 0.55 \end{bmatrix}$$

which we interpreted in Example 1 to mean that channel 1 has 45% of the market and channel 2 has 55%, can also be interpreted to mean that an individual picked at random from the market will be a channel 1 viewer with probability 0.45 and a channel 2 viewer with probability 0.55. ◀

A square matrix, each of whose columns is a probability vector, is called a ***stochastic matrix***. Such matrices commonly occur in formulas that relate successive states of a stochastic process. For example, the state vectors $\mathbf{x}(k+1)$ and $\mathbf{x}(k)$ in (7) are related by an equation of the form $\mathbf{x}(k+1) = P\mathbf{x}(k)$ in which

$$P = \begin{bmatrix} 0.8 & 0.1 \\ 0.2 & 0.9 \end{bmatrix} \qquad (10)$$

is a stochastic matrix. It should not be surprising that the column vectors of P are probability vectors, since the entries in each column provide a breakdown of what happens to each channel's market share over the year—the entries in column 1 convey that each year channel 1 retains 80% of its market share and loses 20%; and the entries in column 2 convey that each year channel 2 retains 90% of its market share and loses 10%. The entries in (10) can also be viewed as probabilities:

$p_{11} = 0.8 =$ probability that a channel 1 viewer remains a channel 1 viewer
$p_{21} = 0.2 =$ probability that a channel 1 viewer becomes a channel 2 viewer
$p_{12} = 0.1 =$ probability that a channel 2 viewer becomes a channel 1 viewer
$p_{22} = 0.9 =$ probability that a channel 2 viewer remains a channel 2 viewer

Example 1 is a special case of a large class of stochastic processes, called *Markov chains*.

Historical Note Markov chains are named in honor of the Russian mathematician A. A. Markov, a lover of poetry, who used them to analyze the alternation of vowels and consonants in the poem *Eugene Onegin* by Pushkin. Markov believed that the only applications of his chains were to the analysis of literary works, so he would be astonished to learn that his discovery is used today in the social sciences, quantum theory, and genetics!

[*Image: wikipedia*]

**Andrei Andreyevich Markov
(1856–1922)**

286 Chapter 4 General Vector Spaces

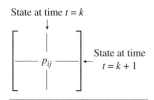

The entry p_{ij} is the probability that the system is in state i at time $t = k+1$ if it is in state j at time $t = k$.

▲ Figure 4.12.2

DEFINITION 1 A *Markov chain* is a dynamical system whose state vectors at a succession of time intervals are probability vectors and for which the state vectors at successive time intervals are related by an equation of the form

$$\mathbf{x}(k+1) = P\mathbf{x}(k)$$

in which $P = [p_{ij}]$ is a stochastic matrix and p_{ij} is the probability that the system will be in state i at time $t = k+1$ if it is in state j at time $t = k$. The matrix P is called the *transition matrix* for the system.

Remark Note that in this definition the row index i corresponds to the later state and the column index j to the earlier state (Figure 4.12.2).

▶ **EXAMPLE 4 Wildlife Migration as a Markov Chain**

Suppose that a tagged lion can migrate over three adjacent game reserves in search of food, reserve 1, reserve 2, and reserve 3. Based on data about the food resources, researchers conclude that the monthly migration pattern of the lion can be modeled by a Markov chain with transition matrix

$$P = \begin{bmatrix} 0.5 & 0.4 & 0.6 \\ 0.2 & 0.2 & 0.3 \\ 0.3 & 0.4 & 0.1 \end{bmatrix} \begin{matrix} 1 \\ 2 \\ 3 \end{matrix} \quad \text{Reserve at time } t = k+1$$

with column headers "Reserve at time $t = k$": 1, 2, 3.

(see Figure 4.12.3). That is,

$p_{11} = 0.5 =$ probability that the lion will stay in reserve 1 when it is in reserve 1
$p_{12} = 0.4 =$ probability that the lion will move from reserve 2 to reserve 1
$p_{13} = 0.6 =$ probability that the lion will move from reserve 3 to reserve 1
$p_{21} = 0.2 =$ probability that the lion will move from reserve 1 to reserve 2
$p_{22} = 0.2 =$ probability that the lion will stay in reserve 2 when it is in reserve 2
$p_{23} = 0.3 =$ probability that the lion will move from reserve 3 to reserve 2
$p_{31} = 0.3 =$ probability that the lion will move from reserve 1 to reserve 3
$p_{32} = 0.4 =$ probability that the lion will move from reserve 2 to reserve 3
$p_{33} = 0.1 =$ probability that the lion will stay in reserve 3 when it is in reserve 3

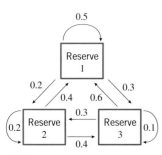

▲ Figure 4.12.3

Assuming that t is in months and the lion is released in reserve 2 at time $t = 0$, track its probable locations over a six-month period.

Solution Let $x_1(k)$, $x_2(k)$, and $x_3(k)$ be the probabilities that the lion is in reserve 1, 2, or 3, respectively, at time $t = k$, and let

$$\mathbf{x}(k) = \begin{bmatrix} x_1(k) \\ x_2(k) \\ x_3(k) \end{bmatrix}$$

be the state vector at that time. Since we know with certainty that the lion is in reserve 2 at time $t = 0$, the initial state vector is

$$\mathbf{x}(0) = \begin{bmatrix} 0 \\ 1 \\ 0 \end{bmatrix}$$

We leave it for you to show that the state vectors over a six-month period are

$$\mathbf{x}(1) = P\mathbf{x}(0) = \begin{bmatrix} 0.400 \\ 0.200 \\ 0.400 \end{bmatrix}, \quad \mathbf{x}(2) = P\mathbf{x}(1) = \begin{bmatrix} 0.520 \\ 0.240 \\ 0.240 \end{bmatrix}, \quad \mathbf{x}(3) = P\mathbf{x}(2) = \begin{bmatrix} 0.500 \\ 0.224 \\ 0.276 \end{bmatrix}$$

$$\mathbf{x}(4) = P\mathbf{x}(3) \approx \begin{bmatrix} 0.505 \\ 0.228 \\ 0.267 \end{bmatrix}, \quad \mathbf{x}(5) = P\mathbf{x}(4) \approx \begin{bmatrix} 0.504 \\ 0.227 \\ 0.269 \end{bmatrix}, \quad \mathbf{x}(6) = P\mathbf{x}(5) \approx \begin{bmatrix} 0.504 \\ 0.227 \\ 0.269 \end{bmatrix}$$

As in Example 2, the state vectors here seem to stabilize over time with a probability of approximately 0.504 that the lion is in reserve 1, a probability of approximately 0.227 that it is in reserve 2, and a probability of approximately 0.269 that it is in reserve 3. ◀

Markov Chains in Terms of Powers of the Transition Matrix

In a Markov chain with an initial state of $\mathbf{x}(0)$, the successive state vectors are

$$\mathbf{x}(1) = P\mathbf{x}(0), \quad \mathbf{x}(2) = P\mathbf{x}(1), \quad \mathbf{x}(3) = P\mathbf{x}(2), \quad \mathbf{x}(4) = P\mathbf{x}(3), \ldots$$

For brevity, it is common to denote $\mathbf{x}(k)$ by $\mathbf{x}_k$, which allows us to write the successive state vectors more briefly as

$$\mathbf{x}_1 = P\mathbf{x}_0, \quad \mathbf{x}_2 = P\mathbf{x}_1, \quad \mathbf{x}_3 = P\mathbf{x}_2, \quad \mathbf{x}_4 = P\mathbf{x}_3, \ldots \tag{11}$$

Alternatively, these state vectors can be expressed in terms of the initial state vector $\mathbf{x}_0$ as

$$\mathbf{x}_1 = P\mathbf{x}_0, \quad \mathbf{x}_2 = P(P\mathbf{x}_0) = P^2\mathbf{x}_0, \quad \mathbf{x}_3 = P(P^2\mathbf{x}_0) = P^3\mathbf{x}_0, \quad \mathbf{x}_4 = P(P^3\mathbf{x}_0) = P^4\mathbf{x}_0, \ldots$$

from which it follows that

$$\mathbf{x}_k = P^k\mathbf{x}_0 \tag{12}$$

Note that Formula (12) makes it possible to compute the state vector $\mathbf{x}_k$ without first computing the earlier state vectors as required in Formula (11).

▶ **EXAMPLE 5 Finding a State Vector Directly from $\mathbf{x}_0$**

Use Formula (12) to find the state vector $\mathbf{x}(3)$ in Example 2.

Solution From (1) and (7), the initial state vector and transition matrix are

$$\mathbf{x}_0 = \mathbf{x}(0) = \begin{bmatrix} 0.5 \\ 0.5 \end{bmatrix} \quad \text{and} \quad P = \begin{bmatrix} 0.8 & 0.1 \\ 0.2 & 0.9 \end{bmatrix}$$

We leave it for you to calculate P^3 and show that

$$\mathbf{x}(3) = \mathbf{x}_3 = P^3\mathbf{x}_0 = \begin{bmatrix} 0.562 & 0.219 \\ 0.438 & 0.781 \end{bmatrix} \begin{bmatrix} 0.5 \\ 0.5 \end{bmatrix} = \begin{bmatrix} 0.3905 \\ 0.6095 \end{bmatrix}$$

which agrees with the result in (8). ◀

Long-Term Behavior of a Markov Chain

We have seen two examples of Markov chains in which the state vectors seem to stabilize after a period of time. Thus, it is reasonable to ask whether all Markov chains have this property. The following example shows that this is not the case.

▶ **EXAMPLE 6 A Markov Chain That Does Not Stabilize**

The matrix

$$P = \begin{bmatrix} 0 & 1 \\ 1 & 0 \end{bmatrix}$$

is stochastic and hence can be regarded as the transition matrix for a Markov chain. A simple calculation shows that $P^2 = I$, from which it follows that

$$I = P^2 = P^4 = P^6 = \cdots \quad \text{and} \quad P = P^3 = P^5 = P^7 = \cdots$$

Thus, the successive states in the Markov chain with initial vector $\mathbf{x}_0$ are

$$\mathbf{x}_0, \quad P\mathbf{x}_0, \quad \mathbf{x}_0, \quad P\mathbf{x}_0, \quad \mathbf{x}_0, \ldots$$

which oscillate between $\mathbf{x}_0$ and $P\mathbf{x}_0$. Thus, the Markov chain does not stabilize unless both components of $\mathbf{x}_0$ are $\frac{1}{2}$ (verify). ◀

A precise definition of what it means for a sequence of numbers or vectors to stabilize is given in calculus; however, that level of precision will not be needed here. Stated informally, we will say that a sequence of vectors

$$\mathbf{x}_1, \quad \mathbf{x}_2, \ldots, \quad \mathbf{x}_k, \ldots$$

approaches a *limit* $\mathbf{q}$ or that it *converges* to $\mathbf{q}$ if all entries in $\mathbf{x}_k$ can be made as close as we like to the corresponding entries in the vector $\mathbf{q}$ by taking k sufficiently large. We denote this by writing $\mathbf{x}_k \to \mathbf{q}$ as $k \to \infty$.

We saw in Example 6 that the state vectors of a Markov chain need not approach a limit in all cases. However, by imposing a mild condition on the transition matrix of a Markov chain, we can guarantee that the state vectors will approach a limit.

DEFINITION 2 A stochastic matrix P is said to be *regular* if P or some positive power of P has all positive entries, and a Markov chain whose transition matrix is regular is said to be a *regular Markov chain*.

▶ **EXAMPLE 7** Regular Stochastic Matrices

The transition matrices in Examples 2 and 4 are regular because their entries are positive. The matrix

$$P = \begin{bmatrix} 0.5 & 1 \\ 0.5 & 0 \end{bmatrix}$$

is regular because

$$P^2 = \begin{bmatrix} 0.75 & 0.5 \\ 0.25 & 0.5 \end{bmatrix}$$

has positive entries. The matrix P in Example 6 is not regular because P and every positive power of P have some zero entries (verify). ◀

The following theorem, which we state without proof, is the fundamental result about the long-term behavior of Markov chains.

THEOREM 4.12.1 *If P is the transition matrix for a regular Markov chain, then:*

(a) *There is a unique probability vector $\mathbf{q}$ such that $P\mathbf{q} = \mathbf{q}$.*

(b) *For any initial probability vector $\mathbf{x}_0$, the sequence of state vectors*

$$\mathbf{x}_0, \quad P\mathbf{x}_0, \ldots, \quad P^k\mathbf{x}_0, \ldots$$

converges to $\mathbf{q}$.

4.12 Dynamical Systems and Markov Chains

The vector $\mathbf{q}$ in this theorem is called the ***steady-state*** vector of the Markov chain. It can be found by rewriting the equation in part (a) as

$$(I - P)\mathbf{q} = \mathbf{0}$$

and then solving this equation for $\mathbf{q}$ subject to the requirement that $\mathbf{q}$ be a probability vector. Here are some examples.

▶ **EXAMPLE 8** **Examples 1 and 2 Revisited**

The transition matrix for the Markov chain in Example 2 is

$$P = \begin{bmatrix} 0.8 & 0.1 \\ 0.2 & 0.9 \end{bmatrix}$$

Since the entries of P are positive, the Markov chain is regular and hence has a unique steady-state vector $\mathbf{q}$. To find $\mathbf{q}$ we will solve the system $(I - P)\mathbf{q} = \mathbf{0}$, which we can write as

$$\begin{bmatrix} 0.2 & -0.1 \\ -0.2 & 0.1 \end{bmatrix} \begin{bmatrix} q_1 \\ q_2 \end{bmatrix} = \begin{bmatrix} 0 \\ 0 \end{bmatrix}$$

The general solution of this system is

$$q_1 = 0.5s, \quad q_2 = s$$

(verify), which we can write in vector form as

$$\mathbf{q} = \begin{bmatrix} q_1 \\ q_2 \end{bmatrix} = \begin{bmatrix} 0.5s \\ s \end{bmatrix} = \begin{bmatrix} \frac{1}{2}s \\ s \end{bmatrix} \quad (13)$$

For $\mathbf{q}$ to be a probability vector, we must have

$$1 = q_1 + q_2 = \tfrac{3}{2}s$$

which implies that $s = \tfrac{2}{3}$. Substituting this value in (13) yields the steady-state vector

$$\mathbf{q} = \begin{bmatrix} \frac{1}{3} \\ \frac{2}{3} \end{bmatrix}$$

which is consistent with the numerical results obtained in (9).

▶ **EXAMPLE 9** **Example 4 Revisited**

The transition matrix for the Markov chain in Example 4 is

$$P = \begin{bmatrix} 0.5 & 0.4 & 0.6 \\ 0.2 & 0.2 & 0.3 \\ 0.3 & 0.4 & 0.1 \end{bmatrix}$$

Since the entries of P are positive, the Markov chain is regular and hence has a unique steady-state vector $\mathbf{q}$. To find $\mathbf{q}$ we will solve the system $(I - P)\mathbf{q} = \mathbf{0}$, which we can write (using fractions) as

$$\begin{bmatrix} \frac{1}{2} & -\frac{2}{5} & -\frac{3}{5} \\ -\frac{1}{5} & \frac{4}{5} & -\frac{3}{10} \\ -\frac{3}{10} & -\frac{2}{5} & \frac{9}{10} \end{bmatrix} \begin{bmatrix} q_1 \\ q_2 \\ q_3 \end{bmatrix} = \begin{bmatrix} 0 \\ 0 \\ 0 \end{bmatrix} \quad (14)$$

(We have converted to fractions to avoid roundoff error in this illustrative example.) We leave it for you to confirm that the reduced row echelon form of the coefficient matrix is

$$\begin{bmatrix} 1 & 0 & -\frac{15}{8} \\ 0 & 1 & -\frac{27}{32} \\ 0 & 0 & 0 \end{bmatrix}$$

and that the general solution of (14) is

$$q_1 = \tfrac{15}{8}s, \quad q_2 = \tfrac{27}{32}s, \quad q_3 = s \tag{15}$$

For **q** to be a probability vector we must have $q_1 + q_2 + q_3 = 1$, from which it follows that $s = \tfrac{32}{119}$ (verify). Substituting this value in (15) yields the steady-state vector

$$\mathbf{q} = \begin{bmatrix} \frac{60}{119} \\ \frac{27}{119} \\ \frac{32}{119} \end{bmatrix} \approx \begin{bmatrix} 0.5042 \\ 0.2269 \\ 0.2689 \end{bmatrix}$$

(verify), which is consistent with the results obtained in Example 4. ◄

Concept Review

- Dynamical system
- State of a variable
- State of a dynamical system
- Stochastic process
- Probability
- Probability vector
- Stochastic matrix
- Markov chain
- Transition matrix
- Regular stochastic matrix
- Regular Markov chain
- Steady-state vector

Skills

- Determine whether a matrix is stochastic.
- Compute the state vectors from a transition matrix and an initial state.
- Determine whether a stochastic matrix is regular.
- Determine whether a Markov chain is regular.
- Find the steady-state vector for a regular transition matrix.

Exercise Set 4.12

▶ In Exercises **1–2**, determine whether A is a stochastic matrix. If A is not stochastic, then explain why not. ◄

1. (a) $A = \begin{bmatrix} 0.2 & 0.8 \\ 0.5 & 0.5 \end{bmatrix}$ (b) $A = \begin{bmatrix} 0.8 & 0.5 \\ 0.2 & 0.5 \end{bmatrix}$

   (c) $A = \begin{bmatrix} 1 & \frac{1}{2} & \frac{1}{3} \\ 0 & 0 & \frac{1}{3} \\ 0 & \frac{1}{2} & \frac{1}{3} \end{bmatrix}$ (d) $A = \begin{bmatrix} \frac{1}{3} & \frac{1}{3} & \frac{1}{2} \\ \frac{1}{6} & \frac{1}{3} & -\frac{1}{2} \\ \frac{1}{2} & \frac{1}{3} & 1 \end{bmatrix}$

2. (a) $A = \begin{bmatrix} \frac{2}{3} & \frac{1}{3} \\ \frac{1}{6} & \frac{2}{3} \end{bmatrix}$ (b) $A = \begin{bmatrix} 0.2 & 0.8 \\ 0.9 & 0.1 \end{bmatrix}$

   (c) $A = \begin{bmatrix} 0 & 0 & 0 \\ \frac{5}{6} & \frac{2}{7} & \frac{1}{3} \\ \frac{1}{6} & \frac{5}{7} & \frac{2}{3} \end{bmatrix}$ (d) $A = \begin{bmatrix} -1 & \frac{1}{3} & \frac{1}{2} \\ 0 & \frac{1}{3} & \frac{1}{2} \\ 2 & \frac{1}{3} & 0 \end{bmatrix}$

▶ In Exercises **3–4**, use Formulas (11) and (12) to compute the state vector $\mathbf{x}_4$ in two different ways. ◄

3. $P = \begin{bmatrix} 0.1 & 0.4 \\ 0.9 & 0.6 \end{bmatrix}$; $\mathbf{x}_0 = \begin{bmatrix} 0.5 \\ 0.5 \end{bmatrix}$

4. $P = \begin{bmatrix} 0.8 & 0.5 \\ 0.2 & 0.5 \end{bmatrix}$; $\mathbf{x}_0 = \begin{bmatrix} 1 \\ 0 \end{bmatrix}$

▶ In Exercises **5–6**, determine whether P is a regular stochastic matrix. ◄

5. (a) $P = \begin{bmatrix} \frac{1}{5} & \frac{1}{7} \\ \frac{4}{5} & \frac{6}{7} \end{bmatrix}$ (b) $P = \begin{bmatrix} \frac{1}{5} & 0 \\ \frac{4}{5} & 1 \end{bmatrix}$ (c) $P = \begin{bmatrix} \frac{1}{5} & 1 \\ \frac{4}{5} & 0 \end{bmatrix}$

6. (a) $P = \begin{bmatrix} \frac{1}{2} & 1 \\ \frac{1}{2} & 0 \end{bmatrix}$ (b) $P = \begin{bmatrix} 1 & \frac{2}{3} \\ 0 & \frac{1}{3} \end{bmatrix}$ (c) $P = \begin{bmatrix} \frac{3}{4} & \frac{1}{3} \\ \frac{1}{4} & \frac{2}{3} \end{bmatrix}$

In Exercises **7–10**, verify that P is a regular stochastic matrix, and find the steady-state vector for the associated Markov chain.

7. $P = \begin{bmatrix} \frac{1}{4} & \frac{2}{3} \\ \frac{3}{4} & \frac{1}{3} \end{bmatrix}$

8. $P = \begin{bmatrix} 0.2 & 0.6 \\ 0.8 & 0.4 \end{bmatrix}$

9. $P = \begin{bmatrix} \frac{1}{3} & 0 & \frac{1}{2} \\ \frac{1}{3} & \frac{1}{2} & \frac{1}{4} \\ \frac{1}{3} & \frac{1}{2} & \frac{1}{4} \end{bmatrix}$

10. $P = \begin{bmatrix} \frac{1}{3} & \frac{1}{4} & \frac{2}{5} \\ 0 & \frac{3}{4} & \frac{2}{5} \\ \frac{2}{3} & 0 & \frac{1}{5} \end{bmatrix}$

11. Consider a Markov process with transition matrix

$$\begin{array}{c} \\ \text{State 1} \\ \text{State 2} \end{array} \begin{array}{cc} \text{State 1} & \text{State 2} \\ \begin{bmatrix} 0.2 & 0.1 \\ 0.8 & 0.9 \end{bmatrix} \end{array}$$

(a) What does the entry 0.2 represent?

(b) What does the entry 0.1 represent?

(c) If the system is in state 1 initially, what is the probability that it will be in state 2 at the next observation?

(d) If the system has a 50% chance of being in state 1 initially, what is the probability that it will be in state 2 at the next observation?

12. Consider a Markov process with transition matrix

$$\begin{array}{c} \\ \text{State 1} \\ \text{State 2} \end{array} \begin{array}{cc} \text{State 1} & \text{State 2} \\ \begin{bmatrix} 0 & \frac{1}{7} \\ 1 & \frac{6}{7} \end{bmatrix} \end{array}$$

(a) What does the entry $\frac{6}{7}$ represent?

(b) What does the entry 0 represent?

(c) If the system is in state 1 initially, what is the probability that it will be in state 1 at the next observation?

(d) If the system has a 50% chance of being in state 1 initially, what is the probability that it will be in state 2 at the next observation?

13. On a given day the air quality in a certain city is either good or bad. Records show that when the air quality is good on one day, then there is a 95% chance that it will be good the next day, and when the air quality is bad on one day, then there is a 45% chance that it will be bad the next day.

(a) Find a transition matrix for this phenomenon.

(b) If the air quality is good today, what is the probability that it will be good two days from now?

(c) If the air quality is bad today, what is the probability that it will be bad three days from now?

(d) If there is a 20% chance that the air quality will be good today, what is the probability that it will be good tomorrow?

14. In a laboratory experiment, a mouse can choose one of two food types each day, type I or type II. Records show that if the mouse chooses type I on a given day, then there is a 75% chance that it will choose type I the next day, and if it chooses type II on one day, then there is a 50% chance that it will choose type II the next day.

(a) Find a transition matrix for this phenomenon.

(b) If the mouse chooses type I today, what is the probability that it will choose type I two days from now?

(c) If the mouse chooses type II today, what is the probability that it will choose type II three days from now?

(d) If there is a 10% chance that the mouse will choose type I today, what is the probability that it will choose type I tomorrow?

15. Suppose that at some initial point in time 100,000 people live in a certain city and 25,000 people live in its suburbs. The Regional Planning Commission determines that each year 5% of the city population moves to the suburbs and 3% of the suburban population moves to the city.

(a) Assuming that the total population remains constant, make a table that shows the populations of the city and its suburbs over a five-year period (round to the nearest integer).

(b) Over the long term, how will the population be distributed between the city and its suburbs?

16. Suppose that two competing television stations, station 1 and station 2, each have 50% of the viewer market at some initial point in time. Assume that over each one-year period station 1 captures 5% of station 2's market share and station 2 captures 10% of station 1's market share.

(a) Make a table that shows the market share of each station over a five-year period.

(b) Over the long term, how will the market share be distributed between the two stations?

17. Suppose that a car rental agency has three locations, numbered 1, 2, and 3. A customer may rent a car from any of the three locations and return it to any of the three locations. Records show that cars are rented and returned in accordance with the following probabilities:

		Rented from Location		
		1	2	3
Returned to Location	1	$\frac{1}{10}$	$\frac{1}{5}$	$\frac{3}{5}$
	2	$\frac{4}{5}$	$\frac{3}{10}$	$\frac{1}{5}$
	3	$\frac{1}{10}$	$\frac{1}{2}$	$\frac{1}{5}$

(a) Assuming that a car is rented from location 1, what is the probability that it will be at location 1 after two rentals?

(b) Assuming that this dynamical system can be modeled as a Markov chain, find the steady-state vector.

(c) If the rental agency owns 120 cars, how many parking spaces should it allocate at each location to be reasonably certain that it will have enough spaces for the cars over the long term? Explain your reasoning.

18. Physical traits are determined by the genes that an offspring receives from its parents. In the simplest case a trait in the offspring is determined by one pair of genes, one member of the pair inherited from the male parent and the other from the female parent. Typically, each gene in a pair can assume one of two forms, called *alleles*, denoted by A and a. This leads to three possible pairings:

$$AA, \quad Aa, \quad aa$$

called *genotypes* (the pairs Aa and aA determine the same trait and hence are not distinguished from one another). It is shown in the study of heredity that if a parent of known genotype is crossed with a random parent of unknown genotype, then the offspring will have the genotype probabilities given in the following table, which can be viewed as a transition matrix for a Markov process:

	Genotype of Parent		
	AA	Aa	aa
AA	$\frac{1}{2}$	$\frac{1}{4}$	0
Aa	$\frac{1}{2}$	$\frac{1}{2}$	$\frac{1}{2}$
aa	0	$\frac{1}{4}$	$\frac{1}{2}$

(Genotype of Offspring)

Thus, for example, the offspring of a parent of genotype AA that is crossed at random with a parent of unknown genotype will have a 50% chance of being AA, a 50% chance of being Aa, and no chance of being aa.

(a) Show that the transition matrix is regular.

(b) Find the steady-state vector, and discuss its physical interpretation.

19. Fill in the missing entries of the stochastic matrix

$$P = \begin{bmatrix} \frac{1}{3} & \frac{1}{12} & \frac{1}{6} \\ * & \frac{1}{6} & * \\ \frac{1}{6} & * & \frac{1}{2} \end{bmatrix}$$

and find its steady-state vector.

20. If P is an $n \times n$ stochastic matrix, and if M is a $1 \times n$ matrix whose entries are all 1's, then $MP = $ _____.

21. If P is a regular stochastic matrix with steady-state vector $\mathbf{q}$, what can you say about the sequence of products

$$P\mathbf{q}, \quad P^2\mathbf{q}, \quad P^3\mathbf{q}, \ldots, \quad P^k\mathbf{q}, \ldots$$

as $k \to \infty$?

22. (a) If P is a regular $n \times n$ stochastic matrix with steady-state vector $\mathbf{q}$, and if $\mathbf{e}_1, \mathbf{e}_2, \ldots, \mathbf{e}_n$ are the standard unit vectors in column form, what can you say about the behavior of the sequence

$$P\mathbf{e}_i, \quad P^2\mathbf{e}_i, \quad P^3\mathbf{e}_i, \ldots, \quad P^k\mathbf{e}_i, \ldots$$

as $k \to \infty$ for each $i = 1, 2, \ldots, n$?

(b) What does this tell you about the behavior of the column vectors of P^k as $k \to \infty$?

23. Prove that the product of two stochastic matrices is a stochastic matrix. [*Hint:* Write each column of the product as a linear combination of the columns of the first factor.]

24. Prove that if P is a stochastic matrix whose entries are all greater than or equal to ρ, then the entries of P^2 are greater than or equal to ρ.

True-False Exercises

In parts (a)–(e) determine whether the statement is true or false, and justify your answer.

(a) The vector $\begin{bmatrix} \frac{1}{3} \\ 0 \\ \frac{2}{3} \end{bmatrix}$ is a probability vector.

(b) The matrix $\begin{bmatrix} 0.2 & 1 \\ 0.8 & 0 \end{bmatrix}$ is a regular stochastic matrix.

(c) The column vectors of a transition matrix are probability vectors.

(d) A steady-state vector for a Markov chain with transition matrix P is any solution of the linear system $(I - P)\mathbf{q} = \mathbf{0}$.

(e) The square of every regular stochastic matrix is stochastic.

Chapter 4 Supplementary Exercises

1. Let V be the set of all ordered pairs of real numbers, and consider the following addition and scalar multiplication operations on $\mathbf{u} = (u_1, u_2, u_3)$ and $\mathbf{v} = (v_1, v_2, v_3)$:

$$\mathbf{u} + \mathbf{v} = (u_1 + v_1, u_2 + v_2, u_3 + v_3), \quad k\mathbf{u} = (ku_1, 0, 0)$$

(a) Compute $\mathbf{u} + \mathbf{v}$ and $k\mathbf{u}$ for $\mathbf{u} = (3, -2, 4)$, $\mathbf{v} = (1, 5, -2)$, and $k = -1$.

(b) In words, explain why V is closed under addition and scalar multiplication.

(c) Since the addition operation on V is the standard addition operation on R^3, certain vector space axioms hold for V because they are known to hold for R^3. Which axioms in Definition 1 of Section 4.1 are they?

(d) Show that Axioms 7, 8, and 9 hold.

(e) Show that Axiom 10 fails for the given operations.

2. In each part, the solution space of the system is a subspace of R^3 and so must be a line through the origin, a plane through the origin, all of R^3, or the origin only. For each system, determine which is the case. If the subspace is a plane, find an equation for it, and if it is a line, find parametric equations.

(a) $0x + 0y + 0z = 0$

(b) $2x - 3y + z = 0$
$6x - 9y + 3z = 0$
$-4x + 6y - 2z = 0$

(c) $x - 2y + 7z = 0$
$-4x + 8y + 5z = 0$
$2x - 4y + 3z = 0$

(d) $x + 4y + 8z = 0$
$2x + 5y + 6z = 0$
$3x + y - 4z = 0$

3. For what values of s is the solution space of

$$x_1 + x_2 + sx_3 = 0$$
$$x_1 + sx_2 + x_3 = 0$$
$$sx_1 + x_2 + x_3 = 0$$

the origin only, a line through the origin, a plane through the origin, or all of R^3?

4. (a) Express $(4a, a - b, a + 2b)$ as a linear combination of $(4, 1, 1)$ and $(0, -1, 2)$.

(b) Express $(3a + b + 3c, -a + 4b - c, 2a + b + 2c)$ as a linear combination of $(3, -1, 2)$ and $(1, 4, 1)$.

(c) Express $(2a - b + 4c, 3a - c, 4b + c)$ as a linear combination of three nonzero vectors.

5. Let W be the space spanned by $\mathbf{f} = \sin x$ and $\mathbf{g} = \cos x$.

(a) Show that for any value of θ, $\mathbf{f}_1 = \sin(x + \theta)$ and $\mathbf{g}_1 = \cos(x + \theta)$ are vectors in W.

(b) Show that $\mathbf{f}_1$ and $\mathbf{g}_1$ form a basis for W.

6. (a) Express $\mathbf{v} = (1, 1)$ as a linear combination of $\mathbf{v}_1 = (1, -1)$, $\mathbf{v}_2 = (3, 0)$, and $\mathbf{v}_3 = (2, 1)$ in two different ways.

(b) Explain why this does not violate Theorem 4.4.1.

7. Let A be an $n \times n$ matrix, and let $\mathbf{v}_1, \mathbf{v}_2, \ldots, \mathbf{v}_n$ be linearly independent vectors in R^n expressed as $n \times 1$ matrices. What must be true about A for $A\mathbf{v}_1, A\mathbf{v}_2, \ldots, A\mathbf{v}_n$ to be linearly independent?

8. Must a basis for P_n contain a polynomial of degree k for each $k = 0, 1, 2, \ldots, n$? Justify your answer.

9. For the purpose of this exercise, let us define a "checkerboard matrix" to be a square matrix $A = [a_{ij}]$ such that

$$a_{ij} = \begin{cases} 1 & \text{if } i + j \text{ is even} \\ 0 & \text{if } i + j \text{ is odd} \end{cases}$$

Find the rank and nullity of the following checkerboard matrices.

(a) The 3×3 checkerboard matrix.

(b) The 4×4 checkerboard matrix.

(c) The $n \times n$ checkerboard matrix.

10. For the purpose of this exercise, let us define an "X-matrix" to be a square matrix with an odd number of rows and columns that has 0's everywhere except on the two diagonals where it has 1's. Find the rank and nullity of the following X-matrices.

(a) $\begin{bmatrix} 1 & 0 & 1 \\ 0 & 1 & 0 \\ 1 & 0 & 1 \end{bmatrix}$

(b) $\begin{bmatrix} 1 & 0 & 0 & 0 & 1 \\ 0 & 1 & 0 & 1 & 0 \\ 0 & 0 & 1 & 0 & 0 \\ 0 & 1 & 0 & 1 & 0 \\ 1 & 0 & 0 & 0 & 1 \end{bmatrix}$

(c) the X-matrix of size $(2n + 1) \times (2n + 1)$

11. In each part, show that the stated set of polynomials is a subspace of P_n and find a basis for it.

(a) All polynomials in P_n such that $p(-x) = p(x)$.

(b) All polynomials in P_n such that $p(0) = 0$.

12. (*Calculus required*) Show that the set of all polynomials in P_n that have a horizontal tangent at $x = 0$ is a subspace of P_n. Find a basis for this subspace.

13. (a) Find a basis for the vector space of all 3×3 symmetric matrices.

(b) Find a basis for the vector space of all 3×3 skew-symmetric matrices.

14. Various advanced texts in linear algebra prove the following determinant criterion for rank: *The rank of a matrix A is r if and only if A has some $r \times r$ submatrix with a nonzero determinant, and all square submatrices of larger size have determinant zero.* [*Note:* A submatrix of A is any matrix obtained by deleting rows or columns of A. The matrix A itself is also considered to be a submatrix of A.] In each part, use this criterion to find the rank of the matrix.

(a) $\begin{bmatrix} 1 & 2 & 0 \\ 2 & 4 & -1 \end{bmatrix}$

(b) $\begin{bmatrix} 1 & 2 & 3 \\ 2 & 4 & 6 \end{bmatrix}$

(c) $\begin{bmatrix} 1 & 0 & 1 \\ 2 & -1 & 3 \\ 3 & -1 & 4 \end{bmatrix}$

(d) $\begin{bmatrix} 1 & -1 & 2 & 0 \\ 3 & 1 & 0 & 0 \\ -1 & 2 & 4 & 0 \end{bmatrix}$

15. Use the result in Exercise 14 above to find the possible ranks for matrices of the form

$$\begin{bmatrix} 0 & 0 & 0 & 0 & 0 & a_{16} \\ 0 & 0 & 0 & 0 & 0 & a_{26} \\ 0 & 0 & 0 & 0 & 0 & a_{36} \\ 0 & 0 & 0 & 0 & 0 & a_{46} \\ a_{51} & a_{52} & a_{53} & a_{54} & a_{55} & a_{56} \end{bmatrix}$$

16. Prove: If S is a basis for a vector space V, then for any vectors $\mathbf{u}$ and $\mathbf{v}$ in V and any scalar k, the following relationships hold.

(a) $(\mathbf{u} + \mathbf{v})_S = (\mathbf{u})_S + (\mathbf{v})_S$

(b) $(k\mathbf{u})_S = k(\mathbf{u})_S$

CHAPTER 5

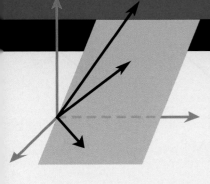

Eigenvalues and Eigenvectors

CHAPTER CONTENTS
5.1 Eigenvalues and Eigenvectors 295
5.2 Diagonalization 305
5.3 Complex Vector Spaces 315
5.4 Differential Equations 327

INTRODUCTION In this chapter we will focus on classes of scalars and vectors known as "eigenvalues" and "eigenvectors," terms derived from the German word *eigen*, meaning "own," "peculiar to," "characteristic," or "individual." The underlying idea first appeared in the study of rotational motion but was later used to classify various kinds of surfaces and to describe solutions of certain differential equations. In the early 1900s it was applied to matrices and matrix transformations, and today it has applications in such diverse fields as computer graphics, mechanical vibrations, heat flow, population dynamics, quantum mechanics, and economics to name just a few.

5.1 Eigenvalues and Eigenvectors

In this section we will define the notions of "eigenvalue" and "eigenvector" and discuss some of their basic properties.

Definition of Eigenvalue and Eigenvector

We begin with the main definition in this section.

> **DEFINITION 1** If A is an $n \times n$ matrix, then a nonzero vector $\mathbf{x}$ in R^n is called an ***eigenvector*** of A (or of the matrix operator T_A) if $A\mathbf{x}$ is a scalar multiple of $\mathbf{x}$; that is,
>
> $$A\mathbf{x} = \lambda \mathbf{x}$$
>
> for some scalar λ. The scalar λ is called an ***eigenvalue*** of A (or of T_A), and $\mathbf{x}$ is said to be an ***eigenvector corresponding to*** λ.

The requirement that an eigenvector be nonzero is imposed to avoid the unimportant case $A\mathbf{0} = \lambda \mathbf{0}$, which holds for every A and λ.

In general, the image of a vector $\mathbf{x}$ under multiplication by a square matrix A differs from $\mathbf{x}$ in both magnitude and direction. However, in the special case where $\mathbf{x}$ is an eigenvector of A, multiplication by A leaves the direction unchanged. For example, in R^2 or R^3 multiplication by A maps each eigenvector $\mathbf{x}$ of A (if any) along the same line through the origin as $\mathbf{x}$. Depending on the sign and magnitude of the eigenvalue λ corresponding to $\mathbf{x}$, the operation $A\mathbf{x} = \lambda\mathbf{x}$ compresses or stretches $\mathbf{x}$ by a factor of λ, with a reversal of direction in the case where λ is negative (Figure 5.1.1).

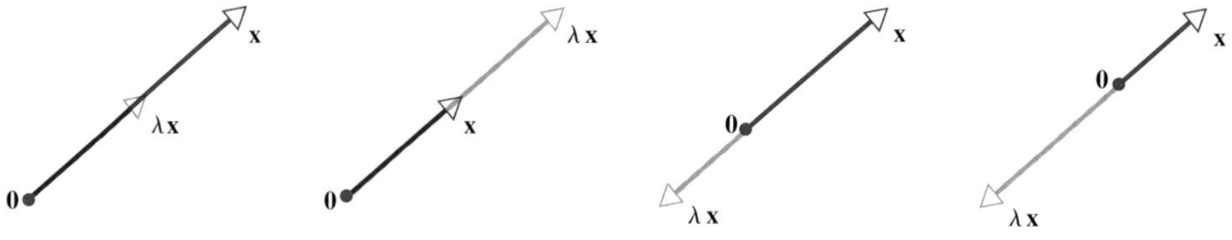

(a) $0 \leq \lambda \leq 1$ (b) $\lambda \leq 1$ (c) $-1 \leq \lambda \leq 0$ (d) $\lambda \leq -1$

▲ Figure 5.1.1

▶ **EXAMPLE 1** **Eigenvector of a 2 × 2 Matrix**

The vector $\mathbf{x} = \begin{bmatrix} 1 \\ 2 \end{bmatrix}$ is an eigenvector of

$$A = \begin{bmatrix} 3 & 0 \\ 8 & -1 \end{bmatrix}$$

corresponding to the eigenvalue $\lambda = 3$, since

$$A\mathbf{x} = \begin{bmatrix} 3 & 0 \\ 8 & -1 \end{bmatrix} \begin{bmatrix} 1 \\ 2 \end{bmatrix} = \begin{bmatrix} 3 \\ 6 \end{bmatrix} = 3\mathbf{x}$$

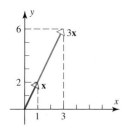

▲ Figure 5.1.2

Geometrically, multiplication by A has stretched the vector $\mathbf{x}$ by a factor of 3 (Figure 5.1.2). ◀

Computing Eigenvalues and Eigenvectors

Our next objective is to obtain a general procedure for finding eigenvalues and eigenvectors of an $n \times n$ matrix A. We will begin with the problem of finding the eigenvalues of A. Note first that the equation $A\mathbf{x} = \lambda\mathbf{x}$ can be rewritten as $A\mathbf{x} = \lambda I\mathbf{x}$, or equivalently, as

$$(\lambda I - A)\mathbf{x} = 0$$

For λ to be an eigenvalue of A this equation must have a nonzero solution for $\mathbf{x}$. But it follows from parts (b) and (g) of Theorem 4.9.4 that this is so if and only if the coefficient matrix $\lambda I - A$ has a zero determinant. Thus, we have the following result.

THEOREM 5.1.1 *If A is an $n \times n$ matrix, then λ is an eigenvalue of A if and only if it satisfies the equation*

$$\det(\lambda I - A) = 0 \qquad (1)$$

*This is called the **characteristic equation** of A.*

▶ **EXAMPLE 2** **Finding Eigenvalues**

In Example 1 we observed that $\lambda = 3$ is an eigenvalue of the matrix

$$A = \begin{bmatrix} 3 & 0 \\ 8 & -1 \end{bmatrix}$$

but we did not explain how we found it. Use the characteristic equation to find all eigenvalues of this matrix.

Solution It follows from Formula (1) that the eigenvalues of A are the solutions of the equation $\det(\lambda I - A) = 0$, which we can write as

$$\begin{vmatrix} \lambda - 3 & 0 \\ -8 & \lambda + 1 \end{vmatrix} = 0$$

from which we obtain
$$(\lambda - 3)(\lambda + 1) = 0 \tag{2}$$

This shows that the eigenvalues of A are $\lambda = 3$ and $\lambda = -1$. Thus, in addition to the eigenvalue $\lambda = 3$ noted in Example 1, we have discovered a second eigenvalue $\lambda = -1$. ◀

When the determinant $\det(\lambda I - A)$ that appears on the left side of (1) is expanded, the result is a polynomial $p(\lambda)$ of degree n that is called the **characteristic polynomial** of A. For example, it follows from (2) that the characteristic polynomial of the 2×2 matrix A in Example 2 is
$$p(\lambda) = (\lambda - 3)(\lambda + 1) = \lambda^2 - 2\lambda - 3$$
which is a polynomial of degree 2. In general, the characteristic polynomial of an $n \times n$ matrix has the form
$$p(\lambda) = \lambda^n + c_1 \lambda^{n-1} + \cdots + c_n$$
in which the coefficient of λ^n is 1 (Exercise 17). Since a polynomial of degree n has at most n distinct roots, it follows that the equation
$$\lambda^n + c_1 \lambda^{n-1} + \cdots + c_n = 0 \tag{3}$$
has at most n distinct solutions and consequently that an $n \times n$ matrix has at most n distinct eigenvalues. Since some of these solutions may be complex numbers, it is possible for a matrix to have complex eigenvalues, even if that matrix itself has real entries. We will discuss this issue in more detail later, but for now we will focus on examples in which the eigenvalues are real numbers.

▶ **EXAMPLE 3 Eigenvalues of a 3 × 3 Matrix**

Find the eigenvalues of
$$A = \begin{bmatrix} 0 & 1 & 0 \\ 0 & 0 & 1 \\ 4 & -17 & 8 \end{bmatrix}$$

Solution The characteristic polynomial of A is
$$\det(\lambda I - A) = \det \begin{bmatrix} \lambda & -1 & 0 \\ 0 & \lambda & -1 \\ -4 & 17 & \lambda - 8 \end{bmatrix} = \lambda^3 - 8\lambda^2 + 17\lambda - 4$$

The eigenvalues of A must therefore satisfy the cubic equation
$$\lambda^3 - 8\lambda^2 + 17\lambda - 4 = 0 \tag{4}$$

To solve this equation, we will begin by searching for integer solutions. This task can be simplified by exploiting the fact that all integer solutions (if there are any) of a polynomial equation with *integer coefficients*
$$\lambda^n + c_1 \lambda^{n-1} + \cdots + c_n = 0$$
must be divisors of the constant term, c_n. Thus, the only possible integer solutions of (4) are the divisors of -4, that is, $\pm 1, \pm 2, \pm 4$. Successively substituting these values in (4) shows that $\lambda = 4$ is an integer solution. As a consequence, $\lambda - 4$ must be a factor of the left side of (4). Dividing $\lambda - 4$ into $\lambda^3 - 8\lambda^2 + 17\lambda - 4$ shows that (4) can be rewritten as
$$(\lambda - 4)(\lambda^2 - 4\lambda + 1) = 0$$

> In applications involving large matrices it is often not feasible to compute the characteristic equation directly so other methods must be used to find eigenvalues. We will consider such methods in Chapter 9.

Thus, the remaining solutions of (4) satisfy the quadratic equation

$$\lambda^2 - 4\lambda + 1 = 0$$

which can be solved by the quadratic formula. Thus the eigenvalues of A are

$$\lambda = 4, \quad \lambda = 2 + \sqrt{3}, \quad \text{and} \quad \lambda = 2 - \sqrt{3}$$

▶ **EXAMPLE 4** **Eigenvalues of an Upper Triangular Matrix**

Find the eigenvalues of the upper triangular matrix

$$A = \begin{bmatrix} a_{11} & a_{12} & a_{13} & a_{14} \\ 0 & a_{22} & a_{23} & a_{24} \\ 0 & 0 & a_{33} & a_{34} \\ 0 & 0 & 0 & a_{44} \end{bmatrix}$$

Solution Recalling that the determinant of a triangular matrix is the product of the entries on the main diagonal (Theorem 2.1.2), we obtain

$$\det(\lambda I - A) = \det \begin{bmatrix} \lambda - a_{11} & -a_{12} & -a_{13} & -a_{14} \\ 0 & \lambda - a_{22} & -a_{23} & -a_{24} \\ 0 & 0 & \lambda - a_{33} & -a_{34} \\ 0 & 0 & 0 & \lambda - a_{44} \end{bmatrix}$$

$$= (\lambda - a_{11})(\lambda - a_{22})(\lambda - a_{33})(\lambda - a_{44})$$

Thus, the characteristic equation is

$$(\lambda - a_{11})(\lambda - a_{22})(\lambda - a_{33})(\lambda - a_{44}) = 0$$

and the eigenvalues are

$$\lambda = a_{11}, \quad \lambda = a_{22}, \quad \lambda = a_{33}, \quad \lambda = a_{44}$$

which are precisely the diagonal entries of A. ◀

The following general theorem should be evident from the computations in the preceding example.

THEOREM 5.1.2 *If A is an $n \times n$ triangular matrix (upper triangular, lower triangular, or diagonal), then the eigenvalues of A are the entries on the main diagonal of A.*

▶ **EXAMPLE 5** **Eigenvalues of a Lower Triangular Matrix**

By inspection, the eigenvalues of the lower triangular matrix

$$A = \begin{bmatrix} \frac{1}{2} & 0 & 0 \\ -1 & \frac{2}{3} & 0 \\ 5 & -8 & -\frac{1}{4} \end{bmatrix}$$

are $\lambda = \frac{1}{2}, \lambda = \frac{2}{3},$ and $\lambda = -\frac{1}{4}$. ◀

Had Theorem 5.1.2 been available earlier, we could have anticipated the result obtained in Example 2.

5.1 Eigenvalues and Eigenvectors

THEOREM 5.1.3 *If A is an n × n matrix, the following statements are equivalent.*

(a) λ *is an eigenvalue of A.*

(b) *The system of equations* $(\lambda I - A)\mathbf{x} = \mathbf{0}$ *has nontrivial solutions.*

(c) *There is a nonzero vector* $\mathbf{x}$ *such that* $A\mathbf{x} = \lambda \mathbf{x}$.

(d) λ *is a solution of the characteristic equation* $\det(\lambda I - A) = 0$.

Finding Eigenvectors and Bases for Eigenspaces

Now that we know how to find the eigenvalues of a matrix, we will consider the problem of finding the corresponding eigenvectors. Since the eigenvectors corresponding to an eigenvalue λ of a matrix A are the nonzero vectors that satisfy the equation

$$(\lambda I - A)\mathbf{x} = \mathbf{0}$$

Notice that $\mathbf{x} = \mathbf{0}$ is in every eigenspace even though it is not an eigenvector. Thus, it is the *nonzero* vectors in an eigenspace that are the eigenvectors.

these eigenvectors are the nonzero vectors in the null space of the matrix $\lambda I - A$. We call this null space the ***eigenspace*** of A corresponding to λ. Stated another way, *the eigenspace of A corresponding to the eigenvalue λ is the solution space of the homogeneous system* $(\lambda I - A)\mathbf{x} = \mathbf{0}$.

▶ **EXAMPLE 6** Bases for Eigenspaces

Find bases for the eigenspaces of the matrix

$$A = \begin{bmatrix} 3 & 0 \\ 8 & -1 \end{bmatrix}$$

Solution In Example 1 we found the characteristic equation of A to be

$$(\lambda - 3)(\lambda + 1) = 0$$

from which we obtained the eigenvalues $\lambda = 3$ and $\lambda = -1$. Thus, there are two eigenspaces of A, one corresponding to each of these eigenvalues.

By definition,

$$\mathbf{x} = \begin{bmatrix} x_1 \\ x_2 \end{bmatrix}$$

is an eigenvector of A corresponding to an eigenvalue λ if and only if $\mathbf{x}$ is a nontrivial solution of $(\lambda I - A)\mathbf{x} = \mathbf{0}$, that is, of

$$\begin{bmatrix} \lambda - 3 & 0 \\ -8 & \lambda + 1 \end{bmatrix} \begin{bmatrix} x_1 \\ x_2 \end{bmatrix} = \begin{bmatrix} 0 \\ 0 \end{bmatrix}$$

If $\lambda = 3$, then this equation becomes

$$\begin{bmatrix} 0 & 0 \\ -8 & 4 \end{bmatrix} \begin{bmatrix} x_1 \\ x_2 \end{bmatrix} = \begin{bmatrix} 0 \\ 0 \end{bmatrix}$$

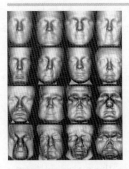

Historical Note Methods of linear algebra are used in the emerging field of computerized face recognition. Researchers are working with the idea that every human face in a racial group is a combination of a few dozen primary shapes. For example, by analyzing three-dimensional scans of many faces, researchers at Rockefeller University have produced both an average head shape in the Caucasian group—dubbed the **meanhead** (top row left in the figure to the left)—and a set of standardized variations from that shape, called **eigenheads** (15 of which are shown in the picture). These are so named because they are eigenvectors of a certain matrix that stores digitized facial information. Face shapes are represented mathematically as linear combinations of the eigenheads.

[Image: Courtesy Dr. Joseph Atick, Dr. Norman Redlich, and Dr. Paul Griffith]

whose general solution is
$$x_1 = \tfrac{1}{2}t, \quad x_2 = t$$
(verify) or in matrix form,
$$\begin{bmatrix} x_1 \\ x_2 \end{bmatrix} = \begin{bmatrix} \tfrac{1}{2}t \\ t \end{bmatrix} = t \begin{bmatrix} \tfrac{1}{2} \\ 1 \end{bmatrix}$$
Thus,
$$\begin{bmatrix} \tfrac{1}{2} \\ 1 \end{bmatrix}$$
is a basis for the eigenspace corresponding to $\lambda = 3$. We leave it as an exercise for you to follow the pattern of these computations and show that
$$\begin{bmatrix} 0 \\ 1 \end{bmatrix}$$
is a basis for the eigenspace corresponding to $\lambda = -1$.

▶ **EXAMPLE 7** **Eigenvectors and Bases for Eigenspaces**

Find bases for the eigenspaces of
$$A = \begin{bmatrix} 0 & 0 & -2 \\ 1 & 2 & 1 \\ 1 & 0 & 3 \end{bmatrix}$$

Solution The characteristic equation of A is $\lambda^3 - 5\lambda^2 + 8\lambda - 4 = 0$, or in factored form, $(\lambda - 1)(\lambda - 2)^2 = 0$ (verify). Thus, the distinct eigenvalues of A are $\lambda = 1$ and $\lambda = 2$, so there are two eigenspaces of A.

By definition,
$$\mathbf{x} = \begin{bmatrix} x_1 \\ x_2 \\ x_3 \end{bmatrix}$$
is an eigenvector of A corresponding to λ if and only if $\mathbf{x}$ is a nontrivial solution of $(\lambda I - A)\mathbf{x} = \mathbf{0}$, or in matrix form,
$$\begin{bmatrix} \lambda & 0 & 2 \\ -1 & \lambda - 2 & -1 \\ -1 & 0 & \lambda - 3 \end{bmatrix} \begin{bmatrix} x_1 \\ x_2 \\ x_3 \end{bmatrix} = \begin{bmatrix} 0 \\ 0 \\ 0 \end{bmatrix} \quad (5)$$

In the case where $\lambda = 2$, Formula (5) becomes
$$\begin{bmatrix} 2 & 0 & 2 \\ -1 & 0 & -1 \\ -1 & 0 & -1 \end{bmatrix} \begin{bmatrix} x_1 \\ x_2 \\ x_3 \end{bmatrix} = \begin{bmatrix} 0 \\ 0 \\ 0 \end{bmatrix}$$

Solving this system using Gaussian elimination yields (verify)
$$x_1 = -s, \quad x_2 = t, \quad x_3 = s$$

Thus, the eigenvectors of A corresponding to $\lambda = 2$ are the nonzero vectors of the form
$$\mathbf{x} = \begin{bmatrix} -s \\ t \\ s \end{bmatrix} = \begin{bmatrix} -s \\ 0 \\ s \end{bmatrix} + \begin{bmatrix} 0 \\ t \\ 0 \end{bmatrix} = s \begin{bmatrix} -1 \\ 0 \\ 1 \end{bmatrix} + t \begin{bmatrix} 0 \\ 1 \\ 0 \end{bmatrix}$$

Since
$$\begin{bmatrix} -1 \\ 0 \\ 1 \end{bmatrix} \quad \text{and} \quad \begin{bmatrix} 0 \\ 1 \\ 0 \end{bmatrix}$$
are linearly independent (why?), these vectors form a basis for the eigenspace corresponding to $\lambda = 2$.

If $\lambda = 1$, then (5) becomes

$$\begin{bmatrix} 1 & 0 & 2 \\ -1 & -1 & -1 \\ -1 & 0 & -2 \end{bmatrix} \begin{bmatrix} x_1 \\ x_2 \\ x_3 \end{bmatrix} = \begin{bmatrix} 0 \\ 0 \\ 0 \end{bmatrix}$$

Solving this system yields (verify)

$$x_1 = -2s, \quad x_2 = s, \quad x_3 = s$$

Thus, the eigenvectors corresponding to $\lambda = 1$ are the nonzero vectors of the form

$$\begin{bmatrix} -2s \\ s \\ s \end{bmatrix} = s \begin{bmatrix} -2 \\ 1 \\ 1 \end{bmatrix} \quad \text{so that} \quad \begin{bmatrix} -2 \\ 1 \\ 1 \end{bmatrix}$$

is a basis for the eigenspace corresponding to $\lambda = 1$. ◄

Powers of a Matrix Once the eigenvalues and eigenvectors of a matrix A are found, it is a simple matter to find the eigenvalues and eigenvectors of any positive integer power of A; for example, if λ is an eigenvalue of A and $\mathbf{x}$ is a corresponding eigenvector, then

$$A^2\mathbf{x} = A(A\mathbf{x}) = A(\lambda \mathbf{x}) = \lambda(A\mathbf{x}) = \lambda(\lambda \mathbf{x}) = \lambda^2 \mathbf{x}$$

which shows that λ^2 is an eigenvalue of A^2 and that $\mathbf{x}$ is a corresponding eigenvector. In general, we have the following result.

THEOREM 5.1.4 *If k is a positive integer, λ is an eigenvalue of a matrix A, and $\mathbf{x}$ is a corresponding eigenvector, then λ^k is an eigenvalue of A^k and $\mathbf{x}$ is a corresponding eigenvector.*

► **EXAMPLE 8** **Powers of a Matrix**

In Example 7 we showed that the eigenvalues of

$$A = \begin{bmatrix} 0 & 0 & -2 \\ 1 & 2 & 1 \\ 1 & 0 & 3 \end{bmatrix}$$

are $\lambda = 2$ and $\lambda = 1$, so from Theorem 5.1.4 both $\lambda = 2^7 = 128$ and $\lambda = 1^7 = 1$ are eigenvalues of A^7. We also showed that

$$\begin{bmatrix} -1 \\ 0 \\ 1 \end{bmatrix} \quad \text{and} \quad \begin{bmatrix} 0 \\ 1 \\ 0 \end{bmatrix}$$

are eigenvectors of A corresponding to the eigenvalue $\lambda = 2$, so from Theorem 5.1.4 they are also eigenvectors of A^7 corresponding to $\lambda = 2^7 = 128$. Similarly, the eigenvector

$$\begin{bmatrix} -2 \\ 1 \\ 1 \end{bmatrix}$$

of A corresponding to the eigenvalue $\lambda = 1$ is also an eigenvector of A^7 corresponding to $\lambda = 1^7 = 1$. ◄

Eigenvalues and Invertibility The next theorem establishes a relationship between eigenvalues and the invertibility of a matrix.

THEOREM 5.1.5 *A square matrix A is invertible if and only if $\lambda = 0$ is not an eigenvalue of A.*

Proof Assume that A is an $n \times n$ matrix and observe first that $\lambda = 0$ is a solution of the characteristic equation

$$\lambda^n + c_1 \lambda^{n-1} + \cdots + c_n = 0$$

if and only if the constant term c_n is zero. Thus, it suffices to prove that A is invertible if and only if $c_n \neq 0$. But

$$\det(\lambda I - A) = \lambda^n + c_1 \lambda^{n-1} + \cdots + c_n$$

or, on setting $\lambda = 0$,

$$\det(-A) = c_n \quad \text{or} \quad (-1)^n \det(A) = c_n$$

It follows from the last equation that $\det(A) = 0$ if and only if $c_n = 0$, and this in turn implies that A is invertible if and only if $c_n \neq 0$. ◄

▶ **EXAMPLE 9** **Eigenvalues and Invertibility**

The matrix A in Example 7 is invertible since it has eigenvalues $\lambda = 1$ and $\lambda = 2$, neither of which is zero. We leave it for you to check this conclusion by showing that $\det(A) \neq 0$. ◄

More on the Equivalence Theorem

As our final result in this section, we will use Theorem 5.1.5 to add one additional part to Theorem 4.10.4.

THEOREM 5.1.6 **Equivalent Statements**

If A is an $n \times n$ matrix, then the following statements are equivalent.

(a) A is invertible.
(b) $A\mathbf{x} = \mathbf{0}$ has only the trivial solution.
(c) The reduced row echelon form of A is I_n.
(d) A is expressible as a product of elementary matrices.
(e) $A\mathbf{x} = \mathbf{b}$ is consistent for every $n \times 1$ matrix $\mathbf{b}$.
(f) $A\mathbf{x} = \mathbf{b}$ has exactly one solution for every $n \times 1$ matrix $\mathbf{b}$.
(g) $\det(A) \neq 0$.
(h) The column vectors of A are linearly independent.
(i) The row vectors of A are linearly independent.
(j) The column vectors of A span R^n.
(k) The row vectors of A span R^n.
(l) The column vectors of A form a basis for R^n.
(m) The row vectors of A form a basis for R^n.
(n) A has rank n.
(o) A has nullity 0.
(p) The orthogonal complement of the null space of A is R^n.
(q) The orthogonal complement of the row space of A is $\{\mathbf{0}\}$.
(r) The range of T_A is R^n.
(s) T_A is one-to-one.
(t) $\lambda = 0$ is not an eigenvalue of A.

This theorem relates all of the major topics we have studied thus far.

5.1 Eigenvalues and Eigenvectors

Concept Review
- Eigenvector
- Eigenvalue
- Characteristic equation
- Characteristic polynomial
- Eigenspace
- Equivalence Theorem

Skills
- Find the eigenvalues of a matrix.
- Find bases for the eigenspaces of a matrix.

Exercise Set 5.1

▶ In Exercises 1–2, confirm by multiplication that **x** is an eigenvector of A, and find the corresponding eigenvalue. ◀

1. $A = \begin{bmatrix} 8 & -9 & 4 \\ 3 & -4 & 3 \\ -3 & 3 & 1 \end{bmatrix}$; $\mathbf{x} = \begin{bmatrix} 1 \\ 2 \\ 3 \end{bmatrix}$

2. $A = \begin{bmatrix} 2 & -1 & -1 \\ -1 & 2 & -1 \\ -1 & -1 & 2 \end{bmatrix}$; $\mathbf{x} = \begin{bmatrix} 1 \\ 1 \\ 1 \end{bmatrix}$

3. Find the characteristic equations of the following matrices:

(a) $\begin{bmatrix} 4 & 3 \\ 0 & -2 \end{bmatrix}$ (b) $\begin{bmatrix} 2 & -1 \\ 10 & -9 \end{bmatrix}$ (c) $\begin{bmatrix} 0 & 3 \\ 4 & 0 \end{bmatrix}$

(d) $\begin{bmatrix} -2 & -7 \\ 1 & 2 \end{bmatrix}$ (e) $\begin{bmatrix} 0 & 0 \\ 0 & 0 \end{bmatrix}$ (f) $\begin{bmatrix} 1 & 0 \\ 0 & 1 \end{bmatrix}$

4. Find the eigenvalues of the matrices in Exercise 3.

5. Find bases for the eigenspaces of the matrices in Exercise 3.

6. Find the characteristic equations of the following matrices:

(a) $\begin{bmatrix} 5 & 1 & 3 \\ 0 & -1 & 0 \\ 0 & 1 & 2 \end{bmatrix}$ (b) $\begin{bmatrix} 0 & 6 & 12 \\ 0 & 3 & 10 \\ 0 & 0 & -2 \end{bmatrix}$

(c) $\begin{bmatrix} -2 & 0 & 1 \\ -6 & -2 & 0 \\ 19 & 5 & -4 \end{bmatrix}$ (d) $\begin{bmatrix} -1 & 0 & 1 \\ -1 & 3 & 0 \\ -4 & 13 & -1 \end{bmatrix}$

(e) $\begin{bmatrix} 5 & 0 & 1 \\ 1 & 1 & 0 \\ -7 & 1 & 0 \end{bmatrix}$ (f) $\begin{bmatrix} 5 & 6 & 2 \\ 0 & -1 & -8 \\ 1 & 0 & -2 \end{bmatrix}$

7. Find the eigenvalues of the matrices in Exercise 6.

8. Find bases for the eigenspaces of the matrices in Exercise 6.

9. Find the characteristic equations of the following matrices:

(a) $\begin{bmatrix} 0 & 0 & 2 & 0 \\ 1 & 0 & 1 & 0 \\ 0 & 1 & -2 & 0 \\ 0 & 0 & 0 & 1 \end{bmatrix}$ (b) $\begin{bmatrix} 10 & -9 & 0 & 0 \\ 4 & -2 & 0 & 0 \\ 0 & 0 & -2 & -7 \\ 0 & 0 & 1 & 2 \end{bmatrix}$

10. Find the eigenvalues of the matrices in Exercise 9.

11. Find bases for the eigenspaces of the matrices in Exercise 9.

12. By inspection, find the eigenvalues of the following matrices:

(a) $\begin{bmatrix} -1 & 6 \\ 0 & 5 \end{bmatrix}$ (b) $\begin{bmatrix} 3 & 0 & 0 \\ -2 & 7 & 0 \\ 4 & 8 & 1 \end{bmatrix}$

(c) $\begin{bmatrix} -\frac{1}{3} & 0 & 0 & 0 \\ 0 & -\frac{1}{3} & 0 & 0 \\ 0 & 0 & 1 & 0 \\ 0 & 0 & 0 & \frac{1}{2} \end{bmatrix}$

13. Find the eigenvalues of A^7 for

$$A = \begin{bmatrix} 2 & 0 & 0 & 0 \\ 3 & -1 & 0 & 0 \\ 8 & 7 & \frac{1}{2} & 0 \\ -1 & 9 & 6 & 0 \end{bmatrix}$$

14. Find the eigenvalues and bases for the eigenspaces of A^{25} for

$$A = \begin{bmatrix} -1 & -2 & -2 \\ 1 & 2 & 1 \\ -1 & -1 & 0 \end{bmatrix}$$

15. Let A be a 2×2 matrix, and call a line through the origin of R^2 *invariant* under A if $A\mathbf{x}$ lies on the line when $\mathbf{x}$ does. Find equations for all lines in R^2, if any, that are invariant under the given matrix.

(a) $A = \begin{bmatrix} 4 & -1 \\ 2 & 1 \end{bmatrix}$ (b) $A = \begin{bmatrix} 0 & 1 \\ -1 & 0 \end{bmatrix}$ (c) $A = \begin{bmatrix} 2 & 3 \\ 0 & 2 \end{bmatrix}$

16. Find $\det(A)$ given that A has $p(\lambda)$ as its characteristic polynomial.

(a) $p(\lambda) = \lambda^3 - 2\lambda^2 + \lambda + 5$

(b) $p(\lambda) = \lambda^4 - \lambda^3 + 7$

[*Hint:* See the proof of Theorem 5.1.5.]

17. Let A be an $n \times n$ matrix.

(a) Prove that the characteristic polynomial of A has degree n.

(b) Prove that the coefficient of λ^n in the characteristic polynomial is 1.

18. Show that the characteristic equation of a 2×2 matrix A can be expressed as $\lambda^2 - \text{tr}(A)\lambda + \det(A) = 0$, where $\text{tr}(A)$ is the trace of A.

19. Use the result in Exercise 18 to show that if
$$A = \begin{bmatrix} a & b \\ c & d \end{bmatrix}$$
then the solutions of the characteristic equation of A are
$$\lambda = \tfrac{1}{2}\left[(a+d) \pm \sqrt{(a-d)^2 + 4bc}\right]$$
Use this result to show that A has

(a) two distinct real eigenvalues if $(a-d)^2 + 4bc > 0$.

(b) two repeated real eigenvalues if $(a-d)^2 + 4bc = 0$.

(c) complex conjugate eigenvalues if $(a-d)^2 + 4bc < 0$.

20. Let A be the matrix in Exercise 19. Show that if $b \neq 0$, then
$$\mathbf{x}_1 = \begin{bmatrix} -b \\ a - \lambda_1 \end{bmatrix} \quad \text{and} \quad \mathbf{x}_2 = \begin{bmatrix} -b \\ a - \lambda_2 \end{bmatrix}$$
are eigenvectors of A that correspond, respectively, to the eigenvalues
$$\lambda_1 = \tfrac{1}{2}\left[(a+d) + \sqrt{(a-d)^2 + 4bc}\right]$$
and
$$\lambda_2 = \tfrac{1}{2}\left[(a+d) - \sqrt{(a-d)^2 + 4bc}\right]$$

21. Use the result of Exercise 18 to prove that if $p(\lambda)$ is the characteristic polynomial of a 2×2 matrix A, then $p(A) = 0$.

22. Prove: If a, b, c, and d are integers such that $a + b = c + d$, then
$$A = \begin{bmatrix} a & b \\ c & d \end{bmatrix}$$
has integer eigenvalues—namely, $\lambda_1 = a + b$ and $\lambda_2 = a - c$.

23. Prove: If λ is an eigenvalue of an invertible matrix A, and $\mathbf{x}$ is a corresponding eigenvector, then $1/\lambda$ is an eigenvalue of A^{-1}, and $\mathbf{x}$ is a corresponding eigenvector.

24. Prove: If λ is an eigenvalue of A, $\mathbf{x}$ is a corresponding eigenvector, and s is a scalar, then $\lambda - s$ is an eigenvalue of $A - sI$, and $\mathbf{x}$ is a corresponding eigenvector.

25. Prove: If λ is an eigenvalue of A and $\mathbf{x}$ is a corresponding eigenvector, then $s\lambda$ is an eigenvalue of sA for every scalar s, and $\mathbf{x}$ is a corresponding eigenvector.

26. Find the eigenvalues and bases for the eigenspaces of
$$A = \begin{bmatrix} -2 & 2 & 3 \\ -2 & 3 & 2 \\ -4 & 2 & 5 \end{bmatrix}$$

and then use Exercises 23 and 24 to find the eigenvalues and bases for the eigenspaces of

(a) A^{-1} (b) $A - 3I$ (c) $A + 2I$

27. (a) Prove that if A is a square matrix, then A and A^T have the same eigenvalues. [*Hint:* Look at the characteristic equation $\det(\lambda I - A) = 0$.]

(b) Show that A and A^T need not have the same eigenspaces. [*Hint:* Use the result in Exercise 20 to find a 2×2 matrix for which A and A^T have different eigenspaces.]

28. Suppose that the characteristic polynomial of some matrix A is found to be $p(\lambda) = \lambda^2(\lambda + 3)^3(\lambda - 4)$. In each part, answer the question and explain your reasoning.

(a) What is the size of A?

(b) Is A invertible?

(c) How many eigenspaces does A have?

29. The eigenvectors that we have been studying are sometimes called **right eigenvectors** to distinguish them from **left eigenvectors**, which are $n \times 1$ column matrices $\mathbf{x}$ that satisfy the equation $\mathbf{x}^T A = \mu \mathbf{x}^T$ for some scalar μ. What is the relationship, if any, between the right eigenvectors and corresponding eigenvalues λ of A and the left eigenvectors and corresponding eigenvalues μ of A?

True-False Exercises

In parts (a)–(g) determine whether the statement is true or false, and justify your answer.

(a) If A is a square matrix and $A\mathbf{x} = \lambda\mathbf{x}$ for some nonzero scalar λ, then $\mathbf{x}$ is an eigenvector of A.

(b) If λ is an eigenvalue of a matrix A, then the linear system $(\lambda I - A)\mathbf{x} = \mathbf{0}$ has only the trivial solution.

(c) If the characteristic polynomial of a matrix A is $p(\lambda) = \lambda^2 + 1$, then A is invertible.

(d) If λ is an eigenvalue of a matrix A, then the eigenspace of A corresponding to λ is the set of eigenvectors of A corresponding to λ.

(e) If 0 is an eigenvalue of a matrix A, then A^2 is singular.

(f) The eigenvalues of a matrix A are the same as the eigenvalues of the reduced row echelon form of A.

(g) If 0 is an eigenvalue of a matrix A, then the set of columns of A is linearly independent.

5.2 Diagonalization

In this section we will be concerned with the problem of finding a basis for R^n that consists of eigenvectors of an $n \times n$ matrix A. Such bases can be used to study geometric properties of A and to simplify various numerical computations. These bases are also of physical significance in a wide variety of applications, some of which will be considered later in this text.

The Matrix Diagonalization Problem

Our first objective in this section is to show that the following two seemingly different problems are equivalent.

> **Problem 1** Given an $n \times n$ matrix A, does there exist an invertible matrix P such that $P^{-1}AP$ is diagonal?
>
> **Problem 2** Given an $n \times n$ matrix A, does A have n linearly independent eigenvectors?

Similarity

The matrix product $P^{-1}AP$ that appears in Problem 1 is called a ***similarity transformation*** of the matrix A. Such products are important in the study of eigenvectors and eigenvalues, so we will begin with some terminology about them.

> **DEFINITION 1** If A and B are square matrices, then we say that ***B is similar to A*** if there is an invertible matrix P such that $B = P^{-1}AP$.

Note that if B is similar to A, then it is also true that A is similar to B, since we can express B as $B = Q^{-1}AQ$ by taking $Q = P^{-1}$. This being the case, we will usually say that A and B are ***similar matrices*** if either is similar to the other.

Similarity Invariants

Similar matrices have many properties in common. For example, if $B = P^{-1}AP$, then it follows that A and B have the same determinant, since

$$\det(B) = \det(P^{-1}AP) = \det(P^{-1})\det(A)\det(P)$$
$$= \frac{1}{\det(P)}\det(A)\det(P) = \det(A)$$

In general, any property that is shared by all similar matrices is called a ***similarity invariant*** or is said to be ***invariant under similarity***. Table 1 lists the most important similarity invariants. The proofs of some of these results are given as exercises.

Expressed in the language of similarity, Problem 1 posed above is equivalent to asking whether the matrix A is similar to a diagonal matrix. If so, the diagonal matrix will have all of the similarity-invariant properties of A, but will have a simpler form, making it easier to analyze and work with. This important idea has some associated terminology.

> **DEFINITION 2** A square matrix A is said to be ***diagonalizable*** if it is similar to some diagonal matrix; that is, if there exists an invertible matrix P such that $P^{-1}AP$ is diagonal. In this case the matrix P is said to ***diagonalize*** A.

Chapter 5 Eigenvalues and Eigenvectors

Table 1 Similarity Invariants

Property	Description
Determinant	A and $P^{-1}AP$ have the same determinant.
Invertibility	A is invertible if and only if $P^{-1}AP$ is invertible.
Rank	A and $P^{-1}AP$ have the same rank.
Nullity	A and $P^{-1}AP$ have the same nullity.
Trace	A and $P^{-1}AP$ have the same trace.
Characteristic polynomial	A and $P^{-1}AP$ have the same characteristic polynomial.
Eigenvalues	A and $P^{-1}AP$ have the same eigenvalues.
Eigenspace dimension	If λ is an eigenvalue of A and hence of $P^{-1}AP$, then the eigenspace of A corresponding to λ and the eigenspace of $P^{-1}AP$ corresponding to λ have the same dimension.

The following theorem shows that Problems 1 and 2 posed above are actually two different forms of the same mathematical problem.

THEOREM 5.2.1 *If A is an $n \times n$ matrix, the following statements are equivalent.*
(a) A is diagonalizable.
(b) A has n linearly independent eigenvectors.

Part (b) of Theorem 5.2.1 is equivalent to saying that there is a *basis* for R^n consisting of eigenvectors of A. Why?

Proof (a) $\Rightarrow$ (b) Since A is assumed to be diagonalizable, it follows that there exists an invertible matrix P and a diagonal matrix D such that $P^{-1}AP = D$ or, equivalently,

$$AP = PD \tag{1}$$

If we denote the column vectors of P by $\mathbf{p}_1, \mathbf{p}_2, \ldots, \mathbf{p}_n$, and if we assume that the diagonal entries of D are $\lambda_1, \lambda_2, \ldots, \lambda_n$, then by Formula (6) of Section 1.3 the left side of (1) can be expressed as

$$AP = A[\mathbf{p}_1 \quad \mathbf{p}_2 \quad \cdots \quad \mathbf{p}_n] = [A\mathbf{p}_1 \quad A\mathbf{p}_2 \quad \cdots \quad A\mathbf{p}_n]$$

and, as noted in the comment following Example 1 of Section 1.7, the right side of (1) can be expressed as

$$PD = [\lambda_1\mathbf{p}_1 \quad \lambda_2\mathbf{p}_2 \quad \cdots \quad \lambda_n\mathbf{p}_n]$$

Thus, it follows from (1) that

$$A\mathbf{p}_1 = \lambda_1\mathbf{p}_1, \quad A\mathbf{p}_2 = \lambda_2\mathbf{p}_2, \ldots, \quad A\mathbf{p}_n = \lambda_n\mathbf{p}_n \tag{2}$$

Since P is invertible, we know from Theorem 5.1.6 that its column vectors $\mathbf{p}_1, \mathbf{p}_2, \ldots, \mathbf{p}_n$ are linearly independent (and hence nonzero). Thus, it follows from (2) that these n column vectors are eigenvectors of A.

Proof (b) $\Rightarrow$ (a) Assume that A has n linearly independent eigenvectors, $\mathbf{p}_1, \mathbf{p}_2, \ldots, \mathbf{p}_n$, and that $\lambda_1, \lambda_2, \ldots, \lambda_n$ are the corresponding eigenvalues. If we let

$$P = [\mathbf{p}_1 \quad \mathbf{p}_2 \quad \cdots \quad \mathbf{p}_n]$$

and if we let D be the diagonal matrix that has $\lambda_1, \lambda_2, \ldots, \lambda_n$ as its successive diagonal entries, then

$$AP = A[\mathbf{p}_1 \quad \mathbf{p}_2 \quad \cdots \quad \mathbf{p}_n] = [A\mathbf{p}_1 \quad A\mathbf{p}_2 \quad \cdots \quad A\mathbf{p}_n]$$
$$= [\lambda_1\mathbf{p}_1 \quad \lambda_2\mathbf{p}_2 \quad \cdots \quad \lambda_n\mathbf{p}_n] = PD$$

5.2 Diagonalization

Since the column vectors of P are linearly independent, it follows from Theorem 5.1.6 that P is invertible, so that this last equation can be rewritten as $P^{-1}AP = D$, which shows that A is diagonalizable. ◀

Procedure for Diagonalizing a Matrix

The preceding theorem guarantees that an $n \times n$ matrix A with n linearly independent eigenvectors is diagonalizable, and the proof suggests the following method for diagonalizing A.

Procedure for Diagonalizing a Matrix

Step 1. Confirm that the matrix is actually diagonalizable by finding n linearly independent eigenvectors. One way to do this is by finding a basis for each eigenspace and merging these basis vectors into a single set S. If this set has fewer than n vectors, then the matrix is not diagonalizable.

Step 2. Form the matrix $P = [\mathbf{p}_1 \ \mathbf{p}_2 \ \cdots \ \mathbf{p}_n]$ that has the vectors in S as its column vectors.

Step 3. The matrix $P^{-1}AP$ will be diagonal and have the eigenvalues $\lambda_1, \lambda_2, \ldots, \lambda_n$ corresponding to the eigenvectors $\mathbf{p}_1, \mathbf{p}_2, \ldots, \mathbf{p}_n$ as its successive diagonal entries.

▶ **EXAMPLE 1** Finding a Matrix P That Diagonalizes a Matrix A

Find a matrix P that diagonalizes

$$A = \begin{bmatrix} 0 & 0 & -2 \\ 1 & 2 & 1 \\ 1 & 0 & 3 \end{bmatrix}$$

Solution In Example 7 of the preceding section we found the characteristic equation of A to be

$$(\lambda - 1)(\lambda - 2)^2 = 0$$

and we found the following bases for the eigenspaces:

$$\lambda = 2: \quad \mathbf{p}_1 = \begin{bmatrix} -1 \\ 0 \\ 1 \end{bmatrix}, \quad \mathbf{p}_2 = \begin{bmatrix} 0 \\ 1 \\ 0 \end{bmatrix}; \quad \lambda = 1: \quad \mathbf{p}_3 = \begin{bmatrix} -2 \\ 1 \\ 1 \end{bmatrix}$$

There are three basis vectors in total, so the matrix

$$P = \begin{bmatrix} -1 & 0 & -2 \\ 0 & 1 & 1 \\ 1 & 0 & 1 \end{bmatrix}$$

diagonalizes A. As a check, you should verify that

$$P^{-1}AP = \begin{bmatrix} 1 & 0 & 2 \\ 1 & 1 & 1 \\ -1 & 0 & -1 \end{bmatrix} \begin{bmatrix} 0 & 0 & -2 \\ 1 & 2 & 1 \\ 21 & 0 & 3 \end{bmatrix} \begin{bmatrix} -1 & 0 & -2 \\ 0 & 1 & 1 \\ 1 & 0 & 1 \end{bmatrix} = \begin{bmatrix} 2 & 0 & 0 \\ 0 & 2 & 0 \\ 0 & 0 & 1 \end{bmatrix} \blacktriangleleft$$

In general, there is no preferred order for the columns of P. Since the ith diagonal entry of $P^{-1}AP$ is an eigenvalue for the ith column vector of P, changing the order of the columns of P just changes the order of the eigenvalues on the diagonal of $P^{-1}AP$. Thus, had we written

$$P = \begin{bmatrix} -1 & -2 & 0 \\ 0 & 1 & 1 \\ 1 & 1 & 0 \end{bmatrix}$$

in the preceding example, we would have obtained

$$P^{-1}AP = \begin{bmatrix} 2 & 0 & 0 \\ 0 & 1 & 0 \\ 0 & 0 & 2 \end{bmatrix}$$

▶ **EXAMPLE 2** **A Matrix That Is Not Diagonalizable**

Find a matrix P that diagonalizes

$$A = \begin{bmatrix} 1 & 0 & 0 \\ 1 & 2 & 0 \\ -3 & 5 & 2 \end{bmatrix}$$

Solution The characteristic polynomial of A is

$$\det(\lambda I - A) = \begin{vmatrix} \lambda - 1 & 0 & 0 \\ -1 & \lambda - 2 & 0 \\ 3 & -5 & \lambda - 2 \end{vmatrix} = (\lambda - 1)(\lambda - 2)^2$$

so the characteristic equation is

$$(\lambda - 1)(\lambda - 2)^2 = 0$$

Thus, the distinct eigenvalues of A are $\lambda = 1$ and $\lambda = 2$. We leave it for you to show that bases for the eigenspaces are

$$\lambda = 1: \quad \mathbf{p}_1 = \begin{bmatrix} \frac{1}{8} \\ -\frac{1}{8} \\ 1 \end{bmatrix}; \quad \lambda = 2: \quad \mathbf{p}_2 = \begin{bmatrix} 0 \\ 0 \\ 1 \end{bmatrix}$$

Since A is a 3×3 matrix and there are only two basis vectors in total, A is not diagonalizable.

Alternative Solution If you are concerned only in determining whether a matrix is diagonalizable and not with actually finding a diagonalizing matrix P, then it is not necessary to compute bases for the eigenspaces—it suffices to find the dimensions of the eigenspaces. For this example, the eigenspace corresponding to $\lambda = 1$ is the solution space of the system

$$\begin{bmatrix} 0 & 0 & 0 \\ -1 & -1 & 0 \\ 3 & -5 & -1 \end{bmatrix} \begin{bmatrix} x_1 \\ x_2 \\ x_3 \end{bmatrix} = \begin{bmatrix} 0 \\ 0 \\ 0 \end{bmatrix}$$

Since the coefficient matrix has rank 2 (verify), the nullity of this matrix is 1 by Theorem 4.8.2, and hence the eigenspace corresponding to $\lambda = 1$ is one-dimensional.

The eigenspace corresponding to $\lambda = 2$ is the solution space of the system

$$\begin{bmatrix} 1 & 0 & 0 \\ -1 & 0 & 0 \\ 3 & -5 & 0 \end{bmatrix} \begin{bmatrix} x_1 \\ x_2 \\ x_3 \end{bmatrix} = \begin{bmatrix} 0 \\ 0 \\ 0 \end{bmatrix}$$

This coefficient matrix also has rank 2 and nullity 1 (verify), so the eigenspace corresponding to $\lambda = 2$ is also one-dimensional. Since the eigenspaces produce a total of two basis vectors, and since three are needed, the matrix A is not diagonalizable. ◀

There is an assumption in Example 1 that the column vectors of P, which are made up of basis vectors from the various eigenspaces of A, are linearly independent. The following theorem, proved at the end of this section, shows that this is so.

THEOREM 5.2.2 *If $\mathbf{v}_1, \mathbf{v}_2, \ldots, \mathbf{v}_k$ are eigenvectors of a matrix A corresponding to distinct eigenvalues, then $\{\mathbf{v}_1, \mathbf{v}_2, \ldots, \mathbf{v}_k\}$ is a linearly independent set.*

Remark Theorem 5.2.2 is a special case of a more general result: Suppose that $\lambda_1, \lambda_2, \ldots, \lambda_k$ are distinct eigenvalues and that we choose a linearly independent set in each of the corresponding eigenspaces. If we then merge all these vectors into a single set, the result will still be a linearly independent set. For example, if we choose three linearly independent vectors from one eigenspace and two linearly independent vectors from another eigenspace, then the five vectors together form a linearly independent set. We omit the proof.

As a consequence of Theorem 5.2.2, we obtain the following important result.

THEOREM 5.2.3 *If an $n \times n$ matrix A has n distinct eigenvalues, then A is diagonalizable.*

Proof If $\mathbf{v}_1, \mathbf{v}_2, \ldots, \mathbf{v}_n$ are eigenvectors corresponding to the distinct eigenvalues $\lambda_1, \lambda_2, \ldots, \lambda_n$, then by Theorem 5.2.2, $\mathbf{v}_1, \mathbf{v}_2, \ldots, \mathbf{v}_n$ are linearly independent. Thus, A is diagonalizable by Theorem 5.2.1. ◄

▶ **EXAMPLE 3** **Using Theorem 5.2.3**

We saw in Example 3 of the preceding section that

$$A = \begin{bmatrix} 0 & 1 & 0 \\ 0 & 0 & 1 \\ 4 & -17 & 8 \end{bmatrix}$$

has three distinct eigenvalues: $\lambda = 4$, $\lambda = 2 + \sqrt{3}$, and $\lambda = 2 - \sqrt{3}$. Therefore, A is diagonalizable and

$$P^{-1}AP = \begin{bmatrix} 4 & 0 & 0 \\ 0 & 2+\sqrt{3} & 0 \\ 0 & 0 & 2-\sqrt{3} \end{bmatrix}$$

for some invertible matrix P. If needed, the matrix P can be found using the method shown in Example 1 of this section.

▶ **EXAMPLE 4** **Diagonalizability of Triangular Matrices**

From Theorem 5.1.2, the eigenvalues of a triangular matrix are the entries on its main diagonal. Thus, a triangular matrix with distinct entries on the main diagonal is diagonalizable. For example,

$$A = \begin{bmatrix} -1 & 2 & 4 & 0 \\ 0 & 3 & 1 & 7 \\ 0 & 0 & 5 & 8 \\ 0 & 0 & 0 & -2 \end{bmatrix}$$

is a diagonalizable matrix with eigenvalues $\lambda_1 = -1$, $\lambda_2 = 3$, $\lambda_3 = 5$, $\lambda_4 = -2$. ◄

310 Chapter 5 Eigenvalues and Eigenvectors

Computing Powers of a Matrix

There are many applications in which it is necessary to compute high powers of a square matrix A. We will show next that if A happens to be diagonalizable, then the computations can be simplified by diagonalizing A.

To start, suppose that A is a diagonalizable $n \times n$ matrix, that P diagonalizes A, and that

$$P^{-1}AP = \begin{bmatrix} \lambda_1 & 0 & \cdots & 0 \\ 0 & \lambda_2 & \cdots & 0 \\ \vdots & \vdots & & \vdots \\ 0 & 0 & \cdots & \lambda_n \end{bmatrix} = D$$

Squaring both sides of this equation yields

$$(P^{-1}AP)^2 = \begin{bmatrix} \lambda_1^2 & 0 & \cdots & 0 \\ 0 & \lambda_2^2 & \cdots & 0 \\ \vdots & \vdots & & \vdots \\ 0 & 0 & \cdots & \lambda_n^2 \end{bmatrix} = D^2$$

We can rewrite the left side of this equation as

$$(P^{-1}AP)^2 = P^{-1}APP^{-1}AP = P^{-1}AIAP = P^{-1}A^2P$$

from which we obtain the relationship $P^{-1}A^2P = D^2$. More generally, if k is a positive integer, then a similar computation will show that

$$P^{-1}A^kP = D^k = \begin{bmatrix} \lambda_1^k & 0 & \cdots & 0 \\ 0 & \lambda_2^k & \cdots & 0 \\ \vdots & \vdots & & \vdots \\ 0 & 0 & \cdots & \lambda_n^k \end{bmatrix}$$

which we can rewrite as

$$A^k = PD^kP^{-1} = P \begin{bmatrix} \lambda_1^k & 0 & \cdots & 0 \\ 0 & \lambda_2^k & \cdots & 0 \\ \vdots & \vdots & & \vdots \\ 0 & 0 & \cdots & \lambda_n^k \end{bmatrix} P^{-1} \quad (3)$$

Formula (3) reveals that raising a diagonalizable matrix A to a positive integer power has the effect of raising its eigenvalues to that power.

Note that computing the right side of this formula involves only three matrix multiplications and the powers of the diagonal entries of D. For matrices of large size and high powers of λ, this involves substantially fewer operations than computing A^k directly.

▶ **EXAMPLE 5** Power of a Matrix

Use (3) to find A^{13}, where

$$A = \begin{bmatrix} 0 & 0 & -2 \\ 1 & 2 & 1 \\ 1 & 0 & 3 \end{bmatrix}$$

Solution We showed in Example 1 that the matrix A is diagonalized by

$$P = \begin{bmatrix} -1 & 0 & -2 \\ 0 & 1 & 1 \\ 1 & 0 & 1 \end{bmatrix}$$

and that

$$D = P^{-1}AP = \begin{bmatrix} 2 & 0 & 0 \\ 0 & 2 & 0 \\ 0 & 0 & 1 \end{bmatrix}$$

Thus, it follows from (3) that

$$A^{13} = PD^{13}P^{-1} = \begin{bmatrix} -1 & 0 & -2 \\ 0 & 1 & 1 \\ 1 & 0 & 1 \end{bmatrix} \begin{bmatrix} 2^{13} & 0 & 0 \\ 0 & 2^{13} & 0 \\ 0 & 0 & 1^{13} \end{bmatrix} \begin{bmatrix} 1 & 0 & 2 \\ 1 & 1 & 1 \\ -1 & 0 & -1 \end{bmatrix} \quad (4)$$

$$= \begin{bmatrix} -8190 & 0 & -16382 \\ 8191 & 8192 & 8191 \\ 8191 & 0 & 16383 \end{bmatrix} \blacktriangleleft$$

Remark With the method in the preceding example, most of the work is in diagonalizing A. Once that work is done, it can be used to compute any power of A. Thus, to compute A^{1000} we need only change the exponents from 13 to 1000 in (4).

Eigenvalues of Powers of a Matrix

Once the eigenvalues and eigenvectors of any square matrix A are found, it is a simple matter to find the eigenvalues and eigenvectors of any positive integer power of A. For example, if λ is an eigenvalue of A and $\mathbf{x}$ is a corresponding eigenvector, then

$$A^2\mathbf{x} = A(A\mathbf{x}) = A(\lambda\mathbf{x}) = \lambda(A\mathbf{x}) = \lambda(\lambda\mathbf{x}) = \lambda^2\mathbf{x}$$

which shows not only that λ^2 is an eigenvalue of A^2 but that $\mathbf{x}$ is a corresponding eigenvector. In general, we have the following result.

Note that diagonalizability is not a requirement in Theorem 5.2.4.

THEOREM 5.2.4 *If λ is an eigenvalue of a square matrix A and $\mathbf{x}$ is a corresponding eigenvector, and if k is any positive integer, then λ^k is an eigenvalue of A^k and $\mathbf{x}$ is a corresponding eigenvector.*

Some problems that use this theorem are given in the exercises.

Geometric and Algebraic Multiplicity

Theorem 5.2.3 does not completely settle the diagonalizability question since it only guarantees that a square matrix with n distinct eigenvalues is diagonalizable, but does not preclude the possibility that there may exist diagonalizable matrices with fewer than n distinct eigenvalues. The following example shows that this is indeed the case.

▶ **EXAMPLE 6** **The Converse of Theorem 5.2.3 Is False**

Consider the matrices

$$I = \begin{bmatrix} 1 & 0 & 0 \\ 0 & 1 & 0 \\ 0 & 0 & 1 \end{bmatrix} \quad \text{and} \quad J = \begin{bmatrix} 1 & 1 & 0 \\ 0 & 1 & 1 \\ 0 & 0 & 1 \end{bmatrix}$$

It follows from Theorem 5.1.2 that both of these matrices have only one distinct eigenvalue, namely $\lambda = 1$, and hence only one eigenspace. We leave it as an exercise for you to solve the characteristic equations

$$(\lambda I - I)\mathbf{x} = \mathbf{0} \quad \text{and} \quad (\lambda J - I)\mathbf{x} = \mathbf{0}$$

with $\lambda = 1$ and show that for I the eigenspace is three-dimensional (all of R^3) and for J it is one-dimensional, consisting of all scalar multiples of

$$\mathbf{x} = \begin{bmatrix} 1 \\ 0 \\ 0 \end{bmatrix}$$

This shows that the converse of Theorem 5.2.3 is false, since we have produced two 3×3 matrices with fewer than three distinct eigenvalues, one of which is diagonalizable and the other of which is not. ◀

A full excursion into the study of diagonalizability is left for more advanced courses, but we will touch on one theorem that is important to a fuller understanding of diagonalizability. It can be proved that if λ_0 is an eigenvalue of A, then the dimension of the eigenspace corresponding to λ_0 cannot exceed the number of times that $\lambda - \lambda_0$ appears as a factor of the characteristic polynomial of A. For example, in Examples 1 and 2 the characteristic polynomial is

$$(\lambda - 1)(\lambda - 2)^2$$

Thus, the eigenspace corresponding to $\lambda = 1$ is at most (hence exactly) one-dimensional, and the eigenspace corresponding to $\lambda = 2$ is at most two-dimensional. In Example 1 the eigenspace corresponding to $\lambda = 2$ actually had dimension 2, resulting in diagonalizability, but in Example 2 the eigenspace corresponding to $\lambda = 2$ had only dimension 1, resulting in nondiagonalizability.

There is some terminology that is related to these ideas. If λ_0 is an eigenvalue of an $n \times n$ matrix A, then the dimension of the eigenspace corresponding to λ_0 is called the **geometric multiplicity** of λ_0, and the number of times that $\lambda - \lambda_0$ appears as a factor in the characteristic polynomial of A is called the **algebraic multiplicity** of λ_0. The following theorem, which we state without proof, summarizes the preceding discussion.

THEOREM 5.2.5 Geometric and Algebraic Multiplicity

If A is a square matrix, then:

(a) For every eigenvalue of A, the geometric multiplicity is less than or equal to the algebraic multiplicity.

(b) A is diagonalizable if and only if the geometric multiplicity of every eigenvalue is equal to the algebraic multiplicity.

OPTIONAL

We will complete this section with an optional proof of Theorem 5.2.2.

Proof of Theorem 5.2.2 Let $\mathbf{v}_1, \mathbf{v}_2, \ldots, \mathbf{v}_k$ be eigenvectors of A corresponding to distinct eigenvalues $\lambda_1, \lambda_2, \ldots, \lambda_k$. We will assume that $\mathbf{v}_1, \mathbf{v}_2, \ldots, \mathbf{v}_k$ are linearly dependent and obtain a contradiction. We can then conclude that $\mathbf{v}_1, \mathbf{v}_2, \ldots, \mathbf{v}_k$ are linearly independent.

Since an eigenvector is nonzero by definition, $\{\mathbf{v}_1\}$ is linearly independent. Let r be the largest integer such that $\{\mathbf{v}_1, \mathbf{v}_2, \ldots, \mathbf{v}_r\}$ is linearly independent. Since we are assuming that $\{\mathbf{v}_1, \mathbf{v}_2, \ldots, \mathbf{v}_k\}$ is linearly dependent, r satisfies $1 \leq r < k$. Moreover, by the definition of r, $\{\mathbf{v}_1, \mathbf{v}_2, \ldots, \mathbf{v}_{r+1}\}$ is linearly dependent. Thus, there are scalars $c_1, c_2, \ldots, c_{r+1}$, not all zero, such that

$$c_1 \mathbf{v}_1 + c_2 \mathbf{v}_2 + \cdots + c_{r+1} \mathbf{v}_{r+1} = \mathbf{0} \tag{5}$$

Multiplying both sides of (5) by A and using the fact that

$$A\mathbf{v}_1 = \lambda_1 \mathbf{v}_1, \quad A\mathbf{v}_2 = \lambda_2 \mathbf{v}_2, \ldots, \quad A\mathbf{v}_{r+1} = \lambda_{r+1} \mathbf{v}_{r+1}$$

we obtain

$$c_1 \lambda_1 \mathbf{v}_1 + c_2 \lambda_2 \mathbf{v}_2 + \cdots + c_{r+1} \lambda_{r+1} \mathbf{v}_{r+1} = \mathbf{0} \tag{6}$$

If we now multiply both sides of (5) by λ_{r+1} and subtract the resulting equation from (6) we obtain

$$c_1(\lambda_1 - \lambda_{r+1})\mathbf{v}_1 + c_2(\lambda_2 - \lambda_{r+1})\mathbf{v}_2 + \cdots + c_r(\lambda_r - \lambda_{r+1})\mathbf{v}_r = \mathbf{0}$$

Since $\{\mathbf{v}_1, \mathbf{v}_2, \ldots, \mathbf{v}_r\}$ is a linearly independent set, this equation implies that

$$c_1(\lambda_1 - \lambda_{r+1}) = c_2(\lambda_2 - \lambda_{r+1}) = \cdots = c_r(\lambda_r - \lambda_{r+1}) = 0$$

and since $\lambda_1, \lambda_2, \ldots, \lambda_{r+1}$ are assumed to be distinct, it follows that

$$c_1 = c_2 = \cdots = c_r = 0 \tag{7}$$

Substituting these values in (5) yields

$$c_{r+1}\mathbf{v}_{r+1} = \mathbf{0}$$

Since the eigenvector $\mathbf{v}_{r+1}$ is nonzero, it follows that

$$c_{r+1} = 0 \tag{8}$$

But equations (7) and (8) contradict the fact that $c_1, c_2, \ldots, c_{r+1}$ are not all zero so the proof is complete. ◀

Concept Review
- Similarity transformation
- Similarity invariant
- Similar matrices
- Diagonalizable matrix
- Geometric multiplicity
- Algebraic multiplicity

Skills
- Determine whether a square matrix A is diagonalizable.
- Diagonalize a square matrix A.
- Find powers of a matrix using similarity.
- Find the geometric multiplicity and the algebraic multiplicity of an eigenvalue.

Exercise Set 5.2

▶ In Exercises 1–4, show that A and B are not similar matrices.

1. $A = \begin{bmatrix} 2 & 1 \\ 3 & 4 \end{bmatrix}$, $B = \begin{bmatrix} 2 & 0 \\ 3 & 3 \end{bmatrix}$

2. $A = \begin{bmatrix} 4 & -1 \\ 2 & 4 \end{bmatrix}$, $B = \begin{bmatrix} 4 & 1 \\ 2 & 4 \end{bmatrix}$

3. $A = \begin{bmatrix} 4 & 2 & 0 \\ 2 & 1 & 0 \\ 1 & 1 & 7 \end{bmatrix}$, $B = \begin{bmatrix} 0 & 3 & 4 \\ 0 & 7 & 2 \\ 0 & 0 & 4 \end{bmatrix}$

4. $A = \begin{bmatrix} 1 & 0 & 1 \\ 2 & 0 & 2 \\ 3 & 0 & 3 \end{bmatrix}$, $B = \begin{bmatrix} 1 & 1 & 0 \\ 2 & 2 & 0 \\ 0 & 1 & 1 \end{bmatrix}$

5. Let A be a 6×6 matrix with characteristic equation $\lambda^2(\lambda - 1)(\lambda - 2)^3 = 0$. What are the possible dimensions for eigenspaces of A?

6. Let
$$A = \begin{bmatrix} 4 & 0 & 1 \\ 2 & 3 & 2 \\ 1 & 0 & 4 \end{bmatrix}$$

(a) Find the eigenvalues of A.

(b) For each eigenvalue λ, find the rank of the matrix $\lambda I - A$.

(c) Is A diagonalizable? Justify your conclusion.

▶ In Exercises 7–11, use the method of Exercise 6 to determine whether the matrix is diagonalizable. ◀

7. $\begin{bmatrix} 4 & 0 \\ -2 & 4 \end{bmatrix}$

8. $\begin{bmatrix} 2 & -3 \\ 1 & -1 \end{bmatrix}$

9. $\begin{bmatrix} 6 & 3 & 1 \\ 0 & 3 & 1 \\ 0 & 0 & 9 \end{bmatrix}$

10. $\begin{bmatrix} -1 & 0 & 1 \\ -1 & 3 & 0 \\ -4 & 13 & -1 \end{bmatrix}$

11. $\begin{bmatrix} 2 & -1 & 0 & 1 \\ 0 & 2 & 1 & -1 \\ 0 & 0 & 3 & 2 \\ 0 & 0 & 0 & 3 \end{bmatrix}$

▶ In Exercises 12–15, find a matrix P that diagonalizes A, and compute $P^{-1}AP$. ◀

12. $A = \begin{bmatrix} -14 & 12 \\ -20 & 17 \end{bmatrix}$

13. $A = \begin{bmatrix} 5 & 7 \\ 0 & -3 \end{bmatrix}$

14. $A = \begin{bmatrix} 1 & 0 & 0 \\ 0 & 1 & 1 \\ 0 & 1 & 1 \end{bmatrix}$

15. $A = \begin{bmatrix} 2 & 0 & -2 \\ 0 & 3 & 0 \\ 0 & 0 & 3 \end{bmatrix}$

In Exercises 16–21, find the geometric and algebraic multiplicity of each eigenvalue of the matrix A, and determine whether A is diagonalizable. If A is diagonalizable, then find a matrix P that diagonalizes A, and find $P^{-1}AP$.

16. $A = \begin{bmatrix} 19 & -9 & -6 \\ 25 & -11 & -9 \\ 17 & -9 & -4 \end{bmatrix}$

17. $A = \begin{bmatrix} 1 & 0 & 9 \\ 0 & 0 & 0 \\ 0 & 0 & 1 \end{bmatrix}$

18. $A = \begin{bmatrix} 5 & 0 & 0 \\ 1 & 5 & 0 \\ 0 & 1 & 5 \end{bmatrix}$

19. $A = \begin{bmatrix} 1 & 2 & -2 \\ -3 & 4 & 0 \\ -3 & 1 & 3 \end{bmatrix}$

20. $A = \begin{bmatrix} -2 & 0 & 0 & 0 \\ 0 & -2 & 0 & 0 \\ 0 & 0 & 3 & 0 \\ 0 & 0 & 1 & 3 \end{bmatrix}$

21. $A = \begin{bmatrix} -2 & 0 & 0 & 0 \\ 0 & -2 & 5 & -5 \\ 0 & 0 & 3 & 0 \\ 0 & 0 & 0 & 3 \end{bmatrix}$

22. Use the method of Example 5 to compute A^{10}, where
$$A = \begin{bmatrix} 2 & 3 \\ 0 & -1 \end{bmatrix}$$

23. Use the method of Example 5 to compute A^{11}, where
$$A = \begin{bmatrix} -1 & 0 & 1 \\ 0 & 2 & 0 \\ 0 & -3 & 1 \end{bmatrix}$$

24. In each part, compute the stated power of
$$A = \begin{bmatrix} 1 & -2 & 8 \\ 0 & -1 & 0 \\ 0 & 0 & -1 \end{bmatrix}$$

   (a) A^{1000} (b) A^{-1000} (c) A^{2301} (d) A^{-2301}

25. Find A^n if n is a positive integer and
$$A = \begin{bmatrix} 3 & -1 & 0 \\ -1 & 2 & -1 \\ 0 & -1 & 3 \end{bmatrix}$$

26. Let
$$A = \begin{bmatrix} a & b \\ c & d \end{bmatrix}$$

   Show that

   (a) A is diagonalizable if $(a-d)^2 + 4bc > 0$.

   (b) A is not diagonalizable if $(a-d)^2 + 4bc < 0$.

   [*Hint:* See Exercise 19 of Section 5.1.]

27. In the case where the matrix A in Exercise 26 is diagonalizable, find a matrix P that diagonalizes A. [*Hint:* See Exercise 20 of Section 5.1.]

28. Prove that similar matrices have the same rank.

29. Prove that similar matrices have the same nullity.

30. Prove that similar matrices have the same trace.

31. Prove that if A is diagonalizable, then so is A^k for every positive integer k.

32. Prove that if A is a diagonalizable matrix, then the rank of A is the number of nonzero eigenvalues of A.

33. Suppose that the characteristic polynomial of some matrix A is found to be $p(\lambda) = (\lambda - 1)(\lambda - 3)^2(\lambda - 4)^3$. In each part, answer the question and explain your reasoning.

   (a) What can you say about the dimensions of the eigenspaces of A?

   (b) What can you say about the dimensions of the eigenspaces if you know that A is diagonalizable?

   (c) If $\{v_1, v_2, v_3\}$ is a linearly independent set of eigenvectors of A all of which correspond to the same eigenvalue of A, what can you say about the eigenvalue?

34. This problem will lead you through a proof of the fact that the algebraic multiplicity of an eigenvalue of an $n \times n$ matrix A is greater than or equal to the geometric multiplicity. For this purpose, assume that λ_0 is an eigenvalue with geometric multiplicity k.

   (a) Prove that there is a basis $B = \{u_1, u_2, \ldots, u_n\}$ for R^n in which the first k vectors of B form a basis for the eigenspace corresponding to λ_0.

   (b) Let P be the matrix having the vectors in B as columns. Prove that the product AP can be expressed as
   $$AP = P \begin{bmatrix} \lambda_0 I_k & X \\ 0 & Y \end{bmatrix}$$
   [*Hint:* Compare the first k column vectors on both sides.]

   (c) Use the result in part (b) to prove that A is similar to
   $$C = \begin{bmatrix} \lambda_0 I_k & X \\ 0 & Y \end{bmatrix}$$
   and hence that A and C have the same characteristic polynomial.

   (d) By considering $\det(\lambda I - C)$, prove that the characteristic polynomial of C (and hence A) contains the factor $(\lambda - \lambda_0)$ at least k times, thereby proving that the algebraic multiplicity of λ_0 is greater than or equal to the geometric multiplicity k.

True-False Exercises

In parts (a)–(h) determine whether the statement is true or false, and justify your answer.

(a) Every square matrix is similar to itself.

(b) If A, B, and C are matrices for which A is similar to B and B is similar to C, then A is similar to C.

(c) If A and B are similar invertible matrices, then A^{-1} and B^{-1} are similar.

(d) If A is diagonalizable, then there is a unique matrix P such that $P^{-1}AP$ is diagonal.

(e) If A is diagonalizable and invertible, then A^{-1} is diagonalizable.

(f) If A is diagonalizable, then A^T is diagonalizable.

(g) If there is a basis for R^n consisting of eigenvectors of an $n \times n$ matrix A, then A is diagonalizable.

(h) If every eigenvalue of a matrix A has algebraic multiplicity 1, then A is diagonalizable.

5.3 Complex Vector Spaces

Because the characteristic equation of any square matrix can have complex solutions, the notions of complex eigenvalues and eigenvectors arise naturally, even within the context of matrices with real entries. In this section we will discuss this idea and apply our results to study symmetric matrices in more detail. A review of the essentials of complex numbers appears in the back of this text.

Review of Complex Numbers

Recall that if $z = a + bi$ is a complex number, then:

- $\text{Re}(z) = a$ and $\text{Im}(z) = b$ are called the ***real part*** of z and the ***imaginary part*** of z, respectively,
- $|z| = \sqrt{a^2 + b^2}$ is called the ***modulus*** (or ***absolute value***) of z,
- $\bar{z} = a - bi$ is called the ***complex conjugate*** of z,
- $z\bar{z} = a^2 + b^2 = |z|^2$,
- the angle ϕ in Figure 5.3.1 is called an ***argument*** of z,
- $\text{Re}(z) = |z| \cos \phi$
- $\text{Im}(z) = |z| \sin \phi$
- $z = |z|(\cos \phi + i \sin \phi)$ is called the ***polar form*** of z.

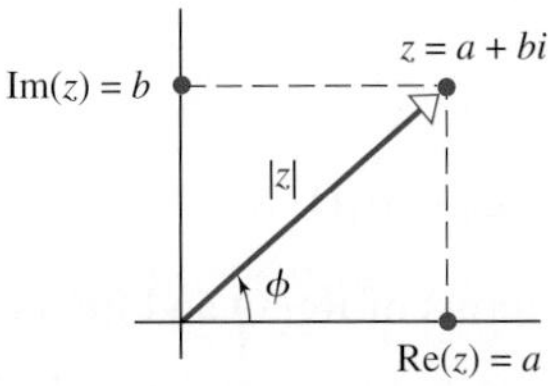

▲ Figure 5.3.1

Complex Eigenvalues

In Formula (3) of Section 5.1 we observed that the characteristic equation of a general $n \times n$ matrix A has the form

$$\lambda^n + c_1 \lambda^{n-1} + \cdots + c_n = 0 \qquad (1)$$

in which the highest power of λ has a coefficient of 1. Up to now we have limited our discussion to matrices in which the solutions of (1) are real numbers. However, it is possible for the characteristic equation of a matrix A with real entries to have imaginary solutions; for example, the characteristic equation of the matrix

$$A = \begin{bmatrix} -2 & -1 \\ 5 & 2 \end{bmatrix}$$

is

$$\begin{vmatrix} \lambda + 2 & 1 \\ -5 & \lambda - 2 \end{vmatrix} = \lambda^2 + 1 = 0$$

which has the imaginary solutions $\lambda = i$ and $\lambda = -i$. To deal with this case we will need to explore the notion of a complex vector space and some related ideas.

Vectors in C^n

A vector space in which scalars are allowed to be complex numbers is called a ***complex vector space***. In this section we will be concerned only with the following complex generalization of the real vector space R^n.

> **DEFINITION 1** If n is a positive integer, then a ***complex n-tuple*** is a sequence of n complex numbers $(v_1, v_2, \ldots, v_n)$. The set of all complex n-tuples is called ***complex n-space*** and is denoted by C^n. Scalars are complex numbers, and the operations of addition, subtraction, and scalar multiplication are performed componentwise.

The terminology used for n-tuples of real numbers applies to complex n-tuples without change. Thus, if $v_1, v_2, \ldots, v_n$ are complex numbers, then we call $\mathbf{v} = (v_1, v_2, \ldots, v_n)$ a ***vector*** in C^n and $v_1, v_2, \ldots, v_n$ its ***components***. Some examples of vectors in C^3 are

$$\mathbf{u} = (1+i, -4i, 3+2i), \quad \mathbf{v} = (0, i, 5), \quad \mathbf{w} = \left(6 - \sqrt{2}i, 9 + \tfrac{1}{2}i, \pi i\right)$$

Every vector

$$\mathbf{v} = (v_1, v_2, \ldots, v_n) = (a_1 + b_1 i, a_2 + b_2 i, \ldots, a_n + b_n i)$$

in C^n can be split into ***real*** and ***imaginary parts*** as

$$\mathbf{v} = (a_1, a_2, \ldots, a_n) + i(b_1, b_2, \ldots, b_n)$$

which we also denote as

$$\mathbf{v} = \text{Re}(\mathbf{v}) + i\,\text{Im}(\mathbf{v})$$

where

$$\text{Re}(\mathbf{v}) = (a_1, a_2, \ldots, a_n) \quad \text{and} \quad \text{Im}(\mathbf{v}) = (b_1, b_2, \ldots, b_n)$$

The vector

$$\bar{\mathbf{v}} = (\bar{v}_1, \bar{v}_2, \ldots, \bar{v}_n) = (a_1 - b_1 i, a_2 - b_2 i, \ldots, a_n - b_n i)$$

is called the ***complex conjugate*** of $\mathbf{v}$ and can be expressed in terms of $\text{Re}(\mathbf{v})$ and $\text{Im}(\mathbf{v})$ as

$$\bar{\mathbf{v}} = (a_1, a_2, \ldots, a_n) - i(b_1, b_2, \ldots, b_n) = \text{Re}(\mathbf{v}) - i\,\text{Im}(\mathbf{v}) \tag{2}$$

It follows that the vectors in R^n can be viewed as those vectors in C^n whose imaginary part is zero; or stated another way, a vector $\mathbf{v}$ in C^n is in R^n if and only if $\bar{\mathbf{v}} = \mathbf{v}$.

In this section we will also need to consider matrices with complex entries, so henceforth we will call a matrix A a ***real matrix*** if its entries are required to be real numbers and a ***complex matrix*** if its entries are allowed to be complex numbers. The standard operations on real matrices carry over to complex matrices without change, and all of the familiar properties of matrices continue to hold.

If A is a complex matrix, then $\text{Re}(A)$ and $\text{Im}(A)$ are the matrices formed from the real and imaginary parts of the entries of A, and $\bar{A}$ is the matrix formed by taking the complex conjugate of each entry in A.

▶ **EXAMPLE 1** Real and Imaginary Parts of Vectors and Matrices

Let

$$\mathbf{v} = (3+i, -2i, 5) \quad \text{and} \quad A = \begin{bmatrix} 1+i & -i \\ 4 & 6-2i \end{bmatrix}$$

Then
$$\bar{\mathbf{v}} = (3 - i, 2i, 5), \quad \text{Re}(\mathbf{v}) = (3, 0, 5), \quad \text{Im}(\mathbf{v}) = (1, -2, 0)$$

$$\bar{A} = \begin{bmatrix} 1-i & i \\ 4 & 6+2i \end{bmatrix}, \quad \text{Re}(A) = \begin{bmatrix} 1 & 0 \\ 4 & 6 \end{bmatrix}, \quad \text{Im}(A) = \begin{bmatrix} 1 & -1 \\ 0 & -2 \end{bmatrix}$$

$$\det(A) = \begin{vmatrix} 1+i & -i \\ 4 & 6-2i \end{vmatrix} = (1+i)(6-2i) - (-i)(4) = 8 + 8i \blacktriangleleft$$

Algebraic Properties of the Complex Conjugate

The next two theorems list some properties of complex vectors and matrices that we will need in this section. Some of the proofs are given as exercises.

THEOREM 5.3.1 *If $\mathbf{u}$ and $\mathbf{v}$ are vectors in C^n, and if k is a scalar, then*:

(a) $\bar{\bar{\mathbf{u}}} = \mathbf{u}$

(b) $\overline{k\mathbf{u}} = \bar{k}\bar{\mathbf{u}}$

(c) $\overline{\mathbf{u} + \mathbf{v}} = \bar{\mathbf{u}} + \bar{\mathbf{v}}$

(d) $\overline{\mathbf{u} - \mathbf{v}} = \bar{\mathbf{u}} - \bar{\mathbf{v}}$

THEOREM 5.3.2 *If A is an $m \times k$ complex matrix and B is a $k \times n$ complex matrix, then*:

(a) $\bar{\bar{A}} = A$

(b) $\overline{(A^T)} = (\bar{A})^T$

(c) $\overline{AB} = \bar{A}\bar{B}$

The Complex Euclidean Inner Product

The following definition extends the notions of dot product and norm to C^n.

DEFINITION 2 If $\mathbf{u} = (u_1, u_2, \ldots, u_n)$ and $\mathbf{v} = (v_1, v_2, \ldots, v_n)$ are vectors in C^n, then the *complex Euclidean inner product* of $\mathbf{u}$ and $\mathbf{v}$ (also called the *complex dot product*) is denoted by $\mathbf{u} \cdot \mathbf{v}$ and is defined as

$$\mathbf{u} \cdot \mathbf{v} = u_1 \bar{v}_1 + u_2 \bar{v}_2 + \cdots + u_n \bar{v}_n \quad (3)$$

We also define the *Euclidean norm* on C^n to be

$$\|\mathbf{v}\| = \sqrt{\mathbf{v} \cdot \mathbf{v}} = \sqrt{|v_1|^2 + |v_2|^2 + \cdots + |v_n|^2} \quad (4)$$

The complex conjugates in (3) ensure that $\|\mathbf{v}\|$ is a real number, for without them the quantity $\mathbf{v} \cdot \mathbf{v}$ in (4) might be imaginary.

As in the real case, we call $\mathbf{v}$ a *unit vector* in C^n if $\|\mathbf{v}\| = 1$, and we say two vectors $\mathbf{u}$ and $\mathbf{v}$ are *orthogonal* if $\mathbf{u} \cdot \mathbf{v} = 0$.

▶ **EXAMPLE 2** Complex Euclidean Inner Product and Norm

Find $\mathbf{u} \cdot \mathbf{v}$, $\mathbf{v} \cdot \mathbf{u}$, $\|\mathbf{u}\|$, and $\|\mathbf{v}\|$ for the vectors

$$\mathbf{u} = (1+i, i, 3-i) \quad \text{and} \quad \mathbf{v} = (1+i, 2, 4i)$$

Solution

$$\mathbf{u} \cdot \mathbf{v} = (1+i)(\overline{1+i}) + i(\overline{2}) + (3-i)(\overline{4i}) = (1+i)(1-i) + 2i + (3-i)(-4i) = -2 - 10i$$
$$\mathbf{v} \cdot \mathbf{u} = (1+i)(\overline{1+i}) + 2(\overline{i}) + (4i)(\overline{3-i}) = (1+i)(1-i) - 2i + 4i(3+i) = -2 + 10i$$
$$\|\mathbf{u}\| = \sqrt{|1+i|^2 + |i|^2 + |3-i|^2} = \sqrt{2 + 1 + 10} = \sqrt{13}$$
$$\|\mathbf{v}\| = \sqrt{|1+i|^2 + |2|^2 + |4i|^2} = \sqrt{2 + 4 + 16} = \sqrt{22} \blacktriangleleft$$

Recall from Table 1 of Section 3.2 that if $\mathbf{u}$ and $\mathbf{v}$ are *column vectors* in R^n, then their dot product can be expressed as

$$\mathbf{u} \cdot \mathbf{v} = \mathbf{u}^T \mathbf{v} = \mathbf{v}^T \mathbf{u}$$

The analogous formulas in C^n are (verify)

$$\mathbf{u} \cdot \mathbf{v} = \mathbf{u}^T \overline{\mathbf{v}} = \overline{\mathbf{v}}^T \mathbf{u} \tag{5}$$

Example 2 reveals a major difference between the dot product on R^n and the complex dot product on C^n. For the dot product on R^n we always have $\mathbf{v} \cdot \mathbf{u} = \mathbf{u} \cdot \mathbf{v}$ (the *symmetry property*), but for the complex dot product the corresponding relationship is given by $\mathbf{u} \cdot \mathbf{v} = \overline{\mathbf{v} \cdot \mathbf{u}}$, which is called its **antisymmetry** property. The following theorem is an analog of Theorem 3.2.2.

THEOREM 5.3.3 *If $\mathbf{u}, \mathbf{v}$, and $\mathbf{w}$ are vectors in C^n, and if k is a scalar, then the complex Euclidean inner product has the following properties:*

(a) $\mathbf{u} \cdot \mathbf{v} = \overline{\mathbf{v} \cdot \mathbf{u}}$ [Antisymmetry property]
(b) $\mathbf{u} \cdot (\mathbf{v} + \mathbf{w}) = \mathbf{u} \cdot \mathbf{v} + \mathbf{u} \cdot \mathbf{w}$ [Distributive property]
(c) $k(\mathbf{u} \cdot \mathbf{v}) = (k\mathbf{u}) \cdot \mathbf{v}$ [Homogeneity property]
(d) $\mathbf{u} \cdot k\mathbf{v} = \overline{k}(\mathbf{u} \cdot \mathbf{v})$ [Antihomogeneity property]
(e) $\mathbf{v} \cdot \mathbf{v} \geq 0$ and $\mathbf{v} \cdot \mathbf{v} = 0$ if and only if $\mathbf{v} = \mathbf{0}$.   [Positivity property]

Parts (c) and (d) of this theorem state that a scalar multiplying a complex Euclidean inner product can be regrouped with the first vector, but to regroup it with the second vector you must first take its complex conjugate. We will prove part (d), and leave the others as exercises.

Proof (d)

$$k(\mathbf{u} \cdot \mathbf{v}) = k(\overline{\mathbf{v} \cdot \mathbf{u}}) = \overline{\overline{k}(\mathbf{v} \cdot \mathbf{u})} = \overline{\overline{k}(\mathbf{v} \cdot \mathbf{u})} = \overline{(\overline{k}\mathbf{v}) \cdot \mathbf{u}} = \mathbf{u} \cdot (\overline{k}\mathbf{v})$$

To complete the proof, substitute $\overline{k}$ for k and use the fact that $\overline{\overline{k}} = k$. $\blacktriangleleft$

Vector Concepts in C^n

Except for the use of complex scalars, the notions of linear combination, linear independence, subspace, spanning, basis, and dimension carry over without change to C^n.

Eigenvalues and eigenvectors are defined for complex matrices exactly as for real matrices. If A is an $n \times n$ matrix with complex entries, then the complex roots of the characteristic equation $\det(\lambda I - A) = 0$ are called **complex eigenvalues** of A. As in the real case, λ is a complex eigenvalue of A if and only if there exists a nonzero vector $\mathbf{x}$ in C^n such that $A\mathbf{x} = \lambda\mathbf{x}$. Each such $\mathbf{x}$ is called a **complex eigenvector** of A corresponding to λ. The complex eigenvectors of A corresponding to λ are the nonzero solutions of the linear system $(\lambda I - A)\mathbf{x} = \mathbf{0}$, and the set of all such solutions is a subspace of C^n, called the **eigenspace** of A corresponding to λ.

Is R^n a subspace of C^n? Explain.

The following theorem states that if a *real matrix* has complex eigenvalues, then those eigenvalues and their corresponding eigenvectors occur in conjugate pairs.

THEOREM 5.3.4 *If λ is an eigenvalue of a real $n \times n$ matrix A, and if $\mathbf{x}$ is a corresponding eigenvector, then $\bar{\lambda}$ is also an eigenvalue of A, and $\bar{\mathbf{x}}$ is a corresponding eigenvector.*

Proof Since λ is an eigenvalue of A and $\mathbf{x}$ is a corresponding eigenvector, we have
$$\overline{A\mathbf{x}} = \overline{\lambda \mathbf{x}} = \bar{\lambda}\bar{\mathbf{x}} \tag{6}$$
However, $\bar{A} = A$, since A has real entries, so it follows from part (c) of Theorem 5.3.2 that
$$\overline{A\mathbf{x}} = \bar{A}\bar{\mathbf{x}} = A\bar{\mathbf{x}} \tag{7}$$
Equations (6) and (7) together imply that
$$A\bar{\mathbf{x}} = \overline{A\mathbf{x}} = \bar{\lambda}\bar{\mathbf{x}}$$
in which $\bar{\mathbf{x}} \neq \mathbf{0}$ (why?); this tells us that $\bar{\lambda}$ is an eigenvalue of A and $\bar{\mathbf{x}}$ is a corresponding eigenvector. ◄

▶ **EXAMPLE 3 Complex Eigenvalues and Eigenvectors**

Find the eigenvalues and bases for the eigenspaces of
$$A = \begin{bmatrix} -2 & -1 \\ 5 & 2 \end{bmatrix}$$

Solution The characteristic polynomial of A is
$$\begin{vmatrix} \lambda + 2 & 1 \\ -5 & \lambda - 2 \end{vmatrix} = \lambda^2 + 1 = (\lambda - i)(\lambda + i)$$
so the eigenvalues of A are $\lambda = i$ and $\lambda = -i$. Note that these eigenvalues are complex conjugates, as guaranteed by Theorem 5.3.4. To find the eigenvectors we must solve the system
$$\begin{bmatrix} \lambda + 2 & 1 \\ -5 & \lambda - 2 \end{bmatrix} \begin{bmatrix} x_1 \\ x_2 \end{bmatrix} = \begin{bmatrix} 0 \\ 0 \end{bmatrix}$$
with $\lambda = i$ and then with $\lambda = -i$. With $\lambda = i$, this system becomes
$$\begin{bmatrix} i + 2 & 1 \\ -5 & i - 2 \end{bmatrix} \begin{bmatrix} x_1 \\ x_2 \end{bmatrix} = \begin{bmatrix} 0 \\ 0 \end{bmatrix} \tag{8}$$
We could solve this system by reducing the augmented matrix
$$\begin{bmatrix} i + 2 & 1 & 0 \\ -5 & i - 2 & 0 \end{bmatrix} \tag{9}$$
to reduced row echelon form by Gauss–Jordan elimination, though the complex arithmetic is somewhat tedious. A simpler procedure here is first to observe that the reduced row echelon form of (9) must have a row of zeros because (8) has nontrivial solutions. This being the case, each row of (9) must be a scalar multiple of the other, and hence the first row can be made into a row of zeros by adding a suitable multiple of the second row to it. Accordingly, we can simply set the entries in the first row to zero, then interchange

the rows, and then multiply the new first row by $-\frac{1}{5}$ to obtain the reduced row echelon form

$$\begin{bmatrix} 1 & \frac{2}{5} - \frac{1}{5}i & 0 \\ 0 & 0 & 0 \end{bmatrix}$$

Thus, a general solution of the system is

$$x_1 = \left(-\tfrac{2}{5} + \tfrac{1}{5}i\right)t, \quad x_2 = t$$

This tells us that the eigenspace corresponding to $\lambda = i$ is one-dimensional and consists of all complex scalar multiples of the basis vector

$$\mathbf{x} = \begin{bmatrix} -\frac{2}{5} + \frac{1}{5}i \\ 1 \end{bmatrix} \tag{10}$$

As a check, let us confirm that $A\mathbf{x} = i\mathbf{x}$. We obtain

$$A\mathbf{x} = \begin{bmatrix} -2 & -1 \\ 5 & 2 \end{bmatrix} \begin{bmatrix} -\frac{2}{5} + \frac{1}{5}i \\ 1 \end{bmatrix} = \begin{bmatrix} -2\left(-\frac{2}{5} + \frac{1}{5}i\right) - 1 \\ 5\left(-\frac{2}{5} + \frac{1}{5}i\right) + 2 \end{bmatrix} = \begin{bmatrix} -\frac{1}{5} - \frac{2}{5}i \\ i \end{bmatrix} = i\mathbf{x}$$

We could find a basis for the eigenspace corresponding to $\lambda = -i$ in a similar way, but the work is unnecessary, since Theorem 5.3.4 implies that

$$\bar{\mathbf{x}} = \begin{bmatrix} -\frac{2}{5} - \frac{1}{5}i \\ 1 \end{bmatrix} \tag{11}$$

must be a basis for this eigenspace. The following computations confirm that $\bar{\mathbf{x}}$ is an eigenvector of A corresponding to $\lambda = -i$:

$$A\bar{\mathbf{x}} = \begin{bmatrix} -2 & -1 \\ 5 & 2 \end{bmatrix} \begin{bmatrix} -\frac{2}{5} - \frac{1}{5}i \\ 1 \end{bmatrix}$$

$$= \begin{bmatrix} -2\left(-\frac{2}{5} - \frac{1}{5}i\right) - 1 \\ 5\left(-\frac{2}{5} - \frac{1}{5}i\right) + 2 \end{bmatrix} = \begin{bmatrix} -\frac{1}{5} + \frac{2}{5}i \\ -i \end{bmatrix} = -i\bar{\mathbf{x}} \blacktriangleleft$$

Since a number of our subsequent examples will involve 2×2 matrices with real entries, it will be useful to discuss some general results about the eigenvalues of such matrices. Observe first that the characteristic polynomial of the matrix

$$A = \begin{bmatrix} a & b \\ c & d \end{bmatrix}$$

Olga Taussky-Todd (1906–1995)

Historical Note Olga Taussky-Todd was one of the pioneering women in matrix analysis and the first woman appointed to the faculty at the California Institute of Technology. She worked at the National Physical Laboratory in London during World War II, where she was assigned to study flutter in supersonic aircraft. While there, she realized that some results about the eigenvalues of a certain 6×6 complex matrix could be used to answer key questions about the flutter problem that would otherwise have required laborious calculation. After World War II Olga Taussky-Todd continued her work on matrix-related subjects and helped to draw many known but disparate results about matrices into the coherent subject that we now call matrix theory.

[Image: Courtesy of the Archives, California Institute of Technology]

is

$$\det(\lambda I - A) = \begin{vmatrix} \lambda - a & -b \\ -c & \lambda - d \end{vmatrix} = (\lambda - a)(\lambda - d) - bc = \lambda^2 - (a+d)\lambda + (ad - bc)$$

We can express this in terms of the trace and determinant of A as

$$\det(\lambda I - A) = \lambda^2 - \text{tr}(A)\lambda + \det(A) \tag{12}$$

from which it follows that the characteristic equation of A is

$$\lambda^2 - \text{tr}(A)\lambda + \det(A) = 0 \tag{13}$$

Now recall from algebra that if $ax^2 + bx + c = 0$ is a quadratic equation with real coefficients, then the **discriminant** $b^2 - 4ac$ determines the nature of the roots:

$$b^2 - 4ac > 0 \quad \text{[Two distinct real roots]}$$
$$b^2 - 4ac = 0 \quad \text{[One repeated real root]}$$
$$b^2 - 4ac < 0 \quad \text{[Two conjugate imaginary roots]}$$

Applying this to (13) with $a = 1$, $b = -\text{tr}(A)$, and $c = \det(A)$ yields the following theorem.

THEOREM 5.3.5 *If A is a 2×2 matrix with real entries, then the characteristic equation of A is $\lambda^2 - \text{tr}(A)\lambda + \det(A) = 0$ and*

(a) *A has two distinct real eigenvalues if $\text{tr}(A)^2 - 4\det(A) > 0$;*

(b) *A has one repeated real eigenvalue if $\text{tr}(A)^2 - 4\det(A) = 0$;*

(c) *A has two complex conjugate eigenvalues if $\text{tr}(A)^2 - 4\det(A) < 0$.*

▶ **EXAMPLE 4 Eigenvalues of a 2 × 2 Matrix**

In each part, use Formula (13) for the characteristic equation to find the eigenvalues of

(a) $A = \begin{bmatrix} 2 & 2 \\ -1 & 5 \end{bmatrix}$ (b) $A = \begin{bmatrix} 0 & -1 \\ 1 & 2 \end{bmatrix}$ (c) $A = \begin{bmatrix} 2 & 3 \\ -3 & 2 \end{bmatrix}$

Solution (a) We have $\text{tr}(A) = 7$ and $\det(A) = 12$, so the characteristic equation of A is

$$\lambda^2 - 7\lambda + 12 = 0$$

Factoring yields $(\lambda - 4)(\lambda - 3) = 0$, so the eigenvalues of A are $\lambda = 4$ and $\lambda = 3$.

Solution (b) We have $\text{tr}(A) = 2$ and $\det(A) = 1$, so the characteristic equation of A is

$$\lambda^2 - 2\lambda + 1 = 0$$

Factoring this equation yields $(\lambda - 1)^2 = 0$, so $\lambda = 1$ is the only eigenvalue of A; it has algebraic multiplicity 2.

Solution (c) We have $\text{tr}(A) = 4$ and $\det(A) = 13$, so the characteristic equation of A is

$$\lambda^2 - 4\lambda + 13 = 0$$

Solving this equation by the quadratic formula yields

$$\lambda = \frac{4 \pm \sqrt{(-4)^2 - 4(13)}}{2} = \frac{4 \pm \sqrt{-36}}{2} = 2 \pm 3i$$

Thus, the eigenvalues of A are $\lambda = 2 + 3i$ and $\lambda = 2 - 3i$. ◀

Symmetric Matrices Have Real Eigenvalues

Our next result, which is concerned with the eigenvalues of real symmetric matrices, is important in a wide variety of applications. The key to its proof is to think of a real symmetric matrix as a complex matrix whose entries have an imaginary part of zero.

THEOREM 5.3.6 *If A is a real symmetric matrix, then A has real eigenvalues.*

Proof Suppose that λ is an eigenvalue of A and $\mathbf{x}$ is a corresponding eigenvector, where we allow for the possibility that λ is complex and $\mathbf{x}$ is in C^n. Thus,

$$A\mathbf{x} = \lambda \mathbf{x}$$

where $\mathbf{x} \neq \mathbf{0}$. If we multiply both sides of this equation by $\overline{\mathbf{x}}^T$ and use the fact that

$$\overline{\mathbf{x}}^T A\mathbf{x} = \overline{\mathbf{x}}^T(\lambda \mathbf{x}) = \lambda(\overline{\mathbf{x}}^T \mathbf{x}) = \lambda(\mathbf{x} \cdot \mathbf{x}) = \lambda \|\mathbf{x}\|^2$$

then we obtain

$$\lambda = \frac{\overline{\mathbf{x}}^T A\mathbf{x}}{\|\mathbf{x}\|^2}$$

Since the denominator in this expression is real, we can prove that λ is real by showing that

$$\overline{\overline{\mathbf{x}}^T A\mathbf{x}} = \overline{\mathbf{x}}^T A\mathbf{x} \tag{14}$$

But, A is symmetric and has real entries, so it follows from the second equality in (14) and properties of the conjugate that

$$\overline{\overline{\mathbf{x}}^T A\mathbf{x}} = \overline{\overline{\mathbf{x}}}^T \overline{A\mathbf{x}} = \mathbf{x}^T \overline{A\mathbf{x}} = (\overline{A\mathbf{x}})^T \mathbf{x} = (A\overline{\mathbf{x}})^T \mathbf{x} = (A\overline{\mathbf{x}})^T \mathbf{x} = \overline{\mathbf{x}}^T A^T \mathbf{x} = \overline{\mathbf{x}}^T A\mathbf{x} \blacktriangleleft$$

A Geometric Interpretation of Complex Eigenvalues

The following theorem is the key to understanding the geometric significance of complex eigenvalues of real 2×2 matrices.

THEOREM 5.3.7 *The eigenvalues of the real matrix*

$$C = \begin{bmatrix} a & -b \\ b & a \end{bmatrix} \tag{15}$$

are $\lambda = a \pm bi$. If a and b are not both zero, then this matrix can be factored as

$$\begin{bmatrix} a & -b \\ b & a \end{bmatrix} = \begin{bmatrix} |\lambda| & 0 \\ 0 & |\lambda| \end{bmatrix} \begin{bmatrix} \cos\phi & -\sin\phi \\ \sin\phi & \cos\phi \end{bmatrix} \tag{16}$$

where ϕ is the angle from the positive x-axis to the ray that joins the origin to the point (a, b) (Figure 5.3.2).

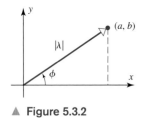

▲ Figure 5.3.2

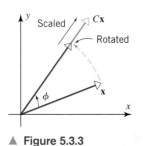

▲ Figure 5.3.3

Geometrically, this theorem states that multiplication by a matrix of form (15) can be viewed as a rotation through the angle ϕ followed by a scaling with factor $|\lambda|$ (Figure 5.3.3).

Proof The characteristic equation of C is $(\lambda - a)^2 + b^2 = 0$ (verify), from which it follows that the eigenvalues of C are $\lambda = a \pm bi$. Assuming that a and b are not both zero, let ϕ be the angle from the positive x-axis to the ray that joins the origin to the point (a, b). The angle ϕ is an argument of the eigenvalue $\lambda = a + bi$, so we see from Figure 5.3.2 that

$$a = |\lambda|\cos\phi \quad \text{and} \quad b = |\lambda|\sin\phi$$

It follows from this that the matrix in (15) can be written as

$$\begin{bmatrix} a & -b \\ b & a \end{bmatrix} = \begin{bmatrix} |\lambda| & 0 \\ 0 & |\lambda| \end{bmatrix} \begin{bmatrix} \dfrac{a}{|\lambda|} & -\dfrac{b}{|\lambda|} \\ \dfrac{b}{|\lambda|} & \dfrac{a}{|\lambda|} \end{bmatrix} = \begin{bmatrix} |\lambda| & 0 \\ 0 & |\lambda| \end{bmatrix} \begin{bmatrix} \cos\phi & -\sin\phi \\ \sin\phi & \cos\phi \end{bmatrix} \blacktriangleleft$$

The following theorem, whose proof is considered in the exercises, shows that every real 2×2 matrix with complex eigenvalues is similar to a matrix of form (15).

THEOREM 5.3.8 *Let A be a real 2×2 matrix with complex eigenvalues $\lambda = a \pm bi$ (where $b \neq 0$). If $\mathbf{x}$ is an eigenvector of A corresponding to $\lambda = a - bi$, then the matrix $P = [\text{Re}(\mathbf{x}) \quad \text{Im}(\mathbf{x})]$ is invertible and*

$$A = P \begin{bmatrix} a & -b \\ b & a \end{bmatrix} P^{-1} \tag{17}$$

▶ **EXAMPLE 5 A Matrix Factorization Using Complex Eigenvalues**

Factor the matrix in Example 3 into form (17) using the eigenvalue $\lambda = -i$ and the corresponding eigenvector that was given in (11).

Solution For consistency with the notation in Theorem 5.3.8, let us denote the eigenvector in (11) that corresponds to $\lambda = -i$ by $\mathbf{x}$ (rather than $\bar{\mathbf{x}}$ as before). For this λ and $\mathbf{x}$ we have

$$a = 0, \quad b = 1, \quad \text{Re}(\mathbf{x}) = \begin{bmatrix} -\tfrac{2}{5} \\ 1 \end{bmatrix}, \quad \text{Im}(\mathbf{x}) = \begin{bmatrix} -\tfrac{1}{5} \\ 0 \end{bmatrix}$$

Thus,

$$P = [\text{Re}(\mathbf{x}) \quad \text{Im}(\mathbf{x})] = \begin{bmatrix} -\tfrac{2}{5} & -\tfrac{1}{5} \\ 1 & 0 \end{bmatrix}$$

so A can be factored in form (17) as

$$\begin{bmatrix} -2 & -1 \\ 5 & 2 \end{bmatrix} = \begin{bmatrix} -\tfrac{2}{5} & -\tfrac{1}{5} \\ 1 & 0 \end{bmatrix} \begin{bmatrix} 0 & -1 \\ 1 & 0 \end{bmatrix} \begin{bmatrix} 0 & 1 \\ -5 & -2 \end{bmatrix}$$

You may want to confirm this by multiplying out the right side. ◀

A Geometric Interpretation of Theorem 5.3.8

To clarify what Theorem 5.3.8 says geometrically, let us denote the matrices on the right side of (16) by S and R_ϕ, respectively, and then use (16) to rewrite (17) as

$$A = PSR_\phi P^{-1} = P \begin{bmatrix} |\lambda| & 0 \\ 0 & |\lambda| \end{bmatrix} \begin{bmatrix} \cos\phi & -\sin\phi \\ \sin\phi & \cos\phi \end{bmatrix} P^{-1} \tag{18}$$

If we now view P as the transition matrix from the basis $B = \{\text{Re}(\mathbf{x}), \text{Im}(\mathbf{x})\}$ to the standard basis, then (18) tells us that computing a product $A\mathbf{x}_0$ can be broken down into a three-step process:

Step 1. Map $\mathbf{x}_0$ from standard coordinates into B-coordinates by forming the product $P^{-1}\mathbf{x}_0$.

Step 2. Rotate and scale the vector $P^{-1}\mathbf{x}_0$ by forming the product $SR_\phi P^{-1}\mathbf{x}_0$.

Step 3. Map the rotated and scaled vector back to standard coordinates to obtain $A\mathbf{x}_0 = PSR_\phi P^{-1}\mathbf{x}_0$.

Power Sequences

There are many problems in which one is interested in how successive applications of a matrix transformation affect a specific vector. For example, if A is the standard matrix for an operator on R^n and $\mathbf{x}_0$ is some fixed vector in R^n, then one might be interested in the behavior of the power sequence

$$\mathbf{x}_0, \quad A\mathbf{x}_0, \quad A^2\mathbf{x}_0, \ldots, \quad A^k\mathbf{x}_0, \ldots$$

For example, if

$$A = \begin{bmatrix} \frac{1}{2} & \frac{3}{4} \\ -\frac{3}{5} & \frac{11}{10} \end{bmatrix} \quad \text{and} \quad \mathbf{x}_0 = \begin{bmatrix} 1 \\ 1 \end{bmatrix}$$

then with the help of a computer or calculator one can show that the first four terms in the power sequence are

$$\mathbf{x}_0 = \begin{bmatrix} 1 \\ 1 \end{bmatrix}, \quad A\mathbf{x}_0 = \begin{bmatrix} 1.25 \\ 0.5 \end{bmatrix}, \quad A^2\mathbf{x}_0 = \begin{bmatrix} 1.0 \\ -0.2 \end{bmatrix}, \quad A^3\mathbf{x}_0 = \begin{bmatrix} 0.35 \\ -0.82 \end{bmatrix}$$

With the help of MATLAB or a computer algebra system one can show that if the first 100 terms are plotted as ordered pairs (x, y), then the points move along the elliptical path shown in Figure 5.3.4a.

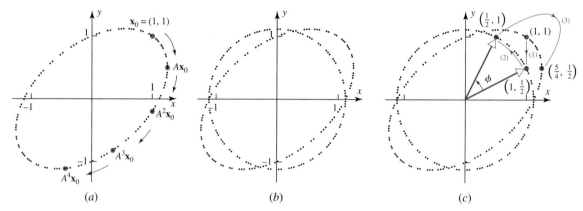

▲ Figure 5.3.4

To understand why the points move along an elliptical path, we will need to examine the eigenvalues and eigenvectors of A. We leave it for you to show that the eigenvalues of A are $\lambda = \frac{4}{5} \pm \frac{3}{5}i$ and that the corresponding eigenvectors are

$$\lambda_1 = \tfrac{4}{5} - \tfrac{3}{5}i: \quad \mathbf{v}_1 = \left(\tfrac{1}{2} + i, 1\right) \quad \text{and} \quad \lambda_2 = \tfrac{4}{5} + \tfrac{3}{5}i: \quad \mathbf{v}_2 = \left(\tfrac{1}{2} - i, 1\right)$$

If we take $\lambda = \lambda_1 = \frac{4}{5} - \frac{3}{5}i$ and $\mathbf{x} = \mathbf{v}_1 = \left(\frac{1}{2} + i, 1\right)$ in (17) and use the fact that $|\lambda| = 1$, then we obtain the factorization

$$\underbrace{\begin{bmatrix} \frac{1}{2} & \frac{3}{4} \\ -\frac{3}{5} & \frac{11}{10} \end{bmatrix}}_{A} = \underbrace{\begin{bmatrix} \frac{1}{2} & 1 \\ 1 & 0 \end{bmatrix}}_{P} \underbrace{\begin{bmatrix} \frac{4}{5} & -\frac{3}{5} \\ \frac{3}{5} & \frac{4}{5} \end{bmatrix}}_{R_\phi} \underbrace{\begin{bmatrix} 0 & 1 \\ 1 & -\frac{1}{2} \end{bmatrix}}_{P^{-1}} \tag{19}$$

where R_ϕ is a rotation about the origin through the angle ϕ whose tangent is

$$\tan \phi = \frac{\sin \phi}{\cos \phi} = \frac{3/5}{4/5} = \frac{3}{4} \quad (\phi = \tan^{-1} \tfrac{3}{4} \approx 36.9°)$$

The matrix P in (19) is the transition matrix from the basis

$$B = \{\operatorname{Re}(\mathbf{x}), \operatorname{Im}(\mathbf{x})\} = \left\{\left(\tfrac{1}{2}, 1\right), (1, 0)\right\}$$

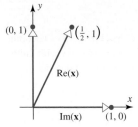

▲ Figure 5.3.5

to the standard basis, and P^{-1} is the transition matrix from the standard basis to the basis B (Figure 5.3.5). Next, observe that if n is a positive integer, then (19) implies that

$$A^n \mathbf{x}_0 = (PR_\phi P^{-1})^n \mathbf{x}_0 = PR_\phi^n P^{-1} \mathbf{x}_0$$

so the product $A^n \mathbf{x}_0$ can be computed by first mapping $\mathbf{x}_0$ into the point $P^{-1}\mathbf{x}_0$ in B-coordinates, then multiplying by R_ϕ^n to rotate this point about the origin through the angle $n\phi$, and then multiplying $R_\phi^n P^{-1} \mathbf{x}_0$ by P to map the resulting point back to standard coordinates. We can now see what is happening geometrically: In B-coordinates each successive multiplication by A causes the point $P^{-1}\mathbf{x}_0$ to advance through an angle ϕ, thereby tracing a circular orbit about the origin. However, the basis B is *skewed* (not orthogonal), so when the points on the circular orbit are transformed back to standard coordinates, the effect is to distort the circular orbit into the elliptical orbit traced by $A^n \mathbf{x}_0$ (Figure 5.3.4b). Here are the computations for the first step (successive steps are illustrated in Figure 5.3.4c):

$$\begin{bmatrix} \frac{1}{2} & \frac{3}{4} \\ -\frac{3}{5} & \frac{11}{10} \end{bmatrix} \begin{bmatrix} 1 \\ 1 \end{bmatrix} = \begin{bmatrix} \frac{1}{2} & 1 \\ 1 & 0 \end{bmatrix} \begin{bmatrix} \frac{4}{5} & -\frac{3}{5} \\ \frac{3}{5} & \frac{4}{5} \end{bmatrix} \begin{bmatrix} 0 & 1 \\ 1 & -\frac{1}{2} \end{bmatrix} \begin{bmatrix} 1 \\ 1 \end{bmatrix}$$

$$= \begin{bmatrix} \frac{1}{2} & 1 \\ 1 & 0 \end{bmatrix} \begin{bmatrix} \frac{4}{5} & -\frac{3}{5} \\ \frac{3}{5} & \frac{4}{5} \end{bmatrix} \begin{bmatrix} 1 \\ \frac{1}{2} \end{bmatrix} \qquad [\mathbf{x}_0 \text{ is mapped to } B\text{-coordinates.}]$$

$$= \begin{bmatrix} \frac{1}{2} & 1 \\ 1 & 0 \end{bmatrix} \begin{bmatrix} \frac{1}{2} \\ 1 \end{bmatrix} \qquad [\text{The point } (1, \tfrac{1}{2}) \text{ is rotated through the angle } \phi.]$$

$$= \begin{bmatrix} \frac{5}{4} \\ \frac{1}{2} \end{bmatrix} \qquad [\text{The point } (\tfrac{1}{2}, 1) \text{ is mapped to standard coordinates.}]$$

Concept Review

- Real part of z
- Imaginary part of z
- Modulus of z
- Complex conjugate of z
- Argument of z
- Polar form of z

- Complex vector space
- Complex n-tuple
- Complex n-space
- Real matrix
- Complex matrix
- Complex Euclidean inner product

- Euclidean norm on C^n
- Antisymmetry property
- Complex eigenvalue
- Complex eigenvector
- Eigenspace in C^n
- Discriminant

Skills

- Find the real part, imaginary part, and complex conjugate of a complex matrix or vector.
- Find the determinant of a complex matrix.
- Find complex inner products and norms of complex vectors.
- Find the eigenvalues and bases for the eigenspaces of complex matrices.
- Factor a 2×2 real matrix with complex eigenvalues into a product of a scaling matrix and a rotation matrix.

Exercise Set 5.3

In Exercises 1–2, find $\bar{\mathbf{u}}$, $\text{Re}(\mathbf{u})$, $\text{Im}(\mathbf{u})$, and $\|\mathbf{u}\|$.

1. $\mathbf{u} = (3i, 1 - 4i, 2 + i)$
2. $\mathbf{u} = (6, 1 + 4i, 6 - 2i)$

In Exercises 3–4, show that $\mathbf{u}$, $\mathbf{v}$, and k satisfy Theorem 5.3.1.

3. $\mathbf{u} = (3 + i, 2 + 2i, -5i)$, $\mathbf{v} = (1 - i, 3, 1 + i)$, $k = 2i$
4. $\mathbf{u} = (6, 1 + 4i, 6 - 2i)$, $\mathbf{v} = (4, 3 + 2i, i - 3)$, $k = -i$

5. Solve the equation $2\mathbf{x} - 3i\mathbf{u} = \bar{\mathbf{v}}$ for $\mathbf{x}$, where $\mathbf{u}$ and $\mathbf{v}$ are the vectors in Exercise 3.

6. Solve the equation $(1 + i)\mathbf{x} + 2\mathbf{u} = \bar{\mathbf{v}}$ for $\mathbf{x}$, where $\mathbf{u}$ and $\mathbf{v}$ are the vectors in Exercise 4.

In Exercises 7–8, find $\bar{A}$, $\text{Re}(A)$, $\text{Im}(A)$, $\det(A)$, and $\text{tr}(A)$.

7. $A = \begin{bmatrix} 1 + 3i & 2 \\ 4 + i & -3i \end{bmatrix}$
8. $A = \begin{bmatrix} 4i & 2 - 3i \\ 2 + 3i & 1 \end{bmatrix}$

9. Let A be the matrix given in Exercise 7. Confirm that if $B = (3i, 2 + i)$ is written in column form, then A and B have the properties stated in Theorem 5.3.2.

10. Let A be the matrix given in Exercise 8, and let B be the matrix

$$B = \begin{bmatrix} 5i \\ 1 - 4i \end{bmatrix}$$

Confirm that these matrices have the properties stated in Theorem 5.3.2.

In Exercises 11–12, compute $\mathbf{u} \cdot \mathbf{v}$, $\mathbf{u} \cdot \mathbf{w}$, and $\mathbf{v} \cdot \mathbf{w}$, and show that the vectors satisfy Formula (5) and parts (a), (b), and (c) of Theorem 5.3.3.

11. $\mathbf{u} = (3i, 2 + 2i, 5)$, $\mathbf{v} = (1 + i, 4 - i, 1 + i)$,
$\mathbf{w} = (3, 5i, -3i)$, $k = 3i$

12. $\mathbf{u} = (1 + i, 4, 3i)$, $\mathbf{v} = (3, -4i, 2 + 3i)$,
$\mathbf{w} = (1 - i, 4i, 4 - 5i)$, $k = 1 + i$

13. Compute $\overline{(\mathbf{u} \cdot \bar{\mathbf{v}})} - \overline{\mathbf{w} \cdot \mathbf{u}}$ for the vectors $\mathbf{u}$, $\mathbf{v}$, and $\mathbf{w}$ in Exercise 11.

14. Compute $\overline{(i\mathbf{u} \cdot \mathbf{w})} + (\|\mathbf{u}\|\mathbf{v}) \cdot \mathbf{u}$ for the vectors $\mathbf{u}$, $\mathbf{v}$, and $\mathbf{w}$ in Exercise 12.

In Exercises 15–18, find the eigenvalues and bases for the eigenspaces of A.

15. $A = \begin{bmatrix} 4 & -5 \\ 1 & 0 \end{bmatrix}$
16. $A = \begin{bmatrix} -1 & -5 \\ 4 & 7 \end{bmatrix}$

17. $A = \begin{bmatrix} 3 & -5 \\ 2 & -3 \end{bmatrix}$
18. $A = \begin{bmatrix} 8 & 6 \\ -3 & 2 \end{bmatrix}$

In Exercises 19–22, each matrix C has form (15). Theorem 5.3.7 implies that C is the product of a scaling matrix with factor $|\lambda|$ and a rotation matrix with angle ϕ. Find $|\lambda|$ and ϕ for which $-\pi < \phi \leq \pi$.

19. $C = \begin{bmatrix} 1 & -1 \\ 1 & 1 \end{bmatrix}$
20. $C = \begin{bmatrix} 0 & 5 \\ -5 & 0 \end{bmatrix}$

21. $C = \begin{bmatrix} 1 & \sqrt{3} \\ -\sqrt{3} & 1 \end{bmatrix}$
22. $C = \begin{bmatrix} \sqrt{2} & \sqrt{2} \\ -\sqrt{2} & \sqrt{2} \end{bmatrix}$

In Exercises 23–26, find an invertible matrix P and a matrix C of form (15) such that $A = PCP^{-1}$.

23. $A = \begin{bmatrix} 7 & 5 \\ -1 & 5 \end{bmatrix}$
24. $A = \begin{bmatrix} 4 & -5 \\ 1 & 0 \end{bmatrix}$

25. $A = \begin{bmatrix} 8 & 6 \\ -3 & 2 \end{bmatrix}$
26. $A = \begin{bmatrix} 5 & -2 \\ 1 & 3 \end{bmatrix}$

27. Find all complex scalars k, if any, for which $\mathbf{u}$ and $\mathbf{v}$ are orthogonal in C^3.

(a) $\mathbf{u} = (3i, 1, i)$, $\mathbf{v} = (-i, 5i, k)$
(b) $\mathbf{u} = (k, k, 1 + i)$, $\mathbf{v} = (1, -1, 1 - i)$

28. Show that if A is a real $n \times n$ matrix and $\mathbf{x}$ is a column vector in C^n, then $\text{Re}(A\mathbf{x}) = A(\text{Re}(\mathbf{x}))$ and $\text{Im}(A\mathbf{x}) = A(\text{Im}(\mathbf{x}))$.

29. The matrices

$$\sigma_1 = \begin{bmatrix} 0 & 1 \\ 1 & 0 \end{bmatrix}, \quad \sigma_2 = \begin{bmatrix} 0 & -i \\ i & 0 \end{bmatrix}, \quad \sigma_3 = \begin{bmatrix} 1 & 0 \\ 0 & -1 \end{bmatrix}$$

called **Pauli spin matrices**, are used in quantum mechanics to study particle spin. The **Dirac matrices**, which are also used in quantum mechanics, are expressed in terms of the Pauli spin matrices and the 2×2 identity matrix I_2 as

$$\beta = \begin{bmatrix} I_2 & 0 \\ 0 & -I_2 \end{bmatrix}, \quad \alpha_x = \begin{bmatrix} 0 & \sigma_1 \\ \sigma_1 & 0 \end{bmatrix},$$

$$\alpha_y = \begin{bmatrix} 0 & \sigma_2 \\ \sigma_2 & 0 \end{bmatrix}, \quad \alpha_z = \begin{bmatrix} 0 & \sigma_3 \\ \sigma_3 & 0 \end{bmatrix}$$

(a) Show that $\beta^2 = \alpha_x^2 = \alpha_y^2 = \alpha_z^2$.

(b) Matrices A and B for which $AB = -BA$ are said to be **anticommutative**. Show that the Dirac matrices are anticommutative.

30. If k is a real scalar and $\mathbf{v}$ is a vector in R^n, then Theorem 3.2.1 states that $\|k\mathbf{v}\| = |k|\|\mathbf{v}\|$. Is this relationship also true if k is a complex scalar and $\mathbf{v}$ is a vector in C^n? Justify your answer.

31. Prove part (c) of Theorem 5.3.1.

32. Prove Theorem 5.3.2.

33. Prove that if $\mathbf{u}$ and $\mathbf{v}$ are vectors in C^n, then

$$\mathbf{u} \cdot \mathbf{v} = \frac{1}{4} \|\mathbf{u} + \mathbf{v}\|^2 - \frac{1}{4} \|\mathbf{u} - \mathbf{v}\|^2$$
$$+ \frac{i}{4} \|\mathbf{u} + i\mathbf{v}\|^2 - \frac{i}{4} \|\mathbf{u} - i\mathbf{v}\|^2$$

34. It follows from Theorem 5.3.7 that the eigenvalues of the rotation matrix

$$R_\phi = \begin{bmatrix} \cos\phi & -\sin\phi \\ \sin\phi & \cos\phi \end{bmatrix}$$

are $\lambda = \cos\phi \pm i \sin\phi$. Prove that if $\mathbf{x}$ is an eigenvector corresponding to either eigenvalue, then $\text{Re}(\mathbf{x})$ and $\text{Im}(\mathbf{x})$ are orthogonal and have the same length. [*Note:* This implies that $P = [\text{Re}(\mathbf{x}) \mid \text{Im}(\mathbf{x})]$ is a real scalar multiple of an orthogonal matrix.]

35. The two parts of this exercise lead you through a proof of Theorem 5.3.8.

    (a) For notational simplicity, let

    $$M = \begin{bmatrix} a & -b \\ b & a \end{bmatrix}$$

    and let $\mathbf{u} = \text{Re}(\mathbf{x})$ and $\mathbf{v} = \text{Im}(\mathbf{x})$, so $P = [\mathbf{u} \mid \mathbf{v}]$. Show that the relationship $A\mathbf{x} = \lambda \mathbf{x}$ implies that

    $$A\mathbf{x} = (a\mathbf{u} + b\mathbf{v}) + i(-b\mathbf{u} + a\mathbf{v})$$

    and then equate real and imaginary parts in this equation to show that

    $$AP = [A\mathbf{u} \mid A\mathbf{v}] = [a\mathbf{u} + b\mathbf{v} \mid -b\mathbf{u} + a\mathbf{v}] = PM$$

    (b) Show that P is invertible, thereby completing the proof, since the result in part (a) implies that $A = PMP^{-1}$. [*Hint:* If P is not invertible, then one of its column vectors is a real scalar multiple of the other, say $\mathbf{v} = c\mathbf{u}$. Substitute this into the equations $A\mathbf{u} = a\mathbf{u} + b\mathbf{v}$ and $A\mathbf{v} = -b\mathbf{u} + a\mathbf{v}$ obtained in part (a), and show that $(1 + c^2)b\mathbf{u} = \mathbf{0}$. Finally, show that this leads to a contradiction, thereby proving that P is invertible.]

36. In this problem you will prove the complex analog of the Cauchy–Schwarz inequality.

    (a) Prove: If k is a complex number, and $\mathbf{u}$ and $\mathbf{v}$ are vectors in C^n, then

    $$(\mathbf{u} - k\mathbf{v}) \cdot (\mathbf{u} - k\mathbf{v}) = \mathbf{u} \cdot \mathbf{u} - \overline{k}(\mathbf{u} \cdot \mathbf{v}) - k\overline{(\mathbf{u} \cdot \mathbf{v})} + k\overline{k}(\mathbf{v} \cdot \mathbf{v})$$

    (b) Use the result in part (a) to prove that

    $$0 \le \mathbf{u} \cdot \mathbf{u} - \overline{k}(\mathbf{u} \cdot \mathbf{v}) - k\overline{(\mathbf{u} \cdot \mathbf{v})} + k\overline{k}(\mathbf{v} \cdot \mathbf{v})$$

    (c) Take $k = (\mathbf{u} \cdot \mathbf{v})/(\mathbf{v} \cdot \mathbf{v})$ in part (b) to prove that

    $$|\mathbf{u} \cdot \mathbf{v}| \le \|\mathbf{u}\| \, \|\mathbf{v}\|$$

True-False Exercises

In parts (a)–(f) determine whether the statement is true or false, and justify your answer.

(a) There is a real 5×5 matrix with no real eigenvalues.

(b) The eigenvalues of a 2×2 complex matrix are the solutions of the equation $\lambda^2 - \text{tr}(A)\lambda + \det(A) = 0$.

(c) Matrices that have the same complex eigenvalues with the same algebraic multiplicities have the same trace.

(d) If λ is a complex eigenvalue of a real matrix A with a corresponding complex eigenvector $\mathbf{v}$, then $\overline{\lambda}$ is a complex eigenvalue of A and $\overline{\mathbf{v}}$ is a complex eigenvector of A corresponding to $\overline{\lambda}$.

(e) Every eigenvalue of a complex symmetric matrix is real.

(f) If a 2×2 real matrix A has complex eigenvalues and $\mathbf{x}_0$ is a vector in R^2, then the vectors $\mathbf{x}_0, A\mathbf{x}_0, A^2\mathbf{x}_0, \ldots, A^n\mathbf{x}_0, \ldots$ lie on an ellipse.

5.4 Differential Equations

Many laws of physics, chemistry, biology, engineering, and economics are described in terms of "differential equations"—that is, equations involving functions and their derivatives. In this section we will illustrate one way in which linear algebra, eigenvalues and eigenvectors can be applied to solving systems of differential equations. Calculus is a prerequisite for this section.

Terminology   Recall from calculus that a **differential equation** is an equation involving unknown functions and their derivatives. The **order** of a differential equation is the order of the highest derivative it contains. The simplest differential equations are the first-order equations of the form

$$y' = ay \tag{1}$$

328 Chapter 5 Eigenvalues and Eigenvectors

where $y = f(x)$ is an unknown differentiable function to be determined, $y' = dy/dx$ is its derivative, and a is a constant. As with most differential equations, this equation has infinitely many solutions; they are the functions of the form

$$y = ce^{ax} \tag{2}$$

where c is an arbitrary constant. That every function of this form is a solution of (1) follows from the computation

$$y' = cae^{ax} = ay$$

and that these are the only solution is shown in the exercises. Accordingly, we call (2) the ***general solution*** of (1). As an example, the general solution of the differential equation $y' = 5y$ is

$$y = ce^{5x} \tag{3}$$

Often, a physical problem that leads to a differential equation imposes some conditions that enable us to isolate one particular solution from the general solution. For example, if we require that solution (3) of the equation $y' = 5y$ satisfy the added condition

$$y(0) = 6 \tag{4}$$

(that is, $y = 6$ when $x = 0$), then on substituting these values in (3), we obtain $6 = ce^0 = c$, from which we conclude that

$$y = 6e^{5x}$$

is the only solution $y' = 5y$ that satisfies (4).

A condition such as (4), which specifies the value of the general solution at a point is called an ***initial condition***, and the problem of solving a differential equation subject to an initial condition is called an ***initial-value problem***.

First-Order Linear Systems   In this section we will be concerned with solving systems of differential equations of the form

$$
\begin{aligned}
y_1' &= a_{11}y_1 + a_{12}y_2 + \cdots + a_{1n}y_n \\
y_2' &= a_{21}y_1 + a_{22}y_2 + \cdots + a_{2n}y_n \\
&\vdots \\
y_n' &= a_{n1}y_1 + a_{n2}y_2 + \cdots + a_{nn}y_n
\end{aligned}
\tag{5}
$$

where $y_1 = f_1(x), y_2 = f_2(x), \ldots, y_n = f_n(x)$ are functions to be determined, and the a_{ij}'s are constants. In matrix notation, (5) can be written as

$$
\begin{bmatrix} y_1' \\ y_2' \\ \vdots \\ y_n' \end{bmatrix} = \begin{bmatrix} a_{11} & a_{12} & \cdots & a_{1n} \\ a_{21} & a_{22} & \cdots & a_{2n} \\ \vdots & \vdots & & \vdots \\ a_{n1} & a_{n2} & \cdots & a_{nn} \end{bmatrix} \begin{bmatrix} y_1 \\ y_2 \\ \vdots \\ y_n \end{bmatrix}
$$

or, more briefly as

$$\mathbf{y'} = A\mathbf{y} \tag{6}$$

where the notation $\mathbf{y}'$ denotes the vector obtained by differentiating each component of $\mathbf{y}$.

A system of differential equations of form (5) is called a ***first-order linear system***.

▶ **EXAMPLE 1** **Solution of a Linear System with Initial Conditions**

(a) Write the following system in matrix form:

$$
\begin{aligned}
y_1' &= 3y_1 \\
y_2' &= -2y_2 \\
y_3' &= 5y_3
\end{aligned}
\tag{7}
$$

(b) Solve the system.
(c) Find a solution of the system that satisfies the initial conditions $y_1(0) = 1$, $y_2(0) = 4$, and $y_3(0) = -2$.

Solution (a)

$$\begin{bmatrix} y_1' \\ y_2' \\ y_3' \end{bmatrix} = \begin{bmatrix} 3 & 0 & 0 \\ 0 & -2 & 0 \\ 0 & 0 & 5 \end{bmatrix} \begin{bmatrix} y_1 \\ y_2 \\ y_3 \end{bmatrix} \quad (8)$$

or

$$\mathbf{y}' = \begin{bmatrix} 3 & 0 & 0 \\ 0 & -2 & 0 \\ 0 & 0 & 5 \end{bmatrix} \mathbf{y} \quad (9)$$

Solution (b) Because each equation in (7) involves only one unknown function, we can solve the equations individually. It follows from (2) that these solutions are

$$y_1 = c_1 e^{3x}$$
$$y_2 = c_2 e^{-2x}$$
$$y_3 = c_3 e^{5x}$$

or, in matrix notation,

$$\mathbf{y} = \begin{bmatrix} y_1 \\ y_2 \\ y_3 \end{bmatrix} = \begin{bmatrix} c_1 e^{3x} \\ c_2 e^{-2x} \\ c_3 e^{5x} \end{bmatrix} \quad (10)$$

Solution (c) From the given initial conditions, we obtain

$$1 = y_1(0) = c_1 e^0 = c_1$$
$$4 = y_2(0) = c_2 e^0 = c_2$$
$$-2 = y_3(0) = c_3 e^0 = c_3$$

so the solution satisfying these conditions are

$$y_1 = e^{3x}, \quad y_2 = 4e^{-2x}, \quad y_3 = -2e^{5x}$$

or, in matrix notation,

$$\mathbf{y} = \begin{bmatrix} y_1 \\ y_2 \\ y_3 \end{bmatrix} = \begin{bmatrix} e^{3x} \\ 4e^{-2x} \\ -2e^{5x} \end{bmatrix} \blacktriangleleft$$

Solution by Diagonalization

What made the system in Example 1 easy to solve was the fact that each equation involved only one of the unknown functions, so its matrix formulation, $\mathbf{y}' = A\mathbf{y}$, had a *diagonal* coefficient matrix A [Formula (9)]. A more complicated situation occurs when some or all of the equations in the system involve more than one of the unknown functions, for in this case the coefficient matrix is not diagonal. Let us now consider how we might solve such a system.

The basic idea for solving a system $\mathbf{y}' = A\mathbf{y}$ whose coefficient matrix A is not diagonal is to introduce a new unknown vector $\mathbf{u}$ that is related to the unknown vector $\mathbf{y}$ by an equation of the form $\mathbf{y} = P\mathbf{u}$ in which P is an invertible matrix that diagonalizes A. Of course, such a matrix may or may not exist, but if it does then we can rewrite the equation $\mathbf{y}' = A\mathbf{y}$ as

$$P\mathbf{u}' = A(P\mathbf{u})$$

or alternatively as
$$\mathbf{u}' = (P^{-1}AP)\mathbf{u}$$
Since P is assumed to diagonalize A, this equation has the form
$$\mathbf{u}' = D\mathbf{u}$$
where D is diagonal. We can now solve this equation for $\mathbf{u}$ using the method of Example 1, and then obtain $\mathbf{y}$ by matrix multiplication using the relationship $\mathbf{y} = P\mathbf{u}$.

In summary, we have the following procedure for solving a system $\mathbf{y}' = A\mathbf{y}$ in the case were A is diagonalizable.

A Procedure for Solving $\mathbf{y}' = A\mathbf{y}$ if A is Diagonalizable

Step 1. Find a matrix P that diagonalizes A.

Step 2. Make the substitutions $\mathbf{y} = P\mathbf{u}$ and $\mathbf{y}' = P\mathbf{u}'$ to obtain a new "diagonal system" $\mathbf{u}' = D\mathbf{u}$, where $D = P^{-1}AP$.

Step 3. Solve $\mathbf{u}' = D\mathbf{u}$.

Step 4. Determine $\mathbf{y}$ from the equation $\mathbf{y} = P\mathbf{u}$.

▶ **EXAMPLE 2** Solution Using Diagonalization

(a) Solve the system
$$\begin{aligned} y_1' &= y_1 + y_2 \\ y_2' &= 4y_1 - 2y_2 \end{aligned}$$

(b) Find the solution that satisfies the initial conditions $y_1(0) = 1$, $y_2(0) = 6$.

Solution (a) The coefficient matrix for the system is
$$A = \begin{bmatrix} 1 & 1 \\ 4 & -2 \end{bmatrix}$$
As discussed in Section 5.2, A will be diagonalized by any matrix P whose columns are linearly independent eigenvectors of A. Since
$$\det(\lambda I - A) = \begin{vmatrix} \lambda - 1 & -1 \\ -4 & \lambda + 2 \end{vmatrix} = \lambda^2 + \lambda - 6 = (\lambda + 3)(\lambda - 2)$$
the eigenvalues of A are $\lambda = 2$ and $\lambda = -3$. By definition,
$$\mathbf{x} = \begin{bmatrix} x_1 \\ x_2 \end{bmatrix}$$
is an eigenvector of A corresponding to λ if and only if $\mathbf{x}$ is a nontrivial solution of
$$\begin{bmatrix} \lambda - 1 & -1 \\ -4 & \lambda + 2 \end{bmatrix} \begin{bmatrix} x_1 \\ x_2 \end{bmatrix} = \begin{bmatrix} 0 \\ 0 \end{bmatrix}$$
If $\lambda = 2$, this system becomes
$$\begin{bmatrix} 1 & -1 \\ -4 & 4 \end{bmatrix} \begin{bmatrix} x_1 \\ x_2 \end{bmatrix} = \begin{bmatrix} 0 \\ 0 \end{bmatrix}$$
Solving this system yields $x_1 = t$, $x_2 = t$, so
$$\begin{bmatrix} x_1 \\ x_2 \end{bmatrix} = \begin{bmatrix} t \\ t \end{bmatrix} = t \begin{bmatrix} 1 \\ 1 \end{bmatrix}$$

Thus,
$$\mathbf{p}_1 = \begin{bmatrix} 1 \\ 1 \end{bmatrix}$$

is a basis for the eigenspace corresponding to $\lambda = 2$. Similarly, you can show that

$$\mathbf{p}_2 = \begin{bmatrix} -\frac{1}{4} \\ 1 \end{bmatrix}$$

is a basis for the eigenspace corresponding to $\lambda = -3$. Thus,

$$P = \begin{bmatrix} 1 & -\frac{1}{4} \\ 1 & 1 \end{bmatrix}$$

diagonalizes A, and

$$D = P^{-1}AP = \begin{bmatrix} 2 & 0 \\ 0 & -3 \end{bmatrix}$$

Thus, as noted in Step 2 of the procedure stated above, the substitution

$$\mathbf{y} = P\mathbf{u} \quad \text{and} \quad \mathbf{y}' = P\mathbf{u}'$$

yields the "diagonal system"

$$\mathbf{u}' = D\mathbf{u} = \begin{bmatrix} 2 & 0 \\ 0 & -3 \end{bmatrix} \mathbf{u} \quad \text{or} \quad \begin{array}{l} u_1' = 2u_1 \\ u_2' = -3u_2 \end{array}$$

From (2) the solution of this system is

$$\begin{array}{l} u_1 = c_1 e^{2x} \\ u_2 = c_2 e^{-3x} \end{array} \quad \text{or} \quad \mathbf{u} = \begin{bmatrix} c_1 e^{2x} \\ c_2 e^{-3x} \end{bmatrix}$$

so the equation $\mathbf{y} = P\mathbf{u}$ yields, as the solution for $\mathbf{y}$,

$$\mathbf{y} = \begin{bmatrix} y_1 \\ y_2 \end{bmatrix} = \begin{bmatrix} 1 & -\frac{1}{4} \\ 1 & 1 \end{bmatrix} \begin{bmatrix} c_1 e^{2x} \\ c_2 e^{-3x} \end{bmatrix} = \begin{bmatrix} c_1 e^{2x} - \frac{1}{4} c_2 e^{-3x} \\ c_1 e^{2x} + c_2 e^{-3x} \end{bmatrix}$$

or

$$\begin{array}{l} y_1 = c_1 e^{2x} - \frac{1}{4} c_2 e^{-3x} \\ y_2 = c_1 e^{2x} + c_2 e^{-3x} \end{array} \tag{11}$$

Solution (b) If we substitute the given initial conditions in (11), we obtain

$$\begin{array}{l} c_1 - \frac{1}{4} c_2 = 1 \\ c_1 + c_2 = 6 \end{array}$$

Solving this system, we obtain $c_1 = 2$, $c_2 = 4$, so it follows from (11) that the solution satisfying the initial conditions is

$$\begin{array}{l} y_1 = 2e^{2x} - e^{-3x} \\ y_2 = 2e^{2x} + 4e^{-3x} \end{array} \blacktriangleleft$$

Remark Keep in mind that the method of Example 2 works because the coefficient matrix of the system can be diagonalized. In cases where this is not so, other methods are required. These are typically discussed in books devoted to differential equations.

Concept Review

- Differential equation
- Order of a differential equation
- General solution
- Particular solution
- Initial condition
- Initial-value problem
- First-order linear system

Skills

- Find the matrix form of a system of linear differential equations.
- Find the general solution of a system of linear differential equations by diagonalization.
- Find the particular solution of a system of linear differential equations satisfying an initial condition.

Exercise Set 5.4

1. (a) Solve the system
$$y_1' = 2y_1 - 3y_2$$
$$y_2' = -4y_1 + y_2$$
   (b) Find the solution that satisfies the initial conditions $y_1(0) = 0$, $y_2(0) = 0$.

2. (a) Solve the system
$$y_1' = y_1$$
$$y_2' = -2y_1 - 4y_2$$
   (b) Find the solution that satisfies the conditions $y_1(0) = 10$, $y_2(0) = 5$.

3. (a) Solve the system
$$y_1' = 4y_1 \quad\quad + y_3$$
$$y_2' = -2y_1 + y_2$$
$$y_3' = -2y_1 \quad\quad + y_3$$
   (b) Find the solution that satisfies the initial conditions $y_1(0) = -1$, $y_2(0) = 1$, $y_3(0) = 0$.

4. Solve the system
$$y_1' = 3y_1 + y_2 + y_3$$
$$y_2' = y_1 + 3y_2 + y_3$$
$$y_3' = y_1 + y_2 + 3y_3$$

5. Show that every solution of $y' = ay$ has the form $y = ce^{ax}$. [*Hint:* Let $y = f(x)$ be a solution of the equation, and show that $f(x)e^{-ax}$ is constant.]

6. Show that if A is diagonalizable and
$$\mathbf{y} = \begin{bmatrix} y_1 \\ y_2 \\ \vdots \\ y_n \end{bmatrix}$$
is a solution of the system $\mathbf{y}' = A\mathbf{y}$, then each y_i is a linear combination of $e^{\lambda_1 x}, e^{\lambda_2 x}, \ldots, e^{\lambda_n x}$, where $\lambda_1, \lambda_2, \ldots, \lambda_n$ are the eigenvalues of A.

7. Sometimes it is possible to solve a single higher-order linear differential equation with constant coefficients by expressing it as a system and applying the methods of this section. For the differential equation $y'' - 7y' + 6y = 0$, show that the substitutions $y_1 = y$ and $y_2 = y'$ lead to the system
$$y_1' = \quad\quad y_2$$
$$y_2' = -6y_1 + 7y_2$$
Solve this system, and use the result to solve the original differential equation.

8. Use the procedure in Exercise 7 to solve $y'' + y' - 12y = 0$.

9. Explain how you might use the procedure in Exercise 7 to solve $y''' - 6y'' + 11y' - 6y = 0$. Use your procedure to solve the equation.

10. (a) By rewriting (11) in matrix form, show that the solution of the system in Example 2 can be expressed as
$$\mathbf{y} = c_1 e^{2x} \begin{bmatrix} 1 \\ 1 \end{bmatrix} + c_2 e^{-3x} \begin{bmatrix} -\frac{1}{4} \\ 1 \end{bmatrix}$$
This is called the ***general solution*** of the system.

   (b) Note that in part (a), the vector in the first term is an eigenvector corresponding to the eigenvalue $\lambda_1 = 2$, and the vector in the second term is an eigenvector corresponding to the eigenvalue $\lambda_2 = -3$. This is a special case of the following general result:

   Theorem. *If the coefficient matrix A of the system $\mathbf{y}' = A\mathbf{y}$ is diagonalizable, then the general solution of the system can be expressed as*
$$\mathbf{y} = c_1 e^{\lambda_1 x} \mathbf{x}_1 + c_2 e^{\lambda_2 x} \mathbf{x}_2 + \cdots + c_n e^{\lambda_n x} \mathbf{x}_n$$
   where $\lambda_1, \lambda_2, \ldots, \lambda_n$ are the eigenvalues of A, and $\mathbf{x}_i$ is an eigenvector of A corresponding to λ_i.

Prove this result by tracing through the four-step procedure preceding Example 2 with

$$D = \begin{bmatrix} \lambda_1 & 0 & \cdots & 0 \\ 0 & \lambda_2 & \cdots & 0 \\ \vdots & \vdots & & \vdots \\ 0 & 0 & \cdots & \lambda_n \end{bmatrix} \quad \text{and} \quad P = [\mathbf{x}_1 \mid \mathbf{x}_2 \mid \cdots \mid \mathbf{x}_n]$$

11. Consider the system of differential equations $\mathbf{y}' = A\mathbf{y}$, where A is a 2×2 matrix. For what values of $a_{11}, a_{12}, a_{21}, a_{22}$ do the component solutions $y_1(t), y_2(t)$ tend to zero as $t \to \infty$? In particular, what must be true about the determinant and the trace of A for this to happen?

12. Solve the nondiagonalizable system

$$y_1' = y_1 + y_2$$
$$y_2' = y_2$$

True-False Exercises

In parts (a)–(e) determine whether the statement is true or false, and justify your answer.

(a) Every system of differential equations $\mathbf{y}' = A\mathbf{y}$ has a solution.

(b) If $\mathbf{x}' = A\mathbf{x}$ and $\mathbf{y}' = A\mathbf{y}$, then $\mathbf{x} = \mathbf{y}$.

(c) If $\mathbf{x}' = A\mathbf{x}$ and $\mathbf{y}' = A\mathbf{y}$, then $(c\mathbf{x} + d\mathbf{y})' = A(c\mathbf{x} + d\mathbf{y})$ for all scalars c and d.

(d) If A is a square matrix with distinct real eigenvalues, then it is possible to solve $\mathbf{x}' = A\mathbf{x}$ by diagonalization.

(e) If A and P are similar matrices, then $\mathbf{y}' = A\mathbf{y}$ and $\mathbf{u}' = P\mathbf{u}$ have the same solutions.

Chapter 5 Supplementary Exercises

1. (a) Show that if $0 < \theta < \pi$, then

$$A = \begin{bmatrix} \cos\theta & -\sin\theta \\ \sin\theta & \cos\theta \end{bmatrix}$$

has no real eigenvalues and consequently no real eigenvectors.

(b) Give a geometric explanation of the result in part (a).

2. Find the eigenvalues of

$$A = \begin{bmatrix} 0 & 1 & 0 \\ 0 & 0 & 1 \\ k^3 & -3k^2 & 3k \end{bmatrix}$$

3. (a) Show that if D is a diagonal matrix with nonnegative entries on the main diagonal, then there is a matrix S such that $S^2 = D$.

(b) Show that if A is a diagonalizable matrix with nonnegative eigenvalues, then there is a matrix S such that $S^2 = A$.

(c) Find a matrix S such that $S^2 = A$, given that

$$A = \begin{bmatrix} 1 & 3 & 1 \\ 0 & 4 & 5 \\ 0 & 0 & 9 \end{bmatrix}$$

4. Prove: If A is a square matrix, then A and A^T have the same characteristic polynomial.

5. Prove: If A is a square matrix and $p(\lambda) = \det(\lambda I - A)$ is the characteristic polynomial of A, then the coefficient of λ^{n-1} in $p(\lambda)$ is the negative of the trace of A.

6. Prove: If $b \neq 0$, then

$$A = \begin{bmatrix} a & b \\ 0 & a \end{bmatrix}$$

is not diagonalizable.

7. In advanced linear algebra, one proves the **Cayley–Hamilton Theorem**, which states that a square matrix A satisfies its characteristic equation; that is, if

$$c_0 + c_1\lambda + c_2\lambda^2 + \cdots + c_{n-1}\lambda^{n-1} + \lambda^n = 0$$

is the characteristic equation of A, then

$$c_0 I + c_1 A + c_2 A^2 + \cdots + c_{n-1} A^{n-1} + A^n = 0$$

Verify this result for

(a) $A = \begin{bmatrix} 3 & 6 \\ 1 & 2 \end{bmatrix}$ (b) $A = \begin{bmatrix} 0 & 1 & 0 \\ 0 & 0 & 1 \\ 1 & -3 & 3 \end{bmatrix}$

▶ In Exercises 8–10, use the Cayley–Hamilton Theorem, stated in Exercise 7. ◀

8. (a) Use Exercise 18 of Section 5.1 to prove the Cayley–Hamilton Theorem for 2×2 matrices.

(b) Prove the Cayley–Hamilton Theorem for $n \times n$ diagonalizable matrices.

9. The Cayley–Hamilton Theorem provides a method for calculating powers of a matrix. For example, if A is a 2×2 matrix with characteristic equation

$$c_0 + c_1\lambda + \lambda^2 = 0$$

then $c_0 I + c_1 A + A^2 = 0$, so

$$A^2 = -c_1 A - c_0 I$$

Multiplying through by A yields $A^3 = -c_1 A^2 - c_0 A$, which expresses A^3 in terms of A^2 and A, and multiplying through by A^2 yields $A^4 = -c_1 A^3 - c_0 A^2$, which expresses A^4 in terms of A^3 and A^2. Continuing in this way, we can calculate successive powers of A by expressing them in terms of lower powers. Use this procedure to calculate A^2, A^3, A^4, and A^5 for

$$A = \begin{bmatrix} 3 & 6 \\ 1 & 2 \end{bmatrix}$$

10. Use the method of the preceding exercise to calculate A^3 and A^4 for

$$A = \begin{bmatrix} 0 & 1 & 0 \\ 0 & 0 & 1 \\ 1 & -3 & 3 \end{bmatrix}$$

11. Find the eigenvalues of the matrix

$$A = \begin{bmatrix} c_1 & c_2 & \cdots & c_n \\ c_1 & c_2 & \cdots & c_n \\ \vdots & \vdots & & \vdots \\ c_1 & c_2 & \cdots & c_n \end{bmatrix}$$

12. (a) It was shown in Exercise 17 of Section 5.1 that if A is an $n \times n$ matrix, then the coefficient of λ^n in the characteristic polynomial of A is 1. (A polynomial with this property is called ***monic***.) Show that the matrix

$$\begin{bmatrix} 0 & 0 & 0 & \cdots & 0 & -c_0 \\ 1 & 0 & 0 & \cdots & 0 & -c_1 \\ 0 & 1 & 0 & \cdots & 0 & -c_2 \\ \vdots & \vdots & \vdots & & \vdots & \vdots \\ 0 & 0 & 0 & \cdots & 1 & -c_{n-1} \end{bmatrix}$$

has characteristic polynomial

$$p(\lambda) = c_0 + c_1\lambda + \cdots + c_{n-1}\lambda^{n-1} + \lambda^n$$

This shows that every monic polynomial is the characteristic polynomial of some matrix. The matrix in this example is called the ***companion matrix*** of $p(\lambda)$. [*Hint:* Evaluate all determinants in the problem by adding a multiple of the second row to the first to introduce a zero at the top of the first column, and then expanding by cofactors along the first column.]

(b) Find a matrix with characteristic polynomial

$$p(\lambda) = 1 - 2\lambda + \lambda^2 + 3\lambda^3 + \lambda^4$$

13. A square matrix A is called ***nilpotent*** if $A^n = 0$ for some positive integer n. What can you say about the eigenvalues of a nilpotent matrix?

14. Prove: If A is an $n \times n$ matrix and n is odd, then A has at least one real eigenvalue.

15. Find a 3×3 matrix A that has eigenvalues $\lambda = 0, 1$, and -1 with corresponding eigenvectors

$$\begin{bmatrix} 0 \\ 1 \\ -1 \end{bmatrix}, \quad \begin{bmatrix} 1 \\ -1 \\ 1 \end{bmatrix}, \quad \begin{bmatrix} 0 \\ 1 \\ 1 \end{bmatrix}$$

respectively.

16. Suppose that a 4×4 matrix A has eigenvalues $\lambda_1 = 1$, $\lambda_2 = -2$, $\lambda_3 = 3$, and $\lambda_4 = -3$.

(a) Use the method of Exercise 16 of Section 5.1 to find $\det(A)$.

(b) Use Exercise 5 above to find $\text{tr}(A)$.

17. Let A be a square matrix such that $A^3 = A$. What can you say about the eigenvalues of A?

18. (a) Solve the system

$$\begin{aligned} y_1' &= y_1 + 3y_2 \\ y_2' &= 2y_1 + 4y_2 \end{aligned}$$

(b) Find the solution satisfying the initial conditions $y_1(0) = 5$ and $y_2(0) = 6$.

CHAPTER 6

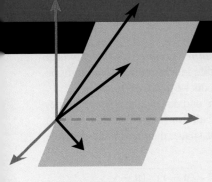

Inner Product Spaces

CHAPTER CONTENTS
6.1 Inner Products 335
6.2 Angle and Orthogonality in Inner Product Spaces 345
6.3 Gram–Schmidt Process; QR-Decomposition 352
6.4 Best Approximation; Least Squares 366
6.5 Least Squares Fitting to Data 376
6.6 Function Approximation; Fourier Series 382

INTRODUCTION In Chapter 3 we defined the dot product of vectors in R^n, and we used that concept to define notions of length, angle, distance, and orthogonality. In this chapter we will generalize those ideas so they are applicable in any vector space, not just R^n. We will also discuss various applications of these ideas.

6.1 Inner Products

In this section we will use the most important properties of the dot product on R^n as axioms, which, if satisfied by the vectors in a vector space V, will enable us to extend the notions of length, distance, angle, and perpendicularity to general vector spaces.

General Inner Products

In Definition 4 of Section 3.2 we defined the dot product of two vectors in R^n, and in Theorem 3.2.2 we listed four fundamental properties of such products. Our first goal in this section is to extend the notion of a dot product to general real vector spaces by using those four properties as axioms. We make the following definition.

Note that Definition 1 applies only to *real* vector spaces. A definition of inner products on *complex* vector spaces is given in the exercises. Since we will have little need for complex vector spaces from this point on, you can assume that all vector spaces under discussion are real, even though some of the theorems are also valid in complex vector spaces.

DEFINITION 1 An ***inner product*** on a real vector space V is a function that associates a real number $\langle \mathbf{u}, \mathbf{v} \rangle$ with each pair of vectors in V in such a way that the following axioms are satisfied for all vectors $\mathbf{u}$, $\mathbf{v}$, and $\mathbf{w}$ in V and all scalars k.

1. $\langle \mathbf{u}, \mathbf{v} \rangle = \langle \mathbf{v}, \mathbf{u} \rangle$ [Symmetry axiom]
2. $\langle \mathbf{u} + \mathbf{v}, \mathbf{w} \rangle = \langle \mathbf{u}, \mathbf{w} \rangle + \langle \mathbf{v}, \mathbf{w} \rangle$ [Additivity axiom]
3. $\langle k\mathbf{u}, \mathbf{v} \rangle = k\langle \mathbf{u}, \mathbf{v} \rangle$ [Homogeneity axiom]
4. $\langle \mathbf{v}, \mathbf{v} \rangle \geq 0$ and $\langle \mathbf{v}, \mathbf{v} \rangle = 0$ if and only if $\mathbf{v} = \mathbf{0}$ [Positivity axiom]

A real vector space with an inner product is called a ***real inner product space***.

Because the axioms for a real inner product space are based on properties of the dot product, these inner product space axioms will be satisfied automatically if we define the inner product of two vectors $\mathbf{u}$ and $\mathbf{v}$ in R^n to be

$$\langle \mathbf{u}, \mathbf{v} \rangle = \mathbf{u} \cdot \mathbf{v} = u_1 v_1 + u_2 v_2 + \cdots + u_n v_n$$

335

This inner product is commonly called the **Euclidean inner product** (or the **standard inner product**) on R^n to distinguish it from other possible inner products that might be defined on R^n. We call R^n with the Euclidean inner product **Euclidean n-space**.

Inner products can be used to define notions of norm and distance in a general inner product space just as we did with dot products in R^n. Recall from Formulas (11) and (19) of Section 3.2 that if **u** and **v** are vectors in Euclidean n-space, then norm and distance can be expressed in terms of the dot product as

$$\|\mathbf{v}\| = \sqrt{\mathbf{v} \cdot \mathbf{v}} \quad \text{and} \quad d(\mathbf{u}, \mathbf{v}) = \|\mathbf{u} - \mathbf{v}\| = \sqrt{(\mathbf{u} - \mathbf{v}) \cdot (\mathbf{u} - \mathbf{v})}$$

Motivated by these formulas we make the following definition.

DEFINITION 2 If V is a real inner product space, then the **norm** (or **length**) of a vector **v** in V is denoted by $\|\mathbf{v}\|$ and is defined by

$$\|\mathbf{v}\| = \sqrt{\langle \mathbf{v}, \mathbf{v} \rangle}$$

and the **distance** between two vectors is denoted by $d(\mathbf{u}, \mathbf{v})$ and is defined by

$$d(\mathbf{u}, \mathbf{v}) = \|\mathbf{u} - \mathbf{v}\| = \sqrt{\langle \mathbf{u} - \mathbf{v}, \mathbf{u} - \mathbf{v} \rangle}$$

A vector of norm 1 is called a **unit vector**.

The following theorem, which we state without proof, shows that norms and distances in real inner product spaces have many of the properties that you might expect.

THEOREM 6.1.1 *If* **u** *and* **v** *are vectors in a real inner product space V, and if k is a scalar, then:*

(a) $\|\mathbf{v}\| \geq 0$ with equality if and only if $\mathbf{v} = \mathbf{0}$.
(b) $\|k\mathbf{v}\| = |k| \|\mathbf{v}\|$.
(c) $d(\mathbf{u}, \mathbf{v}) = d(\mathbf{v}, \mathbf{u})$.
(d) $d(\mathbf{u}, \mathbf{v}) \geq 0$ with equality if and only if $\mathbf{u} = \mathbf{v}$.

Although the Euclidean inner product is the most important inner product on R^n, there are various applications in which it is desirable to modify it by *weighting* each term differently. More precisely, if

$$w_1, w_2, \ldots, w_n$$

are *positive* real numbers, which we will call **weights**, and if $\mathbf{u} = (u_1, u_2, \ldots, u_n)$ and $\mathbf{v} = (v_1, v_2, \ldots, v_n)$ are vectors in R^n, then it can be shown that the formula

$$\langle \mathbf{u}, \mathbf{v} \rangle = w_1 u_1 v_1 + w_2 u_2 v_2 + \cdots + w_n u_n v_n \tag{1}$$

defines an inner product on R^n that we call the **weighted Euclidean inner product with weights** $w_1, w_2, \ldots, w_n$.

Note that the standard Euclidean inner product is the special case of the weighted Euclidean inner product in which all the weights are 1.

▶ **EXAMPLE 1 Weighted Euclidean Inner Product**

Let $\mathbf{u} = (u_1, u_2)$ and $\mathbf{v} = (v_1, v_2)$ be vectors in R^2. Verify that the weighted Euclidean inner product

$$\langle \mathbf{u}, \mathbf{v} \rangle = 3u_1 v_1 + 2u_2 v_2 \tag{2}$$

satisfies the four inner product axioms.

Solution

Axiom 1: Interchanging **u** and **v** in Formula (2) does not change the sum on the right side, so $\langle \mathbf{u}, \mathbf{v} \rangle = \langle \mathbf{v}, \mathbf{u} \rangle$.

Axiom 2: If $\mathbf{w} = (w_1, w_2)$, then

$$\begin{aligned}\langle \mathbf{u} + \mathbf{v}, \mathbf{w} \rangle &= 3(u_1 + v_1)w_1 + 2(u_2 + v_2)w_2 \\ &= 3(u_1 w_1 + v_1 w_1) + 2(u_2 w_2 + v_2 w_2) \\ &= (3u_1 w_1 + 2u_2 w_2) + (3v_1 w_1 + 2v_2 w_2) \\ &= \langle \mathbf{u}, \mathbf{w} \rangle + \langle \mathbf{v}, \mathbf{w} \rangle\end{aligned}$$

Axiom 3: $\begin{aligned}\langle k\mathbf{u}, \mathbf{v} \rangle &= 3(ku_1)v_1 + 2(ku_2)v_2 \\ &= k(3u_1 v_1 + 2u_2 v_2) \\ &= k\langle \mathbf{u}, \mathbf{v} \rangle\end{aligned}$

Axiom 4: $\langle \mathbf{v}, \mathbf{v} \rangle = 3(v_1 v_1) + 2(v_2 v_2) = 3v_1^2 + 2v_2^2 \geq 0$ with equality if and only if $v_1 = v_2 = 0$; that is, if and only if $\mathbf{v} = \mathbf{0}$. ◀

> In Example 1, we are using subscripted w's to denote the components of the vector **w**, not the weights. The weights are the numbers 3 and 2 in Formula (2).

An Application of Weighted Euclidean Inner Products

To illustrate one way in which a weighted Euclidean inner product can arise, suppose that some physical experiment has n possible numerical outcomes

$$x_1, x_2, \ldots, x_n$$

and that a series of m repetitions of the experiment yields these values with various frequencies. Specifically, suppose that x_1 occurs f_1 times, x_2 occurs f_2 times, and so forth. Since there are a total of m repetitions of the experiment, it follows that

$$f_1 + f_2 + \cdots + f_n = m$$

Thus, the **arithmetic average** of the observed numerical values (denoted by $\bar{x}$) is

$$\bar{x} = \frac{f_1 x_1 + f_2 x_2 + \cdots + f_n x_n}{f_1 + f_2 + \cdots + f_n} = \frac{1}{m}(f_1 x_1 + f_2 x_2 + \cdots + f_n x_n) \tag{3}$$

If we let

$$\mathbf{f} = (f_1, f_2, \ldots, f_n)$$
$$\mathbf{x} = (x_1, x_2, \ldots, x_n)$$
$$w_1 = w_2 = \cdots = w_n = 1/m$$

then (3) can be expressed as the weighted Euclidean inner product

$$\bar{x} = \langle \mathbf{f}, \mathbf{x} \rangle = w_1 f_1 x_1 + w_2 f_2 x_2 + \cdots + w_n f_n x_n$$

▶ **EXAMPLE 2** Using a Weighted Euclidean Inner Product

It is important to keep in mind that norm and distance depend on the inner product being used. If the inner product is changed, then the norms and distances between vectors also change. For example, for the vectors $\mathbf{u} = (1, 0)$ and $\mathbf{v} = (0, 1)$ in R^2 with the Euclidean inner product we have

$$\|\mathbf{u}\| = \sqrt{1^2 + 0^2} = 1$$

and

$$d(\mathbf{u}, \mathbf{v}) = \|\mathbf{u} - \mathbf{v}\| = \|(1, -1)\| = \sqrt{1^2 + (-1)^2} = \sqrt{2}$$

but if we change to the weighted Euclidean inner product

$$\langle \mathbf{u}, \mathbf{v} \rangle = 3u_1 v_1 + 2u_2 v_2$$

338 Chapter 6 Inner Product Spaces

we have
$$\|\mathbf{u}\| = \langle \mathbf{u}, \mathbf{u} \rangle^{1/2} = [3(1)(1) + 2(0)(0)]^{1/2} = \sqrt{3}$$

and
$$d(\mathbf{u}, \mathbf{v}) = \|\mathbf{u} - \mathbf{v}\| = \langle (1, -1), (1, -1) \rangle^{1/2}$$
$$= [3(1)(1) + 2(-1)(-1)]^{1/2} = \sqrt{5} \blacktriangleleft$$

Unit Circles and Spheres in Inner Product Spaces

If V is an inner product space, then the set of points in V that satisfy
$$\|\mathbf{u}\| = 1$$
is called the **unit sphere** or sometimes the **unit circle** in V.

▶ **EXAMPLE 3** Unusual Unit Circles in R^2

(a) Sketch the unit circle in an xy-coordinate system in R^2 using the Euclidean inner product $\langle \mathbf{u}, \mathbf{v} \rangle = u_1 v_1 + u_2 v_2$.

(b) Sketch the unit circle in an xy-coordinate system in R^2 using the weighted Euclidean inner product $\langle \mathbf{u}, \mathbf{v} \rangle = \frac{1}{9} u_1 v_1 + \frac{1}{4} u_2 v_2$.

Solution (a) If $\mathbf{u} = (x, y)$, then $\|\mathbf{u}\| = \langle \mathbf{u}, \mathbf{u} \rangle^{1/2} = \sqrt{x^2 + y^2}$, so the equation of the unit circle is $\sqrt{x^2 + y^2} = 1$, or, on squaring both sides,
$$x^2 + y^2 = 1$$
As expected, the graph of this equation is a circle of radius 1 centered at the origin (Figure 6.1.1a).

Solution (b) If $\mathbf{u} = (x, y)$, then $\|\mathbf{u}\| = \langle \mathbf{u}, \mathbf{u} \rangle^{1/2} = \sqrt{\frac{1}{9} x^2 + \frac{1}{4} y^2}$, so the equation of the unit circle is $\sqrt{\frac{1}{9} x^2 + \frac{1}{4} y^2} = 1$, or, on squaring both sides,
$$\frac{x^2}{9} + \frac{y^2}{4} = 1$$
The graph of this equation is the ellipse shown in Figure 6.1.1b. ◀

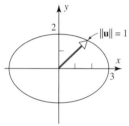

(a) The unit circle using the standard Euclidean inner product.

(b) The unit circle using a weighted Euclidean inner product.

▲ Figure 6.1.1

Remark It may seem odd that the "unit circle" in the second part of the last example turned out to have an elliptical shape. This will make more sense if you think of circles and spheres in general vector spaces *algebraically* ($\|\mathbf{u}\| = 1$) rather than geometrically. The change in geometry occurs because the norm, not being Euclidean, has the effect of distorting the space that we are used to seeing through "Euclidean eyes."

Inner Products Generated by Matrices

The Euclidean inner product and the weighted Euclidean inner products are special cases of a general class of inner products on R^n called **matrix inner products**. To define this class of inner products, let $\mathbf{u}$ and $\mathbf{v}$ be vectors in R^n that are expressed in *column form*, and let A be an *invertible* $n \times n$ matrix. It can be shown (Exercise 31) that if $\mathbf{u} \cdot \mathbf{v}$ is the Euclidean inner product on R^n, then the formula

$$\langle \mathbf{u}, \mathbf{v} \rangle = A\mathbf{u} \cdot A\mathbf{v} \tag{4}$$

also defines an inner product; it is called the **inner product on R^n generated by A**.

Recall from Table 1 of Section 3.2 that if $\mathbf{u}$ and $\mathbf{v}$ are in column form, then $\mathbf{u} \cdot \mathbf{v}$ can be written as $\mathbf{v}^T \mathbf{u}$ from which it follows that (4) can be expressed as

$$\langle \mathbf{u}, \mathbf{v} \rangle = (A\mathbf{v})^T A\mathbf{u}$$

or, equivalently as

$$\langle \mathbf{u}, \mathbf{v} \rangle = \mathbf{v}^T A^T A \mathbf{u} \quad (5)$$

▶ **EXAMPLE 4** Matrices Generating Weighted Euclidean Inner Products

The standard Euclidean and weighted Euclidean inner products are examples of matrix inner products. The standard Euclidean inner product on R^n is generated by the $n \times n$ identity matrix, since setting $A = I$ in Formula (4) yields

$$\langle \mathbf{u}, \mathbf{v} \rangle = I\mathbf{u} \cdot I\mathbf{v} = \mathbf{u} \cdot \mathbf{v}$$

and the weighted Euclidean inner product

$$\langle \mathbf{u}, \mathbf{v} \rangle = w_1 u_1 v_1 + w_2 u_2 v_2 + \cdots + w_n u_n v_n \quad (6)$$

is generated by the matrix

$$A = \begin{bmatrix} \sqrt{w_1} & 0 & 0 & \cdots & 0 \\ 0 & \sqrt{w_2} & 0 & \cdots & 0 \\ \vdots & \vdots & \vdots & & \vdots \\ 0 & 0 & 0 & \cdots & \sqrt{w_n} \end{bmatrix} \quad (7)$$

This can be seen by first observing that $A^T A$ is the $n \times n$ diagonal matrix whose diagonal entries are the weights $w_1, w_2, \ldots, w_n$ and then observing that (5) simplifies to (6) when A is the matrix in Formula (7).

▶ **EXAMPLE 5** Example 1 Revisited

The weighted Euclidean inner product $\langle \mathbf{u}, \mathbf{v} \rangle = 3u_1 v_1 + 2u_2 v_2$ discussed in Example 1 is the inner product on R^2 generated by

$$A = \begin{bmatrix} \sqrt{3} & 0 \\ 0 & \sqrt{2} \end{bmatrix} \quad ◀$$

Every diagonal matrix with positive diagonal entries generates a weighted inner product. Why?

Other Examples of Inner Products

So far, we have only considered examples of inner products on R^n. We will now consider examples of inner products on some of the other kinds of vector spaces that we discussed earlier.

▶ **EXAMPLE 6** An Inner Product on M_{nn}

If U and V are $n \times n$ matrices, then the formula

$$\langle U, V \rangle = \text{tr}(U^T V) \quad (8)$$

defines an inner product on the vector space M_{nn} (see Definition 8 of Section 1.3 for a definition of trace). This can be proved by confirming that the four inner product space axioms are satisfied, but you can visualize why this is so by computing (8) for the 2×2 matrices

$$U = \begin{bmatrix} u_1 & u_2 \\ u_3 & u_4 \end{bmatrix} \quad \text{and} \quad V = \begin{bmatrix} v_1 & v_2 \\ v_3 & v_4 \end{bmatrix}$$

This yields

$$\langle U, V \rangle = \text{tr}(U^T V) = u_1 v_1 + u_2 v_2 + u_3 v_3 + u_4 v_4$$

which is just the dot product of the corresponding entries in the two matrices. For example, if

$$U = \begin{bmatrix} 1 & 2 \\ 3 & 4 \end{bmatrix} \quad \text{and} \quad V = \begin{bmatrix} -1 & 0 \\ 3 & 2 \end{bmatrix}$$

then

$$\langle U, V \rangle = 1(-1) + 2(0) + 3(3) + 4(2) = 16$$

The norm of a matrix U relative to this inner product is

$$\|U\| = \langle U, U \rangle^{1/2} = \sqrt{u_1^2 + u_2^2 + u_3^2 + u_4^2}$$

▶ **EXAMPLE 7** *The Standard Inner Product on P_n*

If

$$\mathbf{p} = a_0 + a_1 x + \cdots + a_n x^n \quad \text{and} \quad \mathbf{q} = b_0 + b_1 x + \cdots + b_n x^n$$

are polynomials in P_n, then the following formula defines an inner product on P_n (verify) that we will call the ***standard inner product*** on this space:

$$\langle \mathbf{p}, \mathbf{q} \rangle = a_0 b_0 + a_1 b_1 + \cdots + a_n b_n \tag{9}$$

The norm of a polynomial $\mathbf{p}$ relative to this inner product is

$$\|\mathbf{p}\| = \sqrt{\langle \mathbf{p}, \mathbf{p} \rangle} = \sqrt{a_0^2 + a_1^2 + \cdots + a_n^2}$$

▶ **EXAMPLE 8** *The Evaluation Inner Product on P_n*

If

$$\mathbf{p} = p(x) = a_0 + a_1 x + \cdots + a_n x^n \quad \text{and} \quad \mathbf{q} = q(x) = b_0 + b_1 x + \cdots + b_n x^n$$

are polynomials in P_n, and if $x_0, x_1, \ldots, x_n$ are distinct real numbers (called ***sample points***), then the formula

$$\langle \mathbf{p}, \mathbf{q} \rangle = p(x_0)q(x_0) + p(x_1)q(x_1) + \cdots + p(x_n)q(x_n) \tag{10}$$

defines an inner product on P_n called the ***evaluation inner product*** at $x_0, x_1, \ldots, x_n$. Algebraically, this can be viewed as the dot product in R^n of the n-tuples

$$\bigl(p(x_0), p(x_1), \ldots, p(x_n)\bigr) \quad \text{and} \quad \bigl(q(x_0), q(x_1), \ldots, q(x_n)\bigr)$$

and hence the first three inner product axioms follow from properties of the dot product. The fourth inner product axiom follows from the fact that

$$\langle \mathbf{p}, \mathbf{p} \rangle = [p(x_0)]^2 + [p(x_1)]^2 + \cdots + [p(x_n)]^2 \geq 0$$

with equality holding if and only if

$$p(x_0) = p(x_1) = \cdots = p(x_n) = 0$$

But a nonzero polynomial of degree n or less can have at most n distinct roots, so it must be that $\mathbf{p} = \mathbf{0}$, which proves that the fourth inner product axiom holds.

The norm of a polynomial $\mathbf{p}$ relative to the evaluation inner product is

$$\|\mathbf{p}\| = \sqrt{\langle \mathbf{p}, \mathbf{p} \rangle} = \sqrt{[p(x_0)]^2 + [p(x_1)]^2 + \cdots + [p(x_n)]^2} \tag{11}$$

▶ **EXAMPLE 9** *Working with the Evaluation Inner Product*

Let P_2 have the evaluation inner product at the points

$$x_0 = -2, \quad x_1 = 0, \quad \text{and} \quad x_2 = 2$$

Compute $\langle \mathbf{p}, \mathbf{q} \rangle$ and $\|\mathbf{p}\|$ for the polynomials $\mathbf{p} = p(x) = x^2$ and $\mathbf{q} = q(x) = 1 + x$.

Solution It follows from (10) and (11) that

$$\langle \mathbf{p}, \mathbf{q} \rangle = p(-2)q(-2) + p(0)q(0) + p(2)q(2) = (4)(-1) + (0)(1) + (4)(3) = 8$$

$$\|\mathbf{p}\| = \sqrt{[p(x_0)]^2 + [p(x_1)]^2 + [p(x_2)]^2} = \sqrt{[p(-2)]^2 + [p(0)]^2 + [p(2)]^2}$$
$$= \sqrt{4^2 + 0^2 + 4^2} = \sqrt{32} = 4\sqrt{2}$$

CALCULUS REQUIRED

▶ **EXAMPLE 10 An Inner Product on C[a, b]**

Let $\mathbf{f} = f(x)$ and $\mathbf{g} = g(x)$ be two functions in $C[a, b]$ and define

$$\langle \mathbf{f}, \mathbf{g} \rangle = \int_a^b f(x)g(x)\, dx \tag{12}$$

We will show that this formula defines an inner product on $C[a, b]$ by verifying the four inner product axioms for functions $\mathbf{f} = f(x)$, $\mathbf{g} = g(x)$, and $\mathbf{h} = h(x)$ in $C[a, b]$:

1. $\langle \mathbf{f}, \mathbf{g} \rangle = \int_a^b f(x)g(x)\, dx = \int_a^b g(x)f(x)\, dx = \langle \mathbf{g}, \mathbf{f} \rangle$

    which proves that Axiom 1 holds.

2. $\langle \mathbf{f} + \mathbf{g}, \mathbf{h} \rangle = \int_a^b (f(x) + g(x))h(x)\, dx$

    $= \int_a^b f(x)h(x)\, dx + \int_a^b g(x)h(x)\, dx$

    $= \langle \mathbf{f}, \mathbf{h} \rangle + \langle \mathbf{g}, \mathbf{h} \rangle$

    which proves that Axiom 2 holds.

3. $\langle k\mathbf{f}, \mathbf{g} \rangle = \int_a^b kf(x)g(x)\, dx = k \int_a^b f(x)g(x)\, dx = k \langle \mathbf{f}, \mathbf{g} \rangle$

    which proves that Axiom 3 holds.

4. If $\mathbf{f} = f(x)$ is any function in $C[a, b]$, then

$$\langle \mathbf{f}, \mathbf{f} \rangle = \int_a^b f^2(x)\, dx \geq 0 \tag{13}$$

since $f^2(x) \geq 0$ for all x in the interval $[a, b]$. Moreover because f is continuous on $[a, b]$, the equality holds in Formula (13) if and only if the function f is identically zero on $[a, b]$, that is, if and only if $\mathbf{f} = \mathbf{0}$; and this proves that Axiom 4 holds.

CALCULUS REQUIRED

▶ **EXAMPLE 11 Norm of a Vector in C[a, b]**

If $C[a, b]$ has the inner product that was defined in Example 10, then the norm of a function $\mathbf{f} = f(x)$ relative to this inner product is

$$\|\mathbf{f}\| = \langle \mathbf{f}, \mathbf{f} \rangle^{1/2} = \sqrt{\int_a^b f^2(x)\, dx} \tag{14}$$

and the unit sphere in this space consists of all functions $\mathbf{f}$ in $C[a, b]$ that satisfy the equation

$$\int_a^b f^2(x)\, dx = 1 \quad ◀$$

Remark Note that the vector space P_n is a subspace of $C[a, b]$ because polynomials are continuous functions. Thus, Formula (12) defines an inner product on P_n.

Remark Recall from calculus that the arc length of a curve $y = f(x)$ over an interval $[a, b]$ is given by the formula

$$L = \int_a^b \sqrt{1 + [f'(x)]^2}\, dx \tag{15}$$

Do not confuse this concept of arc length with $\|\mathbf{f}\|$, which is the length (norm) of $\mathbf{f}$ when $\mathbf{f}$ is viewed as a vector in $C[a, b]$. Formulas (14) and (15) are quite different.

Algebraic Properties of Inner Products

The following theorem lists some of the algebraic properties of inner products that follow from the inner product axioms. This result is a generalization of Theorem 3.2.3, which applied only to the dot product on R^n.

THEOREM 6.1.2 *If $\mathbf{u}$, $\mathbf{v}$, and $\mathbf{w}$ are vectors in a real inner product space V, and if k is a scalar, then:*

(a) $\langle \mathbf{0}, \mathbf{v} \rangle = \langle \mathbf{v}, \mathbf{0} \rangle = 0$
(b) $\langle \mathbf{u}, \mathbf{v} + \mathbf{w} \rangle = \langle \mathbf{u}, \mathbf{v} \rangle + \langle \mathbf{u}, \mathbf{w} \rangle$
(c) $\langle \mathbf{u}, \mathbf{v} - \mathbf{w} \rangle = \langle \mathbf{u}, \mathbf{v} \rangle - \langle \mathbf{u}, \mathbf{w} \rangle$
(d) $\langle \mathbf{u} - \mathbf{v}, \mathbf{w} \rangle = \langle \mathbf{u}, \mathbf{w} \rangle - \langle \mathbf{v}, \mathbf{w} \rangle$
(e) $k \langle \mathbf{u}, \mathbf{v} \rangle = \langle \mathbf{u}, k\mathbf{v} \rangle$

Proof We will prove part (b) and leave the proofs of the remaining parts as exercises.

$$\begin{aligned}
\langle \mathbf{u}, \mathbf{v} + \mathbf{w} \rangle &= \langle \mathbf{v} + \mathbf{w}, \mathbf{u} \rangle && \text{[By symmetry]} \\
&= \langle \mathbf{v}, \mathbf{u} \rangle + \langle \mathbf{w}, \mathbf{u} \rangle && \text{[By additivity]} \\
&= \langle \mathbf{u}, \mathbf{v} \rangle + \langle \mathbf{u}, \mathbf{w} \rangle && \text{[By symmetry]} \blacktriangleleft
\end{aligned}$$

The following example illustrates how Theorem 6.1.2 and the defining properties of inner products can be used to perform algebraic computations with inner products. As you read through the example, you will find it instructive to justify the steps.

▶ **EXAMPLE 12** Calculating with Inner Products

$$\begin{aligned}
\langle \mathbf{u} - 2\mathbf{v}, 3\mathbf{u} + 4\mathbf{v} \rangle &= \langle \mathbf{u}, 3\mathbf{u} + 4\mathbf{v} \rangle - \langle 2\mathbf{v}, 3\mathbf{u} + 4\mathbf{v} \rangle \\
&= \langle \mathbf{u}, 3\mathbf{u} \rangle + \langle \mathbf{u}, 4\mathbf{v} \rangle - \langle 2\mathbf{v}, 3\mathbf{u} \rangle - \langle 2\mathbf{v}, 4\mathbf{v} \rangle \\
&= 3\langle \mathbf{u}, \mathbf{u} \rangle + 4\langle \mathbf{u}, \mathbf{v} \rangle - 6\langle \mathbf{v}, \mathbf{u} \rangle - 8\langle \mathbf{v}, \mathbf{v} \rangle \\
&= 3\|\mathbf{u}\|^2 + 4\langle \mathbf{u}, \mathbf{v} \rangle - 6\langle \mathbf{u}, \mathbf{v} \rangle - 8\|\mathbf{v}\|^2 \\
&= 3\|\mathbf{u}\|^2 - 2\langle \mathbf{u}, \mathbf{v} \rangle - 8\|\mathbf{v}\|^2 \blacktriangleleft
\end{aligned}$$

Concept Review

- Inner product axioms
- Euclidean inner product
- Euclidean n-space
- Weighted Euclidean inner product
- Unit circle (sphere)
- Matrix inner product
- Norm in an inner product space
- Distance between two vectors in an inner product space
- Examples of inner products
- Properties of inner products

Skills

- Compute the inner product of two vectors.
- Find the norm of a vector.
- Find the distance between two vectors.
- Show that a given formula defines an inner product.
- Show that a given formula does not define an inner product by demonstrating that at least one of the inner product space axioms fails.

Exercise Set 6.1

1. Let $\langle \mathbf{u}, \mathbf{v} \rangle$ be the Euclidean inner product on R^2, and let $\mathbf{u} = (1, 1)$, $\mathbf{v} = (3, 2)$, $\mathbf{w} = (-1, 0)$, and $k = 5$. Compute the following.

    (a) $\langle \mathbf{v}, \mathbf{w} \rangle$ (b) $\langle k\mathbf{u}, \mathbf{v} \rangle$ (c) $\langle \mathbf{u} + \mathbf{v}, \mathbf{w} \rangle$
    (d) $\|\mathbf{u}\|$ (e) $d(\mathbf{u}, \mathbf{v})$ (f) $\|\mathbf{u} - k\mathbf{v}\|$

2. Repeat Exercise 1 for the weighted Euclidean inner product $\langle \mathbf{u}, \mathbf{v} \rangle = 2u_1v_1 + 3u_2v_2$.

3. Let $\langle \mathbf{u}, \mathbf{v} \rangle$ be the Euclidean inner product on R^2, and let $\mathbf{u} = (3, -2)$, $\mathbf{v} = (4, 5)$, $\mathbf{w} = (-1, 6)$, and $k = -4$. Verify the following.

    (a) $\langle \mathbf{u}, \mathbf{v} \rangle = \langle \mathbf{v}, \mathbf{u} \rangle$
    (b) $\langle \mathbf{u} + \mathbf{v}, \mathbf{w} \rangle = \langle \mathbf{u}, \mathbf{w} \rangle + \langle \mathbf{v}, \mathbf{w} \rangle$
    (c) $\langle \mathbf{u}, \mathbf{v} + \mathbf{w} \rangle = \langle \mathbf{u}, \mathbf{v} \rangle + \langle \mathbf{u}, \mathbf{w} \rangle$
    (d) $\langle k\mathbf{u}, \mathbf{v} \rangle = k\langle \mathbf{u}, \mathbf{v} \rangle = \langle \mathbf{u}, k\mathbf{v} \rangle$
    (e) $\langle \mathbf{0}, \mathbf{v} \rangle = \langle \mathbf{v}, \mathbf{0} \rangle = 0$

4. Repeat Exercise 3 for the weighted Euclidean inner product $\langle \mathbf{u}, \mathbf{v} \rangle = 4u_1v_1 + 5u_2v_2$.

5. Let $\langle \mathbf{u}, \mathbf{v} \rangle$ be the inner product on R^2 generated by $\begin{bmatrix} 2 & 1 \\ 1 & 1 \end{bmatrix}$, and let $\mathbf{u} = (2, 1)$, $\mathbf{v} = (-1, 1)$, $\mathbf{w} = (0, -1)$. Compute the following.

    (a) $\langle \mathbf{u}, \mathbf{v} \rangle$ (b) $\langle \mathbf{v}, \mathbf{w} \rangle$ (c) $\langle \mathbf{u}, \mathbf{v} + \mathbf{w} \rangle$
    (d) $\|\mathbf{u}\|$ (e) $d(\mathbf{v}, \mathbf{w})$ (f) $\|\mathbf{v} - \mathbf{w}\|^2$

6. Repeat Exercise 5 for the inner product on R^2 generated by $\begin{bmatrix} 0 & -1 \\ 2 & 1 \end{bmatrix}$.

7. Compute $\langle \mathbf{u}, \mathbf{v} \rangle$ using the inner product in Example 6.

    (a) $\mathbf{u} = \begin{bmatrix} 3 & -2 \\ 4 & 8 \end{bmatrix}$, $\mathbf{v} = \begin{bmatrix} -1 & 3 \\ 1 & 1 \end{bmatrix}$

    (b) $\mathbf{u} = \begin{bmatrix} 1 & 2 \\ -3 & 5 \end{bmatrix}$, $\mathbf{v} = \begin{bmatrix} 4 & 6 \\ 0 & 8 \end{bmatrix}$

8. Compute $\langle \mathbf{p}, \mathbf{q} \rangle$ using the inner product in Example 7.

    (a) $\mathbf{p} = 3 - x + 2x^2$, $\mathbf{q} = 2 - 4x^2$
    (b) $\mathbf{p} = -5 + 2x + x^2$, $\mathbf{q} = 3 + 2x - 4x^2$

9. (a) Use Formula (4) to show that $\langle \mathbf{u}, \mathbf{v} \rangle = 9u_1v_1 + 4u_2v_2$ is the inner product on R^2 generated by

    $$A = \begin{bmatrix} 3 & 0 \\ 0 & 2 \end{bmatrix}$$

    (b) Use the inner product in part (a) to compute $\langle \mathbf{u}, \mathbf{v} \rangle$ if $\mathbf{u} = (-3, 2)$ and $\mathbf{v} = (1, 7)$.

10. (a) Use Formula (4) to show that

    $$\langle \mathbf{u}, \mathbf{v} \rangle = 10u_1v_1 - 7u_2v_1 - 7u_1v_2 + 5u_2v_2$$

is the inner product on R^2 generated by

$$A = \begin{bmatrix} 3 & -2 \\ -1 & 1 \end{bmatrix}$$

(b) Use the inner product in part (a) to compute $\langle \mathbf{u}, \mathbf{v} \rangle$ if $\mathbf{u} = (0, -3)$ and $\mathbf{v} = (6, 2)$.

11. Let $\mathbf{u} = (u_1, u_2)$ and $\mathbf{v} = (v_1, v_2)$. In each part, the given expression is an inner product on R^2. Find a matrix that generates it.

    (a) $\langle \mathbf{u}, \mathbf{v} \rangle = 3u_1v_1 + 5u_2v_2$ (b) $\langle \mathbf{u}, \mathbf{v} \rangle = 4u_1v_1 + 6u_2v_2$

12. Let P_2 have the inner product in Example 7. In each part, find $\|\mathbf{p}\|$.

    (a) $\mathbf{p} = -2 + 3x + 2x^2$ (b) $\mathbf{p} = 4 - 3x^2$

13. Let M_{22} have the inner product in Example 6. In each part, find $\|A\|$.

    (a) $A = \begin{bmatrix} 5 & 3 \\ 2 & -6 \end{bmatrix}$ (b) $A = \begin{bmatrix} 0 & 0 \\ 0 & 0 \end{bmatrix}$

14. Let P_2 have the inner product in Example 7. Find $d(\mathbf{p}, \mathbf{q})$ if $\mathbf{p} = x + 3x^2$ and $\mathbf{q} = 2 - x + 4x^2$.

15. Let M_{22} have the inner product in Example 6. Find $d(A, B)$.

    (a) $A = \begin{bmatrix} 2 & 6 \\ 9 & 4 \end{bmatrix}$, $B = \begin{bmatrix} -4 & 7 \\ 1 & 6 \end{bmatrix}$

    (b) $A = \begin{bmatrix} -2 & 4 \\ 1 & 0 \end{bmatrix}$, $B = \begin{bmatrix} -5 & 1 \\ 6 & 2 \end{bmatrix}$

16. Let P_2 have the inner product of Example 9, and let $\mathbf{p} = 1 + x + x^2$ and $\mathbf{q} = 1 - 2x^2$. Compute the following.

    (a) $\langle \mathbf{p}, \mathbf{q} \rangle$ (b) $\|\mathbf{p}\|$ (c) $d(\mathbf{p}, \mathbf{q})$

17. Let P_3 have the evaluation inner product at the sample points

    $$x_0 = 2, \quad x_1 = -1, \quad x_2 = 0, \quad x_3 = 1$$

    Find $\langle \mathbf{p}, \mathbf{q} \rangle$ and $\|\mathbf{p}\|$ for $\mathbf{p} = x + x^3$ and $\mathbf{q} = 1 + x^2$.

18. In each part, use the given inner product on R^2 to find $\|\mathbf{w}\|$, where $\mathbf{w} = (-1, 3)$.

    (a) the Euclidean inner product
    (b) the weighted Euclidean inner product $\langle \mathbf{u}, \mathbf{v} \rangle = 3u_1v_1 + 2u_2v_2$, where $\mathbf{u} = (u_1, u_2)$ and $\mathbf{v} = (v_1, v_2)$
    (c) the inner product generated by the matrix

    $$A = \begin{bmatrix} 1 & 2 \\ -1 & 3 \end{bmatrix}$$

19. Use the inner products in Exercise 18 to find $d(\mathbf{u}, \mathbf{v})$ for $\mathbf{u} = (-1, 2)$ and $\mathbf{v} = (2, 5)$.

20. Suppose that $\mathbf{u}$, $\mathbf{v}$, and $\mathbf{w}$ are vectors such that

$$\langle \mathbf{u}, \mathbf{v} \rangle = 2, \quad \langle \mathbf{v}, \mathbf{w} \rangle = -3, \quad \langle \mathbf{u}, \mathbf{w} \rangle = 5$$
$$\|\mathbf{u}\| = 1, \quad \|\mathbf{v}\| = 2, \quad \|\mathbf{w}\| = 7$$

Evaluate the given expression.

(a) $\langle \mathbf{u} + \mathbf{v}, \mathbf{v} + \mathbf{w} \rangle$
(b) $\langle 2\mathbf{v} - \mathbf{w}, 3\mathbf{u} + 2\mathbf{w} \rangle$
(c) $\langle \mathbf{u} - \mathbf{v} - 2\mathbf{w}, 4\mathbf{u} + \mathbf{v} \rangle$
(d) $\|\mathbf{u} + \mathbf{v}\|$
(e) $\|2\mathbf{w} - \mathbf{v}\|$
(f) $\|\mathbf{u} - 2\mathbf{v} + 4\mathbf{w}\|$

21. Sketch the unit circle in R^2 using the given inner product.

(a) $\langle \mathbf{u}, \mathbf{v} \rangle = \frac{1}{4} u_1 v_1 + \frac{1}{16} u_2 v_2$
(b) $\langle \mathbf{u}, \mathbf{v} \rangle = 2 u_1 v_1 + u_2 v_2$

22. Find a weighted Euclidean inner product on R^2 for which the unit circle is the ellipse shown in the accompanying figure.

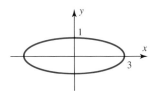

◀ Figure Ex-22

23. Let $\mathbf{u} = (u_1, u_2)$ and $\mathbf{v} = (v_1, v_2)$. Show that the following are inner products on R^2 by verifying that the inner product axioms hold.

(a) $\langle \mathbf{u}, \mathbf{v} \rangle = 3 u_1 v_1 + 5 u_2 v_2$
(b) $\langle \mathbf{u}, \mathbf{v} \rangle = 4 u_1 v_1 + u_2 v_1 + u_1 v_2 + 4 u_2 v_2$

24. Let $\mathbf{u} = (u_1, u_2, u_3)$ and $\mathbf{v} = (v_1, v_2, v_3)$. Determine which of the following are inner products on R^3. For those that are not, list the axioms that do not hold.

(a) $\langle \mathbf{u}, \mathbf{v} \rangle = u_1 v_1 + u_3 v_3$
(b) $\langle \mathbf{u}, \mathbf{v} \rangle = u_1^2 v_1^2 + u_2^2 v_2^2 + u_3^2 v_3^2$
(c) $\langle \mathbf{u}, \mathbf{v} \rangle = 2 u_1 v_1 + u_2 v_2 + 4 u_3 v_3$
(d) $\langle \mathbf{u}, \mathbf{v} \rangle = u_1 v_1 - u_2 v_2 + u_3 v_3$

25. Show that the following identity holds for vectors in any inner product space.

$$\|\mathbf{u} + \mathbf{v}\|^2 + \|\mathbf{u} - \mathbf{v}\|^2 = 2\|\mathbf{u}\|^2 + 2\|\mathbf{v}\|^2$$

26. Show that the following identity holds for vectors in any inner product space.

$$\langle \mathbf{u}, \mathbf{v} \rangle = \frac{1}{4} \|\mathbf{u} + \mathbf{v}\|^2 - \frac{1}{4} \|\mathbf{u} - \mathbf{v}\|^2$$

27. Let $U = \begin{bmatrix} u_1 & u_2 \\ u_3 & u_4 \end{bmatrix}$ and $V = \begin{bmatrix} v_1 & v_2 \\ v_3 & v_4 \end{bmatrix}$.

Show that $\langle U, V \rangle = u_1 v_1 + u_2 v_3 + u_3 v_2 + u_4 v_4$ is *not* an inner product on M_{22}.

28. (*Calculus required*) Let the vector space P_2 have the inner product

$$\langle \mathbf{p}, \mathbf{q} \rangle = \int_{-1}^{1} p(x) q(x) \, dx$$

(a) Find $\|\mathbf{p}\|$ for $\mathbf{p} = 1$, $\mathbf{p} = x$, and $\mathbf{p} = x^2$.
(b) Find $d(\mathbf{p}, \mathbf{q})$ if $\mathbf{p} = 1$ and $\mathbf{q} = x$.

29. (*Calculus required*) Use the inner product

$$\langle \mathbf{p}, \mathbf{q} \rangle = \int_{-1}^{1} p(x) q(x) \, dx$$

on P_3, to compute $\langle \mathbf{p}, \mathbf{q} \rangle$.

(a) $\mathbf{p} = 1 - x^2 + 3x^3$, $\mathbf{q} = 2 - x$
(b) $\mathbf{p} = x - 5x^3$, $\mathbf{q} = 2 + 8x^2$

30. (*Calculus required*) In each part, use the inner product

$$\langle \mathbf{f}, \mathbf{g} \rangle = \int_{0}^{1} f(x) g(x) \, dx$$

on $C[0, 1]$ to compute $\langle \mathbf{f}, \mathbf{g} \rangle$.

(a) $\mathbf{f} = \cos 2\pi x$, $\mathbf{g} = \sin 2\pi x$
(b) $\mathbf{f} = x$, $\mathbf{g} = e^x$
(c) $\mathbf{f} = \tan \frac{\pi}{4} x$, $\mathbf{g} = 1$

31. Prove that Formula (4) defines an inner product on R^n.

32. The definition of a complex vector space was given in the first margin note in Section 4.1. The definition of a *complex inner product* on a complex vector space V is identical to Definition 1 except that scalars are allowed to be complex numbers, and Axiom 1 is replaced by $\langle \mathbf{u}, \mathbf{v} \rangle = \overline{\langle \mathbf{v}, \mathbf{u} \rangle}$. The remaining axioms are unchanged. A complex vector space with a complex inner product is called a *complex inner product space*. Prove that if V is a complex inner product space then $\langle \mathbf{u}, k\mathbf{v} \rangle = \bar{k} \langle \mathbf{u}, \mathbf{v} \rangle$.

True-False Exercises

In parts (a)–(g) determine whether the statement is true or false, and justify your answer.

(a) The dot product on R^2 is an example of a weighted inner product.

(b) The inner product of two vectors cannot be a negative real number.

(c) $\langle \mathbf{u}, \mathbf{v} + \mathbf{w} \rangle = \langle \mathbf{v}, \mathbf{u} \rangle + \langle \mathbf{w}, \mathbf{u} \rangle$.

(d) $\langle k\mathbf{u}, k\mathbf{v} \rangle = k^2 \langle \mathbf{u}, \mathbf{v} \rangle$.

(e) If $\langle \mathbf{u}, \mathbf{v} \rangle = 0$, then $\mathbf{u} = \mathbf{0}$ or $\mathbf{v} = \mathbf{0}$.

(f) If $\|\mathbf{v}\|^2 = 0$, then $\mathbf{v} = \mathbf{0}$.

(g) If A is an $n \times n$ matrix, then $\langle \mathbf{u}, \mathbf{v} \rangle = A\mathbf{u} \cdot A\mathbf{v}$ defines an inner product on R^n.

6.2 Angle and Orthogonality in Inner Product Spaces

In Section 3.2 we defined the notion of "angle" between vectors in R^n. In this section we will extend this idea to general vector spaces. This will enable us to extend the notion of orthogonality as well, thereby setting the groundwork for a variety of new applications.

Cauchy–Schwarz Inequality

Recall from Formula (20) of Section 3.2 that the angle θ between two vectors $\mathbf{u}$ and $\mathbf{v}$ in R^n is

$$\theta = \cos^{-1}\left(\frac{\mathbf{u} \cdot \mathbf{v}}{\|\mathbf{u}\|\|\mathbf{v}\|}\right) \tag{1}$$

We were assured that this formula was valid because it followed from the Cauchy–Schwarz inequality (Theorem 3.2.4) that

$$-1 \leq \frac{\mathbf{u} \cdot \mathbf{v}}{\|\mathbf{u}\|\|\mathbf{v}\|} \leq 1 \tag{2}$$

as required for the inverse cosine to be defined. The following generalization of Theorem 3.2.4 will enable us to define the angle between two vectors in *any* real inner product space.

THEOREM 6.2.1 **Cauchy–Schwarz Inequality**

If $\mathbf{u}$ and $\mathbf{v}$ are vectors in a real inner product space V, then

$$|\langle \mathbf{u}, \mathbf{v} \rangle| \leq \|\mathbf{u}\|\|\mathbf{v}\| \tag{3}$$

Proof We warn you in advance that the proof presented here depends on a clever trick that is not easy to motivate.

In the case where $\mathbf{u} = \mathbf{0}$ the two sides of (3) are equal since $\langle \mathbf{u}, \mathbf{v} \rangle$ and $\|\mathbf{u}\|$ are both zero. Thus, we need only consider the case where $\mathbf{u} \neq \mathbf{0}$. Making this assumption, let

$$a = \langle \mathbf{u}, \mathbf{u} \rangle, \quad b = 2\langle \mathbf{u}, \mathbf{v} \rangle, \quad c = \langle \mathbf{v}, \mathbf{v} \rangle$$

and let t be any real number. Since the positivity axiom states that the inner product of any vector with itself is nonnegative, it follows that

$$0 \leq \langle t\mathbf{u} + \mathbf{v}, t\mathbf{u} + \mathbf{v} \rangle = \langle \mathbf{u}, \mathbf{u} \rangle t^2 + 2\langle \mathbf{u}, \mathbf{v} \rangle t + \langle \mathbf{v}, \mathbf{v} \rangle$$
$$= at^2 + bt + c$$

This inequality implies that the quadratic polynomial $at^2 + bt + c$ has either no real roots or a repeated real root. Therefore, its discriminant must satisfy the inequality $b^2 - 4ac \leq 0$. Expressing the coefficients a, b, and c in terms of the vectors $\mathbf{u}$ and $\mathbf{v}$ gives $4\langle \mathbf{u}, \mathbf{v} \rangle^2 - 4\langle \mathbf{u}, \mathbf{u} \rangle \langle \mathbf{v}, \mathbf{v} \rangle \leq 0$ or, equivalently,

$$\langle \mathbf{u}, \mathbf{v} \rangle^2 \leq \langle \mathbf{u}, \mathbf{u} \rangle \langle \mathbf{v}, \mathbf{v} \rangle$$

Taking square roots of both sides and using the fact that $\langle \mathbf{u}, \mathbf{u} \rangle$ and $\langle \mathbf{v}, \mathbf{v} \rangle$ are nonnegative yields

$$|\langle \mathbf{u}, \mathbf{v} \rangle| \leq \langle \mathbf{u}, \mathbf{u} \rangle^{1/2} \langle \mathbf{v}, \mathbf{v} \rangle^{1/2} \quad \text{or equivalently} \quad |\langle \mathbf{u}, \mathbf{v} \rangle| \leq \|\mathbf{u}\|\|\mathbf{v}\|$$

which completes the proof. ◂

The following two alternative forms of the Cauchy–Schwarz inequality are useful to know:

$$\langle \mathbf{u}, \mathbf{v} \rangle^2 \leq \langle \mathbf{u}, \mathbf{u} \rangle \langle \mathbf{v}, \mathbf{v} \rangle \tag{4}$$

$$\langle \mathbf{u}, \mathbf{v} \rangle^2 \leq \|\mathbf{u}\|^2 \|\mathbf{v}\|^2 \tag{5}$$

The first of these formulas was obtained in the proof of Theorem 6.2.1, and the second is a variation of the first.

Angle Between Vectors

Our next goal is to define what is meant by the "angle" between vectors in a real inner product space. As the first step, we leave it for you to use the Cauchy–Schwarz inequality to show that

$$-1 \leq \frac{\langle \mathbf{u}, \mathbf{v} \rangle}{\|\mathbf{u}\| \|\mathbf{v}\|} \leq 1 \tag{6}$$

This being the case, there is a unique angle θ in radian measure for which

$$\cos \theta = \frac{\langle \mathbf{u}, \mathbf{v} \rangle}{\|\mathbf{u}\| \|\mathbf{v}\|} \quad \text{and} \quad 0 \leq \theta \leq \pi \tag{7}$$

(Figure 6.2.1). This enables us to *define* the **angle θ between u and v** to be

$$\theta = \cos^{-1}\left(\frac{\langle \mathbf{u}, \mathbf{v} \rangle}{\|\mathbf{u}\| \|\mathbf{v}\|} \right) \tag{8}$$

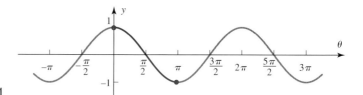

▶ Figure 6.2.1

▶ **EXAMPLE 1** **Cosine of an Angle Between Two Vectors in R^4**

Let R^4 have the Euclidean inner product. Find the cosine of the angle θ between the vectors $\mathbf{u} = (4, 3, 1, -2)$ and $\mathbf{v} = (-2, 1, 2, 3)$.

Solution We leave it for you to verify that

$$\|\mathbf{u}\| = \sqrt{30}, \quad \|\mathbf{v}\| = \sqrt{18}, \quad \text{and} \quad \langle \mathbf{u}, \mathbf{v} \rangle = -9$$

from which it follows that

$$\cos \theta = \frac{\langle \mathbf{u}, \mathbf{v} \rangle}{\|\mathbf{u}\| \|\mathbf{v}\|} = -\frac{9}{\sqrt{30}\sqrt{18}} = -\frac{3}{2\sqrt{15}} \blacktriangleleft$$

Properties of Length and Distance in General Inner Product Spaces

In Section 3.2 we used the dot product to extend the notions of length and distance to R^n, and we showed that various familiar theorems remained valid (see Theorems 3.2.5, 3.2.6, and 3.2.7). By making only minor adjustments to the proofs of those theorems, we can show that they remain valid in any real inner product space. For example, here is the generalization of Theorem 3.2.5 (the triangle inequalities).

THEOREM 6.2.2 *If $\mathbf{u}$, $\mathbf{v}$, and $\mathbf{w}$ are vectors in a real inner product space V, and if k is any scalar, then*:

(a) $\|\mathbf{u} + \mathbf{v}\| \leq \|\mathbf{u}\| + \|\mathbf{v}\|$ [Triangle inequality for vectors]

(b) $d(\mathbf{u}, \mathbf{v}) \leq d(\mathbf{u}, \mathbf{w}) + d(\mathbf{w}, \mathbf{v})$ [Triangle inequality for distances]

Proof (a)

$$\begin{aligned}
\|\mathbf{u}+\mathbf{v}\|^2 &= \langle \mathbf{u}+\mathbf{v}, \mathbf{u}+\mathbf{v}\rangle \\
&= \langle \mathbf{u},\mathbf{u}\rangle + 2\langle \mathbf{u},\mathbf{v}\rangle + \langle \mathbf{v},\mathbf{v}\rangle \\
&\leq \langle \mathbf{u},\mathbf{u}\rangle + 2|\langle \mathbf{u},\mathbf{v}\rangle| + \langle \mathbf{v},\mathbf{v}\rangle \quad \text{[Property of absolute value]}\\
&\leq \langle \mathbf{u},\mathbf{u}\rangle + 2\|\mathbf{u}\|\|\mathbf{v}\| + \langle \mathbf{v},\mathbf{v}\rangle \quad \text{[By (3)]}\\
&= \|\mathbf{u}\|^2 + 2\|\mathbf{u}\|\|\mathbf{v}\| + \|\mathbf{v}\|^2 \\
&= (\|\mathbf{u}\| + \|\mathbf{v}\|)^2
\end{aligned}$$

Taking square roots gives $\|\mathbf{u}+\mathbf{v}\| \leq \|\mathbf{u}\| + \|\mathbf{v}\|$.

Proof (b) Identical to the proof of part (*b*) of Theorem 3.2.5. ◀

Orthogonality

Although Example 1 is a useful mathematical exercise, there is only an occasional need to compute angles in vector spaces other than R^2 and R^3. A problem of more interest in general vector spaces is ascertaining whether the angle between vectors is $\pi/2$. You should be able to see from Formula (8) that if **u** and **v** are *nonzero* vectors, then the angle between them is $\theta = \pi/2$ if and only if $\langle \mathbf{u},\mathbf{v}\rangle = 0$. Accordingly, we make the following definition (which is applicable even if one or both of the vectors is zero).

> **DEFINITION 1** Two vectors **u** and **v** in an inner product space are called ***orthogonal*** if $\langle \mathbf{u},\mathbf{v}\rangle = 0$.

As the following example shows, orthogonality depends on the inner product in the sense that for different inner products two vectors can be orthogonal with respect to one but not the other.

▶ **EXAMPLE 2 Orthogonality Depends on the Inner Product**

The vectors $\mathbf{u}=(1,1)$ and $\mathbf{v}=(1,-1)$ are orthogonal with respect to the Euclidean inner product on R^2, since

$$\mathbf{u}\cdot\mathbf{v} = (1)(1) + (1)(-1) = 0$$

However, they are not orthogonal with respect to the weighted Euclidean inner product $\langle \mathbf{u},\mathbf{v}\rangle = 3u_1v_1 + 2u_2v_2$, since

$$\langle \mathbf{u},\mathbf{v}\rangle = 3(1)(1) + 2(1)(-1) = 1 \neq 0$$

▶ **EXAMPLE 3 Orthogonal Vectors in M_{22}**

If M_{22} has the inner product of Example 6 in the preceding section, then the matrices

$$U = \begin{bmatrix} 1 & 0 \\ 1 & 1 \end{bmatrix} \quad \text{and} \quad V = \begin{bmatrix} 0 & 2 \\ 0 & 0 \end{bmatrix}$$

are orthogonal, since

$$\langle U, V\rangle = 1(0) + 0(2) + 1(0) + 1(0) = 0$$

CALCULUS REQUIRED

▶ **EXAMPLE 4 Orthogonal Vectors in P_2**

Let P_2 have the inner product

$$\langle \mathbf{p},\mathbf{q}\rangle = \int_{-1}^{1} p(x)q(x)\,dx$$

and let $\mathbf{p} = x$ and $\mathbf{q} = x^2$. Then

$$\|\mathbf{p}\| = \langle \mathbf{p}, \mathbf{p} \rangle^{1/2} = \left[\int_{-1}^{1} xx \, dx \right]^{1/2} = \left[\int_{-1}^{1} x^2 \, dx \right]^{1/2} = \sqrt{\frac{2}{3}}$$

$$\|\mathbf{q}\| = \langle \mathbf{q}, \mathbf{q} \rangle^{1/2} = \left[\int_{-1}^{1} x^2 x^2 \, dx \right]^{1/2} = \left[\int_{-1}^{1} x^4 \, dx \right]^{1/2} = \sqrt{\frac{2}{5}}$$

$$\langle \mathbf{p}, \mathbf{q} \rangle = \int_{-1}^{1} xx^2 \, dx = \int_{-1}^{1} x^3 \, dx = 0$$

Because $\langle \mathbf{p}, \mathbf{q} \rangle = 0$, the vectors $\mathbf{p} = x$ and $\mathbf{q} = x^2$ are orthogonal relative to the given inner product. ◀

In Section 3.3 we proved the Theorem of Pythagoras for vectors in Euclidean n-space. The following theorem extends this result to vectors in any real inner product space.

THEOREM 6.2.3 Generalized Theorem of Pythagoras

If $\mathbf{u}$ and $\mathbf{v}$ are orthogonal vectors in an inner product space, then

$$\|\mathbf{u} + \mathbf{v}\|^2 = \|\mathbf{u}\|^2 + \|\mathbf{v}\|^2$$

Proof The orthogonality of $\mathbf{u}$ and $\mathbf{v}$ implies that $\langle \mathbf{u}, \mathbf{v} \rangle = 0$, so

$$\|\mathbf{u} + \mathbf{v}\|^2 = \langle \mathbf{u} + \mathbf{v}, \mathbf{u} + \mathbf{v} \rangle = \|\mathbf{u}\|^2 + 2\langle \mathbf{u}, \mathbf{v} \rangle + \|\mathbf{v}\|^2$$
$$= \|\mathbf{u}\|^2 + \|\mathbf{v}\|^2 \quad \blacktriangleleft$$

CALCULUS REQUIRED

▶ **EXAMPLE 5** Theorem of Pythagoras in P_2

In Example 4 we showed that $\mathbf{p} = x$ and $\mathbf{q} = x^2$ are orthogonal with respect to the inner product

$$\langle \mathbf{p}, \mathbf{q} \rangle = \int_{-1}^{1} p(x)q(x) \, dx$$

on P_2. It follows from Theorem 6.2.3 that

$$\|\mathbf{p} + \mathbf{q}\|^2 = \|\mathbf{p}\|^2 + \|\mathbf{q}\|^2$$

Thus, from the computations in Example 4, we have

$$\|\mathbf{p} + \mathbf{q}\|^2 = \left(\sqrt{\frac{2}{3}} \right)^2 + \left(\sqrt{\frac{2}{5}} \right)^2 = \frac{2}{3} + \frac{2}{5} = \frac{16}{15}$$

We can check this result by direct integration:

$$\|\mathbf{p} + \mathbf{q}\|^2 = \langle \mathbf{p} + \mathbf{q}, \mathbf{p} + \mathbf{q} \rangle = \int_{-1}^{1} (x + x^2)(x + x^2) \, dx$$
$$= \int_{-1}^{1} x^2 \, dx + 2 \int_{-1}^{1} x^3 \, dx + \int_{-1}^{1} x^4 \, dx = \frac{2}{3} + 0 + \frac{2}{5} = \frac{16}{15} \quad \blacktriangleleft$$

Orthogonal Complements   In Section 4.8 we defined the notion of an *orthogonal complement* for subspaces of R^n, and we used that definition to establish a geometric link between the fundamental spaces of a matrix. The following definition extends that idea to general inner product spaces.

6.2 Angle and Orthogonality in Inner Product Spaces

DEFINITION 2 If W is a subspace of an inner product space V, then the set of all vectors in V that are orthogonal to every vector in W is called the **orthogonal complement** of W and is denoted by the symbol $W^\perp$.

In Theorem 4.8.8 we stated three properties of orthogonal complements in R^n. The following theorem generalizes parts (a) and (b) of that theorem to general inner product spaces.

THEOREM 6.2.4 *If W is a subspace of an inner product space V, then:*

(a) $W^\perp$ *is a subspace of V.*

(b) $W \cap W^\perp = \{\mathbf{0}\}$.

Proof (a) The set $W^\perp$ contains at least the zero vector, since $\langle \mathbf{0}, \mathbf{w} \rangle = 0$ for every vector $\mathbf{w}$ in W. Thus, it remains to show that $W^\perp$ is closed under addition and scalar multiplication. To do this, suppose that $\mathbf{u}$ and $\mathbf{v}$ are vectors in $W^\perp$, so that for every vector $\mathbf{w}$ in W we have $\langle \mathbf{u}, \mathbf{w} \rangle = 0$ and $\langle \mathbf{v}, \mathbf{w} \rangle = 0$. It follows from the additivity and homogeneity axioms of inner products that

$$\langle \mathbf{u} + \mathbf{v}, \mathbf{w} \rangle = \langle \mathbf{u}, \mathbf{w} \rangle + \langle \mathbf{v}, \mathbf{w} \rangle = 0 + 0 = 0$$
$$\langle k\mathbf{u}, \mathbf{w} \rangle = k\langle \mathbf{u}, \mathbf{w} \rangle = k(0) = 0$$

which proves that $\mathbf{u} + \mathbf{v}$ and $k\mathbf{u}$ are in $W^\perp$.

Proof (b) If $\mathbf{v}$ is any vector in both W and $W^\perp$, then $\mathbf{v}$ is orthogonal to itself; that is, $\langle \mathbf{v}, \mathbf{v} \rangle = 0$. It follows from the positivity axiom for inner products that $\mathbf{v} = \mathbf{0}$. ◄

The next theorem, which we state without proof, generalizes part (c) of Theorem 4.8.8. Note, however, that this theorem applies only to finite-dimensional inner product spaces, whereas Theorem 6.2.5 does not have this restriction.

Theorem 6.2.5 implies that in a finite-dimensional inner product space orthogonal complements occur in pairs, each being orthogonal to the other (Figure 6.2.2).

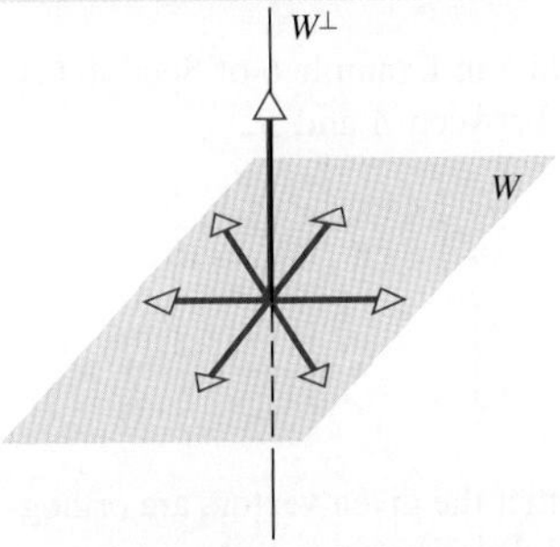

THEOREM 6.2.5 *If W is a subspace of a finite-dimensional inner product space V, then the orthogonal complement of $W^\perp$ is W; that is,*

$$(W^\perp)^\perp = W$$

In our study of the fundamental spaces of a matrix in Section 4.8 we showed that the row space and null space of a matrix are orthogonal complements with respect to the Euclidean inner product on R^n (Theorem 4.8.9). The following example takes advantage of that fact.

▲ Figure 6.2.2 Each vector in W is orthogonal to each vector in $W^\perp$ and conversely.

► **EXAMPLE 6** Basis for an Orthogonal Complement

Let W be the subspace of R^6 spanned by the vectors

$$\mathbf{w}_1 = (1, 3, -2, 0, 2, 0), \quad \mathbf{w}_2 = (2, 6, -5, -2, 4, -3),$$
$$\mathbf{w}_3 = (0, 0, 5, 10, 0, 15), \quad \mathbf{w}_4 = (2, 6, 0, 8, 4, 18)$$

Find a basis for the orthogonal complement of W.

Solution The space W is the same as the row space of the matrix

$$A = \begin{bmatrix} 1 & 3 & -2 & 0 & 2 & 0 \\ 2 & 6 & -5 & -2 & 4 & -3 \\ 0 & 0 & 5 & 10 & 0 & 15 \\ 2 & 6 & 0 & 8 & 4 & 18 \end{bmatrix}$$

Since the row space and null space of A are orthogonal complements, our problem reduces to finding a basis for the null space of this matrix. In Example 4 of Section 4.7 we showed that

$$\mathbf{v}_1 = \begin{bmatrix} -3 \\ 1 \\ 0 \\ 0 \\ 0 \\ 0 \end{bmatrix}, \quad \mathbf{v}_2 = \begin{bmatrix} -4 \\ 0 \\ -2 \\ 1 \\ 0 \\ 0 \end{bmatrix}, \quad \mathbf{v}_3 = \begin{bmatrix} -2 \\ 0 \\ 0 \\ 0 \\ 1 \\ 0 \end{bmatrix}$$

form a basis for this null space. Expressing these vectors in comma-delimited form (to match that of $\mathbf{w}_1$, $\mathbf{w}_2$, $\mathbf{w}_3$, and $\mathbf{w}_4$), we obtain the basis vectors

$$\mathbf{v}_1 = (-3, 1, 0, 0, 0, 0), \quad \mathbf{v}_2 = (-4, 0, -2, 1, 0, 0), \quad \mathbf{v}_3 = (-2, 0, 0, 0, 1, 0)$$

You may want to check that these vectors are orthogonal to $\mathbf{w}_1$, $\mathbf{w}_2$, $\mathbf{w}_3$, and $\mathbf{w}_4$ by computing the necessary dot products. ◀

Concept Review

- Cauchy–Schwarz inequality
- Angle between vectors
- Orthogonal vectors
- Orthogonal complement

Skills

- Find the angle between two vectors in an inner product space.
- Determine whether two vectors in an inner product space are orthogonal.
- Find a basis for the orthogonal complement of a subspace of an inner product space.

Exercise Set 6.2

1. Let R^2, R^3, and R^4 have the Euclidean inner product. In each part, find the cosine of the angle between $\mathbf{u}$ and $\mathbf{v}$.
   (a) $\mathbf{u} = (-2, 1)$, $\mathbf{v} = (3, 1)$
   (b) $\mathbf{u} = (-1, 0)$, $\mathbf{v} = (3, 8)$
   (c) $\mathbf{u} = (1, 2, 3)$, $\mathbf{v} = (4, 4, -4)$
   (d) $\mathbf{u} = (4, 1, 8)$, $\mathbf{v} = (1, 0, -3)$
   (e) $\mathbf{u} = (0, -1, 1, 0)$, $\mathbf{v} = (3, -3, 3, -3)$
   (f) $\mathbf{u} = (2, 1, 7, -1)$, $\mathbf{v} = (4, 0, 0, 0)$

2. Let P_2 have the inner product in Example 7 of Section 6.1. Find the cosine of the angle between $\mathbf{p}$ and $\mathbf{q}$.
   (a) $\mathbf{p} = -1 + 5x + 2x^2$, $\mathbf{q} = 2 + 4x - 9x^2$
   (b) $\mathbf{p} = x - x^2$, $\mathbf{q} = 7 + 3x + 3x^2$

3. Let M_{22} have the inner product in Example 6 of Section 6.1. Find the cosine of the angle between A and B.
   (a) $A = \begin{bmatrix} 2 & 4 \\ 4 & 2 \end{bmatrix}$, $B = \begin{bmatrix} 0 & 1 \\ 0 & -2 \end{bmatrix}$
   (b) $A = \begin{bmatrix} 1 & 2 \\ 3 & 1 \end{bmatrix}$, $B = \begin{bmatrix} 1 & 0 \\ 2 & 2 \end{bmatrix}$

4. In each part, determine whether the given vectors are orthogonal with respect to the Euclidean inner product.
   (a) $\mathbf{u} = (2, 1, 3)$, $\mathbf{v} = (-1, -4, 2)$
   (b) $\mathbf{u} = (0, 1, -3)$, $\mathbf{v} = (3, -1, 0)$
   (c) $\mathbf{u} = (u_1, u_2, u_3)$, $\mathbf{v} = (0, 0, 0)$
   (d) $\mathbf{u} = (4, 1, 2, 1)$, $\mathbf{v} = (-1, 1, 1, 1)$

(e) $\mathbf{u} = (1, 5, 2, -2)$, $\mathbf{v} = (-2, 1, 0, 3)$

(f) $\mathbf{u} = (a, b)$, $\mathbf{v} = (-b, a)$

5. Show that $\mathbf{p} = 1 - x + 2x^2$ and $\mathbf{q} = 2x + x^2$ are orthogonal with respect to the inner product in Exercise 2.

6. Let
$$A = \begin{bmatrix} 2 & 1 \\ -1 & 3 \end{bmatrix}$$

Which of the following matrices are orthogonal to A with respect to the inner product in Exercise 3?

(a) $\begin{bmatrix} -3 & 0 \\ 0 & 2 \end{bmatrix}$
(b) $\begin{bmatrix} 1 & 1 \\ 0 & -1 \end{bmatrix}$

(c) $\begin{bmatrix} 0 & 0 \\ 0 & 0 \end{bmatrix}$
(d) $\begin{bmatrix} 2 & 1 \\ 5 & 2 \end{bmatrix}$

7. Do there exist scalars k and l such that the vectors $\mathbf{u} = (k, 3, 2)$, $\mathbf{v} = (-3, 1, l)$, and $\mathbf{w} = (-5, 5, 1)$ are mutually orthogonal with respect to the Euclidean inner product?

8. Let R^3 have the Euclidean inner product, and suppose that $\mathbf{u} = (1, 1, -1)$ and $\mathbf{v} = (6, 7, -15)$. Find a value of k for which $\|k\mathbf{u} + \mathbf{v}\| = 13$.

9. Let R^3 have the Euclidean inner product. For which values of k are $\mathbf{u}$ and $\mathbf{v}$ orthogonal?

(a) $\mathbf{u} = (1, 4, 2)$, $\mathbf{v} = (3, -2, k)$

(b) $\mathbf{u} = (k, -2, 4)$, $\mathbf{v} = (k, k, -2)$

10. Let R^4 have the Euclidean inner product. Find two unit vectors that are orthogonal to all three of the vectors $\mathbf{u} = (2, 1, -4, 0)$, $\mathbf{v} = (-1, -1, 2, 2)$, and $\mathbf{w} = (3, 2, 5, 4)$.

11. In each part, verify that the Cauchy–Schwarz inequality holds for the given vectors using the Euclidean inner product.

(a) $\mathbf{u} = (3, 2)$, $\mathbf{v} = (4, -1)$

(b) $\mathbf{u} = (-3, 1, 0)$, $\mathbf{v} = (2, -1, 3)$

(c) $\mathbf{u} = (-4, 2, 1)$, $\mathbf{v} = (8, -4, -2)$

(d) $\mathbf{u} = (0, -2, 2, 1)$, $\mathbf{v} = (-1, -1, 1, 1)$

12. In each part, verify that the Cauchy–Schwarz inequality holds for the given vectors.

(a) $\mathbf{u} = (-2, 1)$ and $\mathbf{v} = (1, 0)$ using the inner product of Example 1 of Section 6.1.

(b) $U = \begin{bmatrix} -1 & 2 \\ 6 & 1 \end{bmatrix}$ and $V = \begin{bmatrix} 1 & 0 \\ 3 & 3 \end{bmatrix}$

using the inner product in Example 6 of Section 6.1.

(c) $\mathbf{p} = -1 + 2x + x^2$ and $\mathbf{q} = 2 - 4x^2$ using the inner product given in Example 7 of Section 6.1.

13. Let R^4 have the Euclidean inner product, and let $\mathbf{u} = (-1, 1, 0, 2)$. Determine whether the vector $\mathbf{u}$ is orthogonal to the subspace spanned by the vectors $\mathbf{w}_1 = (0, 0, 0, 0)$, $\mathbf{w}_2 = (1, -1, 3, 0)$, and $\mathbf{w}_3 = (4, 0, 9, 2)$.

▶ In Exercises 14–15, assume that R^n has the Euclidean inner product. ◀

14. Let W be the line in R^2 with equation $y = -x/3$. Find an equation for $W^\perp$.

15. (a) Let W be the plane in R^3 with equation $2x - y - 4z = 0$. Find parametric equations for $W^\perp$.

(b) Let W be the line in R^3 with parametric equations
$$x = 2t, \quad y = -5t, \quad z = 4t$$
Find an equation for $W^\perp$.

(c) Let W be the intersection of the two planes
$$x - y - z = 0 \quad \text{and} \quad x + 2y + z = 0$$
in R^3. Find an equation for $W^\perp$.

16. Find a basis for the orthogonal complement of the subspace of R^n spanned by the vectors.

(a) $\mathbf{v}_1 = (2, 1, 3)$, $\mathbf{v}_2 = (-1, -4, 2)$, $\mathbf{v}_3 = (4, -5, 13)$

(b) $\mathbf{v}_1 = (0, 2, 1)$, $\mathbf{v}_2 = (4, 0, -3)$, $\mathbf{v}_3 = (6, 1, -4)$

(c) $\mathbf{v}_1 = (3, 0, 1, -2)$, $\mathbf{v}_2 = (-1, -2, -2, 1)$, $\mathbf{v}_3 = (4, 2, 3, -3)$

(d) $\mathbf{v}_1 = (1, 4, 5, 6, 9)$, $\mathbf{v}_2 = (3, -2, 1, 4, -1)$, $\mathbf{v}_3 = (-1, 0, -1, -2, -1)$, $\mathbf{v}_4 = (2, 3, 5, 7, 8)$

17. Let V be an inner product space. Show that if $\mathbf{u}$ and $\mathbf{v}$ are orthogonal unit vectors in V, then $\|\mathbf{u} - \mathbf{v}\| = \sqrt{2}$.

18. Let V be an inner product space. Show that if $\mathbf{w}$ is orthogonal to both $\mathbf{u}_1$ and $\mathbf{u}_2$, then it is orthogonal to $k_1 \mathbf{u}_1 + k_2 \mathbf{u}_2$ for all scalars k_1 and k_2. Interpret this result geometrically in the case where V is R^3 with the Euclidean inner product.

19. Let V be an inner product space. Show that if $\mathbf{w}$ is orthogonal to each of the vectors $\mathbf{u}_1, \mathbf{u}_2, \ldots, \mathbf{u}_r$, then it is orthogonal to every vector in $\text{span}\{\mathbf{u}_1, \mathbf{u}_2, \ldots, \mathbf{u}_r\}$.

20. Let $\{\mathbf{v}_1, \mathbf{v}_2, \ldots, \mathbf{v}_r\}$ be a basis for an inner product space V. Show that the zero vector is the only vector in V that is orthogonal to all of the basis vectors.

21. Let $\{\mathbf{w}_1, \mathbf{w}_2, \ldots, \mathbf{w}_k\}$ be a basis for a subspace W of V. Show that $W^\perp$ consists of all vectors in V that are orthogonal to every basis vector.

22. Prove the following generalization of Theorem 6.2.3: If $\mathbf{v}_1, \mathbf{v}_2, \ldots, \mathbf{v}_r$ are pairwise orthogonal vectors in an inner product space V, then
$$\|\mathbf{v}_1 + \mathbf{v}_2 + \cdots + \mathbf{v}_r\|^2 = \|\mathbf{v}_1\|^2 + \|\mathbf{v}_2\|^2 + \cdots + \|\mathbf{v}_r\|^2$$

23. Prove: If $\mathbf{u}$ and $\mathbf{v}$ are $n \times 1$ matrices and A is an $n \times n$ matrix, then
$$(\mathbf{v}^T A^T A \mathbf{u})^2 \leq (\mathbf{u}^T A^T A \mathbf{u})(\mathbf{v}^T A^T A \mathbf{v})$$

24. Use the Cauchy–Schwarz inequality to prove that for all real values of a, b, and θ,
$$(a \cos \theta + b \sin \theta)^2 \leq a^2 + b^2$$

25. Prove: If $w_1, w_2, \ldots, w_n$ are positive real numbers, and if $\mathbf{u} = (u_1, u_2, \ldots, u_n)$ and $\mathbf{v} = (v_1, v_2, \ldots, v_n)$ are any two vectors in R^n, then

$$|w_1 u_1 v_1 + w_2 u_2 v_2 + \cdots + w_n u_n v_n|$$
$$\leq (w_1 u_1^2 + w_2 u_2^2 + \cdots + w_n u_n^2)^{1/2} (w_1 v_1^2 + w_2 v_2^2 + \cdots + w_n v_n^2)^{1/2}$$

26. Show that equality holds in the Cauchy–Schwarz inequality if and only if $\mathbf{u}$ and $\mathbf{v}$ are linearly dependent.

27. Use vector methods to prove that a triangle that is inscribed in a circle so that it has a diameter for a side must be a right triangle. [*Hint:* Express the vectors $\overrightarrow{AB}$ and $\overrightarrow{BC}$ in the accompanying figure in terms of $\mathbf{u}$ and $\mathbf{v}$.]

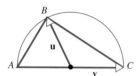

◀ Figure Ex-27

28. As illustrated in the accompanying figure, the vectors $\mathbf{u} = (1, \sqrt{3})$ and $\mathbf{v} = (-1, \sqrt{3})$ have norm 2 and an angle of 60° between them relative to the Euclidean inner product. Find a weighted Euclidean inner product with respect to which $\mathbf{u}$ and $\mathbf{v}$ are orthogonal unit vectors.

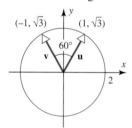

◀ Figure Ex-28

29. (*Calculus required*) Let $f(x)$ and $g(x)$ be continuous functions on $[0, 1]$. Prove:

(a) $\left[\int_0^1 f(x) g(x) \, dx \right]^2 \leq \left[\int_0^1 f^2(x) \, dx \right] \left[\int_0^1 g^2(x) \, dx \right]$

(b) $\left[\int_0^1 [f(x) + g(x)]^2 \, dx \right]^{1/2} \leq \left[\int_0^1 f^2(x) \, dx \right]^{1/2} + \left[\int_0^1 g^2(x) \, dx \right]^{1/2}$

[*Hint:* Use the Cauchy–Schwarz inequality.]

30. (*Calculus required*) Let $C[0, \pi]$ have the inner product

$$\langle \mathbf{f}, \mathbf{g} \rangle = \int_0^\pi f(x) g(x) \, dx$$

and let $\mathbf{f}_n = \cos nx$ ($n = 0, 1, 2, \ldots$). Show that if $k \neq l$, then $\mathbf{f}_k$ and $\mathbf{f}_l$ are orthogonal vectors.

31. (a) Let W be the line $y = x$ in an xy-coordinate system in R^2. Describe the subspace $W^\perp$.

(b) Let W be the y-axis in an xyz-coordinate system in R^3. Describe the subspace $W^\perp$.

(c) Let W be the yz-plane of an xyz-coordinate system in R^3. Describe the subspace $W^\perp$.

32. Prove that Formula (4) holds for all nonzero vectors $\mathbf{u}$ and $\mathbf{v}$ in an inner product space V.

True-False Exercises

In parts (a)–(f) determine whether the statement is true or false, and justify your answer.

(a) If $\mathbf{u}$ is orthogonal to every vector of a subspace W, then $\mathbf{u} = \mathbf{0}$.

(b) If $\mathbf{u}$ is a vector in both W and $W^\perp$, then $\mathbf{u} = \mathbf{0}$.

(c) If $\mathbf{u}$ and $\mathbf{v}$ are vectors in $W^\perp$, then $\mathbf{u} + \mathbf{v}$ is in $W^\perp$.

(d) If $\mathbf{u}$ is a vector in $W^\perp$ and k is a real number, then $k\mathbf{u}$ is in $W^\perp$.

(e) If $\mathbf{u}$ and $\mathbf{v}$ are orthogonal, then $|\langle \mathbf{u}, \mathbf{v} \rangle| = \|\mathbf{u}\| \|\mathbf{v}\|$.

(f) If $\mathbf{u}$ and $\mathbf{v}$ are orthogonal, then $\|\mathbf{u} + \mathbf{v}\| = \|\mathbf{u}\| + \|\mathbf{v}\|$.

6.3 Gram–Schmidt Process; QR-Decomposition

In many problems involving vector spaces, the problem solver is free to choose any basis for the vector space that seems appropriate. In inner product spaces, the solution of a problem is often greatly simplified by choosing a basis in which the vectors are orthogonal to one another. In this section we will show how such bases can be obtained.

Orthogonal and Orthonormal Sets

Recall from Section 6.2 that two vectors in an inner product space are said to be *orthogonal* if their inner product is zero. The following definition extends the notion of orthogonality to sets of vectors in an inner product space.

DEFINITION 1 A set of two or more vectors in a real inner product space is said to be ***orthogonal*** if all pairs of distinct vectors in the set are orthogonal. An orthogonal set in which each vector has norm 1 is said to be ***orthonormal***.

▶ **EXAMPLE 1** An Orthogonal Set in R^3

Let
$$\mathbf{u}_1 = (0, 1, 0), \quad \mathbf{u}_2 = (1, 0, 1), \quad \mathbf{u}_3 = (1, 0, -1)$$
and assume that R^3 has the Euclidean inner product. It follows that the set of vectors $S = \{\mathbf{u}_1, \mathbf{u}_2, \mathbf{u}_3\}$ is orthogonal since $\langle \mathbf{u}_1, \mathbf{u}_2 \rangle = \langle \mathbf{u}_1, \mathbf{u}_3 \rangle = \langle \mathbf{u}_2, \mathbf{u}_3 \rangle = 0$. ◀

If $\mathbf{v}$ is a nonzero vector in an inner product space, then it follows from Theorem 6.1.1b with $k = \|\mathbf{v}\|$ that
$$\left\| \frac{1}{\|\mathbf{v}\|} \mathbf{v} \right\| = \left| \frac{1}{\|\mathbf{v}\|} \right| \|\mathbf{v}\| = \frac{1}{\|\mathbf{v}\|} \|\mathbf{v}\| = 1$$
from which we see that multiplying a nonzero vector by the reciprocal of its norm produces a vector of norm 1. This process is called ***normalizing*** $\mathbf{v}$. It follows that *any orthogonal set of nonzero vectors can be converted to an orthonormal set by normalizing each of its vectors.*

▶ **EXAMPLE 2** Constructing an Orthonormal Set

The Euclidean norms of the vectors in Example 1 are
$$\|\mathbf{u}_1\| = 1, \quad \|\mathbf{u}_2\| = \sqrt{2}, \quad \|\mathbf{u}_3\| = \sqrt{2}$$
Consequently, normalizing $\mathbf{u}_1$, $\mathbf{u}_2$, and $\mathbf{u}_3$ yields
$$\mathbf{v}_1 = \frac{\mathbf{u}_1}{\|\mathbf{u}_1\|} = (0, 1, 0), \quad \mathbf{v}_2 = \frac{\mathbf{u}_2}{\|\mathbf{u}_2\|} = \left(\frac{1}{\sqrt{2}}, 0, \frac{1}{\sqrt{2}} \right),$$
$$\mathbf{v}_3 = \frac{\mathbf{u}_3}{\|\mathbf{u}_3\|} = \left(\frac{1}{\sqrt{2}}, 0, -\frac{1}{\sqrt{2}} \right)$$
We leave it for you to verify that the set $S = \{\mathbf{v}_1, \mathbf{v}_2, \mathbf{v}_3\}$ is orthonormal by showing that
$$\langle \mathbf{v}_1, \mathbf{v}_2 \rangle = \langle \mathbf{v}_1, \mathbf{v}_3 \rangle = \langle \mathbf{v}_2, \mathbf{v}_3 \rangle = 0 \quad \text{and} \quad \|\mathbf{v}_1\| = \|\mathbf{v}_2\| = \|\mathbf{v}_3\| = 1 \quad ◀$$

In R^2 any two nonzero perpendicular vectors are linearly independent because neither is a scalar multiple of the other; and in R^3 any three nonzero mutually perpendicular vectors are linearly independent because no one lies in the plane of the other two (and hence is not expressible as a linear combination of the other two). The following theorem generalizes these observations.

THEOREM 6.3.1 *If $S = \{\mathbf{v}_1, \mathbf{v}_2, \ldots, \mathbf{v}_n\}$ is an orthogonal set of nonzero vectors in an inner product space, then S is linearly independent.*

Proof Assume that
$$k_1 \mathbf{v}_1 + k_2 \mathbf{v}_2 + \cdots + k_n \mathbf{v}_n = \mathbf{0} \tag{1}$$
To demonstrate that $S = \{\mathbf{v}_1, \mathbf{v}_2, \ldots, \mathbf{v}_n\}$ is linearly independent, we must prove that $k_1 = k_2 = \cdots = k_n = 0$.

For each v_i in S, it follows from (1) that

$$\langle k_1 \mathbf{v}_1 + k_2 \mathbf{v}_2 + \cdots + k_n \mathbf{v}_n, \mathbf{v}_i \rangle = \langle \mathbf{0}, \mathbf{v}_i \rangle = 0$$

or, equivalently,

$$k_1 \langle \mathbf{v}_1, \mathbf{v}_i \rangle + k_2 \langle \mathbf{v}_2, \mathbf{v}_i \rangle + \cdots + k_n \langle \mathbf{v}_n, \mathbf{v}_i \rangle = 0$$

From the orthogonality of S it follows that $\langle \mathbf{v}_j, \mathbf{v}_i \rangle = 0$ when $j \neq i$, so this equation reduces to

$$k_i \langle \mathbf{v}_i, \mathbf{v}_i \rangle = 0$$

Since the vectors in S are assumed to be nonzero, it follows from the positivity axiom for inner products that $\langle \mathbf{v}_i, \mathbf{v}_i \rangle \neq 0$. Thus, the preceding equation implies that each k_i in Equation (1) is zero, which is what we wanted to prove. ◄

In an inner product space, a basis consisting of orthonormal vectors is called an ***orthonormal basis***, and a basis consisting of orthogonal vectors is called an ***orthogonal basis***. A familiar example of an orthonormal basis is the standard basis for R^n with the Euclidean inner product:

$$\mathbf{e}_1 = (1, 0, 0, \ldots, 0), \quad \mathbf{e}_2 = (0, 1, 0, \ldots, 0), \ldots, \quad \mathbf{e}_n = (0, 0, 0, \ldots, 1)$$

Since an orthonormal set is orthogonal, and since its vectors are nonzero (norm 1), it follows from Theorem 6.3.1 that every *orthonormal* set is linearly independent.

▶ **EXAMPLE 3** **An Orthonormal Basis**

In Example 2 we showed that the vectors

$$\mathbf{v}_1 = (0, 1, 0), \quad \mathbf{v}_2 = \left(\frac{1}{\sqrt{2}}, 0, \frac{1}{\sqrt{2}} \right), \quad \text{and} \quad \mathbf{v}_3 = \left(\frac{1}{\sqrt{2}}, 0, -\frac{1}{\sqrt{2}} \right)$$

form an orthonormal set with respect to the Euclidean inner product on R^3. By Theorem 6.3.1, these vectors form a linearly independent set, and since R^3 is three-dimensional, it follows from Theorem 4.5.4 that $S = \{\mathbf{v}_1, \mathbf{v}_2, \mathbf{v}_3\}$ is an orthonormal basis for R^3. ◄

Coordinates Relative to Orthonormal Bases

One way to express a vector $\mathbf{u}$ as a linear combination of basis vectors

$$S = \{\mathbf{v}_1, \mathbf{v}_2, \ldots, \mathbf{v}_n\}$$

is to convert the vector equation

$$\mathbf{u} = c_1 \mathbf{v}_1 + c_2 \mathbf{v}_2 + \cdots + c_n \mathbf{v}_n$$

to a linear system and solve for the coefficients $c_1, c_2, \ldots, c_n$. However, if the basis happens to be orthogonal or orthonormal, then the following theorem shows that the coefficients can be obtained more simply by computing appropriate inner products.

THEOREM 6.3.2

(a) *If $S = \{\mathbf{v}_1, \mathbf{v}_2, \ldots, \mathbf{v}_n\}$ is an orthogonal basis for an inner product space V, and if $\mathbf{u}$ is any vector in V, then*

$$\mathbf{u} = \frac{\langle \mathbf{u}, \mathbf{v}_1 \rangle}{\|\mathbf{v}_1\|^2} \mathbf{v}_1 + \frac{\langle \mathbf{u}, \mathbf{v}_2 \rangle}{\|\mathbf{v}_2\|^2} \mathbf{v}_2 + \cdots + \frac{\langle \mathbf{u}, \mathbf{v}_n \rangle}{\|\mathbf{v}_n\|^2} \mathbf{v}_n \quad (2)$$

(b) *If $S = \{\mathbf{v}_1, \mathbf{v}_2, \ldots, \mathbf{v}_n\}$ is an orthonormal basis for an inner product space V, and if $\mathbf{u}$ is any vector in V, then*

$$\mathbf{u} = \langle \mathbf{u}, \mathbf{v}_1 \rangle \mathbf{v}_1 + \langle \mathbf{u}, \mathbf{v}_2 \rangle \mathbf{v}_2 + \cdots + \langle \mathbf{u}, \mathbf{v}_n \rangle \mathbf{v}_n \quad (3)$$

Proof (a) Since $S = \{\mathbf{v}_1, \mathbf{v}_2, \ldots, \mathbf{v}_n\}$ is a basis for V, every vector $\mathbf{u}$ in V can be expressed in the form

$$\mathbf{u} = c_1\mathbf{v}_1 + c_2\mathbf{v}_2 + \cdots + c_n\mathbf{v}_n$$

We will complete the proof by showing that

$$c_i = \frac{\langle \mathbf{u}, \mathbf{v}_i \rangle}{\|\mathbf{v}_i\|^2} \tag{4}$$

for $i = 1, 2, \ldots, n$. To do this, observe first that

$$\langle \mathbf{u}, \mathbf{v}_i \rangle = \langle c_1\mathbf{v}_1 + c_2\mathbf{v}_2 + \cdots + c_n\mathbf{v}_n, \mathbf{v}_i \rangle$$
$$= c_1\langle \mathbf{v}_1, \mathbf{v}_i \rangle + c_2\langle \mathbf{v}_2, \mathbf{v}_i \rangle + \cdots + c_n\langle \mathbf{v}_n, \mathbf{v}_i \rangle$$

Since S is an orthogonal set, all of the inner products in the last equality are zero except the ith, so we have

$$\langle \mathbf{u}, \mathbf{v}_i \rangle = c_i \langle \mathbf{v}_i, \mathbf{v}_i \rangle = c_i \|\mathbf{v}_i\|^2$$

Solving this equation for c_i yields (4), which completes the proof.

Proof (b) In this case, $\|\mathbf{v}_1\| = \|\mathbf{v}_2\| = \cdots = \|\mathbf{v}_n\| = 1$, so Formula (2) simplifies to Formula (3). ◂

Using the terminology and notation from Definition 2 of Section 4.4, it follows from Theorem 6.3.2 that the coordinate vector of a vector $\mathbf{u}$ in V relative to an orthogonal basis $S = \{\mathbf{v}_1, \mathbf{v}_2, \ldots, \mathbf{v}_n\}$ is

$$(\mathbf{u})_S = \left(\frac{\langle \mathbf{u}, \mathbf{v}_1 \rangle}{\|\mathbf{v}_1\|^2}, \frac{\langle \mathbf{u}, \mathbf{v}_2 \rangle}{\|\mathbf{v}_2\|^2}, \ldots, \frac{\langle \mathbf{u}, \mathbf{v}_n \rangle}{\|\mathbf{v}_n\|^2} \right) \tag{5}$$

and relative to an orthonormal basis $S = \{\mathbf{v}_1, \mathbf{v}_2, \ldots, \mathbf{v}_n\}$ is

$$(\mathbf{u})_S = (\langle \mathbf{u}, \mathbf{v}_1 \rangle, \langle \mathbf{u}, \mathbf{v}_2 \rangle, \ldots, \langle \mathbf{u}, \mathbf{v}_n \rangle) \tag{6}$$

▶ **EXAMPLE 4** **A Coordinate Vector Relative to an Orthonormal Basis**

Let

$$\mathbf{v}_1 = (0, 1, 0), \quad \mathbf{v}_2 = \left(-\tfrac{4}{5}, 0, \tfrac{3}{5}\right), \quad \mathbf{v}_3 = \left(\tfrac{3}{5}, 0, \tfrac{4}{5}\right)$$

It is easy to check that $S = \{\mathbf{v}_1, \mathbf{v}_2, \mathbf{v}_3\}$ is an orthonormal basis for R^3 with the Euclidean inner product. Express the vector $\mathbf{u} = (1, 1, 1)$ as a linear combination of the vectors in S, and find the coordinate vector $(\mathbf{u})_S$.

Solution We leave it for you to verify that

$$\langle \mathbf{u}, \mathbf{v}_1 \rangle = 1, \quad \langle \mathbf{u}, \mathbf{v}_2 \rangle = -\tfrac{1}{5}, \quad \text{and} \quad \langle \mathbf{u}, \mathbf{v}_3 \rangle = \tfrac{7}{5}$$

Therefore, by Theorem 6.3.2 we have

$$\mathbf{u} = \mathbf{v}_1 - \tfrac{1}{5}\mathbf{v}_2 + \tfrac{7}{5}\mathbf{v}_3$$

that is,

$$(1, 1, 1) = (0, 1, 0) - \tfrac{1}{5}\left(-\tfrac{4}{5}, 0, \tfrac{3}{5}\right) + \tfrac{7}{5}\left(\tfrac{3}{5}, 0, \tfrac{4}{5}\right)$$

Thus, the coordinate vector of $\mathbf{u}$ relative to S is

$$(\mathbf{u})_S = (\langle \mathbf{u}, \mathbf{v}_1 \rangle, \langle \mathbf{u}, \mathbf{v}_2 \rangle, \langle \mathbf{u}, \mathbf{v}_3 \rangle) = \left(1, -\tfrac{1}{5}, \tfrac{7}{5}\right)$$

▶ **EXAMPLE 5** **An Orthonormal Basis from an Orthogonal Basis**

(a) Show that the vectors
$$\mathbf{w}_1 = (0, 2, 0), \quad \mathbf{w}_2 = (3, 0, 3), \quad \mathbf{w}_3 = (-4, 0, 4)$$
form an orthogonal basis for R^3 with the Euclidean inner product, and use that basis to find an orthonormal basis by normalizing each vector.

(b) Express the vector $\mathbf{u} = (1, 2, 4)$ as a linear combination of the orthonormal basis vectors obtained in part (a).

Solution (a) The given vectors form an orthogonal set since
$$\langle \mathbf{w}_1, \mathbf{w}_2 \rangle = 0, \quad \langle \mathbf{w}_1, \mathbf{w}_3 \rangle = 0, \quad \langle \mathbf{w}_2, \mathbf{w}_3 \rangle = 0$$
It follows from Theorem 6.3.1 that these vectors are linearly independent and hence form a basis for R^3 by Theorem 4.5.4. We leave it for you to calculate the norms of $\mathbf{w}_1$, $\mathbf{w}_2$, and $\mathbf{w}_3$ and then obtain the orthonormal basis
$$\mathbf{v}_1 = \frac{\mathbf{w}_1}{\|\mathbf{w}_1\|} = (0, 1, 0), \quad \mathbf{v}_2 = \frac{\mathbf{w}_2}{\|\mathbf{w}_2\|} = \left(\frac{1}{\sqrt{2}}, 0, \frac{1}{\sqrt{2}}\right),$$
$$\mathbf{v}_3 = \frac{\mathbf{w}_3}{\|\mathbf{w}_3\|} = \left(-\frac{1}{\sqrt{2}}, 0, \frac{1}{\sqrt{2}}\right)$$

Solution (b) It follows from Formula (3) that
$$\mathbf{u} = \langle \mathbf{u}, \mathbf{v}_1 \rangle \mathbf{v}_1 + \langle \mathbf{u}, \mathbf{v}_2 \rangle \mathbf{v}_2 + \langle \mathbf{u}, \mathbf{v}_3 \rangle \mathbf{v}_3$$
We leave it for you to confirm that
$$\langle \mathbf{u}, \mathbf{v}_1 \rangle = (1, 2, 4) \cdot (0, 1, 0) = 2$$
$$\langle \mathbf{u}, \mathbf{v}_2 \rangle = (1, 2, 4) \cdot \left(\frac{1}{\sqrt{2}}, 0, \frac{1}{\sqrt{2}}\right) = \frac{5}{\sqrt{2}}$$
$$\langle \mathbf{u}, \mathbf{v}_3 \rangle = (1, 2, 4) \cdot \left(-\frac{1}{\sqrt{2}}, 0, \frac{1}{\sqrt{2}}\right) = \frac{3}{\sqrt{2}}$$
and hence that
$$(1, 2, 4) = 2(0, 1, 0) + \frac{5}{\sqrt{2}}\left(\frac{1}{\sqrt{2}}, 0, \frac{1}{\sqrt{2}}\right) + \frac{3}{\sqrt{2}}\left(-\frac{1}{\sqrt{2}}, 0, \frac{1}{\sqrt{2}}\right) \blacktriangleleft$$

Orthogonal Projections

Many applied problems are best solved by working with orthogonal or orthonormal basis vectors. Such bases are typically found by starting with some simple basis (say a standard basis) and then converting that basis into an orthogonal or orthonormal basis. To explain exactly how that is done will require some preliminary ideas about orthogonal projections.

In Section 3.3 we proved a result called the *Projection Theorem* (see Theorem 3.3.2) which dealt with the problem of decomposing a vector $\mathbf{u}$ in R^n into a sum of two terms, $\mathbf{w}_1$ and $\mathbf{w}_2$, in which $\mathbf{w}_1$ is the orthogonal projection of $\mathbf{u}$ on some nonzero vector $\mathbf{a}$ and $\mathbf{w}_2$ is orthogonal to $\mathbf{w}_1$ (Figure 3.3.2). That result is a special case of the following more general theorem.

THEOREM 6.3.3 **Projection Theorem**

If W is a finite-dimensional subspace of an inner product space V, then every vector $\mathbf{u}$ in V can be expressed in exactly one way as
$$\mathbf{u} = \mathbf{w}_1 + \mathbf{w}_2 \tag{7}$$
where $\mathbf{w}_1$ is in W and $\mathbf{w}_2$ is in $W^\perp$.

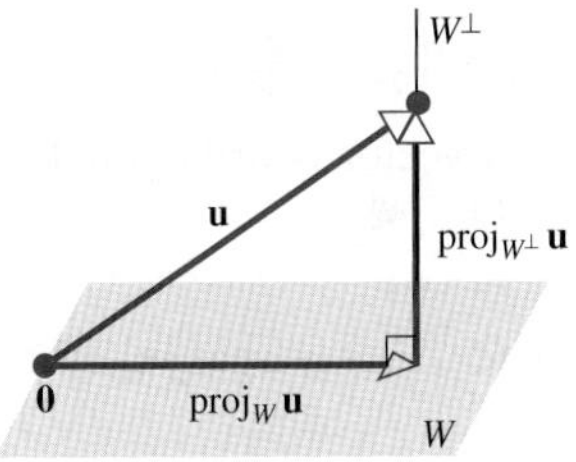

The vectors $\mathbf{w}_1$ and $\mathbf{w}_2$ in Formula (7) are commonly denoted by

$$\mathbf{w}_1 = \text{proj}_W \mathbf{u} \quad \text{and} \quad \mathbf{w}_2 = \text{proj}_{W^\perp} \mathbf{u} \tag{8}$$

They are called the *orthogonal projection of* $\mathbf{u}$ *on* W and the *orthogonal projection of* $\mathbf{u}$ *on* $W^\perp$, respectively. The vector $\mathbf{w}_2$ is also called the *component of* $\mathbf{u}$ *orthogonal to* W. Using the notation in (8), Formula (7) can be expressed as

$$\mathbf{u} = \text{proj}_W \mathbf{u} + \text{proj}_{W^\perp} \mathbf{u} \tag{9}$$

(Figure 6.3.1). Moreover, since $\text{proj}_{W^\perp} \mathbf{u} = \mathbf{u} - \text{proj}_W \mathbf{u}$, we can also express Formula (9) as

$$\mathbf{u} = \text{proj}_W \mathbf{u} + (\mathbf{u} - \text{proj}_W \mathbf{u}) \tag{10}$$

▲ Figure 6.3.1

The following theorem provides formulas for calculating orthogonal projections.

THEOREM 6.3.4 *Let W be a finite-dimensional subspace of an inner product space V.*
(a) *If $\{\mathbf{v}_1, \mathbf{v}_2, \ldots, \mathbf{v}_r\}$ is an orthogonal basis for W, and $\mathbf{u}$ is any vector in V, then*

$$\text{proj}_W \mathbf{u} = \frac{\langle \mathbf{u}, \mathbf{v}_1 \rangle}{\|\mathbf{v}_1\|^2} \mathbf{v}_1 + \frac{\langle \mathbf{u}, \mathbf{v}_2 \rangle}{\|\mathbf{v}_2\|^2} \mathbf{v}_2 + \cdots + \frac{\langle \mathbf{u}, \mathbf{v}_r \rangle}{\|\mathbf{v}_r\|^2} \mathbf{v}_r \tag{11}$$

(b) *If $\{\mathbf{v}_1, \mathbf{v}_2, \ldots, \mathbf{v}_r\}$ is an orthonormal basis for W, and $\mathbf{u}$ is any vector in V, then*

$$\text{proj}_W \mathbf{u} = \langle \mathbf{u}, \mathbf{v}_1 \rangle \mathbf{v}_1 + \langle \mathbf{u}, \mathbf{v}_2 \rangle \mathbf{v}_2 + \cdots + \langle \mathbf{u}, \mathbf{v}_r \rangle \mathbf{v}_r \tag{12}$$

Proof (a) It follows from Theorem 6.3.3 that the vector $\mathbf{u}$ can be expressed in the form $\mathbf{u} = \mathbf{w}_1 + \mathbf{w}_2$, where $\mathbf{w}_1 = \text{proj}_W \mathbf{u}$ is in W and $\mathbf{w}_2$ is in $W^\perp$; and it follows from Theorem 6.3.2 that the component $\text{proj}_W \mathbf{u} = \mathbf{w}_1$ can be expressed in terms of the basis vectors for W as

$$\text{proj}_W \mathbf{u} = \mathbf{w}_1 = \frac{\langle \mathbf{w}_1, \mathbf{v}_1 \rangle}{\|\mathbf{v}_1\|^2} \mathbf{v}_1 + \frac{\langle \mathbf{w}_1, \mathbf{v}_2 \rangle}{\|\mathbf{v}_2\|^2} \mathbf{v}_2 + \cdots + \frac{\langle \mathbf{w}_1, \mathbf{v}_r \rangle}{\|\mathbf{v}_r\|^2} \mathbf{v}_r \tag{13}$$

Since $\mathbf{w}_2$ is orthogonal to W, it follows that

$$\langle \mathbf{w}_2, \mathbf{v}_1 \rangle = \langle \mathbf{w}_2, \mathbf{v}_2 \rangle = \cdots = \langle \mathbf{w}_2, \mathbf{v}_r \rangle = 0$$

so we can rewrite (13) as

$$\text{proj}_W \mathbf{u} = \mathbf{w}_1 = \frac{\langle \mathbf{w}_1 + \mathbf{w}_2, \mathbf{v}_1 \rangle}{\|\mathbf{v}_1\|^2} \mathbf{v}_1 + \frac{\langle \mathbf{w}_1 + \mathbf{w}_2, \mathbf{v}_2 \rangle}{\|\mathbf{v}_2\|^2} \mathbf{v}_2 + \cdots + \frac{\langle \mathbf{w}_1 + \mathbf{w}_2, \mathbf{v}_r \rangle}{\|\mathbf{v}_r\|^2} \mathbf{v}_r$$

or, equivalently, as

$$\text{proj}_W \mathbf{u} = \mathbf{w}_1 = \frac{\langle \mathbf{u}, \mathbf{v}_1 \rangle}{\|\mathbf{v}_1\|^2} \mathbf{v}_1 + \frac{\langle \mathbf{u}, \mathbf{v}_2 \rangle}{\|\mathbf{v}_2\|^2} \mathbf{v}_2 + \cdots + \frac{\langle \mathbf{u}, \mathbf{v}_r \rangle}{\|\mathbf{v}_r\|^2} \mathbf{v}_r$$

Proof (b) In this case, $\|\mathbf{v}_1\| = \|\mathbf{v}_2\| = \cdots = \|\mathbf{v}_r\| = 1$, so Formula (13) simplifies to Formula (12). ◂

▶ **EXAMPLE 6 Calculating Projections**

Let R^3 have the Euclidean inner product, and let W be the subspace spanned by the orthonormal vectors $\mathbf{v}_1 = (0, 1, 0)$ and $\mathbf{v}_2 = \left(-\frac{4}{5}, 0, \frac{3}{5}\right)$. From Formula (12) the orthogonal projection of $\mathbf{u} = (1, 1, 1)$ on W is

$$\text{proj}_W \mathbf{u} = \langle \mathbf{u}, \mathbf{v}_1 \rangle \mathbf{v}_1 + \langle \mathbf{u}, \mathbf{v}_2 \rangle \mathbf{v}_2$$
$$= (1)(0, 1, 0) + \left(-\tfrac{1}{5}\right)\left(-\tfrac{4}{5}, 0, \tfrac{3}{5}\right)$$
$$= \left(\tfrac{4}{25}, 1, -\tfrac{3}{25}\right)$$

The component of **u** orthogonal to W is

$$\text{proj}_{W^\perp} \mathbf{u} = \mathbf{u} - \text{proj}_W \mathbf{u} = (1, 1, 1) - \left(\tfrac{4}{25}, 1, -\tfrac{3}{25}\right) = \left(\tfrac{21}{25}, 0, \tfrac{28}{25}\right)$$

Observe that $\text{proj}_{W^\perp} \mathbf{u}$ is orthogonal to both $\mathbf{v}_1$ and $\mathbf{v}_2$, so this vector is orthogonal to each vector in the space W spanned by $\mathbf{v}_1$ and $\mathbf{v}_2$, as it should be. ◄

A Geometric Interpretation of Orthogonal Projections

If W is a one-dimensional subspace of an inner product space V, say span{**a**}, then Formula (11) has only the one term

$$\text{proj}_W \mathbf{u} = \frac{\langle \mathbf{u}, \mathbf{a} \rangle}{\|\mathbf{a}\|^2} \mathbf{a}$$

In the special case where V is R^3 with the Euclidean inner product, this is exactly Formula (10) of Section 3.3 for the orthogonal projection of **u** along **a**. This suggests that we can think of (11) as the sum of orthogonal projections on "axes" determined by the basis vectors for the subspace W (Figure 6.3.2).

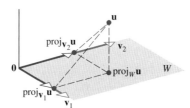

► Figure 6.3.2

The Gram–Schmidt Process

We have seen that orthonormal bases exhibit a variety of useful properties. Our next theorem, which is the main result in this section, shows that every nonzero finite-dimensional vector space has an orthonormal basis. The proof of this result is extremely important, since it provides an algorithm, or method, for converting an arbitrary basis into an orthonormal basis.

THEOREM 6.3.5 *Every nonzero finite-dimensional inner product space has an orthonormal basis.*

Proof Let W be any nonzero finite-dimensional subspace of an inner product space, and suppose that $\{\mathbf{u}_1, \mathbf{u}_2, \ldots, \mathbf{u}_r\}$ is any basis for W. It suffices to show that W has an orthogonal basis, since the vectors in that basis can be normalized to obtain an orthonormal basis. The following sequence of steps will produce an orthogonal basis $\{\mathbf{v}_1, \mathbf{v}_2, \ldots, \mathbf{v}_r\}$ for W:

Step 1. Let $\mathbf{v}_1 = \mathbf{u}_1$.

Step 2. As illustrated in Figure 6.3.3, we can obtain a vector $\mathbf{v}_2$ that is orthogonal to $\mathbf{v}_1$ by computing the component of $\mathbf{u}_2$ that is orthogonal to the space W_1 spanned by $\mathbf{v}_1$. Using Formula (11) to perform this computation we obtain

$$\mathbf{v}_2 = \mathbf{u}_2 - \text{proj}_{W_1} \mathbf{u}_2 = \mathbf{u}_2 - \frac{\langle \mathbf{u}_2, \mathbf{v}_1 \rangle}{\|\mathbf{v}_1\|^2} \mathbf{v}_1$$

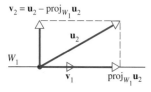

▲ Figure 6.3.3

Of course, if $\mathbf{v}_2 = \mathbf{0}$, then $\mathbf{v}_2$ is not a basis vector. But this cannot happen, since it would then follow from the above formula for $\mathbf{v}_2$ that

$$\mathbf{u}_2 = \frac{\langle \mathbf{u}_2, \mathbf{v}_1 \rangle}{\|\mathbf{v}_1\|^2} \mathbf{v}_1 = \frac{\langle \mathbf{u}_2, \mathbf{v}_1 \rangle}{\|\mathbf{u}_1\|^2} \mathbf{u}_1$$

which implies that $\mathbf{u}_2$ is a multiple of $\mathbf{u}_1$, contradicting the linear independence of the basis $S = \{\mathbf{u}_1, \mathbf{u}_2, \ldots, \mathbf{u}_n\}$.

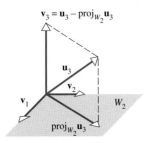

▲ Figure 6.3.4

Step 3. To construct a vector $\mathbf{v}_3$ that is orthogonal to both $\mathbf{v}_1$ and $\mathbf{v}_2$, we compute the component of $\mathbf{u}_3$ orthogonal to the space W_2 spanned by $\mathbf{v}_1$ and $\mathbf{v}_2$ (Figure 6.3.4). Using Formula (11) to perform this computation we obtain

$$\mathbf{v}_3 = \mathbf{u}_3 - \mathrm{proj}_{W_2}\mathbf{u}_3 = \mathbf{u}_3 - \frac{\langle \mathbf{u}_3, \mathbf{v}_1 \rangle}{\|\mathbf{v}_1\|^2}\mathbf{v}_1 - \frac{\langle \mathbf{u}_3, \mathbf{v}_2 \rangle}{\|\mathbf{v}_2\|^2}\mathbf{v}_2$$

As in Step 2, the linear independence of $\{\mathbf{u}_1, \mathbf{u}_2, \ldots, \mathbf{u}_n\}$ ensures that $\mathbf{v}_3 \neq \mathbf{0}$. We leave the details for you.

Step 4. To determine a vector $\mathbf{v}_4$ that is orthogonal to $\mathbf{v}_1$, $\mathbf{v}_2$, and $\mathbf{v}_3$, we compute the component of $\mathbf{u}_4$ orthogonal to the space W_3 spanned by $\mathbf{v}_1$, $\mathbf{v}_2$, and $\mathbf{v}_3$. From (11),

$$\mathbf{v}_4 = \mathbf{u}_4 - \mathrm{proj}_{W_3}\mathbf{u}_4 = \mathbf{u}_4 - \frac{\langle \mathbf{u}_4, \mathbf{v}_1 \rangle}{\|\mathbf{v}_1\|^2}\mathbf{v}_1 - \frac{\langle \mathbf{u}_4, \mathbf{v}_2 \rangle}{\|\mathbf{v}_2\|^2}\mathbf{v}_2 - \frac{\langle \mathbf{u}_4, \mathbf{v}_3 \rangle}{\|\mathbf{v}_3\|^2}\mathbf{v}_3$$

Continuing in this way we will produce an orthogonal set of vectors $\{\mathbf{v}_1, \mathbf{v}_2, \ldots, \mathbf{v}_r\}$ after r steps. Since orthogonal sets are linearly independent, this set will be an orthogonal basis for the r-dimensional space W. By normalizing these basis vectors we can obtain an orthonormal basis. ◀

The step-by-step construction of an orthogonal (or orthonormal) basis given in the foregoing proof is called the *Gram–Schmidt process*. For reference, we provide the following summary of the steps.

The Gram–Schmidt Process

To convert a basis $\{\mathbf{u}_1, \mathbf{u}_2, \ldots, \mathbf{u}_r\}$ into an orthogonal basis $\{\mathbf{v}_1, \mathbf{v}_2, \ldots, \mathbf{v}_r\}$, perform the following computations:

Step 1. $\mathbf{v}_1 = \mathbf{u}_1$

Step 2. $\mathbf{v}_2 = \mathbf{u}_2 - \dfrac{\langle \mathbf{u}_2, \mathbf{v}_1 \rangle}{\|\mathbf{v}_1\|^2}\mathbf{v}_1$

Step 3. $\mathbf{v}_3 = \mathbf{u}_3 - \dfrac{\langle \mathbf{u}_3, \mathbf{v}_1 \rangle}{\|\mathbf{v}_1\|^2}\mathbf{v}_1 - \dfrac{\langle \mathbf{u}_3, \mathbf{v}_2 \rangle}{\|\mathbf{v}_2\|^2}\mathbf{v}_2$

Step 4. $\mathbf{v}_4 = \mathbf{u}_4 - \dfrac{\langle \mathbf{u}_4, \mathbf{v}_1 \rangle}{\|\mathbf{v}_1\|^2}\mathbf{v}_1 - \dfrac{\langle \mathbf{u}_4, \mathbf{v}_2 \rangle}{\|\mathbf{v}_2\|^2}\mathbf{v}_2 - \dfrac{\langle \mathbf{u}_4, \mathbf{v}_3 \rangle}{\|\mathbf{v}_3\|^2}\mathbf{v}_3$

$\vdots$

(continue for r steps)

Optional Step. To convert the orthogonal basis into an orthonormal basis $\{\mathbf{q}_1, \mathbf{q}_2, \ldots, \mathbf{q}_r\}$, normalize the orthogonal basis vectors.

▶ **EXAMPLE 7 Using the Gram–Schmidt Process**

Assume that the vector space R^3 has the Euclidean inner product. Apply the Gram–Schmidt process to transform the basis vectors

$$\mathbf{u}_1 = (1, 1, 1), \quad \mathbf{u}_2 = (0, 1, 1), \quad \mathbf{u}_3 = (0, 0, 1)$$

into an orthogonal basis $\{\mathbf{v}_1, \mathbf{v}_2, \mathbf{v}_3\}$, and then normalize the orthogonal basis vectors to obtain an orthonormal basis $\{\mathbf{q}_1, \mathbf{q}_2, \mathbf{q}_3\}$.

Solution

Step 1. $\mathbf{v}_1 = \mathbf{u}_1 = (1, 1, 1)$

Step 2. $\mathbf{v}_2 = \mathbf{u}_2 - \mathrm{proj}_{W_1}\mathbf{u}_2 = \mathbf{u}_2 - \dfrac{\langle \mathbf{u}_2, \mathbf{v}_1 \rangle}{\|\mathbf{v}_1\|^2}\mathbf{v}_1$

$= (0, 1, 1) - \dfrac{2}{3}(1, 1, 1) = \left(-\dfrac{2}{3}, \dfrac{1}{3}, \dfrac{1}{3}\right)$

Step 3. $\mathbf{v}_3 = \mathbf{u}_3 - \text{proj}_{W_2} \mathbf{u}_3 = \mathbf{u}_3 - \dfrac{\langle \mathbf{u}_3, \mathbf{v}_1 \rangle}{\|\mathbf{v}_1\|^2} \mathbf{v}_1 - \dfrac{\langle \mathbf{u}_3, \mathbf{v}_2 \rangle}{\|\mathbf{v}_2\|^2} \mathbf{v}_2$

$= (0, 0, 1) - \dfrac{1}{3}(1, 1, 1) - \dfrac{1/3}{2/3}\left(-\dfrac{2}{3}, \dfrac{1}{3}, \dfrac{1}{3}\right)$

$= \left(0, -\dfrac{1}{2}, \dfrac{1}{2}\right)$

Thus,

$$\mathbf{v}_1 = (1, 1, 1), \quad \mathbf{v}_2 = \left(-\dfrac{2}{3}, \dfrac{1}{3}, \dfrac{1}{3}\right), \quad \mathbf{v}_3 = \left(0, -\dfrac{1}{2}, \dfrac{1}{2}\right)$$

form an orthogonal basis for R^3. The norms of these vectors are

$$\|\mathbf{v}_1\| = \sqrt{3}, \quad \|\mathbf{v}_2\| = \dfrac{\sqrt{6}}{3}, \quad \|\mathbf{v}_3\| = \dfrac{1}{\sqrt{2}}$$

so an orthonormal basis for R^3 is

$$\mathbf{q}_1 = \dfrac{\mathbf{v}_1}{\|\mathbf{v}_1\|} = \left(\dfrac{1}{\sqrt{3}}, \dfrac{1}{\sqrt{3}}, \dfrac{1}{\sqrt{3}}\right), \quad \mathbf{q}_2 = \dfrac{\mathbf{v}_2}{\|\mathbf{v}_2\|} = \left(-\dfrac{2}{\sqrt{6}}, \dfrac{1}{\sqrt{6}}, \dfrac{1}{\sqrt{6}}\right),$$

$$\mathbf{q}_3 = \dfrac{\mathbf{v}_3}{\|\mathbf{v}_3\|} = \left(0, -\dfrac{1}{\sqrt{2}}, \dfrac{1}{\sqrt{2}}\right) \blacktriangleleft$$

Remark In the last example we normalized at the end to convert the orthogonal basis into an orthonormal basis. Alternatively, we could have normalized each orthogonal basis vector as soon as it was obtained, thereby producing an orthonormal basis step by step. However, that procedure generally has the disadvantage in hand calculation of producing more square roots to manipulate. A more useful variation is to "scale" the orthogonal basis vectors at each step to eliminate some of the fractions. For example, after Step 2 above, we could have multiplied by 3 to produce $(-2, 1, 1)$ as the second orthogonal basis vector, thereby simplifying the calculations in Step 3.

Erhardt Schmidt
(1875–1959)

Jorgen Pederson Gram
(1850–1916)

Historical Note Schmidt was a German mathematician who studied for his doctoral degree at Göttingen University under David Hilbert, one of the giants of modern mathematics. For most of his life he taught at Berlin University where, in addition to making important contributions to many branches of mathematics, he fashioned some of Hilbert's ideas into a general concept, called a *Hilbert space*—a fundamental idea in the study of infinite-dimensional vector spaces. He first described the process that bears his name in a paper on integral equations that he published in 1907.

[*Image: Archives of the Mathematisches Forschungsinst*]

Historical Note Gram was a Danish actuary whose early education was at village schools supplemented by private tutoring. He obtained a doctorate degree in mathematics while working for the Hafnia Life Insurance Company, where he specialized in the mathematics of accident insurance. It was in his dissertation that his contributions to the Gram–Schmidt process were formulated. He eventually became interested in abstract mathematics and received a gold medal from the Royal Danish Society of Sciences and Letters in recognition of his work. His lifelong interest in applied mathematics never wavered, however, and he produced a variety of treatises on Danish forest management.

[*Image: wikipedia*]

CALCULUS REQUIRED

▶ **EXAMPLE 8** **Legendre Polynomials**

Let the vector space P_2 have the inner product

$$\langle \mathbf{p}, \mathbf{q} \rangle = \int_{-1}^{1} p(x)q(x)\, dx$$

Apply the Gram–Schmidt process to transform the standard basis $\{1, x, x^2\}$ for P_2 into an orthogonal basis $\{\phi_1(x), \phi_2(x), \phi_3(x)\}$.

Solution Take $\mathbf{u}_1 = 1$, $\mathbf{u}_2 = x$, and $\mathbf{u}_3 = x^2$.

Step 1. $\mathbf{v}_1 = \mathbf{u}_1 = 1$

Step 2. We have

$$\langle \mathbf{u}_2, \mathbf{v}_1 \rangle = \int_{-1}^{1} x\, dx = 0$$

so

$$\mathbf{v}_2 = \mathbf{u}_2 - \frac{\langle \mathbf{u}_2, \mathbf{v}_1 \rangle}{\|\mathbf{v}_1\|^2} \mathbf{v}_1 = \mathbf{u}_2 = x$$

Step 3. We have

$$\langle \mathbf{u}_3, \mathbf{v}_1 \rangle = \int_{-1}^{1} x^2\, dx = \left.\frac{x^3}{3}\right]_{-1}^{1} = \frac{2}{3}$$

$$\langle \mathbf{u}_3, \mathbf{v}_2 \rangle = \int_{-1}^{1} x^3\, dx = \left.\frac{x^4}{4}\right]_{-1}^{1} = 0$$

$$\|\mathbf{v}_1\|^2 = \langle \mathbf{v}_1, \mathbf{v}_1 \rangle = \int_{-1}^{1} 1\, dx = \left.x\right]_{-1}^{1} = 2$$

so

$$\mathbf{v}_3 = \mathbf{u}_3 - \frac{\langle \mathbf{u}_3, \mathbf{v}_1 \rangle}{\|\mathbf{v}_1\|^2} \mathbf{v}_1 - \frac{\langle \mathbf{u}_3, \mathbf{v}_2 \rangle}{\|\mathbf{v}_2\|^2} \mathbf{v}_2 = x^2 - \frac{1}{3}$$

Thus, we have obtained the orthogonal basis $\{\phi_1(x), \phi_2(x), \phi_3(x)\}$ in which

$$\phi_1(x) = 1, \quad \phi_2(x) = x, \quad \phi_3(x) = x^2 - \frac{1}{3} \quad \blacktriangleleft$$

Remark The orthogonal basis vectors in the foregoing example are often scaled so all three functions have a value of 1 at $x = 1$. The resulting polynomials

$$1, \quad x, \quad \frac{1}{2}(3x^2 - 1)$$

which are known as the first three **Legendre polynomials**, play an important role in a variety of applications. The scaling does not affect the orthogonality.

Extending Orthonormal Sets to Orthonormal Bases

Recall from part (b) of Theorem 4.5.5 that a linearly independent set in a finite-dimensional vector space can be enlarged to a basis by adding appropriate vectors. The following theorem is an analog of that result for orthogonal and orthonormal sets in finite-dimensional inner product spaces.

THEOREM 6.3.6 *If W is a finite-dimensional inner product space, then:*

(a) Every orthogonal set of nonzero vectors in W can be enlarged to an orthogonal basis for W.

(b) Every orthonormal set in W can be enlarged to an orthonormal basis for W.

We will prove part *(b)* and leave part *(a)* as an exercise.

Proof (b) Suppose that $S = \{\mathbf{v}_1, \mathbf{v}_2, \ldots, \mathbf{v}_s\}$ is an orthonormal set of vectors in W. Part *(b)* of Theorem 4.5.5 tells us that we can enlarge S to some basis

$$S' = \{\mathbf{v}_1, \mathbf{v}_2, \ldots, \mathbf{v}_s, \mathbf{v}_{s+1}, \ldots, \mathbf{v}_k\}$$

for W. If we now apply the Gram–Schmidt process to the set S', then the vectors $\mathbf{v}_1, \mathbf{v}_2, \ldots, \mathbf{v}_s$, will not be affected since they are already orthonormal, and the resulting set

$$S'' = \{\mathbf{v}_1, \mathbf{v}_2, \ldots, \mathbf{v}_s, \mathbf{v}_{s+1}, \ldots, \mathbf{v}_k\}$$

will be an orthonormal basis for W. ◀

OPTIONAL
QR-Decomposition

In recent years a numerical algorithm based on the Gram–Schmidt process, and known as **QR-decomposition**, has assumed growing importance as the mathematical foundation for a wide variety of numerical algorithms, including those for computing eigenvalues of large matrices. The technical aspects of such algorithms are discussed in textbooks that specialize in the numerical aspects of linear algebra. However, we will discuss some of the underlying ideas here. We begin by posing the following problem.

Problem If A is an $m \times n$ matrix with linearly independent column vectors, and if Q is the matrix that results by applying the Gram–Schmidt process to the column vectors of A, what relationship, if any, exists between A and Q?

To solve this problem, suppose that the column vectors of A are $\mathbf{u}_1, \mathbf{u}_2, \ldots, \mathbf{u}_n$ and the orthonormal column vectors of Q are $\mathbf{q}_1, \mathbf{q}_2, \ldots, \mathbf{q}_n$. Thus, A and Q can be written in partitioned form as

$$A = [\mathbf{u}_1 \mid \mathbf{u}_2 \mid \cdots \mid \mathbf{u}_n] \quad \text{and} \quad Q = [\mathbf{q}_1 \mid \mathbf{q}_2 \mid \cdots \mid \mathbf{q}_n]$$

It follows from Theorem 6.3.2b that $\mathbf{u}_1, \mathbf{u}_2, \ldots, \mathbf{u}_n$ are expressible in terms of the vectors $\mathbf{q}_1, \mathbf{q}_2, \ldots, \mathbf{q}_n$ as

$$\mathbf{u}_1 = \langle \mathbf{u}_1, \mathbf{q}_1 \rangle \mathbf{q}_1 + \langle \mathbf{u}_1, \mathbf{q}_2 \rangle \mathbf{q}_2 + \cdots + \langle \mathbf{u}_1, \mathbf{q}_n \rangle \mathbf{q}_n$$
$$\mathbf{u}_2 = \langle \mathbf{u}_2, \mathbf{q}_1 \rangle \mathbf{q}_1 + \langle \mathbf{u}_2, \mathbf{q}_2 \rangle \mathbf{q}_2 + \cdots + \langle \mathbf{u}_2, \mathbf{q}_n \rangle \mathbf{q}_n$$
$$\vdots \qquad \vdots \qquad \vdots \qquad \vdots$$
$$\mathbf{u}_n = \langle \mathbf{u}_n, \mathbf{q}_1 \rangle \mathbf{q}_1 + \langle \mathbf{u}_n, \mathbf{q}_2 \rangle \mathbf{q}_2 + \cdots + \langle \mathbf{u}_n, \mathbf{q}_n \rangle \mathbf{q}_n$$

Recalling from Section 1.3 (Example 9) that the jth column vector of a matrix product is a linear combination of the column vectors of the first factor with coefficients coming from the jth column of the second factor, it follows that these relationships can be expressed in matrix form as

$$[\mathbf{u}_1 \mid \mathbf{u}_2 \mid \cdots \mid \mathbf{u}_n] = [\mathbf{q}_1 \mid \mathbf{q}_2 \mid \cdots \mid \mathbf{q}_n] \begin{bmatrix} \langle \mathbf{u}_1, \mathbf{q}_1 \rangle & \langle \mathbf{u}_2, \mathbf{q}_1 \rangle & \cdots & \langle \mathbf{u}_n, \mathbf{q}_1 \rangle \\ \langle \mathbf{u}_1, \mathbf{q}_2 \rangle & \langle \mathbf{u}_2, \mathbf{q}_2 \rangle & \cdots & \langle \mathbf{u}_n, \mathbf{q}_2 \rangle \\ \vdots & \vdots & & \vdots \\ \langle \mathbf{u}_1, \mathbf{q}_n \rangle & \langle \mathbf{u}_2, \mathbf{q}_n \rangle & \cdots & \langle \mathbf{u}_n, \mathbf{q}_n \rangle \end{bmatrix}$$

or more briefly as
$$A = QR \tag{14}$$

where R is the second factor in the product. However, it is a property of the Gram–Schmidt process that for $j \geq 2$, the vector $\mathbf{q}_j$ is orthogonal to $\mathbf{u}_1, \mathbf{u}_2, \ldots, \mathbf{u}_{j-1}$. Thus, all entries below the main diagonal of R are zero, and R has the form

$$R = \begin{bmatrix} \langle \mathbf{u}_1, \mathbf{q}_1 \rangle & \langle \mathbf{u}_2, \mathbf{q}_1 \rangle & \cdots & \langle \mathbf{u}_n, \mathbf{q}_1 \rangle \\ 0 & \langle \mathbf{u}_2, \mathbf{q}_2 \rangle & \cdots & \langle \mathbf{u}_n, \mathbf{q}_2 \rangle \\ \vdots & \vdots & & \vdots \\ 0 & 0 & \cdots & \langle \mathbf{u}_n, \mathbf{q}_n \rangle \end{bmatrix} \tag{15}$$

We leave it for you to show that R is invertible by showing that its diagonal entries are nonzero. Thus, Equation (14) is a factorization of A into the product of a matrix Q with orthonormal column vectors and an invertible upper triangular matrix R. We call Equation (14) the ***QR-decomposition of*** A. In summary, we have the following theorem.

THEOREM 6.3.7 *QR-Decomposition*

If A is an $m \times n$ matrix with linearly independent column vectors, then A can be factored as
$$A = QR$$
where Q is an $m \times n$ matrix with orthonormal column vectors, and R is an $n \times n$ invertible upper triangular matrix.

It is common in numerical linear algebra to say that a matrix with linearly independent columns has **full column rank**.

Recall from Theorem 5.1.6 (the Equivalence Theorem) that a *square* matrix has linearly independent column vectors if and only if it is invertible. Thus, it follows from the foregoing theorem that *every invertible matrix has a QR-decomposition*.

▶ **EXAMPLE 9** *QR-Decomposition of a 3 × 3 Matrix*

Find the QR-decomposition of

$$A = \begin{bmatrix} 1 & 0 & 0 \\ 1 & 1 & 0 \\ 1 & 1 & 1 \end{bmatrix}$$

Solution The column vectors of A are

$$\mathbf{u}_1 = \begin{bmatrix} 1 \\ 1 \\ 1 \end{bmatrix}, \quad \mathbf{u}_2 = \begin{bmatrix} 0 \\ 1 \\ 1 \end{bmatrix}, \quad \mathbf{u}_3 = \begin{bmatrix} 0 \\ 0 \\ 1 \end{bmatrix}$$

Applying the Gram–Schmidt process with normalization to these column vectors yields the orthonormal vectors (see Example 7)

$$\mathbf{q}_1 = \begin{bmatrix} \frac{1}{\sqrt{3}} \\ \frac{1}{\sqrt{3}} \\ \frac{1}{\sqrt{3}} \end{bmatrix}, \quad \mathbf{q}_2 = \begin{bmatrix} -\frac{2}{\sqrt{6}} \\ \frac{1}{\sqrt{6}} \\ \frac{1}{\sqrt{6}} \end{bmatrix}, \quad \mathbf{q}_3 = \begin{bmatrix} 0 \\ -\frac{1}{\sqrt{2}} \\ \frac{1}{\sqrt{2}} \end{bmatrix}$$

Thus, it follows from Formula (15) that R is

$$R = \begin{bmatrix} \langle \mathbf{u}_1, \mathbf{q}_1 \rangle & \langle \mathbf{u}_2, \mathbf{q}_1 \rangle & \langle \mathbf{u}_3, \mathbf{q}_1 \rangle \\ 0 & \langle \mathbf{u}_2, \mathbf{q}_2 \rangle & \langle \mathbf{u}_3, \mathbf{q}_2 \rangle \\ 0 & 0 & \langle \mathbf{u}_3, \mathbf{q}_3 \rangle \end{bmatrix} = \begin{bmatrix} \frac{3}{\sqrt{3}} & \frac{2}{\sqrt{3}} & \frac{1}{\sqrt{3}} \\ 0 & \frac{2}{\sqrt{6}} & \frac{1}{\sqrt{6}} \\ 0 & 0 & \frac{1}{\sqrt{2}} \end{bmatrix}$$

◀ Show that the matrix Q in Example 9 has the property $QQ^T = I$, and show that every $m \times n$ matrix with orthonormal column vectors has this property.

from which it follows that the QR-decomposition of A is

$$\underbrace{\begin{bmatrix} 1 & 0 & 0 \\ 1 & 1 & 0 \\ 1 & 1 & 1 \end{bmatrix}}_{A} = \underbrace{\begin{bmatrix} \frac{1}{\sqrt{3}} & -\frac{2}{\sqrt{6}} & 0 \\ \frac{1}{\sqrt{3}} & \frac{1}{\sqrt{6}} & -\frac{1}{\sqrt{2}} \\ \frac{1}{\sqrt{3}} & \frac{1}{\sqrt{6}} & \frac{1}{\sqrt{2}} \end{bmatrix}}_{Q} \underbrace{\begin{bmatrix} \frac{3}{\sqrt{3}} & \frac{2}{\sqrt{3}} & \frac{1}{\sqrt{3}} \\ 0 & \frac{2}{\sqrt{6}} & \frac{1}{\sqrt{6}} \\ 0 & 0 & \frac{1}{\sqrt{2}} \end{bmatrix}}_{R}$$

Concept Review

- Orthogonal and orthonormal sets
- Normalizing a vector
- Orthogonal projections
- Gram–Schmidt process
- QR-decomposition

Skills

- Determine whether a set of vectors is orthogonal (or orthonormal).
- Compute the coordinates of a vector with respect to an orthogonal (or orthonormal) basis.
- Find the orthogonal projection of a vector onto a subspace.
- Use the Gram–Schmidt process to construct an orthogonal (or orthonormal) basis for an inner product space.
- Find the QR-decomposition of an invertible matrix.

Exercise Set 6.3

1. Which of the following sets of vectors are orthogonal with respect to the Euclidean inner product on R^2?
   (a) $(4, 0)$, $(0, -3)$
   (b) $\left(\frac{1}{\sqrt{2}}, -\frac{1}{\sqrt{2}}\right)$, $\left(\frac{1}{\sqrt{2}}, \frac{1}{\sqrt{2}}\right)$
   (c) $(2, 2)$, $\left(-\frac{1}{2}, -\frac{1}{2}\right)$,
   (d) $(-3, 1)$, $(0, 0)$

2. Which of the sets in Exercise 1 are orthonormal with respect to the Euclidean inner product on R^2?

3. Which of the following sets of vectors are orthogonal with respect to the Euclidean inner product on R^3?
   (a) $\left(0, \frac{1}{\sqrt{2}}, -\frac{1}{\sqrt{2}}\right)$, $\left(-\frac{1}{\sqrt{3}}, \frac{1}{\sqrt{3}}, \frac{1}{\sqrt{3}}\right)$, $\left(0, \frac{1}{\sqrt{2}}, \frac{1}{\sqrt{2}}\right)$
   (b) $\left(\frac{2}{3}, -\frac{2}{3}, \frac{1}{3}\right)$, $\left(\frac{2}{3}, \frac{1}{3}, -\frac{2}{3}\right)$, $\left(\frac{1}{3}, \frac{2}{3}, \frac{2}{3}\right)$
   (c) $(1, 0, 1)$, $\left(-\frac{1}{\sqrt{3}}, \frac{1}{\sqrt{3}}, -\frac{1}{\sqrt{3}}\right)$, $(-1, 0, 1)$
   (d) $\left(\frac{1}{\sqrt{6}}, \frac{1}{\sqrt{6}}, -\frac{2}{\sqrt{6}}\right)$, $\left(\frac{1}{\sqrt{2}}, -\frac{1}{\sqrt{2}}, 0\right)$

4. Which of the sets in Exercise 3 are orthonormal with respect to the Euclidean inner product on R^3?

5. Which of the following sets of polynomials are orthonormal with respect to the inner product on P_2 discussed in Example 7 of Section 6.1?
   (a) $p_1(x) = -\frac{1}{3} + \frac{2}{3}x + \frac{2}{3}x^2$, $p_2(x) = \frac{2}{3} - \frac{1}{3}x + \frac{2}{3}x^2$,
   $p_3(x) = \frac{2}{3} + \frac{2}{3}x + \frac{1}{3}x^2$
   (b) $p_1(x) = 1$, $p_2(x) = \frac{1}{\sqrt{2}}x + \frac{1}{\sqrt{2}}x^2$, $p_3(x) = x^2$

6. Which of the following sets of matrices are orthonormal with respect to the inner product on M_{22} discussed in Example 6 of Section 6.1?
   (a) $\begin{bmatrix} 0 & 0 \\ 0 & 1 \end{bmatrix}$, $\begin{bmatrix} -\frac{1}{3} & \frac{2}{3} \\ -\frac{2}{3} & 0 \end{bmatrix}$, $\begin{bmatrix} -\frac{2}{3} & \frac{1}{3} \\ \frac{2}{3} & 0 \end{bmatrix}$, $\begin{bmatrix} \frac{2}{3} & \frac{2}{3} \\ \frac{1}{3} & 0 \end{bmatrix}$
   (b) $\begin{bmatrix} 1 & 0 \\ 0 & 0 \end{bmatrix}$, $\begin{bmatrix} 0 & 1 \\ 0 & 0 \end{bmatrix}$, $\begin{bmatrix} 0 & 0 \\ 1 & 1 \end{bmatrix}$, $\begin{bmatrix} 0 & 0 \\ 1 & -1 \end{bmatrix}$

7. Verify that the given vectors form an orthogonal set with respect to the Euclidean inner product; then convert it to an

orthonormal set by normalizing the vectors.

(a) $(2, 3)$, $(-6, 4)$

(b) $(1, -1, 0)$, $(3, 3, 0)$, $(0, 0, 2)$

(c) $(\frac{1}{5}, \frac{1}{5}, \frac{1}{5})$, $(-\frac{1}{2}, \frac{1}{2}, 0)$, $(\frac{1}{3}, \frac{1}{3}, -\frac{2}{3})$

8. Verify that the set of vectors $\{(1, 0), (0, 1)\}$ is orthogonal with respect to the inner product $\langle \mathbf{u}, \mathbf{v} \rangle = 4u_1v_1 + u_2v_2$ on R^2; then convert it to an orthonormal set by normalizing the vectors.

9. Verify that the vectors

$$\mathbf{v}_1 = \left(-\tfrac{3}{5}, \tfrac{4}{5}, 0\right), \quad \mathbf{v}_2 = \left(\tfrac{4}{5}, \tfrac{3}{5}, 0\right), \quad \mathbf{v}_3 = (0, 0, 1)$$

form an orthonormal basis for R^3 with the Euclidean inner product; then use Theorem 6.3.2b to express each of the following as linear combinations of $\mathbf{v}_1$, $\mathbf{v}_2$, and $\mathbf{v}_3$.

(a) $(2, 1, -1)$ (b) $(1, 3, 4)$ (c) $\left(\tfrac{1}{7}, -\tfrac{3}{7}, \tfrac{5}{7}\right)$

10. Verify that the vectors

$$\mathbf{v}_1 = (1, -1, 2, -1), \quad \mathbf{v}_2 = (-2, 2, 3, 2),$$
$$\mathbf{v}_3 = (1, 2, 0, -1), \quad \mathbf{v}_4 = (1, 0, 0, 1)$$

form an orthogonal basis for R^4 with the Euclidean inner product; then use Theorem 6.3.2a to express each of the following as linear combinations of $\mathbf{v}_1$, $\mathbf{v}_2$, $\mathbf{v}_3$, and $\mathbf{v}_4$.

(a) $(1, -1, 1, -1)$

(b) $(\sqrt{2}, -3\sqrt{2}, 5\sqrt{2}, -\sqrt{2})$

(c) $\left(-\tfrac{1}{3}, \tfrac{2}{3}, -\tfrac{1}{3}, \tfrac{4}{3}\right)$

11. (a) Show that the vectors

$$\mathbf{v}_1 = (1, -2, 3, -4), \quad \mathbf{v}_2 = (2, 1, -4, -3),$$
$$\mathbf{v}_3 = (-3, 4, 1, -2), \quad \mathbf{v}_4 = (4, 3, 2, 1)$$

form an orthogonal basis for R^4 with the Euclidean inner product.

(b) Use Theorem 6.3.2a to express $\mathbf{u} = (-1, 2, 3, 7)$ as a linear combination of the vectors in part (a).

▶ In Exercises 12–13, an orthonormal basis with respect to the Euclidean inner product is given. Use Theorem 6.3.2b to find the coordinate vector of $\mathbf{w}$ with respect to that basis. ◀

12. (a) $\mathbf{w} = (3, 7)$; $\mathbf{u}_1 = \left(\tfrac{1}{\sqrt{2}}, -\tfrac{1}{\sqrt{2}}\right)$, $\mathbf{u}_2 = \left(\tfrac{1}{\sqrt{2}}, \tfrac{1}{\sqrt{2}}\right)$

(b) $\mathbf{w} = (-1, 0, 2)$; $\mathbf{u}_1 = \left(\tfrac{2}{3}, -\tfrac{2}{3}, \tfrac{1}{3}\right)$,

$\mathbf{u}_2 = \left(\tfrac{2}{3}, \tfrac{1}{3}, -\tfrac{2}{3}\right)$, $\mathbf{u}_3 = \left(\tfrac{1}{3}, \tfrac{2}{3}, \tfrac{2}{3}\right)$

13. (a) $\mathbf{w} = (2, 0, 5)$; $\mathbf{u}_1 = \left(\tfrac{2}{3}, \tfrac{1}{3}, \tfrac{2}{3}\right)$, $\mathbf{u}_2 = \left(\tfrac{1}{3}, \tfrac{2}{3}, -\tfrac{2}{3}\right)$,

$\mathbf{u}_3 = \left(\tfrac{2}{3}, -\tfrac{2}{3}, -\tfrac{1}{3}\right)$

(b) $\mathbf{w} = (-1, 1, 2)$; $\mathbf{u}_1 = \left(\tfrac{3}{\sqrt{11}}, \tfrac{1}{\sqrt{11}}, \tfrac{1}{\sqrt{11}}\right)$,

$\mathbf{u}_2 = \left(-\tfrac{1}{\sqrt{6}}, \tfrac{2}{\sqrt{6}}, \tfrac{1}{\sqrt{6}}\right)$,

$\mathbf{u}_3 = \left(-\tfrac{1}{\sqrt{66}}, -\tfrac{4}{\sqrt{66}}, \tfrac{7}{\sqrt{66}}\right)$

▶ In Exercises 14–15, the given vectors are orthogonal with respect to the Euclidean inner product. Find $\text{proj}_W \mathbf{x}$, where $\mathbf{x} = (1, 2, 0, -2)$ and W is the subspace of R^4 spanned by the vectors. ◀

14. (a) $\mathbf{v}_1 = (1, 1, 1, 1)$, $\mathbf{v}_2 = (1, 1, -1, -1)$

(b) $\mathbf{v}_1 = (0, 1, -4, -1)$, $\mathbf{v}_2 = (3, 5, 1, 1)$

15. (a) $\mathbf{v}_1 = (1, -1, -1, 1)$, $\mathbf{v}_2 = (1, 1, 1, 1)$,
$\mathbf{v}_3 = (1, 1, -1, -1)$

(b) $\mathbf{v}_1 = (0, 1, -4, -1)$, $\mathbf{v}_2 = (3, 5, 1, 1)$,
$\mathbf{v}_3 = (1, 0, 1, -4)$

▶ In Exercises 16–17, the given vectors are orthonormal with respect to the Euclidean inner product. Use Theorem 6.3.4b to find $\text{proj}_W \mathbf{x}$, where $\mathbf{x} = (1, 2, 0, -1)$ and W is the subspace of R^4 spanned by the vectors. ◀

16. (a) $\mathbf{v}_1 = \left(0, \tfrac{1}{\sqrt{18}}, -\tfrac{4}{\sqrt{18}}, -\tfrac{1}{\sqrt{18}}\right)$, $\mathbf{v}_2 = \left(\tfrac{1}{2}, \tfrac{5}{6}, \tfrac{1}{6}, \tfrac{1}{6}\right)$

(b) $\mathbf{v}_1 = \left(\tfrac{1}{2}, \tfrac{1}{2}, \tfrac{1}{2}, \tfrac{1}{2}\right)$, $\mathbf{v}_2 = \left(\tfrac{1}{2}, \tfrac{1}{2}, -\tfrac{1}{2}, -\tfrac{1}{2}\right)$

17. (a) $\mathbf{v}_1 = \left(0, \tfrac{1}{\sqrt{18}}, -\tfrac{4}{\sqrt{18}}, -\tfrac{1}{\sqrt{18}}\right)$, $\mathbf{v}_2 = \left(\tfrac{1}{2}, \tfrac{5}{6}, \tfrac{1}{6}, \tfrac{1}{6}\right)$,

$\mathbf{v}_3 = \left(\tfrac{1}{\sqrt{18}}, 0, \tfrac{1}{\sqrt{18}}, -\tfrac{4}{\sqrt{18}}\right)$

(b) $\mathbf{v}_1 = \left(\tfrac{1}{2}, \tfrac{1}{2}, \tfrac{1}{2}, \tfrac{1}{2}\right)$, $\mathbf{v}_2 = \left(\tfrac{1}{2}, \tfrac{1}{2}, -\tfrac{1}{2}, -\tfrac{1}{2}\right)$,

$\mathbf{v}_3 = \left(\tfrac{1}{2}, -\tfrac{1}{2}, \tfrac{1}{2}, -\tfrac{1}{2}\right)$

18. In Example 6 of Section 4.9 we found the orthogonal projection of the vector $\mathbf{x} = (1, 5)$ onto the line through the origin making an angle of $\pi/6$ radians with the x-axis. Solve that same problem using Theorem 6.3.4.

19. Find the vectors $\mathbf{w}_1$ in W and $\mathbf{w}_2$ in $W^\perp$ such that $\mathbf{x} = \mathbf{w}_1 + \mathbf{w}_2$, where $\mathbf{x}$ and W are as given in

(a) Exercise 14(a). (b) Exercise 15(a).

20. Find the vectors $\mathbf{w}_1$ in W and $\mathbf{w}_2$ in $W^\perp$ such that $\mathbf{x} = \mathbf{w}_1 + \mathbf{w}_2$, where $\mathbf{x}$ and W are as given in

(a) Exercise 16(a). (b) Exercise 17(a).

21. Let R^2 have the Euclidean inner product. Use the Gram–Schmidt process to transform the basis $\{\mathbf{u}_1, \mathbf{u}_2\}$ into an orthonormal basis. Draw both sets of basis vectors in the xy-plane.

(a) $\mathbf{u}_1 = (1, -3)$, $\mathbf{u}_2 = (2, 2)$

(b) $\mathbf{u}_1 = (1, 0)$, $\mathbf{u}_2 = (3, -5)$

22. Let R^3 have the Euclidean inner product. Use the Gram–Schmidt process to transform the basis $\{u_1, u_2, u_3\}$ into an orthonormal basis.

(a) $u_1 = (1, 1, 1)$, $u_2 = (-1, 1, 0)$, $u_3 = (1, 2, 1)$

(b) $u_1 = (1, 0, 0)$, $u_2 = (3, 7, -2)$, $u_3 = (0, 4, 1)$

23. Let R^4 have the Euclidean inner product. Use the Gram–Schmidt process to transform the basis $\{u_1, u_2, u_3, u_4\}$ into an orthonormal basis.

$$u_1 = (0, 2, 1, 0), \quad u_2 = (1, -1, 0, 0),$$
$$u_3 = (1, 2, 0, -1), \quad u_4 = (1, 0, 0, 1)$$

24. Let R^3 have the Euclidean inner product. Find an orthonormal basis for the subspace spanned by $(1, 0, -1)$, $(-1, 1, 3)$, $(0, 1, 2)$.

25. Let R^3 have the inner product
$$\langle u, v \rangle = u_1 v_1 + 2u_2 v_2 + 3u_3 v_3$$
Use the Gram–Schmidt process to transform $u_1 = (1, 1, 1)$, $u_2 = (1, 1, 0)$, $u_3 = (1, 0, 0)$ into an orthonormal basis.

26. Let R^3 have the Euclidean inner product. The subspace of R^3 spanned by the vectors $u_1 = \left(\frac{4}{5}, 0, -\frac{3}{5}\right)$ and $u_2 = (0, 1, 0)$ is a plane passing through the origin. Express $w = (1, 2, 3)$ in the form $w = w_1 + w_2$, where w_1 lies in the plane and w_2 is perpendicular to the plane.

27. Repeat Exercise 26 with $u_1 = (1, 0, -1)$ and $u_2 = (3, 1, 0)$.

28. Let R^4 have the Euclidean inner product. Express the vector $w = (-1, 2, 6, 0)$ in the form $w = w_1 + w_2$, where w_1 is in the space W spanned by $u_1 = (-1, 0, 1, 2)$ and $u_2 = (0, 1, 0, 1)$, and w_2 is orthogonal to W.

29. Find the QR-decomposition of the matrix, where possible.

(a) $\begin{bmatrix} 2 & 1 \\ -1 & 2 \end{bmatrix}$ (b) $\begin{bmatrix} 0 & 1 \\ 2 & 1 \\ 0 & 1 \end{bmatrix}$ (c) $\begin{bmatrix} 1 & 1 \\ -2 & 1 \\ 2 & 1 \end{bmatrix}$

(d) $\begin{bmatrix} 1 & 0 & 2 \\ 0 & 1 & 1 \\ 1 & 2 & 0 \end{bmatrix}$ (e) $\begin{bmatrix} 1 & 2 & 1 \\ 1 & 1 & 1 \\ 0 & 3 & 1 \end{bmatrix}$ (f) $\begin{bmatrix} 1 & 0 & 1 \\ -1 & 1 & 1 \\ 1 & 0 & 1 \\ -1 & 1 & 1 \end{bmatrix}$

30. In Step 3 of the proof of Theorem 6.3.5, it was stated that "the linear independence of $\{u_1, u_2, \ldots, u_n\}$ ensures that $v_3 \neq 0$." Prove this statement.

31. Prove that the diagonal entries of R in Formula (15) are nonzero.

32. (*Calculus required*) Use Theorem 6.3.2a to express the following polynomials as linear combinations of the first three Legendre polynomials (see the Remark following Example 8).

(a) $1 + x + 4x^2$ (b) $2 - 7x^2$ (c) $4 + 3x$

33. (*Calculus required*) Let P_2 have the inner product
$$\langle p, q \rangle = \int_0^1 p(x) q(x)\, dx$$
Apply the Gram–Schmidt process to transform the standard basis $S = \{1, x, x^2\}$ into an orthonormal basis.

34. Find vectors x and y in R^2 that are orthonormal with respect to the inner product $\langle u, v \rangle = 3u_1 v_1 + 2u_2 v_2$ but are not orthonormal with respect to the Euclidean inner product.

True-False Exercises

In parts (a)–(f) determine whether the statement is true or false, and justify your answer.

(a) Every linearly independent set of vectors in an inner product space is orthogonal.

(b) Every orthogonal set of vectors in an inner product space is linearly independent.

(c) Every nontrivial subspace of R^3 has an orthonormal basis with respect to the Euclidean inner product.

(d) Every nonzero finite-dimensional inner product space has an orthonormal basis.

(e) $\text{proj}_W x$ is orthogonal to every vector of W.

(f) If A is an $n \times n$ matrix with a nonzero determinant, then A has a QR-decomposition.

6.4 Best Approximation; Least Squares

In this section we will be concerned with linear systems that cannot be solved exactly and for which an approximate solution is needed. Such systems commonly occur in applications where measurement errors "perturb" the coefficients of a consistent system sufficiently to produce inconsistency.

Least Squares Solutions of Linear Systems

Suppose that $Ax = b$ is an *inconsistent* linear system of m equations in n unknowns in which we suspect the inconsistency to be caused by measurement errors in the coefficients of A. Since no exact solution is possible, we will look for a vector x that comes as "close as possible" to being a solution in the sense that it minimizes $\|b - Ax\|$ with respect

6.4 Best Approximation; Least Squares

to the Euclidean inner product on R^m. You can think of $A\mathbf{x}$ as an approximation to $\mathbf{b}$ and $\|\mathbf{b} - A\mathbf{x}\|$ as the *error* in that approximation—the smaller the error, the better the approximation. This leads to the following problem.

> **Least Squares Problem** Given a linear system $A\mathbf{x} = \mathbf{b}$ of m equations in n unknowns, find a vector $\mathbf{x}$ that minimizes $\|\mathbf{b} - A\mathbf{x}\|$ with respect to the Euclidean inner product on R^m. We call such an $\mathbf{x}$ a *least squares solution* of the system, we call $\mathbf{b} - A\mathbf{x}$ the *least squares error vector*, and we call $\|\mathbf{b} - A\mathbf{x}\|$ the *least squares error*.

To clarify the above terminology, suppose that the matrix form of $\mathbf{b} - A\mathbf{x}$ is

$$\mathbf{b} - A\mathbf{x} = \begin{bmatrix} e_1 \\ e_2 \\ \vdots \\ e_m \end{bmatrix}$$

The term "least squares solution" results from the fact that minimizing $\|\mathbf{b} - A\mathbf{x}\|$ also minimizes $\|\mathbf{b} - A\mathbf{x}\|^2 = e_1^2 + e_2^2 + \cdots + e_m^2$.

Best Approximation

Suppose that $\mathbf{b}$ is a fixed vector in R^3 that we would like to approximate by a vector $\mathbf{w}$ that is required to lie in some subspace W of R^3. Unless $\mathbf{b}$ happens to be in W, then any such approximation will result in an "error vector" $\mathbf{b} - \mathbf{w}$ that cannot be made equal to $\mathbf{0}$ no matter how $\mathbf{w}$ is chosen (Figure 6.4.1a). However, by choosing

$$\mathbf{w} = \operatorname{proj}_W \mathbf{b}$$

we can make the length of the error vector

$$\|\mathbf{b} - \mathbf{w}\| = \|\mathbf{b} - \operatorname{proj}_W \mathbf{b}\|$$

as small as possible (Figure 6.4.1b).

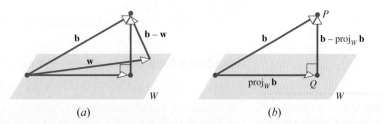

▶ Figure 6.4.1   (a)   (b)

These geometric ideas suggest the following general theorem.

> **THEOREM 6.4.1**   **Best Approximation Theorem**
> *If W is a finite-dimensional subspace of an inner product space V, and if $\mathbf{b}$ is a vector in V, then $\operatorname{proj}_W \mathbf{b}$ is the **best approximation** to $\mathbf{b}$ from W in the sense that*
> $$\|\mathbf{b} - \operatorname{proj}_W \mathbf{b}\| < \|\mathbf{b} - \mathbf{w}\|$$
> *for every vector $\mathbf{w}$ in W that is different from $\operatorname{proj}_W \mathbf{b}$.*

Proof For every vector $\mathbf{w}$ in W, we can write

$$\mathbf{b} - \mathbf{w} = (\mathbf{b} - \operatorname{proj}_W \mathbf{b}) + (\operatorname{proj}_W \mathbf{b} - \mathbf{w}) \tag{1}$$

But $\text{proj}_W \mathbf{b} - \mathbf{w}$ being a difference of vectors in W is itself in W; and since $\mathbf{b} - \text{proj}_W \mathbf{b}$ is orthogonal to W, the two terms on the right side of (1) are orthogonal. Thus, it follows from the Theorem of Pythagoras (Theorem 6.2.3) that

$$\|\mathbf{b} - \mathbf{w}\|^2 = \|\mathbf{b} - \text{proj}_W \mathbf{b}\|^2 + \|\text{proj}_W \mathbf{b} - \mathbf{w}\|^2$$

Since $\mathbf{w} \neq \text{proj}_W \mathbf{b}$, it follows that the second term in this sum is positive, and hence that

$$\|\mathbf{b} - \text{proj}_W \mathbf{b}\|^2 < \|\mathbf{b} - \mathbf{w}\|^2$$

Since norms are nonnegative, it follows (from a property of inequalities) that

$$\|\mathbf{b} - \text{proj}_W \mathbf{b}\| < \|\mathbf{b} - \mathbf{w}\| \blacktriangleleft$$

Least Squares Solutions of Linear Systems

One way to find a least squares solution of $A\mathbf{x} = \mathbf{b}$ is to calculate the orthogonal projection $\text{proj}_W \mathbf{b}$ on the column space W of the matrix A and then solve the equation

$$A\mathbf{x} = \text{proj}_W \mathbf{b} \tag{2}$$

However, we can avoid the need to calculate the projection by rewriting (2) as

$$\mathbf{b} - A\mathbf{x} = \mathbf{b} - \text{proj}_W \mathbf{b}$$

and then multiplying both sides of this equation by A^T to obtain

$$A^T(\mathbf{b} - A\mathbf{x}) = A^T(\mathbf{b} - \text{proj}_W \mathbf{b}) \tag{3}$$

Since $\mathbf{b} - \text{proj}_W \mathbf{b}$ is the component of $\mathbf{b}$ that is orthogonal to the column space of A, it follows from Theorem 4.8.9b that this vector lies in the null space of A^T, and hence that

$$A^T(\mathbf{b} - \text{proj}_W \mathbf{b}) = \mathbf{0}$$

Thus, (3) simplifies to

$$A^T(\mathbf{b} - A\mathbf{x}) = \mathbf{0}$$

which we can rewrite as

$$A^T A \mathbf{x} = A^T \mathbf{b} \tag{4}$$

This is called the ***normal equation*** or the ***normal system*** associated with $A\mathbf{x} = \mathbf{b}$. When viewed as a linear system, the individual equations are called the ***normal equations*** associated with $A\mathbf{x} = \mathbf{b}$.

In summary, we have established the following result.

THEOREM 6.4.2 *For every linear system $A\mathbf{x} = \mathbf{b}$, the associated normal system*

$$A^T A \mathbf{x} = A^T \mathbf{b} \tag{5}$$

is consistent, and all solutions of (5) are least squares solutions of $A\mathbf{x} = \mathbf{b}$. Moreover, if W is the column space of A, and $\mathbf{x}$ is any least squares solution of $A\mathbf{x} = \mathbf{b}$, then the orthogonal projection of $\mathbf{b}$ on W is

$$\text{proj}_W \mathbf{b} = A\mathbf{x} \tag{6}$$

If a linear system is consistent, then its exact solutions are the same as its least squares solutions, in which case the error is zero.

▶ **EXAMPLE 1** **Least Squares Solution**

(a) Find all least squares solutions of the linear system

$$\begin{aligned} x_1 - x_2 &= 4 \\ 3x_1 + 2x_2 &= 1 \\ -2x_1 + 4x_2 &= 3 \end{aligned}$$

(b) Find the error vector and the error.

6.4 Best Approximation; Least Squares

Solution (a) It will be convenient to express the system in the matrix form $A\mathbf{x} = \mathbf{b}$, where

$$A = \begin{bmatrix} 1 & -1 \\ 3 & 2 \\ -2 & 4 \end{bmatrix} \quad \text{and} \quad \mathbf{b} = \begin{bmatrix} 4 \\ 1 \\ 3 \end{bmatrix}$$

It follows that

$$A^T A = \begin{bmatrix} 1 & 3 & -2 \\ -1 & 2 & 4 \end{bmatrix} \begin{bmatrix} 1 & -1 \\ 3 & 2 \\ -2 & 4 \end{bmatrix} = \begin{bmatrix} 14 & -3 \\ -3 & 21 \end{bmatrix}$$

$$A^T \mathbf{b} = \begin{bmatrix} 1 & 3 & -2 \\ -1 & 2 & 4 \end{bmatrix} \begin{bmatrix} 4 \\ 1 \\ 3 \end{bmatrix} = \begin{bmatrix} 1 \\ 10 \end{bmatrix}$$

so the normal system $A^T A \mathbf{x} = A^T \mathbf{b}$ is

$$\begin{bmatrix} 14 & -3 \\ -3 & 21 \end{bmatrix} \begin{bmatrix} x_1 \\ x_2 \end{bmatrix} = \begin{bmatrix} 1 \\ 10 \end{bmatrix}$$

Solving this system yields a unique least squares solution, namely,

$$x_1 = \tfrac{17}{95}, \quad x_2 = \tfrac{143}{285}$$

Solution (b) The error vector is

$$\mathbf{b} - A\mathbf{x} = \begin{bmatrix} 4 \\ 1 \\ 3 \end{bmatrix} - \begin{bmatrix} 1 & -1 \\ 3 & 2 \\ -2 & 4 \end{bmatrix} \begin{bmatrix} \tfrac{17}{95} \\ \tfrac{143}{285} \end{bmatrix} = \begin{bmatrix} 4 \\ 1 \\ 3 \end{bmatrix} - \begin{bmatrix} -\tfrac{92}{285} \\ \tfrac{439}{285} \\ \tfrac{95}{57} \end{bmatrix} = \begin{bmatrix} \tfrac{1232}{285} \\ -\tfrac{154}{285} \\ \tfrac{4}{3} \end{bmatrix}$$

and the error is

$$\|\mathbf{b} - A\mathbf{x}\| \approx 4.556$$

▶ **EXAMPLE 2 Orthogonal Projection on a Subspace**

Find the orthogonal projection of the vector $\mathbf{u} = (-3, -3, 8, 9)$ on the subspace of R^4 spanned by the vectors

$$\mathbf{u}_1 = (3, 1, 0, 1), \quad \mathbf{u}_2 = (1, 2, 1, 1), \quad \mathbf{u}_3 = (-1, 0, 2, -1)$$

Solution We could solve this problem by first using the Gram–Schmidt process to convert $\{\mathbf{u}_1, \mathbf{u}_2, \mathbf{u}_3\}$ into an orthonormal basis and then applying the method used in Example 6 of Section 6.3. However, the following method is more efficient.

The subspace W of R^4 spanned by $\mathbf{u}_1$, $\mathbf{u}_2$, and $\mathbf{u}_3$ is the column space of the matrix

$$A = \begin{bmatrix} 3 & 1 & -1 \\ 1 & 2 & 0 \\ 0 & 1 & 2 \\ 1 & 1 & -1 \end{bmatrix}$$

Thus, if $\mathbf{u}$ is expressed as a column vector, we can find the orthogonal projection of $\mathbf{u}$ on W by finding a least squares solution of the system $A\mathbf{x} = \mathbf{u}$ and then calculating

$\text{proj}_W \mathbf{u} = A\mathbf{x}$ from the least squares solution. The computations are as follows: The system $A\mathbf{x} = \mathbf{u}$ is

$$\begin{bmatrix} 3 & 1 & -1 \\ 1 & 2 & 0 \\ 0 & 1 & 2 \\ 1 & 1 & -1 \end{bmatrix} \begin{bmatrix} x_1 \\ x_2 \\ x_3 \end{bmatrix} = \begin{bmatrix} -3 \\ -3 \\ 8 \\ 9 \end{bmatrix}$$

so

$$A^T A = \begin{bmatrix} 3 & 1 & 0 & 1 \\ 1 & 2 & 1 & 1 \\ -1 & 0 & 2 & -1 \end{bmatrix} \begin{bmatrix} 3 & 1 & -1 \\ 1 & 2 & 0 \\ 0 & 1 & 2 \\ 1 & 1 & -1 \end{bmatrix} = \begin{bmatrix} 11 & 6 & -4 \\ 6 & 7 & 0 \\ -4 & 0 & 6 \end{bmatrix}$$

$$A^T \mathbf{u} = \begin{bmatrix} 3 & 1 & 0 & 1 \\ 1 & 2 & 1 & 1 \\ -1 & 0 & 2 & -1 \end{bmatrix} \begin{bmatrix} -3 \\ -3 \\ 8 \\ 9 \end{bmatrix} = \begin{bmatrix} -3 \\ 8 \\ 10 \end{bmatrix}$$

The normal system $A^T A\mathbf{x} = A^T \mathbf{u}$ in this case is

$$\begin{bmatrix} 11 & 6 & -4 \\ 6 & 7 & 0 \\ -4 & 0 & 6 \end{bmatrix} \begin{bmatrix} x_1 \\ x_2 \\ x_3 \end{bmatrix} = \begin{bmatrix} -3 \\ 8 \\ 10 \end{bmatrix}$$

Solving this system yields

$$\mathbf{x} = \begin{bmatrix} x_1 \\ x_2 \\ x_3 \end{bmatrix} = \begin{bmatrix} -1 \\ 2 \\ 1 \end{bmatrix}$$

as the least squares solution of $A\mathbf{x} = \mathbf{u}$ (verify), so

$$\text{proj}_W \mathbf{u} = A\mathbf{x} = \begin{bmatrix} 3 & 1 & -1 \\ 1 & 2 & 0 \\ 0 & 1 & 2 \\ 1 & 1 & -1 \end{bmatrix} \begin{bmatrix} -1 \\ 2 \\ 1 \end{bmatrix} = \begin{bmatrix} -2 \\ 3 \\ 4 \\ 0 \end{bmatrix}$$

or, in comma-delimited notation, $\text{proj}_W \mathbf{u} = (-2, 3, 4, 0)$. ◄

Uniqueness of Least Squares Solutions

In general, least squares solutions of linear systems are not unique. Although the linear system in Example 1 turned out to have a unique least squares solution, that occurred only because the coefficient matrix of the system happened to satisfy certain conditions that guarantee uniqueness. Our next theorem will show what those conditions are.

THEOREM 6.4.3 *If A is an $m \times n$ matrix, then the following are equivalent.*

(a) *A has linearly independent column vectors.*

(b) *$A^T A$ is invertible.*

Proof We will prove that (a) $\Rightarrow$ (b) and leave the proof that (b) $\Rightarrow$ (a) as an exercise.

(a) $\Rightarrow$ (b) Assume that A has linearly independent column vectors. The matrix $A^T A$ has size $n \times n$, so we can prove that this matrix is invertible by showing that the linear system $A^TA\mathbf{x} = \mathbf{0}$ has only the trivial solution. But if $\mathbf{x}$ is any solution of this system, then $A\mathbf{x}$ is in the null space of A^T and also in the column space of A. By Theorem 4.8.9b these spaces are orthogonal complements, so part (b) of Theorem 6.2.4 implies that $A\mathbf{x} = \mathbf{0}$. But A is assumed to have linearly independent column vectors, so $\mathbf{x} = \mathbf{0}$ by Theorem 1.3.1. ◀

The next theorem, which follows directly from Theorems 6.4.2 and 6.4.3, gives an explicit formula for the least squares solution of a linear system in which the coefficient matrix has linearly independent column vectors.

As an exercise, try using Formula (7) to solve the problem in part (a) of Example 1.

THEOREM 6.4.4 *If A is an $m \times n$ matrix with linearly independent column vectors, then for every $m \times 1$ matrix $\mathbf{b}$, the linear system $A\mathbf{x} = \mathbf{b}$ has a unique least squares solution. This solution is given by*

$$\mathbf{x} = (A^TA)^{-1}A^T\mathbf{b} \quad (7)$$

Moreover, if W is the column space of A, then the orthogonal projection of $\mathbf{b}$ on W is

$$\text{proj}_W \mathbf{b} = A\mathbf{x} = A(A^TA)^{-1}A^T\mathbf{b} \quad (8)$$

OPTIONAL
The Role of QR-Decomposition in Least Squares Problems

Formulas (7) and (8) have theoretical use but are not well suited for numerical computation. In practice, least squares solutions of $A\mathbf{x} = \mathbf{b}$ are typically found by using some variation of Gaussian elimination to solve the normal equations or by using QR-decomposition and the following theorem.

THEOREM 6.4.5 *If A is an $m \times n$ matrix with linearly independent column vectors, and if $A = QR$ is a QR-decomposition of A (see Theorem 6.3.7), then for each $\mathbf{b}$ in R^m the system $A\mathbf{x} = \mathbf{b}$ has a unique least squares solution given by*

$$\mathbf{x} = R^{-1}Q^T\mathbf{b} \quad (9)$$

A proof of this theorem and a discussion of its use can be found in many books on numerical methods of linear algebra. However, you can obtain Formula (9) by making the substitution $A = QR$ in (7) and using the fact that $Q^TQ = I$ to obtain

$$\begin{aligned}\mathbf{x} &= \left((QR)^T(QR)\right)^{-1}(QR)^T\mathbf{b} \\ &= (R^TQ^TQR)^{-1}(QR)^T\mathbf{b} \\ &= R^{-1}(R^T)^{-1}R^TQ^T\mathbf{b} \\ &= R^{-1}Q^T\mathbf{b}\end{aligned}$$

Orthogonal Projections on Subspaces of R^m

In Section 4.8 we showed how to compute orthogonal projections on the coordinate axes of a rectangular coordinate system in R^3 and more generally on lines through the origin of R^3. We will now consider the problem of finding orthogonal projections on *subspaces* of R^m. We begin with the following definition.

DEFINITION 1 If W is a subspace of R^m, then the linear transformation $P: R^m \to W$ that maps each vector $\mathbf{x}$ in R^m into its orthogonal projection $\text{proj}_W \mathbf{x}$ in W is called the ***orthogonal projection of R^m on W***.

It follows from Formula (7) that the standard matrix for the transformation P is

$$[P] = A(A^T A)^{-1} A^T \tag{10}$$

where A is constructed using any basis for W as its column vectors.

▶ **EXAMPLE 3** **The Standard Matrix for an Orthogonal Projection on a Line**

We showed in Formula (16) of Section 4.9 that

$$P_\theta = \begin{bmatrix} \cos^2 \theta & \sin \theta \cos \theta \\ \sin \theta \cos \theta & \sin^2 \theta \end{bmatrix}$$

is the standard matrix for the orthogonal projection on the line W through the origin of R^2 that makes an angle θ with the positive x-axis. Derive this result using Formula (10).

Solution The column vectors of A can be formed from any basis for W. Since W is one-dimensional, we can take $\mathbf{w} = (\cos \theta, \sin \theta)$ as the basis vector (Figure 6.4.2), so

$$A = \begin{bmatrix} \cos \theta \\ \sin \theta \end{bmatrix}$$

We leave it for you to show that $A^T A$ is the 1×1 identity matrix. Thus, Formula (10) simplifies to

$$[P] = A(A^T A)^{-1} A^T = AA^T = \begin{bmatrix} \cos \theta \\ \sin \theta \end{bmatrix} [\cos \theta \quad \sin \theta]$$

$$= \begin{bmatrix} \cos^2 \theta & \sin \theta \cos \theta \\ \sin \theta \cos \theta & \sin^2 \theta \end{bmatrix} = P_\theta \blacktriangleleft$$

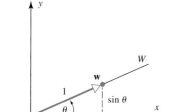

▲ Figure 6.4.2

Another View of Least Squares

Recall from Theorem 4.8.9 that the null space and row space of an $m \times n$ matrix A are orthogonal complements, as are the null space of A^T and the column space of A. Thus, given a linear system $A\mathbf{x} = \mathbf{b}$ in which A is an $m \times n$ matrix, the Projection Theorem (6.3.3) tells us that the vectors $\mathbf{x}$ and $\mathbf{b}$ can each be decomposed into sums of orthogonal terms as

$$\mathbf{x} = \mathbf{x}_{\text{row}(A)} + \mathbf{x}_{\text{null}(A)} \quad \text{and} \quad \mathbf{b} = \mathbf{b}_{\text{null}(A^T)} + \mathbf{b}_{\text{col}(A)}$$

where $\mathbf{x}_{\text{row}(A)}$ and $\mathbf{x}_{\text{null}(A)}$ are the orthogonal projections of $\mathbf{x}$ on the row space of A and the null space of A, and the vectors $\mathbf{b}_{\text{null}(A^T)}$ and $\mathbf{b}_{\text{col}(A)}$ are the orthogonal projections of $\mathbf{b}$ on the null space of A^T and the column space of A.

In Figure 6.4.3 we have represented the fundamental spaces of A by perpendicular lines in R^n and R^m on which we indicated the orthogonal projections of $\mathbf{x}$ and $\mathbf{b}$. (This, of course, is only pictorial since the fundamental spaces need not be one-dimensional.) The figure shows $A\mathbf{x}$ as a point in the column space of A and conveys that $\mathbf{b}_{\text{col}(A)}$ is the point in $\text{col}(A)$ that is closest to $\mathbf{b}$. This illustrates that the least squares solutions of $A\mathbf{x} = \mathbf{b}$ are the exact solutions of the equation $A\mathbf{x} = \mathbf{b}_{\text{col}(A)}$.

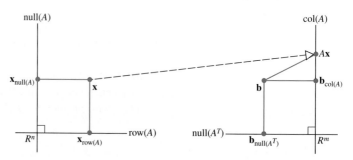

▶ Figure 6.4.3

More on the Equivalence Theorem

As our final result in the main part of this section we will add one additional part to Theorem 5.1.6.

THEOREM 6.4.6 **Equivalent Statements**

If A is an n × n matrix, then the following statements are equivalent.

(a) A is invertible.
(b) $A\mathbf{x} = \mathbf{0}$ has only the trivial solution.
(c) The reduced row echelon form of A is I_n.
(d) A is expressible as a product of elementary matrices.
(e) $A\mathbf{x} = \mathbf{b}$ is consistent for every $n \times 1$ matrix $\mathbf{b}$.
(f) $A\mathbf{x} = \mathbf{b}$ has exactly one solution for every $n \times 1$ matrix $\mathbf{b}$.
(g) $\det(A) \neq 0$.
(h) The column vectors of A are linearly independent.
(i) The row vectors of A are linearly independent.
(j) The column vectors of A span R^n.
(k) The row vectors of A span R^n.
(l) The column vectors of A form a basis for R^n.
(m) The row vectors of A form a basis for R^n.
(n) A has rank n.
(o) A has nullity 0.
(p) The orthogonal complement of the null space of A is R^n.
(q) The orthogonal complement of the row space of A is $\{\mathbf{0}\}$.
(r) The range of T_A is R^n.
(s) T_A is one-to-one.
(t) $\lambda = 0$ is not an eigenvalue of A.
(u) $A^T A$ is invertible.

The proof of part (u) follows from part (h) of this theorem and Theorem 6.4.3 applied to square matrices.

OPTIONAL

We now have all the ingredients needed to prove Theorem 6.3.3 in the special case where V is the vector space R^m.

Proof of Theorem 6.3.3 We will leave the case where $W = \{\mathbf{0}\}$ as an exercise, so assume that $W \neq \{\mathbf{0}\}$. Let $\{\mathbf{v}_1, \mathbf{v}_2, \dots, \mathbf{v}_k\}$ be any basis for W, and form the $m \times k$ matrix M that has these basis vectors as successive columns. This makes W the column space of M and hence $W^\perp$ the null space of M^T. We will complete the proof by showing that every vector $\mathbf{u}$ in R^m can be written in exactly one way as

$$\mathbf{u} = \mathbf{w}_1 + \mathbf{w}_2$$

where $\mathbf{w}_1$ is in the column space of M and $M^T \mathbf{w}_2 = \mathbf{0}$. However, to say that $\mathbf{w}_1$ is in the column space of M is equivalent to saying $\mathbf{w}_1 = M\mathbf{x}$ for some vector $\mathbf{x}$ in R^m, and to say that $M^T \mathbf{w}_2 = \mathbf{0}$ is equivalent to saying that $M^T(\mathbf{u} - \mathbf{w}_1) = \mathbf{0}$. Thus, if we can show that the equation

$$M^T(\mathbf{u} - M\mathbf{x}) = \mathbf{0} \tag{11}$$

Chapter 6 Inner Product Spaces

has a unique solution for $\mathbf{x}$, then $\mathbf{w}_1 = M\mathbf{x}$ and $\mathbf{w}_2 = \mathbf{x} - \mathbf{w}_1$ will be uniquely determined vectors with the required properties. To do this, let us rewrite (11) as

$$M^T M \mathbf{x} = M^T \mathbf{u}$$

Since the matrix M has linearly independent column vectors, the matrix $M^T M$ is invertible by Theorem 6.4.6 and hence the equation has a unique solution as required to complete the proof. ◀

Concept Review

- Least squares problem
- Least squares solution
- Least squares error vector
- Least squares error
- Best approximation
- Normal equation
- Orthogonal projection

Skills

- Find the least squares solution of a linear system.
- Find the error and error vector associated with a least squares solution to a linear system.
- Use the techniques developed in this section to compute orthogonal projections.
- Find the standard matrix of an orthogonal projection.

Exercise Set 6.4

1. Find the normal system associated with the given linear system.

(a) $\begin{bmatrix} 1 & -1 \\ 2 & 3 \\ 4 & 5 \end{bmatrix} \begin{bmatrix} x_1 \\ x_2 \end{bmatrix} = \begin{bmatrix} 2 \\ -1 \\ 5 \end{bmatrix}$

(b) $\begin{bmatrix} 2 & -1 & 0 \\ 3 & 1 & 2 \\ -1 & 4 & 5 \\ 1 & 2 & 4 \end{bmatrix} \begin{bmatrix} x_1 \\ x_2 \\ x_3 \end{bmatrix} = \begin{bmatrix} -1 \\ 0 \\ 1 \\ 2 \end{bmatrix}$

▶ In Exercises 2–4, find the least squares solution of the linear equation $A\mathbf{x} = \mathbf{b}$. ◀

2. (a) $A = \begin{bmatrix} 1 & -1 \\ 2 & 3 \\ 4 & 5 \end{bmatrix}$; $\mathbf{b} = \begin{bmatrix} 2 \\ -1 \\ 5 \end{bmatrix}$

(b) $A = \begin{bmatrix} 2 & -2 \\ 1 & 1 \\ 3 & 1 \end{bmatrix}$; $\mathbf{b} = \begin{bmatrix} 2 \\ -1 \\ 1 \end{bmatrix}$

3. (a) $A = \begin{bmatrix} 1 & 2 \\ 1 & 1 \\ 2 & 0 \end{bmatrix}$; $\mathbf{b} = \begin{bmatrix} 3 \\ 2 \\ 1 \end{bmatrix}$

(b) $A = \begin{bmatrix} 1 & 1 & 0 \\ -2 & 1 & 1 \\ 1 & 0 & 1 \\ -1 & -1 & 1 \end{bmatrix}$; $\mathbf{b} = \begin{bmatrix} 2 \\ 1 \\ 2 \\ 1 \end{bmatrix}$

4. (a) $A = \begin{bmatrix} 3 & 2 & -1 \\ 1 & -4 & 3 \\ 1 & 10 & -7 \end{bmatrix}$; $\mathbf{b} = \begin{bmatrix} 2 \\ -2 \\ 1 \end{bmatrix}$

(b) $A = \begin{bmatrix} 2 & 0 & -1 \\ 1 & -2 & 2 \\ 2 & -1 & 0 \\ 0 & 1 & -1 \end{bmatrix}$; $\mathbf{b} = \begin{bmatrix} 0 \\ 6 \\ 0 \\ 6 \end{bmatrix}$

▶ In Exercises 5–6, find the least squares error vector $\mathbf{e} = \mathbf{b} - A\mathbf{x}$ resulting from the least squares solution $\mathbf{x}$ and verify that it is orthogonal to the column space of A. ◀

5. (a) A and $\mathbf{b}$ are as in Exercise 3(a).

(b) A and $\mathbf{b}$ are as in Exercise 3(b).

6. (a) A and $\mathbf{b}$ are as in Exercise 4(a).

(b) A and $\mathbf{b}$ are as in Exercise 4(b).

7. Find all least squares solutions of $A\mathbf{x} = \mathbf{b}$ and confirm that all of the solutions have the same error vector. Compute the least squares error.

(a) $A = \begin{bmatrix} 1 & -2 \\ 2 & -4 \\ 3 & -6 \end{bmatrix}$; $\mathbf{b} = \begin{bmatrix} 4 \\ 2 \\ 1 \end{bmatrix}$

(b) $A = \begin{bmatrix} 1 & 3 \\ -2 & -6 \\ 3 & 9 \end{bmatrix}$; $\mathbf{b} = \begin{bmatrix} 1 \\ 0 \\ 1 \end{bmatrix}$

(c) $A = \begin{bmatrix} -1 & 3 & 2 \\ 2 & 1 & 3 \\ 0 & 1 & 1 \end{bmatrix}$; $\mathbf{b} = \begin{bmatrix} 7 \\ 0 \\ -7 \end{bmatrix}$

8. Find the orthogonal projection of $\mathbf{u}$ on the subspace of R^3 spanned by the vectors $\mathbf{v}_1$ and $\mathbf{v}_2$.
   (a) $\mathbf{u} = (1, 2, 3);\ \mathbf{v}_1 = (2, 1, 0),\ \mathbf{v}_2 = (-1, 1, 0)$
   (b) $\mathbf{u} = (1, -6, 1);\ \mathbf{v}_1 = (-1, 2, 1),\ \mathbf{v}_2 = (2, 2, 4)$

9. Find the orthogonal projection of $\mathbf{u}$ on the subspace of R^4 spanned by the vectors $\mathbf{v}_1$, $\mathbf{v}_2$, and $\mathbf{v}_3$.
   (a) $\mathbf{u} = (1, -1, 3, 1);\ \mathbf{v}_1 = (1, 2, 1, 1),\ \mathbf{v}_2 = (0, 1, 1, 0),$
       $\mathbf{v}_3 = (2, 1, 2, 1)$
   (b) $\mathbf{u} = (-2, 0, 2, 4);\ \mathbf{v}_1 = (1, 1, 3, 0),$
       $\mathbf{v}_2 = (-2, -1, -2, 1),\ \mathbf{v}_3 = (-3, -1, 1, 3)$

10. Find the orthogonal projection of $\mathbf{u} = (5, 6, 7, 2)$ on the solution space of the homogeneous linear system
$$\begin{aligned} x_1 + x_2 + x_3 &= 0 \\ 2x_2 + x_3 + x_4 &= 0 \end{aligned}$$

11. In each part, find $\det(A^TA)$, and apply Theorem 6.4.3 to determine whether A has linearly independent column vectors.

   (a) $A = \begin{bmatrix} 1 & 0 & -1 \\ 2 & 1 & 2 \\ 3 & 1 & -1 \end{bmatrix}$ (b) $A = \begin{bmatrix} 2 & -1 & 3 \\ 0 & 1 & 1 \\ -1 & 0 & -2 \\ 4 & -5 & 3 \end{bmatrix}$

12. Use Formula (10) and the method of Example 3 to find the standard matrix for the orthogonal projection $P: R^2 \to R^2$ onto
   (a) the x-axis.   (b) the y-axis.
   [Note: Compare your results to Table 3 of Section 4.9.]

13. Use Formula (10) and the method of Example 3 to find the standard matrix for the orthogonal projection $P: R^3 \to R^3$ onto
   (a) the xy-plane.   (b) the yz-plane.
   [Note: Compare your results to Table 4 of Section 4.9.]

14. Show that if $\mathbf{w} = (a, b, c)$ is a nonzero vector, then the standard matrix for the orthogonal projection of R^3 on the line span$\{\mathbf{w}\}$ is
$$P = \frac{1}{a^2 + b^2 + c^2} \begin{bmatrix} a^2 & ab & ac \\ ab & b^2 & bc \\ ac & bc & c^2 \end{bmatrix}$$

15. Let W be the plane with equation $3x - 4y + z = 0$.
   (a) Find a basis for W.
   (b) Use Formula (10) to find the standard matrix for the orthogonal projection on W.
   (c) Use the matrix obtained in part (b) to find the orthogonal projection of a point $P_0(x_0, y_0, z_0)$ on W.
   (d) Find the distance between the point $P_0(2, 1, -1)$ and the plane W, and check your result using Theorem 3.3.4.

16. Let W be the line with parametric equations
$$x = 2t,\quad y = -t,\quad z = 4t$$
   (a) Find a basis for W.
   (b) Use Formula (10) to find the standard matrix for the orthogonal projection on W.
   (c) Use the matrix obtained in part (b) to find the orthogonal projection of a point $P_0(x_0, y_0, z_0)$ on W.
   (d) Find the distance between the point $P_0(2, 1, -3)$ and the line W.

17. In R^3, consider the line l given by the equations
$$x = -1,\quad y = t,\quad z = -t$$
and the line m given by the equations
$$x = 2s,\quad y = 1 + s,\quad z = s$$
Let P be a point on l, and let Q be a point on m. Find the values of t and s that minimize the distance between the lines by minimizing the squared distance $\|P - Q\|^2$.

18. Prove: If A has linearly independent column vectors, and if $A\mathbf{x} = \mathbf{b}$ is consistent, then the least squares solution of $A\mathbf{x} = \mathbf{b}$ and the exact solution of $A\mathbf{x} = \mathbf{b}$ are the same.

19. Prove: If A has linearly independent column vectors, and if $\mathbf{b}$ is orthogonal to the column space of A, then the least squares solution of $A\mathbf{x} = \mathbf{b}$ is $\mathbf{x} = \mathbf{0}$.

20. Let $P: R^m \to W$ be the orthogonal projection of R^m onto a subspace W.
   (a) Prove that $[P]^2 = [P]$.
   (b) What does the result in part (a) imply about the composition $P \circ P$?
   (c) Show that $[P]$ is symmetric.

21. Let A be an $m \times n$ matrix with linearly independent row vectors. Find a standard matrix for the orthogonal projection of R^n onto the row space of A. [Hint: Start with Formula (10).]

22. Prove the implication $(b) \Rightarrow (a)$ of Theorem 6.4.3.

True-False Exercises

In parts (a)–(h) determine whether the statement is true or false, and justify your answer.

(a) If A is an $m \times n$ matrix, then A^TA is a square matrix.

(b) If A^TA is invertible, then A is invertible.

(c) If A is invertible, then A^TA is invertible.

(d) If $A\mathbf{x} = \mathbf{b}$ is a consistent linear system, then $A^TA\mathbf{x} = A^T\mathbf{b}$ is also consistent.

(e) If $A\mathbf{x} = \mathbf{b}$ is an inconsistent linear system, then $A^TA\mathbf{x} = A^T\mathbf{b}$ is also inconsistent.

(f) Every linear system has a least squares solution.

(g) Every linear system has a unique least squares solution.

(h) If A is an $m \times n$ matrix with linearly independent columns and $\mathbf{b}$ is in R^m, then $A\mathbf{x} = \mathbf{b}$ has a unique least squares solution.

6.5 Least Squares Fitting to Data

In this section we will use results about orthogonal projections in inner product spaces to obtain a technique for fitting a line or other polynomial curve to a set of experimentally determined points in the plane.

Fitting a Curve to Data

A common problem in experimental work is to obtain a mathematical relationship $y = f(x)$ between two variables x and y by "fitting" a curve to points in the plane corresponding to various experimentally determined values of x and y, say

$$(x_1, y_1), (x_2, y_2), \ldots, (x_n, y_n)$$

On the basis of theoretical considerations or simply by observing the pattern of the points, the experimenter decides on the general form of the curve $y = f(x)$ to be fitted. Some possibilities are (Figure 6.5.1)

(a) A straight line: $y = a + bx$
(b) A quadratic polynomial: $y = a + bx + cx^2$
(c) A cubic polynomial: $y = a + bx + cx^2 + dx^3$

Because the points are obtained experimentally, there is often some measurement "error" in the data, making it impossible to find a curve of the desired form that passes through all the points. Thus, the idea is to choose the curve (by determining its coefficients) that "best" fits the data. We begin with the simplest and most common case: fitting a straight line to data points.

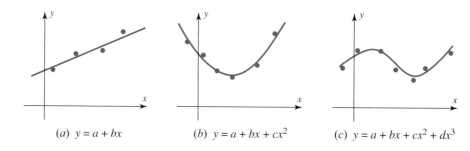

▶ Figure 6.5.1   (a) $y = a + bx$ (b) $y = a + bx + cx^2$ (c) $y = a + bx + cx^2 + dx^3$

Least Squares Fit of a Straight Line

Suppose we want to fit a straight line $y = a + bx$ to the experimentally determined points

$$(x_1, y_1), (x_2, y_2), \ldots, (x_n, y_n)$$

If the data points were collinear, the line would pass through all n points, and the unknown coefficients a and b would satisfy the equations

$$y_1 = a + bx_1$$
$$y_2 = a + bx_2$$
$$\vdots$$
$$y_n = a + bx_n$$

We can write this system in matrix form as

$$\begin{bmatrix} 1 & x_1 \\ 1 & x_2 \\ \vdots & \vdots \\ 1 & x_n \end{bmatrix} \begin{bmatrix} a \\ b \end{bmatrix} = \begin{bmatrix} y_1 \\ y_2 \\ \vdots \\ y_n \end{bmatrix}$$

or more compactly as
$$Mv = y \tag{1}$$
where
$$y = \begin{bmatrix} y_1 \\ y_2 \\ \vdots \\ y_n \end{bmatrix}, \quad M = \begin{bmatrix} 1 & x_1 \\ 1 & x_2 \\ \vdots & \vdots \\ 1 & x_n \end{bmatrix}, \quad v = \begin{bmatrix} a \\ b \end{bmatrix} \tag{2}$$

If the data points are not collinear, then it is impossible to find coefficients a and b that satisfy system (1) exactly; that is, the system is inconsistent. In this case we will look for a least squares solution

$$v = v^* = \begin{bmatrix} a^* \\ b^* \end{bmatrix}$$

We call a line $y = a^* + b^*x$ whose coefficients come from a least squares solution a *regression line* or a *least squares straight line fit* to the data. To explain this terminology, recall that a least squares solution of (1) minimizes

$$\|y - Mv\| \tag{3}$$

If we express the square of (3) in terms of components, we obtain

$$\|y - Mv\|^2 = (y_1 - a - bx_1)^2 + (y_2 - a - bx_2)^2 + \cdots + (y_n - a - bx_n)^2 \tag{4}$$

If we now let

$$d_1 = |y_1 - a - bx_1|, \quad d_2 = |y_2 - a - bx_2|, \ldots, \quad d_n = |y_n - a - bx_n|$$

then (4) can be written as

$$\|y - Mv\|^2 = d_1^2 + d_2^2 + \cdots + d_n^2 \tag{5}$$

As illustrated in Figure 6.5.2, the number d_i can be interpreted as the vertical distance between the line $y = a + bx$ and the data point (x_i, y_i). This distance is a measure of the "error" at the point (x_i, y_i) resulting from the inexact fit of $y = a + bx$ to the data points, the assumption being that the x_i are known exactly and that all the error is in the measurement of the y_i. Since (3) and (5) are minimized by the same vector v^*, the least squares straight line fit minimizes the sum of the squares of the estimated errors d_i, hence the name *least squares straight line fit*.

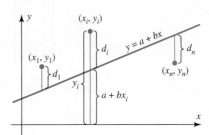

▶ Figure 6.5.2 d_i measures the vertical error in the least squares straight line.

Normal Equations   Recall from Theorem 6.4.2 that the least squares solutions of (1) can be obtained by solving the associated normal system

$$M^T M v = M^T y$$

the equations of which are called the **normal equations**.

In the exercises it will be shown that the column vectors of M are linearly independent if and only if the n data points do not lie on a vertical line in the xy-plane. In this case it follows from Theorem 6.4.4 that the least squares solution is unique and is given by

$$\mathbf{v}^* = (M^T M)^{-1} M^T \mathbf{y}$$

In summary, we have the following theorem.

THEOREM 6.5.1 **Uniqueness of the Least Squares Solution**

Let $(x_1, y_1), (x_2, y_2), \ldots, (x_n, y_n)$ be a set of two or more data points, not all lying on a vertical line, and let

$$M = \begin{bmatrix} 1 & x_1 \\ 1 & x_2 \\ \vdots & \vdots \\ 1 & x_n \end{bmatrix} \quad \text{and} \quad \mathbf{y} = \begin{bmatrix} y_1 \\ y_2 \\ \vdots \\ y_n \end{bmatrix}$$

Then there is a unique least squares straight line fit

$$y = a^* + b^* x$$

to the data points. Moreover,

$$\mathbf{v}^* = \begin{bmatrix} a^* \\ b^* \end{bmatrix}$$

is given by the formula

$$\mathbf{v}^* = (M^T M)^{-1} M^T \mathbf{y} \tag{6}$$

which expresses the fact that $\mathbf{v} = \mathbf{v}^*$ is the unique solution of the normal equations

$$M^T M \mathbf{v} = M^T \mathbf{y} \tag{7}$$

▶ **EXAMPLE 1** **Least Squares Straight Line Fit**

Find the least squares straight line fit to the four points $(0, 1)$, $(1, 3)$, $(2, 4)$, and $(3, 4)$. (See Figure 6.5.3.)

Solution We have

$$M = \begin{bmatrix} 1 & 0 \\ 1 & 1 \\ 1 & 2 \\ 1 & 3 \end{bmatrix}, \quad M^T M = \begin{bmatrix} 4 & 6 \\ 6 & 14 \end{bmatrix}, \quad \text{and} \quad (M^T M)^{-1} = \frac{1}{10} \begin{bmatrix} 7 & -3 \\ -3 & 2 \end{bmatrix}$$

$$\mathbf{v}^* = (M^T M)^{-1} M^T \mathbf{y} = \frac{1}{10} \begin{bmatrix} 7 & -3 \\ -3 & 2 \end{bmatrix} \begin{bmatrix} 1 & 1 & 1 & 1 \\ 0 & 1 & 2 & 3 \end{bmatrix} \begin{bmatrix} 1 \\ 3 \\ 4 \\ 4 \end{bmatrix} = \begin{bmatrix} 1.5 \\ 1 \end{bmatrix}$$

so the desired line is $y = 1.5 + x$.

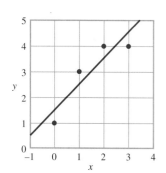

▲ Figure 6.5.3

▶ **EXAMPLE 2** **Spring Constant**

Hooke's law in physics states that the length x of a uniform spring is a linear function of the force y applied to it. If we express this relationship as $y = a + bx$, then the coefficient b is called the **spring constant**. Suppose a particular unstretched spring has a measured length of 6.1 inches (i.e., $x = 6.1$ when $y = 0$). Forces of 2 pounds, 4 pounds,

6.5 Least Squares Fitting to Data

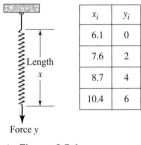

x_i	y_i
6.1	0
7.6	2
8.7	4
10.4	6

▲ Figure 6.5.4

and 6 pounds are then applied to the spring, and the corresponding lengths are found to be 7.6 inches, 8.7 inches, and 10.4 inches (see Figure 6.5.4). Find the spring constant.

Solution We have

$$M = \begin{bmatrix} 1 & 6.1 \\ 1 & 7.6 \\ 1 & 8.7 \\ 1 & 10.4 \end{bmatrix}, \quad \mathbf{y} = \begin{bmatrix} 0 \\ 2 \\ 4 \\ 6 \end{bmatrix},$$

and

$$\mathbf{v}^* = \begin{bmatrix} a^* \\ b^* \end{bmatrix} = (M^T M)^{-1} M^T \mathbf{y} \approx \begin{bmatrix} -8.6 \\ 1.4 \end{bmatrix}$$

where the numerical values have been rounded to one decimal place. Thus, the estimated value of the spring constant is $b^* \approx 1.4$ pounds/inch. ◄

Least Squares Fit of a Polynomial

The technique described for fitting a straight line to data points can be generalized to fitting a polynomial of specified degree to data points. Let us attempt to fit a polynomial of fixed degree m

$$y = a_0 + a_1 x + \cdots + a_m x^m \tag{8}$$

to n points

$$(x_1, y_1), (x_2, y_2), \ldots, (x_n, y_n)$$

Substituting these n values of x and y into (8) yields the n equations

$$\begin{aligned} y_1 &= a_0 + a_1 x_1 + \cdots + a_m x_1^m \\ y_2 &= a_0 + a_1 x_2 + \cdots + a_m x_2^m \\ &\vdots \quad \vdots \quad \vdots \quad \quad \vdots \\ y_n &= a_0 + a_1 x_n + \cdots + a_m x_n^m \end{aligned}$$

or, in matrix form,

$$\mathbf{y} = M\mathbf{v} \tag{9}$$

where

$$\mathbf{y} = \begin{bmatrix} y_1 \\ y_2 \\ \vdots \\ y_n \end{bmatrix}, \quad M = \begin{bmatrix} 1 & x_1 & x_1^2 & \cdots & x_1^m \\ 1 & x_2 & x_2^2 & \cdots & x_2^m \\ \vdots & \vdots & \vdots & & \vdots \\ 1 & x_n & x_n^2 & \cdots & x_n^m \end{bmatrix}, \quad \mathbf{v} = \begin{bmatrix} a_0 \\ a_1 \\ \vdots \\ a_m \end{bmatrix} \tag{10}$$

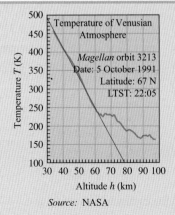

Historical Note On October 5, 1991 the *Magellan* spacecraft entered the atmosphere of Venus and transmitted the temperature T in kelvins (K) versus the altitude h in kilometers (km) until its signal was lost at an altitude of about 34 km. Discounting the initial erratic signal, the data strongly suggested a linear relationship, so a least squares straight line fit was used on the linear part of the data to obtain the equation

$$T = 737.5 - 8.125h$$

By setting $h = 0$ in this equation, the surface temperature of Venus was estimated at $T \approx 737.5$ K.

As before, the solutions of the normal equations

$$M^T M \mathbf{v} = M^T \mathbf{y}$$

determine the coefficients of the polynomial, and the vector $\mathbf{v}$ minimizes

$$\|\mathbf{y} - M\mathbf{v}\|$$

Conditions that guarantee the invertibility of $M^T M$ are discussed in the exercises (Exercise 7). If $M^T M$ is invertible, then the normal equations have a unique solution $\mathbf{v} = \mathbf{v}^*$, which is given by

$$\mathbf{v}^* = (M^T M)^{-1} M^T \mathbf{y} \tag{11}$$

▶ **EXAMPLE 3** **Fitting a Quadratic Curve to Data**

According to Newton's second law of motion, a body near the Earth's surface falls vertically downward according to the equation

$$s = s_0 + v_0 t + \tfrac{1}{2} g t^2 \tag{12}$$

where

$s =$ vertical displacement downward relative to some fixed point
$s_0 =$ initial displacement at time $t = 0$
$v_0 =$ initial velocity at time $t = 0$
$g =$ acceleration of gravity at the Earth's surface

from Equation (12) by releasing a weight with unknown initial displacement and velocity and measuring the distance it has fallen at certain times relative to a fixed reference point. Suppose that a laboratory experiment is performed to evaluate g. Suppose it is found that at times $t = .1, .2, .3, .4,$ and $.5$ seconds the weight has fallen $s = -0.18, 0.31, 1.03, 2.48,$ and 3.73 feet, respectively, from the reference point. Find an approximate value of g using these data.

Solution The mathematical problem is to fit a quadratic curve

$$s = a_0 + a_1 t + a_2 t^2 \tag{13}$$

to the five data points:

$$(.1, -0.18), \quad (.2, 0.31), \quad (.3, 1.03), \quad (.4, 2.48), \quad (.5, 3.73)$$

With the appropriate adjustments in notation, the matrices M and $\mathbf{y}$ in (10) are

$$M = \begin{bmatrix} 1 & t_1 & t_1^2 \\ 1 & t_2 & t_2^2 \\ 1 & t_3 & t_3^2 \\ 1 & t_4 & t_4^2 \\ 1 & t_5 & t_5^2 \end{bmatrix} = \begin{bmatrix} 1 & .1 & .01 \\ 1 & .2 & .04 \\ 1 & .3 & .09 \\ 1 & .4 & .16 \\ 1 & .5 & .25 \end{bmatrix}, \quad \mathbf{y} = \begin{bmatrix} s_1 \\ s_2 \\ s_3 \\ s_4 \\ s_5 \end{bmatrix} = \begin{bmatrix} -0.18 \\ 0.31 \\ 1.03 \\ 2.48 \\ 3.73 \end{bmatrix}$$

Thus, from (11),

$$\mathbf{v}^* = \begin{bmatrix} a_0^* \\ a_1^* \\ a_2^* \end{bmatrix} = (M^T M)^{-1} M^T \mathbf{y} \doteq \begin{bmatrix} -0.40 \\ 0.35 \\ 16.1 \end{bmatrix}$$

From (12) and (13), we have $a_2 = \tfrac{1}{2} g$, so the estimated value of g is

$$g = 2 a_2^* = 2(16.1) = 32.2 \text{ feet/second}^2$$

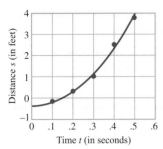

Figure 6.5.5

If desired, we can also estimate the initial displacement and initial velocity of the weight:

$$s_0 = a_0^* = -0.40 \text{ feet}$$
$$v_0 = a_1^* = 0.35 \text{ feet/second}$$

In Figure 6.5.5 we have plotted the five data points and the approximating polynomial. ◀

Concept Review

- Least squares straight line fit
- Regression line
- Least squares polynomial fit

Skills

- Find the least squares straight line fit to a set of data points.
- Find the least squares polynomial fit to a set of data points.
- Use the techniques of this section to solve applied problems.

Exercise Set 6.5

1. Find the least squares straight line fit to the three points $(0, 0)$, $(2, -1)$, and $(3, 4)$.

2. Find the least squares straight line fit to the four points $(0, 1)$, $(2, 0)$, $(3, 1)$, and $(3, 2)$.

3. Find the quadratic polynomial that best fits the four points $(1, 6)$, $(2, 1)$, $(-1, 5)$, and $(-2, 2)$.

4. Find the cubic polynomial that best fits the five points $(-1, -14)$, $(0, -5)$, $(1, -4)$, $(2, 1)$, and $(3, 22)$.

5. Show that the matrix M in Equation (2) has linearly independent columns if and only if at least two of the numbers $x_1, x_2, \ldots, x_n$ are distinct.

6. Show that the columns of the $n \times (m+1)$ matrix M in Equation (10) are linearly independent if $n > m$ and at least $m + 1$ of the numbers $x_1, x_2, \ldots, x_n$ are distinct. [*Hint:* A nonzero polynomial of degree m has at most m distinct roots.]

7. Let M be the matrix in Equation (10). Using Exercise 6, show that a sufficient condition for the matrix $M^T M$ to be invertible is that $n > m$ and that at least $m + 1$ of the numbers $x_1, x_2, \ldots, x_n$ are distinct.

8. The owner of a rapidly expanding business finds that for the first five months of the year the sales (in thousands) are $3.0, $3.5, $5.0, $6.2, and $7.5. The owner plots these figures on a graph and conjectures that for the rest of the year, the sales curve can be approximated by a quadratic polynomial. Find the least squares quadratic polynomial fit to the sales curve, and use it to project the sales for the twelfth month of the year.

9. A corporation obtains the following data relating the number of sales representatives on its staff to annual sales:

Number of Sales Representatives	5	10	15	20	25	30
Annual Sales (millions)	3.4	4.3	5.2	6.1	7.2	8.3

Explain how you might use least squares methods to estimate the annual sales with 45 representatives, and discuss the assumptions that you are making. (You need not perform the actual computations.)

10. *Pathfinder* is an experimental, lightweight, remotely piloted, solar-powered aircraft that was used in a series of experiments by NASA to determine the feasibility of applying solar power for long-duration, high-altitude flight. In August 1997 *Pathfinder* recorded the data in the accompanying table relating altitude H and temperature T. Show that a linear model is reasonable by plotting the data, and then find the least squares line $H = H_0 + kT$ of best fit.

Table Ex-10

Altitude H (thousands of feet)	15	20	25	30	35	40	45
Temperature T (°C)	4.5	−5.9	−16.1	−27.6	−39.8	−50.2	−62.9

11. Find a curve of the form $y = a + (b/x)$ that best fits the data points $(1, 4)$, $(2, 3)$, $(4, 2)$ by making the substitution $X = 1/x$. Draw the curve and plot the data points in the same coordinate system.

True-False Exercises

In parts (a)–(d) determine whether the statement is true or false, and justify your answer.

(a) Every set of data points has a unique least squares straight line fit.

(b) If the data points $(x_1, y_1), (x_2, y_2), \ldots, (x_n, y_n)$ are not collinear, then (1) is an inconsistent system.

(c) If $y = a + bx$ is the least squares line fit to the data points $(x_1, y_1), (x_2, y_2), \ldots, (x_n, y_n)$, then $d_i = |y_i - (a + bx_i)|$ is minimal for every $1 \leq i \leq n$.

(d) If $y = a + bx$ is the least squares line fit to the data points $(x_1, y_1), (x_2, y_2), \ldots, (x_n, y_n)$, then $\sum_{i=1}^{n} |y_i - (a + bx_i)|^2$ is minimal.

6.6 Function Approximation; Fourier Series

In this section we will show how orthogonal projections can be used to approximate certain types of functions by simpler functions that are easier to work with. The ideas explained here have important applications in engineering and science. Calculus is required.

Best Approximations

All of the problems that we will study in this section will be special cases of the following general problem.

> **Approximation Problem** Given a function f that is continuous on an interval $[a, b]$, find the "best possible approximation" to f using only functions from a specified subspace W of $C[a, b]$.

Here are some examples of such problems:

(a) Find the best possible approximation to e^x over $[0, 1]$ by a polynomial of the form $a_0 + a_1 x + a_2 x^2$.

(b) Find the best possible approximation to $\sin \pi x$ over $[-1, 1]$ by a function of the form $a_0 + a_1 e^x + a_2 e^{2x} + a_3 e^{3x}$.

(c) Find the best possible approximation to x over $[0, 2\pi]$ by a function of the form $a_0 + a_1 \sin x + a_2 \sin 2x + b_1 \cos x + b_2 \cos 2x$.

In the first example W is the subspace of $C[0, 1]$ spanned by 1, x, and x^2; in the second example W is the subspace of $C[-1, 1]$ spanned by 1, e^x, e^{2x}, and e^{3x}; and in the third example W is the subspace of $C[0, 2\pi]$ spanned by 1, $\sin x$, $\sin 2x$, $\cos x$, and $\cos 2x$.

Measurements of Error

To solve approximation problems of the preceding types, we first need to make the phrase "best approximation over $[a, b]$" mathematically precise. To do this we will need some way of quantifying the error that results when one continuous function is approximated by another over an interval $[a, b]$. If we were to approximate $f(x)$ by $g(x)$, and if we were concerned only with the error in that approximation at a *single point* x_0, then it would be natural to define the error to be

$$\text{error} = |f(x_0) - g(x_0)|$$

sometimes called the **deviation** between f and g at x_0 (Figure 6.6.1). However, we are not concerned simply with measuring the error at a single point but rather with measuring it over the *entire* interval $[a, b]$. The problem is that an approximation may have small deviations in one part of the interval and large deviations in another. One possible way

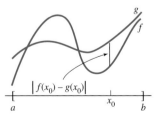

▲ **Figure 6.6.1** The deviation between f and g at x_0.

6.6 Function Approximation; Fourier Series

of accounting for this is to integrate the deviation $|f(x) - g(x)|$ over the interval $[a, b]$ and define the error over the interval to be

$$\text{error} = \int_a^b |f(x) - g(x)|\, dx \tag{1}$$

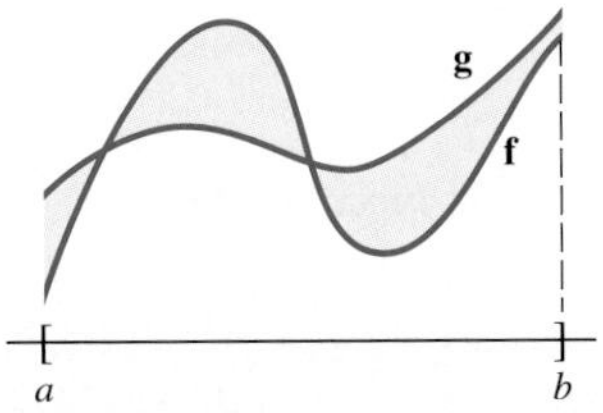

Geometrically, (1) is the area between the graphs of $f(x)$ and $g(x)$ over the interval $[a, b]$ (Figure 6.6.2); the greater the area, the greater the overall error.

Although (1) is natural and appealing geometrically, most mathematicians and scientists generally favor the following alternative measure of error, called the ***mean square error***:

$$\text{mean square error} = \int_a^b [f(x) - g(x)]^2\, dx$$

Figure 6.6.2 The area between the graphs of **f** and **g** over $[a, b]$ measures the error in approximating f by g over $[a, b]$.

Mean square error emphasizes the effect of larger errors because of the squaring and has the added advantage that it allows us to bring to bear the theory of inner product spaces. To see how, suppose that **f** is a continuous function on $[a, b]$ that we want to approximate by a function **g** from a subspace W of $C[a, b]$, and suppose that $C[a, b]$ is given the inner product

$$\langle \mathbf{f}, \mathbf{g} \rangle = \int_a^b f(x)g(x)\, dx$$

It follows that

$$\|\mathbf{f} - \mathbf{g}\|^2 = \langle \mathbf{f} - \mathbf{g}, \mathbf{f} - \mathbf{g} \rangle = \int_a^b [f(x) - g(x)]^2\, dx = \text{mean square error}$$

so minimizing the mean square error is the same as minimizing $\|\mathbf{f} - \mathbf{g}\|^2$. Thus the approximation problem posed informally at the beginning of this section can be restated more precisely as follows.

Least Squares Approximation

Least Squares Approximation Problem Let **f** be a function that is continuous on an interval $[a, b]$, let $C[a, b]$ have the inner product

$$\langle \mathbf{f}, \mathbf{g} \rangle = \int_a^b f(x)g(x)\, dx$$

and let W be a finite-dimensional subspace of $C[a, b]$. Find a function **g** in W that minimizes

$$\|\mathbf{f} - \mathbf{g}\|^2 = \int_a^b [f(x) - g(x)]^2\, dx$$

Since $\|\mathbf{f} - \mathbf{g}\|^2$ and $\|\mathbf{f} - \mathbf{g}\|$ are minimized by the same function **g**, this problem is equivalent to looking for a function **g** in W that is closest to **f**. But we know from Theorem 6.4.1 that $\mathbf{g} = \text{proj}_W \mathbf{f}$ is such a function (Figure 6.6.3).

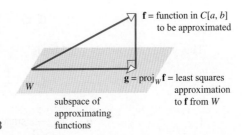

▶ Figure 6.6.3

Thus, we have the following result.

> **THEOREM 6.6.1** *If f is a continuous function on $[a, b]$, and W is a finite-dimensional subspace of $C[a, b]$, then the function g in W that minimizes the mean square error*
> $$\int_a^b [f(x) - g(x)]^2 \, dx$$
> *is $g = \text{proj}_W \mathbf{f}$, where the orthogonal projection is relative to the inner product*
> $$\langle \mathbf{f}, \mathbf{g} \rangle = \int_a^b f(x) g(x) \, dx$$
> *The function $\mathbf{g} = \text{proj}_W \mathbf{f}$ is called the **least squares approximation** to $\mathbf{f}$ from W.*

Fourier Series   A function of the form

$$T(x) = c_0 + c_1 \cos x + c_2 \cos 2x + \cdots + c_n \cos nx \\ + d_1 \sin x + d_2 \sin 2x + \cdots + d_n \sin nx \tag{2}$$

is called a ***trigonometric polynomial***; if c_n and d_n are not both zero, then $T(x)$ is said to have ***order n***. For example,

$$T(x) = 2 + \cos x - 3 \cos 2x + 7 \sin 4x$$

is a trigonometric polynomial of order 4 with

$$c_0 = 2, \quad c_1 = 1, \quad c_2 = -3, \quad c_3 = 0, \quad c_4 = 0, \quad d_1 = 0, \quad d_2 = 0, \quad d_3 = 0, \quad d_4 = 7$$

It is evident from (2) that the trigonometric polynomials of order n or less are the various possible linear combinations of

$$1, \quad \cos x, \quad \cos 2x, \ldots, \quad \cos nx, \qquad \sin x, \quad \sin 2x, \ldots, \quad \sin nx \tag{3}$$

It can be shown that these $2n + 1$ functions are linearly independent and thus form a basis for a $(2n + 1)$-dimensional subspace of $C[a, b]$.

Let us now consider the problem of finding the least squares approximation of a continuous function $f(x)$ over the interval $[0, 2\pi]$ by a trigonometric polynomial of order n or less. As noted above, the least squares approximation to $\mathbf{f}$ from W is the orthogonal projection of $\mathbf{f}$ on W. To find this orthogonal projection, we must find an orthonormal basis $\mathbf{g}_0, \mathbf{g}_1, \ldots, \mathbf{g}_{2n}$ for W, after which we can compute the orthogonal projection on W from the formula

$$\text{proj}_W \mathbf{f} = \langle \mathbf{f}, \mathbf{g}_0 \rangle \mathbf{g}_0 + \langle \mathbf{f}, \mathbf{g}_1 \rangle \mathbf{g}_1 + \cdots + \langle \mathbf{f}, \mathbf{g}_{2n} \rangle \mathbf{g}_{2n} \tag{4}$$

(see Theorem 6.3.4b). An orthonormal basis for W can be obtained by applying the Gram–Schmidt process to the basis vectors in (3) using the inner product

$$\langle \mathbf{f}, \mathbf{g} \rangle = \int_0^{2\pi} f(x) g(x) \, dx$$

This yields the orthonormal basis

$$\mathbf{g}_0 = \frac{1}{\sqrt{2\pi}}, \quad \mathbf{g}_1 = \frac{1}{\sqrt{\pi}} \cos x, \ldots, \quad \mathbf{g}_n = \frac{1}{\sqrt{\pi}} \cos nx,$$
$$\mathbf{g}_{n+1} = \frac{1}{\sqrt{\pi}} \sin x, \ldots, \quad \mathbf{g}_{2n} = \frac{1}{\sqrt{\pi}} \sin nx \tag{5}$$

(see Exercise 6). If we introduce the notation

$$a_0 = \frac{2}{\sqrt{2\pi}}\langle \mathbf{f}, \mathbf{g}_0 \rangle, \quad a_1 = \frac{1}{\sqrt{\pi}}\langle \mathbf{f}, \mathbf{g}_1 \rangle, \ldots, \quad a_n = \frac{1}{\sqrt{\pi}}\langle \mathbf{f}, \mathbf{g}_n \rangle$$

$$b_1 = \frac{1}{\sqrt{\pi}}\langle \mathbf{f}, \mathbf{g}_{n+1} \rangle, \ldots, \quad b_n = \frac{1}{\sqrt{\pi}}\langle \mathbf{f}, \mathbf{g}_{2n} \rangle$$
(6)

then on substituting (5) in (4), we obtain

$$\mathrm{proj}_W \mathbf{f} = \frac{a_0}{2} + [a_1 \cos x + \cdots + a_n \cos nx] + [b_1 \sin x + \cdots + b_n \sin nx] \quad (7)$$

where

$$a_0 = \frac{2}{\sqrt{2\pi}}\langle \mathbf{f}, \mathbf{g}_0 \rangle = \frac{2}{\sqrt{2\pi}}\int_0^{2\pi} f(x)\frac{1}{\sqrt{2\pi}}dx = \frac{1}{\pi}\int_0^{2\pi} f(x)\,dx$$

$$a_1 = \frac{1}{\sqrt{\pi}}\langle \mathbf{f}, \mathbf{g}_1 \rangle = \frac{1}{\sqrt{\pi}}\int_0^{2\pi} f(x)\frac{1}{\sqrt{\pi}}\cos x\,dx = \frac{1}{\pi}\int_0^{2\pi} f(x)\cos x\,dx$$

$$\vdots$$

$$a_n = \frac{1}{\sqrt{\pi}}\langle \mathbf{f}, \mathbf{g}_n \rangle = \frac{1}{\sqrt{\pi}}\int_0^{2\pi} f(x)\frac{1}{\sqrt{\pi}}\cos nx\,dx = \frac{1}{\pi}\int_0^{2\pi} f(x)\cos nx\,dx$$

$$b_1 = \frac{1}{\sqrt{\pi}}\langle \mathbf{f}, \mathbf{g}_{n+1} \rangle = \frac{1}{\sqrt{\pi}}\int_0^{2\pi} f(x)\frac{1}{\sqrt{\pi}}\sin x\,dx = \frac{1}{\pi}\int_0^{2\pi} f(x)\sin x\,dx$$

$$\vdots$$

$$b_n = \frac{1}{\sqrt{\pi}}\langle \mathbf{f}, \mathbf{g}_{2n} \rangle = \frac{1}{\sqrt{\pi}}\int_0^{2\pi} f(x)\frac{1}{\sqrt{\pi}}\sin nx\,dx = \frac{1}{\pi}\int_0^{2\pi} f(x)\sin nx\,dx$$

In short,

$$a_k = \frac{1}{\pi}\int_0^{2\pi} f(x)\cos kx\,dx, \quad b_k = \frac{1}{\pi}\int_0^{2\pi} f(x)\sin kx\,dx \quad (8)$$

The numbers $a_0, a_1, \ldots, a_n, b_1, \ldots, b_n$ are called the *Fourier coefficients* of **f**.

▶ **EXAMPLE 1 Least Squares Approximations**

Find the least squares approximation of $f(x) = x$ on $[0, 2\pi]$ by

(a) a trigonometric polynomial of order 2 or less;
(b) a trigonometric polynomial of order n or less.

Solution (a)

$$a_0 = \frac{1}{\pi}\int_0^{2\pi} f(x)\,dx = \frac{1}{\pi}\int_0^{2\pi} x\,dx = 2\pi \quad (9a)$$

For $k = 1, 2, \ldots$, integration by parts yields (verify)

$$a_k = \frac{1}{\pi}\int_0^{2\pi} f(x)\cos kx\,dx = \frac{1}{\pi}\int_0^{2\pi} x\cos kx\,dx = 0 \quad (9b)$$

$$b_k = \frac{1}{\pi}\int_0^{2\pi} f(x)\sin kx\,dx = \frac{1}{\pi}\int_0^{2\pi} x\sin kx\,dx = -\frac{2}{k} \quad (9c)$$

Thus, the least squares approximation to x on $[0, 2\pi]$ by a trigonometric polynomial of order 2 or less is

$$x \approx \frac{a_0}{2} + a_1 \cos x + a_2 \cos 2x + b_1 \sin x + b_2 \sin 2x$$

or, from (9a), (9b), and (9c),

$$x \approx \pi - 2 \sin x - \sin 2x$$

Solution (b) The least squares approximation to x on $[0, 2\pi]$ by a trigonometric polynomial of order n or less is

$$x \approx \frac{a_0}{2} + [a_1 \cos x + \cdots + a_n \cos nx] + [b_1 \sin x + \cdots + b_n \sin nx]$$

or, from (9a), (9b), and (9c),

$$x \approx \pi - 2\left(\sin x + \frac{\sin 2x}{2} + \frac{\sin 3x}{3} + \cdots + \frac{\sin nx}{n}\right)$$

The graphs of $y = x$ and some of these approximations are shown in Figure 6.6.4.

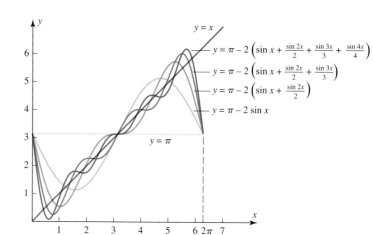

▶ **Figure 6.6.4**

Jean Baptiste Fourier (1768–1830)

Historical Note Fourier was a French mathematician and physicist who discovered the Fourier series and related ideas while working on problems of heat diffusion. This discovery was one of the most influential in the history of mathematics; it is the cornerstone of many fields of mathematical research and a basic tool in many branches of engineering. Fourier, a political activist during the French revolution, spent time in jail for his defense of many victims during the Terror. He later became a favorite of Napoleon and was named a baron.
[Image: The Granger Collection, New York]

It is natural to expect that the mean square error will diminish as the number of terms in the least squares approximation

$$f(x) \approx \frac{a_0}{2} + \sum_{k=1}^{n}(a_k \cos kx + b_k \sin kx)$$

increases. It can be proved that for functions f in $C[0, 2\pi]$, the mean square error approaches zero as $n \to +\infty$; this is denoted by writing

$$f(x) = \frac{a_0}{2} + \sum_{k=1}^{\infty}(a_k \cos kx + b_k \sin kx)$$

The right side of this equation is called the **Fourier series** for f over the interval $[0, 2\pi]$. Such series are of major importance in engineering, science, and mathematics. ◀

Concept Review

- Approximation of functions
- Mean square error
- Least squares approximation
- Trigonometric polynomial
- Fourier coefficients
- Fourier series

Skills

- Find the least squares approximation of a function.
- Find the mean square error of the least squares approximation of a function.
- Compute the Fourier series of a function.

Exercise Set 6.6

1. Find the least squares approximation of $f(x) = 1 - x$ over the interval $[0, 2\pi]$ by
   (a) a trigonometric polynomial of order 2 or less.
   (b) a trigonometric polynomial of order n or less.

2. Find the least squares approximation of $f(x) = x^2$ over the interval $[0, 2\pi]$ by
   (a) a trigonometric polynomial of order 3 or less.
   (b) a trigonometric polynomial of order n or less.

3. (a) Find the least squares approximation of $1 + x$ over the interval $[0, 1]$ by a function of the form $a + be^x$.
   (b) Find the mean square error of the approximation.

4. (a) Find the least squares approximation of e^x over the interval $[0, 1]$ by a polynomial of the form $a_0 + a_1 x$.
   (b) Find the mean square error of the approximation.

5. (a) Find the least squares approximation of $\cos \pi x$ over the interval $[-1, 1]$ by a polynomial of the form $a_0 + a_1 x + a_2 x^2$.
   (b) Find the mean square error of the approximation.

6. Use the Gram–Schmidt process to obtain the orthonormal basis (5) from the basis (3).

7. Carry out the integrations indicated in Formulas (9a), (9b), and (9c).

8. Find the Fourier series of $f(x) = \pi - x$ over the interval $[0, 2\pi]$.

9. Find the Fourier series of $f(x) = 0$, $0 \le x < \pi$ and $f(x) = 1$, $\pi \le x \le 2\pi$ over the interval $[0, 2\pi]$.

10. What is the Fourier series of $\cos(3x)$?

True-False Exercises

In parts (a)–(e) determine whether the statement is true or false, and justify your answer.

(a) If a function $\mathbf{f}$ in $C[a, b]$ is approximated by the function $\mathbf{g}$, then the mean square error is the same as the area between the graphs of $f(x)$ and $g(x)$ over the interval $[a, b]$.

(b) Given a finite-dimensional subspace W of $C[a, b]$, the function $\mathbf{g} = \text{proj}_W \mathbf{f}$ minimizes the mean square error.

(c) $\{1, \cos x, \sin x, \cos 2x, \sin 2x\}$ is an orthogonal subset of the vector space $C[0, 2\pi]$ with respect to the inner product $\langle \mathbf{f}, \mathbf{g} \rangle = \int_0^{2\pi} f(x) g(x) \, dx$.

(d) $\{1, \cos x, \sin x, \cos 2x, \sin 2x\}$ is an orthonormal subset of the vector space $C[0, 2\pi]$ with respect to the inner product $\langle \mathbf{f}, \mathbf{g} \rangle = \int_0^{2\pi} f(x) g(x) \, dx$.

(e) $\{1, \cos x, \sin x, \cos 2x, \sin 2x\}$ is a linearly independent subset of $C[0, 2\pi]$.

Chapter 6 Supplementary Exercises

1. Let R^4 have the Euclidean inner product.
   (a) Find a vector in R^4 that is orthogonal to $\mathbf{u}_1 = (1, 0, 0, 0)$ and $\mathbf{u}_4 = (0, 0, 0, 1)$ and makes equal angles with $\mathbf{u}_2 = (0, 1, 0, 0)$ and $\mathbf{u}_3 = (0, 0, 1, 0)$.
   (b) Find a vector $\mathbf{x} = (x_1, x_2, x_3, x_4)$ of length 1 that is orthogonal to $\mathbf{u}_1$ and $\mathbf{u}_4$ above and such that the cosine of the angle between $\mathbf{x}$ and $\mathbf{u}_2$ is twice the cosine of the angle between $\mathbf{x}$ and $\mathbf{u}_3$.

2. Prove: If $\langle \mathbf{u}, \mathbf{v} \rangle$ is the Euclidean inner product on R^n, and if A is an $n \times n$ matrix, then

$$\langle \mathbf{u}, A\mathbf{v} \rangle = \langle A^T \mathbf{u}, \mathbf{v} \rangle$$

[Hint: Use the fact that $\langle \mathbf{u}, \mathbf{v} \rangle = \mathbf{u} \cdot \mathbf{v} = \mathbf{v}^T \mathbf{u}$.]

3. Let M_{22} have the inner product $\langle U, V \rangle = \text{tr}(U^T V) = \text{tr}(V^T U)$ that was defined in Example 6 of Section 6.1. Describe the orthogonal complement of
   (a) the subspace of all diagonal matrices.
   (b) the subspace of symmetric matrices.

4. Let $A\mathbf{x} = \mathbf{0}$ be a system of m equations in n unknowns. Show that

$$\mathbf{x} = \begin{bmatrix} x_1 \\ x_2 \\ \vdots \\ x_n \end{bmatrix}$$

is a solution of this system if and only if the vector $\mathbf{x} = (x_1, x_2, \ldots, x_n)$ is orthogonal to every row vector of A with respect to the Euclidean inner product on R^n.

5. Use the Cauchy–Schwarz inequality to show that if $a_1, a_2, \ldots, a_n$ are positive real numbers, then
$$(a_1 + a_2 + \cdots + a_n)\left(\frac{1}{a_1} + \frac{1}{a_2} + \cdots + \frac{1}{a_n}\right) \geq n^2$$

6. Show that if $\mathbf{x}$ and $\mathbf{y}$ are vectors in an inner product space and c is any scalar, then
$$\|c\mathbf{x} + \mathbf{y}\|^2 = c^2\|\mathbf{x}\|^2 + 2c\langle \mathbf{x}, \mathbf{y}\rangle + \|\mathbf{y}\|^2$$

7. Let R^3 have the Euclidean inner product. Find two vectors of length 1 that are orthogonal to all three of the vectors $\mathbf{u}_1 = (1, 1, -1)$, $\mathbf{u}_2 = (-2, -1, 2)$, and $\mathbf{u}_3 = (-1, 0, 1)$.

8. Find a weighted Euclidean inner product on R^n such that the vectors
$$\mathbf{v}_1 = (1, 0, 0, \ldots, 0)$$
$$\mathbf{v}_2 = (0, \sqrt{2}, 0, \ldots, 0)$$
$$\mathbf{v}_3 = (0, 0, \sqrt{3}, \ldots, 0)$$
$$\vdots$$
$$\mathbf{v}_n = (0, 0, 0, \ldots, \sqrt{n})$$
form an orthonormal set.

9. Is there a weighted Euclidean inner product on R^2 for which the vectors $(1, 2)$ and $(3, -1)$ form an orthonormal set? Justify your answer.

10. If $\mathbf{u}$ and $\mathbf{v}$ are vectors in an inner product space V, then $\mathbf{u}$, $\mathbf{v}$, and $\mathbf{u} - \mathbf{v}$ can be regarded as sides of a "triangle" in V (see the accompanying figure). Prove that the law of cosines holds for any such triangle; that is,
$$\|\mathbf{u} - \mathbf{v}\|^2 = \|\mathbf{u}\|^2 + \|\mathbf{v}\|^2 - 2\|\mathbf{u}\|\|\mathbf{v}\|\cos\theta$$
where θ is the angle between $\mathbf{u}$ and $\mathbf{v}$.

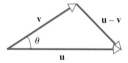

◀ Figure Ex-10

11. (a) As shown in Figure 3.2.6, the vectors $(k, 0, 0)$, $(0, k, 0)$, and $(0, 0, k)$ form the edges of a cube in R^3 with diagonal (k, k, k). Similarly, the vectors
$$(k, 0, 0, \ldots, 0), \quad (0, k, 0, \ldots, 0), \ldots, \quad (0, 0, 0, \ldots, k)$$
can be regarded as edges of a "cube" in R^n with diagonal $(k, k, k, \ldots, k)$. Show that each of the above edges makes an angle of θ with the diagonal, where $\cos\theta = 1/\sqrt{n}$.

(b) *(Calculus required)* What happens to the angle θ in part (a) as the dimension of R^n approaches $+\infty$?

12. Let $\mathbf{u}$ and $\mathbf{v}$ be vectors in an inner product space.

(a) Prove that $\|\mathbf{u}\| = \|\mathbf{v}\|$ if and only if $\mathbf{u} + \mathbf{v}$ and $\mathbf{u} - \mathbf{v}$ are orthogonal.

(b) Give a geometric interpretation of this result in R^2 with the Euclidean inner product.

13. Let $\mathbf{u}$ be a vector in an inner product space V, and let $\{\mathbf{v}_1, \mathbf{v}_2, \ldots, \mathbf{v}_n\}$ be an orthonormal basis for V. Show that if α_i is the angle between $\mathbf{u}$ and $\mathbf{v}_i$, then
$$\cos^2\alpha_1 + \cos^2\alpha_2 + \cdots + \cos^2\alpha_n = 1$$

14. Prove: If $\langle \mathbf{u}, \mathbf{v}\rangle_1$ and $\langle \mathbf{u}, \mathbf{v}\rangle_2$ are two inner products on a vector space V, then the quantity $\langle \mathbf{u}, \mathbf{v}\rangle = \langle \mathbf{u}, \mathbf{v}\rangle_1 + \langle \mathbf{u}, \mathbf{v}\rangle_2$ is also an inner product.

15. Prove Theorem 6.2.5.

16. Prove: If A has linearly independent column vectors, and if $\mathbf{b}$ is orthogonal to the column space of A, then the least squares solution of $A\mathbf{x} = \mathbf{b}$ is $\mathbf{x} = \mathbf{0}$.

17. Is there any value of s for which $x_1 = 1$ and $x_2 = 2$ is the least squares solution of the following linear system?
$$x_1 - x_2 = 1$$
$$2x_1 + 3x_2 = 1$$
$$4x_1 + 5x_2 = s$$
Explain your reasoning.

18. Show that if p and q are distinct positive integers, then the functions $f(x) = \sin px$ and $g(x) = \sin qx$ are orthogonal with respect to the inner product
$$\langle \mathbf{f}, \mathbf{g}\rangle = \int_0^{2\pi} f(x)g(x)\,dx$$

19. Show that if p and q are positive integers, then the functions $f(x) = \cos px$ and $g(x) = \sin qx$ are orthogonal with respect to the inner product
$$\langle \mathbf{f}, \mathbf{g}\rangle = \int_0^{2\pi} f(x)g(x)\,dx$$

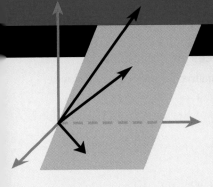

CHAPTER 7

Diagonalization and Quadratic Forms

CHAPTER CONTENTS
7.1 Orthogonal Matrices 389
7.2 Orthogonal Diagonalization 397
7.3 Quadratic Forms 405
7.4 Optimization Using Quadratic Forms 417
7.5 Hermitian, Unitary, and Normal Matrices 424

INTRODUCTION In Section 5.2 we found conditions that guaranteed the diagonalizability of an $n \times n$ matrix, but we did not consider what class or classes of matrices might actually satisfy those conditions. In this chapter we will show that every symmetric matrix is diagonalizable. This is an extremely important result because many applications utilize it in some essential way.

7.1 Orthogonal Matrices

In this section we will discuss the class of matrices whose inverses can be obtained by transposition. Such matrices occur in a variety of applications and arise as well as transition matrices when one orthonormal basis is changed to another.

Orthogonal Matrices   We begin with the following definition.

Recall from Theorem 1.6.3 that if either product in (1) holds, then so does the other. Thus, A is orthogonal if *either* $AA^T = I$ or $A^TA = I$.

DEFINITION 1 A square matrix A is said to be **orthogonal** if its transpose is the same as its inverse, that is, if
$$A^{-1} = A^T$$
or, equivalently, if
$$AA^T = A^TA = I \qquad (1)$$

▶ **EXAMPLE 1 A 3 × 3 Orthogonal Matrix**

The matrix
$$A = \begin{bmatrix} \frac{3}{7} & \frac{2}{7} & \frac{6}{7} \\ -\frac{6}{7} & \frac{3}{7} & \frac{2}{7} \\ \frac{2}{7} & \frac{6}{7} & -\frac{3}{7} \end{bmatrix}$$

is orthogonal since
$$A^TA = \begin{bmatrix} \frac{3}{7} & -\frac{6}{7} & \frac{2}{7} \\ \frac{2}{7} & \frac{3}{7} & \frac{6}{7} \\ \frac{6}{7} & \frac{2}{7} & -\frac{3}{7} \end{bmatrix} \begin{bmatrix} \frac{3}{7} & \frac{2}{7} & \frac{6}{7} \\ -\frac{6}{7} & \frac{3}{7} & \frac{2}{7} \\ \frac{2}{7} & \frac{6}{7} & -\frac{3}{7} \end{bmatrix} = \begin{bmatrix} 1 & 0 & 0 \\ 0 & 1 & 0 \\ 0 & 0 & 1 \end{bmatrix}$$

▶ **EXAMPLE 2** **Rotation and Reflection Matrices are Orthogonal**

Recall from Table 5 of Section 4.9 that the standard matrix for the counterclockwise rotation of R^2 through an angle θ is

$$A = \begin{bmatrix} \cos\theta & -\sin\theta \\ \sin\theta & \cos\theta \end{bmatrix}$$

This matrix is orthogonal for all choices of θ since

$$A^T A = \begin{bmatrix} \cos\theta & \sin\theta \\ -\sin\theta & \cos\theta \end{bmatrix} \begin{bmatrix} \cos\theta & -\sin\theta \\ \sin\theta & \cos\theta \end{bmatrix} = \begin{bmatrix} 1 & 0 \\ 0 & 1 \end{bmatrix}$$

We leave it for you to verify that the reflection matrices in Tables 1 and 2 and the rotation matrices in Table 6 of Section 4.9 are all orthogonal. ◀

Observe that for the orthogonal matrices in Examples 1 and 2, both the row vectors and the column vectors form orthonormal sets with respect to the Euclidean inner product. This is a consequence of the following theorem.

THEOREM 7.1.1 *The following are equivalent for an $n \times n$ matrix A.*

(a) *A is orthogonal.*

(b) *The row vectors of A form an orthonormal set in R^n with the Euclidean inner product.*

(c) *The column vectors of A form an orthonormal set in R^n with the Euclidean inner product.*

Proof We will prove the equivalence of (a) and (b) and leave the equivalence of (a) and (c) as an exercise.

(a) ⇔ (b) The entry in the ith row and jth column of the matrix product AA^T is the dot product of the ith row vector of A and the jth column vector of A^T (see Formula (5) of Section 1.3). But except for a difference in form, the jth column vector of A^T is the jth row vector of A. Thus, if the row vectors of A are $\mathbf{r}_1, \mathbf{r}_2, \ldots, \mathbf{r}_n$, then the matrix product AA^T can be expressed as

$$AA^T = \begin{bmatrix} \mathbf{r}_1 \cdot \mathbf{r}_1 & \mathbf{r}_1 \cdot \mathbf{r}_2 & \cdots & \mathbf{r}_1 \cdot \mathbf{r}_n \\ \mathbf{r}_2 \cdot \mathbf{r}_1 & \mathbf{r}_2 \cdot \mathbf{r}_2 & \cdots & \mathbf{r}_2 \cdot \mathbf{r}_n \\ \vdots & \vdots & & \vdots \\ \mathbf{r}_n \cdot \mathbf{r}_1 & \mathbf{r}_n \cdot \mathbf{r}_2 & \cdots & \mathbf{r}_n \cdot \mathbf{r}_n \end{bmatrix}$$

[see Formula (28) of Section 3.2]. Thus, it follows that $AA^T = I$ if and only if

$$\mathbf{r}_1 \cdot \mathbf{r}_1 = \mathbf{r}_2 \cdot \mathbf{r}_2 = \cdots = \mathbf{r}_n \cdot \mathbf{r}_n = 1$$

and

$$\mathbf{r}_i \cdot \mathbf{r}_j = 0 \quad \text{when } i \neq j$$

which are true if and only if $\{\mathbf{r}_1, \mathbf{r}_2, \ldots, \mathbf{r}_n\}$ is an orthonormal set in R^n. ◀

WARNING Note that an orthogonal matrix is one with *orthonormal* rows and columns—not simply orthogonal rows and columns.

The following theorem lists three more fundamental properties of orthogonal matrices. The proofs are all straightforward and are left as exercises.

THEOREM 7.1.2

(a) *The inverse of an orthogonal matrix is orthogonal.*

(b) *A product of orthogonal matrices is orthogonal.*

(c) *If A is orthogonal, then $\det(A) = 1$ or $\det(A) = -1$.*

▶ **EXAMPLE 3** $\det(A) = \pm 1$ for an Orthogonal Matrix A

The matrix
$$A = \begin{bmatrix} \frac{1}{\sqrt{2}} & \frac{1}{\sqrt{2}} \\ -\frac{1}{\sqrt{2}} & \frac{1}{\sqrt{2}} \end{bmatrix}$$
is orthogonal since its row (and column) vectors form orthonormal sets in R^2 with the Euclidean inner product. We leave it for you to verify that $\det(A) = 1$ and that interchanging the rows produces an orthogonal matrix whose determinant is -1. ◀

Orthogonal Matrices as Linear Operators

We observed in Example 2 that the standard matrices for the basic reflection and rotation operators on R^2 and R^3 are orthogonal. The next theorem will explain why this is so.

THEOREM 7.1.3 *If A is an $n \times n$ matrix, then the following are equivalent.*

(a) *A is orthogonal.*

(b) *$\|A\mathbf{x}\| = \|\mathbf{x}\|$ for all $\mathbf{x}$ in R^n.*

(c) *$A\mathbf{x} \cdot A\mathbf{y} = \mathbf{x} \cdot \mathbf{y}$ for all $\mathbf{x}$ and $\mathbf{y}$ in R^n.*

Proof We will prove the sequence of implications $(a) \Rightarrow (b) \Rightarrow (c) \Rightarrow (a)$.

(a) $\Rightarrow$ (b) Assume that A is orthogonal, so that $A^T A = I$. It follows from Formula (26) of Section 3.2 that
$$\|A\mathbf{x}\| = (A\mathbf{x} \cdot A\mathbf{x})^{1/2} = (\mathbf{x} \cdot A^T A \mathbf{x})^{1/2} = (\mathbf{x} \cdot \mathbf{x})^{1/2} = \|\mathbf{x}\|$$

(b) $\Rightarrow$ (c) Assume that $\|A\mathbf{x}\| = \|\mathbf{x}\|$ for all $\mathbf{x}$ in R^n. From Theorem 3.2.7 we have
$$A\mathbf{x} \cdot A\mathbf{y} = \tfrac{1}{4}\|A\mathbf{x} + A\mathbf{y}\|^2 - \tfrac{1}{4}\|A\mathbf{x} - A\mathbf{y}\|^2 = \tfrac{1}{4}\|A(\mathbf{x}+\mathbf{y})\|^2 - \tfrac{1}{4}\|A(\mathbf{x}-\mathbf{y})\|^2$$
$$= \tfrac{1}{4}\|\mathbf{x}+\mathbf{y}\|^2 - \tfrac{1}{4}\|\mathbf{x}-\mathbf{y}\|^2 = \mathbf{x} \cdot \mathbf{y}$$

(c) $\Rightarrow$ (a) Assume that $A\mathbf{x} \cdot A\mathbf{y} = \mathbf{x} \cdot \mathbf{y}$ for all $\mathbf{x}$ and $\mathbf{y}$ in R^n. It follows from Formula (26) of Section 3.2 that
$$\mathbf{x} \cdot \mathbf{y} = \mathbf{x} \cdot A^T A \mathbf{y}$$
which can be rewritten as $\mathbf{x} \cdot (A^T A \mathbf{y} - \mathbf{y}) = 0$ or as
$$\mathbf{x} \cdot (A^T A - I)\mathbf{y} = 0$$
Since this equation holds for all $\mathbf{x}$ in R^n, it holds in particular if $\mathbf{x} = (A^T A - I)\mathbf{y}$, so
$$(A^T A - I)\mathbf{y} \cdot (A^T A - I)\mathbf{y} = 0$$
Thus, it follows from the positivity axiom for inner products that
$$(A^T A - I)\mathbf{y} = \mathbf{0}$$
Since this equation is satisfied by every vector $\mathbf{y}$ in R^n, it must be that $A^T A - I$ is the zero matrix (why?) and hence that $A^T A = I$. Thus, A is orthogonal. ◀

Theorem 7.1.3 has a useful geometric interpretation when considered from the viewpoint of matrix transformations: If A is an orthogonal matrix and $T_A: R^n \to R^n$ is multiplication by A, then we will call T_A an ***orthogonal operator*** on R^n. It follows from parts (*a*) and (*b*) of Theorem 7.1.3 that the orthogonal operators on R^n are precisely those operators that leave the lengths of all vectors unchanged. This explains why, in Example 2, we found the standard matrices for the basic reflections and rotations of R^2 and R^3 to be orthogonal.

Parts (*a*) and (*c*) of Theorem 7.1.3 imply that orthogonal operators leave the *angle* between two vectors unchanged. Why?

Change of Orthonormal Basis

Orthonormal bases for inner product spaces are convenient because, as the following theorem shows, many familiar formulas hold for such bases. We leave the proof as an exercise.

THEOREM 7.1.4 *If S is an orthonormal basis for an n-dimensional inner product space V, and if*

$$(\mathbf{u})_S = (u_1, u_2, \ldots, u_n) \quad \text{and} \quad (\mathbf{v})_S = (v_1, v_2, \ldots, v_n)$$

then:

(a) $\|\mathbf{u}\| = \sqrt{u_1^2 + u_2^2 + \cdots + u_n^2}$

(b) $d(\mathbf{u}, \mathbf{v}) = \sqrt{(u_1 - v_1)^2 + (u_2 - v_2)^2 + \cdots + (u_n - v_n)^2}$

(c) $\langle \mathbf{u}, \mathbf{v} \rangle = u_1 v_1 + u_2 v_2 + \cdots + u_n v_n$

Remark Note that the three parts of Theorem 7.1.4 can be expressed as

$$\|\mathbf{u}\| = \|(\mathbf{u})_S\| \qquad d(\mathbf{u}, \mathbf{v}) = d\left((\mathbf{u})_S, (\mathbf{v})_S\right) \qquad \langle \mathbf{u}, \mathbf{v} \rangle = \langle (\mathbf{u})_S, (\mathbf{v})_S \rangle$$

where the norm, distance, and inner product on the left sides are relative to the inner product on V and on the right sides are relative to the Euclidean inner product on R^n.

Transitions between orthonormal bases for an inner product space are of special importance in geometry and various applications. The following theorem, whose proof is deferred to the end of this section, is concerned with transitions of this type.

THEOREM 7.1.5 *Let V be a finite-dimensional inner product space. If P is the transition matrix from one orthonormal basis for V to another orthonormal basis for V, then P is an orthogonal matrix.*

▶ **EXAMPLE 4 Rotation of Axes in 2-Space**

In many problems a rectangular xy-coordinate system is given, and a new $x'y'$-coordinate system is obtained by rotating the xy-system counterclockwise about the origin through an angle θ. When this is done, each point Q in the plane has two sets of coordinates—coordinates (x, y) relative to the xy-system and coordinates (x', y') relative to the $x'y'$-system (Figure 7.1.1*a*).

By introducing unit vectors $\mathbf{u}_1$ and $\mathbf{u}_2$ along the positive x- and y-axes and unit vectors $\mathbf{u}'_1$ and $\mathbf{u}'_2$ along the positive x'- and y'-axes, we can regard this rotation as a change from an old basis $B = \{\mathbf{u}_1, \mathbf{u}_2\}$ to a new basis $B' = \{\mathbf{u}'_1, \mathbf{u}'_2\}$ (Figure 7.1.1*b*). Thus, the new coordinates (x', y') and the old coordinates (x, y) of a point Q will be related by

$$\begin{bmatrix} x' \\ y' \end{bmatrix} = P^{-1} \begin{bmatrix} x \\ y \end{bmatrix} \tag{2}$$

7.1 Orthogonal Matrices

where P is the transition from B' to B. To find P we must determine the coordinate matrices of the new basis vectors $\mathbf{u}'_1$ and $\mathbf{u}'_2$ relative to the old basis. As indicated in Figure 7.1.1c, the components of $\mathbf{u}'_1$ in the old basis are $\cos\theta$ and $\sin\theta$, so

$$[\mathbf{u}'_1]_B = \begin{bmatrix} \cos\theta \\ \sin\theta \end{bmatrix}$$

Similarly, from Figure 7.1.1d, we see that the components of $\mathbf{u}'_2$ in the old basis are $\cos(\theta + \pi/2) = -\sin\theta$ and $\sin(\theta + \pi/2) = \cos\theta$, so

$$[\mathbf{u}'_2]_B = \begin{bmatrix} -\sin\theta \\ \cos\theta \end{bmatrix}$$

Thus the transition matrix from B' to B is

$$P = \begin{bmatrix} \cos\theta & -\sin\theta \\ \sin\theta & \cos\theta \end{bmatrix} \quad (3)$$

Observe that P is an orthogonal matrix, as expected, since B and B' are orthonormal bases. Thus

$$P^{-1} = P^T = \begin{bmatrix} \cos\theta & \sin\theta \\ -\sin\theta & \cos\theta \end{bmatrix}$$

so (2) yields

$$\begin{bmatrix} x' \\ y' \end{bmatrix} = \begin{bmatrix} \cos\theta & \sin\theta \\ -\sin\theta & \cos\theta \end{bmatrix} \begin{bmatrix} x \\ y \end{bmatrix} \quad (4)$$

or, equivalently,

$$\begin{aligned} x' &= x\cos\theta + y\sin\theta \\ y' &= -x\sin\theta + y\cos\theta \end{aligned} \quad (5)$$

These are sometimes called the ***rotation equations*** for R^2.

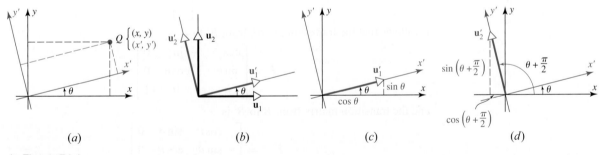

▲ Figure 7.1.1

▶ **EXAMPLE 5** **Rotation of Axes in 2-Space**

Use form (4) of the rotation equations for R^2 to find the new coordinates of the point $Q(2, 1)$ if the coordinate axes of a rectangular coordinate system are rotated through an angle of $\theta = \pi/4$.

Solution Since

$$\sin\frac{\pi}{4} = \cos\frac{\pi}{4} = \frac{1}{\sqrt{2}}$$

the equation in (4) becomes

$$\begin{bmatrix} x' \\ y' \end{bmatrix} = \begin{bmatrix} \frac{1}{\sqrt{2}} & \frac{1}{\sqrt{2}} \\ -\frac{1}{\sqrt{2}} & \frac{1}{\sqrt{2}} \end{bmatrix} \begin{bmatrix} x \\ y \end{bmatrix}$$

Thus, if the old coordinates of a point Q are $(x, y) = (2, -1)$, then

$$\begin{bmatrix} x' \\ y' \end{bmatrix} = \begin{bmatrix} \frac{1}{\sqrt{2}} & \frac{1}{\sqrt{2}} \\ -\frac{1}{\sqrt{2}} & \frac{1}{\sqrt{2}} \end{bmatrix} \begin{bmatrix} 2 \\ -1 \end{bmatrix} = \begin{bmatrix} \frac{1}{\sqrt{2}} \\ -\frac{3}{\sqrt{2}} \end{bmatrix}$$

so the new coordinates of Q are $(x', y') = \left(\frac{1}{\sqrt{2}}, -\frac{3}{\sqrt{2}}\right)$. ◀

Remark Observe that the coefficient matrix in (4) is the same as the standard matrix for the linear operator that rotates the vectors of R^2 through the angle $-\theta$ (see margin note for Table 5 of Section 4.9). This is to be expected since rotating the coordinate axes through the angle θ with the vectors of R^2 kept fixed has the same effect as rotating the vectors in R^2 through the angle $-\theta$ with the axes kept fixed.

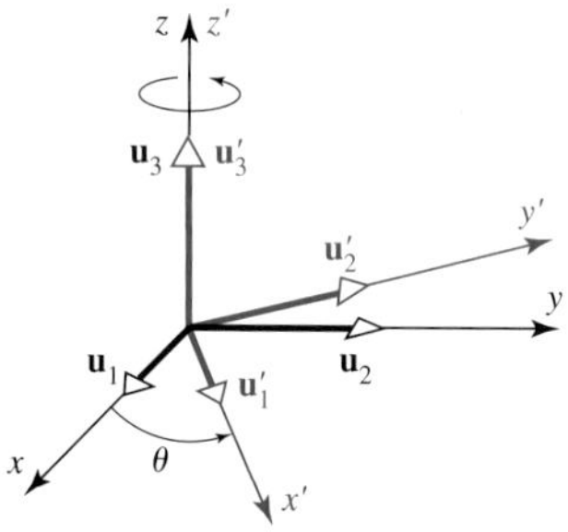

▶ **EXAMPLE 6** Application to Rotation of Axes in 3-Space

Suppose that a rectangular xyz-coordinate system is rotated around its z-axis counterclockwise (looking down the positive z-axis) through an angle θ (Figure 7.1.2). If we introduce unit vectors $\mathbf{u}_1$, $\mathbf{u}_2$, and $\mathbf{u}_3$ along the positive x-, y-, and z-axes and unit vectors $\mathbf{u}'_1$, $\mathbf{u}'_2$, and $\mathbf{u}'_3$ along the positive x'-, y'-, and z'-axes, we can regard the rotation as a change from the old basis $B = \{\mathbf{u}_1, \mathbf{u}_2, \mathbf{u}_3\}$ to the new basis $B' = \{\mathbf{u}'_1, \mathbf{u}'_2, \mathbf{u}'_3\}$. In light of Example 4, it should be evident that

$$[\mathbf{u}'_1]_B = \begin{bmatrix} \cos\theta \\ \sin\theta \\ 0 \end{bmatrix} \quad \text{and} \quad [\mathbf{u}'_2]_B = \begin{bmatrix} -\sin\theta \\ \cos\theta \\ 0 \end{bmatrix}$$

Moreover, since $\mathbf{u}'_3$ extends 1 unit up the positive z'-axis,

$$[\mathbf{u}'_3]_B = \begin{bmatrix} 0 \\ 0 \\ 1 \end{bmatrix}$$

It follows that the transition matrix from B' to B is

$$P = \begin{bmatrix} \cos\theta & -\sin\theta & 0 \\ \sin\theta & \cos\theta & 0 \\ 0 & 0 & 1 \end{bmatrix}$$

and the transition matrix from B to B' is

$$P^{-1} = \begin{bmatrix} \cos\theta & \sin\theta & 0 \\ -\sin\theta & \cos\theta & 0 \\ 0 & 0 & 1 \end{bmatrix}$$

(verify). Thus, the new coordinates (x', y', z') of a point Q can be computed from its old coordinates (x, y, z) by

$$\begin{bmatrix} x' \\ y' \\ z' \end{bmatrix} = \begin{bmatrix} \cos\theta & \sin\theta & 0 \\ -\sin\theta & \cos\theta & 0 \\ 0 & 0 & 1 \end{bmatrix} \begin{bmatrix} x \\ y \\ z \end{bmatrix} \quad ◀$$

▲ Figure 7.1.2

OPTIONAL

We conclude this section with an optional proof of Theorem 7.1.5.

Proof of Theorem 7.1.5 Assume that V is an n-dimensional inner product space and that P is the transition matrix from an orthonormal basis B' to an orthonormal basis B. We will denote the norm relative to the inner product on V by the symbol $\|\ \|_V$ to distinguish it from the norm relative to the Euclidean inner product on R^n, which we will denote by $\|\ \|$.

> Recall that $(\mathbf{u})_S$ denotes a coordinate vector expressed in comma-delimited form whereas $[\mathbf{u}]_S$ denotes a coordinate vector expressed in column form.

To prove that P is orthogonal, we will use Theorem 7.1.3 and show that $\|P\mathbf{x}\| = \|\mathbf{x}\|$ for every vector $\mathbf{x}$ in R^n. As a first step in this direction, recall from Theorem 7.1.4a that for any orthonormal basis for V the norm of any vector $\mathbf{u}$ in V is the same as the norm of its coordinate vector with respect to the Euclidean inner product, that is

$$\|\mathbf{u}\|_V = \|[\mathbf{u}]_{B'}\| = \|[\mathbf{u}]_B\|$$

or

$$\|\mathbf{u}\|_V = \|[\mathbf{u}]_{B'}\| = \|P[\mathbf{u}]_{B'}\| \qquad (6)$$

Now let $\mathbf{x}$ be any vector in R^n, and let $\mathbf{u}$ be the vector in V whose coordinate vector with respect to the basis B' is $\mathbf{x}$; that is, $[\mathbf{u}]_{B'} = \mathbf{x}$. Thus, from (6),

$$\|\mathbf{u}\| = \|\mathbf{x}\| = \|P\mathbf{x}\|$$

which proves that P is orthogonal. ◀

Concept Review

- Orthogonal matrix
- Orthogonal operator
- Properties of orthogonal matrices.
- Geometric properties of an orthogonal operator
- Properties of transition matrices from one orthonormal basis to another.

Skills

- Be able to identify an orthogonal matrix.
- Know the possible values for the determinant of an orthogonal matrix.
- Find the new coordinates of a point resulting from a rotation of axes.

Exercise Set 7.1

1. (a) Show that the matrix

$$A = \begin{bmatrix} \frac{3}{5} & 0 & \frac{4}{5} \\ \frac{4}{5} & 0 & -\frac{3}{5} \\ 0 & 1 & 0 \end{bmatrix}$$

is orthogonal in three ways: by calculating $A^T A$, by using part (b) of Theorem 7.1.1, and by using part (c) of Theorem 7.1.1.

(b) Find the inverse of the matrix A in part (a).

2. (a) Show that the matrix

$$A = \begin{bmatrix} \frac{1}{3} & \frac{2}{3} & \frac{2}{3} \\ \frac{2}{3} & -\frac{2}{3} & \frac{1}{3} \\ -\frac{2}{3} & -\frac{1}{3} & \frac{2}{3} \end{bmatrix}$$

is orthogonal.

(b) Let $T: R^3 \to R^3$ be multiplication by the matrix A in part (a). Find $T(\mathbf{x})$ for the vector $\mathbf{x} = (1, -3, 4)$. Using the Euclidean inner product on R^3, verify that $\|T(\mathbf{x})\| = \|\mathbf{x}\|$.

3. Determine which of the following matrices are orthogonal. For those that are orthogonal, find the inverse.

(a) $\begin{bmatrix} 0 & 1 \\ 1 & 0 \end{bmatrix}$

(b) $\begin{bmatrix} \frac{1}{\sqrt{2}} & -\frac{1}{\sqrt{2}} \\ \frac{1}{\sqrt{2}} & \frac{1}{\sqrt{2}} \end{bmatrix}$

(c) $\begin{bmatrix} \frac{1}{\sqrt{2}} & 0 & \frac{1}{\sqrt{2}} \\ 1 & 0 & 0 \\ 0 & 1 & 0 \end{bmatrix}$

(d) $\begin{bmatrix} \frac{1}{2} & \frac{1}{2} & \frac{1}{2} & \frac{1}{2} \\ -\frac{1}{\sqrt{2}} & \frac{1}{\sqrt{2}} & 0 & 0 \\ 0 & 0 & \frac{1}{\sqrt{2}} & -\frac{1}{\sqrt{2}} \\ -\frac{1}{2} & -\frac{1}{2} & \frac{1}{2} & \frac{1}{2} \end{bmatrix}$

(e) $\begin{bmatrix} \frac{1}{2} & \frac{1}{2} & \frac{1}{2} & \frac{1}{2} \\ \frac{1}{2} & -\frac{5}{6} & \frac{1}{6} & \frac{1}{6} \\ \frac{1}{2} & \frac{1}{6} & \frac{1}{6} & -\frac{5}{6} \\ \frac{1}{2} & \frac{1}{6} & -\frac{5}{6} & \frac{1}{6} \end{bmatrix}$

(f) $\begin{bmatrix} 0 & \frac{1}{\sqrt{2}} & 0 & \frac{1}{\sqrt{2}} \\ 0 & \frac{1}{\sqrt{2}} & 0 & -\frac{1}{\sqrt{2}} \\ \frac{\sqrt{3}}{2} & 0 & \frac{1}{2} & 0 \\ \frac{1}{2} & 0 & -\frac{\sqrt{3}}{2} & 0 \end{bmatrix}$

4. Prove that if A is orthogonal, then A^T is orthogonal.

5. Verify that the reflection matrices in Tables 1 and 2 of Section 4.9 are orthogonal.

6. Let a rectangular $x'y'$-coordinate system be obtained by rotating a rectangular xy-coordinate system counterclockwise through the angle $\theta = 3\pi/4$.

   (a) Find the $x'y'$-coordinates of the point whose xy-coordinates are $(1, -3)$.

   (b) Find the xy-coordinates of the point whose $x'y'$-coordinates are $(2, 4)$.

7. Repeat Exercise 6 with $\theta = \pi/3$.

8. Let a rectangular $x'y'z'$-coordinate system be obtained by rotating a rectangular xyz-coordinate system counterclockwise about the z-axis (looking down the z-axis) through the angle $\theta = \pi/4$.

   (a) Find the $x'y'z'$-coordinates of the point whose xyz-coordinates are $(-1, 2, 5)$.

   (b) Find the xyz-coordinates of the point whose $x'y'z'$-coordinates are $(1, 6, -3)$.

9. Repeat Exercise 8 for a rotation of $\theta = \pi/6$ counterclockwise about the y-axis (looking along the positive y-axis toward the origin).

10. Repeat Exercise 8 for a rotation of $\theta = 3\pi/4$ counterclockwise about the x-axis (looking along the positive x-axis toward the origin).

11. (a) A rectangular $x'y'z'$-coordinate system is obtained by rotating an xyz-coordinate system counterclockwise about the y-axis through an angle θ (looking along the positive y-axis toward the origin). Find a matrix A such that

    $$\begin{bmatrix} x' \\ y' \\ z' \end{bmatrix} = A \begin{bmatrix} x \\ y \\ z \end{bmatrix}$$

    where (x, y, z) and (x', y', z') are the coordinates of the same point in the xyz- and $x'y'z'$-systems, respectively.

    (b) Repeat part (a) for a rotation about the x-axis.

12. A rectangular $x''y''z''$-coordinate system is obtained by first rotating a rectangular xyz-coordinate system 60° counterclockwise about the z-axis (looking down the positive z-axis) to obtain an $x'y'z'$-coordinate system, and then rotating the $x'y'z'$-coordinate system 45° counterclockwise about the y'-axis (looking along the positive y'-axis toward the origin). Find a matrix A such that

    $$\begin{bmatrix} x'' \\ y'' \\ z'' \end{bmatrix} = A \begin{bmatrix} x \\ y \\ z \end{bmatrix}$$

    where (x, y, z) and (x'', y'', z'') are the xyz- and $x''y''z''$-coordinates of the same point.

13. What conditions must a and b satisfy for the matrix

    $$\begin{bmatrix} a+b & b-a \\ a-b & b+a \end{bmatrix}$$

    to be orthogonal?

14. Prove that a 2×2 orthogonal matrix A has only one of two possible forms:

    $$A = \begin{bmatrix} \cos\theta & -\sin\theta \\ \sin\theta & \cos\theta \end{bmatrix} \quad \text{or} \quad A = \begin{bmatrix} \cos\theta & \sin\theta \\ \sin\theta & -\cos\theta \end{bmatrix}$$

    where $0 \leq \theta < 2\pi$. [*Hint:* Start with a general 2×2 matrix $A = (a_{ij})$, and use the fact that the column vectors form an orthonormal set in R^2.]

15. (a) Use the result in Exercise 14 to prove that multiplication by a 2×2 orthogonal matrix is either a rotation or a rotation followed by a reflection about the x-axis.

    (b) Prove that multiplication by A is a rotation if $\det(A) = 1$ and that a rotation followed by a reflection if $\det(A) = -1$.

16. Use the result in Exercise 15 to determine whether multiplication by A is a rotation or a rotation followed by a reflection about the x-axis. Find the angle of rotation in either case.

    (a) $A = \begin{bmatrix} \frac{\sqrt{3}}{2} & \frac{1}{2} \\ -\frac{1}{2} & \frac{\sqrt{3}}{2} \end{bmatrix}$
    (b) $A = \begin{bmatrix} -\frac{1}{\sqrt{2}} & -\frac{1}{\sqrt{2}} \\ -\frac{1}{\sqrt{2}} & \frac{1}{\sqrt{2}} \end{bmatrix}$

17. Find a, b, and c for which the matrix

    $$\begin{bmatrix} a & \frac{1}{\sqrt{2}} & -\frac{1}{\sqrt{2}} \\ b & \frac{1}{\sqrt{6}} & \frac{1}{\sqrt{6}} \\ c & \frac{1}{\sqrt{3}} & \frac{1}{\sqrt{3}} \end{bmatrix}$$

    is orthogonal. Are the values of a, b, and c unique? Explain.

18. The result in Exercise 15 has an analog for 3×3 orthogonal matrices: It can be proved that multiplication by a 3×3 orthogonal matrix A is a rotation about some axis if $\det(A) = 1$ and is a rotation about some axis followed by a reflection about some coordinate plane if $\det(A) = -1$. Determine whether multiplication by A is a rotation or a rotation followed by a reflection.

    (a) $A = \begin{bmatrix} \frac{3}{7} & \frac{2}{7} & \frac{6}{7} \\ -\frac{6}{7} & \frac{3}{7} & \frac{2}{7} \\ \frac{2}{7} & \frac{6}{7} & -\frac{3}{7} \end{bmatrix}$
    (b) $A = \begin{bmatrix} \frac{2}{7} & \frac{3}{7} & \frac{6}{7} \\ \frac{3}{7} & -\frac{6}{7} & \frac{2}{7} \\ \frac{6}{7} & \frac{2}{7} & -\frac{3}{7} \end{bmatrix}$

19. Use the fact stated in Exercise 18 and part (b) of Theorem 7.1.2 to show that a composition of rotations can always be accomplished by a single rotation about some appropriate axis.

20. Prove the equivalence of statements (a) and (c) in Theorem 7.1.1.

21. A linear operator on R^2 is called **rigid** if it does not change the lengths of vectors, and it is called **angle preserving** if it does not change the angle between nonzero vectors.

    (a) Name two different types of linear operators that are rigid.

    (b) Name two different types of linear operators that are angle preserving.

(c) Are there any linear operators on R^2 that are rigid and not angle preserving? Angle preserving and not rigid? Justify your answer.

True-False Exercises

In parts (a)–(h) determine whether the statement is true or false, and justify your answer.

(a) The matrix $\begin{bmatrix} 1 & 0 \\ 0 & 1 \\ 0 & 0 \end{bmatrix}$ is orthogonal.

(b) The matrix $\begin{bmatrix} 1 & -2 \\ 2 & 1 \end{bmatrix}$ is orthogonal.

(c) An $m \times n$ matrix A is orthogonal if $A^T A = I$.

(d) A square matrix whose columns form an orthogonal set is orthogonal.

(e) Every orthogonal matrix is invertible.

(f) If A is an orthogonal matrix, then A^2 is orthogonal and $(\det A)^2 = 1$.

(g) Every eigenvalue of an orthogonal matrix has absolute value 1.

(h) If A is a square matrix and $\|A\mathbf{u}\| = 1$ for all unit vectors $\mathbf{u}$, then A is orthogonal.

7.2 Orthogonal Diagonalization

In this section we will be concerned with the problem of diagonalizing a symmetric matrix A. As we will see, this problem is closely related to that of finding an orthonormal basis for R^n that consists of eigenvectors of A. Problems of this type are important because many of the matrices that arise in applications are symmetric.

The Orthogonal Diagonalization Problem

In Definition 1 of Section 5.2 we defined two square matrices, A and B, to be *similar* if there is an *invertible* matrix P such that $P^{-1}AP = B$. In this section we will be concerned with the special case in which it is possible to find an *orthogonal* matrix P for which this relationship holds.

We begin with the following definition.

> **DEFINITION 1** If A and B are square matrices, then we say that A and B are **orthogonally similar** if there is an orthogonal matrix P such that $P^T A P = B$.

If A is orthogonally similar to some diagonal matrix, say

$$P^T A P = D$$

then we say that A is **orthogonally diagonalizable** and that P **orthogonally diagonalizes** A.

Our first goal in this section is to determine what conditions a matrix must satisfy to be orthogonally diagonalizable. As a first step, observe that there is no hope of orthogonally diagonalizing a matrix that is not symmetric. To see why this is so, suppose that

$$P^T A P = D \tag{1}$$

where P is an orthogonal matrix and D is a diagonal matrix. Multiplying the left side of (1) by P, the right side by P^T, and then using the fact that $PP^T = P^T P = I$, we can rewrite this equation as

$$A = PDP^T \tag{2}$$

Now transposing both sides of this equation and using the fact that a diagonal matrix is the same as its transpose we obtain

$$A^T = (PDP^T)^T = (P^T)^T D^T P^T = PDP^T = A$$

so A must be symmetric.

Conditions for Orthogonal Diagonalizability

The following theorem shows that every symmetric matrix is, in fact, orthogonally diagonalizable. In this theorem, and for the remainder of this section, *orthogonal* will mean orthogonal with respect to the Euclidean inner product on R^n.

THEOREM 7.2.1 *If A is an $n \times n$ matrix, then the following are equivalent.*
(a) A is orthogonally diagonalizable.
(b) A has an orthonormal set of n eigenvectors.
(c) A is symmetric.

Proof (a) $\Rightarrow$ **(b)** Since A is orthogonally diagonalizable, there is an orthogonal matrix P such that $P^{-1}AP$ is diagonal. As shown in the proof of Theorem 5.2.1, the n column vectors of P are eigenvectors of A. Since P is orthogonal, these column vectors are orthonormal, so A has n orthonormal eigenvectors.

(b) $\Rightarrow$ **(a)** Assume that A has an orthonormal set of n eigenvectors $\{\mathbf{p}_1, \mathbf{p}_2, \ldots, \mathbf{p}_n\}$. As shown in the proof of Theorem 5.2.1, the matrix P with these eigenvectors as columns diagonalizes A. Since these eigenvectors are orthonormal, P is orthogonal and thus orthogonally diagonalizes A.

(a) $\Rightarrow$ **(c)** In the proof that $(a) \Rightarrow (b)$ we showed that an orthogonally diagonalizable $n \times n$ matrix A is orthogonally diagonalized by an $n \times n$ matrix P whose columns form an orthonormal set of eigenvectors of A. Let D be the diagonal matrix

$$D = P^T A P$$

from which it follows that

$$A = PDP^T$$

Thus,

$$A^T = (PDP^T)^T = PD^T P^T = PDP^T = A$$

which shows that A is symmetric.

(c) $\Rightarrow$ **(a)** The proof of this part is beyond the scope of this text and will be omitted. ◀

Properties of Symmetric Matrices

Our next goal is to devise a procedure for orthogonally diagonalizing a symmetric matrix, but before we can do so, we need the following critical theorem about eigenvalues and eigenvectors of symmetric matrices.

THEOREM 7.2.2 *If A is a symmetric matrix, then:*
(a) *The eigenvalues of A are all real numbers.*
(b) *Eigenvectors from different eigenspaces are orthogonal.*

Part (a), which requires results about complex vector spaces, will be discussed in Section 7.5.

Proof (b) Let $\mathbf{v}_1$ and $\mathbf{v}_2$ be eigenvectors corresponding to distinct eigenvalues λ_1 and λ_2 of the matrix A. We want to show that $\mathbf{v}_1 \cdot \mathbf{v}_2 = 0$. Our proof of this involves the trick of starting with the expression $A\mathbf{v}_1 \cdot \mathbf{v}_2$. It follows from Formula (26) of Section 3.2 and the symmetry of A that

$$A\mathbf{v}_1 \cdot \mathbf{v}_2 = \mathbf{v}_1 \cdot A^T \mathbf{v}_2 = \mathbf{v}_1 \cdot A\mathbf{v}_2 \qquad (3)$$

But $\mathbf{v}_1$ is an eigenvector of A corresponding to λ_1, and $\mathbf{v}_2$ is an eigenvector of A corresponding to λ_2, so (3) yields the relationship

$$\lambda_1 \mathbf{v}_1 \cdot \mathbf{v}_2 = \mathbf{v}_1 \cdot \lambda_2 \mathbf{v}_2$$

which can be rewritten as

$$(\lambda_1 - \lambda_2)(\mathbf{v}_1 \cdot \mathbf{v}_2) = 0 \qquad (4)$$

But $\lambda_1 - \lambda_2 \neq 0$, since λ_1 and λ_2 were assumed distinct. Thus, it follows from (4) that $\mathbf{v}_1 \cdot \mathbf{v}_2 = 0$. ◀

Theorem 7.2.2 yields the following procedure for orthogonally diagonalizing a symmetric matrix.

Orthogonally Diagonalizing an $n \times n$ Symmetric Matrix

Step 1. Find a basis for each eigenspace of A.

Step 2. Apply the Gram–Schmidt process to each of these bases to obtain an orthonormal basis for each eigenspace.

Step 3. Form the matrix P whose columns are the vectors constructed in Step 2. This matrix will orthogonally diagonalize A, and the eigenvalues on the diagonal of $D = P^T A P$ will be in the same order as their corresponding eigenvectors in P.

Remark The justification of this procedure should be clear: Theorem 7.2.2 ensures that eigenvectors from *different* eigenspaces are orthogonal, and applying the Gram–Schmidt process ensures that the eigenvectors within the *same* eigenspace are orthonormal. It follows that the *entire* set of eigenvectors obtained by this procedure will be orthonormal.

▶ **EXAMPLE 1** **Orthogonally Diagonalizing a Symmetric Matrix**

Find an orthogonal matrix P that diagonalizes

$$A = \begin{bmatrix} 4 & 2 & 2 \\ 2 & 4 & 2 \\ 2 & 2 & 4 \end{bmatrix}$$

Solution We leave it for you to verify that the characteristic equation of A is

$$\det(\lambda I - A) = \det \begin{bmatrix} \lambda - 4 & -2 & -2 \\ -2 & \lambda - 4 & -2 \\ -2 & -2 & \lambda - 4 \end{bmatrix} = (\lambda - 2)^2(\lambda - 8) = 0$$

Thus, the distinct eigenvalues of A are $\lambda = 2$ and $\lambda = 8$. By the method used in Example 5 of Section 7.1, it can be shown that

$$\mathbf{u}_1 = \begin{bmatrix} -1 \\ 1 \\ 0 \end{bmatrix} \quad \text{and} \quad \mathbf{u}_2 = \begin{bmatrix} -1 \\ 0 \\ 1 \end{bmatrix} \qquad (5)$$

form a basis for the eigenspace corresponding to $\lambda = 2$. Applying the Gram–Schmidt process to $\{\mathbf{u}_1, \mathbf{u}_2\}$ yields the following orthonormal eigenvectors (verify):

$$\mathbf{v}_1 = \begin{bmatrix} -\frac{1}{\sqrt{2}} \\ \frac{1}{\sqrt{2}} \\ 0 \end{bmatrix} \quad \text{and} \quad \mathbf{v}_2 = \begin{bmatrix} -\frac{1}{\sqrt{6}} \\ -\frac{1}{\sqrt{6}} \\ \frac{2}{\sqrt{6}} \end{bmatrix} \qquad (6)$$

The eigenspace corresponding to $\lambda = 8$ has

$$\mathbf{u}_3 = \begin{bmatrix} 1 \\ 1 \\ 1 \end{bmatrix}$$

as a basis. Applying the Gram–Schmidt process to $\{\mathbf{u}_3\}$ (i.e., normalizing $\mathbf{u}_3$) yields

$$\mathbf{v}_3 = \begin{bmatrix} \frac{1}{\sqrt{3}} \\ \frac{1}{\sqrt{3}} \\ \frac{1}{\sqrt{3}} \end{bmatrix}$$

Finally, using $\mathbf{v}_1$, $\mathbf{v}_2$, and $\mathbf{v}_3$ as column vectors, we obtain

$$P = \begin{bmatrix} -\frac{1}{\sqrt{2}} & -\frac{1}{\sqrt{6}} & \frac{1}{\sqrt{3}} \\ \frac{1}{\sqrt{2}} & -\frac{1}{\sqrt{6}} & \frac{1}{\sqrt{3}} \\ 0 & \frac{2}{\sqrt{6}} & \frac{1}{\sqrt{3}} \end{bmatrix}$$

which orthogonally diagonalizes A. As a check, we leave it for you to confirm that

$$P^T A P = \begin{bmatrix} -\frac{1}{\sqrt{2}} & \frac{1}{\sqrt{2}} & 0 \\ -\frac{1}{\sqrt{6}} & -\frac{1}{\sqrt{6}} & \frac{2}{\sqrt{6}} \\ \frac{1}{\sqrt{3}} & \frac{1}{\sqrt{3}} & \frac{1}{\sqrt{3}} \end{bmatrix} \begin{bmatrix} 4 & 2 & 2 \\ 2 & 4 & 2 \\ 2 & 2 & 4 \end{bmatrix} \begin{bmatrix} -\frac{1}{\sqrt{2}} & -\frac{1}{\sqrt{6}} & \frac{1}{\sqrt{3}} \\ \frac{1}{\sqrt{2}} & -\frac{1}{\sqrt{6}} & \frac{1}{\sqrt{3}} \\ 0 & \frac{2}{\sqrt{6}} & \frac{1}{\sqrt{3}} \end{bmatrix} = \begin{bmatrix} 2 & 0 & 0 \\ 0 & 2 & 0 \\ 0 & 0 & 8 \end{bmatrix} \blacktriangleleft$$

Spectral Decomposition   If A is a symmetric matrix that is orthogonally diagonalized by

$$P = [\mathbf{u}_1 \quad \mathbf{u}_2 \quad \cdots \quad \mathbf{u}_n]$$

and if $\lambda_1, \lambda_2, \ldots, \lambda_n$ are the eigenvalues of A corresponding to the unit eigenvectors $\mathbf{u}_1, \mathbf{u}_2, \ldots, \mathbf{u}_n$, then we know that $D = P^T A P$, where D is a diagonal matrix with the eigenvalues in the diagonal positions. It follows from this that the matrix A can be expressed as

$$A = PDP^T = [\mathbf{u}_1 \quad \mathbf{u}_2 \quad \cdots \quad \mathbf{u}_n] \begin{bmatrix} \lambda_1 & 0 & \cdots & 0 \\ 0 & \lambda_2 & \cdots & 0 \\ \vdots & \vdots & \ddots & \vdots \\ 0 & 0 & \cdots & \lambda_n \end{bmatrix} \begin{bmatrix} \mathbf{u}_1^T \\ \mathbf{u}_2^T \\ \vdots \\ \mathbf{u}_n^T \end{bmatrix}$$

$$= [\lambda_1 \mathbf{u}_1 \quad \lambda_2 \mathbf{u}_2 \quad \cdots \quad \lambda_n \mathbf{u}_n] \begin{bmatrix} \mathbf{u}_1^T \\ \mathbf{u}_2^T \\ \vdots \\ \mathbf{u}_n^T \end{bmatrix}$$

Multiplying out, we obtain the formula

$$A = \lambda_1 \mathbf{u}_1 \mathbf{u}_1^T + \lambda_2 \mathbf{u}_2 \mathbf{u}_2^T + \cdots + \lambda_n \mathbf{u}_n \mathbf{u}_n^T \tag{7}$$

which is called a ***spectral decomposition of A***.[*]

---

[*]The terminology *spectral decomposition* is derived from the fact that the set of all eigenvalues of a matrix A is sometimes called the **spectrum** of A. The terminology *eigenvalue decomposition* is due to Professor Dan Kalman, who introduced it in an award-winning paper entitled "A Singularly Valuable Decomposition: The SVD of a Matrix," *The College Mathematics Journal*, Vol. 27, No. 1, January 1996.

Note that in each term of the spectral decomposition of A has the form $\lambda \mathbf{u}\mathbf{u}^T$, where $\mathbf{u}$ is a unit eigenvector of A in column form, and λ is an eigenvalue of A corresponding to $\mathbf{u}$. Since $\mathbf{u}$ has size $n \times 1$, it follows that the product $\mathbf{u}\mathbf{u}^T$ has size $n \times n$. It can be proved (though we will not do it) that $\mathbf{u}\mathbf{u}^T$ is the standard matrix for the orthogonal projection of R^n on the subspace spanned by the vector $\mathbf{u}$. Accepting this to be so, the spectral decomposition of A tells that the image of a vector $\mathbf{x}$ under multiplication by a symmetric matrix A can be obtained by projecting $\mathbf{x}$ orthogonally on the lines (one-dimensional subspaces) determined by the eigenvectors of A, then scaling those projections by the eigenvalues, and then adding the scaled projections. Here is an example.

▶ **EXAMPLE 2 A Geometric Interpretation of a Spectral Decomposition**

The matrix

$$A = \begin{bmatrix} 1 & 2 \\ 2 & -2 \end{bmatrix}$$

has eigenvalues $\lambda_1 = -3$ and $\lambda_2 = 2$ with corresponding eigenvectors

$$\mathbf{x}_1 = \begin{bmatrix} 1 \\ -2 \end{bmatrix} \quad \text{and} \quad \mathbf{x}_2 = \begin{bmatrix} 2 \\ 1 \end{bmatrix}$$

(verify). Normalizing these basis vectors yields

$$\mathbf{u}_1 = \frac{\mathbf{x}_1}{\|\mathbf{x}_1\|} = \begin{bmatrix} \frac{1}{\sqrt{5}} \\ -\frac{2}{\sqrt{5}} \end{bmatrix} \quad \text{and} \quad \mathbf{u}_2 = \frac{\mathbf{x}_2}{\|\mathbf{x}_2\|} = \begin{bmatrix} \frac{2}{\sqrt{5}} \\ \frac{1}{\sqrt{5}} \end{bmatrix}$$

so a spectral decomposition of A is

$$\begin{bmatrix} 1 & 2 \\ 2 & -2 \end{bmatrix} = \lambda_1 \mathbf{u}_1 \mathbf{u}_1^T + \lambda_2 \mathbf{u}_2 \mathbf{u}_2^T = (-3) \begin{bmatrix} \frac{1}{\sqrt{5}} \\ -\frac{2}{\sqrt{5}} \end{bmatrix} \begin{bmatrix} \frac{1}{\sqrt{5}} & -\frac{2}{\sqrt{5}} \end{bmatrix} + (2) \begin{bmatrix} \frac{2}{\sqrt{5}} \\ \frac{1}{\sqrt{5}} \end{bmatrix} \begin{bmatrix} \frac{2}{\sqrt{5}} & \frac{1}{\sqrt{5}} \end{bmatrix}$$

$$= (-3) \begin{bmatrix} \frac{1}{5} & -\frac{2}{5} \\ -\frac{2}{5} & \frac{4}{5} \end{bmatrix} + (2) \begin{bmatrix} \frac{4}{5} & \frac{2}{5} \\ \frac{2}{5} & \frac{1}{5} \end{bmatrix} \quad (8)$$

where, as noted above, the 2×2 matrices on the right side of (8) are the standard matrices for the orthogonal projections onto the eigenspaces corresponding to $\lambda_1 = -3$ and $\lambda_2 = 2$, respectively.

Now let us see what this spectral decomposition tells us about the image of the vector $\mathbf{x} = (1, 1)$ under multiplication by A. Writing $\mathbf{x}$ in column form, it follows that

$$A\mathbf{x} = \begin{bmatrix} 1 & 2 \\ 2 & -2 \end{bmatrix} \begin{bmatrix} 1 \\ 1 \end{bmatrix} = \begin{bmatrix} 3 \\ 0 \end{bmatrix} \quad (9)$$

and from (8) that

$$A\mathbf{x} = \begin{bmatrix} 1 & 2 \\ 2 & -2 \end{bmatrix} \begin{bmatrix} 1 \\ 1 \end{bmatrix} = (-3) \begin{bmatrix} \frac{1}{5} & -\frac{2}{5} \\ -\frac{2}{5} & \frac{4}{5} \end{bmatrix} \begin{bmatrix} 1 \\ 1 \end{bmatrix} + (2) \begin{bmatrix} \frac{4}{5} & \frac{2}{5} \\ \frac{2}{5} & \frac{1}{5} \end{bmatrix} \begin{bmatrix} 1 \\ 1 \end{bmatrix}$$

$$= (-3) \begin{bmatrix} -\frac{1}{5} \\ \frac{2}{5} \end{bmatrix} + (2) \begin{bmatrix} \frac{6}{5} \\ \frac{3}{5} \end{bmatrix}$$

$$= \begin{bmatrix} \frac{3}{5} \\ -\frac{6}{5} \end{bmatrix} + \begin{bmatrix} \frac{12}{5} \\ \frac{6}{5} \end{bmatrix} = \begin{bmatrix} 3 \\ 0 \end{bmatrix} \quad (10)$$

Formulas (9) and (10) provide two different ways of viewing the image of the vector $(1, 1)$ under multiplication by A: Formula (9) tells us directly that the image of this vector is $(3, 0)$, whereas Formula (10) tells us that this image can also be obtained by projecting $(1, 1)$ onto the eigenspaces corresponding to $\lambda_1 = -3$ and $\lambda_2 = 2$ to obtain the vectors $\left(-\frac{1}{5}, \frac{2}{5}\right)$ and $\left(\frac{6}{5}, \frac{3}{5}\right)$, then scaling by the eigenvalues to obtain $\left(\frac{3}{5}, -\frac{6}{5}\right)$ and $\left(\frac{12}{5}, \frac{6}{5}\right)$, and then adding these vectors (see Figure 7.2.1). ◀

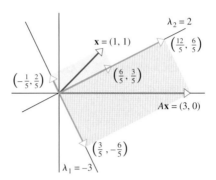

▶ Figure 7.2.1

The Nondiagonalizable Case

If A is an $n \times n$ matrix that is not orthogonally diagonalizable, it may still be possible to achieve considerable simplification in the form of $P^T A P$ by choosing the orthogonal matrix P appropriately. We will consider two theorems (without proof) that illustrate this. The first, due to the German mathematician Isaai Schur, states that every square matrix A is orthogonally similar to an upper triangular matrix that has the eigenvalues of A on the main diagonal.

THEOREM 7.2.3 Schur's Theorem

If A is an $n \times n$ matrix with real entries and real eigenvalues, then there is an orthogonal matrix P such that $P^T A P$ is an upper triangular matrix of the form

$$P^T A P = \begin{bmatrix} \lambda_1 & \times & \times & \cdots & \times \\ 0 & \lambda_2 & \times & \cdots & \times \\ 0 & 0 & \lambda_3 & \cdots & \times \\ \vdots & \vdots & \vdots & \ddots & \vdots \\ 0 & 0 & 0 & \cdots & \lambda_n \end{bmatrix} \tag{11}$$

in which $\lambda_1, \lambda_2, \ldots, \lambda_n$ are the eigenvalues of the matrix A repeated according to multiplicity.

Issai Schur (1875–1941)

Historical Note The life of the German mathematician Issai Schur is a sad reminder of the effect that Nazi policies had on Jewish intellectuals during the 1930s. Schur was a brilliant mathematician and a popular lecturer who attracted many students and researchers to the University of Berlin, where he worked and taught. His lectures sometimes attracted so many students that opera glasses were needed to see him from the back row. Schur's life became increasingly difficult under Nazi rule, and in April of 1933 he was forced to "retire" from the university under a law that prohibited non-Aryans from holding "civil service" positions. There was an outcry from many of his students and colleagues who respected and liked him, but it did not stave off his complete dismissal in 1935. Schur, who thought of himself as a loyal German never understood the persecution and humiliation he received at Nazi hands. He left Germany for Palestine in 1939, a broken man. Lacking in financial resources, he had to sell his beloved mathematics books and lived in poverty until his death in 1941.

[*Image: Courtesy Electronic Publishing Services, Inc., New York City*]

It is common to denote the upper triangular matrix in (11) by S (for Schur), in which case that equation can be rewritten as

$$A = PSP^T \tag{12}$$

which is called a ***Schur decomposition*** of A.

The next theorem, due to the German mathematician and engineer Karl Hessenberg (1904–1959), states that every square matrix with real entries is orthogonally similar to a matrix in which each entry below the first ***subdiagonal*** is zero (Figure 7.2.2). Such a matrix is said to be in ***upper Hessenberg form***.

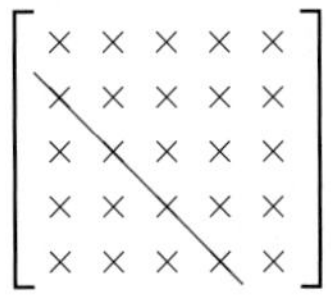

First subdiagonal

▲ Figure 7.2.2

THEOREM 7.2.4 Hessenberg's Theorem

If A is an $n \times n$ matrix, then there is an orthogonal matrix P such that $P^T A P$ is a matrix of the form

$$P^T A P = \begin{bmatrix} \times & \times & \cdots & \times & \times & \times \\ \times & \times & \cdots & \times & \times & \times \\ 0 & \times & \ddots & \times & \times & \times \\ \vdots & \vdots & \ddots & \vdots & \vdots & \vdots \\ 0 & 0 & \cdots & \times & \times & \times \\ 0 & 0 & \cdots & 0 & \times & \times \end{bmatrix} \tag{13}$$

Note that unlike those in (11), the diagonal entries in (13) are usually *not* the eigenvalues of A.

It is common to denote the upper Hessenberg matrix in (13) by H (for Hessenberg), in which case that equation can be rewritten as

$$A = PHP^T \tag{14}$$

which is called an ***upper Hessenberg decomposition*** of A.

Remark In many numerical algorithms the initial matrix is first converted to upper Hessenberg form to reduce the amount of computation in subsequent parts of the algorithm. Many computer packages have built-in commands for finding Schur and Hessenberg decompositions.

Concept Review

- Orthogonally similar matrices
- Orthogonally diagonalizable matrix
- Spectral decomposition (or eigenvalue decomposition)
- Schur decomposition
- Subdiagonal
- Upper Hessenburg form
- Upper Hessenburg decomposition

Skills

- Be able to recognize an orthogonally diagonalizable matrix.
- Know that eigenvalues of symmetric matrices are real numbers.
- Know that for a symmetric matrix eigenvectors from different eigenspaces are orthogonal.
- Be able to orthogonally diagonalize a symmetric matrix.
- Be able to find the spectral decomposition of a symmetric matrix.
- Know the statement of Schur's Theorem.
- Know the statement of Hessenburg's Theorem.

Exercise Set 7.2

1. Find the characteristic equation of the given symmetric matrix, and then by inspection determine the dimensions of the eigenspaces.

(a) $\begin{bmatrix} 2 & 6 \\ 6 & 18 \end{bmatrix}$

(b) $\begin{bmatrix} 1 & -2 & 2 \\ -2 & 1 & -2 \\ 2 & -2 & 1 \end{bmatrix}$

(c) $\begin{bmatrix} 1 & 0 & 1 \\ 0 & 1 & 0 \\ 1 & 0 & 1 \end{bmatrix}$

(d) $\begin{bmatrix} 2 & 2 & 2 \\ 2 & 2 & -2 \\ 2 & -2 & 2 \end{bmatrix}$

(e) $\begin{bmatrix} 0 & 3 & 0 & 0 \\ 3 & 0 & 0 & 0 \\ 0 & 0 & 0 & 3 \\ 0 & 0 & 3 & 0 \end{bmatrix}$

(f) $\begin{bmatrix} 0 & 0 & 2 & -1 \\ 0 & 0 & -1 & 2 \\ 2 & -1 & 0 & 0 \\ -1 & 2 & 0 & 0 \end{bmatrix}$

▶ In Exercises 2–9, find a matrix P that orthogonally diagonalizes A, and determine $P^{-1}AP$. ◀

2. $A = \begin{bmatrix} 4 & 2 \\ 2 & 4 \end{bmatrix}$

3. $A = \begin{bmatrix} 1 & -12 \\ -12 & -6 \end{bmatrix}$

4. $A = \begin{bmatrix} -7 & 24 \\ 24 & 7 \end{bmatrix}$

5. $A = \begin{bmatrix} 3 & 0 & 1 \\ 0 & 2 & 0 \\ 1 & 0 & 3 \end{bmatrix}$

6. $A = \begin{bmatrix} 1 & 1 & 0 \\ 1 & 1 & 0 \\ 0 & 0 & 0 \end{bmatrix}$

7. $A = \begin{bmatrix} 2 & -1 & -1 \\ -1 & 2 & -1 \\ -1 & -1 & 2 \end{bmatrix}$

8. $A = \begin{bmatrix} 4 & 2 & 0 & 0 \\ 2 & 4 & 0 & 0 \\ 0 & 0 & 0 & 0 \\ 0 & 0 & 0 & 0 \end{bmatrix}$

9. $A = \begin{bmatrix} 6 & -2 & 0 & 0 \\ -2 & 3 & 0 & 0 \\ 0 & 0 & 6 & -2 \\ 0 & 0 & -2 & 3 \end{bmatrix}$

10. Assuming that $b \neq 0$, find a matrix that orthogonally diagonalizes

$$\begin{bmatrix} a & b \\ b & a \end{bmatrix}$$

11. Prove that if A is any $m \times n$ matrix, then $A^T A$ has an orthonormal set of n eigenvectors.

12. (a) Show that if $\mathbf{v}$ is any $n \times 1$ matrix and I is the $n \times n$ identity matrix, then $I - \mathbf{v}\mathbf{v}^T$ is orthogonally diagonalizable.

(b) Find a matrix P that orthogonally diagonalizes $I - \mathbf{v}\mathbf{v}^T$ if

$$\mathbf{v} = \begin{bmatrix} 1 \\ 0 \\ 1 \end{bmatrix}$$

13. Use the result in Exercise 19 of Section 5.1 to prove Theorem 7.2.2a for 2×2 symmetric matrices.

14. Does there exist a 3×3 symmetric matrix with eigenvalues $\lambda_1 = -1, \lambda_2 = 3, \lambda_3 = 7$ and corresponding eigenvectors

$$\begin{bmatrix} 0 \\ 1 \\ -1 \end{bmatrix}, \begin{bmatrix} 1 \\ 0 \\ 0 \end{bmatrix}, \begin{bmatrix} 0 \\ 1 \\ 1 \end{bmatrix}?$$

If so, find such a matrix; if not, explain why not.

15. Is the converse of Theorem 7.2.2b true? Explain.

16. Find the spectral decomposition of each matrix.

(a) $\begin{bmatrix} 4 & 2 \\ 2 & 4 \end{bmatrix}$

(b) $\begin{bmatrix} -7 & 24 \\ 24 & 7 \end{bmatrix}$

(c) $\begin{bmatrix} -3 & 1 & 2 \\ 1 & -3 & 2 \\ 2 & 2 & 0 \end{bmatrix}$

(d) $\begin{bmatrix} 3 & 0 & 1 \\ 0 & 2 & 0 \\ 1 & 0 & 3 \end{bmatrix}$

17. Show that if A is a symmetric orthogonal matrix, then 1 and -1 are the only possible eigenvalues.

18. (a) Find a 3×3 symmetric matrix whose eigenvalues are $\lambda_1 = -1, \lambda_2 = 3, \lambda_3 = 7$ and for which the corresponding eigenvectors are $\mathbf{v}_1 = (0, 1, -1)$, $\mathbf{v}_2 = (1, 0, 0)$, $\mathbf{v}_3 = (0, 1, 1)$.

(b) Is there a 3×3 symmetric matrix with eigenvalues $\lambda_1 = -1, \lambda_2 = 3, \lambda_3 = 7$ and corresponding eigenvectors $\mathbf{v}_1 = (0, 1, -1)$, $\mathbf{v}_2 = (1, 0, 0)$, $\mathbf{v}_3 = (1, 1, 1)$? Explain your reasoning.

19. Let A be a diagonalizable matrix with the property that eigenvectors from distinct eigenvalues are orthogonal. Must A be symmetric? Explain you reasoning.

20. Prove: If $\{\mathbf{u}_1, \mathbf{u}_2, \ldots, \mathbf{u}_n\}$ is an orthonormal basis for R^n, and if A can be expressed as

$$A = c_1\mathbf{u}_1\mathbf{u}_1^T + c_2\mathbf{u}_2\mathbf{u}_2^T + \cdots + c_n\mathbf{u}_n\mathbf{u}_n^T$$

then A is symmetric and has eigenvalues $c_1, c_2, \ldots, c_n$.

21. In this exercise we will establish that a matrix A is orthogonally diagonalizable if and only if it is symmetric. We have shown that an orthogonally diagonalizable matrix is symmetric. The harder part is to prove that a symmetric matrix A is orthogonally diagonalizable. We will proceed in two steps: first we will show that A is diagonalizable, and then we will build on that result to show that A is orthogonally diagonalizable.

(a) Assume that A is a symmetric $n \times n$ matrix. One way to prove that A is diagonalizable is to show that for each eigenvalue λ_0 the geometric multiplicity is equal to the algebraic multiplicity. For this purpose, assume that the geometric multiplicity of λ_0 is k, let $B_0 = \{\mathbf{u}_1, \mathbf{u}_2, \ldots, \mathbf{u}_k\}$ be an orthonormal basis for the eigenspace corresponding to λ_0, extend this to an orthonormal basis

$B = \{\mathbf{u}_1, \mathbf{u}_2, \ldots, \mathbf{u}_n\}$ for R^n, and let P be the matrix having the vectors of B as columns. As shown in Exercise 34(b) of Section 5.2, the product AP can be written as

$$AP = P \begin{bmatrix} \lambda_0 I_k & X \\ 0 & Y \end{bmatrix}$$

Use the fact that B is an orthonormal basis to prove that $X = 0$ [a zero matrix of size $n \times (n-k)$].

(b) It follows from part (a) and Exercise 34(c) of Section 5.2 that A has the same characteristic polynomial as

$$C = \begin{bmatrix} \lambda_0 I_k & 0 \\ 0 & Y \end{bmatrix}$$

Use this fact and Exercise 34(d) of Section 5.2 to prove that the algebraic multiplicity of λ_0 is the same as the geometric multiplicity of λ_0. This establishes that A is diagonalizable.

(c) Use Theorem 7.2.2(b) and the fact that A is diagonalizable to prove that A is orthogonally diagonalizable.

True-False Exercises

In parts (a)–(g) determine whether the statement is true or false, and justify your answer.

(a) If A is a square matrix, then AA^T and A^TA are orthogonally diagonalizable.

(b) If $\mathbf{v}_1$ and $\mathbf{v}_2$ are eigenvectors from distinct eigenspaces of a symmetric matrix, then $\|\mathbf{v}_1 + \mathbf{v}_2\|^2 = \|\mathbf{v}_1\|^2 + \|\mathbf{v}_2\|^2$.

(c) Every orthogonal matrix is orthogonally diagonalizable.

(d) If A is both invertible and orthogonally diagonalizable, then A^{-1} is orthogonally diagonalizable.

(e) Every eigenvalue of an orthogonal matrix has absolute value 1.

(f) If A is an $n \times n$ orthogonally diagonalizable matrix, then there exists an orthonormal basis for R^n consisting of eigenvectors of A.

(g) If A is orthogonally diagonalizable, then A has real eigenvalues.

7.3 Quadratic Forms

In this section we will use matrix methods to study real-valued functions of several variables in which each term is either the square of a variable or the product of two variables. Such functions arise in a variety of applications, including geometry, vibrations of mechanical systems, statistics, and electrical engineering.

Definition of a Quadratic Form

Expressions of the form

$$a_1 x_1 + a_2 x_2 + \cdots + a_n x_n$$

occurred in our study of linear equations and linear systems. If $a_1, a_2, \ldots, a_n$ are treated as fixed constants, then this expression is a real-valued function of the n variables $x_1, x_2, \ldots, x_n$ and is called a **linear form** on R^n. All variables in a linear form occur to the first power and there are no products of variables. Here we will be concerned with **quadratic forms** on R^n, which are functions of the form

$$a_1 x_1^2 + a_2 x_2^2 + \cdots + a_n x_n^2 + \text{(all possible terms } a_k x_i x_j \text{ in which } x_i \neq x_j\text{)}$$

The terms of the form $a_k x_i x_j$ are called **cross product terms**. It is common to combine the cross product terms involving $x_i x_j$ with those involving $x_j x_i$ to avoid duplication. Thus, a general quadratic form on R^2 would typically be expressed as

$$a_1 x_1^2 + a_2 x_2^2 + 2a_3 x_1 x_2 \tag{1}$$

and a general quadratic form on R^3 as

$$a_1 x_1^2 + a_2 x_2^2 + a_3 x_3^2 + 2a_4 x_1 x_2 + 2a_5 x_1 x_3 + 2a_6 x_2 x_3 \tag{2}$$

If, as usual, we do not distinguish between the number a and the 1×1 matrix $[a]$, and if we let $\mathbf{x}$ be the column vector of variables, then (1) and (2) can be expressed in matrix form as

$$\begin{bmatrix} x_1 & x_2 \end{bmatrix} \begin{bmatrix} a_1 & a_3 \\ a_3 & a_2 \end{bmatrix} \begin{bmatrix} x_1 \\ x_2 \end{bmatrix} = \mathbf{x}^T A \mathbf{x}$$

$$\begin{bmatrix} x_1 & x_2 & x_3 \end{bmatrix} \begin{bmatrix} a_1 & a_4 & a_5 \\ a_4 & a_2 & a_6 \\ a_5 & a_6 & a_3 \end{bmatrix} \begin{bmatrix} x_1 \\ x_2 \\ x_3 \end{bmatrix} = \mathbf{x}^T A \mathbf{x}$$

(verify). Note that the matrix A in these formulas is symmetric, that its diagonal entries are the coefficients of the squared terms, and its off-diagonal entries are half the coefficients of the cross product terms. In general, if A is a symmetric $n \times n$ matrix and $\mathbf{x}$ is an $n \times 1$ column vector of variables, then we call the function

$$Q_A(\mathbf{x}) = \mathbf{x}^T A \mathbf{x} \tag{3}$$

the *quadratic form associated with A*. When convenient, (3) can be expressed in dot product notation as

$$\mathbf{x}^T A \mathbf{x} = \mathbf{x} \cdot A\mathbf{x} = A\mathbf{x} \cdot \mathbf{x} \tag{4}$$

In the case where A is a diagonal matrix, the quadratic form $\mathbf{x}^T A \mathbf{x}$ has no cross product terms; for example, if A has diagonal entries $\lambda_1, \lambda_2, \ldots, \lambda_n$, then

$$\mathbf{x}^T A \mathbf{x} = [x_1 \; x_2 \; \cdots \; x_n] \begin{bmatrix} \lambda_1 & 0 & \cdots & 0 \\ 0 & \lambda_2 & \cdots & 0 \\ \vdots & \vdots & \ddots & \vdots \\ 0 & 0 & \cdots & \lambda_n \end{bmatrix} \begin{bmatrix} x_1 \\ x_2 \\ \vdots \\ x_n \end{bmatrix} = \lambda_1 x_1^2 + \lambda_2 x_2^2 + \cdots + \lambda_n x_n^2$$

▶ **EXAMPLE 1** Expressing Quadratic Forms in Matrix Notation

In each part, express the quadratic form in the matrix notation $\mathbf{x}^T A \mathbf{x}$, where A is symmetric.

(a) $2x^2 + 6xy - 5y^2$ (b) $x_1^2 + 7x_2^2 - 3x_3^2 + 4x_1 x_2 - 2x_1 x_3 + 8x_2 x_2$

Solution The diagonal entries of A are the coefficients of the squared terms, and the off-diagonal entries are half the coefficients of the cross product terms, so

$$2x^2 + 6xy - 5y^2 = [x \; y] \begin{bmatrix} 2 & 3 \\ 3 & -5 \end{bmatrix} \begin{bmatrix} x \\ y \end{bmatrix}$$

$$x_1^2 + 7x_2^2 - 3x_3^2 + 4x_1 x_2 - 2x_1 x_3 + 8x_2 x_3 = [x_1 \; x_2 \; x_3] \begin{bmatrix} 1 & 2 & -1 \\ 2 & 7 & 4 \\ -1 & 4 & -3 \end{bmatrix} \begin{bmatrix} x_1 \\ x_2 \\ x_3 \end{bmatrix} \; ◀$$

Change of Variable in a Quadratic Form

There are three important kinds of problems that occur in applications of quadratic forms:

> **Problem 1** If $\mathbf{x}^T A \mathbf{x}$ is a quadratic form on R^2 or R^3, what kind of curve or surface is represented by the equation $\mathbf{x}^T A \mathbf{x} = k$?
>
> **Problem 2** If $\mathbf{x}^T A \mathbf{x}$ is a quadratic form on R^n, what conditions must A satisfy for $\mathbf{x}^T A \mathbf{x}$ to have positive values for $\mathbf{x} \neq \mathbf{0}$?
>
> **Problem 3** If $\mathbf{x}^T A \mathbf{x}$ is a quadratic form on R^n, what are its maximum and minimum values if $\mathbf{x}$ is constrained to satisfy $\|\mathbf{x}\| = 1$?

We will consider the first two problems in this section and the third problem in the next section.

Many of the techniques for solving these problems are based on simplifying the quadratic form $\mathbf{x}^T A \mathbf{x}$ by making a substitution

$$\mathbf{x} = P\mathbf{y} \tag{5}$$

that expresses the variables $x_1, x_2, \ldots, x_n$ in terms of new variables $y_1, y_2, \ldots, y_n$. If P is invertible, then we call (5) a *change of variable*, and if P is orthogonal, then we call (5) an *orthogonal change of variable*.

If we make the change of variable $\mathbf{x} = P\mathbf{y}$ in the quadratic form $\mathbf{x}^T A \mathbf{x}$, then we obtain

$$\mathbf{x}^T A \mathbf{x} = (P\mathbf{y})^T A (P\mathbf{y}) = \mathbf{y}^T P^T A P \mathbf{y} = \mathbf{y}^T (P^T A P) \mathbf{y} \quad (6)$$

Since the matrix $B = P^T A P$ is symmetric (verify), the effect of the change of variable is to produce a new quadratic form $\mathbf{y}^T B \mathbf{y}$ in the variables $y_1, y_2, \ldots, y_n$. In particular, if we choose P to orthogonally diagonalize A, then the new quadratic form will be $\mathbf{y}^T D \mathbf{y}$, where D is a diagonal matrix with the eigenvalues of A on the main diagonal; that is,

$$\mathbf{x}^T A \mathbf{x} = \mathbf{y}^T D \mathbf{y} = \begin{bmatrix} y_1 & y_2 & \cdots & y_n \end{bmatrix} \begin{bmatrix} \lambda_1 & 0 & \cdots & 0 \\ 0 & \lambda_2 & \cdots & 0 \\ \vdots & \vdots & \ddots & \vdots \\ 0 & 0 & \cdots & \lambda_n \end{bmatrix} \begin{bmatrix} y_1 \\ y_2 \\ \vdots \\ y_n \end{bmatrix}$$

$$= \lambda_1 y_1^2 + \lambda_2 y_2^2 + \cdots + \lambda_n y_n^2$$

Thus, we have the following result, called the *principal axes theorem*.

THEOREM 7.3.1 **The Principal Axes Theorem**

If A is a symmetric $n \times n$ matrix, then there is an orthogonal change of variable that transforms the quadratic form $\mathbf{x}^T A \mathbf{x}$ into a quadratic form $\mathbf{y}^T D \mathbf{y}$ with no cross product terms. Specifically, if P orthogonally diagonalizes A, then making the change of variable $\mathbf{x} = P\mathbf{y}$ in the quadratic form $\mathbf{x}^T A \mathbf{x}$ yields the quadratic form

$$\mathbf{x}^T A \mathbf{x} = \mathbf{y}^T D \mathbf{y} = \lambda_1 y_1^2 + \lambda_2 y_2^2 + \cdots + \lambda_n y_n^2$$

in which $\lambda_1, \lambda_2, \ldots, \lambda_n$ are the eigenvalues of A corresponding to the eigenvectors that form the successive columns of P.

▶ **EXAMPLE 2** **An Illustration of the Principal Axes Theorem**

Find an orthogonal change of variable that eliminates the cross product terms in the quadratic form $Q = x_1^2 - x_3^2 - 4x_1 x_2 + 4x_2 x_3$, and express Q in terms of the new variables.

Solution The quadratic form can be expressed in matrix notation as

$$Q = \mathbf{x}^T A \mathbf{x} = \begin{bmatrix} x_1 & x_2 & x_3 \end{bmatrix} \begin{bmatrix} 1 & -2 & 0 \\ -2 & 0 & 2 \\ 0 & 2 & -1 \end{bmatrix} \begin{bmatrix} x_1 \\ x_2 \\ x_3 \end{bmatrix}$$

The characteristic equation of the matrix A is

$$\begin{vmatrix} \lambda - 1 & 2 & 0 \\ 2 & \lambda & -2 \\ 0 & -2 & \lambda + 1 \end{vmatrix} = \lambda^3 - 9\lambda = \lambda(\lambda + 3)(\lambda - 3) = 0$$

so the eigenvalues are $\lambda = 0, -3, 3$. We leave it for you to show that orthonormal bases for the three eigenspaces are

$$\lambda = 0: \begin{bmatrix} \frac{2}{3} \\ \frac{1}{3} \\ \frac{2}{3} \end{bmatrix}, \quad \lambda = -3: \begin{bmatrix} -\frac{1}{3} \\ -\frac{2}{3} \\ \frac{2}{3} \end{bmatrix}, \quad \lambda = 3: \begin{bmatrix} -\frac{2}{3} \\ \frac{2}{3} \\ \frac{1}{3} \end{bmatrix}$$

Thus, a substitution $\mathbf{x} = P\mathbf{y}$ that eliminates the cross product terms is

$$\begin{bmatrix} x_1 \\ x_2 \\ x_3 \end{bmatrix} = \begin{bmatrix} \frac{2}{3} & -\frac{1}{3} & -\frac{2}{3} \\ \frac{1}{3} & -\frac{2}{3} & \frac{2}{3} \\ \frac{2}{3} & \frac{2}{3} & \frac{1}{3} \end{bmatrix} \begin{bmatrix} y_1 \\ y_2 \\ y_3 \end{bmatrix}$$

This produces the new quadratic form

$$Q = \mathbf{y}^T(P^TAP)\mathbf{y} = \begin{bmatrix} y_1 & y_2 & y_3 \end{bmatrix} \begin{bmatrix} 0 & 0 & 0 \\ 0 & -3 & 0 \\ 0 & 0 & 3 \end{bmatrix} \begin{bmatrix} y_1 \\ y_2 \\ y_3 \end{bmatrix} = -3y_2^2 + 3y_3^2$$

in which there are no cross product terms. ◂

Remark If A is a symmetric $n \times n$ matrix, then the quadratic form $\mathbf{x}^T A \mathbf{x}$ is a real-valued function whose range is the set of all possible values for $\mathbf{x}^T A \mathbf{x}$ as $\mathbf{x}$ varies over R^n. It can be shown that an orthogonal change of variable $\mathbf{x} = P\mathbf{y}$ does not alter the range of a quadratic form; that is, the set of all values for $\mathbf{x}^T A \mathbf{x}$ as $\mathbf{x}$ varies over R^n is the same as the set of all values for $\mathbf{y}^T(P^TAP)\mathbf{y}$ as $\mathbf{y}$ varies over R^n.

Quadratic Forms in Geometry

Recall that a ***conic section*** or ***conic*** is a curve that results by cutting a double-napped cone with a plane (Figure 7.3.1). The most important conic sections are ellipses, hyperbolas, and parabolas, which result when the cutting plane does not pass through the vertex. Circles are special cases of ellipses that result when the cutting plane is perpendicular to the axis of symmetry of the cone. If the cutting plane passes through the vertex, then the resulting intersection is called a ***degenerate conic***. The possibilities are a point, a pair of intersecting lines, or a single line.

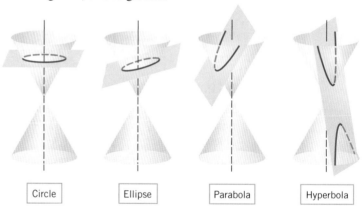

▶ Figure 7.3.1

Quadratic forms in R^2 arise naturally in the study of conic sections. For example, it is shown in analytic geometry that an equation of the form

$$ax^2 + 2bxy + cy^2 + dx + ey + f = 0 \tag{7}$$

in which a, b, and c are not all zero, represents a conic section.* If $d = e = 0$ in (7), then there are no linear terms, so the equation becomes

$$ax^2 + 2bxy + cy^2 + f = 0 \tag{8}$$

---

*We must also allow for the possibility that there are no real values of x and y that satisfy the equation, as with $x^2 + y^2 + 1 = 0$. In such cases we say that the equation has ***no graph*** or has ***an empty graph***.

and is said to represent a ***central conic***. These include circles, ellipses, and hyperbolas, but not parabolas. Furthermore, if $b = 0$ in (8), then there is no cross product term (i.e., term involving xy), and the equation

$$ax^2 + cy^2 + f = 0 \tag{9}$$

is said to represent a ***central conic in standard position***. The most important conics of this type are shown in Table 1.

Table 1

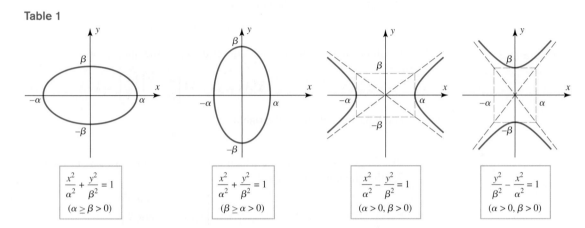

$\dfrac{x^2}{\alpha^2} + \dfrac{y^2}{\beta^2} = 1$	$\dfrac{x^2}{\alpha^2} + \dfrac{y^2}{\beta^2} = 1$	$\dfrac{x^2}{\alpha^2} - \dfrac{y^2}{\beta^2} = 1$	$\dfrac{y^2}{\beta^2} - \dfrac{x^2}{\alpha^2} = 1$
$(\alpha \geq \beta > 0)$	$(\beta \geq \alpha > 0)$	$(\alpha > 0, \beta > 0)$	$(\alpha > 0, \beta > 0)$

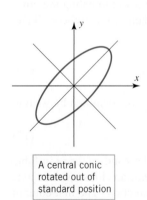

A central conic rotated out of standard position

▲ Figure 7.3.2

If we take the constant f in Equations (8) and (9) to the right side and let $k = -f$, then we can rewrite these equations in matrix form as

$$\begin{bmatrix} x & y \end{bmatrix} \begin{bmatrix} a & b \\ b & c \end{bmatrix} \begin{bmatrix} x \\ y \end{bmatrix} = k \quad \text{and} \quad \begin{bmatrix} x & y \end{bmatrix} \begin{bmatrix} a & 0 \\ 0 & c \end{bmatrix} \begin{bmatrix} x \\ y \end{bmatrix} = k \tag{10}$$

The first of these corresponds to Equation (8) in which there is a cross product term $2bxy$, and the second corresponds to Equation (9) in which there is no cross product term. Geometrically, the existence of a cross product term signals that the graph of the quadratic form is rotated about the origin, as in Figure 7.3.2. The three-dimensional analogs of the equations in (10) are

$$\begin{bmatrix} x & y & z \end{bmatrix} \begin{bmatrix} a & d & e \\ d & b & f \\ e & f & c \end{bmatrix} \begin{bmatrix} x \\ y \\ z \end{bmatrix} = k \quad \text{and} \quad \begin{bmatrix} x & y & z \end{bmatrix} \begin{bmatrix} a & 0 & 0 \\ 0 & b & 0 \\ 0 & 0 & c \end{bmatrix} \begin{bmatrix} x \\ y \\ z \end{bmatrix} = k \tag{11}$$

If a, b, and c are not all zero, then the graphs of these equations in R^3 are called ***central quadrics in standard position***.

Identifying Conic Sections

We are now ready to consider the first of the three problems posed earlier, identifying the curve or surface represented by an equation $\mathbf{x}^T A \mathbf{x} = k$ in two or three variables. We will focus on the two-variable case. We noted above that an equation of the form

$$ax^2 + 2bxy + cy^2 + f = 0 \tag{12}$$

represents a central conic. If $b = 0$, then the conic is in standard position, and if $b \neq 0$, it is rotated. It is an easy matter to identify central conics in standard position by matching the equation with one of the standard forms. For example, the equation

$$9x^2 + 16y^2 - 144 = 0$$

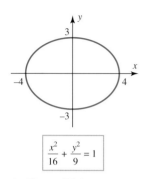

Figure 7.3.3

can be rewritten as

$$\frac{x^2}{16} + \frac{y^2}{9} = 1$$

which, by comparison with Table 1, is the ellipse shown in Figure 7.3.3.

If a central conic is rotated out of standard position, then it can be identified by first rotating the coordinate axes to put it in standard position and then matching the resulting equation with one of the standard forms in Table 1. To find a rotation that eliminates the cross product term in the equation

$$ax^2 + 2bxy + cy^2 = k \tag{13}$$

it will be convenient to express the equation in the matrix form

$$\mathbf{x}^T A \mathbf{x} = \begin{bmatrix} x & y \end{bmatrix} \begin{bmatrix} a & b \\ b & c \end{bmatrix} \begin{bmatrix} x \\ y \end{bmatrix} = k \tag{14}$$

and look for a change of variable

$$\mathbf{x} = P\mathbf{x}'$$

that diagonalizes A and for which $\det(P) = 1$. Since we saw in Example 4 of Section 7.1 that the transition matrix

$$P = \begin{bmatrix} \cos\theta & -\sin\theta \\ \sin\theta & \cos\theta \end{bmatrix} \tag{15}$$

has the effect of rotating the xy-axes of a rectangular coordinate system through an angle θ, our problem reduces to finding θ that diagonalizes A, thereby eliminating the cross product term in (13). If we make this change of variable, then in the $x'y'$-coordinate system, Equation (14) will become

$$\mathbf{x}'^T D \mathbf{x}' = \begin{bmatrix} x' & y' \end{bmatrix} \begin{bmatrix} \lambda_1 & 0 \\ 0 & \lambda_2 \end{bmatrix} \begin{bmatrix} x' \\ y' \end{bmatrix} = k \tag{16}$$

where λ_1 and λ_2 are the eigenvalues of A. The conic can now be identified by writing (16) in the form

$$\lambda_1 x'^2 + \lambda_2 y'^2 = k \tag{17}$$

and performing the necessary algebra to match it with one of the standard forms in Table 1. For example, if λ_1, λ_2, and k are positive, then (17) represents an ellipse with an axis of length $2\sqrt{k/\lambda_1}$ in the x'-direction and $2\sqrt{k/\lambda_2}$ in the y'-direction. The first column vector of P, which is a unit eigenvector corresponding to λ_1, is along the positive x'-axis; and the second column vector of P, which is a unit eigenvector corresponding to λ_2, is a unit vector along the y'-axis. These are called the ***principal axes*** of the ellipse, which explains why Theorem 7.3.1 is called "the principal axes theorem." (See Figure 7.3.4.)

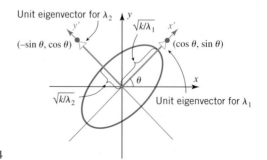

Figure 7.3.4

EXAMPLE 3 Identifying a Conic by Eliminating the Cross Product Term

(a) Identify the conic whose equation is $5x^2 - 4xy + 8y^2 - 36 = 0$ by rotating the xy-axes to put the conic in standard position.

(b) Find the angle θ through which you rotated the xy-axes in part (a).

Solution (a) The given equation can be written in the matrix form

$$\mathbf{x}^T A \mathbf{x} = 36$$

where

$$A = \begin{bmatrix} 5 & -2 \\ -2 & 8 \end{bmatrix}$$

The characteristic polynomial of A is

$$\begin{vmatrix} \lambda - 5 & 2 \\ 2 & \lambda - 8 \end{vmatrix} = (\lambda - 4)(\lambda - 9)$$

so the eigenvalues are $\lambda = 4$ and $\lambda = 9$. We leave it for you to show that orthonormal bases for the eigenspaces are

$$\lambda = 4: \begin{bmatrix} \frac{2}{\sqrt{5}} \\ \frac{1}{\sqrt{5}} \end{bmatrix}, \quad \lambda = 9: \begin{bmatrix} -\frac{1}{\sqrt{5}} \\ \frac{2}{\sqrt{5}} \end{bmatrix}$$

Thus, A is orthogonally diagonalized by

$$P = \begin{bmatrix} \frac{2}{\sqrt{5}} & -\frac{1}{\sqrt{5}} \\ \frac{1}{\sqrt{5}} & \frac{2}{\sqrt{5}} \end{bmatrix} \tag{18}$$

Had it turned out that $\det(P) = -1$, then we would have interchanged the columns to reverse the sign.

Moreover, it happens by chance that $\det(P) = 1$, so we are assured that the substitution $\mathbf{x} = P\mathbf{x}'$ performs a rotation of axes. It follows from (16) that the equation of the conic in the $x'y'$-coordinate system is

$$[x' \ y'] \begin{bmatrix} 4 & 0 \\ 0 & 9 \end{bmatrix} \begin{bmatrix} x' \\ y' \end{bmatrix} = 36$$

which we can write as

$$4x'^2 + 9y'^2 = 36 \quad \text{or} \quad \frac{x'^2}{9} + \frac{y'^2}{4} = 1$$

We can now see from Table 1 that the conic is an ellipse whose axis has length $2\alpha = 6$ in the x'-direction and length $2\beta = 4$ in the y'-direction.

Solution (b) It follows from (15) that

$$P = \begin{bmatrix} \frac{2}{\sqrt{5}} & -\frac{1}{\sqrt{5}} \\ \frac{1}{\sqrt{5}} & \frac{2}{\sqrt{5}} \end{bmatrix} = \begin{bmatrix} \cos\theta & -\sin\theta \\ \sin\theta & \cos\theta \end{bmatrix}$$

which implies that

$$\cos\theta = \frac{2}{\sqrt{5}}, \quad \sin\theta = \frac{1}{\sqrt{5}}, \quad \tan\theta = \frac{\sin\theta}{\cos\theta} = \frac{1}{2}$$

Thus, $\theta = \tan^{-1} \frac{1}{2} \approx 26.6°$ (Figure 7.3.5). ◀

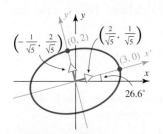

▲ Figure 7.3.5

Remark In the exercises we will ask you to show that if $b \neq 0$, then the cross product term in the equation

$$ax^2 + 2bxy + cy^2 = k$$

412 Chapter 7 Diagonalization and Quadratic Forms

can be eliminated by a rotation through an angle θ that satisfies

$$\cot 2\theta = \frac{a-c}{2b} \tag{19}$$

We leave it for you to confirm that this is consistent with part (b) of the last example.

Positive Definite Quadratic Forms

We will now consider the second of the two problems posed earlier, determining conditions under which $x^TAx > 0$ for all nonzero values of x. We will explain why this is important shortly, but first we introduce some terminology.

> The terminology in Definition 1 also applies to the matrix A; that is, A is positive definite, negative definite, or indefinite in accordance with whether the associated quadratic form has that property.

DEFINITION 1 A quadratic form x^TAx is said to be
- ***positive definite*** if $x^TAx > 0$ for $x \neq 0$
- ***negative definite*** if $x^TAx < 0$ for $x \neq 0$
- ***indefinite*** if x^TAx has both positive and negative values

The following theorem, whose proof is deferred to the end of the section, provides a way of using eigenvalues to determine whether a matrix A and its associated quadratic form x^TAx are positive definite, negative definite, or indefinite.

THEOREM 7.3.2 *If A is a symmetric matrix, then:*
(a) x^TAx *is positive definite if and only if all eigenvalues of A are positive.*
(b) x^TAx *is negative definite if and only if all eigenvalues of A are negative.*
(c) x^TAx *is indefinite if and only if A has at least one positive eigenvalue and at least one negative eigenvalue.*

Remark The three classifications in Definition 1 do not exhaust all of the possibilities. For example, a quadratic form for which $x^TAx \geq 0$ if $x \neq 0$ is called ***positive semidefinite***, and one for which $x^TAx \leq 0$ if $x \neq 0$ is called ***negative semidefinite***. Every positive definite form is positive semidefinite, but not conversely, and every negative definite form is negative semidefinite, but not conversely (why?). By adjusting the proof of Theorem 7.3.2 appropriately, one can prove that x^TAx is positive semidefinite if and only if all eigenvalues of A are nonnegative and is negative semidefinite if and only if all eigenvalues of A are nonpositive.

▶ **EXAMPLE 4** Positive Definite Quadratic Forms

It is not usually possible to tell from the signs of the entries in a symmetric matrix A whether that matrix is positive definite, negative definite, or indefinite. For example, the entries of the matrix

$$A = \begin{bmatrix} 3 & 1 & 1 \\ 1 & 0 & 2 \\ 1 & 2 & 0 \end{bmatrix}$$

are nonnegative, but the matrix is indefinite since its eigenvalues are $\lambda = 1, 4, -2$ (verify). To see this another way, let us write out the quadratic form as

$$x^TAx = \begin{bmatrix} x_1 & x_2 & x_3 \end{bmatrix} \begin{bmatrix} 3 & 1 & 1 \\ 1 & 0 & 2 \\ 1 & 2 & 0 \end{bmatrix} \begin{bmatrix} x_1 \\ x_2 \\ x_3 \end{bmatrix} = 3x_1^2 + 2x_1x_2 + 2x_1x_3 + 4x_2x_3$$

> Positive definite and negative definite matrices are invertible. Why?

We can now see, for example, that

$$\mathbf{x}^T A \mathbf{x} = 4 \quad \text{for} \quad x_1 = 0, \quad x_2 = 1, \quad x_3 = 1$$

and

$$\mathbf{x}^T A \mathbf{x} = -4 \quad \text{for} \quad x_1 = 0, \quad x_2 = 1, \quad x_3 = -1 \blacktriangleleft$$

Classifying Conic Sections Using Eigenvalues

If $\mathbf{x}^T B \mathbf{x} = k$ is the equation of a conic, and if $k \neq 0$, then we can divide through by k and rewrite the equation in the form

$$\mathbf{x}^T A \mathbf{x} = 1 \tag{20}$$

where $A = (1/k)B$. If we now rotate the coordinate axes to eliminate the cross product term (if any) in this equation, then the equation of the conic in the new coordinate system will be of the form

$$\lambda_1 x'^2 + \lambda_2 y'^2 = 1 \tag{21}$$

in which λ_1 and λ_2 are the eigenvalues of A. The particular type of conic represented by this equation will depend on the signs of the eigenvalues λ_1 and λ_2. For example, you should be able to see from (21) that:

- $\mathbf{x}^T A \mathbf{x} = 1$ represents an ellipse if $\lambda_1 > 0$ and $\lambda_2 > 0$.
- $\mathbf{x}^T A \mathbf{x} = 1$ has no graph if $\lambda_1 < 0$ and $\lambda_2 < 0$.
- $\mathbf{x}^T A \mathbf{x} = 1$ represents a hyperbola if λ_1 and λ_2 have opposite signs.

In the case of the ellipse, Equation (21) can be rewritten as

$$\frac{x'^2}{(1/\sqrt{\lambda_1})^2} + \frac{y'^2}{(1/\sqrt{\lambda_2})^2} = 1 \tag{22}$$

so the axes of the ellipse have lengths $2/\sqrt{\lambda_1}$ and $2/\sqrt{\lambda_2}$ (Figure 7.3.6).

The following theorem is an immediate consequence of this discussion and Theorem 7.3.2.

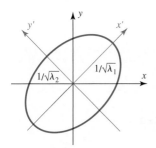

▲ Figure 7.3.6

> **THEOREM 7.3.3** *If A is a symmetric 2×2 matrix, then:*
> *(a)* $\mathbf{x}^T A \mathbf{x} = 1$ *represents an ellipse if A is positive definite.*
> *(b)* $\mathbf{x}^T A \mathbf{x} = 1$ *has no graph if A is negative definite.*
> *(c)* $\mathbf{x}^T A \mathbf{x} = 1$ *represents a hyperbola if A is indefinite.*

In Example 3 we performed a rotation to show that the equation

$$5x^2 - 4xy + 8y^2 - 36 = 0$$

represents an ellipse with a major axis of length 6 and a minor axis of length 4. This conclusion can also be obtained by rewriting the equation in the form

$$\tfrac{5}{36}x^2 - \tfrac{1}{9}xy + \tfrac{2}{9}y^2 = 1$$

and showing that the associated matrix

$$A = \begin{bmatrix} \tfrac{5}{36} & -\tfrac{1}{18} \\ -\tfrac{1}{18} & \tfrac{2}{9} \end{bmatrix}$$

has eigenvalues $\lambda_1 = \tfrac{1}{9}$ and $\lambda_2 = \tfrac{1}{4}$. These eigenvalues are positive, so the matrix A is positive definite and the equation represents an ellipse. Moreover, it follows from (21) that the axes of the ellipse have lengths $2/\sqrt{\lambda_1} = 6$ and $2/\sqrt{\lambda_2} = 4$, which is consistent with Example 3.

Identifying Positive Definite Matrices

Positive definite matrices are the most important symmetric matrices in applications, so it will be useful to learn a little more about them. We already know that a symmetric matrix is positive definite if and only if its eigenvalues are all positive; now we will give a criterion that can be used to determine whether a symmetric matrix is positive definite *without* finding the eigenvalues. For this purpose we define the **kth principal submatrix** of an $n \times n$ matrix A to be the $k \times k$ submatrix consisting of the first k rows and columns of A. For example, here are the principal submatrices of a general 4×4 matrix:

$$\begin{bmatrix} a_{11} & a_{12} & a_{13} & a_{14} \\ a_{21} & a_{22} & a_{23} & a_{24} \\ a_{31} & a_{32} & a_{33} & a_{34} \\ a_{41} & a_{42} & a_{43} & a_{44} \end{bmatrix} \quad \begin{bmatrix} a_{11} & a_{12} & a_{13} & a_{14} \\ a_{21} & a_{22} & a_{23} & a_{24} \\ a_{31} & a_{32} & a_{33} & a_{34} \\ a_{41} & a_{42} & a_{43} & a_{44} \end{bmatrix} \quad \begin{bmatrix} a_{11} & a_{12} & a_{13} & a_{14} \\ a_{21} & a_{22} & a_{23} & a_{24} \\ a_{31} & a_{32} & a_{33} & a_{34} \\ a_{41} & a_{42} & a_{43} & a_{44} \end{bmatrix} \quad \begin{bmatrix} a_{11} & a_{12} & a_{13} & a_{14} \\ a_{21} & a_{22} & a_{23} & a_{24} \\ a_{31} & a_{32} & a_{33} & a_{34} \\ a_{41} & a_{42} & a_{43} & a_{44} \end{bmatrix}$$

First principal submatrix, Second principal submatrix, Third principal submatrix, Fourth principal submatrix $= A$

The following theorem, which we state without proof, provides a determinant test for ascertaining whether a symmetric matrix is positive definite.

THEOREM 7.3.4 *A symmetric matrix A is positive definite if and only if the determinant of every principal submatrix is positive.*

▶ **EXAMPLE 5** Working with Principal Submatrices

The matrix

$$A = \begin{bmatrix} 2 & -1 & -3 \\ -1 & 2 & 4 \\ -3 & 4 & 9 \end{bmatrix}$$

is positive definite since the determinants

$$|2| = 2, \quad \begin{vmatrix} 2 & -1 \\ -1 & 2 \end{vmatrix} = 3, \quad \begin{vmatrix} 2 & -1 & -3 \\ -1 & 2 & 4 \\ -3 & 4 & 9 \end{vmatrix} = 1$$

are all positive. Thus, we are guaranteed that all eigenvalues of A are positive and $\mathbf{x}^T A \mathbf{x} > 0$ for $\mathbf{x} \neq \mathbf{0}$. ◀

OPTIONAL

We conclude this section with an optional proof of Theorem 7.3.2.

Proofs of Theorem 7.3.2(a) and (b) It follows from the principal axes theorem (Theorem 7.3.1) that there is an orthogonal change of variable $\mathbf{x} = P\mathbf{y}$ for which

$$\mathbf{x}^T A \mathbf{x} = \mathbf{y}^T D \mathbf{y} = \lambda_1 y_1^2 + \lambda_2 y_2^2 + \cdots + \lambda_n y_n^2 \tag{23}$$

where the λ's are the eigenvalues of A. Moreover, it follows from the invertibility of P that $\mathbf{y} \neq \mathbf{0}$ if and only if $\mathbf{x} \neq \mathbf{0}$, so the values of $\mathbf{x}^T A \mathbf{x}$ for $\mathbf{x} \neq \mathbf{0}$ are the same as the values of $\mathbf{y}^T D \mathbf{y}$ for $\mathbf{y} \neq \mathbf{0}$. Thus, it follows from (23) that $\mathbf{x}^T A \mathbf{x} > 0$ for $\mathbf{x} \neq \mathbf{0}$ if and only if all of the λ's in that equation are positive, and that $\mathbf{x}^T A \mathbf{x} < 0$ for $\mathbf{x} \neq \mathbf{0}$ if and only if all of the λ's are negative. This proves parts (*a*) and (*b*).

Proof (c) Assume that A has at least one positive eigenvalue and at least one negative eigenvalue, and to be specific, suppose that $\lambda_1 > 0$ and $\lambda_2 < 0$ in (23). Then

$$\mathbf{x}^T A \mathbf{x} > 0 \quad \text{if} \quad y_1 = 1 \text{ and all other } y\text{'s are } 0$$

and

$$\mathbf{x}^T A \mathbf{x} < 0 \quad \text{if} \quad y_2 = 1 \text{ and all other } y\text{'s are } 0$$

which proves that $\mathbf{x}^T A \mathbf{x}$ is indefinite. Conversely, if $\mathbf{x}^T A \mathbf{x} > 0$ for some $\mathbf{x}$, then $\mathbf{y}^T D \mathbf{y} > 0$ for some $\mathbf{y}$, so at least one of the λ's in (23) must be positive. Similarly, if $\mathbf{x}^T A \mathbf{x} < 0$ for some $\mathbf{x}$, then $\mathbf{y}^T D \mathbf{y} < 0$ for some $\mathbf{y}$, so at least one of the λ's in (23) must be negative, which completes the proof. ◄

Concept Review

- Linear form
- Quadratic form
- Cross product term
- Quadratic form associated with a matrix
- Change of variable
- Orthogonal change of variable
- Principal Axes Theorem
- Conic section
- Degenerate conic
- Central conic
- Standard position of a central conic
- Standard form of a central conic
- Central quadric
- Principal axes of an ellipse
- Positive definite quadratic form
- Negative definite quadratic form
- Indefinite quadratic form
- Positive semidefinite quadratic form
- Negative semidefinite quadratic form
- Principal submatrix

Skills

- Express a quadratic form in the matrix notation $\mathbf{x}^T A \mathbf{x}$, where A is a symmetric matrix.
- Find an orthogonal change of variable that eliminates the cross product terms in a quadratic form, and express the quadratic form in terms of the new variable.
- Identify a conic section from an equation by rotating axes to place the conic in standard position, and find the angle of rotation.
- Identify a conic section using eigenvalues.
- Classify matrices and quadratic forms as positive definite, negative definite, indefinite, positive semidefinite or negative semidefinite.

Exercise Set 7.3

▶ In Exercises 1–2, express the quadratic form in the matrix notation $\mathbf{x}^T A \mathbf{x}$, where A is a symmetric matrix. ◄

1. (a) $5x_1^2 - 3x_2^2$ (b) $9x_1^2 + x_2^2 - 4x_1 x_2$
  (c) $6x_1^2 + 4x_2^2 - 7x_3^2 - 2x_1 x_2 + 4x_1 x_3 + x_2 x_3$

2. (a) $5x_1^2 + 5x_1 x_2$ (b) $-7x_1 x_2$
  (c) $x_1^2 + x_2^2 - 3x_3^2 - 5x_1 x_2 + 9x_1 x_3$

▶ In Exercises 3–4, find a formula for the quadratic form that does not use matrices. ◄

3. $\begin{bmatrix} x & y \end{bmatrix} \begin{bmatrix} 4 & -1 \\ -1 & 3 \end{bmatrix} \begin{bmatrix} x \\ y \end{bmatrix}$

4. $\begin{bmatrix} x_1 & x_2 & x_3 \end{bmatrix} \begin{bmatrix} -2 & \frac{7}{2} & 1 \\ \frac{7}{2} & 0 & 6 \\ 1 & 6 & 3 \end{bmatrix} \begin{bmatrix} x_1 \\ x_2 \\ x_3 \end{bmatrix}$

▶ In Exercises 5–8, find an orthogonal change of variables that eliminates the cross product terms in the quadratic form Q, and express Q in terms of the new variables. ◄

5. $Q = -x_1^2 - x_2^2 - 6x_1 x_2$

6. $Q = 5x_1^2 + 2x_2^2 + 4x_3^2 + 4x_1 x_2$

7. $Q = 2x_1^2 + 2x_2^2 + 2x_3^2 + 4x_1 x_2$

8. $Q = 2x_1^2 + 5x_2^2 + 5x_3^2 + 4x_1 x_2 - 4x_1 x_3 - 8x_2 x_3$

▶ In Exercises 9–10, express the quadratic equation in the matrix form $\mathbf{x}^T A \mathbf{x} + K \mathbf{x} + f = 0$, where $\mathbf{x}^T A \mathbf{x}$ is the associated quadratic form and K is an appropriate matrix. ◄

9. (a) $2x^2 + xy + x - 6y + 2 = 0$
  (b) $y^2 + 7x - 8y - 5 = 0$

10. (a) $x^2 - xy + 5x + 8y - 3 = 0$
  (b) $5xy = 8$

▶ In Exercises 11–12, identify the conic section represented by the equation. ◄

11. (a) $2x^2 - 2y^2 = 20$ (b) $5x^2 + 3y^2 - 15 = 0$
  (c) $7y^2 - 2x = 0$ (d) $x^2 + y^2 - 25 = 0$

12. (a) $4x^2 + 9y^2 = 1$ (b) $4x^2 - 5y^2 = 20$
  (c) $-x^2 = 2y$ (d) $x^2 - 3 = -y^2$

In Exercises 13–16, identify the conic section represented by the equation by rotating axes to place the conic in standard position. Find an equation of the conic in the rotated coordinates, and find the angle of rotation.

13. $x^2 - 4xy - 2y^2 + 8 = 0$ 14. $5x^2 + 4xy + 5y^2 = 9$

15. $2x^2 - 12xy - 3y^2 - 7 = 0$

16. $x^2 + xy + y^2 = \frac{1}{2}$

In Exercises 17–18, determine by inspection whether the matrix is positive definite, negative definite, indefinite, positive semidefinite, or negative semidefinite.

17. (a) $\begin{bmatrix} 3 & 0 \\ 0 & -5 \end{bmatrix}$ (b) $\begin{bmatrix} 0 & 0 \\ 0 & -5 \end{bmatrix}$ (c) $\begin{bmatrix} 3 & 0 \\ 0 & 5 \end{bmatrix}$

(d) $\begin{bmatrix} 3 & 0 \\ 0 & 0 \end{bmatrix}$ (e) $\begin{bmatrix} -3 & 0 \\ 0 & -5 \end{bmatrix}$

18. (a) $\begin{bmatrix} 2 & 0 \\ 0 & -5 \end{bmatrix}$ (b) $\begin{bmatrix} -2 & 0 \\ 0 & -5 \end{bmatrix}$ (c) $\begin{bmatrix} 2 & 0 \\ 0 & 5 \end{bmatrix}$

(d) $\begin{bmatrix} 0 & 0 \\ 0 & -5 \end{bmatrix}$ (e) $\begin{bmatrix} 2 & 0 \\ 0 & 0 \end{bmatrix}$

In Exercises 19–24, classify the quadratic form as positive definite, negative definite, indefinite, positive semidefinite, or negative semidefinite.

19. $3x_1^2 - 4x_2^2$ 20. $-x_1^2 - 3x_2^2$ 21. $2x_1^2 + 7x_2^2$

22. $-(x_1 - x_2)^2$ 23. $(4x_1 - x_2)^2$ 24. $x_1 x_2$

In Exercises 25–26, show that the matrix A is positive definite first by using Theorem 7.3.2 and second by using Theorem 7.3.4.

25. (a) $A = \begin{bmatrix} 5 & -2 \\ -2 & 5 \end{bmatrix}$ (b) $A = \begin{bmatrix} 2 & -1 & 0 \\ -1 & 2 & 0 \\ 0 & 0 & 5 \end{bmatrix}$

26. (a) $A = \begin{bmatrix} 2 & 1 \\ 1 & 2 \end{bmatrix}$ (b) $A = \begin{bmatrix} 3 & -1 & 0 \\ -1 & 2 & -1 \\ 0 & -1 & 3 \end{bmatrix}$

In Exercises 27–28, find all values of k for which the quadratic form is positive definite.

27. $5x_1^2 + x_2^2 + kx_3^2 + 4x_1 x_2 - 2x_1 x_3 - 2x_2 x_3$

28. $3x_1^2 + x_2^2 + 2x_3^2 - 2x_1 x_3 + 2kx_2 x_3$

29. Let $\mathbf{x}^T A \mathbf{x}$ be a quadratic form in the variables $x_1, x_2, \ldots, x_n$, and define $T : R^n \to R$ by $T(\mathbf{x}) = \mathbf{x}^T A \mathbf{x}$.

(a) Show that $T(\mathbf{x} + \mathbf{y}) = T(\mathbf{x}) + 2\mathbf{x}^T A \mathbf{y} + T(\mathbf{y})$.

(b) Show that $T(c\mathbf{x}) = c^2 T(\mathbf{x})$.

30. Express the quadratic form $(c_1 x_1 + c_2 x_2 + \cdots + c_n x_n)^2$ in the matrix notation $\mathbf{x}^T A \mathbf{x}$, where A is symmetric.

31. In statistics, the quantities
$$\bar{x} = \frac{1}{n}(x_1 + x_2 + \cdots + x_n)$$
and
$$s_x^2 = \frac{1}{n-1}\left[(x_1 - \bar{x})^2 + (x_2 - \bar{x})^2 + \cdots + (x_n - \bar{x})^2\right]$$
are called, respectively, the *sample mean* and *sample variance* of $\mathbf{x} = (x_1, x_2, \ldots, x_n)$.

(a) Express the quadratic form s_x^2 in the matrix notation $\mathbf{x}^T A \mathbf{x}$, where A is symmetric.

(b) Is s_x^2 a positive definite quadratic form? Explain.

32. The graph in an xyz-coordinate system of an equation of form $ax^2 + by^2 + cz^2 = 1$ in which a, b, and c are positive is a surface called a *central ellipsoid in standard position* (see the accompanying figure). This is the three-dimensional generalization of the ellipse $ax^2 + by^2 = 1$ in the xy-plane. The intersections of the ellipsoid $ax^2 + by^2 + cz^2 = 1$ with the coordinate axes determine three line segments called the *axes* of the ellipsoid. If a central ellipsoid is rotated about the origin so two or more of its axes do not coincide with any of the coordinate axes, then the resulting equation will have one or more cross product terms.

(a) Show that the equation
$$\tfrac{4}{3}x^2 + \tfrac{4}{3}y^2 + \tfrac{4}{3}z^2 + \tfrac{4}{3}xy + \tfrac{4}{3}xz + \tfrac{4}{3}yz = 1$$
represents an ellipsoid, and find the lengths of its axes. [*Suggestion:* Write the equation in the form $\mathbf{x}^T A \mathbf{x} = 1$ and make an orthogonal change of variable to eliminate the cross product terms.]

(b) What property must a symmetric 3×3 matrix have in order for the equation $\mathbf{x}^T A \mathbf{x} = 1$ to represent an ellipsoid?

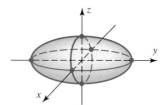

◀ Figure Ex-32

33. What property must a symmetric 2×2 matrix A have for $\mathbf{x}^T A \mathbf{x} = 1$ to represent a circle?

34. Prove: If $b \neq 0$, then the cross product term can be eliminated from the quadratic form $ax^2 + 2bxy + cy^2$ by rotating the coordinate axes through an angle θ that satisfies the equation
$$\cot 2\theta = \frac{a - c}{2b}$$

35. Prove that if A is an $n \times n$ symmetric matrix all of whose eigenvalues are nonnegative, then $\mathbf{x}^T A \mathbf{x} \geq 0$ for all nonzero $\mathbf{x}$ in R^n.

True-False Exercises

In parts (a)–(l) determine whether the statement is true or false, and justify your answer.

(a) A symmetric matrix with positive definite eigenvalues is positive definite.

(b) $x_1^2 - x_2^2 + x_3^2 + 4x_1x_2x_3$ is a quadratic form.

(c) $(x_1 - 3x_2)^2$ is a quadratic form.

(d) A positive definite matrix is invertible.

(e) A symmetric matrix is either positive definite, negative definite, or indefinite.

(f) If A is positive definite, then $-A$ is negative definite.

(g) $\mathbf{x} \cdot \mathbf{x}$ is a quadratic form for all $\mathbf{x}$ in R^n.

(h) If $\mathbf{x}^T A \mathbf{x}$ is a positive definite quadratic form, then so is $\mathbf{x}^T A^{-1} \mathbf{x}$.

(i) If A is a matrix with only positive eigenvalues, then $\mathbf{x}^T A \mathbf{x}$ is a positive definite quadratic form.

(j) If A is a 2×2 symmetric matrix with positive entries and $\det(A) > 0$, then A is positive definite.

(k) If $\mathbf{x}^T A \mathbf{x}$ is a quadratic form with no cross product terms, then A is a diagonal matrix.

(l) If $\mathbf{x}^T A \mathbf{x}$ is a positive definite quadratic form in two variables and $c \neq 0$, then the graph of the equation $\mathbf{x}^T A \mathbf{x} = c$ is an ellipse.

7.4 Optimization Using Quadratic Forms

Quadratic forms arise in various problems in which the maximum or minimum value of some quantity is required. In this section we will discuss some problems of this type.

Constrained Extremum Problems

Our first goal in this section is to consider the problem of finding the maximum and minimum values of a quadratic form $\mathbf{x}^T A \mathbf{x}$ subject of the constraint $\|\mathbf{x}\| = 1$. Problems of this type arise in a wide variety of applications.

To visualize this problem geometrically in the case where $\mathbf{x}^T A \mathbf{x}$ is a quadratic form on R^2, view $z = \mathbf{x}^T A \mathbf{x}$ as the equation of some surface in a rectangular xyz-coordinate system and view $\|\mathbf{x}\| = 1$ as the unit circle centered at the origin of the xy-plane. Geometrically, the problem of finding the maximum and minimum values of $\mathbf{x}^T A \mathbf{x}$ subject to the requirement $\|\mathbf{x}\| = 1$ amounts to finding the highest and lowest points on the intersection of the surface with the right circular cylinder determined by the circle (Figure 7.4.1).

The following theorem, whose proof is deferred to the end of the section, is the key result for solving problems of this type.

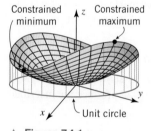

▲ Figure 7.4.1

> **THEOREM 7.4.1** *Constrained Extremum Theorem*
>
> *Let A be a symmetric $n \times n$ matrix whose eigenvalues in order of decreasing size are $\lambda_1 \geq \lambda_2 \geq \cdots \geq \lambda_n$. Then:*
>
> *(a) the quadratic form $\mathbf{x}^T A \mathbf{x}$ attains a maximum value and a minimum value on the set of vectors for which $\|\mathbf{x}\| = 1$;*
>
> *(b) the maximum value attained in part (a) occurs at a unit vector corresponding to the eigenvalue λ_1;*
>
> *(c) the minimum value attained in part (a) occurs at a unit vector corresponding to the eigenvalue λ_n.*

Remark The condition $\|\mathbf{x}\| = 1$ in this theorem is called a ***constraint***, and the maximum or minimum value of $\mathbf{x}^T A \mathbf{x}$ subject to the constraint is called a ***constrained extremum***. This constraint can also be expressed as $\mathbf{x}^T \mathbf{x} = 1$ or as $x_1^2 + x_2^2 + \cdots + x_n^2 = 1$, when convenient.

418 Chapter 7 Diagonalization and Quadratic Forms

▶ **EXAMPLE 1 Finding Constrained Extrema**

Find the maximum and minimum values of the quadratic form
$$z = 5x^2 + 5y^2 + 4xy$$
subject to the constraint $x^2 + y^2 = 1$.

Solution The quadratic form can be expressed in matrix notation as
$$z = 5x^2 + 5y^2 + 4xy = \mathbf{x}^T A \mathbf{x} = [x \ \ y] \begin{bmatrix} 5 & 2 \\ 2 & 5 \end{bmatrix} \begin{bmatrix} x \\ y \end{bmatrix}$$

We leave it for you to show that the eigenvalues of A are $\lambda_1 = 7$ and $\lambda_2 = 3$ and that corresponding eigenvectors are
$$\lambda_1 = 7: \ \begin{bmatrix} 1 \\ 1 \end{bmatrix}, \quad \lambda_2 = 3: \ \begin{bmatrix} -1 \\ 1 \end{bmatrix}$$

Normalizing these eigenvectors yields
$$\lambda_1 = 7: \ \begin{bmatrix} \frac{1}{\sqrt{2}} \\ \frac{1}{\sqrt{2}} \end{bmatrix}, \quad \lambda_2 = 3: \ \begin{bmatrix} -\frac{1}{\sqrt{2}} \\ \frac{1}{\sqrt{2}} \end{bmatrix} \tag{1}$$

Thus, the constrained extrema are

constrained maximum: $z = 7$ at $(x, y) = \left(\frac{1}{\sqrt{2}}, \frac{1}{\sqrt{2}}\right)$

constrained minimum: $z = 3$ at $(x, y) = \left(-\frac{1}{\sqrt{2}}, \frac{1}{\sqrt{2}}\right)$ ◀

Remark Since the negatives of the eigenvectors in (1) are also unit eigenvectors, they too produce the maximum and minimum values of z; that is, the constrained maximum $z = 7$ also occurs at the point $(x, y) = \left(-\frac{1}{\sqrt{2}}, -\frac{1}{\sqrt{2}}\right)$ and the constrained minimum $z = 3$ at $(x, y) = \left(\frac{1}{\sqrt{2}}, -\frac{1}{\sqrt{2}}\right)$.

▶ **EXAMPLE 2 A Constrained Extremum Problem**

A rectangle is to be inscribed in the ellipse $4x^2 + 9y^2 = 36$, as shown in Figure 7.4.2. Use eigenvalue methods to find nonnegative values of x and y that produce the inscribed rectangle with maximum area.

Solution The area z of the inscribed rectangle is given by $z = 4xy$, so the problem is to maximize the quadratic form $z = 4xy$ subject to the constraint $4x^2 + 9y^2 = 36$. In this problem, the graph of the constraint equation is an ellipse rather than the unit circle as required in Theorem 7.4.1, but we can remedy this problem by rewriting the constraint as
$$\left(\frac{x}{3}\right)^2 + \left(\frac{y}{2}\right)^2 = 1$$
and defining new variables, x_1 and y_1, by the equations
$$x = 3x_1 \quad \text{and} \quad y = 2y_1$$
This enables us to reformulate the problem as follows:
$$\text{maximize } z = 4xy = 24x_1 y_1$$
subject to the constraint
$$x_1^2 + y_1^2 = 1$$
To solve this problem, we will write the quadratic form $z = 24x_1 y_1$ as
$$z = \mathbf{x}^T A \mathbf{x} = [x_1 \ \ y_1] \begin{bmatrix} 0 & 12 \\ 12 & 0 \end{bmatrix} \begin{bmatrix} x_1 \\ y_1 \end{bmatrix}$$

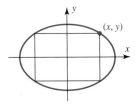

▲ Figure 7.4.2 A rectangle inscribed in the ellipse $4x^2 + 9y^2 = 36$.

We now leave it for you to show that the largest eigenvalue of A is $\lambda = 12$ and that the only corresponding unit eigenvector with nonnegative entries is

$$\mathbf{x} = \begin{bmatrix} x_1 \\ y_1 \end{bmatrix} = \begin{bmatrix} \frac{1}{\sqrt{2}} \\ \frac{1}{\sqrt{2}} \end{bmatrix}$$

Thus, the maximum area is $z = 12$, and this occurs when

$$x = 3x_1 = \frac{3}{\sqrt{2}} \quad \text{and} \quad y = 2y_1 = \frac{2}{\sqrt{2}} \blacktriangleleft$$

Constrained Extrema and Level Curves

A useful way of visualizing the behavior of a function $f(x, y)$ of two variables is to consider the curves in the xy-plane along which $f(x, y)$ is constant. These curves have equations of the form

$$f(x, y) = k$$

and are called the **level curves** of f (Figure 7.4.3). In particular, the level curves of a quadratic form $\mathbf{x}^T A \mathbf{x}$ on R^2 have equations of the form

$$\mathbf{x}^T A \mathbf{x} = k \qquad (2)$$

so the maximum and minimum values of $\mathbf{x}^T A \mathbf{x}$ subject to the constraint $\|\mathbf{x}\| = 1$ are the largest and smallest values of k for which the graph of (2) intersects the unit circle. Typically, such values of k produce level curves that just touch the unit circle (Figure 7.4.4), and the coordinates of the points where the level curves just touch produce the vectors that maximize or minimize $\mathbf{x}^T A \mathbf{x}$ subject to the constraint $\|\mathbf{x}\| = 1$.

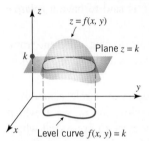

▲ Figure 7.4.3

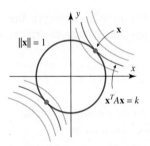

▲ Figure 7.4.4

▶ **EXAMPLE 3 Example 1 Revisited Using Level Curves**

In Example 1 (and its following remark) we found the maximum and minimum values of the quadratic form

$$z = 5x^2 + 5y^2 + 4xy$$

subject to the constraint $x^2 + y^2 = 1$. We showed that the constrained maximum is $z = 7$, and this is attained at the points

$$(x, y) = \left(\frac{1}{\sqrt{2}}, \frac{1}{\sqrt{2}}\right) \quad \text{and} \quad (x, y) = \left(-\frac{1}{\sqrt{2}}, -\frac{1}{\sqrt{2}}\right) \qquad (3)$$

and that the constrained minimum $z = 3$, and this is attained at the points

$$(x, y) = \left(-\frac{1}{\sqrt{2}}, \frac{1}{\sqrt{2}}\right) \quad \text{and} \quad (x, y) = \left(\frac{1}{\sqrt{2}}, -\frac{1}{\sqrt{2}}\right) \qquad (4)$$

Geometrically, this means that the level curve $5x^2 + 5y^2 + 4xy = 7$ should just touch the unit circle at the points in (3), and the level curve $5x^2 + 5y^2 + 4xy = 3$ should just touch it at the points in (4). All of this is consistent with Figure 7.4.5. ◀

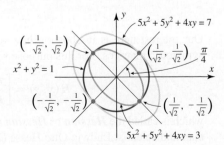

▶ Figure 7.4.5

CALCULUS REQUIRED

Relative Extrema of Functions of Two Variables

We will conclude this section by showing how quadratic forms can be used to study characteristics of real-valued functions of two variables.

Recall that if a function $f(x, y)$ has first-order partial derivatives, then its relative maxima and minima, if any, occur at points where

$$f_x(x, y) = 0 \quad \text{and} \quad f_y(x, y) = 0$$

These are called **critical points** of f. The specific behavior of f at a critical point (x_0, y_0) is determined by the sign of

$$D(x, y) = f(x, y) - f(x_0, y_0) \tag{5}$$

at points (x, y) that are close to, but different from, (x_0, y_0):

- If $D(x, y) > 0$ at points (x, y) that are sufficiently close to, but different from, (x_0, y_0), then $f(x_0, y_0) < f(x, y)$ at such points and f is said to have a **relative minimum** at (x_0, y_0) (Figure 7.4.6a).
- If $D(x, y) < 0$ at points (x, y) that are sufficiently close to, but different from, (x_0, y_0), then $f(x_0, y_0) > f(x, y)$ at such points and f is said to have a **relative maximum** at (x_0, y_0) (Figure 7.4.6b).
- If $D(x, y)$ has both positive and negative values inside *every* circle centered at (x_0, y_0), then there are points (x, y) that are arbitrarily close to (x_0, y_0) at which $f(x_0, y_0) < f(x, y)$ and points (x, y) that are arbitrarily close to (x_0, y_0) at which $f(x_0, y_0) > f(x, y)$. In this case we say that f has a **saddle point** at (x_0, y_0) (Figure 7.4.6c).

In general, it can be difficult to determine the sign of (5) directly. However, the following theorem, which is proved in calculus, makes it possible to analyze critical points using derivatives.

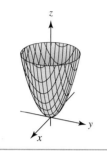

Relative minimum at (0, 0)

(a)

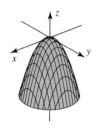

Relative maximum at (0, 0)

(b)

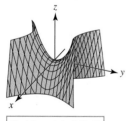

Saddle point at (0, 0)

(c)

▲ Figure 7.4.6

THEOREM 7.4.2 *Second Derivative Test*

Suppose that (x_0, y_0) is a critical point of $f(x, y)$ and that f has continuous second-order partial derivatives in some circular region centered at (x_0, y_0). Then:

(a) *f has a relative minimum at (x_0, y_0) if*

$$f_{xx}(x_0, y_0) f_{yy}(x_0, y_0) - f_{xy}^2(x_0, y_0) > 0 \quad \text{and} \quad f_{xx}(x_0, y_0) > 0$$

(b) *f has a relative maximum at (x_0, y_0) if*

$$f_{xx}(x_0, y_0) f_{yy}(x_0, y_0) - f_{xy}^2(x_0, y_0) > 0 \quad \text{and} \quad f_{xx}(x_0, y_0) < 0$$

(c) *f has a saddle point at (x_0, y_0) if*

$$f_{xx}(x_0, y_0) f_{yy}(x_0, y_0) - f_{xy}^2(x_0, y_0) < 0$$

(d) *The test is inconclusive if*

$$f_{xx}(x_0, y_0) f_{yy}(x_0, y_0) - f_{xy}^2(x_0, y_0) = 0$$

Our interest here is in showing how to reformulate this theorem using properties of symmetric matrices. For this purpose we consider the symmetric matrix

$$H(x, y) = \begin{bmatrix} f_{xx}(x, y) & f_{xy}(x, y) \\ f_{xy}(x, y) & f_{yy}(x, y) \end{bmatrix}$$

which is called the **Hessian** or **Hessian matrix** of f in honor of the German mathematician and scientist Ludwig Otto Hesse (1811–1874). The notation $H(x, y)$ emphasizes that the entries in the matrix depend on x and y. The Hessian is of interest because

$$\det[H(x_0, y_0)] = \begin{vmatrix} f_{xx}(x_0, y_0) & f_{xy}(x_0, y_0) \\ f_{xy}(x_0, y_0) & f_{yy}(x_0, y_0) \end{vmatrix} = f_{xx}(x_0, y_0)f_{yy}(x_0, y_0) - f_{xy}^2(x_0, y_0)$$

is the expression that appears in Theorem 7.4.2. We can now reformulate the second derivative test as follows.

THEOREM 7.4.3 *Hessian Form of the Second Derivative Test*

Suppose that (x_0, y_0) is a critical point of $f(x, y)$ and that f has continuous second-order partial derivatives in some circular region centered at (x_0, y_0). If $H(x_0, y_0)$ is the Hessian of f at (x_0, y_0), then:

(a) *f has a relative minimum at (x_0, y_0) if $H(x_0, y_0)$ is positive definite.*
(b) *f has a relative maximum at (x_0, y_0) if $H(x_0, y_0)$ is negative definite.*
(c) *f has a saddle point at (x_0, y_0) if $H(x_0, y_0)$ is indefinite.*
(d) *The test is inconclusive otherwise.*

We will prove part (*a*). The proofs of the remaining parts will be left as exercises.

Proof (a) If $H(x_0, y_0)$ is positive definite, then Theorem 7.3.4 implies that the principal submatrices of $H(x_0, y_0)$ have positive determinants. Thus,

$$\det[H(x_0, y_0)] = \begin{vmatrix} f_{xx}(x_0, y_0) & f_{xy}(x_0, y_0) \\ f_{xy}(x_0, y_0) & f_{yy}(x_0, y_0) \end{vmatrix} = f_{xx}(x_0, y_0)f_{yy}(x_0, y_0) - f_{xy}^2(x_0, y_0) > 0$$

and

$$\det[f_{xx}(x_0, y_0)] = f_{xx}(x_0, y_0) > 0$$

so f has a relative minimum at (x_0, y_0) by part (*a*) of Theorem 7.4.2. ◀

▶ **EXAMPLE 4** **Using the Hessian to Classify Relative Extrema**

Find the critical points of the function

$$f(x, y) = \tfrac{1}{3}x^3 + xy^2 - 8xy + 3$$

and use the eigenvalues of the Hessian matrix at those points to determine which of them, if any, are relative maxima, relative minima, or saddle points.

Solution To find both the critical points and the Hessian matrix we will need to calculate the first and second partial derivatives of f. These derivatives are

$$f_x(x, y) = x^2 + y^2 - 8y, \quad f_y(x, y) = 2xy - 8x, \quad f_{xy}(x, y) = 2y - 8$$
$$f_{xx}(x, y) = 2x, \quad f_{yy}(x, y) = 2x$$

Thus, the Hessian matrix is

$$H(x, y) = \begin{bmatrix} f_{xx}(x, y) & f_{xy}(x, y) \\ f_{xy}(x, y) & f_{yy}(x, y) \end{bmatrix} = \begin{bmatrix} 2x & 2y - 8 \\ 2y - 8 & 2x \end{bmatrix}$$

To find the critical points we set f_x and f_y equal to zero. This yields the equations

$$f_x(x, y) = x^2 + y^2 - 8y = 0 \quad \text{and} \quad f_y(x, y) = 2xy - 8x = 2x(y - 4) = 0$$

Solving the second equation yields $x = 0$ or $y = 4$. Substituting $x = 0$ in the first equation and solving for y yields $y = 0$ or $y = 8$; and substituting $y = 4$ into the first equation and solving for x yields $x = 4$ or $x = -4$. Thus, we have four critical points:

$$(0, 0), \quad (0, 8), \quad (4, 4), \quad (-4, 4)$$

Evaluating the Hessian matrix at these points yields

$$H(0,0) = \begin{bmatrix} 0 & -8 \\ -8 & 0 \end{bmatrix}, \quad H(0,8) = \begin{bmatrix} 0 & 8 \\ 8 & 0 \end{bmatrix}$$

$$H(4,4) = \begin{bmatrix} 8 & 0 \\ 0 & 8 \end{bmatrix}, \quad H(-4,4) = \begin{bmatrix} -8 & 0 \\ 0 & -8 \end{bmatrix}$$

We leave it for you to find the eigenvalues of these matrices and deduce the following classifications of the stationary points:

Critical Point (x_0, y_0)	λ_1	λ_2	Classification
(0, 0)	8	−8	Saddle point
(0, 8)	8	−8	Saddle point
(4, 4)	8	8	Relative minimum
(−4, 4)	−8	−8	Relative maximum

OPTIONAL

We conclude this section with an optional proof of Theorem 7.4.1.

Proof of Theorem 7.4.1 The first step in the proof is to show that $A\mathbf{x}$ has constrained maximum and minimum values for $\|\mathbf{x}\| = 1$. Since A is symmetric, the principal axes theorem (Theorem 7.3.1) implies that there is an orthogonal change of variable $\mathbf{x} = P\mathbf{y}$ such that

$$\mathbf{x}^T A \mathbf{x} = \lambda_1 y_1^2 + \lambda_2 y_2^2 + \cdots + \lambda_n y_n^2 \tag{6}$$

in which $\lambda_1, \lambda_2, \ldots, \lambda_n$ are the eigenvalues of A. Let us assume that $\|\mathbf{x}\| = 1$ and that the column vectors of P (which are unit eigenvectors of A) have been ordered so that

$$\lambda_1 \geq \lambda_2 \geq \cdots \geq \lambda_n \tag{7}$$

Since the matrix P is orthogonal, multiplication by P is length preserving, so that $\|\mathbf{y}\| = \|\mathbf{x}\| = 1$; that is,

$$y_1^2 + y_2^2 + \cdots + y_n^2 = 1$$

It follows from this equation and (7) that

$$\lambda_n = \lambda_n (y_1^2 + y_2^2 + \cdots + y_n^2) \leq \lambda_1 y_1^2 + \lambda_2 y_2^2 + \cdots + \lambda_n y_n^2$$
$$\leq \lambda_1 (y_1^2 + y_2^2 + \cdots + y_n^2) = \lambda_1$$

and hence from (6) that

$$\lambda_n \leq \mathbf{x}^T A \mathbf{x} \leq \lambda_1$$

This shows that all values of $\mathbf{x}^T A \mathbf{x}$ for which $\|\mathbf{x}\| = 1$ lie between the largest and smallest eigenvalues of A. Now let $\mathbf{x}$ be a unit eigenvector corresponding to λ_1. Then

$$\mathbf{x}^T A \mathbf{x} = \mathbf{x}^T (\lambda_1 \mathbf{x}) = \lambda_1 \mathbf{x}^T \mathbf{x} = \lambda_1 \|\mathbf{x}\|^2 = \lambda_1$$

which shows that $\mathbf{x}^T A \mathbf{x}$ has λ_1 as a constrained maximum and that this maximum occurs if $\mathbf{x}$ is a unit eigenvector of A corresponding to λ_1. Similarly, if $\mathbf{x}$ is a unit eigenvector corresponding to λ_n, then

$$\mathbf{x}^T A \mathbf{x} = \mathbf{x}^T (\lambda_n \mathbf{x}) = \lambda_n \mathbf{x}^T \mathbf{x} = \lambda_n \|\mathbf{x}\|^2 = \lambda_n$$

so $\mathbf{x}^T A \mathbf{x}$ has λ_n as a constrained minimum and this minimum occurs if $\mathbf{x}$ is a unit eigenvector of A corresponding to λ_n. This completes the proof. ◀

Concept Review

- Constraint
- Constrained extremum
- Level curve
- Critical point
- Relative minimum
- Relative maximum
- Saddle point
- Second derivative test
- Hessian matrix

Skills

- Find the maximum and minimum values of a quadratic form subject to a constraint.
- Find the critical points of a real-valued function of two variables, and use the eigenvalues of the Hessian matrix at the critical points to classify them as relative maxima, relative minima, or saddle points.

Exercise Set 7.4

▶ In Exercises 1–4, find the maximum and minimum values of the given quadratic form subject to the constraint $x^2 + y^2 = 1$, and determine the values of x and y at which the maximum and minimum occur. ◀

1. $7x^2 - 3y^2$ 2. xy 3. $4x^2 + 2y^2$ 4. $5x^2 + 5xy$

▶ In Exercises 5–6, find the maximum and minimum values of the given quadratic form subject to the constraint
$$x^2 + y^2 + z^2 = 1$$
and determine the values of x, y, and z at which the maximum and minimum occur. ◀

5. $x^2 - 3y^2 + 8z^2$ 6. $x^2 + 2y^2 + z^2 + 2xy + 2yz$

7. Use the method of Example 2 to find the maximum and minimum values of xy subject to the constraint $4x^2 + 8y^2 = 16$.

8. Use the method of Example 2 to find the maximum and minimum values of $x^2 + 2xy + y^2$ subject to the constraint $x^2 + 4y^2 = 4$.

▶ In Exercises 9–10, draw the unit circle and the level curves corresponding to the given quadratic form. Show that the unit circle intersects each of these curves in exactly two places, label the intersection points, and verify that the constrained extrema occur at those points. ◀

9. $5x^2 - y^2$ 10. xy

11. (a) Show that the function $f(x, y) = x^3 - y^3 + 3xy$ has critical points at $(0, 0)$, and $(1, -1)$.

   (b) Use the Hessian form of the second derivative test to show that f has a saddle point at at $(0, 0)$ and a relative minimum at $(1, -1)$.

12. (a) Show that the function $f(x, y) = x^3 - 6xy - y^3$ has critical points at $(0, 0)$ and $(-2, 2)$.

   (b) Use the Hessian form of the second derivative test to show f has a relative maximum at $(-2, 2)$ and a saddle point at $(0, 0)$.

▶ In Exercises 13–16, find the critical points of f, if any, and classify them as relative maxima, relative minima, or saddle points.

13. $f(x, y) = x^4 - y^4 - 4xy$

14. $f(x, y) = x^3 - 3xy + y^3$

15. $f(x, y) = 6(x + 1)^2 + y^2 - (x + 1)^2 y$

16. $f(x, y) = x^3 + y^3 - 3x - 3y$

17. A rectangle whose center is at the origin and whose sides are parallel to the coordinate axes is to be inscribed in the ellipse $16x^2 + y^2 = 16$. Use the method of Example 2 to find nonnegative values of x and y that produce the inscribed rectangle with maximum area.

18. Suppose that the temperature at a point (x, y) on a metal plate is $T(x, y) = 4x^2 - 4xy + y^2$. An ant, walking on the plate, traverses a circle of radius 5 centered at the origin. What are the highest and lowest temperatures encountered by the ant?

19. (a) Show that the functions
$$f(x, y) = x^4 + y^4 \quad \text{and} \quad g(x, y) = x^4 - y^4$$
have a critical point at $(0, 0)$ but the second derivative test is inconclusive at that point.

   (b) Give a reasonable argument to show that f has a relative minimum at $(0, 0)$ and g has a saddle point at $(0, 0)$.

20. Suppose that the Hessian matrix of a certain quadratic form $f(x, y)$ is
$$H = \begin{bmatrix} 2 & 4 \\ 4 & 2 \end{bmatrix}$$
What can you say about the location and classification of the critical points of f?

21. Suppose that A is an $n \times n$ symmetric matrix and
$$q(\mathbf{x}) = \mathbf{x}^T A \mathbf{x}$$
where $\mathbf{x}$ is a vector in R^n that is expressed in column form. What can you say about the value of q if $\mathbf{x}$ is a unit eigenvector corresponding to an eigenvalue λ of A?

22. Prove: If $\mathbf{x}^T A \mathbf{x}$ is a quadratic form whose minimum and maximum values subject to the constraint $\|\mathbf{x}\| = 1$ are m and M, respectively, then for each number c in the interval $m \leq c \leq M$, there is a unit vector $\mathbf{x}_c$ such that $\mathbf{x}_c^T A \mathbf{x}_c = c$. [*Hint:* In the case where $m < M$, let $\mathbf{u}_m$ and $\mathbf{u}_M$ be unit eigenvectors of A such that $\mathbf{u}_m^T A \mathbf{u}_m = m$ and $\mathbf{u}_M^T A \mathbf{u}_M = M$, and let

$$\mathbf{x}_c = \sqrt{\frac{M-c}{M-m}}\,\mathbf{u}_m + \sqrt{\frac{c-m}{M-m}}\,\mathbf{u}_M$$

Show that $\mathbf{x}_c^T A \mathbf{x}_c = c$.]

True-False Exercises

In parts (a)–(e) determine whether the statement is true or false, and justify your answer.

(a) A quadratic form must have either a maximum or minimum value.

(b) The maximum value of a quadratic form $\mathbf{x}^T A \mathbf{x}$ subject to the constraint $\|\mathbf{x}\| = 1$ occurs at a unit eigenvector corresponding to the largest eigenvalue of A.

(c) The Hessian matrix of a function f with continuous second-order partial derivatives is a symmetric matrix.

(d) If (x_0, y_0) is a critical point of a function f and the Hessian of f at (x_0, y_0) is 0, then f has neither a relative maximum nor a relative minimum at (x_0, y_0).

(e) If A is a symmetric matrix and $\det A < 0$, then the minimum of $\mathbf{x}^T A \mathbf{x}$ subject to the constraint $\|\mathbf{x}\| = 1$ is negative.

7.5 Hermitian, Unitary, and Normal Matrices

We know that every real symmetric matrix is orthogonally diagonalizable and that the real symmetric matrices are the only orthogonally diagonalizable matrices. In this section we will consider the diagonalization problem for complex matrices.

Hermitian and Unitary Matrices

The transpose operation is less important for complex matrices than for real matrices. A more useful operation for complex matrices is given in the following definition.

> **DEFINITION 1** If A is a complex matrix, then the ***conjugate transpose*** of A, denoted by A^*, is defined by
> $$A^* = \overline{A}^T \tag{1}$$

Remark Since part (b) of Theorem 5.3.2 states that $\overline{(A^T)} = (\overline{A})^T$, the order in which the transpose and conjugation operations are performed in computing $A^* = \overline{A}^T$ does not matter. Moreover, in the case where A has real entries we have $A^* = (\overline{A})^T = A^T$, so A^* is the same as A^T for real matrices.

▶ **EXAMPLE 1 Conjugate Transpose**

Find the conjugate transpose A^* of the matrix

$$A = \begin{bmatrix} 1+i & -i & 0 \\ 2 & 3-2i & i \end{bmatrix}$$

Solution We have

$$\overline{A} = \begin{bmatrix} 1-i & i & 0 \\ 2 & 3+2i & -i \end{bmatrix} \quad \text{and hence} \quad A^* = \overline{A}^T = \begin{bmatrix} 1-i & 2 \\ i & 3+2i \\ 0 & -i \end{bmatrix} \blacktriangleleft$$

The following theorem, parts of which are given as exercises, shows that the basic algebraic properties of the conjugate transpose operation are similar to those of the transpose (compare to Theorem 1.4.8).

7.5 Hermitian, Unitary, and Normal Matrices

THEOREM 7.5.1 *If k is a complex scalar, and if A, B, and C are complex matrices whose sizes are such that the stated operations can be performed, then:*
(a) $(A^*)^* = A$
(b) $(A + B)^* = A^* + B^*$
(c) $(A - B)^* = A^* - B^*$
(d) $(kA)^* = \bar{k}A^*$
(e) $(AB)^* = B^*A^*$

Remark Note that the relationship $\mathbf{u} \cdot \mathbf{v} = \bar{\mathbf{v}}^T \mathbf{u}$ in Formula (5) of Section 5.3 can be expressed in terms of the conjugate transpose as

$$\mathbf{u} \cdot \mathbf{v} = \mathbf{v}^* \mathbf{u} \qquad (2)$$

We are now ready to define two new classes of matrices that will be important in our study of diagonalization in C^n.

DEFINITION 2 A square complex matrix A is said to be **unitary** if
$$A^{-1} = A^* \qquad (3)$$
and is said to be **Hermitian**[*] if
$$A^* = A \qquad (4)$$

Note that a unitary matrix can also be defined as a square complex matrix A for which
$$AA^* = A^*A = I$$

If A is a real matrix, then $A^* = A^T$, in which case (3) becomes $A^{-1} = A^T$ and (4) becomes $A^T = A$. Thus, the unitary matrices are complex generalizations of the real orthogonal matrices and Hermitian matrices are complex generalizations of the real symmetric matrices.

▶ **EXAMPLE 2** Recognizing Hermitian Matrices

Hermitian matrices are easy to recognize because their diagonal entries are real (why?), and the entries that are symmetrically positioned across the main diagonal are complex conjugates. Thus, for example, we can tell by inspection that

$$A = \begin{bmatrix} 1 & i & 1+i \\ -i & -5 & 2-i \\ 1-i & 2+i & 3 \end{bmatrix}$$

is Hermitian. ◀

The fact that real symmetric matrices have real eigenvalues is a special case of the following more general result about Hermitian matrices, the proof of which is left for the exercises.

THEOREM 7.5.2 *The eigenvalues of a Hermitian matrix are real numbers.*

The fact that eigenvectors from different eigenspaces of a real symmetric matrix are orthogonal is a special case of the following more general result about Hermitian matrices.

---
[*]In honor of the French mathematician Charles Hermite (1822–1901).

THEOREM 7.5.3 *If A is a Hermitian matrix, then eigenvectors from different eigenspaces are orthogonal.*

Proof Let $\mathbf{v}_1$ and $\mathbf{v}_2$ be eigenvectors of A corresponding to distinct eigenvalues λ_1 and λ_2. Using Formula (2) and the facts that $\lambda_1 = \bar{\lambda}_1$, $\lambda_2 = \bar{\lambda}_2$, and $A = A^*$, we can write

$$\lambda_1(\mathbf{v}_2 \cdot \mathbf{v}_1) = (\lambda_1 \mathbf{v}_1)^* \mathbf{v}_2 = (A\mathbf{v}_1)^* \mathbf{v}_2 = (\mathbf{v}_1^* A^*)\mathbf{v}_2$$
$$= (\mathbf{v}_1^* A)\mathbf{v}_2 = \mathbf{v}_1^*(A\mathbf{v}_2)$$
$$= \mathbf{v}_1^*(\lambda_2 \mathbf{v}_2) = \lambda_2(\mathbf{v}_1^* \mathbf{v}_2) = \lambda_2(\mathbf{v}_2 \cdot \mathbf{v}_1)$$

This implies that $(\lambda_1 - \lambda_2)(\mathbf{v}_2 \cdot \mathbf{v}_1) = 0$ and hence that $\mathbf{v}_2 \cdot \mathbf{v}_1 = 0$ (since $\lambda_1 \neq \lambda_2$). ◀

▶ **EXAMPLE 3** **Eigenvalues and Eigenvectors of a Hermitian Matrix**
Confirm that the Hermitian matrix

$$A = \begin{bmatrix} 2 & 1+i \\ 1-i & 3 \end{bmatrix}$$

has real eigenvalues and that eigenvectors from different eigenspaces are orthogonal.

Solution The characteristic polynomial of A is

$$\det(\lambda I - A) = \begin{vmatrix} \lambda - 2 & -1 - i \\ -1 + i & \lambda - 3 \end{vmatrix}$$
$$= (\lambda - 2)(\lambda - 3) - (-1 - i)(-1 + i)$$
$$= (\lambda^2 - 5\lambda + 6) - 2 = (\lambda - 1)(\lambda - 4)$$

so the eigenvalues of A are $\lambda = 1$ and $\lambda = 4$, which are real. Bases for the eigenspaces of A can be obtained by solving the linear system

$$\begin{bmatrix} \lambda - 2 & -1 - i \\ -1 + i & \lambda - 3 \end{bmatrix} \begin{bmatrix} x_1 \\ x_2 \end{bmatrix} = \begin{bmatrix} 0 \\ 0 \end{bmatrix}$$

with $\lambda = 1$ and with $\lambda = 4$. We leave it for you to do this and to show that the general solutions of these systems are

$$\lambda = 1: \begin{bmatrix} x_1 \\ x_2 \end{bmatrix} = t \begin{bmatrix} -1 - i \\ 1 \end{bmatrix} \quad \text{and} \quad \lambda = 4: \begin{bmatrix} x_1 \\ x_2 \end{bmatrix} = t \begin{bmatrix} \frac{1}{2}(1+i) \\ 1 \end{bmatrix}$$

Thus, bases for these eigenspaces are

$$\lambda = 1: \mathbf{v}_1 = \begin{bmatrix} -1 - i \\ 1 \end{bmatrix} \quad \text{and} \quad \lambda = 4: \mathbf{v}_2 = \begin{bmatrix} \frac{1}{2}(1+i) \\ 1 \end{bmatrix}$$

The vectors $\mathbf{v}_1$ and $\mathbf{v}_2$ are orthogonal since

$$\mathbf{v}_1 \cdot \mathbf{v}_2 = (-1 - i)\overline{\left(\tfrac{1}{2}(1+i)\right)} + (1)(1) = \tfrac{1}{2}(-1 - i)(1 - i) + 1 = 0$$

and hence all scalar multiples of them are also orthogonal. ◀

Unitary matrices are not usually easy to recognize by inspection. However, the following analog of Theorems 7.1.1 and 7.1.3, part of which is proved in the exercises, provides a way of ascertaining whether a matrix is unitary without computing its inverse.

7.5 Hermitian, Unitary, and Normal Matrices

THEOREM 7.5.4 *If A is an $n \times n$ matrix with complex entries, then the following are equivalent.*

(a) *A is unitary.*

(b) *$\|A\mathbf{x}\| = \|\mathbf{x}\|$ for all $\mathbf{x}$ in C^n.*

(c) *$A\mathbf{x} \cdot A\mathbf{y} = \mathbf{x} \cdot \mathbf{y}$ for all $\mathbf{x}$ and $\mathbf{y}$ in C^n.*

(d) *The column vectors of A form an orthonormal set in C^n with respect to the complex Euclidean inner product.*

(e) *The row vectors of A form an orthonormal set in C^n with respect to the complex Euclidean inner product.*

▶ **EXAMPLE 4 A Unitary Matrix**

Use Theorem 7.5.4 to show that

$$A = \begin{bmatrix} \tfrac{1}{2}(1+i) & \tfrac{1}{2}(1+i) \\ \tfrac{1}{2}(1-i) & \tfrac{1}{2}(-1+i) \end{bmatrix}$$

is unitary, and then find A^{-1}.

Solution We will show that the row vectors

$$\mathbf{r}_1 = \begin{bmatrix} \tfrac{1}{2}(1+i) & \tfrac{1}{2}(1+i) \end{bmatrix} \quad \text{and} \quad \mathbf{r}_2 = \begin{bmatrix} \tfrac{1}{2}(1-i) & \tfrac{1}{2}(-1+i) \end{bmatrix}$$

are orthonormal. The relevant computations are

$$\|\mathbf{r}_1\| = \sqrt{\left|\tfrac{1}{2}(1+i)\right|^2 + \left|\tfrac{1}{2}(1+i)\right|^2} = \sqrt{\tfrac{1}{2}+\tfrac{1}{2}} = 1$$

$$\|\mathbf{r}_2\| = \sqrt{\left|\tfrac{1}{2}(1-i)\right|^2 + \left|\tfrac{1}{2}(-1+i)\right|^2} = \sqrt{\tfrac{1}{2}+\tfrac{1}{2}} = 1$$

$$\mathbf{r}_1 \cdot \mathbf{r}_2 = \left(\tfrac{1}{2}(1+i)\right)\overline{\left(\tfrac{1}{2}(1-i)\right)} + \left(\tfrac{1}{2}(1+i)\right)\overline{\left(\tfrac{1}{2}(-1+i)\right)}$$

$$= \left(\tfrac{1}{2}(1+i)\right)\left(\tfrac{1}{2}(1+i)\right) + \left(\tfrac{1}{2}(1+i)\right)\left(\tfrac{1}{2}(-1-i)\right) = \tfrac{1}{2}i - \tfrac{1}{2}i = 0$$

Since we now know that A is unitary, it follows that

$$A^{-1} = A^* = \begin{bmatrix} \tfrac{1}{2}(1-i) & \tfrac{1}{2}(1+i) \\ \tfrac{1}{2}(1-i) & \tfrac{1}{2}(-1-i) \end{bmatrix}$$

You can confirm the validity of this result by showing that $AA^* = A^*A = I$. ◀

Unitary Diagonalizability Since unitary matrices are the complex analogs of the real orthogonal matrices, the following definition is a natural generalization of orthogonal diagonalizability for real matrices.

DEFINITION 3 A square complex matrix is said to be ***unitarily diagonalizable*** if there is a unitary matrix P such that $P^*AP = D$ is a complex diagonal matrix. Any such matrix P is said to ***unitarily diagonalize*** A.

Recall that a real symmetric $n \times n$ matrix A has an orthonormal set of n eigenvectors and is orthogonally diagonalized by any $n \times n$ matrix whose column vectors are an orthonormal set of eigenvectors of A. Here is the complex analog of that result.

THEOREM 7.5.5 *Every $n \times n$ Hermitian matrix A has an orthonormal set of n eigenvectors and is unitarily diagonalized by any $n \times n$ matrix P whose column vectors form an orthonormal set of eigenvectors of A.*

The procedure for unitarily diagonalizing a Hermitian matrix A is exactly the same as that for orthogonally diagonalizing a symmetric matrix:

Unitarily Diagonalizing a Hermitian Matrix

Step 1. Find a basis for each eigenspace of A.

Step 2. Apply the Gram–Schmidt process to each of these bases to obtain orthonormal bases for the eigenspaces.

Step 3. Form the matrix P whose column vectors are the basis vectors obtained in Step 2. This will be a unitary matrix (Theorem 7.5.4) and will unitarily diagonalize A.

▶ **EXAMPLE 5** Unitary Diagonalization of a Hermitian Matrix

Find a matrix P that unitarily diagonalizes the Hermitian matrix

$$A = \begin{bmatrix} 2 & 1+i \\ 1-i & 3 \end{bmatrix}$$

Solution We showed in Example 3 that the eigenvalues of A are $\lambda = 1$ and $\lambda = 4$ and that bases for the corresponding eigenspaces are

$$\lambda = 1: \quad \mathbf{v}_1 = \begin{bmatrix} -1-i \\ 1 \end{bmatrix} \quad \text{and} \quad \lambda = 4: \quad \mathbf{v}_2 = \begin{bmatrix} \frac{1}{2}(1+i) \\ 1 \end{bmatrix}$$

Since each eigenspace has only one basis vector, the Gram–Schmidt process is simply a matter of normalizing these basis vectors. We leave it for you to show that

$$\mathbf{p}_1 = \frac{\mathbf{v}_1}{\|\mathbf{v}_1\|} = \begin{bmatrix} \frac{-1-i}{\sqrt{3}} \\ \frac{1}{\sqrt{3}} \end{bmatrix} \quad \text{and} \quad \mathbf{p}_2 = \frac{\mathbf{v}_2}{\|\mathbf{v}_2\|} = \begin{bmatrix} \frac{1+i}{\sqrt{6}} \\ \frac{2}{\sqrt{6}} \end{bmatrix}$$

Thus, A is unitarily diagonalized by the matrix

$$P = [\mathbf{p}_1 \quad \mathbf{p}_2] = \begin{bmatrix} \frac{-1-i}{\sqrt{3}} & \frac{1+i}{\sqrt{6}} \\ \frac{1}{\sqrt{3}} & \frac{2}{\sqrt{6}} \end{bmatrix}$$

Although it is a little tedious, you may want to check this result by showing that

$$P^*AP = \begin{bmatrix} \frac{-1+i}{\sqrt{3}} & \frac{1}{\sqrt{3}} \\ \frac{1-i}{\sqrt{6}} & \frac{2}{\sqrt{6}} \end{bmatrix} \begin{bmatrix} 2 & 1+i \\ 1-i & 3 \end{bmatrix} \begin{bmatrix} \frac{-1-i}{\sqrt{3}} & \frac{1+i}{\sqrt{6}} \\ \frac{1}{\sqrt{3}} & \frac{2}{\sqrt{6}} \end{bmatrix} = \begin{bmatrix} 1 & 0 \\ 0 & 4 \end{bmatrix} \quad \blacktriangleleft$$

Skew-Symmetric and Skew-Hermitian Matrices

In Exercise 37 of Section 1.7 we defined a square matrix with real entries to be ***skew-symmetric*** if $A^T = -A$. A skew-symmetric matrix must have zeros on the main diagonal (why?), and each entry off the main diagonal must be the negative of its mirror image about the main diagonal. Here is an example.

$$A = \begin{bmatrix} 0 & 1 & -2 \\ -1 & 0 & 4 \\ 2 & -4 & 0 \end{bmatrix} \quad \text{[skew-symmetric]}$$

We leave it for you to confirm that $A^T = -A$.

The complex analogs of the skew-symmetric matrices are the matrices for which $A^* = -A$. Such matrices are said to be **skew-Hermitian**.

Since a skew-Hermitian matrix A has the property

$$A^* = \overline{A}^T = -A$$

it must be that A has zeros or pure imaginary numbers on the main diagonal (why?), and that the complex conjugate of each entry off the main diagonal is the negative of its mirror image about the main diagonal. Here is an example.

$$A = \begin{bmatrix} i & 1-i & 5 \\ -1-i & 2i & i \\ -5 & i & 0 \end{bmatrix} \quad \text{[skew-Hermitian]}$$

Normal Matrices

Hermitian matrices enjoy many, but not all, of the properties of real symmetric matrices. For example, we know that real symmetric matrices are orthogonally diagonalizable and Hermitian matrices are unitarily diagonalizable. However, whereas the real symmetric matrices are the only orthogonally diagonalizable matrices, the Hermitian matrices do not constitute the entire class of unitarily diagonalizable complex matrices; that is, there exist unitarily diagonalizable matrices that are not Hermitian. Specifically, it can be proved that a square complex matrix A is unitarily diagonalizable if and only if

$$AA^* = A^*A \tag{5}$$

Matrices with this property are said to be **normal**. Normal matrices include the Hermitian, skew-Hermitian, and unitary matrices in the complex case and the symmetric, skew-symmetric, and orthogonal matrices in the real case. The nonzero skew-symmetric matrices are particularly interesting because they are examples of real matrices that are not orthogonally diagonalizable but are unitarily diagonalizable.

A Comparison of Eigenvalues

We have seen that Hermitian matrices have real eigenvalues. In the exercises we will ask you to show that the eigenvalues of a skew-Hermitian matrix are either zero or purely imaginary (have real part of zero) and that the eigenvalues of unitary matrices have modulus 1. These ideas are illustrated schematically in Figure 7.5.1.

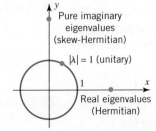

▶ Figure 7.5.1

Concept Review

- Conjugate transpose
- Unitary matrix
- Hermitian matrix
- Unitarily diagonalizable matrix
- Skew-symmetric matrix
- Skew-Hermitian matrix
- Normal matrix

Skills

- Find the conjugate transpose of a matrix.
- Be able to identify Hermitian matrices.
- Find the inverse of a unitary matrix.
- Find a unitary matrix that diagonalizes a Hermitian matrix.

Exercise Set 7.5

▶ In Exercises 1–2, find A^*. ◀

1. $A = \begin{bmatrix} 2+i & 1 \\ 2 & -3i \\ 0 & 1+5i \end{bmatrix}$ **2.** $A = \begin{bmatrix} 2i & 1-i & -1+i \\ 4 & 5-7i & -i \end{bmatrix}$

▶ In Exercises 3–4, substitute numbers for the ×'s so that A is Hermitian. ◀

3. $A = \begin{bmatrix} 3 & \times & \times \\ 3+2i & -2 & \times \\ 7 & 1-5i & 6 \end{bmatrix}$ **4.** $A = \begin{bmatrix} 2 & 0 & 3+5i \\ \times & -4 & -i \\ \times & \times & 6 \end{bmatrix}$

▶ In Exercises 5–6, show that A is not Hermitian for any choice of the ×'s. ◀

5. (a) $A = \begin{bmatrix} 2 & \times & 3-7i \\ \times & 1+i & 2i \\ 3+7i & \times & 0 \end{bmatrix}$

(b) $A = \begin{bmatrix} 2 & 4+i & \times \\ -4+i & -1 & \times \\ \times & \times & \times \end{bmatrix}$

6. (a) $A = \begin{bmatrix} 1 & 1+i & \times \\ 1+i & 7 & \times \\ 6-2i & \times & 0 \end{bmatrix}$

(b) $A = \begin{bmatrix} 1 & \times & 3+5i \\ \times & 3 & 1-i \\ 3-5i & \times & 2+i \end{bmatrix}$

▶ In Exercises 7–8, verify that the eigenvalues of the Hermitian matrix A are real and that eigenvectors from different eigenspaces are orthogonal (see Theorem 7.5.3). ◀

7. $A = \begin{bmatrix} 5 & 1-3i \\ 1+3i & 2 \end{bmatrix}$ **8.** $A = \begin{bmatrix} 0 & 2i \\ -2i & 2 \end{bmatrix}$

▶ In Exercises 9–12, show that A is unitary, and find A^{-1}. ◀

9. $A = \begin{bmatrix} \frac{4}{5} & \frac{3i}{5} \\ -\frac{3i}{5} & -\frac{4}{5} \end{bmatrix}$ **10.** $A = \begin{bmatrix} \frac{1}{\sqrt{3}} & -\frac{1-i}{\sqrt{3}} \\ \frac{1+i}{\sqrt{3}} & \frac{1}{\sqrt{3}} \end{bmatrix}$

11. $A = \begin{bmatrix} \frac{1}{2\sqrt{2}}(\sqrt{3}+i) & \frac{1}{2\sqrt{2}}(1-i\sqrt{3}) \\ \frac{1}{2\sqrt{2}}(1+i\sqrt{3}) & \frac{1}{2\sqrt{2}}(i-\sqrt{3}) \end{bmatrix}$

12. $A = \begin{bmatrix} \frac{6}{7} & \frac{2}{7}+\frac{3i}{7} \\ -\frac{2}{7}+\frac{3i}{7} & \frac{6}{7} \end{bmatrix}$

▶ In Exercises 13–18, find a unitary matrix P that diagonalizes the Hermitian matrix A, and determine $P^{-1}AP$. ◀

13. $A = \begin{bmatrix} 9 & 12i \\ -12i & 16 \end{bmatrix}$ **14.** $A = \begin{bmatrix} 3 & -i \\ i & 3 \end{bmatrix}$

15. $A = \begin{bmatrix} 6 & 2+2i \\ 2-2i & 4 \end{bmatrix}$ **16.** $A = \begin{bmatrix} 0 & 3+i \\ 3-i & -3 \end{bmatrix}$

17. $A = \begin{bmatrix} 5 & 0 & 0 \\ 0 & -1 & -1+i \\ 0 & -1-i & 0 \end{bmatrix}$

18. $A = \begin{bmatrix} 2 & \frac{1}{\sqrt{2}}i & -\frac{1}{\sqrt{2}}i \\ -\frac{1}{\sqrt{2}}i & 2 & 0 \\ \frac{1}{\sqrt{2}}i & 0 & 2 \end{bmatrix}$

▶ In Exercises 19–20, substitute numbers for the ×'s so that A is skew-Hermitian. ◀

19. $A = \begin{bmatrix} -i & \times & \times \\ 2+5i & 0 & \times \\ 1 & 4-6i & 3i \end{bmatrix}$ **20.** $A = \begin{bmatrix} 0 & 0 & 3-5i \\ \times & 0 & -i \\ \times & \times & 0 \end{bmatrix}$

▶ In Exercises 21–22, show that A is not skew-Hermitian for any choice of the ×'s. ◀

21. (a) $A = \begin{bmatrix} 3i & \times & 2-i \\ \times & \times & 4 \\ 2+i & -4 & 0 \end{bmatrix}$

(b) $A = \begin{bmatrix} 0 & 0 & 2-3i \\ 0 & 7 & 1 \\ -2-3i & -1 & \times \end{bmatrix}$

7.5 Hermitian, Unitary, and Normal Matrices

22. (a) $A = \begin{bmatrix} i & \times & 2-3i \\ \times & 0 & 1+i \\ 2+3i & -1-i & \times \end{bmatrix}$

    (b) $A = \begin{bmatrix} 0 & -i & 4+7i \\ \times & 0 & \times \\ -4-7i & \times & 1 \end{bmatrix}$

▶ In Exercises 23–24, verify that the eigenvalues of the skew-Hermitian matrix A are pure imaginary numbers. ◀

23. $A = \begin{bmatrix} i & 1-i \\ -1-i & 0 \end{bmatrix}$ 24. $A = \begin{bmatrix} 0 & 3i \\ 3i & 0 \end{bmatrix}$

▶ In Exercises 25–26, show that A is normal. ◀

25. $A = \begin{bmatrix} 1+2i & 2+i & -2-i \\ 2+i & 1+i & -i \\ -2-i & -i & 1+i \end{bmatrix}$

26. $A = \begin{bmatrix} 2+2i & i & 1-i \\ i & -2i & 1-3i \\ 1-i & 1-3i & -3+8i \end{bmatrix}$

27. Show that the matrix

$$A = \frac{1}{\sqrt{2}} \begin{bmatrix} e^{i\theta} & e^{-i\theta} \\ ie^{i\theta} & -ie^{-i\theta} \end{bmatrix}$$

is unitary for all real values of θ. [*Note:* See Formula (17) in Appendix B for the definition of $e^{i\theta}$.]

28. Prove that each entry on the main diagonal of a skew-Hermitian matrix is either zero or a pure imaginary number.

29. Let A be any $n \times n$ matrix with complex entries, and define the matrices B and C to be

$$B = \frac{1}{2}(A + A^*) \quad \text{and} \quad C = \frac{1}{2i}(A - A^*)$$

    (a) Show that B and C are Hermitian.
    (b) Show that $A = B + iC$ and $A^* = B - iC$.
    (c) What condition must B and C satisfy for A to be normal?

30. Show that if A is an $n \times n$ matrix with complex entries, and if $\mathbf{u}$ and $\mathbf{v}$ are vectors in C^n that are expressed in column form, then

$$A\mathbf{u} \cdot \mathbf{v} = \mathbf{u} \cdot A^*\mathbf{v} \quad \text{and} \quad \mathbf{u} \cdot A\mathbf{v} = A^*\mathbf{u} \cdot \mathbf{v}$$

31. Show that if A is a unitary matrix, then so is A^*.

32. Show that the eigenvalues of a skew-Hermitian matrix are either zero or purely imaginary.

33. Show that the eigenvalues of a unitary matrix have modulus 1.

34. Show that if $\mathbf{u}$ is a nonzero vector in C^n that is expressed in column form, then $P = \mathbf{uu}^*$ is Hermitian.

35. Show that if $\mathbf{u}$ is a unit vector in C^n that is expressed in column form, then $H = I - 2\mathbf{uu}^*$ is Hermitian and unitary.

36. What can you say about the inverse of a matrix A that is both Hermitian and unitary?

37. Find a 2×2 matrix that is both Hermitian and unitary and whose entries are not all real numbers.

38. Under what conditions is the following matrix normal?

$$A = \begin{bmatrix} a & 0 & 0 \\ 0 & 0 & c \\ 0 & b & 0 \end{bmatrix}$$

39. What geometric interpretations might you reasonably give to multiplication by the matrices $P = \mathbf{uu}^*$ and $H = I - 2\mathbf{uu}^*$ in Exercises 34 and 35?

40. Prove that if A is an invertible matrix, then A^* is invertible, and $(A^*)^{-1} = (A^{-1})^*$.

41. (a) Prove that $\det(\overline{A}) = \overline{\det(A)}$.
    (b) Use the result in part (a) and the fact that a square matrix and its transpose have the same determinant to prove that $\det(A^*) = \overline{\det(A)}$.

42. Use part (b) of Exercise 41 to prove:
    (a) If A is Hermitian, then $\det(A)$ is real.
    (b) If A is unitary, then $|\det(A)| = 1$.

43. Use properties of the transpose and complex conjugate to prove parts (*a*) and (*e*) of Theorem 7.5.1.

44. Use properties of the transpose and complex conjugate to prove parts (*b*) and (*d*) of Theorem 7.5.1.

45. Prove that an $n \times n$ matrix with complex entries is unitary if and only if the columns of A form an orthonormal set in C^n.

46. Prove that the eigenvalues of a Hermitian matrix are real.

True-False Exercises

In parts (a)–(e) determine whether the statement is true or false, and justify your answer.

(a) The matrix $\begin{bmatrix} 0 & i \\ i & 2 \end{bmatrix}$ is Hermitian.

(b) The matrix $\begin{bmatrix} -\frac{i}{\sqrt{2}} & \frac{i}{\sqrt{6}} & \frac{i}{\sqrt{3}} \\ 0 & -\frac{i}{\sqrt{6}} & \frac{i}{\sqrt{3}} \\ \frac{i}{\sqrt{2}} & \frac{i}{\sqrt{6}} & \frac{i}{\sqrt{3}} \end{bmatrix}$ is unitary.

(c) The conjugate transpose of a unitary matrix is unitary.

(d) Every unitarily diagonalizable matrix is Hermitian.

(e) A positive integer power of a skew-Hermitian matrix is skew-Hermitian.

Chapter 7 Supplementary Exercises

1. Verify that each matrix is orthogonal, and find its inverse.

   (a) $\begin{bmatrix} \frac{3}{5} & -\frac{4}{5} \\ \frac{4}{5} & \frac{3}{5} \end{bmatrix}$

   (b) $\begin{bmatrix} \frac{4}{5} & 0 & -\frac{3}{5} \\ -\frac{9}{25} & \frac{4}{5} & -\frac{12}{25} \\ \frac{12}{25} & \frac{3}{5} & \frac{16}{25} \end{bmatrix}$

2. Prove: If Q is an orthogonal matrix, then each entry of Q is the same as its cofactor if $\det(Q) = 1$ and is the negative of its cofactor if $\det(Q) = -1$.

3. Prove that if A is a positive definite symmetric matrix, and if $\mathbf{u}$ and $\mathbf{v}$ are vectors in R^n in column form, then
$$\langle \mathbf{u}, \mathbf{v} \rangle = \mathbf{u}^T A \mathbf{v}$$
is an inner product on R^n.

4. Find the characteristic polynomial and the dimensions of the eigenspaces of the symmetric matrix
$$\begin{bmatrix} 3 & 2 & 2 \\ 2 & 3 & 2 \\ 2 & 2 & 3 \end{bmatrix}$$

5. Find a matrix P that orthogonally diagonalizes
$$A = \begin{bmatrix} 1 & 0 & 1 \\ 0 & 1 & 0 \\ 1 & 0 & 1 \end{bmatrix}$$
and determine the diagonal matrix $D = P^T A P$.

6. Express each quadratic form in the matrix notation $\mathbf{x}^T A \mathbf{x}$.
   (a) $-4x_1^2 + 16x_2^2 - 15x_1 x_2$
   (b) $9x_1^2 - x_2^2 + 4x_3^2 + 6x_1 x_2 - 8x_1 x_3 + x_2 x_3$

7. Classify the quadradic form
$$x_1^2 - 3x_1 x_2 + 4x_2^2$$
as positive definite, negative definite, indefinite, positive semidefinite, or negative semidefinite.

8. Find an orthogonal change of variable that eliminates the cross product terms in each quadratic form, and express the quadratic form in terms of the new variables.
   (a) $-3x_1^2 + 5x_2^2 + 2x_1 x_2$
   (b) $-5x_1^2 + x_2^2 - x_3^2 + 6x_1 x_3 + 4x_1 x_2$

9. Identify the type of conic section represented by each equation.
   (a) $y - x^2 = 0$
   (b) $3x - 11y^2 = 0$

10. Find a unitary matrix U that diagonalizes
$$A = \begin{bmatrix} 1 & 1 & 0 \\ 0 & 1 & 1 \\ 1 & 0 & 1 \end{bmatrix}$$
and determine the diagonal matrix $D = U^{-1} A U$.

11. Show that if U is an $n \times n$ unitary matrix and
$$|z_1| = |z_2| = \cdots = |z_n| = 1$$
then the product
$$U \begin{bmatrix} z_1 & 0 & 0 & \cdots & 0 \\ 0 & z_2 & 0 & \cdots & 0 \\ \vdots & \vdots & \vdots & & \vdots \\ 0 & 0 & 0 & \cdots & z_n \end{bmatrix}$$
is also unitary.

12. Suppose that $A^* = -A$.
    (a) Show that iA is Hermitian.
    (b) Show that A is unitarily diagonalizable and has pure imaginary eigenvalues.

CHAPTER 8

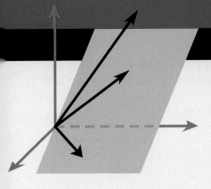

Linear Transformations

CHAPTER CONTENTS
8.1 General Linear Transformations 433
8.2 Isomorphism 445
8.3 Compositions and Inverse Transformations 452
8.4 Matrices for General Linear Transformations 458
8.5 Similarity 468

INTRODUCTION In Sections 4.9 and 4.10 we studied linear transformations from R^n to R^m. In this chapter we will define and study linear transformations from a general vector space V to a general vector space W. The results we obtain here have important applications in physics, engineering, and various branches of mathematics.

8.1 General Linear Transformations

Up to now our study of linear transformations has focused on transformations from R^n to R^m. In this section we will turn our attention to linear transformations involving general vector spaces. We will illustrate ways in which such transformations arise, and we will establish a fundamental relationship between general n-dimensional vector spaces and R^n.

Definitions and Terminology

In Section 4.9 we defined a *matrix transformation* $T_A: R^n \to R^m$ to be a mapping of the form

$$T_A(\mathbf{x}) = A\mathbf{x}$$

in which A is an $m \times n$ matrix. We subsequently established in Theorems 4.10.2 and 4.10.3 that the matrix transformations are precisely the *linear transformations* from R^n to R^m, that is, the transformations with the linearity properties

$$T(\mathbf{u} + \mathbf{v}) = T(\mathbf{u}) + T(\mathbf{v}) \quad \text{and} \quad T(k\mathbf{u}) = kT(\mathbf{u})$$

We will use these two properties as the starting point for defining more general linear transformations.

> **DEFINITION 1** If $T: V \to W$ is a function from a vector space V to a vector space W, then T is called a *linear transformation* from V to W if the following two properties hold for all vectors $\mathbf{u}$ and $\mathbf{v}$ in V and for all scalars k:
>
> (i) $T(k\mathbf{u}) = kT(\mathbf{u})$ [Homogeneity property]
> (ii) $T(\mathbf{u} + \mathbf{v}) = T(\mathbf{u}) + T(\mathbf{v})$ [Additivity property]
>
> In the special case where $V = W$, the linear transformation T is called a *linear operator* on the vector space V.

The homogeneity and additivity properties of a linear transformation $T : V \to W$ can be used in combination to show that if $\mathbf{v}_1$ and $\mathbf{v}_2$ are vectors in V and k_1 and k_2 are any scalars, then

$$T(k_1\mathbf{v}_1 + k_2\mathbf{v}_2) = k_1 T(\mathbf{v}_1) + k_2 T(\mathbf{v}_2)$$

More generally, if $\mathbf{v}_1, \mathbf{v}_2, \ldots, \mathbf{v}_r$ are vectors in V and $k_1, k_2, \ldots, k_r$ are any scalars, then

$$T(k_1\mathbf{v}_1 + k_2\mathbf{v}_2 + \cdots + k_r\mathbf{v}_r) = k_1 T(\mathbf{v}_1) + k_2 T(\mathbf{v}_2) + \cdots + k_r T(\mathbf{v}_r) \quad (1)$$

The following theorem is an analog of parts (*a*) and (*d*) of Theorem 4.9.1.

THEOREM 8.1.1 *If $T : V \to W$ is a linear transformation, then*:
(*a*) $T(\mathbf{0}) = \mathbf{0}$.
(*b*) $T(\mathbf{u} - \mathbf{v}) = T(\mathbf{u}) - T(\mathbf{v})$ *for all $\mathbf{u}$ and $\mathbf{v}$ in V.*

> Use the two parts of Theorem 8.1.1 to prove that
> $$T(-\mathbf{v}) = -\mathbf{v}$$
> for all $\mathbf{v}$ in V.

Proof Let $\mathbf{u}$ be any vector in V. Since $0\mathbf{u} = \mathbf{0}$, it follows from the homogeneity property in Definition 1 that

$$T(\mathbf{0}) = T(0\mathbf{u}) = 0T(\mathbf{u}) = \mathbf{0}$$

which proves (*a*).

We can prove part (*b*) by rewriting $T(\mathbf{u} - \mathbf{v})$ as

$$T(\mathbf{u} - \mathbf{v}) = T(\mathbf{u} + (-1)\mathbf{v})$$
$$= T(\mathbf{u}) + (-1)T(\mathbf{v})$$
$$= T(\mathbf{u}) - T(\mathbf{v})$$

We leave it for you to justify each step. ◀

▶ **EXAMPLE 1** **Matrix Transformations**

Because we have based the definition of a general linear transformation on the homogeneity and additivity properties of *matrix transformations*, it follows that a matrix transformation $T_A : R^n \to R^m$ is also a linear transformation in this more general sense with $V = R^n$ and $W = R^m$.

▶ **EXAMPLE 2** **The Zero Transformation**

Let V and W be any two vector spaces. The mapping $T : V \to W$ such that $T(\mathbf{v}) = \mathbf{0}$ for every $\mathbf{v}$ in V is a linear transformation called the **zero transformation**. To see that T is linear, observe that

$$T(\mathbf{u} + \mathbf{v}) = \mathbf{0}, \quad T(\mathbf{u}) = \mathbf{0}, \quad T(\mathbf{v}) = \mathbf{0}, \quad \text{and} \quad T(k\mathbf{u}) = \mathbf{0}$$

Therefore,

$$T(\mathbf{u} + \mathbf{v}) = T(\mathbf{u}) + T(\mathbf{v}) \quad \text{and} \quad T(k\mathbf{u}) = kT(\mathbf{u})$$

▶ **EXAMPLE 3** **The Identity Operator**

Let V be any vector space. The mapping $I : V \to V$ defined by $I(\mathbf{v}) = \mathbf{v}$ is called the **identity operator** on V. We will leave it for you to verify that I is linear.

8.1 General Linear Transformations

▶ **EXAMPLE 4** Dilation and Contraction Operators

If V is a vector space and k is any scalar, then the mapping $T : V \to V$ given by $T(\mathbf{x}) = k\mathbf{x}$ is a linear operator on V, for if c is any scalar and if $\mathbf{u}$ and $\mathbf{v}$ are any vectors in V, then

$$T(c\mathbf{u}) = k(c\mathbf{u}) = c(k\mathbf{u}) = cT(\mathbf{u})$$
$$T(\mathbf{u} + \mathbf{v}) = k(\mathbf{u} + \mathbf{v}) = k\mathbf{u} + k\mathbf{v} = T(\mathbf{u}) + T(\mathbf{v})$$

If $0 < k < 1$, then T is called the *contraction* of V with factor k, and if $k > 1$, it is called the *dilation* of V with factor k (Figure 8.1.1).

▶ Figure 8.1.1

Dilation of V Contraction of V

▶ **EXAMPLE 5** A Linear Transformation from P_n to P_{n+1}

Let $\mathbf{p} = p(x) = c_0 + c_1 x + \cdots + c_n x^n$ be a polynomial in P_n, and define the transformation $T : P_n \to P_{n+1}$ by

$$T(\mathbf{p}) = T(p(x)) = xp(x) = c_0 x + c_1 x^2 + \cdots + c_n x^{n+1}$$

This transformation is linear because for any scalar k and any polynomials $\mathbf{p}_1$ and $\mathbf{p}_2$ in P_n we have

$$T(k\mathbf{p}) = T(kp(x)) = x(kp(x)) = k(xp(x)) = kT(\mathbf{p})$$

and

$$T(\mathbf{p}_1 + \mathbf{p}_2) = T(p_1(x) + p_2(x)) = x(p_1(x) + p_2(x))$$
$$= xp_1(x) + xp_2(x) = T(\mathbf{p}_1) + T(\mathbf{p}_2)$$

▶ **EXAMPLE 6** A Linear Transformation Using an Inner Product

Let V be an inner product space, let $\mathbf{v}_0$ be any fixed vector in V, and let $T : V \to R$ be the transformation

$$T(\mathbf{x}) = \langle \mathbf{x}, \mathbf{v}_0 \rangle$$

that maps a vector $\mathbf{x}$ into its inner product with $\mathbf{v}_0$. This transformation is linear, for if k is any scalar, and if $\mathbf{u}$ and $\mathbf{v}$ are any vectors in V, then it follows from properties of inner products that

$$T(k\mathbf{u}) = \langle k\mathbf{u}, \mathbf{v}_0 \rangle = k \langle \mathbf{u}, \mathbf{v}_0 \rangle = kT(\mathbf{u})$$
$$T(\mathbf{u} + \mathbf{v}) = \langle \mathbf{u} + \mathbf{v}, \mathbf{v}_0 \rangle = \langle \mathbf{u}, \mathbf{v}_0 \rangle + \langle \mathbf{v}, \mathbf{v}_0 \rangle = T(\mathbf{u}) + T(\mathbf{v})$$

▶ **EXAMPLE 7** Transformations on Matrix Spaces

Let M_{nn} be the vector space of $n \times n$ matrices. In each part determine whether the transformation is linear.

(a) $T_1(A) = A^T$ (b) $T_2(A) = \det(A)$

Solution (a) It follows from parts (b) and (d) of Theorem 1.4.8 that

$$T_1(kA) = (kA)^T = kA^T = kT_1(A)$$
$$T_1(A + B) = (A + B)^T = A^T + B^T = T_1(A) + T_1(B)$$

so T_1 is linear.

Solution (b) It follows from Formula (1) of Section 2.3 that

$$T_2(kA) = \det(kA) = k^n \det(A) = k^n T_2(A)$$

Thus, T_2 is not homogeneous and hence not linear if $n > 1$. Note that additivity also fails because we showed in Example 1 of Section 2.3 that $\det(A + B)$ and $\det(A) + \det(B)$ are not generally equal.

▶ **EXAMPLE 8** Translation Is Not Linear

Part (*a*) of Theorem 8.1.1 states that a linear transformation maps $\mathbf{0}$ to $\mathbf{0}$. This property is useful for identifying transformations that are *not* linear. For example, if $\mathbf{x}_0$ is a fixed nonzero vector in R^2, then the transformation

$$T(\mathbf{x}) = \mathbf{x} + \mathbf{x}_0$$

has the geometric effect of translating each point $\mathbf{x}$ in a direction parallel to $\mathbf{x}_0$ through a distance of $\|\mathbf{x}_0\|$ (Figure 8.1.2). This cannot be a linear transformation since $T(\mathbf{0}) = \mathbf{x}_0$, so T does not map $\mathbf{0}$ to $\mathbf{0}$.

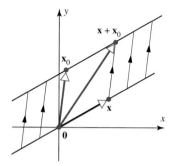

▲ Figure 8.1.2 $T(\mathbf{x}) = \mathbf{x} + \mathbf{x}_0$ translates each point $\mathbf{x}$ along a line parallel to $\mathbf{x}_0$ through a distance $\|\mathbf{x}_0\|$.

▶ **EXAMPLE 9** The Evaluation Transformation

Let V be a subspace of $F(-\infty, \infty)$, let

$$x_1, x_2, \ldots, x_n$$

be distinct real numbers, and let $T : V \to R^n$ be the transformation

$$T(f) = \big(f(x_1), f(x_2), \ldots, f(x_n)\big) \tag{2}$$

that associates with f the n-tuple of function values at $x_1, x_2, \ldots, x_n$. We call this the **evaluation transformation** on V at $x_1, x_2, \ldots, x_n$. Thus, for example, if

$$x_1 = -1, \quad x_2 = 2, \quad x_3 = 4$$

and if $f(x) = x^2 - 1$, then

$$T(f) = \big(f(x_1), f(x_2), f(x_3)\big) = (0, 3, 15)$$

The evaluation transformation in (2) is linear, for if k is any scalar, and if f and g are any functions in V, then

$$\begin{aligned} T(kf) &= \big((kf)(x_1), (kf)(x_2), \ldots, (kf)(x_n)\big) \\ &= \big(kf(x_1), kf(x_2), \ldots, kf(x_n)\big) \\ &= k\big(f(x_1), f(x_2), \ldots, f(x_n)\big) = kT(f) \end{aligned}$$

and

$$\begin{aligned} T(f + g) &= \big((f+g)(x_1), (f+g)(x_2), \ldots, (f+g)(x_n)\big) \\ &= \big(f(x_1) + g(x_1), f(x_2) + g(x_2), \ldots, f(x_n) + g(x_n)\big) \\ &= \big(f(x_1), f(x_2), \ldots, f(x_n)\big) + \big(g(x_1), g(x_2), \ldots, g(x_n)\big) \\ &= T(f) + T(g) \blacktriangleleft \end{aligned}$$

Finding Linear Transformations from Images of Basis Vectors

We saw in Formula (11) of Section 4.10 that if $T: R^n \to R^m$ is a matrix transformation, say multiplication by A, and if $\mathbf{e}_1, \mathbf{e}_2, \ldots, \mathbf{e}_n$ are the standard basis vectors for R^n, then A can be expressed as

$$A = [T(\mathbf{e}_1) \mid T(\mathbf{e}_2) \mid \ldots \mid T(\mathbf{e}_n)]$$

8.1 General Linear Transformations

It follows from this that the image of any vector $\mathbf{v} = (c_1, c_2, \ldots, c_n)$ in R^n under multiplication by A can be expressed as

$$T(\mathbf{v}) = c_1 T(\mathbf{e}_1) + c_2 T(\mathbf{e}_2) + \cdots + c_n T(\mathbf{e}_n)$$

This formula tells us that for a matrix transformation the image of any vector is expressible as a linear combination of the images of the standard basis vectors. This is a special case of the following more general result.

THEOREM 8.1.2 *Let $T : V \to W$ be a linear transformation, where V is finite dimensional. If $S = \{\mathbf{v}_1, \mathbf{v}_2, \ldots, \mathbf{v}_n\}$ is a basis for V, then the image of any vector $\mathbf{v}$ in V can be expressed as*

$$T(\mathbf{v}) = c_1 T(\mathbf{v}_1) + c_2 T(\mathbf{v}_2) + \cdots + c_n T(\mathbf{v}_n) \tag{3}$$

where $c_1, c_2, \ldots, c_n$ are the coefficients required to express $\mathbf{v}$ as a linear combination of the vectors in S.

Proof Express $\mathbf{v}$ as $\mathbf{v} = c_1 \mathbf{v}_1 + c_2 \mathbf{v}_2 + \cdots + c_n \mathbf{v}_n$ and use the linearity of T. ◀

▶ **EXAMPLE 10 Computing with Images of Basis Vectors**

Consider the basis $S = \{\mathbf{v}_1, \mathbf{v}_2, \mathbf{v}_3\}$ for R^3, where

$$\mathbf{v}_1 = (1, 1, 1), \quad \mathbf{v}_2 = (1, 1, 0), \quad \mathbf{v}_3 = (1, 0, 0)$$

Let $T : R^3 \to R^2$ be the linear transformation for which

$$T(\mathbf{v}_1) = (1, 0), \quad T(\mathbf{v}_2) = (2, -1), \quad T(\mathbf{v}_3) = (4, 3)$$

Find a formula for $T(x_1, x_2, x_3)$, and then use that formula to compute $T(2, -3, 5)$.

Solution We first need to express $\mathbf{x} = (x_1, x_2, x_3)$ as a linear combination of $\mathbf{v}_1, \mathbf{v}_2,$ and $\mathbf{v}_3$. If we write

$$(x_1, x_2, x_3) = c_1(1, 1, 1) + c_2(1, 1, 0) + c_3(1, 0, 0)$$

then on equating corresponding components, we obtain

$$\begin{aligned} c_1 + c_2 + c_3 &= x_1 \\ c_1 + c_2 \phantom{{}+ c_3} &= x_2 \\ c_1 \phantom{{}+ c_2 + c_3} &= x_3 \end{aligned}$$

which yields $c_1 = x_3, c_2 = x_2 - x_3, c_3 = x_1 - x_2$, so

$$\begin{aligned}(x_1, x_2, x_3) &= x_3(1, 1, 1) + (x_2 - x_3)(1, 1, 0) + (x_1 - x_2)(1, 0, 0) \\ &= x_3 \mathbf{v}_1 + (x_2 - x_3)\mathbf{v}_2 + (x_1 - x_2)\mathbf{v}_3 \end{aligned}$$

Thus

$$\begin{aligned} T(x_1, x_2, x_3) &= x_3 T(\mathbf{v}_1) + (x_2 - x_3) T(\mathbf{v}_2) + (x_1 - x_2) T(\mathbf{v}_3) \\ &= x_3(1, 0) + (x_2 - x_3)(2, -1) + (x_1 - x_2)(4, 3) \\ &= (4x_1 - 2x_2 - x_3, \, 3x_1 - 4x_2 + x_3) \end{aligned}$$

From this formula, we obtain

$$T(2, -3, 5) = (9, 23)$$

Chapter 8 Linear Transformations

CALCULUS REQUIRED

▶ **EXAMPLE 11** **A Linear Transformation from $C^1(-\infty, \infty)$ to $F(-\infty, \infty)$**

Let $V = C^1(-\infty, \infty)$ be the vector space of functions with continuous first derivatives on $(-\infty, \infty)$, and let $W = F(-\infty, \infty)$ be the vector space of all real-valued functions defined on $(-\infty, \infty)$. Let $D: V \to W$ be the transformation that maps a function $\mathbf{f} = f(x)$ into its derivative—that is,

$$D(\mathbf{f}) = f'(x)$$

From the properties of differentiation, we have

$$D(\mathbf{f} + \mathbf{g}) = D(k\mathbf{f}) = kD(\mathbf{f}) \quad \text{and} \quad D(\mathbf{f}) + D(\mathbf{g})$$

Thus, D is a linear transformation.

CALCULUS REQUIRED

▶ **EXAMPLE 12** **An Integral Transformation**

Let $V = C(-\infty, \infty)$ be the vector space of continuous functions on the interval $(-\infty, \infty)$, let $W = C^1(-\infty, \infty)$ be the vector space of functions with continuous first derivatives on $(-\infty, \infty)$, and let $J: V \to W$ be the transformation that maps a function f in V into

$$J(f) = \int_0^x f(t)\,dt$$

For example, if $f(x) = x^2$, then

$$J(f) = \int_0^x t^2\,dt = \left.\frac{t^3}{3}\right]_0^x = \frac{x^3}{3}$$

The transformation $J: V \to W$ is linear, for if k is any constant, and if f and g are any functions in V, then properties of the integral imply that

$$J(kf) = \int_0^x kf(t)\,dt = k\int_0^x f(t)\,dt = kJ(f)$$

$$J(f + g) = \int_0^x (f(t) + g(t))\,dt = \int_0^x f(t)\,dt + \int_0^x g(t)\,dt = J(f) + J(g) \blacktriangleleft$$

Kernel and Range

Recall that if A is an $m \times n$ matrix, then the null space of A consists of all vectors $\mathbf{x}$ in R^n such that $A\mathbf{x} = \mathbf{0}$, and by Theorem 4.7.1 the column space of A consists of all vectors $\mathbf{b}$ in R^m for which there is at least one vector $\mathbf{x}$ in R^n such that $A\mathbf{x} = \mathbf{b}$. From the viewpoint of matrix transformations, the null space of A consists of all vectors in R^n that multiplication by A maps into $\mathbf{0}$, and the column space of A consists of all vectors in R^m that are images of at least one vector in R^n under multiplication by A. The following definition extends these ideas to general linear transformations.

DEFINITION 2 If $T: V \to W$ is a linear transformation, then the set of vectors in V that T maps into $\mathbf{0}$ is called the **kernel** of T and is denoted by $\ker(T)$. The set of all vectors in W that are images under T of at least one vector in V is called the **range** of T and is denoted by $R(T)$.

▶ **EXAMPLE 13** **Kernel and Range of a Matrix Transformation**

If $T_A: R^n \to R^m$ is multiplication by the $m \times n$ matrix A, then, as discussed above, the kernel of T_A is the null space of A, and the range of T_A is the column space of A.

▶ **EXAMPLE 14** **Kernel and Range of the Zero Transformation**

Let $T: V \to W$ be the zero transformation. Since T maps *every* vector in V into $\mathbf{0}$, it follows that $\ker(T) = V$. Moreover, since $\mathbf{0}$ is the *only* image under T of vectors in V, it follows that $R(T) = \{\mathbf{0}\}$.

▶ **EXAMPLE 15** **Kernel and Range of the Identity Operator**

Let $I: V \to V$ be the identity operator. Since $I(\mathbf{v}) = \mathbf{v}$ for all vectors in V, *every* vector in V is the image of some vector (namely, itself); thus $R(I) = V$. Since the *only* vector that I maps into $\mathbf{0}$ is $\mathbf{0}$, it follows that $\ker(I) = \{\mathbf{0}\}$.

▶ **EXAMPLE 16** **Kernel and Range of an Orthogonal Projection**

As illustrated in Figure 8.1.3a, the points that T maps into $\mathbf{0} = (0, 0, 0)$ are precisely those on the z-axis, so $\ker(T)$ is the set of points of the form $(0, 0, z)$. As illustrated in Figure 8.1.3b, T maps the points in R^3 to the xy-plane, where each point in that plane is the image of each point on the vertical line above it. Thus, $R(T)$ is the set of points of the form $(x, y, 0)$.

▶ **EXAMPLE 17** **Kernel and Range of a Rotation**

Let $T: R^2 \to R^2$ be the linear operator that rotates each vector in the xy-plane through the angle θ (Figure 8.1.4). Since *every* vector in the xy-plane can be obtained by rotating some vector through the angle θ, it follows that $R(T) = R^2$. Moreover, the only vector that rotates into $\mathbf{0}$ is $\mathbf{0}$, so $\ker(T) = \{\mathbf{0}\}$.

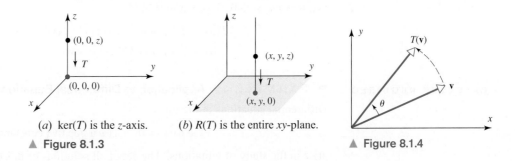

(a) $\ker(T)$ is the z-axis. (b) $R(T)$ is the entire xy-plane.

▲ Figure 8.1.3

▲ Figure 8.1.4

CALCULUS REQUIRED

▶ **EXAMPLE 18** **Kernel of a Differentiation Transformation**

Let $V = C^1(-\infty, \infty)$ be the vector space of functions with continuous first derivatives on $(-\infty, \infty)$, let $W = F(-\infty, \infty)$ be the vector space of all real-valued functions defined on $(-\infty, \infty)$, and let $D: V \to W$ be the differentiation transformation $D(\mathbf{f}) = f'(x)$. The kernel of D is the set of functions in V with derivative zero. From calculus, this is the set of constant functions on $(-\infty, \infty)$. ◀

Properties of Kernel and Range

In all of the preceding examples, $\ker(T)$ and $R(T)$ turned out to be *subspaces*. In Examples 14, 15, and 17 they were either the zero subspace or the entire vector space. In Example 16 the kernel was a line through the origin, and the range was a plane through the origin, both of which are subspaces of R^3. All of this is a consequence of the following general theorem.

THEOREM 8.1.3 *If $T: V \to W$ is a linear transformation, then:*
(a) The kernel of T is a subspace of V.
(b) The range of T is a subspace of W.

Proof (a) To show that $\ker(T)$ is a subspace, we must show that it contains at least one vector and is closed under addition and scalar multiplication. By part (a) of Theorem 8.1.1, the vector $\mathbf{0}$ is in $\ker(T)$, so the kernel contains at least one vector. Let $\mathbf{v}_1$ and $\mathbf{v}_2$ be vectors in $\ker(T)$, and let k be any scalar. Then

$$T(\mathbf{v}_1 + \mathbf{v}_2) = T(\mathbf{v}_1) + T(\mathbf{v}_2) = \mathbf{0} + \mathbf{0} = \mathbf{0}$$

so $\mathbf{v}_1 + \mathbf{v}_2$ is in $\ker(T)$. Also,

$$T(k\mathbf{v}_1) = kT(\mathbf{v}_1) = k\mathbf{0} = \mathbf{0}$$

so $k\mathbf{v}_1$ is in $\ker(T)$.

Proof (b) To show that $R(T)$ is a subspace of W, we must show that it contains at least one vector and is closed under addition and scalar multiplication. However, it contains at least the zero vector of W since $T(\mathbf{0}) = (\mathbf{0})$ by part (a) of Theorem 8.1.1. To prove that it is closed under addition and scalar multiplication, we must show that if $\mathbf{w}_1$ and $\mathbf{w}_2$ are vectors in $R(T)$, and if k is any scalar, then there exist vectors $\mathbf{a}$ and $\mathbf{b}$ in V for which

$$T(\mathbf{a}) = \mathbf{w}_1 + \mathbf{w}_2 \quad \text{and} \quad T(\mathbf{b}) = k\mathbf{w}_1 \tag{4}$$

But the fact $\mathbf{w}_1$ and $\mathbf{w}_2$ are in $R(T)$ tells us that there exist vectors $\mathbf{v}_1$ and $\mathbf{v}_2$ in V such that

$$T(\mathbf{v}_1) = \mathbf{w}_1 \quad \text{and} \quad T(\mathbf{v}_2) = \mathbf{w}_2$$

The following computations complete the proof by showing that the vectors $\mathbf{a} = \mathbf{v}_1 + \mathbf{v}_2$ and $\mathbf{b} = k\mathbf{v}_1$ satisfy the equations in (4):

$$T(\mathbf{a}) = T(\mathbf{v}_1 + \mathbf{v}_2) = T(\mathbf{v}_1) + T(\mathbf{v}_2) = \mathbf{w}_1 + \mathbf{w}_2$$
$$T(\mathbf{b}) = T(k\mathbf{v}_1) = kT(\mathbf{v}_1) = k\mathbf{w}_1 \blacktriangleleft$$

CALCULUS REQUIRED

▶ **EXAMPLE 19** **Application to Differential Equations**

Differential equations of the form

$$y'' + \omega^2 y = 0 \quad (\omega \text{ a positive constant}) \tag{5}$$

arise in the study of vibrations. The set of all solutions of this equation on the interval $(-\infty, \infty)$ is the kernel of the linear transformation $D: C^2(-\infty, \infty) \to C(-\infty, \infty)$, given by

$$D(y) = y'' + \omega^2 y$$

It is proved in standard textbooks on differential equations that the kernel is a two-dimensional subspace of $C^2(-\infty, \infty)$, so that if we can find two linearly independent solutions of (5), then all other solutions can be expressed as linear combinations of those two. We leave it for you to confirm by differentiating that

$$y_1 = \cos \omega x \quad \text{and} \quad y_2 = \sin \omega x$$

are solutions of (5). These functions are linearly independent since neither is a scalar multiple of the other, and thus

$$y = c_1 \cos \omega x + c_2 \sin \omega x \tag{6}$$

is a "general solution" of (5) in the sense that every choice of c_1 and c_2 produces a solution, and every solution is of this form. ◀

8.1 General Linear Transformations

Rank and Nullity of Linear Transformations

In Definition 1 of Section 4.8 we defined the notions of *rank* and *nullity* for an $m \times n$ matrix, and in Theorem 4.8.2, which we called the *Dimension Theorem*, we proved that the sum of the rank and nullity is n. We will show next that this result is a special case of a more general result about linear transformations. We start with the following definition.

> **DEFINITION 3** Let $T: V \to W$ be a linear transformation. If the range of T is finite-dimensional, then its dimension is called the **rank of T**; and if the kernel of T is finite-dimensional, then its dimension is called the **nullity of T**. The rank of T is denoted by $\mathrm{rank}(T)$ and the nullity of T by $\mathrm{nullity}(T)$.

The following theorem, whose proof is optional, generalizes Theorem 4.8.2.

> **THEOREM 8.1.4** *Dimension Theorem for Linear Transformations*
>
> *If $T: V \to W$ is a linear transformation from an n-dimensional vector space V to a vector space W, then*
> $$\mathrm{rank}(T) + \mathrm{nullity}(T) = n \tag{7}$$

In the special case where A is an $m \times n$ matrix and $T_A: R^n \to R^m$ is multiplication by A, the kernel of T_A is the null space of A, and the range of T_A is the column space of A. Thus, it follows from Theorem 8.1.4 that

$$\mathrm{rank}(T_A) + \mathrm{nullity}(T_A) = n$$

OPTIONAL

Proof of Theorem 8.1.4 We must show that

$$\dim(R(T)) + \dim(\ker(T)) = n$$

We will give the proof for the case where $1 \leq \dim(\ker(T)) < n$. The cases where $\dim(\ker(T)) = 0$ and $\dim(\ker(T)) = n$ are left as exercises. Assume $\dim(\ker(T)) = r$, and let $\mathbf{v}_1, \ldots, \mathbf{v}_r$ be a basis for the kernel. Since $\{\mathbf{v}_1, \ldots, \mathbf{v}_r\}$ is linearly independent, Theorem 4.5.5b states that there are $n - r$ vectors, $\mathbf{v}_{r+1}, \ldots, \mathbf{v}_n$, such that the extended set $\{\mathbf{v}_1, \ldots, \mathbf{v}_r, \mathbf{v}_{r+1}, \ldots, \mathbf{v}_n\}$ is a basis for V. To complete the proof, we will show that the $n - r$ vectors in the set $S = \{T(\mathbf{v}_{r+1}), \ldots, T(\mathbf{v}_n)\}$ form a basis for the range of T. It will then follow that

$$\dim(R(T)) + \dim(\ker(T)) = (n - r) + r = n$$

First we show that S spans the range of T. If $\mathbf{b}$ is any vector in the range of T, then $\mathbf{b} = T(\mathbf{v})$ for some vector $\mathbf{v}$ in V. Since $\{\mathbf{v}_1, \ldots, \mathbf{v}_r, \mathbf{v}_{r+1}, \ldots, \mathbf{v}_n\}$ is a basis for V, the vector $\mathbf{v}$ can be written in the form

$$\mathbf{v} = c_1 \mathbf{v}_1 + \cdots + c_r \mathbf{v}_r + c_{r+1} \mathbf{v}_{r+1} + \cdots + c_n \mathbf{v}_n$$

Since $\mathbf{v}_1, \ldots, \mathbf{v}_r$ lie in the kernel of T, we have $T(\mathbf{v}_1) = \cdots = T(\mathbf{v}_r) = \mathbf{0}$, so

$$\mathbf{b} = T(\mathbf{v}) = c_{r+1} T(\mathbf{v}_{r+1}) + \cdots + c_n T(\mathbf{v}_n)$$

Thus S spans the range of T.

Finally, we show that S is a linearly independent set and consequently forms a basis for the range of T. Suppose that some linear combination of the vectors in S is zero; that is,

$$k_{r+1} T(\mathbf{v}_{r+1}) + \cdots + k_n T(\mathbf{v}_n) = \mathbf{0} \tag{8}$$

We must show that $k_{r+1} = \cdots = k_n = 0$. Since T is linear, (8) can be rewritten as

$$T(k_{r+1} \mathbf{v}_{r+1} + \cdots + k_n \mathbf{v}_n) = \mathbf{0}$$

which says that $k_{r+1}\mathbf{v}_{r+1} + \cdots + k_n\mathbf{v}_n$ is in the kernel of T. This vector can therefore be written as a linear combination of the basis vectors $\{\mathbf{v}_1, \ldots, \mathbf{v}_r\}$, say

$$k_{r+1}\mathbf{v}_{r+1} + \cdots + k_n\mathbf{v}_n = k_1\mathbf{v}_1 + \cdots + k_r\mathbf{v}_r$$

Thus,

$$k_1\mathbf{v}_1 + \cdots + k_r\mathbf{v}_r - k_{r+1}\mathbf{v}_{r+1} - \cdots - k_n\mathbf{v}_n = \mathbf{0}$$

Since $\{\mathbf{v}_1, \ldots, \mathbf{v}_n\}$ is linearly independent, all of the k's are zero; in particular, $k_{r+1} = \cdots = k_n = 0$, which completes the proof. ◂

Concept Review

- Linear transformation
- Linear operator
- Zero transformation
- Identity operator
- Contraction
- Dilation
- Evaluation transformation
- Kernel
- Range
- Rank
- Nullity

Skills

- Determine whether a function is a linear transformation.
- Find a formula for a linear transformation $T: V \to W$ given the values of T on a basis for V.
- Find a basis for the kernel of a linear transformation.
- Find a basis for the range of a linear transformation.
- Find the rank of a linear transformation.
- Find the nullity of a linear transformation.

Exercise Set 8.1

1. Use the defintion of a linear operator that was given in this section to show that the function $T: R^2 \to R^2$ given by the formula $T(x_1, x_2) = (4x_1 - x_2, x_1 + 2x_2)$ is a linear operator.

2. Use the defintion of a linear operator given in this section to show that the function $T: R^3 \to R^2$ given by the formula $T(x_1, x_2, x_3) = (x_1 + x_2 - x_3, x_1 - 2x_3)$ is a linear transformation.

▶ In Exercises ??–8, determine whether the function is a linear transformation. Justify your answer. ◂

3. $T: M_{22} \to M_{23}$, where B is a fixed 2×3 matrix and $T(A) = AB$.

4. $T: M_{nn} \to R$, where $T(A) = \text{tr}(A)$.

5. $F: M_{mn} \to M_{nm}$, where $F(A) = A^T$.

6. $T: M_{22} \to R$, where

    (a) $T\left(\begin{bmatrix} a & b \\ c & d \end{bmatrix}\right) = 3a - 4b + c - d$

    (b) $T\left(\begin{bmatrix} a & b \\ c & d \end{bmatrix}\right) = a^2 + b^2$

7. $T: P_2 \to P_2$, where

    (a) $T(a_0 + a_1 x + a_2 x^2) = a_0 + a_1(x + 1) + a_2(x + 1)^2$

    (b) $T(a_0 + a_1 x + a_2 x^2) = (a_0 + 1) + (a_1 + 1)x + (a_2 + 1)x^2$

8. $T: M_{22} \to R$, where

    (a) $T(f(x)) = 1 + f(x)$
    (b) $T(f(x)) = f(x + 1)$

9. Consider the basis $S = \{\mathbf{v}_1, \mathbf{v}_2\}$ for R^2, where $\mathbf{v}_1 = (1, 1)$ and $\mathbf{v}_2 = (1, 0)$, and let $T: R^2 \to R^2$ be the linear operator for which

    $$T(\mathbf{v}_1) = (1, -2) \quad \text{and} \quad T(\mathbf{v}_2) = (-4, 1)$$

    Find a formula for $T(x_1, x_2)$, and use that formula to find $T(5, -3)$.

10. Consider the basis $S = \{\mathbf{v}_1, \mathbf{v}_2\}$ for R^2, where $\mathbf{v}_1 = (-2, 1)$ and $\mathbf{v}_2 = (1, 3)$, and let $T: R^2 \to R^3$ be the linear transformation such that

    $$T(\mathbf{v}_1) = (-1, 2, 0) \quad \text{and} \quad T(\mathbf{v}_2) = (0, -3, 5)$$

    Find a formula for $T(x_1, x_2)$, and use that formula to find $T(2, -3)$.

11. Consider the basis $S = \{\mathbf{v}_1, \mathbf{v}_2\}$ for R^2, where $\mathbf{v}_1 = (1, 0)$ and $\mathbf{v}_2 = (1, 1)$, and let $T: R^2 \to R^2$ be the linear operator such that

    $$T(\mathbf{v}_1) = (-1, 2) \quad \text{and} \quad T(\mathbf{v}_2) = (2, -3)$$

    Find a formula for $T(x_1, x_2)$, and use that formula to find $T(5, -3)$.

12. Consider the basis $S = \{\mathbf{v}_1, \mathbf{v}_2, \mathbf{v}_3\}$ for R^3, where $\mathbf{v}_1 = (1, 2, 1)$, $\mathbf{v}_2 = (2, 9, 0)$, and $\mathbf{v}_3 = (3, 3, 4)$, and let

$T: R^3 \to R^2$ be the linear transformation for which

$$T(\mathbf{v}_1) = (1, 0), \quad T(\mathbf{v}_2) = (-1, 1), \quad T(\mathbf{v}_3) = (0, 1)$$

Find a formula for $T(x_1, x_2, x_3)$, and use that formula to find $T(7, 13, 7)$.

13. Consider the basis $S = \{\mathbf{v}_1, \mathbf{v}_2, \mathbf{v}_3\}$ for R^3, where $\mathbf{v}_1 = (1, 1, 1)$, $\mathbf{v}_2 = (1, 1, 0)$, and $\mathbf{v}_3 = (1, 0, 0)$, and let $T: R^3 \to R^3$ be the linear operator such that

$$T(\mathbf{v}_1) = (-1, 2, 4), \quad T(\mathbf{v}_2) = (0, 3, 2),$$
$$T(\mathbf{v}_3) = (1, 5, -1)$$

Find a formula for $T(x_1, x_2, x_3)$, and use that formula to find $T(2, 4, -1)$.

14. Let $T: R^2 \to R^2$ be the linear operator given by the formula

$$T(x, y) = (2x - y, -8x + 4y)$$

Which of the following vectors are in $R(T)$?

(a) $(1, -4)$ (b) $(5, 0)$ (c) $(-3, 12)$

15. Let $\mathbf{v}_1$, $\mathbf{v}_2$, and $\mathbf{v}_3$ be vectors in a vector space V, and let $T: V \to R^3$ be a linear transformation for which

$$T(\mathbf{v}_1) = (2, -1, 4), \quad T(\mathbf{v}_2) = (-3, 2, 1),$$
$$T(\mathbf{v}_3) = (0, 5, 1)$$

Find $T(3\mathbf{v}_1 - 2\mathbf{v}_2 + \mathbf{v}_3)$.

16. Let $T: R^2 \to R^2$ be the linear operator given by the formula

$$T(x, y) = (x - 3y, -2x + 6y)$$

Which of the following vectors are in $R(T)$?

(a) $(1, -2)$ (b) $(3, 1)$ (c) $(-2, 4)$

17. Let $T: R^2 \to R^2$ be the linear operator in Exercise 16. Which of the following vectors are in $\ker(T)$?

(a) $(1, -3)$ (b) $(3, 1)$ (c) $(-6, -2)$

18. Let $T: P_2 \to P_3$ be the linear transformation defined by $T(p(x)) = xp(x)$. Which of the following are in $\ker(T)$?

(a) x^2 (b) 0 (c) $1 + x$

19. Let $T: P_2 \to P_3$ be the linear transformation in Exercise 18. Which of the following are in $R(T)$?

(a) $x + x^2$ (b) $1 + x$ (c) $3 - x^2$

20. Find a basis for the kernel of
    (a) the linear operator in Exercise 16.
    (b) the linear transformation in Exercise 18.
    (c) the linear transformation in Exercise 19.

21. Find a basis for the range of
    (a) the linear operator in Exercise 16.
    (b) the linear transformation in Exercise 18.
    (c) the linear transformation in Exercise 19.

22. Verify Formula (7) of the dimension theorem for
    (a) the linear operator in Exercise 16.
    (b) the linear transformation in Exercise 18.
    (c) the linear transformation in Exercise 19.

▶ In Exercises 23–26, let T be multiplication by the matrix A. Find

(a) a basis for the range of T.
(b) a basis for the kernel of T.
(c) the rank and nullity of T.
(d) the rank and nullity of A. ◀

23. $A = \begin{bmatrix} 1 & -1 & 3 \\ 5 & 6 & -4 \\ 7 & 4 & 2 \end{bmatrix}$ 24. $A = \begin{bmatrix} 2 & 0 & -1 \\ 4 & 0 & -2 \\ 20 & 0 & 0 \end{bmatrix}$

25. $A = \begin{bmatrix} 1 & 0 & 3 \\ 1 & 2 & 4 \\ 1 & 8 & 25 \end{bmatrix}$

26. $A = \begin{bmatrix} 1 & 4 & 5 & 0 & 9 \\ 3 & -2 & 1 & 0 & -1 \\ -1 & 0 & -1 & 0 & -1 \\ 2 & 3 & 5 & 1 & 8 \end{bmatrix}$

27. Describe the kernel and range of
    (a) the orthogonal projection on the xz-plane.
    (b) the orthogonal projection on the yz-plane.
    (c) the orthogonal projection on the plane defined by the equation $y = x$.

28. Let V be any vector space, and let $T: V \to V$ be defined by $T(\mathbf{v}) = 3\mathbf{v}$.
    (a) What is the kernel of T?
    (b) What is the range of T?

29. In each part, use the given information to find the nullity of the linear transformation T.
    (a) $T: R^5 \to R^7$ has rank 3.
    (b) $T: P_4 \to P_3$ has rank 1.
    (c) The range of $T: R^6 \to R^3$ is R^3.
    (d) $T: M_{22} \to M_{22}$ has rank 3.

30. Let A be a 7×6 matrix such that $A\mathbf{x} = \mathbf{0}$ has only the trivial solution, and let $T: R^6 \to R^7$ be multiplication by A. Find the rank and nullity of T.

31. Let A be a 5×7 matrix with rank 4.
    (a) What is the dimension of the solution space of $A\mathbf{x} = \mathbf{0}$?
    (b) Is $A\mathbf{x} = \mathbf{b}$ consistent for all vectors $\mathbf{b}$ in R^5? Explain.

32. Let $T: R^3 \to W$ be a linear transformation from R^3 to any vector space. Give a geometric description of $\ker(T)$.

33. Let $T: V \to R^3$ be a linear transformation from any vector space to R^3. Give a geometric description of $R(T)$.

34. Let $T: R^3 \to R^3$ be multiplication by
$$\begin{bmatrix} 1 & 3 & 4 \\ 3 & 4 & 7 \\ -2 & 2 & 0 \end{bmatrix}$$

   (a) Show that the kernel of T is a line through the origin, and find parametric equations for it.

   (b) Show that the range of T is a plane through the origin, and find an equation for it.

35. (a) Show that if a_1, a_2, b_1, and b_2 are any scalars, then the formula
$$F(x, y) = (a_1 x + b_1 y, a_2 x + b_2 y)$$
defines a linear operator on R^2.

   (b) Does the formula $F(x, y) = (a_1 x^2 + b_1 y^2, a_2 x^2 + b_2 y^2)$ define a linear operator on R^2? Explain.

36. Let $\{\mathbf{v}_1, \mathbf{v}_2, \ldots, \mathbf{v}_n\}$ be a basis for a vector space V, and let $T: V \to W$ be a linear transformation. Show that if
$$T(\mathbf{v}_1) = T(\mathbf{v}_2) = \cdots = T(\mathbf{v}_n) = \mathbf{0}$$
then T is the zero transformation.

37. Let $\{\mathbf{v}_1, \mathbf{v}_2, \ldots, \mathbf{v}_n\}$ be a basis for a vector space V, and let $T: V \to V$ be a linear operator. Show that if
$$T(\mathbf{v}_1) = \mathbf{v}_1, \quad T(\mathbf{v}_2) = \mathbf{v}_2, \ldots, \quad T(\mathbf{v}_n) = \mathbf{v}_n$$
then T is the identity transformation on V.

38. For a positive integer $n > 1$, let $T: M_{nn} \to R$ be the linear transformation defined by $T(A) = \text{tr}(A)$, where A is an $n \times n$ matrix with real entries. Determine the dimension of $\ker(T)$.

39. Prove: If $\{\mathbf{v}_1, \mathbf{v}_2, \ldots, \mathbf{v}_n\}$ is a basis for V and $\mathbf{w}_1, \mathbf{w}_2, \ldots, \mathbf{w}_n$ are vectors in W, not necessarily distinct, then there exists a linear transformation $T: V \to W$ such that
$$T(\mathbf{v}_1) = \mathbf{w}_1, \quad T(\mathbf{v}_2) = \mathbf{w}_2, \ldots, \quad T(\mathbf{v}_n) = \mathbf{w}_n$$

40. (*Calculus required*) Let $V = C[a, b]$ be the vector space of functions continuous on $[a, b]$, and let $T: V \to V$ be the transformation defined by
$$T(\mathbf{f}) = 5f(x) + 3 \int_a^x f(t)\,dt$$
Is T a linear operator?

41. (*Calculus required*) Let $D: P_3 \to P_2$ be the differentiation transformation $D(\mathbf{p}) = p'(x)$. What is the kernel of D?

42. (*Calculus required*) Let $J: P_1 \to R$ be the integration transformation $J(\mathbf{p}) = \int_{-1}^{1} p(x)\,dx$. What is the kernel of J?

43. (*Calculus required*) Let V be the vector space of real-valued functions with continuous derivatives of all orders on the interval $(-\infty, \infty)$, and let $W = F(-\infty, \infty)$ be the vector space of real-valued functions defined on $(-\infty, \infty)$.

   (a) Find a linear transformation $T: V \to W$ whose kernel is P_3.

   (b) Find a linear transformation $T: V \to W$ whose kernel is P_n.

44. If A is an $m \times n$ matrix, and if the linear system $A\mathbf{x} = \mathbf{b}$ is consistent for every vector $\mathbf{b}$ in R^m, what can you say about the range of $T_A: R^n \to R^m$?

True-False Exercises

In parts (a)–(i) determine whether the statement is true or false, and justify your answer.

(a) If $T(c_1\mathbf{v}_1 + c_2\mathbf{v}_2) = c_1 T(\mathbf{v}_1) + c_2 T(\mathbf{v}_2)$ for all vectors $\mathbf{v}_1$ and $\mathbf{v}_2$ in V and all scalars c_1 and c_2, then T is a linear transformation.

(b) If $\mathbf{v}$ is a nonzero vector in V, then there is exactly one linear transformation $T: V \to W$ such that $T(-\mathbf{v}) = -T(\mathbf{v})$.

(c) There is exactly one linear transformation $T: V \to W$ for which $T(\mathbf{u} + \mathbf{v}) = T(\mathbf{u} - \mathbf{v})$ for all vectors $\mathbf{u}$ and $\mathbf{v}$ in V.

(d) If $\mathbf{v}_0$ is a nonzero vector in V, then the formula $T(\mathbf{v}) = \mathbf{v}_0 + \mathbf{v}$ defines a linear operator on V.

(e) The kernel of a linear transformation is a vector space.

(f) The range of a linear transformation is a vector space.

(g) If $T: P_6 \to M_{22}$ is a linear transformation, then the nullity of T is 3.

(h) The function $T: M_{22} \to R$ defined by $T(A) = \det A$ is a linear transformation.

(i) The linear transformation $T: M_{22} \to M_{22}$ defined by
$$T(A) = \begin{bmatrix} 1 & 3 \\ 2 & 6 \end{bmatrix} A$$
has rank 1.

8.2 Isomorphism

In this section we will establish a fundamental connection between real finite-dimensional vector spaces and the Euclidean space R^n. This connection is not only important theoretically, but it has practical applications in that it allows us to perform vector computations in general vector spaces by working with the vectors in R^n.

One-to-One and Onto Although many of the theorems in this text have been concerned exclusively with the vector space R^n, this is not as limiting as it might seem. As we will show, the vector space R^n is the "mother" of all real n-dimensional vector spaces in the sense that any such space might differ from R^n in the notation used to represent vectors, but not in its algebraic structure. To explain what we mean by this, we will need two definitions, the first of which is a generalization of Definition 1 in Section 4.10. (See Figure 8.2.1).

> **DEFINITION 1** If $T: V \to W$ is a linear transformation from a vector space V to a vector space W, then T is said to be ***one-to-one*** if T maps distinct vectors in V into distinct vectors in W.

> **DEFINITION 2** If $T: V \to W$ is a linear transformation from a vector space V to a vector space W, then T is said to be ***onto*** (or ***onto W***) if every vector in W is the image of at least one vector in V.

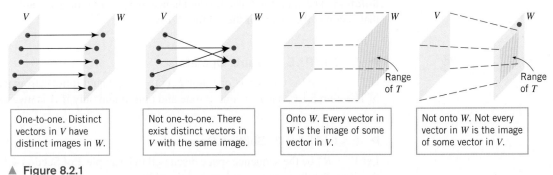

▲ Figure 8.2.1

The following theorem provides a useful way of telling whether a linear transformation is one-to-one by examining its kernel.

> **THEOREM 8.2.1** *If $T: V \to W$ is a linear transformation, then the following statements are equivalent.*
>
> (a) T is one-to-one.
> (b) $\ker(T) = \{\mathbf{0}\}$.

Proof (a) $\Rightarrow$ **(b)** Since T is linear, we know that $T(\mathbf{0}) = \mathbf{0}$ by Theorem 8.1.1a. Since T is one-to-one, there can be no other vectors in V that map into $\mathbf{0}$, so $\ker(T) = \{\mathbf{0}\}$.

(b) $\Rightarrow$ **(a)** Assume that $\ker(T) = \{\mathbf{0}\}$. If $\mathbf{u}$ and $\mathbf{v}$ are distinct vectors in V, then $\mathbf{u} - \mathbf{v} \neq \mathbf{0}$. This implies that $T(\mathbf{u} - \mathbf{v}) \neq \mathbf{0}$, for otherwise $\ker(T)$ would contain a nonzero vector. Since T is linear, it follows that

$$T(\mathbf{u}) - T(\mathbf{v}) = T(\mathbf{u} - \mathbf{v}) \neq \mathbf{0}$$

so T maps distinct vectors in V into distinct vectors in W and hence is one-to-one. ◄

In the special case where V is finite-dimensional and T is a linear operator on V, then we can add a third statement to those in Theorem 8.2.1.

THEOREM 8.2.2 *If V is a finite-dimensional vector space, and if $T : V \to V$ is a linear operator, then the following statements are equivalent.*

(a) *T is one-to-one.*
(b) $\ker(T) = \{\mathbf{0}\}$.
(c) *T is onto* [*i.e., $R(T) = V$*].

Proof We already know that (a) and (b) are equivalent by Theorem 8.2.1, so it suffices to show that (b) and (c) are equivalent. We leave it for you to do this by assuming that $\dim(V) = n$ and applying Theorem 8.1.4. ◄

▶ **EXAMPLE 1** **Dilations and Contractions Are One-to-One and Onto**

Show that if V is a finite-dimensional vector space and c is any *nonzero* scalar, then the linear operator $T : V \to V$ defined by $T(\mathbf{v}) = c\mathbf{v}$ is one-to-one and onto.

Solution The operator T is onto (and hence one-to-one) for if $\mathbf{v}$ is any vector in V then that vector is the image of the vector $(1/c)\mathbf{v}$.

▶ **EXAMPLE 2** **Matrix Operators**

If $T_A : R^n \to R^n$ is the matrix operator $T_A(\mathbf{x}) = A\mathbf{x}$, then it follows from parts (*r*) and (*s*) of Theorem 5.1.6 that T_A is one-to-one and onto if and only if A is invertible.

▶ **EXAMPLE 3** **Shifting Operators**

Let $V = R^\infty$ be the sequence space discussed in Example 3 of Section 4.1, and consider the linear "shifting operators" on V defined by

$$T_1(u_1, u_2, \ldots, u_n, \ldots) = (0, u_1, u_2, \ldots, u_n, \ldots)$$
$$T_2(u_1, u_2, \ldots, u_n, \ldots) = (u_2, u_3, \ldots, u_n, \ldots)$$

(a) Show that T_1 is one-to-one but not onto.
(b) Show that T_2 is onto but not one to one.

Solution (a) The operator T_1 is one-to-one because distinct sequences in R^∞ obviously have distinct images. This operator is not onto because no vector in R^∞ maps into the sequence $(1, 0, 0, \ldots, 0, \ldots)$, for example.

Solution (b) The operator T_2 is not one-to-one because, for example, the vectors $(1, 0, 0, \ldots, 0, \ldots)$ and $(2, 0, 0, \ldots, 0, \ldots)$ both map into $(0, 0, 0, \ldots, 0, \ldots)$. This operator is onto because every possible sequence of real numbers can be obtained with an appropriate choice of the numbers $u_2, u_3, \ldots, u_n, \ldots$.

Why does Example 3 not violate Theorem 8.2.2?

8.2 Isomorphism

▶ **EXAMPLE 4** Basic Transformations That Are One-to-One and Onto

The linear transformations $T_1: P_3 \to R^4$ and $T_2: M_{22} \to R^4$ defined by

$$T_1(a + bx + cx^2 + dx^3) = (a, b, c, d)$$

$$T_2\left(\begin{bmatrix} a & b \\ c & d \end{bmatrix}\right) = (a, b, c, d)$$

are both one-to-one and onto (verify by showing that their kernels contain only the zero vector).

▶ **EXAMPLE 5** A One-to-One Linear Transformation

Let $T: P_n \to P_{n+1}$ be the linear transformation

$$T(\mathbf{p}) = T(p(x)) = xp(x)$$

discussed in Example 5 of Section 8.1. If

$$\mathbf{p} = p(x) = c_0 + c_1 x + \cdots + c_n x^n \quad \text{and} \quad \mathbf{q} = q(x) = d_0 + d_1 x + \cdots + d_n x^n$$

are distinct polynomials, then they differ in at least one coefficient. Thus,

$$T(\mathbf{p}) = c_0 x + c_1 x^2 + \cdots + c_n x^{n+1} \quad \text{and} \quad T(\mathbf{q}) = d_0 x + d_1 x^2 + \cdots + d_n x^{n+1}$$

also differ in at least one coefficient. It follows that T is one-to-one since it maps distinct polynomials $\mathbf{p}$ and $\mathbf{q}$ into distinct polynomials $T(\mathbf{p})$ and $T(\mathbf{q})$.

CALCULUS REQUIRED

▶ **EXAMPLE 6** A Transformation That Is Not One-to-One

Let

$$D: C^1(-\infty, \infty) \to F(-\infty, \infty)$$

be the differentiation transformation discussed in Example 11 of Section 8.1. This linear transformation is *not* one-to-one because it maps functions that differ by a constant into the same function. For example,

$$D(x^2) = D(x^2 + 1) = 2x \quad ◀$$

Dimension and Linear Transformations

In the exercises we will ask you to prove the following two important facts about a linear transformation $T: V \to W$ in the case where V and W are finite-dimensional:

1. If $\dim(W) < \dim(V)$, then T cannot be one-to-one.
2. If $\dim(V) < \dim(W)$, then T cannot be onto.

Stated informally, if a linear transformation maps a "bigger" space to a "smaller" space, then some points in the "bigger" space must have the same image; and if a linear transformation maps a "smaller" space to a "bigger" space, then there must be points in the "bigger" space that are not images of any points in the "smaller" space.

Remark These observations tell us, for example, that any linear transformation from R^3 to R^2 must map some distinct points of R^3 into the *same* point in R^2, and it also tells us that there is no linear transformation that maps R^2 onto *all* of R^3.

Isomorphism

Our next definition paves the way for the main result in this section.

DEFINITION 3 If a linear transformation $T: V \to W$ is both one-to-one and onto, then T is said to be an ***isomorphism***, and the vector spaces V and W are said to be ***isomorphic***.

The word *isomorphic* is derived from the Greek words *iso*, meaning "identical," and *morphe*, meaning "form." This terminology is appropriate because, as we will now explain, isomorphic vector spaces have the same "algebraic form," even though they may consist of different kinds of objects. To illustrate this idea, examine Table 1 in which we have shown how the isomorphism

$$a_0 + a_1 x + a_2 x^2 \xrightarrow{T} (a_0, a_1, a_2)$$

matches up vector operations in P_2 and R^3.

Table 1

Operation in P_2	Operation in R^3
$3(1 - 2x + 3x^2) = 3 - 6x + 9x^2$	$3(1, -2, 3) = (3, -6, 9)$
$(2 + x - x^2) + (1 - x + 5x^2) = 3 + 4x^2$	$(2, 1, -1) + (1, -1, 5) = (3, 0, 4)$
$(4 + 2x + 3x^2) - (2 - 4x + 3x^2) = 2 + 6x$	$(4, 2, 3) - (2, -4, 3) = (2, 6, 0)$

The following theorem, which is one of the most important results in linear algebra, reveals the fundamental importance of the vector space R^n.

THEOREM 8.2.3 *Every real n-dimensional vector space is isomorphic to R^n.*

Theorem 8.2.3 tells us that a real *n*-dimensional vector space may differ from R^n in notation, but its algebraic structure will be the same.

Proof Let V be a real *n*-dimensional vector space. To prove that V is isomorphic to R^n we must find a linear transformation $T : V \rightarrow R^n$ that is one-to-one and onto. For this purpose, let

$$\mathbf{v}_1, \mathbf{v}_2, \ldots, \mathbf{v}_n$$

be any basis for V, let

$$\mathbf{u} = k_1 \mathbf{v}_1 + k_2 \mathbf{v}_2 + \cdots + k_n \mathbf{v}_n \quad (1)$$

be the representation of a vector $\mathbf{u}$ in V as a linear combination of the basis vectors, and define the transformation $T : V \rightarrow R^n$ by

$$T(\mathbf{u}) = (k_1, k_2, \ldots, k_n) \quad (2)$$

We will show that T is an isomorphism (linear, one-to-one, and onto). To prove the linearity, let $\mathbf{u}$ and $\mathbf{v}$ be vectors in V, let c be a scalar, and let

$$\mathbf{u} = k_1 \mathbf{v}_1 + k_2 \mathbf{v}_2 + \cdots + k_n \mathbf{v}_n \quad \text{and} \quad \mathbf{v} = d_1 \mathbf{v}_1 + d_2 \mathbf{v}_2 + \cdots + d_n \mathbf{v}_n \quad (3)$$

be the representations of $\mathbf{u}$ and $\mathbf{v}$ as linear combinations of the basis vectors. Then it follows from (1) that

$$T(c\mathbf{u}) = T(ck_1 \mathbf{v}_1 + ck_2 \mathbf{v}_2 + \cdots + ck_n \mathbf{v}_n)$$
$$= (ck_1, ck_2, \ldots, ck_n)$$
$$= c(k_1, k_2, \ldots, k_n) = cT(\mathbf{u})$$

and it follows from (2) that

$$T(\mathbf{u} + \mathbf{v}) = T\big((k_1 + d_1)\mathbf{v}_1 + (k_2 + d_2)\mathbf{v}_2 + \cdots + (k_n + d_n)\mathbf{v}_n\big)$$
$$= (k_1 + d_1, k_2 + d_2, \ldots, k_n + d_n)$$
$$= (k_1, k_2, \ldots, k_n) + (d_1, d_2, \ldots, d_n)$$
$$= T(\mathbf{u}) + T(\mathbf{v})$$

which shows that T is linear. To show that T is one-to-one, we must show that if $\mathbf{u}$ and $\mathbf{v}$ are distinct vectors in V, then so are their images in R^n. But if $\mathbf{u} \neq \mathbf{v}$, and if the representations of these vectors in terms of the basis vectors are as in (3), then we must have $k_i \neq d_i$ for at least one i. Thus,

$$T(\mathbf{u}) = (k_1, k_2, \ldots, k_n) \neq (d_1, d_2, \ldots, d_n) = T(\mathbf{v})$$

which shows that $\mathbf{u}$ and $\mathbf{v}$ have distinct images under T. Finally, the transformation T is onto, for if

$$\mathbf{w} = (k_1, k_2, \ldots, k_n)$$

is any vector in R^n, then it follows from (2) that $\mathbf{w}$ is the image under T of the vector

$$\mathbf{u} = k_1 \mathbf{v}_1 + k_2 \mathbf{v}_2 + \cdots + k_n \mathbf{v}_n \blacktriangleleft$$

Remark Note that the isomorphism T in Formula (2) of the foregoing proof is the coordinate map

$$\mathbf{u} \xrightarrow{T} (k_1, k_2, \ldots, k_n) = (\mathbf{u})_S$$

that maps $\mathbf{u}$ into its coordinate vector with respect to the basis $S = \{\mathbf{v}_1, \mathbf{v}_2, \ldots, \mathbf{v}_n\}$. Since there are generally many possible bases for a given vector space V, there are generally many possible isomorphisms between V and R^n, one for each different basis.

▶ **EXAMPLE 7** **The Natural Isomorphism from P_{n-1} to R^n**

We leave it for you to verify that the mapping

$$a_0 + a_1 x + \cdots + a_{n-1} x^{n-1} \xrightarrow{T} (a_0, a_1, \ldots, a_{n-1})$$

from P_{n-1} to R^n is one-to-one, onto, and linear. This is called the ***natural isomorphism*** from P_{n-1} to R^n because, as the following computations show, it maps the natural basis $\{1, x, x^2, \ldots, x^{n-1}\}$ for P_{n-1} into the standard basis for R^n:

$$
\begin{aligned}
1 &= 1 + 0x + 0x^2 + \cdots + 0x^{n-1} &\xrightarrow{T}& \quad (1, 0, 0, \ldots, 0) \\
x &= 0 + x + 0x^2 + \cdots + 0x^{n-1} &\xrightarrow{T}& \quad (0, 1, 0, \ldots, 0) \\
&\vdots & \vdots & \quad \vdots \\
x^{n-1} &= 0 + 0x + 0x^2 + \cdots + x^{n-1} &\xrightarrow{T}& \quad (0, 0, 0, \ldots, 1)
\end{aligned}
$$

▶ **EXAMPLE 8** **The Natural Isomorphism from M_{22} to R^4**

The matrices

$$E_1 = \begin{bmatrix} 1 & 0 \\ 0 & 0 \end{bmatrix}, \quad E_2 = \begin{bmatrix} 0 & 1 \\ 0 & 0 \end{bmatrix}, \quad E_3 = \begin{bmatrix} 0 & 0 \\ 1 & 0 \end{bmatrix}, \quad E_4 = \begin{bmatrix} 0 & 0 \\ 0 & 1 \end{bmatrix}$$

form a basis for the vector space M_{22} of 2×2 matrices. An isomorphism $T : M_{22} \to R^4$ can be constructed by first writing a matrix A in M_{22} in terms of the basis vectors as

$$A = \begin{bmatrix} a_1 & a_2 \\ a_3 & a_4 \end{bmatrix} = a_1 \begin{bmatrix} 1 & 0 \\ 0 & 0 \end{bmatrix} + a_2 \begin{bmatrix} 0 & 1 \\ 0 & 0 \end{bmatrix} + a_3 \begin{bmatrix} 0 & 0 \\ 1 & 0 \end{bmatrix} + a_4 \begin{bmatrix} 0 & 0 \\ 0 & 1 \end{bmatrix}$$

and then defining T as

$$T(A) = (a_1, a_2, a_3, a_4)$$

Thus, for example,

$$\begin{bmatrix} 1 & -3 \\ 4 & 6 \end{bmatrix} \xrightarrow{T} (1, -3, 4, 6)$$

More generally, this idea can be used to show that the vector space M_{mn} of $m \times n$ matrices with real entries is isomorphic to R^{mn}.

CALCULUS REQUIRED

▶ **EXAMPLE 9** **Differentiation by Matrix Multiplication**

Consider the differentiation transformation $D: P_3 \to P_2$ on the vector space of polynomials of degree three or less. If we map P_3 and P_2 into R^4 and R^3, respectively, by the natural isomorphisms, then the transformation D produces a corresponding matrix transformation from R^4 to R^3. Specifically, the derivative transformation

$$a_0 + a_1 x + a_2 x^2 + a_3 x^3 \xrightarrow{D} a_1 + 2a_2 x + 3a_3 x^2$$

produces the matrix transformation

$$\begin{bmatrix} 0 & 1 & 0 & 0 \\ 0 & 0 & 2 & 0 \\ 0 & 0 & 0 & 3 \end{bmatrix} \begin{bmatrix} a_0 \\ a_1 \\ a_2 \\ a_3 \end{bmatrix} = \begin{bmatrix} a_1 \\ 2a_2 \\ 3a_3 \end{bmatrix}$$

Thus, for example, the derivative

$$\frac{d}{dx}(2 + x + 4x^2 - x^3) = 1 + 8x - 3x^2$$

can be calculated as the matrix product

$$\begin{bmatrix} 0 & 1 & 0 & 0 \\ 0 & 0 & 2 & 0 \\ 0 & 0 & 0 & 3 \end{bmatrix} \begin{bmatrix} 2 \\ 1 \\ 4 \\ -1 \end{bmatrix} = \begin{bmatrix} 1 \\ 8 \\ -3 \end{bmatrix}$$

This idea is useful for constructing numerical algorithms to perform derivative calculations. ◀

Inner Product Space Isomorphisms

In the case where V is a real n-dimensional inner product space, both V and R^n have, in addition to their algebraic structure, a geometric structure arising from their respective inner products. Thus, it is reasonable to inquire if there exists an isomorphism from V to R^n that preserves the geometric structure as well as the algebraic structure. For example, we would want orthogonal vectors in V to have orthogonal counterparts in R^n, and we would want orthonormal sets in V to correspond to orthonormal sets in R^n.

In order for an isomorphism to preserve geometric structure, it obviously has to preserve inner products, since notions of length, angle, and orthogonality are all based on the inner product. Thus, if V and W are inner product spaces, then we call an isomorphism $T: V \to W$ an ***inner product space isomorphism*** if

$$\langle T(\mathbf{u}), T(\mathbf{v}) \rangle = \langle \mathbf{u}, \mathbf{v} \rangle$$

It can be proved that if V is any real n-dimensional inner product space and R^n has the Euclidean inner product (the dot product), then there exists an inner product space isomorphism from V to R^n. Under such an isomorphism, the inner product space V has the same algebraic and geometric structure as R^n. In this sense, every n-dimensional inner product space is a "carbon copy" of R^n with the Euclidean inner product that differs only in the notation used to represent vectors.

▶ **EXAMPLE 10** **An Inner Product Space Isomorphism**

Let R^n be the vector space of real n-tuples in comma-delimited form, let M_n be the vector space of real $n \times 1$ matrices, let R^n have the Euclidean inner product $\langle \mathbf{u}, \mathbf{v} \rangle = \mathbf{u} \cdot \mathbf{v}$, and let M_n have the inner product $\langle \mathbf{u}, \mathbf{v} \rangle = \mathbf{u}^T\mathbf{v}$ in which $\mathbf{u}$ and $\mathbf{v}$ are expressed in column form. The mapping $T: R^n \to M_n$ defined by

$$(v_1, v_2, \ldots, v_n) \xrightarrow{T} \begin{bmatrix} v_1 \\ v_2 \\ \vdots \\ v_n \end{bmatrix}$$

is an inner product space isomorphism, so the distinction between the inner product space R^n and the inner product space M_n is essentially notational, a fact that we have used many times in this text. ◀

Concept Review

- One-to-one
- Onto
- Isomorphism
- Isomorphic vector spaces
- Natural isomorphism
- Inner product space isomorphism

Skills

- Determine whether a linear transformation is one-to-one.
- Determine whether a linear transformation is onto.
- Determine whether a linear transformation is an isomorphism.

Exercise Set 8.2

1. In each part, find $\ker(T)$, and determine whether the linear transformation T is one-to-one.
   (a) $T: R^2 \to R^2$, where $T(x, y) = (2y, 3x)$
   (b) $T: R^2 \to R^2$, where $T(x, y) = (5x - y, 0)$
   (c) $T: R^3 \to R^2$, where $T(x, y, z) = (x + y, x - z)$
   (d) $T: R^2 \to R^3$, where $T(x, y) = (y, x, x - y)$
   (e) $T: R^2 \to R^3$, where $T(x, y) = (x - y, y - x, 2x - 2y)$
   (f) $T: R^3 \to R^2$, where $T(x, y, z) = (x + y + z, x - y - z)$

2. Which of the transformations in Exercise 1 are onto?

3. In each part, determine whether multiplication by A is a one-to-one linear transformation.
   (a) $A = \begin{bmatrix} 1 & -3 & 2 \\ -2 & 6 & -4 \end{bmatrix}$
   (b) $A = \begin{bmatrix} 1 & 3 & 5 & 7 \\ -1 & 3 & 0 & 0 \\ 1 & 15 & 15 & 21 \end{bmatrix}$
   (c) $A = \begin{bmatrix} 2 & 5 \\ -1 & 3 \\ 2 & 4 \end{bmatrix}$

4. Which of the transformations in Exercise 3 are onto?

5. As indicated in the accompanying figure, let $T: R^2 \to R^2$ be the orthogonal projection on the line $y = x$.
   (a) Find the kernel of T.
   (b) Is T one-to-one? Justify your conclusion.

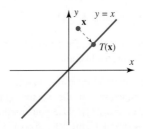

◀ Figure Ex-5

6. As indicated in the accompanying figure, let $T: R^2 \to R^2$ be the linear operator that reflects each point about the y-axis.
   (a) Find the kernel of T.
   (b) Is T one-to-one? Justify your conclusion.

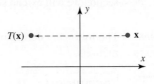

◀ Figure Ex-6

7. In each part, use the given information to determine whether the linear transformation T is one-to-one.
   (a) $T: R^m \to R^m$; nullity$(T) = 1$
   (b) $T: R^n \to R^n$; rank$(T) = n$
   (c) $T: R^m \to R^n$; $n < m$
   (d) $T: R^n \to R^n$; $R(T) = R^n$

8. In each part, determine whether the linear transformation T is one-to-one.

   (a) $T: P_2 \to P_3$, where $T(a_0 + a_1 x + a_2 x^2) = x(a_0 + a_1 x + a_2 x^2)$

   (b) $T: P_2 \to P_2$, where $T(p(x)) = p(x+1)$

9. Prove: If V and W are finite-dimensional vector spaces such that $\dim(W) < \dim(V)$, then there is no one-to-one linear transformation $T: V \to W$.

10. Prove: There can be an onto linear transformation from V to W only if $\dim(V) \geq \dim(W)$.

11. (a) Find an isomorphism between the vector space of all 3×3 symmetric matrices and R^6.

    (b) Find two different isomorphisms between the vector space of all 2×2 matrices and R^4.

    (c) Find an isomorphism between the vector space of all polynomials of degree at most 3 such that $p(0) = 0$ and R^3.

    (d) Find an isomorphism between the vector spaces $\text{span}\{1, \sin(x), \cos(x)\}$ and R^3.

12. (*Calculus required*) Let $J: P_1 \to R$ be the integration transformation $J(\mathbf{p}) = \int_{-1}^{1} p(x)\,dx$. Determine whether J is one-to-one. Justify your conclusion.

13. (*Calculus required*) Let V be the vector space $C^1[0,1]$ and let $T: V \to R$ be defined by

    $$T(\mathbf{f}) = f(0) + 2f'(0) + 3f'(1)$$

    Verify that T is a linear transformation. Determine whether T is one-to-one, and justify your conclusion.

14. (*Calculus required*) Devise a method for using matrix multiplication to differentiate functions in the vector space $\text{span}\{1, \sin(x), \cos(x), \sin(2x), \cos(2x)\}$. Use your method to find the derivative of $3 - 4\sin(x) + \sin(2x) + 5\cos(2x)$.

15. Does the formula $T(a, b, c) = ax^2 + bx + c$ define a one-to-one linear transformation from R^3 to P_2? Explain your reasoning.

16. Let E be a fixed 2×2 elementary matrix. Does the formula $T(A) = EA$ define a one-to-one linear operator on M_{22}? Explain your reasoning.

17. Let $\mathbf{a}$ be a fixed vector in R^3. Does the formula $T(\mathbf{v}) = \mathbf{a} \times \mathbf{v}$ define a one-to-one linear operator on R^3? Explain your reasoning.

18. Prove that an inner product space isomorphism preserves angles and distances—that is, the angle between $\mathbf{u}$ and $\mathbf{v}$ in V is equal to the angle between $T(\mathbf{u})$ and $T(\mathbf{v})$ in W, and $\|\mathbf{u} - \mathbf{v}\|_V = \|T(\mathbf{u}) - T(\mathbf{v})\|_W$.

19. Does an inner product space isomorphism map orthonormal sets to orthonormal sets? Justify your answer.

20. Find an inner product space isomorphism between P_5 and M_{23}.

True-False Exercises

In parts (a)–(f) determine whether the statement is true or false, and justify your answer.

(a) The vector spaces R^2 and P_2 are isomorphic.

(b) If the kernel of a linear transformation $T: P_3 \to P_3$ is $\{\mathbf{0}\}$, then T is an isomorphism.

(c) Every linear transformation from M_{33} to P_9 is an isomorphism.

(d) There is a subspace of M_{23} that is isomorphic to R^4.

(e) There is a 2×2 matrix P such that $T: M_{22} \to M_{22}$ defined by $T(A) = AP - PA$ is an isomorphism.

(f) There is a linear transformation $T: P_4 \to P_4$ such that the kernel of T is isomorphic to the range of T.

8.3 Compositions and Inverse Transformations

In Section 4.10 we discussed compositions and inverses of matrix transformations. In this section we will extend some of those ideas to general linear transformations.

Composition of Linear Transformations

The following definition extends Formula (1) of Section 4.10 to general linear transformations.

Note that the word "with" establishes the order of the operations in a composition. The composition of T_2 with T_1 is

$(T_2 \circ T_1)(\mathbf{u}) = T_2(T_1(\mathbf{u}))$

whereas the composition of T_1 with T_2 is

$(T_1 \circ T_2)(\mathbf{u}) = T_1(T_2(\mathbf{u}))$

DEFINITION 1 If $T_1: U \to V$ and $T_2: V \to W$ are linear transformations, then the **composition of T_2 with T_1**, denoted by $T_2 \circ T_1$ (which is read "T_2 circle T_1"), is the function defined by the formula

$$(T_2 \circ T_1)(\mathbf{u}) = T_2(T_1(\mathbf{u})) \tag{1}$$

where $\mathbf{u}$ is a vector in U.

8.3 Compositions and Inverse Transformations

Remark Observe that this definition requires that the domain of T_2 (which is V) contain the range of T_1. This is essential for the formula $T_2(T_1(\mathbf{u}))$ to make sense (Figure 8.3.1).

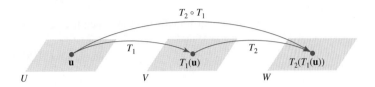

▶ **Figure 8.3.1** The composition of T_2 with T_1.

Our first theorem shows that the composition of two linear transformations is itself a linear transformation.

THEOREM 8.3.1 *If $T_1: U \to V$ and $T_2: V \to W$ are linear transformations, then $(T_2 \circ T_1): U \to W$ is also a linear transformation.*

Proof If $\mathbf{u}$ and $\mathbf{v}$ are vectors in U and c is a scalar, then it follows from (1) and the linearity of T_1 and T_2 that

$$(T_2 \circ T_1)(\mathbf{u} + \mathbf{v}) = T_2(T_1(\mathbf{u}+\mathbf{v})) = T_2(T_1(\mathbf{u}) + T_1(\mathbf{v}))$$
$$= T_2(T_1(\mathbf{u})) + T_2(T_1(\mathbf{v}))$$
$$= (T_2 \circ T_1)(\mathbf{u}) + (T_2 \circ T_1)(\mathbf{v})$$

and

$$(T_2 \circ T_1)(c\mathbf{u}) = T_2(T_1(c\mathbf{u})) = T_2(cT_1(\mathbf{u}))$$
$$= cT_2(T_1(\mathbf{u})) = c(T_2 \circ T_1)(\mathbf{u})$$

Thus, $T_2 \circ T_1$ satisfies the two requirements of a linear transformation. ◂

▶ **EXAMPLE 1** **Composition of Linear Transformations**

Let $T_1: P_1 \to P_2$ and $T_2: P_2 \to P_2$ be the linear transformations given by the formulas

$$T_1(p(x)) = xp(x) \quad \text{and} \quad T_2(p(x)) = p(2x+4)$$

Then the composition $(T_2 \circ T_1): P_1 \to P_2$ is given by the formula

$$(T_2 \circ T_1)(p(x)) = T_2(T_1(p(x))) = T_2(xp(x)) = (2x+4)p(2x+4)$$

In particular, if $p(x) = c_0 + c_1 x$, then

$$(T_2 \circ T_1)(p(x)) = (T_2 \circ T_1)(c_0 + c_1 x) = (2x+4)(c_0 + c_1(2x+4))$$
$$= c_0(2x+4) + c_1(2x+4)^2$$

▶ **EXAMPLE 2** **Composition with the Identity Operator**

If $T: V \to V$ is any linear operator, and if $I: V \to V$ is the identity operator (Example ?? of Section 8.1), then for all vectors $\mathbf{v}$ in V, we have

$$(T \circ I)(\mathbf{v}) = T(I(\mathbf{v})) = T(\mathbf{v})$$
$$(I \circ T)(\mathbf{v}) = I(T(\mathbf{v})) = T(\mathbf{v})$$

It follows that $T \circ I$ and $I \circ T$ are the same as T; that is,

$$T \circ I = T \quad \text{and} \quad I \circ T = T \quad \blacktriangleleft \qquad (2)$$

As illustrated in Figure 8.3.2, compositions can be defined for more than two linear transformations. For example, if

$$T_1: U \to V, \quad T_2: V \to W, \quad \text{and} \quad T_3: W \to Y$$

are linear transformations, then the composition $T_3 \circ T_2 \circ T_1$ is defined by

$$(T_3 \circ T_2 \circ T_1)(\mathbf{u}) = T_3(T_2(T_1(\mathbf{u}))) \tag{3}$$

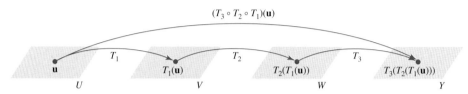

▲ Figure 8.3.2 The composition of three linear transformations.

Inverse Linear Transformations

In Theorem 4.10.1 we showed that a matrix operator $T_A: R^n \to R^n$ is one-to-one if and only if the matrix A is invertible, in which case the inverse operator is $T_{A^{-1}}$. We then showed that if $\mathbf{w}$ is the image of a vector $\mathbf{x}$ under the operator T_A, then $\mathbf{x}$ is the image under $T_{A^{-1}}$ of the vector $\mathbf{w}$ (see Figure 4.10.8). Our next objective is to extend the notion of invertibility to general linear transformations.

Recall that if $T: V \to W$ is a linear transformation, then the range of T, denoted by $R(T)$, is the subspace of W consisting of all images under T of vectors in V. If T is one-to-one, then each vector $\mathbf{w}$ in $R(T)$ is the image of a unique vector $\mathbf{v}$ in V. This uniqueness allows us to define a new function, called the *inverse of* T and denoted by T^{-1}, that maps $\mathbf{w}$ back into $\mathbf{v}$ (Figure 8.3.3).

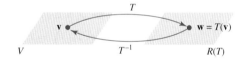

▶ Figure 8.3.3 The inverse of T maps $T(\mathbf{v})$ back into $\mathbf{v}$.

It can be proved (Exercise 19) that $T^{-1}: R(T) \to V$ is a linear transformation. Moreover, it follows from the definition of T^{-1} that

$$T^{-1}(T(\mathbf{v})) = T^{-1}(\mathbf{w}) = \mathbf{v} \tag{4}$$

$$T(T^{-1}(\mathbf{w})) = T(\mathbf{v}) = \mathbf{w} \tag{5}$$

so that T and T^{-1}, when applied in succession in either order, cancel the effect of each other.

Remark It is important to note that if $T: V \to W$ is a one-to-one linear transformation, then the domain of T^{-1} is the *range* of T, where the range may or may not be all of W. However, in the special case where $T: V \to V$ is a one-to-one *linear operator* and V is n-dimensional, then it follows from Theorem 8.2.2 that T must also be onto, so the domain of T^{-1} is *all* of V.

8.3 Compositions and Inverse Transformations

▶ **EXAMPLE 3** An Inverse Transformation

In Example 5 of Section 8.2 we showed that the linear transformation $T: P_n \to P_{n+1}$ given by
$$T(\mathbf{p}) = T(p(x)) = xp(x)$$
is one-to-one; thus, T has an inverse. In this case the range of T is *not* all of P_{n+1} but rather the subspace of P_{n+1} consisting of polynomials with a zero constant term. This is evident from the formula for T:
$$T(c_0 + c_1 x + \cdots + c_n x^n) = c_0 x + c_1 x^2 + \cdots + c_n x^{n+1}$$
It follows that $T^{-1}: R(T) \to P_n$ is given by the formula
$$T^{-1}(c_0 x + c_1 x^2 + \cdots + c_n x^{n+1}) = c_0 + c_1 x + \cdots + c_n x^n$$
For example, in the case where $n \geq 3$,
$$T^{-1}(2x - x^2 + 5x^3 + 3x^4) = 2 - x + 5x^2 + 3x^3$$

▶ **EXAMPLE 4** An Inverse Transformation

Let $T: R^3 \to R^3$ be the linear operator defined by the formula
$$T(x_1, x_2, x_3) = (3x_1 + x_2, -2x_1 - 4x_2 + 3x_3, 5x_1 + 4x_2 - 2x_3)$$
Determine whether T is one-to-one; if so, find $T^{-1}(x_1, x_2, x_3)$.

Solution It follows from Formula (12) of Section 4.9 that the standard matrix for T is
$$[T] = \begin{bmatrix} 3 & 1 & 0 \\ -2 & -4 & 3 \\ 5 & 4 & -2 \end{bmatrix}$$
(verify). This matrix is invertible, and from Formula (7) of Section 4.10 the standard matrix for T^{-1} is
$$[T^{-1}] = [T]^{-1} = \begin{bmatrix} 4 & -2 & -3 \\ -11 & 6 & 9 \\ -12 & 7 & 10 \end{bmatrix}$$
It follows that
$$T^{-1}\left(\begin{bmatrix} x_1 \\ x_2 \\ x_3 \end{bmatrix}\right) = [T^{-1}]\begin{bmatrix} x_1 \\ x_2 \\ x_3 \end{bmatrix} = \begin{bmatrix} 4 & -2 & -3 \\ -11 & 6 & 9 \\ -12 & 7 & 10 \end{bmatrix}\begin{bmatrix} x_1 \\ x_2 \\ x_3 \end{bmatrix} = \begin{bmatrix} 4x_1 - 2x_2 - 3x_3 \\ -11x_1 + 6x_2 + 9x_3 \\ -12x_1 + 7x_2 + 10x_3 \end{bmatrix}$$
Expressing this result in horizontal notation yields
$$T^{-1}(x_1, x_2, x_3) = (4x_1 - 2x_2 - 3x_3, -11x_1 + 6x_2 + 9x_3, -12x_1 + 7x_2 + 10x_3) \blacktriangleleft$$

Composition of One-To-One Linear Transformations

The following theorem shows that a composition of one-to-one linear transformations is one-to-one, and it relates the inverse of a composition to the inverses of its individual linear transformations.

THEOREM 8.3.2 *If $T_1: U \to V$ and $T_2: V \to W$ are one-to-one linear transformations, then*

(a) $T_2 \circ T_1$ *is one-to-one.*
(b) $(T_2 \circ T_1)^{-1} = T_1^{-1} \circ T_2^{-1}$.

Chapter 8 Linear Transformations

Proof (a) We want to show that $T_2 \circ T_1$ maps distinct vectors in U into distinct vectors in W. But if $\mathbf{u}$ and $\mathbf{v}$ are distinct vectors in U, then $T_1(\mathbf{u})$ and $T_1(\mathbf{v})$ are distinct vectors in V since T_1 is one-to-one. This and the fact that T_2 is one-to-one imply that

$$T_2(T_1(\mathbf{u})) \quad \text{and} \quad T_2(T_1(\mathbf{v}))$$

are also distinct vectors. But these expressions can also be written as

$$(T_2 \circ T_1)(\mathbf{u}) \quad \text{and} \quad (T_2 \circ T_1)(\mathbf{v})$$

so $T_2 \circ T_1$ maps $\mathbf{u}$ and $\mathbf{v}$ into distinct vectors in W.

Proof (b) We want to show that

$$(T_2 \circ T_1)^{-1}(\mathbf{w}) = (T_1^{-1} \circ T_2^{-1})(\mathbf{w})$$

for every vector $\mathbf{w}$ in the range of $T_2 \circ T_1$. For this purpose, let

$$\mathbf{u} = (T_2 \circ T_1)^{-1}(\mathbf{w}) \tag{6}$$

so our goal is to show that

$$\mathbf{u} = (T_1^{-1} \circ T_2^{-1})(\mathbf{w})$$

But it follows from (6) that

$$(T_2 \circ T_1)(\mathbf{u}) = \mathbf{w}$$

or, equivalently,

$$T_2(T_1(\mathbf{u})) = \mathbf{w}$$

Now, taking T_2^{-1} of each side of this equation, then taking T_1^{-1} of each side of the result, and then using (4) yields (verify)

$$\mathbf{u} = T_1^{-1}(T_2^{-1}(\mathbf{w}))$$

or, equivalently,

$$\mathbf{u} = (T_1^{-1} \circ T_2^{-1})(\mathbf{w}) \blacktriangleleft$$

In words, part (b) of Theorem 8.3.2 states that *the inverse of a composition is the composition of the inverses in the reverse order.* This result can be extended to compositions of three or more linear transformations; for example,

$$(T_3 \circ T_2 \circ T_1)^{-1} = T_1^{-1} \circ T_2^{-1} \circ T_3^{-1} \tag{7}$$

In the case where T_A, T_B, and T_C are matrix operators on R^n, Formula (7) can be written as

$$(T_C \circ T_B \circ T_A)^{-1} = T_A^{-1} \circ T_B^{-1} \circ T_C^{-1}$$

or alternatively as

$$(T_{CBA})^{-1} = T_{A^{-1}B^{-1}C^{-1}} \tag{8}$$

> Note the order of the subscripts on the two sides of Formula (8).

Concept Review

- Composition of linear transformations
- Inverse of a linear transformation

Skills

- Find the domain and range of the composition of two linear transformations.
- Find the composition of two linear transformations.
- Determine whether a linear transformation has an inverse.
- Find the inverse of a linear transformation.

Exercise Set 8.3

1. For the following, determine if the composition $(T_2 \circ T_1)$ is defined, and if so find the domain and codomain of $(T_2 \circ T_1)$, and find $T_2 \circ T_1$.

   (a) $T_1(x, y) = (3y, 2x)$, $T_2(x, y) = (x - 2y, x + y)$
   (b) $T_1(x, y, z) = (y - z, x)$, $T_2(x, y) = (x, y, x - y)$
   (c) $T_1(x, y) = (x + y, y - x, x - y)$,
       $T_2(x, y) = (y, x + y, 2x)$
   (d) $T_1(x, y) = (4x, -y)$, $T_2(x, y) = (-x, 4y, x - 4y)$

2. Find $(T_3 \circ T_2 \circ T_1)(x, y)$.

   (a) $T_1(x, y) = (-2y, 3x, x - 2y)$, $T_2(x, y, z) = (y, z, x)$,
       $T_3(x, y, z) = (x + z, y - z)$
   (b) $T_1(x, y) = (x + y, y, -x)$,
       $T_2(x, y, z) = (0, x + y + z, 3y)$,
       $T_3(x, y, z) = (3x + 2y, 4z - x - 3y)$

3. Let $T_1: M_{22} \to R$ and $T_2: M_{22} \to M_{22}$ be the linear transformations given by $T_1(A) = \operatorname{tr}(A)$ and $T_2(A) = A^T$.

   (a) Find $(T_1 \circ T_2)(A)$, where $A = \begin{bmatrix} a & b \\ c & d \end{bmatrix}$.
   (b) Can you find $(T_2 \circ T_1)(A)$? Explain.

4. Let $T_1: P_n \to P_n$ and $T_2: P_n \to P_n$ be the linear operators given by $T_1(p(x)) = p(x + 2)$ and $T_2(p(x)) = p(x - 2)$. Find $(T_1 \circ T_2)(p(x))$ and $(T_2 \circ T_1)(p(x))$.

5. Let $T_1: V \to V$ be the dilation $T_1(\mathbf{v}) = 4\mathbf{v}$. Find a linear operator $T_2: V \to V$ such that $T_1 \circ T_2 = I$ and $T_2 \circ T_1 = I$.

6. Suppose that the linear transformations $T_1: P_2 \to P_2$ and $T_2: P_2 \to P_3$ are given by the formulas $T_1(p(x)) = p(x + 1)$ and $T_2(p(x)) = xp(x)$. Find $(T_2 \circ T_1)(a_0 + a_1 x + a_2 x^2)$.

7. Let $q_0(x)$ be a fixed polynomial of degree m, and define a function T with domain P_n by the formula $T(p(x)) = p(q_0(x))$. Show that T is a linear transformation.

8. Use the definition of $T_3 \circ T_2 \circ T_1$ given by Formula (3) to prove that

   (a) $T_3 \circ T_2 \circ T_1$ is a linear transformation.
   (b) $T_3 \circ T_2 \circ T_1 = (T_3 \circ T_2) \circ T_1$.
   (c) $T_3 \circ T_2 \circ T_1 = T_3 \circ (T_2 \circ T_1)$.

9. Let $T: R^3 \to R^3$ be the orthogonal projection of R^3 onto the yz-plane. Show that $T \circ T = T$.

10. In each part, let $T: R^2 \to R^2$ be multiplication by A. Determine whether T has an inverse; if so, find

    $$T^{-1}\left(\begin{bmatrix} x_1 \\ x_2 \end{bmatrix}\right)$$

    (a) $A = \begin{bmatrix} 4 & 8 \\ -3 & -6 \end{bmatrix}$ (b) $A = \begin{bmatrix} 3 & 1 \\ 5 & 2 \end{bmatrix}$ (c) $A = \begin{bmatrix} 2 & 5 \\ 2 & 6 \end{bmatrix}$

11. In each part, let $T: R^3 \to R^3$ be multiplication by A. Determine whether T has an inverse; if so, find

    $$T^{-1}\left(\begin{bmatrix} x_1 \\ x_2 \\ x_3 \end{bmatrix}\right)$$

    (a) $A = \begin{bmatrix} 1 & 2 & 1 \\ 2 & 1 & 2 \\ 1 & 1 & 1 \end{bmatrix}$ (b) $A = \begin{bmatrix} 1 & 3 & 1 \\ 1 & 1 & 1 \\ -1 & 1 & 0 \end{bmatrix}$

    (c) $A = \begin{bmatrix} 2 & 6 & 10 \\ 0 & 1 & -1 \\ 2 & 4 & 6 \end{bmatrix}$ (d) $A = \begin{bmatrix} 1 & -1 & 1 \\ 0 & 2 & -1 \\ 2 & 3 & 0 \end{bmatrix}$

12. In each part, determine whether the linear operator $T: R^n \to R^n$ is one-to-one; if so, find $T^{-1}(x_1, x_2, \ldots, x_n)$.

    (a) $T(x_1, x_2, \ldots, x_n) = (0, x_1, x_2, \ldots, x_{n-1})$
    (b) $T(x_1, x_2, \ldots, x_n) = (x_n, x_{n-1}, \ldots, x_2, x_1)$
    (c) $T(x_1, x_2, \ldots, x_n) = (x_2, x_3, \ldots, x_n, x_1)$

13. Let $T: R^n \to R^n$ be the linear operator defined by the formula

    $$T(x_1, x_2, \ldots, x_n) = (a_1 x_1, a_2 x_2, \ldots, a_n x_n)$$

    where $a_1, \ldots, a_n$ are constants.

    (a) Under what conditions will T have an inverse?
    (b) Assuming that the conditions determined in part (a) are satisfied, find a formula for $T^{-1}(x_1, x_2, \ldots, x_n)$.

14. Let $T_1: R^2 \to R^2$ and $T_2: R^2 \to R^2$ be the linear operators given by the formulas

    $T_1(x, y) = (x + 3y, x - 3y)$ and $T_2(x, y) = (2x - y, 2x + y)$

    (a) Show that T_1 and T_2 are one-to-one.
    (b) Find formulas for

    $$T_1^{-1}(x, y), \quad T_2^{-1}(x, y), \quad (T_2 \circ T_1)^{-1}(x, y)$$

    (c) Verify that $(T_2 \circ T_1)^{-1} = T_1^{-1} \circ T_2^{-1}$.

15. Let $T_1: P_2 \to P_3$ and $T_2: P_3 \to P_3$ be the linear transformations given by the formulas

    $T_1(p(x)) = xp(x)$ and $T_2(p(x)) = p(x + 1)$

    (a) Find formulas for $T_1^{-1}(p(x))$, $T_2^{-1}(p(x))$, and $(T_2 \circ T_1)^{-1}(p(x))$.
    (b) Verify that $(T_2 \circ T_1)^{-1} = T_1^{-1} \circ T_2^{-1}$.

16. Let $T_A: R^3 \to R^3$, $T_B: R^3 \to R^3$, and $T_C: R^3 \to R^3$ be the reflections about the xy-plane, the xz-plane, and the yz-plane, respectively. Verify Formula (8) for these linear operators.

17. Let $T: P_1 \to R^2$ be the function defined by the formula
$$T(p(x)) = (p(0), p(1))$$
   (a) Find $T(2 - x)$.
   (b) Show that T is a linear transformation.
   (c) Show that T is one-to-one.
   (d) Find $T^{-1}(3, 5)$, and sketch its graph.

18. Let $T: R^2 \to R^2$ be the linear operator given by the formula $T(x, y) = (x + ky, -y)$. Show that T is one-to-one and that $T^{-1} = T$ for every real value of k.

19. Prove: If $T: V \to W$ is a one-to-one linear transformation, then $T^{-1}: R(T) \to V$ is a one-to-one linear transformation.

▶ In Exercises 20–21, determine whether $T_1 \circ T_2 = T_2 \circ T_1$. ◀

20. (a) $T_1: R^2 \to R^2$ is the orthogonal projection on the x-axis, and $T_2: R^2 \to R^2$ is the orthogonal projection on the y-axis.
   (b) $T_1: R^2 \to R^2$ is the rotation about the origin through an angle θ_1, and $T_2: R^2 \to R^2$ is the rotation about the origin through an angle θ_2.
   (c) $T_1: R^3 \to R^3$ is the rotation about the x-axis through an angle θ_1, and $T_2: R^3 \to R^3$ is the rotation about the z-axis through an angle θ_2.

21. (a) $T_1: R^2 \to R^2$ is the reflection about the x-axis, and $T_2: R^2 \to R^2$ is the reflection about the y-axis.
   (b) $T_1: R^2 \to R^2$ is the orthogonal projection on the x-axis, and $T_2: R^2 \to R^2$ is the counterclockwise rotation through an angle θ.
   (c) $T_1: R^3 \to R^3$ is a dilation by a factor k, and $T_2: R^3 \to R^3$ is the counterclockwise rotation about the z-axis through an angle θ.

22. (*Calculus required*) Let
$$D(\mathbf{f}) = f'(x) \quad \text{and} \quad J(\mathbf{f}) = \int_0^x f(t)\, dt$$
be the linear transformations in Examples 11 and 12 of Section 8.1. Find $(J \circ D)(\mathbf{f})$ for
   (a) $\mathbf{f}(x) = x^2 + 3x + 2$
   (b) $\mathbf{f}(x) = \sin x$
   (c) $\mathbf{f}(x) = e^x + 3$

23. (*Calculus required*) The Fundamental Theorem of Calculus implies that integration and differentiation reverse the actions of each other. Define a transformation $D: P_n \to P_{n-1}$ by $D(p(x)) = p'(x)$, and define $J: P_{n-1} \to P_n$ by
$$J(p(x)) = \int_0^x p(t)\, dt$$
   (a) Show that D and J are linear transformations.
   (b) Explain why J is not the inverse transformation of D.
   (c) Can the domains and/or codomains of D and J be restricted so they are inverse linear transformations?

True-False Exercises

In parts (a)–(f) determine whether the statement is true or false, and justify your answer.

(a) The composition of two linear transformations is also a linear transformation.

(b) If $T_1: V \to V$ and $T_2: V \to V$ are any two linear operators, then $T_1 \circ T_2 = T_2 \circ T_1$.

(c) The inverse of a linear transformation is a linear transformation.

(d) If a linear transformation T has an inverse, then the kernel of T is the zero subspace.

(e) If $T: R^2 \to R^2$ is the orthogonal projection onto the x-axis, then $T^{-1}: R^2 \to R^2$ maps each point on the x-axis onto a line that is perpendicular to the x-axis.

(f) If $T_1: U \to V$ and $T_2: V \to W$ are linear transformations, and if T_1 is not one-to-one, then neither is $T_2 \circ T_1$.

8.4 Matrices for General Linear Transformations

In this section we will show that a general linear transformation from any n-dimensional vector space V to any m-dimensional vector space W can be performed using an appropriate matrix transformation from R^n to R^m. This idea is used in computer computations since computers are well suited for performing matrix computations.

Matrices of Linear Transformations

Suppose that V is an n-dimensional vector space, W is an m-dimensional vector space, and that $T: V \to W$ is a linear transformation. Suppose further that B is a basis for V, that B' is a basis for W, and that for each vector $\mathbf{x}$ in V, the coordinate matrices for $\mathbf{x}$ and $T(\mathbf{x})$ are $[\mathbf{x}]_B$ and $[T(\mathbf{x})]_{B'}$, respectively (Figure 8.4.1).

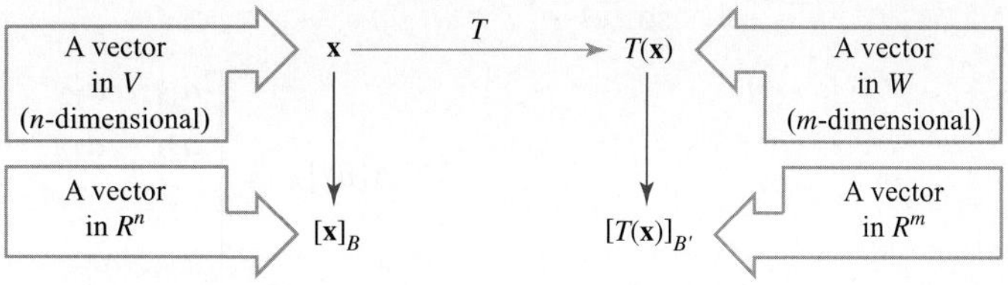

8.4 Matrices for General Linear Transformations

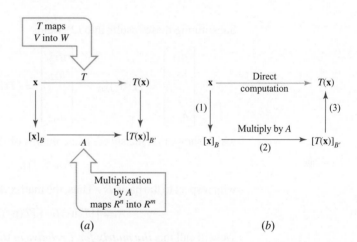

Figure 8.4.1

It will be our goal to find an $m \times n$ matrix A such that multiplication by A maps the vector $[\mathbf{x}]_B$ into the vector $[T(\mathbf{x})]_{B'}$ for each $\mathbf{x}$ in V (Figure 8.4.2a). If we can do so, then, as illustrated in Figure 8.4.2b, we will be able to execute the linear transformation T by using matrix multiplication and the following *indirect* procedure:

Finding $T(\mathbf{x})$ Indirectly

Step 1. Compute the coordinate vector $[\mathbf{x}]_B$.
Step 2. Multiply $[\mathbf{x}]_B$ on the left by A to produce $[T(\mathbf{x})]_{B'}$.
Step 3. Reconstruct $T(\mathbf{x})$ from its coordinate vector $[T(\mathbf{x})]_{B'}$.

Figure 8.4.2

The key to executing this plan is to find an $m \times n$ matrix A with the property that

$$A[\mathbf{x}]_B = [T(\mathbf{x})]_{B'} \tag{1}$$

For this purpose, let $B = \{\mathbf{u}_1, \mathbf{u}_2, \ldots, \mathbf{u}_n\}$ be a basis for the n-dimensional space V and $B' = \{\mathbf{v}_1, \mathbf{v}_2, \ldots, \mathbf{v}_m\}$ a basis for the m-dimensional space W. Since Equation (1) must hold for all vectors in V, it must hold, in particular, for the basis vectors in B; that is,

$$A[\mathbf{u}_1]_B = [T(\mathbf{u}_1)]_{B'}, \quad A[\mathbf{u}_2]_B = [T(\mathbf{u}_2)]_{B'}, \ldots, \quad A[\mathbf{u}_n]_B = [T(\mathbf{u}_n)]_{B'} \tag{2}$$

But

$$[\mathbf{u}_1]_B = \begin{bmatrix} 1 \\ 0 \\ 0 \\ \vdots \\ 0 \end{bmatrix}, \quad [\mathbf{u}_2]_B = \begin{bmatrix} 0 \\ 1 \\ 0 \\ \vdots \\ 0 \end{bmatrix}, \ldots, \quad [\mathbf{u}_n]_B = \begin{bmatrix} 0 \\ 0 \\ 0 \\ \vdots \\ 1 \end{bmatrix}$$

so

$$A[\mathbf{u}_1]_B = \begin{bmatrix} a_{11} & a_{12} & \cdots & a_{1n} \\ a_{21} & a_{22} & \cdots & a_{2n} \\ \vdots & \vdots & & \vdots \\ a_{m1} & a_{m2} & \cdots & a_{mn} \end{bmatrix} \begin{bmatrix} 1 \\ 0 \\ 0 \\ \vdots \\ 0 \end{bmatrix} = \begin{bmatrix} a_{11} \\ a_{21} \\ \vdots \\ a_{m1} \end{bmatrix}$$

$$A[\mathbf{u}_2]_B = \begin{bmatrix} a_{11} & a_{12} & \cdots & a_{1n} \\ a_{21} & a_{22} & \cdots & a_{2n} \\ \vdots & \vdots & & \vdots \\ a_{m1} & a_{m2} & \cdots & a_{mn} \end{bmatrix} \begin{bmatrix} 0 \\ 1 \\ 0 \\ \vdots \\ 0 \end{bmatrix} = \begin{bmatrix} a_{12} \\ a_{22} \\ \vdots \\ a_{m2} \end{bmatrix}$$

$$\vdots$$

$$A[\mathbf{u}_n]_B = \begin{bmatrix} a_{11} & a_{12} & \cdots & a_{1n} \\ a_{21} & a_{22} & \cdots & a_{2n} \\ \vdots & \vdots & & \vdots \\ a_{m1} & a_{m2} & \cdots & a_{mn} \end{bmatrix} \begin{bmatrix} 0 \\ 0 \\ 0 \\ \vdots \\ 1 \end{bmatrix} = \begin{bmatrix} a_{1n} \\ a_{2n} \\ \vdots \\ a_{mn} \end{bmatrix}$$

Substituting these results into (2) yields

$$\begin{bmatrix} a_{11} \\ a_{21} \\ \vdots \\ a_{m1} \end{bmatrix} = [T(\mathbf{u}_1)]_{B'}, \quad \begin{bmatrix} a_{12} \\ a_{22} \\ \vdots \\ a_{m2} \end{bmatrix} = [T(\mathbf{u}_2)]_{B'}, \ldots, \quad \begin{bmatrix} a_{1n} \\ a_{2n} \\ \vdots \\ a_{mn} \end{bmatrix} = [T(\mathbf{u}_n)]_{B'}$$

which shows that the successive columns of A are the coordinate vectors of

$$T(\mathbf{u}_1), T(\mathbf{u}_2), \ldots, T(\mathbf{u}_n)$$

with respect to the basis B'. Thus, the matrix A that completes the link in Figure 8.4.2a is

$$A = \begin{bmatrix} [T(\mathbf{u}_1)]_{B'} \mid [T(\mathbf{u}_2)]_{B'} \mid \cdots \mid [T(\mathbf{u}_n)]_{B'} \end{bmatrix} \tag{3}$$

We will call this *the matrix for T relative to the bases B and B'* and will denote it by the symbol $[T]_{B',B}$. Using this notation, Formula (3) can be written as

$$[T]_{B',B} = \begin{bmatrix} [T(\mathbf{u}_1)]_{B'} \mid [T(\mathbf{u}_2)]_{B'} \mid \cdots \mid [T(\mathbf{u}_n)]_{B'} \end{bmatrix} \tag{4}$$

and from (1), this matrix has the property

$$[T]_{B',B}[\mathbf{x}]_B = [T(\mathbf{x})]_{B'} \tag{5}$$

We leave it as an exercise to show that in the special case where $T_A: R^n \to R^m$ is multiplication by A, and where B and B' are the *standard bases* for R^n and R^m, respectively, then

$$[T]_{B',B} = A \tag{6}$$

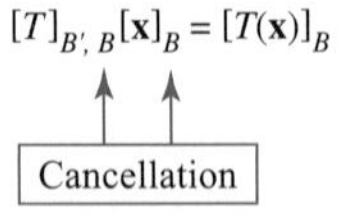

Remark Observe that in the notation $[T]_{B',B}$ the right subscript is a basis for the domain of T, and the left subscript is a basis for the image space of T (Figure 8.4.3). Moreover, observe how the subscript B seems to "cancel out" in Formula (5) (Figure 8.4.4).

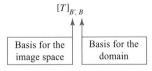

▲ Figure 8.4.3

$[T]_{B',B}[\mathbf{x}]_B = [T(\mathbf{x})]_{B'}$

Cancellation

▲ Figure 8.4.4

8.4 Matrices for General Linear Transformations

▶ **EXAMPLE 1** Matrix for a Linear Transformation

Let $T: P_1 \to P_2$ be the linear transformation defined by

$$T(p(x)) = xp(x)$$

Find the matrix for T with respect to the standard bases

$$B = \{\mathbf{u}_1, \mathbf{u}_2\} \quad \text{and} \quad B' = \{\mathbf{v}_1, \mathbf{v}_2, \mathbf{v}_3\}$$

where

$$\mathbf{u}_1 = 1, \quad \mathbf{u}_2 = x; \quad \mathbf{v}_1 = 1, \quad \mathbf{v}_2 = x, \quad \mathbf{v}_3 = x^2$$

Solution From the given formula for T we obtain

$$T(\mathbf{u}_1) = T(1) = (x)(1) = x$$
$$T(\mathbf{u}_2) = T(x) = (x)(x) = x^2$$

By inspection, the coordinate vectors for $T(\mathbf{u}_1)$ and $T(\mathbf{u}_2)$ relative to B' are

$$[T(\mathbf{u}_1)]_{B'} = \begin{bmatrix} 0 \\ 1 \\ 0 \end{bmatrix}, \quad [T(\mathbf{u}_2)]_{B'} = \begin{bmatrix} 0 \\ 0 \\ 1 \end{bmatrix}$$

Thus, the matrix for T with respect to B and B' is

$$[T]_{B',B} = \big[[T(\mathbf{u}_1)]_{B'} \mid [T(\mathbf{u}_2)]_{B'}\big] = \begin{bmatrix} 0 & 0 \\ 1 & 0 \\ 0 & 1 \end{bmatrix}$$

▶ **EXAMPLE 2** The Three-Step Procedure

Let $T: P_1 \to P_2$ be the linear transformation in Example 1, and use the three-step procedure described in the following figure to perform the computation

$$T(a + bx) = x(a + bx) = ax + bx^2$$

$$\begin{array}{ccc}
\mathbf{x} & \xrightarrow{\text{Direct computation}} & T(\mathbf{x}) \\
{\scriptsize(1)}\downarrow & & \uparrow{\scriptsize(3)} \\
[\mathbf{x}]_B & \xrightarrow[(2)]{\text{Multiply by }[T]_{B',B}} & [T(\mathbf{x})]_{B'}
\end{array}$$

Solution

Step 1. The coordinate matrix for $\mathbf{x} = a + bx$ relative to the basis $B = \{1, x\}$ is

$$[\mathbf{x}]_B = \begin{bmatrix} a \\ b \end{bmatrix}$$

Step 2. Multiplying $[\mathbf{x}]_B$ by the matrix $[T]_{B',B}$ found in Example 1 we obtain

$$[T]_{B',B}[\mathbf{x}]_B = \begin{bmatrix} 0 & 0 \\ 1 & 0 \\ 0 & 1 \end{bmatrix} \begin{bmatrix} a \\ b \end{bmatrix} = \begin{bmatrix} 0 \\ a \\ b \end{bmatrix} = [T(\mathbf{x})]_{B'}$$

Step 3. Reconstructing $T(\mathbf{x}) = T(a + bx)$ from $[T(\mathbf{x})]_{B'}$ we obtain

$$T(a + bx) = 0 + ax + bx^2 = ax + bx^2$$

Although Example 2 is simple, the procedure that it illustrates is applicable to problems of great complexity.

EXAMPLE 3 Matrix for a Linear Transformation

Let $T: R^2 \to R^3$ be the linear transformation defined by

$$T\left(\begin{bmatrix} x_1 \\ x_2 \end{bmatrix}\right) = \begin{bmatrix} x_2 \\ -5x_1 + 13x_2 \\ -7x_1 + 16x_2 \end{bmatrix} = \begin{bmatrix} 0 & 1 \\ -5 & 13 \\ -7 & 16 \end{bmatrix} \begin{bmatrix} x_1 \\ x_2 \end{bmatrix}$$

Find the matrix for the transformation T with respect to the bases $B = \{\mathbf{u}_1, \mathbf{u}_2\}$ for R^2 and $B' = \{\mathbf{v}_1, \mathbf{v}_2, \mathbf{v}_3\}$ for R^3, where

$$\mathbf{u}_1 = \begin{bmatrix} 3 \\ 1 \end{bmatrix}, \quad \mathbf{u}_2 = \begin{bmatrix} 5 \\ 2 \end{bmatrix}; \quad \mathbf{v}_1 = \begin{bmatrix} 1 \\ 0 \\ -1 \end{bmatrix}, \quad \mathbf{v}_2 = \begin{bmatrix} -1 \\ 2 \\ 2 \end{bmatrix}, \quad \mathbf{v}_3 = \begin{bmatrix} 0 \\ 1 \\ 2 \end{bmatrix}$$

Solution From the formula for T,

$$T(\mathbf{u}_1) = \begin{bmatrix} 1 \\ -2 \\ -5 \end{bmatrix}, \quad T(\mathbf{u}_2) = \begin{bmatrix} 2 \\ 1 \\ -3 \end{bmatrix}$$

Expressing these vectors as linear combinations of $\mathbf{v}_1$, $\mathbf{v}_2$, and $\mathbf{v}_3$, we obtain (verify)

$$T(\mathbf{u}_1) = \mathbf{v}_1 - 2\mathbf{v}_3, \quad T(\mathbf{u}_2) = 3\mathbf{v}_1 + \mathbf{v}_2 - \mathbf{v}_3$$

Thus,

$$[T(\mathbf{u}_1)]_{B'} = \begin{bmatrix} 1 \\ 0 \\ -2 \end{bmatrix}, \quad [T(\mathbf{u}_2)]_{B'} = \begin{bmatrix} 3 \\ 1 \\ -1 \end{bmatrix}$$

so

$$[T]_{B',B} = \bigl[[T(\mathbf{u}_1)]_{B'} \mid [T(\mathbf{u}_2)]_{B'}\bigr] = \begin{bmatrix} 1 & 3 \\ 0 & 1 \\ -2 & -1 \end{bmatrix} \blacktriangleleft$$

Remark Example 3 illustrates that a fixed linear transformation generally has multiple representations, each depending on the bases chosen. In this case the matrices

$$[T] = \begin{bmatrix} 0 & 1 \\ -5 & 13 \\ -7 & 16 \end{bmatrix} \quad \text{and} \quad [T]_{B',B} = \begin{bmatrix} 1 & 3 \\ 0 & 1 \\ -2 & -1 \end{bmatrix}$$

both represent the transformation T, the first relative to the standard bases for R^2 and R^3, the second relative to the bases B and B' stated in the example.

Matrices of Linear Operators

In the special case where $V = W$ (so that $T: V \to V$ is a linear operator), it is usual to take $B = B'$ when constructing a matrix for T. In this case the resulting matrix is called the **matrix for T relative to the basis B** and is usually denoted by $[T]_B$ rather than $[T]_{B,B}$. If $B = \{\mathbf{u}_1, \mathbf{u}_2, \ldots, \mathbf{u}_n\}$, then Formulas (4) and (5) become

$$[T]_B = \bigl[[T(\mathbf{u}_1)]_B \mid [T(\mathbf{u}_2)]_B \mid \cdots \mid [T(\mathbf{u}_n)]_B\bigr] \tag{7}$$

$$[T]_B [\mathbf{x}]_B = [T(\mathbf{x})]_B \tag{8}$$

Phrased informally, Formulas (7) and (8) state that *the matrix for T, when multiplied by the coordinate vector for $\mathbf{x}$, produces the coordinate vector for $T(\mathbf{x})$.*

In the special case where $T: R^n \to R^n$ is a matrix operator, say multiplication by A, and B is the standard basis for R^n, then Formula (7) simplifies to

$$[T]_B = A \tag{9}$$

Matrices of Identity Operators

Recall that the identity operator $I: V \to V$ maps every vector in V into itself, that is, $I(\mathbf{x}) = \mathbf{x}$ for every vector $\mathbf{x}$ in V. The following example shows that if V is n-dimensional, then the matrix for I relative to *any* basis B for V is the $n \times n$ identity matrix.

▶ **EXAMPLE 4** **Matrices of Identity Operators**

If $B = \{\mathbf{u}_1, \mathbf{u}_2, \ldots, \mathbf{u}_n\}$ is a basis for a finite-dimensional vector space V, and if $I: V \to V$ is the identity operator on V, then

$$I(\mathbf{u}_1) = \mathbf{u}_1, \quad I(\mathbf{u}_2) = \mathbf{u}_2, \ldots, \quad I(\mathbf{u}_n) = \mathbf{u}_n$$

Therefore,

$$[I]_B = \begin{bmatrix} 1 & 0 & \cdots & 0 \\ 0 & 1 & \cdots & 0 \\ 0 & 0 & \cdots & 0 \\ \vdots & \vdots & & \vdots \\ 0 & 0 & \cdots & 1 \end{bmatrix} = I$$

$\uparrow \qquad \uparrow \qquad \uparrow$
$[I(\mathbf{u}_1)]_B \quad [I(\mathbf{u}_2)]_B \quad [I(\mathbf{u}_n)]_B$

▶ **EXAMPLE 5** **Linear Operator on P_2**

Let $T: P_2 \to P_2$ be the linear operator defined by

$$T(p(x)) = p(3x - 5)$$

that is, $T(c_0 + c_1 x + c_2 x^2) = c_0 + c_1(3x - 5) + c_2(3x - 5)^2$.

(a) Find $[T]_B$ relative to the basis $B = \{1, x, x^2\}$.
(b) Use the indirect procedure to compute $T(1 + 2x + 3x^2)$.
(c) Check the result in (b) by computing $T(1 + 2x + 3x^2)$ directly.

Solution (a) From the formula for T,

$$T(1) = 1, \quad T(x) = 3x - 5, \quad T(x^2) = (3x - 5)^2 = 9x^2 - 30x + 25$$

so

$$[T(1)]_B = \begin{bmatrix} 1 \\ 0 \\ 0 \end{bmatrix}, \quad [T(x)]_B = \begin{bmatrix} -5 \\ 3 \\ 0 \end{bmatrix}, \quad [T(x^2)]_B = \begin{bmatrix} 25 \\ -30 \\ 9 \end{bmatrix}$$

Thus,

$$[T]_B = \begin{bmatrix} 1 & -5 & 25 \\ 0 & 3 & -30 \\ 0 & 0 & 9 \end{bmatrix}$$

Solution (b)

Step 1. The coordinate matrix for $\mathbf{p} = 1 + 2x + 3x^2$ relative to the basis $B = \{1, x, x^2\}$ is

$$[\mathbf{p}]_B = \begin{bmatrix} 1 \\ 2 \\ 3 \end{bmatrix}$$

Step 2. Multiplying $[\mathbf{p}]_B$ by the matrix $[T]_B$ found in part (a) we obtain

$$[T]_B[\mathbf{p}]_B = \begin{bmatrix} 1 & -5 & 25 \\ 0 & 3 & -30 \\ 0 & 0 & 9 \end{bmatrix} \begin{bmatrix} 1 \\ 2 \\ 3 \end{bmatrix} = \begin{bmatrix} 66 \\ -84 \\ 27 \end{bmatrix} = [T(\mathbf{p})]_B$$

Step 3. Reconstructing $T(\mathbf{p}) = T(1 + 2x + 3x^2)$ from $[T(\mathbf{p})]_B$ we obtain

$$T(1 + 2x + 3x^2) = 66 - 84x + 27x^2$$

Solution (c) By direct computation,

$$\begin{aligned} T(1 + 2x + 3x^2) &= 1 + 2(3x - 5) + 3(3x - 5)^2 \\ &= 1 + 6x - 10 + 27x^2 - 90x + 75 \\ &= 66 - 84x + 27x^2 \end{aligned}$$

which agrees with the result in (b). ◀

Matrices of Compositions and Inverse Transformations

We will conclude this section by mentioning two theorems without proof that are generalizations of Formulas (4) and (7) of Section 4.10.

THEOREM 8.4.1 *If $T_1: U \to V$ and $T_2: V \to W$ are linear transformations, and if B, B'', and B' are bases for U, V, and W, respectively, then*

$$[T_2 \circ T_1]_{B', B} = [T_2]_{B', B''}[T_1]_{B'', B} \tag{10}$$

THEOREM 8.4.2 *If $T: V \to V$ is a linear operator, and if B is a basis for V, then the following are equivalent.*

(a) *T is one-to-one.*

(b) *$[T]_B$ is invertible.*

Moreover, when these equivalent conditions hold,

$$[T^{-1}]_B = [T]_B^{-1} \tag{11}$$

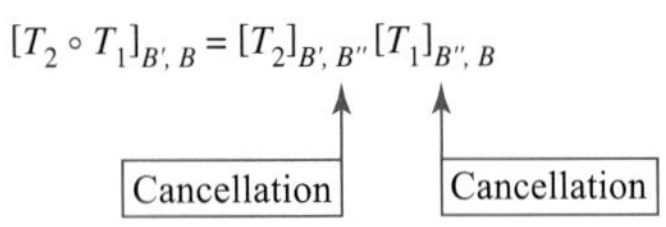

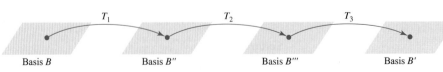

▲ Figure 8.4.5

Remark In (10), observe how the interior subscript B'' (the basis for the intermediate space V) seems to "cancel out," leaving only the bases for the domain and image space of the composition as subscripts (Figure 8.4.5). This cancellation of interior subscripts suggests the following extension of Formula (10) to compositions of three linear transformations (Figure 8.4.6):

$$[T_3 \circ T_2 \circ T_1]_{B', B} = [T_3]_{B', B'''}[T_2]_{B''', B''}[T_1]_{B'', B} \tag{12}$$

▲ Figure 8.4.6

The following example illustrates Theorem 8.4.1.

▶ **EXAMPLE 6 Composition**

Let $T_1: P_1 \to P_2$ be the linear transformation defined by

$$T_1(p(x)) = xp(x)$$

and let $T_2: P_2 \to P_2$ be the linear operator defined by

$$T_2(p(x)) = p(3x - 5)$$

Then the composition $(T_2 \circ T_1): P_1 \to P_2$ is given by

$$(T_2 \circ T_1)(p(x)) = T_2(T_1(p(x))) = T_2(xp(x)) = (3x - 5)p(3x - 5)$$

Thus, if $p(x) = c_0 + c_1 x$, then

$$(T_2 \circ T_1)(c_0 + c_1 x) = (3x - 5)(c_0 + c_1(3x - 5))$$
$$= c_0(3x - 5) + c_1(3x - 5)^2 \tag{13}$$

In this example, P_1 plays the role of U in Theorem 8.4.1, and P_2 plays the roles of both V and W; thus we can take $B' = B''$ in (10) so that the formula simplifies to

$$[T_2 \circ T_1]_{B',B} = [T_2]_{B'}[T_1]_{B',B} \tag{14}$$

Let us choose $B = \{1, x\}$ to be the basis for P_1 and choose $B' = \{1, x, x^2\}$ to be the basis for P_2. We showed in Examples 1 and 5 that

$$[T_1]_{B',B} = \begin{bmatrix} 0 & 0 \\ 1 & 0 \\ 0 & 1 \end{bmatrix} \quad \text{and} \quad [T_2]_{B'} = \begin{bmatrix} 1 & -5 & 25 \\ 0 & 3 & -30 \\ 0 & 0 & 9 \end{bmatrix}$$

Thus, it follows from (14) that

$$[T_2 \circ T_1]_{B',B} = \begin{bmatrix} 1 & -5 & 25 \\ 0 & 3 & -30 \\ 0 & 0 & 9 \end{bmatrix} \begin{bmatrix} 0 & 0 \\ 1 & 0 \\ 0 & 1 \end{bmatrix} = \begin{bmatrix} -5 & 25 \\ 3 & -30 \\ 0 & 9 \end{bmatrix} \tag{15}$$

As a check, we will calculate $[T_2 \circ T_1]_{B',B}$ directly from Formula (4). Since $B = \{1, x\}$, it follows from Formula (4) with $\mathbf{u}_1 = 1$ and $\mathbf{u}_2 = x$ that

$$[T_2 \circ T_1]_{B',B} = \big[[(T_2 \circ T_1)(1)]_{B'} \mid [(T_2 \circ T_1)(x)]_{B'}\big] \tag{16}$$

Using (13) yields

$$(T_2 \circ T_1)(1) = 3x - 5 \quad \text{and} \quad (T_2 \circ T_1)(x) = (3x - 5)^2 = 9x^2 - 30x + 25$$

From this and the fact that $B' = \{1, x, x^2\}$, it follows that

$$[(T_2 \circ T_1)(1)]_{B'} = \begin{bmatrix} -5 \\ 3 \\ 0 \end{bmatrix} \quad \text{and} \quad [(T_2 \circ T_1)(x)]_{B'} = \begin{bmatrix} 25 \\ -30 \\ 9 \end{bmatrix}$$

Substituting in (16) yields

$$[T_2 \circ T_1]_{B',B} = \begin{bmatrix} -5 & 25 \\ 3 & -30 \\ 0 & 9 \end{bmatrix}$$

which agrees with (15). ◀

Concept Review

- Matrix for a linear transformation relative to bases
- Matrix for a linear operator relative to a basis
- The three-step procedure for finding $T(\mathbf{x})$

Skills

- Find the matrix for a linear transformation $T: V \to W$ relative to bases of V and W.
- For a linear transformation $T: V \to W$ find $T(\mathbf{x})$ using the matrix for T relative to bases of V and W.

Exercise Set 8.4

1. Let $T: P_2 \to P_3$ be the linear transformation defined by $T(p(x)) = xp(x)$.

   (a) Find the matrix for T relative to the standard bases
   $$B = \{\mathbf{u}_1, \mathbf{u}_2, \mathbf{u}_3\} \quad \text{and} \quad B' = \{\mathbf{v}_1, \mathbf{v}_2, \mathbf{v}_3, \mathbf{v}_4\}$$
   where
   $$\mathbf{u}_1 = 1, \quad \mathbf{u}_2 = x, \quad \mathbf{u}_3 = x^2$$
   $$\mathbf{v}_1 = 1, \quad \mathbf{v}_2 = x, \quad \mathbf{v}_3 = x^2, \quad \mathbf{v}_4 = x^3$$

   (b) Verify that the matrix $[T]_{B',B}$ obtained in part (a) satisfies Formula (5) for every vector $\mathbf{x} = c_0 + c_1 x + c_2 x^2$ in P_2.

2. Let $T: P_2 \to P_1$ be the linear transformation defined by
   $$T(a_0 + a_1 x + a_2 x^2) = 3a_1 + (2a_0 - a_2)x$$

   (a) Find the matrix for T relative to the standard bases $B = \{1, x, x^2\}$ and $B' = \{1, x\}$ for P_2 and P_1.

   (b) Verify that the matrix $[T]_{B',B}$ obtained in part (a) satisfies Formula (5) for every vector $\mathbf{x} = c_0 + c_1 x + c_2 x^2$ in P_2.

3. Let $T: P_2 \to P_2$ be the linear operator defined by
   $$T(a_0 + a_1 x + a_2 x^2) = a_0 + a_1(x - 1) + a_2(x - 1)^2$$

   (a) Find the matrix for T relative to the standard basis $B = \{1, x, x^2\}$ for P_2.

   (b) Verify that the matrix $[T]_B$ obtained in part (a) satisfies Formula (8) for every vector $\mathbf{x} = a_0 + a_1 x + a_2 x^2$ in P_2.

4. Let $T: R^2 \to R^2$ be the linear operator defined by
   $$T\left(\begin{bmatrix} x_1 \\ x_2 \end{bmatrix}\right) = \begin{bmatrix} x_1 - x_2 \\ x_1 + x_2 \end{bmatrix}$$
   and let $B = \{\mathbf{u}_1, \mathbf{u}_2\}$ be the basis for which
   $$\mathbf{u}_1 = \begin{bmatrix} 1 \\ -1 \end{bmatrix} \quad \text{and} \quad \mathbf{u}_2 = \begin{bmatrix} 0 \\ 1 \end{bmatrix}$$

   (a) Find $[T]_B$.

   (b) Verify that Formula (8) holds for every vector $\mathbf{x}$ in R^2.

5. Let $T: R^2 \to R^3$ be defined by
   $$T\left(\begin{bmatrix} x_1 \\ x_2 \end{bmatrix}\right) = \begin{bmatrix} -x_2 \\ 0 \\ 2x_1 + x_2 \end{bmatrix}$$

   (a) Find the matrix $[T]_{B',B}$ relative to the bases $B = \{\mathbf{u}_1, \mathbf{u}_2\}$ and $B' = \{\mathbf{v}_1, \mathbf{v}_2, \mathbf{v}_3\}$, where
   $$\mathbf{u}_1 = \begin{bmatrix} 1 \\ 2 \end{bmatrix}, \quad \mathbf{u}_2 = \begin{bmatrix} 2 \\ -1 \end{bmatrix}$$
   $$\mathbf{v}_1 = \begin{bmatrix} 1 \\ 1 \\ 1 \end{bmatrix}, \quad \mathbf{v}_2 = \begin{bmatrix} 2 \\ 2 \\ 0 \end{bmatrix}, \quad \mathbf{v}_3 = \begin{bmatrix} 3 \\ 0 \\ 0 \end{bmatrix}$$

   (b) Verify that Formula (5) holds for every vector in R^2.

6. Let $T: R^3 \to R^3$ be the linear operator defined by
   $$T(x_1, x_2, x_3) = (x_2 - x_3, x_1 + x_2, x_3 - x_2)$$

   (a) Find the matrix for T with respect to the basis $B = \{\mathbf{v}_1, \mathbf{v}_2, \mathbf{v}_3\}$, where
   $$\mathbf{v}_1 = (1, 0, 1), \quad \mathbf{v}_2 = (0, 1, 1), \quad \mathbf{v}_3 = (1, 1, 0)$$

   (b) Verify that Formula (8) holds for every vector $\mathbf{x} = (x_1, x_2, x_3)$ in R^3.

   (c) Is T one-to-one? If so, find the matrix of T^{-1} with respect to the basis B.

7. Let $T: P_2 \to P_2$ be the linear operator defined by $T(p(x)) = p(2x + 1)$, that is,
   $$T(c_0 + c_1 x + c_2 x^2) = c_0 + c_1(2x + 1) + c_2(2x + 1)^2$$

   (a) Find $[T]_B$ with respect to the basis $B = \{1, x, x^2\}$.

   (b) Use the three-step procedure illustrated in Example 2 to compute $T(2 - 3x + 4x^2)$.

   (c) Check the result obtained in part (b) by computing $T(2 - 3x + 4x^2)$ directly.

8. Let $T: P_2 \to P_3$ be the linear transformation defined by $T(p(x)) = xp(x - 3)$, that is,
   $$T(c_0 + c_1 x + c_2 x^2) = x(c_0 + c_1(x - 3) + c_2(x - 3)^2)$$

   (a) Find $[T]_{B',B}$ relative to the bases $B = \{1, x, x^2\}$ and $B' = \{1, x, x^2, x^3\}$.

   (b) Use the three-step procedure illustrated in Example 2 to compute $T(1 + x - x^2)$.

   (c) Check the result obtained in part (b) by computing $T(1 + x - x^2)$ directly.

9. Let $\mathbf{v}_1 = \begin{bmatrix} 2 \\ -1 \end{bmatrix}$ and $\mathbf{v}_2 = \begin{bmatrix} 1 \\ -1 \end{bmatrix}$, and let

$$A = \begin{bmatrix} 3 & 2 \\ -2 & 1 \end{bmatrix}$$

be the matrix for $T: R^2 \to R^2$ relative to the basis $B = \{\mathbf{v}_1, \mathbf{v}_2\}$.

(a) Find $[T(\mathbf{v}_1)]_B$ and $[T(\mathbf{v}_2)]_B$.

(b) Find $T(\mathbf{v}_1)$ and $T(\mathbf{v}_2)$.

(c) Find a formula for $T\left(\begin{bmatrix} x_1 \\ x_2 \end{bmatrix}\right)$.

(d) Use the formula obtained in (c) to compute $T\left(\begin{bmatrix} 1 \\ 1 \end{bmatrix}\right)$.

10. Let $A = \begin{bmatrix} 3 & -2 & 1 & 0 \\ 1 & 6 & 2 & 1 \\ -3 & 0 & 7 & 1 \end{bmatrix}$ be the matrix for

$T: R^4 \to R^3$ relative to the bases $B = \{\mathbf{v}_1, \mathbf{v}_2, \mathbf{v}_3, \mathbf{v}_4\}$ and $B' = \{\mathbf{w}_1, \mathbf{w}_2, \mathbf{w}_3\}$, where

$$\mathbf{v}_1 = \begin{bmatrix} 0 \\ 1 \\ 1 \\ 1 \end{bmatrix}, \mathbf{v}_2 = \begin{bmatrix} 2 \\ 1 \\ -1 \\ -1 \end{bmatrix}, \mathbf{v}_3 = \begin{bmatrix} 1 \\ 4 \\ -1 \\ 2 \end{bmatrix}, \mathbf{v}_4 = \begin{bmatrix} 6 \\ 9 \\ 4 \\ 2 \end{bmatrix}$$

$$\mathbf{w}_1 = \begin{bmatrix} 0 \\ 8 \\ 8 \end{bmatrix}, \mathbf{w}_2 = \begin{bmatrix} -7 \\ 8 \\ 1 \end{bmatrix}, \mathbf{w}_3 = \begin{bmatrix} -6 \\ 9 \\ 1 \end{bmatrix}$$

(a) Find $[T(\mathbf{v}_1)]_{B'}$, $[T(\mathbf{v}_2)]_{B'}$, $[T(\mathbf{v}_3)]_{B'}$, and $[T(\mathbf{v}_4)]_{B'}$.

(b) Find $T(\mathbf{v}_1)$, $T(\mathbf{v}_2)$, $T(\mathbf{v}_3)$, and $T(\mathbf{v}_4)$.

(c) Find a formula for $T\left(\begin{bmatrix} x_1 \\ x_2 \\ x_3 \\ x_4 \end{bmatrix}\right)$.

(d) Use the formula obtained in (c) to compute $T\left(\begin{bmatrix} 2 \\ 2 \\ 0 \\ 0 \end{bmatrix}\right)$.

11. Let $A = \begin{bmatrix} 2 & 0 & -1 \\ 1 & 3 & 5 \\ 0 & -2 & 6 \end{bmatrix}$ be the matrix for $T: P_2 \to P_2$ with

respect to the basis $B = \{\mathbf{v}_1, \mathbf{v}_2, \mathbf{v}_3\}$, where $\mathbf{v}_1 = 2 - 2x^2$, $\mathbf{v}_2 = 1 + 2x + 2x^2$, $\mathbf{v}_3 = 3x + x^2$.

(a) Find $[T(\mathbf{v}_1)]_B$, $[T(\mathbf{v}_2)]_B$, and $[T(\mathbf{v}_3)]_B$.

(b) Find $T(\mathbf{v}_1)$, $T(\mathbf{v}_2)$, and $T(\mathbf{v}_3)$.

(c) Find a formula for $T(a_0 + a_1 x + a_2 x^2)$.

(d) Use the formula obtained in (c) to compute $T(1 + x^2)$.

12. Let $T_1: P_1 \to P_2$ be the linear transformation defined by

$$T_1(p(x)) = xp(x)$$

and let $T_2: P_2 \to P_2$ be the linear operator defined by

$$T_2(p(x)) = p(2x + 1)$$

Let $B = \{1, x\}$ and $B' = \{1, x, x^2\}$ be the standard bases for P_1 and P_2.

(a) Find $[T_2 \circ T_1]_{B', B}$, $[T_2]_{B'}$, and $[T_1]_{B', B}$.

(b) State a formula relating the matrices in part (a).

(c) Verify that the matrices in part (a) satisfy the formula you stated in part (b).

13. Let $T_1: P_1 \to P_2$ be the linear transformation defined by

$$T_1(c_0 + c_1 x) = 2c_0 - 3c_1 x$$

and let $T_2: P_2 \to P_3$ be the linear transformation defined by

$$T_2(c_0 + c_1 x + c_2 x^2) = 3c_0 x + 3c_1 x^2 + 3c_2 x^3$$

Let $B = \{1, x\}$, $B'' = \{1, x, x^2\}$, and $B' = \{1, x, x^2, x^3\}$.

(a) Find $[T_2 \circ T_1]_{B', B}$, $[T_2]_{B', B''}$, and $[T_1]_{B'', B}$.

(b) State a formula relating the matrices in part (a).

(c) Verify that the matrices in part (a) satisfy the formula you stated in part (b).

14. Show that if $T: V \to W$ is the zero transformation, then the matrix for T with respect to any bases for V and W is a zero matrix.

15. Show that if $T: V \to V$ is a contraction or a dilation of V (Example 4 of Section 8.1), then the matrix for T relative to any basis for V is a positive scalar multiple of the identity matrix.

16. Let $B = \{\mathbf{v}_1, \mathbf{v}_2, \mathbf{v}_3, \mathbf{v}_4\}$ be a basis for a vector space V. Find the matrix with respect to B of the linear operator $T: V \to V$ defined by $T(\mathbf{v}_1) = \mathbf{v}_2$, $T(\mathbf{v}_2) = \mathbf{v}_3$, $T(\mathbf{v}_3) = \mathbf{v}_4$, $T(\mathbf{v}_4) = \mathbf{v}_1$.

17. Prove that if B and B' are the standard bases for R^n and R^m, respectively, then the matrix for a linear transformation $T: R^n \to R^m$ relative to the bases B and B' is the standard matrix for T.

18. (*Calculus required*) Let $D: P_2 \to P_2$ be the differentiation operator $D(\mathbf{p}) = p'(x)$. In parts (a) and (b), find the matrix of D relative to the basis $B = \{\mathbf{p}_1, \mathbf{p}_2, \mathbf{p}_3\}$.

(a) $\mathbf{p}_1 = 1$, $\mathbf{p}_2 = x$, $\mathbf{p}_3 = x^2$

(b) $\mathbf{p}_1 = 3$, $\mathbf{p}_2 = 3 + 2x$, $\mathbf{p}_3 = 3 + 2x - 6x^2$

(c) Use the matrix in part (a) to compute $D(6 - 6x + 24x^2)$.

(d) Repeat the directions for part (c) for the matrix in part (b).

19. (*Calculus required*) In each part, suppose that $B = \{\mathbf{f}_1, \mathbf{f}_2, \mathbf{f}_3\}$ is a basis for a subspace V of the vector space of real-valued functions defined on the real line. Find the matrix with respect to B for differentiation operator $D: V \to V$.

(a) $\mathbf{f}_1 = 1$, $\mathbf{f}_2 = \sin x$, $\mathbf{f}_3 = \cos x$

(b) $\mathbf{f}_1 = 1$, $\mathbf{f}_2 = e^x$, $\mathbf{f}_3 = e^{2x}$

(c) $\mathbf{f}_1 = e^{2x}$, $\mathbf{f}_2 = xe^{2x}$, $\mathbf{f}_3 = x^2 e^{2x}$

(d) Use the matrix in part (c) to compute $D(3e^{2x} - 4xe^{2x} + 8x^2 e^{2x})$.

20. Let V be a four-dimensional vector space with basis B, let W be a seven-dimensional vector space with basis B', and let $T: V \to W$ be a linear transformation. Identify the four vector spaces that contain the vectors at the corners of the accompanying diagram.

◀ **Figure Ex-20**

21. In each part, fill in the missing part of the equation.

(a) $[T_2 \circ T_1]_{B',B} = [T_2]\_\_?\_\_[T_1]_{B'',B}$

(b) $[T_3 \circ T_2 \circ T_1]_{B',B} = [T_3]\_\_?\_\_[T_2]_{B''',B''}[T_1]_{B'',B}$

(a) If the matrix of a linear transformation $T: V \to W$ relative to some bases of V and W is $\begin{bmatrix} 2 & 4 \\ 0 & 3 \end{bmatrix}$, then there is a nonzero vector $\mathbf{x}$ in V such that $T(\mathbf{x}) = 2\mathbf{x}$.

(b) If the matrix of a linear transformation $T: V \to W$ relative to bases for V and W is $\begin{bmatrix} 2 & 4 \\ 0 & 3 \end{bmatrix}$, then there is a nonzero vector $\mathbf{x}$ in V such that $T(\mathbf{x}) = 4\mathbf{x}$.

(c) If the matrix of a linear transformation $T: V \to W$ relative to certain bases for V and W is $\begin{bmatrix} 1 & 4 \\ 2 & 3 \end{bmatrix}$, then T is one-to-one.

(d) If $S: V \to V$ and $T: V \to V$ are linear operators and B is a basis for V, then the matrix of $S \circ T$ relative to B is $[T]_B[S]_B$.

(e) If $T: V \to V$ is an invertible linear operator and B is a basis for V, then the matrix for T^{-1} relative to B is $[T]_B^{-1}$.

True-False Exercises

In parts (a)–(e) determine whether the statement is true or false, and justify your answer.

8.5 Similarity

The matrix for a linear operator $T: V \to V$ depends on the basis selected for V. One of the fundamental problems of linear algebra is to choose a basis for V that makes the matrix for T as simple as possible—a diagonal or a triangular matrix, for example. In this section we will study this problem.

Simple Matrices for Linear Operators

Standard bases do not necessarily produce the simplest matrices for linear operators. For example, consider the matrix operator $T: R^2 \to R^2$ whose standard matrix is

$$[T] = \begin{bmatrix} 1 & 1 \\ -2 & 4 \end{bmatrix} \quad (1)$$

and view $[T]$ as the matrix for T relative to the standard basis $B = \{\mathbf{e}_1, \mathbf{e}_2\}$ for R^2. Let us compare this to the matrix for T relative to the basis $B' = \{\mathbf{u}'_1, \mathbf{u}'_2\}$ for R^2 in which

$$\mathbf{u}'_1 = \begin{bmatrix} 1 \\ 1 \end{bmatrix}, \quad \mathbf{u}'_2 = \begin{bmatrix} 1 \\ 2 \end{bmatrix} \quad (2)$$

Since

$$T(\mathbf{u}'_1) = \begin{bmatrix} 1 & 1 \\ -2 & 4 \end{bmatrix}\begin{bmatrix} 1 \\ 1 \end{bmatrix} = \begin{bmatrix} 2 \\ 2 \end{bmatrix} = 2\mathbf{u}'_1 \quad \text{and} \quad T(\mathbf{u}'_2) = \begin{bmatrix} 1 & 1 \\ -2 & 4 \end{bmatrix}\begin{bmatrix} 1 \\ 2 \end{bmatrix} = \begin{bmatrix} 3 \\ 6 \end{bmatrix} = 3\mathbf{u}'_2$$

it follows that

$$[T(\mathbf{u}'_1)]_{B'} = \begin{bmatrix} 2 \\ 0 \end{bmatrix} \quad \text{and} \quad [T(\mathbf{u}'_2)]_{B'} = \begin{bmatrix} 0 \\ 3 \end{bmatrix}$$

so the matrix for T relative to the basis B' is

$$[T]_{B'} = \begin{bmatrix} T(\mathbf{u}'_1)_{B'} \mid T(\mathbf{u}'_2)_{B'} \end{bmatrix} = \begin{bmatrix} 2 & 0 \\ 0 & 3 \end{bmatrix}$$

This matrix, being diagonal, has a simpler form than $[T]$ and conveys clearly that the operator T scales $\mathbf{u}'_1$ by a factor of 2 and $\mathbf{u}'_2$ by a factor of 3, information that is not immediately evident from $[T]$.

One of the major themes in more advanced linear algebra courses is to determine the "simplest possible form" that can be obtained for the matrix of a linear operator by choosing the basis appropriately. Sometimes it is possible to obtain a diagonal matrix (as above, for example), whereas other times one must settle for a triangular matrix or some other form. We will only be able to touch on this important topic in this text.

The problem of finding a basis that produces the simplest possible matrix for a linear operator $T: V \to V$ can be attacked by first finding a matrix for T relative to *any* basis, typically a standard basis, where applicable, and then changing the basis in a way that simplifies the matrix. Before pursuing this idea, it will be helpful to revisit some concepts about changing bases.

A New View of Transition Matrices

Recall from Formulas (7) and (8) of Section 4.6 that if $B = \{\mathbf{u}_1, \mathbf{u}_2, \ldots, \mathbf{u}_n\}$ and $B' = \{\mathbf{u}'_1, \mathbf{u}'_2, \ldots, \mathbf{u}'_n\}$ are bases for a vector space V, then the transition matrices from B to B' and from B' to B are

$$P_{B \to B'} = \big[[\mathbf{u}_1]_{B'} \mid [\mathbf{u}_2]_{B'} \mid \cdots \mid [\mathbf{u}_n]_{B'}\big] \tag{3}$$

$$P_{B' \to B} = \big[[\mathbf{u}'_1]_B \mid [\mathbf{u}'_2]_B \mid \cdots \mid [\mathbf{u}'_n]_B\big] \tag{4}$$

where the matrices $P_{B \to B'}$ and $P_{B' \to B}$ are inverses of each other. We also showed in Formulas (9) and (10) of that section that if $\mathbf{v}$ is any vector in V, then

$$P_{B \to B'}[\mathbf{v}]_B = [\mathbf{v}]_{B'} \tag{5}$$

$$P_{B' \to B}[\mathbf{v}]_{B'} = [\mathbf{v}]_B \tag{6}$$

The following theorem shows that transition matrices in Formulas (3) and (4) can be viewed as matrices for identity operators.

THEOREM 8.5.1 *If B and B' are bases for a finite-dimensional vector space V, and if $I: V \to V$ is the identity operator on V, then*

$$P_{B \to B'} = [I]_{B', B} \quad \text{and} \quad P_{B' \to B} = [I]_{B, B'}$$

Proof Suppose that $B = \{\mathbf{u}_1, \mathbf{u}_2, \ldots, \mathbf{u}_n\}$ and $B' = \{\mathbf{u}'_1, \mathbf{u}'_2, \ldots, \mathbf{u}'_n\}$ are bases for V. Using the fact that $I(\mathbf{v}) = \mathbf{v}$ for all $\mathbf{v}$ in V, it follows from Formula (4) of Section 8.4 that

$$\begin{aligned}
[I]_{B', B} &= \big[[I(\mathbf{u}_1)]_{B'} \mid [I(\mathbf{u}_2)]_{B'} \mid \cdots \mid [I(\mathbf{u}_n)]_{B'}\big] \\
&= \big[[\mathbf{u}_1]_{B'} \mid [\mathbf{u}_2]_{B'} \mid \cdots \mid [\mathbf{u}_n]_{B'}\big] \\
&= P_{B \to B'} \quad \text{[Formula (3) above]}
\end{aligned}$$

The proof that $[I]_{B, B'} = P_{B' \to B}$ is similar. ◂

Effect of Changing Bases on Matrices of Linear Operators

We are now ready to consider the main problem in this section.

Problem If B and B' are two bases for a finite-dimensional vector space V, and if $T: V \to V$ is a linear operator, what relationship, if any, exists between the matrices $[T]_B$ and $[T]_{B'}$?

The answer to this question can be obtained by considering the composition of the three linear operators on V pictured in Figure 8.5.1.

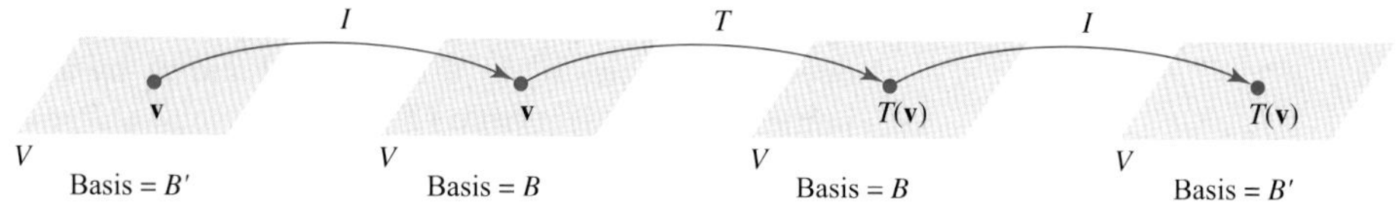

Figure 8.5.1

Basis = B' Basis = B Basis = B Basis = B'

In this figure, **v** is first mapped into itself by the identity operator, then **v** is mapped into $T(\mathbf{v})$ by T, and then $T(\mathbf{v})$ is mapped into itself by the identity operator. All four vector spaces involved in the composition are the same (namely, V), but the bases for the spaces vary. Since the starting vector is **v** and the final vector is $T(\mathbf{v})$, the composition produces the same result as applying T directly; that is,

$$T = I \circ T \circ I \tag{7}$$

If, as illustrated in Figure 8.5.1, if the first and last vector spaces are assigned the basis B' and the middle two spaces are assigned the basis B, then it follows from (7) and Formula (12) of Section 8.4 (with an appropriate adjustment to the names of the bases) that

$$[T]_{B',B'} = [I \circ T \circ I]_{B',B'} = [I]_{B',B}[T]_{B,B}[I]_{B,B'} \tag{8}$$

or, in simpler notation,

$$[T]_{B'} = [I]_{B',B}[T]_B[I]_{B,B'} \tag{9}$$

We can simplify this formula even further by using Theorem 8.5.1 to rewrite it as

$$[T]_{B'} = P_{B \to B'}[T]_B P_{B' \to B} \tag{10}$$

In summary, we have the following theorem.

THEOREM 8.5.2 *Let $T: V \to V$ be a linear operator on a finite-dimensional vector space V, and let B and B' be bases for V. Then*

$$[T]_{B'} = P^{-1}[T]_B P \tag{11}$$

where $P = P_{B' \to B}$ and $P^{-1} = P_{B \to B'}$.

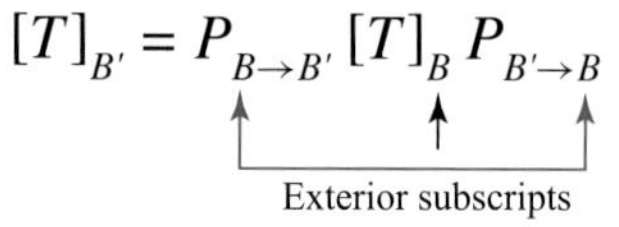

$$[T]_{B'} = P_{B \to B'} [T]_B P_{B' \to B}$$
$$\uparrow \qquad \uparrow \qquad \uparrow$$
Exterior subscripts

▲ **Figure 8.5.2**

Warning When applying Theorem 8.5.2, it is easy to forget whether $P = P_{B' \to B}$ (correct) or $P = P_{B \to B'}$ (incorrect). It may help to use the diagram in Figure 8.5.2 and observe that the *exterior* subscripts of the transition matrices match the subscript of the matrix they enclose.

In the terminology of Definition 1 of Section 5.2, Theorem 8.5.2 tells us that matrices representing the same linear operator relative to different bases must be similar. The following theorem is a rephrasing of Theorem 8.5.2 in the language of similarity.

THEOREM 8.5.3 *Two matrices, A and B, are similar if and only if they represent the same linear operator. Moreover, if $B = P^{-1}AP$, then P is the transition matrix from the basis relative to matrix B to the basis relative to matrix A.*

▶ **EXAMPLE 1** Similar Matrices Represent the Same Linear Operator

We showed at the beginning of this section that the matrices

$$C = \begin{bmatrix} 1 & 1 \\ -2 & 4 \end{bmatrix} \quad \text{and} \quad D = \begin{bmatrix} 2 & 0 \\ 0 & 3 \end{bmatrix}$$

represent the same linear operator $T: R^2 \to R^2$. Verify that these matrices are similar by finding a matrix P for which $D = P^{-1}CP$.

Solution We need to find the transition matrix

$$P = P_{B' \to B} = \begin{bmatrix} [\mathbf{u}'_1]_B \mid [\mathbf{u}'_2]_B \end{bmatrix}$$

where $B' = \{\mathbf{u}'_1, \mathbf{u}'_2\}$ is the basis for R^2 given by (2) and $B = \{\mathbf{e}_1, \mathbf{e}_2\}$ is the standard basis for R^2. We see by inspection that

$$\mathbf{u}'_1 = \mathbf{e}_1 + \mathbf{e}_2$$
$$\mathbf{u}'_2 = \mathbf{e}_1 + 2\mathbf{e}_2$$

from which it follows that

$$[\mathbf{u}'_1]_B = \begin{bmatrix} 1 \\ 1 \end{bmatrix} \quad \text{and} \quad [\mathbf{u}'_2]_B = \begin{bmatrix} 1 \\ 2 \end{bmatrix}$$

Thus,

$$P = P_{B' \to B} = \begin{bmatrix} [\mathbf{u}'_1]_B \mid [\mathbf{u}'_2]_B \end{bmatrix} = \begin{bmatrix} 1 & 1 \\ 1 & 2 \end{bmatrix}$$

We leave it for you to verify that

$$P^{-1} = \begin{bmatrix} 2 & -1 \\ -1 & 1 \end{bmatrix}$$

and hence that

$$\underbrace{\begin{bmatrix} 2 & 0 \\ 0 & 3 \end{bmatrix}}_{D} = \underbrace{\begin{bmatrix} 2 & -1 \\ -1 & 1 \end{bmatrix}}_{P^{-1}} \underbrace{\begin{bmatrix} 1 & 1 \\ -2 & 4 \end{bmatrix}}_{C} \underbrace{\begin{bmatrix} 1 & 1 \\ 1 & 2 \end{bmatrix}}_{P} \blacktriangleleft$$

Similarity Invariants   Recall from Section 5.2 that a property of a square matrix is called a *similarity invariant* if that property is shared by all similar matrices. In Table 1 of that section (table reproduced below), we listed the most important similarity invariants. Since we know from Theorem 8.5.3 that two matrices are similar if and only if they represent the same linear operator $T: V \to V$, it follows that if B and B' are bases for V, then every similarity invariant

Table 1 Similarity Invariants

Property	Description
Determinant	A and $P^{-1}AP$ have the same determinant.
Invertibility	A is invertible if and only if $P^{-1}AP$ is invertible.
Rank	A and $P^{-1}AP$ have the same rank.
Nullity	A and $P^{-1}AP$ have the same nullity.
Trace	A and $P^{-1}AP$ have the same trace.
Characteristic polynomial	A and $P^{-1}AP$ have the same characteristic polynomial.
Eigenvalues	A and $P^{-1}AP$ have the same eigenvalues.
Eigenspace dimension	If λ is an eigenvalue of A and $P^{-1}AP$, then the eigenspace of A corresponding to λ and the eigenspace of $P^{-1}AP$ corresponding to λ have the same dimension.

property of $[T]_B$ is also a similarity invariant property of $[T]_{B'}$ for any other basis B' for V. For example, for any two bases B and B' we must have

$$\det([T]_B) = \det([T]_{B'})$$

It follows from this equation that the value of the determinant depends on T, but not on the particular basis that is used to obtain the matrix for T. Thus, the determinant can be regarded as a property of the linear operator T; indeed, if V is a finite-dimensional vector space, then we can *define* the **determinant of the linear operator** T to be

$$\det(T) = \det([T]_B) \tag{12}$$

where B is *any* basis for V.

▶ **EXAMPLE 2 Determinant of a Linear Operator**

At the beginning of this section we showed that the matrices

$$[T] = \begin{bmatrix} 1 & 1 \\ -2 & 4 \end{bmatrix} \quad \text{and} \quad [T]_{B'} = \begin{bmatrix} 2 & 0 \\ 0 & 3 \end{bmatrix}$$

represent the same linear operator relative to different bases, the first relative to the standard basis $B = \{\mathbf{e}_1, \mathbf{e}_2\}$ for R^2 and the second relative to the basis $B' = \{\mathbf{u}'_1, \mathbf{u}'_2\}$ for which

$$\mathbf{u}'_1 = \begin{bmatrix} 1 \\ 1 \end{bmatrix}, \quad \mathbf{u}'_2 = \begin{bmatrix} 1 \\ 2 \end{bmatrix}$$

This means that $[T]$ and $[T]_{B'}$ must be similar matrices and hence must have the same similarity invariant properties. In particular, they must have the same determinant. We leave it for you to verify that

$$\det[T] = \begin{vmatrix} 1 & 1 \\ -2 & 4 \end{vmatrix} = 6 \quad \text{and} \quad \det[T]_{B'} = \begin{vmatrix} 2 & 0 \\ 0 & 3 \end{vmatrix} = 6$$

▶ **EXAMPLE 3 Eigenvalues and Bases for Eigenspaces**

Find the eigenvalues and bases for the eigenspaces of the linear operator $T : P_2 \to P_2$ defined by

$$T(a + bx + cx^2) = -2c + (a + 2b + c)x + (a + 3c)x^2$$

Solution We leave it for you to show that the matrix for T with respect to the standard basis $B = \{1, x, x^2\}$ is

$$[T]_B = \begin{bmatrix} 0 & 0 & -2 \\ 1 & 2 & 1 \\ 1 & 0 & 3 \end{bmatrix}$$

The eigenvalues of T are $\lambda = 1$ and $\lambda = 2$ (Example 7 of Section 5.1). Also from that example, the eigenspace of $[T]_B$ corresponding to $\lambda = 2$ has the basis $\{\mathbf{u}_1, \mathbf{u}_2\}$, where

$$\mathbf{u}_1 = \begin{bmatrix} -1 \\ 0 \\ 1 \end{bmatrix}, \quad \mathbf{u}_2 = \begin{bmatrix} 0 \\ 1 \\ 0 \end{bmatrix}$$

and the eigenspace of $[T]_B$ corresponding to $\lambda = 1$ has the basis $\{\mathbf{u}_3\}$, where

$$\mathbf{u}_3 = \begin{bmatrix} -2 \\ 1 \\ 1 \end{bmatrix}$$

The matrices $\mathbf{u}_1$, $\mathbf{u}_2$, and $\mathbf{u}_3$ are the coordinate matrices relative to B of

$$\mathbf{p}_1 = -1 + x^2, \quad \mathbf{p}_2 = x, \quad \mathbf{p}_3 = -2 + x + x^2$$

Thus, the eigenspace of T corresponding to $\lambda = 2$ has the basis

$$\{\mathbf{p}_1, \mathbf{p}_2\} = \{-1 + x^2, x\}$$

and that corresponding to $\lambda = 1$ has the basis

$$\{\mathbf{p}_3\} = \{-2 + x + x^2\}$$

As a check, you can use the given formula for T to verify that

$$T(\mathbf{p}_1) = 2\mathbf{p}_1, \quad T(\mathbf{p}_2) = 2\mathbf{p}_2, \quad \text{and} \quad T(\mathbf{p}_3) = \mathbf{p}_3 \blacktriangleleft$$

Concept Review

- Similarity of matrices representing a linear operator
- Similarity invariant
- Determinant of a linear operator

Skills

- Show that two matrices A and B represent the same linear operator, and find a transition matrix P so that $B = P^{-1}AP$.
- Find the eigenvalues and bases for the eigenspaces of a linear operator on a finite-dimensional vector space.

Exercise Set 8.5

▶ In Exercises 1–7, find the matrix for T relative to the basis B, and use Theorem 8.5.2 to compute the matrix for T relative to the basis B'. ◀

1. $T\colon R^2 \to R^2$ is defined by

$$T\left(\begin{bmatrix} x_1 \\ x_2 \end{bmatrix}\right) = \begin{bmatrix} 3x_1 - x_2 \\ x_1 \end{bmatrix}$$

and $B = \{\mathbf{u}_1, \mathbf{u}_2\}$ and $B' = \{\mathbf{v}_1, \mathbf{v}_2\}$, where

$$\mathbf{u}_1 = \begin{bmatrix} 1 \\ 0 \end{bmatrix}, \quad \mathbf{u}_2 = \begin{bmatrix} 0 \\ 1 \end{bmatrix}; \quad \mathbf{v}_1 = \begin{bmatrix} 1 \\ -3 \end{bmatrix}, \quad \mathbf{v}_2 = \begin{bmatrix} 3 \\ 0 \end{bmatrix}$$

2. $T\colon R^2 \to R^2$ is defined by

$$T\left(\begin{bmatrix} x_1 \\ x_2 \end{bmatrix}\right) = \begin{bmatrix} x_1 + 7x_2 \\ 3x_1 - 4x_2 \end{bmatrix}$$

and $B = \{\mathbf{u}_1, \mathbf{u}_2\}$ and $B' = \{\mathbf{v}_1, \mathbf{v}_2\}$, where

$$\mathbf{u}_1 = \begin{bmatrix} 1 \\ 4 \end{bmatrix}, \quad \mathbf{u}_2 = \begin{bmatrix} 2 \\ -2 \end{bmatrix}; \quad \mathbf{v}_1 = \begin{bmatrix} 2 \\ 1 \end{bmatrix}, \quad \mathbf{v}_2 = \begin{bmatrix} -1 \\ 1 \end{bmatrix}$$

3. $T\colon R^2 \to R^2$ is the rotation about the origin through an angle of $45°$; B and B' are the bases in Exercise 1.

4. $T\colon R^3 \to R^3$ is defined by

$$T\left(\begin{bmatrix} x_1 \\ x_2 \\ x_3 \end{bmatrix}\right) = \begin{bmatrix} x_1 + 2x_2 - x_3 \\ -x_2 \\ x_1 + 7x_3 \end{bmatrix}$$

and B is the standard basis for R^3 and $B' = \{\mathbf{v}_1, \mathbf{v}_2, \mathbf{v}_3\}$, where

$$\mathbf{v}_1 = \begin{bmatrix} 1 \\ 0 \\ 0 \end{bmatrix}, \quad \mathbf{v}_2 = \begin{bmatrix} 1 \\ 1 \\ 0 \end{bmatrix}, \quad \mathbf{v}_3 = \begin{bmatrix} 1 \\ 1 \\ 1 \end{bmatrix}$$

5. $T\colon R^3 \to R^3$ is the orthogonal projection on the xy-plane, and B and B' are as in Exercise 4.

6. $T\colon R^2 \to R^2$ is defined by $T(\mathbf{x}) = 4\mathbf{x}$, and B and B' are the bases in Exercise 2.

7. $T\colon P_1 \to P_1$ is defined by $T(a_0 + a_1 x) = a_0 + a_1(x + 1)$, and $B = \{\mathbf{p}_1, \mathbf{p}_2\}$ and $B' = \{\mathbf{q}_1, \mathbf{q}_2\}$, where $\mathbf{p}_1 = 6 + 3x$, $\mathbf{p}_2 = 10 + 2x$, $\mathbf{q}_1 = 2$, $\mathbf{q}_2 = 3 + 2x$.

8. Find $\det(T)$.

   (a) $T\colon R^2 \to R^2$, where
   $T(x_1, x_2) = (3x_1 - 4x_2, -x_1 + 7x_2)$

   (b) $T\colon R^3 \to R^3$, where
   $T(x_1, x_2, x_3) = (x_1 - x_2, x_2 - x_3, x_3 - x_1)$

   (c) $T\colon P_2 \to P_2$, where
   $T(p(x)) = p(x - 1)$

9. Prove that the following are similarity invariants:

   (a) rank   (b) nullity   (c) invertibility

10. Let $T: P_4 \to P_4$ be the linear operator given by the formula $T(p(x)) = p(3x - 2)$.

    (a) Find a matrix for T relative to some convenient basis, and then use it to find the rank and nullity of T.

    (b) Use the result in part (a) to determine whether T is one-to-one.

11. In each part, find a basis for R^2 relative to which the matrix for T is diagonal.

    (a) $T\left(\begin{bmatrix} x_1 \\ x_2 \end{bmatrix}\right) = \begin{bmatrix} 3x_1 - 2x_2 \\ -x_1 + 4x_2 \end{bmatrix}$

    (b) $T\left(\begin{bmatrix} x_1 \\ x_2 \end{bmatrix}\right) = \begin{bmatrix} 2x_1 + 5x_2 \\ x_1 - 2x_2 \end{bmatrix}$

12. In each part, find a basis for R^3 relative to which the matrix for T is diagonal.

    (a) $T\left(\begin{bmatrix} x_1 \\ x_2 \\ x_3 \end{bmatrix}\right) = \begin{bmatrix} -2x_1 + x_2 - x_3 \\ x_1 - 2x_2 - x_3 \\ -x_1 - x_2 - 2x_3 \end{bmatrix}$

    (b) $T\left(\begin{bmatrix} x_1 \\ x_2 \\ x_3 \end{bmatrix}\right) = \begin{bmatrix} -x_2 + x_3 \\ -x_1 + x_3 \\ x_1 + x_2 \end{bmatrix}$

    (c) $T\left(\begin{bmatrix} x_1 \\ x_2 \\ x_3 \end{bmatrix}\right) = \begin{bmatrix} 4x_1 + x_3 \\ 2x_1 + 3x_2 + 2x_3 \\ x_1 + 4x_3 \end{bmatrix}$

13. Let $T: P_2 \to P_2$ be defined by
$$T(a_0 + a_1 x + a_2 x^2) = (2a_0 - a_1 + 3a_2) + (4a_0 - 5a_1)x + (a_1 + 2a_2)x^2$$

    (a) Find the eigenvalues of T.

    (b) Find bases for the eigenspaces of T.

14. Let $T: M_{22} \to M_{22}$ be defined by
$$T\left(\begin{bmatrix} a & b \\ c & d \end{bmatrix}\right) = \begin{bmatrix} 2b & a+b \\ c - 2b & d \end{bmatrix}$$

    (a) Find the eigenvalues of T.

    (b) Find bases for the eigenspaces of T.

15. Let λ be an eigenvalue of a linear operator $T: V \to V$. Prove that the eigenvectors of T corresponding to λ are the nonzero vectors in the kernel of $\lambda I - T$.

16. (a) Prove that if A and B are similar matrices, then A^2 and B^2 are also similar. More generally, prove that A^k and B^k are similar if k is any positive integer.

    (b) If A^2 and B^2 are similar, must A and B be similar? Explain.

17. Let C and D be $m \times n$ matrices, and let $B = \{v_1, v_2, \ldots, v_n\}$ be a basis for a vector space V. Show that if $C[x]_B = D[x]_B$ for all x in V, then $C = D$.

18. Find two nonzero 2×2 matrices that are not similar, and explain why they are not.

19. Complete the proof below by justifying each step.

    Hypothesis: A and B are similar matrices.

    Conclusion: A and B have the same characteristic polynomial.

    Proof: (1) $\det(\lambda I - B) = \det(\lambda I - P^{-1}AP)$

    (2) $\qquad = \det(\lambda P^{-1} P - P^{-1}AP)$

    (3) $\qquad = \det(P^{-1}(\lambda I - A)P)$

    (4) $\qquad = \det(P^{-1})\det(\lambda I - A)\det(P)$

    (5) $\qquad = \det(P^{-1})\det(P)\det(\lambda I - A)$

    (6) $\qquad = \det(\lambda I - A)$

20. If A and B are similar matrices, say $B = P^{-1}AP$, then it follows from Exercise 19 that A and B have the same eigenvalues. Suppose that λ is one of the common eigenvalues and x is a corresponding eigenvector of A. See if you can find an eigenvector of B corresponding to λ (expressed in terms of λ, x, and P).

21. Since the standard basis for R^n is so simple, why would one want to represent a linear operator on R^n in another basis?

22. Prove that trace is a similarity invariant.

True-False Exercises

In parts (a)–(h) determine whether the statement is true or false, and justify your answer.

(a) A matrix cannot be similar to itself.

(b) If A is similar to B, and B is similar to C, then A is similar to C.

(c) If A and B are similar and B is singular, then A is singular.

(d) If A and B are invertible and similar, then A^{-1} and B^{-1} are similar.

(e) If $T_1: R^n \to R^n$ and $T_2: R^n \to R^n$ are linear operators, and if $[T_1]_{B',B} = [T_2]_{B',B}$ with respect to two bases B and B' for R^n, then $T_1(x) = T_2(x)$ for every vector x in R^n.

(f) If $T_1: R^n \to R^n$ is a linear operator, and if $[T_1]_B = [T_1]_{B'}$ with respect to two bases B and B' for R^n, then $B = B'$.

(g) If $T: R^n \to R^n$ is a linear operator, and if $[T]_B = I_n$ with respect to some basis B for R^n, then T is the identity operator on R^n.

(h) If $T: R^n \to R^n$ is a linear operator, and if $[T]_{B',B} = I_n$ with respect to two bases B and B' for R^n, then T is the identity operator on R^n.

Chapter 8 Supplementary Exercises

1. Let A be an $n \times n$ matrix, B a nonzero $n \times 1$ matrix, and $\mathbf{x}$ a vector in R^n expressed in matrix notation. Is $T(\mathbf{x}) = A\mathbf{x} + B$ a linear operator on R^n? Justify your answer.

2. Let
$$A = \begin{bmatrix} \cos\theta & -\sin\theta \\ \sin\theta & \cos\theta \end{bmatrix}$$

   (a) Show that
   $$A^2 = \begin{bmatrix} \cos 2\theta & -\sin 2\theta \\ \sin 2\theta & \cos 2\theta \end{bmatrix} \quad \text{and} \quad A^3 = \begin{bmatrix} \cos 3\theta & -\sin 3\theta \\ \sin 3\theta & \cos 3\theta \end{bmatrix}$$

   (b) Based on your answer to part (a), make a guess at the form of the matrix A^n for any positive integer n.

   (c) By considering the geometric effect of multiplication by A, obtain the result in part (b) geometrically.

3. Let $T: V \to V$ be defined by $T(\mathbf{v}) = \|\mathbf{v}\|\mathbf{v}$. Show that T is not a linear operator on V.

4. Let $\mathbf{v}_1, \mathbf{v}_2, \ldots, \mathbf{v}_m$ be fixed vectors in R^n, and let $T: R^n \to R^m$ be the function defined by $T(\mathbf{x}) = (\mathbf{x} \cdot \mathbf{v}_1, \mathbf{x} \cdot \mathbf{v}_2, \ldots, \mathbf{x} \cdot \mathbf{v}_m)$, where $\mathbf{x} \cdot \mathbf{v}_i$ is the Euclidean inner product on R^n.

   (a) Show that T is a linear transformation.

   (b) Show that the matrix with row vectors $\mathbf{v}_1, \mathbf{v}_2, \ldots, \mathbf{v}_m$ is the standard matrix for T.

5. Let $\{\mathbf{e}_1, \mathbf{e}_2, \mathbf{e}_3, \mathbf{e}_4\}$ be the standard basis for R^4, and let $T: R^4 \to R^3$ be the linear transformation for which
$$T(\mathbf{e}_1) = (1, 2, 1), \quad T(\mathbf{e}_2) = (0, 1, 0),$$
$$T(\mathbf{e}_3) = (1, 3, 0), \quad T(\mathbf{e}_4) = (1, 1, 1)$$

   (a) Find bases for the range and kernel of T.

   (b) Find the rank and nullity of T.

6. Suppose that vectors in R^3 are denoted by 1×3 matrices, and define $T: R^3 \to R^3$ by
$$T([x_1 \; x_2 \; x_3]) = [x_1 \; x_2 \; x_3] \begin{bmatrix} -1 & 2 & 4 \\ 3 & 0 & 1 \\ 2 & 2 & 5 \end{bmatrix}$$

   (a) Find a basis for the kernel of T.

   (b) Find a basis for the range of T.

7. Let $B = \{\mathbf{v}_1, \mathbf{v}_2, \mathbf{v}_3, \mathbf{v}_4\}$ be a basis for a vector space V, and let $T: V \to V$ be the linear operator for which
$$T(\mathbf{v}_1) = \mathbf{v}_1 + \mathbf{v}_2 + \mathbf{v}_3 + 3\mathbf{v}_4$$
$$T(\mathbf{v}_2) = \mathbf{v}_1 - \mathbf{v}_2 + 2\mathbf{v}_3 + 2\mathbf{v}_4$$
$$T(\mathbf{v}_3) = 2\mathbf{v}_1 - 4\mathbf{v}_2 + 5\mathbf{v}_3 + 3\mathbf{v}_4$$
$$T(\mathbf{v}_4) = -2\mathbf{v}_1 + 6\mathbf{v}_2 - 6\mathbf{v}_3 - 2\mathbf{v}_4$$

   (a) Find the rank and nullity of T.

   (b) Determine whether T is one-to-one.

8. Let V and W be vector spaces, let T, T_1, and T_2 be linear transformations from V to W, and let k be a scalar. Define new transformations, $T_1 + T_2$ and kT, by the formulas
$$(T_1 + T_2)(\mathbf{x}) = T_1(\mathbf{x}) + T_2(\mathbf{x})$$
$$(kT)(\mathbf{x}) = k(T(\mathbf{x}))$$

   (a) Show that $(T_1 + T_2): V \to W$ and $kT: V \to W$ are both linear transformations.

   (b) Show that the set of all linear transformations from V to W with the operations in part (a) is a vector space.

9. Let A and B be similar matrices. Prove:

   (a) A^T and B^T are similar.

   (b) If A and B are invertible, then A^{-1} and B^{-1} are similar.

10. (**Fredholm Alternative Theorem**) Let $T: V \to V$ be a linear operator on an n-dimensional vector space. Prove that *exactly one* of the following statements holds:

    (i) The equation $T(\mathbf{x}) = \mathbf{b}$ has a solution for all vectors $\mathbf{b}$ in V.

    (ii) Nullity of $T > 0$.

11. Let $T: M_{22} \to M_{22}$ be the linear operator defined by
$$T(X) = \begin{bmatrix} 1 & 1 \\ 0 & 0 \end{bmatrix} X + X \begin{bmatrix} 0 & 0 \\ 1 & 1 \end{bmatrix}$$

    Find the rank and nullity of T.

12. Prove: If A and B are similar matrices, and if B and C are also similar matrices, then A and C are similar matrices.

13. Let $L: M_{22} \to M_{22}$ be the linear operator that is defined by $L(M) = M^T$. Find the matrix for L with respect to the standard basis for M_{22}.

14. Let $B = \{\mathbf{u}_1, \mathbf{u}_2, \mathbf{u}_3\}$ and $B' = \{\mathbf{v}_1, \mathbf{v}_2, \mathbf{v}_3\}$ be bases for a vector space V, and let
$$P = \begin{bmatrix} 2 & -1 & 3 \\ 1 & 1 & 4 \\ 0 & 1 & 2 \end{bmatrix}$$

    be the transition matrix from B' to B.

    (a) Express $\mathbf{v}_1, \mathbf{v}_2, \mathbf{v}_3$ as linear combinations of $\mathbf{u}_1, \mathbf{u}_2, \mathbf{u}_3$.

    (b) Express $\mathbf{u}_1, \mathbf{u}_2, \mathbf{u}_3$ as linear combinations of $\mathbf{v}_1, \mathbf{v}_2, \mathbf{v}_3$.

15. Let $B = \{\mathbf{u}_1, \mathbf{u}_2, \mathbf{u}_3\}$ be a basis for a vector space V, and let $T: V \to V$ be a linear operator for which
$$[T]_B = \begin{bmatrix} -3 & 4 & 7 \\ 1 & 0 & -2 \\ 0 & 1 & 0 \end{bmatrix}$$

    Find $[T]_{B'}$, where $B' = \{\mathbf{v}_1, \mathbf{v}_2, \mathbf{v}_3\}$ is the basis for V defined by
$$\mathbf{v}_1 = \mathbf{u}_1, \quad \mathbf{v}_2 = \mathbf{u}_1 + \mathbf{u}_2, \quad \mathbf{v}_3 = \mathbf{u}_1 + \mathbf{u}_2 + \mathbf{u}_3$$

16. Show that the matrices
$$\begin{bmatrix} 1 & 1 \\ -1 & 4 \end{bmatrix} \text{ and } \begin{bmatrix} 2 & 1 \\ 1 & 3 \end{bmatrix}$$
are similar but that
$$\begin{bmatrix} 3 & 1 \\ -6 & -2 \end{bmatrix} \text{ and } \begin{bmatrix} -1 & 2 \\ 1 & 0 \end{bmatrix}$$
are not.

17. Suppose that $T: V \to V$ is a linear operator, and B is a basis for V for which
$$[T(\mathbf{x})]_B = \begin{bmatrix} x_1 - x_2 + x_3 \\ x_2 \\ x_1 - x_3 \end{bmatrix} \text{ if } [\mathbf{x}]_B = \begin{bmatrix} x_1 \\ x_2 \\ x_3 \end{bmatrix}$$
Find $[T]_B$.

18. Let $T: V \to V$ be a linear operator. Prove that T is one-to-one if and only if $\det(T) \neq 0$.

19. **(Calculus required)**
   (a) Show that if $\mathbf{f} = f(x)$ is twice differentiable, then the function $D: C^2(-\infty, \infty) \to F(-\infty, \infty)$ defined by $D(\mathbf{f}) = f''(x)$ is a linear transformation.
   (b) Find a basis for the kernel of D.
   (c) Show that the set of functions satisfying the equation $D(\mathbf{f}) = f(x)$ is a two-dimensional subspace of $C^2(-\infty, \infty)$, and find a basis for this subspace.

20. Let $T: P_2 \to R^3$ be the function defined by the formula
$$T(p(x)) = \begin{bmatrix} p(-1) \\ p(0) \\ p(1) \end{bmatrix}$$
   (a) Find $T(x^2 + 5x + 6)$.
   (b) Show that T is a linear transformation.
   (c) Show that T is one-to-one.
   (d) Find $T^{-1}(0, 3, 0)$.
   (e) Sketch the graph of the polynomial in part (d).

21. Let $x_1, x_2,$ and x_3 be distinct real numbers such that
$$x_1 < x_2 < x_3$$
and let $T: P_2 \to R^3$ be the function defined by the formula
$$T(p(x)) = \begin{bmatrix} p(x_1) \\ p(x_2) \\ p(x_3) \end{bmatrix}$$
   (a) Show that T is a linear transformation.
   (b) Show that T is one-to-one.
   (c) Verify that if $a_1, a_2,$ and a_3 are any real numbers, then
$$T^{-1}\left(\begin{bmatrix} a_1 \\ a_2 \\ a_3 \end{bmatrix}\right) = a_1 P_1(x) + a_2 P_2(x) + a_3 P_3(x)$$

where
$$P_1(x) = \frac{(x - x_2)(x - x_3)}{(x_1 - x_2)(x_1 - x_3)}$$
$$P_2(x) = \frac{(x - x_1)(x - x_3)}{(x_2 - x_1)(x_2 - x_3)}$$
$$P_3(x) = \frac{(x - x_1)(x - x_2)}{(x_3 - x_1)(x_3 - x_2)}$$
   (d) What relationship exists between the graph of the function
$$a_1 P_1(x) + a_2 P_2(x) + a_3 P_3(x)$$
   and the points (x_1, a_1), (x_2, a_2), and (x_3, a_3)?

22. **(Calculus required)** Let $p(x)$ and $q(x)$ be continuous functions, and let V be the subspace of $C(-\infty, +\infty)$ consisting of all twice differentiable functions. Define $L: V \to V$ by
$$L(y(x)) = y''(x) + p(x)y'(x) + q(x)y(x)$$
   (a) Show that L is a linear transformation.
   (b) Consider the special case where $p(x) = 0$ and $q(x) = 1$. Show that the function
$$\phi(x) = c_1 \sin x + c_2 \cos x$$
   is in the kernel of L for all real values of c_1 and c_2.

23. **(Calculus required)** Let $D: P_n \to P_n$ be the differentiation operator $D(\mathbf{p}) = \mathbf{p}'$. Show that the matrix for D relative to the basis $B = \{1, x, x^2, \ldots, x^n\}$ is
$$\begin{bmatrix} 0 & 1 & 0 & 0 & \cdots & 0 \\ 0 & 0 & 2 & 0 & \cdots & 0 \\ 0 & 0 & 0 & 3 & \cdots & 0 \\ \vdots & \vdots & \vdots & \vdots & & \vdots \\ 0 & 0 & 0 & 0 & \cdots & n \\ 0 & 0 & 0 & 0 & \cdots & 0 \end{bmatrix}$$

24. **(Calculus required)** It can be shown that for any real number c, the vectors
$$1, \quad x - c, \quad \frac{(x-c)^2}{2!}, \ldots, \frac{(x-c)^n}{n!}$$
form a basis for P_n. Find the matrix for the differentiation operator of Exercise 23 with respect to this basis.

25. **(Calculus required)** Let $J: P_n \to P_{n+1}$ be the integration transformation defined by
$$J(\mathbf{p}) = \int_0^x (a_0 + a_1 t + \cdots + a_n t^n)\, dt$$
$$= a_0 x + \frac{a_1}{2} x^2 + \cdots + \frac{a_n}{n+1} x^{n+1}$$
where $\mathbf{p} = a_0 + a_1 x + \cdots + a_n x^n$. Find the matrix for J with respect to the standard bases for P_n and P_{n+1}.

CHAPTER 9

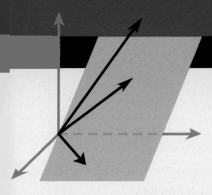

Numerical Methods

CHAPTER CONTENTS
9.1 *LU*-Decompositions 477
9.2 The Power Method 487
9.3 Internet Search Engines 496
9.4 Comparison of Procedures for Solving Linear Systems 501
9.5 Singular Value Decomposition 506
9.6 Data Compression Using Singular Value Decomposition 514

INTRODUCTION This chapter is concerned with "numerical methods" of linear algebra, an area of study that encompasses techniques for solving large-scale linear systems and for finding numerical approximations of various kinds. It is not our objective to discuss algorithms and technical issues in fine detail, since there are many excellent books on the subject. Rather, we will be concerned with introducing some of the basic ideas and exploring important contemporary applications that rely heavily on numerical ideas—singular value decomposition and data compression. A computing utility such as MATLAB, *Mathematica*, or Maple is recommended for Sections 9.2 to 9.6.

9.1 *LU*-Decompositions

Up to now, we have focused on two methods for solving linear systems, Gaussian elimination (reduction to row echelon form) and Gauss–Jordan elimination (reduction to reduced row echelon form). While these methods are fine for the small-scale problems in this text, they are not suitable for large-scale problems in which computer roundoff error, memory usage, and speed are concerns. In this section we will discuss a method for solving a linear system of n equations in n unknowns that is based on factoring its coefficient matrix into a product of lower and upper triangular matrices. This method, called "LU-decomposition," is the basis for many computer algorithms in common use.

Solving Linear Systems by Factoring

Our first goal in this section is to show how to solve a linear system $A\mathbf{x} = \mathbf{b}$ of n equations in n unknowns by factoring the coefficient matrix A into a product

$$A = LU \tag{1}$$

where L is lower triangular and U is upper triangular. Once we understand how to do this, we will discuss how to obtain the factorization itself.

Assuming that we have somehow obtained the factorization in (1), the linear system $A\mathbf{x} = \mathbf{b}$ can be solved by the following procedure, called ***LU-decomposition***.

The Method of *LU*-Decomposition

Step 1. Rewrite the system $A\mathbf{x} = \mathbf{b}$ as

$$LU\mathbf{x} = \mathbf{b} \tag{2}$$

Step 2. Define a new $n \times 1$ matrix $\mathbf{y}$ by

$$U\mathbf{x} = \mathbf{y} \tag{3}$$

Step 3. Use (3) to rewrite (2) as $L\mathbf{y} = \mathbf{b}$ and solve this system for $\mathbf{y}$.

Step 4. Substitute $\mathbf{y}$ in (3) and solve for $\mathbf{x}$.

This procedure, which is illustrated in Figure 9.1.1, replaces the single linear system $A\mathbf{x} = \mathbf{b}$ by a pair of linear systems

$$U\mathbf{x} = \mathbf{y}$$
$$L\mathbf{y} = \mathbf{b}$$

that must be solved in succession. However, since each of these systems has a triangular coefficient matrix, it generally turns out to involve no more computation to solve the two systems than to solve the original system directly.

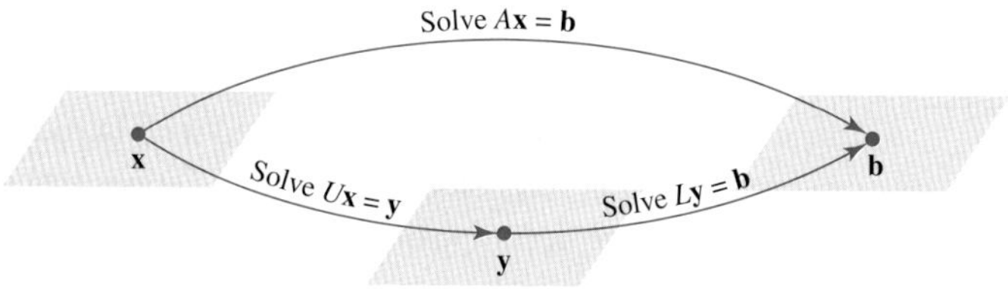

▶ Figure 9.1.1

▶ **EXAMPLE 1** Solving $A\mathbf{x} = \mathbf{b}$ by *LU*-Decomposition

Later in this section we will derive the factorization

$$\underbrace{\begin{bmatrix} 2 & 6 & 2 \\ -3 & -8 & 0 \\ 4 & 9 & 2 \end{bmatrix}}_{A} = \underbrace{\begin{bmatrix} 2 & 0 & 0 \\ -3 & 1 & 0 \\ 4 & -3 & 7 \end{bmatrix}}_{L} \underbrace{\begin{bmatrix} 1 & 3 & 1 \\ 0 & 1 & 3 \\ 0 & 0 & 1 \end{bmatrix}}_{U} \tag{4}$$

Use this result to solve the linear system

$$\underbrace{\begin{bmatrix} 2 & 6 & 2 \\ -3 & -8 & 0 \\ 4 & 9 & 2 \end{bmatrix}}_{A} \underbrace{\begin{bmatrix} x_1 \\ x_2 \\ x_3 \end{bmatrix}}_{\mathbf{x}} = \underbrace{\begin{bmatrix} 2 \\ 2 \\ 3 \end{bmatrix}}_{\mathbf{b}}$$

From (4) we can rewrite this system as

$$\underbrace{\begin{bmatrix} 2 & 0 & 0 \\ -3 & 1 & 0 \\ 4 & -3 & 7 \end{bmatrix}}_{L} \underbrace{\begin{bmatrix} 1 & 3 & 1 \\ 0 & 1 & 3 \\ 0 & 0 & 1 \end{bmatrix}}_{U} \underbrace{\begin{bmatrix} x_1 \\ x_2 \\ x_3 \end{bmatrix}}_{\mathbf{x}} = \underbrace{\begin{bmatrix} 2 \\ 2 \\ 3 \end{bmatrix}}_{\mathbf{b}} \tag{5}$$

Historical Note In 1979 an important library of machine-independent linear algebra programs called LINPACK was developed at Argonne National Laboratories. Many of the programs in that library use the decomposition methods that we will study in this section. Variations of the LINPACK routines are used in many computer programs, including MATLAB, *Mathematica*, and Maple.

As specified in Step 2 above, let us define y_1, y_2, and y_3 by the equation

$$\underbrace{\begin{bmatrix} 1 & 3 & 1 \\ 0 & 1 & 3 \\ 0 & 0 & 1 \end{bmatrix}}_{U} \underbrace{\begin{bmatrix} x_1 \\ x_2 \\ x_3 \end{bmatrix}}_{\mathbf{x}} = \underbrace{\begin{bmatrix} y_1 \\ y_2 \\ y_3 \end{bmatrix}}_{\mathbf{y}} \tag{6}$$

which allows us to rewrite (5) as

$$\underbrace{\begin{bmatrix} 2 & 0 & 0 \\ -3 & 1 & 0 \\ 4 & -3 & 7 \end{bmatrix}}_{L} \underbrace{\begin{bmatrix} y_1 \\ y_2 \\ y_3 \end{bmatrix}}_{\mathbf{y}} = \underbrace{\begin{bmatrix} 2 \\ 2 \\ 3 \end{bmatrix}}_{\mathbf{b}} \tag{7}$$

or equivalently as

$$\begin{aligned} 2y_1 &= 2 \\ -3y_1 + y_2 &= 2 \\ 4y_1 - 3y_2 + 7y_3 &= 3 \end{aligned}$$

This system can be solved by a procedure that is similar to back substitution, except that we solve the equations from the top down instead of from the bottom up. This procedure, called *forward substitution*, yields

$$y_1 = 1, \quad y_2 = 5, \quad y_3 = 2$$

(verify). As indicated in Step 4 above, we substitute these values into (6), which yields the linear system

$$\begin{bmatrix} 1 & 3 & 1 \\ 0 & 1 & 3 \\ 0 & 0 & 1 \end{bmatrix} \begin{bmatrix} x_1 \\ x_2 \\ x_3 \end{bmatrix} = \begin{bmatrix} 1 \\ 5 \\ 2 \end{bmatrix}$$

or, equivalently,

$$\begin{aligned} x_1 + 3x_2 + x_3 &= 1 \\ x_2 + 3x_3 &= 5 \\ x_3 &= 2 \end{aligned}$$

Solving this system by back substitution yields

$$x_1 = 2, \quad x_2 = -1, \quad x_3 = 2$$

(verify). ◀

Alan Mathison Turing (1912–1954)

Historical Note Although the ideas were known earlier, credit for popularizing the matrix formulation of the *LU*-decomposition is often given to the British mathematician Alan Turing for his work on the subject in 1948. Turing, one of the great geniuses of the twentieth century, is the founder of the field of artificial intelligence. Among his many accomplishments in that field, he developed the concept of an internally programmed computer before the practical technology had reached the point where the construction of such a machine was possible. During World War II Turing was secretly recruited by the British government's Code and Cypher School at Bletchley Park to help break the Nazi Enigma codes; it was Turing's statistical approach that provided the breakthrough. In addition to being a brilliant mathematician, Turing was a world-class runner who competed successfully with Olympic-level competition. Sadly, Turing, a homosexual, was tried and convicted of "gross indecency" in 1952, in violation of the then-existing British statutes. Depressed, he committed suicide at age 41 by eating an apple laced with cyanide.

[*Image: Time & Life Pictures/Getty Images, Inc.*]

Finding LU-Decompositions

Example 1 makes it clear that after A is factored into lower and upper triangular matrices, the system $A\mathbf{x} = \mathbf{b}$ can be solved by one forward substitution and one back substitution. We will now show how to obtain such factorizations. We begin with some terminology.

> **DEFINITION 1** A factorization of a square matrix A as $A = LU$, where L is lower triangular and U is upper triangular is called an **LU-decomposition** (or **LU-factorization**) of A.

Not every square matrix has an LU-decomposition. However, we will see that if it is possible to reduce a square matrix A to row echelon form by Gaussian elimination *without performing any row interchanges*, then A will have an LU-decomposition, though it may not be unique. To see why this is so, assume that A has been reduced to a row echelon form U using a sequence of row operations that does not include row interchanges. We know from Theorem 1.5.1 that these operations can be accomplished by multiplying A on the left by an appropriate sequence of elementary matrices; that is, there exist elementary matrices $E_1, E_2, \ldots, E_k$ such that

$$E_k \cdots E_2 E_1 A = U \tag{8}$$

Since elementary matrices are invertible, we can solve (8) for A as

$$A = E_1^{-1} E_2^{-1} \cdots E_k^{-1} U$$

or more briefly as

$$A = LU \tag{9}$$

where

$$L = E_1^{-1} E_2^{-1} \cdots E_k^{-1} \tag{10}$$

We now have all of the ingredients to prove the following result.

> **THEOREM 9.1.1** *If A is a square matrix that can be reduced to a row echelon form U by Gaussian elimination without row interchanges, then A can be factored as $A = LU$, where L is a lower triangular matrix.*

Proof Let L and U be the matrices in Formulas (10) and (8), respectively. The matrix U is upper triangular because it is a row echelon form of a square matrix (so all entries below its main diagonal are zero). To prove that L is lower triangular, it suffices to prove that each factor on the right side of (10) is lower triangular, since Theorem 1.7.1b will then imply that L itself is lower triangular. Since row interchanges are excluded, each E_j results either by adding a scalar multiple of one row of an identity matrix to a row below or by multiplying one row of an identity matrix by a nonzero scalar. In either case, the resulting matrix E_j is lower triangular and hence so is E_j^{-1} by Theorem 1.7.1d. This completes the proof. ◄

▶ **EXAMPLE 2** An *LU*-Decomposition

Find an LU-decomposition of

$$A = \begin{bmatrix} 2 & 6 & 2 \\ -3 & -8 & 0 \\ 4 & 9 & 2 \end{bmatrix}$$

Solution To obtain an LU-decomposition, $A = LU$, we will reduce A to a row echelon form U using Gaussian elimination and then calculate L from (10). The steps are as follows:

	Reduction to Row Echelon Form	Row Operation	Elementary Matrix Corresponding to the Row Operation	Inverse of the Elementary Matrix
	$\begin{bmatrix} 2 & 6 & 2 \\ -3 & -8 & 0 \\ 4 & 9 & 2 \end{bmatrix}$			
Step 1		$\tfrac{1}{2} \times$ row 1	$E_1 = \begin{bmatrix} \tfrac{1}{2} & 0 & 0 \\ 0 & 1 & 0 \\ 0 & 0 & 1 \end{bmatrix}$	$E_1^{-1} = \begin{bmatrix} 2 & 0 & 0 \\ 0 & 1 & 0 \\ 0 & 0 & 1 \end{bmatrix}$
	$\begin{bmatrix} 1 & 3 & 1 \\ -3 & -8 & 0 \\ 4 & 9 & 2 \end{bmatrix}$			
Step 2		$(3 \times$ row 1$) +$ row 2	$E_2 = \begin{bmatrix} 1 & 0 & 0 \\ 3 & 1 & 0 \\ 0 & 0 & 1 \end{bmatrix}$	$E_2^{-1} = \begin{bmatrix} 1 & 0 & 0 \\ -3 & 1 & 0 \\ 0 & 0 & 1 \end{bmatrix}$
	$\begin{bmatrix} 1 & 3 & 1 \\ 0 & 1 & 3 \\ 4 & 9 & 2 \end{bmatrix}$			
Step 3		$(-4 \times$ row 1$) +$ row 3	$E_3 = \begin{bmatrix} 1 & 0 & 0 \\ 0 & 1 & 0 \\ -4 & 0 & 1 \end{bmatrix}$	$E_3^{-1} = \begin{bmatrix} 1 & 0 & 0 \\ 0 & 1 & 0 \\ 4 & 0 & 1 \end{bmatrix}$
	$\begin{bmatrix} 1 & 3 & 1 \\ 0 & 1 & 3 \\ 0 & -3 & -2 \end{bmatrix}$			
Step 4		$(3 \times$ row 2$) +$ row 3	$E_4 = \begin{bmatrix} 1 & 0 & 0 \\ 0 & 1 & 0 \\ 0 & 3 & 1 \end{bmatrix}$	$E_4^{-1} = \begin{bmatrix} 1 & 0 & 0 \\ 0 & 1 & 0 \\ 0 & -3 & 1 \end{bmatrix}$
	$\begin{bmatrix} 1 & 3 & 1 \\ 0 & 1 & 3 \\ 0 & 0 & 7 \end{bmatrix}$			
Step 5		$\tfrac{1}{7} \times$ row 3	$E_5 = \begin{bmatrix} 1 & 0 & 0 \\ 0 & 1 & 0 \\ 0 & 0 & \tfrac{1}{7} \end{bmatrix}$	$E_5^{-1} = \begin{bmatrix} 1 & 0 & 0 \\ 0 & 1 & 0 \\ 0 & 0 & 7 \end{bmatrix}$
	$\begin{bmatrix} 1 & 3 & 1 \\ 0 & 1 & 3 \\ 0 & 0 & 1 \end{bmatrix} = U$			

and, from (10),

$$L = \begin{bmatrix} 2 & 0 & 0 \\ 0 & 1 & 0 \\ 0 & 0 & 1 \end{bmatrix} \begin{bmatrix} 1 & 0 & 0 \\ -3 & 1 & 0 \\ 0 & 0 & 1 \end{bmatrix} \begin{bmatrix} 1 & 0 & 0 \\ 0 & 1 & 0 \\ 4 & 0 & 1 \end{bmatrix} \begin{bmatrix} 1 & 0 & 0 \\ 0 & 1 & 0 \\ 0 & -3 & 1 \end{bmatrix} \begin{bmatrix} 1 & 0 & 0 \\ 0 & 1 & 0 \\ 0 & 0 & 7 \end{bmatrix}$$

$$= \begin{bmatrix} 2 & 0 & 0 \\ -3 & 1 & 0 \\ 4 & -3 & 7 \end{bmatrix}$$

so

$$\begin{bmatrix} 2 & 6 & 2 \\ -3 & -8 & 0 \\ 4 & 9 & 2 \end{bmatrix} = \begin{bmatrix} 2 & 0 & 0 \\ -3 & 1 & 0 \\ 4 & -3 & 7 \end{bmatrix} \begin{bmatrix} 1 & 3 & 1 \\ 0 & 1 & 3 \\ 0 & 0 & 1 \end{bmatrix}$$

is an LU-decomposition of A. ◀

Bookkeeping   As Example 2 shows, most of the work in constructing an LU-decomposition is expended in calculating L. However, *all* this work can be eliminated by some careful bookkeeping of the operations used to reduce A to U.

Because we are assuming that no row interchanges are required to reduce A to U, there are only two types of operations involved—multiplying a row by a nonzero constant, and adding a scalar multiple of one row to another. The first operation is used to introduce the leading 1's and the second to introduce zeros below the leading 1's.

In Example 2, a multiplier of $\frac{1}{2}$ was needed in Step 1 to introduce a leading 1 in the first row, and a multiplier of $\frac{1}{7}$ was needed in Step 5 to introduce a leading 1 in the third row. No actual multiplier was required to introduce a leading 1 in the second row because it was already a 1 at the end of Step 2, but for convenience let us say that the multiplier was 1. Comparing these multipliers with the successive diagonal entries of L, we see that these diagonal entries are precisely the reciprocals of the multipliers used to construct U:

$$L = \begin{bmatrix} ② & 0 & 0 \\ -3 & ① & 0 \\ 4 & -3 & ⑦ \end{bmatrix} \quad (11)$$

Also observe in Example 2 that to introduce zeros below the leading 1 in the first row, we used the operations

add 3 times the first row to the second

add -4 times the first row to the third

and to introduce the zero below the leading 1 in the second row, we used the operation

add 3 times the second row to the third

Now note in (12) that in each position below the main diagonal of L, the entry is the *negative* of the multiplier in the operation that introduced the zero in that position in U:

$$L = \begin{bmatrix} 2 & 0 & 0 \\ -3 & 1 & 0 \\ 4 & -3 & 7 \end{bmatrix} \quad (12)$$

This suggests the following procedure for constructing an LU-decomposition of a square matrix A, assuming that this matrix can be reduced to row echelon form without row interchanges.

Procedure for Constructing an LU-Decomposition

Step 1. Reduce A to a row echelon form U by Gaussian elimination without row interchanges, keeping track of the multipliers used to introduce the leading 1's and the multipliers used to introduce the zeros below the leading 1's.

Step 2. In each position along the main diagonal of L, place the reciprocal of the multiplier that introduced the leading 1 in that position in U.

Step 3. In each position below the main diagonal of L, place the negative of the multiplier used to introduce the zero in that position in U.

Step 4. Form the decomposition $A = LU$.

▶ **EXAMPLE 3 Constructing an LU-Decomposition**

Find an LU-decomposition of

$$A = \begin{bmatrix} 6 & -2 & 0 \\ 9 & -1 & 1 \\ 3 & 7 & 5 \end{bmatrix}$$

Solution We will reduce A to a row echelon form U and at each step we will fill in an entry of L in accordance with the four-step procedure above.

$$A = \begin{bmatrix} 6 & -2 & 0 \\ 9 & -1 & 1 \\ 3 & 7 & 5 \end{bmatrix} \qquad \begin{bmatrix} \bullet & 0 & 0 \\ \bullet & \bullet & 0 \\ \bullet & \bullet & \bullet \end{bmatrix} \quad \bullet \text{ denotes an unknown entry of } L.$$

$$\begin{bmatrix} ① & -\tfrac{1}{3} & 0 \\ 9 & -1 & 1 \\ 3 & 7 & 5 \end{bmatrix} \longleftarrow \text{multiplier} = \tfrac{1}{6} \qquad \begin{bmatrix} 6 & 0 & 0 \\ \bullet & \bullet & 0 \\ \bullet & \bullet & \bullet \end{bmatrix}$$

$$\begin{bmatrix} 1 & -\tfrac{1}{3} & 0 \\ ⓪ & 2 & 1 \\ ⓪ & 8 & 5 \end{bmatrix} \begin{matrix} \\ \longleftarrow \text{multiplier} = -9 \\ \longleftarrow \text{multiplier} = -3 \end{matrix} \qquad \begin{bmatrix} 6 & 0 & 0 \\ 9 & \bullet & 0 \\ 3 & \bullet & \bullet \end{bmatrix}$$

$$\begin{bmatrix} 1 & -\tfrac{1}{3} & 0 \\ 0 & ① & \tfrac{1}{2} \\ 0 & 8 & 5 \end{bmatrix} \longleftarrow \text{multiplier} = \tfrac{1}{2} \qquad \begin{bmatrix} 6 & 0 & 0 \\ 9 & 2 & 0 \\ 3 & \bullet & \bullet \end{bmatrix}$$

$$\begin{bmatrix} 1 & -\tfrac{1}{3} & 0 \\ 0 & 1 & \tfrac{1}{2} \\ 0 & ⓪ & 1 \end{bmatrix} \longleftarrow \text{multiplier} = -8 \qquad \begin{bmatrix} 6 & 0 & 0 \\ 9 & 2 & 0 \\ 3 & 8 & \bullet \end{bmatrix}$$

$$U = \begin{bmatrix} 1 & -\tfrac{1}{3} & 0 \\ 0 & 1 & \tfrac{1}{2} \\ 0 & 0 & ① \end{bmatrix} \longleftarrow \text{multiplier} = 1 \qquad L = \begin{bmatrix} 6 & 0 & 0 \\ 9 & 2 & 0 \\ 3 & 8 & 1 \end{bmatrix} \quad \begin{array}{l} \text{No actual operation is} \\ \text{performed here since} \\ \text{there is already a leading} \\ \text{1 in the third row.} \end{array}$$

Thus, we have constructed the LU-decomposition

$$A = LU = \begin{bmatrix} 6 & 0 & 0 \\ 9 & 2 & 0 \\ 3 & 8 & 1 \end{bmatrix} \begin{bmatrix} 1 & -\frac{1}{3} & 0 \\ 0 & 1 & \frac{1}{2} \\ 0 & 0 & 1 \end{bmatrix}$$

We leave it for you to confirm this end result by multiplying the factors. ◀

LU-Decompositions Are Not Unique

In the absence of restrictions, LU-decompositions are not unique. For example, if

$$A = LU = \begin{bmatrix} l_{11} & 0 & 0 \\ l_{21} & l_{22} & 0 \\ l_{31} & l_{32} & l_{33} \end{bmatrix} \begin{bmatrix} 1 & u_{12} & u_{13} \\ 0 & 1 & u_{23} \\ 0 & 0 & 1 \end{bmatrix}$$

and L has nonzero diagonal entries, then we can shift the diagonal entries from the left factor to the right factor by writing

$$A = \begin{bmatrix} 1 & 0 & 0 \\ l_{21}/l_{11} & 1 & 0 \\ l_{31}/l_{11} & l_{32}/l_{22} & 1 \end{bmatrix} \begin{bmatrix} l_{11} & 0 & 0 \\ 0 & l_{22} & 0 \\ 0 & 0 & l_{33} \end{bmatrix} \begin{bmatrix} 1 & u_{12} & u_{13} \\ 0 & 1 & u_{23} \\ 0 & 0 & 1 \end{bmatrix}$$

$$= \begin{bmatrix} 1 & 0 & 0 \\ l_{21}/l_{11} & 1 & 0 \\ l_{31}/l_{11} & l_{32}/l_{22} & 1 \end{bmatrix} \begin{bmatrix} l_{11} & l_{11}u_{12} & l_{11}u_{13} \\ 0 & l_{22} & l_{22}u_{23} \\ 0 & 0 & l_{33} \end{bmatrix}$$

which is another LU-decomposition of A.

LDU-Decompositions

The method we have described for computing LU-decompositions may result in an "asymmetry" in that the matrix U has 1's on the main diagonal but L need not. However, if it is preferred to have 1's on the main diagonal of the lower triangular factor, then we can "shift" the diagonal entries of L to a diagonal matrix D and write L as

$$L = L'D$$

where L' is a lower triangular matrix with 1's on the main diagonal. For example, a general 3×3 lower triangular matrix with nonzero entries on the main diagonal can be factored as

$$\underbrace{\begin{bmatrix} a_{11} & 0 & 0 \\ a_{21} & a_{22} & 0 \\ a_{31} & a_{32} & a_{33} \end{bmatrix}}_{L} = \underbrace{\begin{bmatrix} 1 & 0 & 0 \\ a_{21}/a_{11} & 1 & 0 \\ a_{31}/a_{11} & a_{32}/a_{22} & 1 \end{bmatrix}}_{L'} \underbrace{\begin{bmatrix} a_{11} & 0 & 0 \\ 0 & a_{22} & 0 \\ 0 & 0 & a_{33} \end{bmatrix}}_{D}$$

Note that the columns of L' are obtained by dividing each entry in the corresponding column of L by the diagonal entry in the column. Thus, for example, we can rewrite (4) as

$$\begin{bmatrix} 2 & 6 & 2 \\ -3 & -8 & 0 \\ 4 & 9 & 2 \end{bmatrix} = \begin{bmatrix} 2 & 0 & 0 \\ -3 & 1 & 0 \\ 4 & -3 & 7 \end{bmatrix} \begin{bmatrix} 1 & 3 & 1 \\ 0 & 1 & 3 \\ 0 & 0 & 1 \end{bmatrix}$$

$$= \begin{bmatrix} 1 & 0 & 0 \\ -\frac{3}{2} & 1 & 0 \\ 2 & -3 & 1 \end{bmatrix} \begin{bmatrix} 2 & 0 & 0 \\ 0 & 1 & 0 \\ 0 & 0 & 7 \end{bmatrix} \begin{bmatrix} 1 & 3 & 1 \\ 0 & 1 & 3 \\ 0 & 0 & 1 \end{bmatrix}$$

One can prove that if A is a square matrix that can be reduced to row echelon form without row interchanges, then A can be factored *uniquely* as

$$A = LDU$$

where L is a lower triangular matrix with 1's on the main diagonal, D is a diagonal matrix, and U is an upper triangular matrix with 1's on the main diagonal. This is called the ***LDU*-decomposition** (or ***LDU*-factorization**) of A.

PLU-Decompositions Many computer algorithms for solving linear systems perform row interchanges to reduce roundoff error, in which case the existence of an LU-decomposition is not guaranteed. However, it is possible to work around this problem by "preprocessing" the coefficient matrix A so that the row interchanges are performed *prior* to computing the LU-decomposition itself. More specifically, the idea is to create a matrix Q (called a ***permutation matrix***) by multiplying, in sequence, those elementary matrices that produce the row interchanges and then execute them by computing the product QA. This product can then be reduced to row echelon form *without* row interchanges, so it is assured to have an LU-decomposition

$$QA = LU \tag{13}$$

Because the matrix Q is invertible (being a product of elementary matrices), the systems $A\mathbf{x} = \mathbf{b}$ and $QA\mathbf{x} = Q\mathbf{b}$ will have the same solutions. But it follows from (13) that the latter system can be rewritten as $LU\mathbf{x} = Q\mathbf{b}$ and hence can be solved using LU-decomposition.

It is common to see Equation (13) expressed as

$$A = PLU \tag{14}$$

in which $P = Q^{-1}$. This is called a ***PLU*-decomposition** or (***PLU*-factorization**) of A.

Concept Review

- LU-decomposition
- LDU-decomposition
- PLU-decomposition

Skills

- Determine whether a square matrix has an LU-decomposition.
- Find an LU-decomposition of a square matrix.
- Use the method of LU-decomposition to solve linear systems.
- Find the LDU-decomposition of a square matrix.
- Find a PLU-decomposition of a square matrix.

Exercise Set 9.1

1. Use the method of Example 1 and the LU-decomposition

$$\begin{bmatrix} 2 & -4 \\ 3 & -2 \end{bmatrix} = \begin{bmatrix} 2 & 0 \\ 3 & 4 \end{bmatrix} \begin{bmatrix} 1 & -2 \\ 0 & 1 \end{bmatrix}$$

to solve the system

$$2x_1 - 4x_2 = -4$$
$$3x_1 - 2x_2 = 2$$

2. Use the method of Example 1 and the LU-decomposition

$$\begin{bmatrix} 3 & -6 & -3 \\ 2 & 0 & 6 \\ -4 & 7 & 4 \end{bmatrix} = \begin{bmatrix} 3 & 0 & 0 \\ 2 & 4 & 0 \\ -4 & -1 & 2 \end{bmatrix} \begin{bmatrix} 1 & -2 & -1 \\ 0 & 1 & 2 \\ 0 & 0 & 1 \end{bmatrix}$$

to solve the system

$$3x_1 - 6x_2 - 3x_3 = -3$$
$$2x_1 + 6x_3 = -22$$
$$-4x_1 + 7x_2 + 4x_3 = 3$$

In Exercises 3–10, find an LU-decomposition of the coefficient matrix, and then use the method of Example 1 to solve the system.

3. $\begin{bmatrix} 5 & 10 \\ 1 & 3 \end{bmatrix} \begin{bmatrix} x_1 \\ x_2 \end{bmatrix} = \begin{bmatrix} -5 \\ -3 \end{bmatrix}$

4. $\begin{bmatrix} -5 & -10 \\ 6 & 5 \end{bmatrix} \begin{bmatrix} x_1 \\ x_2 \end{bmatrix} = \begin{bmatrix} -10 \\ 19 \end{bmatrix}$

5. $\begin{bmatrix} 2 & 4 & -2 \\ 6 & 0 & 3 \\ 4 & 2 & 4 \end{bmatrix} \begin{bmatrix} x_1 \\ x_2 \\ x_3 \end{bmatrix} = \begin{bmatrix} 4 \\ 15 \\ 6 \end{bmatrix}$

6. $\begin{bmatrix} -3 & 12 & -6 \\ 1 & -2 & 2 \\ 0 & 1 & 1 \end{bmatrix} \begin{bmatrix} x_1 \\ x_2 \\ x_3 \end{bmatrix} = \begin{bmatrix} -33 \\ 7 \\ -1 \end{bmatrix}$

7. $\begin{bmatrix} 1 & 4 & 3 \\ -1 & -1 & 3 \\ 2 & 9 & 8 \end{bmatrix} \begin{bmatrix} x_1 \\ x_2 \\ x_3 \end{bmatrix} = \begin{bmatrix} 4 \\ 8 \\ 12 \end{bmatrix}$

8. $\begin{bmatrix} -1 & -3 & -4 \\ 3 & 10 & -10 \\ -2 & -4 & 11 \end{bmatrix} \begin{bmatrix} x_1 \\ x_2 \\ x_3 \end{bmatrix} = \begin{bmatrix} -6 \\ -3 \\ 9 \end{bmatrix}$

9. $\begin{bmatrix} -2 & 0 & -2 & 2 \\ 2 & 1 & 1 & 0 \\ 1 & 2 & 0 & 1 \\ 0 & 1 & -3 & 7 \end{bmatrix} \begin{bmatrix} x_1 \\ x_2 \\ x_3 \\ x_4 \end{bmatrix} = \begin{bmatrix} 4 \\ -6 \\ -3 \\ -10 \end{bmatrix}$

10. $\begin{bmatrix} 2 & -4 & 0 & 0 \\ 1 & 2 & 12 & 0 \\ 0 & -1 & -4 & -5 \\ 0 & 0 & 2 & 11 \end{bmatrix} \begin{bmatrix} x_1 \\ x_2 \\ x_3 \\ x_4 \end{bmatrix} = \begin{bmatrix} 8 \\ 0 \\ 1 \\ 0 \end{bmatrix}$

11. Let
$$A = \begin{bmatrix} 4 & 4 & 0 \\ 8 & 6 & 2 \\ -4 & -10 & 8 \end{bmatrix}$$

(a) Find an LU-decomposition of A.

(b) Express A in the form $A = L_1 D U_1$, where L_1 is lower triangular with 1's along the main diagonal, U_1 is upper triangular, and D is a diagonal matrix.

(c) Express A in the form $A = L_2 U_2$, where L_2 is lower triangular with 1's along the main diagonal and U_2 is upper triangular.

In Exercises 12–13, find the LDU-decomposition of A

12. $A = \begin{bmatrix} 2 & 4 \\ -4 & 1 \end{bmatrix}$

13. $A = \begin{bmatrix} 3 & -12 & 6 \\ 0 & 2 & 0 \\ 6 & -28 & 13 \end{bmatrix}$

14. (a) Show that the matrix
$$\begin{bmatrix} 0 & 1 \\ 1 & 0 \end{bmatrix}$$
has no LU-decomposition.

(b) Find a PLU-decomposition of this matrix.

In Exercises 15–16, use the given PLU-decomposition of A to solve the linear system $A\mathbf{x} = \mathbf{b}$ by rewriting it as $P^{-1}A\mathbf{x} = P^{-1}\mathbf{b}$ and solving this system by LU-decomposition.

15. $\mathbf{b} = \begin{bmatrix} 2 \\ 1 \\ 5 \end{bmatrix}$; $A = \begin{bmatrix} 0 & 1 & 4 \\ 1 & 2 & 2 \\ 3 & 1 & 3 \end{bmatrix}$;

$A = \begin{bmatrix} 0 & 1 & 0 \\ 1 & 0 & 0 \\ 0 & 0 & 1 \end{bmatrix} \begin{bmatrix} 1 & 0 & 0 \\ 0 & 1 & 0 \\ 3 & -5 & 1 \end{bmatrix} \begin{bmatrix} 1 & 2 & 2 \\ 0 & 1 & 4 \\ 0 & 0 & 17 \end{bmatrix} = PLU$

16. $\mathbf{b} = \begin{bmatrix} 3 \\ 0 \\ 6 \end{bmatrix}$; $A = \begin{bmatrix} 4 & 1 & 2 \\ 0 & 2 & 1 \\ 8 & 1 & 8 \end{bmatrix}$;

$A = \begin{bmatrix} 1 & 0 & 0 \\ 0 & 0 & 1 \\ 0 & 1 & 0 \end{bmatrix} \begin{bmatrix} 1 & 0 & 0 \\ 2 & 1 & 0 \\ 0 & -2 & 1 \end{bmatrix} \begin{bmatrix} 4 & 1 & 2 \\ 0 & -1 & 4 \\ 0 & 0 & 9 \end{bmatrix} = PLU$

In Exercises 17–18, find a PLU-decomposition of A, and use it to solve the linear system $A\mathbf{x} = \mathbf{b}$ by the method of Exercises 15 and 16.

17. $A = \begin{bmatrix} 5 & 15 & 0 \\ 2 & 6 & 1 \\ 0 & 5 & 0 \end{bmatrix}$; $\mathbf{b} = \begin{bmatrix} 0 \\ 2 \\ 5 \end{bmatrix}$

18. $A = \begin{bmatrix} 0 & 3 & -2 \\ 1 & 1 & 4 \\ 2 & 2 & 5 \end{bmatrix}$; $\mathbf{b} = \begin{bmatrix} 7 \\ 5 \\ -2 \end{bmatrix}$

19. Let
$$A = \begin{bmatrix} a & b \\ c & d \end{bmatrix}$$

(a) Prove: If $a \neq 0$, then the matrix A has a unique LU-decomposition with 1's along the main diagonal of L.

(b) Find the LU-decomposition described in part (a).

20. Let $A\mathbf{x} = \mathbf{b}$ be a linear system of n equations in n unknowns, and assume that A is an invertible matrix that can be reduced to row-echelon form without row interchanges. How many additions and multiplications are required to solve the system by the method of Example 1?

21. Prove: If A is any $n \times n$ matrix, then A can be factored as $A = PLU$, where L is lower triangular, U is upper triangular, and P can be obtained by interchanging the rows of I_n appropriately. [*Hint:* Let U be a row echelon form of A, and let all row interchanges required in the reduction of A to U be performed first.]

True-False Exercises

In parts (a)–(e) determine whether the statement is true or false, and justify your answer.

(a) Every square matrix has an LU-decomposition.

(b) If a square matrix A is row equivalent to an upper triangular matrix U, then A has an LU-decomposition.

(c) If $L_1, L_2, \ldots, L_k$ are $n \times n$ lower triangular matrices, then the product $L_1 L_2 \cdots L_k$ is lower triangular.

(d) If a square matrix A has an LU-decomposition, then A has a unique LDU-decomposition.

(e) Every square matrix has a PLU-decomposition.

9.2 The Power Method

The eigenvalues of a square matrix can, in theory, be found by solving the characteristic equation. However, this procedure has so many computational difficulties that it is almost never used in applications. In this section we will discuss an algorithm that can be used to approximate the eigenvalue with greatest absolute value and a corresponding eigenvector. This particular eigenvalue and its corresponding eigenvectors are important because they arise naturally in many iterative processes. The methods we will study in this section have recently been used to create Internet search engines such as Google. We will discuss this application in the next section.

The Power Method There are many applications in which some vector $\mathbf{x}_0$ in R^n is multiplied repeatedly by an $n \times n$ matrix A to produce a sequence

$$\mathbf{x}_0, \quad A\mathbf{x}_0, \quad A^2\mathbf{x}_0, \ldots, \quad A^k\mathbf{x}_0, \ldots$$

We call a sequence of this form a ***power sequence generated by*** A. In this section we will be concerned with the convergence of power sequences and how such sequences can be used to approximate eigenvalues and eigenvectors. For this purpose, we make the following definition.

> **DEFINITION 1** If the *distinct* eigenvalues of a matrix A are $\lambda_1, \lambda_2, \ldots, \lambda_k$, and if $|\lambda_1|$ is larger than $|\lambda_2|, \ldots, |\lambda_k|$, then λ_1 is called a ***dominant eigenvalue*** of A. Any eigenvector corresponding to a dominant eigenvalue is called a ***dominant eigenvector*** of A.

▶ **EXAMPLE 1** Dominant Eigenvalues

Some matrices have dominant eigenvalues and some do not. For example, if the distinct eigenvalues of a matrix are

$$\lambda_1 = -4, \quad \lambda_2 = -2, \quad \lambda_3 = 1, \quad \lambda_4 = 3$$

then $\lambda_1 = -4$ is dominant since $|\lambda_1| = 4$ is greater than the absolute values of all the other eigenvalues; but if the distinct eigenvalues of a matrix are

$$\lambda_1 = 7, \quad \lambda_2 = -7, \quad \lambda_3 = -2, \quad \lambda_4 = 5$$

then $|\lambda_1| = |\lambda_2| = 7$, so there is no eigenvalue whose absolute value is greater than the absolute value of all the other eigenvalues. ◀

The most important theorems about convergence of power sequences apply to $n \times n$ matrices with n linearly independent eigenvectors (symmetric matrices, for example), so we will limit our discussion to this case in this section.

488 Chapter 9 Numerical Methods

THEOREM 9.2.1 *Let A be a symmetric $n \times n$ matrix with a positive** *dominant eigenvalue λ. If $\mathbf{x}_0$ is a unit vector in R^n that is not orthogonal to the eigenspace corresponding to λ, then the normalized power sequence*

$$\mathbf{x}_0, \quad \mathbf{x}_1 = \frac{A\mathbf{x}_0}{\|A\mathbf{x}_0\|}, \quad \mathbf{x}_2 = \frac{A\mathbf{x}_1}{\|A\mathbf{x}_1\|}, \ldots, \quad \mathbf{x}_k = \frac{A\mathbf{x}_{k-1}}{\|A\mathbf{x}_{k-1}\|}, \ldots \quad (1)$$

converges to a unit dominant eigenvector, and the sequence

$$A\mathbf{x}_1 \cdot \mathbf{x}_1, \quad A\mathbf{x}_2 \cdot \mathbf{x}_2, \quad A\mathbf{x}_3 \cdot \mathbf{x}_3, \ldots, \quad A\mathbf{x}_k \cdot \mathbf{x}_k, \ldots \quad (2)$$

converges to the dominant eigenvalue λ.

Remark In the exercises we will ask you to show that (1) can also be expressed as

$$\mathbf{x}_0, \quad \mathbf{x}_1 = \frac{A\mathbf{x}_0}{\|A\mathbf{x}_0\|}, \quad \mathbf{x}_2 = \frac{A^2\mathbf{x}_0}{\|A^2\mathbf{x}_0\|}, \ldots, \quad \mathbf{x}_k = \frac{A^k\mathbf{x}_0}{\|A^k\mathbf{x}_0\|}, \ldots \quad (3)$$

This form of the power sequence expresses each iterate in terms of the starting vector $\mathbf{x}_0$, rather than in terms of its predecessor.

We will not prove Theorem 9.2.1, but we can make it plausible geometrically in the 2×2 case where A is a symmetric matrix with distinct positive eigenvalues, λ_1 and λ_2, one of which is dominant. To be specific, assume that λ_1 is dominant and

$$\lambda_1 > \lambda_2 > 0$$

Since we are assuming that A is symmetric and has distinct eigenvalues, it follows from Theorem 7.2.2 that the eigenspaces corresponding to λ_1 and λ_2 are perpendicular lines through the origin. Thus, the assumption that $\mathbf{x}_0$ is a unit vector that is not orthogonal to the eigenspace corresponding to λ_1 implies that $\mathbf{x}_0$ does not lie in the eigenspace corresponding to λ_2. To see the geometric effect of multiplying $\mathbf{x}_0$ by A, it will be useful to split $\mathbf{x}_0$ into the sum

$$\mathbf{x}_0 = \mathbf{v}_0 + \mathbf{w}_0 \quad (4)$$

where $\mathbf{v}_0$ and $\mathbf{w}_0$ are the orthogonal projections of $\mathbf{x}_0$ on the eigenspaces of λ_1 and λ_2, respectively (Figure 9.2.1a).

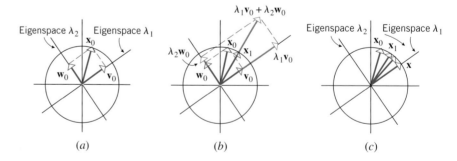

▶ **Figure 9.2.1**   (a)   (b)   (c)

This enables us to express $A\mathbf{x}_0$ as

$$A\mathbf{x}_0 = A\mathbf{v}_0 + A\mathbf{w}_0 = \lambda_1 \mathbf{v}_0 + \lambda_2 \mathbf{w}_0 \quad (5)$$

---

*If the dominant eigenvalue is not positive, sequence (2) will still converge to the dominant eigenvalue, but sequence (1) may not converge to a *specific* dominant eigenvector because of *alternation* (see Exercise 11). Nevertheless, each term of (1) will closely approximate *some* dominant eigenvector for sufficiently large values of k.

which tells us that multiplying $\mathbf{x}_0$ by A "scales" the terms $\mathbf{v}_0$ and $\mathbf{w}_0$ in (4) by λ_1 and λ_2, respectively. However, λ_1 is larger than λ_2, so the scaling is greater in the direction of $\mathbf{v}_0$ than in the direction of $\mathbf{w}_0$. Thus, multiplying $\mathbf{x}_0$ by A "pulls" $\mathbf{x}_0$ toward the eigenspace of λ_1, and normalizing produces a vector $\mathbf{x}_1 = A\mathbf{x}_0/\|A\mathbf{x}_0\|$, which is on the unit circle and is closer to the eigenspace of λ_1 than $\mathbf{x}_0$ (Figure 9.2.1b). Similarly, multiplying $\mathbf{x}_1$ by A and normalizing produces a unit vector $\mathbf{x}_2$ that is closer to the eigenspace of λ_1 than $\mathbf{x}_1$. Thus, it seems reasonable that by repeatedly multiplying by A and normalizing we will produce a sequence of vectors $\mathbf{x}_k$ that lie on the unit circle and converge to a unit vector $\mathbf{x}$ in the eigenspace of λ_1 (Figure 9.2.1c). Moreover, if $\mathbf{x}_k$ converges to $\mathbf{x}$, then it also seems reasonable that $A\mathbf{x}_k \cdot \mathbf{x}_k$ will converge to

$$A\mathbf{x} \cdot \mathbf{x} = \lambda_1 \mathbf{x} \cdot \mathbf{x} = \lambda_1 \|\mathbf{x}\|^2 = \lambda_1$$

which is the dominant eigenvalue of A.

The Power Method with Euclidean Scaling

Theorem 9.2.1 provides us with an algorithm for approximating the dominant eigenvalue and a corresponding unit eigenvector of a symmetric matrix A, provided the dominant eigenvalue is positive. This algorithm, called the ***power method with Euclidean scaling***, is as follows:

The Power Method with Euclidean Scaling

Step 1. Choose an arbitrary nonzero vector and normalize it, if need be, to obtain a unit vector $\mathbf{x}_0$.

Step 2. Compute $A\mathbf{x}_0$ and normalize it to obtain the first approximation $\mathbf{x}_1$ to a dominant unit eigenvector. Compute $A\mathbf{x}_1 \cdot \mathbf{x}_1$ to obtain the first approximation to the dominant eigenvalue.

Step 3. Compute $A\mathbf{x}_1$ and normalize it to obtain the second approximation $\mathbf{x}_2$ to a dominant unit eigenvector. Compute $A\mathbf{x}_2 \cdot \mathbf{x}_2$ to obtain the second approximation to the dominant eigenvalue.

Step 4. Compute $A\mathbf{x}_2$ and normalize it to obtain the third approximation $\mathbf{x}_3$ to a dominant unit eigenvector. Compute $A\mathbf{x}_3 \cdot \mathbf{x}_3$ to obtain the third approximation to the dominant eigenvalue.

Continuing in this way will usually generate a sequence of better and better approximations to the dominant eigenvalue and a corresponding unit eigenvector.[*]

▶ **EXAMPLE 2** **The Power Method with Euclidean Scaling**

Apply the power method with Euclidean scaling to

$$A = \begin{bmatrix} 3 & 2 \\ 2 & 3 \end{bmatrix} \quad \text{with} \quad \mathbf{x}_0 = \begin{bmatrix} 1 \\ 0 \end{bmatrix}$$

Stop at $\mathbf{x}_5$ and compare the resulting approximations to the exact values of the dominant eigenvalue and eigenvector.

---

[*]If the vector $\mathbf{x}_0$ happens to be orthogonal to the eigenspace of the dominant eigenvalue, then the hypotheses of Theorem 9.2.1 will be violated and the method may fail. However, the reality is that computer roundoff errors usually perturb $\mathbf{x}_0$ enough to destroy any orthogonality and make the algorithm work. This is one instance in which errors help to obtain correct results!

Solution We will leave it for you to show that the eigenvalues of A are $\lambda = 1$ and $\lambda = 5$ and that the eigenspace corresponding to the dominant eigenvalue $\lambda = 5$ is the line represented by the parametric equations $x_1 = t$, $x_2 = t$, which we can write in vector form as

$$\mathbf{x} = t \begin{bmatrix} 1 \\ 1 \end{bmatrix} \tag{6}$$

Setting $t = 1/\sqrt{2}$ yields the normalized dominant eigenvector

$$\mathbf{v}_1 = \begin{bmatrix} \frac{1}{\sqrt{2}} \\ \frac{1}{\sqrt{2}} \end{bmatrix} \approx \begin{bmatrix} 0.707106781187\ldots \\ 0.707106781187\ldots \end{bmatrix} \tag{7}$$

Now let us see what happens when we use the power method, starting with the unit vector $\mathbf{x}_0$.

$$A\mathbf{x}_0 = \begin{bmatrix} 3 & 2 \\ 2 & 3 \end{bmatrix} \begin{bmatrix} 1 \\ 0 \end{bmatrix} = \begin{bmatrix} 3 \\ 2 \end{bmatrix} \qquad \mathbf{x}_1 = \frac{A\mathbf{x}_0}{\|A\mathbf{x}_0\|} = \frac{1}{\sqrt{13}} \begin{bmatrix} 3 \\ 2 \end{bmatrix} \approx \frac{1}{3.60555} \begin{bmatrix} 3 \\ 2 \end{bmatrix} \approx \begin{bmatrix} 0.83205 \\ 0.55470 \end{bmatrix}$$

$$A\mathbf{x}_1 \approx \begin{bmatrix} 3 & 2 \\ 2 & 3 \end{bmatrix} \begin{bmatrix} 0.83205 \\ 0.55470 \end{bmatrix} \approx \begin{bmatrix} 3.60555 \\ 3.32820 \end{bmatrix} \qquad \mathbf{x}_2 = \frac{A\mathbf{x}_1}{\|A\mathbf{x}_1\|} \approx \frac{1}{4.90682} \begin{bmatrix} 3.60555 \\ 3.32820 \end{bmatrix} \approx \begin{bmatrix} 0.73480 \\ 0.67828 \end{bmatrix}$$

$$A\mathbf{x}_2 \approx \begin{bmatrix} 3 & 2 \\ 2 & 3 \end{bmatrix} \begin{bmatrix} 0.73480 \\ 0.67828 \end{bmatrix} \approx \begin{bmatrix} 3.56097 \\ 3.50445 \end{bmatrix} \qquad \mathbf{x}_3 = \frac{A\mathbf{x}_2}{\|A\mathbf{x}_2\|} \approx \frac{1}{4.99616} \begin{bmatrix} 3.56097 \\ 3.50445 \end{bmatrix} \approx \begin{bmatrix} 0.71274 \\ 0.70143 \end{bmatrix}$$

$$A\mathbf{x}_3 \approx \begin{bmatrix} 3 & 2 \\ 2 & 3 \end{bmatrix} \begin{bmatrix} 0.71274 \\ 0.70143 \end{bmatrix} \approx \begin{bmatrix} 3.54108 \\ 3.52976 \end{bmatrix} \qquad \mathbf{x}_4 = \frac{A\mathbf{x}_3}{\|A\mathbf{x}_3\|} \approx \frac{1}{4.99985} \begin{bmatrix} 3.54108 \\ 3.52976 \end{bmatrix} \approx \begin{bmatrix} 0.70824 \\ 0.70597 \end{bmatrix}$$

$$A\mathbf{x}_4 \approx \begin{bmatrix} 3 & 2 \\ 2 & 3 \end{bmatrix} \begin{bmatrix} 0.70824 \\ 0.70597 \end{bmatrix} \approx \begin{bmatrix} 3.53666 \\ 3.53440 \end{bmatrix} \qquad \mathbf{x}_5 = \frac{A\mathbf{x}_4}{\|A\mathbf{x}_4\|} \approx \frac{1}{4.99999} \begin{bmatrix} 3.53666 \\ 3.53440 \end{bmatrix} \approx \begin{bmatrix} 0.70733 \\ 0.70688 \end{bmatrix}$$

$$\lambda^{(1)} = (A\mathbf{x}_1) \cdot \mathbf{x}_1 = (A\mathbf{x}_1)^T \mathbf{x}_1 \approx \begin{bmatrix} 3.60555 & 3.32820 \end{bmatrix} \begin{bmatrix} 0.83205 \\ 0.55470 \end{bmatrix} \approx 4.84615$$

$$\lambda^{(2)} = (A\mathbf{x}_2) \cdot \mathbf{x}_2 = (A\mathbf{x}_2)^T \mathbf{x}_2 \approx \begin{bmatrix} 3.56097 & 3.50445 \end{bmatrix} \begin{bmatrix} 0.73480 \\ 0.67828 \end{bmatrix} \approx 4.99361$$

$$\lambda^{(3)} = (A\mathbf{x}_3) \cdot \mathbf{x}_3 = (A\mathbf{x}_3)^T \mathbf{x}_3 \approx \begin{bmatrix} 3.54108 & 3.52976 \end{bmatrix} \begin{bmatrix} 0.71274 \\ 0.70143 \end{bmatrix} \approx 4.99974$$

$$\lambda^{(4)} = (A\mathbf{x}_4) \cdot \mathbf{x}_4 = (A\mathbf{x}_4)^T \mathbf{x}_4 \approx \begin{bmatrix} 3.53666 & 3.53440 \end{bmatrix} \begin{bmatrix} 0.70824 \\ 0.70597 \end{bmatrix} \approx 4.99999$$

$$\lambda^{(5)} = (A\mathbf{x}_5) \cdot \mathbf{x}_5 = (A\mathbf{x}_5)^T \mathbf{x}_5 \approx \begin{bmatrix} 3.53576 & 3.53531 \end{bmatrix} \begin{bmatrix} 0.70733 \\ 0.70688 \end{bmatrix} \approx 5.00000$$

It is accidental that $\lambda^{(5)}$ (the fifth approximation) produced five decimal place accuracy. In general, n iterations need not produce n decimal place accuracy.

Thus, $\lambda^{(5)}$ approximates the dominant eigenvalue to five decimal place accuracy and $\mathbf{x}_5$ approximates the dominant eigenvector in (7) correctly to three decimal place accuracy. ◀

The Power Method with Maximum Entry Scaling

There is a variation of the power method in which the iterates, rather than being normalized at each stage, are scaled to make the maximum entry 1. To describe this method, it will be convenient to denote the maximum *absolute value* of the entries in a vector $\mathbf{x}$ by max(**x**). Thus, for example, if

$$\mathbf{x} = \begin{bmatrix} 5 \\ 3 \\ -7 \\ 2 \end{bmatrix}$$

then max(**x**) = 7. We will need the following variation of Theorem 9.2.1.

THEOREM 9.2.2 *Let A be a symmetric $n \times n$ matrix with a positive dominant[*] eigenvalue λ. If $\mathbf{x}_0$ is a nonzero vector in R^n that is not orthogonal to the eigenspace corresponding to λ, then the sequence*

$$\mathbf{x}_0, \quad \mathbf{x}_1 = \frac{A\mathbf{x}_0}{\max(A\mathbf{x}_0)}, \quad \mathbf{x}_2 = \frac{A\mathbf{x}_1}{\max(A\mathbf{x}_1)}, \ldots, \quad \mathbf{x}_k = \frac{A\mathbf{x}_{k-1}}{\max(A\mathbf{x}_{k-1})}, \ldots \quad (8)$$

converges to an eigenvector corresponding to λ, and the sequence

$$\frac{A\mathbf{x}_1 \cdot \mathbf{x}_1}{\mathbf{x}_1 \cdot \mathbf{x}_1}, \quad \frac{A\mathbf{x}_2 \cdot \mathbf{x}_2}{\mathbf{x}_2 \cdot \mathbf{x}_2}, \quad \frac{A\mathbf{x}_3 \cdot \mathbf{x}_3}{\mathbf{x}_3 \cdot \mathbf{x}_3}, \ldots, \quad \frac{A\mathbf{x}_k \cdot \mathbf{x}_k}{\mathbf{x}_k \cdot \mathbf{x}_k}, \ldots \quad (9)$$

converges to λ.

Remark In the exercises we will ask you to show that (8) can be written in the alternative form

$$\mathbf{x}_0, \quad \mathbf{x}_1 = \frac{A\mathbf{x}_0}{\max(A\mathbf{x}_0)}, \quad \mathbf{x}_2 = \frac{A^2\mathbf{x}_0}{\max(A^2\mathbf{x}_0)}, \ldots, \quad \mathbf{x}_k = \frac{A^k\mathbf{x}_0}{\max(A^k\mathbf{x}_0)}, \ldots \quad (10)$$

which expresses the iterates in terms of the initial vector $\mathbf{x}_0$.

We will omit the proof of this theorem, but if we accept that (8) converges to an eigenvector of A, then it is not hard to see why (9) converges to the dominant eigenvalue. For this purpose we note that each term in (9) is of the form

$$\frac{A\mathbf{x} \cdot \mathbf{x}}{\mathbf{x} \cdot \mathbf{x}} \quad (11)$$

which is called a **Rayleigh quotient** of A. In the case where λ is an eigenvalue of A and **x** is a corresponding eigenvector, the Rayleigh quotient is

$$\frac{A\mathbf{x} \cdot \mathbf{x}}{\mathbf{x} \cdot \mathbf{x}} = \frac{\lambda \mathbf{x} \cdot \mathbf{x}}{\mathbf{x} \cdot \mathbf{x}} = \frac{\lambda(\mathbf{x} \cdot \mathbf{x})}{\mathbf{x} \cdot \mathbf{x}} = \lambda$$

Thus, if $\mathbf{x}_k$ converges to a dominant eigenvector **x**, then it seems reasonable that

$$\frac{A\mathbf{x}_k \cdot \mathbf{x}_k}{\mathbf{x}_k \cdot \mathbf{x}_k} \quad \text{converges to} \quad \frac{A\mathbf{x} \cdot \mathbf{x}}{\mathbf{x} \cdot \mathbf{x}} = \lambda$$

which is the dominant eigenvalue.

Theorem 9.2.2 produces the following algorithm, called the **power method with maximum entry scaling**.

---

[*]As in Theorem 9.2.1, if the dominant eigenvalue is not positive, sequence (9) will still converge to the dominant eigenvalue, but sequence (8) may not converge to a *specific* dominant eigenvector. Nevertheless, each term of (8) will closely approximate *some* dominant eigenvector for sufficiently large values of k.

**John William Strutt Rayleigh
(1842–1919)**

Historical Note The British mathematical physicist John Rayleigh won the Nobel prize in physics in 1904 for his discovery of the inert gas argon. Rayleigh also made fundamental discoveries in acoustics and optics, and his work in wave phenomena enabled him to give the first accurate explanation of why the sky is blue.

[*Image: The Granger Collection, New York*]

The Power Method with Maximum Entry Scaling

Step 1. Choose an arbitrary nonzero vector $\mathbf{x}_0$.

Step 2. Compute $A\mathbf{x}_0$ and multiply it by the factor $1/\max(A\mathbf{x}_0)$ to obtain the first approximation $\mathbf{x}_1$ to a dominant eigenvector. Compute the Rayleigh quotient of $\mathbf{x}_1$ to obtain the first approximation to the dominant eigenvalue.

Step 3. Compute $A\mathbf{x}_1$ and scale it by the factor $1/\max(A\mathbf{x}_1)$ to obtain the second approximation $\mathbf{x}_2$ to a dominant eigenvector. Compute the Rayleigh quotient of $\mathbf{x}_2$ to obtain the second approximation to the dominant eigenvalue.

Step 4. Compute $A\mathbf{x}_2$ and scale it by the factor $1/\max(A\mathbf{x}_2)$ to obtain the third approximation $\mathbf{x}_3$ to a dominant eigenvector. Compute the Rayleigh quotient of $\mathbf{x}_3$ to obtain the third approximation to the dominant eigenvalue.

Continuing in this way will generate a sequence of better and better approximations to the dominant eigenvalue and a corresponding eigenvector.

▶ **EXAMPLE 3** **Example 2 Revisited Using Maximum Entry Scaling**

Apply the power method with maximum entry scaling to

$$A = \begin{bmatrix} 3 & 2 \\ 2 & 3 \end{bmatrix} \quad \text{with} \quad \mathbf{x}_0 = \begin{bmatrix} 1 \\ 0 \end{bmatrix}$$

Stop at $\mathbf{x}_5$ and compare the resulting approximations to the exact values and to the approximations obtained in Example 2.

Solution We leave it for you to confirm that

$$A\mathbf{x}_0 = \begin{bmatrix} 3 & 2 \\ 2 & 3 \end{bmatrix}\begin{bmatrix} 1 \\ 0 \end{bmatrix} = \begin{bmatrix} 3 \\ 2 \end{bmatrix} \qquad \mathbf{x}_1 = \frac{A\mathbf{x}_0}{\max(A\mathbf{x}_0)} = \frac{1}{3}\begin{bmatrix} 3 \\ 2 \end{bmatrix} \approx \begin{bmatrix} 1.00000 \\ 0.66667 \end{bmatrix}$$

$$A\mathbf{x}_1 \approx \begin{bmatrix} 3 & 2 \\ 2 & 3 \end{bmatrix}\begin{bmatrix} 1.00000 \\ 0.66667 \end{bmatrix} \approx \begin{bmatrix} 4.33333 \\ 4.00000 \end{bmatrix} \qquad \mathbf{x}_2 = \frac{A\mathbf{x}_1}{\max(A\mathbf{x}_1)} \approx \frac{1}{4.33333}\begin{bmatrix} 4.33333 \\ 4.00000 \end{bmatrix} \approx \begin{bmatrix} 1.00000 \\ 0.92308 \end{bmatrix}$$

$$A\mathbf{x}_2 \approx \begin{bmatrix} 3 & 2 \\ 2 & 3 \end{bmatrix}\begin{bmatrix} 1.00000 \\ 0.92308 \end{bmatrix} \approx \begin{bmatrix} 4.84615 \\ 4.76923 \end{bmatrix} \qquad \mathbf{x}_3 = \frac{A\mathbf{x}_2}{\max(A\mathbf{x}_2)} \approx \frac{1}{4.84615}\begin{bmatrix} 4.84615 \\ 4.76923 \end{bmatrix} \approx \begin{bmatrix} 1.00000 \\ 0.98413 \end{bmatrix}$$

$$A\mathbf{x}_3 \approx \begin{bmatrix} 3 & 2 \\ 2 & 3 \end{bmatrix}\begin{bmatrix} 1.00000 \\ 0.98413 \end{bmatrix} \approx \begin{bmatrix} 4.96825 \\ 4.95238 \end{bmatrix} \qquad \mathbf{x}_4 = \frac{A\mathbf{x}_3}{\max(A\mathbf{x}_3)} \approx \frac{1}{4.96825}\begin{bmatrix} 4.96825 \\ 4.95238 \end{bmatrix} \approx \begin{bmatrix} 1.00000 \\ 0.99681 \end{bmatrix}$$

$$A\mathbf{x}_4 \approx \begin{bmatrix} 3 & 2 \\ 2 & 3 \end{bmatrix}\begin{bmatrix} 1.00000 \\ 0.99681 \end{bmatrix} \approx \begin{bmatrix} 4.99361 \\ 4.99042 \end{bmatrix} \qquad \mathbf{x}_5 = \frac{A\mathbf{x}_4}{\max(A\mathbf{x}_4)} \approx \frac{1}{4.99361}\begin{bmatrix} 4.99361 \\ 4.99042 \end{bmatrix} \approx \begin{bmatrix} 1.00000 \\ 0.99936 \end{bmatrix}$$

$$\lambda^{(1)} = \frac{A\mathbf{x}_1 \cdot \mathbf{x}_1}{\mathbf{x}_1 \cdot \mathbf{x}_1} = \frac{(A\mathbf{x}_1)^T \mathbf{x}_1}{\mathbf{x}_1^T \mathbf{x}_1} \approx \frac{7.00000}{1.44444} \approx 4.84615$$

$$\lambda^{(2)} = \frac{A\mathbf{x}_2 \cdot \mathbf{x}_2}{\mathbf{x}_2 \cdot \mathbf{x}_2} = \frac{(A\mathbf{x}_2)^T \mathbf{x}_2}{\mathbf{x}_2^T \mathbf{x}_2} \approx \frac{9.24852}{1.85207} \approx 4.99361$$

$$\lambda^{(3)} = \frac{A\mathbf{x}_3 \cdot \mathbf{x}_3}{\mathbf{x}_3 \cdot \mathbf{x}_3} = \frac{(A\mathbf{x}_3)^T \mathbf{x}_3}{\mathbf{x}_3^T \mathbf{x}_3} \approx \frac{9.84203}{1.96851} \approx 4.99974$$

$$\lambda^{(4)} = \frac{A\mathbf{x}_4 \cdot \mathbf{x}_4}{\mathbf{x}_4 \cdot \mathbf{x}_4} = \frac{(A\mathbf{x}_4)^T \mathbf{x}_4}{\mathbf{x}_4^T \mathbf{x}_4} \approx \frac{9.96808}{1.99362} \approx 4.99999$$

$$\lambda^{(5)} = \frac{A\mathbf{x}_5 \cdot \mathbf{x}_5}{\mathbf{x}_5 \cdot \mathbf{x}_5} = \frac{(A\mathbf{x}_5)^T \mathbf{x}_5}{\mathbf{x}_5^T \mathbf{x}_5} \approx \frac{9.99360}{1.99872} \approx 5.00000$$

Whereas the power method with Euclidean scaling produces a sequence that approaches a *unit* dominant eigenvector, maximum entry scaling produces a sequence that approaches an eigenvector whose largest component is 1.

Thus, $\lambda^{(5)}$ approximates the dominant eigenvalue correctly to five decimal places and $\mathbf{x}_5$ closely approximates the dominant eigenvector

$$\mathbf{x} = \begin{bmatrix} 1 \\ 1 \end{bmatrix}$$

that results by taking $t = 1$ in (6). ◀

Rate of Convergence If A is a symmetric matrix whose distinct eigenvalues can be arranged so that

$$|\lambda_1| > |\lambda_2| \geq |\lambda_3| \geq \cdots \geq |\lambda_k|$$

then the "rate" at which the Rayleigh quotients converge to the dominant eigenvalue λ_1 depends on the ratio $|\lambda_1|/|\lambda_2|$; that is, the convergence is slow when this ratio is near 1 and rapid when it is large—the greater the ratio, the more rapid the convergence. For example, if A is a 2×2 symmetric matrix, then the greater the ratio $|\lambda_1|/|\lambda_2|$, the greater the disparity between the scaling effects of λ_1 and λ_2 in Figure 9.2.1, and hence the greater the effect that multiplication by A has on pulling the iterates toward the eigenspace of λ_1. Indeed, the rapid convergence in Example 3 is due to the fact that $|\lambda_1|/|\lambda_2| = 5/1 = 5$, which is considered to be a large ratio. In cases where the ratio is close to 1, the convergence of the power method may be so slow that other methods must be used.

Stopping Procedures If λ is the exact value of the dominant eigenvalue, and if a power method produces the approximation $\lambda^{(k)}$ at the kth iteration, then we call

$$\left| \frac{\lambda - \lambda^{(k)}}{\lambda} \right| \tag{12}$$

the ***relative error*** in $\lambda^{(k)}$. If this is expressed as a percentage, then it is called the ***percentage error*** in $\lambda^{(k)}$. For example, if $\lambda = 5$ and the approximation after three iterations is $\lambda^{(3)} = 5.1$, then

$$\text{relative error in } \lambda^{(3)} = \left| \frac{\lambda - \lambda^{(3)}}{\lambda} \right| = \left| \frac{5 - 5.1}{5} \right| = |-0.02| = 0.02$$

$$\text{percentage error in } \lambda^{(3)} = 0.02 \times 100\% = 2\%$$

In applications one usually knows the relative error E that can be tolerated in the dominant eigenvalue, so the goal is to stop computing iterates once the relative error in the approximation to that eigenvalue is less than E. However, there is a problem in computing the relative error from (12) in that the eigenvalue λ is unknown. To circumvent this problem, it is usual to estimate λ by $\lambda^{(k)}$ and stop the computations when

$$\left| \frac{\lambda^{(k)} - \lambda^{(k-1)}}{\lambda^{(k)}} \right| < E \tag{13}$$

The quantity on the left side of (13) is called the ***estimated relative error*** in $\lambda^{(k)}$ and its percentage form is called the ***estimated percentage error*** in $\lambda^{(k)}$.

▶ **EXAMPLE 4** **Estimated Relative Error**

For the computations in Example 3, find the smallest value of k for which the estimated percentage error in $\lambda^{(k)}$ is less than 0.1%.

Solution The estimated percentage errors in the approximations in Example 3 are as follows:

APPROXIMATION		RELATIVE ERROR	PERCENTAGE ERROR
$\lambda^{(2)}$:	$\left\| \dfrac{\lambda^{(2)} - \lambda^{(1)}}{\lambda^{(2)}} \right\| \approx \left\| \dfrac{4.99361 - 4.84615}{4.99361} \right\|$	≈ 0.02953	$= 2.953\%$
$\lambda^{(3)}$:	$\left\| \dfrac{\lambda^{(3)} - \lambda^{(2)}}{\lambda^{(3)}} \right\| \approx \left\| \dfrac{4.99974 - 4.99361}{4.99974} \right\|$	≈ 0.00123	$= 0.123\%$
$\lambda^{(4)}$:	$\left\| \dfrac{\lambda^{(4)} - \lambda^{(3)}}{\lambda^{(4)}} \right\| \approx \left\| \dfrac{4.99999 - 4.99974}{4.99999} \right\|$	≈ 0.00005	$= 0.005\%$
$\lambda^{(5)}$:	$\left\| \dfrac{\lambda^{(5)} - \lambda^{(4)}}{\lambda^{(5)}} \right\| \approx \left\| \dfrac{5.00000 - 4.99999}{5.00000} \right\|$	≈ 0.00000	$= 0\%$

Thus, $\lambda^{(4)} = 4.99999$ is the first approximation whose estimated percentage error is less than 0.1%. ◀

Remark A rule for deciding when to stop an iterative process is called a ***stopping procedure***. In the exercises, we will discuss stopping procedures for the power method that are based on the dominant eigenvector rather than the dominant eigenvalue.

Concept Review

- Power sequence
- Dominant eigenvalue
- Dominant eigenvector
- Power method with Euclidean scaling
- Rayleigh quotient
- Power method with maximum entry scaling
- Relative error
- Percentage error
- Estimated relative error
- Estimated percentage error
- Stopping procedure

Skills

- Identify the dominant eigenvalue of a matrix.
- Use the power methods described in this section to approximate a dominant eigenvector.
- Find the estimated relative and percentage errors associated with the power methods.

Exercise Set 9.2

▶ In Exercises 1–2, the distinct eigenvalues of a matrix are given. Determine whether A has a dominant eigenvalue, and if so, find it.

1. (a) $\lambda_1 = -3$, $\lambda_2 = -1$, $\lambda_3 = 0$, $\lambda_4 = 2$
(b) $\lambda_1 = 4$, $\lambda_2 = -3$, $\lambda_3 = -4$, $\lambda_4 = 1$

2. (a) $\lambda_1 = 1$, $\lambda_2 = 0$, $\lambda_3 = -3$, $\lambda_4 = 2$
(b) $\lambda_1 = -3$, $\lambda_2 = -2$, $\lambda_3 = -1$, $\lambda_4 = 3$

▶ In Exercises 3–4, apply the power method with Euclidean scaling to the matrix A, starting with $\mathbf{x}_0$ and stopping at $\mathbf{x}_4$. Compare the resulting approximations to the exact values of the dominant eigenvalue and the corresponding unit eigenvector. ◀

3. $A = \begin{bmatrix} -14 & 12 \\ -20 & 17 \end{bmatrix}$; $\mathbf{x}_0 = \begin{bmatrix} 0 \\ 1 \end{bmatrix}$, $\mathbf{x}_1 \approx \begin{bmatrix} 0.5766 \\ 0.8169 \end{bmatrix}$,

$\mathbf{x}_2 \approx \begin{bmatrix} 0.5920 \\ 0.8058 \end{bmatrix}$, $\mathbf{x}_3 = \begin{bmatrix} 0.5965 \\ 0.8025 \end{bmatrix}$, $\mathbf{x}_4 = \begin{bmatrix} 0.5984 \\ 0.8011 \end{bmatrix}$

4. $A = \begin{bmatrix} 7 & -2 & 0 \\ -2 & 6 & -2 \\ 0 & -2 & 5 \end{bmatrix}$; $\mathbf{x}_0 = \begin{bmatrix} 1 \\ 0 \\ 0 \end{bmatrix}$

▶ In Exercises 5–6, apply the power method with maximum entry scaling to the matrix A, starting with $\mathbf{x}_0$ and stopping at $\mathbf{x}_4$. Compare the resulting approximations to the exact values of the dominant eigenvalue and the corresponding scaled eigenvector.

5. $A = \begin{bmatrix} 3 & -1 \\ -1 & 3 \end{bmatrix}$; $\mathbf{x}_0 = \begin{bmatrix} 0 \\ 1 \end{bmatrix}$, $\mathbf{x}_1 = \begin{bmatrix} -0.333 \\ 1 \end{bmatrix}$,

$\mathbf{x}_2 = \begin{bmatrix} -0.6 \\ 1 \end{bmatrix}$, $\mathbf{x}_3 = \begin{bmatrix} -0.778 \\ 1 \end{bmatrix}$, $\mathbf{x}_4 = \begin{bmatrix} -0.882 \\ 1 \end{bmatrix}$

6. $A = \begin{bmatrix} 3 & 2 & 2 \\ 2 & 2 & 0 \\ 2 & 0 & 4 \end{bmatrix}$; $\mathbf{x}_0 = \begin{bmatrix} 1 \\ 1 \\ 1 \end{bmatrix}$

7. Let

$$A = \begin{bmatrix} 2 & -1 \\ -1 & 2 \end{bmatrix}; \quad \mathbf{x}_0 = \begin{bmatrix} 1 \\ 0 \end{bmatrix}$$

(a) Use the power method with maximum entry scaling to approximate a dominant eigenvector of A. Start with $\mathbf{x}_0$, round off all computations to three decimal places, and stop after three iterations.

(b) Use the result in part (a) and the Rayleigh quotient to approximate the dominant eigenvalue of A.

(c) Find the exact values of the eigenvector and eigenvalue approximated in parts (a) and (b).

(d) Find the percentage error in the approximation of the dominant eigenvalue.

8. Repeat the directions of Exercise 7 with

$$A = \begin{bmatrix} 2 & 1 & 0 \\ 1 & 2 & 0 \\ 0 & 0 & 10 \end{bmatrix}; \quad \mathbf{x}_0 = \begin{bmatrix} 1 \\ 1 \\ 1 \end{bmatrix}$$

▶ In Exercises 9–10, a matrix A with a dominant eigenvalue and a sequence $\mathbf{x}_0, A\mathbf{x}_0, \ldots, A^5\mathbf{x}_0$ are given. Use Formulas (9) and (10) to approximate the dominant eigenvalue and a corresponding eigenvector. ◀

9. $A = \begin{bmatrix} 4 & 2 \\ 2 & 4 \end{bmatrix}$; $\mathbf{x}_0 = \begin{bmatrix} 1 \\ 0 \end{bmatrix}$, $A\mathbf{x}_0 = \begin{bmatrix} 4 \\ 2 \end{bmatrix}$, $A^2\mathbf{x}_0 = \begin{bmatrix} 20 \\ 16 \end{bmatrix}$,

$A^3\mathbf{x}_0 = \begin{bmatrix} 112 \\ 104 \end{bmatrix}$, $A^4\mathbf{x}_0 = \begin{bmatrix} 656 \\ 640 \end{bmatrix}$, $A^5\mathbf{x}_0 = \begin{bmatrix} 3904 \\ 3872 \end{bmatrix}$

10. $A = \begin{bmatrix} 4 & 2 \\ 2 & 4 \end{bmatrix}$; $\mathbf{x}_0 = \begin{bmatrix} 0 \\ 1 \end{bmatrix}$, $A\mathbf{x}_0 = \begin{bmatrix} 2 \\ 4 \end{bmatrix}$, $A^2\mathbf{x}_0 = \begin{bmatrix} 16 \\ 20 \end{bmatrix}$,

$A^3\mathbf{x}_0 = \begin{bmatrix} 104 \\ 112 \end{bmatrix}$, $A^4\mathbf{x}_0 = \begin{bmatrix} 640 \\ 656 \end{bmatrix}$, $A^5\mathbf{x}_0 = \begin{bmatrix} 3872 \\ 3904 \end{bmatrix}$

11. Consider matrices

$$A = \begin{bmatrix} -1 & 0 \\ 0 & 0 \end{bmatrix} \quad \text{and} \quad \mathbf{x}_0 = \begin{bmatrix} a \\ b \end{bmatrix}$$

where $\mathbf{x}_0$ is a unit vector and $a \neq 0$. Show that even though the matrix A is symmetric and has a dominant eigenvalue, the power sequence (1) in Theorem 9.2.1 does not converge. This shows that the requirement in that theorem that the dominant eigenvalue be positive is essential.

12. Use the power method with Euclidean scaling to approximate the dominant eigenvalue and a corresponding eigenvector of A. Choose your own starting vector, and stop when the estimated percentage error in the eigenvalue approximation is less than 0.1%.

(a) $\begin{bmatrix} 4 & -3 & 0 \\ -3 & 4 & 0 \\ 0 & 0 & 2 \end{bmatrix}$ (b) $\begin{bmatrix} 1 & 0 & 1 & 1 \\ 0 & 2 & -1 & 1 \\ 1 & -1 & 4 & 1 \\ 1 & 1 & 1 & 8 \end{bmatrix}$

13. Repeat Exercise 12, but this time stop when all corresponding entries in two successive eigenvector approximations differ by less than 0.01 in absolute value.

14. Repeat Exercise 12 using maximum entry scaling.

15. Prove: If A is a nonzero $n \times n$ matrix, then $A^T A$ and $A A^T$ have positive dominant eigenvalues.

16. (*For readers familiar with proof by induction*) Let A be an $n \times n$ matrix, let $\mathbf{x}_0$ be a unit vector in R^n, and define the sequence $\mathbf{x}_1, \mathbf{x}_2, \ldots, \mathbf{x}_k, \ldots$ by

$$\mathbf{x}_1 = \frac{A\mathbf{x}_0}{\|A\mathbf{x}_0\|}, \quad \mathbf{x}_2 = \frac{A\mathbf{x}_1}{\|A\mathbf{x}_1\|}, \ldots, \quad \mathbf{x}_k = \frac{A\mathbf{x}_{k-1}}{\|A\mathbf{x}_{k-1}\|}, \ldots$$

Prove by induction that $\mathbf{x}_k = A^k \mathbf{x}_0 / \|A^k \mathbf{x}_0\|$.

17. (*For readers familiar with proof by induction*) Let A be an $n \times n$ matrix, let $\mathbf{x}_0$ be a nonzero vector in R^n, and define the sequence $\mathbf{x}_1, \mathbf{x}_2, \ldots, \mathbf{x}_k, \ldots$ by

$$\mathbf{x}_1 = \frac{A\mathbf{x}_0}{\max(A\mathbf{x}_0)}, \quad \mathbf{x}_2 = \frac{A\mathbf{x}_1}{\max(A\mathbf{x}_1)}, \ldots,$$

$$\mathbf{x}_k = \frac{A\mathbf{x}_{k-1}}{\max(A\mathbf{x}_{k-1})}, \ldots$$

Prove by induction that

$$\mathbf{x}_k = \frac{A^k \mathbf{x}_0}{\max(A^k \mathbf{x}_0)}$$

9.3 Internet Search Engines

Early search engines on the Internet worked by examining key words and phrases in pages and titles of posted documents. Today's most popular search engines use algorithms based on the power method to analyze hyperlinks (references) between documents. In this section we will discuss one of the ways in which this is done.

Google, the most widely used engine for searching the Internet, was developed in 1996 by Larry Page and Sergey Brin while both were graduate students at Stanford University. Google uses a procedure known as the **PageRank algorithm** to analyze how documents at relevant sites reference one another. It then assigns to each site a **PageRank score**, stores those scores as a matrix, and uses the components of the dominant eigenvector of that matrix to establish the relative importance of the sites to the search.

Google starts by using a standard text-based search engine to find an initial set S_0 of sites containing relevant pages. Since words can have multiple meanings, the set S_0 will typically contain irrelevant sites and miss others of relevance. To compensate for this, the set S_0 is expanded to a larger set S by adjoining all sites referenced by the pages in the sites of S_0. The underlying assumption is that S will contain the most important sites relevant to the search. This process is then repeated a number of times to refine the search information still further.

To be more specific, suppose that the search set S contains n sites, and define the **adjacency matrix** for S to be the $n \times n$ matrix $A = [a_{ij}]$ in which

$$a_{ij} = 1 \text{ if site } i \text{ references site } j$$
$$a_{ij} = 0 \text{ if site } i \text{ does not reference site } j$$

We will assume that no site references itself, so the diagonal entries of A will all be zero.

▶ **EXAMPLE 1 Adjacency Matrices**

Here is a typical adjacency matrix for a search set with four sites:

$$A = \begin{bmatrix} 0 & 0 & 1 & 1 \\ 1 & 0 & 0 & 0 \\ 1 & 0 & 0 & 1 \\ 1 & 1 & 1 & 0 \end{bmatrix} \begin{matrix} 1 \\ 2 \\ 3 \\ 4 \end{matrix} \quad \text{Referencing Site} \quad (1)$$

(Referenced Site: 1 2 3 4)

Thus, Site 1 references Sites 3 and 4, Site 2 references Site 1, and so forth. ◀

There are two basic roles that a site can play in the search process—the site may be a **hub**, meaning that it *references* many other sites, or it may be an **authority**, meaning that it is *referenced by* many other sites. A given site will typically have both hub and authority properties in that it will both reference and be referenced.

Historical Note The term *google* is a variation of the word *googol*, which stands for the number 10^{100} (1 followed by 100 zeros). This term was invented by the American mathematician Edward Kasner (1878–1955) in 1938, and the story goes that it came about when Kasner asked his eight-year-old nephew to give a name to a really big number—he responded with "googol." Kasner then went on to define a *googolplex* to be 10^{googol} (1 followed by googol zeros).

In general, if A is an adjacency matrix for n sites, then the column sums of A measure the authority aspect of the sites and the row sums of A measure their hub aspect. For example, the column sums of the matrix in (1) are 3, 1, 2, and 2, which means that Site 1 is referenced by three other sites, Site 2 is referenced by one other site, and so forth. Similarly, the row sums of the matrix in (1) are 2, 1, 2, and 3, so Site 1 references two other sites, Site 2 references one other site, and so forth.

Accordingly, if A is an adjacency matrix, then we call the vector $\mathbf{h}_0$ of row sums of A the *initial hub vector* of A, and we call the vector $\mathbf{a}_0$ of column sums of A the *initial authority vector* of A. Alternatively, we can think of $\mathbf{a}_0$ as the vector of *row* sums of A^T, which turns out to be more convenient for computations. The entries in the hub vector are called **hub weights** and those in the authority vector **authority weights**.

▶ **EXAMPLE 2** Initial Hub and Authority Vectors of an Adjacency Matrix

Find the initial hub and authority vectors for the adjacency matrix A in Example 1.

Solution The row sums of A yield the initial hub vector

$$\mathbf{h}_0 = \begin{bmatrix} 2 \\ 1 \\ 2 \\ 3 \end{bmatrix} \begin{matrix} \text{Site 1} \\ \text{Site 2} \\ \text{Site 3} \\ \text{Site 4} \end{matrix} \qquad (2)$$

and the row sums of A^T (the column sums of A) yield the initial authority vector

$$\mathbf{a}_0 = \begin{bmatrix} 3 \\ 1 \\ 2 \\ 2 \end{bmatrix} \begin{matrix} \text{Site 1} \\ \text{Site 2} \\ \text{Site 3} \\ \text{Site 4} \end{matrix} \qquad (3)$$ ◀

The link counting in Example 2 suggests that Site 4 is the major hub and Site 1 is the greatest authority. However, counting links does not tell the whole story; for example, it seems reasonable that if Site 1 is to be considered the greatest authority, then more weight should be given to hubs that link to that site, and if Site 4 is to be considered a major hub, then more weight should be given to sites to which it links. Thus, there is an interaction between hubs and authorities that needs to be accounted for in the search process. Accordingly, once the search engine has calculated the initial authority vector $\mathbf{a}_0$, it then uses the information in that vector to create new hub and authority vectors $\mathbf{h}_1$ and $\mathbf{a}_1$ using the formulas

$$\mathbf{h}_1 = \frac{A\mathbf{a}_0}{\|A\mathbf{a}_0\|} \quad \text{and} \quad \mathbf{a}_1 = \frac{A^T\mathbf{h}_1}{\|A^T\mathbf{h}_1\|} \qquad (4)$$

The numerators in these formulas do the weighting, and the normalization serves to control the size of the entries. To understand how the numerators accomplish the weighting, view the product $A\mathbf{a}_0$ as a linear combination of the column vectors of A with coefficients from $\mathbf{a}_0$. For example, with the adjacency matrix in Example 1 and the authority vector calculated in Example 2 we have

$$A\mathbf{a}_0 = \begin{bmatrix} 0 & 0 & 1 & 1 \\ 1 & 0 & 0 & 0 \\ 1 & 0 & 0 & 1 \\ 1 & 1 & 1 & 0 \end{bmatrix} \begin{bmatrix} 3 \\ 1 \\ 2 \\ 2 \end{bmatrix} = 3\begin{bmatrix} 0 \\ 1 \\ 1 \\ 1 \end{bmatrix} + 1\begin{bmatrix} 0 \\ 0 \\ 0 \\ 1 \end{bmatrix} + 2\begin{bmatrix} 1 \\ 0 \\ 0 \\ 1 \end{bmatrix} + 2\begin{bmatrix} 1 \\ 0 \\ 1 \\ 0 \end{bmatrix} = \begin{bmatrix} 4 \\ 3 \\ 5 \\ 6 \end{bmatrix} \begin{matrix} \text{Site 1} \\ \text{Site 2} \\ \text{Site 3} \\ \text{Site 4} \end{matrix}$$

with column labels "Referenced Site" 1, 2, 3, 4 above the matrix.

Thus, we see that the links to each referenced site are weighted by the authority values in $\mathbf{a}_0$. To control the size of the entries, the search engine normalizes $A\mathbf{a}_0$ to produce the updated hub vector

$$\mathbf{h}_1 = \frac{A\mathbf{a}_0}{\|A\mathbf{a}_0\|} = \frac{1}{\sqrt{86}} \begin{bmatrix} 4 \\ 3 \\ 5 \\ 6 \end{bmatrix} \approx \begin{bmatrix} 0.43133 \\ 0.32350 \\ 0.53916 \\ 0.64700 \end{bmatrix} \begin{matrix} \text{Site 1} \\ \text{Site 2} \\ \text{Site 3} \\ \text{Site 4} \end{matrix} \quad \text{New Hub Weights}$$

The new hub vector $\mathbf{h}_1$ can now be used to update the authority vector using Formula (4). The product $A^T\mathbf{h}_1$ performs the weighting, and the normalization controls the size:

$$A^T\mathbf{h}_1 \approx \begin{matrix} \text{Referencing Site} \\ \begin{matrix} 1 & 2 & 3 & 4 \end{matrix} \\ \begin{bmatrix} 0 & 1 & 1 & 1 \\ 0 & 0 & 0 & 1 \\ 1 & 0 & 0 & 1 \\ 1 & 0 & 1 & 0 \end{bmatrix} \end{matrix} \begin{bmatrix} 0.43133 \\ 0.32350 \\ 0.53916 \\ 0.64700 \end{bmatrix} \approx 0.43133 \begin{bmatrix} 0 \\ 0 \\ 1 \\ 1 \end{bmatrix} + 0.32350 \begin{bmatrix} 1 \\ 0 \\ 0 \\ 0 \end{bmatrix} + 0.53916 \begin{bmatrix} 1 \\ 0 \\ 0 \\ 1 \end{bmatrix} + 0.64700 \begin{bmatrix} 1 \\ 1 \\ 1 \\ 0 \end{bmatrix} \approx \begin{bmatrix} 1.50966 \\ 0.64700 \\ 1.07833 \\ 0.97049 \end{bmatrix} \begin{matrix} \text{Site 1} \\ \text{Site 2} \\ \text{Site 3} \\ \text{Site 4} \end{matrix}$$

$$\mathbf{a}_1 = \frac{A^T\mathbf{h}_1}{\|A^T\mathbf{h}_1\|} \approx \frac{1}{2.19142} \begin{bmatrix} 1.50966 \\ 0.64700 \\ 1.07833 \\ 0.97049 \end{bmatrix} \approx \begin{bmatrix} 0.68889 \\ 0.29524 \\ 0.49207 \\ 0.44286 \end{bmatrix} \begin{matrix} \text{Site 1} \\ \text{Site 2} \\ \text{Site 3} \\ \text{Site 4} \end{matrix} \quad \text{New Authority Weights}$$

Once the updated hub and authority vectors, $\mathbf{h}_1$ and $\mathbf{a}_1$, are obtained, the search engine repeats the process and computes a succession of hub and authority vectors, thereby generating the interrelated sequences

$$\mathbf{h}_1 = \frac{A\mathbf{a}_0}{\|A\mathbf{a}_0\|}, \quad \mathbf{h}_2 = \frac{A\mathbf{a}_1}{\|A\mathbf{a}_1\|}, \quad \mathbf{h}_3 = \frac{A\mathbf{a}_2}{\|A\mathbf{a}_2\|}, \ldots, \quad \mathbf{h}_k = \frac{A\mathbf{a}_{k-1}}{\|A\mathbf{a}_{k-1}\|}, \ldots \quad (5)$$

$$\mathbf{a}_0, \quad \mathbf{a}_1 = \frac{A^T\mathbf{h}_1}{\|A^T\mathbf{h}_1\|}, \quad \mathbf{a}_2 = \frac{A^T\mathbf{h}_2}{\|A^T\mathbf{h}_2\|}, \quad \mathbf{a}_3 = \frac{A^T\mathbf{h}_3}{\|A^T\mathbf{h}_3\|}, \ldots, \quad \mathbf{a}_k = \frac{A^T\mathbf{h}_k}{\|A^T\mathbf{h}_k\|}, \ldots \quad (6)$$

However, each of these is a power sequence in disguise. For example, if we substitute the expression for $\mathbf{h}_k$ into the expression for $\mathbf{a}_k$, then we obtain

$$\mathbf{a}_k = \frac{A^T\mathbf{h}_k}{\|A^T\mathbf{h}_k\|} = \frac{A^T\left(\frac{A\mathbf{a}_{k-1}}{\|A\mathbf{a}_{k-1}\|}\right)}{\left\|A^T\left(\frac{A\mathbf{a}_{k-1}}{\|A\mathbf{a}_{k-1}\|}\right)\right\|} = \frac{(A^TA)\mathbf{a}_{k-1}}{\|(A^TA)\mathbf{a}_{k-1}\|}$$

which means that we can rewrite (6) as

$$\mathbf{a}_0, \quad \mathbf{a}_1 = \frac{(A^TA)\mathbf{a}_0}{\|(A^TA)\mathbf{a}_0\|}, \quad \mathbf{a}_2 = \frac{(A^TA)\mathbf{a}_1}{\|(A^TA)\mathbf{a}_1\|}, \ldots, \quad \mathbf{a}_k = \frac{(A^TA)\mathbf{a}_{k-1}}{\|(A^TA)\mathbf{a}_{k-1}\|}, \ldots \quad (7)$$

Similarly, we can rewrite (5) as

$$\mathbf{h}_1 = \frac{A\mathbf{a}_0}{\|A\mathbf{a}_0\|}, \quad \mathbf{h}_2 = \frac{(AA^T)\mathbf{h}_1}{\|(AA^T)\mathbf{h}_1\|}, \ldots, \quad \mathbf{h}_k = \frac{(AA^T)\mathbf{h}_{k-1}}{\|(AA^T)\mathbf{h}_{k-1}\|}, \ldots \quad (8)$$

Remark In Exercise 15 of Section 9.2 you were asked to show that A^TA and AA^T both have positive dominant eigenvalues. That being the case, Theorem 9.2.1 ensures that (7) and (8) converge to the dominant eigenvectors of A^TA and AA^T, respectively. The entries in those eigenvectors are

9.3 Internet Search Engines

the authority and hub weights that Google uses to rank the search sites in order of importance as hubs and authorities.

▶ **EXAMPLE 3** **A Ranking Procedure**

Suppose that a search engine produces 10 Internet sites in its search set and that the adjacency matrix for those sites is

$$A = \begin{bmatrix} 0 & 1 & 0 & 0 & 1 & 0 & 0 & 1 & 0 & 0 \\ 0 & 0 & 0 & 0 & 1 & 0 & 0 & 0 & 0 & 0 \\ 0 & 0 & 0 & 0 & 1 & 0 & 0 & 0 & 0 & 0 \\ 0 & 0 & 0 & 0 & 0 & 1 & 1 & 0 & 0 & 0 \\ 0 & 0 & 0 & 0 & 0 & 0 & 0 & 1 & 0 & 0 \\ 0 & 1 & 1 & 1 & 1 & 0 & 0 & 1 & 0 & 1 \\ 0 & 0 & 0 & 0 & 0 & 0 & 0 & 0 & 1 & 0 \\ 0 & 0 & 0 & 0 & 1 & 0 & 0 & 0 & 0 & 0 \\ 0 & 0 & 0 & 0 & 0 & 1 & 0 & 0 & 0 & 0 \\ 0 & 0 & 0 & 0 & 0 & 1 & 0 & 0 & 0 & 0 \end{bmatrix}$$

where columns are Referenced Sites (1–10) and rows are Referencing Sites (1–10).

Use Formula (7) to rank the sites in decreasing order of authority.

Solution We will take $\mathbf{a}_0$ to be the normalized vector of column sums of A, and then we will compute the iterates in (7) until the authority vectors seem to stabilize. We leave it for you to show that

$$\mathbf{a}_0 = \frac{1}{\sqrt{54}} \begin{bmatrix} 0 \\ 2 \\ 1 \\ 1 \\ 5 \\ 3 \\ 1 \\ 3 \\ 0 \\ 2 \end{bmatrix} \approx \begin{bmatrix} 0 \\ 0.27217 \\ 0.13608 \\ 0.13608 \\ 0.68041 \\ 0.40825 \\ 0.13608 \\ 0.40825 \\ 0 \\ 0.27217 \end{bmatrix}$$

and that

$$(A^T A)\mathbf{a}_0 \approx \begin{bmatrix} 0 & 0 & 0 & 0 & 0 & 0 & 0 & 0 & 0 & 0 \\ 0 & 2 & 1 & 1 & 2 & 0 & 0 & 2 & 0 & 1 \\ 0 & 1 & 1 & 1 & 1 & 0 & 0 & 1 & 0 & 1 \\ 0 & 1 & 1 & 1 & 1 & 0 & 0 & 1 & 0 & 1 \\ 0 & 2 & 1 & 1 & 5 & 0 & 0 & 2 & 0 & 1 \\ 0 & 0 & 0 & 0 & 0 & 3 & 1 & 0 & 0 & 0 \\ 0 & 0 & 0 & 0 & 0 & 1 & 1 & 0 & 0 & 0 \\ 0 & 2 & 1 & 1 & 2 & 0 & 0 & 3 & 0 & 1 \\ 0 & 0 & 0 & 0 & 0 & 0 & 0 & 0 & 0 & 0 \\ 0 & 1 & 1 & 1 & 1 & 0 & 0 & 1 & 0 & 2 \end{bmatrix} \begin{bmatrix} 0 \\ 0.27217 \\ 0.13608 \\ 0.13608 \\ 0.68041 \\ 0.40825 \\ 0.13608 \\ 0.40825 \\ 0 \\ 0.27217 \end{bmatrix} \approx \begin{bmatrix} 0 \\ 3.26599 \\ 1.90516 \\ 1.90516 \\ 5.30723 \\ 1.36083 \\ 0.54433 \\ 3.67423 \\ 0 \\ 2.17732 \end{bmatrix}$$

Thus,

$$\mathbf{a}_1 = \frac{(A^TA)\mathbf{a}_0}{\|(A^TA)\mathbf{a}_0\|} \approx \frac{1}{8.15362} \begin{bmatrix} 0 \\ 3.26599 \\ 1.90516 \\ 1.90516 \\ 5.30723 \\ 1.36083 \\ 0.54433 \\ 3.67423 \\ 0 \\ 2.17732 \end{bmatrix} \approx \begin{bmatrix} 0 \\ 0.40056 \\ 0.23366 \\ 0.23366 \\ 0.65090 \\ 0.16690 \\ 0.06676 \\ 0.45063 \\ 0 \\ 0.26704 \end{bmatrix}$$

Continuing in this way yields the following authority iterates:

$$\mathbf{a}_0 \quad \mathbf{a}_1 = \frac{(A^TA)\mathbf{a}_0}{\|(A^TA)\mathbf{a}_0\|} \quad \mathbf{a}_2 = \frac{(A^TA)\mathbf{a}_1}{\|(A^TA)\mathbf{a}_1\|} \quad \mathbf{a}_3 = \frac{(A^TA)\mathbf{a}_2}{\|(A^TA)\mathbf{a}_2\|} \quad \mathbf{a}_4 = \frac{(A^TA)\mathbf{a}_3}{\|(A^TA)\mathbf{a}_3\|} \quad \cdots \quad \mathbf{a}_9 = \frac{(A^TA)\mathbf{a}_8}{\|(A^TA)\mathbf{a}_8\|} \quad \mathbf{a}_{10} = \frac{(A^TA)\mathbf{a}_9}{\|(A^TA)\mathbf{a}_9\|}$$

$\mathbf{a}_0$	$\mathbf{a}_1$	$\mathbf{a}_2$	$\mathbf{a}_3$	$\mathbf{a}_4$	$\mathbf{a}_9$	$\mathbf{a}_{10}$	
0	0	0	0	0	0	0	Site 1
0.27217	0.40056	0.41652	0.41918	0.41973	0.41990	0.41990	Site 2
0.13608	0.23366	0.24917	0.25233	0.25309	0.25337	0.25337	Site 3
0.13608	0.23366	0.24917	0.25233	0.25309	0.25337	0.25337	Site 4
0.68041	0.65090	0.63407	0.62836	0.62665	0.62597	0.62597	Site 5
0.40825	0.16690	0.06322	0.02372	0.00889	0.00007	0.00002	Site 6
0.13608	0.06676	0.02603	0.00981	0.00368	0.00003	0.00001	Site 7
0.40825	0.45063	0.46672	0.47050	0.47137	0.47165	0.47165	Site 8
0	0	0	0	0	0	0	Site 9
0.27217	0.26704	0.27892	0.28300	0.28416	0.28460	0.28460	Site 10

The small changes between $\mathbf{a}_9$ and $\mathbf{a}_{10}$ suggest that the iterates have stabilized near a dominant eigenvector of A^TA. From the entries in $\mathbf{a}_{10}$ we conclude that Sites 1, 6, 7, and 9 are probably irrelevant to the search and that the remaining sites should be searched in order of decreasing importance as

$$\text{Site 5}, \quad \text{Site 8}, \quad \text{Site 2}, \quad \text{Site 10}, \quad \text{Sites 3 and 4 (a tie)} \blacktriangleleft$$

Concept Review

- Adjacency matrix
- Hub vector
- Authority vector
- Hub weights
- Authority weights

Skills

- Find the initial hub and authority vectors of an adjacency matrix.
- Use the method of Example 3 to rank sites.

Exercise Set 9.3

▶ In Exercises **1–2**, find the initial hub and authority vectors for the given adjacency matrix A. ◀

1. Referenced Site

$$A = \begin{bmatrix} 1 & 0 & 1 \\ 1 & 1 & 0 \\ 0 & 0 & 1 \end{bmatrix} \begin{matrix} 1 \\ 2 \\ 3 \end{matrix} \text{ Referencing Site}$$

with column headers 1 2 3

2. Referenced Site

$$A = \begin{bmatrix} 1 & 1 & 1 & 0 \\ 1 & 0 & 0 & 0 \\ 0 & 1 & 1 & 1 \\ 1 & 0 & 0 & 0 \end{bmatrix} \begin{matrix} 1 \\ 2 \\ 3 \\ 4 \end{matrix} \text{ Referencing Site}$$

with column headers 1 2 3 4

▶ In Exercises 3–4, find the updated hub and authority vectors $\mathbf{h}_1$ and $\mathbf{a}_1$ for the adjacency matrix A. ◀

3. The matrix in Exercise 1.   **4.** The matrix in Exercise 2.

▶ In Exercises 5–8, the adjacency matrix A of an Internet search engine is given. Use the method of Example 3 to rank the sites in decreasing order of authority. ◀

5.   Referenced Site

$$A = \begin{bmatrix} 1 & 1 & 0 & 1 \\ 0 & 0 & 1 & 1 \\ 0 & 1 & 0 & 1 \\ 1 & 1 & 1 & 0 \end{bmatrix} \begin{matrix} 1 \\ 2 \\ 3 \\ 4 \end{matrix}$$ Referencing Site

with columns labeled 1 2 3 4.

6.   Referenced Site

$$A = \begin{bmatrix} 1 & 0 & 0 & 0 \\ 0 & 0 & 0 & 1 \\ 0 & 0 & 0 & 0 \\ 0 & 0 & 1 & 1 \end{bmatrix} \begin{matrix} 1 \\ 2 \\ 3 \\ 4 \end{matrix}$$ Referencing Site

with columns labeled 1 2 3 4.

7.   Referenced Site

$$A = \begin{bmatrix} 0 & 0 & 1 & 1 & 0 \\ 0 & 0 & 1 & 1 & 1 \\ 0 & 1 & 0 & 1 & 0 \\ 0 & 0 & 0 & 1 & 0 \\ 1 & 0 & 0 & 0 & 0 \end{bmatrix} \begin{matrix} 1 \\ 2 \\ 3 \\ 4 \\ 5 \end{matrix}$$ Referencing Site

with columns labeled 1 2 3 4 5.

8.   Referenced Site

$$A = \begin{bmatrix} 0 & 0 & 1 & 1 & 1 & 0 & 0 & 0 & 0 & 0 \\ 0 & 1 & 0 & 0 & 0 & 1 & 1 & 0 & 0 & 0 \\ 1 & 0 & 0 & 1 & 0 & 0 & 0 & 1 & 0 & 0 \\ 1 & 1 & 1 & 1 & 1 & 1 & 1 & 1 & 1 & 1 \\ 0 & 0 & 0 & 0 & 1 & 0 & 0 & 0 & 1 & 0 \\ 1 & 1 & 1 & 0 & 1 & 1 & 0 & 0 & 0 & 1 \\ 0 & 0 & 1 & 0 & 0 & 0 & 1 & 0 & 1 & 1 \\ 0 & 1 & 0 & 1 & 0 & 1 & 0 & 1 & 1 & 1 \\ 0 & 0 & 0 & 0 & 1 & 0 & 1 & 0 & 1 & 1 \\ 0 & 0 & 1 & 1 & 0 & 1 & 1 & 0 & 1 & 0 \end{bmatrix} \begin{matrix} 1 \\ 2 \\ 3 \\ 4 \\ 5 \\ 6 \\ 7 \\ 8 \\ 9 \\ 10 \end{matrix}$$ Referencing Site

with columns labeled 1 2 3 4 5 6 7 8 9 10.

9.4 Comparison of Procedures for Solving Linear Systems

There is an old saying that "time is money." This is especially true in industry where the cost of solving a linear system is generally determined by the time it takes for a computer to perform the required computations. This typically depends both on the speed of the computer processor and on the number of operations required by the algorithm. Thus, choosing the right algorithm has important financial implication in an industrial or research setting. In this section we will discuss some of the factors that affect the choice of algorithms for solving large-scale linear systems.

Flops and the Cost of Solving a Linear System

In computer jargon, an arithmetic operation ($+$, $-$, $*$, $\div$) on two real numbers is called a ***flop***, which is an acronym for "floating-point operation."[*] The total number of flops required to solve a problem, which is called the ***cost*** of the solution, provides a convenient way of choosing between various algorithms for solving the problem. When needed, the cost in flops can be converted to units of time or money if the speed of the computer processor and the financial aspects of its operation are known. For example, many of today's personal computers are capable of performing in excess of 10 gigaflops per second (1 gigaflop $= 10^9$ flops). Thus, an algorithm that costs 1,000,000 flops would be executed in 0.0001 seconds.

---

[*]Real numbers are stored in computers as numerical approximations called ***floating-point numbers***. In base 10, a floating-point number has the form $\pm .d_1 d_2 \cdots d_n \times 10^m$, where m is an integer, called the ***mantissa***, and n is the number of digits to the right of the decimal point. The value of n varies with the computer. In some literature the term *flop* is used as a measure of processing speed and stands for "floating-point operations *per second*." In our usage it is interpreted as a counting unit.

502 Chapter 9 Numerical Methods

To illustrate how costs (in flops) can be computed, let us count the number of flops required to solve a linear system of n equations in n unknowns by Gauss–Jordan elimination. For this purpose we will need the following formulas for the sum of the first n positive integers and the sum of the squares of the first n positive integers:

$$1 + 2 + 3 + \cdots + n = \frac{n(n+1)}{2} \tag{1}$$

$$1^2 + 2^2 + 3^2 + \cdots + n^2 = \frac{n(n+1)(2n+1)}{6} \tag{2}$$

Let $A\mathbf{x} = \mathbf{b}$ be a linear system of n equations in n unknowns to be solved by Gauss–Jordan elimination (or, equivalently, by Gaussian elimination with back substitution). For simplicity, let us assume that A is invertible and that no row interchanges are required to reduce the augmented matrix $[A \mid \mathbf{b}]$ to row echelon form. The diagrams that accompany the following analysis provide a convenient way of counting the operations required to introduce a leading 1 in the first row and then zeros below it. In our operation counts, we will lump divisions and multiplications together as "multiplications," and we will lump additions and subtractions together as "additions."

Step 1. It requires n flops (multiplications) to introduce the leading 1 in the first row.

$$\begin{bmatrix} 1 & \times & \times & \cdots & \times & \times & \vert & \times \\ \bullet & \bullet & \bullet & \cdots & \bullet & \bullet & \vert & \bullet \\ \bullet & \bullet & \bullet & \cdots & \bullet & \bullet & \vert & \bullet \\ \vdots & \vdots & \vdots & & \vdots & \vdots & \vert & \vdots \\ \bullet & \bullet & \bullet & \cdots & \bullet & \bullet & \vert & \bullet \\ \bullet & \bullet & \bullet & \cdots & \bullet & \bullet & \vert & \bullet \end{bmatrix} \quad \begin{bmatrix} \times \text{ denotes a quantity that is being computed.} \\ \bullet \text{ denotes a quantity that is not being computed.} \\ \text{The augmented matrix size is } n \times (n+1). \end{bmatrix}$$

Step 2. It requires n multiplications and n additions to introduce a zero below the leading 1, and there are $n - 1$ rows below the leading 1, so the number of flops required to introduce zeros below the leading 1 is $2n(n - 1)$.

$$\begin{bmatrix} 1 & \bullet & \bullet & \cdots & \bullet & \bullet & \vert & \bullet \\ 0 & \times & \times & \cdots & \times & \times & \vert & \times \\ 0 & \times & \times & \cdots & \times & \times & \vert & \times \\ \vdots & \vdots & \vdots & & \vdots & \vdots & \vert & \vdots \\ 0 & \times & \times & \cdots & \times & \times & \vert & \times \\ 0 & \times & \times & \cdots & \times & \times & \vert & \times \end{bmatrix}$$

Column 1. Combining Steps 1 and 2, the number of flops required for column 1 is

$$n + 2n(n-1) = 2n^2 - n$$

Column 2. The procedure for column 2 is the same as for column 1, except that now we are dealing with one less row and one less column. Thus, the number of flops required to introduce the leading 1 in row 2 and the zeros below it can be obtained by replacing n by $n-1$ in the flop count for the first column. Thus, the number of flops required for column 2 is

$$2(n-1)^2 - (n-1)$$

Column 3. By the argument for column 2, the number of flops required for column 3 is

$$2(n-2)^2 - (n-2)$$

Total for all columns. The pattern should now be clear. The total number of flops required to create the n leading 1's and the associated zeros is

$$(2n^2 - n) + [2(n-1)^2 - (n-1)] + [2(n-2)^2 - (n-2)] + \cdots + (2-1)$$

which we can rewrite as

$$2[n^2 + (n-1)^2 + \cdots + 1] - [n + (n-1) + \cdots + 1]$$

or on applying Formulas (1) and (2) as

$$2\frac{n(n+1)(2n+1)}{6} - \frac{n(n+1)}{2} = \frac{2}{3}n^3 + \frac{1}{2}n^2 - \frac{1}{6}n$$

Next, let us count the number of operations required to complete the backward phase (the back substitution).

Column n. It requires $n-1$ multiplications and $n-1$ additions to introduce zeros above the leading 1 in the nth column, so the total number of flops required for the column is $2(n-1)$.

$$\begin{bmatrix} 1 & \bullet & \bullet & \cdots & \bullet & 0 & | & \times \\ 0 & 1 & \bullet & \cdots & \bullet & 0 & | & \times \\ 0 & 0 & 1 & \cdots & \bullet & 0 & | & \times \\ \vdots & \vdots & \vdots & & \vdots & \vdots & | & \vdots \\ 0 & 0 & 0 & \cdots & 1 & 0 & | & \times \\ 0 & 0 & 0 & \cdots & 0 & 1 & | & \bullet \end{bmatrix}$$

Column $(n-1)$. The procedure is the same as for Step 1, except that now we are dealing with one less row. Thus, the number of flops required for the $(n-1)$st column is $2(n-2)$.

$$\begin{bmatrix} 1 & \bullet & \bullet & \cdots & 0 & 0 & | & \times \\ 0 & 1 & \bullet & \cdots & 0 & 0 & | & \times \\ 0 & 0 & 1 & \cdots & 0 & 0 & | & \times \\ \vdots & \vdots & \vdots & & \vdots & \vdots & | & \vdots \\ 0 & 0 & 0 & \cdots & 1 & 0 & | & \bullet \\ 0 & 0 & 0 & \cdots & 0 & 1 & | & \bullet \end{bmatrix}$$

Column $(n-2)$. By the argument for column $(n-1)$, the number of flops required for column $(n-2)$ is $2(n-3)$.

Total. The pattern should now be clear. The total number of flops to complete the backward phase is

$$2(n-1) + 2(n-2) + 2(n-3) + \cdots + 2(n-n) = 2[n^2 - (1 + 2 + \cdots + n)]$$

which we can rewrite using Formula (1) as

$$2\left(n^2 - \frac{n(n+1)}{2}\right) = n^2 - n$$

In summary, we have shown that for Gauss–Jordan elimination the number of flops required for the forward and backward phases is

$$\text{flops for forward phase} = \tfrac{2}{3}n^3 + \tfrac{1}{2}n^2 - \tfrac{1}{6}n \tag{3}$$

$$\text{flops for backward phase} = n^2 - n \tag{4}$$

Thus, the total cost of solving a linear system by Gauss–Jordan elimination is

$$\text{flops for both phases} = \tfrac{2}{3}n^3 + \tfrac{3}{2}n^2 - \tfrac{7}{6}n \tag{5}$$

Cost Estimates for Solving Large Linear Systems

It is a property of polynomials that for large values of the independent variable the term of highest power makes the major contribution to the value of the polynomial. Thus, for *large* linear systems we can use (3) and (4) to approximate the number of flops in the forward and backward phases as

$$\text{flops for forward phase} \approx \tfrac{2}{3}n^3 \tag{6}$$

$$\text{flops for backward phase} \approx n^2 \tag{7}$$

This shows that it is more costly to execute the forward phase than the backward phase for large linear systems. Indeed, the cost difference between the forward and backward phases can be enormous, as the next example shows.

▶ **EXAMPLE 1 Cost of Solving a Large Linear System**

Approximate the time required to execute the forward and backward phases of Gauss–Jordan elimination for a system of 10,000 ($=10^4$) equations in 10,000 unknowns using a computer that can execute 10 gigaflops per second.

Solution We have $n = 10^4$ for the given system, so from (6) and (7) the number of gigaflops required for the forward and backward phases is

$$\text{gigaflops for forward phase} \approx \tfrac{2}{3}n^3 \times 10^{-9} = \tfrac{2}{3}(10^4)^3 \times 10^{-9} = \tfrac{2}{3} \times 10^3$$

$$\text{gigaflops for backward phase} \approx n^2 \times 10^{-9} = (10^4)^2 \times 10^{-9} = 10^{-1}$$

Thus, at 10 gigaflops/s the execution times for the forward and backward phases are

$$\text{time for forward phase} \approx \left(\tfrac{2}{3} \times 10^3\right) \times 10^{-1} \text{ s} \approx 66.67 \text{ s}$$

$$\text{time for backward phase} \approx (10^{-1}) \times 10^{-1} \text{ s} \approx 0.01 \text{ s} \blacktriangleleft$$

We leave it as an exercise for you to confirm the results in Table 1.

Considerations in Choosing an Algorithm for Solving a Linear System

For a *single* linear system $A\mathbf{x} = \mathbf{b}$ of n equations in n unknowns, the methods of LU-decomposition and Gauss–Jordan elimination differ in bookkeeping but otherwise involve the same number of flops. Thus, neither method has a cost advantage over the other. However, LU-decomposition has other advantages that make it the method of choice:

- Gauss–Jordan elimination and Gaussian elimination both use the augmented matrix $[A \mid \mathbf{b}]$, so $\mathbf{b}$ must be known. In contrast, LU-decomposition uses only the matrix A, so once that decomposition is known it can be used with as many right-hand sides as are required, one at a time.
- The LU-decomposition that is computed to solve $A\mathbf{x} = \mathbf{b}$ can be used to compute A^{-1}, if needed, with little additional work.
- For large linear systems in which computer memory is at a premium, one can dispense with the storage of the 1's and zeros that appear on or below the main diagonal of U, since those entries are known from the form of U. The space that this opens up can then be used to store the entries of L, thereby reducing the amount of memory required to solve the system.
- If A is a large matrix consisting mostly of zeros, and if the nonzero entries are concentrated in a "band" around the main diagonal, then there are techniques that can be used to reduce the cost of LU-decomposition, giving it an advantage over Gauss–Jordan elimination.

Table 1

The cost in flops for Gaussian elimination is the same as that for the forward phase of Gauss–Jordan elimination.

Approximate Cost for an $n \times n$ Matrix A with Large n	
Algorithm	**Cost in Flops**
Gauss–Jordan elimination (forward phase)	$\approx \frac{2}{3}n^3$
Gauss–Jordan elimination (backward phase)	$\approx n^2$
LU-decomposition of A	$\approx \frac{2}{3}n^3$
Forward substitution to solve $L\mathbf{y} = \mathbf{b}$	$\approx n^2$
Backward substitution to solve $U\mathbf{x} = \mathbf{y}$	$\approx n^2$
A^{-1} by reducing $[A \mid I]$ to $[I \mid A^{-1}]$	$\approx 2n^3$
Compute $A^{-1}\mathbf{b}$	$\approx 2n^3$

Concept Review

- Flop
- Formula for the sum of the first n positive integers
- Formula for the sum of the squares of the first n positive integers
- Cost in flops for solving large linear systems by various methods
- Cost in flops for inverting a matrix by row reduction
- Issues to consider when choosing an algorithm to solve a large linear system

Skills

- Compute the cost of solving a linear system by Gauss–Jordan elimination.
- Approximate the time required to execute the forward and backward phases of Gauss–Jordan elimination.
- Approximate the time required to find an LU-decomposition of a matrix.
- Approximate the time required to find the inverse of an invertible matrix.

Exercise Set 9.4

1. A certain computer can execute 10 gigaflops per second. Use Formula (5) to find the time required to solve the system using Gauss–Jordan elimination.
   (a) A system of 1000 equations in 1000 unknowns.
   (b) A system of 10,000 equations in 10,000 unknowns.
   (c) A system of 100,000 equations in 100,000 unknowns.

2. A certain computer can execute 100 gigaflops per second. Use Formula (5) to find the time required to solve the system using Gauss–Jordan elimination.
   (a) A system of 10,000 equations in 10,000 unknowns.
   (b) A system of 100,000 equations in 100,000 unknowns.
   (c) A system of 1,000,000 equations in 1,000,000 unknowns.

3. Today's personal computers can execute 70 gigaflops per second. Use Table 1 to estimate the time required to perform the following operations on the invertible $10,000 \times 10,000$ matrix A.
   (a) Execute the forward phase of Gauss–Jordan elimination.
   (b) Execute the backward phase of Gauss–Jordan elimination.
   (c) LU-decomposition of A.
   (d) Find A^{-1} by reducing $[A \mid I]$ to $[I \mid A^{-1}]$.

4. The IBM Roadrunner computer can operate at speeds in excess of 1 petaflop per second (1 petaflop $= 10^{15}$ flops). Use Table 1 to estimate the time required to perform the following operations of the invertible $100,000 \times 100,000$ matrix A.
   (a) Execute the forward phase of Gauss–Jordan elimination.
   (b) Execute the backward phase of Gauss–Jordan elimination.
   (c) LU-decomposition of A.
   (d) Find A^{-1} by reducing $[A \mid I]$ to $[I \mid A^{-1}]$.

5. (a) Approximate the time required to execute the forward phase of Gauss–Jordan elimination for a system of 100,000 equations in 100,000 unknowns using a computer that can execute 1 gigaflop per second. Do the same for the backward phase. (See Table 1.)
   (b) How many gigaflops per second must a computer be able to execute to find the LU-decomposition of a matrix of size $10,000 \times 10,000$ in less than 0.5 s? (See Table 1.)

6. About how many teraflops per second must a computer be able to execute to find the inverse of a matrix of size $100,000 \times 100,000$ in less than 0.5 s? (1 teraflop $= 10^{12}$ flops.)

▶ In Exercises 7–10, A and B are $n \times n$ matrices and c is a real number. ◀

7. How many flops are required to compute cA?

8. How many flops are required to compute $A + B$?

9. How many flops are required to compute AB?

10. If A is a diagonal matrix and k is a positive integer, how many flops are required to compute A^k?

11. If A and B are upper triangular matrices of size n how many flops are required to compute AB?

9.5 Singular Value Decomposition

In this section we will discuss an extension of the diagonalization theory for $n \times n$ symmetric matrices to general $m \times n$ matrices. The results that we will develop in this section have applications to compression, storage, and transmission of digitized information and form the basis for many of the best computational algorithms that are currently available for solving linear systems.

Decompositions of Square Matrices

We saw in Formula (2) of Section 7.2 that every symmetric matrix A can be expressed as

$$A = PDP^T \tag{1}$$

where P is an $n \times n$ orthogonal matrix of eigenvectors of A, and D is the diagonal matrix whose diagonal entries are the eigenvalues corresponding to the column vectors of P. In this section we will call (1) an *eigenvalue decomposition* of A (abbreviated EVD of A).

If an $n \times n$ matrix A is not symmetric, then it does not have an eigenvalue decomposition, but it does have a **Hessenberg decomposition**

$$A = PHP^T$$

in which P is an orthogonal matrix and H is in upper Hessenberg form (Theorem 7.2.4). Moreover, if A has real eigenvalues, then it has a **Schur decomposition**

$$A = PSP^T$$

in which P is an orthogonal matrix and S is upper triangular (Theorem 7.2.3).

The eigenvalue, Hessenberg, and Schur decompositions are important in numerical algorithms not only because the matrices D, H, and S have simpler forms than A, but also because the orthogonal matrices that appear in these factorizations do not magnify roundoff error. To see why this is so, suppose that $\hat{\mathbf{x}}$ is a column vector whose entries are known exactly and that

$$\mathbf{x} = \hat{\mathbf{x}} + \mathbf{e}$$

is the vector that results when roundoff error is present in the entries of $\hat{\mathbf{x}}$. If P is an orthogonal matrix, then the length-preserving property of orthogonal transformations implies that

$$\|P\mathbf{x} - P\hat{\mathbf{x}}\| = \|\mathbf{x} - \hat{\mathbf{x}}\| = \|\mathbf{e}\|$$

which tells us that the error in approximating $P\hat{\mathbf{x}}$ by $P\mathbf{x}$ has the same magnitude as the error in approximating $\hat{\mathbf{x}}$ by $\mathbf{x}$.

There are two main paths that one might follow in looking for other kinds of decompositions of a general square matrix A: One might look for decompositions of the form

$$A = PJP^{-1}$$

in which P is invertible but not necessarily orthogonal, or one might look for decompositions of the form

$$A = U\Sigma V^T$$

in which U and V are orthogonal but not necessarily the same. The first path leads to decompositions in which J is either diagonal or a certain kind of block diagonal matrix, called a **Jordan canonical form** in honor of the French mathematician Camille Jordan (see p. 510). Jordan canonical forms, which we will not consider in this text, are important theoretically and in certain applications, but they are of lesser importance numerically because of the roundoff difficulties that result from the lack of orthogonality in P. In this section we will focus on the second path.

Singular Values Since matrix products of the form A^TA will play an important role in our work, we will begin with two basic theorems about them.

THEOREM 9.5.1 *If A is an $m \times n$ matrix, then:*

(a) *A and A^TA have the same null space.*

(b) *A and A^TA have the same row space.*

(c) *A^T and A^TA have the same column space.*

(d) *A and A^TA have the same rank.*

We will prove part (a) and leave the remaining proofs for the exercises.

Proof (a) We must show that every solution of $A\mathbf{x} = \mathbf{0}$ is a solution of $A^TA\mathbf{x} = \mathbf{0}$, and conversely. If $\mathbf{x}_0$ is any solution of $A\mathbf{x} = \mathbf{0}$, then $\mathbf{x}_0$ is also a solution of $A^TA\mathbf{x} = \mathbf{0}$ since

$$A^TA\mathbf{x}_0 = A^T(A\mathbf{x}_0) = A^T\mathbf{0} = \mathbf{0}$$

Conversely, if $\mathbf{x}_0$ is any solution of $A^TA\mathbf{x} = \mathbf{0}$, then $\mathbf{x}_0$ is in the null space of A^TA and hence is orthogonal to all vectors in the row space of A^TA by part (*q*) of Theorem 4.8.10.

However, A^TA is symmetric, so $\mathbf{x}_0$ is also orthogonal to every vector in the column space of A^TA. In particular, $\mathbf{x}_0$ must be orthogonal to the vector $(A^TA)\mathbf{x}_0$; that is,

$$\mathbf{x}_0 \cdot (A^TA)\mathbf{x}_0 = 0$$

Using the first formula in Table 1 of Section 3.2 and properties of the transpose operation we can rewrite this as

$$\mathbf{x}_0^T(A^TA)\mathbf{x}_0 = (A\mathbf{x}_0)^T(A\mathbf{x}_0) = (A\mathbf{x}_0) \cdot (A\mathbf{x}_0) = \|A\mathbf{x}_0\|^2 = 0$$

which implies that $A\mathbf{x}_0 = 0$, thereby proving that $\mathbf{x}_0$ is a solution of $A\mathbf{x}_0 = 0$. ◄

THEOREM 9.5.2 *If A is an $m \times n$ matrix, then:*

(a) *A^TA is orthogonally diagonalizable.*

(b) *The eigenvalues of A^TA are nonnegative.*

Proof (a) The matrix A^TA, being symmetric, is orthogonally diagonalizable by Theorem 7.2.1.

Proof (b) Since A^TA is orthogonally diagonalizable, there is an orthonormal basis for R^n consisting of eigenvectors of A^TA, say $\{\mathbf{v}_1, \mathbf{v}_2, \ldots, \mathbf{v}_n\}$. If we let $\lambda_1, \lambda_2, \ldots, \lambda_n$ be the corresponding eigenvalues, then for $1 \leq i \leq n$ we have

$$\|A\mathbf{v}_i\|^2 = A\mathbf{v}_i \cdot A\mathbf{v}_i = \mathbf{v}_i \cdot A^TA\mathbf{v}_i \qquad \text{[Formula (26) of Section 3.2]}$$
$$= \mathbf{v}_i \cdot \lambda_i\mathbf{v}_i = \lambda_i(\mathbf{v}_i \cdot \mathbf{v}_i) = \lambda_i\|\mathbf{v}_i\|^2 = \lambda_i$$

It follows from this relationship that $\lambda_i \geq 0$. ◄

We will assume throughout this section that the eigenvalues of A^TA are named so that

$$\lambda_1 \geq \lambda_2 \geq \cdots \geq \lambda_n \geq 0$$

and hence that

$$\sigma_1 \geq \sigma_2 \geq \cdots \geq \sigma_n \geq 0$$

DEFINITION 1 If A is an $m \times n$ matrix, and if $\lambda_1, \lambda_2, \ldots, \lambda_n$ are the eigenvalues of A^TA, then the numbers

$$\sigma_1 = \sqrt{\lambda_1}, \quad \sigma_2 = \sqrt{\lambda_2}, \ldots, \quad \sigma_n = \sqrt{\lambda_n}$$

are called the **singular values** of A.

▶ **EXAMPLE 1 Singular Values**

Find the singular values of the matrix

$$\begin{bmatrix} 1 & 1 \\ 0 & 1 \\ 1 & 0 \end{bmatrix}$$

Solution The first step is to find the eigenvalues of the matrix

$$A^TA = \begin{bmatrix} 1 & 0 & 1 \\ 1 & 1 & 0 \end{bmatrix} \begin{bmatrix} 1 & 1 \\ 0 & 1 \\ 1 & 0 \end{bmatrix} = \begin{bmatrix} 2 & 1 \\ 1 & 2 \end{bmatrix}$$

The characteristic polynomial of A^TA is

$$\lambda^2 - 4\lambda + 3 = (\lambda - 3)(\lambda - 1)$$

so the eigenvalues of A^TA are $\lambda_1 = 3$ and $\lambda_2 = 1$ and the singular values of A in order of decreasing size are

$$\sigma_1 = \sqrt{\lambda_1} = \sqrt{3}, \quad \sigma_2 = \sqrt{\lambda_2} = 1 \quad ◄$$

Singular Value Decomposition

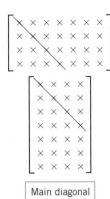

Main diagonal

▲ Figure 9.5.1

Harry Bateman
(1882–1946)

Historical Note The term *singular value* is apparently due to the British-born mathematician Harry Bateman, who used it in a research paper published in 1908. Bateman emigrated to the United States in 1910, teaching at Bryn Mawr College, Johns Hopkins University, and finally at the California Institute of Technology. Interestingly, he was awarded his Ph.D. in 1913 by Johns Hopkins at which point in time he was already an eminent mathematician with 60 publications to his name.
[*Image: Courtesy of the Archives, California Institute of Technology*]

The vectors $\mathbf{u}_1, \mathbf{u}_2, \ldots, \mathbf{u}_k$ are called the **left singular vectors** of A, and the vectors $\mathbf{v}_1, \mathbf{v}_2, \ldots, \mathbf{v}_k$ are called the **right singular vectors** of A.

Before turning to the main result in this section, we will find it useful to extend the notion of a "main diagonal" to matrices that are not square. We define the **main diagonal** of an $m \times n$ matrix to be the line of entries shown in Figure 9.5.1—it starts at the upper left corner and extends diagonally as far as it can go. We will refer to the entries on the main diagonal as the **diagonal entries**.

We are now ready to consider the main result in this section, which is concerned with a specific way of factoring a general $m \times n$ matrix A. This factorization, called **singular value decomposition** (abbreviated SVD) will be given in two forms, a brief form that captures the main idea, and an expanded form that spells out the details. The proof is given at the end of this section.

THEOREM 9.5.3 Singular Value Decomposition

If A is an $m \times n$ matrix, then A can be expressed in the form

$$A = U \Sigma V^T$$

where U and V are orthogonal matrices and Σ is an $m \times n$ matrix whose diagonal entries are the singular values of A and whose other entries are zero.

THEOREM 9.5.4 Singular Value Decomposition (Expanded Form)

If A is an $m \times n$ matrix of rank k, then A can be factored as

$$A = U\Sigma V^T = [\mathbf{u}_1 \; \mathbf{u}_2 \; \cdots \; \mathbf{u}_k \mid \mathbf{u}_{k+1} \; \cdots \; \mathbf{u}_m] \begin{bmatrix} \sigma_1 & 0 & \cdots & 0 & & \\ 0 & \sigma_2 & \cdots & 0 & & 0_{k \times (n-k)} \\ \vdots & \vdots & \ddots & \vdots & & \\ 0 & 0 & \cdots & \sigma_k & & \\ \hline & 0_{(m-k) \times k} & & & 0_{(m-k) \times (n-k)} & \end{bmatrix} \begin{bmatrix} \mathbf{v}_1^T \\ \mathbf{v}_2^T \\ \vdots \\ \mathbf{v}_k^T \\ \mathbf{v}_{k+1}^T \\ \vdots \\ \mathbf{v}_n^T \end{bmatrix}$$

in which U, Σ, and V have sizes $m \times m$, $m \times n$, and $n \times n$, respectively, and in which

(a) $V = [\mathbf{v}_1 \; \mathbf{v}_2 \; \cdots \; \mathbf{v}_n]$ orthogonally diagonalizes $A^T A$.

(b) The nonzero diagonal entries of Σ are $\sigma_1 = \sqrt{\lambda_1}, \sigma_2 = \sqrt{\lambda_2}, \ldots, \sigma_k = \sqrt{\lambda_k}$, where $\lambda_1, \lambda_2, \ldots, \lambda_k$ are the nonzero eigenvalues of $A^T A$ corresponding to the column vectors of V.

(c) The column vectors of V are ordered so that $\sigma_1 \geq \sigma_2 \geq \cdots \geq \sigma_k > 0$.

(d) $\mathbf{u}_i = \dfrac{A\mathbf{v}_i}{\|A\mathbf{v}_i\|} = \dfrac{1}{\sigma_i} A\mathbf{v}_i \quad (i = 1, 2, \ldots, k)$

(e) $\{\mathbf{u}_1, \mathbf{u}_2, \ldots, \mathbf{u}_k\}$ is an orthonormal basis for col(A).

(f) $\{\mathbf{u}_1, \mathbf{u}_2, \ldots, \mathbf{u}_k, \mathbf{u}_{k+1}, \ldots, \mathbf{u}_m\}$ is an extension of $\{\mathbf{u}_1, \mathbf{u}_2, \ldots, \mathbf{u}_k\}$ to an orthonormal basis for R^m.

▶ **EXAMPLE 2** **Singular Value Decomposition if *A* Is Not Square**

Find a singular value decomposition of the matrix

$$A = \begin{bmatrix} 1 & 1 \\ 0 & 1 \\ 1 & 0 \end{bmatrix}$$

Solution We showed in Example 1 that the eigenvalues of $A^T A$ are $\lambda_1 = 3$ and $\lambda_2 = 1$ and that the corresponding singular values of A are $\sigma_1 = \sqrt{3}$ and $\sigma_2 = 1$. We leave it for you to verify that

$$\mathbf{v}_1 = \begin{bmatrix} \frac{\sqrt{2}}{2} \\ \frac{\sqrt{2}}{2} \end{bmatrix} \quad \text{and} \quad \mathbf{v}_2 = \begin{bmatrix} \frac{\sqrt{2}}{2} \\ -\frac{\sqrt{2}}{2} \end{bmatrix}$$

are eigenvectors corresponding to λ_1 and λ_2, respectively, and that $V = [\mathbf{v}_1 \mid \mathbf{v}_2]$ orthogonally diagonalizes $A^T A$. From part (*d*) of Theorem 9.5.4, the vectors

$$\mathbf{u}_1 = \frac{1}{\sigma_1} A \mathbf{v}_1 = \frac{\sqrt{3}}{3} \begin{bmatrix} 1 & 1 \\ 0 & 1 \\ 1 & 0 \end{bmatrix} \begin{bmatrix} \frac{\sqrt{2}}{2} \\ \frac{\sqrt{2}}{2} \end{bmatrix} = \begin{bmatrix} \frac{\sqrt{6}}{3} \\ \frac{\sqrt{6}}{6} \\ \frac{\sqrt{6}}{6} \end{bmatrix}$$

$$\mathbf{u}_2 = \frac{1}{\sigma_2} A \mathbf{v}_2 = (1) \begin{bmatrix} 1 & 1 \\ 0 & 1 \\ 1 & 0 \end{bmatrix} \begin{bmatrix} \frac{\sqrt{2}}{2} \\ -\frac{\sqrt{2}}{2} \end{bmatrix} = \begin{bmatrix} 0 \\ -\frac{\sqrt{2}}{2} \\ \frac{\sqrt{2}}{2} \end{bmatrix}$$

**Eugenio Beltrami
(1835–1900)**

**Camille Jordan
(1838–1922)**

Historical Note The theory of singular value decompositions can be traced back to the work of five people: the Italian mathematician Eugenio Beltrami, the French mathematician Camille Jordan, the English mathematician James Sylvester (see p. 34), and the German mathematicians Erhard Schmidt (see p. 360) and the mathematician Herman Weyl. More recently, the pioneering efforts of the American mathematician Gene Golub produced a stable and efficient algorithm for computing it. Beltrami and Jordan were the progenitors of the decomposition—Beltrami gave a proof of the result for real, invertible matrices with distinct singular values in 1873. Subsequently, Jordan refined the theory and eliminated the unnecessary restrictions imposed by Beltrami. Sylvester, apparently unfamiliar with the work of Beltrami and Jordan, rediscovered the result in 1889 and suggested its importance. Schmidt was the first person to show that the singular value decomposition could be used to approximate a matrix by another matrix with lower rank, and, in so doing, he transformed it from a mathematical curiosity to an important practical tool. Weyl showed how to find the lower rank approximations in the presence of error.

[Images: wikipedia (Beltrami); The Granger Collection, New York (Jordan); Courtesy Electronic Publishing Services, Inc., New York City (Weyl); wikipedia (Golub)]

**Herman Klaus Weyl
(1885–1955)**

**Gene H. Golub
(1932–   )**

are two of the three column vectors of U. Note that $\mathbf{u}_1$ and $\mathbf{u}_2$ are orthonormal, as expected. We could extend the set $\{\mathbf{u}_1, \mathbf{u}_2\}$ to an orthonormal basis for R^3. However, the computations will be easier if we first remove the messy radicals by multiplying $\mathbf{u}_1$ and $\mathbf{u}_2$ by appropriate scalars. Thus, we will look for a unit vector $\mathbf{u}_3$ that is orthogonal to

$$\sqrt{6}\,\mathbf{u}_1 = \begin{bmatrix} 2 \\ 1 \\ 1 \end{bmatrix} \quad \text{and} \quad \sqrt{2}\,\mathbf{u}_2 = \begin{bmatrix} 0 \\ -1 \\ 1 \end{bmatrix}$$

To satisfy these two orthogonality conditions, the vector $\mathbf{u}_3$ must be a solution of the homogeneous linear system

$$\begin{bmatrix} 2 & 1 & 1 \\ 0 & -1 & 1 \end{bmatrix} \begin{bmatrix} x_1 \\ x_2 \\ x_3 \end{bmatrix} = \begin{bmatrix} 0 \\ 0 \end{bmatrix}$$

We leave it for you to show that a general solution of this system is

$$\begin{bmatrix} x_1 \\ x_2 \\ x_3 \end{bmatrix} = t \begin{bmatrix} -1 \\ 1 \\ 1 \end{bmatrix}$$

Normalizing the vector on the right yields

$$\mathbf{u}_3 = \begin{bmatrix} -\frac{1}{\sqrt{3}} \\ \frac{1}{\sqrt{3}} \\ \frac{1}{\sqrt{3}} \end{bmatrix}$$

Thus, the singular value decomposition of A is

$$\begin{bmatrix} 1 & 1 \\ 0 & 1 \\ 1 & 0 \end{bmatrix} = \begin{bmatrix} \frac{\sqrt{6}}{3} & 0 & -\frac{1}{\sqrt{3}} \\ \frac{\sqrt{6}}{6} & -\frac{\sqrt{2}}{2} & \frac{1}{\sqrt{3}} \\ \frac{\sqrt{6}}{6} & \frac{\sqrt{2}}{2} & \frac{1}{\sqrt{3}} \end{bmatrix} \begin{bmatrix} \sqrt{3} & 0 \\ 0 & 1 \\ 0 & 0 \end{bmatrix} \begin{bmatrix} \frac{\sqrt{2}}{2} & \frac{\sqrt{2}}{2} \\ \frac{\sqrt{2}}{2} & -\frac{\sqrt{2}}{2} \end{bmatrix}$$

$$A \quad = \quad\quad\quad U \quad\quad\quad\quad \Sigma \quad\quad\quad V^T$$

You may want to confirm the validity of this equation by multiplying out the matrices on the right side. ◂

OPTIONAL

We conclude this section with an optional proof of Theorem 9.5.4.

Proof of Theorem 9.5.4 For notational simplicity we will prove this theorem in the case where A is an $n \times n$ matrix. To modify the argument for an $m \times n$ matrix you need only make the notational adjustments required to account for the possibility that $m > n$ or $n > m$.

The matrix $A^T A$ is symmetric, so it has an eigenvalue decomposition

$$A^T A = VDV^T$$

in which the column vectors of

$$V = [\mathbf{v}_1 \mid \mathbf{v}_2 \mid \cdots \mid \mathbf{v}_n]$$

are unit eigenvectors of $A^T A$, and D is a diagonal matrix whose successive diagonal entries $\lambda_1, \lambda_2, \ldots, \lambda_n$ are the eigenvalues of $A^T A$ corresponding in succession to the column vectors of V. Since A is assumed to have rank k, it follows from Theorem 9.5.1

that A^TA also has rank k. It follows as well that D has rank k, since it is similar to A^TA and rank is a similarity invariant. Thus, D can be expressed in the form

$$D = \begin{bmatrix} \lambda_1 & & & & & & & 0 \\ & \lambda_2 & & & & & & \\ & & \ddots & & & & & \\ & & & \lambda_k & & & & \\ & & & & 0 & & & \\ & & & & & \ddots & & \\ 0 & & & & & & & 0 \end{bmatrix} \quad (2)$$

where $\lambda_1 \geq \lambda_2 \geq \cdots \geq \lambda_k > 0$. Now let us consider the set of image vectors

$$\{A\mathbf{v}_1, A\mathbf{v}_2, \ldots, A\mathbf{v}_n\} \quad (3)$$

This is an orthogonal set, for if $i \neq j$, then the orthogonality of $\mathbf{v}_i$ and $\mathbf{v}_j$ implies that

$$A\mathbf{v}_i \cdot A\mathbf{v}_j = \mathbf{v}_i \cdot A^TA\mathbf{v}_j = \mathbf{v}_i \cdot \lambda_j \mathbf{v}_j = \lambda_j(\mathbf{v}_i \cdot \mathbf{v}_j) = 0$$

Moreover, the first k vectors in (3) are nonzero since we showed in the proof of Theorem 9.5.2b that $\|A\mathbf{v}_i\|^2 = \lambda_i$ for $i = 1, 2, \ldots, n$, and we have assumed that the first k diagonal entries in (2) are positive. Thus,

$$S = \{A\mathbf{v}_1, A\mathbf{v}_2, \ldots, A\mathbf{v}_k\}$$

is an orthogonal set of *nonzero* vectors in the column space of A. But the column space of A has dimension k since

$$\text{rank}(A) = \text{rank}(A^TA) = k$$

and hence S, being a linearly independent set of k vectors, must be an orthogonal basis for $\text{col}(A)$. If we now normalize the vectors in S, we will obtain an orthonormal basis $\{\mathbf{u}_1, \mathbf{u}_2, \ldots, \mathbf{u}_k\}$ for $\text{col}(A)$ in which

$$\mathbf{u}_i = \frac{A\mathbf{v}_i}{\|A\mathbf{v}_i\|} = \frac{1}{\sqrt{\lambda_i}} A\mathbf{v}_i \quad (1 \leq i \leq k)$$

or, equivalently, in which

$$A\mathbf{v}_1 = \sqrt{\lambda_1}\mathbf{u}_1 = \sigma_1\mathbf{u}_1, \quad A\mathbf{v}_2 = \sqrt{\lambda_2}\mathbf{u}_2 = \sigma_2\mathbf{u}_2, \ldots, \quad A\mathbf{v}_k = \sqrt{\lambda_k}\mathbf{u}_k = \sigma_k\mathbf{u}_k \quad (4)$$

It follows from Theorem 6.3.6 that we can extend this to an orthonormal basis

$$\{\mathbf{u}_1, \mathbf{u}_2, \ldots, \mathbf{u}_k, \mathbf{u}_{k+1}, \ldots, \mathbf{u}_n\}$$

for R^n. Now let U be the orthogonal matrix

$$U = \begin{bmatrix} \mathbf{u}_1 & \mathbf{u}_2 & \cdots & \mathbf{u}_k & \mathbf{u}_{k+1} & \cdots & \mathbf{u}_n \end{bmatrix}$$

and let Σ be the diagonal matrix

$$\Sigma = \begin{bmatrix} \sigma_1 & & & & & & & 0 \\ & \sigma_2 & & & & & & \\ & & \ddots & & & & & \\ & & & \sigma_k & & & & \\ & & & & 0 & & & \\ & & & & & \ddots & & \\ 0 & & & & & & & 0 \end{bmatrix}$$

It follows from (4), and the fact that $A\mathbf{v}_i = 0$ for $i > k$, that

$$U\Sigma = [\sigma_1 \mathbf{u}_1 \quad \sigma_2 \mathbf{u}_2 \quad \cdots \quad \sigma_k \mathbf{u}_k \quad 0 \quad \cdots \quad 0]$$
$$= [A\mathbf{v}_1 \quad A\mathbf{v}_2 \quad \cdots \quad A\mathbf{v}_k \quad A\mathbf{v}_{k+1} \quad \cdots \quad A\mathbf{v}_n]$$
$$= AV$$

which we can rewrite using the orthogonality of V as $A = U\Sigma V^T$. ◀

Concept Review

- Eigenvalue decomposition
- Hessenberg decomposition
- Schur decomposition
- Magnification of roundoff error
- Properties that A and $A^T A$ have in common
- $A^T A$ is orthogonally diagonalizable
- Eigenvalues of $A^T A$ are nonnegative
- Singular values
- Diagonal entries of a matrix that is not square
- Singular value decomposition

Skills

- Find the singular values of an $m \times n$ matrix.
- Find a singular value decomposition of an $m \times n$ matrix.

Exercise Set 9.5

▶ In Exercises 1–4, find the distinct singular values of A. ◀

1. $A = \begin{bmatrix} 4 & 0 & 3 \end{bmatrix}$

2. $A = \begin{bmatrix} 5 & 0 \\ 0 & 2 \end{bmatrix}$

3. $A = \begin{bmatrix} 2 & -1 \\ 1 & 2 \end{bmatrix}$

4. $A = \begin{bmatrix} \sqrt{2} & 0 \\ 1 & \sqrt{2} \end{bmatrix}$

▶ In Exercises 5–12, find a singular value decomposition of A. ◀

5. $A = \begin{bmatrix} 1 & -1 \\ 1 & 1 \end{bmatrix}$

6. $A = \begin{bmatrix} -3 & 0 \\ 0 & -4 \end{bmatrix}$

7. $A = \begin{bmatrix} 2 & -1 \\ -2 & 1 \end{bmatrix}$

8. $A = \begin{bmatrix} 3 & 3 \\ 3 & 3 \end{bmatrix}$

9. $A = \begin{bmatrix} 2 & 1 \\ 4 & 2 \\ 4 & 2 \end{bmatrix}$

10. $A = \begin{bmatrix} 4 & 0 & 3 \\ 0 & 0 & 5 \end{bmatrix}$

11. $A = \begin{bmatrix} 1 & 0 \\ 1 & 1 \\ -1 & 1 \end{bmatrix}$

12. $A = \begin{bmatrix} -1 & 4 \\ -2 & 2 \\ -2 & -4 \end{bmatrix}$

13. Prove: If A is an $m \times n$ matrix, then $A^T A$ and AA^T have the same rank.

14. Prove part (d) of Theorem 9.5.1 by using part (a) of the theorem and the fact that A and $A^T A$ have n columns.

15. (a) Prove part (b) of Theorem 9.5.1 by first showing that row($A^T A$) is a subspace of row(A).

    (b) Prove part (c) of Theorem 9.5.1 by using part (b).

16. Let $T: R^n \to R^m$ be a linear transformation whose standard matrix A has the singular value decomposition $A = U\Sigma V^T$, and let $B = \{\mathbf{v}_1, \mathbf{v}_2, \ldots, \mathbf{v}_n\}$ and $B' = \{\mathbf{u}_1, \mathbf{u}_2, \ldots, \mathbf{u}_m\}$ be the column vectors of V and U, respectively. Show that $\Sigma = [T]_{B', B}$.

17. Show that the singular values of $A^T A$ are the squares of the singular values of A.

18. Show that if $A = U\Sigma V^T$ is a singular value decomposition of A, then U orthogonally diagonalizes AA^T.

True-False Exercises

In parts (a)–(g) determine whether the statement is true or false, and justify your answer.

(a) If A is an $m \times n$ matrix, then $A^T A$ is an $m \times m$ matrix.

(b) If A is an $m \times n$ matrix, then $A^T A$ is a symmetric matrix.

(c) If A is an $m \times n$ matrix, then the eigenvalues of $A^T A$ are positive real numbers.

(d) If A is an $n \times n$ matrix, then A is orthogonally diagonalizable.

(e) If A is an $m \times n$ matrix, then $A^T A$ is orthogonally diagonalizable.

(f) The eigenvalues of $A^T A$ are the singular values of A.

(g) Every $m \times n$ matrix has a singular value decomposition.

9.6 Data Compression Using Singular Value Decomposition

Efficient transmission and storage of large quantities of digital data has become a major problem in our technological world. In this section we will discuss the role that singular value decomposition plays in compressing digital data so that it can be transmitted more rapidly and stored in less space. We assume here that you have read Section 9.5.

Reduced Singular Value Decomposition

Algebraically, the zero rows and columns of the matrix Σ in Theorem 9.5.4 are superfluous and can be eliminated by multiplying out the expression $U\Sigma V^T$ using block multiplication and the partitioning shown in that formula. The products that involve zero blocks as factors drop out, leaving

$$A = [\mathbf{u}_1 \quad \mathbf{u}_2 \quad \cdots \quad \mathbf{u}_k] \begin{bmatrix} \sigma_1 & 0 & \cdots & 0 \\ 0 & \sigma_2 & \cdots & 0 \\ \vdots & \vdots & \ddots & \vdots \\ 0 & 0 & \cdots & \sigma_k \end{bmatrix} \begin{bmatrix} \mathbf{v}_1^T \\ \mathbf{v}_2^T \\ \vdots \\ \mathbf{v}_k^T \end{bmatrix} \qquad (1)$$

which is called a ***reduced singular value decomposition*** of A. In this text we will denote the matrices on the right side of (1) by U_1, Σ_1, and V_1^T, respectively, and we will write this equation as

$$A = U_1 \Sigma_1 V_1^T \qquad (2)$$

Note that the sizes of U_1, Σ_1, and V_1^T are $m \times k$, $k \times k$, and $k \times n$, respectively, and that the matrix Σ_1 is invertible, since its diagonal entries are positive.

If we multiply out on the right side of (1) using the column-row rule, then we obtain

$$A = \sigma_1 \mathbf{u}_1 \mathbf{v}_1^T + \sigma_2 \mathbf{u}_2 \mathbf{v}_2^T + \cdots + \sigma_k \mathbf{u}_k \mathbf{v}_k^T \qquad (3)$$

which is called a ***reduced singular value expansion*** of A. This result applies to *all* matrices, whereas the spectral decomposition [Formula (7) of Section 7.2] applies only to symmetric matrices.

Remark It can be proved that an $m \times n$ matrix M has rank 1 if and only if it can be factored as $M = \mathbf{uv}^T$, where $\mathbf{u}$ is a column vector in R^m and V is a column vector in R^n. Thus, a reduced singular value decomposition expresses a matrix A of rank k as a linear combination of k rank 1 matrices.

▶ **EXAMPLE 1** Reduced Singular Value Decomposition

Find a reduced singular value decomposition and a reduced singular value expansion of the matrix

$$A = \begin{bmatrix} 1 & 1 \\ 0 & 1 \\ 1 & 0 \end{bmatrix}$$

9.6 Data Compression Using Singular Value Decomposition

Solution In Example 2 of Section 9.5 we found the singular value decomposition

$$\begin{bmatrix} 1 & 1 \\ 0 & 1 \\ 1 & 0 \end{bmatrix} = \begin{bmatrix} \frac{\sqrt{6}}{3} & 0 & -\frac{1}{\sqrt{3}} \\ \frac{\sqrt{6}}{6} & -\frac{\sqrt{2}}{2} & \frac{1}{\sqrt{3}} \\ \frac{\sqrt{6}}{6} & \frac{\sqrt{2}}{2} & \frac{1}{\sqrt{3}} \end{bmatrix} \begin{bmatrix} \sqrt{3} & 0 \\ 0 & 1 \\ 0 & 0 \end{bmatrix} \begin{bmatrix} \frac{\sqrt{2}}{2} & \frac{\sqrt{2}}{2} \\ \frac{\sqrt{2}}{2} & -\frac{\sqrt{2}}{2} \end{bmatrix} \quad (4)$$

$$A \quad = \quad U \quad\quad \Sigma \quad\quad V^T$$

Since A has rank 2 (verify), it follows from (1) with $k = 2$ that the reduced singular value decomposition of A corresponding to (4) is

$$\begin{bmatrix} 1 & 1 \\ 0 & 1 \\ 1 & 0 \end{bmatrix} = \begin{bmatrix} \frac{\sqrt{6}}{3} & 0 \\ \frac{\sqrt{6}}{6} & -\frac{\sqrt{2}}{2} \\ \frac{\sqrt{6}}{6} & \frac{\sqrt{2}}{2} \end{bmatrix} \begin{bmatrix} \sqrt{3} & 0 \\ 0 & 1 \end{bmatrix} \begin{bmatrix} \frac{\sqrt{2}}{2} & \frac{\sqrt{2}}{2} \\ \frac{\sqrt{2}}{2} & -\frac{\sqrt{2}}{2} \end{bmatrix}$$

This yields the reduced singular value expansion

$$\begin{bmatrix} 1 & 1 \\ 0 & 1 \\ 1 & 0 \end{bmatrix} = \sigma_1 \mathbf{u}_1 \mathbf{v}_1^T + \sigma_2 \mathbf{u}_2 \mathbf{v}_2^T = \sqrt{3} \begin{bmatrix} \frac{\sqrt{6}}{3} \\ \frac{\sqrt{6}}{6} \\ \frac{\sqrt{6}}{6} \end{bmatrix} \begin{bmatrix} \frac{\sqrt{2}}{2} & \frac{\sqrt{2}}{2} \end{bmatrix} + (1) \begin{bmatrix} 0 \\ -\frac{\sqrt{2}}{2} \\ \frac{\sqrt{2}}{2} \end{bmatrix} \begin{bmatrix} \frac{\sqrt{2}}{2} & -\frac{\sqrt{2}}{2} \end{bmatrix}$$

$$= \sqrt{3} \begin{bmatrix} \frac{\sqrt{3}}{3} & \frac{\sqrt{3}}{3} \\ \frac{\sqrt{3}}{6} & \frac{\sqrt{3}}{6} \\ \frac{\sqrt{3}}{6} & \frac{\sqrt{3}}{6} \end{bmatrix} + (1) \begin{bmatrix} 0 & 0 \\ -\frac{1}{2} & \frac{1}{2} \\ \frac{1}{2} & -\frac{1}{2} \end{bmatrix}$$

Note that the matrices in the expansion have rank 1, as expected. ◀

Data Compression and Image Processing

Singular value decompositions can be used to "compress" visual information for the purpose of reducing its required storage space and speeding up its electronic transmission. The first step in compressing a visual image is to represent it as a numerical matrix from which the visual image can be recovered when needed.

For example, a black and white photograph might be scanned as a rectangular array of pixels (points) and then stored as a matrix A by assigning each pixel a numerical value in accordance with its gray level. If 256 different gray levels are used (0 = white to 255 = black), then the entries in the matrix would be integers between 0 and 255. The image can be recovered from the matrix A by printing or displaying the pixels with their assigned gray levels.

Original   Reconstruction

Historical Note In 1924 the U.S. Federal Bureau of Investigation (FBI) began collecting fingerprints and handprints and now has more than 30 million such prints in its files. To reduce the storage cost, the FBI began working with the Los Alamos National Laboratory, the National Bureau of Standards, and other groups in 1993 to devise rank based compression methods for storing prints in digital form. The following figure shows an original fingerprint and a reconstruction from digital data that was compressed at a ratio of 26:1.

If the matrix A has size $m \times n$, then one might store each of its mn entries individually. An alternative procedure is to compute the reduced singular value decomposition

$$A = \sigma_1 \mathbf{u}_1 \mathbf{v}_1^T + \sigma_2 \mathbf{u}_2 \mathbf{v}_2^T + \cdots + \sigma_k \mathbf{u}_k \mathbf{v}_k^T \tag{5}$$

in which $\sigma_1 \geq \sigma_2 \geq \cdots \geq \sigma_k$, and store the σ's, the $\mathbf{u}$'s, and the $\mathbf{v}$'s. When needed, the matrix A (and hence the image it represents) can be reconstructed from (5). Since each $\mathbf{u}_j$ has m entries and each $\mathbf{v}_j$ has n entries, this method requires storage space for

$$km + kn + k = k(m + n + 1)$$

numbers. Suppose, however, that the singular values $\sigma_{r+1}, \ldots, \sigma_k$ are sufficiently small that dropping the corresponding terms in (5) produces an acceptable approximation

$$A_r = \sigma_1 \mathbf{u}_1 \mathbf{v}_1^T + \sigma_2 \mathbf{u}_2 \mathbf{v}_2^T + \cdots + \sigma_r \mathbf{u}_r \mathbf{v}_r^T \tag{6}$$

to A and the image that it represents. We call (6) the **rank r approximation of A**. This matrix requires storage space for only

$$rm + rn + r = r(m + n + 1)$$

numbers, compared to mn numbers required for entry-by-entry storage of A. For example, the rank 100 approximation of a 1000×1000 matrix A requires storage for only

$$100(1000 + 1000 + 1) = 200{,}100$$

numbers, compared to the 1,000,000 numbers required for entry-by-entry storage of A—a compression of almost 80%.

Figure 9.6.1 shows some approximations of a digitized mandrill image obtained using (6).

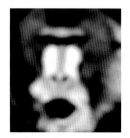 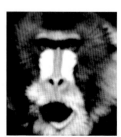   

Rank 4　　　　　Rank 10　　　　　Rank 20　　　　　Rank 50　　　　　Rank 128

▲ **Figure 9.6.1**

Concept Review

- Reduced singular value decomposition
- Reduced singular value expansion
- Rank of an approximation

Skills

- Find the reduced singular value decomposition of an $m \times n$ matrix.
- Find the reduced singular value expansion of an $m \times n$ matrix.

Exercise Set 9.6

▶ In Exercises 1–4, find a reduced singular value decomposition of A. [*Note:* Each matrix appears in Exercise Set 9.5, where you were asked to find its (unreduced) singular value decomposition.]

1. $A = \begin{bmatrix} 2 & 1 \\ 4 & 2 \\ 4 & 2 \end{bmatrix}$

2. $A = \begin{bmatrix} 4 & 0 & 3 \\ 0 & 0 & 5 \end{bmatrix}$

3. $A = \begin{bmatrix} 1 & 0 \\ 1 & 1 \\ -1 & 1 \end{bmatrix}$

4. $A = \begin{bmatrix} -1 & 4 \\ -2 & 2 \\ -2 & -4 \end{bmatrix}$

▶ In Exercises 5–8, find a reduced singular value expansion of A.

5. The matrix A in Exercise 1.

6. The matrix A in Exercise 2.

7. The matrix A in Exercise 3.

8. The matrix A in Exercise 4.

9. Suppose A is a 200×500 matrix. How many numbers must be stored in the rank 100 approximation of A? Compare this with the number of entries of A.

True-False Exercises

In parts (a)–(c) determine whether the statement is true or false, and justify your answer. Assume that $U_1 \Sigma_1 V_1^T$ is a reduced singular value decomposition of an $m \times n$ matrix of rank k.

(a) U_1 has size $m \times k$.

(b) Σ_1 has size $k \times k$.

(c) V_1 has size $k \times n$.

Chapter 9 Supplementary Exercises

1. Find an LU-decomposition of $A = \begin{bmatrix} -6 & 2 \\ 6 & 0 \end{bmatrix}$.

2. Find the LDU-decomposition of the matrix A in Exercise 1.

3. Find an LU-decomposition of $A = \begin{bmatrix} 2 & 4 & 6 \\ 1 & 4 & 7 \\ 1 & 3 & 7 \end{bmatrix}$.

4. Find the LDU-decomposition of the matrix A in Exercise 3.

5. Let $A = \begin{bmatrix} 2 & 1 \\ 1 & 2 \end{bmatrix}$ and $\mathbf{x}_0 = \begin{bmatrix} 1 \\ 0 \end{bmatrix}$.

   (a) Identify the dominant eigenvalue of A and then find the corresponding dominant unit eigenvector $\mathbf{v}$ with *positive* entries.

   (b) Apply the power method with Euclidean scaling to A and $\mathbf{x}_0$, stopping at $\mathbf{x}_5$. Compare your value of $\mathbf{x}_5$ to the eigenvector $\mathbf{v}$ found in part (a).

   (c) Apply the power method with maximum entry scaling to A and $\mathbf{x}_0$, stopping at $\mathbf{x}_5$. Compare your result with the eigenvector $\begin{bmatrix} 1 \\ 1 \end{bmatrix}$.

6. Consider the symmetric matrix

   $$A = \begin{bmatrix} 0 & 1 \\ 1 & 0 \end{bmatrix}$$

   Discuss the behavior of the power sequence

   $$\mathbf{x}_0, \ \mathbf{x}_1, \ldots, \ \mathbf{x}_k, \ldots$$

   with Euclidean scaling for a general *nonzero* vector $\mathbf{x}_0$. What is it about the matrix that causes the observed behavior?

7. Suppose that a symmetric matrix A has distinct eigenvalues $\lambda_1 = 8$, $\lambda_2 = 1.4$, $\lambda_3 = 2.3$, and $\lambda_4 = -8.1$. What can you say about the convergence of the Rayleigh quotients?

8. Find a singular value decomposition of $A = \begin{bmatrix} 1 & 1 \\ 1 & 1 \end{bmatrix}$.

9. Find a singular value decomposition of $A = \begin{bmatrix} 1 & 1 \\ 0 & 0 \\ 1 & 1 \end{bmatrix}$.

10. Find a reduced singular value decomposition and a reduced singular value expansion of the matrix A in Exercise 9.

11. Find the reduced singular value decomposition of the matrix whose singular value decomposition is

$$A = \begin{bmatrix} \tfrac{1}{2} & \tfrac{1}{2} & \tfrac{1}{2} & \tfrac{1}{2} \\ \tfrac{1}{2} & -\tfrac{1}{2} & -\tfrac{1}{2} & \tfrac{1}{2} \\ \tfrac{1}{2} & -\tfrac{1}{2} & \tfrac{1}{2} & -\tfrac{1}{2} \\ \tfrac{1}{2} & \tfrac{1}{2} & -\tfrac{1}{2} & -\tfrac{1}{2} \end{bmatrix} \begin{bmatrix} 24 & 0 & 0 \\ 0 & 12 & 0 \\ 0 & 0 & 0 \\ 0 & 0 & 0 \end{bmatrix} \begin{bmatrix} \tfrac{2}{3} & -\tfrac{1}{3} & \tfrac{2}{3} \\ \tfrac{2}{3} & \tfrac{2}{3} & -\tfrac{1}{3} \\ -\tfrac{1}{3} & \tfrac{2}{3} & \tfrac{2}{3} \end{bmatrix}$$

12. Do orthogonally similar matrices have the same singular values? Justify your answer.

13. If P is the standard matrix for the orthogonal projection of R^n onto a subspace W, what can you say about the singular values of P?

CHAPTER 10

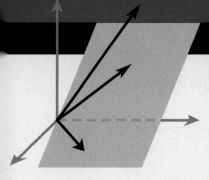

Applications of Linear Algebra

CHAPTER CONTENTS

10.1 Constructing Curves and Surfaces Through Specified Points 520
10.2 Geometric Linear Programming 525
10.3 The Earliest Applications of Linear Algebra 536
10.4 Cubic Spline Interpolation 543
10.5 Markov Chains 553
10.6 Graph Theory 563
10.7 Games of Strategy 572
10.8 Leontief Economic Models 581
10.9 Forest Management 590
10.10 Computer Graphics 597
10.11 Equilibrium Temperature Distributions 605
10.12 Computed Tomography 615
10.13 Fractals 626
10.14 Chaos 641
10.15 Cryptography 654
10.16 Genetics 665
10.17 Age-Specific Population Growth 676
10.18 Harvesting of Animal Populations 686
10.19 A Least Squares Model for Human Hearing 693
10.20 Warps and Morphs 700

INTRODUCTION This chapter consists of 20 applications of linear algebra. With one clearly marked exception, each application is in its own independent section, so sections can be deleted or permuted as desired. Each topic begins with a list of linear algebra prerequisites.

Because our primary objective in this chapter is to present applications of linear algebra, proofs are often omitted. Whenever results from other fields are needed, they are stated precisely, with motivation where possible, but usually without proof.

10.1 Constructing Curves and Surfaces Through Specified Points

In this section we describe a technique that uses determinants to construct lines, circles, and general conic sections through specified points in the plane. The procedure is also used to pass planes and spheres in 3-space through fixed points.

> **PREREQUISITES:** Linear Systems
> Determinants
> Analytic Geometry

The following theorem follows from Theorem 2.3.8.

> **THEOREM 10.1.1** *A homogeneous linear system with as many equations as unknowns has a nontrivial solution if and only if the determinant of the coefficient matrix is zero.*

We will now show how this result can be used to determine equations of various curves and surfaces through specified points.

A Line Through Two Points

Suppose that (x_1, y_1) and (x_2, y_2) are two distinct points in the plane. There exists a unique line

$$c_1 x + c_2 y + c_3 = 0 \tag{1}$$

that passes through these two points (Figure 10.1.1). Note that c_1, c_2, and c_3 are not all zero and that these coefficients are unique only up to a multiplicative constant. Because (x_1, y_1) and (x_2, y_2) lie on the line, substituting them in (1) gives the two equations

$$c_1 x_1 + c_2 y_1 + c_3 = 0 \tag{2}$$
$$c_1 x_2 + c_2 y_2 + c_3 = 0 \tag{3}$$

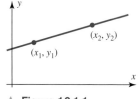

▲ Figure 10.1.1

The three equations, (1), (2), and (3), can be grouped together and rewritten as

$$x c_1 + y c_2 + c_3 = 0$$
$$x_1 c_1 + y_1 c_2 + c_3 = 0$$
$$x_2 c_1 + y_2 c_2 + c_3 = 0$$

which is a homogeneous linear system of three equations for c_1, c_2, and c_3. Because c_1, c_2, and c_3 are not all zero, this system has a nontrivial solution, so the determinant of the coefficient matrix of the system must be zero. That is,

$$\begin{vmatrix} x & y & 1 \\ x_1 & y_1 & 1 \\ x_2 & y_2 & 1 \end{vmatrix} = 0 \tag{4}$$

Consequently, every point (x, y) on the line satisfies (4); conversely, it can be shown that every point (x, y) that satisfies (4) lies on the line.

10.1 Constructing Curves and Surfaces Through Specified Points

▶ **EXAMPLE 1** **Equation of a Line**

Find the equation of the line that passes through the two points (2, 1) and (3, 7).

Solution Substituting the coordinates of the two points into Equation (4) gives

$$\begin{vmatrix} x & y & 1 \\ 2 & 1 & 1 \\ 3 & 7 & 1 \end{vmatrix} = 0$$

The cofactor expansion of this determinant along the first row then gives

$$-6x + y + 11 = 0 \quad \blacktriangleleft$$

A Circle Through Three Points

Suppose that there are three distinct points in the plane, (x_1, y_1), (x_2, y_2), and (x_3, y_3), not all lying on a straight line. From analytic geometry we know that there is a unique circle, say,

$$c_1(x^2 + y^2) + c_2 x + c_3 y + c_4 = 0 \tag{5}$$

that passes through them (Figure 10.1.2). Substituting the coordinates of the three points into this equation gives

$$c_1(x_1^2 + y_1^2) + c_2 x_1 + c_3 y_1 + c_4 = 0 \tag{6}$$
$$c_1(x_2^2 + y_2^2) + c_2 x_2 + c_3 y_2 + c_4 = 0 \tag{7}$$
$$c_1(x_3^2 + y_3^2) + c_2 x_3 + c_3 y_3 + c_4 = 0 \tag{8}$$

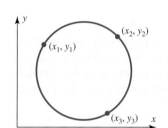

▲ Figure 10.1.2

As before, Equations (5) through (8) form a homogeneous linear system with a nontrivial solution for c_1, c_2, c_3, and c_4. Thus the determinant of the coefficient matrix is zero:

$$\begin{vmatrix} x^2 + y^2 & x & y & 1 \\ x_1^2 + y_1^2 & x_1 & y_1 & 1 \\ x_2^2 + y_2^2 & x_2 & y_2 & 1 \\ x_3^2 + y_3^2 & x_3 & y_3 & 1 \end{vmatrix} = 0 \tag{9}$$

This is a determinant form for the equation of the circle.

▶ **EXAMPLE 2** **Equation of a Circle**

Find the equation of the circle that passes through the three points (1, 7), (6, 2), and (4, 6).

Solution Substituting the coordinates of the three points into Equation (9) gives

$$\begin{vmatrix} x^2 + y^2 & x & y & 1 \\ 50 & 1 & 7 & 1 \\ 40 & 6 & 2 & 1 \\ 52 & 4 & 6 & 1 \end{vmatrix} = 0$$

which reduces to

$$10(x^2 + y^2) - 20x - 40y - 200 = 0$$

In standard form this is

$$(x - 1)^2 + (y - 2)^2 = 5^2$$

Thus the circle has center (1, 2) and radius 5. ◀

A General Conic Section Through Five Points

In his momumental work *Principia Mathematica*, Issac Newton posed and solved the following problem (Book I, Proposition 22, Problem 14): "To describe a conic that shall pass through five given points." Newton solved this problem geometrically, as shown in Figure 10.1.3, in which he passed an ellipse through the points A, B, D, P, C; however, the methods of this section can also be applied.

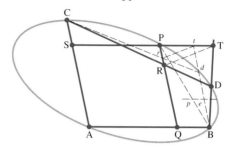

▶ Figure 10.1.3

The general equation of a conic section in the plane (a parabola, hyperbola, or ellipse, or degenerate forms of these curves) is given by

$$c_1 x^2 + c_2 xy + c_3 y^2 + c_4 x + c_5 y + c_6 = 0$$

This equation contains six coefficients, but we can reduce the number to five if we divide through by any one of them that is not zero. Thus only five coefficients must be determined, so five distinct points in the plane are sufficient to determine the equation of the conic section (Figure 10.1.4). As before, the equation can be put in determinant form (see Exercise 7):

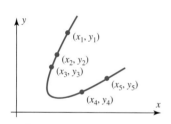

▲ Figure 10.1.4

$$\begin{vmatrix} x^2 & xy & y^2 & x & y & 1 \\ x_1^2 & x_1 y_1 & y_1^2 & x_1 & y_1 & 1 \\ x_2^2 & x_2 y_2 & y_2^2 & x_2 & y_2 & 1 \\ x_3^2 & x_3 y_3 & y_3^2 & x_3 & y_3 & 1 \\ x_4^2 & x_4 y_4 & y_4^2 & x_4 & y_4 & 1 \\ x_5^2 & x_5 y_5 & y_5^2 & x_5 & y_5 & 1 \end{vmatrix} = 0 \qquad (10)$$

▶ **EXAMPLE 3 Equation of an Orbit**

An astronomer who wants to determine the orbit of an asteroid about the Sun sets up a Cartesian coordinate system in the plane of the orbit with the Sun at the origin. Astronomical units of measurement are used along the axes (1 astronomical unit = mean distance of Earth to Sun = 93 million miles). By Kepler's first law, the orbit must be an ellipse, so the astronomer makes five observations of the asteroid at five different times and finds five points along the orbit to be

$(8.025, 8.310), (10.170, 6.355), (11.202, 3.212), (10.736, 0.375), (9.092, -2.267)$

Find the equation of the orbit.

Solution Substituting the coordinates of the five given points into (10) and rounding to three decimal places give

$$\begin{vmatrix} x^2 & xy & y^2 & x & y & 1 \\ 64.401 & 66.688 & 69.056 & 8.025 & 8.310 & 1 \\ 103.429 & 64.630 & 40.386 & 10.170 & 6.355 & 1 \\ 125.485 & 35.981 & 10.317 & 11.202 & 3.212 & 1 \\ 115.262 & 4.026 & 0.141 & 10.736 & 0.375 & 1 \\ 82.664 & -20.612 & 5.139 & 9.092 & -2.267 & 1 \end{vmatrix} = 0$$

The cofactor expansion of this determinant along the first row yields

$$386.802x^2 - 102.895xy + 446.029y^2 - 2476.443x - 1427.998y - 17109.375 = 0$$

Figure 10.1.5 is an accurate diagram of the orbit, together with the five given points. ◄

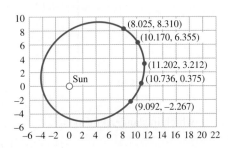

▶ Figure 10.1.5

A Plane Through Three Points

In Exercise 8 we ask you to show the following: The plane in 3-space with equation

$$c_1 x + c_2 y + c_3 z + c_4 = 0$$

that passes through three noncollinear points (x_1, y_1, z_1), (x_2, y_2, z_2), and (x_3, y_3, z_3) is given by the determinant equation

$$\begin{vmatrix} x & y & z & 1 \\ x_1 & y_1 & z_1 & 1 \\ x_2 & y_2 & z_2 & 1 \\ x_3 & y_3 & z_3 & 1 \end{vmatrix} = 0 \tag{11}$$

▶ **EXAMPLE 4** Equation of a Plane

The equation of the plane that passes through the three noncollinear points $(1, 1, 0)$, $(2, 0, -1)$, and $(2, 9, 2)$ is

$$\begin{vmatrix} x & y & z & 1 \\ 1 & 1 & 0 & 1 \\ 2 & 0 & -1 & 1 \\ 2 & 9 & 2 & 1 \end{vmatrix} = 0$$

which reduces to

$$2x - y + 3z - 1 = 0 \blacktriangleleft$$

A Sphere Through Four Points

In Exercise 9 we ask you to show the following: The sphere in 3-space with equation

$$c_1(x^2 + y^2 + z^2) + c_2 x + c_3 y + c_4 z + c_5 = 0$$

that passes through four noncoplanar points (x_1, y_1, z_1), (x_2, y_2, z_2), (x_3, y_3, z_3), and (x_4, y_4, z_4) is given by the following determinant equation:

$$\begin{vmatrix} x^2 + y^2 + z^2 & x & y & z & 1 \\ x_1^2 + y_1^2 + z_1^2 & x_1 & y_1 & z_1 & 1 \\ x_2^2 + y_2^2 + z_2^2 & x_2 & y_2 & z_2 & 1 \\ x_3^2 + y_3^2 + z_3^2 & x_3 & y_3 & z_3 & 1 \\ x_4^2 + y_4^2 + z_4^2 & x_4 & y_4 & z_4 & 1 \end{vmatrix} = 0 \tag{12}$$

▶ **EXAMPLE 5** **Equation of a Sphere**

The equation of the sphere that passes through the four points $(0, 3, 2)$, $(1, -1, 1)$, $(2, 1, 0)$, and $(5, 1, 3)$ is

$$\begin{vmatrix} x^2 + y^2 + z^2 & x & y & z & 1 \\ 13 & 0 & 3 & 2 & 1 \\ 3 & 1 & -1 & 1 & 1 \\ 5 & 2 & 1 & 0 & 1 \\ 35 & 5 & 1 & 3 & 1 \end{vmatrix} = 0$$

This reduces to

$$x^2 + y^2 + z^2 - 4x - 2y - 6z + 5 = 0$$

which in standard form is

$$(x - 2)^2 + (y - 1)^2 + (z - 3)^2 = 9 \quad \blacktriangleleft$$

Exercise Set 10.1

1. Find the equations of the lines that pass through the following points:

   (a) $(1, -1)$, $(2, 2)$ (b) $(0, 1)$, $(1, -1)$

2. Find the equations of the circles that pass through the following points:

   (a) $(2, 6)$, $(2, 0)$, $(5, 3)$ (b) $(2, -2)$, $(3, 5)$, $(-4, 6)$

3. Find the equation of the conic section that passes through the points $(0, 0)$, $(0, -1)$, $(2, 0)$, $(2, -5)$, and $(4, -1)$.

4. Find the equations of the planes in 3-space that pass through the following points:

   (a) $(1, 1, -3)$, $(1, -1, 1)$, $(0, -1, 2)$

   (b) $(2, 3, 1)$, $(2, -1, -1)$, $(1, 2, 1)$

5. (a) Alter Equation (11) so that it determines the plane that passes through the origin and is parallel to the plane that passes through three specified noncollinear points.

   (b) Find the two planes described in part (a) corresponding to the triplets of points in Exercises 4(a) and 4(b).

6. Find the equations of the spheres in 3-space that pass through the following points:

   (a) $(1, 2, 3)$, $(-1, 2, 1)$, $(1, 0, 1)$, $(1, 2, -1)$

   (b) $(0, 1, -2)$, $(1, 3, 1)$, $(2, -1, 0)$, $(3, 1, -1)$

7. Show that Equation (10) is the equation of the conic section that passes through five given distinct points in the plane.

8. Show that Equation (11) is the equation of the plane in 3-space that passes through three given noncollinear points.

9. Show that Equation (12) is the equation of the sphere in 3-space that passes through four given noncoplanar points.

10. Find a determinant equation for the parabola of the form

    $$c_1 y + c_2 x^2 + c_3 x + c_4 = 0$$

    that passes through three given noncollinear points in the plane.

11. What does Equation (9) become if the three distinct points are collinear?

12. What does Equation (11) become if the three distinct points are collinear?

13. What does Equation (12) become if the four points are coplanar?

Section 10.1 Technology Exercises

The following exercises are designed to be solved using a technology utility. Typically, this will be MATLAB, *Mathematica*, Maple, Derive, or Mathcad, but it may also be some other type of linear algebra software or a scientific calculator with some linear algebra capabilities. For each exercise you will need to read the relevant documentation for the particular utility you are using. The goal of these exercises is to provide you with a basic proficiency with your technology utility. Once you have mastered the techniques in these exercises, you will be able to use your technology utility to solve many of the problems in the regular exercise sets.

T1. The general equation of a quadric surface is given by

$$a_1 x^2 + a_2 y^2 + a_3 z^2 + a_4 xy + a_5 xz + a_6 yz + a_7 x + a_8 y + a_9 z + a_{10} = 0$$

Given nine points on this surface, it may be possible to determine its equation.

(a) Show that if the nine points (x_i, y_i) for $i = 1, 2, 3, \ldots, 9$ lie on this surface, and if they determine uniquely the equation of this surface, then its equation can be written in determinant form as

$$\begin{vmatrix} x^2 & y^2 & z^2 & xy & xz & yz & x & y & z & 1 \\ x_1^2 & y_1^2 & z_1^2 & x_1 y_1 & x_1 z_1 & y_1 z_1 & x_1 & y_1 & z_1 & 1 \\ x_2^2 & y_2^2 & z_2^2 & x_2 y_2 & x_2 z_2 & y_2 z_2 & x_2 & y_2 & z_2 & 1 \\ x_3^2 & y_3^2 & z_3^2 & x_3 y_3 & x_3 z_3 & y_3 z_3 & x_3 & y_3 & z_3 & 1 \\ x_4^2 & y_4^2 & z_4^2 & x_4 y_4 & x_4 z_4 & y_4 z_4 & x_4 & y_4 & z_4 & 1 \\ x_5^2 & y_5^2 & z_5^2 & x_5 y_5 & x_5 z_5 & y_5 z_5 & x_5 & y_5 & z_5 & 1 \\ x_6^2 & y_6^2 & z_6^2 & x_6 y_6 & x_6 z_6 & y_6 z_6 & x_6 & y_6 & z_6 & 1 \\ x_7^2 & y_7^2 & z_7^2 & x_7 y_7 & x_7 z_7 & y_7 z_7 & x_7 & y_7 & z_7 & 1 \\ x_8^2 & y_8^2 & z_8^2 & x_8 y_8 & x_8 z_8 & y_8 z_8 & x_8 & y_8 & z_8 & 1 \\ x_9^2 & y_9^2 & z_9^2 & x_9 y_9 & x_9 z_9 & y_9 z_9 & x_9 & y_9 & z_9 & 1 \end{vmatrix} = 0$$

(b) Use the result in part (a) to determine the equation of the quadric surface that passes through the points $(1, 2, 3)$, $(2, 1, 7)$, $(0, 4, 6)$, $(3, -1, 4)$, $(3, 0, 11)$, $(-1, 5, 8)$, $(9, -8, 3)$, $(4, 5, 3)$, and $(-2, 6, 10)$.

T2. (a) A hyperplane in the n-dimensional Euclidean space R^n has an equation of the form

$$a_1 x_1 + a_2 x_2 + a_3 x_3 + \cdots + a_n x_n + a_{n+1} = 0$$

where a_i, $i = 1, 2, 3, \ldots, n+1$, are constants, not all zero, and x_i, $i = 1, 2, 3, \ldots, n$, are variables for which

$$(x_1, x_2, x_3, \ldots, x_n) \in R^n$$

A point

$$(x_{10}, x_{20}, x_{30}, \ldots, x_{n0}) \in R^n$$

lies on this hyperplane if

$$a_1 x_{10} + a_2 x_{20} + a_3 x_{30} + \cdots + a_n x_{n0} + a_{n+1} = 0$$

Given that the n points $(x_{1i}, x_{2i}, x_{3i}, \ldots, x_{ni})$, $i = 1, 2, 3, \ldots, n$, lie on this hyperplane and that they uniquely determine the equation of the hyperplane, show that the equation of the hyperplane can be written in determinant form as

$$\begin{vmatrix} x_1 & x_2 & x_3 & \cdots & x_n & 1 \\ x_{11} & x_{21} & x_{31} & \cdots & x_{n1} & 1 \\ x_{12} & x_{22} & x_{32} & \cdots & x_{n2} & 1 \\ x_{13} & x_{23} & x_{33} & \cdots & x_{n3} & 1 \\ \vdots & \vdots & \vdots & \ddots & \vdots & \vdots \\ x_{1n} & x_{2n} & x_{3n} & \cdots & x_{nn} & 1 \end{vmatrix} = 0$$

(b) Determine the equation of the hyperplane in R^9 that goes through the following nine points:

$(1, 2, 3, 4, 5, 6, 7, 8, 9)$ $(2, 3, 4, 5, 6, 7, 8, 9, 1)$
$(3, 4, 5, 6, 7, 8, 9, 1, 2)$ $(4, 5, 6, 7, 8, 9, 1, 2, 3)$
$(5, 6, 7, 8, 9, 1, 2, 3, 4)$ $(6, 7, 8, 9, 1, 2, 3, 4, 5)$
$(7, 8, 9, 1, 2, 3, 4, 5, 6)$ $(8, 9, 1, 2, 3, 4, 5, 6, 7)$
$(9, 1, 2, 3, 4, 5, 6, 7, 8)$

10.2 Geometric Linear Programming

In this section we describe a geometric technique for maximizing or minimizing a linear expression in two variables subject to a set of linear constraints.

> **PREREQUISITES:** Linear Systems
> Linear Inequalities

Linear Programming   The study of linear programming theory has expanded greatly since the pioneering work of George Dantzig in the late 1940s. Today, linear programming is applied to a wide variety of problems in industry and science. In this section we present a geometric approach to the solution of simple linear programming problems. Let us begin with some examples.

▶ **EXAMPLE 1** **Maximizing Sales Revenue**

A candy manufacturer has 130 pounds of chocolate-covered cherries and 170 pounds of chocolate-covered mints in stock. He decides to sell them in the form of two different mixtures. One mixture will contain half cherries and half mints by weight and will sell for $2.00 per pound. The other mixture will contain one-third cherries and two-thirds mints by weight and will sell for $1.25 per pound. How many pounds of each mixture should the candy manufacturer prepare in order to maximize his sales revenue?

Mathematical Formulation Let the mixture of half cherries and half mints be called mix A, and let x_1 be the number of pounds of this mixture to be prepared. Let the mixture of one-third cherries and two-thirds mints be called mix B, and let x_2 be the number of pounds of this mixture to be prepared. Since mix A sells for $2.00 per pound and mix B sells for $1.25 per pound, the total sales z (in dollars) will be

$$z = 2.00x_1 + 1.25x_2$$

Since each pound of mix A contains $\frac{1}{2}$ pound of cherries and each pound of mix B contains $\frac{1}{3}$ pound of cherries, the total number of pounds of cherries used in both mixtures is

$$\tfrac{1}{2}x_1 + \tfrac{1}{3}x_2$$

Similarly, since each pound of mix A contains $\frac{1}{2}$ pound of mints and each pound of mix B contains $\frac{2}{3}$ pound of mints, the total number of pounds of mints used in both mixtures is

$$\tfrac{1}{2}x_1 + \tfrac{2}{3}x_2$$

Because the manufacturer can use at most 130 pounds of cherries and 170 pounds of mints, we must have

$$\tfrac{1}{2}x_1 + \tfrac{1}{3}x_2 \le 130$$
$$\tfrac{1}{2}x_1 + \tfrac{2}{3}x_2 \le 170$$

Furthermore, since x_1 and x_2 cannot be negative numbers, we must have

$$x_1 \ge 0 \quad \text{and} \quad x_2 \ge 0$$

The problem can therefore be formulated mathematically as follows: Find values of x_1 and x_2 that maximize

$$z = 2.00x_1 + 1.25x_2$$

subject to

$$\tfrac{1}{2}x_1 + \tfrac{1}{3}x_2 \le 130$$
$$\tfrac{1}{2}x_1 + \tfrac{2}{3}x_2 \le 170$$
$$x_1 \ge 0$$
$$x_2 \ge 0$$

Later in this section we will show how to solve this type of mathematical problem geometrically.

▶ **EXAMPLE 2** **Maximizing Annual Yield**

A woman has up to $10,000 to invest. Her broker suggests investing in two bonds, A and B. Bond A is a rather risky bond with an annual yield of 10%, and bond B is a rather safe bond with an annual yield of 7%. After some consideration, she decides to invest at most $6000 in bond A, to invest at least $2000 in bond B, and to invest at least as much in bond A as in bond B. How should she invest her money in order to maximize her annual yield?

Mathematical Formulation Let x_1 be the number of dollars to be invested in bond A, and let x_2 be the number of dollars to be invested in bond B. Since each dollar invested in bond A earns $.10 per year and each dollar invested in bond B earns $.07 per year, the total dollar amount z earned each year by both bonds is

$$z = .10x_1 + .07x_2$$

The constraints imposed can be formulated mathematically as follows:

Invest no more than $10,000:	$x_1 + x_2 \leq 10{,}000$
Invest at most $6000 in bond A:	$x_1 \leq 6000$
Invest at least $2000 in bond B:	$x_2 \geq 2000$
Invest at least as much in bond A as in bond B:	$x_1 \geq x_2$

We also have the implicit assumption that x_1 and x_2 are nonnegative:

$$x_1 \geq 0 \quad \text{and} \quad x_2 \geq 0$$

Thus the complete mathematical formulation of the problem is as follows: Find values of x_1 and x_2 that maximize

$$z = .10x_1 + .07x_2$$

subject to

$$x_1 + x_2 \leq 10{,}000$$
$$x_1 \leq 6000$$
$$x_2 \geq 2000$$
$$x_1 - x_2 \geq 0$$
$$x_1 \geq 0$$
$$x_2 \geq 0$$

▶ **EXAMPLE 3 Minimizing Cost**

A student desires to design a breakfast of cornflakes and milk that is as economical as possible. On the basis of what he eats during his other meals, he decides that his breakfast should supply him with at least 9 grams of protein, at least $\frac{1}{3}$ the recommended daily allowance (RDA) of vitamin D, and at least $\frac{1}{4}$ the RDA of calcium. He finds the following nutrition and cost information on the milk and cornflakes containers:

	Milk ($\frac{1}{2}$ cup)	Cornflakes (1 ounce)
Cost	7.5 cents	5.0 cents
Protein	4 grams	2 grams
Vitamin D	$\frac{1}{8}$ of RDA	$\frac{1}{10}$ of RDA
Calcium	$\frac{1}{6}$ of RDA	None

In order not to have his mixture too soggy or too dry, the student decides to limit himself to mixtures that contain 1 to 3 ounces of cornflakes per cup of milk, inclusive. What quantities of milk and cornflakes should he use to minimize the cost of his breakfast?

Mathematical Formulation Let x_1 be the quantity of milk used (measured in $\frac{1}{2}$-cup units), and let x_2 be the quantity of cornflakes used (measured in 1-ounce units). Then if z is the cost of the breakfast in cents, we may write the following.

Cost of breakfast: $\quad z = 7.5x_1 + 5.0x_2$

At least 9 grams protein: $\quad 4x_1 + 2x_2 \geq 9$

At least $\frac{1}{3}$ RDA vitamin D: $\quad \frac{1}{8}x_1 + \frac{1}{10}x_2 \geq \frac{1}{3}$

At least $\frac{1}{4}$ RDA calcium: $\quad \frac{1}{6}x_1 \geq \frac{1}{4}$

At least 1 ounce cornflakes per cup (two $\frac{1}{2}$-cups) of milk: $\quad \frac{x_2}{x_1} \geq \frac{1}{2}$ (or $x_1 - 2x_2 \leq 0$)

At most 3 ounces cornflakes per cup (two $\frac{1}{2}$-cups) of milk: $\quad \frac{x_2}{x_1} \leq \frac{3}{2}$ (or $3x_1 - 2x_2 \geq 0$)

As before, we also have the implicit assumption that $x_1 \geq 0$ and $x_2 \geq 0$. Thus the complete mathematical formulation of the problem is as follows: Find values of x_1 and x_2 that minimize

$$z = 7.5x_1 + 5.0x_2$$

subject to

$$4x_1 + 2x_2 \geq 9$$
$$\tfrac{1}{8}x_1 + \tfrac{1}{10}x_2 \geq \tfrac{1}{3}$$
$$\tfrac{1}{6}x_1 \geq \tfrac{1}{4}$$
$$x_1 - 2x_2 \leq 0$$
$$3x_1 - 2x_2 \geq 0$$
$$x_1 \geq 0$$
$$x_2 \geq 0 \quad \blacktriangleleft$$

Geometric Solution of Linear Programming Problems

Each of the preceding three examples is a special case of the following problem.

Problem Find values of x_1 and x_2 that either maximize or minimize

$$z = c_1 x_1 + c_2 x_2 \tag{1}$$

subject to

$$\begin{aligned} a_{11}x_1 + a_{12}x_2 \; (\leq)(\geq)(=) \; b_1 \\ a_{21}x_1 + a_{22}x_2 \; (\leq)(\geq)(=) \; b_2 \\ \vdots \qquad \vdots \qquad \vdots \\ a_{m1}x_1 + a_{m2}x_2 \; (\leq)(\geq)(=) \; b_m \end{aligned} \tag{2}$$

and

$$x_1 \geq 0, \quad x_2 \geq 0 \tag{3}$$

In each of the m conditions of (2), any one of the symbols $\leq, \geq$, and $=$ may be used.

The problem above is called the **general linear programming problem** in two variables. The linear function z in (1) is called the **objective function**. Equations (2) and (3) are called the **constraints**; in particular, the equations in (3) are called the **nonnegativity constraints** on the variables x_1 and x_2.

We will now show how to solve a linear programming problem in two variables graphically. A pair of values (x_1, x_2) that satisfy all of the constraints is called a *feasible solution*. The set of all feasible solutions determines a subset of the $x_1 x_2$-plane called the *feasible region*. Our desire is to find a feasible solution that maximizes the objective function. Such a solution is called an *optimal solution*.

To examine the feasible region of a linear programming problem, let us note that each constraint of the form

$$a_{i1}x_1 + a_{i2}x_2 = b_i$$

defines a line in the x_1x_2-plane, whereas each constraint of the form

$$a_{i1}x_1 + a_{i2}x_2 \leq b_i \quad \text{or} \quad a_{i1}x_1 + a_{i2}x_2 \geq b_i$$

defines a half-plane that includes its boundary line

$$a_{i1}x_1 + a_{i2}x_2 = b_i$$

Thus the feasible region is always an intersection of finitely many lines and half-planes. For example, the four constraints

$$\tfrac{1}{2}x_1 + \tfrac{1}{3}x_2 \leq 130$$
$$\tfrac{1}{2}x_1 + \tfrac{2}{3}x_2 \leq 170$$
$$x_1 \geq 0$$
$$x_2 \geq 0$$

of Example 1 define the half-planes illustrated in parts (*a*), (*b*), (*c*), and (*d*) of Figure 10.2.1. The feasible region of this problem is thus the intersection of these four half-planes, which is illustrated in Figure 10.2.1*e*.

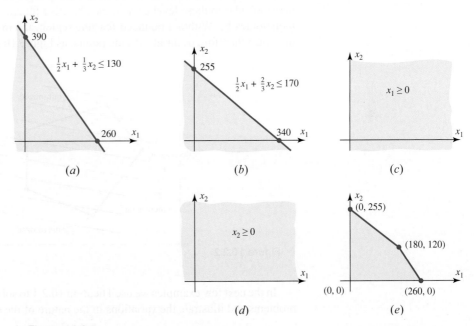

▲ Figure 10.2.1

It can be shown that the feasible region of a linear programming problem has a boundary consisting of a finite number of straight line segments. If the feasible region can be enclosed in a sufficiently large circle, it is called **bounded** (Figure 10.2.1*e*); otherwise, it is called **unbounded** (see Figure 10.2.5). If the feasible region is *empty* (contains no points), then the constraints are inconsistent and the linear programming problem has no solution (see Figure 10.2.6).

Those boundary points of a feasible region that are intersections of two of the straight line boundary segments are called **extreme points**. (They are also called *corner points* and *vertex points*.) For example, in Figure 10.2.1e, we see that the feasible region of Example 1 has four extreme points:

$$(0, 0), \quad (0, 255), \quad (180, 120), \quad (260, 0) \tag{4}$$

The importance of the extreme points of a feasible region is shown by the following theorem.

THEOREM 10.2.1 **Maximum and Minimum Values**

If the feasible region of a linear programming problem is nonempty and bounded, then the objective function attains both a maximum and a minimum value, and these occur at extreme points of the feasible region. If the feasible region is unbounded, then the objective function may or may not attain a maximum or minimum value; however, if it attains a maximum or minimum value, it does so at an extreme point.

Figure 10.2.2 suggests the idea behind the proof of this theorem. Since the objective function

$$z = c_1 x_1 + c_2 x_2$$

of a linear programming problem is a linear function of x_1 and x_2, its level curves (the curves along which z has constant values) are straight lines. As we move in a direction perpendicular to these level curves, the objective function either increases or decreases monotonically. Within a bounded feasible region, the maximum and minimum values of z must therefore occur at extreme points, as Figure 10.2.2 indicates.

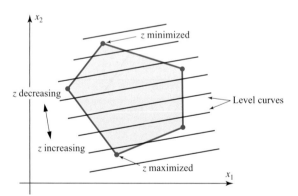

▶ Figure 10.2.2

In the next few examples we use Theorem 10.2.1 to solve several linear programming problems and illustrate the variations in the nature of the solutions that may occur.

▶ **EXAMPLE 4** **Example 1 Revisited**

Figure 10.2.1e shows that the feasible region of Example 1 is bounded. Consequently, from Theorem 10.2.1 the objective function

$$z = 2.00 x_1 + 1.25 x_2$$

attains both its minimum and maximum values at extreme points. The four extreme points and the corresponding values of z are given in the following table.

10.2 Geometric Linear Programming

Extreme Point (x_1, x_2)	Value of $z = 2.00x_1 + 1.25x_2$
(0, 0)	0
(0, 255)	318.75
(180, 120)	510.00
(260, 0)	520.00

We see that the largest value of z is 520.00 and the corresponding optimal solution is (260, 0). Thus the candy manufacturer attains maximum sales of $520 when he produces 260 pounds of mixture A and none of mixture B.

▶ **EXAMPLE 5** Using Theorem 10.2.1

Find values of x_1 and x_2 that maximize

$$z = x_1 + 3x_2$$

subject to

$$2x_1 + 3x_2 \leq 24$$
$$x_1 - x_2 \leq 7$$
$$x_2 \leq 6$$
$$x_1 \geq 0$$
$$x_2 \geq 0$$

Solution In Figure 10.2.3 we have drawn the feasible region of this problem. Since it is bounded, the maximum value of z is attained at one of the five extreme points. The values of the objective function at the five extreme points are given in the following table.

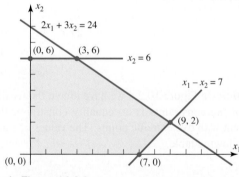

▲ Figure 10.2.3

Extreme Point (x_1, x_2)	Value of $z = x_1 + 3x_2$
(0, 6)	18
(3, 6)	21
(9, 2)	15
(7, 0)	7
(0, 0)	0

From this table, the maximum value of z is 21, which is attained at $x_1 = 3$ and $x_2 = 6$.

▶ **EXAMPLE 6** Using Theorem 10.2.1

Find values of x_1 and x_2 that maximize

$$z = 4x_1 + 6x_2$$

subject to

$$2x_1 + 3x_2 \leq 24$$
$$x_1 - x_2 \leq 7$$
$$x_2 \leq 6$$
$$x_1 \geq 0$$
$$x_2 \geq 0$$

Solution The constraints in this problem are identical to the constraints in Example 5, so the feasible region of this problem is also given by Figure 10.2.3. The values of the objective function at the extreme points are given in the following table.

Extreme Point (x_1, x_2)	Value of $z = 4x_1 + 6x_2$
(0, 6)	36
(3, 6)	48
(9, 2)	48
(7, 0)	28
(0, 0)	0

We see that the objective function attains a maximum value of 48 at two adjacent extreme points, (3, 6) and (9, 2). This shows that an optimal solution to a linear programming problem need not be unique. As we ask you to show in Exercise 10, if the objective function has the same value at two adjacent extreme points, it has the same value at all points on the straight line boundary segment connecting the two extreme points. Thus, in this example the maximum value of z is attained at all points on the straight line segment connecting the extreme points (3, 6) and (9, 2).

▶ **EXAMPLE 7** **The Feasible Region Is a Line Segment**

Find values of x_1 and x_2 that minimize

$$z = 2x_1 - x_2$$

subject to

$$2x_1 + 3x_2 = 12$$
$$2x_1 - 3x_2 \geq 0$$
$$x_1 \geq 0$$
$$x_2 \geq 0$$

Solution In Figure 10.2.4 we have drawn the feasible region of this problem. Because one of the constraints is an equality constraint, the feasible region is a straight line segment with two extreme points. The values of z at the two extreme points are given in the following table.

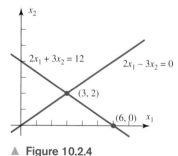

▲ Figure 10.2.4

Extreme Point (x_1, x_2)	Value of $z = 2x_1 - x_2$
(3, 2)	4
(6, 0)	12

The minimum value of z is thus 4 and is attained at $x_1 = 3$ and $x_2 = 2$.

▶ **EXAMPLE 8** Using Theorem 10.2.1

Find values of x_1 and x_2 that maximize

$$z = 2x_1 + 5x_2$$

subject to

$$\begin{aligned} 2x_1 + x_2 &\geq 8 \\ -4x_1 + x_2 &\leq 2 \\ 2x_1 - 3x_2 &\leq 0 \\ x_1 &\geq 0 \\ x_2 &\geq 0 \end{aligned}$$

Solution The feasible region of this linear programming problem is illustrated in Figure 10.2.5. Since it is unbounded, we are not assured by Theorem 10.2.1 that the objective function attains a maximum value. In fact, it is easily seen that since the feasible region contains points for which both x_1 and x_2 are arbitrarily large and positive, the objective function

$$z = 2x_1 + 5x_2$$

can be made arbitrarily large and positive. This problem has no optimal solution. Instead, we say the problem has an *unbounded solution*.

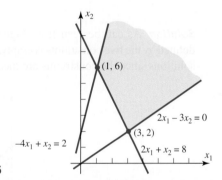

▶ Figure 10.2.5

▶ **EXAMPLE 9** Using Theorem 10.2.1

Find values of x_1 and x_2 that maximize

$$z = -5x_1 + x_2$$

subject to

$$\begin{aligned} 2x_1 + x_2 &\geq 8 \\ -4x_1 + x_2 &\leq 2 \\ 2x_1 - 3x_2 &\leq 0 \\ x_1 &\geq 0 \\ x_2 &\geq 0 \end{aligned}$$

Solution The above constraints are the same as those in Example 8, so the feasible region of this problem is also given by Figure 10.2.5. In Exercise 11 we ask you to show that the objective function of this problem attains a maximum within the feasible region.

By Theorem 10.2.1, this maximum must be attained at an extreme point. The values of z at the two extreme points of the feasible region are given in the following table.

Extreme Point (x_1, x_2)	Value of $z = -5x_1 + x_2$
(1, 6)	1
(3, 2)	−13

The maximum value of z is thus 1 and is attained at the extreme point $x_1 = 1$, $x_2 = 6$.

▶ **EXAMPLE 10 Inconsistent Constraints**

Find values of x_1 and x_2 that minimize

$$z = 3x_1 - 8x_2$$

subject to

$$2x_1 - x_2 \leq 4$$
$$3x_1 + 11x_2 \leq 33$$
$$3x_1 + 4x_2 \geq 24$$
$$x_1 \geq 0$$
$$x_2 \geq 0$$

Solution As can be seen from Figure 10.2.6, the intersection of the five half-planes defined by the five constraints is empty. This linear programming problem has no feasible solutions since the constraints are inconsistent. ◀

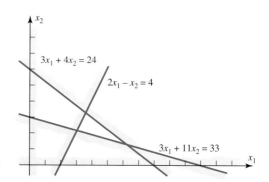

▶ **Figure 10.2.6** There are no points common to all five shaded half-planes.

Exercise Set 10.2

1. Find values of x_1 and x_2 that maximize

$$z = 3x_1 + 2x_2$$

subject to

$$2x_1 + 3x_2 \leq 6$$
$$2x_1 - x_2 \geq 0$$
$$x_1 \leq 2$$
$$x_2 \leq 1$$
$$x_1 \geq 0$$
$$x_2 \geq 0$$

2. Find values of x_1 and x_2 that minimize

$$z = 3x_1 - 5x_2$$

subject to

$$2x_1 - x_2 \leq -2$$
$$4x_1 - x_2 \geq 0$$
$$x_2 \leq 3$$
$$x_1 \geq 0$$
$$x_2 \geq 0$$

3. Find values of x_1 and x_2 that minimize
$$z = -3x_1 + 2x_2$$
subject to
$$\begin{aligned} 3x_1 - x_2 &\geq -5 \\ -x_1 + x_2 &\geq 1 \\ 2x_1 + 4x_2 &\geq 12 \\ x_1 &\geq 0 \\ x_2 &\geq 0 \end{aligned}$$

4. Solve the linear programming problem posed in Example 2.

5. Solve the linear programming problem posed in Example 3.

6. In Example 5 the constraint $x_1 - x_2 \leq 7$ is said to be *nonbinding* because it can be removed from the problem without affecting the solution. Likewise, the constraint $x_2 \leq 6$ is said to be *binding* because removing it will change the solution.

   (a) Which of the remaining constraints are nonbinding and which are binding?

   (b) For what values of the right-hand side of the nonbinding constraint $x_1 - x_2 \leq 7$ will this constraint become binding? For what values will the resulting feasible set be empty?

   (c) For what values of the right-hand side of the binding constraints $x_2 \leq 6$ will this constraint become nonbinding? For what values will the resulting feasible set be empty?

7. A trucking firm ships the containers of two companies, A and B. Each container from company A weighs 40 pounds and is 2 cubic feet in volume. Each container from company B weighs 50 pounds and is 3 cubic feet in volume. The trucking firm charges company A $2.20 for each container shipped and charges company B $3.00 for each container shipped. If one of the firm's trucks cannot carry more than 37,000 pounds and cannot hold more than 2000 cubic feet, how many containers from companies A and B should a truck carry to maximize the shipping charges?

8. Repeat Exercise 7 if the trucking firm raises its price for shipping a container from company A to $2.50.

9. A manufacturer produces sacks of chicken feed from two ingredients, A and B. Each sack is to contain at least 10 ounces of nutrient N_1, at least 8 ounces of nutrient N_2, and at least 12 ounces of nutrient N_3. Each pound of ingredient A contains 2 ounces of nutrient N_1, 2 ounces of nutrient N_2, and 6 ounces of nutrient N_3. Each pound of ingredient B contains 5 ounces of nutrient N_1, 3 ounces of nutrient N_2, and 4 ounces of nutrient N_3. If ingredient A costs 8 cents per pound and ingredient B costs 9 cents per pound, how much of each ingredient should the manufacturer use in each sack of feed to minimize his costs?

10. If the objective function of a linear programming problem has the same value at two adjacent extreme points, show that it has the same value at all points on the straight line segment connecting the two extreme points. [*Hint:* If (x_1', x_2') and (x_1'', x_2'') are any two points in the plane, a point (x_1, x_2) lies on the straight line segment connecting them if
$$x_1 = tx_1' + (1-t)x_1''$$
and
$$x_2 = tx_2' + (1-t)x_2''$$
where t is a number in the interval $[0, 1]$.]

11. Show that the objective function in Example 9 attains a maximum value in the feasible set. [*Hint:* Examine the level curves of the objective function.]

Section 10.2 Technology Exercises

The following exercises are designed to be solved using a technology utility. Typically, this will be MATLAB, *Mathematica*, Maple, Derive, or Mathcad, but it may also be some other type of linear algebra software or a scientific calculator with some linear algebra capabilities. For each exercise you will need to read the relevant documentation for the particular utility you are using. The goal of these exercises is to provide you with a basic proficiency with your technology utility. Once you have mastered the techniques in these exercises, you will be able to use your technology utility to solve many of the problems in the regular exercise sets.

T1. Consider the feasible region consisting of $0 \leq x, 0 \leq y$ along with the set of inequalities
$$x \cos\left(\frac{(2k+1)\pi}{4n}\right) + y \sin\left(\frac{(2k+1)\pi}{4n}\right) \leq \cos\left(\frac{\pi}{4n}\right)$$
for $k = 0, 1, 2, \ldots, n-1$. Maximize the objective function
$$z = 3x + 4y$$
assuming that (a) $n = 1$, (b) $n = 2$, (c) $n = 3$, (d) $n = 4$, (e) $n = 5$, (f) $n = 6$, (g) $n = 7$, (h) $n = 8$, (i) $n = 9$, (j) $n = 10$, and (k) $n = 11$. (l) Next, maximize this objective function using the nonlinear feasible region, $0 \leq x, 0 \leq y$, and
$$x^2 + y^2 \leq 1$$
(m) Let the results of parts (a) through (k) begin a sequence of values for $z_{\max}$. Do these values approach the value determined in part (l)? Explain.

T2. Repeat Exercise T1 using the objective function $z = x + y$.

10.3 The Earliest Applications of Linear Algebra

Linear systems can be found in the earliest writings of many ancient civilizations. In this section we give some examples of the types of problems that they used to solve.

> **PREREQUISITES:** Linear Systems

The practical problems of early civilizations included the measurement of land, the distribution of goods, the tracking of resources such as wheat and cattle, and taxation and inheritance calculations. In many cases, these problems led to linear systems of equations since linearity is one of the simplest relationships that can exist among variables. In this section we present examples from five diverse ancient cultures illustrating how they used and solved systems of linear equations. We restrict ourselves to examples before A.D. 500. These examples consequently predate the development of the field of algebra by Islamic/Arab mathematicians, a field that ultimately led in the nineteenth century to the branch of mathematics now called linear algebra.

▶ **EXAMPLE 1** **Egypt (about 1650 B.C.)**

Problem 40 of the Ahmes Papyrus

The Ahmes (or Rhind) Papyrus is the source of most of our information about ancient Egyptian mathematics. This 5-meter-long papyrus contains 84 short mathematical problems, together with their solutions, and dates from about 1650 B.C. Problem 40 in this papyrus is the following:

> *Divide 100 hekats of barley among five men in arithmetic progression so that the sum of the two smallest is one-seventh the sum of the three largest.*

Let a be the least amount that any man obtains, and let d be the common difference of the terms in the arithmetic progression. Then the other four men receive $a + d, a + 2d$, $a + 3d$, and $a + 4d$ hekats. The two conditions of the problem require that

$$a + (a + d) + (a + 2d) + (a + 3d) + (a + 4d) = 100$$
$$\tfrac{1}{7}[(a + 2d) + (a + 3d) + (a + 4d)] = a + (a + d)$$

These equations reduce to the following system of two equations in two unknowns:

$$\begin{aligned} 5a + 10d &= 100 \\ 11a - 2d &= 0 \end{aligned} \qquad (1)$$

The solution technique described in the papyrus is known as the method of false position or false assumption. It begins by assuming some convenient value of a (in our case $a = 1$), substituting that value into the second equation, and obtaining $d = 11/2$. Substituting $a = 1$ and $d = 11/2$ into the left-hand side of the first equation gives 60, whereas the right-hand side is 100. Adjusting the initial guess for a by multiplying it by 100/60 leads to the correct value $a = 5/3$. Substituting $a = 5/3$ into the second equation then gives $d = 55/6$, so the quantities of barley received by the five men are 10/6, 65/6, 120/6, 175/6, and 230/6 hekats. This technique of guessing a value of an unknown and later adjusting it has been used by many cultures throughout the ages.

▶ **EXAMPLE 2 Babylonia (1900–1600 B.C.)**

The Old Babylonian Empire flourished in Mesopotamia between 1900 and 1600 B.C. Many clay tablets containing mathematical tables and problems survive from that period, one of which (designated Ca MLA 1950) contains the next problem. The statement of the problem is a bit muddled because of the condition of the tablet, but the diagram and the solution on the tablet indicate that the problem is as follows:

Babylonian clay tablet Ca MLA 1950

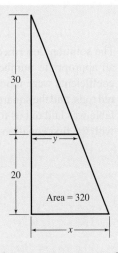

A trapezoid with an area of 320 square units is cut off from a right triangle by a line parallel to one of its sides. The other side has length 50 units, and the height of the trapezoid is 20 units. What are the upper and the lower widths of the trapezoid?

Let x be the lower width of the trapezoid and y its upper width. The area of the trapezoid is its height times its average width, so $20\left(\frac{x+y}{2}\right) = 320$. Using similar triangles, we also have $\frac{x}{50} = \frac{y}{30}$. The solution on the tablet uses these relations to generate the linear system

$$\frac{1}{2}(x + y) = 16$$
$$\frac{1}{2}(x - y) = 4 \qquad (2)$$

Adding and subtracting these two equations then gives the solution $x = 20$ and $y = 12$.

▶ **EXAMPLE 3 China (A.D. 263)**

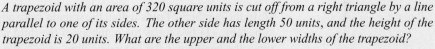

Chiu Chang Suan Shu in Chinese characters

The most important treatise in the history of Chinese mathematics is the Chiu Chang Suan Shu, or "The Nine Chapters of the Mathematical Art." This treatise, which is a collection of 246 problems and their solutions, was assembled in its final form by Liu Hui in A.D. 263. Its contents, however, go back to at least the beginning of the Han dynasty in the second century B.C. The eighth of its nine chapters, entitled "The Way of Calculating by Arrays," contains 18 word problems that lead to linear systems in three to six unknowns. The general solution procedure described is almost identical to the

Gaussian elimination technique developed in Europe in the nineteenth century by Carl Friedrich Gauss (see page 15). The first problem in the eighth chapter is the following:

> *There are three classes of corn, of which three bundles of the first class, two of the second, and one of the third make 39 measures. Two of the first, three of the second, and one of the third make 34 measures. And one of the first, two of the second, and three of the third make 26 measures. How many measures of grain are contained in one bundle of each class?*

Let x, y, and z be the measures of the first, second, and third classes of corn. Then the conditions of the problem lead to the following linear system of three equations in three unknowns:

$$3x + 2y + z = 39$$
$$2x + 3y + z = 34 \qquad (3)$$
$$x + 2y + 3z = 26$$

The solution described in the treatise represented the coefficients of each equation by an appropriate number of rods placed within squares on a counting table. Positive coefficients were represented by black rods, negative coefficients were represented by red rods, and the squares corresponding to zero coefficients were left empty. The counting table was laid out as follows so that the coefficients of each equation appear in columns with the first equation in the rightmost column:

1	2	3
2	3	2
3	1	1
26	34	39

Next, the numbers of rods within the squares were adjusted to accomplish the following two steps: (1) two times the numbers of the third column were subtracted from three times the numbers in the second column and (2) the numbers in the third column were subtracted from three times the numbers in the first column. The result was the following array:

		3
4	5	2
8	1	1
39	24	39

In this array, four times the numbers in the second column were subtracted from five times the numbers in the first column, yielding

		3
	5	2
36	1	1
99	24	39

This last array is equivalent to the linear system

$$3x + 2y + z = 39$$
$$5y + z = 24$$
$$36z = 99$$

This triangular system was solved by a method equivalent to back substitution to obtain $x = 37/4$, $y = 17/4$, and $z = 11/4$.

▶ **EXAMPLE 4** Greece (third century B.C.)

Perhaps the most famous system of linear equations from antiquity is the one associated with the first part of Archimedes' celebrated Cattle Problem. This problem supposedly was posed by Archimedes as a challenge to his colleague Eratosthenes. No solution has come down to us from ancient times, so that it is not known how, or even whether, either of these two geometers solved it.

Archimedes c. 287–212 B.C.

> *If thou art diligent and wise, O stranger, compute the number of cattle of the Sun, who once upon a time grazed on the fields of the Thrinacian isle of Sicily, divided into four herds of different colors, one milk white, another glossy black, a third yellow, and the last dappled. In each herd were bulls, mighty in number according to these proportions: Understand, stranger, that the white bulls were equal to a half and a third of the black together with the whole of the yellow, while the black were equal to the fourth part of the dappled and a fifth, together with, once more, the whole of the yellow. Observe further that the remaining bulls, the dappled, were equal to a sixth part of the white and a seventh, together with all of the yellow. These were the proportions of the cows: The white were precisely equal to the third part and a fourth of the whole herd of the black; while the black were equal to the fourth part once more of the dappled and with it a fifth part, when all, including the bulls, went to pasture together. Now the dappled in four parts were equal in number to a fifth part and a sixth of the yellow herd. Finally the yellow were in number equal to a sixth part and a seventh of the white herd. If thou canst accurately tell, O stranger, the number of cattle of the Sun, giving separately the number of well-fed bulls and again the number of females according to each color, thou wouldst not be called unskilled or ignorant of numbers, but not yet shalt thou be numbered among the wise.*

The conventional designation of the eight variables in this problem is

W = number of white bulls
B = number of black bulls
Y = number of yellow bulls
D = number of dappled bulls
w = number of white cows
b = number of black cows
y = number of yellow cows
d = number of dappled cows

The problem can now be stated as the following seven homogeneous equations in eight unknowns:

1. $W = \left(\frac{1}{2} + \frac{1}{3}\right) B + Y$ (The white bulls were equal to a half and a third of the black [bulls] together with the whole of the yellow [bulls].)

2. $B = \left(\frac{1}{4} + \frac{1}{5}\right) D + Y$ (The black [bulls] were equal to the fourth part of the dappled [bulls] and a fifth, together with, once more, the whole of the yellow [bulls].)

3. $D = \left(\frac{1}{6} + \frac{1}{7}\right) W + Y$ (The remaining bulls, the dappled, were equal to a sixth part of the white [bulls] and a seventh, together with all of the yellow [bulls].)

4. $w = \left(\frac{1}{3} + \frac{1}{4}\right) (B + b)$ (The white [cows] were precisely equal to the third part and a fourth of the whole herd of the black.)

5. $b = \left(\frac{1}{4} + \frac{1}{5}\right) (D + d)$ (The black [cows] were equal to the fourth part once more of the dappled and with it a fifth part, when all, including the bulls, went to pasture together.)

6. $d = \left(\frac{1}{5} + \frac{1}{6}\right) (Y + y)$ (The dappled [cows] in four parts [that is, in totality] were equal in number to a fifth part and a sixth of the yellow herd.)

7. $y = \left(\frac{1}{6} + \frac{1}{7}\right) (W + w)$ (The yellow [cows] were in number equal to a sixth part and a seventh of the white herd.)

As we ask you to show in the exercises, this system has infinitely many solutions of the form

$$\begin{aligned} W &= 10{,}366{,}482k \\ B &= 7{,}460{,}514k \\ Y &= 4{,}149{,}387k \\ D &= 7{,}358{,}060k \\ w &= 7{,}206{,}360k \\ b &= 4{,}893{,}246k \\ y &= 5{,}439{,}213k \\ d &= 3{,}515{,}820k \end{aligned} \quad (4)$$

where k is any real number. The values $k = 1, 2, \ldots$ give infinitely many positive integer solutions to the problem, with $k = 1$ giving the smallest solution.

▶ **EXAMPLE 5 India (fourth century A.D.)**

The Bakhshali Manuscript is an ancient work of Indian/Hindu mathematics dating from around the fourth century A.D., although some of its materials undoubtedly come from many centuries before. It consists of about 70 leaves or sheets of birch bark containing mathematical problems and their solutions. Many of its problems are so-called equalization problems that lead to systems of linear equations. One such problem on the fragment shown is the following:

Fragment III-5-3v of the Bakhshali Manuscript

> *One merchant has seven asava horses, a second has nine haya horses, and a third has ten camels. They are equally well off in the value of their animals if each gives two animals, one to each of the others. Find the price of each animal and the total value of the animals possessed by each merchant.*

Let x be the price of an asava horse, let y be the price of a haya horse, let z be the price of a camel, and the let K be the total value of the animals possessed by each merchant. Then the conditions of the problem lead to the following system of equations:

$$5x + y + z = K$$
$$x + 7y + z = K \qquad (5)$$
$$x + y + 8z = K$$

The method of solution described in the manuscript begins by subtracting the quantity $(x + y + z)$ from both sides of the three equations to obtain $4x = 6y = 7z = K - (x + y + z)$. This shows that if the prices x, y, and z are to be integers, then the quantity $K - (x + y + z)$ must be an integer that is divisible by 4, 6, and 7. The manuscript takes the product of these three numbers, or 168, for the value of $K - (x + y + z)$, which yields $x = 42$, $y = 28$, and $z = 24$ for the prices and $K = 262$ for the total value. (See Exercise 6 for more solutions to this problem.) ◀

Exercise Set 10.3

1. The following lines from Book 12 of Homer's *Odyssey* relate a precursor of Archimedes' Cattle Problem:

   > Thou shalt ascend the isle triangular,
   > Where many oxen of the Sun are fed,
   > And fatted flocks. Of oxen fifty head
   > In every herd feed, and their herds are seven;
   > And of his fat flocks is their number even.

   The last line means that there are as many sheep in all the flocks as there are oxen in all the herds. What is the total number of oxen and sheep that belong to the god of the Sun? (This was a difficult problem in Homer's day.)

2. Solve the following problems from the Bakhshali Manuscript.
   (a) B possesses two times as much as A; C has three times as much as A and B together; D has four times as much as A, B, and C together. Their total possessions are 300. What is the possession of A?
   (b) B gives 2 times as much as A; C gives 3 times as much as B; D gives 4 times as much as C. Their total gift is 132. What is the gift of A?

3. A problem on a Babylonian tablet requires finding the length and width of a rectangle given that the length and the width add up to 10, while the length and one-fourth of the width add up to 7. The solution provided on the tablet consists of the following four statements:

   > Multiply 7 by 4 to obtain 28.
   > Take away 10 from 28 to obtain 18.
   > Take one-third of 18 to obtain 6, the length.
   > Take away 6 from 10 to obtain 4, the width.

   Explain how these steps lead to the answer.

4. The following two problems are from "The Nine Chapters of the Mathematical Art." Solve them using the array technique described in Example 3.
   (a) Five oxen and two sheep are worth 10 units and two oxen and five sheep are worth 8 units. What is the value of each ox and sheep?
   (b) There are three kinds of corn. The grains contained in two, three, and four bundles, respectively, of these three classes of corn, are not sufficient to make a whole measure. However, if we added to them one bundle of the second, third, and first classes, respectively, then the grains would become on full measure in each case. How many measures of grain does each bundle of the different classes contain?

5. This problem in part (a) is known as the "Flower of Thymaridas," named after a Pythagorean of the fourth century B.C.
   (a) Given the n numbers $a_1, a_2, \ldots, a_n$, solve for $x_1, x_2, \ldots, x_n$ in the following linear system:

   $$x_1 + x_2 + \cdots + x_n = a_1$$
   $$x_1 + x_2 = a_2$$
   $$x_1 + x_3 = a_3$$
   $$\vdots$$
   $$x_1 + x_n = a_n$$

   (b) Identify a problem in this exercise set that fits the pattern in part (a), and solve it using your general solution.

6. For Example 5 from the Bakhshali Manuscript:
   (a) Express Equations (5) as a homogeneous linear system of three equations in four unknowns (x, y, z, and K) and show that the solution set has one arbitrary parameter.
   (b) Find the smallest solution for which all four variables are positive integers.
   (c) Show that the solution given in Example 5 is included among your solutions.

7. Solve the problems posed in the following three epigrams, which appear in a collection entitled "The Greek Anthology," compiled in part by a scholar named Metrodorus around A.D. 500. Some of its 46 mathematical problems are believed to date as far back as 600 B.C. [*Note:* Before solving parts (a) and (c), you will have to formulate the question.]

   (a) I desire my two sons to receive the thousand staters of which I am possessed, but let the fifth part of the legitimate one's share exceed by ten the fourth part of what falls to the illegitimate one.

   (b) Make me a crown weighing sixty minae, mixing gold and brass, and with them tin and much-wrought iron. Let the gold and brass together form two-thirds, the gold and tin together three-fourths, and the gold and iron three-fifths. Tell me how much gold you must put in, how much brass, how much tin, and how much iron, so as to make the whole crown weigh sixty minae.

   (c) First person: I have what the second has and the third of what the third has. Second person: I have what the third has and the third of what the first has. Third person: And I have ten minae and the third of what the second has.

 ## Section 10.3 Technology Exercises

The following exercises are designed to be solved using a technology utility. Typically, this will be MATLAB, *Mathematica*, Maple, Derive, or Mathcad, but it may also be some other type of linear algebra software or a scientific calculator with some linear algebra capabilities. For each exercise you will need to read the relevant documentation for the particular utility you are using. The goal of these exercises is to provide you with a basic proficiency with your technology utility. Once you have mastered the techniques in these exercises, you will be able to use your technology utility to solve many of the problems in the regular exercise sets.

T1. (a) Solve Archimedes' Cattle Problem using a symbolic algebra program.

(b) The Cattle Problem has a second part in which two additional conditions are imposed. The first of these states that "When the white bulls mingled their number with the black, they stood firm, equal in depth and breadth." This requires that $W + B$ be a square number, that is, 1, 4, 9, 16, 25, and so on. Show that this requires that the values of k in Eq. (4) be restricted as follows:

$$k = 4{,}456{,}749 r^2, \quad r = 1, 2, 3, \ldots$$

and find the smallest total number of cattle that satisfies this second condition.

Remark The second condition imposed in the second part of the Cattle Problem states that "When the yellow and the dappled bulls were gathered into one herd, they stood in such a manner that their number, beginning from one, grew slowly greater 'til it completed a triangular figure." This requires that the quantity $Y + D$ be a triangular number—that is, a number of the form $1, 1 + 2, 1 + 2 + 3, 1 + 2 + 3 + 4, \ldots$. This final part of the problem was not completely solved until 1965 when all 206,545 digits of the smallest number of cattle that satisfies this condition were found using a computer.

T2. The following problem is from "The Nine Chapters of the Mathematical Art" and determines a homogeneous linear system of five equations in six unknowns. Show that the system has infinitely many solutions, and find the one for which the depth of the well and the lengths of the five ropes are the smallest possible positive integers.

Suppose that five families share a well. Suppose further that

2 of A's ropes are short of the well's depth by one of B's ropes.
3 of B's ropes are short of the well's depth by one of C's ropes.
4 of C's ropes are short of the well's depth by one of D's ropes.
5 of D's ropes are short of the well's depth by one of E's ropes.
6 of E's ropes are short of the well's depth by one of A's ropes.

10.4 Cubic Spline Interpolation

In this section an artist's drafting aid is used as a physical model for the mathematical problem of finding a curve that passes through specified points in the plane. The parameters of the curve are determined by solving a linear system of equations.

> **PREREQUISITES:** Linear Systems
> Matrix Algebra
> Differential Calculus

Curve Fitting

Fitting a curve through specified points in the plane is a common problem encountered in analyzing experimental data, in ascertaining the relations among variables, and in design work. A ubiquitous application is in the design and description of computer and printer fonts, such as PostScript^{TM} and TrueType^{TM} fonts (Figure 10.4.1). In Figure 10.4.2 seven points in the xy-plane are displayed, and in Figure 10.4.4 a smooth curve has been drawn that passes through them. A curve that passes through a set of points in the plane is said to *interpolate* those points, and the curve is called an *interpolating curve* for those points. The interpolating curve in Figure 10.4.4 was drawn with the aid of a *drafting spline* (Figure 10.4.3). This drafting aid consists of a thin, flexible strip of wood or other material that is bent to pass through the points to be interpolated. Attached sliding weights hold the spline in position while the artist draws the interpolating curve. The drafting spline will serve as the physical model for a mathematical theory of interpolation that we will discuss in this section.

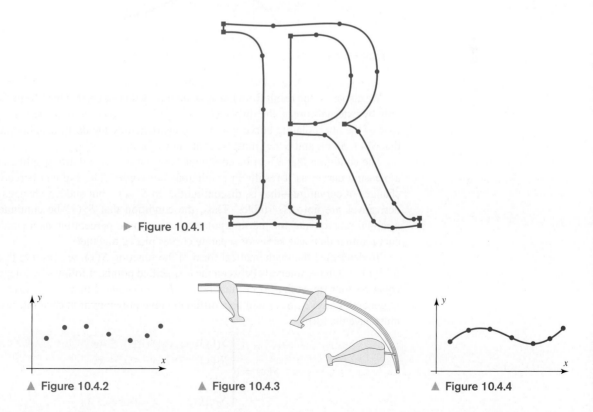

▶ Figure 10.4.1

▲ Figure 10.4.2   ▲ Figure 10.4.3   ▲ Figure 10.4.4

Chapter 10 Applications of Linear Algebra

Statement of the Problem Suppose that we are given n points in the xy-plane,

$$(x_1, y_1), (x_2, y_2), \ldots, (x_n, y_n)$$

which we wish to interpolate with a "well-behaved" curve (Figure 10.4.5). For convenience, we take the points to be equally spaced in the x-direction, although our results can easily be extended to the case of unequally spaced points. If we let the common distance between the x-coordinates of the points be h, then we have

$$x_2 - x_1 = x_3 - x_2 = \cdots = x_n - x_{n-1} = h$$

Let $y = S(x)$, $x_1 \leq x \leq x_n$ denote the interpolating curve that we seek. We assume that this curve describes the displacement of a drafting spline that interpolates the n points when the weights holding down the spline are situated precisely at the n points. It is known from linear beam theory that for small displacements, the fourth derivative of the displacement of a beam is zero along any interval of the x-axis that contains no external forces acting on the beam. If we treat our drafting spline as a thin beam and realize that the only external forces acting on it arise from the weights at the n specified points, then it follows that

$$S^{(iv)}(x) \equiv 0 \tag{1}$$

for values of x lying in the $n - 1$ open intervals

$$(x_1, x_2), (x_2, x_3), \ldots, (x_{n-1}, x_n)$$

between the n points.

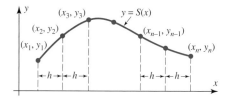

▶ **Figure 10.4.5**

We also need the result from linear beam theory that states that for a beam acted upon only by external forces, the displacement must have two continuous derivatives. In the case of the interpolating curve $y = S(x)$ constructed by the drafting spline, this means that $S(x)$, $S'(x)$, and $S''(x)$ must be continuous for $x_1 \leq x \leq x_n$.

The condition that $S''(x)$ be continuous is what causes a drafting spline to produce a pleasing curve, as it results in continuous *curvature*. The eye can perceive sudden changes in curvature—that is, discontinuities in $S''(x)$—but sudden changes in higher derivatives are not discernible. Thus, the condition that $S''(x)$ be continuous is the minimal prerequisite for the interpolating curve to be perceptible as a single smooth curve, rather than as a series of separate curves pieced together.

To determine the mathematical form of the function $S(x)$, we observe that because $S^{(iv)}(x) \equiv 0$ in the intervals between the n specified points, it follows by integrating this equation four times that $S(x)$ must be a *cubic polynomial* in x in each such interval. In general, however, $S(x)$ will be a different cubic polynomial in each interval, so $S(x)$ must have the form

$$S(x) = \begin{cases} S_1(x), & x_1 \leq x \leq x_2 \\ S_2(x), & x_2 \leq x \leq x_3 \\ \vdots \\ S_{n-1}(x), & x_{n-1} \leq x \leq x_n \end{cases} \tag{2}$$

10.4 Cubic Spline Interpolation

where $S_1(x), S_2(x), \ldots, S_{n-1}(x)$ are cubic polynomials. For convenience, we will write these in the form

$$S_1(x) = a_1(x - x_1)^3 + b_1(x - x_1)^2 + c_1(x - x_1) + d_1, \qquad x_1 \leq x \leq x_2$$
$$S_2(x) = a_2(x - x_2)^3 + b_2(x - x_2)^2 + c_2(x - x_2) + d_2, \qquad x_2 \leq x \leq x_3$$
$$\vdots$$
$$S_{n-1}(x) = a_{n-1}(x - x_{n-1})^3 + b_{n-1}(x - x_{n-1})^2$$
$$\qquad + c_{n-1}(x - x_{n-1}) + d_{n-1}, \qquad x_{n-1} \leq x \leq x_n \tag{3}$$

The a_i's, b_i's, c_i's, and d_i's constitute a total of $4n - 4$ coefficients that we must determine to specify $S(x)$ completely. If we choose these coefficients so that $S(x)$ interpolates the n specified points in the plane and $S(x)$, $S'(x)$, and $S''(x)$ are continuous, then the resulting interpolating curve is called a **cubic spline**.

Derivation of the Formula of a Cubic Spline

From Equations (2) and (3), we have

$$S(x) = S_1(x) = a_1(x - x_1)^3 + b_1(x - x_1)^2 + c_1(x - x_1) + d_1, \qquad x_1 \leq x \leq x_2$$
$$S(x) = S_2(x) = a_2(x - x_2)^3 + b_2(x - x_2)^2 + c_2(x - x_2) + d_2, \qquad x_2 \leq x \leq x_3$$
$$\vdots$$
$$S(x) = S_{n-1}(x) = a_{n-1}(x - x_{n-1})^3 + b_{n-1}(x - x_{n-1})^2$$
$$\qquad + c_{n-1}(x - x_{n-1}) + d_{n-1}, \qquad x_{n-1} \leq x \leq x_n \tag{4}$$

so

$$S'(x) = S'_1(x) = 3a_1(x - x_1)^2 + 2b_1(x - x_1) + c_1, \qquad x_1 \leq x \leq x_2$$
$$S'(x) = S'_2(x) = 3a_2(x - x_2)^3 + 2b_2(x - x_2) + c_2, \qquad x_2 \leq x \leq x_3$$
$$\vdots$$
$$S'(x) = S'_{n-1}(x) = 3a_{n-1}(x - x_{n-1})^2 + 2b_{n-1}(x - x_{n-1}) + c_{n-1}, \qquad x_{n-1} \leq x \leq x_n \tag{5}$$

and

$$S''(x) = S''_1(x) = 6a_1(x - x_1) + 2b_1, \qquad x_1 \leq x \leq x_2$$
$$S''(x) = S''_2(x) = 6a_2(x - x_2) + 2b_2, \qquad x_2 \leq x \leq x_3$$
$$\vdots$$
$$S''(x) = S''_{n-1}(x) = 6a_{n-1}(x - x_{n-1}) + 2b_{n-1}, \qquad x_{n-1} \leq x \leq x_n \tag{6}$$

We will now use these equations and the four properties of cubic splines stated below to express the unknown coefficients a_i, b_i, c_i, d_i, $i = 1, 2, \ldots, n - 1$, in terms of the known coordinates $y_1, y_2, \ldots, y_n$.

1. $S(x)$ interpolates the points (x_i, y_i), $i = 1, 2, \ldots, n$.

   Because $S(x)$ interpolates the points (x_i, y_i), $i = 1, 2, \ldots, n$, we have

   $$S(x_1) = y_1, \quad S(x_2) = y_2, \ldots, \quad S(x_n) = y_n \tag{7}$$

   From the first $n - 1$ of these equations and (4), we obtain

   $$d_1 = y_1$$
   $$d_2 = y_2$$
   $$\vdots$$
   $$d_{n-1} = y_{n-1} \tag{8}$$

From the last equation in (7), the last equation in (4), and the fact that $x_n - x_{n-1} = h$, we obtain

$$a_{n-1}h^3 + b_{n-1}h^2 + c_{n-1}h + d_{n-1} = y_n \tag{9}$$

2. $S(x)$ is continuous on $[x_1, x_n]$.

   Because $S(x)$ is continuous for $x_1 \leq x \leq x_n$, it follows that at each point x_i in the set $x_2, x_3, \ldots, x_{n-1}$ we must have

   $$S_{i-1}(x_i) = S_i(x_i), \qquad i = 2, 3, \ldots, n-1 \tag{10}$$

   Otherwise, the graphs of $S_{i-1}(x)$ and $S_i(x)$ would not join together to form a continuous curve at x_i. When we apply the interpolating property $S_i(x_i) = y_i$, it follows from (10) that $S_{i-1}(x_i) = y_i$, $i = 2, 3, \ldots, n-1$, or from (4) that

   $$\begin{aligned} a_1 h^3 + b_1 h^2 + c_1 h + d_1 &= y_2 \\ a_2 h^3 + b_2 h^2 + c_2 h + d_2 &= y_3 \\ &\vdots \\ a_{n-2} h^3 + b_{n-2} h^2 + c_{n-2} h + d_{n-2} &= y_{n-1} \end{aligned} \tag{11}$$

3. $S'(x)$ is continuous on $[x_1, x_n]$.

   Because $S'(x)$ is continuous for $x_1 \leq x \leq x_n$, it follows that

   $$S'_{i-1}(x_i) = S'_i(x_i), \qquad i = 2, 3, \ldots, n-1$$

   or, from (5),

   $$\begin{aligned} 3a_1 h^2 + 2b_1 h + c_1 &= c_2 \\ 3a_2 h^2 + 2b_2 h + c_2 &= c_3 \\ &\vdots \\ 3a_{n-2} h^2 + 2b_{n-2} h + c_{n-2} &= c_{n-1} \end{aligned} \tag{12}$$

4. $S''(x)$ is continuous on $[x_1, x_2]$.

   Because $S''(x)$ is continuous for $x_1 \leq x \leq x_n$, it follows that

   $$S''_{i-1}(x_i) = S''_i(x_i), \qquad i = 2, 3, \ldots, n-1$$

   or, from (6),

   $$\begin{aligned} 6a_1 h + 2b_1 &= 2b_2 \\ 6a_2 h + 2b_2 &= 2b_3 \\ &\vdots \\ 6a_{n-2} h + 2b_{n-2} &= 2b_{n-1} \end{aligned} \tag{13}$$

Equations (8), (9), (11), (12), and (13) constitute a system of $4n - 6$ linear equations in the $4n - 4$ unknown coefficients a_i, b_i, c_i, d_i, $i = 1, 2, \ldots, n-1$. Consequently, we need two more equations to determine these coefficients uniquely. Before obtaining these additional equations, however, we can simplify our existing system by expressing the unknowns a_i, b_i, c_i, and d_i in terms of new unknown quantities

$$M_1 = S''(x_1), \quad M_2 = S''(x_2), \ldots, \quad M_n = S''(x_n)$$

and the known quantities

$$y_1, y_2, \ldots, y_n$$

For example, from (6) it follows that

$$\begin{aligned} M_1 &= 2b_1 \\ M_2 &= 2b_2 \\ &\vdots \\ M_{n-1} &= 2b_{n-1} \end{aligned}$$

so
$$b_1 = \tfrac{1}{2}M_1, \quad b_2 = \tfrac{1}{2}M_2, \ldots, \quad b_{n-1} = \tfrac{1}{2}M_{n-1}$$

Moreover, we already know from (8) that
$$d_1 = y_1, \quad d_2 = y_2, \ldots, \quad d_{n-1} = y_{n-1}$$

We leave it as an exercise for you to derive the expressions for the a_i's and c_i's in terms of the M_i's and y_i's. The final result is as follows:

THEOREM 10.4.1 Cubic Spline Interpolation

Given n points $(x_1, y_1), (x_2, y_2), \ldots, (x_n, y_n)$ with $x_{i+1} - x_i = h$, $i = 1, 2, \ldots, n-1$, the cubic spline

$$S(x) = \begin{cases} a_1(x - x_1)^3 + b_1(x - x_1)^2 + c_1(x - x_1) + d_1, & x_1 \le x \le x_2 \\ a_2(x - x_2)^3 + b_2(x - x_2)^2 + c_2(x - x_2) + d_2, & x_2 \le x \le x_3 \\ \quad\vdots \\ a_{n-1}(x - x_{n-1})^3 + b_{n-1}(x - x_{n-1})^2 \\ \quad + c_{n-1}(x - x_{n-1}) + d_{n-1}, & x_{n-1} \le x \le x_n \end{cases}$$

that interpolates these points has coefficients given by

$$\begin{aligned} a_i &= (M_{i+1} - M_i)/6h \\ b_i &= M_i/2 \\ c_i &= (y_{i+1} - y_i)/h - [(M_{i+1} + 2M_i)h/6] \\ d_i &= y_i \end{aligned} \qquad (14)$$

for $i = 1, 2, \ldots, n-1$, where $M_i = S''(x_i)$, $i = 1, 2, \ldots, n$.

From this result, we see that the quantities $M_1, M_2, \ldots, M_n$ uniquely determine the cubic spline. To find these quantities, we substitute the expressions for a_i, b_i, and c_i given in (14) into (12). After some algebraic simplification, we obtain

$$\begin{aligned} M_1 + 4M_2 + M_3 &= 6(y_1 - 2y_2 + y_3)/h^2 \\ M_2 + 4M_3 + M_4 &= 6(y_2 - 2y_3 + y_4)/h^2 \\ &\vdots \\ M_{n-2} + 4M_{n-1} + M_n &= 6(y_{n-2} - 2y_{n-1} + y_n)/h^2 \end{aligned} \qquad (15)$$

or, in matrix form,

$$\begin{bmatrix} 1 & 4 & 1 & 0 & \cdots & 0 & 0 & 0 & 0 \\ 0 & 1 & 4 & 1 & \cdots & 0 & 0 & 0 & 0 \\ 0 & 0 & 1 & 4 & \cdots & 0 & 0 & 0 & 0 \\ \vdots & \vdots & \vdots & \vdots & & \vdots & \vdots & \vdots & \vdots \\ 0 & 0 & 0 & 0 & \cdots & 4 & 1 & 0 & 0 \\ 0 & 0 & 0 & 0 & \cdots & 1 & 4 & 1 & 0 \\ 0 & 0 & 0 & 0 & \cdots & 0 & 1 & 4 & 1 \end{bmatrix} \begin{bmatrix} M_1 \\ M_2 \\ M_3 \\ M_4 \\ \vdots \\ M_{n-3} \\ M_{n-2} \\ M_{n-1} \\ M_n \end{bmatrix} = \frac{6}{h^2} \begin{bmatrix} y_1 - 2y_2 + y_3 \\ y_2 - 2y_3 + y_4 \\ y_3 - 2y_4 + y_5 \\ \vdots \\ y_{n-4} - 2y_{n-3} + y_{n-2} \\ y_{n-3} - 2y_{n-2} + y_{n-1} \\ y_{n-2} - 2y_{n-1} + y_n \end{bmatrix}$$

This is a linear system of $n - 2$ equations for the n unknowns $M_1, M_2, \ldots, M_n$. Thus, we still need two additional equations to determine $M_1, M_2, \ldots, M_n$ uniquely. The reason for this is that there are infinitely many cubic splines that interpolate the given points, so we simply do not have enough conditions to determine a unique cubic spline passing through the points. We discuss below three possible ways of specifying the two additional conditions required to obtain a unique cubic spline through the points. (The exercises present two more.) They are summarized in Table 1.

Table 1

Natural Spline	The second derivative of the spline is zero at the endpoints.	$M_1 = 0$ $M_n = 0$	$\begin{bmatrix} 4 & 1 & 0 & \cdots & 0 & 0 & 0 \\ 1 & 4 & 1 & \cdots & 0 & 0 & 0 \\ \vdots & \vdots & \vdots & & \vdots & \vdots & \vdots \\ 0 & 0 & 0 & \cdots & 1 & 4 & 1 \\ 0 & 0 & 0 & \cdots & 0 & 1 & 4 \end{bmatrix} \begin{bmatrix} M_2 \\ M_3 \\ \vdots \\ M_{n-2} \\ M_{n-1} \end{bmatrix} = \dfrac{6}{h^2}$	$\begin{bmatrix} y_1 - 2y_2 + y_3 \\ y_2 - 2y_3 + y_4 \\ \vdots \\ y_{n-2} - 2y_{n-1} + y_n \end{bmatrix}$
Parabolic Runout Spline	The spline reduces to a parabolic curve on the first and last intervals.	$M_1 = M_2$ $M_n = M_{n-1}$	$\begin{bmatrix} 5 & 1 & 0 & \cdots & 0 & 0 & 0 \\ 1 & 4 & 1 & \cdots & 0 & 0 & 0 \\ \vdots & \vdots & \vdots & & \vdots & \vdots & \vdots \\ 0 & 0 & 0 & \cdots & 1 & 4 & 1 \\ 0 & 0 & 0 & \cdots & 0 & 1 & 5 \end{bmatrix} \begin{bmatrix} M_2 \\ M_3 \\ \vdots \\ M_{n-2} \\ M_{n-1} \end{bmatrix} = \dfrac{6}{h^2}$	$\begin{bmatrix} y_1 - 2y_2 + y_3 \\ y_2 - 2y_3 + y_4 \\ \vdots \\ y_{n-2} - 2y_{n-1} + y_n \end{bmatrix}$
Cubic Runout Spline	The spline is a single cubic curve on the first two and last two intervals.	$M_1 = 2M_2 - M_3$ $M_n = 2M_{n-1} - M_{n-2}$	$\begin{bmatrix} 6 & 0 & 0 & \cdots & 0 & 0 & 0 \\ 1 & 4 & 1 & \cdots & 0 & 0 & 0 \\ \vdots & \vdots & \vdots & & \vdots & \vdots & \vdots \\ 0 & 0 & 0 & \cdots & 1 & 4 & 1 \\ 0 & 0 & 0 & \cdots & 0 & 0 & 6 \end{bmatrix} \begin{bmatrix} M_2 \\ M_3 \\ \vdots \\ M_{n-2} \\ M_{n-1} \end{bmatrix} = \dfrac{6}{h^2}$	$\begin{bmatrix} y_1 - 2y_2 + y_3 \\ y_2 - 2y_3 + y_4 \\ \vdots \\ y_{n-2} - 2y_{n-1} + y_n \end{bmatrix}$

The Natural Spline The two simplest mathematical conditions we can impose are

$$M_1 = M_n = 0$$

These conditions together with (15) result in an $n \times n$ linear system for $M_1, M_2, \ldots, M_n$, which can be written in matrix form as

$$\begin{bmatrix} 1 & 0 & 0 & 0 & \cdots & 0 & 0 & 0 \\ 1 & 4 & 1 & 0 & \cdots & 0 & 0 & 0 \\ 0 & 1 & 4 & 1 & \cdots & 0 & 0 & 0 \\ \vdots & \vdots & \vdots & \vdots & & \vdots & \vdots & \vdots \\ 0 & 0 & 0 & 0 & \cdots & 1 & 4 & 1 \\ 0 & 0 & 0 & 0 & \cdots & 0 & 0 & 1 \end{bmatrix} \begin{bmatrix} M_1 \\ M_2 \\ M_3 \\ \vdots \\ M_{n-1} \\ M_n \end{bmatrix} = \frac{6}{h^2} \begin{bmatrix} 0 \\ y_1 - 2y_2 + y_3 \\ y_2 - 2y_3 + y_4 \\ \vdots \\ y_{n-2} - 2y_{n-1} + y_n \\ 0 \end{bmatrix}$$

For numerical calculations it is more convenient to eliminate M_1 and M_n from this system and write

$$\begin{bmatrix} 4 & 1 & 0 & 0 & \cdots & 0 & 0 & 0 \\ 1 & 4 & 1 & 0 & \cdots & 0 & 0 & 0 \\ 0 & 1 & 4 & 1 & \cdots & 0 & 0 & 0 \\ \vdots & \vdots & \vdots & \vdots & & \vdots & \vdots & \vdots \\ 0 & 0 & 0 & 0 & \cdots & 1 & 4 & 1 \\ 0 & 0 & 0 & 0 & \cdots & 0 & 1 & 4 \end{bmatrix} \begin{bmatrix} M_2 \\ M_3 \\ M_4 \\ \vdots \\ M_{n-2} \\ M_{n-1} \end{bmatrix} = \frac{6}{h^2} \begin{bmatrix} y_1 - 2y_2 + y_3 \\ y_2 - 2y_3 + y_4 \\ y_3 - 2y_4 + y_5 \\ \vdots \\ y_{n-3} - 2y_{n-2} + y_{n-1} \\ y_{n-2} - 2y_{n-1} + y_n \end{bmatrix} \quad (16)$$

10.4 Cubic Spline Interpolation

together with

$$M_1 = 0 \quad (17)$$
$$M_n = 0 \quad (18)$$

Thus, the $(n-2) \times (n-2)$ linear system can be solved for the $n-2$ coefficients $M_2, M_3, \ldots, M_{n-1}$, and M_1 and M_n are determined by (17) and (18).

Physically, the natural spline results when the ends of a drafting spline extend freely beyond the interpolating points without constraint. The end portions of the spline outside the interpolating points will fall on straight line paths, causing $S''(x)$ to vanish at the endpoints x_1 and x_n and resulting in the mathematical conditions $M_1 = M_n = 0$.

The natural spline tends to flatten the interpolating curve at the endpoints, which may be undesirable. Of course, if it is required that $S''(x)$ vanish at the endpoints, then the natural spline must be used.

The Parabolic Runout Spline The two additional constraints imposed for this type of spline are

$$M_1 = M_2 \quad (19)$$
$$M_n = M_{n-1} \quad (20)$$

If we use the preceding two equations to eliminate M_1 and M_n from (15), we obtain the $(n-2) \times (n-2)$ linear system

$$\begin{bmatrix} 5 & 1 & 0 & 0 & \cdots & 0 & 0 & 0 \\ 1 & 4 & 1 & 0 & \cdots & 0 & 0 & 0 \\ 0 & 1 & 4 & 1 & \cdots & 0 & 0 & 0 \\ \vdots & \vdots & \vdots & \vdots & & \vdots & \vdots & \vdots \\ 0 & 0 & 0 & 0 & \cdots & 1 & 4 & 1 \\ 0 & 0 & 0 & 0 & \cdots & 0 & 1 & 5 \end{bmatrix} \begin{bmatrix} M_2 \\ M_3 \\ M_4 \\ \vdots \\ M_{n-2} \\ M_{n-1} \end{bmatrix} = \frac{6}{h^2} \begin{bmatrix} y_1 - 2y_2 + y_3 \\ y_2 - 2y_3 + y_4 \\ y_3 - 2y_4 + y_5 \\ \vdots \\ y_{n-3} - 2y_{n-2} + y_{n-1} \\ y_{n-2} - 2y_{n-1} + y_n \end{bmatrix} \quad (21)$$

for $M_2, M_3, \ldots, M_{n-1}$. Once these $n-2$ values have been determined, M_1 and M_n are determined from (19) and (20).

From (14) we see that $M_1 = M_2$ implies that $a_1 = 0$, and $M_n = M_{n-1}$ implies that $a_{n-1} = 0$. Thus, from (3) there are no cubic terms in the formula for the spline over the end intervals $[x_1, x_2]$ and $[x_{n-1}, x_n]$. Hence, as the name suggests, the parabolic runout spline reduces to a parabolic curve over these end intervals.

The Cubic Runout Spline For this type of spline, we impose the two additional conditions

$$M_1 = 2M_2 - M_3 \quad (22)$$
$$M_n = 2M_{n-1} - M_{n-2} \quad (23)$$

Using these two equations to eliminate M_1 and M_n from (15) results in the following $(n-2) \times (n-2)$ linear system for $M_2, M_3, \ldots, M_{n-1}$:

$$\begin{bmatrix} 6 & 0 & 0 & 0 & \cdots & 0 & 0 & 0 \\ 1 & 4 & 1 & 0 & \cdots & 0 & 0 & 0 \\ 0 & 1 & 4 & 1 & \cdots & 0 & 0 & 0 \\ \vdots & \vdots & \vdots & \vdots & & \vdots & \vdots & \vdots \\ 0 & 0 & 0 & 0 & \cdots & 1 & 4 & 1 \\ 0 & 0 & 0 & 0 & \cdots & 0 & 0 & 6 \end{bmatrix} \begin{bmatrix} M_2 \\ M_3 \\ M_4 \\ \vdots \\ M_{n-2} \\ M_{n-1} \end{bmatrix} = \frac{6}{h^2} \begin{bmatrix} y_1 - 2y_2 + y_3 \\ y_2 - 2y_3 + y_4 \\ y_3 - 2y_4 + y_5 \\ \vdots \\ y_{n-3} - 2y_{n-2} + y_{n-1} \\ y_{n-2} - 2y_{n-1} + y_n \end{bmatrix} \quad (24)$$

After we solve this linear system for $M_2, M_3, \ldots, M_{n-1}$, we can use (22) and (23) to determine M_1 and M_n.

If we rewrite (22) as
$$M_2 - M_1 = M_3 - M_2$$
it follows from (14) that $a_1 = a_2$. Because $S'''(x) = 6a_1$ on $[x_1, x_2]$ and $S'''(x) = 6a_2$ on $[x_2, x_3]$, we see that $S'''(x)$ is constant over the entire interval $[x_1, x_3]$. Consequently, $S(x)$ consists of a *single* cubic curve over the interval $[x_1, x_3]$ rather than two different cubic curves pieced together at x_2. [To see this, integrate $S'''(x)$ three times.] A similar analysis shows that $S(x)$ consists of a single cubic curve over the last two intervals.

Whereas the natural spline tends to produce an interpolating curve that is flat at the endpoints, the cubic runout spline has the opposite tendency: it produces a curve with pronounced curvature at the endpoints. If neither behavior is desired, the parabolic runout spline is a reasonable compromise.

▶ **EXAMPLE 1 Using a Parabolic Runout Spline**

The density of water is well known to reach a maximum at a temperature slightly above freezing. Table 2, from the *Handbook of Chemistry and Physics* (CRC Press, 2009), gives the density of water in grams per cubic centimeter for five equally spaced temperatures from $-10°C$ to $30°C$. We will interpolate these five temperature–density measurements with a parabolic runout spline and attempt to find the maximum density of water in this range by finding the maximum value on this cubic spline. In the exercises we ask you to perform similar calculations using a natural spline and a cubic runout spline to interpolate the data points.

Set
$$x_1 = -10, \quad y_1 = .99815$$
$$x_2 = 0, \quad y_2 = .99987$$
$$x_3 = 10, \quad y_3 = .99973$$
$$x_4 = 20, \quad y_4 = .99823$$
$$x_5 = 30, \quad y_5 = .99567$$

Then
$$6[y_1 - 2y_2 + y_3]/h^2 = -.0001116$$
$$6[y_2 - 2y_3 + y_4]/h^2 = -.0000816$$
$$6[y_3 - 2y_4 + y_5]/h^2 = -.0000636$$

and the linear system (21) for the parabolic runout spline becomes
$$\begin{bmatrix} 5 & 1 & 0 \\ 1 & 4 & 1 \\ 0 & 1 & 5 \end{bmatrix} \begin{bmatrix} M_2 \\ M_3 \\ M_4 \end{bmatrix} = \begin{bmatrix} -.0001116 \\ -.0000816 \\ -.0000636 \end{bmatrix}$$

Solving this system yields
$$M_2 = -.00001973$$
$$M_3 = -.00001293$$
$$M_4 = -.00001013$$

From (19) and (20), we have
$$M_1 = M_2 = -.00001973$$
$$M_5 = M_4 = -.00001013$$

10.4 Cubic Spline Interpolation

Table 2

Temperature (°C)	Density (g/cm³)
−10	.99815
0	.99987
10	.99973
20	.99823
30	.99567

Solving for the a_i's, b_i's, c_i's, and d_i's in (14), we obtain the following expression for the interpolating parabolic runout spline:

$$S(x) = \begin{cases} -.00000987(x+10)^2 + .0002707(x+10) + .99815, & -10 \leq x \leq 0 \\ .000000113(x-0)^3 - .00000987(x-0)^2 + .0000733(x-0) + .99987, & 0 \leq x \leq 10 \\ .000000047(x-10)^3 - .00000647(x-10)^2 - .0000900(x-10) + .99973, & 10 \leq x \leq 20 \\ -.00000507(x-20)^2 - .0002053(x-20) + .99823, & 20 \leq x \leq 30 \end{cases}$$

This spline is plotted in Figure 10.4.6. From that figure we see that the maximum is attained in the interval $[0, 10]$. To find this maximum, we set $S'(x)$ equal to zero in the interval $[0, 10]$:

$$S'(x) = .000000339x^2 - .0000197x + .0000733 = 0$$

To three significant digits the root of this quadratic in the interval $[0, 10]$ is $x = 3.99$, and for this value of x, $S(3.99) = 1.00001$. Thus, according to our interpolated estimate, the maximum density of water is 1.00001 g/cm³ attained at 3.99°C. This agrees well with the experimental maximum density of 1.00000 g/cm³ attained at 3.98°C. (In the original metric system, the gram was *defined* as the mass of one cubic centimeter of water at its maximum density.) ◀

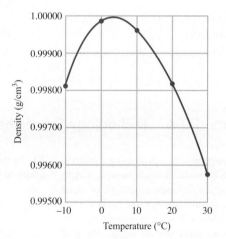

▶ Figure 10.4.6

Closing Remarks In addition to producing excellent interpolating curves, cubic splines and their generalizations are useful for numerical integration and differentiation, for the numerical solution of differential and integral equations, and in optimization theory.

Exercise Set 10.4

1. Derive the expressions for a_i and c_i in Equations (14) of Theorem 10.4.1.

2. The six points

$$(0, .00000), \quad (.2, .19867), \quad (.4, .38942),$$
$$(.6, .56464), \quad (.8, .71736), \quad (1.0, .84147)$$

   lie on the graph of $y = \sin x$, where x is in radians.

   (a) Find the portion of the parabolic runout spline that interpolates these six points for $.4 \leq x \leq .6$. Maintain an accuracy of five decimal places in your calculations.

   (b) Calculate $S(.5)$ for the spline you found in part (a). What is the percentage error of $S(.5)$ with respect to the "exact" value of $\sin(.5) = .47943$?

3. The following five points

$$(0, 1), \quad (1, 7), \quad (2, 27), \quad (3, 79), \quad (4, 181)$$

   lie on a single cubic curve.

   (a) Which of the three types of cubic splines (natural, parabolic runout, or cubic runout) would agree exactly with the single cubic curve on which the five points lie?

   (b) Determine the cubic spline you chose in part (a), and verify that it is a single cubic curve that interpolates the five points.

4. Repeat the calculations in Example 1 using a natural spline to interpolate the five data points.

5. Repeat the calculations in Example 1 using a cubic runout spline to interpolate the five data points.

6. Consider the five points $(0, 0)$, $(.5, 1)$, $(1, 0)$, $(1.5, -1)$, and $(2, 0)$ on the graph of $y = \sin(\pi x)$.

   (a) Use a natural spline to interpolate the data points $(0, 0)$, $(.5, 1)$, and $(1, 0)$.

   (b) Use a natural spline to interpolate the data points $(.5, 1)$, $(1, 0)$, and $(1.5, -1)$.

   (c) Explain the unusual nature of your result in part (b).

7. **(The Periodic Spline)** If it is known or if it is desired that the n points $(x_1, y_1), (x_2, y_2), \ldots, (x_n, y_n)$ to be interpolated lie on a single cycle of a periodic curve with period $x_n - x_1$, then an interpolating cubic spline $S(x)$ must satisfy

$$S(x_1) = S(x_n)$$
$$S'(x_1) = S'(x_n)$$
$$S''(x_1) = S''(x_n)$$

   (a) Show that these three periodicity conditions require that

$$y_1 = y_n$$
$$M_1 = M_n$$
$$4M_1 + M_2 + M_{n-1} = 6(y_{n-1} - 2y_1 + y_2)/h^2$$

   (b) Using the three equations in part (a) and Equations (15), construct an $(n-1) \times (n-1)$ linear system for $M_1, M_2, \ldots, M_{n-1}$ in matrix form.

8. **(The Clamped Spline)** Suppose that, in addition to the n points to be interpolated, we are given specific values y_1' and y_n' for the slopes $S'(x_1)$ and $S'(x_n)$ of the interpolating cubic spline at the endpoints x_1 and x_n.

   (a) Show that

$$2M_1 + M_2 = 6(y_2 - y_1 - hy_1')/h^2$$
$$2M_n + M_{n-1} = 6(y_{n-1} - y_n + hy_n')/h^2$$

   (b) Using the equations in part (a) and Equations (15), construct an $n \times n$ linear system for $M_1, M_2, \ldots, M_n$ in matrix form.

   Remark The clamped spline described in this exercise is the most accurate type of spline for interpolation work if the slopes at the endpoints are known or can be estimated.

Section 10.4   Technology Exercises

The following exercises are designed to be solved using a technology utility. Typically, this will be MATLAB, *Mathematica*, Maple, Derive, or Mathcad, but it may also be some other type of linear algebra software or a scientific calculator with some linear algebra capabilities. For each exercise you will need to read the relevant documentation for the particular utility you are using. The goal of these exercises is to provide you with a basic proficiency with your technology utility. Once you have mastered the techniques in these exercises, you will be able to use your technology utility to solve many of the problems in the regular exercise sets.

T1. In the solution of the natural cubic spline problem, it is necessary to solve a system of equations having coefficient matrix

$$A_n = \begin{bmatrix} 4 & 1 & 0 & \cdots & 0 & 0 & 0 \\ 1 & 4 & 1 & \cdots & 0 & 0 & 0 \\ \vdots & \vdots & \vdots & \ddots & \vdots & \vdots & \vdots \\ 0 & 0 & 0 & \cdots & 1 & 4 & 1 \\ 0 & 0 & 0 & \cdots & 0 & 1 & 4 \end{bmatrix}$$

If we can present a formula for the inverse of this matrix, then the solution for the natural cubic spline problem can be easily obtained. In this exercise and the next, we use a computer to discover this formula. Toward this end, we first determine an expression for the determinant of A_n, denoted by the symbol D_n. Given that

$$A_1 = [4] \quad \text{and} \quad A_2 = \begin{bmatrix} 4 & 1 \\ 1 & 4 \end{bmatrix}$$

we see that
$$D_1 = \det(A_1) = \det[4] = 4$$
and
$$D_2 = \det(A_2) = \det\begin{bmatrix} 4 & 1 \\ 1 & 4 \end{bmatrix} = 15$$

(a) Use the cofactor expansion of determinants to show that
$$D_n = 4D_{n-1} - D_{n-2}$$
for $n = 3, 4, 5, \ldots$. This says, for example, that
$$D_3 = 4D_2 - D_1 = 4(15) - 4 = 56$$
$$D_4 = 4D_3 - D_2 = 4(56) - 15 = 209$$
and so on. Using a computer, check this result for $5 \leq n \leq 10$.

(b) By writing
$$D_n = 4D_{n-1} - D_{n-2}$$
and the identity, $D_{n-1} = D_{n-1}$, in matrix form,
$$\begin{bmatrix} D_n \\ D_{n-1} \end{bmatrix} = \begin{bmatrix} 4 & -1 \\ 1 & 0 \end{bmatrix} \begin{bmatrix} D_{n-1} \\ D_{n-2} \end{bmatrix}$$
show that
$$\begin{bmatrix} D_n \\ D_{n-1} \end{bmatrix} = \begin{bmatrix} 4 & -1 \\ 1 & 0 \end{bmatrix}^{n-2} \begin{bmatrix} D_2 \\ D_1 \end{bmatrix} = \begin{bmatrix} 4 & -1 \\ 1 & 0 \end{bmatrix}^{n-2} \begin{bmatrix} 15 \\ 4 \end{bmatrix}$$

(c) Use the methods in Section 5.2 and a computer to show that

$$\begin{bmatrix} 4 & -1 \\ 1 & 0 \end{bmatrix}^{n-2} = \frac{\begin{bmatrix} (2+\sqrt{3})^{n-1} - (2-\sqrt{3})^{n-1} & (2-\sqrt{3})^{n-2} - (2+\sqrt{3})^{n-2} \\ (2+\sqrt{3})^{n-2} - (2-\sqrt{3})^{n-2} & (2-\sqrt{3})^{n-3} - (2+\sqrt{3})^{n-3} \end{bmatrix}}{2\sqrt{3}}$$

and hence
$$D_n = \frac{(2+\sqrt{3})^{n+1} - (2-\sqrt{3})^{n+1}}{2\sqrt{3}}$$
for $n = 1, 2, 3, \ldots$.

(d) Using a computer, check this result for $1 \leq n \leq 10$.

T2. In this exercise, we determine a formula for calculating A_n^{-1} from D_k for $k = 0, 1, 2, 3, \ldots, n$, assuming that D_0 is defined to be 1.

(a) Use a computer to compute A_k^{-1} for $k = 1, 2, 3, 4$, and 5.

(b) From your results in part (a), discover the conjecture that
$$A_n^{-1} = [\alpha_{ij}]$$
where $\alpha_{ij} = \alpha_{ji}$ and
$$\alpha_{ij} = (-1)^{i+j}\left(\frac{D_{n-j}D_{i-1}}{D_n}\right)$$
for $i \leq j$.

(c) Use the result in part (b) to compute A_7^{-1} and compare it to the result obtained using the computer.

10.5 Markov Chains

In this section we describe a general model of a system that changes from state to state. We then apply the model to several concrete problems.

> **PREREQUISITES:** Linear Systems
> Matrices
> Intuitive Understanding of Limits

A Markov Process Suppose a physical or mathematical system undergoes a process of change such that at any moment it can occupy one of a finite number of states. For example, the weather in a certain city could be in one of three possible states: sunny, cloudy, or rainy. Or an individual could be in one of four possible emotional states: happy, sad, angry, or apprehensive. Suppose that such a system changes with time from one state to another and at scheduled times the state of the system is observed. If the state of the system at any observation cannot be predicted with certainty, but the probability that a given state occurs can be predicted by just knowing the state of the system at the preceding observation, then the process of change is called a **Markov chain** or **Markov process**.

DEFINITION 1 If a Markov chain has k possible states, which we label as $1, 2, \ldots, k$, then the probability that the system is in state i at any observation after it was in state j at the preceding observation is denoted by p_{ij} and is called the ***transition probability*** from state j to state i. The matrix $P = [p_{ij}]$ is called the ***transition matrix of the Markov chain***.

For example, in a three-state Markov chain, the transition matrix has the form

$$\begin{array}{c} \text{Preceding State} \\ \begin{array}{ccc} 1 & 2 & 3 \end{array} \\ \begin{bmatrix} p_{11} & p_{12} & p_{13} \\ p_{21} & p_{22} & p_{23} \\ p_{31} & p_{32} & p_{33} \end{bmatrix} \begin{array}{l} 1 \\ 2 \\ 3 \end{array} \text{New State} \end{array}$$

In this matrix, p_{32} is the probability that the system will change from state 2 to state 3, p_{11} is the probability that the system will still be in state 1 if it was previously in state 1, and so forth.

▶ **EXAMPLE 1 Transition Matrix of the Markov Chain**

A car rental agency has three rental locations, denoted by 1, 2, and 3. A customer may rent a car from any of the three locations and return the car to any of the three locations. The manager finds that customers return the cars to the various locations according to the following probabilities:

$$\begin{array}{c} \text{Rented from Location} \\ \begin{array}{ccc} 1 & 2 & 3 \end{array} \\ \begin{bmatrix} .8 & .3 & .2 \\ .1 & .2 & .6 \\ .1 & .5 & .2 \end{bmatrix} \begin{array}{l} 1 \\ 2 \\ 3 \end{array} \begin{array}{l} \text{Returned} \\ \text{to} \\ \text{Location} \end{array} \end{array}$$

This matrix is the transition matrix of the system considered as a Markov chain. From this matrix, the probability is .6 that a car rented from location 3 will be returned to location 2, the probability is .8 that a car rented from location 1 will be returned to location 1, and so forth.

▶ **EXAMPLE 2 Transition Matrix of the Markov Chain**

By reviewing its donation records, the alumni office of a college finds that 80% of its alumni who contribute to the annual fund one year will also contribute the next year, and 30% of those who do not contribute one year will contribute the next. This can be viewed as a Markov chain with two states: state 1 corresponds to an alumnus giving a donation in any one year, and state 2 corresponds to the alumnus not giving a donation in that year. The transition matrix is

$$P = \begin{bmatrix} .8 & .3 \\ .2 & .7 \end{bmatrix} \blacktriangleleft$$

In the examples above, the transition matrices of the Markov chains have the property that the entries in any column sum to 1. This is not accidental. If $P = [p_{ij}]$ is the transition matrix of any Markov chain with k states, then for each j we must have

$$p_{1j} + p_{2j} + \cdots + p_{kj} = 1 \tag{1}$$

because if the system is in state j at one observation, it is certain to be in one of the k possible states at the next observation.

A matrix with property (1) is called a ***stochastic matrix***, a ***probability matrix***, or a ***Markov matrix***. From the preceding discussion, it follows that the transition matrix for a Markov chain must be a stochastic matrix.

In a Markov chain, the state of the system at any observation time cannot generally be determined with certainty. The best one can usually do is specify probabilities for each of the possible states. For example, in a Markov chain with three states, we might describe the possible state of the system at some observation time by a column vector

$$\mathbf{x} = \begin{bmatrix} x_1 \\ x_2 \\ x_3 \end{bmatrix}$$

in which x_1 is the probability that the system is in state 1, x_2 the probability that it is in state 2, and x_3 the probability that it is in state 3. In general we make the following definition.

> **DEFINITION 2** The ***state vector*** for an observation of a Markov chain with k states is a column vector $\mathbf{x}$ whose ith component x_i is the probability that the system is in the ith state at that time.

Observe that the entries in any state vector for a Markov chain are nonnegative and have a sum of 1. (Why?) A column vector that has this property is called a ***probability vector***.

Let us suppose now that we know the state vector $\mathbf{x}^{(0)}$ for a Markov chain at some initial observation. The following theorem will enable us to determine the state vectors

$$\mathbf{x}^{(1)}, \mathbf{x}^{(2)}, \ldots, \mathbf{x}^{(n)}, \ldots$$

at the subsequent observation times.

> **THEOREM 10.5.1** *If P is the transition matrix of a Markov chain and $\mathbf{x}^{(n)}$ is the state vector at the nth observation, then $\mathbf{x}^{(n+1)} = P\mathbf{x}^{(n)}$.*

The proof of this theorem involves ideas from probability theory and will not be given here. From this theorem, it follows that

$$\mathbf{x}^{(1)} = P\mathbf{x}^{(0)}$$
$$\mathbf{x}^{(2)} = P\mathbf{x}^{(1)} = P^2\mathbf{x}^{(0)}$$
$$\mathbf{x}^{(3)} = P\mathbf{x}^{(2)} = P^3\mathbf{x}^{(0)}$$
$$\vdots$$
$$\mathbf{x}^{(n)} = P\mathbf{x}^{(n-1)} = P^n\mathbf{x}^{(0)}$$

In this way, the initial state vector $\mathbf{x}^{(0)}$ and the transition matrix P determine $\mathbf{x}^{(n)}$ for $n = 1, 2, \ldots$.

▶ **EXAMPLE 3 Example 2 Revisited**

The transition matrix in Example 2 was

$$P = \begin{bmatrix} .8 & .3 \\ .2 & .7 \end{bmatrix}$$

We now construct the probable future donation record of a new graduate who did not give a donation in the initial year after graduation. For such a graduate the system is initially in state 2 with certainty, so the initial state vector is

$$\mathbf{x}^{(0)} = \begin{bmatrix} 0 \\ 1 \end{bmatrix}$$

From Theorem 10.5.1 we then have

$$\mathbf{x}^{(1)} = P\mathbf{x}^{(0)} = \begin{bmatrix} .8 & .3 \\ .2 & .7 \end{bmatrix} \begin{bmatrix} 0 \\ 1 \end{bmatrix} = \begin{bmatrix} .3 \\ .7 \end{bmatrix}$$

$$\mathbf{x}^{(2)} = P\mathbf{x}^{(1)} = \begin{bmatrix} .8 & .3 \\ .2 & .7 \end{bmatrix} \begin{bmatrix} .3 \\ .7 \end{bmatrix} = \begin{bmatrix} .45 \\ .55 \end{bmatrix}$$

$$\mathbf{x}^{(3)} = P\mathbf{x}^{(2)} = \begin{bmatrix} .8 & .3 \\ .2 & .7 \end{bmatrix} \begin{bmatrix} .45 \\ .55 \end{bmatrix} = \begin{bmatrix} .525 \\ .475 \end{bmatrix}$$

Thus, after three years the alumnus can be expected to make a donation with probability .525. Beyond three years, we find the following state vectors (to three decimal places):

$$\mathbf{x}^{(4)} = \begin{bmatrix} .563 \\ .438 \end{bmatrix}, \quad \mathbf{x}^{(5)} = \begin{bmatrix} .581 \\ .419 \end{bmatrix}, \quad \mathbf{x}^{(6)} = \begin{bmatrix} .591 \\ .409 \end{bmatrix}, \quad \mathbf{x}^{(7)} = \begin{bmatrix} .595 \\ .405 \end{bmatrix}$$

$$\mathbf{x}^{(8)} = \begin{bmatrix} .598 \\ .402 \end{bmatrix}, \quad \mathbf{x}^{(9)} = \begin{bmatrix} .599 \\ .401 \end{bmatrix}, \quad \mathbf{x}^{(10)} = \begin{bmatrix} .599 \\ .401 \end{bmatrix}, \quad \mathbf{x}^{(11)} = \begin{bmatrix} .600 \\ .400 \end{bmatrix}$$

For all n beyond 11, we have

$$\mathbf{x}^{(n)} = \begin{bmatrix} .600 \\ .400 \end{bmatrix}$$

to three decimal places. In other words, the state vectors converge to a fixed vector as the number of observations increases. (We will discuss this further below.)

▶ **EXAMPLE 4** **Example 1 Revisited**

The transition matrix in Example 1 was

$$\begin{bmatrix} .8 & .3 & .2 \\ .1 & .2 & .6 \\ .1 & .5 & .2 \end{bmatrix}$$

If a car is rented initially from location 2, then the initial state vector is

$$\mathbf{x}^{(0)} = \begin{bmatrix} 0 \\ 1 \\ 0 \end{bmatrix}$$

Using this vector and Theorem 10.5.1, one obtains the later state vectors listed in Table 1.

Table 1

n \ $x^{(n)}$	0	1	2	3	4	5	6	7	8	9	10	11
$x_1^{(n)}$	0	.300	.400	.477	.511	.533	.544	.550	.553	.555	.556	.557
$x_2^{(n)}$	1	.200	.370	.252	.261	.240	.238	.233	.232	.231	.230	.230
$x_3^{(n)}$	0	.500	.230	.271	.228	.227	.219	.217	.215	.214	.214	.213

For all values of n greater than 11, all state vectors are equal to $\mathbf{x}^{(11)}$ to three decimal places.

Two things should be observed in this example. First, it was not necessary to know how long a customer kept the car. That is, in a Markov process the time period between observations need not be regular. Second, the state vectors approach a fixed vector as n increases, just as in the first example. ◄

▶ **EXAMPLE 5** Using Theorem 10.5.1

A traffic officer is assigned to control the traffic at the eight intersections indicated in Figure 10.5.1. She is instructed to remain at each intersection for an hour and then to either remain at the same intersection or move to a neighboring intersection. To avoid establishing a pattern, she is told to choose her new intersection on a random basis, with each possible choice equally likely. For example, if she is at intersection 5, her next intersection can be 2, 4, 5, or 8, each with probability $\tfrac{1}{4}$. Every day she starts at the location where she stopped the day before. The transition matrix for this Markov chain is

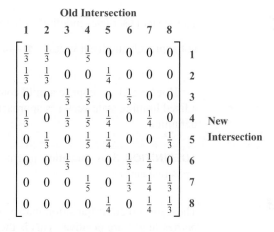

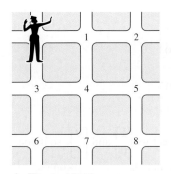

▲ Figure 10.5.1

If the traffic officer begins at intersection 5, her probable locations, hour by hour, are given by the state vectors given in Table 2. For all values of n greater than 22, all state vectors are equal to $\mathbf{x}^{(22)}$ to three decimal places. Thus, as with the first two examples, the state vectors approach a fixed vector as n increases. ◄

Table 2

$\mathbf{x}^{(n)}$ \ n	0	1	2	3	4	5	10	15	20	22
$x_1^{(n)}$	0	.000	.133	.116	.130	.123	.113	.109	.108	.107
$x_2^{(n)}$	0	.250	.146	.163	.140	.138	.115	.109	.108	.107
$x_3^{(n)}$	0	.000	.050	.039	.067	.073	.100	.106	.107	.107
$x_4^{(n)}$	0	.250	.113	.187	.162	.178	.178	.179	.179	.179
$x_5^{(n)}$	1	.250	.279	.190	.190	.168	.149	.144	.143	.143
$x_6^{(n)}$	0	.000	.000	.050	.056	.074	.099	.105	.107	.107
$x_7^{(n)}$	0	.000	.133	.104	.131	.125	.138	.142	.143	.143
$x_8^{(n)}$	0	.250	.146	.152	.124	.121	.108	.107	.107	.107

Limiting Behavior of the State Vectors

In our examples we saw that the state vectors approached some fixed vector as the number of observations increased. We now ask whether the state vectors always approach a fixed vector in a Markov chain. A simple example shows that this is not the case.

▶ **EXAMPLE 6** **System Oscillates Between Two State Vectors**

Let
$$P = \begin{bmatrix} 0 & 1 \\ 1 & 0 \end{bmatrix} \quad \text{and} \quad \mathbf{x}^{(0)} = \begin{bmatrix} 1 \\ 0 \end{bmatrix}$$

Then, because $P^2 = I$ and $P^3 = P$, we have that
$$\mathbf{x}^{(0)} = \mathbf{x}^{(2)} = \mathbf{x}^{(4)} = \cdots = \begin{bmatrix} 1 \\ 0 \end{bmatrix}$$

and
$$\mathbf{x}^{(1)} = \mathbf{x}^{(3)} = \mathbf{x}^{(5)} = \cdots = \begin{bmatrix} 0 \\ 1 \end{bmatrix}$$

This system oscillates indefinitely between the two state vectors $\begin{bmatrix} 1 \\ 0 \end{bmatrix}$ and $\begin{bmatrix} 0 \\ 1 \end{bmatrix}$, so it does not approach any fixed vector. ◀

However, if we impose a mild condition on the transition matrix, we can show that a fixed limiting state vector is approached. This condition is described by the following definition.

> **DEFINITION 3** A transition matrix is ***regular*** if some integer power of it has all positive entries.

Thus, for a regular transition matrix P, there is some positive integer m such that all entries of P^m are positive. This is the case with the transition matrices of Examples 1 and 2 for $m = 1$. In Example 5 it turns out that P^4 has all positive entries. Consequently, in all three examples the transition matrices are regular.

A Markov chain that is governed by a regular transition matrix is called a ***regular Markov chain***. We will see that every regular Markov chain has a fixed state vector $\mathbf{q}$ such that $P^n \mathbf{x}^{(0)}$ approaches $\mathbf{q}$ as n increases for any choice of $\mathbf{x}^{(0)}$. This result is of major importance in the theory of Markov chains. It is based on the following theorem.

> **THEOREM 10.5.2** **Behavior of P^n as $n \to \infty$**
>
> If P is a regular transition matrix, then as $n \to \infty$,
> $$P^n \to \begin{bmatrix} q_1 & q_1 & \cdots & q_1 \\ q_2 & q_2 & \cdots & q_2 \\ \vdots & \vdots & & \vdots \\ q_k & q_k & \cdots & q_k \end{bmatrix}$$
> where the q_i are positive numbers such that $q_1 + q_2 + \cdots + q_k = 1$.

We will not prove this theorem here. We refer you to a more specialized text, such as J. Kemeny and J. Snell, *Finite Markov Chains* (New York: Springer-Verlag, 1976).

Let us set

$$Q = \begin{bmatrix} q_1 & q_1 & \cdots & q_1 \\ q_2 & q_2 & \cdots & q_2 \\ \vdots & \vdots & & \vdots \\ q_k & q_k & \cdots & q_k \end{bmatrix} \quad \text{and} \quad \mathbf{q} = \begin{bmatrix} q_1 \\ q_2 \\ \vdots \\ q_k \end{bmatrix}$$

Thus, Q is a transition matrix, all of whose columns are equal to the probability vector $\mathbf{q}$. Q has the property that if $\mathbf{x}$ is any probability vector, then

$$Q\mathbf{x} = \begin{bmatrix} q_1 & q_1 & \cdots & q_1 \\ q_2 & q_2 & \cdots & q_2 \\ \vdots & \vdots & & \vdots \\ q_k & q_k & \cdots & q_k \end{bmatrix} \begin{bmatrix} x_1 \\ x_2 \\ \vdots \\ x_k \end{bmatrix} = \begin{bmatrix} q_1 x_1 + q_1 x_2 + \cdots + q_1 x_k \\ q_2 x_1 + q_2 x_2 + \cdots + q_2 x_k \\ \vdots & \vdots & & \vdots \\ q_k x_1 + q_k x_2 + \cdots + q_k x_k \end{bmatrix}$$

$$= (x_1 + x_2 + \cdots + x_k) \begin{bmatrix} q_1 \\ q_2 \\ \vdots \\ q_k \end{bmatrix} = (1)\mathbf{q} = \mathbf{q}$$

That is, Q transforms any probability vector $\mathbf{x}$ into the fixed probability vector $\mathbf{q}$. This result leads to the following theorem.

THEOREM 10.5.3 Behavior of $P^n\mathbf{x}$ as $n \to \infty$

If P is a regular transition matrix and $\mathbf{x}$ is any probability vector, then as $n \to \infty$,

$$P^n \mathbf{x} \to \begin{bmatrix} q_1 \\ q_2 \\ \vdots \\ q_k \end{bmatrix} = \mathbf{q}$$

where $\mathbf{q}$ is a fixed probability vector, independent of n, all of whose entries are positive.

This result holds since Theorem 10.5.2 implies that $P^n \to Q$ as $n \to \infty$. This in turn implies that $P^n \mathbf{x} \to Q\mathbf{x} = \mathbf{q}$ as $n \to \infty$. Thus, for a regular Markov chain, the system eventually approaches a fixed state vector $\mathbf{q}$. The vector $\mathbf{q}$ is called the **steady-state vector** of the regular Markov chain.

For systems with many states, usually the most efficient technique of computing the steady-state vector $\mathbf{q}$ is simply to calculate $P^n \mathbf{x}$ for some large n. Our examples illustrate this procedure. Each is a regular Markov process, so that convergence to a steady-state vector is ensured. Another way of computing the steady-state vector is to make use of the following theorem.

THEOREM 10.5.4 Steady-State Vector

The steady-state vector $\mathbf{q}$ of a regular transition matrix P is the unique probability vector that satisfies the equation $P\mathbf{q} = \mathbf{q}$.

To see this, consider the matrix identity $PP^n = P^{n+1}$. By Theorem 10.5.2, both P^n and P^{n+1} approach Q as $n \to \infty$. Thus, we have $PQ = Q$. Any one column of this matrix equation gives $P\mathbf{q} = \mathbf{q}$. To show that $\mathbf{q}$ is the only probability vector that satisfies this equation, suppose $\mathbf{r}$ is another probability vector such that $P\mathbf{r} = \mathbf{r}$. Then also $P^n\mathbf{r} = \mathbf{r}$ for $n = 1, 2, \ldots$. When we let $n \to \infty$, Theorem 10.5.3 leads to $\mathbf{q} = \mathbf{r}$.

Theorem 10.5.4 can also be expressed by the statement that the homogeneous linear system

$$(I - P)\mathbf{q} = \mathbf{0}$$

has a unique solution vector $\mathbf{q}$ with nonnegative entries that satisfy the condition $q_1 + q_2 + \cdots + q_k = 1$. We can apply this technique to the computation of the steady-state vectors for our examples.

▶ **EXAMPLE 7** **Example 2 Revisited**

In Example 2 the transition matrix was

$$P = \begin{bmatrix} .8 & .3 \\ .2 & .7 \end{bmatrix}$$

so the linear system $(I - P)\mathbf{q} = \mathbf{0}$ is

$$\begin{bmatrix} .2 & -.3 \\ -.2 & .3 \end{bmatrix} \begin{bmatrix} q_1 \\ q_2 \end{bmatrix} = \begin{bmatrix} 0 \\ 0 \end{bmatrix} \tag{2}$$

This leads to the single independent equation

$$.2q_1 - .3q_2 = 0$$

or

$$q_1 = 1.5q_2$$

Thus, when we set $q_2 = s$, any solution of (2) is of the form

$$\mathbf{q} = s \begin{bmatrix} 1.5 \\ 1 \end{bmatrix}$$

where s is an arbitrary constant. To make the vector $\mathbf{q}$ a probability vector, we set $s = 1/(1.5 + 1) = .4$. Consequently,

$$\mathbf{q} = \begin{bmatrix} .6 \\ .4 \end{bmatrix}$$

is the steady-state vector of this regular Markov chain. This means that over the long run, 60% of the alumni will give a donation in any one year, and 40% will not. Observe that this agrees with the result obtained numerically in Example 3.

▶ **EXAMPLE 8** **Example 1 Revisited**

In Example 1 the transition matrix was

$$P = \begin{bmatrix} .8 & .3 & .2 \\ .1 & .2 & .6 \\ .1 & .5 & .2 \end{bmatrix}$$

so the linear system $(I - P)\mathbf{q} = \mathbf{0}$ is

$$\begin{bmatrix} .2 & -.3 & -.2 \\ -.1 & .8 & -.6 \\ -.1 & -.5 & .8 \end{bmatrix} \begin{bmatrix} q_1 \\ q_2 \\ q_3 \end{bmatrix} = \begin{bmatrix} 0 \\ 0 \\ 0 \end{bmatrix}$$

The reduced row echelon form of the coefficient matrix is (verify)

$$\begin{bmatrix} 1 & 0 & -\frac{34}{13} \\ 0 & 1 & -\frac{14}{13} \\ 0 & 0 & 0 \end{bmatrix}$$

so the original linear system is equivalent to the system

$$q_1 = \left(\frac{34}{13}\right)q_3$$
$$q_2 = \left(\frac{14}{13}\right)q_3$$

When we set $q_3 = s$, any solution of the linear system is of the form

$$\mathbf{q} = s \begin{bmatrix} \frac{34}{13} \\ \frac{14}{13} \\ 1 \end{bmatrix}$$

To make this a probability vector, we set

$$s = \frac{1}{\frac{34}{13} + \frac{14}{13} + 1} = \frac{13}{61}$$

Thus, the steady-state vector of the system is

$$\mathbf{q} = \begin{bmatrix} \frac{34}{61} \\ \frac{14}{61} \\ \frac{13}{61} \end{bmatrix} = \begin{bmatrix} .5573\ldots \\ .2295\ldots \\ .2131\ldots \end{bmatrix}$$

This agrees with the result obtained numerically in Table 1. The entries of $\mathbf{q}$ give the long-run probabilities that any one car will be returned to location 1, 2, or 3, respectively. If the car rental agency has a fleet of 1000 cars, it should design its facilities so that there are at least 558 spaces at location 1, at least 230 spaces at location 2, and at least 214 spaces at location 3.

▶ **EXAMPLE 9** **Example 5 Revisited**

We will not give the details of the calculations but simply state that the unique probability vector solution of the linear system $(I - P)\mathbf{q} = \mathbf{0}$ is

$$\mathbf{q} = \begin{bmatrix} \frac{3}{28} \\ \frac{3}{28} \\ \frac{3}{28} \\ \frac{5}{28} \\ \frac{4}{28} \\ \frac{3}{28} \\ \frac{4}{28} \\ \frac{3}{28} \end{bmatrix} = \begin{bmatrix} .1071\ldots \\ .1071\ldots \\ .1071\ldots \\ .1785\ldots \\ .1428\ldots \\ .1071\ldots \\ .1428\ldots \\ .1071\ldots \end{bmatrix}$$

The entries in this vector indicate the proportion of time the traffic officer spends at each intersection over the long term. Thus, if the objective is for her to spend the same proportion of time at each intersection, then the strategy of random movement with equal probabilities from one intersection to another is not a good one. (See Exercise 5.) ◀

Exercise Set 10.5

1. Consider the transition matrix
$$P = \begin{bmatrix} .4 & .5 \\ .6 & .5 \end{bmatrix}$$

   (a) Calculate $\mathbf{x}^{(n)}$ for $n = 1, 2, 3, 4, 5$ if $\mathbf{x}^{(0)} = \begin{bmatrix} 1 \\ 0 \end{bmatrix}$.

   (b) State why P is regular and find its steady-state vector.

2. Consider the transition matrix
$$P = \begin{bmatrix} .2 & .1 & .7 \\ .6 & .4 & .2 \\ .2 & .5 & .1 \end{bmatrix}$$

   (a) Calculate $\mathbf{x}^{(1)}$, $\mathbf{x}^{(2)}$, and $\mathbf{x}^{(3)}$ to three decimal places if
$$\mathbf{x}^{(0)} = \begin{bmatrix} 0 \\ 0 \\ 1 \end{bmatrix}$$

   (b) State why P is regular and find its steady-state vector.

3. Find the steady-state vectors of the following regular transition matrices:

   (a) $\begin{bmatrix} \frac{1}{3} & \frac{3}{4} \\ \frac{2}{3} & \frac{1}{4} \end{bmatrix}$ (b) $\begin{bmatrix} .81 & .26 \\ .19 & .74 \end{bmatrix}$ (c) $\begin{bmatrix} \frac{1}{3} & \frac{1}{2} & 0 \\ \frac{1}{3} & 0 & \frac{1}{4} \\ \frac{1}{3} & \frac{1}{2} & \frac{3}{4} \end{bmatrix}$

4. Let P be the transition matrix
$$\begin{bmatrix} \frac{1}{2} & 0 \\ \frac{1}{2} & 1 \end{bmatrix}$$

   (a) Show that P is not regular.

   (b) Show that as n increases, $P^n\mathbf{x}^{(0)}$ approaches $\begin{bmatrix} 0 \\ 1 \end{bmatrix}$ for any initial state vector $\mathbf{x}^{(0)}$.

   (c) What conclusion of Theorem 10.5.3 is not valid for the steady state of this transition matrix?

5. Verify that if P is a $k \times k$ regular transition matrix all of whose row sums are equal to 1, then the entries of its steady-state vector are all equal to $1/k$.

6. Show that the transition matrix
$$P = \begin{bmatrix} 0 & \frac{1}{2} & \frac{1}{2} \\ \frac{1}{2} & \frac{1}{2} & 0 \\ \frac{1}{2} & 0 & \frac{1}{2} \end{bmatrix}$$

   is regular, and use Exercise 5 to find its steady-state vector.

7. John is either happy or sad. If he is happy one day, then he is happy the next day four times out of five. If he is sad one day, then he is sad the next day one time out of three. Over the long term, what are the chances that John is happy on any given day?

8. A country is divided into three demographic regions. It is found that each year 5% of the residents of region 1 move to region 2, and 5% move to region 3. Of the residents of region 2, 15% move to region 1 and 10% move to region 3. And of the residents of region 3, 10% move to region 1 and 5% move to region 2. What percentage of the population resides in each of the three regions after a long period of time?

Section 10.5 Technology Exercises

The following exercises are designed to be solved using a technology utility. Typically, this will be MATLAB, Mathematica, Maple, Derive, or Mathcad, but it may also be some other type of linear algebra software or a scientific calculator with some linear algebra capabilities. For each exercise you will need to read the relevant documentation for the particular utility you are using. The goal of these exercises is to provide you with a basic proficiency with your technology utility. Once you have mastered the techniques in these exercises, you will be able to use your technology utility to solve many of the problems in the regular exercise sets.

T1. Consider the sequence of transition matrices
$$\{P_2, P_3, P_4, \ldots\}$$

with

$$P_2 = \begin{bmatrix} 0 & \frac{1}{2} \\ 1 & \frac{1}{2} \end{bmatrix}, \quad P_3 = \begin{bmatrix} 0 & 0 & \frac{1}{3} \\ 0 & \frac{1}{2} & \frac{1}{3} \\ 1 & \frac{1}{2} & \frac{1}{3} \end{bmatrix},$$

$$P_4 = \begin{bmatrix} 0 & 0 & 0 & \frac{1}{4} \\ 0 & 0 & \frac{1}{3} & \frac{1}{4} \\ 0 & \frac{1}{2} & \frac{1}{3} & \frac{1}{4} \\ 1 & \frac{1}{2} & \frac{1}{3} & \frac{1}{4} \end{bmatrix}, \quad P_5 = \begin{bmatrix} 0 & 0 & 0 & 0 & \frac{1}{5} \\ 0 & 0 & 0 & \frac{1}{4} & \frac{1}{5} \\ 0 & 0 & \frac{1}{3} & \frac{1}{4} & \frac{1}{5} \\ 0 & \frac{1}{2} & \frac{1}{3} & \frac{1}{4} & \frac{1}{5} \\ 1 & \frac{1}{2} & \frac{1}{3} & \frac{1}{4} & \frac{1}{5} \end{bmatrix},$$

and so on.

(a) Use a computer to show that each of these four matrices is regular by computing their squares.

(b) Verify Theorem 10.5.2 by computing the 100th power of P_k for $k = 2, 3, 4, 5$. Then make a conjecture as to the limiting value of P_k^n as $n \to \infty$ for all $k = 2, 3, 4, \ldots$.

(c) Verify that the common column $\mathbf{q}_k$ of the limiting matrix you found in part (b) satisfies the equation $P_k \mathbf{q}_k = \mathbf{q}_k$, as required by Theorem 10.5.4.

T2. A mouse is placed in a box with nine rooms as shown in the accompanying figure. Assume that it is equally likely that the mouse goes through any door in the room or stays in the room.

(a) Construct the 9×9 transition matrix for this problem and show that it is regular.

(b) Determine the steady-state vector for the matrix.

(c) Use a symmetry argument to show that this problem may be solved using only a 3×3 matrix.

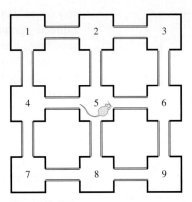

Figure Ex-T2

10.6 Graph Theory

In this section we introduce matrix representations of relations among members of a set. We use matrix arithmetic to analyze these relationships.

PREREQUISITES: Matrix Addition and Multiplication

Relations Among Members of a Set

There are countless examples of sets with finitely many members in which some relation exists among members of the set. For example, the set could consist of a collection of people, animals, countries, companies, sports teams, or cities; and the relation between two members, A and B, of such a set could be that person A dominates person B, animal A feeds on animal B, country A militarily supports country B, company A sells its product to company B, sports team A consistently beats sports team B, or city A has a direct airline flight to city B.

We will now show how the theory of *directed graphs* can be used to mathematically model relations such as those in the preceding examples.

Directed Graphs

A ***directed graph*** is a finite set of elements, $\{P_1, P_2, \ldots, P_n\}$, together with a finite collection of ordered pairs (P_i, P_j) of distinct elements of this set, with no ordered pair being repeated. The elements of the set are called ***vertices***, and the ordered pairs are called ***directed edges***, of the directed graph. We use the notation $P_i \to P_j$ (which is read "P_i is connected to P_j") to indicate that the directed edge (P_i, P_j) belongs to the directed graph. Geometrically, we can visualize a directed graph (Figure 10.6.1) by representing the vertices as points in the plane and representing the directed edge $P_i \to P_j$ by drawing a line or arc from vertex P_i to vertex P_j, with an arrow pointing from P_i to P_j. If both $P_i \to P_j$ and $P_j \to P_i$ hold (denoted $P_i \leftrightarrow P_j$), we draw a single line between P_i and P_j with two oppositely pointing arrows (as with P_2 and P_3 in the figure).

As in Figure 10.6.1, for example, a directed graph may have separate "components" of vertices that are connected only among themselves; and some vertices, such as P_5,

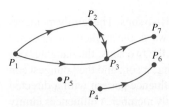

▲ **Figure 10.6.1**

564 Chapter 10 Applications of Linear Algebra

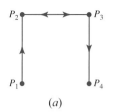
(a)

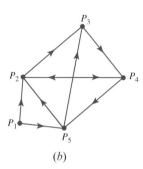
(b)

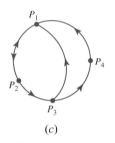
(c)

▲ Figure 10.6.2

may not be connected with any other vertex. Also, because $P_i \to P_i$ is not permitted in a directed graph, a vertex cannot be connected with itself by a single arc that does not pass through any other vertex.

Figure 10.6.2 shows diagrams representing three more examples of directed graphs. With a directed graph having n vertices, we may associate an $n \times n$ matrix $M = [m_{ij}]$, called the **vertex matrix** of the directed graph. Its elements are defined by

$$m_{ij} = \begin{cases} 1, & \text{if } P_i \to P_j \\ 0, & \text{otherwise} \end{cases}$$

for $i, j = 1, 2, \ldots, n$. For the three directed graphs in Figure 10.6.2, the corresponding vertex matrices are

Figure 10.6.2a: $M = \begin{bmatrix} 0 & 1 & 0 & 0 \\ 0 & 0 & 1 & 0 \\ 0 & 1 & 0 & 1 \\ 0 & 0 & 0 & 0 \end{bmatrix}$

Figure 10.6.2b: $M = \begin{bmatrix} 0 & 1 & 0 & 0 & 1 \\ 0 & 0 & 1 & 1 & 0 \\ 0 & 0 & 0 & 1 & 0 \\ 0 & 1 & 0 & 0 & 1 \\ 0 & 1 & 1 & 0 & 0 \end{bmatrix}$

Figure 10.6.2c: $M = \begin{bmatrix} 0 & 1 & 0 & 0 \\ 1 & 0 & 1 & 0 \\ 1 & 0 & 0 & 1 \\ 1 & 0 & 0 & 0 \end{bmatrix}$

By their definition, vertex matrices have the following two properties:

(i) All entries are either 0 or 1.
(ii) All diagonal entries are 0.

Conversely, any matrix with these two properties determines a unique directed graph having the given matrix as its vertex matrix. For example, the matrix

$$M = \begin{bmatrix} 0 & 1 & 1 & 0 \\ 0 & 0 & 1 & 0 \\ 1 & 0 & 0 & 1 \\ 0 & 0 & 0 & 0 \end{bmatrix}$$

determines the directed graph in Figure 10.6.3.

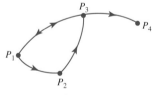
▲ Figure 10.6.3

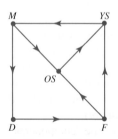
▲ Figure 10.6.4

▶ **EXAMPLE 1 Influences Within a Family**

A certain family consists of a mother, father, daughter, and two sons. The family members have influence, or power, over each other in the following ways: the mother can influence the daughter and the oldest son; the father can influence the two sons; the daughter can influence the father; the oldest son can influence the youngest son; and the youngest son can influence the mother. We may model this family influence pattern with a directed graph whose vertices are the five family members. If family member A influences family member B, we write $A \to B$. Figure 10.6.4 is the resulting directed graph, where we have used obvious letter designations for the five family members. The vertex matrix of this directed graph is

$$\begin{array}{c} \\ M \\ F \\ D \\ OS \\ YS \end{array} \begin{array}{c} M F D OS YS \\ \begin{bmatrix} 0 & 0 & 1 & 1 & 0 \\ 0 & 0 & 0 & 1 & 1 \\ 0 & 1 & 0 & 0 & 0 \\ 0 & 0 & 0 & 0 & 1 \\ 1 & 0 & 0 & 0 & 0 \end{bmatrix} \end{array}$$

▶ **EXAMPLE 2** **Vertex Matrix: Moves on a Chessboard**

In chess the knight moves in an "L"-shaped pattern about the chessboard. For the board in Figure 10.6.5 it may move horizontally two squares and then vertically one square, or it may move vertically two squares and then horizontally one square. Thus, from the center square in the figure, the knight may move to any of the eight marked shaded squares. Suppose that the knight is restricted to the nine numbered squares in Figure 10.6.6. If by $i \to j$ we mean that the knight may move from square i to square j, the directed graph in Figure 10.6.7 illustrates all possible moves that the knight may make among these nine squares. In Figure 10.6.8 we have "unraveled" Figure 10.6.7 to make the pattern of possible moves clearer.

The vertex matrix of this directed graph is given by

$$M = \begin{bmatrix} 0 & 0 & 0 & 0 & 0 & 1 & 0 & 1 & 0 \\ 0 & 0 & 0 & 0 & 0 & 0 & 1 & 0 & 1 \\ 0 & 0 & 0 & 1 & 0 & 0 & 0 & 1 & 0 \\ 0 & 0 & 1 & 0 & 0 & 0 & 0 & 0 & 1 \\ 0 & 0 & 0 & 0 & 0 & 0 & 0 & 0 & 0 \\ 1 & 0 & 0 & 0 & 0 & 0 & 1 & 0 & 0 \\ 0 & 1 & 0 & 0 & 0 & 1 & 0 & 0 & 0 \\ 1 & 0 & 1 & 0 & 0 & 0 & 0 & 0 & 0 \\ 0 & 1 & 0 & 1 & 0 & 0 & 0 & 0 & 0 \end{bmatrix} ◀$$

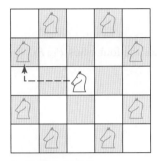

▲ Figure 10.6.5

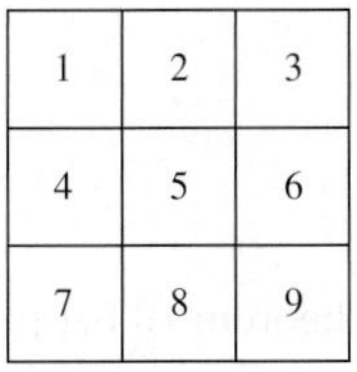

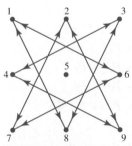

▲ Figure 10.6.6

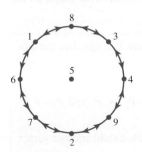

▲ Figure 10.6.7

▲ Figure 10.6.8

In Example 1 the father cannot directly influence the mother; that is, $F \to M$ is not true. But he can influence the youngest son, who can then influence the mother. We write this as $F \to YS \to M$ and call it a **2-step connection** from F to M. Analogously, we call $M \to D$ a **1-step connection**, $F \to OS \to YS \to M$ a **3-step connection**, and so forth. Let us now consider a technique for finding the number of all possible r-step connections ($r = 1, 2, \ldots$) from one vertex P_i to another vertex P_j of an arbitrary directed graph. (This will include the case when P_i and P_j are the same vertex.) The number of 1-step connections from P_i to P_j is simply m_{ij}. That is, there is either zero or one 1-step connection from P_i to P_j, depending on whether m_{ij} is zero or one. For the number of 2-step connections, we consider the square of the vertex matrix. If we let $m_{ij}^{(2)}$ be the (i, j)-th element of M^2, we have

$$m_{ij}^{(2)} = m_{i1}m_{1j} + m_{i2}m_{2j} + \cdots + m_{in}m_{nj} \tag{1}$$

Now, if $m_{i1} = m_{1j} = 1$, there is a 2-step connection $P_i \to P_1 \to P_j$ from P_i to P_j. But if either m_{i1} or m_{1j} is zero, such a 2-step connection is not possible. Thus $P_i \to P_1 \to P_j$ is a 2-step connection if and only if $m_{i1}m_{1j} = 1$. Similarly, for any $k = 1, 2, \ldots, n$,

566 Chapter 10 Applications of Linear Algebra

$P_i \to P_k \to P_j$ is a 2-step connection from P_i to P_j if and only if the term $m_{ik}m_{kj}$ on the right side of (1) is one; otherwise, the term is zero. Thus, the right side of (1) is the total number of two 2-step connections from P_i to P_j.

A similar argument will work for finding the number of 3-, 4-, ..., r-step connections from P_i to P_j. In general, we have the following result.

> **THEOREM 10.6.1** Let M be the vertex matrix of a directed graph and let $m_{ij}^{(r)}$ be the (i, j)-th element of M^r. Then $m_{ij}^{(r)}$ is equal to the number of r-step connections from P_i to P_j.

▶ **EXAMPLE 3 Using Theorem 10.6.1**

Figure 10.6.9 is the route map of a small airline that services the four cities P_1, P_2, P_3, P_4. As a directed graph, its vertex matrix is

$$M = \begin{bmatrix} 0 & 1 & 1 & 0 \\ 1 & 0 & 1 & 0 \\ 1 & 0 & 0 & 1 \\ 0 & 1 & 1 & 0 \end{bmatrix}$$

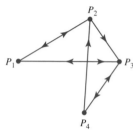

▲ Figure 10.6.9

We have that

$$M^2 = \begin{bmatrix} 2 & 0 & 1 & 1 \\ 1 & 1 & 1 & 1 \\ 0 & 2 & 2 & 0 \\ 2 & 0 & 1 & 1 \end{bmatrix} \quad \text{and} \quad M^3 = \begin{bmatrix} 1 & 3 & 3 & 1 \\ 2 & 2 & 3 & 1 \\ 4 & 0 & 2 & 2 \\ 1 & 3 & 3 & 1 \end{bmatrix}$$

If we are interested in connections from city P_4 to city P_3, we may use Theorem 10.6.1 to find their number. Because $m_{43} = 1$, there is one 1-step connection; because $m_{43}^{(2)} = 1$, there is one 2-step connection; and because $m_{43}^{(3)} = 3$, there are three 3-step connections. To verify this, from Figure 10.6.9 we find

$$\begin{aligned}
\text{1-step connections from } P_4 \text{ to } P_3: \quad & P_4 \to P_3 \\
\text{2-step connections from } P_4 \text{ to } P_3: \quad & P_4 \to P_2 \to P_3 \\
\text{3-step connections from } P_4 \text{ to } P_3: \quad & P_4 \to P_3 \to P_4 \to P_3 \\
& P_4 \to P_2 \to P_1 \to P_3 \\
& P_4 \to P_3 \to P_1 \to P_3 \quad \blacktriangleleft
\end{aligned}$$

Cliques

In everyday language a "clique" is a closely knit group of people (usually three or more) that tends to communicate within itself and has no place for outsiders. In graph theory this concept is given a more precise meaning.

> **DEFINITION 1** A subset of a directed graph is called a ***clique*** if it satisfies the following three conditions:
>
> (i) The subset contains at least three vertices.
> (ii) For each pair of vertices P_i and P_j in the subset, both $P_i \to P_j$ and $P_j \to P_i$ are true.
> (iii) The subset is as large as possible; that is, it is not possible to add another vertex to the subset and still satisfy condition (ii).

10.6 Graph Theory

This definition suggests that cliques are maximal subsets that are in perfect "communication" with each other. For example, if the vertices represent cities, and $P_i \to P_j$ means that there is a direct airline flight from city P_i to city P_j, then there is a direct flight between any two cities within a clique in either direction.

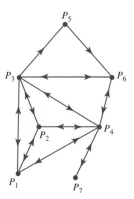

▲ Figure 10.6.10

▶ **EXAMPLE 4** **A Directed Graph with Two Cliques**

The directed graph illustrated in Figure 10.6.10 (which might represent the route map of an airline) has two cliques:

$$\{P_1, P_2, P_3, P_4\} \quad \text{and} \quad \{P_3, P_4, P_6\}$$

This example shows that a directed graph may contain several cliques and that a vertex may simultaneously belong to more than one clique. ◀

For simple directed graphs, cliques can be found by inspection. But for large directed graphs, it would be desirable to have a systematic procedure for detecting cliques. For this purpose, it will be helpful to define a matrix $S = [s_{ij}]$ related to a given directed graph as follows:

$$s_{ij} = \begin{cases} 1, & \text{if } P_i \leftrightarrow P_j \\ 0, & \text{otherwise} \end{cases}$$

The matrix S determines a directed graph that is the same as the given directed graph, with the exception that the directed edges with only one arrow are deleted. For example, if the original directed graph is given by Figure 10.6.11a, the directed graph that has S as its vertex matrix is given in Figure 10.6.11b. The matrix S may be obtained from the vertex matrix M of the original directed graph by setting $s_{ij} = 1$ if $m_{ij} = m_{ji} = 1$ and setting $s_{ij} = 0$ otherwise.

The following theorem, which uses the matrix S, is helpful for identifying cliques.

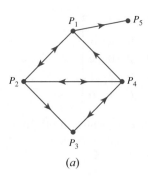

(a)

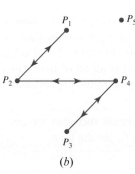

(b)

▲ Figure 10.6.11

THEOREM 10.6.2 **Identifying Cliques**

Let $s_{ij}^{(3)}$ be the (i, j)-th element of S^3. Then a vertex P_i belongs to some clique if and only if $s_{ii}^{(3)} \neq 0$.

Proof If $s_{ii}^{(3)} \neq 0$, then there is at least one 3-step connection from P_i to itself in the modified directed graph determined by S. Suppose it is $P_i \to P_j \to P_k \to P_i$. In the modified directed graph, all directed relations are two-way, so we also have the connections $P_i \leftrightarrow P_j \leftrightarrow P_k \leftrightarrow P_i$. But this means that $\{P_i, P_j, P_k\}$ is either a clique or a subset of a clique. In either case, P_i must belong to some clique. The converse statement, "if P_i belongs to a clique, then $s_{ii}^{(3)} \neq 0$," follows in a similar manner. ◀

▶ **EXAMPLE 5** **Using Theorem 10.6.2**

Suppose that a directed graph has as its vertex matrix

$$M = \begin{bmatrix} 0 & 1 & 1 & 1 \\ 1 & 0 & 1 & 0 \\ 0 & 1 & 0 & 1 \\ 1 & 0 & 0 & 0 \end{bmatrix}$$

Then

$$S = \begin{bmatrix} 0 & 1 & 0 & 1 \\ 1 & 0 & 1 & 0 \\ 0 & 1 & 0 & 0 \\ 1 & 0 & 0 & 0 \end{bmatrix} \quad \text{and} \quad S^3 = \begin{bmatrix} 0 & 3 & 0 & 2 \\ 3 & 0 & 2 & 0 \\ 0 & 2 & 0 & 1 \\ 2 & 0 & 1 & 0 \end{bmatrix}$$

Because all diagonal entries of S^3 are zero, it follows from Theorem 10.6.2 that the directed graph has no cliques.

▶ **EXAMPLE 6 Using Theorem 10.6.2**

Suppose that a directed graph has as its vertex matrix

$$M = \begin{bmatrix} 0 & 1 & 0 & 1 & 1 \\ 1 & 0 & 0 & 1 & 0 \\ 1 & 1 & 0 & 1 & 0 \\ 1 & 1 & 0 & 0 & 0 \\ 1 & 0 & 0 & 1 & 0 \end{bmatrix}$$

Then

$$S = \begin{bmatrix} 0 & 1 & 0 & 1 & 1 \\ 1 & 0 & 0 & 1 & 0 \\ 0 & 0 & 0 & 0 & 0 \\ 1 & 1 & 0 & 0 & 0 \\ 1 & 0 & 0 & 0 & 0 \end{bmatrix} \quad \text{and} \quad S^3 = \begin{bmatrix} 2 & 4 & 0 & 4 & 3 \\ 4 & 2 & 0 & 3 & 1 \\ 0 & 0 & 0 & 0 & 0 \\ 4 & 3 & 0 & 2 & 1 \\ 3 & 1 & 0 & 1 & 0 \end{bmatrix}$$

The nonzero diagonal entries of S^3 are $s_{11}^{(3)}$, $s_{22}^{(3)}$, and $s_{44}^{(3)}$. Consequently, in the given directed graph, P_1, P_2, and P_4 belong to cliques. Because a clique must contain at least three vertices, the directed graph has only one clique, $\{P_1, P_2, P_4\}$. ◀

Dominance-Directed Graphs

In many groups of individuals or animals, there is a definite "pecking order" or dominance relation between any two members of the group. That is, given any two individuals A and B, either A dominates B or B dominates A, but not both. In terms of a directed graph in which $P_i \to P_j$ means P_i dominates P_j, this means that for all distinct pairs, either $P_i \to P_j$ or $P_j \to P_i$, but not both. In general, we have the following definition.

> **DEFINITION 2** A *dominance-directed graph* is a directed graph such that for any distinct pair of vertices P_i and P_j, either $P_i \to P_j$ or $P_j \to P_i$, but not both.

An example of a directed graph satisfying this definition is a league of n sports teams that play each other exactly one time, as in one round of a round-robin tournament in which no ties are allowed. If $P_i \to P_j$ means that team P_i beat team P_j in their single match, it is easy to see that the definition of a dominance-directed group is satisfied. For this reason, dominance-directed graphs are sometimes called *tournaments*.

Figure 10.6.12 illustrates some dominance-directed graphs with three, four, and five vertices, respectively. In these three graphs, the circled vertices have the following interesting property: from each one there is either a 1-step or a 2-step connection to any other vertex in its graph. In a sports tournament, these vertices would correspond to the most "powerful" teams in the sense that these teams either beat any given team or beat some other team that beat the given team. We can now state and prove a theorem that guarantees that any dominance-directed graph has at least one vertex with this property.

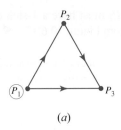

(a)

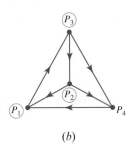

(b)

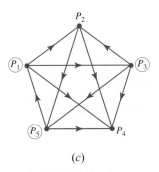

(c)

▲ Figure 10.6.12

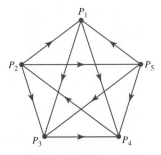

▲ Figure 10.6.13

THEOREM 10.6.3 Connections in Dominance-Directed Graphs

In any dominance-directed graph, there is at least one vertex from which there is a 1-step or 2-step connection to any other vertex.

Proof Consider a vertex (there may be several) with the largest total number of 1-step and 2-step connections to other vertices in the graph. By renumbering the vertices, we may assume that P_1 is such a vertex. Suppose there is some vertex P_i such that there is no 1-step or 2-step connection from P_1 to P_i. Then, in particular, $P_1 \to P_i$ is not true, so that by definition of a dominance-directed graph, it must be that $P_i \to P_1$. Next, let P_k be any vertex such that $P_1 \to P_k$ is true. Then we cannot have $P_k \to P_i$, as then $P_1 \to P_k \to P_i$ would be a 2-step connection from P_1 to P_i. Thus, it must be that $P_i \to P_k$. That is, P_i has 1-step connections to all the vertices to which P_1 has 1-step connections. The vertex P_i must then also have 2-step connections to all the vertices to which P_1 has 2-step connections. But because, in addition, we have that $P_i \to P_1$, this means that P_i has more 1-step and 2-step connections to other vertices than does P_1. However, this contradicts the way in which P_1 was chosen. Hence, there can be no vertex P_i to which P_1 has no 1-step or 2-step connection. ◀

This proof shows that a vertex with the largest total number of 1-step and 2-step connections to other vertices has the property stated in the theorem. There is a simple way of finding such vertices using the vertex matrix M and its square M^2. The sum of the entries in the ith row of M is the total number of 1-step connections from P_i to other vertices, and the sum of the entries of the ith row of M^2 is the total number of 2-step connections from P_i to other vertices. Consequently, the sum of the entries of the ith row of the matrix $A = M + M^2$ is the total number of 1-step and 2-step connections from P_i to other vertices. In other words, a row of $A = M + M^2$ with the largest row sum identifies a vertex having the property stated in Theorem 10.6.3.

▶ **EXAMPLE 7 Using Theorem 10.6.3**

Suppose that five baseball teams play each other exactly once, and the results are as indicated in the dominance-directed graph of Figure 10.6.13. The vertex matrix of the graph is

$$M = \begin{bmatrix} 0 & 0 & 1 & 1 & 0 \\ 1 & 0 & 1 & 0 & 1 \\ 0 & 0 & 0 & 1 & 0 \\ 0 & 1 & 0 & 0 & 0 \\ 1 & 0 & 1 & 1 & 0 \end{bmatrix}$$

so

$$A = M + M^2 = \begin{bmatrix} 0 & 0 & 1 & 1 & 0 \\ 1 & 0 & 1 & 0 & 1 \\ 0 & 0 & 0 & 1 & 0 \\ 0 & 1 & 0 & 0 & 0 \\ 1 & 0 & 1 & 1 & 0 \end{bmatrix} + \begin{bmatrix} 0 & 1 & 0 & 1 & 0 \\ 1 & 0 & 2 & 3 & 0 \\ 0 & 1 & 0 & 0 & 0 \\ 1 & 0 & 1 & 0 & 1 \\ 0 & 1 & 1 & 2 & 0 \end{bmatrix} = \begin{bmatrix} 0 & 1 & 1 & 2 & 0 \\ 2 & 0 & 3 & 3 & 1 \\ 0 & 1 & 0 & 1 & 0 \\ 1 & 1 & 1 & 0 & 1 \\ 1 & 1 & 2 & 3 & 0 \end{bmatrix}$$

The row sums of A are

1st row sum $= 4$

2nd row sum $= 9$

3rd row sum $= 2$

4th row sum $= 4$

5th row sum $= 7$

Because the second row has the largest row sum, the vertex P_2 must have a 1-step or 2-step connection to any other vertex. This is easily verified from Figure 10.6.13. ◀

We have informally suggested that a vertex with the largest number of 1-step and 2-step connections to other vertices is a "powerful" vertex. We can formalize this concept with the following definition.

> **DEFINITION 3** The *power* of a vertex of a dominance-directed graph is the total number of 1-step and 2-step connections from it to other vertices. Alternatively, the power of a vertex P_i is the sum of the entries of the ith row of the matrix $A = M + M^2$, where M is the vertex matrix of the directed graph.

▶ **EXAMPLE 8** **Example 7 Revisited**

Let us rank the five baseball teams in Example 7 according to their powers. From the calculations for the row sums in that example, we have

$$\text{Power of team } P_1 = 4$$
$$\text{Power of team } P_2 = 9$$
$$\text{Power of team } P_3 = 2$$
$$\text{Power of team } P_4 = 4$$
$$\text{Power of team } P_5 = 7$$

Hence, the ranking of the teams according to their powers would be

$$P_2 \text{ (first)}, \quad P_5 \text{ (second)}, \quad P_1 \text{ and } P_4 \text{ (tied for third)}, \quad P_3 \text{ (last)} \blacktriangleleft$$

Exercise Set 10.6

1. Construct the vertex matrix for each of the directed graphs illustrated in Figure Ex-1.

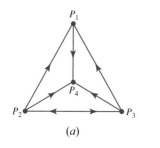

(a)

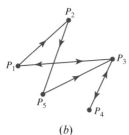

(b)

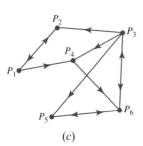

(c)

▲ Figure Ex-1

2. Draw a diagram of the directed graph corresponding to each of the following vertex matrices.

(a) $\begin{bmatrix} 0 & 1 & 1 & 0 \\ 1 & 0 & 0 & 0 \\ 0 & 0 & 0 & 1 \\ 1 & 0 & 1 & 0 \end{bmatrix}$
(b) $\begin{bmatrix} 0 & 0 & 1 & 0 & 0 \\ 1 & 0 & 0 & 0 & 1 \\ 0 & 1 & 0 & 1 & 1 \\ 0 & 0 & 0 & 0 & 0 \\ 1 & 1 & 1 & 0 & 0 \end{bmatrix}$

(c) $\begin{bmatrix} 0 & 1 & 0 & 1 & 0 & 1 \\ 1 & 0 & 0 & 0 & 1 & 0 \\ 0 & 0 & 0 & 0 & 0 & 0 \\ 1 & 1 & 0 & 0 & 1 & 0 \\ 0 & 0 & 0 & 1 & 0 & 1 \\ 0 & 1 & 0 & 0 & 1 & 0 \end{bmatrix}$

3. Let M be the following vertex matrix of a directed graph:

$$\begin{bmatrix} 0 & 1 & 1 & 1 \\ 1 & 0 & 0 & 0 \\ 0 & 1 & 0 & 1 \\ 0 & 1 & 1 & 0 \end{bmatrix}$$

(a) Draw a diagram of the directed graph.
(b) Use Theorem 10.6.1 to find the number of 1-, 2-, and 3-step connections from the vertex P_1 to the vertex P_2. Verify your answer by listing the various connections as in Example 3.
(c) Repeat part (b) for the 1-, 2-, and 3-step connections from P_1 to P_4.

4. (a) Compute the matrix product M^TM for the vertex matrix M in Example 1.

   (b) Verify that the kth diagonal entry of M^TM is the number of family members who influence the kth family member. Why is this true?

   (c) Find a similar interpretation for the values of the nondiagonal entries of M^TM.

5. By inspection, locate all cliques in each of the directed graphs illustrated in Figure Ex-5.

6. For each of the following vertex matrices, use Theorem 10.6.2 to find all cliques in the corresponding directed graphs.

(a) $\begin{bmatrix} 0 & 1 & 0 & 1 & 0 \\ 1 & 0 & 1 & 0 & 1 \\ 0 & 1 & 0 & 1 & 1 \\ 1 & 0 & 0 & 0 & 1 \\ 1 & 0 & 1 & 1 & 0 \end{bmatrix}$ (b) $\begin{bmatrix} 0 & 1 & 0 & 1 & 1 & 0 \\ 1 & 0 & 1 & 0 & 1 & 1 \\ 0 & 1 & 0 & 1 & 0 & 1 \\ 1 & 0 & 1 & 0 & 1 & 1 \\ 0 & 1 & 0 & 1 & 0 & 0 \\ 0 & 0 & 1 & 1 & 1 & 0 \end{bmatrix}$

7. For the dominance-directed graph illustrated in Figure Ex-7 construct the vertex matrix and find the power of each vertex.

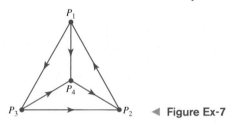

◀ Figure Ex-7

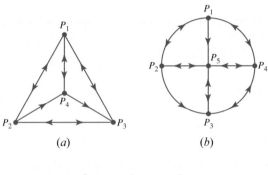

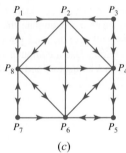

▲ Figure Ex-5

8. Five baseball teams play each other one time with the following results:

A beats B, C, D
B beats C, E
C beats D, E
D beats B
E beats A, D

Rank the five baseball teams in accordance with the powers of the vertices they correspond to in the dominance-directed graph representing the outcomes of the games.

Section 10.6 Technology Exercises

The following exercises are designed to be solved using a technology utility. Typically, this will be MATLAB, *Mathematica*, Maple, Derive, or Mathcad, but it may also be some other type of linear algebra software or a scientific calculator with some linear algebra capabilities. For each exercise you will need to read the relevant documentation for the particular utility you are using. The goal of these exercises is to provide you with a basic proficiency with your technology utility. Once you have mastered the techniques in these exercises, you will be able to use your technology utility to solve many of the problems in the regular exercise sets.

T1. A graph having n vertices such that every vertex is connected to every other vertex has a vertex matrix given by

$$M_n = \begin{bmatrix} 0 & 1 & 1 & 1 & 1 & \cdots & 1 \\ 1 & 0 & 1 & 1 & 1 & \cdots & 1 \\ 1 & 1 & 0 & 1 & 1 & \cdots & 1 \\ 1 & 1 & 1 & 0 & 1 & \cdots & 1 \\ 1 & 1 & 1 & 1 & 0 & \cdots & 1 \\ \vdots & \vdots & \vdots & \vdots & \vdots & \ddots & \vdots \\ 1 & 1 & 1 & 1 & 1 & \cdots & 0 \end{bmatrix}$$

In this problem we develop a formula for M_n^k whose (i, j)-th entry equals the number of k-step connections from P_i to P_j.

(a) Use a computer to compute the eight matrices M_n^k for $n = 2, 3$ and for $k = 2, 3, 4, 5$.

(b) Use the results in part (a) and symmetry arguments to show that M_n^k can be written as

$$M_n^k = \begin{bmatrix} 0 & 1 & 1 & 1 & 1 & \cdots & 1 \\ 1 & 0 & 1 & 1 & 1 & \cdots & 1 \\ 1 & 1 & 0 & 1 & 1 & \cdots & 1 \\ 1 & 1 & 1 & 0 & 1 & \cdots & 1 \\ 1 & 1 & 1 & 1 & 0 & \cdots & 1 \\ \vdots & \vdots & \vdots & \vdots & \vdots & \ddots & \vdots \\ 1 & 1 & 1 & 1 & 1 & \cdots & 0 \end{bmatrix}^k$$

$$= \begin{bmatrix} \alpha_k & \beta_k & \beta_k & \beta_k & \beta_k & \cdots & \beta_k \\ \beta_k & \alpha_k & \beta_k & \beta_k & \beta_k & \cdots & \beta_k \\ \beta_k & \beta_k & \alpha_k & \beta_k & \beta_k & \cdots & \beta_k \\ \beta_k & \beta_k & \beta_k & \alpha_k & \beta_k & \cdots & \beta_k \\ \beta_k & \beta_k & \beta_k & \beta_k & \alpha_k & \cdots & \beta_k \\ \vdots & \vdots & \vdots & \vdots & \vdots & \ddots & \vdots \\ \beta_k & \beta_k & \beta_k & \beta_k & \beta_k & \cdots & \alpha_k \end{bmatrix}$$

(c) Using the fact that $M_n^k = M_n M_n^{k-1}$, show that

$$\begin{bmatrix} \alpha_k \\ \beta_k \end{bmatrix} = \begin{bmatrix} 0 & n-1 \\ 1 & n-2 \end{bmatrix} \begin{bmatrix} \alpha_{k-1} \\ \beta_{k-1} \end{bmatrix}$$

with

$$\begin{bmatrix} \alpha_1 \\ \beta_1 \end{bmatrix} = \begin{bmatrix} 0 \\ 1 \end{bmatrix}$$

(d) Using part (c), show that

$$\begin{bmatrix} \alpha_k \\ \beta_k \end{bmatrix} = \begin{bmatrix} 0 & n-1 \\ 1 & n-2 \end{bmatrix}^{k-1} \begin{bmatrix} 0 \\ 1 \end{bmatrix}$$

(e) Use the methods of Section 5.2 to compute

$$\begin{bmatrix} 0 & n-1 \\ 1 & n-2 \end{bmatrix}^{k-1}$$

and thereby obtain expressions for α_k and β_k, and eventually show that

$$M_n^k = \left(\frac{(n-1)^k - (-1)^k}{n} \right) U_n + (-1)^k I_n$$

where U_n is the $n \times n$ matrix all of whose entries are ones and I_n is the $n \times n$ identity matrix.

(f) Show that for $n > 2$, all vertices for these directed graphs belong to cliques.

T2. Consider a round-robin tournament among n players (labeled $a_1, a_2, a_3, \ldots, a_n$) where a_1 beats a_2, a_2 beats a_3, a_3 beats a_4, $\ldots$, a_{n-1} beats a_n, and a_n beats a_1. Compute the "power" of each player, showing that they all have the same power; then determine that common power. [*Hint:* Use a computer to study the cases $n = 3, 4, 5, 6$; then make a conjecture and prove your conjecture to be true.]

10.7 Games of Strategy

In this section we discuss a general game in which two competing players choose separate strategies to reach opposing objectives. The optimal strategy of each player is found in certain cases with the use of matrix techniques.

> **PREREQUISITES:** Matrix Multiplication
> Basic Probability Concepts

Game Theory   To introduce the basic concepts in the theory of games, we will consider the following carnival-type game that two people agree to play. We will call the participants in the game *player R* and *player C*. Each player has a stationary wheel with a movable pointer on it as in Figure 10.7.1. For reasons that will become clear, we will call player R's wheel the *row-wheel* and player C's wheel the *column-wheel*. The row-wheel is divided into three sectors numbered 1, 2, and 3, and the column-wheel is divided into four sectors numbered 1, 2, 3, and 4. The fractions of the area occupied by the various sectors are indicated in the figure. To play the game, each player spins the pointer of his or her wheel and lets it come to rest at random. The number of the sector in which each pointer comes

10.7 Games of Strategy

to rest is called the *move* of that player. Thus, player R has three possible moves and player C has four possible moves. Depending on the move each player makes, player C then makes a payment of money to player R according to Table 1.

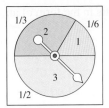

Row-wheel of player R

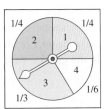

Column-wheel of player C

▲ Figure 10.7.1

Table 1 Payment to Player R

		Player C's Move			
		1	2	3	4
Player R's Move	1	\$3	\$5	−\$2	−\$1
	2	−\$2	\$4	−\$3	−\$4
	3	\$6	−\$5	\$0	\$3

For example, if the row-wheel pointer comes to rest in sector 1 (player R makes move 1), and the column-wheel pointer comes to rest in sector 2 (player C makes move 2), then player C must pay player R the sum of \$5. Some of the entries in this table are negative, indicating that player C makes a negative payment to player R. By this we mean that player R makes a positive payment to player C. For example, if the row-wheel shows 2 and the column-wheel shows 4, then player R pays player C the sum of \$4, because the corresponding entry in the table is −\$4. In this way the positive entries of the table are the gains of player R and the losses of player C, and the negative entries are the gains of player C and the losses of player R.

In this game the players have no control over their moves; each move is determined by chance. However, if each player can decide whether he or she wants to play, then each would want to know how much he or she can expect to win or lose over the long term if he or she chooses to play. (Later in the section we will discuss this question and also consider a more complicated situation in which the players can exercise some control over their moves by varying the sectors of their wheels.)

Two-Person Zero-Sum Matrix Games

The game described above is an example of a **two-person zero-sum matrix game**. The term *zero-sum* means that in each play of the game, the positive gain of one player is equal to the negative gain (loss) of the other player. That is, the sum of the two gains is zero. The term *matrix game* is used to describe a two-person game in which each player has only a finite number of moves, so that all possible outcomes of each play, and the corresponding gains of the players, can be displayed in tabular or matrix form, as in Table 1.

In a general game of this type, let player R have m possible moves and let player C have n possible moves. In a play of the game, each player makes one of his or her possible moves, and then a *payoff* is made from player C to player R, depending on the moves. For $i = 1, 2, \ldots, m$, and $j = 1, 2, \ldots, n$, let us set

$$a_{ij} = \text{payoff that player } C \text{ makes to player } R \text{ if player } R$$
$$\text{makes move } i \text{ and player } C \text{ makes move } j$$

This payoff need not be money; it can be any type of commodity to which we can attach a numerical value. As before, if an entry a_{ij} is negative, we mean that player C receives a payoff of $|a_{ij}|$ from player R. We arrange these mn possible payoffs in the form of an $m \times n$ matrix

$$A = \begin{bmatrix} a_{11} & a_{12} & \cdots & a_{1n} \\ a_{21} & a_{22} & \cdots & a_{2n} \\ \vdots & \vdots & & \vdots \\ a_{m1} & a_{m2} & \cdots & a_{mn} \end{bmatrix}$$

which we will call the *payoff matrix* of the game.

Each player is to make his or her moves on a probabilistic basis. For example, for the game discussed in the introduction, the ratio of the area of a sector to the area of the wheel would be the probability that the player makes the move corresponding to that sector. Thus, from Figure 10.7.1, we see that player R would make move 2 with probability $\frac{1}{3}$, and player C would make move 2 with probability $\frac{1}{4}$. In the general case we make the following definitions:

p_i = probability that player R makes move i $(i = 1, 2, \ldots, m)$

q_j = probability that player C makes move j $(j = 1, 2, \ldots, n)$

It follows from these definitions that

$$p_1 + p_2 + \cdots + p_m = 1$$

and

$$q_1 + q_2 + \cdots + q_n = 1$$

With the probabilities p_i and q_j we form two vectors:

$$\mathbf{p} = \begin{bmatrix} p_1 & p_2 & \cdots & p_m \end{bmatrix} \quad \text{and} \quad \mathbf{q} = \begin{bmatrix} q_1 \\ q_2 \\ \vdots \\ q_n \end{bmatrix}$$

We call the row vector $\mathbf{p}$ the *strategy of player R* and the column vector $\mathbf{q}$ the *strategy of player C*. For example, from Figure 10.7.1 we have

$$\mathbf{p} = \begin{bmatrix} \tfrac{1}{6} & \tfrac{1}{3} & \tfrac{1}{2} \end{bmatrix} \quad \text{and} \quad \mathbf{q} = \begin{bmatrix} \tfrac{1}{4} \\ \tfrac{1}{4} \\ \tfrac{1}{3} \\ \tfrac{1}{6} \end{bmatrix}$$

for the carnival game described earlier.

From the theory of probability, if the probability that player R makes move i is p_i, and independently the probability that player C makes move j is q_j, then $p_i q_j$ is the probability that for any one play of the game, player R makes move i *and* player C makes move j. The payoff to player R for such a pair of moves is a_{ij}. If we multiply each possible payoff by its corresponding probability and sum over all possible payoffs, we obtain the expression

$$a_{11} p_1 q_1 + a_{12} p_1 q_2 + \cdots + a_{1n} p_1 q_n + a_{21} p_2 q_1 + \cdots + a_{mn} p_m q_n \tag{1}$$

Equation (1) is a weighted average of the payoffs to player R; each payoff is weighted according to the probability of its occurrence. In the theory of probability, this weighted average is called the *expected payoff* to player R. It can be shown that if the game is played many times, the long-term average payoff per play to player R is given by this expression. We denote this expected payoff by $E(\mathbf{p}, \mathbf{q})$ to emphasize the fact that it depends on the strategies of the two players. From the definition of the payoff matrix A

and the strategies **p** and **q**, it can be verified that we may express the expected payoff in matrix notation as

$$E(\mathbf{p}, \mathbf{q}) = [p_1 \quad p_2 \quad \cdots \quad p_m] \begin{bmatrix} a_{11} & a_{12} & \cdots & a_{1n} \\ a_{21} & a_{22} & \cdots & a_{2n} \\ \vdots & \vdots & & \vdots \\ a_{m1} & a_{m2} & \cdots & a_{mn} \end{bmatrix} \begin{bmatrix} q_1 \\ q_2 \\ \vdots \\ q_n \end{bmatrix} = \mathbf{p}A\mathbf{q} \quad (2)$$

Because $E(\mathbf{p}, \mathbf{q})$ is the expected payoff to player R, it follows that $-E(\mathbf{p}, \mathbf{q})$ is the expected payoff to player C.

▶ **EXAMPLE 1** **Expected Payoff to Player R**

For the carnival game described earlier, we have

$$E(\mathbf{p}, \mathbf{q}) = \mathbf{p}A\mathbf{q} = \begin{bmatrix} \frac{1}{6} & \frac{1}{3} & \frac{1}{2} \end{bmatrix} \begin{bmatrix} 3 & 5 & -2 & -1 \\ -2 & 4 & -3 & -4 \\ 6 & -5 & 0 & 3 \end{bmatrix} \begin{bmatrix} \frac{1}{4} \\ \frac{1}{4} \\ \frac{1}{3} \\ \frac{1}{6} \end{bmatrix} = \frac{13}{72} = .1805\ldots$$

Thus, in the long run, player R can expect to receive an average of about 18 cents from player C in each play of the game. ◀

So far we have been discussing the situation in which each player has a predetermined strategy. We will now consider the more difficult situation in which both players can change their strategies independently. For example, in the game described in the introduction, we would allow both players to alter the areas of the sectors of their wheels and thereby control the probabilities of their respective moves. This qualitatively changes the nature of the problem and puts us firmly in the field of true game theory. It is understood that neither player knows what strategy the other will choose. It is also assumed that each player will make the best possible choice of strategy and that the other player knows this. Thus, player R attempts to choose a strategy **p** such that $E(\mathbf{p}, \mathbf{q})$ is as large as possible for the best strategy **q** that player C can choose; and similarly, player C attempts to choose a strategy **q** such that $E(\mathbf{p}, \mathbf{q})$ is as small as possible for the best strategy **p** that player R can choose. To see that such choices are actually possible, we will need the following theorem, called the **Fundamental Theorem of Two-Person Zero-Sum Games**. (The general proof, which involves ideas from the theory of linear programming, will be omitted. However, below we will prove this theorem for what are called strictly determined games and 2×2 matrix games.)

THEOREM 10.7.1 **Fundamental Theorem of Zero-Sum Games**

There exist strategies $\mathbf{p}^*$ *and* $\mathbf{q}^*$ *such that*

$$E(\mathbf{p}^*, \mathbf{q}) \geq E(\mathbf{p}^*, \mathbf{q}^*) \geq E(\mathbf{p}, \mathbf{q}^*) \quad (3)$$

for all strategies **p** *and* **q**.

The strategies $\mathbf{p}^*$ and $\mathbf{q}^*$ in this theorem are the best possible strategies for players R and C, respectively. To see why this is so, let $v = E(\mathbf{p}^*, \mathbf{q}^*)$. The left-hand inequality of Equation (3) then reads

$$E(\mathbf{p}^*, \mathbf{q}) \geq v \quad \text{for all strategies } \mathbf{q}$$

This means that if player R chooses the strategy $\mathbf{p}^*$, then no matter what strategy $\mathbf{q}$ player C chooses, the expected payoff to player R will never be below v. Moreover, it is not possible for player R to achieve an expected payoff greater than v. To see why, suppose there is some strategy $\mathbf{p}^{**}$ that player R can choose such that

$$E(\mathbf{p}^{**}, \mathbf{q}) > v \quad \text{for all strategies } \mathbf{q}$$

Then, in particular,

$$E(\mathbf{p}^{**}, \mathbf{q}^*) > v$$

But this contradicts the right-hand inequality of Equation (3), which requires that $v \geq E(\mathbf{p}^{**}, \mathbf{q}^*)$. Consequently, the best player R can do is prevent his or her expected payoff from falling below the value v. Similarly, the best player C can do is ensure that player R's expected payoff does not exceed v, and this can be achieved by using strategy $\mathbf{q}^*$.

On the basis of this discussion, we arrive at the following definitions.

DEFINITION 1 If $\mathbf{p}^*$ and $\mathbf{q}^*$ are strategies such that

$$E(\mathbf{p}^*, \mathbf{q}) \geq E(\mathbf{p}^*, \mathbf{q}^*) \geq E(\mathbf{p}, \mathbf{q}^*) \tag{4}$$

for all strategies $\mathbf{p}$ and $\mathbf{q}$, then

(i) $\mathbf{p}^*$ is called an ***optimal strategy for player R***.
(ii) $\mathbf{q}^*$ is called an ***optimal strategy for player C***.
(iii) $v = E(\mathbf{p}^*, \mathbf{q}^*)$ is called the ***value*** of the game.

The wording in this definition suggests that optimal strategies are not necessarily unique. This is indeed the case, and in Exercise 2 we ask you to show this. However, it can be proved that any two sets of optimal strategies always result in the same value v of the game. That is, if $\mathbf{p}^*, \mathbf{q}^*$ and $\mathbf{p}^{**}, \mathbf{q}^{**}$ are optimal strategies, then

$$E(\mathbf{p}^*, \mathbf{q}^*) = E(\mathbf{p}^{**}, \mathbf{q}^{**}) \tag{5}$$

The value of a game is thus the expected payoff to player R when both players choose any possible optimal strategies.

To find optimal strategies, we must find vectors $\mathbf{p}^*$ and $\mathbf{q}^*$ that satisfy Equation (4). This is generally done by using linear programming techniques. Next, we discuss special cases for which optimal strategies may be found by more elementary techniques.

We now introduce the following definition.

DEFINITION 2 An entry a_{rs} in a payoff matrix A is called a ***saddle point*** if

(i) a_{rs} is the smallest entry in its row, and
(ii) a_{rs} is the largest entry in its column.

A game whose payoff matrix has a saddle point is called ***strictly determined***.

For example, the shaded element in each of the following payoff matrices is a saddle point:

$$\begin{bmatrix} 3 & 1 \\ -4 & 0 \end{bmatrix}, \quad \begin{bmatrix} 30 & -50 & -5 \\ 60 & 90 & 75 \\ -10 & 60 & -30 \end{bmatrix}, \quad \begin{bmatrix} 0 & -3 & 5 & -9 \\ 15 & -8 & -2 & 10 \\ 7 & 10 & 6 & 9 \\ 6 & 11 & -3 & 2 \end{bmatrix}$$

If a matrix has a saddle point a_{rs}, it turns out that the following strategies are optimal strategies for the two players:

$$\mathbf{p}^* = [0 \ 0 \ \cdots \ \underset{r\text{th entry}}{1} \ \cdots \ 0], \qquad \mathbf{q}^* = \begin{bmatrix} 0 \\ 0 \\ \vdots \\ 1 \\ \vdots \\ 0 \end{bmatrix} \leftarrow s\text{th entry}$$

That is, an optimal strategy for player R is to always make the rth move, and an optimal strategy for player C is to always make the sth move. Such strategies for which only one move is possible are called **pure strategies**. Strategies for which more than one move is possible are called **mixed strategies**. To show that the above pure strategies are optimal, you can verify the following three equations (see Exercise 6):

$$E(\mathbf{p}^*, \mathbf{q}^*) = \mathbf{p}^* A \mathbf{q}^* = a_{rs} \qquad (6)$$

$$E(\mathbf{p}^*, \mathbf{q}) = \mathbf{p}^* A \mathbf{q} \geq a_{rs} \quad \text{for any strategy } \mathbf{q} \qquad (7)$$

$$E(\mathbf{p}, \mathbf{q}^*) = \mathbf{p} A \mathbf{q}^* \leq a_{rs} \quad \text{for any strategy } \mathbf{p} \qquad (8)$$

Together, these three equations imply that

$$E(\mathbf{p}^*, \mathbf{q}) \geq E(\mathbf{p}^*, \mathbf{q}^*) \geq E(\mathbf{p}, \mathbf{q}^*)$$

for all strategies $\mathbf{p}$ and $\mathbf{q}$. Because this is exactly Equation (4), it follows that $\mathbf{p}^*$ and $\mathbf{q}^*$ are optimal strategies.

From Equation (6) the value of a strictly determined game is simply the numerical value of a saddle point a_{rs}. It is possible for a payoff matrix to have several saddle points, but then the uniqueness of the value of a game guarantees that the numerical values of all saddle points are the same.

▶ **EXAMPLE 2 Optimal Strategies to Maximize a Viewing Audience**

Two competing television networks, R and C, are scheduling one-hour programs in the same time period. Network R can schedule one of three possible programs, and network C can schedule one of four possible programs. Neither network knows which program the other will schedule. Both networks ask the same outside polling agency to give them an estimate of how all possible pairings of the programs will divide the viewing audience. The agency gives them each Table 2, whose (i, j)-th entry is the percentage of the viewing audience that will watch network R if network R's program i is paired against network C's program j. What program should each network schedule in order to maximize its viewing audience?

Table 2 Audience Percentage for Network R

		Network C's Program			
		1	2	3	4
Network R's Program	1	60	20	30	55
	2	50	75	45	60
	3	70	45	35	30

Solution Subtract 50 from each entry in Table 2 to construct the following matrix:

$$\begin{bmatrix} 10 & -30 & -20 & 5 \\ 0 & 25 & -5 & 10 \\ 20 & -5 & -15 & -20 \end{bmatrix}$$

This is the payoff matrix of the two-person zero-sum game in which each network is considered to start with 50% of the audience, and the (i, j)-th entry of the matrix is the percentage of the viewing audience that network C loses to network R if programs i and j are paired against each other. It is easy to see that the entry

$$a_{23} = -5$$

is a saddle point of the payoff matrix. Hence, the optimal strategy of network R is to schedule program 2, and the optimal strategy of network C is to schedule program 3. This will result in network R's receiving 45% of the audience and network C's receiving 55% of the audience. ◀

2 × 2 Matrix Games

Another case in which the optimal strategies can be found by elementary means occurs when each player has only two possible moves. In this case, the payoff matrix is a 2 × 2 matrix

$$A = \begin{bmatrix} a_{11} & a_{12} \\ a_{21} & a_{22} \end{bmatrix}$$

If the game is strictly determined, at least one of the four entries of A is a saddle point, and the techniques discussed above can then be applied to determine optimal strategies for the two players. If the game is not strictly determined, we first compute the expected payoff for arbitrary strategies $\mathbf{p}$ and $\mathbf{q}$:

$$E(\mathbf{p}, \mathbf{q}) = \mathbf{p}A\mathbf{q} = [p_1 \quad p_2] \begin{bmatrix} a_{11} & a_{12} \\ a_{21} & a_{22} \end{bmatrix} \begin{bmatrix} q_1 \\ q_2 \end{bmatrix}$$

$$= a_{11}p_1q_1 + a_{12}p_1q_2 + a_{21}p_2q_1 + a_{22}p_2q_2 \qquad (9)$$

Because

$$p_1 + p_2 = 1 \quad \text{and} \quad q_1 + q_2 = 1 \qquad (10)$$

we may substitute $p_2 = 1 - p_1$ and $q_2 = 1 - q_1$ into (9) to obtain

$$E(\mathbf{p}, \mathbf{q}) = a_{11}p_1q_1 + a_{12}p_1(1-q_1) + a_{21}(1-p_1)q_1 + a_{22}(1-p_1)(1-q_1) \qquad (11)$$

If we rearrange the terms in Equation (11), we can write

$$E(\mathbf{p}, \mathbf{q}) = [(a_{11} + a_{22} - a_{12} - a_{21})p_1 - (a_{22} - a_{21})]q_1 + (a_{12} - a_{22})p_1 + a_{22} \qquad (12)$$

By examining the coefficient of the q_1 term in (12), we see that if we set

$$p_1 = p_1^* = \frac{a_{22} - a_{21}}{a_{11} + a_{22} - a_{12} - a_{21}} \qquad (13)$$

then that coefficient is zero, and (12) reduces to

$$E(\mathbf{p}^*, \mathbf{q}) = \frac{a_{11}a_{22} - a_{12}a_{21}}{a_{11} + a_{22} - a_{12} - a_{21}} \qquad (14)$$

Equation (14) is independent of $\mathbf{q}$; that is, if player R chooses the strategy determined by (13), player C cannot change the expected payoff by varying his or her strategy.

In a similar manner, it can be verified that if player C chooses the strategy determined by

$$q_1 = q_1^* = \frac{a_{22} - a_{12}}{a_{11} + a_{22} - a_{12} - a_{21}} \qquad (15)$$

then substituting in (12) gives

$$E(\mathbf{p}, \mathbf{q}^*) = \frac{a_{11}a_{22} - a_{12}a_{21}}{a_{11} + a_{22} - a_{12} - a_{21}} \quad (16)$$

Equations (14) and (16) show that

$$E(\mathbf{p}^*, \mathbf{q}) = E(\mathbf{p}^*, \mathbf{q}^*) = E(\mathbf{p}, \mathbf{q}^*) \quad (17)$$

for all strategies $\mathbf{p}$ and $\mathbf{q}$. Thus, the strategies determined by (13), (15), and (10) are optimal strategies for players R and C, respectively, and so we have the following result.

THEOREM 10.7.2 **Optimal Strategies for a 2 × 2 Matrix Game**

For a 2 × 2 game that is not strictly determined, optimal strategies for players R and C are

$$\mathbf{p}^* = \begin{bmatrix} \dfrac{a_{22} - a_{21}}{a_{11} + a_{22} - a_{12} - a_{21}} & \dfrac{a_{11} - a_{12}}{a_{11} + a_{22} - a_{12} - a_{21}} \end{bmatrix}$$

and

$$\mathbf{q}^* = \begin{bmatrix} \dfrac{a_{22} - a_{12}}{a_{11} + a_{22} - a_{12} - a_{21}} \\ \dfrac{a_{11} - a_{21}}{a_{11} + a_{22} - a_{12} - a_{21}} \end{bmatrix}$$

The value of the game is

$$v = \frac{a_{11}a_{22} - a_{12}a_{21}}{a_{11} + a_{22} - a_{12} - a_{21}}$$

In order to be complete, we must show that the entries in the vectors $\mathbf{p}^*$ and $\mathbf{q}^*$ are numbers strictly between 0 and 1. In Exercise 8 we ask you to show that this is the case as long as the game is not strictly determined.

Equation (17) is interesting in that it implies that either player can force the expected payoff to be the value of the game by choosing his or her optimal strategy, regardless of which strategy the other player chooses. This is not true, in general, for games in which either player has more than two moves.

▶ **EXAMPLE 3** **Using Theorem 10.7.2**

The federal government desires to inoculate its citizens against a certain flu virus. The virus has two strains, and the proportions in which the two strains occur in the virus population is not known. Two vaccines have been developed and each citizen is given only one of them. Vaccine 1 is 85% effective against strain 1 and 70% effective against strain 2. Vaccine 2 is 60% effective against strain 1 and 90% effective against strain 2. What inoculation policy should the government adopt?

Solution We can consider this a two-person game in which player R (the government) desires to make the payoff (the fraction of citizens resistant to the virus) as large as possible, and player C (the virus) desires to make the payoff as small as possible. The payoff matrix is

$$\text{Vaccine} \begin{array}{c} \\ 1 \\ 2 \end{array} \overset{\text{Strain}}{\begin{bmatrix} 1 & 2 \\ .85 & .70 \\ .60 & .90 \end{bmatrix}}$$

This matrix has no saddle points, so Theorem 10.7.2 is applicable. Consequently,

$$p_1^* = \frac{a_{22} - a_{21}}{a_{11} + a_{22} - a_{12} - a_{21}} = \frac{.90 - .60}{.85 + .90 - .70 - .60} = \frac{.30}{.45} = \frac{2}{3}$$

$$p_2^* = 1 - p_1^* = 1 - \frac{2}{3} = \frac{1}{3}$$

$$q_1^* = \frac{a_{22} - a_{12}}{a_{11} + a_{22} - a_{12} - a_{21}} = \frac{.90 - .70}{.85 + .90 - .70 - .60} = \frac{.20}{.45} = \frac{4}{9}$$

$$q_2^* = 1 - q_1^* = 1 - \frac{4}{9} = \frac{5}{9}$$

$$v = \frac{a_{11}a_{22} - a_{12}a_{21}}{a_{11} + a_{22} - a_{12} - a_{21}} = \frac{(.85)(.90) - (.70)(.60)}{.85 + .90 - .70 - .60} = \frac{.345}{.45} = .7666\ldots$$

Thus, the optimal strategy for the government is to inoculate $\frac{2}{3}$ of the citizens with vaccine 1 and $\frac{1}{3}$ of the citizens with vaccine 2. This will guarantee that about 76.7% of the citizens will be resistant to a virus attack regardless of the distribution of the two strains.

In contrast, a virus distribution of $\frac{4}{9}$ of strain 1 and $\frac{5}{9}$ of strain 2 will result in the same 76.7% of resistant citizens, regardless of the inoculation strategy adopted by the government (see Exercise 7). ◀

Exercise Set 10.7

1. Suppose that a game has a payoff matrix

$$A = \begin{bmatrix} -4 & 6 & -4 & 1 \\ 5 & -7 & 3 & 8 \\ -8 & 0 & 6 & -2 \end{bmatrix}$$

   (a) If players R and C use strategies

   $$\mathbf{p} = \begin{bmatrix} \frac{1}{2} & 0 & \frac{1}{2} \end{bmatrix} \quad \text{and} \quad \mathbf{q} = \begin{bmatrix} \frac{1}{4} \\ \frac{1}{4} \\ \frac{1}{4} \\ \frac{1}{4} \end{bmatrix}$$

   respectively, what is the expected payoff of the game?

   (b) If player C keeps his strategy fixed as in part (a), what strategy should player R choose to maximize his expected payoff?

   (c) If player R keeps her strategy fixed as in part (a), what strategy should player C choose to minimize the expected payoff to player R?

2. Construct a simple example to show that optimal strategies are not necessarily unique. For example, find a payoff matrix with several equal saddle points.

3. For the strictly determined games with the following payoff matrices, find optimal strategies for the two players, and find the values of the games.

   (a) $\begin{bmatrix} 5 & 2 \\ 7 & 3 \end{bmatrix}$

   (b) $\begin{bmatrix} -3 & -2 \\ 2 & 4 \\ -4 & 1 \end{bmatrix}$

   (c) $\begin{bmatrix} 2 & -2 & 0 \\ -6 & 0 & -5 \\ 5 & 2 & 3 \end{bmatrix}$

   (d) $\begin{bmatrix} -3 & 2 & -1 \\ -2 & -1 & 5 \\ -4 & 1 & 0 \\ -3 & 4 & 6 \end{bmatrix}$

4. For the 2 × 2 games with the following payoff matrices, find optimal strategies for the two players, and find the values of the games.

   (a) $\begin{bmatrix} 6 & 3 \\ -1 & 4 \end{bmatrix}$

   (b) $\begin{bmatrix} 40 & 20 \\ -10 & 30 \end{bmatrix}$

   (c) $\begin{bmatrix} 3 & 7 \\ -5 & 4 \end{bmatrix}$

   (d) $\begin{bmatrix} 3 & 5 \\ 5 & 2 \end{bmatrix}$

   (e) $\begin{bmatrix} 7 & -3 \\ -5 & -2 \end{bmatrix}$

5. Player R has two playing cards: a black ace and a red four. Player C also has two cards: a black two and a red three. Each player secretly selects one of his or her cards. If both selected cards are the same color, player C pays player R the sum of the face values in dollars. If the cards are different colors, player R pays player C the sum of the face values. What are optimal strategies for both players, and what is the value of the game?

6. Verify Equations (6), (7), and (8).

7. Verify the statement in the last paragraph of Example 3.

8. Show that the entries of the optimal strategies $\mathbf{p}^*$ and $\mathbf{q}^*$ given in Theorem 10.7.2 are numbers strictly between zero and one.

Section 10.7 Technology Exercises

The following exercises are designed to be solved using a technology utility. Typically, this will be MATLAB, *Mathematica*, Maple, Derive, or Mathcad, but it may also be some other type of linear algebra software or a scientific calculator with some linear algebra capabilities. For each exercise you will need to read the relevant documentation for the particular utility you are using. The goal of these exercises is to provide you with a basic proficiency with your technology utility. Once you have mastered the techniques in these exercises, you will be able to use your technology utility to solve many of the problems in the regular exercise sets.

T1. Consider a game between two players where each player can make up to n different moves ($n > 1$). If the ith move of player R and the jth move of player C are such that $i + j$ is even, then C pays R \$1. If $i + j$ is odd, then R pays C \$1. Assume that both players have the same strategy—that is, $\mathbf{p}_n = [\rho_i]_{1 \times n}$ and $\mathbf{q}_n = [\rho_i]_{n \times 1}$, where $\rho_1 + \rho_2 + \rho_3 + \cdots + \rho_n = 1$. Use a computer to show that

$$E(\mathbf{p}_2, \mathbf{q}_2) = (\rho_1 - \rho_2)^2$$
$$E(\mathbf{p}_3, \mathbf{q}_3) = (\rho_1 - \rho_2 + \rho_3)^2$$
$$E(\mathbf{p}_4, \mathbf{q}_4) = (\rho_1 - \rho_2 + \rho_3 - \rho_4)^2$$
$$E(\mathbf{p}_5, \mathbf{q}_5) = (\rho_1 - \rho_2 + \rho_3 - \rho_4 + \rho_5)^2$$

Using these results as a guide, prove in general that the expected payoff to player R is

$$E(\mathbf{p}_n, \mathbf{q}_n) = \left(\sum_{j=1}^{n} (-1)^{j+1} \rho_j \right)^2 \geq 0$$

which shows that in the long run, player R *will not lose* in this game.

T2. Consider a game between two players where each player can make up to n different moves ($n > 1$). If both players make the same move, then player C pays player R \$($n - 1$). However, if both players make different moves, then player R pays player C \$1. Assume that both players have the same strategy—that is, $\mathbf{p}_n = [\rho_i]_{1 \times n}$ and $\mathbf{q}_n = [\rho_i]_{n \times 1}$, where $\rho_1 + \rho_2 + \rho_3 + \cdots + \rho_n = 1$. Use a computer to show that

$$E(\mathbf{p}_2, \mathbf{q}_2) = \tfrac{1}{2}(\rho_1 - \rho_1)^2 + \tfrac{1}{2}(\rho_1 - \rho_2)^2 + \tfrac{1}{2}(\rho_2 - \rho_1)^2$$
$$+ \tfrac{1}{2}(\rho_2 - \rho_2)^2$$
$$E(\mathbf{p}_3, \mathbf{q}_3) = \tfrac{1}{2}(\rho_1 - \rho_1)^2 + \tfrac{1}{2}(\rho_1 - \rho_2)^2 + \tfrac{1}{2}(\rho_1 - \rho_3)^2$$
$$+ \tfrac{1}{2}(\rho_2 - \rho_1)^2 + \tfrac{1}{2}(\rho_2 - \rho_2)^2 + \tfrac{1}{2}(\rho_2 - \rho_3)^2$$
$$+ \tfrac{1}{2}(\rho_3 - \rho_1)^2 + \tfrac{1}{2}(\rho_3 - \rho_2)^2 + \tfrac{1}{2}(\rho_3 - \rho_3)^2$$
$$E(\mathbf{p}_4, \mathbf{q}_4) = \tfrac{1}{2}(\rho_1 - \rho_1)^2 + \tfrac{1}{2}(\rho_1 - \rho_2)^2 + \tfrac{1}{2}(\rho_1 - \rho_3)^2$$
$$+ \tfrac{1}{2}(\rho_1 - \rho_4)^2 + \tfrac{1}{2}(\rho_2 - \rho_1)^2 + \tfrac{1}{2}(\rho_2 - \rho_2)^2$$
$$+ \tfrac{1}{2}(\rho_2 - \rho_3)^2 + \tfrac{1}{2}(\rho_2 - \rho_4)^2 + \tfrac{1}{2}(\rho_3 - \rho_1)^2$$
$$+ \tfrac{1}{2}(\rho_3 - \rho_2)^2 + \tfrac{1}{2}(\rho_3 - \rho_3)^2 + \tfrac{1}{2}(\rho_3 - \rho_4)^2$$
$$+ \tfrac{1}{2}(\rho_4 - \rho_1)^2 + \tfrac{1}{2}(\rho_4 - \rho_2)^2 + \tfrac{1}{2}(\rho_4 - \rho_3)^2$$
$$+ \tfrac{1}{2}(\rho_4 - \rho_4)^2$$

Using these results as a guide, prove in general that the expected payoff to player R is

$$E(\mathbf{p}_n, \mathbf{q}_n) = \frac{1}{2} \sum_{i=1}^{n} \sum_{j=1}^{n} (\rho_i - \rho_j)^2 \geq 0$$

which shows that in the long run, player R *will not lose* in this game.

10.8 Leontief Economic Models

In this section we discuss two linear models for economic systems. Some results about nonnegative matrices are applied to determine equilibrium price structures and outputs necessary to satisfy demand.

> **PREREQUISITES:** Linear Systems
> Matrices

Economic Systems   Matrix theory has been very successful in describing the interrelations among prices, outputs, and demands in economic systems. In this section we discuss some simple models based on the ideas of Nobel laureate Wassily Leontief. We examine two different but related models: the closed or input-output model, and the open or production model. In each, we are given certain economic parameters that describe the interrelations between the "industries" in the economy under consideration. Using matrix theory, we then evaluate certain other parameters, such as prices or output levels, in order to satisfy a desired economic objective. We begin with the closed model.

Chapter 10 Applications of Linear Algebra

Leontief Closed (Input-Output) Model

First we present a simple example; then we proceed to the general theory of the model.

▶ **EXAMPLE 1** **An Input-Output Model**

Three homeowners—a carpenter, an electrician, and a plumber—agree to make repairs in their three homes. They agree to work a total of 10 days each according to the following schedule:

	Work Performed by		
	Carpenter	Electrician	Plumber
Days of Work in Home of Carpenter	2	1	6
Days of Work in Home of Electrician	4	5	1
Days of Work in Home of Plumber	4	4	3

For tax purposes, they must report and pay each other a reasonable daily wage, even for the work each does on his or her own home. Their normal daily wages are about $100, but they agree to adjust their respective daily wages so that each homeowner will come out even—that is, so that the total amount paid out by each is the same as the total amount each receives. We can set

$$p_1 = \text{daily wage of carpenter}$$
$$p_2 = \text{daily wage of electrician}$$
$$p_3 = \text{daily wage of plumber}$$

To satisfy the "equilibrium" condition that each homeowner comes out even, we require that

$$\text{total expenditures} = \text{total income}$$

for each of the homeowners for the 10-day period. For example, the carpenter pays a total of $2p_1 + p_2 + 6p_3$ for the repairs in his own home and receives a total income of $10p_1$ for the repairs that he performs on all three homes. Equating these two expressions then gives the first of the following three equations:

$$2p_1 + p_2 + 6p_3 = 10p_1$$
$$4p_1 + 5p_2 + p_3 = 10p_2$$
$$4p_1 + 4p_2 + 3p_3 = 10p_3$$

The remaining two equations are the equilibrium equations for the electrician and the plumber. Dividing these equations by 10 and rewriting them in matrix form yields

$$\begin{bmatrix} .2 & .1 & .6 \\ .4 & .5 & .1 \\ .4 & .4 & .3 \end{bmatrix} \begin{bmatrix} p_1 \\ p_2 \\ p_3 \end{bmatrix} = \begin{bmatrix} p_1 \\ p_2 \\ p_3 \end{bmatrix} \quad (1)$$

Equation (1) can be rewritten as a homogeneous system by subtracting the left side from the right side to obtain

$$\begin{bmatrix} .8 & -.1 & -.6 \\ -.4 & .5 & -.1 \\ -.4 & -.4 & .7 \end{bmatrix} \begin{bmatrix} p_1 \\ p_2 \\ p_3 \end{bmatrix} = \begin{bmatrix} 0 \\ 0 \\ 0 \end{bmatrix}$$

The solution of this homogeneous system is found to be (verify)

$$\begin{bmatrix} p_1 \\ p_2 \\ p_3 \end{bmatrix} = s \begin{bmatrix} 31 \\ 32 \\ 36 \end{bmatrix}$$

where s is an arbitrary constant. This constant is a scale factor, which the homeowners may choose for their convenience. For example, they can set $s = 3$ so that the corresponding daily wages—\$93, \$96, and \$108—are about \$100. ◀

This example illustrates the salient features of the Leontief input-output model of a closed economy. In the basic Equation (1), each column sum of the coefficient matrix is 1, corresponding to the fact that each of the homeowners' "output" of labor is completely distributed among these same homeowners in the proportions given by the entries in the column. Our problem is to determine suitable "prices" for these outputs so as to put the system in equilibrium—that is, so that each homeowner's total expenditures equal his or her total income.

In the general model we have an economic system consisting of a finite number of "industries," which we number as industries $1, 2, \ldots, k$. Over some fixed period of time, each industry produces an "output" of some good or service that is completely utilized in a predetermined manner by the k industries. An important problem is to find suitable "prices" to be charged for these k outputs so that for each industry, total expenditures equal total income. Such a price structure represents an equilibrium position for the economy.

For the fixed time period in question, let us set

p_i = price charged by the ith industry for its total output

e_{ij} = fraction of the total output of the jth industry purchased by the ith industry

for $i, j = 1, 2, \ldots, k$. By definition, we have

(i) $p_i \geq 0$, $i = 1, 2, \ldots, k$
(ii) $e_{ij} \geq 0$, $i, j = 1, 2, \ldots, k$
(iii) $e_{1j} + e_{2j} + \cdots + e_{kj} = 1$, $j = 1, 2, \ldots, k$

With these quantities, we form the ***price vector***

$$\mathbf{p} = \begin{bmatrix} p_1 \\ p_2 \\ \vdots \\ p_k \end{bmatrix}$$

and the ***exchange matrix*** or ***input-output matrix***

$$E = \begin{bmatrix} e_{11} & e_{12} & \cdots & e_{1k} \\ e_{21} & e_{22} & \cdots & e_{2k} \\ \vdots & \vdots & & \vdots \\ e_{k1} & e_{k2} & \cdots & e_{kk} \end{bmatrix}$$

Condition (iii) expresses the fact that all the column sums of the exchange matrix are 1.

As in the example, in order that the expenditures of each industry be equal to its income, the following matrix equation must be satisfied [see (1)]:

$$E\mathbf{p} = \mathbf{p} \qquad (2)$$

or

$$(I - E)\mathbf{p} = \mathbf{0} \qquad (3)$$

Equation (3) is a homogeneous linear system for the price vector **p**. It will have a nontrivial solution if and only if the determinant of its coefficient matrix $I - E$ is zero. In Exercise 7 we ask you to show that this is the case for any exchange matrix E. Thus, (3) always has nontrivial solutions for the price vector **p**.

Actually, for our economic model to make sense, we need more than just the fact that (3) has nontrivial solutions for **p**. We also need the prices p_i of the k outputs to be nonnegative numbers. We express this condition as $\mathbf{p} \geq 0$. (In general, if A is any vector or matrix, the notation $A \geq 0$ means that every entry of A is nonnegative, and the notation $A > 0$ means that every entry of A is positive. Similarly, $A \geq B$ means $A - B \geq 0$, and $A > B$ means $A - B > 0$.) To show that (3) has a nontrivial solution for which $\mathbf{p} \geq 0$ is a bit more difficult than showing merely that some nontrivial solution exists. But it is true, and we state this fact without proof in the following theorem.

THEOREM 10.8.1 *If E is an exchange matrix, then $E\mathbf{p} = \mathbf{p}$ always has a nontrivial solution **p** whose entries are nonnegative.*

Let us consider a few simple examples of this theorem.

▶ **EXAMPLE 2** **Using Theorem 10.8.1**

Let
$$E = \begin{bmatrix} \frac{1}{2} & 0 \\ \frac{1}{2} & 1 \end{bmatrix}$$

Then $(I - E)\mathbf{p} = \mathbf{0}$ is
$$\begin{bmatrix} \frac{1}{2} & 0 \\ -\frac{1}{2} & 0 \end{bmatrix} \begin{bmatrix} p_1 \\ p_2 \end{bmatrix} = \begin{bmatrix} 0 \\ 0 \end{bmatrix}$$

which has the general solution
$$\mathbf{p} = s \begin{bmatrix} 0 \\ 1 \end{bmatrix}$$

where s is an arbitrary constant. We then have nontrivial solutions $\mathbf{p} \geq 0$ for any $s > 0$.

▶ **EXAMPLE 3** **Using Theorem 10.8.1**

Let
$$E = \begin{bmatrix} 1 & 0 \\ 0 & 1 \end{bmatrix}$$

Then $(I - E)\mathbf{p} = \mathbf{0}$ has the general solution
$$\mathbf{p} = s \begin{bmatrix} 1 \\ 0 \end{bmatrix} + t \begin{bmatrix} 0 \\ 1 \end{bmatrix}$$

where s and t are independent arbitrary constants. Nontrivial solutions $\mathbf{p} \geq 0$ then result from any $s \geq 0$ and $t \geq 0$, not both zero. ◀

Example 2 indicates that in some situations one of the prices must be zero in order to satisfy the equilibrium condition. Example 3 indicates that there may be several linearly independent price structures available. Neither of these situations describes a truly interdependent economic structure. The following theorem gives sufficient conditions for both cases to be excluded.

THEOREM 10.8.2 *Let E be an exchange matrix such that for some positive integer m all the entries of E^m are positive. Then there is exactly one linearly independent solution of $(I - E)\mathbf{p} = \mathbf{0}$, and it may be chosen so that all its entries are positive.*

We will not give a proof of this theorem. If you have read Section 10.5 on Markov chains, observe that this theorem is essentially the same as Theorem 10.5.4. What we are calling exchange matrices in this section were called stochastic or Markov matrices in Section 10.5.

▶ **EXAMPLE 4** Using Theorem 10.8.2

The exchange matrix in Example 1 was

$$E = \begin{bmatrix} .2 & .1 & .6 \\ .4 & .5 & .1 \\ .4 & .4 & .3 \end{bmatrix}$$

Because $E > 0$, the condition $E^m > 0$ in Theorem 10.8.2 is satisfied for $m = 1$. Consequently, we are guaranteed that there is exactly one linearly independent solution of $(I - E)\mathbf{p} = \mathbf{0}$, and it can be chosen so that $\mathbf{p} > 0$. In that example, we found that

$$\mathbf{p} = \begin{bmatrix} 31 \\ 32 \\ 36 \end{bmatrix}$$

is such a solution. ◀

Leontief Open (Production) Model

In contrast with the closed model, in which the outputs of k industries are distributed only among themselves, the open model attempts to satisfy an outside demand for the outputs. Portions of these outputs can still be distributed among the industries themselves, to keep them operating, but there is to be some excess, some net production, with which to satisfy the outside demand. In the closed model the outputs of the industries are fixed, and our objective is to determine prices for these outputs so that the equilibrium condition, that expenditures equal incomes, is satisfied. In the open model it is the prices that are fixed, and our objective is to determine levels of the outputs of the industries needed to satisfy the outside demand. We will measure the levels of the outputs in terms of their economic values using the fixed prices. To be precise, over some fixed period of time, let

x_i = monetary value of the total output of the ith industry

d_i = monetary value of the output of the ith industry needed to satisfy the outside demand

c_{ij} = monetary value of the output of the ith industry needed by the jth industry to produce one unit of monetary value of its own output

With these quantities, we define the **production vector**

$$\mathbf{x} = \begin{bmatrix} x_1 \\ x_2 \\ \vdots \\ x_k \end{bmatrix}$$

the ***demand vector***

$$\mathbf{d} = \begin{bmatrix} d_1 \\ d_2 \\ \vdots \\ d_k \end{bmatrix}$$

and the ***consumption matrix***

$$C = \begin{bmatrix} c_{11} & c_{12} & \cdots & c_{1k} \\ c_{21} & c_{22} & \cdots & c_{2k} \\ \vdots & \vdots & & \vdots \\ c_{k1} & c_{k2} & \cdots & c_{kk} \end{bmatrix}$$

By their nature, we have that

$$\mathbf{x} \geq 0, \quad \mathbf{d} \geq 0, \quad \text{and} \quad C \geq 0$$

From the definition of c_{ij} and x_j, it can be seen that the quantity

$$c_{i1}x_1 + c_{i2}x_2 + \cdots + c_{ik}x_k$$

is the value of the output of the ith industry needed by all k industries to produce a total output specified by the production vector $\mathbf{x}$. Because this quantity is simply the ith entry of the column vector $C\mathbf{x}$, we can say further that the ith entry of the column vector

$$\mathbf{x} - C\mathbf{x}$$

is the value of the excess output of the ith industry available to satisfy the outside demand. The value of the outside demand for the output of the ith industry is the ith entry of the demand vector $\mathbf{d}$. Consequently, we are led to the following equation

$$\mathbf{x} - C\mathbf{x} = \mathbf{d}$$

or

$$(I - C)\mathbf{x} = \mathbf{d} \tag{4}$$

for the demand to be exactly met, without any surpluses or shortages. Thus, given C and $\mathbf{d}$, our objective is to find a production vector $\mathbf{x} \geq 0$ that satisfies Equation (4).

▶ **EXAMPLE 5** **Production Vector for a Town**

A town has three main industries: a coal-mining operation, an electric power-generating plant, and a local railroad. To mine \$1 of coal, the mining operation must purchase \$.25 of electricity to run its equipment and \$.25 of transportation for its shipping needs. To produce \$1 of electricity, the generating plant requires \$.65 of coal for fuel, \$.05 of its own electricity to run auxiliary equipment, and \$.05 of transportation. To provide \$1 of transportation, the railroad requires \$.55 of coal for fuel and \$.10 of electricity for its auxiliary equipment. In a certain week the coal-mining operation receives orders for \$50,000 of coal from outside the town, and the generating plant receives orders for \$25,000 of electricity from outside. There is no outside demand for the local railroad. How much must each of the three industries produce in that week to exactly satisfy their own demand and the outside demand?

Solution For the one-week period let

$$x_1 = \text{value of total output of coal-mining operation}$$
$$x_2 = \text{value of total output of power-generating plant}$$
$$x_3 = \text{value of total output of local railroad}$$

From the information supplied, the consumption matrix of the system is

$$C = \begin{bmatrix} 0 & .65 & .55 \\ .25 & .05 & .10 \\ .25 & .05 & 0 \end{bmatrix}$$

The linear system $(I - C)\mathbf{x} = \mathbf{d}$ is then

$$\begin{bmatrix} 1.00 & -.65 & -.55 \\ -.25 & .95 & -.10 \\ -.25 & -.05 & 1.00 \end{bmatrix} \begin{bmatrix} x_1 \\ x_2 \\ x_3 \end{bmatrix} = \begin{bmatrix} 50,000 \\ 25,000 \\ 0 \end{bmatrix}$$

The coefficient matrix on the left is invertible, and the solution is given by

$$\mathbf{x} = (I - C)^{-1}\mathbf{d} = \frac{1}{503} \begin{bmatrix} 756 & 542 & 470 \\ 220 & 690 & 190 \\ 200 & 170 & 630 \end{bmatrix} \begin{bmatrix} 50,000 \\ 25,000 \\ 0 \end{bmatrix} = \begin{bmatrix} 102,087 \\ 56,163 \\ 28,330 \end{bmatrix}$$

Thus, the total output of the coal-mining operation should be \$102,087, the total output of the power-generating plant should be \$56,163, and the total output of the railroad should be \$28,330. ◀

Let us reconsider Equation (4):

$$(I - C)\mathbf{x} = \mathbf{d}$$

If the square matrix $I - C$ is invertible, we can write

$$\mathbf{x} = (I - C)^{-1}\mathbf{d} \tag{5}$$

In addition, if the matrix $(I - C)^{-1}$ has only nonnegative entries, then we are guaranteed that for any $\mathbf{d} \geq 0$, Equation (5) has a unique nonnegative solution for $\mathbf{x}$. This is a particularly desirable situation, as it means that any outside demand can be met. The terminology used to describe this case is given in the following definition.

> **DEFINITION 1** A consumption matrix C is said to be **productive** if $(I - C)^{-1}$ exists and
>
> $$(I - C)^{-1} \geq 0$$

We will now consider some simple criteria that guarantee that a consumption matrix is productive. The first is given in the following theorem.

> **THEOREM 10.8.3 Productive Consumption Matrix**
>
> A consumption matrix C is productive if and only if there is some production vector $\mathbf{x} \geq 0$ such that $\mathbf{x} > C\mathbf{x}$.

(The proof is outlined in Exercise 9.) The condition $\mathbf{x} > C\mathbf{x}$ means that there is some production schedule possible such that each industry produces more than it consumes.

Theorem 10.8.3 has two interesting corollaries. Suppose that all the row sums of C are less than 1. If

$$\mathbf{x} = \begin{bmatrix} 1 \\ 1 \\ \vdots \\ 1 \end{bmatrix}$$

then $C\mathbf{x}$ is a column vector whose entries are these row sums. Therefore, $\mathbf{x} > C\mathbf{x}$, and the condition of Theorem 10.8.3 is satisfied. Thus, we arrive at the following corollary:

COROLLARY 10.8.4 *A consumption matrix is productive if each of its row sums is less than 1.*

As we ask you to show in Exercise 8, this corollary leads to the following:

COROLLARY 10.8.5 *A consumption matrix is productive if each of its column sums is less than 1.*

Recalling the definition of the entries of the consumption matrix C, we see that the jth column sum of C is the total value of the outputs of all k industries needed to produce one unit of value of output of the jth industry. The jth industry is thus said to be **profitable** if that jth column sum is less than 1. In other words, Corollary 10.8.5 says that a consumption matrix is productive if all k industries in the economic system are profitable.

▶ **EXAMPLE 6 Using Corollary 10.8.5**

The consumption matrix in Example 5 was

$$C = \begin{bmatrix} 0 & .65 & .55 \\ .25 & .05 & .10 \\ .25 & .05 & 0 \end{bmatrix}$$

All three column sums in this matrix are less than 1, so all three industries are profitable. Consequently, by Corollary 10.8.5, the consumption matrix C is productive. This can also be seen in the calculations in Example 5, as $(I - C)^{-1}$ is nonnegative. ◀

Exercise Set 10.8

1. For the following exchange matrices, find nonnegative price vectors that satisfy the equilibrium condition (3).

   (a) $\begin{bmatrix} \frac{1}{2} & \frac{1}{3} \\ \frac{1}{2} & \frac{2}{3} \end{bmatrix}$

   (b) $\begin{bmatrix} \frac{1}{2} & 0 & \frac{1}{2} \\ \frac{1}{3} & 0 & \frac{1}{2} \\ \frac{1}{6} & 1 & 0 \end{bmatrix}$

   (c) $\begin{bmatrix} .35 & .50 & .30 \\ .25 & .20 & .30 \\ .40 & .30 & .40 \end{bmatrix}$

2. Using Theorem 10.8.3 and its corollaries, show that each of the following consumption matrices is productive.

   (a) $\begin{bmatrix} .8 & .1 \\ .3 & .6 \end{bmatrix}$

   (b) $\begin{bmatrix} .70 & .30 & .25 \\ .20 & .40 & .25 \\ .05 & .15 & .25 \end{bmatrix}$

   (c) $\begin{bmatrix} .7 & .3 & .2 \\ .1 & .4 & .3 \\ .2 & .4 & .1 \end{bmatrix}$

3. Using Theorem 10.8.2, show that there is only one linearly independent price vector for the closed economic system with exchange matrix

   $$E = \begin{bmatrix} 0 & .2 & .5 \\ 1 & .2 & .5 \\ 0 & .6 & 0 \end{bmatrix}$$

4. Three neighbors have backyard vegetable gardens. Neighbor A grows tomatoes, neighbor B grows corn, and neighbor C grows lettuce. They agree to divide their crops among themselves as follows: A gets $\frac{1}{2}$ of the tomatoes, $\frac{1}{3}$ of the corn, and $\frac{1}{4}$ of the lettuce. B gets $\frac{1}{3}$ of the tomatoes, $\frac{1}{3}$ of the corn, and $\frac{1}{4}$ of the lettuce. C gets $\frac{1}{6}$ of the tomatoes, $\frac{1}{3}$ of the corn, $\frac{1}{2}$ of the lettuce. What prices should the neighbors assign to their respective crops if the equilibrium condition of a closed economy is to be satisfied, and if the lowest-priced crop is to have a price of $100?

5. Three engineers—a civil engineer (CE), an electrical engineer (EE), and a mechanical engineer (ME)—each have a consulting firm. The consulting they do is of a multidisciplinary nature,

so they buy a portion of each others' services. For each $1 of consulting the CE does, she buys $.10 of the EE's services and $.30 of the ME's services. For each $1 of consulting the EE does, she buys $.20 of the CE's services and $.40 of the ME's services. And for each $1 of consulting the ME does, she buys $.30 of the CE's services and $.40 of the EE's services. In a certain week the CE receives outside consulting orders of $500, the EE receives outside consulting orders of $700, and the ME receives outside consulting orders of $600. What dollar amount of consulting does each engineer perform in that week?

6. (a) Suppose that the demand d_i for the output of the ith industry increases by one unit. Explain why the ith column of the matrix $(I - C)^{-1}$ is the increase that must be made to the production vector $\mathbf{x}$ to satisfy this additional demand.

   (b) Referring to Example 5, use the result in part (a) to determine the increase in the value of the output of the coal-mining operation needed to satisfy a demand of one additional unit in the value of the output of the power-generating plant.

7. Using the fact that the column sums of an exchange matrix E are all 1, show that the column sums of $I - E$ are zero. From this, show that $I - E$ has zero determinant, and so $(I - E)\mathbf{p} = \mathbf{0}$ has nontrivial solutions for $\mathbf{p}$.

8. Show that Corollary 10.8.5 follows from Corollary 10.8.4. [*Hint:* Use the fact that $(A^T)^{-1} = (A^{-1})^T$ for any invertible matrix A.]

9. (*Calculus required*) Prove Theorem 10.8.3 as follows:

   (a) Prove the "only if" part of the theorem; that is, show that if C is a productive consumption matrix, then there is a vector $\mathbf{x} \geq 0$ such that $\mathbf{x} > C\mathbf{x}$.

   (b) Prove the "if" part of the theorem as follows:

   Step 1. Show that if there is a vector $\mathbf{x}^* \geq 0$ such that $C\mathbf{x}^* < \mathbf{x}^*$, then $\mathbf{x}^* > 0$.

   Step 2. Show that there is a number λ such that $0 < \lambda < 1$ and $C\mathbf{x}^* < \lambda \mathbf{x}^*$.

   Step 3. Show that $C^n \mathbf{x}^* < \lambda^n \mathbf{x}^*$ for $n = 1, 2, \ldots$.

   Step 4. Show that $C^n \to 0$ as $n \to \infty$.

   Step 5. By multiplying out, show that

   $$(I - C)(I + C + C^2 + \cdots + C^{n-1}) = I - C^n$$

   for $n = 1, 2, \ldots$.

   Step 6. By letting $n \to \infty$ in Step 5, show that the matrix infinite sum

   $$S = I + C + C^2 + \cdots$$

   exists and that $(I - C)S = I$.

   Step 7. Show that $S \geq 0$ and that $S = (I - C)^{-1}$.

   Step 8. Show that C is a productive consumption matrix.

Section 10.8 Technology Exercises

The following exercises are designed to be solved using a technology utility. Typically, this will be MATLAB, *Mathematica*, Maple, Derive, or Mathcad, but it may also be some other type of linear algebra software or a scientific calculator with some linear algebra capabilities. For each exercise you will need to read the relevant documentation for the particular utility you are using. The goal of these exercises is to provide you with a basic proficiency with your technology utility. Once you have mastered the techniques in these exercises, you will be able to use your technology utility to solve many of the problems in the regular exercise sets.

T1. Consider a sequence of exchange matrices $\{E_2, E_3, E_4, E_5, \ldots, E_n\}$, where

$$E_2 = \begin{bmatrix} 0 & \frac{1}{2} \\ 1 & \frac{1}{2} \end{bmatrix}, \quad E_3 = \begin{bmatrix} 0 & \frac{1}{2} & \frac{1}{3} \\ 1 & 0 & \frac{1}{3} \\ 0 & \frac{1}{2} & \frac{1}{3} \end{bmatrix},$$

$$E_4 = \begin{bmatrix} 0 & \frac{1}{2} & \frac{1}{3} & \frac{1}{4} \\ 1 & 0 & \frac{1}{3} & \frac{1}{4} \\ 0 & \frac{1}{2} & 0 & \frac{1}{4} \\ 0 & 0 & \frac{1}{3} & \frac{1}{4} \end{bmatrix}, \quad E_5 = \begin{bmatrix} 0 & \frac{1}{2} & \frac{1}{3} & \frac{1}{4} & \frac{1}{5} \\ 1 & 0 & \frac{1}{3} & \frac{1}{4} & \frac{1}{5} \\ 0 & \frac{1}{2} & 0 & \frac{1}{4} & \frac{1}{5} \\ 0 & 0 & \frac{1}{3} & 0 & \frac{1}{5} \\ 0 & 0 & 0 & \frac{1}{4} & \frac{1}{5} \end{bmatrix}$$

and so on. Use a computer to show that $E_2^2 > 0_2$, $E_3^3 > 0_3$, $E_4^4 > 0_4$, $E_5^5 > 0_5$, and make the conjecture that although $E_n^n > 0_n$ is true, $E_n^k > 0_n$ is not true for $k = 1, 2, 3, \ldots, n - 1$. Next, use a computer to determine the vectors $\mathbf{p}_n$ such that $E_n \mathbf{p}_n = \mathbf{p}_n$ (for $n = 2, 3, 4, 5, 6$), and then see if you can discover a pattern that would allow you to compute $\mathbf{p}_{n+1}$ easily from $\mathbf{p}_n$. Test your discovery by first constructing $\mathbf{p}_8$ from

$$\mathbf{p}_7 = \begin{bmatrix} 2520 \\ 3360 \\ 1890 \\ 672 \\ 175 \\ 36 \\ 7 \end{bmatrix}$$

and then checking to see whether $E_8 \mathbf{p}_8 = \mathbf{p}_8$.

T2. Consider an open production model having n industries with $n > 1$. In order to produce $1 of its own output, the jth industry must spend $(1/n)$ for the output of the ith industry (for all $i \neq j$), but the jth industry (for all $j = 1, 2, 3, \ldots, n$) spends nothing for its own output. Construct the consumption matrix

C_n, show that it is productive, and determine an expression for $(I_n - C_n)^{-1}$. In determining an expression for $(I_n - C_n)^{-1}$, use a computer to study the cases when $n = 2, 3, 4,$ and 5; then make a conjecture and prove your conjecture to be true. [*Hint:* If $F_n = [1]_{n \times n}$ (i.e., the $n \times n$ matrix with every entry equal to 1), first show that

$$F_n^2 = n F_n$$

and then express your value of $(I_n - C_n)^{-1}$ in terms of n, I_n, and F_n.]

10.9 Forest Management

In this section we discuss a matrix model for the management of a forest where trees are grouped into classes according to height. The optimal sustainable yield of a periodic harvest is calculated when the trees of different height classes can have different economic values.

PREREQUISITES: Matrix Operations

Optimal Sustainable Yield Our objective is to introduce a simplified model for the sustainable harvesting of a forest whose trees are classified by height. The height of a tree is assumed to determine its economic value when it is cut down and sold. Initially, there is a distribution of trees of various heights. The forest is then allowed to grow for a certain period of time, after which some of the trees of various heights are harvested. The trees left unharvested are to be of the same height configuration as the original forest, so that the harvest is sustainable. As we will see, there are many such sustainable harvesting procedures. We want to find one for which the total economic value of all the trees removed is as large as possible. This determines the **optimal sustainable yield** of the forest and is the largest yield that can be attained continually without depleting the forest.

The Model Suppose that a harvester has a forest of Douglas fir trees that are to be sold as Christmas trees year after year. Every December the harvester cuts down some of the trees to be sold. For each tree cut down, a seedling is planted in its place. In this way the total number of trees in the forest is always the same. (In this simplified model, we will not take into account trees that die between harvests. We assume that every seedling planted survives and grows until it is harvested.)

In the marketplace, trees of different heights have different economic values. Suppose that there are n different price classes corresponding to certain height intervals, as shown in Table 1 and Figure 10.9.1. The first class consists of seedlings with heights in the interval $[0, h_1)$, and these seedlings are of no economic value. The nth class consists of trees with heights greater than or equal to h_{n-1}.

Table 1

Class	Value (dollars)	Height Interval
1 (seedlings)	None	$[0, h_1)$
2	p_2	$[h_1, h_2)$
3	p_3	$[h_2, h_3)$
$\vdots$	$\vdots$	$\vdots$
$n-1$	p_{n-1}	$[h_{n-2}, h_{n-1})$
n	p_n	$[h_{n-1},)$

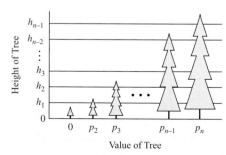

▲ Figure 10.9.1

Let x_i ($i = 1, 2, \ldots, n$) be the number of trees within the ith class that remain after each harvest. We form a column vector with the numbers and call it the **nonharvest vector**:

$$\mathbf{x} = \begin{bmatrix} x_1 \\ x_2 \\ \vdots \\ x_n \end{bmatrix}$$

For a sustainable harvesting policy, the forest is to be returned after each harvest to the fixed configuration given by the nonharvest vector $\mathbf{x}$. Part of our problem is to find those nonharvest vectors $\mathbf{x}$ for which sustainable harvesting is possible.

Because the total number of trees in the forest is fixed, we can set

$$x_1 + x_2 + \cdots + x_n = s \qquad (1)$$

where s is predetermined by the amount of land available and the amount of space each tree requires. Referring to Figure 10.9.2, we have the following situation. The forest configuration is given by the vector $\mathbf{x}$ after each harvest. Between harvests the trees grow and produce a new forest configuration before each harvest. A certain number of trees are removed from each class at the harvest. Finally, a seedling is planted in place of each tree removed, to return the forest again to the configuration $\mathbf{x}$.

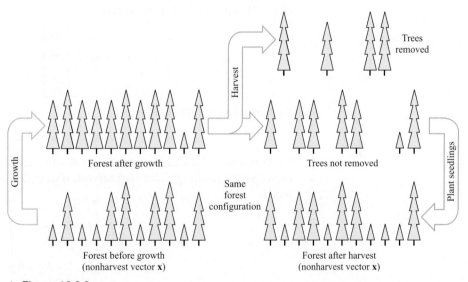

▲ Figure 10.9.2

Consider first the growth of the forest between harvests. During this period a tree in the ith class may grow and move up to a higher height class. Or its growth may be retarded for some reason, and it will remain in the same class. We consequently define the following growth parameters g_i for $i = 1, 2, \ldots, n - 1$:

g_i = the fraction of trees in the ith class that grow into
the $(i + 1)$-st class during a growth period

For simplicity we assume that a tree can move at most one height class upward in one growth period. With this assumption, we have

$1 - g_i$ = the fraction of trees in the ith class that remain in
the ith class during a growth period

With these $n-1$ growth parameters, we form the following $n \times n$ **growth matrix**:

$$G = \begin{bmatrix} 1-g_1 & 0 & 0 & \cdots & 0 \\ g_1 & 1-g_2 & 0 & \cdots & 0 \\ 0 & g_2 & 1-g_3 & \cdots & 0 \\ \vdots & \vdots & \vdots & & \vdots \\ 0 & 0 & 0 & \cdots & 1-g_{n-1} & 0 \\ 0 & 0 & 0 & \cdots & g_{n-1} & 1 \end{bmatrix} \quad (2)$$

Because the entries of the vector **x** are the numbers of trees in the n classes before the growth period, you can verify that the entries of the vector

$$G\mathbf{x} = \begin{bmatrix} (1-g_1)x_1 \\ g_1 x_1 + (1-g_2)x_2 \\ g_2 x_2 + (1-g_3)x_3 \\ \vdots \\ g_{n-2} x_{n-2} + (1-g_{n-1})x_{n-1} \\ g_{n-1} x_{n-1} + x_n \end{bmatrix} \quad (3)$$

are the numbers of trees in the n classes after the growth period.

Suppose that during the harvest we remove y_i ($i = 1, 2, \ldots, n$) trees from the ith class. We will call the column vector

$$\mathbf{y} = \begin{bmatrix} y_1 \\ y_2 \\ \vdots \\ y_n \end{bmatrix}$$

the **harvest vector**. Thus, a total of

$$y_1 + y_2 + \cdots + y_n$$

trees are removed at each harvest. This is also the total number of trees added to the first class (the new seedlings) after each harvest. If we define the following $n \times n$ **replacement matrix**

$$R = \begin{bmatrix} 1 & 1 & \cdots & 1 \\ 0 & 0 & \cdots & 0 \\ \vdots & \vdots & & \vdots \\ 0 & 0 & \cdots & 0 \end{bmatrix} \quad (4)$$

then the column vector

$$R\mathbf{y} = \begin{bmatrix} y_1 + y_2 + \cdots + y_n \\ 0 \\ 0 \\ \vdots \\ 0 \end{bmatrix} \quad (5)$$

specifies the configuration of trees planted after each harvest.

At this point we are ready to write the following equation, which characterizes a sustainable harvesting policy:

$$\begin{bmatrix} \text{configuration} \\ \text{at end of} \\ \text{growth period} \end{bmatrix} - [\text{harvest}] + \begin{bmatrix} \text{new seedling} \\ \text{replacement} \end{bmatrix} = \begin{bmatrix} \text{configuration} \\ \text{at beginning of} \\ \text{growth period} \end{bmatrix}$$

or mathematically,
$$Gx - y + Ry = x$$
This equation can be rewritten as
$$(I - R)y = (G - I)x \tag{6}$$
or more comprehensively as

$$\begin{bmatrix} 0 & -1 & -1 & \cdots & -1 & -1 \\ 0 & 1 & 0 & \cdots & 0 & 0 \\ 0 & 0 & 1 & \cdots & 0 & 0 \\ \vdots & \vdots & \vdots & & \vdots & \vdots \\ 0 & 0 & 0 & \cdots & 1 & 0 \\ 0 & 0 & 0 & \cdots & 0 & 1 \end{bmatrix} \begin{bmatrix} y_1 \\ y_2 \\ y_3 \\ \vdots \\ y_{n-1} \\ y_n \end{bmatrix}$$

$$= \begin{bmatrix} -g_1 & 0 & 0 & \cdots & 0 & 0 \\ g_1 & -g_2 & 0 & \cdots & 0 & 0 \\ 0 & g_2 & -g_3 & \cdots & 0 & 0 \\ \vdots & \vdots & \vdots & & \vdots & \vdots \\ 0 & 0 & 0 & \cdots & -g_{n-1} & 0 \\ 0 & 0 & 0 & \cdots & g_{n-1} & 0 \end{bmatrix} \begin{bmatrix} x_1 \\ x_2 \\ x_3 \\ \vdots \\ x_{n-1} \\ x_n \end{bmatrix}$$

We will refer to Equation (6) as the **sustainable harvesting condition**. Any vectors x and y with nonnegative entries, and such that $x_1 + x_2 + \cdots + x_n = s$, which satisfy this matrix equation, determine a sustainable harvesting policy for the forest. Note that if $y_1 > 0$, then the harvester is removing seedlings of no economic value and replacing them with new seedlings. Because there is no point in doing this, we assume that

$$y_1 = 0 \tag{7}$$

With this assumption, it can be verified that (6) is the matrix form of the following set of equations:

$$\begin{aligned} y_2 + y_3 + \cdots + y_n &= g_1 x_1 \\ y_2 &= g_1 x_1 - g_2 x_2 \\ y_3 &= g_2 x_2 - g_3 x_3 \\ &\vdots \\ y_{n-1} &= g_{n-2} x_{n-2} - g_{n-1} x_{n-1} \\ y_n &= g_{n-1} x_{n-1} \end{aligned} \tag{8}$$

Note that the first equation in (8) is the sum of the remaining $n - 1$ equations.

Because we must have $y_i \geq 0$ for $i = 2, 3, \ldots, n$, Equations (8) require that

$$g_1 x_1 \geq g_2 x_2 \geq \cdots \geq g_{n-1} x_{n-1} \geq 0 \tag{9}$$

Conversely, if x is a column vector with nonnegative entries that satisfy Equation (9), then (7) and (8) define a column vector y with nonnegative entries. Furthermore, x and y then satisfy the sustainable harvesting condition (6). In other words, a necessary and sufficient condition for a nonnegative column vector x to determine a forest configuration that is capable of sustainable harvesting is that its entries satisfy (9).

Optimal Sustainable Yield   Because we remove y_i trees from the ith class ($i = 2, 3, \ldots, n$) and each tree in the ith class has an economic value of p_i, the total yield of the harvest, Yld, is given by

$$Yld = p_2 y_2 + p_3 y_3 + \cdots + p_n y_n \tag{10}$$

594 Chapter 10 Applications of Linear Algebra

Using (8), we may substitute for the y_i's in (10) to obtain

$$Yld = p_2 g_1 x_1 + (p_3 - p_2) g_2 x_2 + \cdots + (p_n - p_{n-1}) g_{n-1} x_{n-1} \tag{11}$$

Combining (11), (1), and (9), we can now state the problem of maximizing the yield of the forest over all possible sustainable harvesting policies as follows:

Problem Find nonnegative numbers $x_1, x_2, \ldots, x_n$ that maximize

$$Yld = p_2 g_1 x_1 + (p_3 - p_2) g_2 x_2 + \cdots + (p_n - p_{n-1}) g_{n-1} x_{n-1}$$

subject to

$$x_1 + x_2 + \cdots + x_n = s$$

and

$$g_1 x_1 \geq g_2 x_2 \geq \cdots \geq g_{n-1} x_{n-1} \geq 0$$

As formulated above, this problem belongs to the field of linear programming. However, we will illustrate the following result, without linear programming theory, by actually exhibiting a sustainable harvesting policy.

THEOREM 10.9.1 Optimal Sustainable Yield

The optimal sustainable yield is achieved by harvesting all the trees from one particular height class and none of the trees from any other height class.

Let us first set

$$Yld_k = \text{yield obtained by harvesting all of the } k\text{th}$$
$$\text{class and none of the other classes}$$

The largest value of Yld_k for $k = 2, 3, \ldots, n$ will then be the optimal sustainable yield, and the corresponding value of k will be the class that should be completely harvested to attain the optimal sustainable yield. Because no class but the kth is harvested, we have

$$y_2 = y_3 = \cdots = y_{k-1} = y_{k+1} = \cdots = y_n = 0 \tag{12}$$

In addition, because all of the kth class is harvested, no trees are ever present in the height classes above the kth class. Thus,

$$x_k = x_{k+1} = \cdots = x_n = 0 \tag{13}$$

Substituting (12) and (13) into the sustainable harvesting condition (8) gives

$$\begin{aligned} y_k &= g_1 x_1 \\ 0 &= g_1 x_1 - g_2 x_2 \\ 0 &= g_2 x_2 - g_3 x_3 \\ &\vdots \\ 0 &= g_{k-2} x_{k-2} - g_{k-1} x_{k-1} \\ y_k &= g_{k-1} x_{k-1} \end{aligned} \tag{14}$$

Equations (14) can also be written as

$$y_k = g_1 x_1 = g_2 x_2 = \cdots = g_{k-1} x_{k-1} \tag{15}$$

from which it follows that

$$
\begin{aligned}
x_2 &= g_1 x_1 / g_2 \\
x_3 &= g_1 x_1 / g_3 \\
&\vdots \\
x_{k-1} &= g_1 x_1 / g_{k-1}
\end{aligned}
\tag{16}
$$

If we substitute Equations (13) and (16) into

$$x_1 + x_2 + \cdots + x_n = s$$

[which is Equation (1)], we can solve for x_1 and obtain

$$x_1 = \frac{s}{1 + \dfrac{g_1}{g_2} + \dfrac{g_1}{g_3} + \cdots + \dfrac{g_1}{g_{k-1}}} \tag{17}$$

For the yield Yld_k, we combine (10), (12), (15), and (17) to obtain

$$
\begin{aligned}
Yld_k &= p_2 y_2 + p_3 y_3 + \cdots + p_n y_n \\
&= p_k y_k \\
&= p_k g_1 x_1 \\
&= \frac{p_k s}{\dfrac{1}{g_1} + \dfrac{1}{g_2} + \cdots + \dfrac{1}{g_{k-1}}}
\end{aligned}
\tag{18}
$$

Equation (18) determines Yld_k in terms of the known growth and economic parameters for any $k = 2, 3, \ldots, n$. Thus, the optimal sustainable yield is found as follows.

> **THEOREM 10.9.2** **Finding the Optimal Sustainable Yield**
>
> *The optimal sustainable yield is the largest value of*
>
> $$\frac{p_k s}{\dfrac{1}{g_1} + \dfrac{1}{g_2} + \cdots + \dfrac{1}{g_{k-1}}}$$
>
> *for $k = 2, 3, \ldots, n$. The corresponding value of k is the number of the class that is completely harvested.*

In Exercise 4 we ask you to show that the nonharvest vector $\mathbf{x}$ for the optimal sustainable yield is

$$\mathbf{x} = \frac{s}{\dfrac{1}{g_1} + \dfrac{1}{g_2} + \cdots + \dfrac{1}{g_{k-1}}} \begin{bmatrix} 1/g_1 \\ 1/g_2 \\ \vdots \\ 1/g_{k-1} \\ 0 \\ 0 \\ \vdots \\ 0 \end{bmatrix} \tag{19}$$

Theorem 10.9.2 implies that it is not necessarily the highest-priced class of trees that should be totally cropped. The growth parameters g_i must also be taken into account to determine the optimal sustainable yield.

EXAMPLE 1 Using Theorem 10.9.2

For a Scots pine forest in Scotland with a growth period of six years, the following growth matrix was found (see M. B. Usher, "A Matrix Approach to the Management of Renewable Resources, with Special Reference to Selection Forests," *Journal of Applied Ecology*, vol. 3, 1966, pp. 355–367):

$$G = \begin{bmatrix} .72 & 0 & 0 & 0 & 0 & 0 \\ .28 & .69 & 0 & 0 & 0 & 0 \\ 0 & .31 & .75 & 0 & 0 & 0 \\ 0 & 0 & .25 & .77 & 0 & 0 \\ 0 & 0 & 0 & .23 & .63 & 0 \\ 0 & 0 & 0 & 0 & .37 & 1.00 \end{bmatrix}$$

Suppose that the prices of trees in the five tallest height classes are

$$p_2 = \$50, \quad p_3 = \$100, \quad p_4 = \$150, \quad p_5 = \$200, \quad p_6 = \$250$$

Which class should be completely harvested to obtain the optimal sustainable yield, and what is that yield?

Solution From matrix G we have that

$$g_1 = .28, \quad g_2 = .31, \quad g_3 = .25, \quad g_4 = .23, \quad g_5 = .37$$

Equation (18) then gives

$$Yld_2 = 50s/(.28^{-1}) = 14.0s$$
$$Yld_3 = 100s/(.28^{-1} + .31^{-1}) = 14.7s$$
$$Yld_4 = 150s/(.28^{-1} + .31^{-1} + .25^{-1}) = 13.9s$$
$$Yld_5 = 200s/(.28^{-1} + .31^{-1} + .25^{-1} + .23^{-1}) = 13.2s$$
$$Yld_6 = 250s/(.28^{-1} + .31^{-1} + .25^{-1} + .23^{-1} + .37^{-1}) = 14.0s$$

We see that Yld_3 is the largest of these five quantities, so from Theorem 10.9.2 the third class should be completely harvested every six years to maximize the sustainable yield. The corresponding optimal sustainable yield is $\$14.7s$, where s is the total number of trees in the forest. ◀

Exercise Set 10.9

1. A certain forest is divided into three height classes and has a growth matrix between harvests given by

$$G = \begin{bmatrix} \frac{1}{2} & 0 & 0 \\ \frac{1}{2} & \frac{1}{3} & 0 \\ 0 & \frac{2}{3} & 1 \end{bmatrix}$$

If the price of trees in the second class is $30 and the price of trees in the third class is $50, which class should be completely harvested to attain the optimal sustainable yield? What is the optimal yield if there are 1000 trees in the forest?

2. In Example 1, to what level must the price of trees in the fifth class rise so that the fifth class is the one to harvest completely in order to attain the optimal sustainable yield?

3. In Example 1, what must the ratio of the prices $p_2: p_3: p_4: p_5: p_6$ be in order that the yields Yld_k, $k = 2, 3, 4, 5, 6$, all be the same? (In this case, any sustainable harvesting policy will produce the same optimal sustainable yield.)

4. Derive Equation (19) for the nonharvest vector $\mathbf{x}$ corresponding to the optimal sustainable harvesting policy described in Theorem 10.9.2.

5. For the optimal sustainable harvesting policy described in Theorem 10.9.2, how many trees are removed from the forest during each harvest?

6. If all the growth parameters $g_1, g_2, \ldots, g_{n-1}$ in the growth matrix G are equal, what should the ratio of the prices $p_2: p_3: \ldots: p_n$ be in order that any sustainable harvesting policy be an optimal sustainable harvesting policy? (See Exercise 3.)

Section 10.9 Technology Exercises

The following exercises are designed to be solved using a technology utility. Typically, this will be MATLAB, *Mathematica*, Maple, Derive, or Mathcad, but it may also be some other type of linear algebra software or a scientific calculator with some linear algebra capabilities. For each exercise you will need to read the relevant documentation for the particular utility you are using. The goal of these exercises is to provide you with a basic proficiency with your technology utility. Once you have mastered the techniques in these exercises, you will be able to use your technology utility to solve many of the problems in the regular exercise sets.

T1. A particular forest has growth parameters given by

$$g_i = \frac{1}{i}$$

for $i = 1, 2, 3, \ldots, n-1$, where n (the total number of height classes) can be chosen as large as needed. Suppose that the value of a tree in the kth height interval is given by

$$p_k = a(k-1)^\rho$$

where a is a constant (in dollars) and ρ is a parameter satisfying $1 \leq \rho \leq 2$.

(a) Show that the yield Yld_k is given by

$$Yld_k = \frac{2a(k-1)^{\rho-1}s}{k}$$

(b) For

$$\rho = 1.0, \ 1.1, \ 1.2, \ 1.3, \ 1.4, \ 1.5, \ 1.6, \ 1.7, \ 1.8, \ 1.9$$

use a computer to determine the class number that should be completely harvested, and determine the optimal sustainable yield in each case. Make sure that you allow k to take on only integer values in your calculations.

(c) Repeat the calculations in part (b) using

$$\rho = 1.91, \ 1.92, \ 1.93, \ 1.94, \ 1.95, \\ 1.96, \ 1.97, \ 1.98, \ 1.99$$

(d) Show that if $\rho = 2$, then the optimal sustainable yield can never be larger than $2as$.

(e) Compare the values of k determined in parts (b) and (c) to $1/(2-\rho)$, and use some calculus to explain why

$$k \simeq \frac{1}{2-\rho}$$

T2. A particular forest has growth parameters given by

$$g_i = \frac{1}{2^i}$$

for $i = 1, 2, 3, \ldots, n-1$, where n (the total number of height classes) can be chosen as large as needed. Suppose that the value of a tree in the kth height interval is given by

$$p_k = a(k-1)^\rho$$

where a is a constant (in dollars) and ρ is a parameter satisfying $1 \leq \rho$.

(a) Show that the yield Yld_k is given by

$$Yld_k = \frac{a(k-1)^\rho s}{2^k - 2}$$

(b) For

$$\rho = 1, \ 2, \ 3, \ 4, \ 5, \ 6, \ 7, \ 8, \ 9, \ 10$$

use a computer to determine the class number that should be completely harvested in order to obtain an optimal yield, and determine the optimal sustainable yield in each case. Make sure that you allow k to take on only integer values in your calculations.

(c) Compare the values of k determined in part (b) to $1 + \rho/\ln(2)$ and use some calculus to explain why

$$k \simeq 1 + \frac{\rho}{\ln(2)}$$

10.10 Computer Graphics

In this section we assume that a view of a three-dimensional object is displayed on a video screen and show how matrix algebra can be used to obtain new views of the object by rotation, translation, and scaling.

> **PREREQUISITES:** Matrix Algebra
> Analytic Geometry

Visualization of a Three-Dimensional Object

Suppose that we want to visualize a three-dimensional object by displaying various views of it on a video screen. The object we have in mind to display is to be determined by a finite number of straight line segments. As an example, consider the truncated right pyramid with hexagonal base illustrated in Figure 10.10.1. We first introduce an xyz-coordinate system in which to embed the object. As in Figure 10.10.1, we orient the coordinate system so that its origin is at the center of the video screen and the xy-plane coincides with the plane of the screen. Consequently, an observer will see only the projection of the view of the three-dimensional object onto the two-dimensional xy-plane.

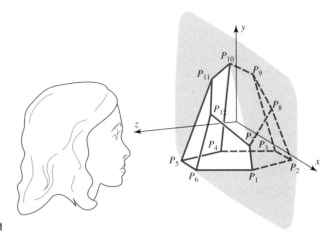

▶ **Figure 10.10.1**

In the xyz-coordinate system, the endpoints $P_1, P_2, \ldots, P_n$ of the straight line segments that determine the view of the object will have certain coordinates—say,

$$(x_1, y_1, z_1), \quad (x_2, y_2, z_2), \ldots, \quad (x_n, y_n, z_n)$$

These coordinates, together with a specification of which pairs are to be connected by straight line segments, are to be stored in the memory of the video display system. For example, assume that the 12 vertices of the truncated pyramid in Figure 10.10.1 have the following coordinates (the screen is 4 units wide by 3 units high):

$P_1: (1.000, -.800, .000)$, $P_2: (.500, -.800, -.866)$,
$P_3: (-.500, -.800, -.866)$, $P_4: (-1.000, -.800, .000)$,
$P_5: (-.500, -.800, .866)$, $P_6: (.500, -.800, .866)$,
$P_7: (.840, -.400, .000)$, $P_8: (.315, .125, -.546)$,
$P_9: (-.210, .650, -.364)$, $P_{10}: (-.360, .800, .000)$,
$P_{11}: (-.210, .650, .364)$, $P_{12}: (.315, .125, .546)$

These 12 vertices are connected pairwise by 18 straight line segments as follows, where $P_i \leftrightarrow P_j$ denotes that point P_i is connected to point P_j:

$P_1 \leftrightarrow P_2$, $P_2 \leftrightarrow P_3$, $P_3 \leftrightarrow P_4$, $P_4 \leftrightarrow P_5$, $P_5 \leftrightarrow P_6$, $P_6 \leftrightarrow P_1$,
$P_7 \leftrightarrow P_8$, $P_8 \leftrightarrow P_9$, $P_9 \leftrightarrow P_{10}$, $P_{10} \leftrightarrow P_{11}$, $P_{11} \leftrightarrow P_{12}$, $P_{12} \leftrightarrow P_7$,
$P_1 \leftrightarrow P_7$, $P_2 \leftrightarrow P_8$, $P_3 \leftrightarrow P_9$, $P_4 \leftrightarrow P_{10}$, $P_5 \leftrightarrow P_{11}$, $P_6 \leftrightarrow P_{12}$

In View 1 these 18 straight line segments are shown as they would appear on the video screen. It should be noticed that only the x- and y-coordinates of the vertices are needed by the video display system to draw the view, because only the projection of the object onto the xy-plane is displayed. However, we must keep track of the z-coordinates to carry out certain transformations discussed later.

We now show how to form new views of the object by scaling, translating, or rotating the initial view. We first construct a $3 \times n$ matrix P, referred to as the *coordinate matrix of the view*, whose columns are the coordinates of the n points of a view:

$$P = \begin{bmatrix} x_1 & x_2 & \cdots & x_n \\ y_1 & y_2 & \cdots & y_n \\ z_1 & z_2 & \cdots & z_n \end{bmatrix}$$

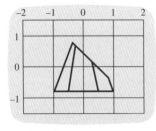

▶ **View 1**

For example, the coordinate matrix P corresponding to View 1 is the 3×12 matrix

$$\begin{bmatrix} 1.000 & .500 & -.500 & -1.000 & -.500 & .500 & .840 & .315 & -.210 & -.360 & -.210 & .315 \\ -.800 & -.800 & -.800 & -.800 & -.800 & -.800 & -.400 & .125 & .650 & .800 & .650 & .125 \\ .000 & -.866 & -.866 & .000 & .866 & .866 & .000 & -.546 & -.364 & .000 & .364 & .546 \end{bmatrix}$$

We will show below how to transform the coordinate matrix P of a view to a new coordinate matrix P' corresponding to a new view of the object. The straight line segments connecting the various points move with the points as they are transformed. In this way, each view is uniquely determined by its coordinate matrix once we have specified which pairs of points in the original view are to be connected by straight lines.

Scaling The first type of transformation we consider consists of scaling a view along the x, y, and z directions by factors of α, β, and γ, respectively. By this we mean that if a point P_i has coordinates (x_i, y_i, z_i) in the original view, it is to move to a new point P'_i with coordinates $(\alpha x_i, \beta y_i, \gamma z_i)$ in the new view. This has the effect of transforming a unit cube in the original view to a rectangular parallelepiped of dimensions $\alpha \times \beta \times \gamma$ (Figure 10.10.2). Mathematically, this may be accomplished with matrix multiplication as follows. Define a 3×3 diagonal matrix

$$S = \begin{bmatrix} \alpha & 0 & 0 \\ 0 & \beta & 0 \\ 0 & 0 & \gamma \end{bmatrix}$$

Then, if a point P_i in the original view is represented by the column vector

$$\begin{bmatrix} x_i \\ y_i \\ z_i \end{bmatrix}$$

then the transformed point P'_i is represented by the column vector

$$\begin{bmatrix} x'_i \\ y'_i \\ z'_i \end{bmatrix} = \begin{bmatrix} \alpha & 0 & 0 \\ 0 & \beta & 0 \\ 0 & 0 & \gamma \end{bmatrix} \begin{bmatrix} x_i \\ y_i \\ z_i \end{bmatrix}$$

Using the coordinate matrix P, which contains the coordinates of all n points of the original view as its columns, we can transform these n points simultaneously to produce the coordinate matrix P' of the scaled view, as follows:

$$SP = \begin{bmatrix} \alpha & 0 & 0 \\ 0 & \beta & 0 \\ 0 & 0 & \gamma \end{bmatrix} \begin{bmatrix} x_1 & x_2 & \cdots & x_n \\ y_1 & y_2 & \cdots & y_n \\ z_1 & z_2 & \cdots & z_n \end{bmatrix}$$

$$= \begin{bmatrix} \alpha x_1 & \alpha x_2 & \cdots & \alpha x_n \\ \beta y_1 & \beta y_2 & \cdots & \beta y_n \\ \gamma z_1 & \gamma z_2 & \cdots & \gamma z_n \end{bmatrix} = P'$$

The new coordinate matrix can then be entered into the video display system to produce the new view of the object. As an example, View 2 is View 1 scaled by setting $\alpha = 1.8$, $\beta = 0.5$, and $\gamma = 3.0$. Note that the scaling $\gamma = 3.0$ along the z-axis is not visible in View 2, since we see only the projection of the object onto the xy-plane.

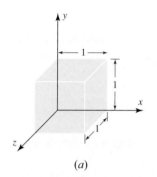

(a)

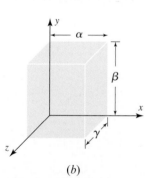

(b)

▲ Figure 10.10.2

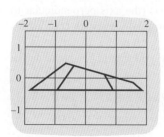

▲ View 2 View 1 scaled by $\alpha = 1.8$, $\beta = 0.5$, $\gamma = 3.0$.

Translation We next consider the transformation of translating or displacing an object to a new position on the screen. Referring to Figure 10.10.3, suppose we desire to change an existing view so that each point P_i with coordinates (x_i, y_i, z_i) moves to a new point P_i' with coordinates $(x_i + x_0, y_i + y_0, z_i + z_0)$. The vector

$$\begin{bmatrix} x_0 \\ y_0 \\ z_0 \end{bmatrix}$$

is called the **translation vector** of the transformation. By defining a $3 \times n$ matrix T as

$$T = \begin{bmatrix} x_0 & x_0 & \cdots & x_0 \\ y_0 & y_0 & \cdots & y_0 \\ z_0 & z_0 & \cdots & z_0 \end{bmatrix}$$

we can translate all n points of the view determined by the coordinate matrix P by matrix addition via the equation

$$P' = P + T$$

The coordinate matrix P' then specifies the new coordinates of the n points. For example, if we wish to translate View 1 according to the translation vector

$$\begin{bmatrix} 1.2 \\ 0.4 \\ 1.7 \end{bmatrix}$$

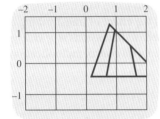

▲ **View 3** View 1 translated by $x_0 = 1.2$, $y_0 = 0.4$, $z_0 = 1.7$.

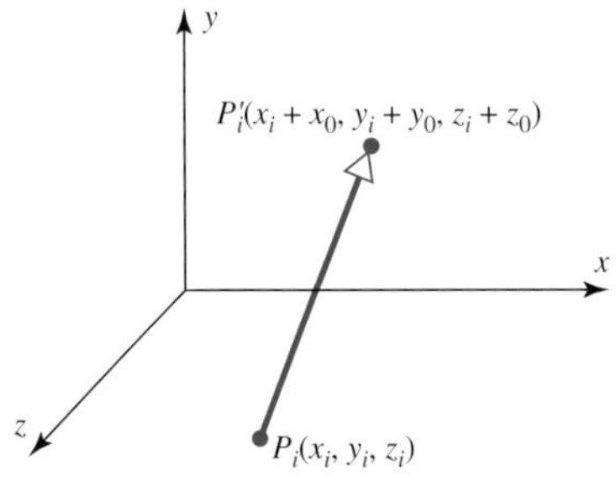

the result is View 3. Note, again, that the translation $z_0 = 1.7$ along the z-axis does not show up explicitly in View 3.

In Exercise 7, a technique of performing translations by matrix multiplication rather than by matrix addition is explained.

▶ **Figure 10.10.3**

▲ **Figure 10.10.4**

Rotation A more complicated type of transformation is a rotation of a view about one of the three coordinate axes. We begin with a rotation about the z-axis (the axis perpendicular to the screen) through an angle θ. Given a point P_i in the original view with coordinates (x_i, y_i, z_i), we wish to compute the new coordinates (x_i', y_i', z_i') of the rotated point P_i'. Referring to Figure 10.10.4 and using a little trigonometry, you should be able to derive the following:

$$x_i' = \rho \cos(\phi + \theta) = \rho \cos\phi \cos\theta - \rho \sin\phi \sin\theta = x_i \cos\theta - y_i \sin\theta$$
$$y_i' = \rho \sin(\phi + \theta) = \rho \cos\phi \sin\theta + \rho \sin\phi \cos\theta = x_i \sin\theta + y_i \cos\theta$$
$$z_i' = z_i$$

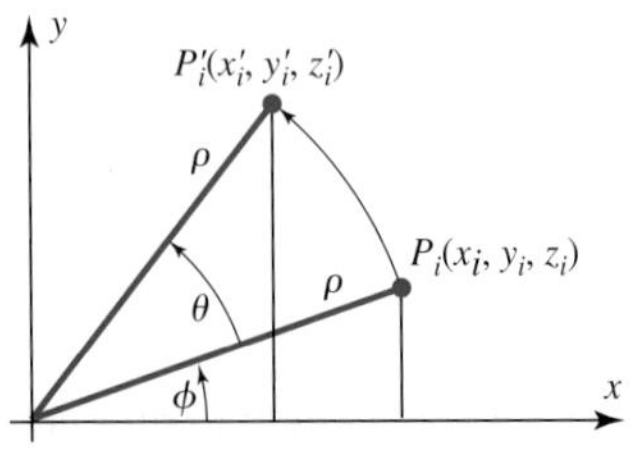

These equations can be written in matrix form as

$$\begin{bmatrix} x'_i \\ y'_i \\ z'_i \end{bmatrix} = \begin{bmatrix} \cos\theta & -\sin\theta & 0 \\ \sin\theta & \cos\theta & 0 \\ 0 & 0 & 1 \end{bmatrix} \begin{bmatrix} x_i \\ y_i \\ z_i \end{bmatrix}$$

If we let R denote the 3×3 matrix in this equation, all n points can be rotated by the matrix product

$$P' = RP$$

to yield the coordinate matrix P' of the rotated view.

Rotations about the x- and y-axes can be accomplished analogously, and the resulting rotation matrices are given with Views 4, 5, and 6. These three new views of the truncated pyramid correspond to rotations of View 1 about the x-, y-, and z-axes, respectively, each through an angle of $90°$.

Rotation about the x-axis

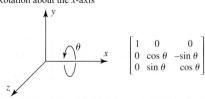
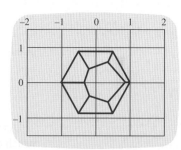

▲ **View 4** View 1 rotated $90°$ about the x-axis.

Rotation about the y-axis

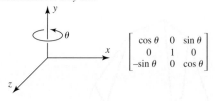
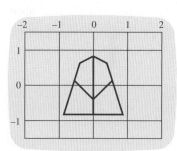

▲ **View 5** View 1 rotated $90°$ about the y-axis.

Rotation about the z-axis

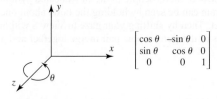
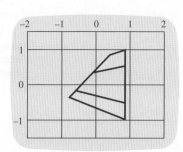

▲ **View 6** View 1 rotated $90°$ about the z-axis.

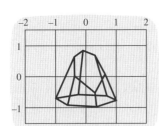

View 7 Oblique view of truncated pyramid.

Rotations about three coordinate axes may be combined to give oblique views of an object. For example, View 7 is View 1 rotated first about the x-axis through $30°$, then about the y-axis through $-70°$, and finally about the z-axis through $-27°$. Mathematically, these three successive rotations can be embodied in the single transformation equation $P' = RP$, where R is the product of three individual rotation matrices:

$$R_1 = \begin{bmatrix} 1 & 0 & 0 \\ 0 & \cos(30°) & -\sin(30°) \\ 0 & \sin(30°) & \cos(30°) \end{bmatrix}$$

$$R_2 = \begin{bmatrix} \cos(-70°) & 0 & \sin(-70°) \\ 0 & 1 & 0 \\ -\sin(-70°) & 0 & \cos(-70°) \end{bmatrix}$$

$$R_3 = \begin{bmatrix} \cos(-27°) & -\sin(-27°) & 0 \\ \sin(-27°) & \cos(-27°) & 0 \\ 0 & 0 & 1 \end{bmatrix}$$

in the order

$$R = R_3 R_2 R_1 = \begin{bmatrix} .305 & -.025 & -.952 \\ -.155 & .985 & -.076 \\ .940 & .171 & .296 \end{bmatrix}$$

As a final illustration, in View 8 we have two separate views of the truncated pyramid, which constitute a stereoscopic pair. They were produced by first rotating View 7 about the y-axis through an angle of $-3°$ and translating it to the right, then rotating the same View 7 about the y-axis through an angle of $+3°$ and translating it to the left. The translation distances were chosen so that the stereoscopic views are about $2\frac{1}{2}$ inches apart—the approximate distance between a pair of eyes.

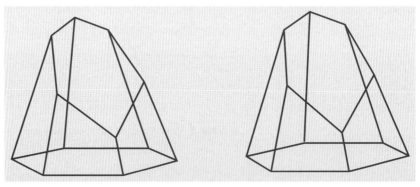

View 8 Stereoscopic figure of truncated pyramid. The three-dimensionality of the diagram can be seen by holding the book about one foot away and focusing on a distant object. Then by shifting your gaze to View 8 without refocusing, you can make the two views of the stereoscopic pair merge together and produce the desired effect.

Exercise Set 10.10

1. View 9 is a view of a square with vertices $(0, 0, 0)$, $(1, 0, 0)$, $(1, 1, 0)$, and $(0, 1, 0)$.

   (a) What is the coordinate matrix of View 9?

   (b) What is the coordinate matrix of View 9 after it is scaled by a factor $1\frac{1}{2}$ in the x-direction and $\frac{1}{2}$ in the y-direction? Draw a sketch of the scaled view.

   (c) What is the coordinate matrix of View 9 after it is translated by the following vector?
   $$\begin{bmatrix} -2 \\ -1 \\ 3 \end{bmatrix}$$
   Draw a sketch of the translated view.

   (d) What is the coordinate matrix of View 9 after it is rotated through an angle of $-30°$ about the z-axis? Draw a sketch of the rotated view.

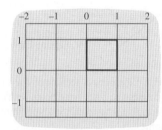

▲ **View 9** Square with vertices $(0, 0, 0)$, $(1, 0, 0)$, $(1, 1, 0)$, and $(0, 1, 0)$ (Exercises 1 and 2).

2. (a) If the coordinate matrix of View 9 is multiplied by the matrix
   $$\begin{bmatrix} 1 & \frac{1}{2} & 0 \\ 0 & 1 & 0 \\ 0 & 0 & 1 \end{bmatrix}$$
   the result is the coordinate matrix of View 10. Such a transformation is called a *shear in the x-direction with factor* $\frac{1}{2}$ *with respect to the y-coordinate*. Show that under such a transformation, a point with coordinates (x_i, y_i, z_i) has new coordinates $(x_i + \frac{1}{2}y_i, y_i, z_i)$.

   (b) What are the coordinates of the four vertices of the shear square in View 10?

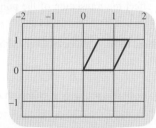

▲ **View 10** View 9 sheared along the x-axis by $\frac{1}{2}$ with respect to the y-coordinate (Exercise 2).

   (c) The matrix
   $$\begin{bmatrix} 1 & 0 & 0 \\ .6 & 1 & 0 \\ 0 & 0 & 1 \end{bmatrix}$$
   determines a *shear in the y-direction with factor .6 with respect to the x-coordinate* (an example appears in View 11). Sketch a view of the square in View 9 after such a shearing transformation, and find the new coordinates of its four vertices.

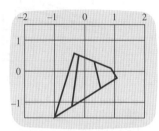

▲ **View 11** View 1 sheared along the y-axis by .6 with respect to the x-coordinate (Exercise 2).

3. (a) The *reflection about the xz-plane* is defined as the transformation that takes a point (x_i, y_i, z_i) to the point $(x_i, -y_i, z_i)$ (e.g., View 12). If P and P' are the coordinate matrices of a view and its reflection about the xz-plane, respectively, find a matrix M such that $P' = MP$.

   (b) Analogous to part (a), define the *reflection about the yz-plane* and construct the corresponding transformation matrix. Draw a sketch of View 1 reflected about the yz-plane.

   (c) Analogous to part (a), define the *reflection about the xy-plane* and construct the corresponding transformation matrix. Draw a sketch of View 1 reflected about the xy-plane.

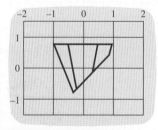

▲ **View 12** View 1 reflected about the xz-plane (Exercise 3).

4. (a) View 13 is View 1 subject to the following five transformations:

   1. Scale by a factor of $\frac{1}{2}$ in the x-direction, 2 in the y-direction, and $\frac{1}{3}$ in the z-direction.
   2. Translate $\frac{1}{2}$ unit in the x-direction.

3. Rotate 20° about the x-axis.
4. Rotate $-45°$ about the y-axis.
5. Rotate 90° about the z-axis.

Construct the five matrices M_1, M_2, M_3, M_4, and M_5 associated with these five transformations.

(b) If P is the coordinate matrix of View 1 and P' is the coordinate matrix of View 13, express P' in terms of M_1, M_2, M_3, M_4, M_5, and P.

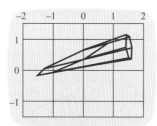

▲ **View 13** View 1 scaled, translated, and rotated (Exercise 4).

5. (a) View 14 is View 1 subject to the following seven transformations:

1. Scale by a factor of .3 in the x-direction and by a factor of .5 in the y-direction.
2. Rotate 45° about the x-axis.
3. Translate 1 unit in the x-direction.
4. Rotate 35° about the y-axis.
5. Rotate $-45°$ about the z-axis.
6. Translate 1 unit in the z-direction.
7. Scale by a factor of 2 in the x-direction.

Construct the matrices $M_1, M_2, \ldots, M_7$ associated with these seven transformations.

(b) If P is the coordinate matrix of View 1 and P' is the coordinate matrix of View 14, express P' in terms of $M_1, M_2, \ldots, M_7$, and P.

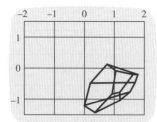

▲ **View 14** View 1 scaled, translated, and rotated (Exercise 5).

6. Suppose that a view with coordinate matrix P is to be rotated through an angle θ about an axis through the origin and specified by two angles α and β (see Figure Ex-6). If P' is the coordinate matrix of the rotated view, find rotation matrices R_1, R_2, R_3, R_4, and R_5 such that

$$P' = R_5 R_4 R_3 R_2 R_1 P$$

[*Hint:* The desired rotation can be accomplished in the following five steps:

1. Rotate through an angle of β about the y-axis.
2. Rotate through an angle of α about the z-axis.
3. Rotate through an angle of θ about the y-axis.
4. Rotate through an angle of $-\alpha$ about the z-axis.
5. Rotate through an angle of $-\beta$ about the y-axis.]

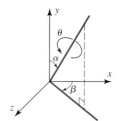

◀ **Figure Ex-6**

7. This exercise illustrates a technique for translating a point with coordinates (x_i, y_i, z_i) to a point with coordinates $(x_i + x_0, y_i + y_0, z_i + z_0)$ by matrix multiplication rather than matrix addition.

(a) Let the point (x_i, y_i, z_i) be associated with the column vector

$$\mathbf{v}_i = \begin{bmatrix} x_i \\ y_i \\ z_i \\ 1 \end{bmatrix}$$

and let the point $(x_i + x_0, y_i + y_0, z_i + z_0)$ be associated with the column vector

$$\mathbf{v}'_i = \begin{bmatrix} x_i + x_0 \\ y_i + y_0 \\ z_i + z_0 \\ 1 \end{bmatrix}$$

Find a 4×4 matrix M such that $\mathbf{v}'_i = M\mathbf{v}_i$.

(b) Find the specific 4×4 matrix of the above form that will effect the translation of the point $(4, -2, 3)$ to the point $(-1, 7, 0)$.

8. For the three rotation matrices given with Views 4, 5, and 6, show that

$$R^{-1} = R^T$$

(A matrix with this property is called an ***orthogonal matrix***. See Section 7.1.)

Section 10.10 Technology Exercises

The following exercises are designed to be solved using a technology utility. Typically, this will be MATLAB, *Mathematica*, Maple, Derive, or Mathcad, but it may also be some other type of linear algebra software or a scientific calculator with some linear algebra capabilities. For each exercise you will need to read the relevant documentation for the particular utility you are using. The goal of these exercises is to provide you with a basic proficiency with your technology utility. Once you have mastered the techniques in these exercises, you will be able to use your technology utility to solve many of the problems in the regular exercise sets.

T1. Let (a, b, c) be a unit vector normal to the plane $ax + by + cz = 0$, and let $\mathbf{r} = (x, y, z)$ be a vector. It can be shown that the mirror image of the vector $\mathbf{r}$ through the above plane has coordinates $\mathbf{r}_m = (x_m, y_m, z_m)$, where

$$\begin{bmatrix} x_m \\ y_m \\ z_m \end{bmatrix} = M \begin{bmatrix} x \\ y \\ z \end{bmatrix}$$

with

$$M = I - 2\mathbf{n}\mathbf{n}^T = \begin{bmatrix} 1 & 0 & 0 \\ 0 & 1 & 0 \\ 0 & 0 & 1 \end{bmatrix} - 2 \begin{bmatrix} a \\ b \\ c \end{bmatrix} [a \ b \ c]$$

(a) Show that $M^2 = I$ and give a physical reason why this must be so. [*Hint:* Use the fact that (a, b, c) is a unit vector to show that $\mathbf{n}^T\mathbf{n} = 1$.]

(b) Use a computer to show that $\det(M) = -1$.

(c) The eigenvectors of M satisfy the equation

$$\begin{bmatrix} x_m \\ y_m \\ z_m \end{bmatrix} = M \begin{bmatrix} x \\ y \\ z \end{bmatrix} = \lambda \begin{bmatrix} x \\ y \\ z \end{bmatrix}$$

and therefore correspond to those vectors whose direction is not affected by a reflection through the plane. Use a computer to determine the eigenvectors and eigenvalues of M, and then give a physical argument to support your answer.

T2. A vector $\mathbf{v} = (x, y, z)$ is rotated by an angle θ about an axis having unit vector (a, b, c), thereby forming the rotated vector $\mathbf{v}_R = (x_R, y_R, z_R)$. It can be shown that

$$\begin{bmatrix} x_R \\ y_R \\ z_R \end{bmatrix} = R(\theta) \begin{bmatrix} x \\ y \\ z \end{bmatrix}$$

with

$$R(\theta) = \cos(\theta) \begin{bmatrix} 1 & 0 & 0 \\ 0 & 1 & 0 \\ 0 & 0 & 1 \end{bmatrix} + (1 - \cos(\theta)) \begin{bmatrix} a \\ b \\ c \end{bmatrix} [a \ b \ c]$$

$$+ \sin(\theta) \begin{bmatrix} 0 & -c & b \\ c & 0 & -a \\ -b & a & 0 \end{bmatrix}$$

(a) Use a computer to show that $R(\theta)R(\varphi) = R(\theta + \varphi)$, and then give a physical reason why this must be so. Depending on the sophistication of the computer you are using, you may have to experiment using different values of a, b, and

$$c = \sqrt{1 - a^2 - b^2}$$

(b) Show also that $R^{-1}(\theta) = R(-\theta)$ and give a physical reason why this must be so.

(c) Use a computer to show that $\det(R(\theta)) = +1$.

10.11 Equilibrium Temperature Distributions

In this section we will see that the equilibrium temperature distribution within a trapezoidal plate can be found when the temperatures around the edges of the plate are specified. The problem is reduced to solving a system of linear equations. Also, an iterative technique for solving the problem and a "random walk" approach to the problem are described.

> **PREREQUISITES:** Linear Systems
> Matrices
> Intuitive Understanding of Limits

Boundary Data Suppose that the two faces of the thin trapezoidal plate shown in Figure 10.11.1*a* are insulated from heat. Suppose that we are also given the temperature along the four edges of the plate. For example, let the temperature be constant on each edge with values of

0°, 0°, 1°, and 2°, as in the figure. After a period of time, the temperature inside the plate will stabilize. Our objective in this section is to determine this equilibrium temperature distribution at the points inside the plate. As we will see, the interior equilibrium temperature is completely determined by the **boundary data**—that is, the temperature along the edges of the plate.

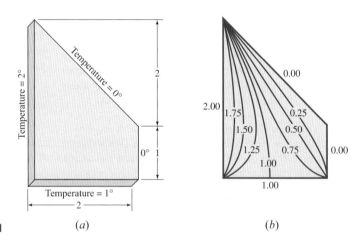

▶ Figure 10.11.1   (a)   (b)

The equilibrium temperature distribution can be visualized by the use of curves that connect points of equal temperature. Such curves are called *isotherms* of the temperature distribution. In Figure 10.11.1b we have sketched a few isotherms, using information we derive later in the chapter.

Although all our calculations will be for the trapezoidal plate illustrated, our techniques generalize easily to a plate of any practical shape. They also generalize to the problem of finding the temperature within a three-dimensional body. In fact, our "plate" could be the cross section of some solid object if the flow of heat perpendicular to the cross section is negligible. For example, Figure 10.11.1 could represent the cross section of a long dam. The dam is exposed to three different temperatures: the temperature of the ground at its base, the temperature of the water on one side, and the temperature of the air on the other side. A knowledge of the temperature distribution inside the dam is necessary to determine the thermal stresses to which it is subjected.

Next we will consider a certain thermodynamic principle that characterizes the temperature distribution we are seeking.

The Mean-Value Property

There are many different ways to obtain a mathematical model for our problem. The approach we use is based on the following property of equilibrium temperature distributions.

> **THEOREM 10.11.1** **The Mean-Value Property**
>
> *Let a plate be in thermal equilibrium and let P be a point inside the plate. Then if C is any circle with center at P that is completely contained in the plate, the temperature at P is the average value of the temperature on the circle (Figure 10.11.2).*

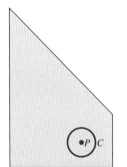

▲ Figure 10.11.2

This property is a consequence of certain basic laws of molecular motion, and we will not attempt to derive it. Basically, this property states that in equilibrium, thermal energy tends to distribute itself as evenly as possible consistent with the boundary conditions.

10.11 Equilibrium Temperature Distributions

It can be shown that the mean-value property uniquely determines the equilibrium temperature distribution of a plate.

Unfortunately, determining the equilibrium temperature distribution from the mean-value property is not an easy matter. However, if we restrict ourselves to finding the temperature only at a finite set of points within the plate, the problem can be reduced to solving a linear system. We pursue this idea next.

Discrete Formulation of the Problem

We can overlay our trapezoidal plate with a succession of finer and finer square nets or meshes (Figure 10.11.3). In (*a*) we have a rather coarse net; in (*b*) we have a net with half the spacing as in (*a*); and in (*c*) we have a net with the spacing again reduced by half. The points of intersection of the net lines are called *mesh points*. We classify them as **boundary mesh points** if they fall on the boundary of the plate or as **interior mesh points** if they lie in the interior of the plate. For the three net spacings we have chosen, there are 1, 9, and 49 interior mesh points, respectively.

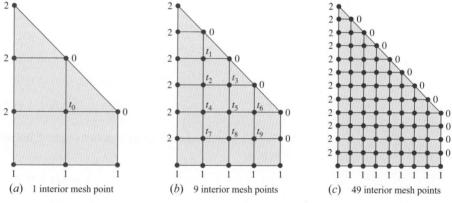

▲ **Figure 10.11.3**

In the discrete formulation of our problem, we try to find the temperature only at the interior mesh points of some particular net. For a rather fine net, as in (*c*), this will provide an excellent picture of the temperature distribution throughout the entire plate.

At the boundary mesh points, the temperature is given by the boundary data. (In Figure 10.11.3 we have labeled all the boundary mesh points with their corresponding temperatures.) At the interior mesh points, we will apply the following discrete version of the mean-value property.

> **THEOREM 10.11.2** *Discrete Mean-Value Property*
>
> *At each interior mesh point, the temperature is approximately the average of the temperatures at the four neighboring mesh points.*

This discrete version is a reasonable approximation to the true mean-value property. But because it is only an approximation, it will provide only an approximation to the true temperatures at the interior mesh points. However, the approximations will get better as the mesh spacing decreases. In fact, as the mesh spacing approaches zero, the approximations approach the exact temperature distribution, a fact proved in advanced courses in numerical analysis. We will illustrate this convergence by computing the approximate temperatures at the mesh points for the three mesh spacings given in Figure 10.11.3.

Case (*a*) of Figure 10.11.3 is simple, for there is only one interior mesh point. If we let t_0 be the temperature at this mesh point, the discrete mean-value property immediately gives

$$t_0 = \tfrac{1}{4}(2+1+0+0) = 0.75$$

In case (*b*) we can label the temperatures at the nine interior mesh points $t_1, t_2, \ldots, t_9$, as in Figure 10.11.3*b*. (The particular ordering is not important.) By applying the discrete mean-value property successively to each of these nine mesh points, we obtain the following nine equations:

$$\begin{aligned}
t_1 &= \tfrac{1}{4}(t_2 + 2 + 0 + 0) \\
t_2 &= \tfrac{1}{4}(t_1 + t_3 + t_4 + 2) \\
t_3 &= \tfrac{1}{4}(t_2 + t_5 + 0 + 0) \\
t_4 &= \tfrac{1}{4}(t_2 + t_5 + t_7 + 2) \\
t_5 &= \tfrac{1}{4}(t_3 + t_4 + t_6 + t_8) \\
t_6 &= \tfrac{1}{4}(t_5 + t_9 + 0 + 0) \\
t_7 &= \tfrac{1}{4}(t_4 + t_8 + 1 + 2) \\
t_8 &= \tfrac{1}{4}(t_5 + t_7 + t_9 + 1) \\
t_9 &= \tfrac{1}{4}(t_6 + t_8 + 1 + 0)
\end{aligned} \tag{1}$$

This is a system of nine linear equations in nine unknowns. We can rewrite it in matrix form as

$$\mathbf{t} = M\mathbf{t} + \mathbf{b} \tag{2}$$

where

$$\mathbf{t} = \begin{bmatrix} t_1 \\ t_2 \\ t_3 \\ t_4 \\ t_5 \\ t_6 \\ t_7 \\ t_8 \\ t_9 \end{bmatrix}, \quad M = \begin{bmatrix} 0 & \tfrac{1}{4} & 0 & 0 & 0 & 0 & 0 & 0 & 0 \\ \tfrac{1}{4} & 0 & \tfrac{1}{4} & \tfrac{1}{4} & 0 & 0 & 0 & 0 & 0 \\ 0 & \tfrac{1}{4} & 0 & 0 & \tfrac{1}{4} & 0 & 0 & 0 & 0 \\ 0 & \tfrac{1}{4} & 0 & 0 & \tfrac{1}{4} & 0 & \tfrac{1}{4} & 0 & 0 \\ 0 & 0 & \tfrac{1}{4} & \tfrac{1}{4} & 0 & \tfrac{1}{4} & 0 & \tfrac{1}{4} & 0 \\ 0 & 0 & 0 & 0 & \tfrac{1}{4} & 0 & 0 & 0 & \tfrac{1}{4} \\ 0 & 0 & 0 & \tfrac{1}{4} & 0 & 0 & 0 & \tfrac{1}{4} & 0 \\ 0 & 0 & 0 & 0 & \tfrac{1}{4} & 0 & \tfrac{1}{4} & 0 & \tfrac{1}{4} \\ 0 & 0 & 0 & 0 & 0 & \tfrac{1}{4} & 0 & \tfrac{1}{4} & 0 \end{bmatrix}, \quad \mathbf{b} = \begin{bmatrix} \tfrac{1}{2} \\ \tfrac{1}{2} \\ 0 \\ \tfrac{1}{2} \\ 0 \\ 0 \\ \tfrac{3}{4} \\ \tfrac{1}{4} \\ \tfrac{1}{4} \end{bmatrix}$$

To solve Equation (2), we write it as

$$(I - M)\mathbf{t} = \mathbf{b}$$

The solution for **t** is thus

$$\mathbf{t} = (I - M)^{-1}\mathbf{b} \tag{3}$$

as long as the matrix $(I - M)$ is invertible. This is indeed the case, and the solution for

t as calculated by (3) is

$$\mathbf{t} = \begin{bmatrix} 0.7846 \\ 1.1383 \\ 0.4719 \\ 1.2967 \\ 0.7491 \\ 0.3265 \\ 1.2995 \\ 0.9014 \\ 0.5570 \end{bmatrix} \quad (4)$$

Figure 10.11.4 is a diagram of the plate with the nine interior mesh points labeled with their temperatures as given by this solution.

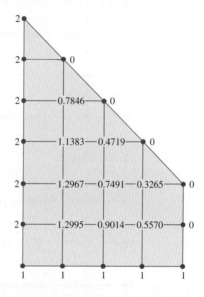

▶ Figure 10.11.4

For case (c) of Figure 10.11.3, we repeat this same procedure. We label the temperatures at the 49 interior mesh points as $t_1, t_2, \ldots, t_{49}$ in some manner. For example, we may begin at the top of the plate and proceed from left to right along each row of mesh points. Applying the discrete mean-value property to each mesh point gives a system of 49 linear equations in 49 unknowns:

$$\begin{aligned} t_1 &= \tfrac{1}{4}(t_2 + 2 + 0 + 0) \\ t_2 &= \tfrac{1}{4}(t_1 + t_3 + t_4 + 2) \\ &\vdots \\ t_{48} &= \tfrac{1}{4}(t_{41} + t_{47} + t_{49} + 1) \\ t_{49} &= \tfrac{1}{4}(t_{42} + t_{48} + 0 + 1) \end{aligned} \quad (5)$$

In matrix form, Equations (5) are

$$\mathbf{t} = M\mathbf{t} + \mathbf{b}$$

where **t** and **b** are column vectors with 49 entries, and M is a 49×49 matrix. As in (3), the solution for **t** is

$$\mathbf{t} = (I - M)^{-1}\mathbf{b} \quad (6)$$

In Figure 10.11.5 we display the temperatures at the 49 mesh points found by Equation (6). The nine unshaded temperatures in this figure fall on the mesh points of Figure 10.11.4.

610 Chapter 10 Applications of Linear Algebra

In Table 1 we compare the temperatures at these nine common mesh points for the three different mesh spacings used.

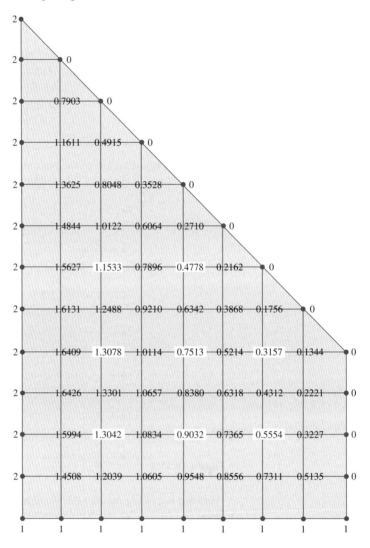

▶ Figure 10.11.5

Table 1

	Temperatures at Common Mesh Points		
	Case (*a*)	Case (*b*)	Case (*c*)
t_1	—	0.7846	0.8048
t_2	—	1.1383	1.1533
t_3	—	0.4719	0.4778
t_4	—	1.2967	1.3078
t_5	0.7500	0.7491	0.7513
t_6	—	0.3265	0.3157
t_7	—	1.2995	1.3042
t_8	—	0.9014	0.9032
t_9	—	0.5570	0.5554

10.11 Equilibrium Temperature Distributions

Knowing that the temperatures of the discrete problem approach the exact temperatures as the mesh spacing decreases, we may surmise that the nine temperatures obtained in case (c) are closer to the exact values than those in case (b).

A Numerical Technique   To obtain the 49 temperatures in case (c) of Figure 10.11.3, it was necessary to solve a linear system with 49 unknowns. A finer net might involve a linear system with hundreds or even thousands of unknowns. Exact algorithms for the solutions of such large systems are impractical, and for this reason we now discuss a numerical technique for the practical solution of these systems.

To describe this technique, we look again at Equation (2):

$$\mathbf{t} = M\mathbf{t} + \mathbf{b} \tag{7}$$

The vector **t** we are seeking appears on both sides of this equation. We consider a way of generating better and better approximations to the vector solution **t**. For the initial approximation $\mathbf{t}^{(0)}$ we can take $\mathbf{t}^{(0)} = \mathbf{0}$ if no better choice is available. If we substitute $\mathbf{t}^{(0)}$ into the right side of (7) and label the resulting left side as $\mathbf{t}^{(1)}$, we have

$$\mathbf{t}^{(1)} = M\mathbf{t}^{(0)} + \mathbf{b} \tag{8}$$

If we substitute $\mathbf{t}^{(1)}$ into the right side of (7), we generate another approximation, which we label $\mathbf{t}^{(2)}$:

$$\mathbf{t}^{(2)} = M\mathbf{t}^{(1)} + \mathbf{b} \tag{9}$$

Continuing in this way, we generate a sequence of approximations as follows:

$$\begin{aligned} \mathbf{t}^{(1)} &= M\mathbf{t}^{(0)} + \mathbf{b} \\ \mathbf{t}^{(2)} &= M\mathbf{t}^{(1)} + \mathbf{b} \\ \mathbf{t}^{(3)} &= M\mathbf{t}^{(2)} + \mathbf{b} \\ &\vdots \\ \mathbf{t}^{(n)} &= M\mathbf{t}^{(n-1)} + \mathbf{b} \\ &\vdots \end{aligned} \tag{10}$$

One would hope that this sequence of approximations $\mathbf{t}^{(0)}, \mathbf{t}^{(1)}, \mathbf{t}^{(2)}, \ldots$ converges to the exact solution of (7). We do not have the space here to go into the theoretical considerations necessary to show this. Suffice it to say that for the particular problem we are considering, the sequence converges to the exact solution for any mesh size and for any initial approximation $\mathbf{t}^{(0)}$.

This technique of generating successive approximations to the solution of (7) is a variation of a technique called **Jacobi iteration**; the approximations themselves are called *iterates*. As a numerical example, let us apply Jacobi iteration to the calculation of the nine mesh point temperatures of case (b). Setting $\mathbf{t}^{(0)} = \mathbf{0}$, we have, from Equation (2),

$$\mathbf{t}^{(1)} = M\mathbf{t}^{(0)} + \mathbf{b} = M\mathbf{0} + \mathbf{b} = \mathbf{b} = \begin{bmatrix} .5000 \\ .5000 \\ .0000 \\ .5000 \\ .0000 \\ .0000 \\ .7500 \\ .2500 \\ .2500 \end{bmatrix}$$

$$\mathbf{t}^{(2)} = M\mathbf{t}^{(1)} + \mathbf{b}$$

$$= \begin{bmatrix} 0 & \frac{1}{4} & 0 & 0 & 0 & 0 & 0 & 0 & 0 \\ \frac{1}{4} & 0 & \frac{1}{4} & \frac{1}{4} & 0 & 0 & 0 & 0 & 0 \\ 0 & \frac{1}{4} & 0 & 0 & \frac{1}{4} & 0 & 0 & 0 & 0 \\ 0 & \frac{1}{4} & 0 & 0 & \frac{1}{4} & 0 & \frac{1}{4} & 0 & 0 \\ 0 & 0 & \frac{1}{4} & \frac{1}{4} & 0 & \frac{1}{4} & 0 & \frac{1}{4} & 0 \\ 0 & 0 & 0 & 0 & \frac{1}{4} & 0 & 0 & 0 & \frac{1}{4} \\ 0 & 0 & 0 & \frac{1}{4} & 0 & 0 & 0 & \frac{1}{4} & 0 \\ 0 & 0 & 0 & 0 & \frac{1}{4} & 0 & \frac{1}{4} & 0 & \frac{1}{4} \\ 0 & 0 & 0 & 0 & 0 & \frac{1}{4} & 0 & \frac{1}{4} & 0 \end{bmatrix} \begin{bmatrix} .5000 \\ .5000 \\ .0000 \\ .5000 \\ .0000 \\ .0000 \\ .7500 \\ .2500 \\ .2500 \end{bmatrix} + \begin{bmatrix} .5000 \\ .5000 \\ .0000 \\ .5000 \\ .0000 \\ .0000 \\ .7500 \\ .2500 \\ .2500 \end{bmatrix} = \begin{bmatrix} .6250 \\ .7500 \\ .1250 \\ .8125 \\ .1875 \\ .0625 \\ .9375 \\ .5000 \\ .3125 \end{bmatrix}$$

Some additional iterates are

$$\mathbf{t}^{(3)} = \begin{bmatrix} 0.6875 \\ 0.8906 \\ 0.2344 \\ 0.9688 \\ 0.3750 \\ 0.1250 \\ 1.0781 \\ 0.6094 \\ 0.3906 \end{bmatrix}, \quad \mathbf{t}^{(10)} = \begin{bmatrix} 0.7791 \\ 1.1230 \\ 0.4573 \\ 1.2770 \\ 0.7236 \\ 0.3131 \\ 1.2848 \\ 0.8827 \\ 0.5446 \end{bmatrix}, \quad \mathbf{t}^{(20)} = \begin{bmatrix} 0.7845 \\ 1.1380 \\ 0.4716 \\ 1.2963 \\ 0.7486 \\ 0.3263 \\ 1.2992 \\ 0.9010 \\ 0.5567 \end{bmatrix}, \quad \mathbf{t}^{(30)} = \begin{bmatrix} 0.7846 \\ 1.1383 \\ 0.4719 \\ 1.2967 \\ 0.7491 \\ 0.3265 \\ 1.2995 \\ 0.9014 \\ 0.5570 \end{bmatrix}$$

All iterates beginning with the thirtieth are equal to $\mathbf{t}^{(30)}$ to four decimal places. Consequently, $\mathbf{t}^{(30)}$ is the exact solution to four decimal places. This agrees with our previous result given in Equation (4).

The Jacobi iteration scheme applied to the linear system (5) with 49 unknowns produces iterates that begin repeating to four decimal places after 119 iterations. Thus, $\mathbf{t}^{(119)}$ would provide the 49 temperatures of case (c) correct to four decimal places.

A Monte Carlo Technique

In this section we describe a so-called *Monte Carlo technique* for computing the temperature at a single interior mesh point of the discrete problem without having to compute the temperatures at the remaining interior mesh points. First we define a *discrete random walk* along the net. By this we mean a directed path along the net lines (Figure 10.11.6) that joins a succession of mesh points such that the direction of departure from each mesh point is chosen at random. Each of the four possible directions of departure from each mesh point along the path is to be equally probable.

By the use of random walks, we can compute the temperature at a specified interior mesh point on the basis of the following property.

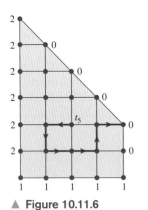

▲ Figure 10.11.6

THEOREM 10.11.3 Random Walk Property

Let $W_1, W_2, \ldots, W_n$ be a succession of random walks, all of which begin at a specified interior mesh point. Let $t_1^*, t_2^*, \ldots, t_n^*$ be the temperatures at the boundary mesh points first encountered along each of these random walks. Then the average value $(t_1^* + t_2^* + \cdots + t_n^*)/n$ of these boundary temperatures approaches the temperature at the specified interior mesh point as the number of random walks n increases without bound.

10.11 Equilibrium Temperature Distributions

This property is a consequence of the discrete mean-value property that the mesh point temperatures satisfy. The proof of the random walk property involves elementary concepts from probability theory, and we will not give it here.

In Table 2 we display the results of a large number of computer-generated random walks for the evaluation of the temperature t_5 of the nine-point mesh of case (b) in Figure 10.11.6. The first column lists the number n of the random walk. The second column lists the temperature t_n^* of the boundary point first encountered along the corresponding random walk. The last column contains the cumulative average of the boundary temperatures encountered along the n random walks. Thus, after 1000 random walks we have the approximation $t_5 \simeq .7550$. This compares with the exact value $t_5 = .7491$ that we had previously evaluated. As can be seen, the convergence to the exact value is not too rapid.

Table 2

n	t_n^*	$(t_1^* + \cdots + t_n^*)/n$	n	t_n^*	$(t_1^* + \cdots + t_n^*)/n$
1	1	1.0000	20	1	0.9500
2	2	1.5000	30	0	0.8000
3	1	1.3333	40	0	0.8250
4	0	1.0000	50	2	0.8400
5	2	1.2000	100	0	0.8300
6	0	1.0000	150	1	0.8000
7	2	1.1429	200	0	0.8050
8	0	1.0000	250	1	0.8240
9	2	1.1111	500	1	0.7860
10	0	1.0000	1000	0	0.7550

Exercise Set 10.11

1. A plate in the form of a circular disk has boundary temperatures of 0° on the left of its circumference and 1° on the right half of its circumference. A net with four interior mesh points is overlaid on the disk (see Figure Ex-1).

   (a) Using the discrete mean-value property, write the 4×4 linear system $\mathbf{t} = M\mathbf{t} + \mathbf{b}$ that determines the approximate temperatures at the four interior mesh points.

   (b) Solve the linear system in part (a).

   (c) Use the Jacobi iteration scheme with $\mathbf{t}^{(0)} = \mathbf{0}$ to generate the iterates $\mathbf{t}^{(1)}, \mathbf{t}^{(2)}, \mathbf{t}^{(3)}, \mathbf{t}^{(4)}$, and $\mathbf{t}^{(5)}$ for the linear system in part (a). What is the "error vector" $\mathbf{t}^{(5)} - \mathbf{t}$, where $\mathbf{t}$ is the solution found in part (b)?

   (d) By certain advanced methods, it can be determined that the exact temperatures to four decimal places at the four mesh points are $t_1 = t_3 = .2871$ and $t_2 = t_4 = .7129$. What are the percentage errors in the values found in part (b)?

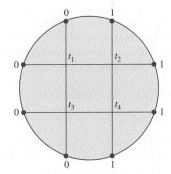

◀ Figure Ex-1

2. Use Theorem 10.11.1 to find the exact equilibrium temperature at the center of the disk in Exercise 1.

3. Calculate the first two iterates $\mathbf{t}^{(1)}$ and $\mathbf{t}^{(2)}$ for case (b) of Figure 10.11.3 with nine interior mesh points [Equation (2)] when the

initial iterate is chosen as

$$\mathbf{t}^{(0)} = [1 \ 1 \ 1 \ 1 \ 1 \ 1 \ 1 \ 1 \ 1]^T$$

4. The random walk illustrated in Figure Ex-4a can be described by six arrows

$$\leftarrow \downarrow \rightarrow \rightarrow \uparrow \rightarrow$$

that specify the directions of departure from the successive mesh points along the path. Figure Ex-4b is an array of 100 computer-generated, randomly oriented arrows arranged in a 10×10 array. Use these arrows to determine random walks to approximate the temperature t_5, as in Table 2. Proceed as follows:

1. Take the last two digits of your telephone number. Use the last digit to specify a row and the other to specify a column.
2. Go to the arrow in the array with that row and column number.
3. Using this arrow as a starting point, move through the array of arrows as you would read a book (left to right and top to bottom). Beginning at the point labeled t_5 in Figure Ex-4a and using this sequence of arrows to specify a sequence of directions, move from mesh point to mesh point until you reach a boundary mesh point. This completes your first random walk. Record the temperature at the boundary mesh point. (If you reach the end of the arrow array, continue with the arrow in the upper left corner.)
4. Return to the interior mesh point labeled t_5 and begin where you left off in the arrow array; generate your next random walk. Repeat this process until you have completed 10 random walks and have recorded 10 boundary temperatures.
5. Calculate the average of the 10 boundary temperatures recorded. (The exact value is $t_5 = .7491$.)

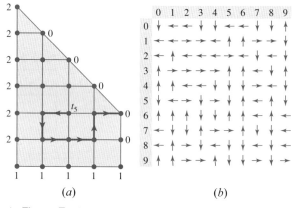

▲ Figure Ex-4

Section 10.11 Technology Exercises

The following exercises are designed to be solved using a technology utility. Typically, this will be MATLAB, *Mathematica*, Maple, Derive, or Mathcad, but it may also be some other type of linear algebra software or a scientific calculator with some linear algebra capabilities. For each exercise you will need to read the relevant documentation for the particular utility you are using. The goal of these exercises is to provide you with a basic proficiency with your technology utility. Once you have mastered the techniques in these exercises, you will be able to use your technology utility to solve many of the problems in the regular exercise sets.

T1. Suppose that we have the square region described by

$$R = \{(x, y) \mid 0 \le x \le 1, 0 \le y \le 1\}$$

and suppose that the equilibrium temperature distribution $u(x, y)$ along the boundary is given by $u(x, 0) = T_B$, $u(x, 1) = T_T$, $u(0, y) = T_L$, and $u(1, y) = T_R$. Suppose next that this region is partitioned into an $(n+1) \times (n+1)$ mesh using

$$x_i = \frac{i}{n} \quad \text{and} \quad y_j = \frac{j}{n}$$

for $i = 0, 1, 2, \ldots, n$ and $j = 0, 1, 2, \ldots, n$. If the temperatures of the interior mesh points are labeled by

$$u_{i,j} = u(x_i, y_i) = u(i/n, j/n)$$

then show that

$$u_{i,j} = \tfrac{1}{4}(u_{i-1,j} + u_{i+1,j} + u_{i,j-1} + u_{i,j+1})$$

for $i = 1, 2, 3, \ldots, n-1$ and $j = 1, 2, 3, \ldots, n-1$. To handle the boundary points, define

$$u_{0,j} = T_L, \quad u_{n,j} = T_R, \quad u_{i,0} = T_B, \quad \text{and} \quad u_{i,n} = T_T$$

for $i = 1, 2, 3, \ldots, n-1$ and $j = 1, 2, 3, \ldots, n-1$. Next let

$$F_{n+1} = \begin{bmatrix} 0 & I_n \\ 1 & 0 \end{bmatrix}$$

be the $(n+1) \times (n+1)$ matrix with the $n \times n$ identity matrix in the upper right-hand corner, a one in the lower left-hand corner, and zeros everywhere else. For example,

$$F_2 = \begin{bmatrix} 0 & 1 \\ 1 & 0 \end{bmatrix}, \quad F_3 = \begin{bmatrix} 0 & 1 & 0 \\ 0 & 0 & 1 \\ 1 & 0 & 0 \end{bmatrix},$$

$$F_4 = \begin{bmatrix} 0 & 1 & 0 & 0 \\ 0 & 0 & 1 & 0 \\ 0 & 0 & 0 & 1 \\ 1 & 0 & 0 & 0 \end{bmatrix}, \quad F_5 = \begin{bmatrix} 0 & 1 & 0 & 0 & 0 \\ 0 & 0 & 1 & 0 & 0 \\ 0 & 0 & 0 & 1 & 0 \\ 0 & 0 & 0 & 0 & 1 \\ 1 & 0 & 0 & 0 & 0 \end{bmatrix}$$

and so on. By defining the $(n+1) \times (n+1)$ matrix

$$M_{n+1} = F_{n+1} + F_{n+1}^T = \begin{bmatrix} 0 & I_n \\ 1 & 0 \end{bmatrix} + \begin{bmatrix} 0 & I_n \\ 1 & 0 \end{bmatrix}^T$$

show that if U_{n+1} is the $(n+1) \times (n+1)$ matrix with entries u_{ij}, then the set of equations

$$u_{i,j} = \tfrac{1}{4}(u_{i-1,j} + u_{i+1,j} + u_{i,j-1} + u_{i,j+1})$$

for $i = 1, 2, 3, \ldots, n-1$ and $j = 1, 2, 3, \ldots, n-1$ can be written as the matrix equation

$$U_{n+1} = \tfrac{1}{4}(M_{n+1} U_{n+1} + U_{n+1} M_{n+1})$$

where we consider only those elements of U_{n+1} with $i = 1, 2, 3, \ldots, n-1$ and $j = 1, 2, 3, \ldots, n-1$.

T2. The results of the preceding exercise and the discussion in the text suggest the following algorithm for solving for the equilibrium temperature in the square region

$$\mathcal{R} = \{(x, y) \mid 0 \leq x \leq 1, 0 \leq y \leq 1\}$$

given the boundary conditions

$$u(x, 0) = T_B, \quad u(x, 1) = T_T,$$
$$u(0, y) = T_L, \quad u(1, y) = T_R$$

1. Choose a value for n, and then choose an initial guess, say

$$\mathbf{U}_{n+1}^{(0)} = \begin{bmatrix} 0 & T_L & \cdots & T_L & 0 \\ T_B & 0 & \cdots & 0 & T_T \\ \vdots & \vdots & & \vdots & \vdots \\ T_B & 0 & \cdots & 0 & T_T \\ 0 & T_R & \cdots & T_R & 0 \end{bmatrix}$$

2. For each value of $k = 0, 1, 2, 3, \ldots$, compute $U_{n+1}^{(k+1)}$ using

$$U_{n+1}^{(k+1)} = \tfrac{1}{4}(M_{n+1} U_{n+1}^{(k)} + U_{n+1}^{(k)} M_{n+1})$$

where M_{n+1} is as defined in Exercise T1. Then adjust $U_{n+1}^{(k+1)}$ by replacing all edge entries by the initial edge entries in $U_{n+1}^{(0)}$. [*Note:* The edge entries of a matrix are the entries in the first and last columns and first and last rows.]

3. Continue this process until $U_{n+1}^{(k+1)} - U_{n+1}^{(k)}$ is approximately the zero matrix. This suggests that

$$U_{n+1} = \lim_{k \to \infty} U_{n+1}^{(k)}$$

Use a computer and this algorithm to solve for $u(x, y)$ given that

$$u(x, 0) = 0, \quad u(x, 1) = 0, \quad u(0, y) = 0, \quad u(1, y) = 2$$

Choose $n = 6$ and compute up to $U_{n+1}^{(30)}$. The exact solution can be expressed as

$$u(x, y) = \frac{8}{\pi} \sum_{m=1}^{\infty} \frac{\sinh[(2m-1)\pi x] \sin[(2m-1)\pi y]}{(2m-1) \sinh[(2m-1)\pi]}$$

Use a computer to compute $u(i/6, j/6)$ for $i, j = 0, 1, 2, 3, 4, 5, 6$, and then compare your results to the values of $u(i/6, j/6)$ in $U_{n+1}^{(30)}$.

T3. Using the exact solution $u(x, y)$ for the temperature distribution described in Exercise T2, use a graphing program to do the following:

(a) Plot the surface $z = u(x, y)$ in three-dimensional xyz-space in which z is the temperature at the point (x, y) in the square region.

(b) Plot several isotherms of the temperature distribution (curves in the xy-plane over which the temperature is a constant).

(c) Plot several curves of the temperature as a function of x with y held constant.

(d) Plot several curves of the temperature as a function of y with x held constant.

10.12 Computed Tomography

In this section we will see how constructing a cross-sectional view of a human body by analyzing X-ray scans leads to an inconsistent linear system. We present an iteration technique that provides an "approximate solution" of the linear system.

> **PREREQUISITES:** Linear Systems
> Natural Logarithms
> Euclidean Space R^n

The basic problem of computed tomography is to construct an image of a cross section of the human body using data collected from many individual beams of X rays that are passed through the cross section. These data are processed by a computer, and the computed cross section is displayed on a video monitor. Figure 10.12.1 is a diagram of

616   Chapter 10 Applications of Linear Algebra

General Electric's CT system showing a patient prepared to have a cross section of his head scanned by X-ray beams.

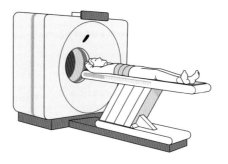

▶ Figure 10.12.1

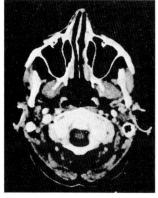

▲ Figure 10.12.2

Such a system is also known as a **CAT scanner**, for *C*omputer-*A*ided *T*omography scanner. Figure 10.12.2 shows a typical cross section of a human head produced by the system.

The first commercial system of computed tomography for medical use was developed in 1971 by G. N. Hounsfield of EMI, Ltd., in England. In 1979, Houndsfield and A. M. Cormack were awarded the Nobel Prize for their pioneering work in the field. As we will see in this section, the construction of a cross section, or tomograph, requires the solution of a large linear system of equations. Certain algorithms, called algebraic reconstruction techniques (ARTs), can be used to solve these linear systems, whose solutions yield the cross sections in digital form.

Scanning Modes

Unlike conventional X-ray pictures that are formed by X rays that are projected *perpendicular* to the plane of the picture, tomographs are constructed from thousands of individual, hairline-thin X-ray beams that *lie in the plane* of the cross section. After they pass through the cross section, the intensities of the X-ray beams are measured by an X-ray detector, and these measurements are relayed to a computer where they are processed. Figures 10.12.3 and 10.12.4 illustrate two possible modes of scanning the cross section: the *parallel mode* and the *fan-beam mode*. In the parallel mode a single X-ray source and X-ray detector pair are translated across the field of view containing the cross section, and many measurements of the parallel beams are recorded. Then the source and detector pair are rotated through a small angle, and another set of measurements is taken. This is repeated until the desired number of beam measurements is completed. For example, in the original 1971 machine, 160 parallel measurements

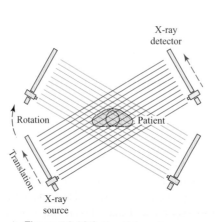

▲ Figure 10.12.3 Parallel mode.

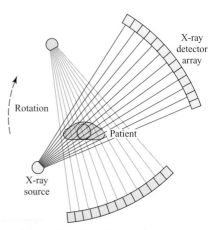

▲ Figure 10.12.4 Fan-beam mode.

were taken through 180 angles spaced 1° apart: a total of 160 × 180 = 28,800 beam measurements. Each such scan took approximately $5\frac{1}{2}$ minutes.

In the fan-beam mode of scanning, a single X-ray tube generates a fan of collimated beams whose intensities are measured simultaneously by an array of detectors on the other side of the field of view. The X-ray tube and detector array are rotated through many angles, and a set of measurements is taken at each angle until the scan is completed. In the General Electric CT system, which uses the fan-beam mode, each scan takes 1 second.

Derivation of Equations

To see how the cross section is reconstructed from the many individual beam measurements, refer to Figure 10.12.5. Here the field of view in which the cross section is situated has been divided into many square *pixels* (picture elements) numbered 1 through N as indicated. It is our desire to determine the X-ray density of each pixel. In the EMI system, 6400 pixels were used, arranged in a square 80 × 80 array. The G.E. CT system uses 262,144 pixels in a 512 × 512 array, each pixel being about 1 mm on a side. After the densities of the pixels are determined by the method we will describe, they are reproduced on a video monitor, with each pixel shaded a level of gray proportional to its X-ray density. Because different tissues within the human body have different X-ray densities, the video display clearly distinguishes the various tissues and organs within the cross section.

Figure 10.12.6 shows a single pixel with an X-ray beam of roughly the same width as the pixel passing squarely through it. The photons constituting the X-ray beam are absorbed by the tissue within the pixel at a rate proportional to the X-ray density of the tissue. Quantitatively, the X-ray density of the jth pixel is denoted by x_j and is defined by

$$x_j = \ln\left(\frac{\text{number of photons entering the } j\text{th pixel}}{\text{number of photons leaving the } j\text{th pixel}}\right)$$

where "ln" denotes the natural logarithmic function. Using the logarithm property $\ln(a/b) = -\ln(b/a)$, we also have

$$x_j = -\ln\left(\begin{array}{c}\text{fraction of photons that pass through}\\ \text{the } j\text{th pixel without being absorbed}\end{array}\right)$$

If the X-ray beam passes through an entire row of pixels (Figure 10.12.7), then the number of photons leaving one pixel is equal to the number of photons entering the next pixel in the row. If the pixels are numbered 1, 2, ..., n, then the additive property of the

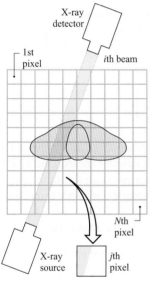

▲ Figure 10.12.5

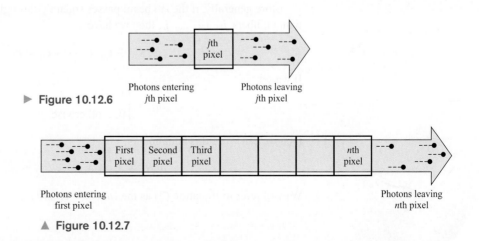

► Figure 10.12.6

▲ Figure 10.12.7

logarithmic function gives

$$x_1 + x_2 + \cdots + x_n = \ln\left(\frac{\text{number of photons entering the first pixel}}{\text{number of photons leaving the nth pixel}}\right)$$

$$= -\ln\left(\begin{array}{c}\text{fraction of photons that pass}\\ \text{through the row of } n \text{ pixels}\\ \text{without being absorbed}\end{array}\right) \quad (1)$$

Thus, to determine the total X-ray density of a row of pixels, we simply sum the individual pixel densities.

Next, consider the X-ray beam in Figure 10.12.5. By the **beam density** of the ith beam of a scan, denoted by b_i, we mean

$$b_i = \ln\left(\frac{\begin{array}{c}\text{number of photons of the } i\text{th beam entering the detector}\\ \text{without the cross section in the field of view}\end{array}}{\begin{array}{c}\text{number of photons of the } i\text{th beam entering the detector}\\ \text{with the cross section in the field of view}\end{array}}\right)$$

$$= -\ln\left(\begin{array}{c}\text{fraction of photons of the } i\text{th beam that}\\ \text{pass through the cross section without}\\ \text{being absorbed}\end{array}\right) \quad (2)$$

The numerator in the first expression for b_i is obtained by performing a calibration scan without the cross section in the field of view. The resulting detector measurements are stored within the computer's memory. Then a clinical scan is performed with the cross section in the field of view, the b_i's of all the beams constituting the scan are computed, and the values are stored for further processing.

For each beam that passes squarely through a row of pixels, we must have

$$\left(\begin{array}{c}\text{fraction of photons of the}\\ \text{beam that pass through the}\\ \text{row of pixels without being}\\ \text{absorbed}\end{array}\right) = \left(\begin{array}{c}\text{fraction of photons of the}\\ \text{beam that pass through the}\\ \text{cross section without being}\\ \text{absorbed}\end{array}\right)$$

Thus, if the ith beam passes squarely through a row of n pixels, then it follows from Equations (1) and (2) that

$$x_1 + x_2 + \cdots + x_n = b_i$$

In this equation, b_i is known from the clinical and calibration measurements, and $x_1, x_2, \ldots, x_n$ are unknown pixel densities that must be determined.

More generally, if the ith beam passes squarely through a row (or column) of pixels with numbers $j_1, j_2, \ldots, j_i$, then we have

$$x_{j_1} + x_{j_2} + \cdots + x_{j_i} = b_i$$

If we set

$$a_{ij} = \begin{cases} 1, & \text{if } j = j_1, j_2, \ldots, j_i \\ 0, & \text{otherwise} \end{cases}$$

then we can write this equation as

$$a_{i1}x_1 + a_{i2}x_2 + \cdots + a_{iN}x_N = b_i \quad (3)$$

We will refer to Equation (3) as the ith beam equation.

Referring to Figure 10.12.5, however, we see that the beams of a scan do not necessarily pass through a row or column of pixels squarely. Instead, a typical beam passes diagonally through each pixel in its path. There are many ways to take this into account. In Figure 10.12.8 we outline three methods of defining the quantities a_{ij} that appear in Equation (3), each of which reduces to our previous definition when the beam passes squarely through a row or column of pixels. Reading down the figure, each method is more exact than its predecessor, but with successively more computational difficulty.

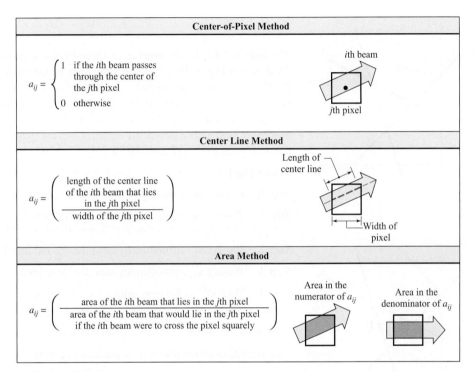

▲ Figure 10.12.8

Using any one of the three methods to define the a_{ij}'s in the ith beam equation, we can write the set of M beam equations in a complete scan as

$$\begin{aligned} a_{11}x_1 + a_{12}x_2 + \cdots + a_{1N}x_N &= b_1 \\ a_{21}x_1 + a_{22}x_2 + \cdots + a_{2N}x_N &= b_2 \\ \vdots \qquad \vdots \qquad \qquad \vdots \qquad &\quad \vdots \\ a_{M1}x_1 + a_{M2}x_2 + \cdots + a_{MN}x_N &= b_M \end{aligned} \tag{4}$$

In this way we have a linear system of M equations (the M beam equations) in N unknowns (the N pixel densities).

Depending on the number of beams and pixels used, we can have $M > N$, $M = N$, or $M < N$. We will consider only the case $M > N$, the so-called *overdetermined case*, in which there are more beams in the scan than pixels in the field of view. Because of inherent modeling and experimental errors in the problem, we should not expect our linear system to have an exact mathematical solution for the pixel densities. In the next section we attempt to find an "approximate" solution to this linear system.

Algebraic Reconstruction Techniques

There have been many mathematical algorithms devised to treat the overdetermined linear system (4). The one we will describe belongs to the class of so-called *Algebraic Reconstruction Techniques* (ARTs). This method, which can be traced to an iterative technique originally introduced by S. Kaczmarz in 1937, was the one used in the first commercial machine. To introduce this technique, consider the following system of three equations in two unknowns:

$$
\begin{aligned}
L_1: & \quad x_1 + x_2 = 2 \\
L_2: & \quad x_1 - 2x_2 = -2 \\
L_3: & \quad 3x_1 - x_2 = 3
\end{aligned}
\tag{5}
$$

The lines L_1, L_2, L_3 determined by these three equations are plotted in the x_1x_2-plane. As shown in Figure 10.12.9a, the three lines do not have a common intersection, and so the three equations do not have an exact solution. However, the points (x_1, x_2) on the shaded triangle formed by the three lines are all situated "near" these three lines and can be thought of as constituting "approximate" solutions to our system. The following iterative procedure describes a geometric construction for generating points on the boundary of that triangular region (Figure 10.12.9b):

Algorithm 1

Step 0. Choose an arbitrary starting point $\mathbf{x}_0$ in the x_1x_2-plane.

Step 1. Project $\mathbf{x}_0$ orthogonally onto the first line L_1 and call the projection $\mathbf{x}_1^{(1)}$. The superscript (1) indicates that this is the first of several cycles through the steps.

Step 2. Project $\mathbf{x}_1^{(1)}$ orthogonally onto the second line L_2 and call the projection $\mathbf{x}_2^{(1)}$.

Step 3. Project $\mathbf{x}_2^{(1)}$ orthogonally onto the third line L_3 and call the projection $\mathbf{x}_3^{(1)}$.

Step 4. Take $\mathbf{x}_3^{(1)}$ as the new value of $\mathbf{x}_0$ and cycle through Steps 1 through 3 again. In the second cycle, label the projected points $\mathbf{x}_1^{(2)}$, $\mathbf{x}_2^{(2)}$, $\mathbf{x}_3^{(2)}$; in the third cycle, label the projected points $\mathbf{x}_1^{(3)}$, $\mathbf{x}_2^{(3)}$, $\mathbf{x}_3^{(3)}$; and so forth.

This algorithm generates three sequences of points

$$
\begin{aligned}
L_1: & \quad \mathbf{x}_1^{(1)}, \mathbf{x}_1^{(2)}, \mathbf{x}_1^{(3)}, \ldots \\
L_2: & \quad \mathbf{x}_2^{(1)}, \mathbf{x}_2^{(2)}, \mathbf{x}_2^{(3)}, \ldots \\
L_3: & \quad \mathbf{x}_3^{(1)}, \mathbf{x}_3^{(2)}, \mathbf{x}_3^{(3)}, \ldots
\end{aligned}
$$

that lie on the three lines L_1, L_2, and L_3, respectively. It can be shown that as long as the three lines are not all parallel, then the first sequence converges to a point $\mathbf{x}_1^*$ on L_1, the second sequence converges to a point $\mathbf{x}_2^*$ on L_2, and the third sequence converges to a point $\mathbf{x}_3^*$ on L_3 (Figure 10.12.9c). These three limit points form what is called the *limit cycle* of the iterative process. It can be shown that the limit cycle is independent of the starting point $\mathbf{x}_0$.

Next we discuss the specific formulas needed to effect the orthogonal projections in Algorithm 1. First, because the equation of a line in x_1x_2-space is

$$a_1x_1 + a_2x_2 = b$$

we can express it in vector form as

$$\mathbf{a}^T\mathbf{x} = b$$

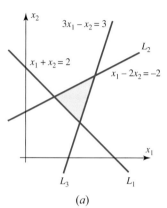

(a)

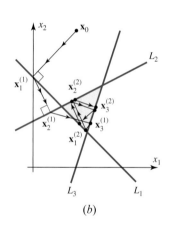

(b)

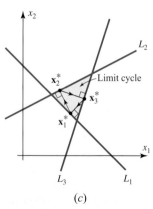

(c)

▲ Figure 10.12.9

10.12 Computed Tomography

where

$$\mathbf{a} = \begin{bmatrix} a_1 \\ a_2 \end{bmatrix} \quad \text{and} \quad \mathbf{x} = \begin{bmatrix} x_1 \\ x_2 \end{bmatrix}$$

The following theorem gives the necessary projection formula (Exercise 5).

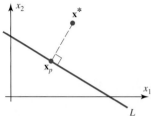

▲ Figure 10.12.10

THEOREM 10.12.1 Orthogonal Projection Formula

Let L be a line in R^2 with equation $\mathbf{a}^T \mathbf{x} = b$, and let $\mathbf{x}^*$ be any point in R^2 (Figure 10.12.10). Then the orthogonal projection, $\mathbf{x}_p$, of $\mathbf{x}^*$ onto L is given by

$$\mathbf{x}_p = \mathbf{x}^* + \frac{(b - \mathbf{a}^T \mathbf{x}^*)}{\mathbf{a}^T \mathbf{a}} \mathbf{a}$$

▶ **EXAMPLE 1 Using Algorithm 1**

We can use Algorithm 1 to find an approximate solution of the linear system given in (5) and illustrated in Figure 10.12.9. If we write the equations of the three lines as

$$L_1: \quad \mathbf{a}_1^T \mathbf{x} = b_1$$
$$L_2: \quad \mathbf{a}_2^T \mathbf{x} = b_2$$
$$L_3: \quad \mathbf{a}_3^T \mathbf{x} = b_3$$

where

$$\mathbf{x} = \begin{bmatrix} x_1 \\ x_2 \end{bmatrix}, \quad \mathbf{a}_1 = \begin{bmatrix} 1 \\ 1 \end{bmatrix}, \quad \mathbf{a}_2 = \begin{bmatrix} 1 \\ -2 \end{bmatrix}, \quad \mathbf{a}_3 = \begin{bmatrix} 3 \\ -1 \end{bmatrix},$$
$$b_1 = 2, \quad b_2 = -2, \quad b_3 = 3$$

then, using Theorem 10.12.1, we can express the iteration scheme in Algorithm 1 as

$$\mathbf{x}_k^{(p)} = \mathbf{x}_{k-1}^{(p)} + \frac{(b_k - \mathbf{a}_k^T \mathbf{x}_{k-1}^{(p)})}{\mathbf{a}_k^T \mathbf{a}_k} \mathbf{a}_k, \quad k = 1, 2, 3$$

where $p = 1$ for the first cycle of iterates, $p = 2$ for the second cycle of iterates, and so forth. After each cycle of iterates (i.e., after $\mathbf{x}_3^{(p)}$ is computed), the next cycle of iterates is begun with $\mathbf{x}_0^{(p+1)}$ set equal to $\mathbf{x}_3^{(p)}$.

Table 1 gives the numerical results of six cycles of iterations starting with the initial point $\mathbf{x}_0 = (1, 3)$.

Using certain techniques that are impractical for large linear systems, we can show the exact values of the points of the limit cycle in this example to be

$$\mathbf{x}_1^* = \left(\tfrac{12}{11}, \tfrac{10}{11}\right) = (1.09090\ldots, .90909\ldots)$$
$$\mathbf{x}_2^* = \left(\tfrac{46}{55}, \tfrac{78}{55}\right) = (.83636\ldots, 1.41818\ldots)$$
$$\mathbf{x}_3^* = \left(\tfrac{31}{22}, \tfrac{27}{22}\right) = (1.40909\ldots, 1.22727\ldots)$$

It can be seen that the sixth cycle of iterates provides an excellent approximation to the limit cycle. Any one of the three iterates $\mathbf{x}_1^{(6)}, \mathbf{x}_2^{(6)}, \mathbf{x}_3^{(6)}$ can be used as an approximate solution of the linear system. (The large discrepancies in the values of $\mathbf{x}_1^{(6)}, \mathbf{x}_2^{(6)}$, and $\mathbf{x}_3^{(6)}$ are due to the artificial nature of this illustrative example. In practical problems, these discrepancies would be much smaller.) ◀

Table 1

	x_1	x_2
$\mathbf{x}_0$	1.00000	3.00000
$\mathbf{x}_1^{(1)}$	.00000	2.00000
$\mathbf{x}_2^{(1)}$	.40000	1.20000
$\mathbf{x}_3^{(1)}$	1.30000	.90000
$\mathbf{x}_1^{(2)}$	1.20000	.80000
$\mathbf{x}_2^{(2)}$	.88000	1.44000
$\mathbf{x}_3^{(2)}$	1.42000	1.26000
$\mathbf{x}_1^{(3)}$	1.08000	.92000
$\mathbf{x}_2^{(3)}$	.83200	1.41600
$\mathbf{x}_3^{(3)}$	1.40800	1.22400
$\mathbf{x}_1^{(4)}$	1.09200	.90800
$\mathbf{x}_2^{(4)}$	.83680	1.41840
$\mathbf{x}_3^{(4)}$	1.40920	1.22760
$\mathbf{x}_1^{(5)}$	1.09080	.90920
$\mathbf{x}_2^{(5)}$	.83632	1.41816
$\mathbf{x}_3^{(5)}$	1.40908	1.22724
$\mathbf{x}_1^{(6)}$	1.09092	.90908
$\mathbf{x}_2^{(6)}$	.83637	1.41818
$\mathbf{x}_3^{(6)}$	1.40909	1.22728

To generalize Algorithm 1 so that it applies to an overdetermined system of M equations in N unknowns,

$$\begin{aligned} a_{11}x_1 + a_{12}x_2 + \cdots + a_{1N}x_N &= b_1 \\ a_{21}x_1 + a_{22}x_2 + \cdots + a_{2N}x_N &= b_2 \\ \vdots \quad\quad \vdots \quad\quad\quad \vdots \quad\quad\quad \vdots& \\ a_{M1}x_1 + a_{M2}x_2 + \cdots + a_{MN}x_N &= b_M \end{aligned} \qquad (6)$$

we introduce column vectors $\mathbf{x}$ and $\mathbf{a}_i$ as follows:

$$\mathbf{x} = \begin{bmatrix} x_1 \\ x_2 \\ \vdots \\ x_N \end{bmatrix}, \quad \mathbf{a}_i = \begin{bmatrix} a_{i1} \\ a_{i2} \\ \vdots \\ a_{iN} \end{bmatrix}, \quad i = 1, 2, \ldots, M$$

With these vectors, the M equations constituting our linear system (6) can be written in vector form as

$$\mathbf{a}_i^T \mathbf{x} = b_i, \quad i = 1, 2, \ldots, M$$

Each of these M equations defines what is called a ***hyperplane*** in the N-dimensional Euclidean space R^N. In general these M hyperplanes have no common intersection, and so we seek instead some point in R^N that is reasonably "close" to all of them. Such a point will constitute an approximate solution of the linear system, and its N entries will determine approximate pixel densities with which to form the desired cross section.

As in the two-dimensional case, we will introduce an iterative process that generates cycles of successive orthogonal projections onto the M hyperplanes beginning with some arbitrary initial point in R^N. Our notation for these successive iterates is

$$\mathbf{x}_k^{(p)} = \begin{pmatrix} \text{the iterate lying on the } k\text{th hyperplane} \\ \text{generated during the } p\text{th cycle of iterations} \end{pmatrix}$$

The algorithm is as follows:

Algorithm 2

Step 0. Choose any point in R^N and label it $\mathbf{x}_0$.

Step 1. For the first cycle of iterates, set $p = 1$.

Step 2. For $k = 1, 2, \ldots, M$, compute

$$\mathbf{x}_k^{(p)} = \mathbf{x}_{k-1}^{(p)} + \frac{(b_k - \mathbf{a}_k^T \mathbf{x}_{k-1}^{(p)})}{\mathbf{a}_k^T \mathbf{a}_k} \mathbf{a}_k$$

Step 3. Set $\mathbf{x}_0^{(p+1)} = \mathbf{x}_M^{(p)}$.

Step 4. Increase the cycle number p by 1 and return to Step 2.

In Step 2 the iterate $\mathbf{x}_k^{(p)}$ is called the ***orthogonal projection*** of $\mathbf{x}_{k-1}^{(p)}$ onto the hyperplane $\mathbf{a}_k^T \mathbf{x} = b_k$. Consequently, as in the two-dimensional case, this algorithm determines a sequence of orthogonal projections from one hyperplane onto the next in which we cycle back to the first hyperplane after each projection onto the last hyperplane.

It can be shown that if the vectors $\mathbf{a}_1, \mathbf{a}_2, \ldots, \mathbf{a}_M$ span R^N, then the iterates $\mathbf{x}_M^{(1)}, \mathbf{x}_M^{(2)}, \mathbf{x}_M^{(3)}, \ldots$ lying on the Mth hyperplane will converge to a point $\mathbf{x}_M^*$ on that hyperplane which does not depend on the choice of the initial point $\mathbf{x}_0$. In computed tomography, one of the iterates $\mathbf{x}_M^{(p)}$ for p sufficiently large is taken as an approximate solution of the linear system for the pixel densities.

Note that for the center-of-pixel method, the scalar quantity $\mathbf{a}_k^T \mathbf{a}_k$ appearing in the equation in Step 2 of the algorithm is simply the number of pixels in which the kth beam passes through the center. Similarly, note that the scalar quantity

$$b_k - \mathbf{a}_k^T \mathbf{x}_{k-1}^{(p)}$$

in that same equation can be interpreted as the *excess kth beam density* that results if the pixel densities are set equal to the entries of $\mathbf{x}_{k-1}^{(p)}$. This provides the following interpretation of our ART iteration scheme for the center-of-pixel method: *Generate the pixel densities of each iterate by distributing the excess beam density of successive beams in the scan evenly among those pixels in which the beam passes through the center. When the last beam in the scan has been reached, return to the first beam and continue.*

▶ **EXAMPLE 2** Using Algorithm 2

We can use Algorithm 2 to find the unknown pixel densities of the 9 pixels arranged in the 3×3 array illustrated in Figure 10.12.11. These 9 pixels are scanned using the parallel mode with 12 beams whose measured beam densities are indicated in the figure. We choose the center-of-pixel method to set up the 12 beam equations. (In Exercises 7 and 8, you are asked to set up the beam equations using the center line and area methods.) As you can verify, the beam equations are

$$
\begin{array}{ll}
x_7 + x_8 + x_9 = 13.00 & x_3 + x_6 + x_9 = 18.00 \\
x_4 + x_5 + x_6 = 15.00 & x_2 + x_5 + x_8 = 12.00 \\
x_1 + x_2 + x_3 = 8.00 & x_1 + x_4 + x_7 = 6.00 \\
x_6 + x_8 + x_9 = 14.79 & x_2 + x_3 + x_6 = 10.51 \\
x_3 + x_5 + x_7 = 14.31 & x_1 + x_5 + x_9 = 16.13 \\
x_1 + x_2 + x_4 = 3.81 & x_4 + x_7 + x_8 = 7.04
\end{array}
$$

Table 2 illustrates the results of the iteration scheme starting with an initial iterate $\mathbf{x}_0 = \mathbf{0}$. The table gives the values of each of the first cycle of iterates, $\mathbf{x}_1^{(1)}$ through $\mathbf{x}_{12}^{(1)}$, but thereafter gives the iterates $\mathbf{x}_{12}^{(p)}$ only for various values of p. The iterates $\mathbf{x}_{12}^{(p)}$ start repeating to two decimal places for $p \geq 45$, and so we take the entries of $\mathbf{x}_{12}^{(45)}$ as approximate values of the 9 pixel densities. ◀

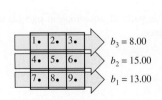
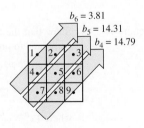
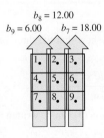
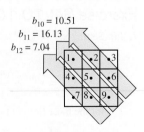

▲ Figure 10.12.11

We close this section by noting that the field of computed tomography is presently a very active research area. In fact, the ART scheme discussed here has been replaced in commercial systems by more sophisticated techniques that are faster and provide a more accurate view of the cross section. However, all the new techniques address the same basic mathematical problem: finding a good approximate solution of a large overdetermined inconsistent linear system of equations.

Table 2

		Pixel Densities								
		x_1	x_2	x_3	x_4	x_5	x_6	x_7	x_8	x_9
	$\mathbf{x}_0$	.00	.00	.00	.00	.00	.00	.00	.00	.00
	$\mathbf{x}_1^{(1)}$	.00	.00	.00	.00	.00	.00	4.33	4.33	4.33
	$\mathbf{x}_2^{(1)}$	.00	.00	.00	5.00	5.00	5.00	4.33	4.33	4.33
	$\mathbf{x}_3^{(1)}$	2.67	2.67	2.67	5.00	5.00	5.00	4.33	4.33	4.33
First Cycle of Iterates	$\mathbf{x}_4^{(1)}$	2.67	2.67	2.67	5.00	5.00	5.37	4.33	4.71	4.71
	$\mathbf{x}_5^{(1)}$	2.67	2.67	3.44	5.00	5.77	5.37	5.10	4.71	4.71
	$\mathbf{x}_6^{(1)}$	.49	.49	3.44	2.83	5.77	5.37	5.10	4.71	4.71
	$\mathbf{x}_7^{(1)}$	.49	.49	4.93	2.83	5.77	6.87	5.10	4.71	6.20
	$\mathbf{x}_8^{(1)}$	.49	.84	4.93	2.83	6.11	6.87	5.10	5.05	6.20
	$\mathbf{x}_9^{(1)}$	−.31	.84	4.93	2.02	6.11	6.87	4.30	5.05	6.20
	$\mathbf{x}_{10}^{(1)}$	−.31	.13	4.22	2.02	6.11	6.16	4.30	5.05	6.20
	$\mathbf{x}_{11}^{(1)}$	1.06	.13	4.22	2.02	7.49	6.16	4.30	5.05	7.58
	$\mathbf{x}_{12}^{(1)}$	1.06	.13	4.22	.58	7.49	6.16	2.85	3.61	7.58
	$\mathbf{x}_{12}^{(2)}$	2.03	.69	4.42	1.34	7.49	5.39	2.65	3.04	6.61
	$\mathbf{x}_{12}^{(3)}$	1.78	.51	4.52	1.26	7.49	5.48	2.56	3.22	6.86
	$\mathbf{x}_{12}^{(4)}$	1.82	.52	4.62	1.37	7.49	5.37	2.45	3.22	6.82
	$\mathbf{x}_{12}^{(5)}$	1.79	.49	4.71	1.43	7.49	5.31	2.37	3.25	6.85
	$\mathbf{x}_{12}^{(10)}$	1.68	.44	5.03	1.70	7.49	5.03	2.04	3.29	6.96
	$\mathbf{x}_{12}^{(20)}$	1.49	.48	5.29	2.00	7.49	4.73	1.79	3.25	7.15
	$\mathbf{x}_{12}^{(30)}$	1.38	.55	5.34	2.11	7.49	4.62	1.74	3.19	7.26
	$\mathbf{x}_{12}^{(40)}$	1.33	.59	5.33	2.14	7.49	4.59	1.75	3.15	7.31
	$\mathbf{x}_{12}^{(45)}$	1.32	.60	5.32	2.15	7.49	4.59	1.76	3.14	7.32

Exercise Set 10.12

1. (a) Setting $\mathbf{x}_k^{(p)} = (x_{k1}^{(p)}, x_{k2}^{(p)})$, show that the three projection equations

$$\mathbf{x}_k^{(p)} = \mathbf{x}_{k-1}^{(p)} + \frac{(b_k - \mathbf{a}_k^T \mathbf{x}_{k-1}^{(p)})}{\mathbf{a}_k^T \mathbf{a}_k} \mathbf{a}_k, \quad k = 1, 2, 3$$

for the three lines in Equation (5) can be written as

$k = 1$:
$$x_{11}^{(p)} = \tfrac{1}{2}[2 + x_{01}^{(p)} - x_{02}^{(p)}]$$
$$x_{12}^{(p)} = \tfrac{1}{2}[2 - x_{01}^{(p)} + x_{02}^{(p)}]$$

$k = 2$:
$$x_{21}^{(p)} = \tfrac{1}{5}[-2 + 4x_{11}^{(p)} + 2x_{12}^{(p)}]$$
$$x_{22}^{(p)} = \tfrac{1}{5}[4 + 2x_{11}^{(p)} + x_{12}^{(p)}]$$

$k = 3$:
$$x_{31}^{(p)} = \tfrac{1}{10}[9 + x_{21}^{(p)} + 3x_{22}^{(p)}]$$
$$x_{32}^{(p)} = \tfrac{1}{10}[-3 + 3x_{21}^{(p)} + 9x_{22}^{(p)}]$$

where $(x_{01}^{(p+1)}, x_{02}^{(p+1)}) = (x_{31}^{(p)}, x_{32}^{(p)})$ for $p = 1, 2, \ldots$.

(b) Show that the three pairs of equations in part (a) can be combined to produce

$$x_{31}^{(p)} = \tfrac{1}{20}[28 + x_{31}^{(p-1)} - x_{32}^{(p-1)}]$$
$$x_{32}^{(p)} = \tfrac{1}{20}[24 + 3x_{31}^{(p-1)} - 3x_{32}^{(p-1)}]$$
$p = 1, 2, \ldots$

where $(x_{31}^{(0)}, x_{32}^{(0)}) = (x_{01}^{(1)}, x_{02}^{(1)}) = \mathbf{x}_0^{(1)}$. [*Note:* Using this pair of equations, we can perform one complete cycle of three orthogonal projections in a single step.]

(c) Because $\mathbf{x}_3^{(p)}$ tends to the limit point $\mathbf{x}_3^*$ as $p \to \infty$, the equations in part (b) become

$$x_{31}^* = \tfrac{1}{20}[28 + x_{31}^* - x_{32}^*]$$
$$x_{32}^* = \tfrac{1}{20}[24 + 3x_{31}^* - 3x_{32}^*]$$

as $p \to \infty$. Solve this linear system for $\mathbf{x}_3^* = (x_{31}^*, x_{32}^*)$. [*Note:* The simplifications of the ART formulas described in this exercise are impractical for the large linear systems that arise in realistic computed tomography problems.]

2. Use the result of Exercise 1(b) to find $\mathbf{x}_3^{(1)}, \mathbf{x}_3^{(2)}, \ldots, \mathbf{x}_3^{(6)}$ to five decimal places in Example 1 using the following initial points:
   (a) $\mathbf{x}_0 = (0, 0)$
   (b) $\mathbf{x}_0 = (1, 1)$
   (c) $\mathbf{x}_0 = (148, -15)$

3. (a) Show directly that the points of the limit cycle in Example 1,
   $$\mathbf{x}_1^* = \left(\tfrac{12}{11}, \tfrac{10}{11}\right), \quad \mathbf{x}_2^* = \left(\tfrac{46}{55}, \tfrac{78}{55}\right), \quad \mathbf{x}_3^* = \left(\tfrac{31}{22}, \tfrac{27}{22}\right)$$
   form a triangle whose vertices lie on the lines L_1, L_2, and L_3 and whose sides are perpendicular to these lines (Figure 10.12.9c).

   (b) Using the equations derived in Exercise 1(a), show that if $\mathbf{x}_0^{(1)} = \mathbf{x}_3^* = \left(\tfrac{31}{22}, \tfrac{27}{22}\right)$, then
   $$\mathbf{x}_1^{(1)} = \mathbf{x}_1^* = \left(\tfrac{12}{11}, \tfrac{10}{11}\right)$$
   $$\mathbf{x}_2^{(1)} = \mathbf{x}_2^* = \left(\tfrac{46}{55}, \tfrac{78}{55}\right)$$
   $$\mathbf{x}_3^{(1)} = \mathbf{x}_3^* = \left(\tfrac{31}{22}, \tfrac{27}{22}\right)$$

   [Note: Either part of this exercise shows that successive orthogonal projections of any point on the limit cycle will move around the limit cycle indefinitely.]

4. The following three lines in the $x_1 x_2$-plane,
   $$L_1: \quad x_2 = 1$$
   $$L_2: \quad x_1 - x_2 = 2$$
   $$L_3: \quad x_1 - x_2 = 0$$
   do not have a common intersection. Draw an accurate sketch of the three lines and graphically perform several cycles of the orthogonal projections described in Algorithm 1, beginning with the initial point $\mathbf{x}_0 = (0, 0)$. On the basis of your sketch, determine the three points of the limit cycle.

5. Prove Theorem 10.12.1 by verifying that
   (a) the point $\mathbf{x}_p$ as defined in the theorem lies on the line $\mathbf{a}^T \mathbf{x} = b$ (i.e., $\mathbf{a}^T \mathbf{x}_p = b$).
   (b) the vector $\mathbf{x}_p - \mathbf{x}^*$ is orthogonal to the line $\mathbf{a}^T \mathbf{x} = b$ (i.e., $\mathbf{x}_p - \mathbf{x}^*$ is parallel to $\mathbf{a}$).

6. As stated in the text, the iterates $\mathbf{x}_M^{(1)}, \mathbf{x}_M^{(2)}, \mathbf{x}_M^{(3)}, \ldots$ defined in Algorithm 2 will converge to a unique limit point $\mathbf{x}_M^*$ if the vectors $\mathbf{a}_1, \mathbf{a}_2, \ldots, \mathbf{a}_M$ span R^N. Show that if this is the case and if the center-of-pixel method is used, then the center of each of the N pixels in the field of view is crossed by at least one of the M beams in the scan.

7. Construct the 12 beam equations in Example 2 using the center line method. Assume that the distance between the center lines of adjacent beams is equal to the width of a single pixel.

8. Construct the 12 beam equations in Example 2 using the area method. Assume that the width of each beam is equal to the width of a single pixel and that the distance between the center lines of adjacent beams is also equal to the width of a single pixel.

Section 10.12 Technology Exercises

The following exercises are designed to be solved using a technology utility. Typically, this will be MATLAB, *Mathematica*, Maple, Derive, or Mathcad, but it may also be some other type of linear algebra software or a scientific calculator with some linear algebra capabilities. For each exercise you will need to read the relevant documentation for the particular utility you are using. The goal of these exercises is to provide you with a basic proficiency with your technology utility. Once you have mastered the techniques in these exercises, you will be able to use your technology utility to solve many of the problems in the regular exercise sets.

T1. Given the set of equations
$$a_k x + b_k y = c_k$$
for $k = 1, 2, 3, \ldots, n$ (with $n > 2$), let us consider the following algorithm for obtaining an approximate solution to the system.

1. Solve all possible pairs of equations
   $$a_i x + b_i y = c_i \quad \text{and} \quad a_j x + b_j y = c_j$$
   for $i, j = 1, 2, 3, \ldots, n$ and $i < j$ for their unique solutions.

This leads to
$$\tfrac{1}{2} n(n-1)$$
solutions, which we label as
$$(x_{ij}, y_{ij})$$
for $i, j = 1, 2, 3, \ldots, n$ and $i < j$.

2. Construct the geometric center of these points defined by
$$(x_C, y_C) = \left(\frac{2}{n(n-1)} \sum_{i=1}^{n-1} \sum_{j=i+1}^{n} x_{ij}, \ \frac{2}{n(n-1)} \sum_{i=1}^{n-1} \sum_{j=i+1}^{n} y_{ij} \right)$$
and use this as the approximate solution to the original system.

Use this algorithm to approximate the solution to the system
$$x + y = 2$$
$$x - 2y = -2$$
$$3x - y = 3$$
and compare your results to those in this section.

T2. (*Calculus required*) Given the set of equations

$$a_k x + b_k y = c_k$$

for $k = 1, 2, 3, \ldots, n$ (with $n > 2$), let us consider the following least squares algorithm for obtaining an approximate solution (x^*, y^*) to the system. Given a point (α, β) and the line $a_i x + b_i y = c_i$, the distance from this point to the line is given by

$$\frac{|a_i \alpha + b_i \beta - c_i|}{\sqrt{a_i^2 + b_i^2}}$$

If we define a function $f(x, y)$ by

$$f(x, y) = \sum_{i=1}^{n} \frac{(a_i x + b_i y - c_i)^2}{a_i^2 + b_i^2}$$

and then determine the point (x^*, y^*) that minimizes this function, we will determine the point that is *closest* to each of these lines in a summed least squares sense. Show that x^* and y^* are solutions to the system

$$\left(\sum_{i=1}^{n} \frac{a_i^2}{a_i^2 + b_i^2}\right) x^* + \left(\sum_{i=1}^{n} \frac{a_i b_i}{a_i^2 + b_i^2}\right) y^* = \sum_{i=1}^{n} \frac{a_i c_i}{a_i^2 + b_i^2}$$

and

$$\left(\sum_{i=1}^{n} \frac{a_i b_i}{a_i^2 + b_i^2}\right) x^* + \left(\sum_{i=1}^{n} \frac{b_i^2}{a_i^2 + b_i^2}\right) y^* = \sum_{i=1}^{n} \frac{b_i c_i}{a_i^2 + b_i^2}$$

Apply this algorithm to the system

$$\begin{aligned} x + y &= 2 \\ x - 2y &= -2 \\ 3x - y &= 3 \end{aligned}$$

and compare your results to those in this section.

10.13 Fractals

In this section we will use certain classes of linear transformations to describe and generate intricate sets in the Euclidean plane. These sets, called fractals, are currently the focus of much mathematical and scientific research.

> **PREREQUISITES:** Geometry of Linear Operators on R^2 (Section 4.11)
> Euclidean Space R^n
> Natural Logarithms
> Intuitive Understanding of Limits

Fractals in the Euclidean Plane

At the end of the nineteenth century and the beginning of the twentieth century, various bizarre and wild sets of points in the Euclidean plane began appearing in mathematics. Although they were initially mathematical curiosities, these sets, called *fractals*, are rapidly growing in importance. It is now recognized that they reveal a regularity in physical and biological phenomena previously dismissed as "random," "noisy," or "chaotic." For example, fractals are all around us in the shapes of clouds, mountains, coastlines, trees, and ferns.

In this section we give a brief description of certain types of fractals in the Euclidean plane R^2. Much of this description is an outgrowth of the work of two mathematicians, Benoit B. Mandelbrot and Michael Barnsley, who are both active researchers in the field.

Self-Similar Sets

To begin our study of fractals, we need to introduce some terminology about sets in R^2. We will call a set in R^2 **bounded** if it can be enclosed by a suitably large circle (Figure 10.13.1) and **closed** if it contains all of its boundary points (Figure 10.13.2). Two sets in R^2 will be called **congruent** if they can be made to coincide exactly by translating and rotating them appropriately within R^2 (Figure 10.13.3). We will also rely on your intuitive concept of **overlapping** and **nonoverlapping sets**, as illustrated in Figure 10.13.4.

If $T: R^2 \to R^2$ is the linear operator that scales by a factor of s (see Table 7 of Section 4.9), and if Q is a set in R^2, then the set $T(Q)$ (the set of images of points in Q under T) is called a **dilation** of the set Q if $s > 1$ and a **contraction** of Q if $0 < s < 1$ (Figure 10.13.5). In either case we say that $T(Q)$ is the set Q **scaled by the factor** s.

10.13 Fractals

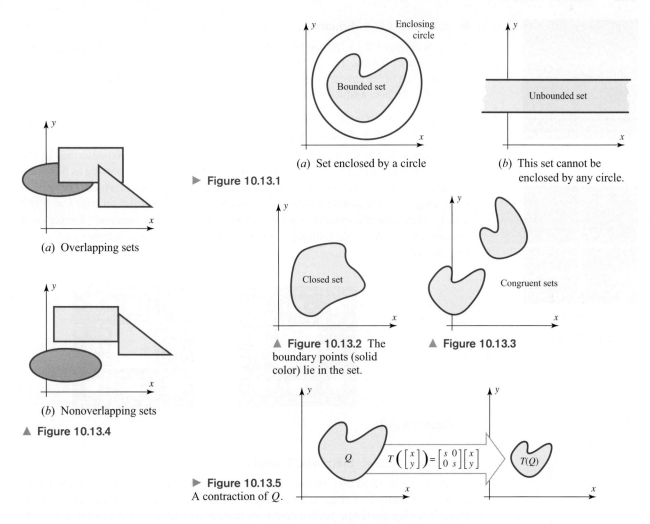

(a) Overlapping sets

(b) Nonoverlapping sets

▲ Figure 10.13.4

▶ Figure 10.13.1

▲ Figure 10.13.2 The boundary points (solid color) lie in the set.

▲ Figure 10.13.3

▶ Figure 10.13.5 A contraction of Q.

The types of fractals we will consider first are called *self-similar*. In general, we define a self-similar set in R^2 as follows:

DEFINITION 1 A closed and bounded subset of the Euclidean plane R^2 is said to be ***self-similar*** if it can be expressed in the form

$$S = S_1 \cup S_2 \cup S_3 \cup \cdots \cup S_k \tag{1}$$

where $S_1, S_2, S_3, \ldots, S_k$ are nonoverlapping sets, each of which is congruent to S scaled by the same factor s $(0 < s < 1)$.

If S is a self-similar set, then (1) is sometimes called a ***decomposition*** of S into nonoverlapping congruent sets.

▶ EXAMPLE 1 Line Segment

A line segment in R^2 (Figure 10.13.6a) can be expressed as the union of two nonoverlapping congruent line segments (Figure 10.13.6b). In Figure 10.13.6b we have separated the two line segments slightly so that they can be seen more easily. Each of these two smaller line segments is congruent to the original line segment scaled by a factor of $\frac{1}{2}$. Hence, a line segment is a self-similar set with $k = 2$ and $s = \frac{1}{2}$.

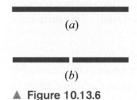

▲ Figure 10.13.6

(a)

(b)

▲ Figure 10.13.7

▶ EXAMPLE 2 Square

A square (Figure 10.13.7a) can be expressed as the union of four nonoverlapping congruent squares (Figure 10.13.7b), where we have again separated the smaller squares slightly. Each of the four smaller squares is congruent to the original square scaled by a factor of $\frac{1}{2}$. Hence, a square is a self-similar set with $k = 4$ and $s = \frac{1}{2}$.

▶ EXAMPLE 3 Sierpinski Carpet

The set suggested by Figure 10.13.8a, the Sierpinski "carpet," was first described by the Polish mathematician Waclaw Sierpinski (1882–1969). It can be expressed as the union of eight nonoverlapping congruent subsets (Figure 10.13.8b), each of which is congruent to the original set scaled by a factor of $\frac{1}{3}$. Hence, it is a self-similar set with $k = 8$ and $s = \frac{1}{3}$. Note that the intricate square-within-a-square pattern continues forever on a smaller and smaller scale (although this can only be suggested in a figure such as the one shown).

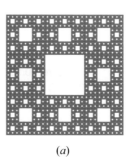 

▶ Figure 10.13.8   (a)   (b)

▶ EXAMPLE 4 Sierpinski Triangle

Figure 10.13.9a illustrates another set described by Sierpinski. It is a self-similar set with $k = 3$ and $s = \frac{1}{2}$ (Figure 10.13.9b). As with the Sierpinski carpet, the intricate triangle-within-a-triangle pattern continues forever on a smaller and smaller scale. ◀

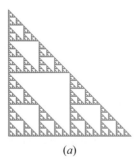 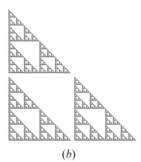

▶ Figure 10.13.9   (a)   (b)

The Sierpinski carpet and triangle have a more intricate structure than the line segment and the square in that they exhibit a pattern that is repeated indefinitely. This difference will be explored later in this section.

Topological Dimension of a Set

In Section 4.5 we defined the dimension of a *subspace* of a vector space to be the number of vectors in a basis, and we found that definition to coincide with our intuitive sense of dimension. For example, the origin of R^2 is zero-dimensional, lines through the origin are one-dimensional, and R^2 itself is two-dimensional. This definition of dimension is a special case of a more general concept called **topological dimension**, which is applicable to sets in R^n that are not necessarily subspaces. A precise definition of this concept is studied in a branch of mathematics called *topology*. Although that definition is beyond the scope of this text, we can state informally that

- a point in R^2 has topological dimension zero;
- a curve in R^2 has topological dimension one;
- a region in R^2 has topological dimension two.

It can be proved that the topological dimension of a set in R^n must be an integer between 0 and n, inclusive. In this text we will denote the topological dimension of a set S by $d_T(S)$.

Table 1

Set S	$d_T(S)$
Line segment	1
Square	2
Sierpinski carpet	1
Sierpinski triangle	1

▶ **EXAMPLE 5** Topological Dimensions of Sets

Table 1 gives the topological dimensions of the sets studied in our earlier examples. The first two results in this table are intuitively obvious; however, the last two are not. Informally stated, the Sierpinski carpet and triangle both contain so many "holes" that those sets resemble web-like networks of lines rather than regions. Hence they have topological dimension one. The proofs are quite difficult. ◀

Hausdorff Dimension of a Self-Similar Set

In 1919 the German mathematician Felix Hausdorff (1868–1942) gave an alternative definition for the dimension of an arbitrary set in R^n. His definition is quite complicated, but for a self-similar set, it reduces to something rather simple:

DEFINITION 2 The ***Hausdorff dimension*** of a self-similar set S of form (1) is denoted by $d_H(S)$ and is defined by

$$d_H(S) = \frac{\ln k}{\ln(1/s)} \qquad (2)$$

In this definition, "ln" denotes the natural logarithm function. Equation (2) can also be expressed as

$$s^{d_H(S)} = \frac{1}{k} \qquad (3)$$

in which the Hausdorff dimension $d_H(S)$ appears as an exponent. Formula (3) is more helpful for interpreting the concept of Hausdorff dimension; it states, for example, that if you scale a self-similar set by a factor of $s = \frac{1}{2}$, then its area (or more properly its *measure*) decreases by a factor of $\left(\frac{1}{2}\right)^{d_H(S)}$. Thus, scaling a line segment by a factor of $\frac{1}{2}$ reduces its measure (length) by a factor of $\left(\frac{1}{2}\right)^1 = \frac{1}{2}$, and scaling a square region by a factor of $\frac{1}{2}$ reduces its measure (area) by a factor of $\left(\frac{1}{2}\right)^2 = \frac{1}{4}$.

Before proceeding to some examples, we should note a few facts about the Hausdorff dimension of a set:

- The topological dimension and Hausdorff dimension of a set need not be the same.
- The Hausdorff dimension of a set need not be an integer.
- The topological dimension of a set is less than or equal to its Hausdorff dimension; that is, $d_T(S) \leq d_H(S)$.

▶ **EXAMPLE 6** **Hausdorff Dimensions of Sets**

Table 2 lists the Hausdorff dimensions of the sets studied in our earlier examples.

Table 2

Set S	s	k	$d_H(S) = \dfrac{\ln k}{\ln (1/s)}$
Line segment	$\tfrac{1}{2}$	2	$\ln 2/\ln 2 = 1$
Square	$\tfrac{1}{2}$	4	$\ln 4/\ln 2 = 2$
Sierpinski carpet	$\tfrac{1}{3}$	8	$\ln 8/\ln 3 = 1.892\ldots$
Sierpinski triangle	$\tfrac{1}{2}$	3	$\ln 3/\ln 2 = 1.584\ldots$

◀

Fractals Comparing Tables 1 and 2, we see that the Hausdorff and topological dimensions are equal for both the line segment and square but are unequal for the Sierpinski carpet and triangle. In 1977 Benoit B. Mandelbrot suggested that sets for which the topological and Hausdorff dimensions differ must be quite complicated (as Hausdorff had earlier suggested in 1919). Mandelbrot proposed calling such sets ***fractals***, and he offered the following definition.

> **DEFINITION 3** A ***fractal*** is a subset of a Euclidean space whose Hausdorff dimension and topological dimension are not equal.

According to this definition, the Sierpinski carpet and Sierpinski triangle are fractals, whereas the line segment and square are not.

It follows from the preceding definition that a set whose Hausdorff dimension is not an integer must be a fractal (why?). However, we will see later that the converse is not true; that is, it is possible for a fractal to have an integer Hausdorff dimension.

Similitudes We will now show how some techniques from linear algebra can be used to generate fractals. This linear algebra approach also leads to algorithms that can be exploited to draw fractals on a computer. We begin with a definition.

> **DEFINITION 4** A ***similitude*** with scale factor s is a mapping of R^2 into R^2 of the form
> $$T\left(\begin{bmatrix} x \\ y \end{bmatrix}\right) = s \begin{bmatrix} \cos\theta & -\sin\theta \\ \sin\theta & \cos\theta \end{bmatrix} \begin{bmatrix} x \\ y \end{bmatrix} + \begin{bmatrix} e \\ f \end{bmatrix}$$
> where s, θ, e, and f are scalars.

Geometrically, a similitude is a composition of three simpler mappings: a scaling by a factor of s, a rotation about the origin through an angle θ, and a translation (e units in the x-direction and f units in the y-direction). Figure 10.13.10 illustrates the effect of a similitude on the unit square U.

For our application to fractals, we will need only similitudes that are ***contractions***, by which we mean that the scale factor s is restricted to the range $0 < s < 1$. Consequently, when we refer to similitudes we will always mean similitudes subject to this restriction.

10.13 Fractals

Figure 10.13.10

(a) Unit square

(b) Unit square after similitude

Similitudes are important in the study of fractals because of the following fact:

> If $T: R^2 \to R^2$ is a similitude with scale factor s and if S is a closed and bounded set in R^2, then the image $T(S)$ of the set S under T is congruent to S scaled by s.

Recall from the definition of a self-similar set in R^2 that a closed and bounded set S in R^2 is self-similar if it can be expressed in the form

$$S = S_1 \cup S_2 \cup S_3 \cup \cdots \cup S_k$$

where $S_1, S_2, S_3, \ldots, S_k$ are nonoverlapping sets each of which is congruent to S scaled by the same factor s ($0 < s < 1$) [see (1)]. In the following examples, we will find similitudes that produce the sets $S_1, S_2, S_3, \ldots, S_k$ from S for the line segment, square, Sierpinski carpet, and Sierpinski triangle.

► EXAMPLE 7 Line Segment

We will take as our line segment the line segment S connecting the points $(0, 0)$ and $(1, 0)$ in the xy-plane (Figure 10.13.11a). Consider the two similitudes

$$T_1\left(\begin{bmatrix} x \\ y \end{bmatrix}\right) = \frac{1}{2}\begin{bmatrix} 1 & 0 \\ 0 & 1 \end{bmatrix}\begin{bmatrix} x \\ y \end{bmatrix}$$

$$T_2\left(\begin{bmatrix} x \\ y \end{bmatrix}\right) = \frac{1}{2}\begin{bmatrix} 1 & 0 \\ 0 & 1 \end{bmatrix}\begin{bmatrix} x \\ y \end{bmatrix} + \begin{bmatrix} \frac{1}{2} \\ 0 \end{bmatrix}$$

(4)

both of which have $s = \frac{1}{2}$ and $\theta = 0$. In Figure 10.13.11b we show how these two similitudes map the unit square U. The similitude T_1 maps U onto the smaller square $T_1(U)$, and the similitude T_2 maps U onto the smaller square $T_2(U)$. At the same time, T_1 maps the line segment S onto the smaller line segment $T_1(S)$, and T_2 maps S onto the smaller nonoverlapping line segment $T_2(S)$. The union of these two smaller nonoverlapping line segments is precisely the original line segment S; that is,

$$S = T_1(S) \cup T_2(S) \tag{5}$$

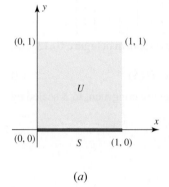

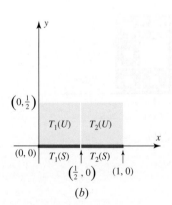

▲ **Figure 10.13.11**

► EXAMPLE 8 Square

Let us consider the unit square U in the xy-plane (Figure 10.13.12a) and the following four similitudes, all having $s = \frac{1}{2}$ and $\theta = 0$:

$$T_1\left(\begin{bmatrix} x \\ y \end{bmatrix}\right) = \frac{1}{2}\begin{bmatrix} 1 & 0 \\ 0 & 1 \end{bmatrix}\begin{bmatrix} x \\ y \end{bmatrix} \qquad T_2\left(\begin{bmatrix} x \\ y \end{bmatrix}\right) = \frac{1}{2}\begin{bmatrix} 1 & 0 \\ 0 & 1 \end{bmatrix}\begin{bmatrix} x \\ y \end{bmatrix} + \begin{bmatrix} \frac{1}{2} \\ 0 \end{bmatrix}$$

$$T_3\left(\begin{bmatrix} x \\ y \end{bmatrix}\right) = \frac{1}{2}\begin{bmatrix} 1 & 0 \\ 0 & 1 \end{bmatrix}\begin{bmatrix} x \\ y \end{bmatrix} + \begin{bmatrix} 0 \\ \frac{1}{2} \end{bmatrix} \qquad T_4\left(\begin{bmatrix} x \\ y \end{bmatrix}\right) = \frac{1}{2}\begin{bmatrix} 1 & 0 \\ 0 & 1 \end{bmatrix}\begin{bmatrix} x \\ y \end{bmatrix} + \begin{bmatrix} \frac{1}{2} \\ \frac{1}{2} \end{bmatrix}$$

(6)

632 Chapter 10 Applications of Linear Algebra

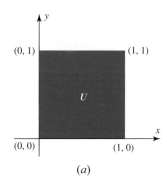

(a)

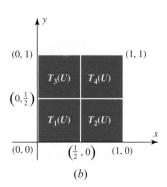

(b)

▲ Figure 10.13.12

The images of the unit square U under these four similitudes are the four squares shown in Figure 10.13.12b. Thus,

$$U = T_1(U) \cup T_2(U) \cup T_3(U) \cup T_4(U) \tag{7}$$

is a decomposition of U into four nonoverlapping squares that are congruent to U scaled by the same scale factor $\left(s = \frac{1}{2}\right)$.

▶ **EXAMPLE 9** Sierpinski Carpet

Let us consider a Sierpinski carpet S over the unit square U of the xy-plane (Figure 10.13.13a) and the following eight similitudes, all having $s = \frac{1}{3}$ and $\theta = 0$:

$$T_i\left(\begin{bmatrix} x \\ y \end{bmatrix}\right) = \frac{1}{3}\begin{bmatrix} 1 & 0 \\ 0 & 1 \end{bmatrix}\begin{bmatrix} x \\ y \end{bmatrix} + \begin{bmatrix} e_i \\ f_i \end{bmatrix}, \quad i = 1, 2, 3, \ldots, 8 \tag{8}$$

where the eight values of $\begin{bmatrix} e_i \\ f_i \end{bmatrix}$ are

$$\begin{bmatrix} 0 \\ 0 \end{bmatrix}, \begin{bmatrix} \frac{1}{3} \\ 0 \end{bmatrix}, \begin{bmatrix} \frac{2}{3} \\ 0 \end{bmatrix}, \begin{bmatrix} 0 \\ \frac{1}{3} \end{bmatrix}, \begin{bmatrix} \frac{2}{3} \\ \frac{1}{3} \end{bmatrix}, \begin{bmatrix} 0 \\ \frac{2}{3} \end{bmatrix}, \begin{bmatrix} \frac{1}{3} \\ \frac{2}{3} \end{bmatrix}, \begin{bmatrix} \frac{2}{3} \\ \frac{2}{3} \end{bmatrix}$$

The images of S under these eight similitudes are the eight sets shown in Figure 10.13.13b. Thus,

$$S = T_1(S) \cup T_2(S) \cup T_3(S) \cup \cdots \cup T_8(S) \tag{9}$$

is a decomposition of S into eight nonoverlapping sets that are congruent to S scaled by the same scale factor $\left(s = \frac{1}{3}\right)$.

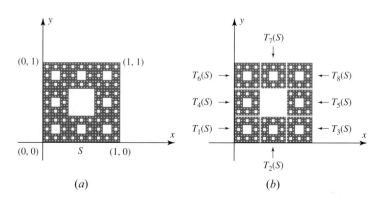

▶ Figure 10.13.13   (a)   (b)

▶ **EXAMPLE 10** Sierpinski Triangle

Let us consider a Sierpinski triangle S fitted inside the unit square U of the xy-plane, as shown in Figure 10.13.14a, and the following three similitudes, all having $s = \frac{1}{2}$ and $\theta = 0$:

$$T_1\left(\begin{bmatrix} x \\ y \end{bmatrix}\right) = \frac{1}{2}\begin{bmatrix} 1 & 0 \\ 0 & 1 \end{bmatrix}\begin{bmatrix} x \\ y \end{bmatrix}$$

$$T_2\left(\begin{bmatrix} x \\ y \end{bmatrix}\right) = \frac{1}{2}\begin{bmatrix} 1 & 0 \\ 0 & 1 \end{bmatrix}\begin{bmatrix} x \\ y \end{bmatrix} + \begin{bmatrix} \frac{1}{2} \\ 0 \end{bmatrix} \tag{10}$$

$$T_3\left(\begin{bmatrix} x \\ y \end{bmatrix}\right) = \frac{1}{2}\begin{bmatrix} 1 & 0 \\ 0 & 1 \end{bmatrix}\begin{bmatrix} x \\ y \end{bmatrix} + \begin{bmatrix} 0 \\ \frac{1}{2} \end{bmatrix}$$

The images of S under these three similitudes are the three sets in Figure 10.13.14b. Thus,

$$S = T_1(S) \cup T_2(S) \cup T_3(S) \tag{11}$$

is a decomposition of S into three nonoverlapping sets that are congruent to S scaled by the same scale factor $\left(s = \frac{1}{2}\right)$. ◄

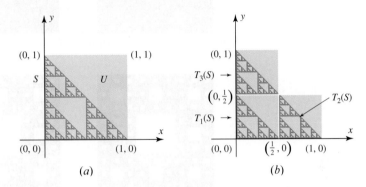

► Figure 10.13.14

In the preceding examples we started with a specific set S and showed that it was self-similar by finding similitudes $T_1, T_2, T_3, \ldots, T_k$ with the same scale factor such that $T_1(S), T_2(S), T_3(S), \ldots, T_k(S)$ were nonoverlapping sets and such that

$$S = T_1(S) \cup T_2(S) \cup T_3(S) \cup \cdots \cup T_k(S) \tag{12}$$

The following theorem addresses the converse problem of determining a self-similar set from a collection of similitudes.

THEOREM 10.13.1 *If $T_1, T_2, T_3, \ldots, T_k$ are contracting similitudes with the same scale factor, then there is a unique nonempty closed and bounded set S in the Euclidean plane such that*

$$S = T_1(S) \cup T_2(S) \cup T_3(S) \cup \cdots \cup T_k(S)$$

Furthermore, if the sets $T_1(S), T_2(S), T_3(S), \ldots, T_k(S)$ are nonoverlapping, then S is self-similar.

Algorithms for Generating Fractals

In general, there is no simple way to obtain the set S in the preceding theorem directly. We now describe an iterative procedure that will determine S from the similitudes that define it. We first give an example of the procedure and then give an algorithm for the general case.

► **EXAMPLE 11** **Sierpinski Carpet**

Figure 10.13.15 shows the unit square region S_0 in the xy-plane, which will serve as an "initial" set for an iterative procedure for the construction of the Sierpinski carpet. The set S_1 in the figure is the result of mapping S_0 with each of the eight similitudes T_i ($i = 1, 2, \ldots, 8$) in (8) that determine the Sierpinski carpet. It consists of eight square regions, each of side length $\frac{1}{3}$, surrounding an empty middle square. Next we apply the eight similitudes to S_1 and arrive at the set S_2. Similarly, applying the eight similitudes to S_2 results in the set S_3. It we continue this process indefinitely, the sequence of sets $S_1, S_2, S_3, \ldots$ will "converge" to a set S, which is the Sierpinski carpet. ◄

634 Chapter 10 Applications of Linear Algebra

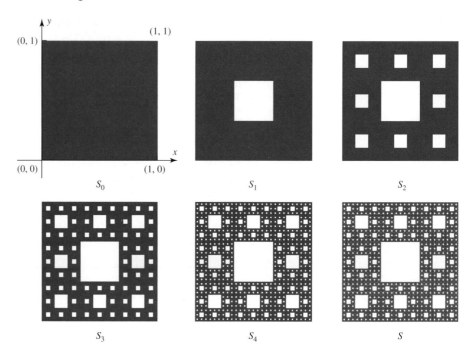

▶ Figure 10.13.15

Remark Although we should properly give a definition of what it means for a sequence of sets to "converge" to a given set, an intuitive interpretation will suffice in this introductory treatment.

Although we started in Figure 10.13.15 with the unit square region to arrive at the Sierpinski carpet, we could have started with any nonempty set S_0. The only restriction is that the set S_0 be closed and bounded. For example, if we start with the particular set S_0 shown in Figure 10.13.16, then S_1 is the set obtained by applying each of the eight similitudes in (8). Applying the eight similitudes to S_1 results in the set S_2. As before, applying the eight similitudes indefinitely yields the Sierpinski carpet S as the limiting set.

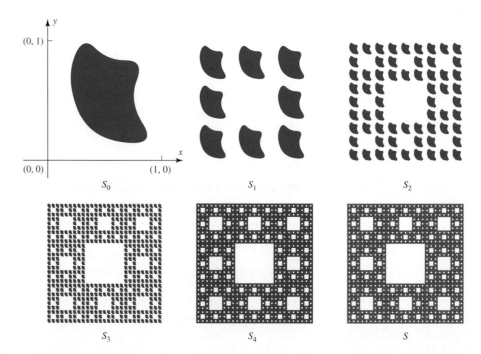

▶ Figure 10.13.16

The general algorithm illustrated in the preceding example is as follows: Let T_1, T_2, $T_3, \ldots, T_k$ be contracting similitudes with the same scale factor, and for an arbitrary set Q in R^2, define the set $\mathfrak{J}(Q)$ by

$$\mathfrak{J}(Q) = T_1(Q) \cup T_2(Q) \cup T_3(Q) \cup \cdots \cup T_k(Q)$$

The following algorithm generates a sequence of sets $S_0, S_1, \ldots, S_n, \ldots$ that converges to the set S in Theorem 10.13.1.

Algorithm 1

Step 0. Choose an arbitrary nonempty closed and bounded set S_0 in R^2.
Step 1. Compute $S_1 = \mathfrak{J}(S_0)$.
Step 2. Compute $S_2 = \mathfrak{J}(S_1)$.
Step 3. Compute $S_3 = \mathfrak{J}(S_2)$.
$\vdots$
Step n. Compute $S_n = \mathfrak{J}(S_{n-1})$.
$\vdots$

▶ **EXAMPLE 12** Sierpinski Triangle

Let us construct the Sierpinski triangle determined by the three similitudes given in (10). The corresponding set mapping is $\mathfrak{J}(Q) = T_1(Q) \cup T_2(Q) \cup T_3(Q)$. Figure 10.13.17 shows an arbitrary closed and bounded set S_0; the first four iterates S_1, S_2, S_3, S_4; and the limiting set S (the Sierpinski triangle).

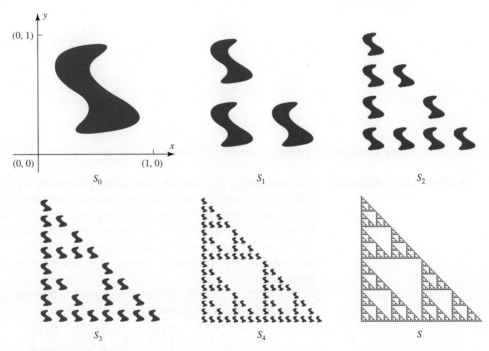

▲ Figure 10.13.17

► EXAMPLE 13 Using Algorithm 1

Consider the following two similitudes:

$$T_1\left(\begin{bmatrix}x\\y\end{bmatrix}\right) = \frac{1}{2}\begin{bmatrix}1 & 0\\0 & 1\end{bmatrix}$$

$$T_2\left(\begin{bmatrix}x\\y\end{bmatrix}\right) = \frac{1}{2}\begin{bmatrix}\cos\theta & -\sin\theta\\\sin\theta & \cos\theta\end{bmatrix}\begin{bmatrix}x\\y\end{bmatrix} + \begin{bmatrix}.3\\.3\end{bmatrix}$$

The actions of these two similitudes on the unit square U are illustrated in Figure 10.13.18. Here, the rotation angle θ is a parameter that we will vary to generate different self-similar sets. The self-similar sets determined by these two similitudes are shown in Figure 10.13.19 for various values of θ. For simplicity, we have not drawn the xy-axes, but in each case the origin is the lower left point of the set. These sets were generated on a computer using Algorithm 1 for the various values of θ. Because $k = 2$ and $s = \frac{1}{2}$, it follows from (2) that the Hausdorff dimension of these sets for any value of θ is 1. It can be shown that the topological dimension of these sets is 1 for $\theta = 0$ and 0 for all other values of θ. It follows that the self-similar set for $\theta = 0$ is not a fractal [it is the straight line segment from $(0, 0)$ to $(.6, .6)$], while the self-similar sets for all other values of θ are fractals. In particular, they are examples of fractals with integer Hausdorff dimension. ◄

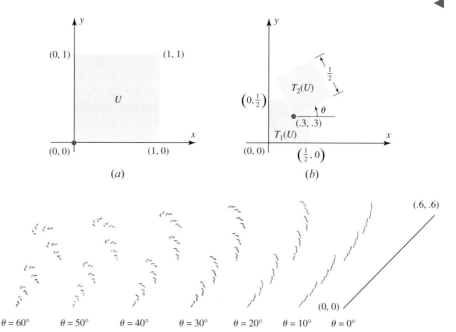

► Figure 10.13.18

► Figure 10.13.19 $\theta = 60°$ $\theta = 50°$ $\theta = 40°$ $\theta = 30°$ $\theta = 20°$ $\theta = 10°$ $\theta = 0°$

A Monte Carlo Approach   The set-mapping approach of constructing self-similar sets described in Algorithm 1 is rather time-consuming on a computer because the similitudes involved must be applied to each of the many computer screen pixels in the successive iterated sets. In 1985 Michael Barnsley described an alternative, more practical method of generating a self-similar set defined through its similitudes. It is a so-called **Monte Carlo method** that takes advantage of probability theory. Barnsley refers to it as the **Random Iteration Algorithm**.

Let $T_1, T_2, T_3, \ldots, T_k$ be contracting similitudes with the same scale factor. The following algorithm generates a sequence of points

$$\begin{bmatrix}x_0\\y_0\end{bmatrix}, \begin{bmatrix}x_1\\y_1\end{bmatrix}, \ldots, \begin{bmatrix}x_n\\y_n\end{bmatrix}, \ldots$$

that collectively converge to the set S in Theorem 10.13.1.

10.13 Fractals

Algorithm 2

Step 0. Choose an arbitrary point $\begin{bmatrix} x_0 \\ y_0 \end{bmatrix}$ in S.

Step 1. Choose one of the k similitudes at random, say T_{k_1}, and compute
$$\begin{bmatrix} x_1 \\ y_1 \end{bmatrix} = T_{k_1}\left(\begin{bmatrix} x_0 \\ y_0 \end{bmatrix}\right)$$

Step 2. Choose one of the k similitudes at random, say T_{k_2}, and compute
$$\begin{bmatrix} x_2 \\ y_2 \end{bmatrix} = T_{k_2}\left(\begin{bmatrix} x_1 \\ y_1 \end{bmatrix}\right)$$

$\vdots$

Step n. Choose one of the k similitudes at random, say T_{k_n}, and compute
$$\begin{bmatrix} x_n \\ y_n \end{bmatrix} = T_{k_n}\left(\begin{bmatrix} x_{n-1} \\ y_{n-1} \end{bmatrix}\right)$$

$\vdots$

On a computer screen the pixels corresponding to the points generated by this algorithm will fill out the pixel representation of the limiting set S.

Figure 10.13.20 shows four stages of the Random Iteration Algorithm that generate the Sierpinski carpet, starting with the initial point $\begin{bmatrix} 0 \\ 0 \end{bmatrix}$.

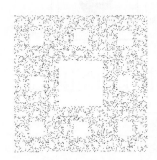 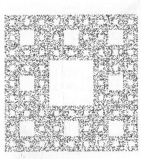 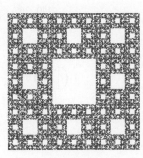 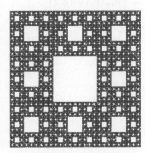

5000 iterations   15,000 iterations   45,000 iterations   100,000 iterations

▲ **Figure 10.13.20**

Remark Although Step 0 in the preceding algorithm requires the selection of an initial point in the set S, which may not be known in advance, this is not a serious problem. In practice, one can usually start with any point in R^2 and after a few iterations (say ten or so), the point generated will be sufficiently close to S that the algorithm will work correctly from that point on.

More General Fractals   So far, we have discussed fractals that are self-similar sets according to the definition of a self-similar set in R^2. However, Theorem 10.13.1 remains true if the similitudes $T_1, T_2, \ldots, T_k$ are replaced by more general transformations, called *contracting affine transformations*. An affine transformation is defined as follows:

DEFINITION 5 An ***affine transformation*** is a mapping of R^2 into R^2 of the form
$$T\left(\begin{bmatrix} x \\ y \end{bmatrix}\right) = \begin{bmatrix} a & b \\ c & d \end{bmatrix}\begin{bmatrix} x \\ y \end{bmatrix} + \begin{bmatrix} e \\ f \end{bmatrix}$$
where a, b, c, d, e, and f are scalars.

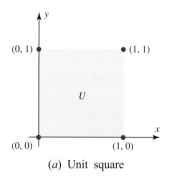

(a) Unit square

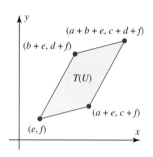

(b) Unit square after affine transformation

▲ Figure 10.13.21

Figure 10.13.21 shows how an affine transformation maps the unit square U onto a parallelogram $T(U)$. An affine transformation is said to be **contracting** if the Euclidean distance between any two points in the plane is strictly decreased after the two points are mapped by the transformation. It can be shown that any k contracting affine transformations $T_1, T_2, \ldots, T_k$ determine a unique closed and bounded set S satisfying the equation

$$S = T_1(S) \cup T_2(S) \cup T_3(S) \cup \cdots \cup T_k(S) \tag{13}$$

Equation (13) has the same form as Equation (12), which we used to find self-similar sets. Although Equation (13), which uses contracting affine transformations, does not determine a self-similar set S, the set it does determine has many of the features of self-similar sets. For example, Figure 10.13.22 shows how a set in the plane resembling a fern (an example made famous by Barnsley) can be generated through four contracting affine transformations. Note that the middle fern is the slightly overlapping union of the four smaller affine-image ferns surrounding it. Note also how T_3, because the determinant of its matrix part is zero, maps the entire fern onto the small straight line segment between the points (.50, 0) and (.50, .16). Figure 10.13.22 contains a wealth of information and should be studied carefully.

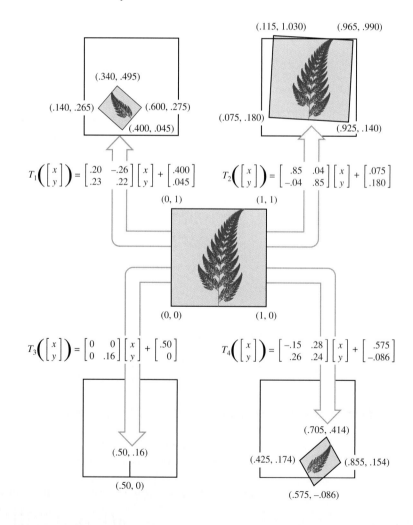

▶ Figure 10.13.22

Michael Barnsley has applied the above theory to the field of data compression and transmission. The fern, for example, is completely determined by the four affine transformations T_1, T_2, T_3, T_4. These four transformations, in turn, are determined by the

24 numbers given in Figure 10.13.22 defining their corresponding values of a, b, c, d, e, and f. In other words, these 24 numbers completely *encode* the picture of the fern. Storing these 24 numbers in a computer requires considerably less memory space than storing a pixel-by-pixel description of the fern. In principle, any picture represented by a pixel map on a computer screen can be described through a finite number of affine transformations, although it is not easy to determine which transformations to use. Nevertheless, once encoded, the affine transformations generally require several orders of magnitude less computer memory than a pixel-by-pixel description of the pixel map.

FURTHER READINGS

Readers interested in learning more about fractals are referred to the following books, the first of which elaborates on the linear transformation approach of this section.

1. MICHAEL BARNSLEY, *Fractals Everywhere* (New York: Academic Press, 1993).
2. BENOIT B. MANDELBROT, *The Fractal Geometry of Nature* (New York: W. H. Freeman, 1982).
3. HEINZ-OTTO PEITGEN AND P. H. RICHTER, *The Beauty of Fractals* (New York: Springer-Verlag, 1986).
4. HEINZ-OTTO PEITGEN AND DIETMAR SAUPE, *The Science of Fractal Images* (New York: Springer-Verlag, 1988).

Exercise Set 10.13

1. The self-similar set in Figure Ex-1 has the sizes indicated. Given that its lower left corner is situated at the origin of the xy-plane, find the similitudes that determine the set. What is its Hausdorff dimension? Is it a fractal?

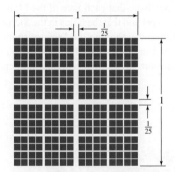

◀ Figure Ex-1

2. Find the Hausdorff dimension of the self-similar set shown in Figure Ex-2. Use a ruler to measure the figure and determine an approximate value of the scale factor s. What are the rotation angles of the similitudes determining this set?

◀ Figure Ex-2

3. Each of the 12 self-similar sets in Figure Ex-3 results from three similitudes with scale factor of $\frac{1}{2}$, and so all have Hausdorff dimension $\ln 3 / \ln 2 = 1.584 \ldots$. The rotation angles of the three similitudes are all multiples of $90°$. Find these rotation angles for each set and express them as a triplet of integers (n_1, n_2, n_3), where n_i is the corresponding integer multiple of $90°$ in the order upper right, lower left, lower right. For example, the first set (the Sierpinski triangle) generates the triplet $(0, 0, 0)$.

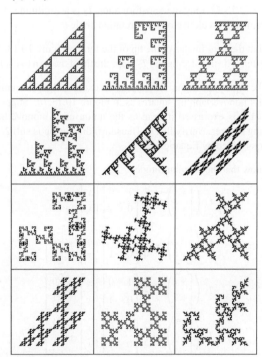

▲ Figure Ex-3

640 Chapter 10 Applications of Linear Algebra

4. For each of the self-similar sets in Figure Ex-4, find: (i) the scale factor s of the similitudes describing the set; (ii) the rotation angles θ of all similitudes describing the set (all rotation angles are multiples of 90°); and (iii) the Hausdorff dimension of the set. Which of the sets are fractals and why?

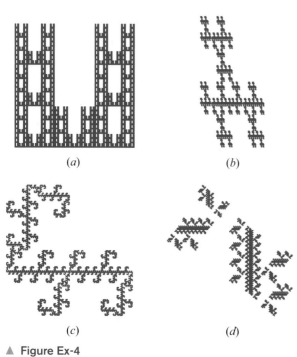

(a) (b)

(c) (d)

▲ Figure Ex-4

5. Show that of the four affine transformations shown in Figure 10.13.22, only the transformation T_2 is a similitude. Determine its scale factor s and rotation angle θ.

6. Find the coordinates of the tip of the fern in Figure 10.13.22. [*Hint:* The transformation T_2 maps the tip of the fern to itself.]

7. The square in Figure 10.13.7*a* was expressed as the union of 4 nonoverlapping squares as in Figure 10.13.7*b*. Suppose that it is expressed instead as the union of 16 nonoverlapping squares. Verify that its Hausdorff dimension is still 2, as determined by Equation (2).

8. Show that the four similitudes

$$T_1\left(\begin{bmatrix} x \\ y \end{bmatrix}\right) = \frac{3}{4}\begin{bmatrix} 1 & 0 \\ 0 & 1 \end{bmatrix}\begin{bmatrix} x \\ y \end{bmatrix}$$

$$T_2\left(\begin{bmatrix} x \\ y \end{bmatrix}\right) = \frac{3}{4}\begin{bmatrix} 1 & 0 \\ 0 & 1 \end{bmatrix}\begin{bmatrix} x \\ y \end{bmatrix} + \begin{bmatrix} \frac{1}{4} \\ 0 \end{bmatrix}$$

$$T_3\left(\begin{bmatrix} x \\ y \end{bmatrix}\right) = \frac{3}{4}\begin{bmatrix} 1 & 0 \\ 0 & 1 \end{bmatrix}\begin{bmatrix} x \\ y \end{bmatrix} + \begin{bmatrix} 0 \\ \frac{1}{4} \end{bmatrix}$$

$$T_4\left(\begin{bmatrix} x \\ y \end{bmatrix}\right) = \frac{3}{4}\begin{bmatrix} 1 & 0 \\ 0 & 1 \end{bmatrix}\begin{bmatrix} x \\ y \end{bmatrix} + \begin{bmatrix} \frac{1}{4} \\ \frac{1}{4} \end{bmatrix}$$

express the unit square as the union of four *overlapping* squares. Evaluate the right-hand side of Equation (2) for the values of k and s determined by these similitudes, and show that the result is not the correct value of the Hausdorff dimension of the unit square. [*Note:* This exercise shows the necessity of the nonoverlapping condition in the definition of a self-similar set and its Hausdorff dimension.]

9. All of the results in this section can be extended to R^n. Compute the Hausdorff dimension of the unit cube in R^3 (see Figure Ex-9). Given that the topological dimension of the unit cube is 3, determine whether it is a fractal. [*Hint:* Express the unit cube as the union of eight smaller congruent nonoverlapping cubes.]

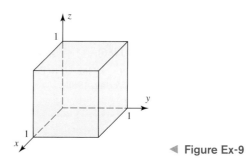

◀ Figure Ex-9

10. The set in R^3 in Figure Ex-10 is called the **Menger sponge**. It is a self-similar set obtained by drilling out certain square holes from the unit cube. Note that each face of the Menger sponge is a Sierpinski carpet and that the holes in the Sierpinski carpet now run all the way through the Menger sponge. Determine the values of k and s for the Menger sponge and find its Hausdorff dimension. Is the Menger sponge a fractal?

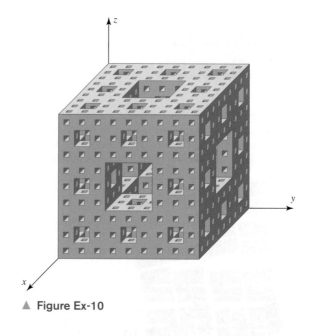

▲ Figure Ex-10

11. The two similitudes

$$T_1\left(\begin{bmatrix} x \\ y \end{bmatrix}\right) = \frac{1}{3}\begin{bmatrix} 1 & 0 \\ 0 & 1 \end{bmatrix}\begin{bmatrix} x \\ y \end{bmatrix}$$

and

$$T_2\left(\begin{bmatrix} x \\ y \end{bmatrix}\right) = \frac{1}{3}\begin{bmatrix} 1 & 0 \\ 0 & 1 \end{bmatrix}\begin{bmatrix} x \\ y \end{bmatrix} + \begin{bmatrix} \frac{2}{3} \\ 0 \end{bmatrix}$$

determine a fractal known as the **Cantor set**. Starting with the unit square region U as an initial set, sketch the first four sets that Algorithm 1 determines. Also, find the Hausdorff dimension of the Cantor set. (This famous set was the first example that Hausdorff gave in his 1919 paper of a set whose Hausdorff dimension is not equal to its topological dimension.)

12. Compute the areas of the sets S_0, S_1, S_2, S_3, and S_4 in Figure 10.13.15.

Section 10.13 Technology Exercises

The following exercises are designed to be solved using a technology utility. Typically, this will be MATLAB, *Mathematica*, Maple, Derive, or Mathcad, but it may also be some other type of linear algebra software or a scientific calculator with some linear algebra capabilities. For each exercise you will need to read the relevant documentation for the particular utility you are using. The goal of these exercises is to provide you with a basic proficiency with your technology utility. Once you have mastered the techniques in these exercises, you will be able to use your technology utility to solve many of the problems in the regular exercise sets.

T1. Use similitudes of the form

$$T_i\left(\begin{bmatrix} x \\ y \\ z \end{bmatrix}\right) = \frac{1}{3}\begin{bmatrix} 1 & 0 & 0 \\ 0 & 1 & 0 \\ 0 & 0 & 1 \end{bmatrix}\begin{bmatrix} x \\ y \\ z \end{bmatrix} + \begin{bmatrix} a_i \\ b_i \\ c_i \end{bmatrix}$$

to show that the Menger sponge (see Exercise 10) is the set S satisfying

$$S = \bigcup_{i=1}^{20} T_i(S)$$

for appropriately chosen similitudes T_i (for $i = 1, 2, 3, \ldots, 20$). Determine these similitudes by determining the collection of 3×1 matrices

$$\left\{ \begin{bmatrix} a_i \\ b_i \\ c_i \end{bmatrix} \text{ for } i = 1, 2, 3, \ldots, 20 \right\}$$

T2. Generalize the ideas involved in the Cantor set (in R^1), the Sierpinski carpet (in R^2), and the Menger sponge (in R^3) to R^n by considering the set S satisfying

$$S = \bigcup_{i=1}^{m_n} T_i(S)$$

with

$$T_i\left(\begin{bmatrix} x_1 \\ x_2 \\ x_3 \\ \vdots \\ x_n \end{bmatrix}\right) = \frac{1}{3}\begin{bmatrix} 1 & 0 & 0 & \cdots & 0 \\ 0 & 1 & 0 & \cdots & 0 \\ 0 & 0 & 1 & \cdots & 0 \\ \vdots & \vdots & \vdots & \ddots & \vdots \\ 0 & 0 & 0 & \cdots & 1 \end{bmatrix}\begin{bmatrix} x_1 \\ x_2 \\ x_3 \\ \vdots \\ x_n \end{bmatrix} + \begin{bmatrix} a_{1i} \\ a_{2i} \\ a_{3i} \\ \vdots \\ a_{ni} \end{bmatrix}$$

where each a_{ki} equals 0, $\frac{1}{3}$, or $\frac{2}{3}$, and no two of them ever equal $\frac{1}{3}$ at the same time. Use a computer to construct the set

$$\left\{ \begin{bmatrix} a_{1i} \\ a_{2i} \\ a_{3i} \\ \vdots \\ a_{ni} \end{bmatrix} \text{ for } i = 1, 2, 3, \ldots, m_n \right\}$$

thereby determining the value of m_n for $n = 2, 3, 4$. Then develop an expression for m_n.

10.14 Chaos

In this section we use a map of the unit square in the xy-plane onto itself to describe the concept of a chaotic mapping.

> **PREREQUISITES:** Geometry of Linear Operators on R^2 (Section 4.11)
> Eigenvalues and Eigenvectors
> Intuitive Understanding of Limits and Continuity

Chaos The word **chaos** was first used in a mathematical sense in 1975 by Tien-Yien Li and James Yorke in a paper entitled "Period Three Implies Chaos." The term is now used to describe the behavior of certain mathematical mappings and physical phenomena that

642 Chapter 10 Applications of Linear Algebra

at first glance seem to behave in a random or disorderly fashion but actually have an underlying element of order (examples include random-number generation, shuffling cards, cardiac arrhythmia, fluttering airplane wings, changes in the red spot of Jupiter, and deviations in the orbit of Pluto). In this section we discuss a particular chaotic mapping called ***Arnold's cat map***, after the Russian mathematician Vladimir I. Arnold who first described it using a diagram of a cat.

Arnold's Cat Map   To describe Arnold's cat map, we need a few ideas about ***modular arithmetic***. If x is a real number, then the notation $x \bmod 1$ denotes the unique number in the interval $[0, 1)$ that differs from x by an integer. For example,

$$2.3 \bmod 1 = 0.3, \quad 0.9 \bmod 1 = 0.9, \quad -3.7 \bmod 1 = 0.3, \quad 2.0 \bmod 1 = 0$$

Note that if x is a nonnegative number, then $x \bmod 1$ is simply the fractional part of x. If (x, y) is an ordered pair of real numbers, then the notation $(x, y) \bmod 1$ denotes $(x \bmod 1, y \bmod 1)$. For example,

$$(2.3, -7.9) \bmod 1 = (0.3, 0.1)$$

Observe that for every real number x, the point $x \bmod 1$ lies in the unit interval $[0, 1)$ and that for every ordered pair (x, y), the point $(x, y) \bmod 1$ lies in the unit square

$$S = \{(x, y) \mid 0 \leq x < 1, 0 \leq y < 1\}$$

Also observe that the upper boundary and the right-hand boundary of the square are not included in S.

Arnold's cat map is the transformation $\Gamma \colon R^2 \to R^2$ defined by the formula

$$\Gamma \colon (x, y) \to (x + y, x + 2y) \bmod 1$$

or, in matrix notation,

$$\Gamma\left(\begin{bmatrix} x \\ y \end{bmatrix}\right) = \begin{bmatrix} 1 & 1 \\ 1 & 2 \end{bmatrix} \begin{bmatrix} x \\ y \end{bmatrix} \bmod 1 \tag{1}$$

To understand the geometry of Arnold's cat map, it is helpful to write (1) in the factored form

$$\Gamma\left(\begin{bmatrix} x \\ y \end{bmatrix}\right) = \begin{bmatrix} 1 & 0 \\ 1 & 1 \end{bmatrix} \begin{bmatrix} 1 & 1 \\ 0 & 1 \end{bmatrix} \begin{bmatrix} x \\ y \end{bmatrix} \bmod 1$$

which expresses Arnold's cat map as the composition of a shear in the x-direction with factor 1, followed by a shear in the y-direction with factor 1. Because the computations are performed mod 1, Γ maps all points of R^2 into the unit square S.

We will illustrate the effect of Arnold's cat map on the unit square S, which is shaded in Figure 10.14.1a and contains a picture of a cat. It can be shown that it does not matter whether the mod 1 computations are carried out after each shear or at the very end. We will discuss both methods, first performing them at the end. The steps are as follows:

Step 1.   Shear in the x-direction with factor 1 (Figure 10.14.1b):

$$(x, y) \to (x + y, y)$$

or in matrix notation

$$\begin{bmatrix} 1 & 1 \\ 0 & 1 \end{bmatrix} \begin{bmatrix} x \\ y \end{bmatrix} = \begin{bmatrix} x + y \\ y \end{bmatrix}$$

Step 2.   Shear in the y-direction with factor 1 (Figure 10.14.1c):

$$(x, y) \to (x, x + y)$$

or, in matrix notation,

$$\begin{bmatrix} 1 & 0 \\ 1 & 1 \end{bmatrix} \begin{bmatrix} x \\ y \end{bmatrix} = \begin{bmatrix} x \\ x + y \end{bmatrix}$$

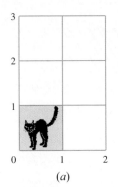 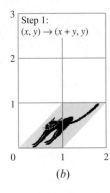 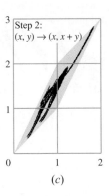 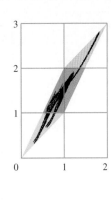 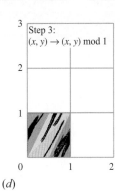

▲ Figure 10.14.1

Step 3. Reassembly into S (Figure 10.14.1d):

$$(x, y) \to (x, y) \bmod 1$$

The geometric effect of the mod 1 arithmetic is to break up the parallelogram in Figure 10.14.1c and reassemble the pieces of S as shown in Figure 10.14.1d.

For computer implementation, it is more convenient to perform the mod 1 arithmetic at each step, rather than at the end. With this approach there is a reassembly at each step, but the net effect is the same. The steps are as follows:

Step 1. Shear in the x-direction with factor 1, followed by a reassembly into S (Figure 10.14.2b):

$$(x, y) \to (x + y, y) \bmod 1$$

Step 2. Shear in the y-direction with factor 1, followed by a reassembly into S (Figure 10.14.2c):

$$(x, y) \to (x, x + y) \bmod 1$$

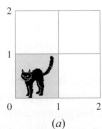 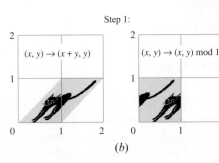 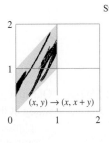 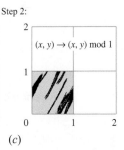

▲ Figure 10.14.2

Repeated Mappings Chaotic mappings such as Arnold's cat map usually arise in physical models in which an operation is performed repeatedly. For example, cards are mixed by repeated shuffles, paint is mixed by repeated stirs, water in a tidal basin is mixed by repeated tidal changes, and so forth. Thus, we are interested in examining the effect on S of repeated applications (or *iterations*) of Arnold's cat map. Figure 10.14.3, which was generated on a computer, shows the effect of 25 iterations of Arnold's cat map on the cat in the unit square S. Two interesting phenomena occur:

- The cat returns to its original form at the 25th iteration.
- At some of the intermediate iterations, the cat is decomposed into streaks that seem to have a specific direction.

Much of the remainder of this section is devoted to explaining these phenomena.

644 Chapter 10 Applications of Linear Algebra

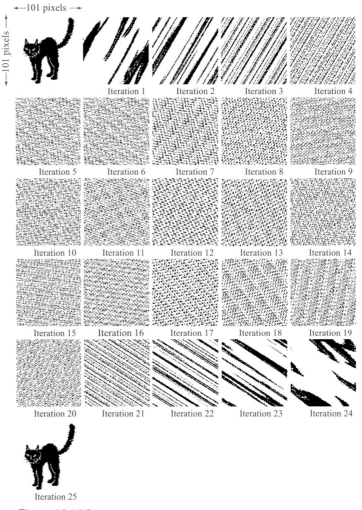

▲ Figure 10.14.3

Periodic Points   Our first goal is to explain why the cat in Figure 10.14.3 returns to its original configuration at the 25th iteration. For this purpose it will be helpful to think of a ***picture*** in the xy-plane as an assignment of colors to the points in the plane. For pictures generated on a computer screen or other digital device, hardware limitations require that a picture be broken up into discrete squares, called ***pixels***. For example, in the computer-generated pictures in Figure 10.14.3 the unit square S is divided into a grid with 101 pixels on a side for a total of 10,201 pixels, each of which is black or white (Figure 10.14.4). An assignment of colors to pixels to create a picture is called a ***pixel map***.

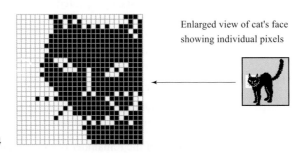

▶ Figure 10.14.4

As shown in Figure 10.14.5, each pixel in S can be assigned a unique pair of coordinates of the form $(m/101, n/101)$ that identifies its lower left-hand corner, where m and n are integers in the range $0, 1, 2, \ldots, 100$. We call these points *pixel points* because each such point identifies a unique pixel. Instead of restricting the discussion to the case where S is subdivided into an array with 101 pixels on a side, let us consider the more general case where there are p pixels per side. Thus, each pixel map in S consists of p^2 pixels uniformly spaced $1/p$ units apart in both the x- and the y-directions. The pixel points in S have coordinates of the form $(m/p, n/p)$, where m and n are integers ranging from 0 to $p-1$.

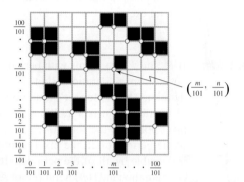

▶ **Figure 10.14.5**

Under Arnold's cat map each pixel point of S is transformed into another pixel point of S. To see why this is so, observe that the image of the pixel point $(m/p, n/p)$ under Γ is given in matrix form by

$$\Gamma\left(\begin{bmatrix} \dfrac{m}{p} \\ \dfrac{n}{p} \end{bmatrix}\right) = \begin{bmatrix} 1 & 1 \\ 1 & 2 \end{bmatrix} \begin{bmatrix} \dfrac{m}{p} \\ \dfrac{n}{p} \end{bmatrix} \bmod 1 = \begin{bmatrix} \dfrac{m+n}{p} \\ \dfrac{m+2n}{p} \end{bmatrix} \bmod 1 \quad (2)$$

The ordered pair $((m+n)/p, (m+2n)/p) \bmod 1$ is of the form $(m'/p, n'/p)$, where m' and n' lie in the range $0, 1, 2, \ldots, p-1$. Specifically, m' and n' are the remainders when $m+n$ and $m+2n$ are divided by p, respectively. Consequently, each point in S of the form $(m/p, n/p)$ is mapped onto another point of the same form.

Because Arnold's cat map transforms every pixel point of S into another pixel point of S, and because there are only p^2 different pixel points in S, it follows that any given pixel point must return to its original position after at most p^2 iterations of Arnold's cat map.

▶ **EXAMPLE 1** **Using Formula (2)**

If $p = 76$, then (2) becomes

$$\Gamma\left(\begin{bmatrix} \dfrac{m}{76} \\ \dfrac{n}{76} \end{bmatrix}\right) = \begin{bmatrix} \dfrac{m+n}{76} \\ \dfrac{m+2n}{76} \end{bmatrix} \bmod 1$$

In this case the successive iterates of the point $\left(\dfrac{27}{76}, \dfrac{58}{76}\right)$ are

$$\overset{0}{\begin{bmatrix} \tfrac{27}{76} \\ \tfrac{58}{76} \end{bmatrix}} \to \overset{1}{\begin{bmatrix} \tfrac{9}{76} \\ \tfrac{67}{76} \end{bmatrix}} \to \overset{2}{\begin{bmatrix} \tfrac{0}{76} \\ \tfrac{67}{76} \end{bmatrix}} \to \overset{3}{\begin{bmatrix} \tfrac{67}{76} \\ \tfrac{58}{76} \end{bmatrix}} \to \overset{4}{\begin{bmatrix} \tfrac{49}{76} \\ \tfrac{31}{76} \end{bmatrix}} \to \overset{5}{\begin{bmatrix} \tfrac{4}{76} \\ \tfrac{35}{76} \end{bmatrix}} \to \overset{6}{\begin{bmatrix} \tfrac{39}{76} \\ \tfrac{74}{76} \end{bmatrix}} \to \overset{7}{\begin{bmatrix} \tfrac{37}{76} \\ \tfrac{35}{76} \end{bmatrix}} \to \overset{8}{\begin{bmatrix} \tfrac{72}{76} \\ \tfrac{31}{76} \end{bmatrix}}$$

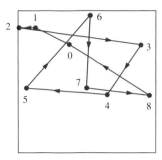

▲ Figure 10.14.6

(verify). Because the point returns to its initial position on the ninth application of Arnold's cat map (but no sooner), the point is said to have period 9, and the set of nine distinct iterates of the point is called a 9-cycle. Figure 10.14.6 shows this 9-cycle with the initial point labeled 0 and its successive iterates labeled accordingly. ◄

In general, a point that returns to its initial position after n applications of Arnold's cat map, but does not return with fewer than n applications, is said to have **period n**, and its set of n distinct iterates is called an **n-cycle**. Arnold's cat map maps $(0, 0)$ into $(0, 0)$, so this point has period 1. Points with period 1 are also called **fixed points**. We leave it as an exercise (Exercise 11) to show that $(0, 0)$ is the only fixed point of Arnold's cat map.

Period Versus Pixel Width

If P_1 and P_2 are points with periods q_1 and q_2, respectively, then P_1 returns to its initial position in q_1 iterations (but no sooner), and P_2 returns to its initial position in q_2 iterations (but no sooner); thus, both points return to their initial positions in any number of iterations that is a multiple of both q_1 and q_2. In general, for a pixel map with p^2 pixel points of the form $(m/p, n/p)$, we let $\Pi(p)$ denote the least common multiple of the periods of all the pixel points in the map [i.e., $\Pi(p)$ is the smallest integer that is divisible by all of the periods]. It follows that the pixel map will return to its initial configuration in $\Pi(p)$ iterations of Arnold's cat map (but no sooner). For this reason, we call $\Pi(p)$ the **period of the pixel map**. In Exercise 4 we ask you to show that if $p = 101$, then all pixel points have period 1, 5, or 25, so $\Pi(101) = 25$. This explains why the cat in Figure 10.14.3 returned to its initial configuration in 25 iterations.

Figure 10.14.7 shows how the period of a pixel map varies with p. Although the general tendency is for the period to increase as p increases, there is a surprising amount of irregularity in the graph. Indeed, there is no simple function that specifies this relationship (see Exercise 1).

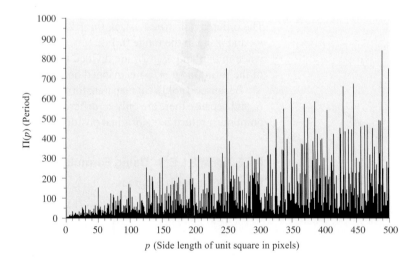

▶ Figure 10.14.7

Although a pixel map with p pixels on a side does not return to its initial configuration until $\Pi(p)$ iterations have occurred, various unexpected things can occur at intermediate iterations. For example, Figure 10.14.8 shows a pixel map with $p = 250$ of the famous Hungarian-American mathematician John von Neumann. It can be shown that $\Pi(250) = 750$; hence, the pixel map will return to its initial configuration after 750 iterations of

Arnold's cat map (but no sooner). However, after 375 iterations the pixel map is turned upside down, and after another 375 iterations (for a total of 750) the pixel map is returned to its initial configuration. Moreover, there are so many pixel points with periods that divide 750 that multiple ghostlike images of the original likeness occur at intermediate iterations; at 195 iterations numerous miniatures of the original likeness occur in diagonal rows.

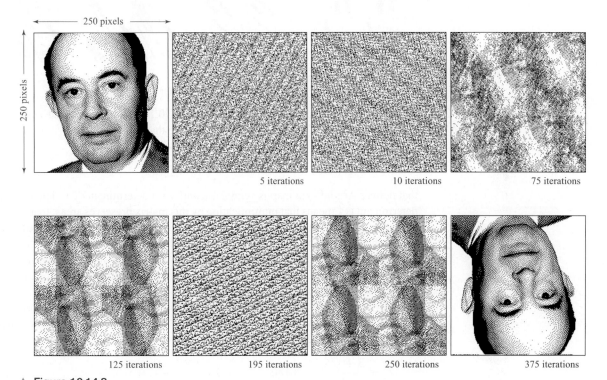

▲ Figure 10.14.8

The Tiled Plane

Our next objective is to explain the cause of the linear streaks that occur in Figure 10.14.3. For this purpose it will be helpful to view Arnold's cat map another way. As defined, Arnold's cat map is not a linear transformation because of the mod 1 arithmetic. However, there is an alternative way of defining Arnold's cat map that avoids the mod 1 arithmetic and results in a linear transformation. For this purpose, imagine that the unit square S with its picture of the cat is a "tile," and suppose that the entire plane is covered with such tiles, as in Figure 10.14.9. We say that the xy-plane has been **tiled** with the unit square. If we apply the matrix transformation in (1) to the entire tiled plane without performing the mod 1 arithmetic, then it can be shown that the portion of the image within S will be identical to the image that we obtained using the mod 1 arithmetic (Figure 10.14.9). In short, the tiling results in the same pixel map in S as the mod 1 arithmetic, but in the tiled case Arnold's cat map is a linear transformation.

It is important to understand, however, that tiling and mod 1 arithmetic produce periodicity in different ways. If a pixel map in S has period n, then in the case of mod 1 arithmetic, each point returns to its original position at the end of n iterations. In the case of tiling, points need not return to their original positions; rather, each point is replaced by a point of the same color at the end of n iterations.

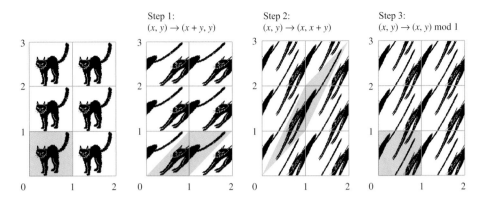

▶ Figure 10.14.9

Properties of Arnold's Cat Map

To understand the cause of the streaks in Figure 10.14.3, think of Arnold's cat map as a linear transformation on the tiled plane. Observe that the matrix

$$C = \begin{bmatrix} 1 & 1 \\ 1 & 2 \end{bmatrix}$$

that defines Arnold's cat map is symmetric and has a determinant of 1. The fact that the determinant is 1 means that multiplication by this matrix preserves areas; that is, the area of any figure in the plane and the area of its image are the same. This is also true for figures in S in the case of mod 1 arithmetic, since the effect of the mod 1 arithmetic is to cut up the figure and reassemble the pieces without any overlap, as shown in Figure 10.14.1d. Thus, in Figure 10.14.3 the area of the cat (whatever it is) is the same as the total area of the blotches in each iteration.

The fact that the matrix is symmetric means that its eigenvalues are real and the corresponding eigenvectors are perpendicular. We leave it for you to show that the eigenvalues and corresponding eigenvectors of C are

$$\lambda_1 = \frac{3 + \sqrt{5}}{2} = 2.6180\ldots, \qquad \lambda_2 = \frac{3 - \sqrt{5}}{2} = 0.3819\ldots,$$

$$\mathbf{v}_1 = \begin{bmatrix} 1 \\ \frac{1 + \sqrt{5}}{2} \end{bmatrix} = \begin{bmatrix} 1 \\ 1.6180\ldots \end{bmatrix}, \qquad \mathbf{v}_2 = \begin{bmatrix} \frac{-1 - \sqrt{5}}{2} \\ 1 \end{bmatrix} = \begin{bmatrix} -1.6180\ldots \\ 1 \end{bmatrix}$$

For each application of Arnold's cat map, the eigenvalue λ_1 causes a stretching in the direction of the eigenvector $\mathbf{v}_1$ by a factor of $2.6180\ldots$, and the eigenvalue λ_2 causes a compression in the direction of the eigenvector $\mathbf{v}_2$ by a factor of $0.3819\ldots$. Figure 10.14.10 shows a square centered at the origin whose sides are parallel to the two eigenvector directions. Under the above mapping, this square is deformed into the rectangle whose sides are also parallel to the two eigenvector directions. The area of the square and rectangle are the same.

To explain the cause of the streaks in Figure 10.14.3, consider S to be part of the tiled plane, and let $\mathbf{p}$ be a point of S with period n. Because we are considering tiling, there is a point $\mathbf{q}$ in the plane with the same color as $\mathbf{p}$ that on successive iterations moves toward the position initially occupied by $\mathbf{p}$, reaching that position on the nth iteration. This point is $\mathbf{q} = (A^{-1})^n \mathbf{p} = A^{-n}\mathbf{p}$, since

$$A^n \mathbf{q} = A^n(A^{-n}\mathbf{p}) = \mathbf{p}$$

Thus, with successive iterations, points of S flow away from their initial positions, while at the same time other points in the plane (with corresponding colors) flow toward those initial positions, completing their trip on the final iteration of the cycle. Figure 10.14.11

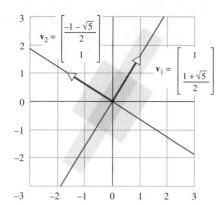

▶ **Figure 10.14.10**

illustrates this in the case where $n = 4$, $\mathbf{q} = \left(-\frac{8}{3}, \frac{5}{3}\right)$, and $\mathbf{p} = A^4\mathbf{q} = \left(\frac{1}{3}, \frac{2}{3}\right)$. Note that $\mathbf{p} \bmod 1 = \mathbf{q} \bmod 1 = \left(\frac{1}{3}, \frac{2}{3}\right)$, so both points occupy the same positions on their respective tiles. The outgoing point moves in the general direction of the eigenvector $\mathbf{v}_1$, as indicated by the arrows in Figure 10.14.11, and the incoming point moves in the general direction of eigenvector $\mathbf{v}_2$. It is the "flow lines" in the general directions of the eigenvectors that form the streaks in Figure 10.14.3.

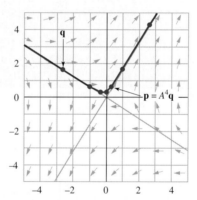

▶ **Figure 10.14.11**

Nonperiodic Points

Thus far we have considered the effect of Arnold's cat map on pixel points of the form $(m/p, n/p)$ for an arbitrary positive integer p. We know that all such points are periodic. We now consider the effect of Arnold's cat map on an arbitrary point (a, b) in S. We classify such points as rational if the coordinates a and b are both rational numbers, and irrational if at least one of the coordinates is irrational. Every rational point is periodic, since it is a pixel point for a suitable choice of p. For example, the rational point $(r_1/s_1, r_2/s_2)$ can be written as $(r_1 s_2/s_1 s_2, r_2 s_1/s_1 s_2)$, so it is a pixel point with $p = s_1 s_2$. It can be shown (Exercise 13) that the converse is also true: Every periodic point must be a rational point.

It follows from the preceding discussion that the irrational points in S are nonperiodic, so that successive iterates of an irrational point (x_0, y_0) in S must all be distinct points in S. Figure 10.14.12, which was computer generated, shows an irrational point and selected iterates up to 100,000. For the particular irrational point that we selected, the iterates do not seem to cluster in any particular region of S; rather, they appear to be spread throughout S, becoming denser with successive iterations.

The behavior of the iterates in Figure 10.14.12 is sufficiently important that there is some terminology associated with it. We say that a set D of points in S is **dense in S**

if every circle centered at any point of S encloses points of D, no matter how small the radius of the circle is taken (Figure 10.14.13). It can be shown that the rational points are dense in S and the iterates of most (but not all) of the irrational points are dense in S.

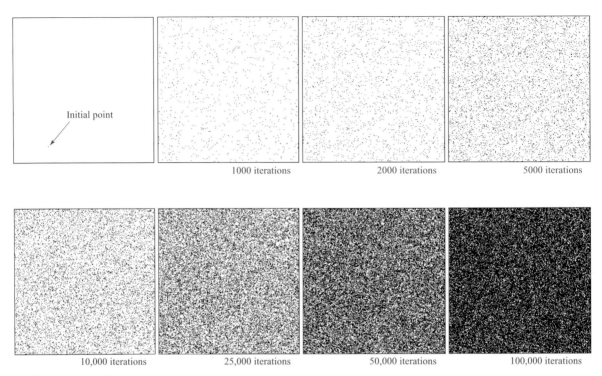

▲ Figure 10.14.12

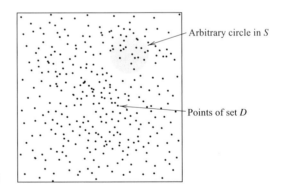

▶ Figure 10.14.13

Definition of Chaos   We know that under Arnold's cat map, the rational points of S are periodic and dense in S and that some but not all of the irrational points have iterates that are dense in S. These are the basic ingredients of chaos. There are several definitions of chaos in current use, but the following one, which is an outgrowth of a definition introduced by Robert L. Devaney in 1986 in his book *An Introduction to Chaotic Dynamical Systems* (Benjamin/Cummings Publishing Company), is most closely related to our work.

> **DEFINITION 1** A mapping T of S onto itself is said to be **chaotic** if:
> (i) S contains a dense set of periodic points of the mapping T.
> (ii)   There is a point in S whose iterates under T are dense in S.

10.14 Chaos

Thus Arnold's cat map satisfies the definition of a chaotic mapping. What is noteworthy about this definition is that a chaotic mapping exhibits an element of order and an element of disorder—the periodic points move regularly in cycles, but the points with dense iterates move irregularly, often obscuring the regularity of the periodic points. This fusion of order and disorder characterizes chaotic mappings.

Dynamical Systems

Chaotic mappings arise in the study of **dynamical systems**. Informally stated, a dynamical system can be viewed as a system that has a specific state or configuration at each point of time but that changes its state with time. Chemical systems, ecological systems, electrical systems, biological systems, economic systems, and so forth can be looked at in this way. In a **discrete-time dynamical system**, the state changes at discrete points of time rather than at each instant. In a **discrete-time chaotic dynamical system**, each state results from a chaotic mapping of the preceding state. For example, if one imagines that Arnold's cat map is applied at discrete points of time, then the pixel maps in Figure 10.14.3 can be viewed as the evolution of a discrete-time chaotic dynamical system from some initial set of states (each point of the cat is a single initial state) to successive sets of states.

One of the fundamental problems in the study of dynamical systems is to predict future states of the system from a known initial state. In practice, however, the exact initial state is rarely known because of errors in the devices used to measure the initial state. It was believed at one time that if the measuring devices were sufficiently accurate and the computers used to perform the iteration were sufficiently powerful, then one could predict the future states of the system to any degree of accuracy. But the discovery of chaotic systems shattered this belief because it was found that for such systems the slightest error in measuring the initial state or in the computation of the iterates becomes magnified exponentially, thereby preventing an accurate prediction of future states. Let us demonstrate this **sensitivity to initial conditions** with Arnold's cat map.

Suppose that P_0 is a point in the xy-plane whose exact coordinates are (0.77837, 0.70904). A measurement error of 0.00001 is made in the y-coordinate, such that the point is thought to be located at (0.77837, 0.70905), which we denote by Q_0. Both P_0 and Q_0 are pixel points with $p = 100{,}000$ (why?), and thus, since $\Pi(100{,}000) = 75{,}000$, both return to their initial positions after 75,000 iterations. In Figure 10.14.14 we show the first 50 iterates of P_0 under Arnold's cat map as crosses and the first 50 iterates of Q_0 as circles. Although P_0 and Q_0 are close enough that their symbols overlap initially, only their first eight iterates have overlapping symbols; from the ninth iteration on their iterates follow divergent paths.

It is possible to quantify the growth of the error from the eigenvalues and eigenvectors of Arnold's cat map. For this purpose we will think of Arnold's cat map as a linear transformation on the tiled plane. Recall from Figure 10.14.10 and the related discussion that the projected distance between two points in S in the direction of the eigenvector $\mathbf{v}_1$ increases by a factor of $2.6180\ldots (= \lambda_1)$ with each iteration (Figure 10.14.15). After nine iterations this projected distance increases by a factor of $(2.6180\ldots)^9 = 5777.99\ldots$, and with an initial error of roughly $1/100{,}000$ in the direction of $\mathbf{v}_1$, this distance is $0.05777\ldots$, or about $\frac{1}{17}$ the width of the unit square S. After 12 iterations this small initial error grows to $(2.6180\ldots)^{12}/100{,}000 = 1.0368\ldots$, which is greater than the width of S. Thus, we completely lose track of the true iterates within S after 12 iterations because of the exponential growth of the initial error.

Although sensitivity to initial conditions limits the ability to predict the future evolution of dynamical systems, new techniques are presently being investigated to describe this future evolution in alternative ways.

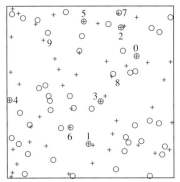

▲ Figure 10.14.14

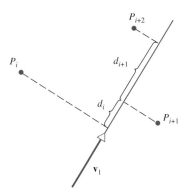

▲ Figure 10.14.15

Exercise Set 10.14

1. In a journal article [F. J. Dyson and H. Falk, "Period of a Discrete Cat Mapping," *The American Mathematical Monthly*, 99 (August–September 1992), pp. 603–614] the following results concerning the nature of the function $\Pi(p)$ were established:

   (i) $\Pi(p) = 3p$ if and only if $p = 2 \cdot 5^k$ for $k = 1, 2, \ldots$.

   (ii) $\Pi(p) = 2p$ if and only if $p = 5^k$ for $k = 1, 2, \ldots$ or $p = 6 \cdot 5^k$ for $k = 0, 1, 2, \ldots$.

   (iii) $\Pi(p) \le 12p/7$ for all other choices of p.

   Find $\Pi(250)$, $\Pi(25)$, $\Pi(125)$, $\Pi(30)$, $\Pi(10)$, $\Pi(50)$, $\Pi(3750)$, $\Pi(6)$, and $\Pi(5)$.

2. Find all the n-cycles that are subsets of the 36 points in S of the form $(m/6, n/6)$ with m and n in the range 0, 1, 2, 3, 4, 5. Then find $\Pi(6)$.

3. (*Fibonacci Shift-Register Random-Number Generator*) A well-known method of generating a sequence of "pseudorandom" integers $x_0, x_1, x_2, x_3, \ldots$ in the interval from 0 to $p - 1$ is based on the following algorithm:

   (i) Pick any two integers x_0 and x_1 from the range $0, 1, 2, \ldots, p - 1$.

   (ii) Set $x_{n+1} = (x_n + x_{n-1}) \bmod p$ for $n = 1, 2, \ldots$.

   Here $x \bmod p$ denotes the number in the interval from 0 to $p - 1$ that differs from x by a multiple of p. For example, 35 mod $9 = 8$ (because $8 = 35 - 3 \cdot 9$); 36 mod $9 = 0$ (because $0 = 36 - 4 \cdot 9$); and -3 mod $9 = 6$ (because $6 = -3 + 1 \cdot 9$).

   (a) Generate the sequence of pseudorandom numbers that results from the choices $p = 15$, $x_0 = 3$, and $x_1 = 7$ until the sequence starts repeating.

   (b) Show that the following formula is equivalent to step (ii) of the algorithm:
   $$\begin{bmatrix} x_{n+1} \\ x_{n+2} \end{bmatrix} = \begin{bmatrix} 1 & 1 \\ 1 & 2 \end{bmatrix} \begin{bmatrix} x_{n-1} \\ x_n \end{bmatrix} \bmod p \quad \text{for } n = 1, 2, 3, \ldots$$

   (c) Use the formula in part (b) to generate the sequence of vectors for the choices $p = 21$, $x_0 = 5$, and $x_1 = 5$ until the sequence starts repeating.

   Remark If we take $p = 1$ and pick x_0 and x_1 from the interval [0, 1), then the above random-number generator produces pseudorandom numbers in the interval [0, 1). The resulting scheme is precisely Arnold's cat map. Furthermore, if we eliminate the modular arithmetic in the algorithm and take $x_0 = x_1 = 1$, then the resulting sequence of integers is the famous Fibonacci sequence, 1, 1, 2, 3, 5, 8, 13, 21, 34, 55, 89, ..., in which each number after the first two is the sum of the preceding two numbers.

4. For $C = \begin{bmatrix} 1 & 1 \\ 1 & 2 \end{bmatrix}$, it can be verified that
   $$C^{25} = \begin{bmatrix} 7{,}778{,}742{,}049 & 12{,}586{,}269{,}025 \\ 12{,}586{,}269{,}025 & 20{,}365{,}011{,}074 \end{bmatrix}$$

   It can also be verified that 12,586,269,025 is divisible by 101 and that when 7,778,742,049 and 20,365,011,074 are divided by 101, the remainder is 1.

   (a) Show that every point in S of the form $(m/101, n/101)$ returns to its starting position after 25 iterations under Arnold's cat map.

   (b) Show that every point in S of the form $(m/101, n/101)$ has period 1, 5, or 25.

   (c) Show that the point $\left(\frac{1}{101}, 0\right)$ has period greater than 5 by iterating it five times.

   (d) Show that $\Pi(101) = 25$.

5. Show that for the mapping $T: S \to S$ defined by $T(x, y) = \left(x + \frac{5}{12}, y\right) \bmod 1$, every point in S is a periodic point. Why does this show that the mapping is not chaotic?

6. An *Anosov automorphism* on R^2 is a mapping from the unit square S onto S of the form
   $$\begin{bmatrix} x \\ y \end{bmatrix} \to \begin{bmatrix} a & b \\ c & d \end{bmatrix} \begin{bmatrix} x \\ y \end{bmatrix} \bmod 1$$
   in which (i) a, b, c, and d are integers, (ii) the determinant of the matrix is ± 1, and (iii) the eigenvalues of the matrix

do not have magnitude 1. It can be shown that all Anosov automorphisms are chaotic mappings.

(a) Show that Arnold's cat map is an Anosov automorphism.

(b) Which of the following are the matrices of an Anosov automorphism?

$$\begin{bmatrix} 0 & 1 \\ 1 & 0 \end{bmatrix}, \quad \begin{bmatrix} 3 & 2 \\ 1 & 1 \end{bmatrix}, \quad \begin{bmatrix} 1 & 0 \\ 0 & 1 \end{bmatrix},$$

$$\begin{bmatrix} 5 & 7 \\ 2 & 3 \end{bmatrix}, \quad \begin{bmatrix} 6 & 2 \\ 5 & 2 \end{bmatrix}$$

(c) Show that the following mapping of S onto S is not an Anosov automorphism.

$$\begin{bmatrix} x \\ y \end{bmatrix} \rightarrow \begin{bmatrix} 0 & 1 \\ -1 & 0 \end{bmatrix} \begin{bmatrix} x \\ y \end{bmatrix} \bmod 1$$

What is the geometric effect of this transformation on S? Use your observation to show that the mapping is not a chaotic mapping by showing that all points in S are periodic points.

7. Show that Arnold's cat map is one-to-one over the unit square S and that its range is S.

8. Show that the inverse of Arnold's cat map is given by

$$\Gamma^{-1}(x, y) = (2x - y, -x + y) \bmod 1$$

9. Show that the unit square S can be partitioned into four triangular regions on each of which Arnold's cat map is a transformation of the form

$$\begin{bmatrix} x \\ y \end{bmatrix} \rightarrow \begin{bmatrix} 1 & 1 \\ 1 & 2 \end{bmatrix} \begin{bmatrix} x \\ y \end{bmatrix} + \begin{bmatrix} a \\ b \end{bmatrix}$$

where a and b need not be the same for each region. [Hint: Find the regions in S that map onto the four shaded regions of the parallelogram in Figure 10.14.1d.]

10. If (x_0, y_0) is a point in S and (x_n, y_n) is its nth iterate under Arnold's cat map, show that

$$\begin{bmatrix} x_n \\ y_n \end{bmatrix} = \begin{bmatrix} 1 & 1 \\ 1 & 2 \end{bmatrix}^n \begin{bmatrix} x_0 \\ y_0 \end{bmatrix} \bmod 1$$

This result implies that the modular arithmetic need only be performed once rather than after each iteration.

11. Show that $(0, 0)$ is the only fixed point of Arnold's cat map by showing that the only solution of the equation

$$\begin{bmatrix} x_0 \\ y_0 \end{bmatrix} = \begin{bmatrix} 1 & 1 \\ 1 & 2 \end{bmatrix} \begin{bmatrix} x_0 \\ y_0 \end{bmatrix} \bmod 1$$

with $0 \leq x_0 < 1$ and $0 \leq y_0 < 1$ is $x_0 = y_0 = 0$. [Hint: For appropriate nonnegative integers, r and s, we can write

$$\begin{bmatrix} x_0 \\ y_0 \end{bmatrix} = \begin{bmatrix} 1 & 1 \\ 1 & 2 \end{bmatrix} \begin{bmatrix} x_0 \\ y_0 \end{bmatrix} - \begin{bmatrix} r \\ s \end{bmatrix}$$

for the preceding equation.]

12. Find all 2-cycles of Arnold's cat map by finding all solutions of the equation

$$\begin{bmatrix} x_0 \\ y_0 \end{bmatrix} = \begin{bmatrix} 1 & 1 \\ 1 & 2 \end{bmatrix}^2 \begin{bmatrix} x_0 \\ y_0 \end{bmatrix} \bmod 1$$

with $0 \leq x_0 < 1$ and $0 \leq y_0 < 1$. [Hint: For appropriate nonnegative integers, r and s, we can write

$$\begin{bmatrix} x_0 \\ y_0 \end{bmatrix} = \begin{bmatrix} 2 & 3 \\ 3 & 5 \end{bmatrix} \begin{bmatrix} x_0 \\ y_0 \end{bmatrix} - \begin{bmatrix} r \\ s \end{bmatrix}$$

for the preceding equation.]

13. Show that every periodic point of Arnold's cat map must be a rational point by showing that for all solutions of the equation

$$\begin{bmatrix} x_0 \\ y_0 \end{bmatrix} = \begin{bmatrix} 1 & 1 \\ 1 & 2 \end{bmatrix}^n \begin{bmatrix} x_0 \\ y_0 \end{bmatrix} \bmod 1$$

the numbers x_0 and y_0 are quotients of integers.

14. Let T be the Arnold's cat map applied five times in a row; that is, $T = \Gamma^5$. Figure Ex-14 represents four successive mappings of T on the first image, each image having a resolution of 101×101 pixels. The fifth mapping returns to the first image because this cat map has a period of 25. Explain how you might generate this particular sequence of images.

▲ Figure Ex-14

Section 10.14 Technology Exercises

The following exercises are designed to be solved using a technology utility. Typically, this will be MATLAB, *Mathematica*, Maple, Derive, or Mathcad, but it may also be some other type of linear algebra software or a scientific calculator with some linear algebra capabilities. For each exercise you will need to read the relevant documentation for the particular utility you are using. The goal of these exercises is to provide you with a basic proficiency with your technology utility. Once you have mastered the techniques in these exercises, you will be able to use your technology utility to solve many of the problems in the regular exercise sets.

T1. The methods of Exercise 4 show that for the cat map, $\Pi(p)$ is the smallest integer satisfying the equation

$$\begin{bmatrix} 1 & 1 \\ 1 & 2 \end{bmatrix}^{\Pi(p)} \mod p = \begin{bmatrix} 1 & 0 \\ 0 & 1 \end{bmatrix}$$

This suggests that one way to determine $\Pi(p)$ is to compute

$$\begin{bmatrix} 1 & 1 \\ 1 & 2 \end{bmatrix}^{n} \mod p$$

starting with $n = 1$ and stopping when this produces the identity matrix. Use this idea to compute $\Pi(p)$ for $p = 2, 3, \ldots, 10$. Compare your results to the formulas given in Exercise 1, if they apply. What can you conjecture about

$$\begin{bmatrix} 1 & 1 \\ 1 & 2 \end{bmatrix}^{\frac{1}{2}\Pi(p)} \mod p$$

when $\Pi(p)$ is even?

T2. The eigenvalues and eigenvectors for the cat map matrix

$$C = \begin{bmatrix} 1 & 1 \\ 1 & 2 \end{bmatrix}$$

are

$$\lambda_1 = \frac{3 + \sqrt{5}}{2}, \quad \lambda_2 = \frac{3 - \sqrt{5}}{2},$$

$$\mathbf{v}_1 = \begin{bmatrix} 1 \\ \frac{1+\sqrt{5}}{2} \end{bmatrix}, \quad \mathbf{v}_2 = \begin{bmatrix} 1 \\ \frac{1-\sqrt{5}}{2} \end{bmatrix}$$

Using these eigenvalues and eigenvectors, we can define

$$D = \begin{bmatrix} \dfrac{3 + \sqrt{5}}{2} & 0 \\ 0 & \dfrac{3 - \sqrt{5}}{2} \end{bmatrix} \quad \text{and} \quad P = \begin{bmatrix} 1 & 1 \\ \dfrac{1+\sqrt{5}}{2} & \dfrac{1-\sqrt{5}}{2} \end{bmatrix}$$

and write $C = PDP^{-1}$; hence, $C^n = PD^nP^{-1}$. Use a computer to show that

$$C^n = \begin{bmatrix} c_{11}^{(n)} & c_{12}^{(n)} \\ c_{21}^{(n)} & c_{22}^{(n)} \end{bmatrix}$$

where

$$c_{11}^{(n)} = \left(\frac{1+\sqrt{5}}{2\sqrt{5}}\right)\left(\frac{3-\sqrt{5}}{2}\right)^n - \left(\frac{1-\sqrt{5}}{2\sqrt{5}}\right)\left(\frac{3+\sqrt{5}}{2}\right)^n$$

$$c_{22}^{(n)} = \left(\frac{1+\sqrt{5}}{2\sqrt{5}}\right)\left(\frac{3+\sqrt{5}}{2}\right)^n - \left(\frac{1-\sqrt{5}}{2\sqrt{5}}\right)\left(\frac{3-\sqrt{5}}{2}\right)^n$$

and

$$c_{12}^{(n)} = c_{21}^{(n)} = \frac{1}{\sqrt{5}}\left\{\left(\frac{3+\sqrt{5}}{2}\right)^n - \left(\frac{3-\sqrt{5}}{2}\right)^n\right\}$$

How can you use these results and your conclusions in Exercise T1 to simplify the method for computing $\Pi(p)$?

10.15 Cryptography

In this section we present a method of encoding and decoding messages. We also examine modular arithmetic and show how Gaussian elimination can sometimes be used to break an opponent's code.

> **PREREQUISITES:** Matrices
> Gaussian Elimination
> Matrix Operations
> Linear Independence
> Linear Transformations (Section 4.9)

Ciphers   The study of encoding and decoding secret messages is called **cryptography**. Although secret codes date to the earliest days of written communication, there has been a recent surge of interest in the subject because of the need to maintain the privacy of information transmitted over public lines of communication. In the language of cryptography, codes are called **ciphers**, uncoded messages are called **plaintext**, and coded messages are called **ciphertext**. The process of converting from plaintext to ciphertext is called **enciphering**, and the reverse process of converting from ciphertext to plaintext is called **deciphering**.

The simplest ciphers, called **substitution ciphers**, are those that replace each letter of the alphabet by a different letter. For example, in the substitution cipher

Plain   A B C D E F G H I J K L M N O P Q R S T U V W X Y Z
Cipher   D E F G H I J K L M N O P Q R S T U V W X Y Z A B C

the plaintext letter A is replaced by D, the plaintext letter B by E, and so forth. With this cipher the plaintext message

ROME WAS NOT BUILT IN A DAY

becomes

URPH ZDV QRW EXLOW LQ D GDB

Hill Ciphers   A disadvantage of substitution ciphers is that they preserve the frequencies of individual letters, making it relatively easy to break the code by statistical methods. One way to overcome this problem is to divide the plaintext into groups of letters and encipher the plaintext group by group, rather than one letter at a time. A system of cryptography in which the plaintext is divided into sets of n letters, each of which is replaced by a set of n cipher letters, is called a **polygraphic system**. In this section we will study a class of polygraphic systems based on matrix transformations. [The ciphers that we will discuss are called **Hill ciphers** after Lester S. Hill, who introduced them in two papers: "Cryptography in an Algebraic Alphabet," *American Mathematical Monthly, 36* (June–July 1929), pp. 306–312; and "Concerning Certain Linear Transformation Apparatus of Cryptography," *American Mathematical Monthly, 38* (March 1931), pp. 135–154.]

In the discussion to follow, we assume that each plaintext and ciphertext letter except Z is assigned the numerical value that specifies its position in the standard alphabet (Table 1). For reasons that will become clear later, Z is assigned a value of zero.

Table 1

A	B	C	D	E	F	G	H	I	J	K	L	M	N	O	P	Q	R	S	T	U	V	W	X	Y	Z
1	2	3	4	5	6	7	8	9	10	11	12	13	14	15	16	17	18	19	20	21	22	23	24	25	0

In the simplest Hill ciphers, successive *pairs* of plaintext are transformed into ciphertext by the following procedure:

Step 1. Choose a 2×2 matrix with integer entries

$$A = \begin{bmatrix} a_{11} & a_{12} \\ a_{21} & a_{22} \end{bmatrix}$$

to perform the encoding. Certain additional conditions on A will be imposed later.

Step 2. Group successive plaintext letters into pairs, adding an arbitrary "dummy" letter to fill out the last pair if the plaintext has an odd number of letters, and replace each plaintext letter by its numerical value.

Step 3. Successively convert each plaintext pair $p_1 p_2$ into a column vector

$$\mathbf{p} = \begin{bmatrix} p_1 \\ p_2 \end{bmatrix}$$

and form the product $A\mathbf{p}$. We will call $\mathbf{p}$ a *plaintext vector* and $A\mathbf{p}$ the corresponding *ciphertext vector*.

Step 4. Convert each ciphertext vector into its alphabetic equivalent.

► EXAMPLE 1 Hill Cipher of a Message

Use the matrix
$$\begin{bmatrix} 1 & 2 \\ 0 & 3 \end{bmatrix}$$
to obtain the Hill cipher for the plaintext message

$$I\ AM\ HIDING$$

Solution If we group the plaintext into pairs and add the dummy letter G to fill out the last pair, we obtain

$$IA \quad MH \quad ID \quad IN \quad GG$$

or, equivalently, from Table 1,

$$9\ 1 \quad 13\ 8 \quad 9\ 4 \quad 9\ 14 \quad 7\ 7$$

To encipher the pair IA, we form the matrix product

$$\begin{bmatrix} 1 & 2 \\ 0 & 3 \end{bmatrix} \begin{bmatrix} 9 \\ 1 \end{bmatrix} = \begin{bmatrix} 11 \\ 3 \end{bmatrix}$$

which, from Table 1, yields the ciphertext KC.

To encipher the pair MH, we form the product

$$\begin{bmatrix} 1 & 2 \\ 0 & 3 \end{bmatrix} \begin{bmatrix} 13 \\ 8 \end{bmatrix} = \begin{bmatrix} 29 \\ 24 \end{bmatrix} \tag{1}$$

However, there is a problem here, because the number 29 has no alphabet equivalent (Table 1). To resolve this problem, we make the following agreement:

> *Whenever an integer greater than* 25 *occurs, it will be replaced by the remainder that results when this integer is divided by* 26.

Because the remainder after division by 26 is one of the integers $0, 1, 2, \ldots, 25$, this procedure will always yield an integer with an alphabet equivalent.

Thus, in (1) we replace 29 by 3, which is the remainder after dividing 29 by 26. It now follows from Table 1 that the ciphertext for the pair MH is CX.

The computations for the remaining ciphertext vectors are

$$\begin{bmatrix} 1 & 2 \\ 0 & 3 \end{bmatrix} \begin{bmatrix} 9 \\ 4 \end{bmatrix} = \begin{bmatrix} 17 \\ 12 \end{bmatrix}$$

$$\begin{bmatrix} 1 & 2 \\ 0 & 3 \end{bmatrix} \begin{bmatrix} 9 \\ 14 \end{bmatrix} = \begin{bmatrix} 37 \\ 42 \end{bmatrix} \quad \text{or} \quad \begin{bmatrix} 11 \\ 16 \end{bmatrix}$$

$$\begin{bmatrix} 1 & 2 \\ 0 & 3 \end{bmatrix} \begin{bmatrix} 7 \\ 7 \end{bmatrix} = \begin{bmatrix} 21 \\ 21 \end{bmatrix}$$

These correspond to the ciphertext pairs QL, KP, and UU, respectively. In summary, the entire ciphertext message is

$$KC \quad CX \quad QL \quad KP \quad UU$$

which would usually be transmitted as a single string without spaces:

$$KCCXQLKPUU \quad \blacktriangleleft$$

Because the plaintext was grouped in pairs and enciphered by a 2×2 matrix, the Hill cipher in Example 1 is referred to as a **Hill 2-cipher**. It is obviously also possible to group the plaintext in triples and encipher by a 3 × 3 matrix with integer entries; this is called a **Hill 3-cipher**. In general, for a **Hill n-cipher**, plaintext is grouped into sets of n letters and enciphered by an $n \times n$ matrix with integer entries.

Modular Arithmetic

In Example 1, integers greater than 25 were replaced by their remainders after division by 26. This technique of working with remainders is at the core of a body of mathematics called *modular arithmetic*. Because of its importance in cryptography, we will digress for a moment to touch on some of the main ideas in this area.

In modular arithmetic we are given a positive integer m, called the **modulus**, and any two integers whose difference is an integer multiple of the modulus are regarded as "equal" or "equivalent" with respect to the modulus. More precisely, we make the following definition.

> **DEFINITION 1** If m is a positive integer and a and b are any integers, then we say that a is **equivalent** to b modulo m, written
>
> $$a = b \pmod{m}$$
>
> if $a - b$ is an integer multiple of m.

▶ **EXAMPLE 2** Various Equivalences

$$7 = 2 \pmod{5}$$
$$19 = 3 \pmod{2}$$
$$-1 = 25 \pmod{26}$$
$$12 = 0 \pmod{4}$$ ◀

For any modulus m it can be proved that every integer a is equivalent, modulo m, to exactly one of the integers

$$0, 1, 2, \ldots, m - 1$$

We call this integer the **residue** of a modulo m, and we write

$$Z_m = \{0, 1, 2, \ldots, m - 1\}$$

to denote the set of residues modulo m.

If a is a *nonnegative* integer, then its residue modulo m is simply the remainder that results when a is divided by m. For an arbitrary integer a, the residue can be found using the following theorem.

> **THEOREM 10.15.1** *For any integer a and modulus m, let*
>
> $$R = \text{remainder of } \frac{|a|}{m}$$
>
> *Then the residue r of a modulo m is given by*
>
> $$r = \begin{cases} R & \text{if } a \geq 0 \\ m - R & \text{if } a < 0 \text{ and } R \neq 0 \\ 0 & \text{if } a < 0 \text{ and } R = 0 \end{cases}$$

▶ **EXAMPLE 3 Residues mod 26**

Find the residue modulo 26 of (a) 87, (b) −38, and (c) −26.

Solution (a) Dividing $|87| = 87$ by 26 yields a remainder of $R = 9$, so $r = 9$. Thus,
$$87 = 9 \quad (\text{mod } 26)$$

Solution (b) Dividing $|-38| = 38$ by 26 yields a remainder of $R = 12$, so $r = 26 - 12 = 14$. Thus,
$$-38 = 14 \quad (\text{mod } 26)$$

Solution (c) Dividing $|-26| = 26$ by 26 yields a remainder of $R = 0$. Thus,
$$-26 = 0 \quad (\text{mod } 26) \quad \blacktriangleleft$$

In ordinary arithmetic every nonzero number a has a *reciprocal* or *multiplicative inverse*, denoted by a^{-1}, such that
$$aa^{-1} = a^{-1}a = 1$$

In modular arithmetic we have the following corresponding concept:

> **DEFINITION 2** If a is a number in Z_m, then a number a^{-1} in Z_m is called a **reciprocal** or **multiplicative inverse** of a modulo m if $aa^{-1} = a^{-1}a = 1 \ (\text{mod } m)$.

It can be proved that if a and m have no common prime factors, then a has a unique reciprocal modulo m; conversely, if a and m have a common prime factor, then a has no reciprocal modulo m.

▶ **EXAMPLE 4 Reciprocal of 3 mod 26**

The number 3 has a reciprocal modulo 26 because 3 and 26 have no common prime factors. This reciprocal can be obtained by finding the number x in Z_{26} that satisfies the modular equation
$$3x = 1 \quad (\text{mod } 26)$$

Although there are general methods for solving such modular equations, it would take us too far afield to study them. However, because 26 is relatively small, this equation can be solved by trying the possible solutions, 0 to 25, one at a time. With this approach we find that $x = 9$ is the solution, because
$$3 \cdot 9 = 27 = 1 \quad (\text{mod } 26)$$

Thus,
$$3^{-1} = 9 \quad (\text{mod } 26)$$

▶ **EXAMPLE 5 A Number with No Reciprocal mod 26**

The number 4 has no reciprocal modulo 26, because 4 and 26 have 2 as a common prime factor (see Exercise 8). ◀

For future reference, in Table 2 we provide the following reciprocals modulo 26:

Table 2 Reciprocals Modulo 26

a	1	3	5	7	9	11	15	17	19	21	23	25
a^{-1}	1	9	21	15	3	19	7	23	11	5	17	25

Deciphering   Every useful cipher must have a procedure for decipherment. In the case of a Hill cipher, decipherment uses the inverse (mod 26) of the enciphering matrix. To be precise, if m is a positive integer, then a square matrix A with entries in Z_m is said to be **invertible modulo m** if there is a matrix B with entries in Z_m such that

$$AB = BA = I \pmod{m}$$

Suppose now that

$$A = \begin{bmatrix} a_{11} & a_{12} \\ a_{21} & a_{22} \end{bmatrix}$$

is invertible modulo 26 and this matrix is used in a Hill 2-cipher. If

$$\mathbf{p} = \begin{bmatrix} p_1 \\ p_2 \end{bmatrix}$$

is a plaintext vector, then

$$\mathbf{c} = A\mathbf{p} \pmod{26}$$

is the corresponding ciphertext vector and

$$\mathbf{p} = A^{-1}\mathbf{c} \pmod{26}$$

Thus, each plaintext vector can be recovered from the corresponding ciphertext vector by multiplying it on the left by A^{-1} (mod 26).

In cryptography it is important to know which matrices are invertible modulo 26 and how to obtain their inverses. We now investigate these questions.

In ordinary arithmetic, a square matrix A is invertible if and only if $\det(A) \neq 0$, or, equivalently, if and only if $\det(A)$ has a reciprocal. The following theorem is the analog of this result in modular arithmetic.

THEOREM 10.15.2 *A square matrix A with entries in Z_m is invertible modulo m if and only if the residue of $\det(A)$ modulo m has a reciprocal modulo m.*

Because the residue of $\det(A)$ modulo m will have a reciprocal modulo m if and only if this residue and m have no common prime factors, we have the following corollary.

COROLLARY 10.15.3 *A square matrix A with entries in Z_m is invertible modulo m if and only if m and the residue of $\det(A)$ modulo m have no common prime factors.*

Because the only prime factors of $m = 26$ are 2 and 13, we have the following corollary, which is useful in cryptography.

COROLLARY 10.15.4 *A square matrix A with entries in Z_{26} is invertible modulo 26 if and only if the residue of $\det(A)$ modulo 26 is not divisible by 2 or 13.*

We leave it for you to verify that if

$$A = \begin{bmatrix} a & b \\ c & d \end{bmatrix}$$

has entries in Z_{26} and the residue of $\det(A) = ad - bc$ modulo 26 is not divisible by 2 or 13, then the inverse of A (mod 26) is given by

$$A^{-1} = (ad - bc)^{-1} \begin{bmatrix} d & -b \\ -c & a \end{bmatrix} \pmod{26} \tag{2}$$

where $(ad - bc)^{-1}$ is the reciprocal of the residue of $ad - bc$ (mod 26).

► EXAMPLE 6 Inverse of a Matrix mod 26

Find the inverse of

$$A = \begin{bmatrix} 5 & 6 \\ 2 & 3 \end{bmatrix}$$

modulo 26.

Solution

$$\det(A) = ad - bc = 5 \cdot 3 - 6 \cdot 2 = 3$$

so from Table 2,

$$(ad - bc)^{-1} = 3^{-1} = 9 \quad (\text{mod } 26)$$

Thus, from (2),

$$A^{-1} = 9 \begin{bmatrix} 3 & -6 \\ -2 & 5 \end{bmatrix} = \begin{bmatrix} 27 & -54 \\ -18 & 45 \end{bmatrix} = \begin{bmatrix} 1 & 24 \\ 8 & 19 \end{bmatrix} \quad (\text{mod } 26)$$

As a check,

$$AA^{-1} = \begin{bmatrix} 5 & 6 \\ 2 & 3 \end{bmatrix} \begin{bmatrix} 1 & 24 \\ 8 & 19 \end{bmatrix} = \begin{bmatrix} 53 & 234 \\ 26 & 105 \end{bmatrix} = \begin{bmatrix} 1 & 0 \\ 0 & 1 \end{bmatrix} \quad (\text{mod } 26)$$

Similarly, $A^{-1}A = I$.

► EXAMPLE 7 Decoding a Hill 2-Cipher

Decode the following Hill 2-cipher, which was enciphered by the matrix in Example 6:

GTNKGKDUSK

Solution From Table 1 the numerical equivalent of this ciphertext is

7 20   14 11   7 11   4 21   19 11

To obtain the plaintext pairs, we multiply each ciphertext vector by the inverse of A (obtained in Example 6):

$$\begin{bmatrix} 1 & 24 \\ 8 & 19 \end{bmatrix} \begin{bmatrix} 7 \\ 20 \end{bmatrix} = \begin{bmatrix} 487 \\ 436 \end{bmatrix} = \begin{bmatrix} 19 \\ 20 \end{bmatrix} \quad (\text{mod } 26)$$

$$\begin{bmatrix} 1 & 24 \\ 8 & 19 \end{bmatrix} \begin{bmatrix} 14 \\ 11 \end{bmatrix} = \begin{bmatrix} 278 \\ 321 \end{bmatrix} = \begin{bmatrix} 18 \\ 9 \end{bmatrix} \quad (\text{mod } 26)$$

$$\begin{bmatrix} 1 & 24 \\ 8 & 19 \end{bmatrix} \begin{bmatrix} 7 \\ 11 \end{bmatrix} = \begin{bmatrix} 271 \\ 265 \end{bmatrix} = \begin{bmatrix} 11 \\ 5 \end{bmatrix} \quad (\text{mod } 26)$$

$$\begin{bmatrix} 1 & 24 \\ 8 & 19 \end{bmatrix} \begin{bmatrix} 4 \\ 21 \end{bmatrix} = \begin{bmatrix} 508 \\ 431 \end{bmatrix} = \begin{bmatrix} 14 \\ 15 \end{bmatrix} \quad (\text{mod } 26)$$

$$\begin{bmatrix} 1 & 24 \\ 8 & 19 \end{bmatrix} \begin{bmatrix} 19 \\ 11 \end{bmatrix} = \begin{bmatrix} 283 \\ 361 \end{bmatrix} = \begin{bmatrix} 23 \\ 23 \end{bmatrix} \quad (\text{mod } 26)$$

From Table 1, the alphabet equivalents of these vectors are

ST RI KE NO WW

which yields the message

STRIKE NOW ◄

Breaking a Hill Cipher   Because the purpose of enciphering messages and information is to prevent "opponents" from learning their contents, cryptographers are concerned with the *security* of their ciphers—that is, how readily they can be broken (deciphered by their opponents). We will conclude this section by discussing one technique for breaking Hill ciphers.

Suppose that you are able to obtain some corresponding plaintext and ciphertext from an opponent's message. For example, on examining some intercepted ciphertext, you may be able to deduce that the message is a letter that begins *DEAR SIR*. We will show that with a small amount of such data, it may be possible to determine the deciphering matrix of a Hill code and consequently obtain access to the rest of the message.

It is a basic result in linear algebra that a linear transformation is completely determined by its values at a basis. This principle suggests that if we have a Hill n-cipher, and if

$$\mathbf{p}_1, \mathbf{p}_2, \ldots, \mathbf{p}_n$$

are linearly independent plaintext vectors whose corresponding ciphertext vectors

$$A\mathbf{p}_1, A\mathbf{p}_2, \ldots, A\mathbf{p}_n$$

are known, then there is enough information available to determine the matrix A and hence A^{-1} (mod m).

The following theorem, whose proof is discussed in the exercises, provides a way to do this.

THEOREM 10.15.5 **Determining the Deciphering Matrix**

Let $\mathbf{p}_1, \mathbf{p}_2, \ldots, \mathbf{p}_n$ be linearly independent plaintext vectors, and let $\mathbf{c}_1, \mathbf{c}_2, \ldots, \mathbf{c}_n$ be the corresponding ciphertext vectors in a Hill n-cipher. If

$$P = \begin{bmatrix} \mathbf{p}_1^T \\ \mathbf{p}_2^T \\ \vdots \\ \mathbf{p}_n^T \end{bmatrix}$$

is the $n \times n$ matrix with row vectors $\mathbf{p}_1^T, \mathbf{p}_2^T, \ldots, \mathbf{p}_n^T$ and if

$$C = \begin{bmatrix} \mathbf{c}_1^T \\ \mathbf{c}_2^T \\ \vdots \\ \mathbf{c}_n^T \end{bmatrix}$$

is the $n \times n$ matrix with row vectors $\mathbf{c}_1^T, \mathbf{c}_2^T, \ldots, \mathbf{c}_n^T$, then the sequence of elementary row operations that reduces C to I transforms P to $(A^{-1})^T$.

This theorem tells us that to find the transpose of the deciphering matrix A^{-1}, we must find a sequence of row operations that reduces C to I and then perform this same sequence of operations on P. The following example illustrates a simple algorithm for doing this.

▶ **EXAMPLE 8** **Using Theorem 10.15.5**

The following Hill 2-cipher is intercepted:

$$IOSBTGXESPXHOPDE$$

Decipher the message, given that it starts with the word *DEAR*.

Solution From Table 1, the numerical equivalent of the known plaintext is

$$\begin{array}{cc} DE & AR \\ 4\ 5 & 1\ 18 \end{array}$$

and the numerical equivalent of the corresponding ciphertext is

$$\begin{array}{cc} IO & SB \\ 9\ 15 & 19\ 2 \end{array}$$

so the corresponding plaintext and ciphertext vectors are

$$\mathbf{p}_1 = \begin{bmatrix} 4 \\ 5 \end{bmatrix} \leftrightarrow \mathbf{c}_1 = \begin{bmatrix} 9 \\ 15 \end{bmatrix}$$

$$\mathbf{p}_2 = \begin{bmatrix} 1 \\ 18 \end{bmatrix} \leftrightarrow \mathbf{c}_2 = \begin{bmatrix} 19 \\ 2 \end{bmatrix}$$

We want to reduce

$$C = \begin{bmatrix} \mathbf{c}_1^T \\ \mathbf{c}_2^T \end{bmatrix} = \begin{bmatrix} 9 & 15 \\ 19 & 2 \end{bmatrix}$$

to I by elementary row operations and simultaneously apply these operations to

$$P = \begin{bmatrix} \mathbf{p}_1^T \\ \mathbf{p}_2^T \end{bmatrix} = \begin{bmatrix} 4 & 5 \\ 1 & 18 \end{bmatrix}$$

to obtain $(A^{-1})^T$ (the transpose of the deciphering matrix). This can be accomplished by adjoining P to the right of C and applying row operations to the resulting matrix $[C \mid P]$ until the left side is reduced to I. The final matrix will then have the form $[I \mid (A^{-1})^T]$. The computations can be carried out as follows:

$$\begin{bmatrix} 9 & 15 & | & 4 & 5 \\ 19 & 2 & | & 1 & 18 \end{bmatrix}$$
⟵ We formed the matrix $[C \mid P]$.

$$\begin{bmatrix} 1 & 45 & | & 12 & 15 \\ 19 & 2 & | & 1 & 18 \end{bmatrix}$$
⟵ We multiplied the first row by $9^{-1} = 3$.

$$\begin{bmatrix} 1 & 19 & | & 12 & 15 \\ 19 & 2 & | & 1 & 18 \end{bmatrix}$$
⟵ We replaced 45 by its residue modulo 26.

$$\begin{bmatrix} 1 & 19 & | & 12 & 15 \\ 0 & -359 & | & -227 & -267 \end{bmatrix}$$
⟵ We added -19 times the first row to the second.

$$\begin{bmatrix} 1 & 19 & | & 12 & 15 \\ 0 & 5 & | & 7 & 19 \end{bmatrix}$$
⟵ We replaced the entries in the second row by their residues modulo 26.

$$\begin{bmatrix} 1 & 19 & | & 12 & 15 \\ 0 & 1 & | & 147 & 399 \end{bmatrix}$$
⟵ We multiplied the second row by $5^{-1} = 21$.

$$\begin{bmatrix} 1 & 19 & | & 12 & 15 \\ 0 & 1 & | & 17 & 9 \end{bmatrix}$$
⟵ We replaced the entries in the second row by their residues modulo 26.

$$\begin{bmatrix} 1 & 0 & | & -311 & -156 \\ 0 & 1 & | & 17 & 9 \end{bmatrix}$$
⟵ We added -19 times the second row to the first.

$$\begin{bmatrix} 1 & 0 & | & 1 & 0 \\ 0 & 1 & | & 17 & 9 \end{bmatrix}$$
⟵ We replaced the entries in the first row by their residues modulo 26.

Thus,
$$(A^{-1})^T = \begin{bmatrix} 1 & 0 \\ 17 & 9 \end{bmatrix}$$

so the deciphering matrix is
$$A^{-1} = \begin{bmatrix} 1 & 17 \\ 0 & 9 \end{bmatrix}$$

To decipher the message, we first group the ciphertext into pairs and find the numerical equivalent of each letter:

IO	SB	TG	XE	SP	XH	OP	DE
9 15	19 2	20 7	24 5	19 16	24 8	15 16	4 5

Next, we multiply successive ciphertext vectors on the left by A^{-1} and find the alphabet equivalents of the resulting plaintext pairs:

$$\begin{bmatrix} 1 & 17 \\ 0 & 9 \end{bmatrix} \begin{bmatrix} 9 \\ 15 \end{bmatrix} = \begin{bmatrix} 4 \\ 5 \end{bmatrix} \quad \begin{matrix} D \\ E \end{matrix}$$

$$\begin{bmatrix} 1 & 17 \\ 0 & 9 \end{bmatrix} \begin{bmatrix} 19 \\ 2 \end{bmatrix} = \begin{bmatrix} 1 \\ 18 \end{bmatrix} \quad \begin{matrix} A \\ R \end{matrix}$$

$$\begin{bmatrix} 1 & 17 \\ 0 & 9 \end{bmatrix} \begin{bmatrix} 20 \\ 7 \end{bmatrix} = \begin{bmatrix} 9 \\ 11 \end{bmatrix} \quad \begin{matrix} I \\ K \end{matrix}$$

$$\begin{bmatrix} 1 & 17 \\ 0 & 9 \end{bmatrix} \begin{bmatrix} 24 \\ 5 \end{bmatrix} = \begin{bmatrix} 5 \\ 19 \end{bmatrix} \quad \begin{matrix} E \\ S \end{matrix} \quad (\text{mod } 26)$$

$$\begin{bmatrix} 1 & 17 \\ 0 & 9 \end{bmatrix} \begin{bmatrix} 19 \\ 16 \end{bmatrix} = \begin{bmatrix} 5 \\ 14 \end{bmatrix} \quad \begin{matrix} E \\ N \end{matrix}$$

$$\begin{bmatrix} 1 & 17 \\ 0 & 9 \end{bmatrix} \begin{bmatrix} 24 \\ 8 \end{bmatrix} = \begin{bmatrix} 4 \\ 20 \end{bmatrix} \quad \begin{matrix} D \\ T \end{matrix}$$

$$\begin{bmatrix} 1 & 17 \\ 0 & 9 \end{bmatrix} \begin{bmatrix} 15 \\ 16 \end{bmatrix} = \begin{bmatrix} 1 \\ 14 \end{bmatrix} \quad \begin{matrix} A \\ N \end{matrix}$$

$$\begin{bmatrix} 1 & 17 \\ 0 & 9 \end{bmatrix} \begin{bmatrix} 4 \\ 5 \end{bmatrix} = \begin{bmatrix} 11 \\ 19 \end{bmatrix} \quad \begin{matrix} K \\ S \end{matrix}$$

Finally, we construct the message from the plaintext pairs:

DE AR IK ES EN DT AN KS

DEAR IKE SEND TANKS ◀

Further Readings

Readers interested in learning more about mathematical cryptography are referred to the following books, the first of which is elementary and the second more advanced.

1. ABRAHAM SINKOV, *Elementary Cryptanalysis, a Mathematical Approach* (Mathematical Association of America, 2009).
2. ALAN G. KONHEIM, *Cryptography, a Primer* (New York: Wiley-Interscience, 1981).

Exercise Set 10.15

1. Obtain the Hill cipher of the message

   DARK NIGHT

   for each of the following enciphering matrices:

   (a) $\begin{bmatrix} 1 & 3 \\ 2 & 1 \end{bmatrix}$ (b) $\begin{bmatrix} 4 & 3 \\ 1 & 2 \end{bmatrix}$

2. In each part determine whether the matrix is invertible modulo 26. If so, find its inverse modulo 26 and check your work by verifying that $AA^{-1} = A^{-1}A = I \pmod{26}$.

   (a) $A = \begin{bmatrix} 9 & 1 \\ 7 & 2 \end{bmatrix}$ (b) $A = \begin{bmatrix} 3 & 1 \\ 5 & 3 \end{bmatrix}$ (c) $A = \begin{bmatrix} 8 & 11 \\ 1 & 9 \end{bmatrix}$

   (d) $A = \begin{bmatrix} 2 & 1 \\ 1 & 7 \end{bmatrix}$ (e) $A = \begin{bmatrix} 3 & 1 \\ 6 & 2 \end{bmatrix}$ (f) $A = \begin{bmatrix} 1 & 8 \\ 1 & 3 \end{bmatrix}$

3. Decode the message

   SAKNOXAOJX

   given that it is a Hill cipher with enciphering matrix

   $\begin{bmatrix} 4 & 1 \\ 3 & 2 \end{bmatrix}$

4. A Hill 2-cipher is intercepted that starts with the pairs

   SL   HK

   Find the deciphering and enciphering matrices, given that the plaintext is known to start with the word *ARMY*.

5. Decode the following Hill 2-cipher if the last four plaintext letters are known to be *ATOM*.

   LNGIHGYBVRENJYQO

6. Decode the following Hill 3-cipher if the first nine plaintext letters are *IHAVECOME*:

   HPAFQGGDUGDDHPGODYNOR

7. All of the results of this section can be generalized to the case where the plaintext is a binary message; that is, it is a sequence of 0's and 1's. In this case we do all of our modular arithmetic using modulus 2 rather than modulus 26. Thus, for example, $1 + 1 = 0 \pmod 2$. Suppose we want to encrypt the message 110101111. Let us first break it into triplets to form the three vectors $\begin{bmatrix} 1 \\ 1 \\ 0 \end{bmatrix}, \begin{bmatrix} 1 \\ 0 \\ 1 \end{bmatrix}, \begin{bmatrix} 1 \\ 1 \\ 1 \end{bmatrix}$, and let us take $\begin{bmatrix} 1 & 1 & 0 \\ 0 & 1 & 1 \\ 1 & 1 & 1 \end{bmatrix}$ as our enciphering matrix.

   (a) Find the encoded message.

   (b) Find the inverse modulo 2 of the enciphering matrix, and verify that it decodes your encoded message.

8. If, in addition to the standard alphabet, a period, comma, and question mark were allowed, then 29 plaintext and ciphertext symbols would be available and all matrix arithmetic would be done modulo 29. Under what conditions would a matrix with entries in Z_{29} be invertible modulo 29?

9. Show that the modular equation $4x = 1 \pmod{26}$ has no solution in Z_{26} by successively substituting the values $x = 0, 1, 2, \ldots, 25$.

10. (a) Let P and C be the matrices in Theorem 10.15.5. Show that $P = C(A^{-1})^T$.

    (b) To prove Theorem 10.15.5, let $E_1, E_2, \ldots, E_n$ be the elementary matrices that correspond to the row operations that reduce C to I, so

    $$E_n \cdots E_2 E_1 C = I$$

    Show that

    $$E_n \cdots E_2 E_1 P = (A^{-1})^T$$

    from which it follows that the same sequence of row operations that reduces C to I converts P to $(A^{-1})^T$.

11. (a) If A is the enciphering matrix of a Hill n-cipher, show that

    $$A^{-1} = (C^{-1}P)^T \pmod{26}$$

    where C and P are the matrices defined in Theorem 10.15.5.

    (b) Instead of using Theorem 10.15.5 as in the text, find the deciphering matrix A^{-1} of Example 8 by using the result in part (a) and Equation (2) to compute C^{-1}. [*Note:* Although this method is practical for Hill 2-ciphers, Theorem 10.15.5 is more efficient for Hill n-ciphers with $n > 2$.]

Section 10.15 Technology Exercises

The following exercises are designed to be solved using a technology utility. Typically, this will be MATLAB, *Mathematica*, Maple, Derive, or Mathcad, but it may also be some other type of linear algebra software or a scientific calculator with some linear algebra capabilities. For each exercise you will need to read the relevant documentation for the particular utility you are using. The goal of these exercises is to provide you with a basic proficiency with your technology utility. Once you have mastered the techniques in these exercises, you will be able to use your technology utility to solve many of the problems in the regular exercise sets.

T1. Two integers that have no common factors (except 1) are said to be relatively prime. Given a positive integer n, let $S_n = \{a_1, a_2, a_3, \ldots, a_m\}$, where $a_1 < a_2 < a_3 < \cdots < a_m$, be the set of all positive integers less than n and relatively prime to n. For example, if $n = 9$, then

$$S_9 = \{a_1, a_2, a_3, \ldots, a_6\} = \{1, 2, 4, 5, 7, 8\}$$

(a) Construct a table consisting of n and S_n for $n = 2, 3, \ldots, 15$, and then compute

$$\sum_{k=1}^{m} a_k \text{ and } \left(\sum_{k=1}^{m} a_k\right) \pmod{n}$$

in each case. Draw a conjecture for $n > 15$ and prove your conjecture to be true. [*Hint:* Use the fact that if a is relatively prime to n, then $n - a$ is also relatively prime to n.]

(b) Given a positive integer n and the set S_n, let P_n be the $m \times m$ matrix

$$P_n = \begin{bmatrix} a_1 & a_2 & a_3 & \cdots & a_{m-1} & a_m \\ a_2 & a_3 & a_4 & \cdots & a_m & a_1 \\ a_3 & a_4 & a_5 & \cdots & a_1 & a_2 \\ \vdots & \vdots & \vdots & \ddots & \vdots & \vdots \\ a_{m-1} & a_m & a_1 & \cdots & a_{m-3} & a_{m-2} \\ a_m & a_1 & a_2 & \cdots & a_{m-2} & a_{m-1} \end{bmatrix}$$

so that, for example,

$$P_9 = \begin{bmatrix} 1 & 2 & 4 & 5 & 7 & 8 \\ 2 & 4 & 5 & 7 & 8 & 1 \\ 4 & 5 & 7 & 8 & 1 & 2 \\ 5 & 7 & 8 & 1 & 2 & 4 \\ 7 & 8 & 1 & 2 & 4 & 5 \\ 8 & 1 & 2 & 4 & 5 & 7 \end{bmatrix}$$

Use a computer to compute $\det(P_n)$ and $\det(P_n) \pmod{n}$ for $n = 2, 3, \ldots, 15$, and then use these results to construct a conjecture.

(c) Use the results of part (a) to prove your conjecture to be true. [*Hint:* Add the first $m - 1$ rows of P_n to its last row and then use Theorem 2.2.3.] What do these results imply about the inverse of $P_n \pmod{n}$?

T2. Given a positive integer n greater than 1, the number of positive integers less than n and relatively prime to n is called the *Euler phi function* of n and is denoted by $\varphi(n)$. For example, $\varphi(6) = 2$ since only two positive integers (1 and 5) are less than 6 and have no common factor with 6.

(a) Using a computer, for each value of $n = 2, 3, \ldots, 25$ compute and print out all positive integers that are less than n and relatively prime to n. Then use these integers to determine the values of $\varphi(n)$ for $n = 2, 3, \ldots, 25$. Can you discover a pattern in the results?

(b) It can be shown that if $\{p_1, p_2, p_3, \ldots, p_m\}$ are all the distinct prime factors of n, then

$$\varphi(n) = n\left(1 - \frac{1}{p_1}\right)\left(1 - \frac{1}{p_2}\right)\left(1 - \frac{1}{p_3}\right)\cdots\left(1 - \frac{1}{p_m}\right)$$

For example, since $\{2, 3\}$ are the distinct prime factors of 12, we have

$$\varphi(12) = 12\left(1 - \frac{1}{2}\right)\left(1 - \frac{1}{3}\right) = 4$$

which agrees with the fact that $\{1, 5, 7, 11\}$ are the only positive integers less than 12 and relatively prime to 12. Using a computer, print out all the prime factors of n for $n = 2, 3, \ldots, 25$. Then compute $\varphi(n)$ using the formula above and compare it to your results in part (a).

10.16 Genetics

In this section we investigate the propagation of an inherited trait in successive generations by computing powers of a matrix.

> **PREREQUISITES:** Eigenvalues and Eigenvectors
> Diagonalization of a Matrix
> Intuitive Understanding of Limits

Inheritance Traits In this section we examine the inheritance of traits in animals or plants. The inherited trait under consideration is assumed to be governed by a set of two genes, which we designate by A and a. Under **autosomal inheritance** each individual in the population of either gender possesses two of these genes, the possible pairings being designated AA, Aa, and aa. This pair of genes is called the individual's **genotype**, and it determines how the trait controlled by the genes is manifested in the individual. For example, in snapdragons a set of two genes determines the color of the flower. Genotype AA produces red flowers, genotype Aa produces pink flowers, and genotype aa produces white flowers. In humans, eye coloration is controlled through autosomal inheritance. Genotypes AA and Aa have brown eyes, and genotype aa has blue eyes. In this case we say that gene A **dominates** gene a, or that gene a is **recessive** to gene A, because genotype Aa has the same outward trait as genotype AA.

In addition to autosomal inheritance we will also discuss **X-linked inheritance**. In this type of inheritance, the male of the species possesses only one of the two possible genes (A or a), and the female possesses a pair of the two genes (AA, Aa, or aa). In humans, color blindness, hereditary baldness, hemophilia, and muscular dystrophy, to name a few, are traits controlled by X-linked inheritance.

Below we explain the manner in which the genes of the parents are passed on to their offspring for the two types of inheritance. We construct matrix models that give the probable genotypes of the offspring in terms of the genotypes of the parents, and we use these matrix models to follow the genotype distribution of a population through successive generations.

Autosomal Inheritance In autosomal inheritance an individual inherits one gene from each of its parents' pairs of genes to form its own particular pair. As far as we know, it is a matter of chance which of the two genes a parent passes on to the offspring. Thus, if one parent is of genotype Aa, it is equally likely that the offspring will inherit the A gene or the a gene from that parent. If one parent is of genotype aa and the other parent is of genotype Aa, the offspring will always receive an a gene from the aa parent and will receive either an A gene or an a gene, with equal probability, from the Aa parent. Consequently, each of the offspring has equal probability of being genotype aa or Aa. In Table 1 we list the probabilities of the possible genotypes of the offspring for all possible combinations of the genotypes of the parents.

Table 1

Genotype of Offspring	Genotypes of Parents					
	AA–AA	AA–Aa	AA–aa	Aa–Aa	Aa–aa	aa–aa
AA	1	$\frac{1}{2}$	0	$\frac{1}{4}$	0	0
Aa	0	$\frac{1}{2}$	1	$\frac{1}{2}$	$\frac{1}{2}$	0
aa	0	0	0	$\frac{1}{4}$	$\frac{1}{2}$	1

▶ **EXAMPLE 1 Distribution of Genotypes in a Population**

Suppose that a farmer has a large population of plants consisting of some distribution of all three possible genotypes AA, Aa, and aa. The farmer desires to undertake a breeding program in which each plant in the population is always fertilized with a plant of genotype AA and is then replaced by one of its offspring. We want to derive an expression for the distribution of the three possible genotypes in the population after any number of generations.

For $n = 0, 1, 2, \ldots$, let us set

$a_n =$ fraction of plants of genotype AA in nth generation

$b_n =$ fraction of plants of genotype Aa in nth generation

$c_n =$ fraction of plants of genotype aa in nth generation

Thus a_0, b_0, and c_0 specify the initial distribution of the genotypes. We also have that

$$a_n + b_n + c_n = 1 \quad \text{for } n = 0, 1, 2, \ldots$$

From Table 1 we can determine the genotype distribution of each generation from the genotype distribution of the preceding generation by the following equations:

$$\begin{aligned} a_n &= a_{n-1} + \tfrac{1}{2} b_{n-1} \\ b_n &= c_{n-1} + \tfrac{1}{2} b_{n-1} \quad n = 1, 2, \ldots \\ c_n &= 0 \end{aligned} \tag{1}$$

For example, the first of these three equations states that all the offspring of a plant of genotype AA will be of genotype AA under this breeding program and that half of the offspring of a plant of genotype Aa will be of genotype AA.

Equations (1) can be written in matrix notation as

$$\mathbf{x}^{(n)} = M \mathbf{x}^{(n-1)}, \quad n = 1, 2, \ldots \tag{2}$$

where

$$\mathbf{x}^{(n)} = \begin{bmatrix} a_n \\ b_n \\ c_n \end{bmatrix}, \quad \mathbf{x}^{(n-1)} = \begin{bmatrix} a_{n-1} \\ b_{n-1} \\ c_{n-1} \end{bmatrix}, \quad \text{and} \quad M = \begin{bmatrix} 1 & \tfrac{1}{2} & 0 \\ 0 & \tfrac{1}{2} & 1 \\ 0 & 0 & 0 \end{bmatrix}$$

Note that the three columns of the matrix M are the same as the first three columns of Table 1.

From Equation (2) it follows that

$$\mathbf{x}^{(n)} = M \mathbf{x}^{(n-1)} = M^2 \mathbf{x}^{(n-2)} = \cdots = M^n \mathbf{x}^{(0)} \tag{3}$$

Consequently, if we can find an explicit expression for M^n, we can use (3) to obtain an explicit expression for $\mathbf{x}^{(n)}$. To find an explicit expression for M^n, we first diagonalize M. That is, we find an invertible matrix P and a diagonal matrix D such that

$$M = PDP^{-1} \tag{4}$$

With such a diagonalization, we then have (see Exercise 1)

$$M^n = PD^n P^{-1} \quad \text{for } n = 1, 2, \ldots$$

where

$$D^n = \begin{bmatrix} \lambda_1 & 0 & 0 & \cdots & 0 \\ 0 & \lambda_2 & 0 & \cdots & 0 \\ \vdots & \vdots & \vdots & & \vdots \\ 0 & 0 & 0 & \cdots & \lambda_k \end{bmatrix}^n = \begin{bmatrix} \lambda_1^n & 0 & 0 & \cdots & 0 \\ 0 & \lambda_2^n & 0 & \cdots & 0 \\ \vdots & \vdots & \vdots & & \vdots \\ 0 & 0 & 0 & \cdots & \lambda_k^n \end{bmatrix}$$

The diagonalization of M is accomplished by finding its eigenvalues and corresponding eigenvectors. These are as follows (verify):

Eigenvalues: $\quad \lambda_1 = 1, \quad \lambda_2 = \tfrac{1}{2}, \quad \lambda_3 = 0$

Corresponding eigenvectors: $\quad \mathbf{v}_1 = \begin{bmatrix} 1 \\ 0 \\ 0 \end{bmatrix}, \quad \mathbf{v}_2 = \begin{bmatrix} 1 \\ -1 \\ 0 \end{bmatrix}, \quad \mathbf{v}_3 = \begin{bmatrix} 1 \\ -2 \\ 1 \end{bmatrix}$

Thus, in Equation (4) we have

$$D = \begin{bmatrix} \lambda_1 & 0 & 0 \\ 0 & \lambda_2 & 0 \\ 0 & 0 & \lambda_3 \end{bmatrix} = \begin{bmatrix} 1 & 0 & 0 \\ 0 & \tfrac{1}{2} & 0 \\ 0 & 0 & 0 \end{bmatrix}$$

and
$$P = [\mathbf{v}_1 \mid \mathbf{v}_2 \mid \mathbf{v}_3] = \begin{bmatrix} 1 & 1 & 1 \\ 0 & -1 & -2 \\ 0 & 0 & 1 \end{bmatrix}$$

Therefore,
$$\mathbf{x}^{(n)} = PD^n P^{-1} \mathbf{x}^{(0)} = \begin{bmatrix} 1 & 1 & 1 \\ 0 & -1 & -2 \\ 0 & 0 & 1 \end{bmatrix} \begin{bmatrix} 1 & 0 & 0 \\ 0 & \left(\tfrac{1}{2}\right)^n & 0 \\ 0 & 0 & 0 \end{bmatrix} \begin{bmatrix} 1 & 1 & 1 \\ 0 & -1 & -2 \\ 0 & 0 & 1 \end{bmatrix} \begin{bmatrix} a_0 \\ b_0 \\ c_0 \end{bmatrix}$$

or
$$\mathbf{x}^{(n)} = \begin{bmatrix} a_n \\ b_n \\ c_n \end{bmatrix} = \begin{bmatrix} 1 & 1 - \left(\tfrac{1}{2}\right)^n & 1 - \left(\tfrac{1}{2}\right)^{n-1} \\ 0 & \left(\tfrac{1}{2}\right)^n & \left(\tfrac{1}{2}\right)^{n-1} \\ 0 & 0 & 0 \end{bmatrix} \begin{bmatrix} a_0 \\ b_0 \\ c_0 \end{bmatrix}$$
$$= \begin{bmatrix} a_0 + b_0 + c_0 - \left(\tfrac{1}{2}\right)^n b_0 - \left(\tfrac{1}{2}\right)^{n-1} c_0 \\ \left(\tfrac{1}{2}\right)^n b_0 + \left(\tfrac{1}{2}\right)^{n-1} c_0 \\ 0 \end{bmatrix}$$

Using the fact that $a_0 + b_0 + c_0 = 1$, we thus have
$$\begin{aligned} a_n &= 1 - \left(\tfrac{1}{2}\right)^n b_0 - \left(\tfrac{1}{2}\right)^{n-1} c_0 \\ b_n &= \left(\tfrac{1}{2}\right)^n b_0 + \left(\tfrac{1}{2}\right)^{n-1} c_0 \qquad n = 1, 2, \ldots \\ c_n &= 0 \end{aligned} \tag{5}$$

These are explicit formulas for the fractions of the three genotypes in the nth generation of plants in terms of the initial genotype fractions.

Because $\left(\tfrac{1}{2}\right)^n$ tends to zero as n approaches infinity, it follows from these equations that
$$\begin{aligned} a_n &\to 1 \\ b_n &\to 0 \\ c_n &= 0 \end{aligned}$$

as n approaches infinity. That is, in the limit all plants in the population will be genotype AA.

▶ **EXAMPLE 2** **Modifying Example 1**

We can modify Example 1 so that instead of each plant being fertilized with one of genotype AA, each plant is fertilized with a plant of its own genotype. Using the same notation as in Example 1, we then find
$$\mathbf{x}^{(n)} = M^n \mathbf{x}^{(0)}$$
where
$$M = \begin{bmatrix} 1 & \tfrac{1}{4} & 0 \\ 0 & \tfrac{1}{2} & 0 \\ 0 & \tfrac{1}{4} & 1 \end{bmatrix}$$

The columns of this new matrix M are the same as the columns of Table 1 corresponding to parents with genotypes AA–AA, Aa–Aa, and aa–aa.

The eigenvalues of M are (verify)

$$\lambda_1 = 1, \qquad \lambda_2 = 1, \qquad \lambda_3 = \tfrac{1}{2}$$

The eigenvalue $\lambda_1 = 1$ has multiplicity two and its corresponding eigenspace is two-dimensional. Picking two linearly independent eigenvectors $\mathbf{v}_1$ and $\mathbf{v}_2$ in that eigenspace, and a single eigenvector $\mathbf{v}_3$ for the simple eigenvalue $\lambda_3 = \tfrac{1}{2}$, we have (verify)

$$\mathbf{v}_1 = \begin{bmatrix} 1 \\ 0 \\ 0 \end{bmatrix}, \qquad \mathbf{v}_2 = \begin{bmatrix} 0 \\ 0 \\ 1 \end{bmatrix}, \qquad \mathbf{v}_3 = \begin{bmatrix} 1 \\ -2 \\ 1 \end{bmatrix}$$

The calculations for $\mathbf{x}^{(n)}$ are then

$$\mathbf{x}^{(n)} = M^n \mathbf{x}^{(0)} = PD^n P^{-1} \mathbf{x}^{(0)}$$

$$= \begin{bmatrix} 1 & 0 & 1 \\ 0 & 0 & -2 \\ 0 & 1 & 1 \end{bmatrix} \begin{bmatrix} 1 & 0 & 0 \\ 0 & 1 & 0 \\ 0 & 0 & \left(\tfrac{1}{2}\right)^n \end{bmatrix} \begin{bmatrix} 1 & \tfrac{1}{2} & 0 \\ 0 & \tfrac{1}{2} & 1 \\ 0 & -\tfrac{1}{2} & 0 \end{bmatrix} \begin{bmatrix} a_0 \\ b_0 \\ c_0 \end{bmatrix}$$

$$= \begin{bmatrix} 1 & \tfrac{1}{2} - \left(\tfrac{1}{2}\right)^{n+1} & 0 \\ 0 & \left(\tfrac{1}{2}\right)^n & 0 \\ 0 & \tfrac{1}{2} - \left(\tfrac{1}{2}\right)^{n+1} & 1 \end{bmatrix} \begin{bmatrix} a_0 \\ b_0 \\ c_0 \end{bmatrix}$$

Thus,

$$\begin{aligned} a_n &= a_0 + \left[\tfrac{1}{2} - \left(\tfrac{1}{2}\right)^{n+1}\right] b_0 \\ b_n &= \left(\tfrac{1}{2}\right)^n b_0 \qquad\qquad n = 1, 2, \ldots \\ c_n &= c_0 + \left[\tfrac{1}{2} - \left(\tfrac{1}{2}\right)^{n+1}\right] b_0 \end{aligned} \qquad (6)$$

In the limit, as n tends to infinity, $\left(\tfrac{1}{2}\right)^n \to 0$ and $\left(\tfrac{1}{2}\right)^{n+1} \to 0$, so

$$\begin{aligned} a_n &\to a_0 + \tfrac{1}{2} b_0 \\ b_n &\to 0 \\ c_n &\to c_0 + \tfrac{1}{2} b_0 \end{aligned}$$

Thus, fertilization of each plant with one of its own genotype produces a population that in the limit contains only genotypes AA and aa. ◂

Autosomal Recessive Diseases

There are many genetic diseases governed by autosomal inheritance in which a normal gene A dominates an abnormal gene a. Genotype AA is a normal individual; genotype Aa is a carrier of the disease but is not afflicted with the disease; and genotype aa is afflicted with the disease. In humans such genetic diseases are often associated with a particular racial group—for instance, cystic fibrosis (predominant among Caucasians), sickle-cell anemia (predominant among people of African origin), Cooley's anemia (predominant among people of Mediterranean origin), and Tay-Sachs disease (predominant among Eastern European Jews).

Suppose that an animal breeder has a population of animals that carries an autosomal recessive disease. Suppose further that those animals afflicted with the disease do not survive to maturity. One possible way to control such a disease is for the breeder to always mate a female, regardless of her genotype, with a normal male. In this way, all future offspring will either have a normal father and a normal mother (AA–AA matings) or a normal father and a carrier mother (AA–Aa matings). There can be no AA–aa matings since animals of genotype aa do not survive to maturity. Under this type of mating program no future offspring will be afflicted with the disease, although there will still be carriers in future generations. Let us now determine the fraction of carriers in future generations. We set

$$\mathbf{x}^{(n)} = \begin{bmatrix} a_n \\ b_n \end{bmatrix}, \quad n = 1, 2, \ldots$$

where

a_n = fraction of population of genotype AA in nth generation

b_n = fraction of population of genotype Aa (carriers) in nth generation

Because each offspring has at least one normal parent, we may consider the controlled mating program as one of continual mating with genotype AA, as in Example 1. Thus, the transition of genotype distributions from one generation to the next is governed by the equation

$$\mathbf{x}^{(n)} = M\mathbf{x}^{(n-1)}, \quad n = 1, 2, \ldots$$

where

$$M = \begin{bmatrix} 1 & \frac{1}{2} \\ 0 & \frac{1}{2} \end{bmatrix}$$

Because we know the initial distribution $\mathbf{x}^{(0)}$, the distribution of genotypes in the nth generation is thus given by

$$\mathbf{x}^{(n)} = M^n \mathbf{x}^{(0)}, \quad n = 1, 2, \ldots$$

The diagonalization of M is easily carried out (see Exercise 4) and leads to

$$\mathbf{x}^{(n)} = PD^n P^{-1} \mathbf{x}^{(0)} = \begin{bmatrix} 1 & 1 \\ 0 & -1 \end{bmatrix} \begin{bmatrix} 1 & 0 \\ 0 & \left(\frac{1}{2}\right)^n \end{bmatrix} \begin{bmatrix} 1 & 1 \\ 0 & -1 \end{bmatrix} \begin{bmatrix} a_0 \\ b_0 \end{bmatrix}$$

$$= \begin{bmatrix} 1 & 1 - \left(\frac{1}{2}\right)^n \\ 0 & \left(\frac{1}{2}\right)^n \end{bmatrix} \begin{bmatrix} a_0 \\ b_0 \end{bmatrix} = \begin{bmatrix} a_0 + b_0 - \left(\frac{1}{2}\right)^n b_0 \\ \left(\frac{1}{2}\right)^n b_0 \end{bmatrix}$$

Because $a_0 + b_0 = 1$, we have

$$\begin{aligned} a_n &= 1 - \left(\tfrac{1}{2}\right)^n b_0 \\ b_n &= \left(\tfrac{1}{2}\right)^n b_0 \end{aligned} \quad n = 1, 2, \ldots \tag{7}$$

Thus, as n tends to infinity, we have

$$\begin{aligned} a_n &\to 1 \\ b_n &\to 0 \end{aligned}$$

so in the limit there will be no carriers in the population.

From (7) we see that

$$b_n = \tfrac{1}{2} b_{n-1}, \quad n = 1, 2, \ldots \tag{8}$$

That is, the fraction of carriers in each generation is one-half the fraction of carriers in the preceding generation. It would be of interest also to investigate the propagation of carriers under random mating, when two animals mate without regard to their genotypes. Unfortunately, such random mating leads to nonlinear equations, and the techniques of this section are not applicable. However, by other techniques it can be shown that under random mating, Equation (8) is replaced by

$$b_n = \frac{b_{n-1}}{1 + \tfrac{1}{2} b_{n-1}}, \quad n = 1, 2, \ldots \tag{9}$$

As a numerical example, suppose that the breeder starts with a population in which 10% of the animals are carriers. Under the controlled-mating program governed by Equation (8), the percentage of carriers can be reduced to 5% in one generation. But under random mating, Equation (9) predicts that 9.5% of the population will be carriers after one generation ($b_n = .095$ if $b_{n-1} = .10$). In addition, under controlled mating no offspring will ever be afflicted with the disease, but with random mating it can be shown that about 1 in 400 offspring will be born with the disease when 10% of the population are carriers.

X-Linked Inheritance

As mentioned in the introduction, in X-linked inheritance the male possesses one gene (A or a) and the female possesses two genes (AA, Aa, or aa). The term *X-linked* is used because such genes are found on the X-chromosome, of which the male has one and the female has two. The inheritance of such genes is as follows: A male offspring receives one of his mother's two genes with equal probability, and a female offspring receives the one gene of her father and one of her mother's two genes with equal probability. Readers familiar with basic probability can verify that this type of inheritance leads to the genotype probabilities in Table 2.

Table 2

			Genotypes of Parents (Father, Mother)					
			(A, AA)	(A, Aa)	(A, aa)	(a, AA)	(a, Aa)	(a, aa)
Offspring	Male	A	1	$\frac{1}{2}$	0	1	$\frac{1}{2}$	0
		a	0	$\frac{1}{2}$	1	0	$\frac{1}{2}$	1
	Female	AA	1	$\frac{1}{2}$	0	0	0	0
		Aa	0	$\frac{1}{2}$	1	1	$\frac{1}{2}$	0
		aa	0	0	0	0	$\frac{1}{2}$	1

We will discuss a program of inbreeding in connection with X-linked inheritance. We begin with a male and female; select two of their offspring at random, one of each gender, and mate them; select two of the resulting offspring and mate them; and so forth. Such inbreeding is commonly performed with animals. (Among humans, such brother-sister marriages were used by the rulers of ancient Egypt to keep the royal line pure.)

The original male-female pair can be one of the six types, corresponding to the six columns of Table 2:

$$(A, AA), \quad (A, Aa), \quad (A, aa), \quad (a, AA), \quad (a, Aa), \quad (a, aa)$$

The sibling pairs mated in each successive generation have certain probabilities of being one of these six types. To compute these probabilities, for $n = 0, 1, 2, \ldots$, let us set

a_n = probability sibling-pair mated in nth generation is type (A, AA)
b_n = probability sibling-pair mated in nth generation is type (A, Aa)
c_n = probability sibling-pair mated in nth generation is type (A, aa)
d_n = probability sibling-pair mated in nth generation is type (a, AA)
e_n = probability sibling-pair mated in nth generation is type (a, Aa)
f_n = probability sibling-pair mated in nth generation is type (a, aa)

With these probabilities we form a column vector

$$\mathbf{x}^{(n)} = \begin{bmatrix} a_n \\ b_n \\ c_n \\ d_n \\ e_n \\ f_n \end{bmatrix}, \quad n = 0, 1, 2, \ldots$$

From Table 2 it follows that

$$\mathbf{x}^{(n)} = M\mathbf{x}^{(n-1)}, \quad n = 1, 2, \ldots \tag{10}$$

where

$$M = \begin{array}{c} \begin{array}{cccccc} (A,AA) & (A,Aa) & (A,aa) & (a,AA) & (a,Aa) & (a,aa) \end{array} \\ \begin{bmatrix} 1 & \frac{1}{4} & 0 & 0 & 0 & 0 \\ 0 & \frac{1}{4} & 0 & 1 & \frac{1}{4} & 0 \\ 0 & 0 & 0 & 0 & \frac{1}{4} & 0 \\ 0 & \frac{1}{4} & 0 & 0 & 0 & 0 \\ 0 & \frac{1}{4} & 1 & 0 & \frac{1}{4} & 0 \\ 0 & 0 & 0 & 0 & \frac{1}{4} & 1 \end{bmatrix} \begin{array}{l} (A,AA) \\ (A,Aa) \\ (A,aa) \\ (a,AA) \\ (a,Aa) \\ (a,aa) \end{array} \end{array}$$

For example, suppose that in the $(n-1)$-st generation, the sibling pair mated is type (A, Aa). Then their male offspring will be genotype A or a with equal probability, and their female offspring will be genotype AA or Aa with equal probability. Because one of the male offspring and one of the female offspring are chosen at random for mating, the next sibling pair will be one of type (A, AA), (A, Aa), (a, AA), or (a, Aa) with equal probability. Thus, the second column of M contains "$\frac{1}{4}$" in each of the four rows corresponding to these four sibling pairs. (See Exercise 9 for the remaining columns.)

As in our previous examples, it follows from (10) that

$$\mathbf{x}^{(n)} = M^n \mathbf{x}^{(0)}, \quad n = 1, 2, \ldots \tag{11}$$

After lengthy calculations, the eigenvalues and eigenvectors of M turn out to be

$$\lambda_1 = 1, \quad \lambda_2 = 1, \quad \lambda_3 = \tfrac{1}{2}, \quad \lambda_4 = -\tfrac{1}{2}, \quad \lambda_5 = \tfrac{1}{4}(1+\sqrt{5}), \quad \lambda_6 = \tfrac{1}{4}(1-\sqrt{5})$$

$$\mathbf{v}_1 = \begin{bmatrix} 1 \\ 0 \\ 0 \\ 0 \\ 0 \\ 0 \end{bmatrix}, \quad \mathbf{v}_2 = \begin{bmatrix} 0 \\ 0 \\ 0 \\ 0 \\ 0 \\ 1 \end{bmatrix}, \quad \mathbf{v}_3 = \begin{bmatrix} -1 \\ 2 \\ -1 \\ 1 \\ -2 \\ 1 \end{bmatrix}, \quad \mathbf{v}_4 = \begin{bmatrix} 1 \\ -6 \\ -3 \\ 3 \\ 6 \\ -1 \end{bmatrix},$$

$$\mathbf{v}_5 = \begin{bmatrix} \tfrac{1}{4}(-3-\sqrt{5}) \\ 1 \\ \tfrac{1}{4}(-1+\sqrt{5}) \\ \tfrac{1}{4}(-1+\sqrt{5}) \\ 1 \\ \tfrac{1}{4}(-3-\sqrt{5}) \end{bmatrix}, \quad \mathbf{v}_6 = \begin{bmatrix} \tfrac{1}{4}(-3+\sqrt{5}) \\ 1 \\ \tfrac{1}{4}(-1-\sqrt{5}) \\ \tfrac{1}{4}(-1-\sqrt{5}) \\ 1 \\ \tfrac{1}{4}(-3+\sqrt{5}) \end{bmatrix}$$

The diagonalization of M then leads to

$$\mathbf{x}^{(n)} = PD^n P^{-1} \mathbf{x}^{(0)}, \quad n = 1, 2, \ldots \tag{12}$$

where

$$P = \begin{bmatrix} 1 & 0 & -1 & 1 & \frac{1}{4}(-3-\sqrt{5}) & \frac{1}{4}(-3+\sqrt{5}) \\ 0 & 0 & 2 & -6 & 1 & 1 \\ 0 & 0 & -1 & -3 & \frac{1}{4}(-1+\sqrt{5}) & \frac{1}{4}(-1-\sqrt{5}) \\ 0 & 0 & 1 & 3 & \frac{1}{4}(-1+\sqrt{5}) & \frac{1}{4}(-1-\sqrt{5}) \\ 0 & 0 & -2 & 6 & 1 & 1 \\ 0 & 1 & 1 & -1 & \frac{1}{4}(-3-\sqrt{5}) & \frac{1}{4}(-3+\sqrt{5}) \end{bmatrix}$$

$$D^n = \begin{bmatrix} 1 & 0 & 0 & 0 & 0 & 0 \\ 0 & 1 & 0 & 0 & 0 & 0 \\ 0 & 0 & \left(\frac{1}{2}\right)^n & 0 & 0 & 0 \\ 0 & 0 & 0 & \left(-\frac{1}{2}\right)^n & 0 & 0 \\ 0 & 0 & 0 & 0 & \left[\frac{1}{4}(1+\sqrt{5})\right]^n & 0 \\ 0 & 0 & 0 & 0 & 0 & \left[\frac{1}{4}(1-\sqrt{5})\right]^n \end{bmatrix}$$

$$P^{-1} = \begin{bmatrix} 1 & \frac{2}{3} & \frac{1}{3} & \frac{2}{3} & \frac{1}{3} & 0 \\ 0 & \frac{1}{3} & \frac{2}{3} & \frac{1}{3} & \frac{2}{3} & 1 \\ 0 & \frac{1}{8} & -\frac{1}{4} & \frac{1}{4} & -\frac{1}{8} & 0 \\ 0 & -\frac{1}{24} & -\frac{1}{12} & \frac{1}{12} & \frac{1}{24} & 0 \\ 0 & \frac{1}{20}(5+\sqrt{5}) & \frac{1}{5}\sqrt{5} & \frac{1}{5}\sqrt{5} & \frac{1}{20}(5+\sqrt{5}) & 0 \\ 0 & \frac{1}{20}(5-\sqrt{5}) & -\frac{1}{5}\sqrt{5} & -\frac{1}{5}\sqrt{5} & \frac{1}{20}(5-\sqrt{5}) & 0 \end{bmatrix}$$

We will not write out the matrix product in (12), as it is rather unwieldy. However, if a specific vector $\mathbf{x}^{(0)}$ is given, the calculation for $\mathbf{x}^{(n)}$ is not too cumbersome (see Exercise 6).

Because the absolute values of the last four diagonal entries of D are less than 1, we see that as n tends to infinity,

$$D^n \to \begin{bmatrix} 1 & 0 & 0 & 0 & 0 & 0 \\ 0 & 1 & 0 & 0 & 0 & 0 \\ 0 & 0 & 0 & 0 & 0 & 0 \\ 0 & 0 & 0 & 0 & 0 & 0 \\ 0 & 0 & 0 & 0 & 0 & 0 \\ 0 & 0 & 0 & 0 & 0 & 0 \end{bmatrix}$$

And so, from Equation (12),

$$\mathbf{x}^{(n)} \to P \begin{bmatrix} 1 & 0 & 0 & 0 & 0 & 0 \\ 0 & 1 & 0 & 0 & 0 & 0 \\ 0 & 0 & 0 & 0 & 0 & 0 \\ 0 & 0 & 0 & 0 & 0 & 0 \\ 0 & 0 & 0 & 0 & 0 & 0 \\ 0 & 0 & 0 & 0 & 0 & 0 \end{bmatrix} P^{-1} \mathbf{x}^{(0)}$$

Performing the matrix multiplication on the right, we obtain (verify)

$$\mathbf{x}^{(n)} \to \begin{bmatrix} a_0 + \tfrac{2}{3}b_0 + \tfrac{1}{3}c_0 + \tfrac{2}{3}d_0 + \tfrac{1}{3}e_0 \\ 0 \\ 0 \\ 0 \\ 0 \\ f_0 + \tfrac{1}{3}b_0 + \tfrac{2}{3}c_0 + \tfrac{1}{3}d_0 + \tfrac{2}{3}e_0 \end{bmatrix} \quad (13)$$

That is, in the limit all sibling pairs will be either type (A, AA) or type (a, aa). For example, if the initial parents are type (A, Aa) (that is, $b_0 = 1$ and $a_0 = c_0 = d_0 = e_0 = f_0 = 0$), then as n tends to infinity,

$$\mathbf{x}^{(n)} \to \begin{bmatrix} \tfrac{2}{3} \\ 0 \\ 0 \\ 0 \\ 0 \\ \tfrac{1}{3} \end{bmatrix}$$

Thus, in the limit there is probability $\tfrac{2}{3}$ that the sibling pairs will be (A, AA), and probability $\tfrac{1}{3}$ that they will be (a, aa).

Exercise Set 10.16

1. Show that if $M = PDP^{-1}$, then $M^n = PD^nP^{-1}$ for $n = 1, 2, \ldots$.

2. In Example 1 suppose that the plants are always fertilized with a plant of genotype Aa rather than one of genotype AA. Derive formulas for the fractions of the plants of genotypes AA, Aa, and aa in the nth generation. Also, find the limiting genotype distribution as n tends to infinity.

3. In Example 1 suppose that the initial plants are fertilized with genotype AA, the first generation is fertilized with genotype Aa, the second generation is fertilized with genotype AA, and this alternating pattern of fertilization is kept up. Find formulas for the fractions of the plants of genotypes AA, Aa, and aa in the nth generation.

4. In the section on autosomal recessive diseases, find the eigenvalues and eigenvectors of the matrix M and verify Equation (7).

5. Suppose that a breeder has an animal population in which 25% of the population are carriers of an autosomal recessive disease. If the breeder allows the animals to mate irrespective of their genotype, use Equation (9) to calculate the number of generations required for the percentage of carriers to fall from 25% to 10%. If the breeder instead implements the controlled-mating program determined by Equation (8), what will the percentage of carriers be after the same number of generations?

6. In the section on X-linked inheritance, suppose that the initial parents are equally likely to be of any of the six possible genotype parents; that is,

$$\mathbf{x}^{(0)} = \begin{bmatrix} \tfrac{1}{6} \\ \tfrac{1}{6} \\ \tfrac{1}{6} \\ \tfrac{1}{6} \\ \tfrac{1}{6} \\ \tfrac{1}{6} \end{bmatrix}$$

Using Equation (12), calculate $\mathbf{x}^{(n)}$ and also calculate the limit of $\mathbf{x}^{(n)}$ as n tends to infinity.

7. From (13) show that under X-linked inheritance with inbreeding, the probability that the limiting sibling pairs will be of type (A, AA) is the same as the proportion of A genes in the initial population.

8. In X-linked inheritance suppose that none of the females of genotype Aa survive to maturity. Under inbreeding the possible sibling pairs are then

$$(A, AA), \quad (A, aa), \quad (a, AA), \quad \text{and} \quad (a, aa)$$

Find the transition matrix that describes how the genotype distribution changes in one generation.

9. Derive the matrix M in Equation (10) from Table 2.

Section 10.16 Technology Exercises

The following exercises are designed to be solved using a technology utility. Typically, this will be MATLAB, *Mathematica*, Maple, Derive, or Mathcad, but it may also be some other type of linear algebra software or a scientific calculator with some linear algebra capabilities. For each exercise you will need to read the relevant documentation for the particular utility you are using. The goal of these exercises is to provide you with a basic proficiency with your technology utility. Once you have mastered the techniques in these exercises, you will be able to use your technology utility to solve many of the problems in the regular exercise sets.

T1. (a) Use a computer to verify that the eigenvalues and eigenvectors of

$$M = \begin{bmatrix} 1 & \frac{1}{4} & 0 & 0 & 0 & 0 \\ 0 & \frac{1}{4} & 0 & 1 & \frac{1}{4} & 0 \\ 0 & 0 & 0 & 0 & \frac{1}{4} & 0 \\ 0 & \frac{1}{4} & 0 & 0 & 0 & 0 \\ 0 & \frac{1}{4} & 1 & 0 & \frac{1}{4} & 0 \\ 0 & 0 & 0 & 0 & \frac{1}{4} & 1 \end{bmatrix}$$

as given in the text are correct.

(b) Starting with $\mathbf{x}^{(n)} = M\mathbf{x}^{(n-1)}$ and the assumption that

$$\lim_{n \to \infty} \mathbf{x}^{(n)} = \mathbf{x}$$

exists, we must have

$$\lim_{n \to \infty} \mathbf{x}^{(n)} = M \lim_{n \to \infty} \mathbf{x}^{(n-1)} \quad \text{or} \quad \mathbf{x} = M\mathbf{x}$$

This suggests that $\mathbf{x}$ can be solved directly using the equation $(M - I)\mathbf{x} = \mathbf{0}$. Use a computer to solve the equation $\mathbf{x} = M\mathbf{x}$, where

$$\mathbf{x} = \begin{bmatrix} a \\ b \\ c \\ d \\ e \\ f \end{bmatrix}$$

and $a + b + c + d + e + f = 1$; compare your results to Equation (13). Explain why the solution to $(M - I)\mathbf{x} = \mathbf{0}$ along with $a + b + c + d + e + f = 1$ is not specific enough to determine $\lim_{n \to \infty} \mathbf{x}^{(n)}$.

T2. (a) Given

$$P = \begin{bmatrix} 1 & 0 & -1 & 1 & \frac{1}{4}(-3-\sqrt{5}) & \frac{1}{4}(-3+\sqrt{5}) \\ 0 & 0 & 2 & -6 & 1 & 1 \\ 0 & 0 & -1 & -3 & \frac{1}{4}(-1+\sqrt{5}) & \frac{1}{4}(-1-\sqrt{5}) \\ 0 & 0 & 1 & 3 & \frac{1}{4}(-1+\sqrt{5}) & \frac{1}{4}(-1-\sqrt{5}) \\ 0 & 0 & -2 & 6 & 1 & 1 \\ 0 & 1 & 1 & -1 & \frac{1}{4}(-3-\sqrt{5}) & \frac{1}{4}(-3+\sqrt{5}) \end{bmatrix}$$

from Equation (12) and

$$\lim_{n \to \infty} D^n = \begin{bmatrix} 1 & 0 & 0 & 0 & 0 & 0 \\ 0 & 1 & 0 & 0 & 0 & 0 \\ 0 & 0 & 0 & 0 & 0 & 0 \\ 0 & 0 & 0 & 0 & 0 & 0 \\ 0 & 0 & 0 & 0 & 0 & 0 \\ 0 & 0 & 0 & 0 & 0 & 0 \end{bmatrix}$$

use a computer to show that

$$\lim_{n \to \infty} M^n = \begin{bmatrix} 1 & \frac{2}{3} & \frac{1}{3} & \frac{2}{3} & \frac{1}{3} & 0 \\ 0 & 0 & 0 & 0 & 0 & 0 \\ 0 & 0 & 0 & 0 & 0 & 0 \\ 0 & 0 & 0 & 0 & 0 & 0 \\ 0 & 0 & 0 & 0 & 0 & 0 \\ 0 & \frac{1}{3} & \frac{2}{3} & \frac{1}{3} & \frac{2}{3} & 1 \end{bmatrix}$$

(b) Use a computer to calculate M^n for $n = 10, 20, 30, 40, 50, 60, 70$, and then compare your results to the limit in part (a).

10.17 Age-Specific Population Growth

In this section we investigate, using the Leslie matrix model, the growth over time of a female population that is divided into age classes. We then determine the limiting age distribution and growth rate of the population.

> **PREREQUISITES:** Eigenvalues and Eigenvectors
> Diagonalization of a Matrix
> Intuitive Understanding of Limits

One of the most common models of population growth used by demographers is the so-called Leslie model developed in the 1940s. This model describes the growth of the female portion of a human or animal population. In this model the females are divided into age classes of equal duration. To be specific, suppose that the maximum age attained by any female in the population is L years (or some other time unit) and we divide the population into n age classes. Then each class is L/n years in duration. We label the age classes according to Table 1.

Table 1

Age Class	Age Interval
1	$[0, L/n)$
2	$[L/n, 2L/n)$
3	$[2L/n, 3L/n)$
$\vdots$	$\vdots$
$n-1$	$[(n-2)L/n, (n-1)L/n)$
n	$[(n-1)L/n, L]$

Suppose that we know the number of females in each of the n classes at time $t = 0$. In particular, let there be $x_1^{(0)}$ females in the first class, $x_2^{(0)}$ females in the second class, and so forth. With these n numbers we form a column vector:

$$\mathbf{x}^{(0)} = \begin{bmatrix} x_1^{(0)} \\ x_2^{(0)} \\ \vdots \\ x_n^{(0)} \end{bmatrix}$$

We call this vector the ***initial age distribution vector***.

As time progresses, the number of females within each of the n classes changes because of three biological processes: birth, death, and aging. By describing these three processes quantitatively, we will see how to project the initial age distribution vector into the future.

The easiest way to study the aging process is to observe the population at discrete times—say, $t_0, t_1, t_2, \ldots, t_k, \ldots$. The Leslie model requires that the duration between any two successive observation times be the same as the duration of the age intervals.

10.17 Age-Specific Population Growth

Therefore, we set

$$t_0 = 0$$
$$t_1 = L/n$$
$$t_2 = 2L/n$$
$$\vdots$$
$$t_k = kL/n$$
$$\vdots$$

With this assumption, all females in the $(i+1)$-st class at time t_{k+1} were in the ith class at time t_k.

The birth and death processes between two successive observation times can be described by means of the following demographic parameters:

a_i $(i = 1, 2, \ldots, n)$	The average number of daughters born to each female during the time she is in the ith age class
b_i $(i = 1, 2, \ldots, n-1)$	The fraction of females in the ith age class that can be expected to survive and pass into the $(i+1)$-st age class

By their definitions, we have that

(i) $a_i \geq 0$ for $i = 1, 2, \ldots, n$
(ii) $0 < b_i \leq 1$ for $i = 1, 2, \ldots, n-1$

Note that we do not allow any b_i to equal zero, because then no females would survive beyond the ith age class. We also assume that at least one a_i is positive so that some births occur. Any age class for which the corresponding value of a_i is positive is called a ***fertile age class***.

We next define the age distribution vector $\mathbf{x}^{(k)}$ at time t_k by

$$\mathbf{x}^{(k)} = \begin{bmatrix} x_1^{(k)} \\ x_2^{(k)} \\ \vdots \\ x_n^{(k)} \end{bmatrix}$$

where $x_i^{(k)}$ is the number of females in the ith age class at time t_k. Now, at time t_k, the females in the first age class are just those daughters born between times t_{k-1} and t_k. Thus, we can write

$$\left\{\begin{array}{c}\text{number of}\\\text{females}\\\text{in class 1}\\\text{at time } t_k\end{array}\right\} = \left\{\begin{array}{c}\text{number of}\\\text{daughters}\\\text{born to}\\\text{females in}\\\text{class 1}\\\text{between times}\\t_{k-1}\text{ and }t_k\end{array}\right\} + \left\{\begin{array}{c}\text{number of}\\\text{daughters}\\\text{born to}\\\text{females in}\\\text{class 2}\\\text{between times}\\t_{k-1}\text{ and }t_k\end{array}\right\} + \cdots + \left\{\begin{array}{c}\text{number of}\\\text{daughters}\\\text{born to}\\\text{females in}\\\text{class }n\\\text{between times}\\t_{k-1}\text{ and }t_k\end{array}\right\}$$

or, mathematically,

$$x_1^{(k)} = a_1 x_1^{(k-1)} + a_2 x_2^{(k-1)} + \cdots + a_n x_n^{(k-1)} \qquad (1)$$

The females in the $(i+1)$-st age class $(i = 1, 2, \ldots, n-1)$ at time t_k are those females in the ith class at time t_{k-1} who are still alive at time t_k. Thus,

$$\left\{ \begin{array}{l} \text{number of} \\ \text{females in} \\ \text{class } i+1 \\ \text{at time } t_k \end{array} \right\} = \left\{ \begin{array}{l} \text{fraction of} \\ \text{females in} \\ \text{class } i \\ \text{who survive} \\ \text{and pass into} \\ \text{class } i+1 \end{array} \right\} \left\{ \begin{array}{l} \text{number of} \\ \text{females in} \\ \text{class } i \\ \text{at time } t_{k-1} \end{array} \right\}$$

or, mathematically,

$$x_{i+1}^{(k)} = b_i x_i^{(k-1)}, \qquad i = 1, 2, \ldots, n-1 \qquad (2)$$

Using matrix notation, we can write Equations (1) and (2) as

$$\begin{bmatrix} x_1^{(k)} \\ x_2^{(k)} \\ x_3^{(k)} \\ \vdots \\ x_n^{(k)} \end{bmatrix} = \begin{bmatrix} a_1 & a_2 & a_3 & \cdots & a_{n-1} & a_n \\ b_1 & 0 & 0 & \cdots & 0 & 0 \\ 0 & b_2 & 0 & \cdots & 0 & 0 \\ \vdots & \vdots & \vdots & & \vdots & \vdots \\ 0 & 0 & 0 & \cdots & b_{n-1} & 0 \end{bmatrix} \begin{bmatrix} x_1^{(k-1)} \\ x_2^{(k-1)} \\ x_3^{(k-1)} \\ \vdots \\ x_n^{(k-1)} \end{bmatrix}$$

or more compactly as

$$\mathbf{x}^{(k)} = L \mathbf{x}^{(k-1)}, \qquad k = 1, 2, \ldots \qquad (3)$$

where L is the **Leslie matrix**

$$L = \begin{bmatrix} a_1 & a_2 & a_3 & \cdots & a_{n-1} & a_n \\ b_1 & 0 & 0 & \cdots & 0 & 0 \\ 0 & b_2 & 0 & \cdots & 0 & 0 \\ \vdots & \vdots & \vdots & & \vdots & \vdots \\ 0 & 0 & 0 & \cdots & b_{n-1} & 0 \end{bmatrix} \qquad (4)$$

From Equation (3) it follows that

$$\begin{aligned} \mathbf{x}^{(1)} &= L \mathbf{x}^{(0)} \\ \mathbf{x}^{(2)} &= L \mathbf{x}^{(1)} = L^2 \mathbf{x}^{(0)} \\ \mathbf{x}^{(3)} &= L \mathbf{x}^{(2)} = L^3 \mathbf{x}^{(0)} \\ &\vdots \\ \mathbf{x}^{(k)} &= L \mathbf{x}^{(k-1)} = L^k \mathbf{x}^{(0)} \end{aligned} \qquad (5)$$

Thus, if we know the initial age distribution $\mathbf{x}^{(0)}$ and the Leslie matrix L, we can determine the female age distribution at any later time.

▶ **EXAMPLE 1 Female Age Distribution for Animals**

Suppose that the oldest age attained by the females in a certain animal population is 15 years and we divide the population into three age classes with equal durations of five years. Let the Leslie matrix for this population be

$$L = \begin{bmatrix} 0 & 4 & 3 \\ \frac{1}{2} & 0 & 0 \\ 0 & \frac{1}{4} & 0 \end{bmatrix}$$

10.17 Age-Specific Population Growth

If there are initially 1000 females in each of the three age classes, then from Equation (3) we have

$$\mathbf{x}^{(0)} = \begin{bmatrix} 1{,}000 \\ 1{,}000 \\ 1{,}000 \end{bmatrix}$$

$$\mathbf{x}^{(1)} = L\mathbf{x}^{(0)} = \begin{bmatrix} 0 & 4 & 3 \\ \frac{1}{2} & 0 & 0 \\ 0 & \frac{1}{4} & 0 \end{bmatrix} \begin{bmatrix} 1{,}000 \\ 1{,}000 \\ 1{,}000 \end{bmatrix} = \begin{bmatrix} 7{,}000 \\ 500 \\ 250 \end{bmatrix}$$

$$\mathbf{x}^{(2)} = L\mathbf{x}^{(1)} = \begin{bmatrix} 0 & 4 & 3 \\ \frac{1}{2} & 0 & 0 \\ 0 & \frac{1}{4} & 0 \end{bmatrix} \begin{bmatrix} 7{,}000 \\ 500 \\ 250 \end{bmatrix} = \begin{bmatrix} 2{,}750 \\ 3{,}500 \\ 125 \end{bmatrix}$$

$$\mathbf{x}^{(3)} = L\mathbf{x}^{(2)} = \begin{bmatrix} 0 & 4 & 3 \\ \frac{1}{2} & 0 & 0 \\ 0 & \frac{1}{4} & 0 \end{bmatrix} \begin{bmatrix} 2{,}750 \\ 3{,}500 \\ 125 \end{bmatrix} = \begin{bmatrix} 14{,}375 \\ 1{,}375 \\ 875 \end{bmatrix}$$

Thus, after 15 years there are 14,375 females between 0 and 5 years of age, 1375 females between 5 and 10 years of age, and 875 females between 10 and 15 years of age. ◀

Limiting Behavior

Although Equation (5) gives the age distribution of the population at any time, it does not immediately give a general picture of the dynamics of the growth process. For this we need to investigate the eigenvalues and eigenvectors of the Leslie matrix. The eigenvalues of L are the roots of its characteristic polynomial. As we ask you to verify in Exercise 2, this characteristic polynomial is

$$p(\lambda) = |\lambda I - L| \\ = \lambda^n - a_1 \lambda^{n-1} - a_2 b_1 \lambda^{n-2} - a_3 b_1 b_2 \lambda^{n-3} - \cdots - a_n b_1 b_2 \cdots b_{n-1}$$

To analyze the roots of this polynomial, it will be convenient to introduce the function

$$q(\lambda) = \frac{a_1}{\lambda} + \frac{a_2 b_1}{\lambda^2} + \frac{a_3 b_1 b_2}{\lambda^3} + \cdots + \frac{a_n b_1 b_2 \cdots b_{n-1}}{\lambda^n} \tag{6}$$

Using this function, the characteristic equation $p(\lambda) = 0$ can be written (verify)

$$q(\lambda) = 1 \quad \text{for } \lambda \neq 0 \tag{7}$$

Because all the a_i and b_i are nonnegative, we see that $q(\lambda)$ is monotonically decreasing for λ greater than zero. Furthermore, $q(\lambda)$ has a vertical asymptote at $\lambda = 0$ and approaches zero as $\lambda \to \infty$. Consequently, as Figure 10.17.1 indicates, there is a unique λ, say $\lambda = \lambda_1$, such that $q(\lambda_1) = 1$. That is, the matrix L has a unique positive eigenvalue. It can also be shown (see Exercise 3) that λ_1 has multiplicity 1; that is, λ_1 is not a repeated root of the characteristic equation. Although we omit the computational details, you can verify that an eigenvector corresponding to λ_1 is

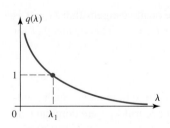

▲ Figure 10.17.1

$$\mathbf{x}_1 = \begin{bmatrix} 1 \\ b_1/\lambda_1 \\ b_1 b_2/\lambda_1^2 \\ b_1 b_2 b_3/\lambda_1^3 \\ \vdots \\ b_1 b_2 \cdots b_{n-1}/\lambda_1^{n-1} \end{bmatrix} \tag{8}$$

Because λ_1 has multiplicity 1, its corresponding eigenspace has dimension 1 (Exercise 3), and so any eigenvector corresponding to it is some multiple of $\mathbf{x}_1$. We can summarize these results in the following theorem.

> **THEOREM 10.17.1 Existence of a Positive Eigenvalue**
>
> *A Leslie matrix L has a unique positive eigenvalue λ_1. This eigenvalue has multiplicity 1 and an eigenvector $\mathbf{x}_1$ all of whose entries are positive.*

We will now show that the long-term behavior of the age distribution of the population is determined by the positive eigenvalue λ_1 and its eigenvector $\mathbf{x}_1$.

In Exercise 9 we ask you to prove the following result.

> **THEOREM 10.17.2 Eigenvalues of a Leslie Matrix**
>
> *If λ_1 is the unique positive eigenvalue of a Leslie matrix L, and λ_k is any other real or complex eigenvalue of L, then $|\lambda_k| \leq \lambda_1$.*

For our purposes the conclusion in Theorem 10.17.2 is not strong enough; we need λ_1 to satisfy $|\lambda_k| < \lambda_1$. In this case λ_1 would be called the ***dominant eigenvalue*** of L. However, as the following example shows, not all Leslie matrices satisfy this condition.

▶ **EXAMPLE 2 Leslie Matrix with No Dominant Eigenvalue**

Let
$$L = \begin{bmatrix} 0 & 0 & 6 \\ \frac{1}{2} & 0 & 0 \\ 0 & \frac{1}{3} & 0 \end{bmatrix}$$

Then the characteristic polynomial of L is
$$p(\lambda) = |\lambda I - L| = \lambda^3 - 1$$

The eigenvalues of L are thus the solutions of $\lambda^3 = 1$—namely,
$$\lambda = 1, \quad -\frac{1}{2} + \frac{\sqrt{3}}{2}i, \quad -\frac{1}{2} - \frac{\sqrt{3}}{2}i$$

All three eigenvalues have absolute value 1, so the unique positive eigenvalue $\lambda_1 = 1$ is not dominant. Note that this matrix has the property that $L^3 = I$. This means that for any choice of the initial age distribution $\mathbf{x}^{(0)}$, we have
$$\mathbf{x}^{(0)} = \mathbf{x}^{(3)} = \mathbf{x}^{(6)} = \cdots = \mathbf{x}^{(3k)} = \cdots$$

The age distribution vector thus oscillates with a period of three time units. Such oscillations (or ***population waves***, as they are called) could not occur if λ_1 were dominant, as we will see below. ◀

It is beyond the scope of this book to discuss necessary and sufficient conditions for λ_1 to be a dominant eigenvalue. However, we will state the following sufficient condition without proof.

> **THEOREM 10.17.3** *Dominant Eigenvalue*
>
> *If two successive entries a_i and a_{i+1} in the first row of a Leslie matrix L are nonzero, then the positive eigenvalue of L is dominant.*

Thus, if the female population has two successive fertile age classes, then its Leslie matrix has a dominant eigenvalue. This is always the case for realistic populations if the duration of the age classes is sufficiently small. Note that in Example 2 there is only one fertile age class (the third), so the condition of Theorem 10.17.3 is not satisfied. In what follows, we always assume that the condition of Theorem 10.17.3 is satisfied.

Let us assume that L is diagonalizable. This is not really necessary for the conclusions we will draw, but it does simplify the arguments. In this case, L has n eigenvalues, $\lambda_1, \lambda_2, \ldots, \lambda_n$, not necessarily distinct, and n linearly independent eigenvectors, $\mathbf{x}_1, \mathbf{x}_2, \ldots, \mathbf{x}_n$, corresponding to them. In this listing we place the dominant eigenvalue λ_1 first. We construct a matrix P whose columns are the eigenvectors of L:

$$P = [\mathbf{x}_1 \mid \mathbf{x}_2 \mid \mathbf{x}_3 \mid \cdots \mid \mathbf{x}_n]$$

The diagonalization of L is then given by the equation

$$L = P \begin{bmatrix} \lambda_1 & 0 & 0 & \cdots & 0 \\ 0 & \lambda_2 & 0 & \cdots & 0 \\ \vdots & \vdots & \vdots & & \vdots \\ 0 & 0 & 0 & \cdots & \lambda_n \end{bmatrix} P^{-1}$$

From this it follows that

$$L^k = P \begin{bmatrix} \lambda_1^k & 0 & 0 & \cdots & 0 \\ 0 & \lambda_2^k & 0 & \cdots & 0 \\ \vdots & \vdots & \vdots & & \vdots \\ 0 & 0 & 0 & \cdots & \lambda_n^k \end{bmatrix} P^{-1}$$

for $k = 1, 2, \ldots$. For any initial age distribution vector $\mathbf{x}^{(0)}$, we then have

$$L^k \mathbf{x}^{(0)} = P \begin{bmatrix} \lambda_1^k & 0 & 0 & \cdots & 0 \\ 0 & \lambda_2^k & 0 & \cdots & 0 \\ \vdots & \vdots & \vdots & & \vdots \\ 0 & 0 & 0 & \cdots & \lambda_n^k \end{bmatrix} P^{-1} \mathbf{x}^{(0)}$$

for $k = 1, 2, \ldots$. Dividing both sides of this equation by λ_1^k and using the fact that $\mathbf{x}^{(k)} = L^k \mathbf{x}^{(0)}$, we have

$$\frac{1}{\lambda_1^k} \mathbf{x}^{(k)} = P \begin{bmatrix} 1 & 0 & 0 & \cdots & 0 \\ 0 & \left(\frac{\lambda_2}{\lambda_1}\right)^k & 0 & \cdots & 0 \\ \vdots & \vdots & \vdots & & \vdots \\ 0 & 0 & 0 & \cdots & \left(\frac{\lambda_n}{\lambda_1}\right)^k \end{bmatrix} P^{-1} \mathbf{x}^{(0)} \qquad (9)$$

Because λ_1 is the dominant eigenvalue, we have $|\lambda_i/\lambda_1| < 1$ for $i = 2, 3, \ldots, n$. It follows that

$$(\lambda_i/\lambda_1)^k \to 0 \text{ as } k \to \infty \qquad \text{for } i = 2, 3, \ldots, n$$

Using this fact, we can take the limit of both sides of (9) to obtain

$$\lim_{k \to \infty} \left\{ \frac{1}{\lambda_1^k} \mathbf{x}^{(k)} \right\} = P \begin{bmatrix} 1 & 0 & 0 & \cdots & 0 \\ 0 & 0 & 0 & \cdots & 0 \\ \vdots & \vdots & \vdots & & \vdots \\ 0 & 0 & 0 & \cdots & 0 \end{bmatrix} P^{-1} \mathbf{x}^{(0)} \qquad (10)$$

Let us denote the first entry of the column vector $P^{-1}\mathbf{x}^{(0)}$ by the constant c. As we ask you to show in Exercise 4, the right side of (10) can be written as $c\mathbf{x}_1$, where c is a positive constant that depends only on the initial age distribution vector $\mathbf{x}^{(0)}$. Thus, (10) becomes

$$\lim_{k \to \infty} \left\{ \frac{1}{\lambda_1^k} \mathbf{x}^{(k)} \right\} = c\mathbf{x}_1 \qquad (11)$$

Equation (11) gives us the approximation

$$\mathbf{x}^{(k)} \simeq c\lambda_1^k \mathbf{x}_1 \qquad (12)$$

for large values of k. From (12) we also have

$$\mathbf{x}^{(k-1)} \simeq c\lambda_1^{k-1} \mathbf{x}_1 \qquad (13)$$

Comparing Equations (12) and (13), we see that

$$\mathbf{x}^{(k)} \simeq \lambda_1 \mathbf{x}^{(k-1)} \qquad (14)$$

for large values of k. This means that for large values of time, each age distribution vector is a scalar multiple of the preceding age distribution vector, the scalar being the positive eigenvalue of the Leslie matrix. Consequently, the *proportion* of females in each of the age classes becomes constant. As we will see in the following example, these limiting proportions can be determined from the eigenvector $\mathbf{x}_1$.

▶ **EXAMPLE 3** **Example 1 Revisited**

The Leslie matrix in Example 1 was

$$L = \begin{bmatrix} 0 & 4 & 3 \\ \frac{1}{2} & 0 & 0 \\ 0 & \frac{1}{4} & 0 \end{bmatrix}$$

Its characteristic polynomial is $p(\lambda) = \lambda^3 - 2\lambda - \frac{3}{8}$, and you can verify that the positive eigenvalue is $\lambda_1 = \frac{3}{2}$. From (8) the corresponding eigenvector $\mathbf{x}_1$ is

$$\mathbf{x}_1 = \begin{bmatrix} 1 \\ b_1/\lambda_1 \\ b_1 b_2/\lambda_1^2 \end{bmatrix} = \begin{bmatrix} 1 \\ \frac{\frac{1}{2}}{\frac{3}{2}} \\ \frac{(\frac{1}{2})(\frac{1}{4})}{(\frac{3}{2})^2} \end{bmatrix} = \begin{bmatrix} 1 \\ \frac{1}{3} \\ \frac{1}{18} \end{bmatrix}$$

From (14) we have

$$\mathbf{x}^{(k)} \simeq \tfrac{3}{2} \mathbf{x}^{(k-1)}$$

for large values of k. Hence, every five years the number of females in each of the three classes will increase by about 50%, as will the total number of females in the population.

From (12) we have

$$\mathbf{x}^{(k)} \simeq c \left(\tfrac{3}{2}\right)^k \begin{bmatrix} 1 \\ \tfrac{1}{3} \\ \tfrac{1}{18} \end{bmatrix}$$

Consequently, eventually the females will be distributed among the three age classes in the ratios $1:\tfrac{1}{3}:\tfrac{1}{18}$. This corresponds to a distribution of 72% of the females in the first age class, 24% of the females in the second age class, and 4% of the females in the third age class.

▶ **EXAMPLE 4** **Female Age Distribution for Humans**

In this example we use birth and death parameters from the year 1965 for Canadian females. Because few women over 50 years of age bear children, we restrict ourselves to the portion of the female population between 0 and 50 years of age. The data are for 5-year age classes, so there are a total of 10 age classes. Rather than writing out the 10×10 Leslie matrix in full, we list the birth and death parameters as follows:

Age Interval	a_i	b_i
[0, 5)	0.00000	0.99651
[5, 10)	0.00024	0.99820
[10, 15)	0.05861	0.99802
[15, 20)	0.28608	0.99729
[20, 25)	0.44791	0.99694
[25, 30)	0.36399	0.99621
[30, 35)	0.22259	0.99460
[35, 40)	0.10457	0.99184
[40, 45)	0.02826	0.98700
[45, 50)	0.00240	—

Using numerical techniques, we can approximate the positive eigenvalue and corresponding eigenvector by

$$\lambda_1 = 1.07622 \quad \text{and} \quad \mathbf{x}_1 = \begin{bmatrix} 1.00000 \\ 0.92594 \\ 0.85881 \\ 0.79641 \\ 0.73800 \\ 0.68364 \\ 0.63281 \\ 0.58482 \\ 0.53897 \\ 0.49429 \end{bmatrix}$$

Thus, if Canadian women continued to reproduce and die as they did in 1965, eventually every 5 years their numbers would increase by 7.622%. From the eigenvector $\mathbf{x}_1$, we see that, in the limit, for every 100,000 females between 0 and 5 years of age, there will be 92,594 females between 5 and 10 years of age, 85,881 females between 10 and 15 years of age, and so forth. ◀

Let us look again at Equation (12), which gives the age distribution vector of the population for large times:

$$\mathbf{x}^{(k)} \simeq c\lambda_1^k \mathbf{x}_1 \qquad (15)$$

Three cases arise according to the value of the positive eigenvalue λ_1:

(i) The population is eventually increasing if $\lambda_1 > 1$.
(ii) The population is eventually decreasing if $\lambda_1 < 1$.
(iii) The population eventually stabilizes if $\lambda_1 = 1$.

The case $\lambda_1 = 1$ is particularly interesting because it determines a population that has **zero population growth**. For any initial age distribution, the population approaches a limiting age distribution that is some multiple of the eigenvector $\mathbf{x}_1$. From Equations (6) and (7), we see that $\lambda_1 = 1$ is an eigenvalue if and only if

$$a_1 + a_2 b_1 + a_3 b_1 b_2 + \cdots + a_n b_1 b_2 \cdots b_{n-1} = 1 \qquad (16)$$

The expression

$$R = a_1 + a_2 b_1 + a_3 b_1 b_2 + \cdots + a_n b_1 b_2 \cdots b_{n-1} \qquad (17)$$

is called the **net reproduction rate** of the population. (See Exercise 5 for a demographic interpretation of R.) Thus, we can say that a population has zero population growth if and only if its net reproduction rate is 1.

Exercise Set 10.17

1. Suppose that a certain animal population is divided into two age classes and has a Leslie matrix

$$L = \begin{bmatrix} 1 & \frac{3}{2} \\ \frac{1}{2} & 0 \end{bmatrix}$$

   (a) Calculate the positive eigenvalue λ_1 of L and the corresponding eigenvector $\mathbf{x}_1$.

   (b) Beginning with the initial age distribution vector

$$\mathbf{x}^{(0)} = \begin{bmatrix} 100 \\ 0 \end{bmatrix}$$

   calculate $\mathbf{x}^{(1)}$, $\mathbf{x}^{(2)}$, $\mathbf{x}^{(3)}$, $\mathbf{x}^{(4)}$, and $\mathbf{x}^{(5)}$, rounding off to the nearest integer when necessary.

   (c) Calculate $\mathbf{x}^{(6)}$ using the exact formula $\mathbf{x}^{(6)} = L\mathbf{x}^{(5)}$ and using the approximation formula $\mathbf{x}^{(6)} \simeq \lambda_1 \mathbf{x}^{(5)}$.

2. Find the characteristic polynomial of a general Leslie matrix given by Equation (4).

3. (a) Show that the positive eigenvalue λ_1 of a Leslie matrix is always simple. Recall that a root λ_0 of a polynomial $q(\lambda)$ is simple if and only if $q'(\lambda_0) \neq 0$.

   (b) Show that the eigenspace corresponding to λ_1 has dimension 1.

4. Show that the right side of Equation (10) is $c\mathbf{x}_1$, where c is the first entry of the column vector $P^{-1}\mathbf{x}^{(0)}$.

5. Show that the net reproduction rate R, defined by (17), can be interpreted as the average number of daughters born to a single female during her expected lifetime.

6. Show that a population is eventually decreasing if and only if its net reproduction rate is less than 1. Similarly, show that a population is eventually increasing if and only if its net reproduction rate is greater than 1.

7. Calculate the net reproduction rate of the animal population in Example 1.

8. (*For readers with a hand calculator*) Calculate the net reproduction rate of the Canadian female population in Example 4.

9. (*For readers who have read Sections 10.1–10.3*) Prove Theorem 10.17.2. [*Hint:* Write $\lambda_k = re^{i\theta}$, substitute into (7), take the real parts of both sides, and show that $r \leq \lambda_1$.]

Section 10.17 Technology Exercises

The following exercises are designed to be solved using a technology utility. Typically, this will be MATLAB, *Mathematica*, Maple, Derive, or Mathcad, but it may also be some other type of linear algebra software or a scientific calculator with some linear algebra capabilities. For each exercise you will need to read the relevant documentation for the particular utility you are using. The goal of these exercises is to provide you with a basic proficiency with your technology utility. Once you have mastered the techniques in these exercises, you will be able to use your technology utility to solve many of the problems in the regular exercise sets.

T1. Consider the sequence of Leslie matrices

$$L_2 = \begin{bmatrix} 0 & a \\ b_1 & 0 \end{bmatrix}, \qquad L_3 = \begin{bmatrix} 0 & 0 & a \\ b_1 & 0 & 0 \\ 0 & b_2 & 0 \end{bmatrix},$$

$$L_4 = \begin{bmatrix} 0 & 0 & 0 & a \\ b_1 & 0 & 0 & 0 \\ 0 & b_2 & 0 & 0 \\ 0 & 0 & b_3 & 0 \end{bmatrix}, \quad L_5 = \begin{bmatrix} 0 & 0 & 0 & 0 & a \\ b_1 & 0 & 0 & 0 & 0 \\ 0 & b_2 & 0 & 0 & 0 \\ 0 & 0 & b_3 & 0 & 0 \\ 0 & 0 & 0 & b_4 & 0 \end{bmatrix}, \ldots$$

(a) Use a computer to show that

$$L_2^2 = I_2, \qquad L_3^3 = I_3, \qquad L_4^4 = I_4, \qquad L_5^5 = I_5, \ldots$$

for a suitable choice of a in terms of $b_1, b_2, \ldots, b_{n-1}$.

(b) From your results in part (a), conjecture a relationship between a and $b_1, b_2, \ldots, b_{n-1}$ that will make $L_n^n = I_n$, where

$$L_n = \begin{bmatrix} 0 & 0 & 0 & \cdots & 0 & a \\ b_1 & 0 & 0 & \cdots & 0 & 0 \\ 0 & b_2 & 0 & \cdots & 0 & 0 \\ 0 & 0 & b_3 & \cdots & 0 & 0 \\ \vdots & \vdots & \vdots & \ddots & \vdots & \vdots \\ 0 & 0 & 0 & \cdots & b_{n-1} & 0 \end{bmatrix}$$

(c) Determine an expression for $p_n(\lambda) = |\lambda I_n - L_n|$ and use it to show that all eigenvalues of L_n satisfy $|\lambda| = 1$ when a and $b_1, b_2, \ldots, b_{n-1}$ are related by the equation determined in part (b).

T2. Consider the sequence of Leslie matrices

$$L_2 = \begin{bmatrix} a & ap \\ b & 0 \end{bmatrix}, \qquad L_3 = \begin{bmatrix} a & ap & ap^2 \\ b & 0 & 0 \\ 0 & b & 0 \end{bmatrix},$$

$$L_4 = \begin{bmatrix} a & ap & ap^2 & ap^3 \\ b & 0 & 0 & 0 \\ 0 & b & 0 & 0 \\ 0 & 0 & b & 0 \end{bmatrix},$$

$$L_5 = \begin{bmatrix} a & ap & ap^2 & ap^3 & ap^4 \\ b & 0 & 0 & 0 & 0 \\ 0 & b & 0 & 0 & 0 \\ 0 & 0 & b & 0 & 0 \\ 0 & 0 & 0 & b & 0 \end{bmatrix}, \ldots$$

$$L_n = \begin{bmatrix} a & ap & ap^2 & \cdots & ap^{n-2} & ap^{n-1} \\ b & 0 & 0 & \cdots & 0 & 0 \\ 0 & b & 0 & \cdots & 0 & 0 \\ 0 & 0 & b & \cdots & 0 & 0 \\ \vdots & \vdots & \vdots & \ddots & \vdots & \vdots \\ 0 & 0 & 0 & \cdots & b & 0 \end{bmatrix}$$

where $0 < p < 1$, $0 < b < 1$, and $1 < a$.

(a) Choose a value for n (say, $n = 8$). For various values of a, b, and p, use a computer to determine the dominant eigenvalue of L_n, and then compare your results to the value of $a + bp$.

(b) Show that

$$p_n(\lambda) = |\lambda I_n - L_n| = \lambda^n - a\left(\frac{\lambda^n - (bp)^n}{\lambda - bp}\right)$$

which means that the eigenvalues of L_n must satisfy

$$\lambda^{n+1} - (a + bp)\lambda^n + a(bp)^n = 0$$

(c) Can you now provide a rough proof to explain the fact that $\lambda_1 \simeq a + bp$?

T3. Suppose that a population of mice has a Leslie matrix L over a 1-month period and an initial age distribution vector $\mathbf{x}^{(0)}$ given by

$$L = \begin{bmatrix} 0 & 0 & \frac{1}{2} & \frac{4}{5} & \frac{3}{10} & 0 \\ \frac{4}{5} & 0 & 0 & 0 & 0 & 0 \\ 0 & \frac{9}{10} & 0 & 0 & 0 & 0 \\ 0 & 0 & \frac{9}{10} & 0 & 0 & 0 \\ 0 & 0 & 0 & \frac{4}{5} & 0 & 0 \\ 0 & 0 & 0 & 0 & \frac{3}{10} & 0 \end{bmatrix} \quad \text{and} \quad \mathbf{x}^{(0)} = \begin{bmatrix} 50 \\ 40 \\ 30 \\ 20 \\ 10 \\ 5 \end{bmatrix}$$

(a) Compute the net reproduction rate of the population.

(b) Compute the age distribution vector after 100 months and 101 months, and show that the vector after 101 weeks is approximately a scalar multiple of the vector after 100 months.

(c) Compute the dominant eigenvalue of L and its corresponding eigenvector. How are they related to your results in part (b)?

(d) Suppose you wish to control the mouse population by feeding it a substance that decreases its age-specific birthrates (the entries in the first row of L) by a constant fraction. What range of fractions would cause the population eventually to decrease?

10.18 Harvesting of Animal Populations

In this section we employ the Leslie matrix model of population growth to model the sustainable harvesting of an animal population. We also examine the effect of harvesting different fractions of different age groups.

> **PREREQUISITES:** Age-Specific Population Growth (Section 10.17)

Harvesting

In Section 10.17 we used the Leslie matrix model to examine the growth of a female population that was divided into discrete age classes. In this section, we investigate the effects of harvesting an animal population growing according to such a model. By **harvesting** we mean the removal of animals from the population. (The word *harvesting* is not necessarily a euphemism for "slaughtering"; the animals may be removed from the population for other purposes.)

In this section we restrict ourselves to *sustainable harvesting policies*. By this we mean the following:

> **DEFINITION 1** A harvesting policy in which an animal population is periodically harvested is said to be **sustainable** if the yield of each harvest is the same and the age distribution of the population remaining after each harvest is the same.

Thus, the animal population is not depleted by a sustainable harvesting policy; only the excess growth is removed.

As in Section 10.17, we will discuss only the females of the population. If the number of males in each age class is equal to the number of females—a reasonable assumption for many populations—then our harvesting policies will also apply to the male portion of the population.

The Harvesting Model

Figure 10.18.1 illustrates the basic idea of the model. We begin with a population having a particular age distribution. It undergoes a growth period that will be described by the Leslie matrix. At the end of the growth period, a certain fraction of each age class is harvested in such a way that the unharvested population has the same age distribution as the original population. This cycle repeats after each harvest so that the yield is sustainable. The duration of the harvest is assumed to be short in comparison with the growth period so that any growth or change in the population during the harvest period can be neglected.

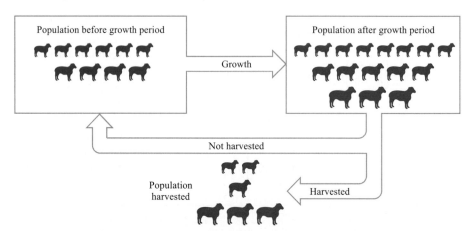

▶ Figure 10.18.1

10.18 Harvesting of Animal Populations

To describe this harvesting model mathematically, let

$$\mathbf{x} = \begin{bmatrix} x_1 \\ x_2 \\ \vdots \\ x_n \end{bmatrix}$$

be the age distribution vector of the population at the beginning of the growth period. Thus x_i is the number of females in the ith class left unharvested. As in Section 10.17, we require that the duration of each age class be identical with the duration of the growth period. For example, if the population is harvested once a year, then the population is divided into 1-year age classes.

If L is the Leslie matrix describing the growth of the population, then the vector $L\mathbf{x}$ is the age distribution vector of the population at the end of the growth period, immediately before the periodic harvest. Let h_i, for $i = 1, 2, \ldots, n$, be the fraction of females from the ith class that is harvested. We use these n numbers to form an $n \times n$ diagonal matrix

$$H = \begin{bmatrix} h_1 & 0 & 0 & \cdots & 0 \\ 0 & h_2 & 0 & \cdots & 0 \\ 0 & 0 & h_3 & \cdots & 0 \\ \vdots & \vdots & \vdots & & \vdots \\ 0 & 0 & 0 & \cdots & h_n \end{bmatrix}$$

which we will call the ***harvesting matrix***. By definition, we have

$$0 \le h_i \le 1 \quad (i = 1, 2, \ldots, n)$$

That is, we can harvest none ($h_i = 0$), all ($h_i = 1$), or some fraction ($0 < h_i < 1$) of each of the n classes. Because the number of females in the ith class immediately before each harvest is the ith entry $(L\mathbf{x})_i$ of the vector $L\mathbf{x}$, the ith entry of the column vector

$$HL\mathbf{x} = \begin{bmatrix} h_1(L\mathbf{x})_1 \\ h_2(L\mathbf{x})_2 \\ \vdots \\ h_n(L\mathbf{x})_n \end{bmatrix}$$

is the number of females harvested from the ith class.

From the definition of a sustainable harvesting policy, we have

$$\begin{bmatrix} \text{age distribution} \\ \text{at end of} \\ \text{growth period} \end{bmatrix} - [\text{harvest}] = \begin{bmatrix} \text{age distribution} \\ \text{at beginning of} \\ \text{growth period} \end{bmatrix}$$

or, mathematically,

$$L\mathbf{x} - HL\mathbf{x} = \mathbf{x} \tag{1}$$

If we write Equation (1) in the form

$$(I - H)L\mathbf{x} = \mathbf{x} \tag{2}$$

we see that $\mathbf{x}$ must be an eigenvector of the matrix $(I - H)L$ corresponding to the eigenvalue 1. As we will now show, this places certain restrictions on the values of h_i and $\mathbf{x}$.

Suppose that the Leslie matrix of the population is

$$L = \begin{bmatrix} a_1 & a_2 & a_3 & \cdots & a_{n-1} & a_n \\ b_1 & 0 & 0 & \cdots & 0 & 0 \\ 0 & b_2 & 0 & \cdots & 0 & 0 \\ \vdots & \vdots & \vdots & & \vdots & \vdots \\ 0 & 0 & 0 & \cdots & b_{n-1} & 0 \end{bmatrix} \tag{3}$$

688 Chapter 10 Applications of Linear Algebra

Then the matrix $(I - H)L$ is (verify)

$$(I - H)L = \begin{bmatrix} (1-h_1)a_1 & (1-h_1)a_2 & (1-h_1)a_3 & \cdots & (1-h_1)a_{n-1} & (1-h_1)a_n \\ (1-h_2)b_1 & 0 & 0 & \cdots & 0 & 0 \\ 0 & (1-h_3)b_2 & 0 & \cdots & 0 & 0 \\ \vdots & \vdots & \vdots & & \vdots & \vdots \\ 0 & 0 & 0 & \cdots & (1-h_n)b_{n-1} & 0 \end{bmatrix}$$

Thus, we see that $(I - H)L$ is a matrix with the same mathematical form as a Leslie matrix. In Section 10.17 we showed that a necessary and sufficient condition for a Leslie matrix to have 1 as an eigenvalue is that its net reproduction rate also be 1 [see Eq. (16) of Section 10.17]. Calculating the net reproduction rate of $(I - H)L$ and setting it equal to 1, we obtain (verify)

$$(1-h_1)[a_1 + a_2 b_1(1-h_2) + a_3 b_1 b_2(1-h_2)(1-h_3) + \cdots \\ + a_n b_1 b_2 \cdots b_{n-1}(1-h_2)(1-h_3) \cdots (1-h_n)] = 1 \quad (4)$$

This equation places a restriction on the allowable harvesting fractions. Only those values of $h_1, h_2, \ldots, h_n$ that satisfy (4) and that lie in the interval $[0, 1]$ can produce a sustainable yield.

If $h_1, h_2, \ldots, h_n$ do satisfy (4), then the matrix $(I - H)L$ has the desired eigenvalue $\lambda_1 = 1$. Furthermore, this eigenvalue has multiplicity 1, because the positive eigenvalue of a Leslie matrix always has multiplicity 1 (Theorem 10.17.1). This means that there is only one linearly independent eigenvector $\mathbf{x}$ satisfying Equation (2). [See Exercise 3(b) of Section 10.17.] One possible choice for $\mathbf{x}$ is the following normalized eigenvector:

$$\mathbf{x}_1 = \begin{bmatrix} 1 \\ b_1(1-h_2) \\ b_1 b_2 (1-h_2)(1-h_3) \\ b_1 b_2 b_3 (1-h_2)(1-h_3)(1-h_4) \\ \vdots \\ b_1 b_2 b_3 \cdots b_{n-1}(1-h_2)(1-h_3)\cdots(1-h_n) \end{bmatrix} \quad (5)$$

Any other solution $\mathbf{x}$ of (2) is a multiple of $\mathbf{x}_1$. Thus, the vector $\mathbf{x}_1$ determines the proportion of females within each of the n classes after a harvest under a sustainable harvesting policy. But there is an ambiguity in the total number of females in the population after each harvest. This can be determined by some auxiliary condition, such as an ecological or economic constraint. For example, for a population economically supported by the harvester, the largest population the harvester can afford to raise between harvests would determine the particular constant that $\mathbf{x}_1$ is multiplied by to produce the appropriate vector $\mathbf{x}$ in Equation (2). For a wild population, the natural habitat of the population would determine how large the total population could be between harvests.

Summarizing our results so far, we see that there is a wide choice in the values of $h_1, h_2, \ldots, h_n$ that will produce a sustainable yield. But once these values are selected, the proportional age distribution of the population after each harvest is uniquely determined by the normalized eigenvector $\mathbf{x}_1$ defined by Equation (5). We now consider a few particular harvesting strategies of this type.

Uniform Harvesting   With many populations it is difficult to distinguish or catch animals of specific ages. If animals are caught at random, we can reasonably assume that the same fraction of each age class is harvested. We therefore set

$$h = h_1 = h_2 = \cdots = h_n$$

Equation (2) then reduces to (verify)

$$L\mathbf{x} = \left(\frac{1}{1-h}\right)\mathbf{x}$$

Hence, $1/(1-h)$ must be the unique positive eigenvalue λ_1 of the Leslie growth matrix L. That is,

$$\lambda_1 = \frac{1}{1-h}$$

Solving for the harvesting fraction h, we obtain

$$h = 1 - (1/\lambda_1) \tag{6}$$

The vector $\mathbf{x}_1$, in this case, is the same as the eigenvector of L corresponding to the eigenvalue λ_1. From Equation (8) of Section 10.17, this is

$$\mathbf{x}_1 = \begin{bmatrix} 1 \\ b_1/\lambda_1 \\ b_1 b_2/\lambda_1^2 \\ b_1 b_2 b_3/\lambda_1^3 \\ \vdots \\ b_1 b_2 \cdots b_{n-1}/\lambda_1^{n-1} \end{bmatrix} \tag{7}$$

From (6) we can see that the larger λ_1 is, the larger is the fraction of animals we can harvest without depleting the population. Note that we need $\lambda_1 > 1$ in order for the harvesting fraction h to lie in the interval $(0, 1)$. This is to be expected, because $\lambda_1 > 1$ is the condition that the population be increasing.

▶ **EXAMPLE 1 Harvesting Sheep**

For a certain species of domestic sheep in New Zealand with a growth period of 1 year, the following Leslie matrix was found (see G. Caughley, "Parameters for Seasonally Breeding Populations," *Ecology*, 48, 1967, pp. 834–839).

$$L = \begin{bmatrix} .000 & .045 & .391 & .472 & .484 & .546 & .543 & .502 & .468 & .459 & .433 & .421 \\ .845 & 0 & 0 & 0 & 0 & 0 & 0 & 0 & 0 & 0 & 0 & 0 \\ 0 & .975 & 0 & 0 & 0 & 0 & 0 & 0 & 0 & 0 & 0 & 0 \\ 0 & 0 & .965 & 0 & 0 & 0 & 0 & 0 & 0 & 0 & 0 & 0 \\ 0 & 0 & 0 & .950 & 0 & 0 & 0 & 0 & 0 & 0 & 0 & 0 \\ 0 & 0 & 0 & 0 & .926 & 0 & 0 & 0 & 0 & 0 & 0 & 0 \\ 0 & 0 & 0 & 0 & 0 & .895 & 0 & 0 & 0 & 0 & 0 & 0 \\ 0 & 0 & 0 & 0 & 0 & 0 & .850 & 0 & 0 & 0 & 0 & 0 \\ 0 & 0 & 0 & 0 & 0 & 0 & 0 & .786 & 0 & 0 & 0 & 0 \\ 0 & 0 & 0 & 0 & 0 & 0 & 0 & 0 & .691 & 0 & 0 & 0 \\ 0 & 0 & 0 & 0 & 0 & 0 & 0 & 0 & 0 & .561 & 0 & 0 \\ 0 & 0 & 0 & 0 & 0 & 0 & 0 & 0 & 0 & 0 & .370 & 0 \end{bmatrix}$$

The sheep have a lifespan of 12 years, so they are divided into 12 age classes of duration 1 year each. By the use of numerical techniques, the unique positive eigenvalue of L can be found to be

$$\lambda_1 = 1.176$$

From Equation (6), the harvesting fraction h is

$$h = 1 - (1/\lambda_1) = 1 - (1/1.176) = .150$$

Thus, the uniform harvesting policy is one in which 15.0% of the sheep from each of the 12 age classes is harvested every year. From (7) the age distribution vector of the sheep after each harvest is proportional to

$$\mathbf{x}_1 = \begin{bmatrix} 1.000 \\ 0.719 \\ 0.596 \\ 0.489 \\ 0.395 \\ 0.311 \\ 0.237 \\ 0.171 \\ 0.114 \\ 0.067 \\ 0.032 \\ 0.010 \end{bmatrix} \tag{8}$$

From (8) we see that for every 1000 sheep between 0 and 1 year of age that are not harvested, there are 719 sheep between 1 and 2 years of age, 596 sheep between 2 and 3 years of age, and so forth. ◄

Harvesting Only the Youngest Age Class

In some populations only the youngest females are of any economic value, so the harvester seeks to harvest only the females from the youngest age class. Accordingly, let us set

$$h_1 = h$$
$$h_2 = h_3 = \cdots = h_n = 0$$

Equation (4) then reduces to

$$(1-h)(a_1 + a_2 b_1 + a_3 b_1 b_2 + \cdots + a_n b_1 b_2 \cdots b_{n-1}) = 1$$

or

$$(1-h)R = 1$$

where R is the net reproduction rate of the population. [See Equation (17) of Section 10.17.] Solving for h, we obtain

$$h = 1 - (1/R) \tag{9}$$

Note from this equation that a sustainable harvesting policy is possible only if $R > 1$. This is reasonable because only if $R > 1$ is the population increasing. From Equation (5), the age distribution vector after each harvest is proportional to the vector

$$\mathbf{x}_1 = \begin{bmatrix} 1 \\ b_1 \\ b_1 b_2 \\ b_1 b_2 b_3 \\ \vdots \\ b_1 b_2 b_3 \cdots b_{n-1} \end{bmatrix} \tag{10}$$

▶ **EXAMPLE 2** **Sustainable Harvesting Policy**

Let us apply this type of sustainable harvesting policy to the sheep population in Example 1. For the net reproduction rate of the population we find

$$R = a_1 + a_2 b_1 + a_3 b_1 b_2 + \cdots + a_n b_1 b_2 \cdots b_{n-1}$$
$$= (.000) + (.045)(.845) + \cdots + (.421)(.845)(.975) \cdots (.370)$$
$$= 2.514$$

From Equation (9), the fraction of the first age class harvested is

$$h = 1 - (1/R) = 1 - (1/2.514) = .602$$

From Equation (10), the age distribution of the sheep population after the harvest is proportional to the vector

$$\mathbf{x}_1 = \begin{bmatrix} 1.000 \\ .845 \\ (.845)(.975) \\ (.845)(.975)(.965) \\ \vdots \\ (.845)(.975)\cdots(.370) \end{bmatrix} = \begin{bmatrix} 1.000 \\ 0.845 \\ 0.824 \\ 0.795 \\ 0.755 \\ 0.699 \\ 0.626 \\ 0.532 \\ 0.418 \\ 0.289 \\ 0.162 \\ 0.060 \end{bmatrix} \quad (11)$$

A direct calculation gives us the following (see also Exercise 3):

$$L\mathbf{x}_1 = \begin{bmatrix} 2.514 \\ 0.845 \\ 0.824 \\ 0.795 \\ 0.755 \\ 0.699 \\ 0.626 \\ 0.532 \\ 0.418 \\ 0.289 \\ 0.162 \\ 0.060 \end{bmatrix} \quad (12)$$

The vector $L\mathbf{x}_1$ is the age distribution vector immediately before the harvest. The total of all entries in $L\mathbf{x}_1$ is 8.520, so the first entry 2.514 is 29.5% of the total. This means that immediately before each harvest, 29.5% of the population is in the youngest age class. Since 60.2% of this class is harvested, it follows that 17.8% (= 60.2% of 29.5%) of the entire sheep population is harvested each year. This can be compared with the uniform harvesting policy of Example 1, in which 15.0% of the sheep population is harvested each year. ◀

Optimal Sustainable Yield   We saw in Example 1 that a sustainable harvesting policy in which the same fraction of each age class is harvested produces a yield of 15.0% of the sheep population. In Example 2 we saw that if only the youngest age class is harvested, the resulting yield is 17.8% of the population. There are many other possible sustainable harvesting policies, and each generally provides a different yield. It would be of interest to find a sustainable harvesting policy that produces the largest possible yield. Such a policy is called an ***optimal sustainable harvesting policy***, and the resulting yield is called the ***optimal sustainable yield***. However, determining the optimal sustainable yield requires linear programming theory, which we will not discuss here. We refer you to the following result, which appears in J. R. Beddington and D. B. Taylor, "Optimum Age Specific Harvesting of a Population," *Biometrics*, 29, 1973, pp. 801–809.

> **THEOREM 10.18.1 Optimal Sustainable Yield**
>
> *An optimal sustainable harvesting policy is one in which either one or two age classes are harvested. If two age classes are harvested, then the older age class is completely harvested.*

As an illustration, it can be shown that the optimal sustainable yield of the sheep population is attained when

$$h_1 = 0.522$$
$$h_9 = 1.000 \tag{13}$$

and all other values of h_i are zero. Thus, 52.2% of the sheep between 0 and 1 year of age and all the sheep between 8 and 9 years of age are harvested. As we ask you to show in Exercise 2, the resulting optimal sustainable yield is 19.9% of the population.

Exercise Set 10.18

1. Let a certain animal population be divided into three 1-year age classes and have as its Leslie matrix

$$L = \begin{bmatrix} 0 & 4 & 3 \\ \frac{1}{2} & 0 & 0 \\ 0 & \frac{1}{4} & 0 \end{bmatrix}$$

   (a) Find the yield and the age distribution vector after each harvest if the same fraction of each of the three age classes is harvested every year.

   (b) Find the yield and the age distribution vector after each harvest if only the youngest age class is harvested every year. Also, find the fraction of the youngest age class that is harvested.

2. For the optimal sustainable harvesting policy described by Equations (13), find the vector $\mathbf{x}_1$ that specifies the age distribution of the population after each harvest. Also calculate the vector $L\mathbf{x}_1$ and verify that the optimal sustainable yield is 19.9% of the population.

3. Use Equation (10) to show that if only the first age class of an animal population is harvested,

$$L\mathbf{x}_1 - \mathbf{x}_1 = \begin{bmatrix} R - 1 \\ 0 \\ 0 \\ \vdots \\ 0 \end{bmatrix}$$

   where R is the net reproduction rate of the population.

4. If only the Ith class of an animal population is to be periodically harvested ($I = 1, 2, \ldots, n$), find the corresponding harvesting fraction h_I.

5. Suppose that all of the Jth class and a certain fraction h_I of the Ith class of an animal population is to be periodically harvested ($1 \leq I < J \leq n$). Calculate h_I.

Section 10.18 Technology Exercises

The following exercises are designed to be solved using a technology utility. Typically, this will be MATLAB, *Mathematica*, Maple, Derive, or Mathcad, but it may also be some other type of linear algebra software or a scientific calculator with some linear algebra capabilities. For each exercise you will need to read the relevant documentation for the particular utility you are using. The goal of these exercises is to provide you with a basic proficiency with your technology utility. Once you have mastered the techniques in these exercises, you will be able to use your technology utility to solve many of the problems in the regular exercise sets.

T1. The results of Theorem 10.18.1 suggest the following algorithm for determining the optimal sustainable yield.

(i) For each value of $i = 1, 2, \ldots, n$, set $h_i = h$ and $h_k = 0$ for $k \neq i$ and calculate the respective yields. These n calculations give the one-age-class results. Of course, any calculation leading to a value of h not between 0 and 1 is rejected.

(ii) For each value of $i = 1, 2, \ldots, n-1$ and $j = i+1, i+2, \ldots, n$, set $h_i = h$, $h_j = 1$, and $h_k = 0$ for $k \neq i, j$ and calculate the respective yields. These $\frac{1}{2}n(n-1)$ calculations give the two-age-class results. Of course, any calculation leading to a value of h not between 0 and 1 is again rejected.

(iii) Of the yields calculated in parts (i) and (ii), the largest is the optimal sustainable yield. Note that there will be at most

$$n + \tfrac{1}{2}n(n-1) = \tfrac{1}{2}n(n+1)$$

calculations in all. Once again, some of these may lead to a value of h not between 0 and 1 and must therefore be rejected.

If we use this algorithm for the sheep example in the text, there will be at most $\frac{1}{2}(12)(12+1) = 78$ calculations to consider. Use a computer to do the two-age-class calculations for $h_1 = h$, $h_j = 1$, and $h_k = 0$ for $k \neq 1$ or j for $j = 2, 3, \ldots, 12$. Construct a summary table consisting of the values of h_1 and the percentage yields using $j = 2, 3, \ldots, 12$, which will show that the largest of these yields occurs when $j = 9$.

T2. Using the algorithm in Exercise T1, do the one-age-class calculations for $h_i = h$ and $h_k = 0$ for $k \neq i$ for $i = 1, 2, \ldots, 12$. Construct a summary table consisting of the values of h_i and the percentage yields using $i = 1, 2, \ldots, 12$, which will show that the largest of these yields occurs when $i = 9$.

T3. Referring to the mouse population in Exercise T3 of Section 10.17, suppose that reducing the birthrates is not practical, so you instead decide to control the population by uniformly harvesting all of the age classes monthly.

(a) What fraction of the population must be harvested monthly to bring the mouse population to equilibrium eventually?

(b) What is the equilibrium age distribution vector under this uniform harvesting policy?

(c) The total number of mice in the original mouse population was 155. What would be the total number of mice after 5, 10, and 200 months under your uniform harvesting policy?

10.19 A Least Squares Model for Human Hearing

In this section we apply the method of least squares approximation to a model for human hearing. The use of this method is motivated by energy considerations.

> **PREREQUISITES:** Inner Product Spaces
> Orthogonal Projection
> Fourier Series (Section 6.6)

Anatomy of the Ear

We begin with a brief discussion of the nature of sound and human hearing. Figure 10.19.1 is a schematic diagram of the ear showing its three main components: the outer ear, middle ear, and inner ear. Sound waves enter the outer ear where they are channeled to the eardrum, causing it to vibrate. Three tiny bones in the middle ear mechanically link the eardrum with the snail-shaped cochlea within the inner ear. These bones pass on the vibrations of the eardrum to a fluid within the cochlea. The cochlea contains thousands of minute hairs that oscillate with the fluid. Those near the entrance of the cochlea are stimulated by high frequencies, and those near the tip are stimulated by low frequencies. The movements of these hairs activate nerve cells that send signals along various neural pathways to the brain, where the signals are interpreted as sound.

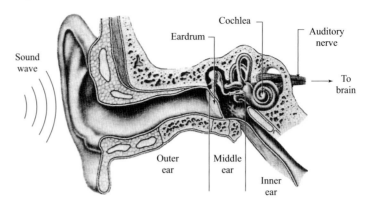

▶ Figure 10.19.1

The sound waves themselves are variations in time of the air pressure. For the auditory system, the most elementary type of sound wave is a sinusoidal variation in the air pressure. This type of sound wave stimulates the hairs within the cochlea in such a way that a nerve impulse along a single neural pathway is produced (Figure 10.19.2). A sinusoidal sound wave can be described by a function of time

$$q(t) = A_0 + A\sin(\omega t - \delta) \tag{1}$$

where $q(t)$ is the atmospheric pressure at the eardrum, A_0 is the normal atmospheric pressure, A is the maximum deviation of the pressure from the normal atmospheric pressure, $\omega/2\pi$ is the frequency of the wave in cycles per second, and δ is the phase angle of the wave. To be perceived as sound, such sinusoidal waves must have frequencies within a certain range. For humans this range is roughly 20 cycles per second (cps) to 20,000 cps. Frequencies outside this range will not stimulate the hairs within the cochlea enough to produce nerve signals.

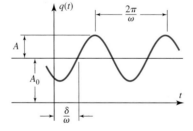
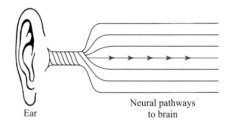

▶ Figure 10.19.2

To a reasonable degree of accuracy, the ear is a linear system. This means that if a complex sound wave is a finite sum of sinusoidal components of different amplitudes, frequencies, and phase angles, say,

$$q(t) = A_0 + A_1 \sin(\omega_1 t - \delta_1) + A_2 \sin(\omega_2 t - \delta_2) + \cdots + A_n \sin(\omega_n t - \delta_n) \tag{2}$$

then the response of the ear consists of nerve impulses along the same neural pathways that would be stimulated by the individual components (Figure 10.19.3).

10.19 A Least Squares Model for Human Hearing

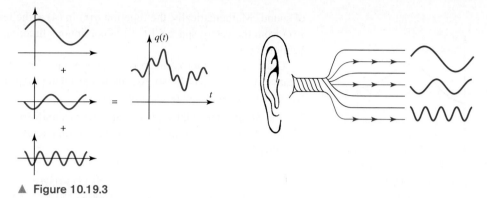

▲ Figure 10.19.3

Let us now consider some periodic sound wave $p(t)$ with period T [i.e., $p(t) \equiv p(t+T)$] that is *not* a finite sum of sinusoidal waves. If we examine the response of the ear to such a periodic wave, we find that it is the same as the response to some wave that is the sum of sinusoidal waves. That is, there is some sound wave $q(t)$ as given by Equation (2) that produces the same response as $p(t)$, even though $p(t)$ and $q(t)$ are different functions of time.

We now want to determine the frequencies, amplitudes, and phase angles of the sinusoidal components of $q(t)$. Because $q(t)$ produces the same response as the periodic wave $p(t)$, it is reasonable to expect that $q(t)$ has the same period T as $p(t)$. This requires that each sinusoidal term in $q(t)$ have period T. Consequently, the frequencies of the sinusoidal components must be integer multiples of the basic frequency $1/T$ of the function $p(t)$. Thus, the ω_k in Equation (2) must be of the form

$$\omega_k = 2k\pi/T, \qquad k = 1, 2, \ldots$$

But because the ear cannot perceive sinusoidal waves with frequencies greater than 20,000 cps, we may omit those values of k for which $\omega_k/2\pi = k/T$ is greater than 20,000. Thus, $q(t)$ is of the form

$$q(t) = A_0 + A_1 \sin\left(\frac{2\pi t}{T} - \delta_1\right) + \cdots + A_n \sin\left(\frac{2n\pi t}{T} - \delta_n\right) \qquad (3)$$

where n is the largest integer such that n/T is not greater than 20,000.

We now turn our attention to the values of the amplitudes $A_0, A_1, \ldots, A_n$ and the phase angles $\delta_1, \delta_2, \ldots, \delta_n$ that appear in Equation (3). There is some criterion by which the auditory system "picks" these values so that $q(t)$ produces the same response as $p(t)$. To examine this criterion, let us set

$$e(t) = p(t) - q(t)$$

If we consider $q(t)$ as an approximation to $p(t)$, then $e(t)$ is the error in this approximation, an error that the ear cannot perceive. In terms of $e(t)$, the criterion for the determination of the amplitudes and the phase angles is that the quantity

$$\int_0^T [e(t)]^2 \, dt = \int_0^T [p(t) - q(t)]^2 \, dt \qquad (4)$$

be as small as possible. We cannot go into the physiological reasons for this, but we note that this expression is proportional to the *acoustic energy* of the error wave $e(t)$ over one period. In other words, it is the energy of the difference between the two sound waves $p(t)$ and $q(t)$ that determines whether the ear perceives any difference between them. If this energy is as small as possible, then the two waves produce the same sensation of sound. Mathematically, the function $q(t)$ in (4) is the least squares approximation to $p(t)$ from the vector space $C[0, T]$ of continuous functions on the interval $[0, T]$. (See Section 6.6.)

Least squares approximations by continuous functions arise in a wide variety of engineering and scientific approximation problems. Apart from the acoustics problem just discussed, some other examples follow.

1. Let $S(x)$ be the axial strain distribution in a uniform rod lying along the x-axis from $x = 0$ to $x = l$ (Figure 10.19.4). The strain energy in the rod is proportional to the integral

$$\int_0^l [S(x)]^2 \, dx$$

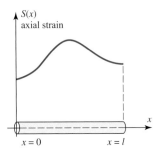

▲ Figure 10.19.4

The closeness of an approximation $q(x)$ to $S(x)$ can be judged according to the strain energy of the difference of the two strain distributions. That energy is proportional to

$$\int_0^l [S(x) - q(x)]^2 \, dx$$

which is a least squares criterion.

2. Let $E(t)$ be a periodic voltage across a resistor in an electrical circuit (Figure 10.19.5). The electrical energy transferred to the resistor during one period T is proportional to

$$\int_0^T [E(t)]^2 \, dt$$

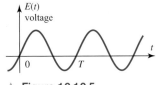

▲ Figure 10.19.5

If $q(t)$ has the same period as $E(t)$ and is to be an approximation to $E(t)$, then the criterion of closeness might be taken as the energy of the difference voltage. This is proportional to

$$\int_0^T [E(t) - q(t)]^2 \, dt$$

which is again a least squares criterion.

3. Let $y(x)$ be the vertical displacement of a uniform flexible string whose equilibrium position is along the x-axis from $x = 0$ to $x = l$ (Figure 10.19.6). The elastic potential energy of the string is proportional to

$$\int_0^l [y(x)]^2 \, dx$$

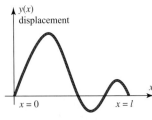

▲ Figure 10.19.6

If $q(x)$ is to be an approximation to the displacement, then as before, the energy integral

$$\int_0^l [y(x) - q(x)]^2 \, dx$$

determines a least squares criterion for the closeness of the approximation.

Least squares approximation is also used in situations where there is no a priori justification for its use, such as for approximating business cycles, population growth curves,

10.19 A Least Squares Model for Human Hearing

sales curves, and so forth. It is used in these cases because of its mathematical simplicity. In general, if no other error criterion is immediately apparent for an approximation problem, the least squares criterion is the one most often chosen.

The following result was obtained in Section 6.6.

THEOREM 10.19.1 Minimizing Mean Square Error on $[0, 2\pi]$

If $f(t)$ is continuous on $[0, 2\pi]$, then the trigonometric function $g(t)$ of the form

$$g(t) = \tfrac{1}{2}a_0 + a_1 \cos t + \cdots + a_n \cos nt + b_1 \sin t + \cdots + b_n \sin nt$$

that minimizes the mean square error

$$\int_0^{2\pi} [f(t) - g(t)]^2 \, dt$$

has coefficients

$$a_k = \frac{1}{\pi} \int_0^{2\pi} f(t) \cos kt \, dt, \qquad k = 0, 1, 2, \ldots, n$$

$$b_k = \frac{1}{\pi} \int_0^{2\pi} f(t) \sin kt \, dt, \qquad k = 1, 2, \ldots, n$$

If the original function $f(t)$ is defined over the interval $[0, T]$ instead of $[0, 2\pi]$, a change of scale will yield the following result (see Exercise 8):

THEOREM 10.19.2 Minimizing Mean Square Error on $[0, T]$

If $f(t)$ is continuous on $[0, T]$, then the trigonometric function $g(t)$ of the form

$$g(t) = \frac{1}{2}a_0 + a_1 \cos \frac{2\pi}{T} t + \cdots + a_n \cos \frac{2n\pi}{T} t + b_1 \sin \frac{2\pi}{T} t + \cdots + b_n \sin \frac{2n\pi}{T} t$$

that minimizes the mean square error

$$\int_0^T [f(t) - g(t)]^2 \, dt$$

has coefficients

$$a_k = \frac{2}{T} \int_0^T f(t) \cos \frac{2k\pi t}{T} \, dt, \qquad k = 0, 1, 2, \ldots, n$$

$$b_k = \frac{2}{T} \int_0^T f(t) \sin \frac{2k\pi t}{T} \, dt, \qquad k = 1, 2, \ldots, n$$

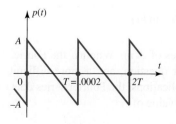

▲ Figure 10.19.7

▶ **EXAMPLE 1** Least Squares Approximation to a Sound Wave

Let a sound wave $p(t)$ have a saw-tooth pattern with a basic frequency of 5000 cps (Figure 10.19.7). Assume units are chosen so that the normal atmospheric pressure is at the zero level and the maximum amplitude of the wave is A. The basic period of the wave is $T = 1/5000 = .0002$ second. From $t = 0$ to $t = T$, the function $p(t)$ has the equation

$$p(t) = \frac{2A}{T}\left(\frac{T}{2} - t\right)$$

Theorem 10.19.2 then yields the following (verify):

$$a_0 = \frac{2}{T}\int_0^T p(t)\,dt = \frac{2}{T}\int_0^T \frac{2A}{T}\left(\frac{T}{2} - t\right)dt = 0$$

$$a_k = \frac{2}{T}\int_0^T p(t)\cos\frac{2k\pi t}{T}\,dt = \frac{2}{T}\int_0^T \frac{2A}{T}\left(\frac{T}{2} - t\right)\cos\frac{2k\pi t}{T}\,dt = 0, \quad k = 1, 2, \ldots$$

$$b_k = \frac{2}{T}\int_0^T p(t)\sin\frac{2k\pi t}{T}\,dt = \frac{2}{T}\int_0^T \frac{2A}{T}\left(\frac{T}{2} - t\right)\sin\frac{2k\pi t}{T}\,dt = \frac{2A}{k\pi}, \quad k = 1, 2, \ldots$$

We can now investigate how the sound wave $p(t)$ is perceived by the human ear. We note that $4/T = 20{,}000$ cps, so we need only go up to $k = 4$ in the formulas above. The least squares approximation to $p(t)$ is then

$$q(t) = \frac{2A}{\pi}\left[\sin\frac{2\pi}{T}t + \frac{1}{2}\sin\frac{4\pi}{T}t + \frac{1}{3}\sin\frac{6\pi}{T}t + \frac{1}{4}\sin\frac{8\pi}{T}t\right]$$

The four sinusoidal terms have frequencies of 5000, 10,000, 15,000, and 20,000 cps, respectively. In Figure 10.19.8 we have plotted $p(t)$ and $q(t)$ over one period. Although $q(t)$ is not a very good point-by-point approximation to $p(t)$, to the ear, both $p(t)$ and $q(t)$ produce the same sensation of sound. ◀

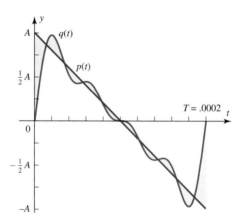

▶ **Figure 10.19.8**

As discussed in Section 6.6, the least squares approximation becomes better as the number of terms in the approximating trigonometric polynomial becomes larger. More precisely,

$$\int_0^{2\pi}\left[f(t) - \frac{1}{2}a_0 - \sum_{k=1}^n (a_k\cos kt + b_k\sin kt)\right]^2 dt$$

tends to zero as n approaches infinity. We denote this by writing

$$f(t) \sim \frac{1}{2}a_0 + \sum_{k=1}^\infty (a_k\cos kt + b_k\sin kt)$$

where the right side of this equation is the Fourier series of $f(t)$. Whether the Fourier series of $f(t)$ converges to $f(t)$ for each t is another question, and a more difficult one. For most continuous functions encountered in applications, the Fourier series does indeed converge to its corresponding function for each value of t.

Exercise Set 10.19

1. Find the trigonometric polynomial of order 3 that is the least squares approximation to the function $f(t) = (t - \pi)^2$ over the interval $[0, 2\pi]$.

2. Find the trigonometric polynomial of order 4 that is the least squares approximation to the function $f(t) = t^2$ over the interval $[0, T]$.

3. Find the trigonometric polynomial of order 4 that is the least squares approximation to the function $f(t)$ over the interval $[0, 2\pi]$, where
$$f(t) = \begin{cases} \sin t, & 0 \leq t \leq \pi \\ 0, & \pi < t \leq 2\pi \end{cases}$$

4. Find the trigonometric polynomial of arbitrary order n that is the least squares approximation to the function $f(t) = \sin \frac{1}{2}t$ over the interval $[0, 2\pi]$.

5. Find the trigonometric polynomial of arbitrary order n that is the least squares approximation to the function $f(t)$ over the interval $[0, T]$, where
$$f(t) = \begin{cases} t, & 0 \leq t \leq \frac{1}{2}T \\ T - t, & \frac{1}{2}T < t \leq T \end{cases}$$

6. For the inner product
$$\langle \mathbf{u}, \mathbf{v} \rangle = \int_0^{2\pi} u(t)v(t)\,dt$$
show that
   (a) $\|1\| = \sqrt{2\pi}$
   (b) $\|\cos kt\| = \sqrt{\pi}$ for $k = 1, 2, \ldots$
   (c) $\|\sin kt\| = \sqrt{\pi}$ for $k = 1, 2, \ldots$

7. Show that the $2n + 1$ functions
$$1, \cos t, \cos 2t, \ldots, \cos nt, \sin t, \sin 2t, \ldots, \sin nt$$
are orthogonal over the interval $[0, 2\pi]$ relative to the inner product $\langle \mathbf{u}, \mathbf{v} \rangle$ defined in Exercise 6.

8. If $f(t)$ is defined and continuous on the interval $[0, T]$, show that $f(T\tau/2\pi)$ is defined and continuous for τ in the interval $[0, 2\pi]$. Use this fact to show how Theorem 10.19.2 follows from Theorem 10.19.1.

Section 10.19 Technology Exercises

The following exercises are designed to be solved using a technology utility. Typically, this will be MATLAB, *Mathematica*, Maple, Derive, or Mathcad, but it may also be some other type of linear algebra software or a scientific calculator with some linear algebra capabilities. For each exercise you will need to read the relevant documentation for the particular utility you are using. The goal of these exercises is to provide you with a basic proficiency with your technology utility. Once you have mastered the techniques in these exercises, you will be able to use your technology utility to solve many of the problems in the regular exercise sets.

T1. Let g be the function
$$g(t) = \frac{3 + 4\sin t}{5 - 4\cos t}$$
for $0 \leq t \leq 2\pi$. Use a computer to determine the Fourier coefficients
$$\begin{Bmatrix} a_k \\ b_k \end{Bmatrix} = \frac{1}{\pi} \int_0^{2\pi} \left(\frac{3 + 4\sin t}{5 - 4\cos t} \right) \begin{Bmatrix} \cos kt \\ \sin kt \end{Bmatrix} dt$$
for $k = 0, 1, 2, 3, 4, 5$. From your results, make a conjecture about the general expressions for a_k and b_k. Test your conjecture by calculating
$$\frac{1}{2}a_0 + \sum_{k=1}^{\infty}(a_k \cos kt + b_k \sin kt)$$
on the computer and see whether it converges to $g(t)$.

T2. Let g be the function
$$g(t) = e^{\cos t}[\cos(\sin t) + \sin(\sin t)]$$
for $0 \leq t \leq 2\pi$. Use a computer to determine the Fourier coefficients
$$\begin{Bmatrix} a_k \\ b_k \end{Bmatrix} = \frac{1}{\pi} \int_0^{2\pi} g(t) \begin{Bmatrix} \cos kt \\ \sin kt \end{Bmatrix} dt$$
for $k = 0, 1, 2, 3, 4, 5$. From your results, make a conjecture about the general expressions for a_k and b_k. Test your conjecture by calculating
$$\frac{1}{2}a_0 + \sum_{k=1}^{\infty}(a_k \cos kt + b_k \sin kt)$$
on the computer and see whether it converges to $g(t)$.

10.20 Warps and Morphs

Among the more interesting image-manipulation techniques available for computer graphics are warps and morphs. In this section we show how linear transformations can be used to distort a single picture to produce a warp, or to distort and blend two pictures to produce a morph.

> **PREREQUISITES:** Geometry of Linear Operators on R^2 (Section 4.11)
> Linear Independence
> Bases in R^2

Computer graphics software enables you to manipulate an image in various ways, such as by scaling, rotating, or slanting the image. Distorting an image by separately moving the corners of a rectangle containing the image is another basic image-manipulation technique. Distorting various pieces of an image in different ways is a more complicated procedure that results in a *warp* of the picture. In addition, warping two different images in complementary ways and blending the warps results in a *morph* of the two pictures (from the Greek root meaning "shape" or "form"). An example is Figure 10.20.1 in which four photographs of a woman taken over a 50-year period (the four diagonal pictures from top left to bottom right) have been pairwise morphed by different amounts to suggest the gradual aging of the woman.

▶ Figure 10.20.1

10.20 Warps and Morphs

The most visible application of warping and morphing images has been the production of special effects in motion pictures and television. However, many scientific and technological applications of such techniques have also arisen—for example, studying the evolution, growth, and development of living organisms, assisting in reconstructive and cosmetic surgery, exploring various designs of a product, and "aging" photographs of missing persons or police suspects.

Warps We begin by describing a simple warp of a triangular region in the plane. Let the three vertices of a triangle be given by the three noncollinear points $\mathbf{v}_1$, $\mathbf{v}_2$, and $\mathbf{v}_3$ (Figure 10.20.2a). We will call this triangle the **begin-triangle**. If $\mathbf{v}$ is any point in the begin-triangle, then there are unique constants c_1 and c_2 such that

$$\mathbf{v} - \mathbf{v}_3 = c_1(\mathbf{v}_1 - \mathbf{v}_3) + c_2(\mathbf{v}_2 - \mathbf{v}_3) \tag{1}$$

Equation (1) expresses the vector $\mathbf{v} - \mathbf{v}_3$ as a (unique) linear combination of the two linearly independent vectors $\mathbf{v}_1 - \mathbf{v}_3$ and $\mathbf{v}_2 - \mathbf{v}_3$ with respect to an origin at $\mathbf{v}_3$. If we set $c_3 = 1 - c_1 - c_2$, then we can rewrite (1) as

$$\mathbf{v} = c_1\mathbf{v}_1 + c_2\mathbf{v}_2 + c_3\mathbf{v}_3 \tag{2}$$

where

$$c_1 + c_2 + c_3 = 1 \tag{3}$$

from the definition of c_3. We say that $\mathbf{v}$ is a **convex combination** of the vectors $\mathbf{v}_1$, $\mathbf{v}_2$, and $\mathbf{v}_3$ if (2) and (3) are satisfied and, in addition, the coefficients c_1, c_2, and c_3 are nonnegative. It can be shown (Exercise 6) that $\mathbf{v}$ lies in the triangle determined by $\mathbf{v}_1$, $\mathbf{v}_2$, and $\mathbf{v}_3$ if and only if it is a convex combination of those three vectors.

Next, given three noncollinear points $\mathbf{w}_1$, $\mathbf{w}_2$, and $\mathbf{w}_3$ of an **end-triangle** (Figure 10.20.2b), there is a unique **affine transformation** that maps $\mathbf{v}_1$ to $\mathbf{w}_1$, $\mathbf{v}_2$ to $\mathbf{w}_2$, and $\mathbf{v}_3$ to $\mathbf{w}_3$. That is, there is a unique 2×2 invertible matrix M and a unique vector $\mathbf{b}$ such that

$$\mathbf{w}_i = M\mathbf{v}_i + \mathbf{b} \quad \text{for } i = 1, 2, 3 \tag{4}$$

(See Exercise 5 for the evaluation of M and $\mathbf{b}$.) Moreover, it can be shown (Exercise 3) that the image $\mathbf{w}$ of the vector $\mathbf{v}$ in (2) under this affine transformation is

$$\mathbf{w} = c_1\mathbf{w}_1 + c_2\mathbf{w}_2 + c_3\mathbf{w}_3 \tag{5}$$

This is a basic property of affine transformations: They map a convex combination of vectors to the same convex combination of the images of the vectors.

Now suppose that the begin-triangle contains a picture within it (Figure 10.20.3a). That is, to each point in the begin-triangle we assign a gray level, say 0 for white and 100 for black, with any other gray level lying between 0 and 100. In particular, let a scalar-valued function ρ_0, called the **picture-density** of the begin-triangle, be defined so that $\rho_0(\mathbf{v})$ is the gray level at the point $\mathbf{v}$ in the begin-triangle. We can now define a picture in the end-triangle, called a **warp** of the original picture, with a picture-density ρ_1 by defining the gray level at the point $\mathbf{w}$ within the end-triangle to be the gray level of the point $\mathbf{v}$ in the begin-triangle that maps onto $\mathbf{w}$. In equation form, the picture-density ρ_1 is determined by

$$\rho_1(\mathbf{w}) = \rho_0(c_1\mathbf{v}_1 + c_2\mathbf{v}_2 + c_3\mathbf{v}_3) \tag{6}$$

In this way, as c_1, c_2, and c_3 vary over all nonnegative values that add to one, (5) generates all points $\mathbf{w}$ in the end-triangle, and (6) generates the gray levels $\rho_1(\mathbf{w})$ of the warped picture at those points (Figure 10.20.3b).

Equation (6) determines a very simple warp of a picture within a single triangle. More generally, we can break up a picture into many triangular regions and warp each

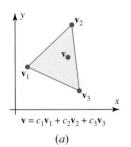

▲ Figure 10.20.2

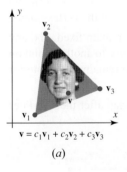

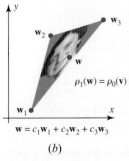

▲ Figure 10.20.3

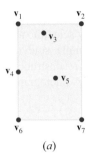

(a)

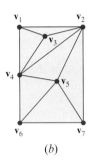

(b)

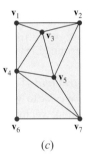

(c)

▲ Figure 10.20.4

triangular region differently. This gives us much freedom in designing a warp through our choice of triangular regions and how we change them. To this end, suppose we are given a picture contained within some rectangular region of the plane. We choose n points $\mathbf{v}_1, \mathbf{v}_2, \ldots, \mathbf{v}_n$ within the rectangle, which we call ***vertex points***, so that they fall on key elements or features of the picture we wish to warp (Figure 10.20.4a). Once the vertex points are chosen, we complete a ***triangulation*** of the rectangular region; that is, we draw line segments between the vertex points in such a way that we have the following conditions (Figure 10.20.4b):

1. The line segments form the sides of a set of triangles.
2. The line segments do not intersect.
3. Each vertex point is the vertex of at least one triangle.
4. The union of the triangles is the rectangle.
5. The set of triangles is maximal (i.e., no more vertices can be connected).

Note that condition 4 requires that each corner of the rectangle containing the picture be a vertex point.

One can always form a triangulation from any n vertex points, but the triangulation is not necessarily unique. For example, Figures 10.20.4b and 10.20.4c are two different triangulations of the set of vertex points in Figure 10.20.4a. Since there are various computer algorithms that perform triangulations very quickly, it is not necessary to perform the tiresome triangulation task by hand; one need only specify the desired vertex points and let a computer generate a triangulation from them. If n is the number of vertex points chosen, it can be shown that the number of triangles m of any triangulation of those points is given by

$$m = 2n - 2 - k \tag{7}$$

where k is the number of vertex points lying on the boundary of the rectangle, including the four situated at the corner points.

The warp is specified by moving the n vertex points $\mathbf{v}_1, \mathbf{v}_2, \ldots, \mathbf{v}_n$ to new locations $\mathbf{w}_1, \mathbf{w}_2, \ldots, \mathbf{w}_n$ according to the changes we desire in the picture (Figures 10.20.5a and 10.20.5b). However, we impose two restrictions on the movements of the vertex points:

1. The four vertex points at the corners of the rectangle are to remain fixed, and any vertex point on a side of the rectangle is to remain fixed or move to another point on the same side of the rectangle. All other vertex points are to remain in the interior of the rectangle.
2. The triangles determined by the triangulation are not to overlap after their vertices have been moved.

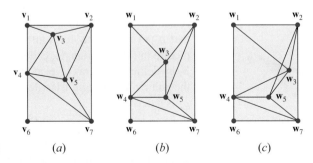

▶ Figure 10.20.5   (a)   (b)   (c)

The first restriction guarantees that the rectangular shape of the begin-picture is preserved. The second restriction guarantees that the displaced vertex points still form a triangulation of the rectangle and that the new triangulation is similar to the original one. For example, Figure 10.20.5c is not an allowable movement of the vertex points shown in Figure 10.20.5a. Although a violation of this condition can be handled mathematically without too much additional effort, the resulting warps usually produce unnatural results and we will not consider them here.

Figure 10.20.6 is a warp of a photograph of a woman using a triangulation with 94 vertex points and 179 triangles. Note that the vertex points in the begin-triangulation are chosen to lie along key features of the picture (hairline, eyes, lips, etc.). These vertex points were moved to final positions corresponding to those same features in a picture of the woman taken 20 years after the begin-picture. Thus, the warped picture represents the woman forced into her older shape but using her younger gray levels.

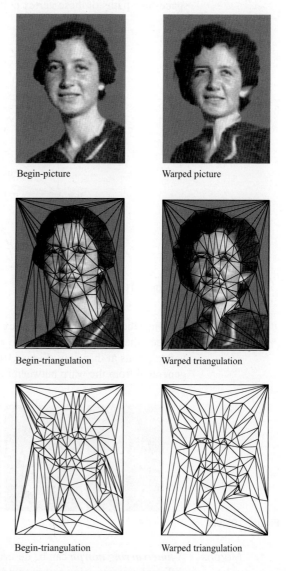

▶ Figure 10.20.6

Begin-picture   Warped picture

Begin-triangulation   Warped triangulation

Begin-triangulation   Warped triangulation

Time-Varying Warps   A ***time-varying warp*** is the set of warps generated when the vertex points of the begin-picture are moved continually in time from their original positions to specified final positions. This gives us a motion picture in which the begin-picture is continually warped to a final warp. Let us choose time units so that $t = 0$ corresponds to our begin-picture

704 Chapter 10 Applications of Linear Algebra

and $t = 1$ corresponds to our final warp. The simplest way of moving the vertex points from time 0 to time 1 is with constant velocity along straight-line paths from their initial positions to their final positions.

To describe such a motion, let $\mathbf{u}_i(t)$ denote the position of the ith vertex point at any time t between 0 and 1. Thus $\mathbf{u}_i(0) = \mathbf{v}_i$ (its given position in the begin-picture) and $\mathbf{u}_i(1) = \mathbf{w}_i$ (its given position in the final warp). In between, we determine its position by

$$\mathbf{u}_i(t) = (1 - t)\mathbf{v}_i + t\mathbf{w}_i \tag{8}$$

Note that (8) expresses $\mathbf{u}_i(t)$ as a convex combination of $\mathbf{v}_i$ and $\mathbf{w}_i$ for each t in $[0, 1]$. Figure 10.20.7 illustrates a time-varying triangulation of a plain rectangular region with six vertex points. The lines connecting the vertex points at the different times are the space-time paths of these vertex points in this space-time diagram.

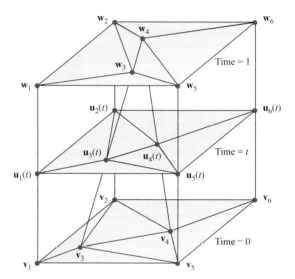

▶ Figure 10.20.7

Once the positions of the vertex points are computed at time t, a warp is performed between the begin-picture and the triangulation at time t determined by the displaced vertex points at that time. Figure 10.20.8 shows a time-varying warp at five values of t generated from the warp between $t = 0$ and $t = 1$ shown in Figure 10.20.6.

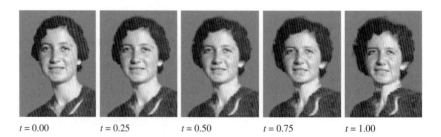

▶ Figure 10.20.8 $t = 0.00$ $t = 0.25$ $t = 0.50$ $t = 0.75$ $t = 1.00$

Morphs   A *time-varying morph* can be described as a blending of two time-varying warps of two different pictures using two triangulations that match corresponding features in the two pictures. One of the two pictures is designated as the begin-picture and the other as the end-picture. First, a time-varying warp from $t = 0$ to $t = 1$ is generated in which the begin-picture is warped into the shape of the end-picture. Then a time-varying warp from $t = 1$ to $t = 0$ is generated in which the end-picture is warped into the shape of the begin-picture. Finally, a weighted average of the gray levels of the two warps at each time t is produced to generate the morph of the two images at time t.

Figure 10.20.9 shows two photographs of a woman taken 20 years apart. Below the pictures are two corresponding triangulations in which corresponding features of the two photographs are matched. The time-varying morph between these two pictures for five values of t between 0 and 1 is shown in Figure 10.20.10.

Begin-picture   End-picture

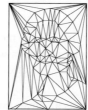 

▶ **Figure 10.20.9** Begin-triangulation   End-triangulation

▶ **Figure 10.20.10** $t = 0.00$ $t = 0.25$ $t = 0.50$ $t = 0.75$ $t = 1.00$

The procedure for producing such a morph is outlined in the following nine steps (Figure 10.20.11):

Step 1. Given a begin-picture with picture-density ρ_0 and an end-picture with picture-density ρ_1, position n vertex points $\mathbf{v}_1, \mathbf{v}_2, \ldots, \mathbf{v}_n$ in the begin-picture at key features of that picture.

Step 2. Position n corresponding vertex points $\mathbf{w}_1, \mathbf{w}_2, \ldots, \mathbf{w}_n$ in the end-picture at the corresponding key features of that picture.

Step 3. Triangulate the begin- and end-pictures in similar ways by drawing lines between corresponding vertex points in both pictures.

Step 4. For any time t between 0 and 1, find the vertex points $\mathbf{u}_1(t), \mathbf{u}_2(t), \ldots, \mathbf{u}_n(t)$ in the morph picture at that time, using the formula

$$\mathbf{u}_i(t) = (1 - t)\mathbf{v}_i + t\mathbf{w}_i, \qquad i = 1, 2, \ldots, n \tag{9}$$

Step 5. Triangulate the morph picture at time t similar to the begin- and end-picture triangulations.

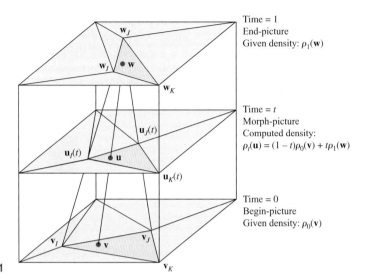

▶ Figure 10.20.11

Step 6. For any point **u** in the morph picture at time t, find the triangle in the triangulation of the morph picture in which it lies and the vertices $\mathbf{u}_I(t)$, $\mathbf{u}_J(t)$, and $\mathbf{u}_K(t)$ of that triangle. (See Exercise 1 to determine whether a given point lies in a given triangle.)

Step 7. Express **u** as a convex combination of $\mathbf{u}_I(t)$, $\mathbf{u}_J(t)$, and $\mathbf{u}_K(t)$ by finding the constants c_I, c_J, and c_K such that

$$\mathbf{u} = c_I \mathbf{u}_I(t) + c_J \mathbf{u}_J(t) + c_K \mathbf{u}_K(t) \tag{10}$$

and

$$c_I + c_J + c_K = 1 \tag{11}$$

Step 8. Determine the locations of the point **u** in the begin- and end-pictures using

$$\mathbf{v} = c_I \mathbf{v}_I + c_J \mathbf{v}_J + c_K \mathbf{v}_K \quad \text{(in the begin-picture)} \tag{12}$$

and

$$\mathbf{w} = c_I \mathbf{w}_I + c_J \mathbf{w}_J + c_K \mathbf{w}_K \quad \text{(in the end-picture)} \tag{13}$$

Step 9. Finally, determine the picture-density $\rho_t(\mathbf{u})$ of the morph-picture at the point **u** using

$$\rho_t(\mathbf{u}) = (1-t)\rho_0(\mathbf{v}) + t\rho_1(\mathbf{w}) \tag{14}$$

Step 9 is the key step in distinguishing a warp from a morph. Equation (14) takes weighted averages of the gray levels of the begin- and end-pictures to produce the gray levels of the morph-picture. The weights depend on the fraction of the distances that the vertex points have moved from their beginning positions to their ending positions. For example, if the vertex points have moved one-fourth of the way to their destinations (i.e., if $t = 0.25$), then we use one-fourth of the gray levels of the end-picture and three-fourths of the gray levels of the begin-picture. Thus, as time progresses, not only does the shape of the begin-picture gradually change into the shape of the end-picture (as in a warp) but the gray levels of the begin-picture also gradually change into the gray levels of the end-picture.

The procedure described above to generate a morph is cumbersome to perform by hand, but it is the kind of dull, repetitive procedure at which computers excel. A successful morph demands good preparation and requires more artistic ability than mathematical ability. (The software designer is required to have the mathematical ability.) The two photographs to be morphed should be carefully chosen so that they have matching features, and the vertex points in the two photographs also should be carefully chosen so that the triangles in the two resulting triangulations contain similar features of the two pictures. When the procedure is done correctly, each frame of the morph should look just as "real" as the begin- and end-pictures.

The techniques we have discussed in this section can be generalized in numerous ways to produce much more elaborate warps and morphs. For example:

1. If the pictures are in color, the three components of the picture colors (red, green, and blue) can be morphed separately to produce a color morph.
2. Rather than following straight-line paths to their destinations, the vertices of a triangulation can be directed separately along more complicated paths to produce a variety of results.
3. Rather than travel with constant speeds along their paths, the vertices of a triangulation can be directed to have different speeds at different times. For example, in a morph between two faces, the hairline can be made to change first, then the nose, and so forth.
4. Similarly, the gray-level mixing of the begin-picture and end-picture at different times and different vertices can be varied in a more complicated way than that in Equation (14).
5. One can morph two surfaces in three-dimensional space (representing two complete heads, for example) by triangulating the surfaces and using the techniques in this section.
6. One can morph two solids in three-dimensional space (for example, two three-dimensional tomographs of a beating human heart at two different times) by dividing the two solids into corresponding tetrahedral regions.
7. Two film strips can be morphed frame by frame by different amounts between each pair of frames to produce a morphed film strip in which, say, an actor walking along a set is gradually morphed into an ape walking along the set.
8. Instead of using straight lines to triangulate two pictures to be morphed, more complicated curves, such as spline curves, can be matched between the two pictures.
9. Three or more pictures can be morphed together by generalizing the formulas given in this section.

These and other generalizations have made warping and morphing two of the most active areas in computer graphics.

Exercise Set 10.20

1. Determine whether the vector $\mathbf{v}$ is a convex combination of the vectors $\mathbf{v}_1$, $\mathbf{v}_2$, and $\mathbf{v}_3$. Do this by solving Equations (1) and (3) for c_1, c_2, and c_3 and ascertaining whether these coefficients are nonnegative.

   (a) $\mathbf{v} = \begin{bmatrix} 3 \\ 3 \end{bmatrix}$, $\mathbf{v}_1 = \begin{bmatrix} 1 \\ 1 \end{bmatrix}$, $\mathbf{v}_2 = \begin{bmatrix} 3 \\ 5 \end{bmatrix}$, $\mathbf{v}_3 = \begin{bmatrix} 4 \\ 2 \end{bmatrix}$

   (b) $\mathbf{v} = \begin{bmatrix} 2 \\ 4 \end{bmatrix}$, $\mathbf{v}_1 = \begin{bmatrix} 1 \\ 1 \end{bmatrix}$, $\mathbf{v}_2 = \begin{bmatrix} 3 \\ 5 \end{bmatrix}$, $\mathbf{v}_3 = \begin{bmatrix} 4 \\ 2 \end{bmatrix}$

   (c) $\mathbf{v} = \begin{bmatrix} 0 \\ 0 \end{bmatrix}$, $\mathbf{v}_1 = \begin{bmatrix} 3 \\ 3 \end{bmatrix}$, $\mathbf{v}_2 = \begin{bmatrix} -2 \\ -2 \end{bmatrix}$, $\mathbf{v}_3 = \begin{bmatrix} 3 \\ 0 \end{bmatrix}$

   (d) $\mathbf{v} = \begin{bmatrix} 1 \\ 0 \end{bmatrix}$, $\mathbf{v}_1 = \begin{bmatrix} 3 \\ 3 \end{bmatrix}$, $\mathbf{v}_2 = \begin{bmatrix} -2 \\ -2 \end{bmatrix}$, $\mathbf{v}_3 = \begin{bmatrix} 3 \\ 0 \end{bmatrix}$

2. Verify Equation (7) for the two triangulations given in Figure 10.20.4.

3. Let an affine transformation be given by a 2×2 matrix M and a two-dimensional vector $\mathbf{b}$. Let $\mathbf{v} = c_1\mathbf{v}_1 + c_2\mathbf{v}_2 + c_3\mathbf{v}_3$, where $c_1 + c_2 + c_3 = 1$; let $\mathbf{w} = M\mathbf{v} + \mathbf{b}$; and let $\mathbf{w}_i = M\mathbf{v}_i + \mathbf{b}$ for $i = 1, 2, 3$. Show that $\mathbf{w} = c_1\mathbf{w}_1 + c_2\mathbf{w}_2 + c_3\mathbf{w}_3$. (This shows that an affine transformation maps a convex combination of vectors to the same convex combination of the images of the vectors.)

4. (a) Exhibit a triangulation of the points in Figure 10.20.4 in which the points $\mathbf{v}_3$, $\mathbf{v}_5$, and $\mathbf{v}_6$ form the vertices of a single triangle.

   (b) Exhibit a triangulation of the points in Figure 10.20.4 in which the points $\mathbf{v}_2$, $\mathbf{v}_5$, and $\mathbf{v}_7$ do *not* form the vertices of a single triangle.

5. Find the 2 × 2 matrix M and two-dimensional vector $\mathbf{b}$ that define the affine transformation that maps the three vectors $\mathbf{v}_1$, $\mathbf{v}_2$, and $\mathbf{v}_3$ to the three vectors $\mathbf{w}_1$, $\mathbf{w}_2$, and $\mathbf{w}_3$. Do this by setting up a system of six linear equations for the four entries of the matrix M and the two entries of the vector $\mathbf{b}$.

(a) $\mathbf{v}_1 = \begin{bmatrix} 1 \\ 1 \end{bmatrix}$, $\mathbf{v}_2 = \begin{bmatrix} 2 \\ 3 \end{bmatrix}$, $\mathbf{v}_3 = \begin{bmatrix} 2 \\ 1 \end{bmatrix}$,

$\mathbf{w}_1 = \begin{bmatrix} 4 \\ 3 \end{bmatrix}$, $\mathbf{w}_2 = \begin{bmatrix} 9 \\ 5 \end{bmatrix}$, $\mathbf{w}_3 = \begin{bmatrix} 5 \\ 3 \end{bmatrix}$,

(b) $\mathbf{v}_1 = \begin{bmatrix} -2 \\ 2 \end{bmatrix}$, $\mathbf{v}_2 = \begin{bmatrix} 0 \\ 0 \end{bmatrix}$, $\mathbf{v}_3 = \begin{bmatrix} 2 \\ 1 \end{bmatrix}$,

$\mathbf{w}_1 = \begin{bmatrix} -8 \\ 1 \end{bmatrix}$, $\mathbf{w}_2 = \begin{bmatrix} 0 \\ 1 \end{bmatrix}$, $\mathbf{w}_3 = \begin{bmatrix} 5 \\ 4 \end{bmatrix}$,

(c) $\mathbf{v}_1 = \begin{bmatrix} -2 \\ 1 \end{bmatrix}$, $\mathbf{v}_2 = \begin{bmatrix} 3 \\ 5 \end{bmatrix}$, $\mathbf{v}_3 = \begin{bmatrix} 1 \\ 0 \end{bmatrix}$,

$\mathbf{w}_1 = \begin{bmatrix} 0 \\ -2 \end{bmatrix}$, $\mathbf{w}_2 = \begin{bmatrix} 5 \\ 2 \end{bmatrix}$, $\mathbf{w}_3 = \begin{bmatrix} 3 \\ -3 \end{bmatrix}$,

(d) $\mathbf{v}_1 = \begin{bmatrix} 0 \\ 2 \end{bmatrix}$, $\mathbf{v}_2 = \begin{bmatrix} 2 \\ 2 \end{bmatrix}$, $\mathbf{v}_3 = \begin{bmatrix} -4 \\ -2 \end{bmatrix}$,

$\mathbf{w}_1 = \begin{bmatrix} \frac{5}{2} \\ -1 \end{bmatrix}$, $\mathbf{w}_2 = \begin{bmatrix} \frac{7}{2} \\ 3 \end{bmatrix}$, $\mathbf{w}_3 = \begin{bmatrix} -\frac{7}{2} \\ -9 \end{bmatrix}$

6. (a) Let $\mathbf{a}$ and $\mathbf{b}$ be linearly independent vectors in the plane. Show that if c_1 and c_2 are nonnegative numbers such that $c_1 + c_2 = 1$, then the vector $c_1 \mathbf{a} + c_2 \mathbf{b}$ lies on the line segment connecting the tips of the vectors $\mathbf{a}$ and $\mathbf{b}$.

(b) Let $\mathbf{a}$ and $\mathbf{b}$ be linearly independent vectors in the plane. Show that if c_1 and c_2 are nonnegative numbers such that $c_1 + c_2 \leq 1$, then the vector $c_1 \mathbf{a} + c_2 \mathbf{b}$ lies in the triangle connecting the origin and the tips of the vectors $\mathbf{a}$ and $\mathbf{b}$. [*Hint:* First examine the vector $c_1 \mathbf{a} + c_2 \mathbf{b}$ multiplied by the scale factor $1/(c_1 + c_2)$.]

(c) Let $\mathbf{v}_1$, $\mathbf{v}_2$, and $\mathbf{v}_3$ be noncollinear points in the plane. Show that if c_1, c_2, and c_3 are nonnegative numbers such that $c_1 + c_2 + c_3 = 1$, then the vector $c_1 \mathbf{v}_1 + c_2 \mathbf{v}_2 + c_3 \mathbf{v}_3$ lies in the triangle connecting the tips of the three vectors. [*Hint:* Let $\mathbf{a} = \mathbf{v}_1 - \mathbf{v}_3$ and $\mathbf{b} = \mathbf{v}_2 - \mathbf{v}_3$, and then use Equation (1) and part (b) of this exercise.]

7. (a) What can you say about the coefficients c_1, c_2, and c_3 that determine a convex combination $\mathbf{v} = c_1 \mathbf{v}_1 + c_2 \mathbf{v}_2 + c_3 \mathbf{v}_3$ if $\mathbf{v}$ lies on one of the three vertices of the triangle determined by the three vectors $\mathbf{v}_1$, $\mathbf{v}_2$, and $\mathbf{v}_3$?

(b) What can you say about the coefficients c_1, c_2, and c_3 that determine a convex combination $\mathbf{v} = c_1 \mathbf{v}_1 + c_2 \mathbf{v}_2 + c_3 \mathbf{v}_3$ if $\mathbf{v}$ lies on one of the three sides of the triangle determined by the three vectors $\mathbf{v}_1$, $\mathbf{v}_2$, and $\mathbf{v}_3$?

(c) What can you say about the coefficients c_1, c_2, and c_3 that determine a convex combination $\mathbf{v} = c_1 \mathbf{v}_1 + c_2 \mathbf{v}_2 + c_3 \mathbf{v}_3$ if $\mathbf{v}$ lies in the interior of the triangle determined by the three vectors $\mathbf{v}_1$, $\mathbf{v}_2$, and $\mathbf{v}_3$?

8. (a) The centroid of a triangle lies on the line segment connecting any one of the three vertices of the triangle with the midpoint of the opposite side. Its location on this line segment is two-thirds of the distance from the vertex. If the three vertices are given by the vectors $\mathbf{v}_1$, $\mathbf{v}_2$, and $\mathbf{v}_3$, write the centroid as a convex combination of these three vectors.

(b) Use your result in part (a) to find the vector defining the centroid of the triangle with the three vertices $\begin{bmatrix} 2 \\ 3 \end{bmatrix}$, $\begin{bmatrix} 5 \\ 2 \end{bmatrix}$, and $\begin{bmatrix} 1 \\ 1 \end{bmatrix}$.

Section 10.20 Technology Exercises

The following exercises are designed to be solved using a technology utility. Typically, this will be MATLAB, *Mathematica*, Maple, Derive, or Mathcad, but it may also be some other type of linear algebra software or a scientific calculator with some linear algebra capabilities. For each exercise you will need to read the relevant documentation for the particular utility you are using. The goal of these exercises is to provide you with a basic proficiency with your technology utility. Once you have mastered the techniques in these exercises, you will be able to use your technology utility to solve many of the problems in the regular exercise sets.

T1. To warp or morph a surface in R^3 we must be able to triangulate the surface. Let $\mathbf{v}_1 = \begin{bmatrix} v_{11} \\ v_{12} \\ v_{13} \end{bmatrix}$, $\mathbf{v}_2 = \begin{bmatrix} v_{21} \\ v_{22} \\ v_{23} \end{bmatrix}$, and $\mathbf{v}_3 = \begin{bmatrix} v_{31} \\ v_{32} \\ v_{33} \end{bmatrix}$ be three noncollinear vectors on the surface. Then a vector $\mathbf{v} = \begin{bmatrix} v_1 \\ v_2 \\ v_3 \end{bmatrix}$ lies in the triangle formed by these three vectors if and only if $\mathbf{v}$ is a convex combination of the three vectors; that is, $\mathbf{v} = c_1 \mathbf{v}_1 + c_2 \mathbf{v}_2 + c_3 \mathbf{v}_3$ for some nonnegative coefficients c_1, c_2, and c_3 whose sum is 1.

(a) Show that in this case, c_1, c_2, and c_3 are solutions of the following linear system:

$$\begin{bmatrix} v_{11} & v_{21} & v_{31} \\ v_{12} & v_{22} & v_{32} \\ v_{13} & v_{23} & v_{33} \\ 1 & 1 & 1 \end{bmatrix} \begin{bmatrix} c_1 \\ c_2 \\ c_3 \end{bmatrix} = \begin{bmatrix} v_1 \\ v_2 \\ v_3 \\ 1 \end{bmatrix}$$

In parts (b)–(d) determine whether the vector $\mathbf{v}$ is a convex combination of the vectors $\mathbf{v}_1 = \begin{bmatrix} 2 \\ 7 \\ -5 \end{bmatrix}$, $\mathbf{v}_2 = \begin{bmatrix} 3 \\ 0 \\ 9 \end{bmatrix}$, and $\mathbf{v}_3 = \begin{bmatrix} 2 \\ 2 \\ -4 \end{bmatrix}$.

(b) $\mathbf{v} = \dfrac{1}{4}\begin{bmatrix} 9 \\ 9 \\ 9 \end{bmatrix}$ (c) $\mathbf{v} = \dfrac{1}{4}\begin{bmatrix} 10 \\ 9 \\ 9 \end{bmatrix}$ (d) $\mathbf{v} = \dfrac{1}{4}\begin{bmatrix} 13 \\ -7 \\ 50 \end{bmatrix}$

T2. To warp or morph a solid object in R^3 we first partition the object into disjoint tetrahedrons. Let $\mathbf{v}_1 = \begin{bmatrix} v_{11} \\ v_{12} \\ v_{13} \end{bmatrix}$, $\mathbf{v}_2 = \begin{bmatrix} v_{21} \\ v_{22} \\ v_{23} \end{bmatrix}$, $\mathbf{v}_3 = \begin{bmatrix} v_{31} \\ v_{32} \\ v_{33} \end{bmatrix}$, and $\mathbf{v}_4 = \begin{bmatrix} v_{41} \\ v_{42} \\ v_{43} \end{bmatrix}$ be four noncoplanar vectors. Then a vector $\mathbf{v} = \begin{bmatrix} v_1 \\ v_2 \\ v_3 \end{bmatrix}$ lies in the solid tetrahedron formed by these four vectors if and only if $\mathbf{v}$ is a convex combination of the three vectors; that is, $\mathbf{v} = c_1\mathbf{v}_1 + c_2\mathbf{v}_2 + c_3\mathbf{v}_3 + c_4\mathbf{v}_4$ for some nonnegative coefficients c_1, c_2, c_3, and c_4 whose sum is one.

(a) Show that in this case, c_1, c_2, c_3, and c_4 are solutions of the following linear system:
$$\begin{bmatrix} v_{11} & v_{21} & v_{31} & v_{41} \\ v_{12} & v_{22} & v_{32} & v_{42} \\ v_{13} & v_{23} & v_{33} & v_{43} \\ 1 & 1 & 1 & 1 \end{bmatrix} \begin{bmatrix} c_1 \\ c_2 \\ c_3 \\ c_4 \end{bmatrix} = \begin{bmatrix} v_1 \\ v_2 \\ v_3 \\ 1 \end{bmatrix}$$

In parts (b)–(d) determine whether the vector $\mathbf{v}$ is a convex combination of the vectors $\mathbf{v}_1 = \begin{bmatrix} 2 \\ -6 \\ 1 \end{bmatrix}$, $\mathbf{v}_2 = \begin{bmatrix} -3 \\ 4 \\ 2 \end{bmatrix}$, $\mathbf{v}_3 = \begin{bmatrix} 7 \\ 2 \\ 3 \end{bmatrix}$, and $\mathbf{v}_4 = \begin{bmatrix} -1 \\ 3 \\ 2 \end{bmatrix}$.

(b) $\mathbf{v} = \begin{bmatrix} 5 \\ 0 \\ 7 \end{bmatrix}$ (c) $\mathbf{v} = \begin{bmatrix} 1 \\ 1 \\ 2 \end{bmatrix}$ (d) $\mathbf{v} = \begin{bmatrix} 1 \\ 2 \\ 2 \end{bmatrix}$

APPENDIX A HOW TO READ THEOREMS

Since many of the most important concepts in linear algebra occur as theorem statements, it is important to be familiar with the various ways in which theorems can be structured. This appendix will help you to do that.

Contrapositive Form of a Theorem

The simplest theorems are of the form

$$\text{If } H \text{ is true, then } C \text{ is true.} \tag{1}$$

where H is a statement, called the **hypothesis**, and C is a statement, called the **conclusion**. The theorem is true if the conclusion is true whenever the hypothesis is true, and the theorem is false if there is some case where the hypothesis is true but the conclusion is false. It is common to denote a theorem of form (1) as

$$H \Rightarrow C \tag{2}$$

(read, "H implies C"). As an example, the theorem

$$\text{If } a \text{ and } b \text{ are both positive numbers, then } ab \text{ is a positive number.} \tag{3}$$

is of form (2), where

$$H = a \text{ and } b \text{ are both positive numbers} \tag{4}$$

$$C = ab \text{ is a positive number} \tag{5}$$

Sometimes it is desirable to phrase theorems in a *negative* way. For example, the theorem in (3) can be rephrased equivalently as

$$\text{If } ab \text{ is not a positive number, then } a \text{ and } b \text{ are not both positive numbers.} \tag{6}$$

If we write $\sim H$ to mean that (4) is false and $\sim C$ to mean that (5) is false, then the structure of the theorem in (6) is

$$\sim C \Rightarrow \sim H \tag{7}$$

In general, any theorem of form (2) can be rephrased in form (7), which is called the **contrapositive** of (2). If a theorem is true, then so is its contrapositive, and vice versa.

Converse of a Theorem

The **converse** of a theorem is the statement that results when the hypothesis and conclusion are interchanged. Thus, the converse of the theorem $H \Rightarrow C$ is the statement $C \Rightarrow H$. Whereas the contrapositive of a true theorem must itself be a true theorem, the converse of a true theorem may or may not be true. For example, the converse of (3) is the *false* statement

$$\text{If } ab \text{ is a positive number, then } a \text{ and } b \text{ are both positive numbers.}$$

but the converse of the true theorem

$$\text{If } a > b, \text{ then } 2a > 2b. \tag{8}$$

is the *true* theorem

$$\text{If } 2a > 2b, \text{ then } a > b. \tag{9}$$

Equivalent Statements

If a theorem $H \Rightarrow C$ and its converse $C \Rightarrow H$ are both true, then we say that H and C are **equivalent** statements, which we denote by writing

$$H \Leftrightarrow C \tag{10}$$

(read, "H and C are equivalent"). There are various ways of phrasing equivalent statements as a single theorem. Here are three ways in which (8) and (9) can be combined into a single theorem.

> **Form 1** If $a > b$, then $2a > 2b$, and conversely, if $2a > 2b$, then $a > b$.

> **Form 2** $a > b$ if and only if $2a > 2b$.

> **Form 3** The following statements are equivalent.
> (i) $a > b$
> (ii) $2a > 2b$

Theorems Involving Three or More Statements

Sometimes two true theorems will give you a third true theorem for free. Specifically, if $H \Rightarrow C$ is a true theorem, and $C \Rightarrow D$ is a true theorem, then $H \Rightarrow D$ must also be a true theorem. For example, the theorems

> If opposite sides of a quadrilateral are parallel, then the quadrilateral is a parallelogram.

and

> Opposite sides of a parallelogram have equal lengths.

imply the third theorem

> If opposite sides of a quadrilateral are parallel, then they have equal lengths.

Sometimes three theorems yield equivalent statements for free. For example, if

$$H \Rightarrow C, \quad C \Rightarrow D, \quad D \Rightarrow H \tag{11}$$

then we have the *implication loop* in Figure A.1 from which we can conclude that

$$C \Rightarrow H, \quad D \Rightarrow C, \quad H \Rightarrow D \tag{12}$$

Combining this with (11) we obtain

$$H \Leftrightarrow C, \quad C \Leftrightarrow D, \quad D \Leftrightarrow H \tag{13}$$

In summary, if you want to prove the three equivalences in (13), you need only prove the three implications in (11).

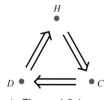

▲ Figure 1.0.1

APPENDIX B COMPLEX NUMBERS

Complex numbers arise naturally in the course of solving polynomial equations. For example, the solutions of the quadratic equation $ax^2 + bx + c = 0$, which are given by the quadratic formula

$$x = \frac{-b \pm \sqrt{b^2 - 4ac}}{2a}$$

are complex numbers if the expression inside the radical is negative. In this appendix we will review some of the basic ideas about complex numbers that are used in this text.

Complex Numbers To deal with the problem that the equation $x^2 = -1$ has no real solutions, mathematicians of the eighteenth century invented the "imaginary" number

$$i = \sqrt{-1}$$

which is assumed to have the property

$$i^2 = (\sqrt{-1})^2 = -1$$

but which otherwise has the algebraic properties of a real number. An expression of the form

$$a + bi \quad \text{or} \quad a + ib$$

in which a and b are *real* numbers is called a **complex number**. Sometimes it will be convenient to use a single letter, typically z, to denote a complex number, in which case we write

$$z = a + bi \quad \text{or} \quad z = a + ib$$

The number a is called the **real part** of z and is denoted by $\text{Re}(z)$, and the number b is called the **imaginary part** of z and is denoted by $\text{Im}(z)$. Thus,

$$\text{Re}(3 + 2i) = 3, \qquad \text{Im}(3 + 2i) = 2$$
$$\text{Re}(1 - 5i) = 1, \qquad \text{Im}(1 - 5i) = \text{Im}(1 + (-5)i) = -5$$
$$\text{Re}(7i) = \text{Re}(0 + 7i) = 0, \quad \text{Im}(7i) = 7$$
$$\text{Re}(4) = 4, \qquad \text{Im}(4) = \text{Im}(4 + 0i) = 0$$

Two complex numbers are considered **equal** if and only if their real parts are equal and their imaginary parts are equal; that is,

$$a + bi = c + di \quad \text{if and only if} \quad a = c \text{ and } b = d$$

A complex number $z = bi$ whose real part is zero is said to be **pure imaginary**. A complex number $z = a$ whose imaginary part is zero is a real number, so the real numbers can be viewed as a subset of the complex numbers.

Complex numbers are added, subtracted, and multiplied in accordance with the standard rules of algebra but with $i^2 = -1$:

$$(a + bi) + (c + di) = (a + c) + (b + d)i \tag{1}$$
$$(a + bi) - (c + di) = (a - c) + (b - d)i \tag{2}$$
$$(a + bi)(c + di) = (ac - bd) + (ad + bc)i \tag{3}$$

The multiplication formula is obtained by expanding the left side and using the fact that $i^2 = -1$. Also note that if $b = 0$, then the multiplication formula simplifies to

$$a(c + di) = ac + adi \tag{4}$$

The set of complex numbers with these operations is commonly denoted by the symbol C and is called the ***complex number system***.

▶ **EXAMPLE 1** **Multiplying Complex Numbers**

As a practical matter, it is usually more convenient to compute products of complex numbers by expansion, rather than substituting in (3). For example,

$$(3 - 2i)(4 + 5i) = 12 + 15i - 8i - 10i^2 = (12 + 10) + 7i = 22 + 7i$$ ◀

The Complex Plane

A complex number $z = a + bi$ can be associated with the ordered pair (a, b) of real numbers and represented geometrically by a point or a vector in the xy-plane (Figure B.1). We call this the ***complex plane***. Points on the x-axis have an imaginary part of zero and hence correspond to real numbers, whereas points on the y-axis have a real part of zero and correspond to pure imaginary numbers. Accordingly, we call the x-axis the ***real axis*** and the y-axis the ***imaginary axis*** (Figure B.2).

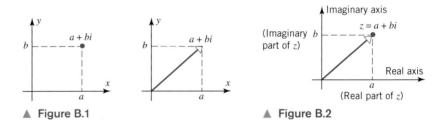

▲ Figure B.1   ▲ Figure B.2

Complex numbers can be added, subtracted, or multiplied by real numbers geometrically by performing these operations on their associated vectors (Figure B.3, for example). In this sense the complex number system C is closely related to R^2, the main difference being that complex numbers can be multiplied to produce other complex numbers, whereas there is no multiplication operation on R^2 that produces other vectors in R^2 (the dot product produces a scalar, not a vector in R^2).

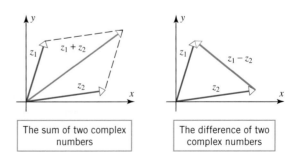

▶ Figure B.3

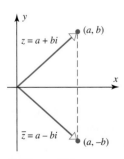

▲ Figure B.4

If $z = a + bi$ is a complex number, then the ***complex conjugate*** of z, or more simply, the ***conjugate*** of z, is denoted by $\bar{z}$ (read, "z bar") and is defined by

$$\bar{z} = a - bi \tag{5}$$

Numerically, $\bar{z}$ is obtained from z by reversing the sign of the imaginary part, and geometrically it is obtained by reflecting the vector for z about the real axis (Figure B.4).

▶ **EXAMPLE 2** **Some Complex Conjugates**

$$z = 3 + 4i \qquad \bar{z} = 3 - 4i$$
$$z = -2 - 5i \qquad \bar{z} = -2 + 5i$$
$$z = i \qquad \bar{z} = -i$$
$$z = 7 \qquad \bar{z} = 7 \blacktriangleleft$$

Remark The last computation in this example illustrates the fact that a real number is equal to its complex conjugate. More generally, $z = \bar{z}$ if and only if z is a real number.

The following computation shows that the product of a complex number $z = a + bi$ and its conjugate $\bar{z} = a - bi$ is a nonnegative real number:

$$z\bar{z} = (a + bi)(a - bi) = a^2 - abi + bai - b^2i^2 = a^2 + b^2 \qquad (6)$$

You will recognize that

$$\sqrt{z\bar{z}} = \sqrt{a^2 + b^2}$$

is the length of the vector corresponding to z (Figure B.5); we call this length the ***modulus*** (or ***absolute value*** of z) and denote it by $|z|$. Thus,

$$|z| = \sqrt{z\bar{z}} = \sqrt{a^2 + b^2} \qquad (7)$$

Note that if $b = 0$, then $z = a$ is a real number and $|z| = \sqrt{a^2} = |a|$, which tells us that the modulus of a real number is the same as its absolute value as defined in beginning algebra.

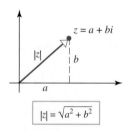

▲ Figure B.5

▶ **EXAMPLE 3** **Some Modulus Computations**

$$z = 3 + 4i \qquad |z| = \sqrt{3^2 + 4^2} = 5$$
$$z = -4 - 5i \qquad |z| = \sqrt{(-4)^2 + (-5)^2} = \sqrt{41}$$
$$z = i \qquad |z| = \sqrt{0^2 + 1^2} = 1 \blacktriangleleft$$

Reciprocals and Division

If $z \neq 0$, then the ***reciprocal*** (or ***multiplicative inverse***) of z is denoted by $1/z$ (or z^{-1}) and is defined by the property

$$\left(\frac{1}{z}\right) z = 1$$

This equation has a unique solution for $1/z$, which we can obtain by multiplying both sides by $\bar{z}$ and using the fact that $z\bar{z} = |z|^2$ [see (7)]. This yields

$$\frac{1}{z} = \frac{\bar{z}}{|z|^2} \qquad (8)$$

If $z_2 \neq 0$, then the ***quotient*** z_1/z_2 is defined to be the product of z_1 and $1/z_2$. This yields the formula

$$\frac{z_1}{z_2} = \frac{\bar{z}_2}{|z_2|^2} z_1 = \frac{z_1 \bar{z}_2}{|z_2|^2} \qquad (9)$$

Observe that the expression on the right side of (9) results if the numerator and denominator of z_1/z_2 are multiplied by $\bar{z}_2$. As a practical matter, this is often the best way to perform divisions of complex numbers.

▶ **EXAMPLE 4** **Division of Complex Numbers**

Let $z_1 = 3 + 4i$ and $z_2 = 1 - 2i$. Express z_1/z_2 in the form $a + bi$.

Solution We will multiply the numerator and denominator of z_1/z_2 by $\bar{z}_2$. This yields

$$\frac{z_1}{z_2} = \frac{z_1 \bar{z}_2}{z_2 \bar{z}_2} = \frac{3+4i}{1-2i} \cdot \frac{1+2i}{1+2i}$$

$$= \frac{3 + 6i + 4i + 8i^2}{1 - 4i^2}$$

$$= \frac{-5 + 10i}{5}$$

$$= -1 + 2i \blacktriangleleft$$

The following theorems list some useful properties of the modulus and conjugate operations.

THEOREM 2.0.1 *The following results hold for any complex numbers z, z_1, and z_2.*
(a) $\overline{z_1 + z_2} = \bar{z}_1 + \bar{z}_2$
(b) $\overline{z_1 - z_2} = \bar{z}_1 - \bar{z}_2$
(c) $\overline{z_1 z_2} = \bar{z}_1 \bar{z}_2$
(d) $\overline{z_1/z_2} = \bar{z}_1/\bar{z}_2$
(e) $\bar{\bar{z}} = z$

THEOREM 2.0.2 *The following results hold for any complex numbers z, z_1, and z_2.*
(a) $|\bar{z}| = |z|$
(b) $|z_1 z_2| = |z_1||z_2|$
(c) $|z_1/z_2| = |z_1|/|z_2|$
(d) $|z_1 + z_2| \leq |z_1| + |z_2|$

Polar Form of a Complex Number

If $z = a + bi$ is a nonzero complex number, and if ϕ is an angle from the real axis to the vector z, then, as suggested in Figure B.6, the real and imaginary parts of z can be expressed as

$$a = |z|\cos\phi \quad \text{and} \quad b = |z|\sin\phi \tag{10}$$

Thus, the complex number $z = a + bi$ can be expressed as

$$z = |z|(\cos\phi + i\sin\phi) \tag{11}$$

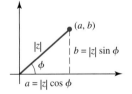

▲ Figure B.6

which is called a *polar form* of z. The angle ϕ in this formula is called an **argument** of z. The argument of z is not unique because we can add or subtract any multiple of 2π to it to obtain a different argument of z. However, there is only one argument whose radian measure satisfies

$$-\pi < \phi \leq \pi \tag{12}$$

This is called the **principal argument** of z.

EXAMPLE 5 Polar Form of a Complex Number

Express $z = 1 - \sqrt{3}i$ in polar form using the principal argument.

Solution The modulus of z is

$$|z| = \sqrt{1^2 + (-\sqrt{3})^2} = \sqrt{4} = 2$$

Thus, it follows from (10) with $a = 1$ and $b = -\sqrt{3}$ that

$$1 = 2\cos\phi \quad \text{and} \quad -\sqrt{3} = 2\sin\phi$$

and this implies that

$$\cos\phi = \frac{1}{2} \quad \text{and} \quad \sin\phi = -\frac{\sqrt{3}}{2}$$

The unique angle ϕ that satisfies these equations and whose radian measure satisfies (12) is $\phi = -\pi/3$ (Figure B.7). Thus, a polar form of z is

$$z = 2\left(\cos\left(-\frac{\pi}{3}\right) + i\sin\left(-\frac{\pi}{3}\right)\right) = 2\left(\cos\frac{\pi}{3} - i\sin\frac{\pi}{3}\right) \blacktriangleleft$$

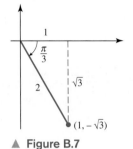

▲ Figure B.7

Geometric Interpretation of Multiplication and Division of Complex Numbers

We now show how polar forms of complex numbers provide geometric interpretations of multiplication and division. Let

$$z_1 = |z_1|(\cos\phi_1 + i\sin\phi_1) \quad \text{and} \quad z_2 = |z_2|(\cos\phi_2 + i\sin\phi_2)$$

be polar forms of the nonzero complex numbers z_1 and z_2. Multiplying, we obtain

$$z_1 z_2 = |z_1||z_2|[(\cos\phi_1\cos\phi_2 - \sin\phi_1\sin\phi_2) + i(\sin\phi_1\cos\phi_2 + \cos\phi_1\sin\phi_2)]$$

Now applying the trigonometric identities

$$\cos(\phi_1 + \phi_2) = \cos\phi_1\cos\phi_2 - \sin\phi_1\sin\phi_2$$
$$\sin(\phi_1 + \phi_2) = \sin\phi_1\cos\phi_2 + \cos\phi_1\sin\phi_2$$

yields

$$z_1 z_2 = |z_1||z_2|[\cos(\phi_1 + \phi_2) + i\sin(\phi_1 + \phi_2)] \quad (13)$$

which is a polar form of the complex number with modulus $|z_1||z_2|$ and argument $\phi_1 + \phi_2$. Thus, we have shown that *multiplying two complex numbers has the geometric effect of multiplying their moduli and adding their arguments* (Figure B.8).

Similar kinds of computations show that

$$\frac{z_1}{z_2} = \frac{|z_1|}{|z_2|}[\cos(\phi_1 - \phi_2) + i\sin(\phi_1 - \phi_2)] \quad (14)$$

which tells us that *dividing complex numbers has the geometric effect of dividing their moduli and subtracting their arguments* (both in the appropriate order).

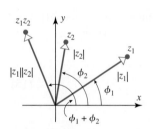

▲ Figure B.8

EXAMPLE 6 Multiplying and Dividing in Polar Form

Use polar forms of the complex numbers $z_1 = 1 + \sqrt{3}i$ and $z_2 = \sqrt{3} + i$ to compute $z_1 z_2$ and z_1/z_2.

Solution Polar forms of these complex numbers are

$$z_1 = 2\left(\cos\frac{\pi}{3} + i\sin\frac{\pi}{3}\right) \quad \text{and} \quad z_2 = 2\left(\cos\frac{\pi}{6} + i\sin\frac{\pi}{6}\right)$$

(verify). Thus, it follows from (13) that

$$z_1 z_2 = 4\left[\cos\left(\frac{\pi}{3}+\frac{\pi}{6}\right) + i\sin\left(\frac{\pi}{3}+\frac{\pi}{6}\right)\right] = 4\left[\cos\left(\frac{\pi}{2}\right) + i\sin\left(\frac{\pi}{2}\right)\right] = 4i$$

and from (14) that

$$\frac{z_1}{z_2} = 1 \cdot \left[\cos\left(\frac{\pi}{3}-\frac{\pi}{6}\right) + i\sin\left(\frac{\pi}{3}-\frac{\pi}{6}\right)\right] = \cos\left(\frac{\pi}{6}\right) + i\sin\left(\frac{\pi}{6}\right) = \frac{\sqrt{3}}{2} + \frac{1}{2}i$$

As a check, let us calculate $z_1 z_2$ and z_1/z_2 directly:

$$z_1 z_2 = (1+\sqrt{3}i)(\sqrt{3}+i) = \sqrt{3} + i + 3i + \sqrt{3}i^2 = 4i$$

$$\frac{z_1}{z_2} = \frac{1+\sqrt{3}i}{\sqrt{3}+i} = \frac{1+\sqrt{3}i}{\sqrt{3}+i} \cdot \frac{\sqrt{3}-i}{\sqrt{3}-i} = \frac{\sqrt{3}-i+3i-\sqrt{3}i^2}{3-i^2} = \frac{2\sqrt{3}+2i}{4} = \frac{\sqrt{3}}{2} + \frac{1}{2}i$$

which agrees with the results obtained using polar forms. ◀

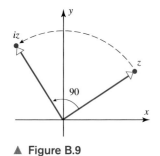

▲ Figure B.9

Remark The complex number i has a modulus of 1 and a principal argument of $\pi/2$. Thus, if z is a complex number, then iz has the same modulus as z but its argument is greater by $\pi/2$ ($= 90°$); that is, multiplication by i has the geometric effect of rotating the vector z counterclockwise by $90°$ (Figure B.9).

DeMoivre's Formula If n is a positive integer, and if z is a nonzero complex number with polar form

$$z = |z|(\cos\phi + i\sin\phi)$$

then raising z to the nth power yields

$$z^n = \underbrace{z \cdot z \cdot \cdots \cdot z}_{n \text{ factors}} = |z|^n[\cos(\underbrace{\phi+\phi+\cdots+\phi}_{n \text{ terms}})] + i[\sin(\underbrace{\phi+\phi+\cdots+\phi}_{n \text{ terms}})]$$

which we can write more succinctly as

$$z^n = |z|^n(\cos n\phi + i\sin n\phi) \quad (15)$$

In the special case where $|z| = 1$ this formula simplifies to

$$z^n = \cos n\phi + i\sin n\phi$$

which, using the polar form for z, becomes

$$\boxed{(\cos\phi + i\sin\phi)^n = \cos n\phi + i\sin n\phi} \quad (16)$$

This result is called ***DeMoivre's formula***.

Euler's Formula If θ is a real number, say the radian measure of some angle, then the ***complex exponential*** function $e^{i\theta}$ is defined to be

$$\boxed{e^{i\theta} = \cos\theta + i\sin\theta} \quad (17)$$

which is sometimes called ***Euler's formula***. One motivation for this formula comes from the Maclaurin series in calculus. Readers who have studied infinite series in calculus can deduce (17) by formally substituting $i\theta$ for x in the Maclaurin series for e^x and writing

$$\begin{aligned}
e^{i\theta} &= 1 + i\theta + \frac{(i\theta)^2}{2!} + \frac{(i\theta)^3}{3!} + \frac{(i\theta)^4}{4!} + \frac{(i\theta)^5}{5!} + \frac{(i\theta)^6}{6!} + \cdots \\
&= 1 + i\theta - \frac{\theta^2}{2!} - i\frac{\theta^3}{3!} + \frac{\theta^4}{4!} + i\frac{\theta^5}{5!} - \frac{\theta^6}{6!} + \cdots \\
&= \left(1 - \frac{\theta^2}{2!} + \frac{\theta^4}{4!} - \frac{\theta^6}{6!} + \cdots\right) + i\left(\theta - \frac{\theta^3}{3!} + \frac{\theta^5}{5!} - \cdots\right) \\
&= \cos\theta + i\sin\theta
\end{aligned}$$

where the last step follows from the Maclaurin series for $\cos\theta$ and $\sin\theta$.

If $z = a + bi$ is any complex number, then the ***complex exponential*** e^z is defined to be

$$e^z = e^{a+bi} = e^a e^{ib} = e^a(\cos b + i\sin b) \tag{18}$$

It can be proved that complex exponentials satisfy the standard laws of exponents. Thus, for example,

$$e^{z_1} e^{z_2} = e^{z_1 + z_2}, \quad \frac{e^{z_1}}{e^{z_2}} = e^{z_1 - z_2}, \quad \frac{1}{e^z} = e^{-z}$$

ANSWERS TO EXERCISES

Exercise Set 1.1 (page 9)

1. (a), (c), and (f) are linear equations; (b), (d) and (e) are not linear equations   3. (a) Not linear   (b) Linear

5. (a) delete
   ((c) and (d) unchanged)

7. (a) first equation is not satisfied
   (b) all the equations are satisfied
   (c) second equation is not satisfied
   (d) all the equations are satisfied
   (e) second equation is not satisfied

9. (a) $x = t$, $y = \frac{3}{4} - \frac{1}{2}t$.
   (b) $x_1 = 3 + \frac{5}{3}r - \frac{1}{3}s - \frac{4}{3}t$, $x_2 = r$, $x_3 = s$, $x_4 = t$.

11. (a) $\begin{aligned} 2x + 5y &= 6 \\ y &= 2 \\ -x &= 0 \end{aligned}$ (b) $\begin{aligned} 3x_1 - 2x_3 &= 5 \\ 7x_1 + x_2 + 4x_3 &= -3 \\ -2x_2 + x_3 &= 7 \end{aligned}$ (c) $\begin{aligned} x_1 + 5x_2 + 7x_3 - x_4 &= 3 \\ 2x_1 + 2x_2 + x_3 + x_4 &= 0 \end{aligned}$ (d) $\begin{aligned} x_1 &= 7 \\ x_2 &= -2 \\ x_3 &= 3 \\ x_4 &= 4 \end{aligned}$

13. (a) $\begin{bmatrix} -2 & 6 \\ 3 & 8 \\ 9 & -3 \end{bmatrix}$ (b) $\begin{bmatrix} 3 & 0 & -1 & 6 & 0 \\ 0 & 2 & -1 & -5 & -2 \end{bmatrix}$ (c) $\begin{bmatrix} 0 & 2 & 0 & -3 & 1 & 0 \\ -3 & -1 & 1 & 0 & 0 & -1 \\ 6 & 2 & -1 & 2 & -3 & 6 \end{bmatrix}$

   (d) $\begin{bmatrix} 1 & 0 & -1 & 0 & 4 \\ 0 & 1 & 0 & 1 & 9 \end{bmatrix}$

18. (a) False.   (e) True.

True/False 1.1

(a) True   (b) False   (c) True   (d) True   (e) False   (f) False   (g) True   (h) False

Exercise Set 1.2 (page 22)

1. (a) Both   (b) Both   (c) Both   (d) Both   (e) Both   (f) Both   (g) Row echelon

3. (a) $x_1 = -37$, $x_2 = -8$, $x_3 = 5$ (b) $x_1 = 13t - 10$, $x_2 = 13t - 5$, $x_3 = -t + 2$, $x_4 = t$
   (c) $x_1 = -7s + 2t - 11$, $x_2 = s$, $x_3 = -3t - 4$, $x_4 = -3t + 9$, $x_5 = t$ (d) Inconsistent

5. $x_1 = \frac{17}{5}$, $x_2 = \frac{-7}{5}$, $x_3 = \frac{-9}{5}$.   7. $x = 4 - 4t$, $y = -t$, $z = 1 + 4t$, $w = t$.

9. $x_1 = \frac{17}{5}$, $x_2 = \frac{-7}{5}$, $x_3 = \frac{-9}{5}$.   11. $x = 4 - 4t$, $y = -t$, $z = 1 + 4t$, $w = t$.

13. Has nontrivial solutions   15. Has nontrivial solutions   17. $x_1 = 0$, $x_2 = 0$, $x_3 = 0$

19. $x_1 = -\frac{8}{3}s - \frac{7}{3}t$, $x_2 = \frac{13}{3}s + \frac{5}{3}t$, $x_3 = s$, $x_4 = t$.

21. $w = t$, $x = -t$, $y = t$, $z = 0$ 23. $I_1 = -1$, $I_2 = 0$, $I_3 = 1$, $I_4 = 2$

25. No solutions if $a = -1$; unique solution if $a \neq -1$

27. inconsistent system if $a = -2$; infinitely many solutions if $a = 2$; unique solutions for any other value of a

29. $x = \frac{2a}{3} - \frac{b}{9}$, $y = -\frac{a}{3} + \frac{2b}{9}$ 31. $\begin{bmatrix} 1 & 3 \\ 0 & 1 \end{bmatrix}$ and $\begin{bmatrix} 1 & 0 \\ 0 & 1 \end{bmatrix}$ are possible answers.   35. $x = \pm 1$, $y = \pm\sqrt{3}$, $z = \pm\sqrt{2}$

37. $a = 1$, $b = -6$, $c = 2$, $d = 10$ 39. The nonhomogeneous system will have exactly one solution.

True/False 1.2

(a) True   (b) False   (c) False   (d) True   (e) True   (f) False   (g) True   (h) False   (i) False

Exercise Set 1.3 (page 35)

1. (a) Undefined (b) 4×2 (c) Undefined (d) Undefined (e) 5×5 (f) 5×2 (g) Undefined (h) 5×2

3. (a) $\begin{bmatrix} -2 & 4 & 8 \\ -2 & 1 & 3 \\ 11 & 0 & 5 \end{bmatrix}$ (b) $\begin{bmatrix} -2 & -2 & 8 \\ 8 & -1 & 1 \\ -3 & -12 & 1 \end{bmatrix}$ (c) $\begin{bmatrix} 10 & 0 \\ -20 & 30 \end{bmatrix}$ (d) $\begin{bmatrix} 18 & -9 & -72 \\ -27 & 0 & -18 \\ -36 & 54 & -27 \end{bmatrix}$

   (e) Undefined

   (f) $\begin{bmatrix} 6 & 18 & -24 \\ -44 & 7 & 1 \\ 37 & 60 & 5 \end{bmatrix}$ (g) $\begin{bmatrix} -4 & 32 & 16 \\ -44 & 10 & 14 \\ 78 & 48 & 26 \end{bmatrix}$ (h) $\begin{bmatrix} 0 & 0 & 0 \\ 0 & 0 & 0 \end{bmatrix}$ (i) $-2 + 0 + 3 = 1$ (j) -2

   (k) Undefined (l) 8

5. (a) $\begin{bmatrix} 2 & -14 & 4 \\ 26 & 46 & -8 \end{bmatrix}$ (b) Undefined (c) $\begin{bmatrix} 27 & 0 & 18 \\ 51 & -33 & -105 \\ 36 & -15 & 222 \end{bmatrix}$ (d) $= \begin{bmatrix} 58 & 22 \\ -50 & 226 \end{bmatrix}$ (e) $= \begin{bmatrix} 58 & 22 \\ -50 & 226 \end{bmatrix}$

   (f) $= \begin{bmatrix} 97 & -12 & 17 \\ -12 & 9 & -6 \\ 17 & -6 & 5 \end{bmatrix}$ (g) $\begin{bmatrix} 5 & 16 & 40 \\ -10 & 29 & 39 \end{bmatrix}$

   (h) Not defined. (i) 143 (j) 11 (k) Not defined. (l) 18

7. (a) $[67\ 41\ 41]$ (b) $[63\ 67\ 57]$ (c) $\begin{bmatrix} 41 \\ 21 \\ 67 \end{bmatrix}$ (d) $\begin{bmatrix} 6 \\ 6 \\ 63 \end{bmatrix}$ (e) $[24\ 56\ 97]$ (f) $\begin{bmatrix} 76 \\ 98 \\ 97 \end{bmatrix}$

9. (a) $\begin{bmatrix} -3 \\ 48 \\ 24 \end{bmatrix} = 3\begin{bmatrix} 3 \\ 6 \\ 0 \end{bmatrix} + 6\begin{bmatrix} -2 \\ 5 \\ 4 \end{bmatrix}$; $\begin{bmatrix} 12 \\ 29 \\ 56 \end{bmatrix} = -2\begin{bmatrix} 3 \\ 6 \\ 0 \end{bmatrix} + 5\begin{bmatrix} -2 \\ 5 \\ 4 \end{bmatrix} + 4\begin{bmatrix} 7 \\ 4 \\ 9 \end{bmatrix}$;

   $\begin{bmatrix} 76 \\ 98 \\ 97 \end{bmatrix} = 7\begin{bmatrix} 3 \\ 6 \\ 0 \end{bmatrix} + 4\begin{bmatrix} -2 \\ 5 \\ 4 \end{bmatrix} + 9\begin{bmatrix} 7 \\ 4 \\ 9 \end{bmatrix}$

   (b) $\begin{bmatrix} 64 \\ 21 \\ 77 \end{bmatrix} = 6\begin{bmatrix} 6 \\ 0 \\ 7 \end{bmatrix} + 7\begin{bmatrix} 4 \\ 3 \\ 5 \end{bmatrix}$; $\begin{bmatrix} 14 \\ 22 \\ 28 \end{bmatrix} = -2\begin{bmatrix} 6 \\ 0 \\ 7 \end{bmatrix} + \begin{bmatrix} -2 \\ 1 \\ 7 \end{bmatrix} + 7\begin{bmatrix} 4 \\ 3 \\ 5 \end{bmatrix}$;

   $\begin{bmatrix} 38 \\ 18 \\ 74 \end{bmatrix} = 4\begin{bmatrix} 6 \\ 0 \\ 7 \end{bmatrix} + 3\begin{bmatrix} -2 \\ 1 \\ 7 \end{bmatrix} + 5\begin{bmatrix} 4 \\ 3 \\ 5 \end{bmatrix}$

11. (a) $\begin{bmatrix} 5 & 1 & 1 \\ 2 & 0 & 3 \\ 1 & 2 & 0 \end{bmatrix} \begin{bmatrix} x \\ y \\ z \end{bmatrix} = \begin{bmatrix} 2 \\ 1 \\ 0 \end{bmatrix}$

    (b) $\begin{bmatrix} 1 & 1 & -1 & -7 \\ 0 & -1 & 4 & 1 \\ 4 & 2 & 1 & 8 \end{bmatrix} \begin{bmatrix} x_1 \\ x_2 \\ x_3 \\ x_4 \end{bmatrix} = \begin{bmatrix} 6 \\ 1 \\ 0 \end{bmatrix}$

13. (a) $\begin{aligned} 5x_1 + 6x_2 - 7x_3 &= 2 \\ -x_1 - 2x_2 + 3x_3 &= 0 \\ 4x_2 - x_3 &= 3 \end{aligned}$ (b) $\begin{aligned} x_1 + x_2 + x_3 &= 2 \\ 2x_1 + 3x_2 &= 2 \\ 5x_1 - 3x_2 - 6x_3 &= -9 \end{aligned}$ 15. -1

17. $a = -3, b = 3, c = -1, d = 1$

23. (a) $\begin{bmatrix} a_{11} & 0 & 0 & 0 & 0 & 0 \\ 0 & a_{22} & 0 & 0 & 0 & 0 \\ 0 & 0 & a_{33} & 0 & 0 & 0 \\ 0 & 0 & 0 & a_{44} & 0 & 0 \\ 0 & 0 & 0 & 0 & a_{55} & 0 \\ 0 & 0 & 0 & 0 & 0 & a_{66} \end{bmatrix}$ (b) $\begin{bmatrix} a_{11} & a_{12} & a_{13} & a_{14} & a_{15} & a_{16} \\ 0 & a_{22} & a_{23} & a_{24} & a_{25} & a_{26} \\ 0 & 0 & a_{33} & a_{34} & a_{35} & a_{36} \\ 0 & 0 & 0 & a_{44} & a_{45} & a_{46} \\ 0 & 0 & 0 & 0 & a_{55} & a_{56} \\ 0 & 0 & 0 & 0 & 0 & a_{66} \end{bmatrix}$

(c) $\begin{bmatrix} a_{11} & 0 & 0 & 0 & 0 & 0 \\ a_{21} & a_{22} & 0 & 0 & 0 & 0 \\ a_{31} & a_{32} & a_{33} & 0 & 0 & 0 \\ a_{41} & a_{42} & a_{43} & a_{44} & 0 & 0 \\ a_{51} & a_{52} & a_{53} & a_{54} & a_{55} & 0 \\ a_{61} & a_{62} & a_{63} & a_{64} & a_{65} & a_{66} \end{bmatrix}$ (d) $\begin{bmatrix} a_{11} & a_{12} & 0 & 0 & 0 & 0 \\ a_{21} & a_{22} & a_{23} & 0 & 0 & 0 \\ 0 & a_{32} & a_{33} & a_{34} & 0 & 0 \\ 0 & 0 & a_{43} & a_{44} & a_{45} & 0 \\ 0 & 0 & 0 & a_{54} & a_{55} & a_{56} \\ 0 & 0 & 0 & 0 & a_{65} & a_{66} \end{bmatrix}$

25. $f\begin{bmatrix} x_1 \\ x_2 \end{bmatrix} = \begin{bmatrix} x_1 + x_2 \\ x_2 \end{bmatrix}$

(a) $f\begin{bmatrix} 1 \\ 1 \end{bmatrix} = \begin{bmatrix} 2 \\ 1 \end{bmatrix}$ (b) $f\begin{bmatrix} 2 \\ 0 \end{bmatrix} = \begin{bmatrix} 2 \\ 0 \end{bmatrix}$ (c) $f\begin{bmatrix} 4 \\ 3 \end{bmatrix} = \begin{bmatrix} 7 \\ 3 \end{bmatrix}$ (d) $f\begin{bmatrix} 2 \\ -2 \end{bmatrix} = \begin{bmatrix} 0 \\ -2 \end{bmatrix}$

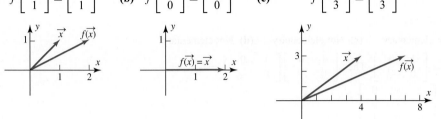

27. One; namely, $A = \begin{bmatrix} 1 & 1 & 0 \\ 1 & -1 & 0 \\ 0 & 0 & 0 \end{bmatrix}$

29. (a) $\begin{bmatrix} 1 & 1 \\ 1 & 1 \end{bmatrix}$ and $\begin{bmatrix} -1 & -1 \\ -1 & -1 \end{bmatrix}$ (b) Four; $\begin{bmatrix} \sqrt{5} & 0 \\ 0 & 3 \end{bmatrix}, \begin{bmatrix} -\sqrt{5} & 0 \\ 0 & 3 \end{bmatrix}, \begin{bmatrix} \sqrt{5} & 0 \\ 0 & -3 \end{bmatrix}, \begin{bmatrix} -\sqrt{5} & 0 \\ 0 & -3 \end{bmatrix}$

True/False 1.3

(a) True (b) False (c) False (d) False (e) True (f) False (g) False (h) True (i) True (j) True
(k) True (l) False (m) True (n) True (o) False

Exercise Set 1.4 (page 49)

5. $B^{-1} = \frac{1}{3}\begin{bmatrix} -2 & -3 \\ 5 & 6 \end{bmatrix}$. 7. $D^{-1} = \frac{1}{21}\begin{bmatrix} 2 & 3 \\ -7 & 0 \end{bmatrix}$ 9. $\begin{bmatrix} \frac{1}{2}(e^x + e^{-x}) & -\frac{1}{2}(e^x - e^{-x}) \\ -\frac{1}{2}(e^x - e^{-x}) & \frac{1}{2}(e^x + e^{-x}) \end{bmatrix}$

15. $A = \frac{1}{50}\begin{bmatrix} 3 & -2 \\ -1 & 4 \end{bmatrix}$ 17. $\begin{bmatrix} -\frac{9}{13} & \frac{1}{13} \\ \frac{2}{13} & -\frac{6}{13} \end{bmatrix}$

19. (a) $\begin{bmatrix} 11 & -15 \\ -30 & 41 \end{bmatrix}$ (b) $\begin{bmatrix} 41 & 15 \\ 30 & 11 \end{bmatrix}$ (c) $\begin{bmatrix} 2 & -2 \\ -4 & 6 \end{bmatrix}$
(d) $\begin{bmatrix} -1 & -1 \\ -2 & 1 \end{bmatrix}$ (e) $\begin{bmatrix} 6 & -7 \\ -14 & 20 \end{bmatrix}$ (f) $\begin{bmatrix} 13 & -13 \\ -26 & 39 \end{bmatrix}$

21. (a) $\begin{bmatrix} 27 & 0 & 0 \\ 0 & 26 & -18 \\ 0 & 18 & 26 \end{bmatrix}$ (b) $\begin{bmatrix} \frac{1}{27} & 0 & 0 \\ 0 & 0.026 & 0.018 \\ 0 & -0.018 & 0.026 \end{bmatrix}$ (c) $\begin{bmatrix} 4 & 0 & 0 \\ 0 & -5 & -12 \\ 0 & 12 & -5 \end{bmatrix}$ (d) $\begin{bmatrix} 1 & 0 & 0 \\ 0 & -3 & 3 \\ 0 & -3 & -3 \end{bmatrix}$
(e) $\begin{bmatrix} 16 & 0 & 0 \\ 0 & -14 & -15 \\ 0 & 15 & -14 \end{bmatrix}$ (f) $\begin{bmatrix} 25 & 0 & 0 \\ 0 & 32 & -24 \\ 0 & 24 & 32 \end{bmatrix}$

27. $\begin{bmatrix} \frac{1}{a_{11}} & 0 & \cdots & 0 \\ 0 & \frac{1}{a_{22}} & \cdots & 0 \\ \vdots & \vdots & & \vdots \\ 0 & 0 & \cdots & \frac{1}{a_{nn}} \end{bmatrix}$ 31. $D = CA^{-1}B^{-1}A^{-2}BC^2(B^T)^{-1}A^2$ 33. $D^{-2}C$ 35. $A^{-1} = \begin{bmatrix} \frac{1}{2} & \frac{1}{2} & -\frac{1}{2} \\ -\frac{1}{2} & \frac{1}{2} & \frac{1}{2} \\ \frac{1}{2} & -\frac{1}{2} & \frac{1}{2} \end{bmatrix}$

724   Answers to Exercises

37. $\begin{bmatrix} \frac{1}{2} & 0 & \frac{1}{2} \\ 0 & 1 & 0 \\ \frac{1}{2} & 0 & -\frac{1}{2} \end{bmatrix}$ 39. $x_1 = \frac{1}{23}$, $x_2 = \frac{13}{23}$

True/False 1.4

(a) False   (b) False   (c) False   (d) False   (e) False   (f) True   (g) True   (h) True   (i) False   (j) True
(k) False

Exercise Set 1.5 (page 58)

1. (a) Elementary   (b) Not elementary   (c) Not elementary   (d) Not elementary

3. (a) Multiply R_2 by $\frac{-1}{4}$, $E = \begin{bmatrix} 1 & 0 \\ 0 & \frac{-1}{4} \end{bmatrix}$. Then $\begin{bmatrix} 1 & 0 \\ 0 & \frac{-1}{4} \end{bmatrix}\begin{bmatrix} 1 & 0 \\ 0 & -4 \end{bmatrix} = I_2$.

(b) Add $-9R_2$ to R_1. $E = \begin{bmatrix} 1 & -9 & 0 \\ 0 & 1 & 0 \\ 0 & 0 & 1 \end{bmatrix}$.

(c) Interchage R_2 and R_3. $E = \begin{bmatrix} 1 & 0 & 0 \\ 0 & 0 & 1 \\ 0 & 1 & 0 \end{bmatrix}$

(d) Add R_4 to R_2. $E = \begin{bmatrix} 1 & 0 & 0 & 0 \\ 0 & 1 & 0 & 1 \\ 0 & 0 & 1 & 0 \\ 0 & 0 & 0 & 1 \end{bmatrix}$

5. (a) Swap rows 1 and 2: $EA = \begin{bmatrix} 3 & -6 & -6 & -6 \\ -1 & -2 & 5 & -1 \end{bmatrix}$

(b) Add -3 times row 2 to row 3: $EA = \begin{bmatrix} 2 & -1 & 0 & -4 & -4 \\ 1 & -3 & -1 & 5 & 3 \\ -1 & 9 & 4 & -12 & -10 \end{bmatrix}$

(c) Add 4 times row 3 to row 1: $EA = \begin{bmatrix} 13 & 28 \\ 2 & 5 \\ 3 & 6 \end{bmatrix}$

7. (a) $E = \begin{bmatrix} 1 & 0 & 0 \\ 0 & 1 & 0 \\ -2 & 0 & 1 \end{bmatrix}$ (b) $E = \begin{bmatrix} 1 & 0 & 0 \\ 0 & 1 & 0 \\ 2 & 0 & 1 \end{bmatrix}$ (c) $E = \begin{bmatrix} \frac{1}{2} & 0 & 0 \\ 0 & 1 & 0 \\ 0 & 0 & 1 \end{bmatrix}$ (d) $E = \begin{bmatrix} 2 & 0 & 0 \\ 0 & 1 & 0 \\ 0 & 0 & 1 \end{bmatrix}$

9. $\begin{bmatrix} -7 & 4 \\ 2 & -1 \end{bmatrix}$ 11. No inverse   13. $A^{-1} = \begin{bmatrix} \frac{1}{2} & 0 & \frac{1}{4} \\ 0 & \frac{1}{10} & -\frac{1}{5} \\ 0 & \frac{1}{10} & \frac{3}{10} \end{bmatrix}$ 15. No inverse

17. $\begin{bmatrix} \frac{1}{2} & -\frac{1}{2} & \frac{1}{2} \\ -\frac{1}{2} & \frac{1}{2} & \frac{1}{2} \\ \frac{1}{2} & \frac{1}{2} & -\frac{1}{2} \end{bmatrix}$ 19. $A^{-1} = \begin{bmatrix} -2 & 1 & 2 \\ \frac{1}{2} & -\frac{1}{2} & 0 \\ \frac{1}{4} & \frac{1}{4} & -\frac{1}{2} \end{bmatrix}$

21. $\begin{bmatrix} \frac{1}{4} & \frac{1}{2} & -3 & 0 \\ -\frac{1}{8} & \frac{1}{4} & -\frac{3}{2} & 0 \\ 0 & 0 & \frac{1}{2} & 0 \\ \frac{1}{40} & -\frac{1}{20} & -\frac{1}{10} & -\frac{1}{5} \end{bmatrix}$ 23. $\begin{bmatrix} -\frac{7}{12} & \frac{5}{24} & \frac{5}{8} & -\frac{1}{4} \\ \frac{5}{6} & \frac{5}{12} & \frac{1}{4} & -\frac{1}{2} \\ \frac{5}{12} & \frac{5}{24} & \frac{5}{8} & -\frac{1}{4} \\ -\frac{1}{12} & -\frac{1}{24} & -\frac{1}{8} & \frac{1}{4} \end{bmatrix}$

25. (a) $\begin{bmatrix} \frac{1}{k_1} & 0 & 0 & 0 \\ 0 & \frac{1}{k_2} & 0 & 0 \\ 0 & 0 & \frac{1}{k_3} & 0 \\ 0 & 0 & 0 & \frac{1}{k_4} \end{bmatrix}$ (b) $\begin{bmatrix} \frac{1}{k} & -\frac{1}{k} & 0 & 0 \\ 0 & 1 & 0 & 0 \\ 0 & 0 & \frac{1}{k} & -\frac{1}{k} \\ 0 & 0 & 0 & 1 \end{bmatrix}$ 27. $c \neq 0, -1$

29. $\begin{pmatrix} -2 & 3 \\ 1 & 0 \end{pmatrix} = \begin{pmatrix} 1 & -2 \\ 0 & 1 \end{pmatrix} \begin{pmatrix} 3 & 0 \\ 0 & 1 \end{pmatrix} \begin{pmatrix} 0 & 1 \\ 1 & 0 \end{pmatrix}$

31. $\begin{bmatrix} 1 & 0 & -2 \\ 0 & 4 & 3 \\ 0 & 0 & 1 \end{bmatrix} = \begin{bmatrix} 1 & 0 & -2 \\ 0 & 1 & 0 \\ 0 & 0 & 1 \end{bmatrix} \begin{bmatrix} 1 & 0 & 0 \\ 0 & 1 & 3 \\ 0 & 0 & 1 \end{bmatrix} \begin{bmatrix} 1 & 0 & 0 \\ 0 & 4 & 0 \\ 0 & 0 & 1 \end{bmatrix}$

33. $\begin{pmatrix} -2 & 3 \\ 1 & 0 \end{pmatrix}^{-1} = \begin{pmatrix} 0 & 1 \\ 1 & 0 \end{pmatrix} \begin{pmatrix} \frac{1}{3} & 0 \\ 0 & 1 \end{pmatrix} \begin{pmatrix} 1 & 2 \\ 0 & 1 \end{pmatrix}$

35. $\begin{bmatrix} 1 & 0 & 2 \\ 0 & \frac{1}{4} & -\frac{3}{4} \\ 0 & 0 & 1 \end{bmatrix} = \begin{bmatrix} 1 & 0 & 0 \\ 0 & \frac{1}{4} & 0 \\ 0 & 0 & 1 \end{bmatrix} \begin{bmatrix} 1 & 0 & 0 \\ 0 & 1 & -3 \\ 0 & 0 & 1 \end{bmatrix} \begin{bmatrix} 1 & 0 & 2 \\ 0 & 1 & 0 \\ 0 & 0 & 1 \end{bmatrix}$

37. Keeping track of the elementary matrices used to row reduce A to B, we see that
$\begin{pmatrix} 1 & 0 & 0 \\ 0 & 0 & 1 \\ 0 & 1 & 0 \end{pmatrix} \begin{pmatrix} 1 & 2 & 0 \\ 0 & 1 & 0 \\ 0 & 0 & 1 \end{pmatrix} \begin{pmatrix} -3 & -11 & -18 \\ 5 & 6 & 8 \\ -1 & 3 & 4 \end{pmatrix} = \begin{pmatrix} 7 & 1 & -2 \\ -1 & 3 & 4 \\ 5 & 6 & 8 \end{pmatrix}$, and so

$\begin{pmatrix} 1 & -2 & 0 \\ 0 & 1 & 0 \\ 0 & 0 & 1 \end{pmatrix} \begin{pmatrix} 1 & 0 & 0 \\ 0 & 0 & 1 \\ 0 & 1 & 0 \end{pmatrix} \begin{pmatrix} 7 & 1 & -2 \\ -1 & 3 & 4 \\ 5 & 6 & 8 \end{pmatrix} = \begin{pmatrix} -3 & -11 & -18 \\ 5 & 6 & 8 \\ -1 & 3 & 4 \end{pmatrix}$

True/False 1.5
(a) False (b) True (c) True (d) True (e) True (f) True (g) False

Exercise Set 1.6 (page 65)

1. $\begin{pmatrix} -19 \\ 11 \end{pmatrix}$. 3. $x_1 = -1$, $x_2 = 4$, $x_3 = -7$ 5. $x = \begin{pmatrix} -3 \\ 9 \\ -9 \end{pmatrix}$

7. $\begin{pmatrix} 6b_1 - b_2 \\ -5b_1 + b_2 \end{pmatrix}$ 9. (i) $x_1 = \frac{22}{17}$, $x_2 = \frac{1}{17}$ (ii) $x_1 = \frac{21}{17}$, $x_2 = \frac{11}{17}$

11. (a) $x = \begin{pmatrix} 5 \\ -6 \end{pmatrix}$ (b) $x = \begin{pmatrix} 46 \\ -56 \end{pmatrix}$ (c) $x = \begin{pmatrix} 19 \\ -23 \end{pmatrix}$ (d) $x = \begin{pmatrix} 25 \\ -31 \end{pmatrix}$

13. $-4b_1 + b_2 = 0$, or $b_2 = 4b_1$. 15. $b_3 = b_1 - b_2$.

17. $b_1 - 2b_3 + b_4 = 0$, or $b_1 = 2b_3 - b_4$. 19. $X = \begin{pmatrix} -83 & -186 & 264 & 113 & -56 \\ 27 & 59 & -85 & -37 & 19 \\ 10 & 24 & -32 & -13 & 7 \end{pmatrix}$.

True/False 1.6
(a) True (b) True (c) True (d) True (e) True (f) True (g) True

Exercise Set 1.7 (page 71)

1. invertible 3. $\begin{bmatrix} -1 & 0 & 0 \\ 0 & \frac{1}{2} & 0 \\ 0 & 0 & 3 \end{bmatrix}$ 5. $\begin{bmatrix} 6 & 3 \\ 4 & -1 \\ 4 & 10 \end{bmatrix}$ 7. $\begin{bmatrix} -15 & 10 & 0 & 20 & -20 \\ 2 & -10 & 6 & 0 & 6 \\ 18 & -6 & -6 & -6 & -6 \end{bmatrix}$

9. $A^2 = \begin{pmatrix} 4 & 0 \\ 0 & 1 \end{pmatrix}$, $A^{-2} = \begin{pmatrix} \frac{1}{4} & 0 \\ 0 & 1 \end{pmatrix}$, $A^{-k} = \begin{pmatrix} \frac{1}{2^k} & 0 \\ 0 & (-1)^k \end{pmatrix}$

11. $A^2 = \begin{pmatrix} 9 & 0 & 0 \\ 0 & \frac{1}{4} & 0 \\ 0 & 0 & \frac{1}{25} \end{pmatrix}$, $A^{-2} = \begin{pmatrix} \frac{1}{9} & 0 & 0 \\ 0 & 4 & 0 \\ 0 & 0 & 25 \end{pmatrix}$, $A^{-k} = \begin{pmatrix} \frac{1}{3^k} & 0 & 0 \\ 0 & 2^k & 0 \\ 0 & 0 & 25^k \end{pmatrix}$.

13. Symmetric 15. Symmetric 17. Symmetric 19. Not symmetric 21. Invertible

23. a^2 must equal 4, so $a = 2, a = -2$.

25. $x \neq 2, x \neq -3$, and $x \neq 0$, the matrix will be invertible

27. $\begin{bmatrix} 1 & 0 & 0 \\ 0 & -1 & 0 \\ 0 & 0 & -1 \end{bmatrix}$

35. **(a)** Yes **(b)** No (unless $n = 1$) **(c)** Yes **(d)** No (unless $n = 1$)

39. $\begin{bmatrix} 0 & 0 & -8 \\ 0 & 0 & -4 \\ 8 & 4 & 0 \end{bmatrix}$ 43. $A = \begin{pmatrix} 2 & 0 \\ 3 & -1 \end{pmatrix}$

True/False 1.7

(a) True (b) False (c) False (d) True (e) True (f) False (g) False (h) True (i) True (j) False
(k) False (l) False (m) True

Exercise Set 1.8 (page 84)

1.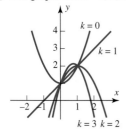

3. **(a)** $x_3 - x_4 = -500$, $-x_1 + x_4 = 100$, $x_1 - x_2 = 300$, $x_2 - x_3 = 100$
   (b) $x_1 = -100 + t$, $x_2 = -400 + t$, $x_3 = -500 + t$, $x_4 = t$
   (c) For all rates to be nonnegative, we need $t = 500$ cars per hour, so $x_1 = 400$, $x_2 = 100$, $x_3 = 0$, $x_4 = 500$

5. $I_1 = \frac{13}{5}A$, $I_2 = -\frac{2}{5}A$, $I_3 = \frac{11}{5}A$ 7. $I_1 = I_4 = I_5 = I_6 = \frac{1}{2}A$, $I_2 = I_3 = 0A$

9. $x_1 = 1$, $x_2 = 5$, $x_3 = 3$, and $x_4 = 4$; the balanced equation is $C_3H_8 + 5O_2 \rightarrow 3CO_2 + 4H_2O$

11. $x_1 = x_2 = x_3 = x_4 = t$; the balanced equation is $CH_3COF + H_2O \rightarrow CH_3COOH + HF$

13. $y = 2x^2 + x - 1$ 15. $p(x) = 1 + \frac{13}{6}x - \frac{1}{6}x^3$

17. **(a)** Using $a_1 = k$ as a parameter, $p(x) = 1 + kx + (1-k)x^2$ where $- < k <$.
    (b) The graphs for k = 0, 1, 2, and 3 are shown.

True/False 1.8

(a) True (b) False (c) True (d) False (e) False

Exercise Set 1.9 (page 90)

1. **(a)** $C = \begin{bmatrix} .1 & .2 \\ .3 & .1 \end{bmatrix}$ **(b)** $\begin{bmatrix} \$12{,}400 \\ \$10{,}800 \end{bmatrix}$ 3. **(a)** $C = \begin{bmatrix} .2 & .6 & .5 \\ .4 & .2 & .2 \\ .2 & .1 & .2 \end{bmatrix}$ **(b)** $\begin{pmatrix} 1 & 0 & 0 & \$35{,}277.80 \\ 0 & 1 & 0 & \$25{,}000.00 \\ 0 & 0 & 1 & \$14{,}444.40 \end{pmatrix}$

5. $\begin{bmatrix} 123.08 \\ 202.56 \end{bmatrix}$

True/False 1.9

(a) False (b) True (c) False (d) True (e) True

Chapter 1 Supplementary Exercises (page 91)

1. $\begin{aligned} 3x_1 - x_2 + x_4 &= 1 \\ 2x_1 + 3x_3 + 3x_4 &= -1 \end{aligned}$
   $x_1 = -\frac{3}{2}s - \frac{3}{2}t - \frac{1}{2},\ x_2 = -\frac{9}{2}s - \frac{1}{2}t - \frac{5}{2},\ x_3 = s,\ x_4 = t$

3. $\begin{aligned} 2x_1 - 4x_2 + x_3 &= 6 \\ -4x_1 + 3x_3 &= -1 \\ x_2 - x_3 &= 3 \end{aligned}$
   $x_1 = -\frac{17}{2},\ x_2 = -\frac{26}{3},\ x_3 = -\frac{35}{3}$

5. $x' = \frac{3}{5}x + \frac{4}{5}y,\ y' = -\frac{4}{5}x + \frac{3}{5}y$ 7. $x = 4,\ y = 2,\ z = 3$

9. (a) $a \neq 0,\ b \neq 2$ (b) $a \neq 0,\ b = 2$ (c) $a = 0,\ b = 2$ (d) $a = 0,\ b \neq 2$

11. $K = \begin{bmatrix} 0 & 2 \\ 1 & 1 \end{bmatrix}$ 13. (a) $X = \begin{bmatrix} -1 & 3 & -1 \\ 6 & 0 & 1 \end{bmatrix}$ (b) $X = \begin{bmatrix} 1 & -2 \\ 3 & 1 \end{bmatrix}$ (c) $X = \begin{bmatrix} -\frac{113}{37} & -\frac{160}{37} \\ -\frac{20}{37} & -\frac{46}{37} \end{bmatrix}$

15. $a = 1,\ b = -2,\ c = 3$

Exercise Set 2.1 (page 98)

1. $M_{11} = 29,\ C_{11} = 29$
   $M_{12} = 21,\ C_{12} = -21$
   $M_{13} = 27,\ C_{13} = 27$
   $M_{21} = -11,\ C_{21} = 11$
   $M_{22} = 13,\ C_{22} = 13$
   $M_{23} = -5,\ C_{23} = 5$
   $M_{31} = -19,\ C_{31} = -19$
   $M_{32} = -19,\ C_{32} = 19$
   $M_{33} = 19,\ C_{33} = 19$

3. (a) $M_{11} = -27,\ C_{11} = -27$
   (b) $M_{32} = 24,$ and $C_{32} = -24$
   (c) $M_{12} = -108$ and $C_{12} = 108$
   (d) $M_{43} = -24$ and $C_{43} = 24$

5. $= 4;\ \frac{1}{4}\begin{pmatrix} 3 & -4 \\ -2 & 4 \end{pmatrix}$. 7. $59;\ \begin{bmatrix} -\frac{2}{59} & -\frac{7}{59} \\ \frac{7}{59} & -\frac{5}{59} \end{bmatrix}$ 9. $a^2 - 5a + 21$ 11. 40 13. -123 15. $\lambda = -4$ or $\lambda = -2$

17. $\lambda = \pm\sqrt{8} = \pm 2\sqrt{2}$ 19. (all parts) -123 21. -40 23. 0 25. -248

27. -1 29. 0 31. 6 33. The determinant is $\sin^2\theta + \cos^2\theta = 1$. 35. $d_2 = d_1 + \lambda$

True/False 2.1

(a) False (b) False (c) True (d) True (e) True (f) True (g) False (h) False (i) False (j) True

---

Exercise Set 2.2 (page 105)

5. 1 7. -2 9. 1 11. 5 13. 33

15. 6 17. -2 19. Exercise 14: 39; Exercise 15: 6; Exercise 16: $-\frac{1}{6}$; Exercise 17: -2 21. -6

23. 60 25. -6 27. -36

True/False 2.2

(a) True (b) True (c) False (d) False (e) True (f) True

Exercise Set 2.3 (page 115)

7. invertible
9. invertible 11. Not invertible 13. Invertible 15. $k \neq \frac{5 \pm \sqrt{17}}{2}$ 17. $k \neq -1$

19. $A^{-1} = \begin{bmatrix} 3 & -5 & -5 \\ -3 & 4 & 5 \\ 2 & -2 & -3 \end{bmatrix}$ 21. $A^{-1} = \begin{bmatrix} \frac{1}{2} & \frac{3}{2} & 1 \\ 0 & 1 & \frac{3}{2} \\ 0 & 0 & \frac{1}{2} \end{bmatrix}$ 23. $A^{-1} = \begin{bmatrix} -4 & 3 & 0 & -1 \\ 2 & -1 & 0 & 0 \\ -7 & 0 & -1 & 8 \\ 6 & 0 & 1 & -7 \end{bmatrix}$

25. $(-1, 2, 3)$ 27. $x_1 = -\frac{30}{11}$, $x_2 = -\frac{38}{11}$, $x_3 = -\frac{40}{11}$ 29. Cramer's rule does not apply. 31. $y = 0$

35. **(a)** 320 **(b)** $\frac{-1}{5}$ **(c)** $\frac{-27}{5}$ **(d)** $\frac{-1}{135}$ **(e)** 5

37. **(a)** 189 **(b)** $\frac{1}{7}$ **(c)** $\frac{8}{7}$ **(d)** $\frac{1}{56}$

True/False 2.3

(a) False **(b)** False **(c)** True **(d)** False **(e)** True **(f)** True **(g)** True **(h)** True **(i)** True **(j)** True
(k) True **(l)** False

Chapter 2 Supplementary Exercises (page 117)

1. -18 3. 24 5. -10 7. 329 9. Exercise 3: 24; Exercise 4: 0; Exercise 5: -10; Exercise 6: -48
11. The matrices in Exercises 1–3 are invertible, the matrix in Exercise 4 is not.

13. $b^2 - 5b + 21$ 15. -120 17. $\begin{bmatrix} -\frac{1}{6} & \frac{1}{9} \\ \frac{1}{6} & \frac{2}{9} \end{bmatrix}$ 19. $\begin{bmatrix} \frac{1}{8} & -\frac{1}{8} & -\frac{3}{8} \\ \frac{1}{8} & \frac{5}{24} & -\frac{1}{24} \\ \frac{1}{4} & -\frac{7}{12} & -\frac{1}{12} \end{bmatrix}$ 21. $\begin{bmatrix} \frac{1}{5} & \frac{2}{5} & -\frac{1}{10} \\ \frac{1}{5} & -\frac{3}{5} & \frac{2}{5} \\ -\frac{2}{5} & \frac{6}{5} & -\frac{3}{10} \end{bmatrix}$

23. $\begin{bmatrix} \frac{10}{329} & -\frac{2}{329} & \frac{52}{329} & -\frac{27}{329} \\ \frac{55}{329} & -\frac{11}{329} & \frac{43}{329} & \frac{16}{329} \\ -\frac{3}{47} & \frac{10}{47} & -\frac{25}{47} & \frac{6}{47} \\ -\frac{31}{329} & \frac{72}{329} & \frac{102}{329} & -\frac{15}{329} \end{bmatrix}$ 25. $x' = \frac{3}{5}x + \frac{4}{5}y$, $y' = -\frac{4}{5}x + \frac{3}{5}y$ 29. **(b)** $\cos \beta = \frac{c^2 + a^2 - b^2}{2ac}$, $\cos \gamma = \frac{a^2 + b^2 - c^2}{2ab}$

Exercise Set 3.1 (page 128)

1. **(a)** **(b)** z $(-3, 4, 5)$ **(c)** **(d)** **(e)** $(-3, -4, 5)$ **(f)**

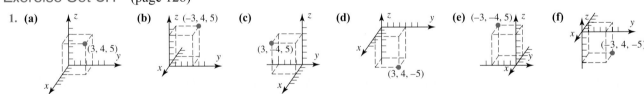

3. **(a)** $v_1 = (2, 5)$ **(b)** $v_2 = (-3, 2)$ **(c)** $v_3 = -4, -3)$

5. (a) and (b)

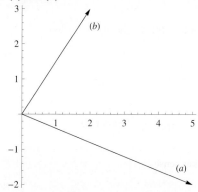

7. (a) $\overrightarrow{P_1P_2} = (-1, 3)$ (b) $\overrightarrow{P_1P_2} = (-3, 6, 1)$

9. (a) terminal point $(8, 4)$.
(b) terminal point $(4, 1, 2)$.

11. (a) $\mathbf{u} = (-1, 2, -4)$ is one possible answer. (b) $\mathbf{u} = (7, -2, -6)$ is one possible answer.

13. (a) $\mathbf{u} + \mathbf{w} = (1, 2)$
(b) $\mathbf{v} - 3\mathbf{u} = (-8, 6)$
(c) $2(\mathbf{u} - 5\mathbf{w}) = (26, -44)$
(d) $3\mathbf{v} - 2(\mathbf{u} + 2\mathbf{w}) = (5, -12)$
(e) $-3(\mathbf{w} - 2\mathbf{u} + \mathbf{v}) = (21, -24)$
(f) $(-2\mathbf{u} - \mathbf{v}) - 5(\mathbf{v} + 3\mathbf{w}) = (18, -56)$

15. (a) $(-1, 9, -11, 1)$ (b) $(22, 53, -19, 14)$ (c) $(-13, 13, -36, -2)$ (d) $(-90, -114, 60, -36)$ (e) $(-9, -5, -5, -3)$
(f) $(27, 29, -27, 9)$

17. (a) $\mathbf{u} - \mathbf{v} = (6, 0, -7, 1, -3)$
(b) $3\mathbf{v} - 2\mathbf{w} = (5, -7, 27, 16, -4)$
(c) $5(\mathbf{u} + 2\mathbf{w}) - 3\mathbf{v} = (-12, 18, -51, -41, 5)$

19. (a) $\mathbf{u} - \mathbf{v} = (-7, 1, 10, 3, 2)$
(b) $2\mathbf{v} + 3\mathbf{w} = (26, -3, -28, 11, -11)$
(c) $(3\mathbf{u} + 4\mathbf{v}) - (7\mathbf{w} + 3\mathbf{u}) = 4\mathbf{v} - 7\mathbf{w} = (-26, 7, -4, -17, 43)$

21. $\mathbf{x} = \left(-\frac{8}{5}, \frac{3}{5}, 0, \frac{6}{5}, 3\right)$

23. (a) Not parallel (b) Parallel (c) Parallel

25. $a = 1, b = -2$ **27.** $c_1 = 2, c_2 = -1, c_3 = 5$ **29.** $c_1 = 2, c_2 = -1, c_3 = 2, c_4 = 3$

33. (a) $\left(\frac{9}{2}, -\frac{1}{2}, -\frac{1}{2}\right)$ (b) $\left(\frac{23}{4}, -\frac{9}{4}, \frac{1}{4}\right)$

True/False 3.1

(a) False (b) False (c) False (d) True (e) True (f) False (g) False (h) True (i) False
(j) True (k) False

Exercise Set 3.2 (page 141)

1. (a) $\|\mathbf{v}\| = \sqrt{34}$, $\frac{1}{\|\mathbf{v}\|}\mathbf{v} = \left(\frac{3}{\sqrt{34}}, \frac{-5}{\sqrt{34}}\right), \left(\frac{-3}{\sqrt{34}}, \frac{5}{\sqrt{34}}\right)$.
(b) $\|\mathbf{v}\| = \sqrt{19}$. $\frac{1}{\|\mathbf{v}\|}\mathbf{v} = \left(\frac{3}{\sqrt{19}}, \frac{3}{\sqrt{19}}, \frac{1}{\sqrt{19}}\right), \left(\frac{-3}{\sqrt{19}}, \frac{-3}{\sqrt{19}}, \frac{-1}{\sqrt{19}}\right)$.
(c) $\|\mathbf{v}\| = \sqrt{42}$. $\frac{1}{\|\mathbf{v}\|}\mathbf{v} = \left(0, \frac{1}{\sqrt{42}}, -\frac{1}{\sqrt{42}}, \frac{2}{\sqrt{42}}, \frac{6}{\sqrt{42}}\right), \left(0, \frac{-1}{\sqrt{42}}, \frac{1}{\sqrt{42}}, \frac{-2}{\sqrt{42}}, \frac{-6}{\sqrt{42}}\right)$.

3. (a) $\|\mathbf{u} - \mathbf{v}\| = \sqrt{3}$
(b) $\|\mathbf{u}\| - \|\mathbf{v}\| = \sqrt{26}$

(c) $\|3\mathbf{u} + 3\mathbf{w}\| = \sqrt{378}$
(d) $\|2\mathbf{u} - 4\mathbf{v} + \mathbf{w}\| = \sqrt{529}$.

5. (a) $\|3\mathbf{u} - 5\mathbf{v} + \mathbf{w}\| = \sqrt{2570}$ (b) $\|3\mathbf{u}\| - 5\|\mathbf{v}\| + \|\mathbf{w}\| = 3\sqrt{46} - 10\sqrt{21} + \sqrt{42}$ (c) $\|-\|\mathbf{u}\|\mathbf{v}\| = 2\sqrt{966}$

7. $k = 2, k = -2$

9. (a) $\mathbf{u} \cdot \mathbf{v} = -18, \mathbf{u} \cdot \mathbf{u} = 43$
   (b) $\mathbf{u} \cdot \mathbf{v} = 9, \mathbf{u} \cdot \mathbf{u} = 11$

11. (a) $\|\mathbf{u} - \mathbf{v}\| = \sqrt{14}$ (b) $\|\mathbf{u} - \mathbf{v}\| = \sqrt{59}$ (c) $\|\mathbf{u} - \mathbf{v}\| = \sqrt{677}$

13. (a) $\cos\theta = \frac{15}{\sqrt{27}\sqrt{17}}$; θ is acute (b) $\cos\theta = -\frac{4}{\sqrt{6}\sqrt{45}}$; θ is obtuse (c) $\cos\theta = -\frac{136}{\sqrt{225}\sqrt{180}}$; θ is obtuse

15. $\mathbf{a} \cdot \mathbf{b} = 6$

17. (a) $\mathbf{u} \cdot (\mathbf{v} \cdot \mathbf{w})$ does not make sense because $\mathbf{v} \cdot \mathbf{w}$ is a scalar. (b) $\mathbf{u} \cdot (\mathbf{v} + \mathbf{w})$ makes sense.
    (c) $\|\mathbf{u} \cdot \mathbf{v}\|$ does not make sense because the quantity inside the norm is a scalar.
    (d) $(\mathbf{u} \cdot \mathbf{v}) - \|\mathbf{u}\|$ makes sense since the terms are both scalars.

19. (a) $\left(\frac{3}{5}, \frac{-4}{5}\right)$
    (b) $\left(\frac{1}{\sqrt{2}}, \frac{1}{\sqrt{2}}\right)$
    (c) $\left(\frac{3}{7}, \frac{6}{7}, \frac{-2}{7}\right)$
    (d) $\left(\frac{1}{\sqrt{39}}, \frac{3}{\sqrt{39}}, \frac{5}{\sqrt{39}}, \frac{-2}{\sqrt{39}}\right)$

23. (a) $\cos\theta = 60°$
    (b) $\cos\theta = 46.91°$
    (c) $\cos\theta = 117.81°$
    (d) $\cos\theta = 105.48°$

25. (a) $|\mathbf{u} \cdot \mathbf{v}| = 11, \|\mathbf{u}\|\|\mathbf{v}\| = 31.02, |\mathbf{u} \cdot \mathbf{v}| = 11 < \|\mathbf{u}\|\|\mathbf{v}\| = 31.02$
    (b) $|\mathbf{u} \cdot \mathbf{v}| = 0 < \|\mathbf{u}\|\|\mathbf{v}\| = \sqrt{42}\sqrt{27}$.

27. A sphere of radius 1 centered at (x_0, y_0, z_0).

True/False 3.2
(a) True   (b) True   (c) False   (d) True   (e) True   (f) False   (g) False   (h) False   (i) True   (j) True

Exercise Set 3.3 (page 150)

1. (a) orthogonal
   (b) not orthogonal
   (c) orthogonal
   (d) not orthogonal

3. (a) orthogonal set
   (b) not orthogonal set
   (c) orthogonal set
   (d) not orthogonal set

5. $\left(\frac{1}{\sqrt{3}}, \frac{1}{\sqrt{3}}, \frac{-1}{\sqrt{3}}\right)$ 7. Yes   9. $(x - 2) - (y - 3) + 2(z + 4) = 0$ 11. $2(x - 1) = 0$ 13. Not parallel

15. Parallel   17. Perpendicular   19. (a) $\frac{6}{\sqrt{2}}$ (b) $\frac{2}{\sqrt{5}}$

21. $(0, 0), (6, 2)$ 23. $\mathbf{u} - \text{proj}_\mathbf{a}\mathbf{u} = \mathbf{u} = (3, -1, 2)$.

25. $\left(0, \frac{2}{5}, -\frac{1}{5}\right)$, $\left(1, \frac{3}{5}, \frac{6}{5}\right)$ 27. $\left(\frac{1}{5}, -\frac{1}{5}, \frac{1}{10}, -\frac{1}{10}\right)$, $\left(\frac{9}{5}, \frac{6}{5}, \frac{9}{10}, \frac{21}{10}\right)$ 29. 1   31. $\frac{1}{\sqrt{17}}$

33. $\frac{3}{\sqrt{14}}$ 35. $\frac{1}{\sqrt{29}}$ 37. $\frac{11}{\sqrt{6}}$ 39. 0 (The planes coincide.)   41. (b) $\cos\beta = \frac{b}{\|\mathbf{v}\|}$, $\cos\gamma = \frac{c}{\|\mathbf{v}\|}$

True/False 3.3
(a) True   (b) True   (c) True   (d) True   (e) True   (f) False   (g) False

Exercise Set 3.4 (page 159)

1. Vector equation: $(3, 2) + t(-1, 0)$, parametric equation: $x = 3 - t, y = 2$
3. Vector equation: $(x, y, z) = t(-3, 0, 1)$;
   parametric equations: $x = -3t, y = 0, z = t$
5. point $(-2, 3)$; parallel to the vector $(4, -1)$
7. point $(2, -2)$; parallel to the vector $(-1, 6)$
9. $\mathbf{x} = (1, -2, 0) + t(0, 2, 4) + s(-1, 3, 2)$.
11. Vector equation: $(x, y, z) = (-1, 1, 4) + t_1(6, -1, 0) + t_2(-1, 3, 1)$;
    parametric equations: $x = -1 + 6t_1 - t_2, y = 1 - t_1 + 3t_2, z = 4 + t_2$
13. Vector equation: $\mathbf{x} = (0, 0) + t(1, 3) = t(1, 3)$; parametric equations: $x = t, y = 3t$
15. A possible answer is vector equation: $(x, y, z) = t_1(0, 1, 0) + t_2(5, 0, 4)$;
    parametric equations: $x + 5t_2, y = t_1, z = 4t_2$
17. $(2, 1, -1) \cdot (0, t, t) = 0, (4, 2, -2) \cdot (0, t, t) = 0$, and $(1, 3, -3) \cdot (0, t, t) = 0$.
19. $x_1 = \frac{3}{7}r - \frac{19}{4}s - \frac{8}{7}t, \; x_2 = -\frac{2}{7}r + \frac{1}{7}s + \frac{3}{7}t, \; x_3 = r, \; x_4 = s, \; x_5 = t$
21. (a) $(1, 0, 0) + s(-1, 1, 0) + t(-1, 0, 1)$ (b) a plane in R^3 passing through $P(1, 0, 0)$ and parallel to $(-1, 1, 0)$ and $(-1, 0, 1)$
23. (a) $\begin{aligned} x + y + z &= 0 \\ -2x + 3y &= 0 \end{aligned}$ (b) a line through the origin in R^3 (c) $x = -\frac{3}{5}t, \; y = -\frac{2}{5}t, \; z = t$
25. (a) $x_1 = \frac{-1}{2}s + \frac{3}{2}t, x_2 = s, x_3 = t$
    (b) $\begin{bmatrix} 2 & 1 & -3 \\ 6 & 3 & -9 \\ -2 & -1 & 3 \end{bmatrix} \begin{bmatrix} 1 \\ -2 \\ 1 \end{bmatrix} = \begin{bmatrix} -3 \\ -9 \\ 3 \end{bmatrix}$
    (c) $\begin{bmatrix} 1 \\ -2 \\ 1 \end{bmatrix} + \begin{pmatrix} \frac{-1}{2} \\ 1 \\ 0 \end{pmatrix} s + \begin{bmatrix} \frac{3}{2} \\ 0 \\ 1 \end{bmatrix} t$.
    (d) $x_1 = \frac{-3}{2} + \frac{-1}{2}s + \frac{3}{2}t, x_2 = s, x_3 = t$
27. $x_1 = -\frac{3}{4}t, x_2 = t, x_3 = 0, x_4 = 0$; The general solution to the non-homogeneous system is $x = \begin{bmatrix} 1 \\ 0 \\ -2 \\ 1 \end{bmatrix} + \begin{bmatrix} \frac{3}{4} \\ 1 \\ 0 \\ 0 \end{bmatrix} t$. A particular solution to the non-homogeneous system is $x = \begin{bmatrix} 1 \\ 0 \\ -2 \\ 1 \end{bmatrix}$.

True/False 3.4

(a) True (b) False (c) True (d) True (e) False (f) True

Exercise Set 3.5 (page 168)

1. (a) $(-4, 9, 6)$ (b) $(-27, -40, 42)$ (c) $(-41, -60, -40)$
3. $(10, -10, -10)$ 5. $(6, -2, -2)$ 7. $= 2\sqrt{131}$ 9. $\sqrt{101}$ 11. 2 13. $\frac{9}{2}$ 15. $\frac{\sqrt{374}}{2}$ 17. 16
19. Vectors do not lie in the same plane 21. 8 23. abc 25. (a) 4 (b) -4 (c) -4
27. (a) $\frac{\sqrt{26}}{2}$ (b) $\frac{\sqrt{26}}{3}$ 29. $2(\mathbf{v} \times \mathbf{u})$ 37. (a) $\frac{17}{6}$ (b) $\frac{1}{2}$

True/False 3.5

(a) True (b) True (c) False (d) True (e) False (f) False

Chapter 3 Supplementary Exercises (page 170)

1. (a) $3\mathbf{v} - 2\mathbf{u} = (13, -3, 10)$ (b) $\|\mathbf{u} + \mathbf{v} + \mathbf{w}\| = \sqrt{70}$ (c) $\sqrt{774}$ (d) $\text{proj}_\mathbf{w}\mathbf{u} = -\frac{12}{27}(2, -5, -5)$
   (e) $\mathbf{u} \cdot (\mathbf{v} \times \mathbf{w}) = -122$ (f) $(-5\mathbf{v} + \mathbf{w}) \times ((\mathbf{u} \cdot \mathbf{v})\mathbf{w}) = (-3150, -2430, 1170)$

3. (a) $3\mathbf{v} - 2\mathbf{u} = (-5, -12, 20, -2)$ (b) $\|\mathbf{u} + \mathbf{v} + \mathbf{w}\| = \sqrt{106}$ (c) $\sqrt{2810}$ (d) $\text{proj}_\mathbf{w}\mathbf{u} = -\frac{15}{77}(9, 1, -6, -6)$

5. Not an orthogonal set

7. (a) A line through the origin, perpendicular to the given vector.
   (b) A plane through the origin, perpendicular to the given vector.   (c) $\{\mathbf{0}\}$ (the origin)
   (d) A line through the origin, perpendicular to the plane containing the two noncollinear vectors.

9. True   11. $S(-1, -1, 5)$ 13. $\sqrt{\frac{14}{17}}$ 15. $\frac{11}{\sqrt{35}}$

17. Vector equation: $(x, y, z) = (-2, 1, 3) + t_1(1, -2, -2) + t_2(5, -1, -5)$;
    parametric equations: $x = -2 + t_1 + 5t_2$, $y = 1 - 2t_1 - t_2$, $z = 3 - 2t_1 - 5t_2$

19. Vector equation: $(x, y) = (0, -3) + t(8, -1)$;
    parametric equations: $x = 8t$, $y = -3 - t$

21. A possible answer is vector equation: $(x, y) = (0, -5) + t(1, 3)$; parametric equations: $x = t$, $y = -5 + 3t$

23. $3(x + 1) + 6(y - 5) + 2(z - 6) = 0$ 25. $-18(x - 9) - 51y - 24(z - 4) = 0$ 29. A plane

Exercise Set 4.1 (page 178)

1. (a) $\mathbf{u} + \mathbf{v} = (3, 1)$ $k\mathbf{u} = (0, 20)$
   (c) Axioms 1–5

3. The set is a vector space with the given operations.

5. Not a vector space, Axioms 5 and 6 fail.   7. Not a vector space. Axiom 8 fails.

9. The set is a vector space with the given operations.   11. The set is not a vector space.

True/False 4.1

(a) False   (b) False   (c) True   (d) False   (e) False

Exercise Set 4.2 (page 188)

1. (a), (d), (e)   3. (a), (b)   5. (a), (c), (d)   7. (a), (c), (d)   9. (a), (c)

11. (a) Vectors span
    (b) Vectors do not span
    (c) Vectors span
    (d) Vectors do not span

13. The polynomials do not span

15. (a) Parametric equations $x_1 = 2s - 6t$, $x_2 = s$, $x_3 = t$. equation of a plane: $x - 2y + 6z = 0$.
    (c) Origin
    (e) Parametric equations $x_1 = 4t$, $x_2 = t$, $x_3 = 0$.

True/False 4.2

(a) True   (b) True   (c) False   (d) False   (e) False   (f) True   (g) True   (h) False   (i) False
(j) True   (k) False

Exercise Set 4.3 (page 199)

1. **(a)** $u_2 = (6, -2) = 2u_1 = 2(3, -1)$
   (b) $3u_1 + 3u_2 = 3(-2, 0, 1) + 3(4, -2, 0) = u_3 = (6, -6, 3)$
   (c) $A = \begin{bmatrix} 0 & 1 \\ 2 & 3 \end{bmatrix} = -B = -\begin{bmatrix} 0 & -1 \\ -2 & -3 \end{bmatrix}$
   (d) $-2p_1 = -2(2 + x - 3x^2) = p_2 = -4 - 2x + 6x^2$

3. (a), (c)

5. **(a)** Do not lie in the same plane
   (b) Lie in the same plane

7. **(a)** $2v_1 - 2v_2 + v_3 = 2(2, 0, -2, 1) - 2(3, 1, -5, 0) + (2, 2, -6, -2) = 0$,
   (b) $-2v_1 + 2v_2 = v_3$, $v_1 = v_2 - \frac{1}{2}v_3$, $v_1 + \frac{1}{2}v_3 = v_2$

9. $\lambda = -\frac{1}{2}$, $\lambda = 1$

19. **(a)** They are linearly independent since v_1, v_2, and v_3 do not lie in the same plane when they are placed with their initial points at the origin.
    (b) They are not linearly independent since v_1, v_2, and v_3 line in the same plane when they are placed with their initial points at the origin.

21. $W(x) = -x \sin x - \cos x \neq 0$ for some x. 23. **(a)** $W(x) = e^x \neq 0$ **(b)** $W(x) = 2 \neq 0$
25. $W(x) = 2 \sin x \neq 0$ for some x.

True/False 4.3
(a) False **(b)** True **(c)** False **(d)** True **(e)** True **(f)** False **(g)** True **(h)** False

Exercise Set 4.4 (page 207)

1. **(a)** $u_1 - u_2 = (3, 2, 1) - (-2, 1, 0) = (5, 1, 1) = u_3$; vectors do not form a linearly independent set
   (b) Three vectors in R^2 cannot be linearly independent
   (c) Vectors do not span P_2. no scalars c_1, c_2
   (d) The vectors do not span M_{22}

3. (a), (b)   7. **(a)** $w_s = (-3, 5)$ **(b)** $w_s = \left(\frac{-1}{5}, \frac{8}{5}\right)$ 9. **(a)** $v_S = (3, -2, -2)$ 11. $A_S = (3, 0, 0, 1)$
13. $A = A_1 - A_2 + A_3 - A_4$
15. $p = 7p_1 - 8p_2 + 3p_3$ 17. **(a)** $(2, 0)$ **(b)** $\left(\frac{2}{\sqrt{3}}, -\frac{1}{\sqrt{3}}\right)$ **(c)** $(0, 1)$ **(d)** $\left(\frac{2}{\sqrt{3}}a, b - \frac{a}{\sqrt{3}}\right)$

True/False 4.4
(a) False **(b)** False **(c)** True **(d)** True **(e)** False

Exercise Set 4.5 (page 216)

1. Basis is the empty set dimension is 0

3. Basis is $\left\{ \begin{bmatrix} \frac{1}{3} \\ 1 \\ 0 \\ 0 \end{bmatrix}, \begin{bmatrix} \frac{1}{6} \\ 0 \\ \frac{-1}{4} \\ 1 \end{bmatrix} \right\}$; dimension is 2

5. No basis; dimension = 0
7. **(a)** $\{(3, 0, 2), (-2, 1, 0)\}$ **(b)** $\{(7, 0, 0), (0, 1, -1)\}$ **(c)** $\{(4, 2, -1)\}$ **(d)** $\{(1, 0, 1), (0, 1, -1)\}$
9. **(a)** n **(b)** $\frac{n(n+1)}{2}$ **(c)** $\frac{n(n+1)}{2}$
13. Any two of $(0, 1, 0, 0)$, $(0, 0, 1, 0)$, and $(0, 0, 0, 1)$ can be used.   15. $v_3 = (a, b, c)$ with $9a - 3b - 5c \neq 0$

True/False 4.5

(a) True (b) True (c) False (d) True (e) True (f) True (g) True (h) True (i) True (j) False

Exercise Set 4.6 (page 222)

1. (a) $\mathbf{w}_S = (-4, 3)$ (b) $\mathbf{w}_S = (\frac{1}{7}, \frac{10}{7})$ (c) $\mathbf{w}_S = (\frac{1}{5}(a+2b), \frac{1}{5}(b-2a))$

3. (a) $\mathbf{p}_S = (1, -1, 3)$ (b) $\mathbf{p}_S = (0, 1, 2)$

5. (a) $w = (16, 10, 12)$ (b) $q = 15 + 4x^2$ (c) $B = \begin{bmatrix} 15 & -1 \\ 6 & 3 \end{bmatrix}$

7. (a) $P_{B' \to B} = \begin{pmatrix} 5 & -7 \\ -7 & 10 \end{pmatrix}$ (b) $P_{B \to B'} = \begin{pmatrix} 10 & 7 \\ 7 & 5 \end{pmatrix}$ (c) $\mathbf{w}_{B'} = \begin{pmatrix} 19 \\ 11 \end{pmatrix}$ (d) $\mathbf{w}_{B'} = \begin{pmatrix} 19 \\ 11 \end{pmatrix}$

9. (a) $\begin{bmatrix} 3 & 2 & \frac{5}{2} \\ -2 & -3 & -\frac{1}{2} \\ 5 & 1 & 6 \end{bmatrix}$ (b) $[\mathbf{w}]_B = \begin{bmatrix} 9 \\ -9 \\ 5 \end{bmatrix}$, $[\mathbf{w}]_{B'} = \begin{bmatrix} -\frac{7}{2} \\ \frac{23}{2} \\ 6 \end{bmatrix}$

11. (b) $\begin{bmatrix} 2 & 0 \\ 1 & 3 \end{bmatrix}$ (c) $\begin{bmatrix} \frac{1}{2} & 0 \\ -\frac{1}{6} & \frac{1}{3} \end{bmatrix}$ (d) $[\mathbf{h}]_B = \begin{bmatrix} 2 \\ -5 \end{bmatrix}$, $[\mathbf{h}]_{B'} = \begin{bmatrix} 1 \\ -2 \end{bmatrix}$

13. (a) $\begin{bmatrix} 1 & 2 & 3 \\ 2 & 5 & 3 \\ 1 & 0 & 8 \end{bmatrix}$ (b) $\begin{bmatrix} -40 & 16 & 9 \\ 13 & -5 & -3 \\ 5 & -2 & -1 \end{bmatrix}$ (d) $[\mathbf{w}]_B = \begin{bmatrix} -239 \\ 77 \\ 30 \end{bmatrix}$, $[\mathbf{w}]_S = \begin{bmatrix} 5 \\ -3 \\ 1 \end{bmatrix}$

(e) $[\mathbf{w}]_S = \begin{bmatrix} 3 \\ -5 \\ 0 \end{bmatrix}$, $[\mathbf{w}]_B = \begin{bmatrix} -200 \\ 64 \\ 25 \end{bmatrix}$

15. (a) $\begin{bmatrix} 3 & 5 \\ -1 & -2 \end{bmatrix}$ (b) $\begin{bmatrix} 2 & 5 \\ -1 & -3 \end{bmatrix}$ (d) $\mathbf{w}_{B_2} = \begin{pmatrix} 2 & 5 \\ -1 & -3 \end{pmatrix}\begin{pmatrix} -3 \\ 2 \end{pmatrix} = \begin{pmatrix} 4 \\ -3 \end{pmatrix}$

(e) $\mathbf{w}_{B_1} = \begin{pmatrix} 3 & 5 \\ -1 & -2 \end{pmatrix}\begin{pmatrix} 15 \\ -12 \end{pmatrix} = \begin{pmatrix} -15 \\ 9 \end{pmatrix}$

17. (a) $\begin{bmatrix} 3 & 2 & \frac{5}{2} \\ -2 & -3 & -\frac{1}{2} \\ 5 & 1 & 6 \end{bmatrix}$ (b) $[\mathbf{w}]_{B_1} = \begin{bmatrix} 9 \\ -9 \\ -5 \end{bmatrix}$, $[\mathbf{w}]_{B_2} = \begin{bmatrix} -\frac{7}{2} \\ \frac{23}{2} \\ 6 \end{bmatrix}$ 19. (a) $\begin{bmatrix} \cos 2\theta & \sin 2\theta \\ \sin 2\theta & -\cos 2\theta \end{bmatrix}$

23. (a) $B = \{(1, 1, 0), (1, 0, 2), (0, 2, 1)\}$ (b) $B = \{(\frac{4}{5}, \frac{1}{5}, -\frac{2}{5}), (\frac{1}{5}, -\frac{1}{5}, \frac{2}{5}), (-\frac{2}{5}, \frac{2}{5}, \frac{1}{5})\}$

True/False 4.6

(a) True (b) True (c) True (d) True (e) False (f) False

Exercise Set 4.7 (page 235)

1. $\mathbf{r}_1 = (2, -1, 0, 1)$, $\mathbf{r}_2 = (3, 5, 7, -1)$, $\mathbf{r}_3 = (1, 4, 2, 7)$;

$\mathbf{c}_1 = \begin{bmatrix} 2 \\ 3 \\ 1 \end{bmatrix}$, $\mathbf{c}_2 = \begin{bmatrix} -1 \\ 5 \\ 4 \end{bmatrix}$, $\mathbf{c}_3 = \begin{bmatrix} 0 \\ 7 \\ 2 \end{bmatrix}$, $\mathbf{c}_4 = \begin{bmatrix} 1 \\ -1 \\ 7 \end{bmatrix}$

3. (a) $b = \begin{pmatrix} 1 \\ 0 \end{pmatrix} = \frac{5}{26}\begin{pmatrix} 5 \\ -1 \end{pmatrix} + \frac{1}{26}\begin{pmatrix} 1 \\ 5 \end{pmatrix}$

(b) b is not in the column space of A (c) $\begin{bmatrix} 1 \\ 9 \\ 1 \end{bmatrix} - 3\begin{bmatrix} -1 \\ 3 \\ 1 \end{bmatrix} + \begin{bmatrix} 1 \\ 1 \\ 1 \end{bmatrix} = \begin{bmatrix} 5 \\ 1 \\ -1 \end{bmatrix}$

(d) $\begin{bmatrix} 2 \\ 0 \\ 0 \end{bmatrix} = \begin{bmatrix} 1 \\ 1 \\ -1 \end{bmatrix} + (t-1)\begin{bmatrix} -1 \\ 1 \\ -1 \end{bmatrix} + t\begin{bmatrix} 1 \\ -1 \\ 1 \end{bmatrix}$ **(e)** $\begin{bmatrix} 4 \\ 3 \\ 5 \\ 7 \end{bmatrix} = -26\begin{bmatrix} 1 \\ 0 \\ 1 \\ 0 \end{bmatrix} + 13\begin{bmatrix} 2 \\ 1 \\ 2 \\ 1 \end{bmatrix} - 7\begin{bmatrix} 0 \\ 2 \\ 1 \\ 2 \end{bmatrix} + 4\begin{bmatrix} 1 \\ 1 \\ 3 \\ 2 \end{bmatrix}$

5. (a) $\mathbf{x} = t\begin{pmatrix} \frac{-1}{3} \\ 1 \end{pmatrix}$ **(b)** $\begin{bmatrix} -2 \\ 7 \\ 0 \end{bmatrix} + t\begin{bmatrix} -1 \\ -1 \\ 1 \end{bmatrix}; t\begin{bmatrix} -1 \\ -1 \\ 1 \end{bmatrix}$ **(c)** $\mathbf{x} = t\begin{pmatrix} -12 \\ -3 \\ 5 \\ 1 \end{pmatrix}$

7. (a) $\mathbf{r}_1 = [1\ 0\ 2],\ \mathbf{r}_2 = [0\ 0\ 1],\ \mathbf{c}_1 = \begin{bmatrix} 1 \\ 0 \\ 0 \end{bmatrix},\ \mathbf{c}_2 = \begin{bmatrix} 2 \\ 1 \\ 0 \end{bmatrix}$

(b) $\mathbf{r}_1 = [1\ -3\ 0\ 0],\ \mathbf{r}_2 = [0\ 1\ 0\ 0],\ \mathbf{c}_1 = \begin{bmatrix} 1 \\ 0 \\ 0 \\ 0 \end{bmatrix},\ \mathbf{c}_2 = \begin{bmatrix} -3 \\ 1 \\ 0 \\ 0 \end{bmatrix}$

(c) $\mathbf{r}_1 = [1\ 2\ 4\ 5],\ \mathbf{r}_2 = [0\ 1\ -3\ 0],\ \mathbf{r}_3 = [0\ 0\ 1\ -3],\ \mathbf{r}_4 = [0\ 0\ 0\ 1],$

$\mathbf{c}_1 = \begin{bmatrix} 1 \\ 0 \\ 0 \\ 0 \\ 0 \end{bmatrix},\ \mathbf{c}_2 = \begin{bmatrix} 2 \\ 1 \\ 0 \\ 0 \\ 0 \end{bmatrix},\ \mathbf{c}_3 = \begin{bmatrix} 4 \\ -3 \\ 1 \\ 0 \\ 0 \end{bmatrix},\ \mathbf{c}_4 = \begin{bmatrix} 5 \\ 0 \\ -3 \\ 1 \\ 0 \end{bmatrix}$

(d) $\mathbf{r}_1 = [1\ 2\ -1\ 5],\ \mathbf{r}_2 = [0\ 1\ 4\ 3],\ \mathbf{r}_3 = [0\ 0\ 1\ -7],\ \mathbf{r}_4 = [0\ 0\ 0\ 1]$

$\mathbf{c}_1 = \begin{bmatrix} 1 \\ 0 \\ 0 \\ 0 \end{bmatrix},\ \mathbf{c}_2 = \begin{bmatrix} 2 \\ 1 \\ 0 \\ 0 \end{bmatrix},\ \mathbf{c}_3 = \begin{bmatrix} -1 \\ 4 \\ 1 \\ 0 \end{bmatrix},\ \mathbf{c}_4 = \begin{bmatrix} 5 \\ 3 \\ -7 \\ 1 \end{bmatrix}$

9. (a) $\mathbf{r}_1 = [\ 1\ \ 0\ \ 2\ ];\ \mathbf{r}_2 = [\ 0\ \ 0\ \ 1\ ];\ \mathbf{c}_1 = \begin{bmatrix} 1 \\ 0 \\ 0 \end{bmatrix};\ \mathbf{c}_2 = \begin{bmatrix} 2 \\ 1 \\ 0 \end{bmatrix}$

(b) $\mathbf{r}_1 = [\ 1\ \ -3\ \ 0\ \ 0\ ];\ \mathbf{r}_2 = [\ 0\ \ 1\ \ 0\ \ 0\ ];\ \mathbf{c}_1 = \begin{bmatrix} 1 \\ 0 \\ 0 \\ 0 \end{bmatrix};\ \mathbf{c}_2 = \begin{bmatrix} -3 \\ 1 \\ 0 \\ 0 \end{bmatrix}$

(c) $\mathbf{r}_1 = [\ 1\ \ 2\ \ 4\ \ 5\ ];\ \mathbf{r}_2 = [\ 0\ \ 1\ \ -3\ \ 0\ ];\ \mathbf{r}_3 = [\ 0\ \ 0\ \ 1\ \ -3\ ];$

$\mathbf{r}_4 = [\ 0\ \ 0\ \ 0\ \ 1\ ];\ \mathbf{c}_1 = \begin{bmatrix} 1 \\ 0 \\ 0 \\ 0 \\ 0 \end{bmatrix};\ \mathbf{c}_2 = \begin{bmatrix} 2 \\ 1 \\ 0 \\ 0 \\ 0 \end{bmatrix};\ \mathbf{c}_3 = \begin{bmatrix} 4 \\ -3 \\ 1 \\ 0 \\ 0 \end{bmatrix};\ \mathbf{c}_4 = \begin{bmatrix} 5 \\ 0 \\ -3 \\ 1 \\ 0 \end{bmatrix}$

(d) $\mathbf{r}_1 = [\ 1\ \ 2\ \ -1\ \ 5\ ];\ \mathbf{r}_2 = [\ 0\ \ 1\ \ 4\ \ 3\ ];\ \mathbf{r}_3 = [\ 0\ \ 0\ \ 1\ \ -7\ ];$

$\mathbf{r}_4 = [\ 0\ \ 0\ \ 0\ \ 1\ ];\ \mathbf{c}_1 = \begin{bmatrix} 1 \\ 0 \\ 0 \\ 0 \end{bmatrix};\ \mathbf{c}_2 = \begin{bmatrix} 2 \\ 1 \\ 0 \\ 0 \end{bmatrix};\ \mathbf{c}_3 = \begin{bmatrix} -1 \\ 4 \\ 1 \\ 0 \end{bmatrix};\ \mathbf{c}_4 = \begin{bmatrix} 5 \\ 3 \\ -7 \\ 1 \end{bmatrix}$

11. (a) $\{(1, 0, -1, \frac{5}{2}), (0, 1, 0, \frac{-1}{2})\}$ **(b)** $(1, -1, 2, 0), (0, 1, 0, 0), (0, 0, 1, -\frac{1}{6})$
(c) $(1, 1, 0, 0), (0, 1, 1, 1), (0, 0, 1, 1), (0, 0, 0, 1)$

15. (b) $\begin{bmatrix} 0 & 0 & 0 \\ 0 & 1 & 0 \\ 0 & 0 & 1 \end{bmatrix}$

17. **(a)** rows of A must be orthogonal to the vector $\begin{pmatrix} 2t \\ -3t \end{pmatrix}$. multiples of the vector $(3, 2)$.

(b) Since A and B are invertible, their null spaces are the origin. The null space of C is the line $3x + y = 0$. The null space of D is the entire xy-plane.

True/False 4.7
(a) True **(b)** False **(c)** False **(d)** False **(e)** False **(f)** True **(g)** True **(h)** False **(i)** True **(j)** False

Exercise Set 4.8 (page 246)

1. $\text{Rank}(A) = \text{Rank}(A^T) = 2$
3. **(a)** 2; 1 **(b)** 1; 2 **(c)** 2; 2 **(d)** 2; 3 **(e)** 3; 2
5. **(a)** rank is 4; nullity is at least 2
   (b) rank is 5; nullity is at least 0
   (c) rank is 4; nullity is at least 0
7. **(a)** Yes, 0 **(b)** No **(c)** Yes, 2 **(d)** Yes, 7 **(e)** No **(f)** Yes, 4 **(g)** Yes, 0
9. $b_1 = \frac{1}{5}b_4 - \frac{1}{5}b_5$
   $b_2 = \frac{1}{10}b_4 + \frac{2}{5}b_5$
   $b_3 = \frac{2}{5}b_4 - \frac{7}{5}b_5$
11. No
13. Rank is 2 if $r = 2$ and $s = 1$; the rank is never 1.
17. **(a)** 3 **(b)** 5 **(c)** 3 **(d)** 3 19. $A = \begin{bmatrix} 0 & 1 \\ 0 & 0 \end{bmatrix}$; $B = \begin{bmatrix} 1 & 2 \\ 2 & 4 \end{bmatrix}$

True/False 4.8
(a) False **(b)** True **(c)** False **(d)** False **(e)** True **(f)** False **(g)** False **(h)** False **(i)** True **(j)** False

Exercise Set 4.9 (page 260)

1. **(a)** Domain: R^2; codomain: R^3 **(b)** Domain: R^3; codomain: R^2 **(c)** Domain: R^3; codomain: R^3
   (d) Domain: R^6; codomain: R^1
3. R^2, R^3, $(3, 2, -5)$
5. **(a)** R^3, R^2, linear **(b)** R^3, R^2, not linear **(c)** Linear; $R^3 \to R^3$ **(d)** Nonlinear; $R^4 \to R^2$
7. **(a)** matrix transformation
   (b) not transformation
   (c) matrix transformation
   (d) not a matrix transformation
   (e) matrix transformation
9. $[T] = \begin{pmatrix} 4 & -3 & 1 \\ 2 & -1 & 5 \\ 1 & 2 & -2 \end{pmatrix}$, $[T]\begin{pmatrix} -1 \\ 2 \\ 4 \end{pmatrix} = \begin{pmatrix} -6 \\ 16 \\ -5 \end{pmatrix}$

11. **(a)** $\begin{bmatrix} 0 & 1 \\ -1 & 0 \\ 1 & 3 \\ 1 & -1 \end{bmatrix}$ **(b)** $\begin{bmatrix} 7 & 2 & -1 & 1 \\ 0 & 1 & 1 & 0 \\ -1 & 0 & 0 & 0 \end{bmatrix}$ **(c)** $\begin{bmatrix} 0 & 0 & 0 \\ 0 & 0 & 0 \\ 0 & 0 & 0 \\ 0 & 0 & 0 \\ 0 & 0 & 0 \end{bmatrix}$ **(d)** $\begin{bmatrix} 0 & 0 & 0 & 1 \\ 1 & 0 & 0 & 0 \\ 0 & 0 & 1 & 0 \\ 0 & 1 & 0 & 0 \\ 1 & 0 & -1 & 0 \end{bmatrix}$

13. **(a)** $T(-1, 4) = (19, -1)$
    (b) $T(2, 1, -3) = (-3, 0, 3)$

15. (a) $(2, -5, -3)$ (b) $(2, 5, 3)$ (c) $(-2, -5, 3)$

17. (a) $(-2, 1, 0)$ (b) $(-2, 0, 3)$ (c) $(0, 1, 3)$ **19.** (a) $\left(-2, \frac{\sqrt{3}-2}{2}, \frac{1+2\sqrt{3}}{2}\right)$ (b) $\left(0, 1, 2\sqrt{2}\right)$ (c) $(-1, -2, 2)$

21. (a) $\left(-2, \frac{\sqrt{3}+2}{2}, \frac{-1+2\sqrt{3}}{2}\right)$ (b) $\left(-2\sqrt{2}, 1, 0\right)$ (c) $(1, 2, 2)$

25. $\begin{bmatrix} -\frac{1}{9} & \frac{8}{9} & \frac{4}{9} \\ \frac{8}{9} & -\frac{1}{9} & \frac{4}{9} \\ \frac{4}{9} & \frac{4}{9} & -\frac{7}{9} \end{bmatrix}$

29. (a) orthogonal projection onto the y-axis with a dilation by -3
(b) expansion in the y-direction by a factor of 4

31. Rotation through the angle 2θ.

33. Rotation through the angle θ and translation by $\mathbf{x}_0$; not a matrix transformation since $\mathbf{x}_0$ is nonzero. **35.** A line in R^n.

True/False 4.9
(a) False (b) False (c) False (d) True (e) False (f) True (g) False (h) False (i) True

Exercise Set 4.10 (page 271)

1. $T_B \circ T_A = \begin{bmatrix} 5 & -1 & 21 \\ 10 & -8 & 4 \\ 45 & 3 & 25 \end{bmatrix}$, $T_A \circ T_B = \begin{bmatrix} -8 & -3 & 1 \\ -5 & -15 & -8 \\ 44 & -11 & 45 \end{bmatrix}$

3. (a) $T_1 = \begin{bmatrix} 1 & 1 \\ 1 & -1 \end{bmatrix}$, $T_2 = \begin{bmatrix} 3 & 0 \\ 2 & 4 \end{bmatrix}$ (b) $T_2 \circ T_1 = \begin{bmatrix} 3 & 3 \\ 6 & -2 \end{bmatrix}$, $T_1 \circ T_2 = \begin{bmatrix} 5 & 4 \\ 1 & -4 \end{bmatrix}$
(c) $T_2(T_1(x_1, x_2)) = (3x_1 + 3x_2, 6x_1 - 2x_2)$,
$T_1(T_2(x_1, x_2)) = (5x_1 + 4x_2, x_1 - 4x_2)$

5. (a) $\begin{pmatrix} -\frac{\sqrt{3}}{2} & \frac{1}{2} \\ \frac{1}{2} & \frac{\sqrt{3}}{2} \end{pmatrix}$ (b) $\begin{pmatrix} 0 & 0 \\ 0 & \frac{1}{3} \end{pmatrix}$ (c) $\begin{pmatrix} -4 & 0 \\ 0 & 4 \end{pmatrix}$

7. (a) $\begin{bmatrix} -1 & 0 & 0 \\ 0 & 0 & 0 \\ 0 & 0 & 1 \end{bmatrix}$ (b) $\begin{bmatrix} 1 & 0 & 1 \\ 0 & \sqrt{2} & 0 \\ -1 & 0 & 1 \end{bmatrix}$ (c) $\begin{bmatrix} -1 & 0 & 0 \\ 0 & 1 & 0 \\ 0 & 0 & 0 \end{bmatrix}$

9. (a) $T_1 \circ T_2 = T_2 \circ T_1$ (b) $T_1 \circ T_2 = T_2 \circ T_1$ (c) $T_1 \circ T_2 \neq T_2 \circ T_1$

11. (a) Not one-to-one (b) One-to-one (c) One-to-one (d) One-to-one (e) One-to-one (f) One-to-one
(g) One-to-one

13. (a) $\begin{pmatrix} 0 & -1 \\ \frac{1}{2} & 0 \end{pmatrix}$, $T^{-1}\begin{matrix} x_1 = -w_2 \\ x_2 = \frac{1}{2}w_1 \end{matrix}$
(b) 1-1, $\begin{matrix} x_1 = \frac{-1}{73}(-7w_1 - 5w_2) \\ x_2 = \frac{-1}{73}(-2w_1 + 9w_2) \end{matrix}$
(c) One-to-one; $\begin{bmatrix} 0 & -1 \\ -1 & 0 \end{bmatrix}$; $T^{-1}(w_1, w_2) = (-w_2, -w_1)$
(d) Not one-to-one

15. (a) Reflection about the x-axis (b) Rotation through the angle $-\frac{\pi}{4}$ (c) Contraction by a factor of $\frac{1}{3}$
(d) Reflection about the yz-plane (e) Dilation by a factor of 5

17. (a) not a matrix transformation (b) matrix transformation (c) matrix transformation
(d) matrix transformation

19. (a) matrix transformation
(b) matrix transformation

21. (a) $\begin{bmatrix} -1 & 0 \\ 0 & 0 \end{bmatrix}$ (b) $\begin{bmatrix} 0 & 1 \\ -1 & 0 \end{bmatrix}$ (c) $\begin{bmatrix} 0 & 0 \\ 3 & 0 \end{bmatrix}$

23. (a) $T_A(e_1) = \begin{pmatrix} 4 \\ 5 \\ 3 \end{pmatrix}$, $T_A(e_2) = \begin{pmatrix} -1 \\ 1 \\ 6 \end{pmatrix}$, $T_A(e_3) = \begin{pmatrix} 2 \\ 2 \\ -4 \end{pmatrix}$

(b) $T_A(e_1 + e_2 + e_3) = \begin{pmatrix} 5 \\ 8 \\ 5 \end{pmatrix}$

(c) $T_A(7e_3) = \begin{pmatrix} 14 \\ 14 \\ -28 \end{pmatrix}$

25. (a) Yes (b) Yes **27.** (b) $T(x_1, x_2) = (x_1^2 + x_2^2, x_1 x_2)$

29. (a) The range of T is a proper subset of R^n. (b) T must map infinitely many vectors to **0**.

True/False 4.10

(a) False (b) True (c) True (d) False (e) False (f) False

Exercise Set 4.11 (page 280)

1. (a) $\begin{bmatrix} 0 & -1 \\ -1 & 0 \end{bmatrix}$ (b) $\begin{bmatrix} -1 & 0 \\ 0 & -1 \end{bmatrix}$ (c) $\begin{bmatrix} 1 & 0 \\ 0 & 0 \end{bmatrix}$ (d) $\begin{bmatrix} 0 & 0 \\ 0 & 1 \end{bmatrix}$

3. (a) $\begin{bmatrix} 1 & 0 & 0 \\ 0 & 1 & 0 \\ 0 & 0 & -1 \end{bmatrix}$ (b) $\begin{bmatrix} 1 & 0 & 0 \\ 0 & -1 & 0 \\ 0 & 0 & 1 \end{bmatrix}$ (c) $\begin{bmatrix} -1 & 0 & 0 \\ 0 & 1 & 0 \\ 0 & 0 & 1 \end{bmatrix}$

5. (a) $\begin{bmatrix} 0 & -1 & 0 \\ 1 & 0 & 0 \\ 0 & 0 & 1 \end{bmatrix}$ (b) $\begin{bmatrix} 1 & 0 & 0 \\ 0 & 0 & -1 \\ 0 & 1 & 0 \end{bmatrix}$ (c) $\begin{bmatrix} 0 & 0 & 1 \\ 0 & 1 & 0 \\ -1 & 0 & 0 \end{bmatrix}$

7.

9. (a) $\begin{bmatrix} 1 & 0 \\ 4 & 1 \end{bmatrix}$ (b) $\begin{bmatrix} 1 & -2 \\ 0 & 1 \end{bmatrix}$

11. (a) compression in the y-direction by a factor of $\frac{1}{3}$
(b) reflection about the x-axis and expansion by a factor of 6 in the x-direction
(c) reflected the line $y = x$, then expanded by a factor of 4 in the y-direction, sheared in the x-direction by a factor of $\frac{1}{2}$

13. (a) $\begin{bmatrix} \frac{1}{2} & 0 \\ 0 & 5 \end{bmatrix}$ (b) $\begin{bmatrix} 1 & 0 \\ 2 & 5 \end{bmatrix}$ (c) $\begin{bmatrix} 0 & -1 \\ -1 & 0 \end{bmatrix}$

17. (a) $y = \frac{2}{7}x$ (b) $y = x$ (c) $y = \frac{1}{2}x$ (d) $y = -2x$ (e) $y = -\frac{8+5\sqrt{3}}{11}x$

19. (a) Since column space of $A = \begin{bmatrix} 4 & 2 \\ 2 & 1 \end{bmatrix}$ is spanned by vector $\begin{bmatrix} 2 \\ 1 \end{bmatrix}$, multiplication maps every point in the plane onto the line $t\begin{bmatrix} 2 \\ 1 \end{bmatrix}$, which is the same as $y = \frac{1}{2}x$. (b) No

23. **(a)** $\begin{bmatrix} 1 & 0 & k \\ 0 & 1 & k \\ 0 & 0 & 1 \end{bmatrix}$

(b) Shear in the xz-direction with factor k maps (x, y, z) to $(x + ky, y, z + ky)$: $\begin{bmatrix} 1 & k & 0 \\ 0 & 1 & 0 \\ 0 & k & 1 \end{bmatrix}$.

Shear in the yz-direction with factor k maps (x, y, z) to $(x, y + kx, z + kx)$: $\begin{bmatrix} 1 & 0 & 0 \\ k & 1 & 0 \\ k & 0 & 1 \end{bmatrix}$.

True/False 4.11
(a) False   **(b)** True   **(c)** True   **(d)** True   **(e)** False   **(f)** False   **(g)** True

Exercise Set 4.12 (page 290)

1. **(a)** Not a stochastic matrix since the columns do not sum to 1
   (b) Stochastic matrix since the columns (with nonnegative entries) sum to 1
   (c) Stochastic
   (d) Not stochastic

3. $\begin{bmatrix} 0.30925 \\ 0.69075 \end{bmatrix}$ 5. **(a)** Regular   **(b)** Not regular   **(c)** Regular   7. $\begin{bmatrix} \frac{8}{17} \\ \frac{9}{17} \end{bmatrix}$ 9. $\begin{bmatrix} \frac{3}{11} \\ \frac{4}{11} \\ \frac{4}{11} \end{bmatrix}$

11. **(a)** Probability that something in state 1 stays in state 1   **(b)** Probability that something in state 2 moves to state 1
    (c) 0.8   **(d)** 0.85

13. **(a)** $\begin{bmatrix} 0.95 & 0.55 \\ 0.05 & 0.45 \end{bmatrix}$ **(b)** 0.93   **(c)** 0.142   **(d)** 0.63

15. **(a)**

Year	1	2	3	4	5
City	95,750	91,840	88,243	84,933	81,889
Suburbs	29,250	33,160	36,757	40,067	43,111

(b)

City	46,875
Suburbs	78,125

17. **(a)** $\frac{23}{100}$ **(b)** $\begin{bmatrix} \frac{46}{159} \\ \frac{22}{53} \\ \frac{47}{159} \end{bmatrix}$ **(c)** 35, 50, 35

19. $\begin{bmatrix} \frac{1}{3} & \frac{1}{12} & \frac{1}{6} \\ \frac{1}{2} & \frac{1}{6} & \frac{1}{3} \\ \frac{1}{6} & \frac{9}{12} & \frac{1}{2} \end{bmatrix}$ $\mathbf{q} = \begin{bmatrix} \frac{12}{71} \\ \frac{22}{71} \\ \frac{37}{71} \end{bmatrix}$

21. $P^k \mathbf{q} = \mathbf{q}$ for every positive integer k

True/False 4.12
(a) True   **(b)** True   **(c)** True   **(d)** False   **(e)** True

Chapter 4 Supplementary Exercises (page 292)

1. **(a)** $\mathbf{u} + \mathbf{v} = (4, 3, 2)$, $-\mathbf{u} = (-3, 0, 0)$ **(c)** Axioms 1–5
3. If $s \neq 1, -2$, the solution space is the origin. If $s = 1$, the solution space is a plane through the origin. If $s = -2$, the solution space is a line through the origin.
7. A must be invertible
9. **(a)** Rank = 2, nullity = 1   **(b)** Rank = 2, nullity = 2   **(c)** Rank = 2, nullity = $n - 2$

740 Answers to Exercises

11. **(a)** $\{1, x^2, x^4, \ldots, x^{2m}\}$ where $2m = n$ if n is even and $2m = n - 1$ if n is odd. **(b)** $\{x, x^2, x^3, \ldots, x^n\}$

13. **(a)** $\left\{\begin{bmatrix} 1 & 0 & 0 \\ 0 & 0 & 0 \\ 0 & 0 & 0 \end{bmatrix}, \begin{bmatrix} 0 & 1 & 0 \\ 1 & 0 & 0 \\ 0 & 0 & 0 \end{bmatrix}, \begin{bmatrix} 0 & 0 & 1 \\ 0 & 0 & 0 \\ 1 & 0 & 0 \end{bmatrix}, \begin{bmatrix} 0 & 0 & 0 \\ 0 & 1 & 0 \\ 0 & 0 & 0 \end{bmatrix}, \begin{bmatrix} 0 & 0 & 0 \\ 0 & 0 & 1 \\ 0 & 1 & 0 \end{bmatrix}, \begin{bmatrix} 0 & 0 & 0 \\ 0 & 0 & 0 \\ 0 & 0 & 1 \end{bmatrix}\right\}$

(b) $\left\{\begin{bmatrix} 0 & 1 & 0 \\ -1 & 0 & 0 \\ 0 & 0 & 0 \end{bmatrix}, \begin{bmatrix} 0 & 0 & 1 \\ 0 & 0 & 0 \\ -1 & 0 & 0 \end{bmatrix}, \begin{bmatrix} 0 & 0 & 0 \\ 0 & 0 & 1 \\ 0 & -1 & 0 \end{bmatrix}\right\}$

15. Possible ranks are 2, 1, and 0.

Exercise Set 5.1 (page 303)

1. 2.   3. **(a)** $(\lambda - 4)(\lambda + 2) = 0$.   **(b)** $\lambda^2 + 7\lambda - 8 = 0$.

5. **(a)** basis for the eigenspace corresponding to $\lambda = 4$ is $\left\{\begin{pmatrix} 1 \\ 0 \end{pmatrix}\right\}$.

basis for the eigenspace corresponding to $\lambda = -2$ is $\left\{\begin{pmatrix} 1 \\ -2 \end{pmatrix}\right\}$.

(b) basis for the eigenspace corresponding to $\lambda = -8$ is $\left\{\begin{pmatrix} 1 \\ 10 \end{pmatrix}\right\}$.

basis for the eigenspace corresponding to $\lambda = 1$ is $\left\{\begin{pmatrix} 1 \\ 1 \end{pmatrix}\right\}$.

7. **(a)** $-1, 2, 5$ **(b)** $0, 3, $ or -2 **(c)** -8 **(d)** 2   **(e)** 2   **(f)** $-4, 3$

9. **(a)** $\lambda^4 + \lambda^3 - 3\lambda^2 - \lambda + 2 = 0$ **(b)** $\lambda^4 - 8\lambda^3 + 19\lambda^2 - 24\lambda + 48 = 0$

11. **(a)** $\lambda = 1$: basis $\begin{bmatrix} 2 \\ 3 \\ 1 \\ 0 \end{bmatrix}, \begin{bmatrix} 0 \\ 0 \\ 0 \\ 1 \end{bmatrix}$; $\lambda = -2$: basis $\begin{bmatrix} -1 \\ 0 \\ 1 \\ 0 \end{bmatrix}$; $\lambda = -1$: basis $\begin{bmatrix} -2 \\ 1 \\ 1 \\ 0 \end{bmatrix}$ **(b)** $\lambda = 4$: basis $\begin{bmatrix} \frac{3}{2} \\ 1 \\ 0 \\ 0 \end{bmatrix}$

13. $2^7, -1, \frac{1}{2^7}$, and 0   15. **(a)** $y = x$ and $y = 2x$ **(b)** No lines   **(c)** $y = 0$

True/False 5.1

(a) False   **(b)** False   **(c)** True   **(d)** False   **(e)** True   **(f)** False   **(g)** False

Exercise Set 5.2 (page 313)

1. cannot be similar   3. cannot be similar   5. $\lambda = 0$: 1 or 2; $\lambda = 1$: 1; $\lambda = 2$: 1, 2, or 3   7. Not diagonalizable
9. Diagonalizable   11. Not diagonalizable

13. $P = \begin{pmatrix} 1 & 7 \\ 0 & -8 \end{pmatrix}$. $P^{-1}AP = \begin{pmatrix} 5 & 0 \\ 0 & -3 \end{pmatrix}$ 15. $P = \begin{bmatrix} -2 & 0 & 1 \\ 0 & 1 & 0 \\ 1 & 0 & 0 \end{bmatrix}$; $P^{-1}AP = \begin{bmatrix} 3 & 0 & 0 \\ 0 & 3 & 0 \\ 0 & 0 & 2 \end{bmatrix}$

17. $P = \begin{pmatrix} 0 & 1 & 0 \\ 1 & 0 & 0 \\ 0 & 0 & 1 \end{pmatrix}, P^{-1}AP = \begin{pmatrix} 0 & 0 & 0 \\ 0 & 1 & 9 \\ 0 & 0 & 1 \end{pmatrix}$ 19. $P = \begin{pmatrix} 1 & 1 & 0 \\ 1 & 3 & 1 \\ 1 & 2 & 1 \end{pmatrix}, P^{-1}AP = \begin{pmatrix} 1 & 0 & 0 \\ 0 & 3 & 0 \\ 0 & 0 & 4 \end{pmatrix}$

21. $P = \begin{bmatrix} 1 & 0 & 0 & 0 \\ 0 & 1 & 1 & -1 \\ 0 & 0 & 1 & 0 \\ 0 & 0 & 0 & 1 \end{bmatrix}$; $P^{-1}AP = \begin{bmatrix} -2 & 0 & 0 & 0 \\ 0 & -2 & 0 & 0 \\ 0 & 0 & 3 & 0 \\ 0 & 0 & 0 & 3 \end{bmatrix}$ 23. $\begin{pmatrix} -1 & -2046 & 1 \\ 0 & 2048 & 0 \\ 0 & -6141 & 1 \end{pmatrix}$

25. $A^n = PD^nP^{-1} = \begin{bmatrix} 1 & 1 & 1 \\ 2 & 0 & -1 \\ 1 & -1 & 1 \end{bmatrix} \begin{bmatrix} 1^n & 0 & 0 \\ 0 & 3^n & 0 \\ 0 & 0 & 4^n \end{bmatrix} \begin{bmatrix} \frac{1}{6} & \frac{1}{3} & \frac{1}{6} \\ \frac{1}{2} & 0 & -\frac{1}{2} \\ \frac{1}{3} & -\frac{1}{3} & \frac{1}{3} \end{bmatrix}$

27. On possibility is $P = \begin{bmatrix} -b & -b \\ a - \lambda_1 & a - \lambda_2 \end{bmatrix}$ where λ_1 and λ_2 are as in Exercise 20 of Section 5.1.

33. (a) $\lambda = 1$: dimension $= 1$; $\lambda = 3$: dimension ≤ 2; $\lambda = 4$: dimension ≤ 3
    (b) Dimensions will be exactly 1, 2, and 3.
    (c) $\lambda = 4$

True/False 5.2

(a) True (b) True (c) True (d) False (e) True (f) True (g) True (h) True

Exercise Set 5.3 (page 326)

1. $\bar{\mathbf{u}} = (-3i, 1 + 4i, 2 - i)$, $\Re(\mathbf{u}) = (0, 1, 2)$, $\Im(\mathbf{u}) = (-3, 4, -1)$, $\|\mathbf{u}\| = \sqrt{31}$

5. $\mathbf{x} = \frac{1}{2}(-2 + 10i, -3 + 6i, 16 - i)$

7. $\bar{A} = \begin{pmatrix} 1 - 3i & 2 \\ 4 - i & 3i \end{pmatrix}$, $\Re(A) = \begin{pmatrix} 1 & 2 \\ 4 & 0 \end{pmatrix}$, $\Im(A) = \begin{pmatrix} 3 & 0 \\ 1 & -3 \end{pmatrix}$,
   $\det A = 1 - 5i$, and $\text{tr}(A) = 1$

11. $\mathbf{u} \cdot \mathbf{v} = 14 + 8i$, $\mathbf{u} \cdot \mathbf{w} = 10 + 14i$, $\mathbf{v} \cdot \mathbf{w} = -5 - 14i$

13. $-11 - 14i$

15. $\lambda_1 = 2 - i$, $\mathbf{x}_1 = \begin{bmatrix} 2 - i \\ 1 \end{bmatrix}$; $\lambda_2 = 2 + i$, $\mathbf{x}_1 = \begin{bmatrix} 2 + i \\ 1 \end{bmatrix}$

17. $\begin{pmatrix} \frac{3}{2} + \frac{1}{2}i \\ 1 \end{pmatrix}, \begin{pmatrix} \frac{3}{2} - \frac{1}{2}i \\ 1 \end{pmatrix}$

19. $|\lambda| = \sqrt{2}$, $\phi = \frac{\pi}{4}$ 21. $|\lambda| = 2$, $\phi = -\frac{\pi}{3}$

23. $P = \begin{pmatrix} -1 - 2i & -1 + 2i \\ 1 & 1 \end{pmatrix}, C = \begin{pmatrix} 6 + 2i & 0 \\ 0 & 6 - 2i \end{pmatrix}$

25. $P = \begin{bmatrix} 1 & -1 \\ 1 & 0 \end{bmatrix}, C = \begin{bmatrix} 5 & -3 \\ 3 & 5 \end{bmatrix}$

27. (a) $k = 8i$ (b) None

True/False 5.3

(a) False (b) True (c) False (d) True (e) False (f) False

Exercise Set 5.4 (page 332)

1. (a) $\begin{matrix} -c_1 + 3c_2 = 0 \\ c_1 + 4c_2 = 0 \end{matrix}$; $y_1 = 0, y_2 = 0$.   (b) $\begin{matrix} y_1 = 0 \\ y_2 = 0 \end{matrix}$ 3. (a) $y_1 = -c_2 e^{2x} + c_3 e^{3x}$ (b) $y_1 = e^{2x} - 2e^{3x}$
   $y_2 = c_1 e^x + 2c_2 e^{2x} - c_3 e^{3x}$ $y_2 = e^x - 2e^{2x} + 2e^{3x}$
   $y_3 = 2c_2 e^{2x} - c_3 e^{3x}$ $y_3 = -2e^{2x} + 2e^{3x}$

7. $y = c_1 e^{3x} + c_2 e^{-2x}$ 9. $y = c_1 e^x + c_2 e^{2x} + c_3 e^{3x}$

True/False 5.4

(a) False (b) False (c) True (d) True (e) False

Chapter 5 Supplementary Exercises (page 333)

1. **(b)** The transformation rotates vectors through the angle θ; therefore, if $0 < \theta < \pi$, then no nonzero vector is transformed into a vector in the same or opposite direction.

3. **(c)** $\begin{bmatrix} 1 & 1 & 0 \\ 0 & 2 & 1 \\ 0 & 0 & 3 \end{bmatrix}$ 9. $A^2 = \begin{bmatrix} 15 & 30 \\ 5 & 10 \end{bmatrix}$, $A^3 = \begin{bmatrix} 75 & 150 \\ 25 & 50 \end{bmatrix}$, $A^4 = \begin{bmatrix} 375 & 750 \\ 125 & 250 \end{bmatrix}$, $A^5 = \begin{bmatrix} 1875 & 3750 \\ 625 & 1250 \end{bmatrix}$

11. 0, $\text{tr}(A)$ 13. They are all 0.   15. $\begin{bmatrix} 1 & 0 & 0 \\ -1 & -\frac{1}{2} & -\frac{1}{2} \\ 1 & -\frac{1}{2} & -\frac{1}{2} \end{bmatrix}$ 17. They are all 0, 1, or -1.

Exercise Set 6.1 (page 343)

1. **(a)** -3 **(b)** 25 **(c)** -4 **(d)** $\sqrt{2}$ **(e)** $\sqrt{5}$ **(f)** $\sqrt{89}$
3. **(a)** 2 **(b)** 11 **(c)** -13 **(d)** -8 **(e)** 0
5. **(c)** -13 **(d)** $\sqrt{34}$ **(a)** -5 **(b)** 1
7. **(a)** 3 **(b)** 56 9. **(b)** 29
11. **(a)** $\begin{bmatrix} \sqrt{3} & 0 \\ 0 & \sqrt{5} \end{bmatrix}$ **(b)** $\begin{bmatrix} 2 & 0 \\ 0 & \sqrt{6} \end{bmatrix}$ 13. **(a)** $\sqrt{74}$ **(b)** 0
15. **(a)** $\sqrt{105}$ **(b)** $\sqrt{47}$
17. $\langle \mathbf{p}, \mathbf{q} \rangle = 50$
    $\|\mathbf{p}\| = \sqrt{108}$
19. **(a)** $3\sqrt{2}$ **(b)** $3\sqrt{5}$ **(c)** $3\sqrt{13}$
21. **(a)** [graph of ellipse with x-intercepts at ± 2 and y-intercepts at ± 4]   **(b)** [graph of ellipse with x-intercepts at $\pm\frac{1}{\sqrt{2}}$ and y-intercepts at ± 1]

27. For $V = \begin{bmatrix} 0 & 1 \\ -1 & 0 \end{bmatrix}$, then $\langle V, V \rangle = -2 < 0$, so Axiom 4 fails.   29. **(a)** $\langle \mathbf{p}, \mathbf{q} \rangle = \frac{22}{15}$ **(b)** 0

True/False 6.1
(a) True   **(b)** False   **(c)** True   **(d)** True   **(e)** False   **(f)** True   **(g)** False

Exercise Set 6.2 (page 350)

1. **(a)** $\frac{3\pi}{4}$ **(b)** $-\frac{3}{\sqrt{73}}$ **(c)** $\frac{\pi}{2}$ **(d)** $-\frac{20}{9\sqrt{10}}$ **(e)** $\frac{\pi}{4}$ **(f)** $\frac{2}{\sqrt{55}}$ 3. **(a)** $\frac{\pi}{2}$ **(b)** .6847   7. none
9. **(a)** $k = \frac{5}{2}$
   (b) $k = 4, -2$
13. No
15. **(a)** $2x - y - 4z = 0$, $(-2, 1, 4)$
    (b) $2x - 5y + 4z = 0$ **(c)** $x - 2y + 3z = 0$.
31. **(a)** The line $y = -x$ **(b)** The xz-plane   **(c)** The x-axis

True/False 6.2

(a) False (b) True (c) True (d) True (e) False (f) False

Exercise Set 6.3 (page 364)

1. (a), (b), (d)

3. (b), (d)

5. none

7. (a) $\left\{\left(\frac{2}{\sqrt{13}}, \frac{3}{\sqrt{13}}\right), \left(\frac{-6}{\sqrt{52}}, \frac{4}{\sqrt{52}}\right)\right\}$.
   (b) $\left\{\left(\frac{1}{2}, -\frac{1}{2}, 0\right), \left(\frac{1}{\sqrt{2}}, \frac{1}{\sqrt{2}}, 0\right), (0, 0, 1)\right\}$

9. (a) $\frac{-2}{5}\left(\frac{-3}{5}, \frac{4}{5}, 0\right) + \frac{11}{5}\left(\frac{4}{5}, \frac{3}{5}, 0\right) - (0, 0, 1)$
   (b) $\frac{9}{5}\left(\frac{-3}{5}, \frac{4}{5}, 0\right) + \frac{13}{5}\left(\frac{4}{5}, \frac{3}{5}, 0\right) + 4(0, 0, 1)$ (c) $-\frac{3}{7}\mathbf{v}_1 - \frac{1}{7}\mathbf{v}_2 + \frac{5}{7}\mathbf{v}_3$

11. (b) $\mathbf{u} = -\frac{4}{5}\mathbf{v}_1 - \frac{11}{10}\mathbf{v}_2 + 0\mathbf{v}_3 + \frac{1}{2}\mathbf{v}_4$

13. (a) $\mathbf{w} = \frac{14}{3}\mathbf{u}_1 - \frac{8}{3}\mathbf{u}_2 - \frac{1}{3}\mathbf{u}_3$ (b) $\mathbf{w} = \frac{5}{\sqrt{6}}\mathbf{u}_2 + \frac{11}{\sqrt{66}}\mathbf{u}_3$

15. (a) $\text{proj}_W \mathbf{x} = \frac{-3}{4}\mathbf{v}_1 + \frac{1}{4}\mathbf{v}_2 + \frac{5}{4}\mathbf{v}_3$
    (b) $\left(\frac{17}{12}, \frac{7}{4}, -\frac{1}{12}, -\frac{23}{12}\right)$

17. (a) $\left(\frac{23}{18}, \frac{11}{6}, -\frac{1}{18}, -\frac{17}{18}\right)$ (b) $\left(\frac{3}{2}, \frac{3}{2}, -\frac{1}{2}, -\frac{1}{2}\right)$

19. (a) $\mathbf{w}_1 = \left(\frac{3}{2}, \frac{3}{2}, -1, -1\right)$, $\mathbf{w}_2 = \left(-\frac{1}{2}, \frac{1}{2}, 1, -1\right)$ (b) $\mathbf{w}_1 = \left(\frac{7}{4}, \frac{5}{4}, -\frac{3}{4}, -\frac{5}{4}\right)$, $\mathbf{w}_2 = \left(-\frac{3}{4}, \frac{3}{4}, \frac{3}{4}, -\frac{3}{4}\right)$

21. (a) $\mathbf{q}_1 = \left(\frac{1}{\sqrt{10}}, -\frac{3}{\sqrt{10}}\right)$, $\mathbf{q}_2 = \left(\frac{3}{\sqrt{10}}, \frac{1}{\sqrt{10}}\right)$ (b) $\mathbf{q}_1 = (1, 0)$, $\mathbf{q}_2 = (0, -1)$

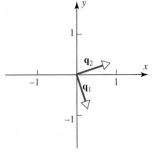
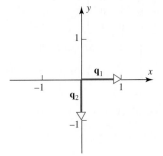

23. $\mathbf{q}_1 = \left(0, \frac{2}{\sqrt{5}}, \frac{1}{\sqrt{5}}, 0\right)$, $\mathbf{q}_2 = \left(\frac{5}{\sqrt{30}}, -\frac{1}{\sqrt{30}}, \frac{2}{\sqrt{30}}, 0\right)$,
    $\mathbf{q}_3 = \left(\frac{1}{\sqrt{10}}, \frac{1}{\sqrt{10}}, -\frac{2}{\sqrt{10}}, -\frac{2}{\sqrt{10}}\right)$, $\mathbf{q}_4 = \left(\frac{1}{\sqrt{15}}, \frac{1}{\sqrt{15}}, -\frac{2}{\sqrt{15}}, \frac{3}{\sqrt{15}}\right)$

25. $\mathbf{q}_1 = \left(\frac{1}{\sqrt{6}}, \frac{1}{\sqrt{6}}, \frac{1}{\sqrt{6}}\right)$, $\mathbf{q}_2 = \left(\frac{1}{\sqrt{6}}, \frac{1}{\sqrt{6}}, -\frac{1}{\sqrt{6}}\right)$, $\mathbf{q}_3 = \left(\frac{2}{\sqrt{6}}, -\frac{1}{\sqrt{6}}, 0\right)$

27. $\mathbf{w}_1 = \frac{1}{11}(13, 16, 35)$. $\mathbf{w}_2 = \frac{1}{11}(-2, 6, -2)$.

29. (a) $QR = \begin{bmatrix} \frac{2}{\sqrt{5}} & \frac{1}{\sqrt{5}} \\ -\frac{1}{\sqrt{5}} & \frac{2}{\sqrt{5}} \end{bmatrix} \begin{bmatrix} \sqrt{5} & 0 \\ 0 & \sqrt{5} \end{bmatrix}$ (b) $\begin{bmatrix} 0 & \frac{1}{\sqrt{2}} \\ 1 & 0 \\ 0 & \frac{1}{\sqrt{2}} \end{bmatrix} \begin{bmatrix} 2 & 1 \\ 0 & \sqrt{2} \end{bmatrix}$ (c) $\begin{bmatrix} \frac{1}{3} & \frac{8}{\sqrt{234}} \\ -\frac{2}{3} & \frac{11}{\sqrt{234}} \\ \frac{2}{3} & \frac{7}{\sqrt{234}} \end{bmatrix} \begin{bmatrix} 3 & \frac{1}{3} \\ 0 & \frac{\sqrt{26}}{3} \end{bmatrix}$

(d) $\begin{bmatrix} \frac{1}{\sqrt{2}} & -\frac{1}{\sqrt{3}} & \frac{1}{\sqrt{6}} \\ 0 & \frac{1}{\sqrt{3}} & \frac{2}{\sqrt{6}} \\ \frac{1}{\sqrt{2}} & \frac{1}{\sqrt{3}} & -\frac{1}{\sqrt{6}} \end{bmatrix} \begin{bmatrix} \sqrt{2} & \sqrt{2} & \sqrt{2} \\ 0 & \sqrt{3} & -\frac{1}{\sqrt{3}} \\ 0 & 0 & \frac{4}{\sqrt{6}} \end{bmatrix}$ (e) $\begin{bmatrix} \frac{1}{\sqrt{2}} & \frac{\sqrt{2}}{2\sqrt{19}} & -\frac{3}{\sqrt{19}} \\ \frac{1}{\sqrt{2}} & -\frac{\sqrt{2}}{2\sqrt{19}} & \frac{3}{\sqrt{19}} \\ 0 & \frac{3\sqrt{2}}{\sqrt{19}} & \frac{1}{\sqrt{19}} \end{bmatrix} \begin{bmatrix} \sqrt{2} & \frac{3}{\sqrt{2}} & \sqrt{2} \\ 0 & \frac{\sqrt{19}}{\sqrt{2}} & \frac{3\sqrt{2}}{\sqrt{19}} \\ 0 & 0 & \frac{1}{\sqrt{19}} \end{bmatrix}$

(f) Columns not linearly independent

33. $\mathbf{v}_1 = 1$, $\mathbf{v}_2 = \sqrt{3}(2x - 1)$, $\mathbf{v}_3 = \sqrt{5}(6x^2 - 6x + 1)$

True/False 6.3
(a) False (b) False (c) True (d) True (e) False (f) True

Exercise Set 6.4 (page 374)

1. (a) $\begin{bmatrix} 21 & 25 \\ 25 & 35 \end{bmatrix} \begin{bmatrix} x_1 \\ x_2 \end{bmatrix} = \begin{bmatrix} 20 \\ 20 \end{bmatrix}$ (b) $\begin{bmatrix} 15 & -1 & 5 \\ -1 & 22 & 30 \\ 5 & 30 & 45 \end{bmatrix} \begin{bmatrix} x_1 \\ x_2 \\ x_3 \end{bmatrix} = \begin{bmatrix} -1 \\ 9 \\ 13 \end{bmatrix}$

3. (a) $\mathbf{x} = \frac{1}{21}\begin{bmatrix} 11 \\ 27 \end{bmatrix}$ (b) $\mathbf{x} = \begin{bmatrix} \frac{11}{17} \\ \frac{2}{3} \\ \frac{30}{17} \end{bmatrix}$ 5. (a) $\mathbf{e} = \begin{bmatrix} \frac{3}{2} \\ \frac{9}{2} \\ -3 \end{bmatrix}$ (b) $\mathbf{e} = \begin{bmatrix} 3 \\ -3 \\ 0 \\ 3 \end{bmatrix}$

7. (a) $\begin{pmatrix} \frac{11}{14} + 2t \\ t \end{pmatrix}$
   (b) Solution: $\mathbf{x} = \left(\frac{2}{7}, 0\right) + t(-3, 1)$ (t a real number); least squares error: $\frac{1}{7}\sqrt{42}$
   (c) Solution: $\mathbf{x} = \left(-\frac{7}{6}, \frac{7}{6}, 0\right) + t(-1, -1, 1)$ (t a real number); least squares error: $\frac{1}{2}\sqrt{294}$

9. (a) $\begin{pmatrix} -\frac{5}{3} \\ 1 \\ \frac{5}{3} \end{pmatrix} \cdot \begin{pmatrix} \frac{5}{3} \\ -\frac{2}{3} \\ \frac{8}{3} \\ 0 \end{pmatrix}$ (b) $\left(-\frac{12}{5}, -\frac{4}{5}, \frac{12}{5}, \frac{16}{5}\right)$

11. (a) $\det(A^T A) = 0$; A does not have linearly independent column vectors.
    (b) $\det(A^T A) = 0$; A does not have linearly independent column vectors.

13. (a) $P = \begin{pmatrix} 1 & 0 & 0 \\ 0 & 1 & 0 \\ 0 & 0 & 0 \end{pmatrix}$ (b) $[P] = \begin{bmatrix} 0 & 0 & 0 \\ 0 & 1 & 0 \\ 0 & 0 & 1 \end{bmatrix}$

15. (a) $\left\{ \begin{bmatrix} 5 \\ 3 \\ -3 \end{bmatrix}, \begin{pmatrix} 4 \\ 2 \\ -4 \end{pmatrix} \right\}$. (b) $P = \begin{bmatrix} \frac{17}{26} & \frac{6}{13} & -\frac{3}{26} \\ \frac{6}{13} & \frac{5}{13} & \frac{2}{13} \\ -\frac{3}{26} & \frac{2}{13} & \frac{25}{26} \end{bmatrix}$. (c) $\begin{pmatrix} \frac{17x}{26} + \frac{6y}{13} - \frac{3z}{26} \\ \frac{6x}{13} + \frac{5y}{13} + \frac{2z}{13} \\ -\frac{3x}{26} + \frac{2y}{13} + \frac{25z}{26} \end{pmatrix}$ (d) $\frac{1}{\sqrt{26}}$

17. t and s distance is 0. 21. $[P] = A^T(AA^T)^{-1}A$

True/False 6.4
(a) True (b) False (c) True (d) True (e) False (f) True (g) False (h) True

Exercise Set 6.5 (page 381)

1. $y = -\frac{11}{14} + \frac{15}{14}x$ 3. $y = \frac{41}{6} - \frac{1}{10}x - \frac{4}{3}x^2$
11. $y = \frac{3}{2} + \frac{18}{7}\frac{1}{x}$

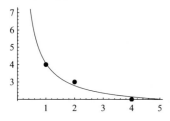

True/False 6.5
(a) False (b) True (c) False (d) True

Exercise Set 6.6 (page 387)

1. (a) $(1 - \pi) + 2\sin(x) + \sin(2x)$
   (b) $(1 - \pi) + 2\sin(x) + \sin(2x) + \frac{2}{3}\sin(3x) + \frac{2}{4}\sin(4x) + \frac{2}{5}\sin(5x) + \ldots + \frac{2}{n}\sin(nx)$
3. (a) $\frac{19-7e}{12-12e}$ (b) $\frac{13}{12} + \frac{1+e}{2(1-e)}$
5. (a) $-\frac{3\sqrt{10}}{\pi^2}x^2$.
   (b) The mean square error is $\int_{-1}^{1} \left(\cos(\pi x) + \frac{3\sqrt{10}}{\pi^2}x^2\right)^2 dx = \frac{36-24\sqrt{10}+\pi^4}{\pi^4} = 0.59$
9. $\frac{1}{2} + \sum_{k=1}^{\infty} \frac{1}{k\pi}\left[(-1)^k - 1\right]\sin(kx)\frac{1}{2} + \sum_{k=1}^{\infty} \frac{-2}{(2k+1)\pi}\sin((2k+1)x)$

True/False 6.6
(a) False (b) True (c) True (d) False (e) True

Chapter 6 Supplementary Exercises (page 387)

1. (a) $(0, a, a, 0)$ with $a \neq 0$ (b) $\pm\left(0, \frac{2}{\sqrt{5}}, \frac{1}{\sqrt{5}}, 0\right)$
3. (a) The subspace of all matrices in M_{22} with only zeros on the diagonal.
   (b) The subspace of all skew-symmetric matrices in M_{22}.
7. $\pm\left(\frac{1}{\sqrt{2}}, 0, \frac{1}{\sqrt{2}}\right)$ 9. No 11. (b) θ approaches $\frac{\pi}{2}$ 17. No

Exercise Set 7.1 (page 395)

1. (b) $A^{-1} = A^T$

3. (a) $\begin{pmatrix} 0 & 1 \\ 1 & 0 \end{pmatrix}$ (c) $\begin{pmatrix} \frac{1}{\sqrt{2}} & 0 & \frac{1}{\sqrt{2}} \\ 1 & 0 & 0 \\ 0 & 1 & 0 \end{pmatrix}$ (d) $\begin{pmatrix} 1 & 0 & 0 & 0 \\ 0 & 1 & 0 & 0 \\ 0 & 0 & 1 & 0 \\ 0 & 0 & 0 & 1 \end{pmatrix}$ (f) $\begin{pmatrix} 1 & 0 & 0 & 0 \\ 0 & 1 & 0 & 0 \\ 0 & 0 & 1 & 0 \\ 0 & 0 & 0 & 1 \end{pmatrix}$

7. (a) $\begin{pmatrix} \frac{1}{2} & -\frac{3\sqrt{3}}{2} \\ -\frac{3}{2} & -\frac{\sqrt{3}}{2} \end{pmatrix}$ (b) $\begin{pmatrix} 1-2\sqrt{3} \\ 2+\sqrt{3} \end{pmatrix}$ 9. (a) $\begin{pmatrix} \frac{5}{2} & -\frac{\sqrt{3}}{2} \\ 2 \\ \frac{1}{2} + \frac{5\sqrt{3}}{2} \end{pmatrix}$ (b) $\begin{pmatrix} \frac{3}{2} + \frac{\sqrt{3}}{2} \\ 6 \\ \frac{1}{2} - \frac{3\sqrt{3}}{2} \end{pmatrix}$

11. (a) $A = \begin{bmatrix} \cos\theta & 0 & -\sin\theta \\ 0 & 1 & 0 \\ \sin\theta & 0 & \cos\theta \end{bmatrix}$ (b) $A = \begin{bmatrix} 1 & 0 & 0 \\ 0 & \cos\theta & \sin\theta \\ 0 & -\sin\theta & \cos\theta \end{bmatrix}$ 13. $a^2 + b^2 = \frac{1}{2}$

17. The only possibilities are $a = 0$, $b = -\frac{2}{\sqrt{6}}$, $c = \frac{1}{\sqrt{3}}$ or $a = 0$, $b = \frac{2}{\sqrt{6}}$, $c = -\frac{1}{\sqrt{3}}$.
21. (a) Rotations about the origin, reflections about any line through the origin, and any combination of these
    (b) Rotation about the origin, dilations, contractions, reflections about lines through the origin, and combinations of these
    (c) No; dilations and contractions

True/False 7.1
(a) False (b) False (c) False (d) False (e) True (f) True (g) True (h) True

Exercise Set 7.2 (page 404)

1. (a) $\lambda^2 - 20\lambda = \lambda(\lambda - 20)$. dimension will be 1.
   (b) $(-5 + \lambda)(1 + \lambda)^2$. dimension 1.

Verify that $\begin{pmatrix} -1 \\ 0 \\ 1 \end{pmatrix}$ and $\begin{pmatrix} 1 \\ 1 \\ 0 \end{pmatrix}$ are both eigenvectors corresponding to $\lambda = -1$, so the dimension of this eigenspace is 2.

(c) $-(-2+\lambda)(-1+\lambda)\lambda$. dimension 1.
(d) $-(-4+\lambda)^2(2+\lambda)$. dimension 1.
(e) $(-3+\lambda)^2(3+\lambda)^2$. dimension 2.
(f) $(-3+\lambda)(-1+\lambda)(1+\lambda)(3+\lambda)$, dimension 1.

3. $P = \begin{pmatrix} \frac{3}{5} & \frac{-4}{5} \\ \frac{4}{5} & \frac{3}{5} \end{pmatrix}$.

$P^{-1}AP = \begin{pmatrix} -15 & 0 \\ 0 & 10 \end{pmatrix}$.

5. $P = \begin{pmatrix} \frac{1}{\sqrt{2}} & -\frac{1}{\sqrt{2}} & 0 \\ 0 & 0 & 1 \\ \frac{1}{\sqrt{2}} & \frac{1}{\sqrt{2}} & 0 \end{pmatrix}$. $P^{-1}AP = \begin{pmatrix} 4 & 0 & 0 \\ 0 & 2 & 0 \\ 0 & 0 & 2 \end{pmatrix}$ 7. $P = \begin{bmatrix} \frac{1}{\sqrt{3}} & -\frac{1}{\sqrt{2}} & -\frac{1}{\sqrt{6}} \\ \frac{1}{\sqrt{3}} & \frac{1}{\sqrt{2}} & -\frac{1}{\sqrt{6}} \\ \frac{1}{\sqrt{3}} & 0 & \frac{2}{\sqrt{6}} \end{bmatrix}$; $P^{-1}AP = \begin{bmatrix} 0 & 0 & 0 \\ 0 & 3 & 0 \\ 0 & 0 & 3 \end{bmatrix}$

9. $P = \begin{pmatrix} 0 & \frac{-2}{\sqrt{5}} & 0 & \frac{1}{\sqrt{5}} \\ 0 & \frac{1}{\sqrt{5}} & 0 & \frac{2}{\sqrt{5}} \\ \frac{-2}{\sqrt{5}} & 0 & \frac{1}{\sqrt{5}} & 0 \\ \frac{1}{\sqrt{5}} & 0 & \frac{2}{\sqrt{5}} & 0 \end{pmatrix}$. $P^{-1}AP = \begin{pmatrix} 7 & 0 & 0 & 0 \\ 0 & 7 & 0 & 0 \\ 0 & 0 & 2 & 0 \\ 0 & 0 & 0 & 2 \end{pmatrix}$ 15. No   19. Yes

True/False 7.2

(a) True   (b) True   (c) False   (d) True   (e) True   (f) False   (g) True

---

Exercise Set 7.3 (page 415)

1. (a) $(x_1 x_2) \begin{pmatrix} 5 & 0 \\ 0 & -3 \end{pmatrix} \begin{pmatrix} x_1 \\ x_2 \end{pmatrix}$

   (b) $(x_1 x_2) \begin{pmatrix} 9 & -2 \\ -2 & 1 \end{pmatrix} \begin{pmatrix} x_1 \\ x_2 \end{pmatrix}$

   (c) $(x_1 x_2 x_3) \begin{pmatrix} 6 & -1 & 2 \\ -1 & 4 & \frac{1}{2} \\ 2 & \frac{1}{2} & -7 \end{pmatrix} \begin{pmatrix} x_1 \\ x_2 \end{pmatrix}$

3. $4x^2 + 3y^2 - 2xy$

5. $\begin{pmatrix} x_1 \\ x_2 \end{pmatrix} = \begin{pmatrix} \frac{1}{\sqrt{2}} & \frac{-1}{\sqrt{2}} \\ \frac{1}{\sqrt{2}} & \frac{1}{\sqrt{2}} \end{pmatrix} \begin{pmatrix} y_1 \\ y_2 \end{pmatrix}$

   $-4y_1^2 + 2y_2^2$

7. $\begin{pmatrix} x_1 \\ x_2 \\ x_3 \end{pmatrix} = \begin{pmatrix} \frac{1}{\sqrt{2}} & 0 & \frac{-1}{\sqrt{2}} \\ \frac{1}{\sqrt{2}} & 0 & \frac{1}{\sqrt{2}} \\ 0 & 1 & 0 \end{pmatrix} \begin{pmatrix} y_1 \\ y_2 \\ y_3 \end{pmatrix}$

   $4y_1^2 + 2y_2^2$

9. (a) $[x\ y] \begin{bmatrix} 2 & \frac{1}{2} \\ \frac{1}{2} & 0 \end{bmatrix} \begin{bmatrix} x \\ y \end{bmatrix} + [-1\ 6] \begin{bmatrix} x \\ y \end{bmatrix} + 2 = 0$ (b) $[x\ y] \begin{bmatrix} 0 & 0 \\ 0 & 1 \end{bmatrix} \begin{bmatrix} x \\ y \end{bmatrix} + [7\ -8] \begin{bmatrix} x \\ y \end{bmatrix} - 5 = 0$

11. (a) hyperbola   (b) ellipse   (c) parabola   (d) circle

13. hyperbola $-3(x')^2 + 2(y')^2 + 8 = 0$.   15. hyperbola $-7(x')^2 + 6(y')^2 - 7 = 0$.

17. (a) indefinite   (b) semi-definite   (c) positive definite.   (d) positive semi-definite.   (e) negative definite.

19. indefinite   21. positive definite.   23. positive semi-definite.   27. $k > 2$

31. (a) $A = \begin{bmatrix} \frac{1}{n} & -\frac{1}{n(n-1)} & \cdots & -\frac{1}{n(n-1)} \\ -\frac{1}{n(n-1)} & \frac{1}{n} & \cdots & -\frac{1}{n(n-1)} \\ \vdots & \vdots & \ddots & \vdots \\ -\frac{1}{n(n-1)} & -\frac{1}{n(n-1)} & \cdots & \frac{1}{n} \end{bmatrix}$ (b) Yes   33. A must have a positive eigenvalue of multiplicity 2.

True/False 7.3

(a) True   (b) False   (c) True   (d) True   (e) False   (f) True   (g) True   (h) True   (i) False   (j) True
(k) False   (l) False

Exercise Set 7.4   (page 423)

1. Maximum of 7 at the point $(1, 0)$ and minimum of -3 at the point $(0, 1)$
3. Maximum of 4 at the point $(1, 0)$ and minimum of 2 at the point $(0, 1)$
5. Maximum of 8 at the point $(0, 0, 1)$ minimum of -3 at the point $(0, 1, 0)$
7. Maximum: $z = 4\sqrt{2}$ at $(x, y) = \left(2\sqrt{2}, 2\right)$ and $\left(-2\sqrt{2}, -2\right)$;
   minimum: $z = -4\sqrt{2}$ at $(x, y) = \left(-2\sqrt{2}, 2\right)$ and $\left(2\sqrt{2}, -2\right)$
9.

13. critical point is $(0,0)$. saddle at $(0, 0)$
15. critical points are $(-1, 0)$, relative minimum at $(-1, 0)$. saddle points
17. $x = \frac{1}{\sqrt{2}}$ and $y = \frac{4}{\sqrt{2}}$

True/False 7.4

(a) False   (b) True   (c) True   (d) False   (e) True

Exercise Set 7.5   (page 430)

1. $A^* = \begin{pmatrix} 2-i & 2 & 0 \\ 1 & 3i & 1-5i \end{pmatrix}$ 3. $A = \begin{pmatrix} 3 & 3-2i & 7 \\ 3+2i & -2 & 1+5i \\ 7 & 1-5i & 6 \end{pmatrix}$

5. (a) complex entry on the diagonal
   (b) $a_{12} \neq a_{21}{}^*$

9. $\begin{pmatrix} 1 & 0 \\ 0 & 1 \end{pmatrix}$, $A^{-1} = A^*$.

11. $A^* = A^{-1} = \begin{bmatrix} \frac{-i+\sqrt{3}}{2\sqrt{2}} & \frac{1-i\sqrt{3}}{2\sqrt{2}} \\ \frac{1+i\sqrt{3}}{2\sqrt{2}} & \frac{-i-\sqrt{3}}{2\sqrt{2}} \end{bmatrix}$

13. $P = \begin{pmatrix} \frac{3i}{5} & -\frac{4i}{5} \\ \frac{4}{5} & \frac{3}{5} \end{pmatrix}$, gives $P^{-1}AP = \begin{pmatrix} 25 & 0 \\ 0 & 0 \end{pmatrix}$.

15. $P = \begin{bmatrix} \frac{-1-i}{\sqrt{6}} & \frac{1+i}{\sqrt{3}} \\ \frac{2}{\sqrt{6}} & \frac{1}{\sqrt{3}} \end{bmatrix}$, $D = \begin{bmatrix} 2 & 0 \\ 0 & 8 \end{bmatrix}$ **17.** $P = \begin{bmatrix} 0 & 0 & 1 \\ \frac{2}{\sqrt{6}} & \frac{-1+i}{\sqrt{6}} & 0 \\ \frac{1+i}{\sqrt{6}} & \frac{2}{\sqrt{6}} & 0 \end{bmatrix}$, $D = \begin{bmatrix} -2 & 0 & 0 \\ 0 & 1 & 0 \\ 0 & 0 & 5 \end{bmatrix}$

19. $A = \begin{pmatrix} -i & -2+5i & -1 \\ 2+5i & 0 & -4-6i \\ 1 & 4-6i & 3i \end{pmatrix}$

21. (a) $a_{31} = -2 + i$.
(b) real diagonal entry so cannot be skew-Hermitian.

29. (c) B and C must commute.

37. $\begin{bmatrix} \frac{1}{\sqrt{2}} & -\frac{i}{\sqrt{2}} \\ \frac{i}{\sqrt{2}} & -\frac{1}{\sqrt{2}} \end{bmatrix}$

39. Multiplication of $\mathbf{x}$ by P corresponds to $\|\mathbf{u}\|^2$ times the orthogonal projection of $\mathbf{x}$ onto $W = \text{span}\{\mathbf{u}\}$. If $\|\mathbf{u}\| = 1$, then multiplications of $\mathbf{x}$ by $H = I - 2\mathbf{u}\mathbf{u}^*$ corresponds to reflection of $\mathbf{x}$ about the hyperplane $\mathbf{u}^\perp$.

True/False 7.5
(a) False **(b)** False **(c)** True **(d)** False **(e)** False

Chapter 7 Supplementary Exercises (page 432)

1. (a) $\begin{bmatrix} \frac{3}{5} & -\frac{4}{5} \\ \frac{4}{5} & \frac{3}{5} \end{bmatrix}^{-1} = \begin{bmatrix} \frac{3}{5} & \frac{4}{5} \\ -\frac{4}{5} & \frac{3}{5} \end{bmatrix}$ **(b)** $\begin{bmatrix} \frac{4}{5} & 0 & -\frac{3}{5} \\ -\frac{9}{25} & \frac{4}{5} & -\frac{12}{25} \\ \frac{12}{25} & \frac{3}{5} & \frac{16}{25} \end{bmatrix}^{-1} = \begin{bmatrix} \frac{4}{5} & -\frac{9}{25} & \frac{12}{25} \\ 0 & \frac{4}{5} & \frac{3}{5} \\ -\frac{3}{5} & -\frac{12}{25} & \frac{16}{25} \end{bmatrix}$

5. $P = \begin{bmatrix} -\frac{1}{\sqrt{2}} & \frac{1}{\sqrt{2}} & 0 \\ 0 & 0 & 1 \\ \frac{1}{\sqrt{2}} & \frac{1}{\sqrt{2}} & 0 \end{bmatrix}$; $P^T A P = \begin{bmatrix} 0 & 0 & 0 \\ 0 & 2 & 0 \\ 0 & 0 & 1 \end{bmatrix}$

7. positive definite **9. (a)** parabola **(b)** parabola

Exercise Set 8.1 (page 442)

1. Linear **3.** Nonlinear **5.** Linear **7.** Linear
9. (a) Linear **(b)** Nonlinear
11. $T(x_1, x_2) = (-x_1 + 3x_2, 2x_1 - 5x_2)$. $T(5, -3) = (-14, 25)$
13. $T(x_1, x_2, x_3) = (x_1 - x_2 - x_3, 5x_1 - 2x_2 - x_3, -x_1 + 3x_2 + 2x_3)$. $T(2, 4, -1) = (-1, 3, 8)$.
15. $T(3v_1 - 2v_2 + v^3) = (12, -2, 11)$ **17. (b)** and **(c)**
19. (a) **21. (a)** $(1, -4)$ **(b)** $(4, 2, 6)$, $(1, 1, 0)$, $(-3, -4, 9)$ **(c)** x, x^2, x^3
23. (a) basis for the range of T is $\{(1,-2)\}$. **(b)** $\begin{bmatrix} -14 \\ 19 \\ 11 \end{bmatrix}$ **(c)** $\text{Rank}(T) = 2$, $\text{nullity}(T) = 1$

25. (a) $\left\{ \begin{pmatrix} 1 \\ -1 \\ -1 \end{pmatrix}, \begin{pmatrix} 0 \\ 2 \\ 8 \end{pmatrix} \right\}$ **(b)** $\left\{ \begin{pmatrix} -3 \\ -7/2 \\ 1 \end{pmatrix} \right\}$.

(c) The rank of $T = 2$ and the nullity of $T = 1$. **(d)** The rank of $A = 2$ and the nullity of $A = 1$.

27. (a) Kernel: y-axis; range: xz-plane **(b)** Kernel: x-axis; range: yz-plane
(c) Kernel: the line through the origin perpendicular to the plane $y = x$; range: plane $y = x$

29. (a) Nullity$(T) = 2$ (b) Nullity$(T) = 4$ (c) Nullity$(T) = 3$ (d) Nullity$(T) = 1$
31. (a) 3 (b) No **33.** A line through the origin, a plane through the origin, the origin only, or all of R^3
35. (b) No **41.** ker(D) consists of all constant polynomials. **43.** (a) $T(f(x)) = f^{(4)}(x)$ (b) $T(f(x)) = f^{(n+1)}(x)$

True/False 8.1
(a) True (b) False (c) True (d) False (e) True (f) True (g) False (h) False (i) False

Exercise Set 8.2 (page 451)

1. (a) kernel of T $\{(0,0)\}$, T is one-to-one. (b) kernel is $\{(1,5)\}$. T is not one-to-one.
  (c) kernel $\begin{pmatrix} 1 \\ -1 \\ 1 \end{pmatrix}$, T is not one-to-one. (d) kernel of T is $\{(0,0)\}$, T is one-to-one.
  (e) ker$(T) = \{k(1,1)\}$; T is not one-to-one
  (f) ker$(T) = \{k(0,1,-1)\}$; T is not one-to-one
3. (a) A is not 1-1. (b) A is not 1-1. (c) A is 1-1.
5. (a) ker$(T) = \{k(-1,1)\}$ (b) T is not one-to-one since ker$(T) \neq \{\mathbf{0}\}$.
7. (a) If nullity $(T) = 1$, T cannot be 1-1. (b) T is 1-1. (c) T is not one-to-one (d) T is one-to-one

11. (a) $T\left(\begin{bmatrix} a & b & c \\ b & d & e \\ c & e & f \end{bmatrix}\right) = \begin{bmatrix} a \\ b \\ c \\ d \\ e \\ f \end{bmatrix}$ (b) $T\left(\begin{bmatrix} a & b \\ c & d \end{bmatrix}\right) = \begin{bmatrix} a \\ b \\ c \\ d \end{bmatrix}$; $T\left(\begin{bmatrix} a & b \\ c & d \end{bmatrix}\right) = \begin{bmatrix} a \\ c \\ b \\ d \end{bmatrix}$

  (c) $T(ax^3 + bx^2 + cx) = \begin{bmatrix} a \\ b \\ c \end{bmatrix}$ (d) $T(a + b\sin(x) + c\cos(x)) = \begin{bmatrix} a \\ b \\ c \end{bmatrix}$

13. T is not one-to-one since, for example, $f(x) = x^2(x-1)^2$ is in its kernel. **15.** Yes; it is one-to-one
17. T is not one-to-one since, for example **a** is in its kernel. **19.** Yes

True/False 8.2
(a) False (b) True (c) False (d) True (e) False (f) False

Exercise Set 8.3 (page 457)

1. (a) $T_2 \circ T_1(x, y) = (3y - 4x, 3y + 2x)$.
  (b) $T_2 \circ T_1(x, y, z) = (y - z, x, y - z - x)$.
  (c) not defined.
  (d) $T_2 \circ T_1(x, y) = (-4x, -4y, 4x + 4y)$.
3. (a) $a + d$ (b) $(T_2 \circ T_1)(A)$ does not exist since $T_1(A)$ is not a 2×2 matrix.
5. $T_2(\mathbf{v}) = \frac{1}{4}\mathbf{v}$
11. (a) T does not have an inverse.
  (b) $T^{-1}\left(\begin{pmatrix} x_1 \\ x_2 \\ x_3 \end{pmatrix}\right) = \begin{pmatrix} \frac{x_1}{2} - \frac{x_2}{2} - x_3 \\ \frac{x_1}{2} - \frac{x_2}{2} \\ -x_1 + 2x_2 + x_3 \end{pmatrix}$.
  (c) $T^{-1}\left(\begin{pmatrix} x_1 \\ x_2 \\ x_3 \end{pmatrix}\right) = \frac{1}{6}\begin{pmatrix} -5x_1 - 2x_2 + 8x_3 \\ x_1 + 4x_2 - x_3 \\ x_1 - 2x_2 - x_3 \end{pmatrix}$.

(d) $T^{-1}\begin{bmatrix} x_1 \\ x_2 \\ x_3 \end{bmatrix} = \begin{bmatrix} 3x_1 + 3x_2 - x_3 \\ -2x_1 - 2x_2 + x_3 \\ -4x_1 - 5x_2 + 2x_3 \end{bmatrix}$

13. (a) $a_i \neq 0$ for $i = 1, 2, 3, \ldots, n$ **(b)** $T^{-1}(x_1, x_2, x_3, \ldots, x_n) = \left(\frac{1}{a_1}x_1, \frac{1}{a_2}x_2, \frac{1}{a_3}x_3, \ldots, \frac{1}{a_n}x_n\right)$

15. (a) $T_1^{-1}(p(x)) = \frac{p(x)}{x}$; $T_2^{-1}(p(x)) = p(x-1)$; $(T_2 \circ T_1)^{-1}(p(x)) = \frac{1}{x}p(x-1)$

17. (a) $(2, 1)$
 (d) $T^{-1}(3, 5) = 3 + 2x$

21. (a) $T_1 \circ T_2 = T_2 \circ T_1$ **(b)** $T_1 \circ T_2 \neq T_2 \circ T_1$ **(c)** $T_1 \circ T_2 = T_2 \circ T_1$

True/False 8.3

(a) True **(b)** False **(c)** False **(d)** True **(e)** False **(f)** True

Exercise Set 8.4 (page 466)

1. (a) $\begin{bmatrix} 0 & 0 & 0 \\ 1 & 0 & 0 \\ 0 & 1 & 0 \\ 0 & 0 & 1 \end{bmatrix}$ **3. (a)** $\begin{bmatrix} 1 & -1 & 1 \\ 0 & 1 & -2 \\ 0 & 0 & 1 \end{bmatrix}$ **5. (a)** $\begin{pmatrix} 4 & 3 \\ -2 & \frac{-3}{2} \\ \frac{-2}{3} & \frac{1}{3} \end{pmatrix}$

7. (a) $\begin{bmatrix} 1 & 1 & 1 \\ 0 & 2 & 4 \\ 0 & 0 & 4 \end{bmatrix}$ **(b)** $3 + 10x + 16x^2$

9. (a) $[T(\mathbf{v}_1)]_B = \begin{pmatrix} 3 \\ -2 \end{pmatrix}$ $[T(\mathbf{v}_2)]_B = \begin{pmatrix} 2 \\ 1 \end{pmatrix}$. **(b)** $T(\mathbf{v}_1) = \begin{pmatrix} 4 \\ -1 \end{pmatrix}$ $T(\mathbf{v}_2) = \begin{pmatrix} 5 \\ -3 \end{pmatrix}$.
 (c) $T\left(\begin{pmatrix} x_1 \\ x_2 \end{pmatrix}\right) = \begin{pmatrix} -x_1 - 6x_2 \\ 2x_1 + 5x_2 \end{pmatrix}$ **(d)** $\begin{pmatrix} -7 \\ 7 \end{pmatrix}$.

11. (a) $[T(\mathbf{v}_1)]_B = \begin{pmatrix} 2 \\ 1 \\ 0 \end{pmatrix}$, $[T(\mathbf{v}_2)]_B = \begin{pmatrix} 0 \\ 3 \\ -2 \end{pmatrix}$, $[T(\mathbf{v}_3)]_B = \begin{pmatrix} -1 \\ 5 \\ 6 \end{pmatrix}$,
 (b) $T(\mathbf{v}_1) = 5 + 2x - 2x^2$, $T(\mathbf{v}_2) = 3 + 4x^2$, $T(\mathbf{v}_1) = 3 + 19x + 20x^2$
 (c) $T(a_0 + a_1x + a_2x^2) = (26 - 68x - 64x^2)a_0 + (17 + 116x + 110x^2)a_1 - (9 + 82x + 50x^2)a_2$
 (d) $T(1 + x^2) = 17 - 150x - 114x^2$

13. (a) $[T_2 \circ T_1]_{B',B} = \begin{bmatrix} 0 & 0 \\ 6 & 0 \\ 0 & -9 \\ 0 & 0 \end{bmatrix}$, $[T_2]_{B',B''} = \begin{bmatrix} 0 & 0 & 0 \\ 3 & 0 & 0 \\ 0 & 3 & 0 \\ 0 & 0 & 3 \end{bmatrix}$, $[T_1]_{B'',B} = \begin{bmatrix} 2 & 0 \\ 0 & -3 \\ 0 & 0 \end{bmatrix}$
 (b) $[T_2 \circ T_1]_{B',B} = [T_2]_{B',B''}[T_1]_{B'',B}$

19. (a) $\begin{bmatrix} 0 & 0 & 0 \\ 0 & 0 & -1 \\ 0 & 1 & 0 \end{bmatrix}$ **(b)** $\begin{bmatrix} 0 & 0 & 0 \\ 0 & 1 & 0 \\ 0 & 0 & 2 \end{bmatrix}$ **(c)** $\begin{bmatrix} 2 & 1 & 0 \\ 0 & 2 & 2 \\ 0 & 0 & 2 \end{bmatrix}$

 (d) $D(3e^{2x} - 4xe^{2x} + 8x^2e^{2x}) = \left[\begin{pmatrix} 2 & 1 & 0 \\ 0 & 2 & 2 \\ 0 & 0 & 2 \end{pmatrix}\begin{pmatrix} 3 \\ -4 \\ 8 \end{pmatrix}\right]_B = 2e^{2x} + 8xe^{2x} + 16x^2e^{2x}$.

21. (a) B', B'' **(b)** B', B'''

True/False 8.4

(a) False **(b)** False **(c)** True **(d)** False **(e)** True

Exercise Set 8.5 (page 473)

1. $[T]_B = \begin{pmatrix} 3 & -1 \\ 1 & 0 \end{pmatrix}$ $[T]_{B'} = \frac{1}{9}\begin{pmatrix} -3 & -9 \\ 19 & 30 \end{pmatrix}$.

3. $[T]_B = \begin{pmatrix} \frac{\sqrt{2}}{2} & -\frac{\sqrt{2}}{2} \\ \frac{\sqrt{2}}{2} & \frac{\sqrt{2}}{2} \end{pmatrix}$ $[T]_{B'} = \begin{pmatrix} \frac{\sqrt{2}}{3} & -\frac{1}{\sqrt{2}} \\ \frac{5\sqrt{2}}{9} & \frac{2\sqrt{2}}{3} \end{pmatrix}$.

5. $[T]_B = \begin{bmatrix} 1 & 0 & 0 \\ 0 & 1 & 0 \\ 0 & 0 & 0 \end{bmatrix}$, $[T]_{B'} = \begin{bmatrix} 1 & 0 & 0 \\ 0 & 1 & 1 \\ 0 & 0 & 0 \end{bmatrix}$

7. $[T]_B = \begin{bmatrix} \frac{2}{3} & -\frac{2}{9} \\ \frac{1}{2} & \frac{4}{3} \end{bmatrix}$, $[T]_{B'} = \begin{bmatrix} 1 & 1 \\ 0 & 1 \end{bmatrix}$

11. (a) $B = \left\{ \begin{pmatrix} -1 \\ 1 \end{pmatrix}, \begin{pmatrix} 2 \\ 1 \end{pmatrix} \right\}, [T]_B = \begin{pmatrix} 5 & 0 \\ 0 & 2 \end{pmatrix}$.

    (b) $B = \left\{ \begin{pmatrix} -1 \\ 1 \end{pmatrix}, \begin{pmatrix} 5 \\ 1 \end{pmatrix} \right\}, [T]_B = \begin{pmatrix} -3 & 0 \\ 0 & 3 \end{pmatrix}$.

13. (a) $\lambda = -4$, $\lambda = 3$, and $\lambda = 0$.

    (b) $\begin{pmatrix} -3 \\ -12 \\ 2 \end{pmatrix}, \begin{pmatrix} 2 \\ 1 \\ 1 \end{pmatrix}$, and $\begin{pmatrix} -5 \\ -4 \\ 2 \end{pmatrix}$.

21. The choice of an appropriate basis can yield a better understanding of the linear operator.

True/False 8.5

(a) False (b) True (c) True (d) True (e) True (f) False (g) True (h) False

Chapter 8 Supplementary Exercises (page 475)

1. No. $T(\mathbf{x}_1 + \mathbf{x}_2) = A(\mathbf{x}_1 + \mathbf{x}_2) + B \neq (A\mathbf{x}_1 + B) + (A\mathbf{x}_2 + B) = T(\mathbf{x}_1) + T(\mathbf{x}_2)$, and if $c \neq 1$, then $T(c\mathbf{x}) = cA\mathbf{x} + B \neq c(A\mathbf{x} + B) = cT(\mathbf{x})$.

5. (a) $T(\mathbf{e}_3)$ and any two of $T(\mathbf{e}_1)$, $T(\mathbf{e}_2)$, and $T(\mathbf{e}_4)$ form bases for the range; $(-1, 1, 0, 1)$ is a basis for the kernel.

   (b) Rank $= 3$, nullity $= 1$

7. (a) Rank$(T) = 2$ and nullity$(T) = 2$ (b) T is not one-to-one.

11. Rank $= 3$, nullity $= 1$ 13. $\begin{bmatrix} 1 & 0 & 0 & 0 \\ 0 & 0 & 1 & 0 \\ 0 & 1 & 0 & 0 \\ 0 & 0 & 0 & 1 \end{bmatrix}$

15. $[T]_{B'} = \begin{bmatrix} -4 & 0 & 9 \\ 1 & 0 & -2 \\ 0 & 1 & 1 \end{bmatrix}$ 17. $[T]_B = \begin{bmatrix} 1 & -1 & 1 \\ 0 & 1 & 0 \\ 1 & 0 & -1 \end{bmatrix}$

19. (b) $f(x) = x$, $g(x) = 1$
    (c) $f(x) = e^x$, $g(x) = e^{-x}$

21. (d) The points are on the graph.   25. $\begin{bmatrix} 0 & 0 & 0 & \cdots & 0 \\ 1 & 0 & 0 & \cdots & 0 \\ 0 & \frac{1}{2} & 0 & \cdots & 0 \\ 0 & 0 & \frac{1}{3} & \cdots & 0 \\ \vdots & \vdots & \vdots & & \vdots \\ 0 & 0 & 0 & \cdots & \frac{1}{n+1} \end{bmatrix}$

Exercise Set 9.1 (page 485)

1. $\mathbf{x} = \begin{pmatrix} 2 \\ 2 \end{pmatrix}$.   3. $\mathbf{y} = \begin{pmatrix} -1 \\ -2 \end{pmatrix}$. $\mathbf{x} = \begin{pmatrix} 3 \\ -2 \end{pmatrix}$.   5. $x_3 = -1, x_2 = -1, x_1 = 3$.

7. $\begin{pmatrix} x_1 \\ x_2 \\ x_3 \end{pmatrix} = \begin{pmatrix} -12 \\ 4 \\ 0 \end{pmatrix} + t \begin{pmatrix} 5 \\ -2 \\ 1 \end{pmatrix}$.   9. $\mathbf{y} = \begin{pmatrix} -2 \\ -2 \\ 3 \\ -2 \end{pmatrix}$. $\mathbf{x} = \begin{pmatrix} -3 \\ 1 \\ -1 \\ -2 \end{pmatrix}$.

11. (a) $A = \begin{pmatrix} 4 & 0 & 0 \\ 8 & -2 & 0 \\ -4 & -6 & 2 \end{pmatrix} \begin{pmatrix} 1 & 1 & 0 \\ 0 & 1 & -1 \\ 0 & 0 & 1 \end{pmatrix}$

(b) $A = \begin{pmatrix} 1 & 0 & 0 \\ 2 & 1 & 0 \\ -1 & 3 & 1 \end{pmatrix} \begin{pmatrix} 4 & 0 & 0 \\ 0 & -2 & 0 \\ 0 & 0 & 2 \end{pmatrix} \begin{pmatrix} 1 & 1 & 0 \\ 0 & 1 & -1 \\ 0 & 0 & 1 \end{pmatrix}$

(c) $A = A = \begin{pmatrix} 1 & 0 & 0 \\ 2 & 1 & 0 \\ -1 & 3 & 1 \end{pmatrix} \begin{pmatrix} 4 & 4 & 0 \\ 0 & -2 & 2 \\ 0 & 0 & 2 \end{pmatrix}$

13. $A = \begin{bmatrix} 1 & 0 & 0 \\ 0 & 1 & 0 \\ 2 & -2 & 1 \end{bmatrix} \begin{bmatrix} 3 & 0 & 0 \\ 0 & 2 & 0 \\ 0 & 0 & 1 \end{bmatrix} \begin{bmatrix} 1 & -4 & 2 \\ 0 & 1 & 0 \\ 0 & 0 & 1 \end{bmatrix}$ 15. $x_1 = \frac{21}{17}$, $x_2 = -\frac{14}{17}$, $x_3 = \frac{12}{17}$

17. $A = \begin{pmatrix} 5 & 0 & 0 \\ 0 & 5 & 0 \\ 2 & 0 & 1 \end{pmatrix} \begin{pmatrix} 1 & 3 & 0 \\ 0 & 1 & 0 \\ 0 & 0 & 1 \end{pmatrix} = LU$.   19. (b) $\begin{bmatrix} a & b \\ c & d \end{bmatrix} = \begin{bmatrix} 1 & 0 \\ \frac{c}{a} & 1 \end{bmatrix} \begin{bmatrix} a & b \\ 0 & \frac{ad-bc}{a} \end{bmatrix}$

True/False 9.1
(a) False   (b) False   (c) True   (d) True   (e) True

Exercise Set 9.2 (page 494)

1. (a) dominant eigenvalue is $\lambda_1 = -3$.
   (b) no dominant eigenvalue

3. $\lambda^{(1)} = (A\mathbf{x}_1) \cdot \mathbf{x}_1 = 2.92179$, $\lambda^{(2)} = (A\mathbf{x}_2) \cdot \mathbf{x}_2 = 2.31557$, $\lambda^{(3)} = (A\mathbf{x}_3) \cdot \mathbf{x}_3 = 2.1372$, $\lambda^{(4)} = (A\mathbf{x}_4) \cdot \mathbf{x}_4 = 2.06176$.

5. $\lambda^{(1)} = \frac{(A\mathbf{x}_1)^T \mathbf{x}_1}{\mathbf{x}_1^T \mathbf{x}_1} = 3.59952$, $\lambda^{(2)} = \frac{(A\mathbf{x}_2)^T \mathbf{x}_2}{\mathbf{x}_2^T \mathbf{x}_2} = 3.88235$, $\lambda^{(3)} = \frac{(A\mathbf{x}_3)^T \mathbf{x}_3}{\mathbf{x}_3^T \mathbf{x}_3} = 3.9693$, $\lambda^{(4)} = \frac{(A\mathbf{x}_4)^T \mathbf{x}_4}{\mathbf{x}_4^T \mathbf{x}_4} = 3.99217$.

7. (a) $\mathbf{x}_1 = \begin{bmatrix} 1 \\ -0.5 \end{bmatrix}$, $\mathbf{x}_2 = \begin{bmatrix} 1 \\ -0.8 \end{bmatrix}$, $\mathbf{x}_3 \approx \begin{bmatrix} 1 \\ -0.929 \end{bmatrix}$ (b) $\lambda^{(1)} = 2.8$, $\lambda^{(2)} \approx 2.976$, $\lambda^{(3)} \approx 2.997$

(c) Dominant eigenvalue: $\lambda = 3$; dominant eigenvector: $\begin{bmatrix} 1 \\ -1 \end{bmatrix}$ (d) 0.1%

9. $\mathbf{x}_1 = \frac{A\mathbf{x}_0}{\max(A\mathbf{x}_0)} = \begin{pmatrix} 1 \\ \frac{1}{2} \end{pmatrix}$, $\mathbf{x}_2 = \frac{A^2\mathbf{x}_0}{\max(A^2\mathbf{x}_0)} = \begin{pmatrix} 1 \\ \frac{4}{5} \end{pmatrix}$, similarly $\mathbf{x}_3 = \begin{pmatrix} 1 \\ .92857 \end{pmatrix}$, $\mathbf{x}_4 = \begin{pmatrix} 1 \\ .97561 \end{pmatrix}$, $\mathbf{x}_5 = \begin{pmatrix} 1 \\ .99180 \end{pmatrix}$.

eigenvalue must be $\lambda = 6$.

13. (a) Starting with $\begin{bmatrix} 1 \\ 0 \\ 0 \end{bmatrix}$, it takes 8 iterations.   (b) Starting with $\begin{bmatrix} 1 \\ 0 \\ 0 \\ 0 \end{bmatrix}$, it takes 8 iterations.

Answers to Exercises

Exercise Set 9.3 (page 500)

1. hub vector is $\begin{pmatrix} 2 \\ 2 \\ 1 \end{pmatrix}$ and authority vector is $\begin{pmatrix} 2 \\ 1 \\ 2 \end{pmatrix}$.

2. hub vector is $\begin{pmatrix} 3 \\ 1 \\ 3 \\ 1 \end{pmatrix}$ and authority vector is $\begin{pmatrix} 3 \\ 2 \\ 2 \\ 1 \end{pmatrix}$.

3. $\mathbf{h}_1 = \dfrac{A\mathbf{a}_0}{\|A\mathbf{a}_0\|} = \begin{pmatrix} .721995 \\ .618853 \\ .309426 \end{pmatrix}$, $\mathbf{a}_1 = \dfrac{A^T\mathbf{h}_1}{\|A^T\mathbf{h}_1\|} \begin{pmatrix} .721995 \\ .309426 \\ .618853 \end{pmatrix}$.

5. site 2, site 4, site 1, site 3.

7. site 4, site 3, site 5, site 2, with site 1 probably irrelevant.

Exercise Set 9.4 (page 506)

1. (a) ≈ 0.067 second   (b) ≈ 66.68 seconds   (c) $\approx 66{,}668$ seconds, or about 18.5 hours

3. (a) ≈ 9.52 seconds   (b) ≈ 0.0014 second   (c) ≈ 9.52 seconds   (d) ≈ 28.6 seconds

4. (a) 0.380952 seconds   (b) 5.71429×10^{-6} seconds   (c) 0.380952 seconds   (d) 1.14286 seconds

5. (a) 6.67×10^5 s for forward phase, 10 s for backward phase   (b) 1334   7. n^2 flops   9. $2n^3 - n^2$ flops

Exercise Set 9.5 (page 513)

1. 5 and 0.   3. $\sqrt{5}$.

5. $A = \begin{bmatrix} \frac{1}{\sqrt{2}} & -\frac{1}{\sqrt{2}} \\ \frac{1}{\sqrt{2}} & \frac{1}{\sqrt{2}} \end{bmatrix} \begin{bmatrix} \sqrt{2} & 0 \\ 0 & \sqrt{2} \end{bmatrix} \begin{bmatrix} 1 & 0 \\ 0 & 1 \end{bmatrix}$

7. singular value decomposition of A is $\begin{pmatrix} -\frac{1}{\sqrt{2}} & \frac{1}{\sqrt{2}} \\ \frac{1}{\sqrt{2}} & \frac{1}{\sqrt{2}} \end{pmatrix} \begin{pmatrix} \sqrt{10} & 0 \\ 0 & 0 \end{pmatrix} \begin{pmatrix} -\frac{2}{\sqrt{5}} & \frac{1}{\sqrt{5}} \\ \frac{1}{\sqrt{5}} & \frac{2}{\sqrt{5}} \end{pmatrix}$.

9. $\begin{pmatrix} \frac{1}{3} & \frac{2}{3} & -\frac{2}{3} \\ \frac{2}{3} & -\frac{2}{3} & -\frac{1}{3} \\ \frac{2}{3} & \frac{1}{3} & \frac{2}{3} \end{pmatrix} \begin{pmatrix} 3\sqrt{5} & 0 \\ 0 & 0 \\ 0 & 0 \end{pmatrix} \begin{pmatrix} \frac{2}{\sqrt{5}} & \frac{1}{\sqrt{5}} \\ \frac{-1}{\sqrt{5}} & \frac{2}{\sqrt{5}} \end{pmatrix}$.

11. $A = \begin{bmatrix} \frac{1}{\sqrt{3}} & 0 & \frac{2}{\sqrt{6}} \\ \frac{1}{\sqrt{3}} & \frac{1}{\sqrt{2}} & -\frac{1}{\sqrt{6}} \\ -\frac{1}{\sqrt{3}} & \frac{1}{\sqrt{2}} & \frac{1}{\sqrt{6}} \end{bmatrix} \begin{bmatrix} \sqrt{3} & 0 \\ 0 & \sqrt{2} \\ 0 & 0 \end{bmatrix} \begin{bmatrix} 1 & 0 \\ 0 & 1 \end{bmatrix}$

True/False 9.5

(a) False   (b) True   (c) False   (d) False   (e) True   (f) False   (g) True

Exercise Set 9.6 (page 517)

1. $\begin{pmatrix} \frac{1}{3} \\ \frac{2}{3} \\ \frac{2}{3} \end{pmatrix} (3\sqrt{5}) \begin{pmatrix} \frac{2}{\sqrt{5}} & \frac{1}{\sqrt{5}} \end{pmatrix}$.   3. $\begin{bmatrix} \frac{1}{\sqrt{3}} & 0 \\ \frac{1}{\sqrt{3}} & \frac{1}{\sqrt{2}} \\ -\frac{1}{\sqrt{3}} & \frac{1}{\sqrt{2}} \end{bmatrix} \begin{bmatrix} \sqrt{3} & 0 \\ 0 & \sqrt{2} \end{bmatrix} \begin{bmatrix} 1 & 0 \\ 0 & 1 \end{bmatrix}$

754 Answers to Exercises

5. $3\sqrt{5}\begin{pmatrix}\frac{1}{3}\\ \frac{2}{3}\\ \frac{2}{3}\end{pmatrix}\begin{pmatrix}\frac{2}{\sqrt{5}} & \frac{1}{\sqrt{5}}\end{pmatrix}$ 7. $\sqrt{3}\begin{bmatrix}\frac{1}{\sqrt{3}}\\ \frac{1}{\sqrt{3}}\\ -\frac{1}{\sqrt{3}}\end{bmatrix}[1\ 0] + \sqrt{2}\begin{bmatrix}0\\ \frac{1}{\sqrt{2}}\\ \frac{1}{\sqrt{2}}\end{bmatrix}[0\ 1]$

9. 70,100 numbers must be stored; A has 100,000 entries

True/False 9.6

(a) True   (b) True   (c) False

Chapter 9 Supplementary Exercises (page 517)

1. $\begin{bmatrix}2 & 0\\ -2 & 1\end{bmatrix}\begin{bmatrix}-3 & 1\\ 0 & 2\end{bmatrix}$ 3. $\begin{bmatrix}2 & 0 & 0\\ 1 & 2 & 0\\ 1 & 1 & 2\end{bmatrix}\begin{bmatrix}1 & 2 & 3\\ 0 & 1 & 2\\ 0 & 0 & 1\end{bmatrix}$

5. (a) $\lambda = 3$, $\mathbf{v} = \begin{bmatrix}\frac{1}{\sqrt{2}}\\ \frac{1}{\sqrt{2}}\end{bmatrix}$ (b) $\mathbf{x}_5 \approx \begin{bmatrix}0.7100\\ 0.7041\end{bmatrix}$, $\mathbf{v} \approx \begin{bmatrix}0.7071\\ 0.7071\end{bmatrix}$ (c) $\mathbf{x}_5 \approx \begin{bmatrix}1\\ 0.9918\end{bmatrix}$

9. $\begin{bmatrix}-\frac{1}{\sqrt{2}} & 0 & \frac{1}{\sqrt{2}}\\ 0 & 1 & 0\\ -\frac{1}{\sqrt{2}} & 0 & -\frac{1}{\sqrt{2}}\end{bmatrix}\begin{bmatrix}2 & 0\\ 0 & 0\\ 0 & 0\end{bmatrix}\begin{bmatrix}-\frac{1}{\sqrt{2}} & -\frac{1}{\sqrt{2}}\\ -\frac{1}{\sqrt{2}} & \frac{1}{\sqrt{2}}\end{bmatrix}$

11. $\begin{bmatrix}12 & 0 & 6\\ 4 & -8 & 10\\ 4 & -8 & 10\\ 12 & 0 & 6\end{bmatrix} = \begin{bmatrix}\frac{1}{2} & \frac{1}{2}\\ \frac{1}{2} & -\frac{1}{2}\\ \frac{1}{2} & -\frac{1}{2}\\ \frac{1}{2} & \frac{1}{2}\end{bmatrix}\begin{bmatrix}24 & 0\\ 0 & 12\end{bmatrix}\begin{bmatrix}\frac{2}{3} & -\frac{1}{3} & \frac{2}{3}\\ \frac{2}{3} & \frac{2}{3} & -\frac{1}{3}\end{bmatrix}$

Exercise Set 10.1 (page 524)

1. (a) $y = 3x - 4$ (b) $y = -2x + 1$
2. (a) $x^2 + y^2 - 4x - 6y + 4 = 0$ or $(x-2)^2 + (y-3)^2 = 9$ (b) $x^2 + y^2 + 2x - 4y - 20 = 0$ or $(x+1)^2 + (y-2)^2 = 25$
3. $x^2 + 2xy + y^2 - 2x + y = 0$ (a parabola)   4. (a) $x + 2y + z = 0$ (b) $-x + y - 2z + 1 = 0$

5. (a) $\begin{vmatrix}x & y & z & 0\\ x_1 & y_1 & z_1 & 1\\ x_2 & y_2 & z_2 & 1\\ x_3 & y_3 & z_3 & 1\end{vmatrix} = 0$ (b) $x + 2y + z = 0$; $-x + y - 2z = 0$

6. (a) $x^2 + y^2 + z^2 - 2x - 4y - 2z = -2$ or $(x-1)^2 + (y-2)^2 + (z-1)^2 = 4$
   (b) $x^2 + y^2 + z^2 - 2x - 2y = 3$ or $(x-1)^2 + (y-1)^2 + z^2 = 5$

10. $\begin{vmatrix}y & x^2 & x & 1\\ y_1 & x_1^2 & x_1 & 1\\ y_2 & x_2^2 & x_2 & 1\\ y_3 & x_3^2 & x_3 & 1\end{vmatrix} = 0$ 11. The equation of the line through the three collinear points   12. $0 = 0$

13. The equation of the plane through the four coplanar points

Exercise Set 10.2 (page 534)

1. $x_1 = 2$, $x_2 = \frac{2}{3}$; maximum value of $z = \frac{22}{3}$ 2. No feasible solutions   3. Unbounded solution
4. Invest $6000 in bond A and $4000 in bond B; the annual yield is $880.

5. $\frac{7}{9}$ cup of milk, $\frac{25}{18}$ ounces of corn flakes; minimum cost $= \frac{335}{18} = 18.6$¢

6. (a) $x_1 \geq 0$ and $x_2 \geq 0$ are nonbinding; $2x_1 + 3x_2 \leq 24$ is binding
   (b) $x_1 - x_2 \leq v$ for $v < -3$ is binding and for $v < -6$ yields the empty set.
   (c) $x_2 \leq v$ for $v < 8$ is nonbinding and for $v < 0$ yields the empty set.

7. 550 containers from company A and 300 containers from company B; maximum shipping charges $= \$2110$

8. 925 containers from company A and no containers from company B; maximum shipping charges $= \$2312.50$

9. 0.4 pound of ingredient A and 2.4 pounds of ingredient B; minimum cost $= 24.8$¢

Exercise Set 10.3 (page 541)

1. 700   2. (a) 5   (b) 4

4. (a) Ox, $\frac{34}{21}$ units; sheep, $\frac{20}{21}$ unit
   (b) First kind, $\frac{9}{25}$ measure; second kind, $\frac{7}{25}$ measure; third kind, $\frac{4}{25}$ measure

5. (a) $x_1 = \dfrac{(a_2 + a_3 + \cdots + a_n) - a_1}{n - 2}$, $x_i = a_i - x_1$, $i = 2, 3, \ldots, n$
   (b) Exercise 7(b); gold, $30\frac{1}{2}$ minae; brass, $9\frac{1}{2}$ minae; tin, $14\frac{1}{2}$ minae; iron, $5\frac{1}{2}$ minae

6. (a) $5x + y + z - K = 0$
       $x + 7y + z - K = 0$
       $x + y + 8z - K = 0$
   $x = \dfrac{21t}{131}$, $y = \dfrac{14t}{131}$, $z = \dfrac{12t}{131}$, $K = t$ where t is an arbitrary number
   (b) Take $t = 131$, so that $x = 21$, $y = 14$, $z = 12$, $K = 131$.
   (c) Take $t = 262$, so that $x = 42$, $y = 28$, $z = 24$, $K = 262$.

7. (a) Legitimate son, $577\frac{7}{9}$ staters; illegitimate son, $422\frac{2}{9}$ staters
   (b) Gold, $30\frac{1}{2}$ minae; brass, $9\frac{1}{2}$ minae; tin, $14\frac{1}{2}$ minae; iron, $5\frac{1}{2}$ minae
   (c) First person, 45; second person, $37\frac{1}{2}$; third person, $22\frac{1}{2}$

Exercise Set 10.4 (page 552)

2. (a) $S(x) = -.12643(x - .4)^3 - .20211(x - .4)^2 + .92158(x - .4) + .38942$
   (b) $S(.5) = .47943$; error $= 0\%$

3. (a) The cubic runout spline   (b) $S(x) = 3x^3 - 2x^2 + 5x + 1$

4. $S(x) = \begin{cases} -.00000042(x + 10)^3 & + .000214(x + 10) + .99815, & -10 \leq x \leq 0 \\ .00000024(x)^3 - .0000126(x)^2 & + .000088(x) + .99987, & 0 \leq x \leq 10 \\ -.00000004(x - 10)^3 - .0000054(x - 10)^2 - .000092(x - 10) + .99973, & 10 \leq x \leq 20 \\ .00000022(x - 20)^3 - .0000066(x - 20)^2 - .000212(x - 20) + .99823, & 20 \leq x \leq 30 \end{cases}$
   Maximum at $(x, S(x)) = (3.93, 1.00004)$

5. $S(x) = \begin{cases} .00000009(x + 10)^3 - .0000121(x + 10)^2 + .000282(x + 10) + .99815, & -10 \leq x \leq 0 \\ .00000009(x)^3 - .0000093(x)^2 + .000070(x) + .99987, & 0 \leq x \leq 10 \\ .00000004(x - 10)^3 - .0000066(x - 10)^2 - .000087(x - 10) + .99973, & 10 \leq x \leq 20 \\ .00000004(x - 20)^3 - .0000053(x - 20)^2 - .000207(x - 20) + .99823, & 20 \leq x \leq 30 \end{cases}$
   Maximum at $(x, S(x)) = (4.00, 1.00001)$

6. (a) $S(x) = \begin{cases} -4x^3 + 3x & 0 \leq x \leq 0.5 \\ 4x^3 - 12x^2 + 9x - 1 & 0.5 \leq x \leq 1 \end{cases}$
   (b) $S(x) = \begin{cases} 2 - 2x & 0.5 \leq x \leq 1 \\ 2 - 2x & 1 \leq x \leq 1.5 \end{cases}$
   (c) The three data points are collinear.

756 Answers to Exercises

7. (b) $\begin{bmatrix} 4 & 1 & 0 & 0 & \cdots & 0 & 0 & 0 & 1 \\ 1 & 4 & 1 & 0 & \cdots & 0 & 0 & 0 & 0 \\ 0 & 1 & 4 & 1 & \cdots & 0 & 0 & 0 & 0 \\ \vdots & \vdots & \vdots & \vdots & & \vdots & \vdots & \vdots & \vdots \\ 0 & 0 & 0 & 0 & \cdots & 0 & 1 & 4 & 1 \\ 1 & 0 & 0 & 0 & \cdots & 0 & 0 & 1 & 4 \end{bmatrix} \begin{bmatrix} M_1 \\ M_2 \\ M_3 \\ \vdots \\ M_{n-2} \\ M_{n-1} \end{bmatrix} = \frac{6}{h^2} \begin{bmatrix} y_{n-1} - 2y_1 + y_2 \\ y_1 - 2y_2 + y_3 \\ y_2 - 2y_3 + y_4 \\ \vdots \\ y_{n-3} - 2y_{n-2} + y_{n-1} \\ y_{n-2} - 2y_{n-1} + y_1 \end{bmatrix}$

8. (b) $\begin{bmatrix} 2 & 1 & 0 & 0 & \cdots & 0 & 0 & 0 & 1 \\ 1 & 4 & 1 & 0 & \cdots & 0 & 0 & 0 & 0 \\ 0 & 1 & 4 & 1 & \cdots & 0 & 0 & 0 & 0 \\ \vdots & \vdots & \vdots & \vdots & & \vdots & \vdots & \vdots & \vdots \\ 0 & 0 & 0 & 0 & \cdots & 0 & 0 & 4 & 1 \\ 0 & 0 & 0 & 0 & \cdots & 0 & 1 & 1 & 2 \end{bmatrix} \begin{bmatrix} M_1 \\ M_2 \\ M_3 \\ \vdots \\ M_{n-1} \\ M_n \end{bmatrix} = \frac{6}{h^2} \begin{bmatrix} -hy_1' - y_1 + y_2 \\ y_1 - 2y_2 + y_3 \\ y_2 - 2y_3 + y_4 \\ \vdots \\ y_{n-2} - 2y_{n-1} + y_n \\ y_{n-1} - y_n + hy_n' \end{bmatrix}$

Exercise Set 10.5 (page 562)

1. (a) $\mathbf{x}^{(1)} = \begin{bmatrix} .4 \\ .6 \end{bmatrix}$, $\mathbf{x}^{(2)} = \begin{bmatrix} .46 \\ .54 \end{bmatrix}$, $\mathbf{x}^{(3)} = \begin{bmatrix} .454 \\ .546 \end{bmatrix}$, $\mathbf{x}^{(4)} = \begin{bmatrix} .4546 \\ .5454 \end{bmatrix}$, $\mathbf{x}^{(5)} = \begin{bmatrix} .45454 \\ .54546 \end{bmatrix}$

   (b) P is regular since all entries of P are positive; $\mathbf{q} = \begin{bmatrix} \frac{5}{11} \\ \frac{6}{11} \end{bmatrix}$

2. (a) $\mathbf{x}^{(1)} = \begin{bmatrix} .7 \\ .2 \\ .1 \end{bmatrix}$, $\mathbf{x}^{(2)} = \begin{bmatrix} .23 \\ .52 \\ .25 \end{bmatrix}$, $\mathbf{x}^{(3)} = \begin{bmatrix} .273 \\ .396 \\ .331 \end{bmatrix}$ (b) P is regular, since all entries of P are positive: $\mathbf{q} = \begin{bmatrix} \frac{22}{72} \\ \frac{29}{72} \\ \frac{21}{72} \end{bmatrix}$

3. (a) $\begin{bmatrix} \frac{9}{17} \\ \frac{8}{17} \end{bmatrix}$ (b) $\begin{bmatrix} \frac{26}{45} \\ \frac{19}{45} \end{bmatrix}$ (c) $\begin{bmatrix} \frac{3}{19} \\ \frac{4}{19} \\ \frac{12}{19} \end{bmatrix}$

4. (a) $P^n = \begin{bmatrix} \left(\frac{1}{2}\right)^n & 0 \\ 1 - \left(\frac{1}{2}\right)^n & 1 \end{bmatrix}$, $n = 1, 2, \ldots$.   Thus, no integer power of P has all positive entries.

   (b) $P^n \to \begin{bmatrix} 0 & 0 \\ 1 & 1 \end{bmatrix}$ as n increases, so $P^n \mathbf{x}^{(0)} \to \begin{bmatrix} 0 \\ 1 \end{bmatrix}$ for any $\mathbf{x}^{(0)}$ as n increases.

   (c) The entries of the limiting vector $\begin{bmatrix} 0 \\ 1 \end{bmatrix}$ are not all positive.

6. $P^2 = \begin{bmatrix} \frac{1}{2} & \frac{1}{4} & \frac{1}{4} \\ \frac{1}{4} & \frac{1}{2} & \frac{1}{4} \\ \frac{1}{4} & \frac{1}{4} & \frac{1}{2} \end{bmatrix}$ has all positive entries; $\mathbf{q} = \begin{bmatrix} \frac{1}{3} \\ \frac{1}{3} \\ \frac{1}{3} \end{bmatrix}$ 7. $\frac{10}{13}$

8. $54\frac{1}{6}\%$ in region 1, $16\frac{2}{3}\%$ in region 2, and $29\frac{1}{6}\%$ in region 3

Answers to Exercises

Exercise Set 10.6 (page 570)

1. (a) $\begin{bmatrix} 0 & 0 & 0 & 1 \\ 1 & 0 & 1 & 1 \\ 1 & 1 & 0 & 1 \\ 0 & 0 & 0 & 0 \end{bmatrix}$ (b) $\begin{bmatrix} 0 & 1 & 1 & 0 & 0 \\ 0 & 0 & 0 & 0 & 1 \\ 1 & 0 & 0 & 1 & 0 \\ 0 & 0 & 1 & 0 & 0 \\ 0 & 0 & 1 & 0 & 0 \end{bmatrix}$ (c) $\begin{bmatrix} 0 & 1 & 0 & 1 & 0 & 0 \\ 1 & 0 & 0 & 0 & 0 & 0 \\ 0 & 1 & 0 & 1 & 1 & 1 \\ 0 & 0 & 0 & 0 & 0 & 1 \\ 0 & 0 & 0 & 0 & 0 & 1 \\ 0 & 0 & 1 & 0 & 1 & 0 \end{bmatrix}$

2. (a)   (b) 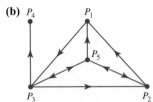 (c)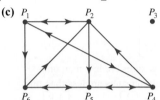

3. (a) 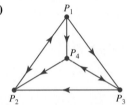 (b) 1-step: $P_1 \to P_2$
2-step: $P_1 \to P_4 \to P_2$
$P_1 \to P_3 \to P_2$
3-step: $P_1 \to P_2 \to P_1 \to P_2$
$P_1 \to P_3 \to P_4 \to P_2$
$P_1 \to P_4 \to P_3 \to P_2$ (c) 1-step: $P_1 \to P_4$
2-step: $P_1 \to P_3 \to P_4$
3-step: $P_1 \to P_2 \to P_1 \to P_4$
$P_1 \to P_4 \to P_3 \to P_4$

4. (a) $\begin{bmatrix} 1 & 0 & 0 & 0 & 0 \\ 0 & 1 & 0 & 0 & 0 \\ 0 & 0 & 1 & 1 & 0 \\ 0 & 0 & 1 & 2 & 1 \\ 0 & 0 & 0 & 1 & 2 \end{bmatrix}$

(c) The ijth entry is the number of family members who influence both the ith and jth family members.

5. (a) $\{P_1, P_2, P_3\}$ (b) $\{P_3, P_4, P_5\}$ (c) $\{P_2, P_4, P_6, P_8\}$ and $\{P_4, P_5, P_6\}$ 6. (a) None (b) $\{P_3, P_4, P_6\}$

7. $\begin{bmatrix} 0 & 0 & 1 & 1 \\ 1 & 0 & 0 & 0 \\ 0 & 1 & 0 & 1 \\ 0 & 1 & 0 & 0 \end{bmatrix}$ Power of $P_1 = 5$
Power of $P_2 = 3$
Power of $P_3 = 4$
Power of $P_4 = 2$

8. First, A; second, B and E (tie); fourth, C; fifth, D

Exercise Set 10.7 (page 580)

1. (a) $-5/8$ (b) $[0 \ 1 \ 0]$ (c) $[1 \ 0 \ 0 \ 0]^T$ 2. Let $A = \begin{bmatrix} 1 & 1 \\ 1 & 1 \end{bmatrix}$, for example.

3. (a) $\mathbf{p}^* = [0 \ 1]$, $\mathbf{q}^* = \begin{bmatrix} 0 \\ 1 \end{bmatrix}$, $v = 3$ (b) $\mathbf{p}^* = [0 \ 1 \ 0]$, $\mathbf{q}^* = \begin{bmatrix} 1 \\ 0 \end{bmatrix}$, $v = 2$

(c) $\mathbf{p}^* = [0 \ 0 \ 1]$, $\mathbf{q}^* = \begin{bmatrix} 0 \\ 1 \\ 0 \end{bmatrix}$, $v = 2$ (d) $\mathbf{p}^* = [0 \ 1 \ 0 \ 0]$, $\mathbf{q}^* = \begin{bmatrix} 1 \\ 0 \\ 0 \\ 0 \end{bmatrix}$, $v = -2$

4. (a) $\mathbf{p}^* = [\frac{5}{8} \ \frac{3}{8}]$, $\mathbf{q}^* = \begin{bmatrix} \frac{1}{8} \\ \frac{7}{8} \end{bmatrix}$, $v = \frac{27}{8}$ (b) $\mathbf{p}^* = [\frac{2}{3} \ \frac{1}{3}]$, $\mathbf{q}^* = \begin{bmatrix} \frac{1}{6} \\ \frac{5}{6} \end{bmatrix}$, $v = \frac{70}{3}$

(c) $\mathbf{p}^* = [1 \ 0]$, $\mathbf{q}^* = \begin{bmatrix} 1 \\ 0 \end{bmatrix}$, $v = 3$ (d) $\mathbf{p}^* = [\frac{3}{5} \ \frac{2}{5}]$, $\mathbf{q}^* = \begin{bmatrix} \frac{3}{5} \\ \frac{2}{5} \end{bmatrix}$, $v = \frac{19}{5}$

(e) $\mathbf{p}^* = \begin{bmatrix} \frac{3}{13} & \frac{10}{13} \end{bmatrix}$, $\mathbf{q}^* = \begin{bmatrix} \frac{1}{13} \\ \frac{12}{13} \end{bmatrix}$, $v = -\frac{29}{13}$

5. $\mathbf{p}^* = \begin{bmatrix} \frac{13}{20} & \frac{7}{20} \end{bmatrix}$, $\mathbf{q}^* = \begin{bmatrix} \frac{11}{20} \\ \frac{9}{20} \end{bmatrix}$, $v = -\frac{3}{20}$

Exercise Set 10.8 (page 588)

1. (a) $\begin{bmatrix} 2 \\ 3 \end{bmatrix}$ **(b)** $\begin{bmatrix} 6 \\ 5 \\ 6 \end{bmatrix}$ **(c)** $\begin{bmatrix} 78 \\ 54 \\ 79 \end{bmatrix}$

2. (a) Use Corollary 10.8.4; all row sums are less than one.
(b) Use Corollary 10.8.5; all column sums are less than one.

(c) Use Theorem 10.8.3, with $\mathbf{x} = \begin{bmatrix} 2 \\ 1 \\ 1 \end{bmatrix} > C\mathbf{x} = \begin{bmatrix} 1.9 \\ .9 \\ .9 \end{bmatrix}$.

3. E^2 has all positive entries. **4.** Price of tomatoes, $120.00; price of corn, $100.00; price of lettuce, $106.67
5. $1256 for the CE, $1448 for the EE, $1556 for the ME **6. (b)** $\frac{542}{503}$

Exercise Set 10.9 (page 596)

1. The second class; $15,000 **2.** $223 **3.** $1 : 1.90 : 3.02 : 4.24 : 5.00$
5. $s/(g_1^{-1} + g_2^{-1} + \cdots + g_{k-1}^{-1})$ **6.** $1 : 2 : 3 : \cdots : n - 1$

Exercise Set 10.10 (page 603)

1. (a) $\begin{bmatrix} 0 & 1 & 1 & 0 \\ 0 & 0 & 1 & 1 \\ 0 & 0 & 0 & 0 \end{bmatrix}$ **(b)** $\begin{bmatrix} 0 & \frac{3}{2} & \frac{3}{2} & 0 \\ 0 & 0 & \frac{1}{2} & \frac{1}{2} \\ 0 & 0 & 0 & 0 \end{bmatrix}$ **(c)** $\begin{bmatrix} -2 & -1 & -1 & -2 \\ -1 & -1 & 0 & 0 \\ 3 & 3 & 3 & 3 \end{bmatrix}$

(d) $\begin{bmatrix} 0 & .866 & 1.366 & .500 \\ 0 & -.500 & .366 & .866 \\ 0 & 0 & 0 & 0 \end{bmatrix}$

2. (b) $(0, 0, 0)$, $(1, 0, 0)$, $\left(1\frac{1}{2}, 1, 0\right)$, and $\left(\frac{1}{2}, 1, 0\right)$
(c) $(0, 0, 0)$, $(1, .6, 0)$, $(1, 1.6, 0)$, $(0, 1, 0)$

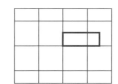
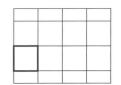
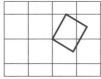
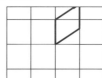

3. (a) $\begin{bmatrix} 1 & 0 & 0 \\ 0 & -1 & 0 \\ 0 & 0 & 1 \end{bmatrix}$ **(b)** $\begin{bmatrix} -1 & 0 & 0 \\ 0 & 1 & 0 \\ 0 & 0 & 1 \end{bmatrix}$ **(c)** $\begin{bmatrix} 1 & 0 & 0 \\ 0 & 1 & 0 \\ 0 & 0 & -1 \end{bmatrix}$

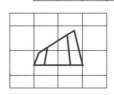
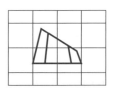

4. (a) $M_1 = \begin{bmatrix} \frac{1}{2} & 0 & 0 \\ 0 & 2 & 0 \\ 0 & 0 & \frac{1}{3} \end{bmatrix}$, $M_2 = \begin{bmatrix} \frac{1}{2} & \frac{1}{2} & \cdots & \frac{1}{2} \\ 0 & 0 & \cdots & 0 \\ 0 & 0 & \cdots & 0 \end{bmatrix}$, $M_3 = \begin{bmatrix} 1 & 0 & 0 \\ 0 & \cos 20° & -\sin 20° \\ 0 & \sin 20° & \cos 20° \end{bmatrix}$,

$M_4 = \begin{bmatrix} \cos(-45°) & 0 & \sin(-45°) \\ 0 & 1 & 0 \\ -\sin(-45°) & 0 & \cos(-45°) \end{bmatrix}$, $M_5 = \begin{bmatrix} 0 & -1 & 0 \\ 1 & 0 & 0 \\ 0 & 0 & 1 \end{bmatrix}$

(b) $P' = M_5 M_4 M_3 (M_1 P + M_2)$

5. (a) $M_1 = \begin{bmatrix} .3 & 0 & 0 \\ 0 & .5 & 0 \\ 0 & 0 & 1 \end{bmatrix}$, $M_2 = \begin{bmatrix} 1 & 0 & 0 \\ 0 & \cos 45° & -\sin 45° \\ 0 & \sin 45° & \cos 45° \end{bmatrix}$, $M_3 = \begin{bmatrix} 1 & 1 & \cdots & 1 \\ 0 & 0 & \cdots & 0 \\ 0 & 0 & \cdots & 0 \end{bmatrix}$,

$M_4 = \begin{bmatrix} \cos 35° & 0 & \sin 35° \\ 0 & 1 & 0 \\ -\sin 35° & 0 & \cos 35° \end{bmatrix}$, $M_5 = \begin{bmatrix} \cos(-45°) & -\sin(-45°) & 0 \\ \sin(-45°) & \cos(-45°) & 0 \\ 0 & 0 & 1 \end{bmatrix}$,

$M_6 = \begin{bmatrix} 0 & 0 & \cdots & 0 \\ 0 & 0 & \cdots & 0 \\ 1 & 1 & \cdots & 1 \end{bmatrix}$, $M_7 = \begin{bmatrix} 2 & 0 & 0 \\ 0 & 1 & 0 \\ 0 & 0 & 1 \end{bmatrix}$

(b) $P' = M_7(M_5 M_4 (M_2 M_1 P + M_3) + M_6)$

6. $R_1 = \begin{bmatrix} \cos \beta & 0 & \sin \beta \\ 0 & 1 & 0 \\ -\sin \beta & 0 & \cos \beta \end{bmatrix}$, $R_2 = \begin{bmatrix} \cos \alpha & -\sin \alpha & 0 \\ \sin \alpha & \cos \alpha & 0 \\ 0 & 0 & 1 \end{bmatrix}$, **7. (a)** $M = \begin{bmatrix} 1 & 0 & 0 & x_0 \\ 0 & 1 & 0 & y_0 \\ 0 & 0 & 1 & z_0 \\ 0 & 0 & 0 & 1 \end{bmatrix}$ **(b)** $\begin{bmatrix} 1 & 0 & 0 & -5 \\ 0 & 1 & 0 & 9 \\ 0 & 0 & 1 & -3 \\ 0 & 0 & 0 & 1 \end{bmatrix}$

$R_3 = \begin{bmatrix} \cos \theta & 0 & \sin \theta \\ 0 & 1 & 0 \\ -\sin \theta & 0 & \cos \theta \end{bmatrix}$, $R_4 = \begin{bmatrix} \cos \alpha & \sin \alpha & 0 \\ -\sin \alpha & \cos \alpha & 0 \\ 0 & 0 & 1 \end{bmatrix}$,

$R_5 = \begin{bmatrix} \cos \beta & 0 & -\sin \beta \\ 0 & 1 & 0 \\ \sin \beta & 0 & \cos \beta \end{bmatrix}$

Exercise Set 10.11 (page 613)

1. (a) $\begin{bmatrix} t_1 \\ t_2 \\ t_3 \\ t_4 \end{bmatrix} = \begin{bmatrix} 0 & \frac{1}{4} & \frac{1}{4} & 0 \\ \frac{1}{4} & 0 & 0 & \frac{1}{4} \\ \frac{1}{4} & 0 & 0 & \frac{1}{4} \\ 0 & \frac{1}{4} & \frac{1}{4} & 0 \end{bmatrix} \begin{bmatrix} t_1 \\ t_2 \\ t_3 \\ t_4 \end{bmatrix} + \begin{bmatrix} 0 \\ \frac{1}{2} \\ 0 \\ \frac{1}{2} \end{bmatrix}$ **(b)** $\mathbf{t} = \begin{bmatrix} \frac{1}{4} \\ \frac{3}{4} \\ \frac{1}{4} \\ \frac{3}{4} \end{bmatrix}$

(c) $\mathbf{t}^{(1)} = \begin{bmatrix} 0 \\ \frac{1}{2} \\ 0 \\ \frac{1}{2} \end{bmatrix}$, $\mathbf{t}^{(2)} = \begin{bmatrix} \frac{1}{8} \\ \frac{5}{8} \\ \frac{1}{8} \\ \frac{5}{8} \end{bmatrix}$, $\mathbf{t}^{(3)} = \begin{bmatrix} \frac{3}{16} \\ \frac{11}{16} \\ \frac{3}{16} \\ \frac{11}{16} \end{bmatrix}$, $\mathbf{t}^{(4)} = \begin{bmatrix} \frac{7}{32} \\ \frac{23}{32} \\ \frac{7}{32} \\ \frac{23}{32} \end{bmatrix}$, $\mathbf{t}^{(5)} = \begin{bmatrix} \frac{15}{64} \\ \frac{47}{64} \\ \frac{15}{64} \\ \frac{47}{64} \end{bmatrix}$, $\mathbf{t}^{(5)} - \mathbf{t} = \begin{bmatrix} -\frac{1}{64} \\ -\frac{1}{64} \\ -\frac{1}{64} \\ -\frac{1}{64} \end{bmatrix}$

(d) for t_1 and t_3, -12.9%; for t_2 and t_4, 5.2%

2. $\frac{1}{2}$ **3.** $\mathbf{t}^{(1)} = \begin{bmatrix} \frac{3}{4} & \frac{5}{4} & \frac{2}{4} & \frac{5}{4} & \frac{4}{4} & \frac{2}{4} & \frac{5}{4} & \frac{4}{4} & \frac{3}{4} \end{bmatrix}^T$

$\mathbf{t}^{(2)} = \begin{bmatrix} \frac{13}{16} & \frac{18}{16} & \frac{9}{16} & \frac{22}{16} & \frac{13}{16} & \frac{7}{16} & \frac{21}{16} & \frac{16}{16} & \frac{10}{16} \end{bmatrix}^T$

Exercise Set 10.12 (page 624)

1. (c) $x_3^* = \left(\frac{31}{22}, \frac{27}{22}\right)$

2. (a) $\mathbf{x}_3^{(1)} = (1.40000, 1.20000)$ (b) Same as part (a) (c) $\mathbf{x}_3^{(1)} = (9.55000, 25.65000)$
   $\mathbf{x}_3^{(2)} = (1.41000, 1.23000)$ $\mathbf{x}_3^{(2)} = (.59500, -1.21500)$
   $\mathbf{x}_3^{(3)} = (1.40900, 1.22700)$ $\mathbf{x}_3^{(3)} = (1.49050, 1.47150)$
   $\mathbf{x}_3^{(4)} = (1.40910, 1.22730)$ $\mathbf{x}_3^{(4)} = (1.40095, 1.20285)$
   $\mathbf{x}_3^{(5)} = (1.40909, 1.22727)$ $\mathbf{x}_3^{(5)} = (1.40991, 1.22972)$
   $\mathbf{x}_3^{(6)} = (1.40909, 1.22727)$ $\mathbf{x}_3^{(6)} = (1.40901, 1.22703)$

4. $\mathbf{x}_1^* = (1, 1), \mathbf{x}_2^* = (2, 0), \mathbf{x}_3^* = (1, 1)$

7.
$$x_7 + x_8 + x_9 = 13.00$$
$$x_4 + x_5 + x_6 = 15.00$$
$$x_1 + x_2 + x_3 = 8.00$$
$$.82843(x_6 + x_8) + .58579x_9 = 14.79$$
$$1.41421(x_3 + x_5 + x_7) = 14.31$$
$$.82843(x_2 + x_4) + .58579x_1 = 3.81$$
$$x_3 + x_6 + x_9 = 18.00$$
$$x_2 + x_5 + x_8 = 12.00$$
$$x_1 + x_4 + x_7 = 6.00$$
$$.82843(x_2 + x_6) + .58579x_3 = 10.51$$
$$1.41421(x_1 + x_5 + x_9) = 16.13$$
$$.82843(x_4 + x_8) + .58579x_7 = 7.04$$

8.
$$x_7 + x_8 + x_9 = 13.00$$
$$x_4 + x_5 + x_6 = 15.00$$
$$x_1 + x_2 + x_3 = 8.00$$
$$.04289(x_3 + x_5 + x_7) + .75000(x_6 + x_8) + .61396x_9 = 14.79$$
$$.91421(x_3 + x_5 + x_7) + .25000(x_2 + x_4 + x_6 + x_8) = 14.31$$
$$.04289(x_3 + x_5 + x_7) + .75000(x_2 + x_4) + .61396x_1 = 3.81$$
$$x_3 + x_6 + x_9 = 18.00$$
$$x_2 + x_5 + x_8 = 12.00$$
$$x_1 + x_4 + x_7 = 6.00$$
$$.04289(x_1 + x_5 + x_9) + .75000(x_2 + x_6) + .61396x_3 = 10.51$$
$$.91421(x_1 + x_5 + x_9) + .25000(x_2 + x_4 + x_6 + x_8) = 16.13$$
$$.04289(x_1 + x_5 + x_9) + .75000(x_4 + x_8) + .61396x_7 = 7.04$$

Exercise Set 10.13 (page 639)

1. $T_i\left(\begin{bmatrix} x \\ y \end{bmatrix}\right) = \frac{12}{25}\begin{bmatrix} 1 & 0 \\ 0 & 1 \end{bmatrix}\begin{bmatrix} x \\ y \end{bmatrix} + \begin{bmatrix} e_i \\ f_i \end{bmatrix}$, $i = 1, 2, 3, 4$, where the four values of $\begin{bmatrix} e_i \\ f_i \end{bmatrix}$ are $\begin{bmatrix} 0 \\ 0 \end{bmatrix}$, $\begin{bmatrix} \frac{13}{25} \\ 0 \end{bmatrix}$, $\begin{bmatrix} 0 \\ \frac{13}{25} \end{bmatrix}$, and $\begin{bmatrix} \frac{13}{25} \\ \frac{13}{25} \end{bmatrix}$; $d_H(S) = \ln(4)/\ln\left(\frac{25}{12}\right) = 1.888\ldots$

2. $s \approx .47$; $d_H(S) \approx \ln(4)/\ln(1/.47) = 1.8\ldots$. Rotation angles: $0°$ (upper left); $-90°$ (upper right); $180°$ (lower left); $180°$ (lower right)

3. $(0, 0, 0), (1, 0, 0), (2, 0, 0), (3, 0, 0), (0, 0, 1), (0, 0, 2), (1, 2, 0), (2, 1, 3), (2, 0, 1), (2, 0, 2), (2, 2, 0), (0, 3, 3)$

4. (a) (i) $s = \frac{1}{3}$; (ii) all rotation angles are $0°$; (iii) $d_H(S) = \ln(7)/\ln(3) = 1.771\ldots$.
   This set is a fractal.

   (b) (i) $s = \frac{1}{2}$; (ii) all rotation angles are $180°$; (iii) $d_H(S) = \ln(3)/\ln(2) = 1.584\ldots$.
   This set is a fractal.

   (c) (i) $s = \frac{1}{2}$; (ii) rotation angles: $-90°$ (top); $180°$ (lower left); $180°$ (lower right);
   (iii) $d_H(S) = \ln(3)/\ln(2) = 1.584\ldots$. This set is a fractal.

   (d) (i) $s = \frac{1}{2}$; (ii) rotation angles: $90°$ (upper left); $180°$ (upper right); $180°$ (lower right);
   (iii) $d_H(S) = \ln(3)/\ln(2) = 1.584\ldots$. This set is a fractal.

5. $s = .8509\ldots$, $\theta = -2.69°\ldots$ 6. $(0.766, 0.996)$ rounded to three decimal places 7. $d_H(S) = \ln(16)/\ln(4) = 2$

8. $\ln(4)/\ln\left(\frac{4}{3}\right) = 4.818\ldots$ 9. $d_H(S) = \ln(8)/\ln(2) = 3$; the cube is not a fractal.

10. $k = 20$; $s = \frac{1}{3}$; $d_H(S) = \ln(20)/\ln(3) = 2.726\ldots$; the set is a fractal.

11.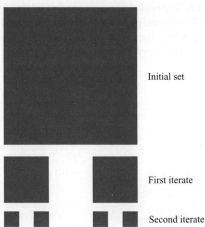

Initial set

First iterate

Second iterate

Third iterate

Fourth iterate

$d_H(S) = \ln(2)/\ln(3) = 0.6309\ldots$

12. Area of $S_0 = 1$; area of $S_1 = \frac{8}{9} = 0.888\ldots$; area of $S_2 = \left(\frac{8}{9}\right)^2 = 0.790\ldots$; area of $S_3 = \left(\frac{8}{9}\right)^3 = 0.702\ldots$; area of $S_4 = \left(\frac{8}{9}\right)^4 = 0.624\ldots$

Exercise Set 10.14 (page 652)

1. $\Pi(250) = 750$, $\Pi(25) = 50$, $\Pi(125) = 250$, $\Pi(30) = 60$, $\Pi(10) = 30$, $\Pi(50) = 150$, $\Pi(3750) = 7500$, $\Pi(6) = 12$, $\Pi(5) = 10$

2. One 1-cycle: $\{(0,0)\}$; one 3-cycle: $\{\left(\frac{3}{6}, 0\right), \left(\frac{3}{6}, \frac{3}{6}\right), \left(0, \frac{3}{6}\right)\}$;
   two 4-cycles: $\{\left(\frac{4}{6}, 0\right), \left(\frac{4}{6}, \frac{4}{6}\right), \left(\frac{2}{6}, 0\right), \left(\frac{2}{6}, \frac{2}{6}\right)\}$ and $\{\left(0, \frac{2}{6}\right), \left(\frac{2}{6}, \frac{4}{6}\right), \left(0, \frac{4}{6}\right), \left(\frac{4}{6}, \frac{2}{6}\right)\}$;
   two 12-cycles: $\{\left(0, \frac{1}{6}\right), \left(\frac{1}{6}, \frac{2}{6}\right), \left(\frac{3}{6}, \frac{5}{6}\right), \left(\frac{2}{6}, \frac{1}{6}\right), \left(\frac{3}{6}, \frac{4}{6}\right), \left(\frac{1}{6}, \frac{5}{6}\right), \left(0, \frac{5}{6}\right), \left(\frac{5}{6}, \frac{4}{6}\right), \left(\frac{3}{6}, \frac{1}{6}\right),$ $\left(\frac{4}{6}, \frac{5}{6}\right), \left(\frac{3}{6}, \frac{2}{6}\right), \left(\frac{5}{6}, \frac{1}{6}\right)\}$ and $\{\left(\frac{1}{6}, 0\right), \left(\frac{1}{6}, \frac{1}{6}\right), \left(\frac{2}{6}, \frac{3}{6}\right), \left(\frac{5}{6}, \frac{2}{6}\right), \left(\frac{1}{6}, \frac{3}{6}\right), \left(\frac{4}{6}, \frac{1}{6}\right), \left(\frac{5}{6}, 0\right),$ $\left(\frac{5}{6}, \frac{5}{6}\right), \left(\frac{4}{6}, \frac{3}{6}\right), \left(\frac{1}{6}, \frac{4}{6}\right), \left(\frac{5}{6}, \frac{3}{6}\right), \left(\frac{2}{6}, \frac{5}{6}\right)\}$. $\Pi(6) = 12$

3. (a) 3, 7, 10, 2, 12, 14, 11, 10, 6, 1, 7, 8, 0, 8, 8, 1, 9, 10, 4, 14, 3, 2, 5, 7, 12, 4, 1, 5, 6, 11, 2, 13, 0, 13, 13, 11, 9, 5, 14, 4, 3, 7, …
   (c) (5, 5), (10, 15), (4, 19), (2, 0), (2, 2), (4, 6), (10, 16), (5, 0), (5, 5), …

4. (c) The first five iterates of $\left(\frac{1}{101}, 0\right)$ are $\left(\frac{1}{101}, \frac{1}{101}\right)$, $\left(\frac{2}{101}, \frac{3}{101}\right)$, $\left(\frac{5}{101}, \frac{8}{101}\right)$, $\left(\frac{13}{101}, \frac{21}{101}\right)$, and $\left(\frac{34}{101}, \frac{55}{101}\right)$.

6. (b) The matrices of Anosov automorphisms are $\begin{bmatrix} 3 & 2 \\ 1 & 1 \end{bmatrix}$ and $\begin{bmatrix} 5 & 7 \\ 2 & 3 \end{bmatrix}$.
   (c) The transformation affects a rotation of S through 90° in the clockwise direction.

9.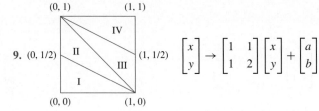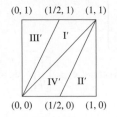

In region I: $\begin{bmatrix} a \\ b \end{bmatrix} = \begin{bmatrix} 0 \\ 0 \end{bmatrix}$; in region II: $\begin{bmatrix} a \\ b \end{bmatrix} = \begin{bmatrix} 0 \\ -1 \end{bmatrix}$;

in region III: $\begin{bmatrix} a \\ b \end{bmatrix} = \begin{bmatrix} -1 \\ -1 \end{bmatrix}$; in region IV: $\begin{bmatrix} a \\ b \end{bmatrix} = \begin{bmatrix} -1 \\ -2 \end{bmatrix}$.

12. $\left(\frac{1}{5}, \frac{3}{5}\right)$ and $\left(\frac{4}{5}, \frac{2}{5}\right)$ form one 2-cycle, and $\left(\frac{2}{5}, \frac{1}{5}\right)$ and $\left(\frac{3}{5}, \frac{4}{5}\right)$ form another 2-cycle.

14. Begin with a 101 × 101 array of white pixels and add the letter 'A' in black pixels to it. Apply the mapping to this image, which will scatter the black pixels throughout the image. Then superimpose the letter 'B' in black pixels onto this image. Apply the mapping again and then superimpose the letter 'C' in black pixels onto the resulting image. Repeat this procedure with the letters 'D' and 'E'. The next application of the mapping will return you to the letter 'A' with the pixels for the letters 'B' through 'E' scattered in the background.

Exercise Set 10.15 (page 664)

1. (a) *GIYUOKEVBH* (b) *SFANEFZWJH*

2. (a) $A^{-1} = \begin{bmatrix} 12 & 7 \\ 23 & 15 \end{bmatrix}$ (b) Not invertible (c) $A^{-1} = \begin{bmatrix} 1 & 19 \\ 23 & 24 \end{bmatrix}$ (d) Not invertible (e) Not invertible

   (f) $A^{-1} = \begin{bmatrix} 15 & 12 \\ 21 & 5 \end{bmatrix}$

3. *WE LOVE MATH*   4. Deciphering matrix $= \begin{bmatrix} 7 & 15 \\ 6 & 5 \end{bmatrix}$; enciphering matrix $= \begin{bmatrix} 7 & 5 \\ 2 & 15 \end{bmatrix}$

5. *THEY SPLIT THE ATOM*   6. *I HAVE COME TO BURY CAESAR*   7. (a) 010110001   (b) $\begin{bmatrix} 0 & 1 & 1 \\ 1 & 1 & 1 \\ 1 & 0 & 1 \end{bmatrix}$

8. A is invertible modulo 29 if and only if $\det(A) \neq 0 \pmod{29}$.

Exercise Set 10.16 (page 674)

2. $\left.\begin{aligned} a_n &= \tfrac{1}{4} + \left(\tfrac{1}{2}\right)^{n+1}(a_0 - c_0) \\ b_n &= \tfrac{1}{2} \\ c_n &= \tfrac{1}{4} - \left(\tfrac{1}{2}\right)^{n+1}(a_0 - c_0) \end{aligned}\right\} n = 1, 2, \ldots$ $\left.\begin{aligned} a_n &\to \tfrac{1}{4} \\ b_n &= \tfrac{1}{2} \\ c_n &\to \tfrac{1}{4} \end{aligned}\right\}$ as $n \to$

3. $\left.\begin{aligned} a_{2n+1} &= \tfrac{2}{3} + \tfrac{1}{6(4)^n}(2a_0 - b_0 - 4c_0) \\ b_{2n+1} &= \tfrac{1}{3} - \tfrac{1}{6(4)^n}(2a_0 - b_0 - 4c_0) \\ c_{2n+1} &= 0 \end{aligned}\right\} n = 0, 1, 2, \ldots$

   $\left.\begin{aligned} a_{2n} &= \tfrac{5}{12} + \tfrac{1}{6(4)^n}(2a_0 - b_0 - 4c_0) \\ b_{2n} &= \tfrac{1}{2} \\ c_{2n} &= \tfrac{1}{12} - \tfrac{1}{6(4)^n}(2a_0 - b_0 - 4c_0) \end{aligned}\right\} n = 1, 2, \ldots$

4. Eigenvalues: $\lambda_1 = 1$, $\lambda_2 = \tfrac{1}{2}$; eigenvectors: $\mathbf{e}_1 = \begin{bmatrix} 1 \\ 0 \end{bmatrix}$, $\mathbf{e}_2 = \begin{bmatrix} 1 \\ -1 \end{bmatrix}$

5. 12 generations; .006%

6. $\mathbf{x}^{(n)} = \begin{bmatrix} \frac{1}{2} + \frac{1}{3} \cdot \frac{1}{4^{n+2}}[(-3-\sqrt{5})(1+\sqrt{5})^{n+1} + (-3+\sqrt{5})(1-\sqrt{5})^{n+1}] \\ \frac{1}{3} \cdot \frac{1}{4^{n+1}}[(1+\sqrt{5})^{n+1} + (1-\sqrt{5})^{n+1}] \\ \frac{1}{3} \cdot \frac{1}{4^{n+1}}[(1+\sqrt{5})^{n} + (1-\sqrt{5})^{n}] \\ \frac{1}{3} \cdot \frac{1}{4^{n+1}}[(1+\sqrt{5})^{n} + (1-\sqrt{5})^{n}] \\ \frac{1}{3} \cdot \frac{1}{4^{n+1}}[(1+\sqrt{5})^{n+1} + (1-\sqrt{5})^{n+1}] \\ \frac{1}{2} + \frac{1}{3} \cdot \frac{1}{4^{n+2}}[(-3-\sqrt{5})(1+\sqrt{5})^{n+1} + (-3+\sqrt{5})(1-\sqrt{5})^{n+1}] \end{bmatrix}$; $\mathbf{x}^{(n)} \to \begin{bmatrix} \frac{1}{2} \\ 0 \\ 0 \\ 0 \\ 0 \\ \frac{1}{2} \end{bmatrix}$ as $n \to$ 8. $\begin{bmatrix} 1 & 0 & 0 & 0 \\ 0 & 0 & 0 & 0 \\ 0 & 0 & 0 & 0 \\ 0 & 0 & 0 & 1 \end{bmatrix}$

Exercise Set 10.17 (page 684)

1. (a) $\lambda_1 = \frac{3}{2}$, $\mathbf{x}_1 = \begin{bmatrix} 1 \\ \frac{1}{3} \end{bmatrix}$ (b) $\mathbf{x}^{(1)} = \begin{bmatrix} 100 \\ 50 \end{bmatrix}$, $\mathbf{x}^{(2)} = \begin{bmatrix} 175 \\ 50 \end{bmatrix}$, $\mathbf{x}^{(3)} = \begin{bmatrix} 250 \\ 88 \end{bmatrix}$, $\mathbf{x}^{(4)} = \begin{bmatrix} 382 \\ 125 \end{bmatrix}$, $\mathbf{x}^{(5)} = \begin{bmatrix} 570 \\ 191 \end{bmatrix}$

(c) $\mathbf{x}^{(6)} = L\mathbf{x}^{(5)} = \begin{bmatrix} 857 \\ 285 \end{bmatrix}$, $\mathbf{x}^{(6)} \simeq \lambda_1 \mathbf{x}^{(5)} = \begin{bmatrix} 855 \\ 287 \end{bmatrix}$

7. 2.375 8. 1.49611

Exercise Set 10.18 (page 692)

1. (a) Yield = $33\frac{1}{3}\%$ of population; $\mathbf{x}_1 = \begin{bmatrix} 1 \\ \frac{1}{3} \\ \frac{1}{18} \end{bmatrix}$

(b) Yield = 45.8% of population; $\mathbf{x}_1 = \begin{bmatrix} 1 \\ \frac{1}{2} \\ \frac{1}{8} \end{bmatrix}$; harvest 57.9% of youngest age class

2. $\mathbf{x}_1 = \begin{bmatrix} 1.000 \\ .845 \\ .824 \\ .795 \\ .755 \\ .699 \\ .626 \\ .532 \\ 0 \\ 0 \\ 0 \\ 0 \end{bmatrix}$, $L\mathbf{x}_1 = \begin{bmatrix} 2.090 \\ .845 \\ .824 \\ .795 \\ .755 \\ .699 \\ .626 \\ .532 \\ .418 \\ 0 \\ 0 \\ 0 \end{bmatrix}$, $\frac{1.090 + .418}{7.584} = .199$ 4. $h_I = (R-1)/(a_I b_1 b_2 \cdots b_{I-1} + \cdots + a_n b_1 b_2 \cdots b_{n-1})$

5. $h_I = \dfrac{a_1 + a_2 b_1 + \cdots + (a_{J-1} b_1 b_2 \cdots b_{J-2}) - 1}{a_I b_1 b_2 \cdots b_{I-1} + \cdots + a_{J-1} b_1 b_2 \cdots b_{J-2}}$

Exercise Set 10.19 (page 699)

1. $\dfrac{\pi^2}{3} + 4\cos t + \cos 2t + \dfrac{4}{9}\cos 3t$

2. $\dfrac{T^2}{3} + \dfrac{T^2}{\pi^2}\left(\cos\dfrac{2\pi}{T}t + \dfrac{1}{2^2}\cos\dfrac{4\pi}{T}t + \dfrac{1}{3^2}\cos\dfrac{6\pi}{T}t + \dfrac{1}{4^2}\cos\dfrac{8\pi}{T}t\right)$
   $-\dfrac{T^2}{\pi}\left(\sin\dfrac{2\pi}{T}t + \dfrac{1}{2}\sin\dfrac{4\pi}{T}t + \dfrac{1}{3}\sin\dfrac{6\pi}{T}t + \dfrac{1}{4}\sin\dfrac{8\pi}{T}t\right)$

3. $\dfrac{1}{\pi} + \dfrac{1}{2}\sin t - \dfrac{2}{3\pi}\cos 2t - \dfrac{2}{15\pi}\cos 4t$
4. $\dfrac{4}{\pi}\left(\dfrac{1}{2} - \dfrac{1}{1\cdot 3}\cos t - \dfrac{1}{3\cdot 5}\cos 2t - \dfrac{1}{5\cdot 7}\cos 3t - \cdots - \dfrac{1}{(2n-1)(2n+1)}\cos nt\right)$

5. $\dfrac{T}{4} - \dfrac{8T}{\pi^2}\left(\dfrac{1}{2^2}\cos\dfrac{2\pi t}{T} + \dfrac{1}{6^2}\cos\dfrac{6\pi t}{T} + \dfrac{1}{10^2}\cos\dfrac{10\pi t}{T} + \cdots + \dfrac{1}{(2n)^2}\cos\dfrac{2n\pi t}{T}\right)$

Exercise Set 10.20 (page 707)

1. (a) Yes; $\mathbf{v} = \tfrac{1}{5}\mathbf{v}_1 + \tfrac{2}{5}\mathbf{v}_2 + \tfrac{2}{5}\mathbf{v}_3$ (b) No; $\mathbf{v} = \tfrac{2}{5}\mathbf{v}_1 + \tfrac{4}{5}\mathbf{v}_2 - \tfrac{1}{5}\mathbf{v}_3$ (c) Yes; $\mathbf{v} = \tfrac{2}{5}\mathbf{v}_1 + \tfrac{3}{5}\mathbf{v}_2 + 0\mathbf{v}_3$
   (d) Yes; $\mathbf{v} = \tfrac{4}{15}\mathbf{v}_1 + \tfrac{6}{15}\mathbf{v}_2 + \tfrac{5}{15}\mathbf{v}_3$

2. $m = $ number of triangles $= 7$, $n = $ number of vertex points $= 7$,
   $k = $ number of boundary vertex points $= 5$; Equation (7) is $7 = 2(7) - 2 - 5$.

3. $\mathbf{w} = M\mathbf{v} + \mathbf{b} = M(c_1\mathbf{v}_1 + c_2\mathbf{v}_2 + c_3\mathbf{v}_3) + (c_1 + c_2 + c_3)\mathbf{b}$
   $= c_1(M\mathbf{v}_1 + \mathbf{b}) + c_2(M\mathbf{v}_2 + \mathbf{b}) + c_3(M\mathbf{v}_3 + \mathbf{b}) = c_1\mathbf{w}_1 + c_2\mathbf{w}_2 + c_3\mathbf{w}_3$

4. (a) 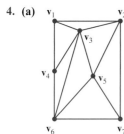   (b)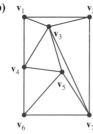

5. (a) $M = \begin{bmatrix} 1 & 2 \\ 0 & 1 \end{bmatrix}$, $\mathbf{b} = \begin{bmatrix} 1 \\ 2 \end{bmatrix}$ (b) $M = \begin{bmatrix} 3 & -1 \\ 1 & 1 \end{bmatrix}$, $\mathbf{b} = \begin{bmatrix} 0 \\ 1 \end{bmatrix}$
   (c) $M = \begin{bmatrix} 1 & 0 \\ 0 & 1 \end{bmatrix}$, $\mathbf{b} = \begin{bmatrix} 2 \\ -3 \end{bmatrix}$ (d) $M = \begin{bmatrix} \tfrac{1}{2} & 1 \\ 2 & 0 \end{bmatrix}$, $\mathbf{b} = \begin{bmatrix} \tfrac{1}{2} \\ -1 \end{bmatrix}$

7. (a) Two of the coefficients are zero.   (b) At least one of the coefficients is zero.
   (c) None of the coefficients are zero.

8. (a) $\tfrac{1}{3}\mathbf{v}_1 + \tfrac{1}{3}\mathbf{v}_2 + \tfrac{1}{3}\mathbf{v}_3$ (b) $\begin{bmatrix} 8/3 \\ 2 \end{bmatrix}$

INDEX

A

Absolute value, 523
  of complex number, 315
  of determinant, 166
Addition
  associative law for, 38, 122
  of complex numbers, 522
  by scalars, 172
  of vectors in R^2 and R^3, 120, 122
  of vectors in R^n, 126
Additivity property
  of linear transformation, 434
  of matrix transformation, 249, 269, 270
Adjacency matrices, 496–497
Adjoint, of a matrix, 110–111
Affine transformations, 637–639
  contracting, 637–638
  with warps, 701
Aeronautics, yaw, pitch, and roll, 256
Age-specific population growth, 676–684
  female age distribution of animals, 678–679
  female age distribution of humans, 683–684
  Leslie matrix, 678, 680–683
  limiting behavior, 679–684
Ahmes Papyrus, 536
Algebraic multiplicity, 312
Algebraic operations, using vector components, 126–127
Algebraic properties of matrices, 38–48
Algebraic properties of vectors, dot product, 135–136
Algebraic Reconstruction Techniques (ARTs), 616, 620–624
Alleles, 292
Amps (unit), 76
Angle
  in R^n, 137, 143
  between vectors, 134–135, 137, 346
Angle of rotation, 254
Animal population harvesting, 686–692
  model for, 686–688
  only in youngest age class, 690–691
  optimal sustainable yield, 692
  uniform, 688–690
Annual yield, maximizing, 526–527
Anosov automorphism, 652–653
Anticommutativity, 326
Antihomogeneity property, of complex Euclidean inner product, 318
Antisymmetry property
  of complex Euclidean inner product, 318

of dot product, 318
Approximate integration, 83
Approximation problems, 382–384
Approximations, best approximations, 367
Area
  of parallelogram, 165
  of triangle, 154
Archimedes, 539
Argument, of complex number, 315, 525
Arithmetic average, 337
Arithmetic operations
  complex numbers, 522–525
  matrices, 27–34, 38–42
  vectors in R^2 and R^3, 120–122
  vectors in R^n, 125
Arnold, Vladimir I., 642
Arnold's cat map, 642–644, 648–650
Artificial intelligence, 479
ARTs. *see* Algebraic Reconstruction Techniques
Associative law for addition, 38, 122
Associative law for matrix multiplication, 38, 39–40
Astronautics, yaw, pitch, and roll, 256
Augmented matrices, 6–7, 11, 12, 18, 25, 33
Authority, 496
Authority vectors, 497
Authority weights, 497
Autosomal inheritance, 665–669
Autosomal recessive diseases, 669–671
Axes
  rotation of in 2–space, 392–394
  rotation of in 3–space, 394–395
Axis of rotation, 254

B

Babylonia, early applications in, 537
Back-substitution, 19–20
Backward phase, 15
Bakhshali Manuscript, 540–541
Balancing (of chemical equation), 79
Barnsley, Michael, 626, 636, 638
Basis, 209–211
  change of, 217–222, 469–470
  coordinate system for vector space, 201–202
  for eigenspaces, 299–301, 472
  finite basis, 201
  by inspection, 212
  linear combinations and, 233–234
  number of vectors in, 209, 211
  ordered basis, 205
  orthogonal basis, 354

  for orthogonal complement, 349–350
  orthonormal basis, 354–355
  for row and column spaces, 229–230, 231
  by row reduction, 230, 231
  for row space of a matrix, 232–233
  standard basis, 202–203, 205–206
  transition matrix, 219
  uniqueness of basis representation, 204
  for vector space using row operations, 232
Basis vectors, 201, 436–437
Bateman, Harry, 509
Battery, 76
Beam density, computed tomography, 618
Begin-triangle, warps, 701
Beltrami, Eugenio, 510
Best approximation theorem, 367
Bôcher, Maxime, 7, 184
Books, ISBN number of, 141
Boundary data, temperature distribution, 605–606
Boundary mesh points, 607
Bounded feasible regions, 529, 530
Bounded sets, 626–627
Branches (network), 73
Brightness, graphical images, 124
Brin, Sergey, 496
Bunyakovsky, Viktor Yakovlevich, 137

C

C^n, 318–321
Calculus of variations, 162
Cancellation law, 41
Cantor set, 641
Carroll, Lewis, 96
Cat map (Arnold's), 642–644, 648–650
CAT scanner, 616, 617
Cattle Problem, 539–540
Cauchy, Augustin, 109, 137, 172
Cauchy-Schwarz inequality, 137, 345–346
Cayley, Arthur, 29, 34, 43
Central conic, 409
Central conic in standard position, 409
Central ellipsoid in standard position, 416
Central quadrics in standard position, 409
Change-of-basis problem, 217–218, 469
Change of variable, 406–407
Chaos, 641–652
  Arnold's cat map, 642–644, 648–650
  defined, 650
  dynamical systems, 651–652
  nonperiodic points, 649–650
  periodic points, 644–646

period vs. pixel width, 646–647
repeated mappings, 643–644
tiled planes, 647–648
Characteristic equation, 296, 308
Characteristic polynomial, 297, 308
Chemical equations, balancing with linear systems, 78–80
Chemical formulas, 78
Chessboard moves, 565–566
China, early applications in, 537–539
Chiu Chang Suan Shu, 537–538
Ciphers, 654–657. *See also* Cryptography
Ciphertext, 654
Ciphertext vector, 655
Circle, through three points, 521
Clamped splines, 552
Cliques, directed graphs, 566–568
Clockwise closed-loop convention, 76
Closed economies, 86
Closed Leontief model, 582–585
Closed sets, 626–627
Closure under addition, 172
Closure under scalar, 172
Codomain, 247
Coefficient matrices, 33, 308, 477
Coefficients
of linear combination of matrices, 31
of linear combination of vectors, 127, 183
literal, 44
Cofactor, 94
Cofactor expansion
determinants by, 93–98
elementary row operations and, 104
of 2×2 matrices, 95
Collinear vectors, 121–122
Column matrices, 26
Column-matrix form of vectors, 128, 225
Column space, 225, 226, 228, 243
basis for, 229–230, 231
equal dimensions of row and column space, 237
Column vectors, 26, 27, 39
Column-wheel, 572
Columns, cofactor expansion and choice of column, 97
Combustion, linear systems to analyze combustion equation for methane, 78–79
Comma-delimited form of vectors, 128, 225
Common initial point, 122
Commutative law for addition, 38
Commutative law for multiplication, 40, 46
Complete reaction (chemical), 78

Complex conjugates
of complex numbers, 315, 522–523
of vectors, 316, 317–318
Complex dot product, 317
Complex eigenvalues, 318, 319, 322–324
Complex eigenvectors, 318, 319
Complex Euclidean inner product, 317–318
Complex exponential, complex numbers, 527
Complex exponential function, 526
Complex inner product space, 344
Complex inner products, 344
Complex matrices, 316
Complex n-space, 316
Complex n-tuples, 316
Complex number system, 522
Complex numbers, 315–316, 521–527
complex conjugates, 522–523
defined, 521
DeMoivre's formula, 526
division of, 524, 525–526
multiplication of, 521, 522, 525–526
polar form of, 315, 524–525
reciprocals, 523
Complex plane, 522
Complex vector spaces, 172, 315–325
Components (of a vector)
algebraic operations using, 126–127
calculating dot products using, 35
complex n-tuples, 316
finding, 123–124
in R^2 and R^3, 122–123
vector components of **u** along **a**, 147–148
Composition
with identity operator, 453–454
of linear transformations, 452–453, 455–456
matrices of, 464, 465
of matrix transformations, 263–267
non-commutative nature of, 264
of one-to-one linear transformations, 455–456
of reflections, 265, 277
of rotations, 264, 275
of three transformations, 265–267
Compression operator, 257, 277
Computed tomography, 615–624
Algebraic Reconstruction Techniques, 620–624
derivation of equations, 617–619
scanning modes, 616–617
Computer graphics, 597–602
morphs, 700, 704–707
rotation, 600–602

scaling, 599
translation, 600
visualization of three-dimensional object, 597–599
warps, 700–704
Computer programs, LU-decomposition and, 478
Computers
Internet search engines, 496–500
LINPACK, 478
Condensation, 96
Congruent set, 626
Conic sections (conics), 408–412
through five points, 522–523
Conjugate transpose, 424
Conjugates, of complex numbers, 522
Consistency, determining by elimination, 64–65
Consistent linear system, 3–4, 227
Constrained extremum, 417, 418
Constrained extremum problems, 417–419
Constrained extremum theorem, 417
Constraint, 417, 528, 534
Consumption matrix, 86, 586
Consumption vectors, 87, 88
Continuous derivatives, functions with, 182
Contracting affine transformation, 637–638
Contraction, 256, 257, 435, 446
Contraction operators, 435
fractals, 626, 627, 630–631
Contrapositive, 519
Contrapositive form of theorem, 519
Convergence
of power sequences, 487
rate of, 493
Converse, 519
Convex combination, 701
Coordinate map, 217
Coordinate matrices, 205
three-dimensional computer graphic views, 604
Coordinate systems, 200
"basis vectors" for, 201
units of measurement, 201
Coordinate vectors, 205
computing, 220
relative to orthonormal basis, 355
relative to standard bases, 206
Coordinates, 205
of generalized point, 124
in R^3, 206–207
relative to standard basis for R^n, 205–206

Index 767

Cormack, A. M., 616
Corner points, 530
Corresponding linear systems, 158
Cost, minimizing, 527–528
Cramer, Gabriel, 113
Cramer's rule, 112–113
Critical points, 420
Cross product, 161–163
  calculating, 161
  determinant form of, 164
  geometric interpretation of, 164–165
  notation, 161
  properties of, 163
Cross product terms, 405, 411
Cryptography, 654–663
  breaking Hill ciphers, 661–663
  ciphers, 654–657
  deciphering, 659–660
  Hill ciphers, 655–657, 660–663
  modular arithmetic, 657–658
CT. *see* Computed tomography
Cubic runout spline, 548–551
Cubic spline, 545–548
Cubic spline interpolation, 543–551
  cubic runout spline, 548–551
  curve fitting, 543
  derivation of formula of cubic spline, 545–548
  natural spline, 548–549
  parabolic runout spline, 548, 549
  statement of problem, 544–545
Current (electrical), 76
Curve fitting, cubic spline interpolation, 543

D

Dantzig, George, 525
Data compression, singular value decomposition, 514–516
Deciphering matrix, 661
Decomposition
  eigenvalue decomposition, 506, 507
  Hessenberg decomposition, 506, 507
  LDU-decomposition, 484–485
  LU-decompositions, 477–484, 504–505
  PLU-decomposition, 485
  Schur decomposition, 507
  self-similar sets, 627
  singular value decomposition, 509–511, 514–516
  of square matrices, 506–507
Degenerate conic, 408
Degrees of freedom, 209
Demand vector, 586
DeMoivre, Abraham, 526

DeMoivre's formula, 526
Dense sets, in chaos theory, 649–650
Dependency equations, 234
Determinants, 43, 93–110
  by cofactor expansion, 93–98
  defined, 93
  of elementary matrices, 102
  equivalence theorem, 114
  evaluating by row reduction, 100–104
  general determinant, 95
  geometric interpretation of, 166–168
  of linear operator, 472
  of matrix product, 107–108
  properties of, 106–112
  sums of, 107
  of 3×3 matrices, 97–98
  of 2×2 matrices, 97–98
  of upper triangular matrix, 97
Devaney, Robert L., 650
Deviation, 382
Diagonal coefficient matrices, 329
Diagonal entries, 509
Diagonal matrices, 66–68, 274–275
Diagonalizability
  nondiagonalizability of $n \times n$ matrix, 402
  orthogonal diagonalizability, 427–428
  of triangular matrices, 309–310
Diagonalization
  matrices, 305–313
  orthogonal diagonalization, 397
  solution of linear system by, 329–331
Dickson, Leonard Eugene, 111
Difference
  of complex numbers, 522
  matrices, 27
  vectors, 121, 126
Differential equations, 327–331, 440
Differentiation, by matrix multiplication, 450
Differentiation transformation, 439
Digital communications, matrix form and, 245
Dilation, 256, 257, 435, 446
Dilation operators, 435, 626
Dimension theorem, for linear transformations, 441–442
Dimensions
  of spans, 209–210
  and transformations, 447
  of vector spaces, 209
Dirac matrices, 326
Directed edges, 563
Directed graphs, 563–568
  cliques, 566–568
  dominance-directed, 568–570

Direct product, 134
Discrete mean-value property, 607
Discrete random walk, 612
Discrete-time chaotic dynamical systems, 651
Discrete-time dynamical systems, 651
Discriminant, 321
Distance, 336
  between a point and a plane, 149–150
  general inner product spaces, 346–347
  orthogonal projections for, 148–149
  between parallel planes, 150
  real inner product spaces, 336
  in R^n, 132–133
  triangle inequality for, 138
Distinct eigenvalues, 487
Distributive property
  of complex Euclidean inner product, 318
  of dot product, 136
Division, of complex numbers, 524, 525–526
Dodgson, Charles Lutwidge, 96
Domain, 247
Dominant eigenvalue, 487–489
Dominance-directed graphs, 568–570
Dominant eigenvalue, of Leslie matrix, 680
Dominant genes, 666
Dot product, 133–136
  algebraic properties of, 135–136
  antisymmetry property of, 318
  application of, 141
  calculating with, 136–137
  complex dot product, 317
  cross product and, 162
  dot product form of linear systems, 156–157
  as matrix multiplication, 139–140
  relationships involving, 162
  symmetry property of, 136, 318
  of vectors, 139
Drafting spline, 543
Dynamical system, 282–284, 651–652

E

Ear
  anatomy of, 693–694
  least squares hearing model, 693–698
Echelon forms, 11–12, 21
Economic modeling, Leontief economic analysis with, 85–89, 581–588
Economic sectors, 85
Economics, n-tuples and, 124
Egypt, early applications in, 536–537

Eigenspaces, 299, 308, 318
  bases for, 299–301
  of real symmetric matrix, 425, 426
Eigenvalue decomposition (EVD), 506, 507
Eigenvalues, 295–296, 308, 319
  complex eigenvalues, 318, 319
  conic sections classified by using, 413
  dominant eigenvalues, 487–489
  of Hermitian, 425, 426
  of Hermitian matrices, 429
  invertibility and, 301–302
  of Leslie matrix, 680–683
  of linear operators, 472
  of square matrix, 311
  of symmetric matrices, 398
  of 3×3 matrix, 297
  of triangular matrices, 298–299
  of 2×2 matrix, 321
Eigenvectors, 295–296, 299
  of 2×2 vector, 296
  bases for eigenspaces and, 300–301
  complex eigenvectors, 318, 319
  left/right eigenvectors, 304
  of real symmetric matrix, 425, 426
  of square matrix, 311
  of symmetric matrices, 398
Einstein, Albert, 123, 124
Eisenstein, Gotthold, 29
Electrical circuits
  n-tuples and, 124
  network analysis with linear systems, 76–78
Electrical current, 76
Electrical potential, 76
Electrical resistance, 76
Elementary matrices, 51
  determinants, 102
  and homogeneous linear systems, 57
  invertibility, 53
  matrix operators corresponding to, 277
Elementary row operations, 7–8, 51, 228
  cofactor expansion and, 104
  determinants and, 100–104
  and inverse operations, 52–55
  and inverse row operations, 52–55
  for inverting matrices, 55–56
  matrix multiplication, 52
  row reduction and determinants, 100–104
Elimination methods, 14–16, 64–65
Ellipse, principal axes of, 410
Empty set, 209
Enciphering, 654
End-triangle, warps, 701
Entries, 26

Equal complex numbers, 521
Equal matrices, 27–28, 39
Equal vectors, 120, 125
Equilibrium temperature distribution, 605–613
  boundary data, 605–606
  discrete formulation of problem, 607–611
  mean-value property, 606–607
  Monte Carlo technique for, 612–613
  numerical technique for, 611–612
Equivalence theorem, 373
  determinants, 114
  invertibility, 53–54, 302
  $n \times n$ matrix, 240–241
Equivalent statements, 519–520
Equivalent vectors, 120, 125
Error vector, 369
Errors
  approximation problems, 384–385
  least squares error, 367
  mean square error, 383
  measurements of, 382–383
  percentage error, 493
  relative error, 493
  roundoff errors, 21–22
Estimated percentage error, 493
Estimated relative error, 493–494
Euclidean inner product, 336–339
  complex Euclidean inner product, 317–318
  of vectors in R^2 or R^3, 133
Euclidean n-space, 336
Euclidean norm, 317
Euclidean scaling, power method with, 489–490
Euler phi functions, 665
Euler's formula, 527
Evaluation inner product, 340
Evaluation transformation, 436–437
EVD (eigenvalue decomposition), 506, 507
Exchange matrix, 583
Expansion operator, 257, 277
Expected payoff, matrix games, 574–575
Exponents, matrix laws, 46
Extreme points, 530

F
Factoring, 447
Factorization, 480
Family influence, 564–565
Fan-beam mode scanning, computed tomography, 616, 617
Feasible region, 529–534
Feasible solution, 528

Fertile age class, 677
Fibonacci shift-register random-number generator, 652
Fingerprint storage, 515
Finite basis, 201
Finite-dimensional inner product space, 349, 362
Finite-dimensional vector space, 204, 212–213, 217
First-order linear system, 328
Fixed points, 646
Floating-point numbers, 501
Floating-point operation, 501
Flops, 501–503
Flow conservation, in networks, 74
Forest management, 590–596
Forward phase, 15
Forward substitution, 479
4×6 matrix, rank and nullity of, 238
Fourier, Jean Baptiste, 386
Fourier coefficients, 385
Fourier series, 384–385, 386
Fractals, 626–639
  algorithms for generating, 633–636
  defined, 630
  in Euclidean plane, 626
  Hausdorff dimension of self-similar sets, 629–630
  Monte Carlo approach for, 636–637
  self-similar sets, 626–628
  similitudes, 630–633
  topological dimension of sets, 628–629
Free variable theorem for homogeneous systems, 19
Free variables, 13, 239
Full column rank, 363
Function spaces, 182–183
Functions
  with continuous derivatives, 182
  defined, 247–248
  linear dependence of, 196–197
Fundamental spaces, 243, 244
Fundamental Theorem of Two-Person Zero-Sum Games, 575–576

G
Games of strategy
  game theory, 572–573
  2×2 matrix games, 578–580
  two-person zero-sum games, 573–578
Game theory, 572–573
Gauss, Carl Friedrich, 15, 29, 94, 538
Gauss-Jordan elimination
  of augmented matrix, 319, 505
  described, 15
  for homogeneous system, 18

polynomial interpolation by, 82
  roundoff errors, 21–22
  using, 44, 504, 505
Gaussian elimination, 11–16, 505
  defined, 16
  roundoff errors, 21–22
General determinant, 95
General Electric CT system, 616, 617
General linear programming problem, 528–530
General solution, 13, 227, 240, 328
Generalized Theorem of Pythagoras, 348
Genes, dominant and recessive, 666
Genetic diseases, 669–671
Genetics, 665–674
  autosomal inheritance, 666–669
  autosomal recessive diseases, 669–671
  inheritance traits, 665–666
  X-linked inheritance, 671–674
Genotypes, 292, 665–666
  defined, 665
  distribution in population, 666–668
Geometric linear programming, 525–534
Geometric multiplicity, 312
Geometric vectors, 119
Geometry
  of linear systems, 152–159
  quadratic forms in, 408–409
  in R^n, 138
Gibbs, Josiah Willard, 134, 161
Golub, Gene H., 510
Google
  algorithms used by, 496
  origin of term, 496
Googol, 496
Gram, Jorgen Pederson, 360
Gram-Schmidt process, 356, 358–360, 361, 362, 363, 384
Graphic images
  images of lines under matrix operators, 279–280
  n-tuples and, 124
  RGB color model, 127
Graph theory, 563–570
  cliques, 566–568
  directed graphs, 563–568
  dominance-directed graphs, 568–570
  relations among members of sets, 563
Grassmann, H.G., 172
Greece, early applications in, 539–540
Growth matrix, forest management model, 592

H

Harvesting
  animal populations, 686–692
  forests, 590–596
Harvesting matrix (animals), 687–688
Harvest vector (forests), 592
Hausdorff, Felix, 629
Hausdorff dimension, 629–630
Hearing, least squares model for, 693–698
Hermite, Charles, 426
Hermitian matrices, 425–429
Hesse, Ludwig Otto, 420
Hessenberg decomposition, 506, 507
Hessenberg's theorem, 403
Hessian matrices, 420, 421
Hilbert, David, 360
Hilbert space, 360
Hill, George William, 184
Hill, Lester S., 655
Hill 2-cipher, 656, 660
Hill 3-cipher, 657
Hill ciphers, 655–657, 660–663
Hill n-cipher, 657
Homogeneity property
  of complex Euclidean inner product, 318
  of dot product, 136
  of linear transformation, 434
  of matrix transformation, 249, 269, 270
Homogeneous equations, 145, 156
Homogeneous linear equations, 2
Homogeneous linear systems, 17–19, 227
  dimensions of solution space, 210
  and elementary matrices, 57
  free variable theorem for, 19
  solutions of, 186–187
Homogeneous systems, solutions spaces of, 187–188
Hooke's law, 378
Houndsfield, G. N., 616
Hub, 496
Hub vectors, 497
Hub weights, 497
Hue, graphical images, 124
Human hearing, least squares model for, 693–698
Hyperplane, 622

I

Idempotency, 50
Identity matrices, 41–42
Identity operators
  about, 251, 434
  composition with, 453–454
  kernel and range of, 439
  matrices of, 463–464
Image processing, data compression and, 515

Images, 247. *See also* Graphic images
  of a line, 279–280
  of a square, 279
Images of basis vectors, 436–437
Imaginary axis, 522
Imaginary numbers, 521. *See also* Complex numbers
Imaginary part
  of complex numbers, 315, 521
  of vectors and matrices, 316
Implication loop, 520
Inconsistent linear system, 3
Indefinite quadratic forms, 412
India, early applications in, 540–541
Infinite-dimensional vector space, 204
Inheritance, 665–666
  autosomal, 665–669
  X-linked, 666, 671–674
Initial age distribution vector, 676
Initial authority vector, 497
Initial condition, 328
Initial hub vector, 497
Initial point, 119
Initial-value problem, 328
Inner product
  algebraic properties of, 342
  calculating, 342
  complex inner products, 344
  Euclidean inner product, 133, 317–318, 336–339
  evaluation inner product, 340
  examples of, 336–341
  linear transformation using, 435
  matrix inner products, 338
  on M_{nn}, 339–340
  on real vector space, 335
  on R^n, 336–339
  standard inner products, 336, 340
Inner product space, 435
  complex inner product space, 344
  unit circle, 338
  unit sphere, 338
Inner product space isomorphism, 450–451
Input-output analysis, 85
Input-output matrix, 583
Inputs, in economics, 85
Instability, 22
Integer coefficients, 297
Integral transformation, 438
Integration, approximate, 83
Interior mesh points, 607
Intermediate demand vector, 87
Internet search engines, 496–500
Interpolating curves, 543
Interpolating polynomial, 80

Interpolation, 543
Invariant under similarity, 305, 471–472
Inverse
   of a product, 45–46
   of diagonal matrices, 67
   of matrix using its adjoint, 111–112
   of 2×2 matrices, 44
Inverse linear transformations, 454–455
Inverse matrices, 41–45
Inverse operations, 52–55
Inverse row operations, 52–55
Inverse transformations, 464
Inversion, solving linear systems by, 44–45, 60–62
Inversion algorithm, 55
Invertibility
   determinant test for, 108–110
   eigenvalues and, 301–302
   of elementary matrices, 53
   equivalence theorem, 53–54
   matrix transformation and, 266–267
   test for determinant, 108–110
   of transition matrices, 220
   of triangular matrices, 68
Invertibility test for determinant, 108–110
Invertible matrices, 41–45, 62–65
   modulo m, 659–660
ISBN (books), 141
Isomorphism, 447–451
Isotherms, 606
Iterates (Jacobi iteration), 611–612
Iterations
   of Arnold's cat map, 643
   Jacobi, 611–612

J
Jacobi iteration, 611–612
Jordan, Camille, 507, 510
Jordan, Wilhelm, 15
Jordan canonical form, 407
Junctions (network), 73, 76

K
Kaczmarz, S., 620
Kalman, Dan, 400
Kasner, Edward, 496
Kernel, 438, 439–440, 445
Kirchhoff, Gustav, 77
Kirchhoff's current law, 76
Kirchhoff's voltage law, 76
kth principal submatrix, 414

L
Lagrange, Joseph Louis, 162
   LDU-decomposition, 484–485
   LDU-factorization, 485

Leading 1!, 11
Leading variables, 13, 239
Least squares, curve fitting, 376
Least squares approximation, 382–385
   in human hearing model, 693–698
Least squares error, 367
Least squares error vector, 367
Least squares fit
   of polynomial, 379–380
   of quadratic curve to data, 380–381
   straight line fit, 376–377, 378
Least squares polynomial fit, 379–380
Least squares solutions, 378
   of linear systems, 366–371, 372
   QR-decomposition and, 371
   straight line fit, 376–377, 378
Least squares straight line fit, 377
Left distributive law, 38
Left eigenvectors, 304
Legendre polynomials, 361
Length, 130, 336, 346
Leontief, Wassily, 85, 86, 581
Leontief economic models, 581–588
   closed (input-output) model, 582–585
   economic systems, 581
   open (production) model, 585–588
Leontief equation, 87
Leontief input-output models, 85–89
Leontief matrices, 87
Leslie matrix age-specific population growth, 678, 680–683
   animal population harvesting, 687–688
   eigenvalues, 680–683
Leslie model, of population growth, 676–684
Level curves, 419
Limit cycle, 620
Line. *See* Lines
Line segment, from one point to another in R^2, 156
Linear algebra, 1. *See also* Linear equations; Linear systems
   coordinate systems, 200
   earliest applications of, 536–541
Linear beam theory, 544
Linear combinations
   basis and, 233–234
   history of term, 184
   of matrices, 31, 32–33
   of vectors, 127, 132–133, 183, 185–186
Linear dependence, 184, 193
Linear equations, 2–3, 156. *See also* Linear systems
Linear form, 405
Linear independence, 184, 190–198, 214
   examples of, 194

   of polynomials, 193
   of sets, 191–194
   of standard unit vectors in R^3, 191–192
   of standard unit vectors in R^4, 192
   of standard unit vectors in R^n, 191
   of two functions, 195
   use of terms, 193
   using the Wronskian, 197–198
Linear operators
   determinants of, 472
   matrices of, 462, 468–469
   orthogonal matrices as, 391–392
   on P^2, 463
Linear programming, geometric, 525–534
Linear systems, 2–3. *See also* Homogeneous linear systems
   with a common coefficient matrix, 61
   applications, 73–83
   augmented matrices, 6–7, 11, 12, 18, 25, 33
   for balancing chemical equations, 78–80
   coefficient matrix, 33
   comparison of procedures for solving, 501–505
   computer solution, 1
   corresponding linear systems, 158
   cost estimate for solving, 501–504
   dot product form of, 156–157
   first-order linear system, 328
   general solution, 13
   geometry of, 152–159
   with infinitely many solutions, 5–7
   least squares solutions of, 366–371, 372
   network analysis with, 73–78
   with no solutions, 5
   non-homogeneous, 19
   number of solutions, 60
   overdetermined/underdetermined, 241
   polynomial interpolation, 80–82
   solution methods, 3, 4–7
   solutions, 3, 11
   solving by elimination row operations, 7–8
   solving by Gaussian elimination, 11–16, 21–22, 505
   solving by matrix inversion, 44–45, 60–62
   solving with Cramer's rule, 112–113
   in three unknowns, 12–13
Linear transformations, 270
   composition of, 452–453, 455–456
   defined, 433
   dimension theorem for, 441–442
   examples of, 435, 438
   inverse linear transformations, 454–455

matrices of, 458–462
one-to-one, 445
onto, 445
from P_n to P_{n+1}, 435
rank and nullity in, 441
using inner product, 435
Linearity conditions, 270
Linearly dependent set, 191
Linearly independent set, 191, 192–193
Lines
  image of, 279–280
  line segment from one point to another in R^2, 156
  orthogonal projection on, 147
  orthogonal projection on lines through the origin, 258–259
  point-normal equations, 144–145
  through origin as subspaces, 180–181
  through two points, 520-521
  through two points in R^3, 155–156
  vector and parametric equations in R^2 and R^3, 152–153, 154
  vector and parametric equations of in R^4, 155
  vector form of, 145, 154
  vectors orthogonal to, 145
LINPACK, 478
Literal coefficients, 44
Liu Hui, 537
Lower triangular matrices, 68, 298
LU-decompositions, 477–484, 504–505
  constructing, 483–484
  examples of, 480–483
  finding, 480
  method, 478
LU-factorization, 480

M

M_{mn}. See $m \times n$ matrices
M_{nn}
  inner products on, 339–340
  standard basis, 203
  subspaces of, 181
$m \times n$ matrices, real vector spaces, 174–175
Maclaurin series, 527
Magnitude (norm), 130
Main diagonal, 27, 509
Mandelbrot, Benoit B., 626, 630
Mantissa, 501
Mapping, 248
Markov, Andrei Andreyevich, 285
Markov chain, 285, 286–288, 553–561
  limiting behavior of state vectors, 558–561

steady-state vector of, 289
transition matrix for, 289–290, 554–557
Markov matrix, 555
MATLAB, 478
Matrices. See also Matrices of specific size, such as 2×2 matrices
  adjacency matrices, 496–497
  adjoint of, 110
  algebraic properties of, 38–48
  arithmetic operations with, 27–34
  coefficient matrices, 33, 308, 477
  column matrices, 26
  complex matrices, 316
  compositions of, 464, 465
  coordinate matrices, 205
  defined, 1, 6, 26
  determinants, 93–110
  diagonal coefficient matrices, 329
  diagonal matrices, 66–68, 274–275
  diagonalization, 305–313
  dimension theorem for matrices, 239
  elementary matrices, 51, 53, 57, 102, 277
  entries, 26
  equality of, 27–28, 39
  examples of, 26–27
  fundamental spaces, 243, 244
  Hermitian matrices, 425–427, 429
  Hessian matrices, 420, 421
  identity matrices, 41–42
  of identity operators, 463–464
  inner products generated by, 338–339
  inverse matrices, 41–45
  of inverse transformations, 464
  invertibility, 53–54, 68, 108–110, 220
  invertible matrices, 41–45, 62–65
  inverting, 55–57
  Leontief economic analysis with, 85–89
  linear combination, 31, 32–33
  of linear operators, 462, 468–469
  of linear transformations, 458–462
  normal matrices, 429
  notation and terminology, 25–27, 33
  orthogonal matrices, 389–395
  orthogonally diagonalizable matrices, 397
  partitioned, 30–31
  permutation matrices, 485
  positive definite matrices, 414
  powers of, 45–46, 301–302, 310–311
  with proportional rows or columns, 102–103
  rank of, 239
  real and imaginary parts of, 316–317
  real matrices, 316, 322
  redundancy in, 245

reflection matrices, 390
rotation matrices, 254, 390
row equivalents, 51
row matrices, 26
scalar multiples, 28
similar matrices, 305
singular/nonsingular matrices, 42
size of, 26, 39
skew-Hermitian matrices, 429
skew-symmetric matrices, 428–429
square matrices, 27, 34, 35, 42, 66, 68, 100–104, 311, 389, 506–507
standard matrices, 248, 251, 268–269, 274, 372
stochastic matrices, 288–289
submatrices, 30, 414
symmetric matrices, 69–70, 299, 322, 398, 420
trace, 35
transition matrices, 219–221, 469
transpose, 33–34
triangular matrices, 68–69, 298–299, 309–310
unitary matrices, 425, 426, 427
upper triangular matrices, 68, 97, 298
zero matrices, 40
Matrix factorization, 323
Matrix games
  defined, 573
  two-person zero-sum, 573–578
Matrix inner products, 338
Matrix multiplication. See Multiplication (matrices)
Matrix notation, 25–27, 33, 406
Matrix operators, 248, 446
  geometric effect of, 278
  graphics images of lines under matrix operators, 279–280
  on R^2, 273–280
Matrix polynomials, 46–47
Matrix spaces, transformations on, 435
Matrix transformations, 248, 434
  composition of, 263–267
  defined, 433
  kernel and range of, 438–439
  notation, 249
  properties of, 249–250
  from R^4 to R^3, 249
  standard matrix for, 251
  zero transformations, 250, 434, 439
Maximization problems
  linear programming, 525–534
  two-person zero-sum games, 577–578
Maximum entry scaling, power method with, 490–493
Mean square error, 383

Mean-value property, 606–607
Mechanical systems, n-tuples and, 125
Menger sponge, 640
Mesh points, 607–611
Methane, linear systems to analyze combustion equation, 78–79
Minimization problems, linear programming, 527–528, 530
Minor, 94
Mixed strategies, of players in matrix games, 577
Modular arithmetic, 642, 657–658
Modulus, 657
  of complex numbers, 315, 323
Monte Carlo technique
  fractal generation, 636–637
  temperature distribution determination, 612–613
Morphs, 700, 704–707
Multiplication (complex numbers), 521, 522, 525–526
Multiplication (matrices), 28–31, 249–250. *See also* Product (of matrices)
  associative law for, 38, 39–40
  by columns and by rows, 30–31
  differentiation by, 450
  dot products as, 139–140
  elementary row operations, 52
  by invertible matrix, 278
  order and, 40
Multiplication (vectors). *See also* Cross product; Euclidean inner product; Inner product; Product (of vectors)
  in R^2 and R^3, 121
  by scalars, 172
Multiplication by A, 248
Multiplicative inverse, 523
  of a modulo m, 658

N

n-cycle, 646
N-dimensional vector space, 212
n-space, 123, 124. *See also* R^n
$n \times n$ matrices
  equivalent statements, 240–241
  Hessenberg's theorem, 403
  nondiagonalizability of, 402
Natural spline, 548–549
Natural isomorphism, 449
Negative, of vector, 121
Negative definite quadratic forms, 412
Negative pole, 76
Negative semidefinite quadratic forms, 412
Net reproduction rate, 684

Network analysis, with linear systems, 73–78
Networks, defined, 73
Newton, Isaac, 522
Nodes (network), 73, 76
Nonharvest vector (forests), 591
Non-homogeneous linear systems, 19
Nonnegativity constraints, 528
Nonoverlapping sets, 626, 627
Nonperiodic pixel points, 649–650
Nonsingular matrices, 42
Nontrivial solution, 17
Nonzero vectors, 188
Norm (length), 130, 148, 336
  calculating, 131
  complex Euclidean inner product and, 317–318
  Euclidean norm, 317
  real inner product spaces, 336
  of vector in $C[a, b]$, 341
Normal, 144
Normal equations, 368, 377–378
Normal matrices, 429
Normal system, 368
Normalization, 132, 364
Null space, 225, 228–229
Nullity, 441
  of 4×6 matrix, 238
  sum of, 239–240
Numerical analysis, 11
Numerical coefficients, 44

O

Objective function, 528
Ohm's law, 76
Ohms (unit), 76
One-to-one linear transformations, 445, 455–456
1-Step connection, directed graphs, 565, 568–569
Onto linear transformations, 445
Open economies, Leontief analysis of, 86–89
Open Leontief model, 585–588
Open sectors, 86
Operators, 248, 435, 446. *See also* Linear operators
Optimal solution, 529
Optimal strategies
  2×2 matrix games, 579–580
  two-person zero-sum games, 576–578
Optimal sustainable harvesting policy, 692
Optimal sustainable yield
  animal harvesting, 692
  forest harvesting, 590, 593–596

Optimization, using quadratic forms, 417–422
Orbits, 522, 523
Order
  of differential equation, 327
  matrix multiplication and, 40
Order n, 384
Ordered basis, 205
Ordered n-tuple, 3, 124
Ordered pair, 3
Ordered triple, 3
Orthogonal basis, 354, 356, 361
Orthogonal change of variable, 407
Orthogonal complement, 243–244, 348–350
Orthogonal diagonalizability, 427–428
Orthogonal diagonalization, 397
Orthogonal matrices, 389–395
Orthogonal operators, 392
Orthogonal projection operators, 251
Orthogonal projections, 146–147, 357
  with Algebraic Reconstruction Technique, 620-622
  on a subspace, 369–370
  geometric interpretation of, 358
  kernel and range of, 439
  on lines through the origin, 258–259
  for standard matrix, 372
  on subspaces of R^m, 371–372
Orthogonal sets, 143, 353
Orthogonal vectors, 143–145, 317
  in M_{22}, 347
  in P_2, 347–348
Orthogonality
  defined, 352–353
  inner product and, 347
  of row vectors and solution vectors, 157–158
Orthogonally diagonalizable matrices, 397
Orthonormal basis, 354–355, 358, 384
  change of, 392
  coordinate vectors relative to, 355
  from orthogonal basis, 356
  orthonormal sets extended to, 361–362
Orthonormal sets, 143, 354
  constructing, 353–354
  extended to orthonormal bases, 361–362
Orthonormality, 353
Outputs, in economics, 85
Outside demand vector, 87, 88
Overdetermined linear system, 241
Overlapping sets, 626, 627

P

P_n. *See* Polynomials
P_2
 linear operators on, 463
 orthogonal vectors in, 347–348
 Theorem of Pythagoras in, 348
Page, Larry, 496
PageRank algorithm, 496
PageRank score, 496
Parabolic runout spline, 548, 549
Parallel mode scanning, computed tomography, 616–617
Parallel planes, distance between, 150
Parallel vectors, 121–122
Parallelogram, area of, 164
Parallelogram equation for vectors, 138
Parallelogram rule for vector addition, 120
Parameters, 5, 13, 152, 153
Parametric equations, 6
 of lines and planes in R^4, 155
 of lines in R^2 and R^3, 152–153, 154
 of planes in R^3, 153, 154–155
Particular solution, 227
Partitioned matrices, 30–31
Pauli spin matrices, 326
Payoff, matrix games, 573
Payoff matrix, 574, 576
Percentage error, 493
Period, of a pixel map, 646
Periodic splines, 552
Permutation matrices, 485
Perpendicular vectors, 143
Photographs, data compression and image processing, 515
Piazzi, Giuseppe, 15
Picture, 644
Picture-density, of begin-triangle, 701
Pine forest growth, 596
Pitch (aircraft), 256
Pivot column, 21
Pivot position, 21
Pixel maps, 644–647
Pixel points, 645
 nonperiodic, 649–650
Pixels
 data compression and image processing, 515
 defined, 644
Plaintext, 654
Plaintext vector, 655
Planes
 distance between a point and a plane, 149–150
 distance between parallel planes, 150
 point-normal equations, 144–145
 through origin as subspaces, 181
 through three points, 523
 tiled, 647–648
 vector and parametric equations in R^3, 153, 154–155
 vector and parametric equations of in R^4, 155
 vector form of, 145, 154
 vectors orthogonal to, 145
PLU-decomposition, 485
PLU-factorization, 485
Plus-minus theorem, 211–212
Point-normal equations, 144–145
Points
 constructing curves and surfaces through, 520–524
 distance between a point and a plane, 149–150
Polar form, of complex numbers, 315, 524–525
Poles (battery), 76
Polygraphic system, 655
Polynomial interpolation, 80–82
Polynomials (P_n), 46–47
 characteristic polynomial, 297, 308
 cubic, 544–551
 least squares fit of, 379–380
 Legendre polynomials, 361
 linear independence of, 193
 linear transformation, 435
 linearly independent set in, 192–193
 spanning set for, 185
 standard basis for, 202
 standard inner product on, 340
 subspaces of, 182
 trigonometric polynomial, 384–385
Population growth, age-specific, 676–684
Population waves, 680
Positive definite matrices, 414
Positive definite quadratic forms, 412–414
Positive pole, 76
Positive semidefinite quadratic forms, 412
Positivity property
 of complex Euclidean inner product, 318
 of dot product, 136
Power, of vertex of dominance-directed graph, 570
Power method, 487–494
 with Euclidean scaling, 489–490
 with maximum entry scaling, 490–493
 for search engine algorithms, 496–500
 stopping procedures, 494
Power sequence generated by A, 487
Powers of a matrix, 45–46, 67, 301–302, 310–311
Price vector, 583
Principal argument, of complex number, 525
Principal axes, 410
Principal axes theorem, 407–408, 410
Principal submatrices, 414
Probability, 284
Probability (Markov) matrix, 555
Probability vector, 284, 555
Product (of matrices), 28
 determinants of, 107–108
 inverse of, 45–46
 as linear combination, 31
 of lower triangular matrices, 68
 of symmetric matrices, 70
 transpose of, 47
Product (of vectors)
 cross product, 161–163
 scalar multiple in R^2 and R^3, 121
Production vector, 87, 88, 585
Productive consumption matrix, 587–588
Productive open economies, 88–89
Products (in chemical equation), 78
Profitable industries, in Leontief model, 588
Profitable sectors, 89
Projection operators, 251, 253, 267–268
Projection theorem, 146–147, 356–357
Pure imaginary complex numbers, 521
Pure strategies, of players in matrix games, 577

Q

QR-decomposition, 362–364, 371
Quadratic curve, of least squares fit, 380–381
Quadratic form associated with A, 406
Quadratic forms, 405–409
 applications of, 406–409
 change of variable, 406–407
 conic sections, 408–409
 expressing in matrix notation, 406
 indefinite quadratic forms, 412
 negative definite quadratic forms, 412
 negative semidefinite quadratic forms, 412
 optimization using, 417–422
 positive definite quadratic forms, 412–414
 positive semidefinite quadratic forms, 412
 principal axes theorem, 407–408
Quotient, complex number division, 524

R

R^m, orthogonal projections on subspaces of, 371–372

R^n
- coordinates relative to standard basis for, 205–206
- distance in, 132–133
- Euclidean inner product, 336–339
- geometry in, 138
- linear independence of standard unit vectors in, 191
- norm of a vector, 131
- span in standard unit vector, 184
- spanning in, 184
- standard basis for, 202
- standard unit vectors in, 132
- Theorem of Pythagoras in, 148
- transition matrices for, 220–222
- two-point vector equations in, 155
- vector forms of lines and planes in, 154
- as vector space, 173
- vectors in, 123–125

R^2
- Anosov automorphism, 652–653
- basic matrix operators on, 258
- dot product of vectors in, 133
- line segment from one point to another in, 156
- lines through origin are subspaces of, 180–181
- matrix operators on, 273–280
- norm of a vector, 131
- parametric equations, of lines in, 152–153, 154
- self-similar sets in, 626–627
- shears in, 258
- spanning in, 184–185
- unit circles in, 338
- vector addition in, 120, 122
- vectors in, 119–129

R^3
- coordinates in, 206–207
- dot product of vectors in, 133
- linear independence of standard unit vectors in, 191–192
- lines through origin are subspaces of, 180–181
- lines through two points in, 155–156
- matrix transformation from R^4 to R^3, 249
- norm of a vector, 131
- orthogonal set in, 353
- rotations in, 254–256
- spanning in, 184–185
- standard basis for, 202–203
- vector addition in, 120, 122
- vector and parametric equations of lines in, 152–153, 154
- vector and parametric equations of planes in, 153, 154–155
- vectors in, 119–129

R^4
- cosine of angle between two vectors in, 346
- linear independence of standard unit vectors in, 192
- matrix transformation from R^4 to R^3, 249
- Theorem of Pythagoras in, 148
- vector and parametric equations of lines and planes in, 155

Random iteration algorithm, 636
Range, 247, 438, 439–440
Rank, 441
- of 4×6 matrix, 238
- dimension theorem for matrices, 239
- maximum value for, 239
- redundancy in a matrix and, 245
- sum of, 239–240

Rank of an approximation, 516
Rate of convergence, 493
Rayleigh, John William Strutt, 492
Rayleigh quotient, 491
Reactants (in chemical equation), 78
Real inner product space, 335, 345–346
Real line, 123
Real matrices, 316, 322
Real part
- of complex numbers, 315, 521
- of vectors and matrices, 316–317

Real-valued functions, vector space of, 175
Real vector space, 171, 172, 335
Recessive genes, 666
Reciprocals
- complex numbers, 523
- of modulo m, 658–659

Rectangular coordinate systems, 200
Reduced row echelon forms, 11–12, 21, 319
Reduced singular value decomposition, 514–515
Reduced singular value expansion, 514
Redundancy, in matrices, 245
Reflection matrices, 390
Reflection operators, 251, 252, 259–260
Reflections, composition of, 265, 277
Regression line, 377
Regular Markov chain, 288, 558
Regular stochastic matrices, 288–289
Regular transition matrix, 558
Relative error, 493

Relative maximum, 420, 421
Relative minimum, 420, 421
Repeated mappings, of Arnold's cat map, 643–644
Replacement matrix, forest management model, 592
Residue, of a modulo m, 657–658
Resistance (electrical), 76
Resistor, 76
Revection transformation, computer graphics, 603
RGB color cube, 127
RGB color model, 127
RGB space, 127
Rhind Papyrus, 536
Right distributive law, 38
Right eigenvectors, 304
Right-hand rule, 164
Roll (aircraft), 256
Rotation equations, 254, 393
Rotation matrices, 254, 390
Rotation of axes
- in 2-space, 392
- in 3-space, 394–395

Rotation operator, 253–255
- properties of, 267
- on R^3, 255

Rotations
- composition of, 264, 275
- kernel and range of, 439
- in R^3, 254–256

Rotation transformation
- computer graphics, 600–602
- self-similar sets, 630

Roundoff errors, 21–22
Row-column method, 30–31
Row echelon form, 11–12, 14–15, 21, 229
Row equivalents, 51
Row matrices, 26
Row-matrix form of vectors, 128, 225
Row operations. *See* Elementary row operations
Row reduction
- basis by, 230, 231
- evaluating determinants by, 100–104

Row space, 225, 228–229, 243
- basis by row reduction, 230
- basis for, 229–230, 232–233
- equal dimensions of row and column space, 237

Row vectors, 26, 27, 39, 157, 225
Row-wheel, 572
Rows, cofactor expansion and choice of row, 97
Runout splines, 548–551

S

Saddle points, 420, 421, 576
Sales revenue, maximizing, 526
Sample points, 340
Saturation, graphical images, 124
Scalar multiples, 28, 172
Scalar multiplication, 121, 172
Scalar triple product, 165–166
Scalars, 26, 119, 121
  from vector multiple, 161
  vector space scalars, 172
Scaling
  Euclidean scaling, 489–490
  maximum entry scaling, 490–493
Scaling transformation
  computer graphics, 599
  self-similar sets, 626, 630–631
Schmidt, Erhardt, 360, 510
Schur, Issai, 402
Schur decomposition, 403, 507
Schur's theorem, 402
Schwarz, Hermann Amandus, 137
Search engines, 496–500
Second derivative test, 420, 421
Sectors (economic), 85
Self-similar sets, 626–630
Sensitivity to initial conditions, dynamical systems, 651
Sets
  linear independence, 191–194
  relations among members of, 563
  self-similar sets, 626–630
Shear operators, 258, 277
Shear transformation, computer graphics, 603
Sheep harvesting, 689–690
Shifting operators, 446
Sierpinski, Waclaw, 628
Sierpinski carpet, 628, 630, 632–635, 637, 640
Sierpinski triangle, 628, 630, 632–633, 635–636
Similar matrices, 305
Similarity invariants, 305, 306, 471–472
Similarity transformation, 305
Similitudes, 630–633
Singular matrices, 42, 43
Singular value decomposition (SVD), 509–511, 514–516
Singular values, 507–508
Skew-Hermitian matrices, 429
Skew product, 161
Skew-symmetric matrices, 428–429
Solution vectors, 157
Solutions
  best approximations, 367
  comparison of procedures for solving linear systems, 501–505
  cost of, 501–504
  factoring, 477
  flops and, 501–504
  Gauss-Jordan elimination, 15, 18, 21–22, 44, 82, 319, 504, 505
  Gaussian elimination, 11–16, 21–22, 505
  general solution, 13, 227, 240, 328
  of homogeneous linear systems, 186–187
  least squares solutions, 366–371, 372
  of linear system by diagonalization, 329–331
  of linear systems, 3, 11
  of linear systems by factoring, 477
  of linear systems with initial conditions, 328–329
  particular solution, 227
  power method, 487–494
  trivial/nontrivial solutions, 17, 191
Solutions spaces, of homogeneous systems, 187–188
Sound waves, in human ear, 693–698
Spacecraft, yaw, pitch, and roll, 256
Spanning
  in R^2 and R^3, 184–185
  in R^n, 184
  testing for, 186
Spanning sets, 185, 188, 204
Spans, 184, 209–210
Spectral decomposition of A, 400, 401
Sphere, through four points, 523–524
Spline interpolation, cubic, 543–551
Springs, constant, 378–379
Square, image of, 279
Square matrices, 42, 66, 68, 389
  decompositions of, 506–507
  determinants of, 100–104
  eigenvalues of, 311
  of order n, 27
  trace, 35
  transpose, 34
Standard basis
  coordinate vectors relative to, 206
  coordinates relative to standard basis for R^n, 205–206
  for M_{nn}, 203
  for polynomials, 202
  for R^3, 202–203
  for R^n, 202
Standard inner product, 336
  on polynomials, 340
Standard matrices, 248, 274
  for matrix transformation, 251
  for orthogonal projection, 372
  for T^{-1}, 268–269
Standard unit vectors, 132, 163
  linear independence in R^3, 191–192
  linear independence in R^4, 192
  linear independence in R^n, 191
  in span R^n, 184
State of a particle system, 125
State of the dynamical system, 272
State of the variable, 282
State vector, 287
  of Markov chains, 555, 558–561
Steady-state vector, of Markov chain, 289, 559–560
Stochastic matrices, 288–289, 555
Stochastic processes, 284
Stopping procedures, 494
Strategies, of players in matrix games, 574, 576–578
Strictly determined games, 576
String theory, 123, 124
Subdiagonal, 403
Submatrices, 30, 414
Subspaces, 179–188, 439
  creating, 183
  defined, 179
  examples of, 180–188
  of M_{nn}, 181
  orthogonal projections on, 369–370
  orthogonal projections on subspaces of R^m, 371–372
  of polynomials, 182
  of polynomials (P_n), 182
  of R^2 and R^3, 180–181
  zero subspace, 180
Substitution ciphers, 655
Subtraction
  of complex numbers, 522
  of vectors in R^2 and R^3, 121
  of vectors in R^n, 126
Sum
  of complex numbers, 522
  matrices, 27, 46
  of rank and nullity, 239–240
  of vectors in R^2 and R^3, 120, 122
  of vectors in R^n, 126
Strategies, of players in matrix games, 574, 576–578
Strictly determined games, 576
SVD (singular value decomposition), 509–511, 514–516
Sylvester, James, 34, 94, 510
Symmetric matrices, 69–70, 322
  eigenvalues of, 398
  Hessian matrices, 420
  orthogonally diagonalizing, 299

Symmetry property, of dot product, 136, 318

T

T^{-1}, standard matrix for, 268–269
Taussky-Todd, Olga, 320
Technology Matrix, 86
Television, market share as dynamical system, 282–283
Temperature distribution, at equilibrium. *see* Equilibrium temperature
Terminal point, 119
Theorem
 contrapositive form of, 519
 converse of, 519
Theorem of Pythagoras
 generalized Theorem of Pythagoras, 348
 in R^4, 148
 in R^n, 148
Three-dimensional object visualization, 597–599
3-Step connection, directed graphs, 565
Three-Step Procedure, 461
3×3 matrices
 adjoint, 111
 orthogonal matrix, 389–390
 QR-decomposition of, 363–364
 determinants, 97
 eigenvalues, 297
3–space, 119
 cross product, 161–163
 scalar triple product, 165–166
3-tuples, 123
Tien-Yien Li, 641
Tiled planes, 647–648
Time, as fourth dimension, 123
Time-varying morphs, 704–707
Time-varying warps, 703–704
Topological dimensions, 628–629
Topology, 628–629
Tournaments, 568
Trace, square matrices, 35
Traffic flow, network analysis with linear systems, 74–75
Transformations, 248. *See also* Linear transformation; Matrix transformations
 differentiation transformation, 439
 dimensions and, 447
 evaluation transformation, 436–437
 integral transformation, 438
 inverse transformations, 464
 on matrix spaces, 435
 one-to-one linear transformation, 447
Transition matrices, 219–221, 469
 invertibility of, 220

Markov chains, 554–557
 for R^n, 220–222
Transition probability, Markov chains, 554
Translation, 120, 436
Translation transformation, computer graphics, 600
Transpose, 33–34
 determinant of, 101
 invertibility, 48
 of lower triangular matrix, 68
 properties, 47–48
 vector spaces, 242
Triangle
 area of, 165
 Sierpinski, 628, 630, 632–633, 635–636
Triangle inequalities
 for distances, 138, 346
 for vectors, 138, 346
Triangle rule for vector addition, 120
Triangular matrices, 68–69
 diagonalizability of, 309–310
 eigenvalues of, 298–299
Triangulation, 702–703
Trigonometric polynomial, 384
Trivial solution, 17, 191
Turing, Alan Mathison, 479
Two-person zero-sum games, 573–578
Two-point vector equations, in R^n, 155
2×2 matrices
 cofactor expansions of, 95
 determinants, 97–98
 eigenvalues of, 321, 322
 games, 578–580
 inverse of, 44
 vector space, 174
2×2 vector, eigenvectors, 296
2–space, 119
2-Step connection, directed graphs, 565–566, 568–569
2–tuples, 123

U

Unbounded feasible regions, 529, 530
Unbounded solution, 533
Underdetermined linear system, 241
Unified field theory, 124
Unit circle, 338
Unit sphere, 338
Unit vectors, 131–132, 317, 336
Unitary diagonalization, of Hermitian matrices, 427–428
Unitary matrices, 425, 426, 427
Units of measurement, 201
Unknowns, 2
Unstable algorithms, 22

Upper Hessenberg decomposition, 403
Upper Hessenberg form, 403
Upper triangular matrices, 68, 97, 298

V

Vaccine distribution, 579–580
Value, 247
Vector addition
 matrix games, 576
 parallelogram rule for, 120
 in R^2 and R^3, 120, 122
 triangle rule for, 120
Vector equations
 of lines and planes in R^4, 155
 of lines in R^2 and R^3, 152–153, 154
 of planes in R^3, 153, 154–155
 two-point vector equations in R^n, 155
Vector forms, 154
Vector space, 171
 axioms, 171–172
 basis for using row operations, 232
 complex vector spaces, 172, 315–325
 dimensions of, 209
 examples of, 173–177, 204
 finite-dimensional vector space, 204, 212–213
 infinite-dimensional/finite-dimensional, 204
 of infinite sequences of real numbers, 173
 isomorphic, 447
 of $m \times n$ matrices, 174–175
 n-dimensional vector space, 212
 of real-valued functions, 175
 real vector space, 171, 172
 subspaces, 179–188, 439
 transposes, 242–243
 of 2×2 matrices, 174
 zero vector space, 173, 209
Vector space scalars, 172
Vector subtraction, in R^2 and R^3, 121
Vectors, 119
 angle between, 134–135, 137, 346
 arithmetic operations, 120–122, 125
 "basis vectors," 201
 collinear vectors, 121–122
 column-matrix form of, 128, 225
 comma-delimited form of, 128, 225
 components of, 122–123
 in coordinate systems, 122–123
 coordinate vector, 205
 coordinate vectors, 206
 dot product, 133–137, 139–140
 equality, 120, 125
 equivalence, 120, 125
 geometric vectors, 119

linear combination of, 127, 132–133, 183, 185–186
linear independence, 184, 190–198
nonzero vectors, 188
norm of, 148
normalizing, 132
notation, 119, 128
orthogonal vectors, 143–145, 317
parallel vectors, 121–122
parallelogram equation for, 138
perpendicular vectors, 143
probability vector, 284
in R^2 and R^3, 119–129
real and imaginary parts of, 316–317
in R^n, 123–125
row-matrix form of, 128, 225
row vectors, 26, 27, 39, 157, 225
solution vectors, 157
standard unit vectors, 132, 163, 184, 191–192
state vector, 287
triangle inequality for, 138
unit vectors, 131–132, 317, 336
zero vector, 120, 125
Vertex matrix, 564, 565
Vertex points
   linear programming, 530
   warps, 702–703
Vertices, graphs, 563–564
Viewing audience maximization, 577–578
Visualization, of three-dimensional objects, 597–599
Voltage rises/drops, 76, 77
Volts (units), 76
von Neumann, John, 646

W

Warps, 700–704
   time-varying, 703–704
Weight, 336
Weighted Euclidean inner products, 336–338, 339
Weyl, Herman Klaus, 510
Wildlife migration, as Markov chain, 286–287
Wilson, Edwin, 161
Wronski, Jozef Hoene de, 194
Wronskian, 197–198

X

X-linked inheritance, 666, 671–674
X-ray computed tomography, 615–624

Y

Yaw, 256
Yorke, James, 641

Z

Zero matrices, 40
Zero population growth, 684
Zero subspace, 180
Zero-sum matrix games, two-person, 573–578
Zero transformations, 250, 434, 439
Zero vector space, 173, 209
Zero vectors, 120, 125